TRIGONOMETRIC FORMS

27. $\int \sin^2 u\, du = \frac{1}{2}u - \frac{1}{4}\sin 2u + c$

28. $\int \cos^2 u\, du = \frac{1}{2}u + \frac{1}{4}\sin 2u + c$

29. $\int \sin^3 u\, du = -\frac{1}{3}(2 + \sin^2 u)\cos u + c$

30. $\int \cos^3 u\, du = \frac{1}{3}(2 + \cos^2 u)\sin u + c$

31. $\int \tan^2 u\, du = \tan u - u + c$

32. $\int \cot^2 u\, du = -$

33. $\int \tan^3 u\, du = \frac{1}{2}$ ... $\ln|\cos u| + c$

34. $\int \cot^3 u\, du = -\frac{1}{2}\cot^2 u - \ln|\sin u| + c$

35. $\int \sec^3 u\, du = \frac{1}{2}\tan^2 u + \ln|\cos u| + c$

36. $\int \csc^3 u\, du = -\frac{1}{2}\csc u \cot u + \frac{1}{2}\ln|\csc u - \cot u| + c$

REDUCTION FORMULAS

37. $\int \frac{du}{(u^2+a^2)^m} = \frac{1}{2(m-1)a^2}\frac{u}{(u^2+a^2)^{m-1}} + \frac{2m-3}{2(m-1)a^2}\int \frac{du}{(u^2+a^2)^{m-1}}$

38. $\int \sin^n u\, du = -\frac{1}{n}\sin^{n-1} u \cos u + \frac{n-1}{n}\int \sin^{n-2} u\, du$

39. $\int \cos^n u\, du = \frac{1}{n}\cos^{n-1} u \sin u + \frac{n-1}{n}\int \cos^{n-2} u\, du$

40. $\int \tan^n u\, du = \frac{\tan^{n-1} u}{n-1} - \int \tan^{n-2} u\, du$

41. $\int \cot^n u\, du = -\frac{\cot^{n-1} u}{n-1} - \int \cot^{n-2} u\, du$

42. $\int \sec^n u\, du = \frac{\sec^{n-2} u \tan u}{n-1} + \frac{n-2}{n-1}\int \sec^{n-2} u\, du$

43. $\int \csc^n u\, du = \frac{\csc^{n-2} u \cot u}{n-1} + \frac{n-2}{n-1}\int \csc^{n-2} u\, du$

USEFUL LIMITS

44. $\lim_{x\to 0} \frac{\sin x}{x} = 1$

45. $\lim_{x\to 0} \frac{1-\cos x}{x} = 0$

46. $\lim_{x\to\infty} x^n e^{-x} = 0$

47. $\lim_{x\to\infty} \frac{\ln x}{x^n} = 0$

48. $\lim_{n\to\infty} x^n = 0 \quad (|x| < 1)$

49. $\lim_{n\to\infty} \left(1 + \frac{x}{n}\right)^n = e^x$

50. $\lim_{n\to\infty} \frac{x^n}{n!} = 0$

51. $\lim_{x\to a} \frac{f(x) - f(a)}{x - a} = f'(a)$

CALCULUS

CALCULUS

GERALD J. JANUSZ

WCB Wm. C. Brown Publishers
Dubuque, Iowa • Melbourne, Australia • Oxford, England

Book Team

Editor *Paula-Christy Heighton*
Developmental Editor *Theresa Grutz*
Publishing Services Coordinator *Julie Avery Kennedy*

Wm. C. Brown Publishers
A Division of Wm. C. Brown Communications, Inc.

Vice President and General Manager *Beverly Kolz*
Vice President, Publisher *Earl McPeek*
Vice President, Director of Sales and Marketing *Virginia S. Moffat*
National Sales Manager *Douglas J. DiNardo*
Marketing Manager *Julie Joyce Keck*
Advertising Manager *Janelle Keeffer*
Director of Production *Colleen A. Yonda*
Publishing Services Manager *Karen J. Slaght*
Permissions/Records Manager *Connie Allendorf*

Wm. C. Brown Communications, Inc.

President and Chief Executive Officer *G. Franklin Lewis*
Corporate Senior Vice President, President of WCB Manufacturing *Roger Meyer*
Corporate Senior Vice President and Chief Financial Officer *Robert Chesterman*

Cover art by Dr. Rudiger Gebauer

Cover design by Tara Bazata

A Times Mirror Company

Library of Congress Catalog Card Number: 93–72119

ISBN 0–697–15374–6

Printed in the United States of America by Wm. C. Brown Communications, Inc., 2460 Kerper Boulevard, Dubuque, IA 52001

10 9 8 7 6 5 4 3 2 1

To Sue,

whose patience, understanding and love
are so important to me.

Contents

12. Vectors in Two and Three Dimensions

13. Vector Functions

14. Partial Derivatives

15. Integration in Higher Dimensions

16. Two Theorems in Vector Calculus

Appendix A

Index

Preface

Calculus is the basic mathematics course studied by all students of science as well as many other disciplines. The applications of calculus to solve practical and theoretical problems started with Isaac Newton who first applied its great computational power to describe the motion of the planets. Applications of calculus are found today in areas ranging from high energy physics to molecular biology and from ecology and the environment to economics and finance; it is generally regarded as the introduction to the study of higher mathematics. With such a wide variety of interests in the students of today, it is impossible to produce a text that caters to the special needs of every interest group. Instead this text provides a solid foundation for understanding the subject, the broad range of its applicability and the techniques necessary to apply the subject to specific problems.

APPROACH

The selection of topics is dictated by several considerations. There is an attempt to present the most powerful calculating tools that calculus can provide—differentiation, integration on the line, in the plane and three-dimensional space and line integrals, power series and approximations by polynomials. These ideas are of value to students of all the sciences. There is the desire to present a solid foundation for understanding and applying these tools. This is of special value to students of the sciences as well as those planning further study of mathematics. There is enough explanation to ensure the students understanding. It is rather common in most calculus texts to state theorems and properties without proof or explanation and then apply them to solve problems. This style frequently leaves students and instructors with many unanswered questions. This text contains very thorough explanations; most theorems are proved completely. There is no development of the real number system so the point of view is that the real numbers are familiar objects. The Completeness Axiom is stated and used as the starting point. I believe that one good way to understand a theorem is to understand its proof. Toward this end, careful explanations are given to lead the reader through the proofs of most theorems.

I have tried to present a text that is interesting for the student and instructor alike. The instructor might find here a thorough treatment of some topic that is traditionally given without explanation. For example, a careful definition of the constant π is given and used to obtain the limit of $\sin x/x$ as x approaches zero. Arc length is treated with some care showing clearly how the smoothness of the function is used. The treatment of error terms is given with great care; the estimates of the difference between the values of a function and its Taylor polynomial approximations serves as the basis for the estimates of the approximations of definite integrals given in chapter 11. Power series are treated before infinite series of constants as a natural follow-up to the study of Taylor polynomials. A novel treatment is given for testing convergence of power series. The Bounded Terms Test is proved and exploited following the suggestions of Professors Jerry Uhl and Horacio Porta. This arrangement of material makes it is possible to omit much of the material on infinite series of constants if one is willing to ignore convergence at the endpoints of the interval of convergence. Although there is a complete treatment, some instructors may wish to omit parts of chapter 10 dealing with series of constants to save time.

Chapter 11 on numerical methods contains the interval bisection method as a first method of finding an approximate solution to an equation. Although it is not an efficient method, it is easily understood and serves to introduce the student to the idea of successive approximation and estimation of the error. Newton's method for finding approximate solutions of equations is treated rather carefully—giving a proof of convergence under suitable hypothesis and mentioning the quadratic convergence as justification for application of the method. There is an extended discussion of various numerical methods for approximating a definite integral. The basis in all cases is the use of Taylor polynomials to approximate a function and the use of the error terms developed in chapter 10 to control the errors in the integral approximations.

It is becomming commonplace for today's students to have a computer or calculator to graph functions and do calculations. The term *accurate sketch* is used in the text to suggest that graphs be generated by machine. The presentation dwells less on the sketching of graphs and devotes more attention to the interpretation of graphs. The presentation of methods of integration is somewhat more extensive than in most texts. The method of partial fractions is treated rather thoroughly because that is the method used by symbolic manipulation programs. The use of the method of partial fractions enables the proof of existence theorems asserting that all functions of certain types can be integrated in closed form. Some of the traditional trigonometric substitutions are given but may be omitted because alternate methods are provided. One benefit of this approach is that the integral of $\sec x$ is easily computed as an integral of a rational function of $\cos x$ for which a standard method is given. This seems preferrable to the more usual, unmotivated method that makes sense only when one already knows the value of the integral.

Parametric equations are emphasized and used quite frequently. This seems the proper course to follow because graphs generated by computer are very often obtained most easily using parametric equations.

I have not included a chapter devoted to differential equations; instead subsections and exercises have been inserted in chapters 5, 7, 10, 13 and 14 devoted to elementary aspects of the solution of differential equations. The sections on exponential growth in chapter 7 and series solutions in chapter 10 seem particularly useful for illustrating some elementary techniques.

EXERCISES

The exercises are an important part of every calculus text. The exercises in this text are generally one of three types. The early exercises in a set include some routine drill questions that ask for direct application of techniques described in the text. There usually follows some slightly more challenging exercises that require a variation of the techniques presented in the text or some specific application of an idea that occurred in the course of an explanation of some general concept. Some exercises present extensions of the text material or related, but different, ideas. There are a small number of exercises that should be regarded as projects that might serve as the basis of a group discussion. The solutions in these exercises might require approximations or data generated solutions. These are indicated in the text by some appropriate comment following the exercise. Some material is presented in the exercises that could, arguably, be made part of the text material. It is recommended that the student read all the exercises even though not all will be worked.

ACKNOWLEDGEMENTS

Many of the ideas in this text were discussed with colleagues at the University of Illinois over many years. Their generosity in sharing ideas is sincerely appreciated. Professor Uhl and Porta were especially unselfish in discussing material. I must thank the reviewers (listed below) for their many excellent suggestions on the first draft; all of their recommendations were seriously considered. Special thanks goes to Jeff Vaaler, who read, and commented on, the entire manuscript, and Donna Szott for checking the mathematical accuracy of many of the examples, statements and proofs and exercises. They found many slips I made in the first draft so that corrections could be made to the final product.

Thanks also to the editors and staff at William C. Brown Publishers who devoted a great deal of time and energy to bring this project to a successful completion. Thanks to Earl McPeek who encouraged me at the beginning to undertake this project and who provided reassurance during some early periods of doubt. Theresa Grutz has been unfailing in her good cheer and encouragement during the later stages of producing the manuscript. Thanks to the many other employees have made contributions to the final form of the manuscript.

Most importantly, I must thank my wife Sue, who had to live through long periods of time when it seemed that the book was the only important thing in my life. Her patience, caring and understanding were critical to my sanity during this time.

REVIEWERS

Daniel Anderson
University of Iowa

Wayne Andrepont
University of Southwestern Louisiana

Duane Blumberg
University of Southwestern Louisiana

Edward B. Burger
Williams College

Daniel Drucker
Wayne State University

Jim Henderson
Colorado College

John Milcetich
University of the District of Columbia

Daniel A. Moran
Michigan State University

Vincent P. Schielack, Jr.
Texan A & M University

Walter Seaman
University of Iowa

Asok Sen
Indiana University–Purdue University

Michael Slack
Stanford University

Mary Jane Sterling
Bradley University

Alvin Swimmer
Arizona State University

Jeffrey D. Vaaler
University of Texas

PRODUCTION

The text was produced from camera-ready pages that I prepared using plain TEX for the typesetting. The macros implementing the design were prepared with the assistance of Dan Latterner, to whom I express my deep appreciation and admiration. Dan's skill helped remove the tedium from thousands of details. Most of the illustrations were prepared using the graphics capability of the *Mathmatica* program; a few were prepared directly in PostScript.

ART

The art on the cover and at the chapter openings is the creation of Professor George Francis of the University of Illinois. The cover piece shows a triangulation of the sphere that was drawn partially by computer and partially by hand. The pieces at the chapter openings are taken from a series of mathematical drawings in the collection of the artist.

GJJ-January 1994

1 Preparations for Calculus

In this chapter we review some basic properties of the real number system, some basic properties of graphs of equations, and other topics necessary in the study of calculus. Many of these properties will be familiar to the reader, but it seems useful to examine the specific properties that occur frequently in the material to follow. Most of this chapter serves as a review of basic material. The treatment is rather brief and touches only on some topics by providing definitions and a minimal number of examples. The notation and language of set theory is introduced. This provides the language that is used throughout the text. Later sections contain a discussion of the mathematical usage of language and grammar that serves to acquaint the reader with some of the peculiarities of mathematical expressions. The study of mathematics will be, for some, like a journey into a foreign country. To enjoy the trip, it helps to first learn the language.

1.1 Real numbers and sets

The *natural numbers* are familiar objects to the reader. These are the so-called counting numbers:

$$\{\text{natural numbers}\} = \{1, 2, 3, \ldots\}.$$

The natural numbers are a subset of the *integers*; the set of integers consists of all the natural numbers, zero, and the negatives of the natural numbers:

$$\{\text{integers}\} = \{\ldots, -2, -1, 0, 1, 2, \ldots\}.$$

We have used set notation to describe the natural numbers and the integers. We use the curly braces { and } to enclose the elements of a set. If the set has only a finite number of elements, we may simply list all the elements within the braces. For example, the set of abbreviations for the days of the week is denoted as {M T W Th F S Su}. When the set has infinitely many elements, it is not possible to list all of them. To describe the infinite set of natural numbers, we gave an indication of the elements in the set 1, 2, 3, and then used the three dots $\cdots$ to indicate that the list continues. This is not absolutely precise because no rule was given to indicate how the process continues. Because the natural numbers are familiar to the reader, this was deemed an adequate description. The description of most sets having an infinite number of elements requires a more precise approach.

Consider the set of all *rational numbers*. This is the set of fractions or quotients of integers. To describe this using set notation, we write

$$\{\text{rational numbers}\} = \left\{\frac{a}{b} : a, b \text{ are integers}, \quad b \neq 0\right\}.$$

Here the infinite set of rational numbers is described by specifying its elements as all quotients a/b, where the symbols a and b stand for arbitrary integers except that b is not allowed to equal 0. One reads the set on the right of the equal sign as "the set of all a over b such that a and b are integers and b is not equal to 0". Every rational number can be represented as a decimal; for example,

$$\frac{3}{5} = 0.600 \qquad \frac{1}{8} = 0.125 \qquad \frac{1}{3} = 0.33333\ldots.$$

The decimal representation of a rational number is found by long division. Not every decimal number is a rational number, however. The decimal representation of a rational number has a special property. One can show that every rational number has a decimal representation that is eventually repeating. This means that from some point onward, the decimal representation has some string of consecutive digits that is repeated infinitely often. This is to be interpreted as allowing the possibility of an infinite string of zeros so that 0.6 and 0.125 are eventually repeating because we regard them as having an infinite number of zeros; $0.6 = 0.6000000\ldots.$ We will verify that every eventually repeating decimal is a rational number after studying more about infinite processes in later chapters. (See chapter 9 for further discussion of this topic.)

The set of rational numbers is not "large" enough to solve many problems. Even rather simple problems involve nonrational numbers. For example consider a square with sides of length 1.

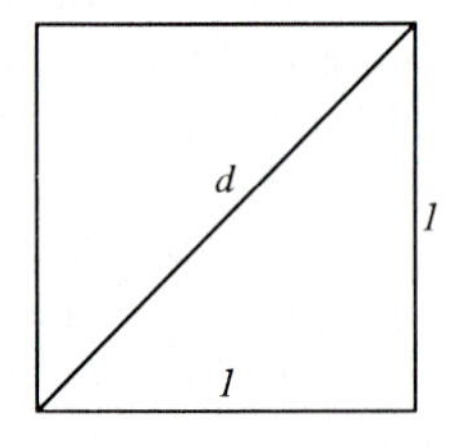

FIGURE 1.1
$d = \sqrt{2}$

The length of each side is a rational number, a natural number even, but the diagonal of the square has length $\sqrt{2}$. By studying some properties of the multiplication of natural numbers, one can prove that $\sqrt{2}$ is *not* a rational number. Part of the decimal representation of $\sqrt{2}$ is $1.41421356237\ldots.$ Here the ... indicates that the sequence continues but no description given for the unspecified decimal digits. This use of the "dots" is slightly different from the use encountered in the definition of the natural numbers. There it was assumed the reader could see how the sequence was to continue; here no such presumption is made. When we obtain a finite number of digits after the decimal point, 11 in this example, we have obtained an approximate value of $\sqrt{2}$. We write

$$\sqrt{2} \approx 1.41421356237$$

to indicate that the right side is approximately equal to the left side. The exact value of $\sqrt{2}$ is an infinite decimal that does not become a repeating decimal after any number of decimal places (because it is not rational).

To have a satisfactory system of numbers we consider the set of *real numbers*.

We take the point of view that the set $\mathcal{R}$ of *real numbers* is the set of all decimals. A *decimal* is a finite string of digits followed by the decimal point followed by a finite or infinite string of digits. A digit is an integer in the set $\{1, 2, 3, 4, 5, 6, 7, 8, 9, 0\}$. Thus a real number has the form

$$a_k a_{k-1} \cdots a_1 a_0 . b_1 b_2 \cdots$$

where the a_i and b_j are digits; there may be infinitely many b's but only finitely many a's.

There is a slight possibility of confusion which we will clear up later. Some real numbers may have more than one representation as a decimal. This can be seen in the examples $1.0000\ldots = 0.9999999\ldots$ and $8.39999999\ldots = 8.4$. This fact will be proved later and should not cause any serious complications for now. We can give a line of reasoning that may be convincing. The fraction $1/3$ has the decimal representation $0.3333\ldots$ which is obtained by the rules of long division; just divide 3 into 1. Now write this equality and multiply both sides by 3:

$$1 = 3\left(\frac{1}{3}\right) = 3(0.33333\ldots) = 0.99999\ldots.$$

In any case, we can avoid these multiple representations if we agree not to write any decimals using a final infinite string of 9s.

We will also assume a number of other important properties of the real number system; namely the fact that two real numbers can be added, subtracted or multiplied to produce another real number. Also we can divide any real number by a nonzero real number to obtain another real number. The fact that division by zero is not allowed will appear in various disguises later. Here is a situation where this consideration must be made.

Suppose we are given two real numbers a, b and we want to describe the set of real numbers x such that $ax = b$. There is no difficulty if $a \neq 0$, because then division by a is allowed. We find $x = b/a$. Suppose that $a = 0$. Then $ax = 0x = 0$ for any real number x whatsoever. The original equation can be true only if $b = 0$, and in this case $ax = b$ holds for every real number x. This analysis of the equation $ax = b$ is summarized by saying that there are three possibilities:

i. If $a \neq 0$, then there is exactly one real number x such that $ax = b$,
ii. If $a = 0$ and $b = 0$, then every real x satisfies $ax = b$,
iii. If $a = 0$ and $b \neq 0$ then there is no real number x with $ax = b$.

If we write

$$B = \{\, x \, : \, ax = b \,\}$$

then the set B depends on the choice of a, b; either

i. B contains exactly one element,
ii. B is the set of all real numbers, or
iii. B is the empty set (that is, B contains no elements).

This is a simple example of a case analysis. The original problem is broken into three cases, each having some additional information with which the problem is easily solved.

It is necessary to describe sets precisely. It is important to be precise with the choice of words. For example, the set of natural numbers defined above does not include the integer 0. We define the set of *nonnegative integers* to be the set of natural numbers with 0 adjoined. We can write this as

$$N = \{\text{nonnegative integers}\} = \{0, 1, 2, \ldots\}.$$

If n is a nonnegative integer, then n belongs to N; this is written as $n \in N$. Here is an example of a set whose definition is slightly more complicated. Let

$$O = \{x \, : \, x = 2n + 1, \text{ with } n \in N\}.$$

To describe the set O without using any mathematical symbols we consider the possible elements of O by first trying some values allowed for n: n can be 0, 1, 2, or 3 so this means $x = 2n + 1$ can be 1, 3, 5 or 7. After a moment's thought, we see that O consists of all the positive, odd integers. To give a slightly more complete statement, we say: if x is a positive odd integer, then $x - 1$ is either 0 or a positive even integer. Every positive even integer is 2 times a natural number; 0 is 2 times 0. Thus $x - 1$ is 2 times a nonnegative integer. That is $x - 1 = 2n$ for some $n \in N$ or $x = 2n + 1$. This shows that O equals the set of positive odd integers.

Exercises 1.1

Describe in words the contents of each of the following sets of real numbers.

1. $\{x : 0 \leq x < 5\}$

2. $\{w : -3 \leq w < -1\}$

3. $\{t : t < 5 \text{ or } t > 25\}$

4. $\{x : 0 \leq 3x + 4 < 12\}$

5. $\{v : -3v < -12 \text{ and } v^2 < 16\}$

6. $\{x : 1 \leq x < 10, x \neq 3, 4, 5\}$

7. $\{x : 9 \leq x^2 < 25 \text{ or } 4x > 200\}$

8. $\{z : z^3 \leq 1000 \text{ and } z^2 < 400\}$

Use set notation to describe the following sets of numbers.

9. The set of numbers no larger than 15 and whose square is less than 100.

10. The set of integers that are multiples of 3 but not multiples of 12.

11. The set of positive integers whose square is at most 10,000.

12. The set of positive rational numbers whose numerator is less than the square of the denominator.

13. The set of even integers whose reciprocals are less than 0.002.

1.2 Inequalities and absolute values

In this section we study the *order* properties of the real numbers. We take for granted that the reader is aware of the notion of positive and negative real numbers and from some of these basic concepts we define the order relation and prove some properties.

We say $a < b$ is true if and only if $b - a$ is positive. The properties of the order can be derived from some well-known properties of the positive real numbers. For example if a and b are positive, then $a + b$ and ab are also positive. There is the *trichotomy law* that states that for any real number x, we have exactly one of the possibilities x is negative, x is 0, x is positive. These facts can be used to prove some simple properties of the ordering. For example we have the following statements.

THEOREM 1.1

Properties of the Ordering

Let a, b, c, d be real numbers. Then,

i. $a < b$ and $b < c$ imply $a < c$,
ii. $a < b$ and $c < d$ imply $a + c < b + d$,
iii. $a < b$ and $0 < c$ imply $ac < bc$,
iv. $c \neq 0$ implies $0 < c^2$,
v. $a < b$ implies $-b < -a$,
vi. $0 < c$ implies $0 < 1/c$.

Proof

i. The inequalities $a < b$ and $b < c$ imply $b - a$ and $c - b$ are positive. Now the sum of two positive numbers is positive so

$$(b - a) + (c - b) = c - a$$

is positive and thus $a < c$.

ii. The two inequalities $a < b$ and $c < d$ imply both $b - a$ and $d - c$ are positive. Hence their sum is positive; that is, $(b-a)+(d-c) = (b+d)-(a+c)$ is positive. It follows that $a + c < b + d$.

iii. If $a < b$, then $b - a$ is positive; if $0 < c$ then c is positive. Because the product of two positive numbers is positive, we have $(b - a)c = bc - ac$ is positive. Thus $ac < bc$. In words, this statement says that both sides of an inequality can be multiplied by a positive number and the inequality will still be correct.

iv. We prove this by a case consideration. Suppose that $0 < c$. Multiply both sides of this inequality by the positive number c to get $0 < c^2$. Now suppose that c is negative. Then $-c$ is positive so that by the case just completed we have

$$0 < (-c)^2 = c^2$$

so property iv is true in both cases. The only other possibility is $c = 0$ but that was ruled out by the hypothesis that $c \neq 0$.

v. Suppose $a < b$ so that $b - a$ is positive. Then $b - a = (-a) - (-b)$ is positive and so $-b < -a$ by definition of the ordering.

vi. There are exactly three possibilities for $1/c$; either $1/c < 0$, $1/c = 0$, or $0 < 1/c$. We are given that $0 < c$ so we may multiply any true inequality by c and obtain another true inequality. Multiply the three possible inequalities for $1/c$ by c to get either $1 < 0$, $1 = 0$ or $0 < 1$. Clearly only the third of these is correct (because $0 < 1^2 = 1$ by property iv) and so $0 < 1/c$ must be correct. □

This gives only a sample of the many statements about inequalities that are probably familiar to the reader. A few more statements are given in the exercises.

The geometric interpretation of $a < b$, namely that the point of the real line corresponding to a is to the left of the point corresponding to b, is given in the next section.

Variations of Notation

If $0 < a$, then we also write $a > 0$. If we know that either a is positive or a is 0, we write $0 \leq a$ or $a \geq 0$. We leave it for the reader to translate the statements involving $<$ of the previous theorem into true statements involving $\leq$. The only statement that may produce an exception is statement vi in which division is not allowed when $c = 0$.

Absolute Values

For any real number x the *absolute value* of x, written as $|x|$, is defined by the conditions

$$|x| = \begin{cases} x & \text{if } x > 0 \\ 0 & \text{if } x = 0 \\ -x & \text{if } x < 0. \end{cases}$$

Thus we have either $|x| = 0$ when $x = 0$ or $|x| > 0$ when $x \neq 0$. In particular we have the statements

$$|16| = 16, \qquad |-12| = 12, \qquad |y^2| = y^2$$

for any number y.

Note that $|x| = x$ whenever $x \geq 0$. Here are some basic properties of the absolute value that are used frequently.

THEOREM 1.2

Properties of Absolute Value

For any numbers x and y, the following are true statements:

A. $|x| \geq 0$,
B. $|x| = |-x|$,
C. $|xy| = |x| \cdot |y|$,
D. $\sqrt{x^2} = |x|$.

Proof

A. If $x > 0$ then $|x| = x > 0$. If $x < 0$ then $-x > 0$ and $|x| = -x > 0$. The remaining case is $x = 0$ where we have $x| = x \geq 0$. Thus property A is true in all cases.

B. If $x = 0$ then $|x| = |-x| = 0$. If $x \neq 0$ then the definition of absolute value implies that both $|x|$ and $|-x|$ are equal to whichever of x and $-x$ is positive.

C. If either x or y is 0, then statement C is true because both sides of the equation equal 0. Assume then that neither x nor y is 0. We have that $|xy|$ is the positive element in the set $\{xy, -xy\}$; that positive element is clearly also equal to $|x| \cdot |y|$.

D. The number $\sqrt{x^2}$ is the nonnegative number whose square is x^2. But then $|x|$ is a nonnegative number whose square is x^2 (as is seen by considering the two possible values of $|x|$). □

In the next theorem we show how inequalities involving an absolute value can be rewritten as pairs of inequalities without absolute values.

THEOREM 1.3

Removal of the Absolute Value Signs

A. The statement $a < |x|$ is true if and only if one of the inequalities $a < x$ or $a < -x$ is true.
B. The statement $|x| < a$ is true if and only if both of the inequalities $-a < x$ and $x < a$ are true.

Proof

A. Suppose that $a < |x|$ holds. If $|x| = x$ then $a < x$ holds. If $|x| = -x$ then $a < -x$ holds. Thus at least one of the two inequalities $a < x$ or $a < -x$ is true. Now suppose that one of the two inequalities $a < x$ or $a < -x$ is true. Because $x \leq |x|$ and $-x \leq |x|$ it follows that $a < |x|$ regardless of which of the two inequalities holds.

B. Suppose $|x| < a$. Because $x \leq |x|$ and $-x \leq |x|$ we have both

$x < a$ and $-x < a$. The second of these is equivalent to $-a < x$ so the two given inequalities are true. If the two inequalities $x < a$ and $-x < a$ hold, then $|x| < a$ because $|x|$ equals one of x or $-x$. □

A slightly more succinct way to express statement B is

$$|x| < a \quad \text{if and only if} \quad -a < x < a.$$

A somewhat less obvious property of the absolute value is the *Triangle Inequality,* the name of which is motivated by the fact that the sum of the lengths of two sides of a triangle is greater than the length of the third side. (We admit that this does not seem like a very good reason for the name but after the study of vectors, the name will have more meaning.)

THEOREM 1.4

Triangle Inequality

For any two real numbers x and y, we have

$$|x + y| \leq |x| + |y|.$$

Proof

We use the idea mentioned just before the statement of the theorem. For any x and y, we have

$$x \leq |x| \quad \text{and} \quad y \leq |y|.$$

Adding these two inequalities gives

$$x + y \leq |x| + |y|.$$

Next exploit the fact that $|-a| = |a|$ to obtain

$$-x \leq |-x| = |x| \quad \text{and} \quad -y \leq |-y| = |y|.$$

Adding these together gives

$$-(x + y) \leq |x| + |y|.$$

Because $|x+y|$ equals either $(x+y)$ or $-(x+y)$ it follows that $|x+y| \leq |x|+|y|$. □

Exercises 1.2

A closed interval is the set of all x satisfying an inequality of the form $a \leq x \leq b$ for some $a < b$. An open interval is the set of x satisfying $a < x < b$. Express the sets below as intervals, if possible, or as the union of several intervals, where necessary.

1. $|x + 1| \leq 4$

2. $|2x - 3| \leq 7$

3. $4 \leq |x - 2| \leq 10$

4. $9 < |4 - x^2| < 29$

Describe the following sets of numbers x without the use of absolute values.

5. $|2x - 3| > 6$

6. $|x^2 - 100| > 25$

7. $|3 - x| \leq 8$

8. $|4 - 16x| < |x|$

9. $9x \leq |3x^2|$

10. $|x - 1| < 5$

11. $2|x - 1| < |x + 2|$

12. $|x^2 - 9| > |x + 3|$

13. Prove that the inequality $x^2 + 1 \geq 2x$ is valid for all real numbers x.

14. Use the Triangle Inequality to prove the following inequality for any numbers x and y:

$$|x - y| \geq \Big| |x| - |y| \Big|.$$

[Hint: Apply the triangle inequality to $x = (x - y) + y$

and again to $y = (y - x) + x$.]

15. Let a and b be nonnegative numbers. Prove the Arithmetic-Geometric Mean Inequality that asserts $\frac{a+b}{2} \geq \sqrt{ab}$. [The more general inequality $(a_1 + \cdots + a_n)/n \geq (a_1 \cdots a_n)^{1/n}$ is known by the same name.]

1.3 Coordinates

In this section we introduce geometry into the study of the real numbers by defining coordinates on a line. We begin with a horizontal line and some designated point, which we take as the origin, the point associated with the real number 0. We then associate with each point of the line a real number subject to the conditions that when $a < b$, the point associated with a is to the left of the point associated with b. If the real number a is associated with the point P, we say that a is the coordinate of P.

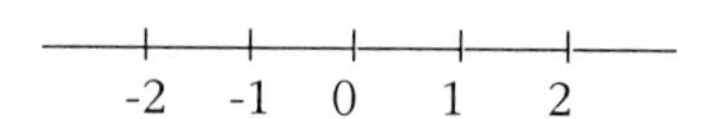

Distances

If the point P has coordinate a and Q has coordinate b, then the *distance* between P and Q is $|a - b|$. The *directed distance* from P to Q is $b - a$. The directed distance can be positive, negative or zero.

Intervals

It is convenient to refer to points on the line by their coordinates. It saves extra words to refer to the point 3, rather than the point with coordinate 3. We adopt this convention whenever it is convenient.

If a and b are points with $a < b$, the set of points lying between a and b is called an *interval.* If both endpoints are included in the set, we denote it as $[a, b]$ and call this the *closed interval* from a to b. If neither endpoint is included in the set, we denote it by (a, b) and call it the *open interval* from a to b. If we want to include one, but not both endpoints, we refer to it as a *half-open interval*; the notation $[a, b)$ denotes the half-open interval that includes a but not b. Similarly $(a, b]$ is the half-open interval including b but not a.

We can use set notation to describe intervals. For example, we have

$$[2, 3] = \{x : 2 \leq x \leq 3\} \quad \text{and} \quad (-3, 5] = \{x : -3 < x \leq 5\}.$$

In many cases the absolute value function is useful for describing intervals such as

$$[-1, 1] = \{x : |x| \leq 1\}.$$

More generally the interval $[-a, a]$ is the set $\{x : |x| \leq a\}$ provided $a > 0$. A more general interval can be described by referring to the distance from its midpoint. The midpoint of the interval $[a, b]$ is $c = (a + b)/2$. The interval $[a, b]$ is the set of points whose distance from c is at most half the length of the interval, namely $(b-a)/2$. Thus

$$[a,b]=\left\{x:\left|x-\frac{a+b}{2}\right|\leq\frac{b-a}{2}\right\}.$$

EXAMPLE 1 Describe the set of all points on the interval $(4,8]$ that are at a distance no more than 10 from -3.

Solution: The points at a distance at most 10 from -3 lie on the interval

$$\{x:|x-(-3)|\leq 10\}=\{x:-10\leq x+3\leq 10\}$$
$$=\{x:-13\leq x\leq 7\}=[-13,7].$$

The points on $(4,8]$ that are also on $[-13,7]$ lie on the interval $(4,7]$. ■

EXAMPLE 2 If x and y are known to have distance no more than 6 from 11, how large can the number $|x-y|$ be?

Solution: The set of numbers at a distance no more than 6 from 11 is the interval $[5,17]$. Thus x and y are points on $[5,17]$ so the largest possible distance between x and y occurs when $|x-y|=17-5=12$. ■

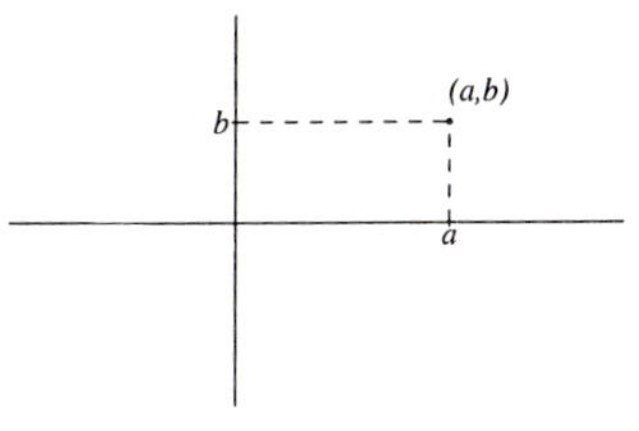

FIGURE 1.2

The Coordinate Plane

Now that we have discussed coordinates for the line, we pass to two dimensions and deal with coordinates for the plane. The points in the plane are given coordinates relative to a pair of coordinate axes; these are two coordinate lines that intersect at right angles at the point with coordinate 0 on each line. It is conventional to refer to the horizontal axis as the x-axis and the vertical axis as the y-axis.

A point P in the plane is assigned coordinates according to the following rule. Through P we draw a line parallel to the y-axis and let it intersect the x-axis at a; through P we draw a line parallel to the x-axis and let it intersect the y-axis at b. Then the coordinate assigned to P is the ordered pair (a,b). This assignment of coordinates to points is referred to as the *rectangular coordinate system* and (a,b) are the rectangular coordinates of P.

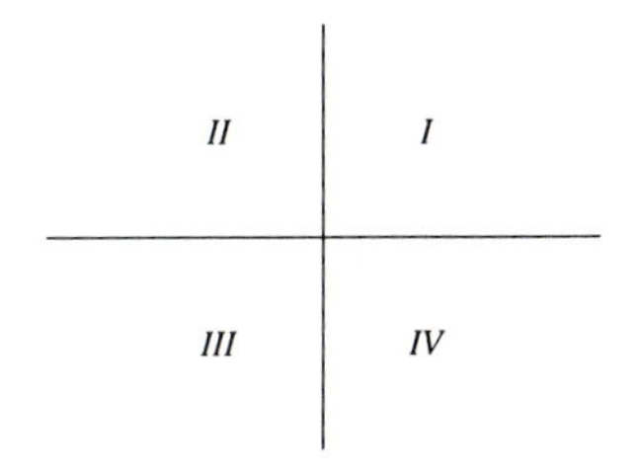

FIGURE 1.3

Quadrants

The region of the plane consisting of points with coordinate (a,b) and $a\geq 0$, $b\geq 0$ is the *first quadrant.* The remaining quadrants are numbered in the counterclockwise direction as indicated in figure 1.4.

EXAMPLE 3 Identify the quadrants containing the four points $(\pm 2,\pm 1)$.

Solution: The points are shown in the figure:

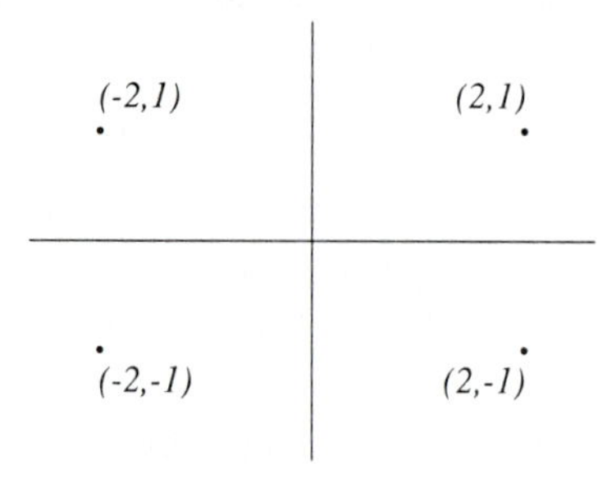

FIGURE 1.4
Four quadrants

Some Regions of the Plane

We consider how some special subsets of the plane can be described by sets of inequalities that restrict the coordinates.

For example, the *upper half plane* is the set of all points (x, y) for which $y \geq 0$. If we exclude the x-axis from this set, it is described by the inequality $y > 0$.

The strip indicated in figure 1.5 by the shaded region is described as the set of all (x, y) such that $-2 \leq x \leq 4$.

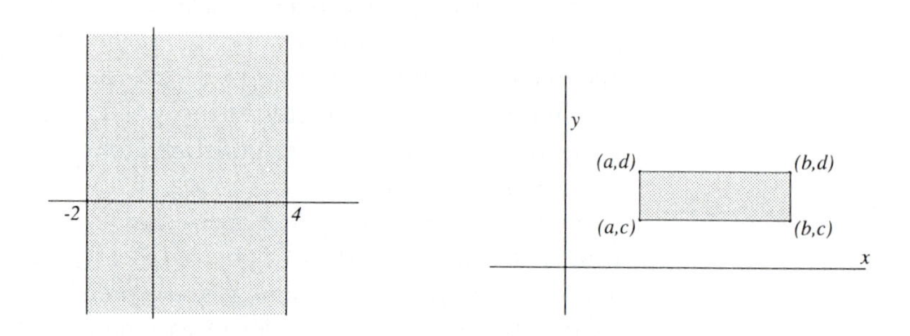

FIGURE 1.5 FIGURE 1.6

If we restrict both coordinates by a set of inequalities of the form

$$a \leq x \leq b \quad \text{and} \quad c \leq y \leq d,$$

the region of points whose coordinates satisfy the inequalities is a rectangle as indicated in figure 1.6.

The Distance Formula

If P has coordinates (a, b) and Q has coordinates (c, d), we can express the distance between P and Q in terms of the coordinates. The line from P to Q is the hypotenuse of the right triangle PQR where R has coordinates (c, b). The length of PR is $|a - c|$ and the length of RQ is $|b - d|$. The Pythagorean Theorem (or the Right Triangle Law) states that

$$PQ^2 = PR^2 + QR^2 = (a-c)^2 + (b-d)^2.$$

Thus the distance D from (a, b) to (c, d) is

$$D = \sqrt{(a-c)^2 + (b-d)^2}.$$

We refer to this formula as the *distance formula.*

EXAMPLE 4 Which of the two points $(1, -2)$ and $(-2, -1)$ is nearer to $(6, 9)$?

Solution: We compute the distances using the distance formula: The distance from $(1, -2)$ to $(6, 9)$ is $\sqrt{(6-1)^2 + (9-(-2))^2} = \sqrt{146}$. The distance from $(-2, -1)$ to $(6, 9)$ is $\sqrt{(6-(-2))^2 + (9-(-1))^2} = \sqrt{164}$. Thus $(1, -2)$ is nearer $(6, 9)$. ■

EXAMPLE 5 What condition must be satisfied by the coordinates of the point (x, y) if its distance from $(2, 3)$ is 10?

Solution: The condition is found from the distance formula; it reads

$$10 = \sqrt{(x-2)^2 + (y-3)^2},$$

or more simply

$$100 = (x-2)^2 + (y-3)^2.$$

Any point (x, y) satisfying this last equation is at a distance 10 from $(2, 3)$. ■

EXAMPLE 6 If x and y are two numbers known to be no more than two units apart, and $|x| \leq 3$, how far apart could $1 - x^2$ be from $1 - y^2$?

Solution: The distance we must estimate is

$$|(1-x^2) - (1-y^2)| = |y^2 - x^2| = |y + x| \cdot |y - x|.$$

We are given that $|y - x| \leq 2$ because x and y are no more than two units apart. Thus it is necessary to estimate the maximum possible value of $|x + y|$. We are given that $|x| \leq 3$, which we write as $-3 \leq x \leq 3$. Because y is no more than two units away from x, we have $x - 2 \leq y \leq x + 2$ and so $2x - 2 \leq y + x \leq 2x + 2$. Taking the extreme values for x gives $x = 3$, $y = 5$ and $x + y = 8$ or $x = -3$, $y = -5$ and $x + y = -8$. Hence we find

$$|(1-x^2) - (1-y^2)| = |y^2 - x^2| = |y + x| \cdot |y - x| \leq 8 \cdot 2 = 16,$$

and the extreme difference is attained with $(x, y) = (3, 5)$ or $(x, y) = (-3, -5)$. ■

Exercises 1.3

Sketch the regions of the plane consisting of all points (x, y) whose coordinates satisfy the following conditions.

1. $1 \leq x \leq 7$, $-1 \leq y \leq 1$

2. $-3 \leq x < 0$, $2 < y < 4$

3. $|x| \leq 1$ and $y > 0$

4. $|x| > 1$ and $y > 0$

5. $|x| > 1$ and $|y| \leq 2$

6. $|x| < 1$ and $|y| < 3$

7. $|x - 1| < 3$ and $y \geq -2$

8. $|x + 12| < 30$ and $y < 5$

9. $|x - y| \leq 1$ and $|x| \leq 1$

10. $|x - y| \leq 1$ and $|y| \leq 1$

11. $x^2 + y^2 \leq 2x$

12. $0 \leq 4y \leq x + 2 \leq 8$

13. $\dfrac{1}{x} < y$

14. $\dfrac{x}{x-3} \geq 0$

15. $\dfrac{2x^2 - 4}{x^2 + 4} \leq 1$

16. $\dfrac{x^2 - 1}{x^2 + 1} > 1$

17. Find all points (x, y) having distance from the origin equal to the distance from $(2, 0)$.

18. Find all points having equal distances from $(1, 1)$

and (3, 1).

19. If x and y are numbers such that $|x - y| \leq 1$ and $|y| \leq 5$, what is the largest possible value of x?

20. If x and y are numbers such that $|x + y| \leq 10$ and $|x| \leq 2$ how large can $x^2 - y^2$ be?

21. If $|u - v| \leq 0.5$ and $|v| < 0.1$, how large can $|\frac{4}{u} - \frac{4}{v}|$ be?

22. If $|a - b| > 1$ and $|b| < 0.1$, what is the least possible value of

$$\sqrt{1 + a^2} + \sqrt{1 + b^2}\,?$$

23. Prove that the inequality $a < u < v < b$ implies $|v - u| < |b - a|$.

1.4 Lines and slopes

Straight lines are described by equations that the coordinates of the points on the line must satisfy. There are several forms that the equation may take depending on the data given that describes the line.

Two-Point Form

Given two points in the plane, there is one and only one straight line passing through the two points. Let P_1 have coordinates (x_1, y_1) and let P_2 have coordinates (x_2, y_2). Give conditions on x and y to ensure that (x, y) is on the line through P_1 and P_2. We first do an easy case, that in which $y_1 = y_2$. Then the two points lie on the vertical line through P_1 so the points on this line are precisely the points (x, y_1), where x can be any real number. The condition can be succinctly described by the equation $y = y_1$, meaning that the y coordinate must equal y_1 and there is no restriction on the x coordinate of the point (x, y).

Now assume that $y_1 \neq y_2$ and that the point P with coordinates (x, y) is on the line through P_1 and P_2. Consider the triangles $P_1P_2Q_1$ and P_1PQ, where $Q_1 = (x_2, y_1)$ and $Q = (x, y_1)$. The two triangles are similar so the ratio of the lengths of the two sides in each right triangle are the same. Thus

$$\frac{y_2 - y_1}{x_2 - x_1} = \frac{y - y_1}{x - x_1}. \qquad \textbf{(1.1)}$$

This is an equation involving x and y and certain constants depending on the two points.

Slope-Intercept Form

If the two points (x_1, y_1) and (x_2, y_2) have different first coordinates, then the line through them is not vertical. The number

$$m = \frac{y_2 - y_1}{x_2 - x_1}$$

is called the *slope of the line.* For a vertical line, the slope is undefined because the definition of m would involve division by zero. For a given nonvertical line, any pair of points on the line can be used to compute the slope. A similar triangle argument will show that the ratio is the same for all pairs on a given line, and thus the slope depends only on the line and not the particular pair of points on the line. We can rewrite the equation of the line given in (1.1) to have the form

$$y = mx + b \tag{1.2}$$

where m is the slope just defined and $b = y_1 - mx_1$. Note that b is the y coordinate of the point where the line crosses the y-axis. This point is called the *y intercept* and the equation (1.2) is called the *slope-intercept form* of the equation of the straight line. Thus the line with equation $y = -2x + 4$ has slope -2 and crosses the y-axis at $(0, 4)$.

Point-Slope Form

The line passing through a point (p, q) and having slope m has equation

$$y = m(x - p) + q = mx - mp + q.$$

The easy way to remember this form is to recognize that when p is substituted for x, the term $(x - p)$ becomes 0 and so the y coordinate of the point on the line is the constant q.

EXAMPLE 1 Find the slope and y intercept of the line through the points $(1, 5)$ and $(-2, 3)$. Write the equation of the line in each of the two-point form, the point-slope form and the slope-intercept form.

Solution: We directly apply the data from the two points to get the two-point form as

$$\frac{y - 5}{x - 1} = \frac{3 - 5}{-2 - 1} = \frac{2}{3}.$$

This is the two-point form and we read the slope as $m = 2/3$. Clear the fractions and get

$$y - 5 = \frac{2}{3}(x - 1) \qquad \text{or} \qquad y = \left(\frac{2}{3}\right) x + \frac{13}{3}.$$

The y intercept is $13/3$ and the slope-intercept form is given by the last equation. To write the equation in the point-slope form, we select any point on the line, say $(1, 5)$. Then $y = (2/3)(x - 1) + 5$ is the point-slope form. If we select a different point, say $(-2, 3)$ the point-slope form is $y = (2/3)(x + 2) + 3$. Written in this form, the two equations appear different. After multiplying across the parenthesis and collecting terms, one finds the equations are exactly the same. ■

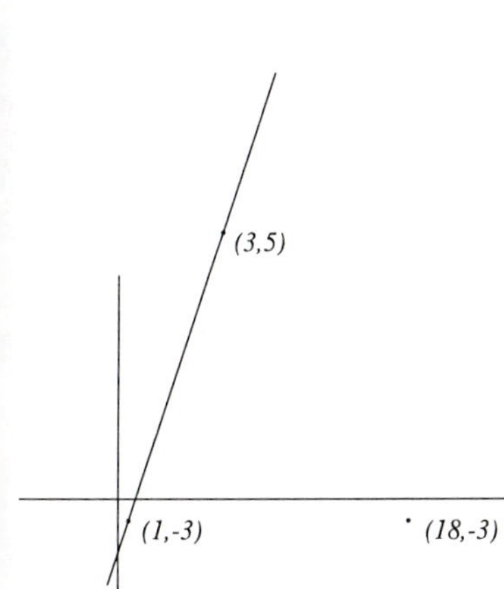

FIGURE 1.7
Two points at distance 17

EXAMPLE 2 Find all points on the line $y = 4x - 7$ that lie at a distance 17 from the point $(18, -3)$.

Solution: A point on the line has coordinates $(x, y) = (x, 4x - 7)$. The distance D from this point to $(18, -3)$ is given by

$$D = \sqrt{(x - 18)^2 + (4x - 7 + 3)^2} = \sqrt{17x^2 - 68x + 340}.$$

If $D = 17$, then after squaring the previous equation we find that x is a solution of

$$17^2 = 17x^2 - 68x + 340 \quad \text{or} \quad 17(x^2 - 4x + 3) = 0.$$ ■

The solutions are $x = 1$ and $x = 3$. The corresponding points on the line are $(1, -3)$ and $(3, 5)$.

Parallel and Perpendicular Lines

Two lines are parallel if and only if they have the same slope. Thus the lines with equations $y = -2x + 5$ and $y = -2x + 13$ are parallel because the slope is the same ($m = -2$ in each case) for both lines. The equations $y = -2x + 5$ and $y = 2x + 5$ represent nonparallel lines because the slopes are different. We know that parallel, but unequal, lines have no points in common whereas nonparallel lines have exactly one point in common. If $y = m_1x + b_1$ and $y = m_2x + b_2$ are equations of two lines then (a, c) will be a point of intersection if we have both

$$a = m_1c + b_1 \qquad \text{and} \qquad a = m_2c + b_2.$$

Subtract the first equation from the second to get $0 = (m_2 - m_1)c + (b_2 - b_1)$. This equation is valid if $m_1 = m_2$ and $b_1 = b_2$, in which case the lines are the same, or if $m_1 \neq m_2$ and $c = (b_1 - b_2)/(m_2 - m_1)$.

By the same sort of computation we may conclude that $m_1 \neq m_2$ implies there is a point satisfying both equations; that is there is a point common to both lines. Hence lines with different slopes have a common point and hence are not parallel.

The slope of a line can be used to test two lines to determine if they are perpendicular. Two lines (neither vertical) are perpendicular if and only if the product of their slopes equals -1. Thus the two lines $y = 2x + 4$ and $y = -(1/2)x - 13$ are perpendicular because the product of their slopes is $(2)(-1/2) = -1$. If m is the slope of a line, then the negative reciprocal $-1/m$ is the slope of a line perpendicular to the given line.

EXAMPLE 3 Find the perpendicular bisector of the line segment having endpoints $(1, 3)$ and $(7, 1)$.

***Solution*:** The slope of the given line segment is

$$m = \frac{3-1}{1-7} = -\frac{1}{3}.$$

Thus the line perpendicular to the segment has slope 3 and must pass through the midpoint of the segment. The point midway between $(1, 3)$ and $(7, 1)$ has coordinates equal to the average of the coordinates of the given points, namely $(4, 2)$. The line we want has slope 3 and passes through $(4, 2)$. The equation that satisfies these conditions is $y = 3(x - 4) + 2$. ■

Exercises 1.4

Find the equation of the straight line satisfying the given conditions.

1. The line passing through $(1, 2)$ and $(7, -2)$.
2. The line passing through $(4, 1)$ and $(-2, -2)$.
3. The line passing through $(-2, -2)$ and perpendicular to the line passing through the origin and $(-4, 2)$.
4. The line passing through $(1, 1)$ and perpendicular to the line passing through $(1, 1)$ and $(3, -5)$.

Find the coordinates of the points described by the following conditions.

5. The points on the line $y = 3x - 1$ and at a distance 2 from the origin.
6. The points on the line $y = x - 1$ and at a distance 3 from $(2, 2)$.
7. The points on the perpendicular bisector of the segment with endpoints $(1, 1)$ and $(11, 7)$ and having distance no more than 2 from the origin.

Two straight lines that have different slopes must intersect at exactly one point. That point is found by solving the equations of the two lines simultaneously. Find the point common to each pair of lines. Sketch the lines and label the common point.

8. $y = 2x - 1$ and $y = x + 4$

9. $2x + y = 3$ and $-2x + 3y = 5$

10. $4x + 3y = 7$ and $3x + 4y = 7$

11. $6x - 9y = 14$ and $11y + 7x = -3$

1.5 Circles and parabolas

In this section we begin an algebraic description of some conic curves and examine graphs associated with them. The four curves, circle, ellipse, parabola and hyperbola, are called *conic curves* because they are the curves traced by intersecting a plane with a (double) cone. Because this description involves a three-dimensional cone, we will not use this interpretation until we take up some three-dimensional topics. Instead, each of the curves will be described as the locus of all points in the plane satisfying some conditions.

Circles

A circle is the set of all points in the plane that lie at a fixed distance from a given point, the center. This verbal definition is translated into an algebraic equation as follows.

Let the center be (h, k) and the radius (the fixed distance from the center) be r ($r \geq 0$). Then the point (x, y) lies on the circle with center (h, k) and radius r if and only if

$$\sqrt{(x-h)^2 + (y-k)^2} = r.$$

This is equivalent to the equation

$$(x-h)^2 + (y-k)^2 = r^2 \qquad \textbf{(1.3)}$$

which we call the *general equation of the circle.*

The set of all points (x, y) satisfying equation (1.3) is called the *graph* of the equation.

EXAMPLE 1 Sketch the graphs of the equations

1. $x^2 + y^2 = 4,$
2. $(x-2)^2 + (y+3)^2 = 36.$

The graph of equation (1) is a circle with center $(0, 0)$ and radius 2. The graph of equation (2) is a circle with center $(2, -3)$ and radius 6.

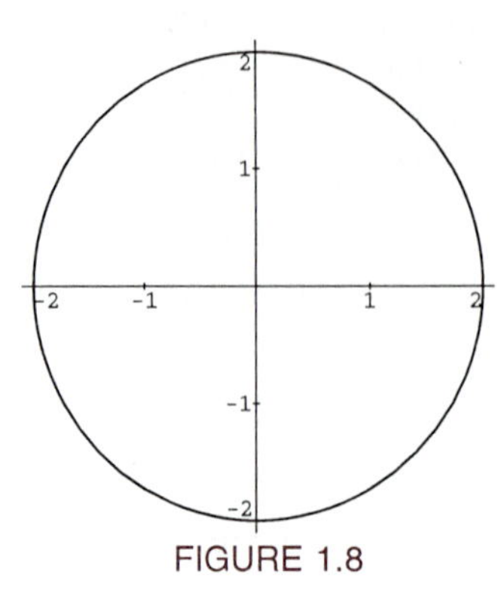

FIGURE 1.8

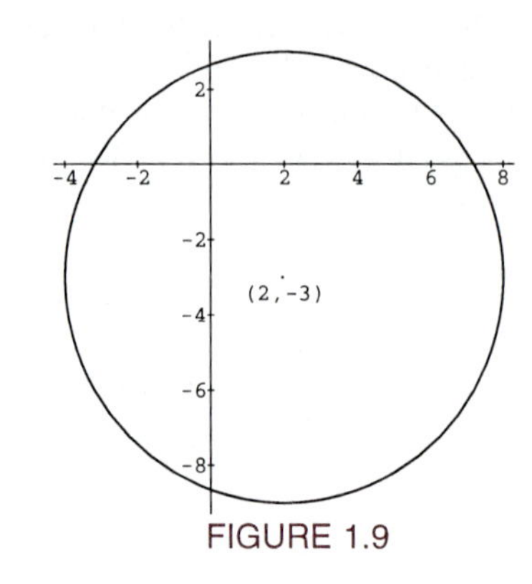

FIGURE 1.9

An equation of the form

$$x^2 + ax + y^2 + by = c \tag{1.4}$$

might be the equation of a circle. To find out if it is, we complete the squares; that is, we add constant terms to both sides of the equation to obtain a sum of perfect squares on the left side:

$$\begin{aligned} x^2 + ax + \frac{a^2}{4} + y^2 + by + \frac{b^2}{4} &= c + \frac{a^2}{4} + \frac{b^2}{4} \\ \left(x + \frac{a}{2}\right)^2 + \left(y + \frac{b}{2}\right)^2 &= c + \frac{a^2}{4} + \frac{b^2}{4} \end{aligned} \tag{1.5}$$

If the term on the right is nonnegative, then there is a real number r such that

$$r^2 = c + \frac{a^2}{4} + \frac{b^2}{4};$$

the equation represents a circle with center $(-a/2, -b/2)$ and radius r.

If $c + \frac{a^2}{4} + \frac{b^2}{4} < 0$, then there are no real numbers x, y that satisfy equation (1.5) because the left side is never negative.

EXAMPLE 2 Describe the graphs of the three equations

1. $x^2 + 6x + y^2 - 4y = 12$,
2. $x^2 + 6x + y^2 - 4y = -13$,
3. $x^2 + 6x + y^2 - 4y = -16$.

Solution: We complete the square by adding the square of half the coefficient of x and the square of half the coefficient of y to both sides; do equation (1) first to get

$$\begin{aligned} x^2 + 6x + 9 + y^2 - 4y + 4 &= 12 + 9 + 4 \\ (x + 3)^2 + (y - 2)^2 &= 25 = 5^2. \end{aligned}$$

Thus equation (1) is the equation of a circle with center $(-3, 2)$ and radius 5.

The same procedure applied to equation (2) yields the equivalent equation

$$(x + 3)^2 + (y - 2)^2 = -13 + 9 + 4 = 0.$$

Thus the graph of equation (2) consists of a single point $(-3, 2)$.

Repeating this procedure one more time with equation (3) transforms it into the equivalent equation

$$(x + 3)^2 + (y - 2)^2 = -16 + 9 + 4 = -3.$$

There are no real numbers that satisfy this equation and so the graph of the equation contains no points. ■

EXAMPLE 3 Sketch the graph of the equation $y = 2 + \sqrt{-5 - 6x - x^2}$.

Solution: At first, the sight of all the negative signs under the radical might be a bit disturbing because no negative number has a real square root. However, for certain values of x, the expression $-5 - 6x - x^2$ is nonnegative. To make this more clear, we complete the square:

$$-5 - 6x - x^2 = -[x^2 + 6x + 5] = -[(x^2 + 6x + 9) - 9 + 5] = -[(x + 3)^2 - 4].$$

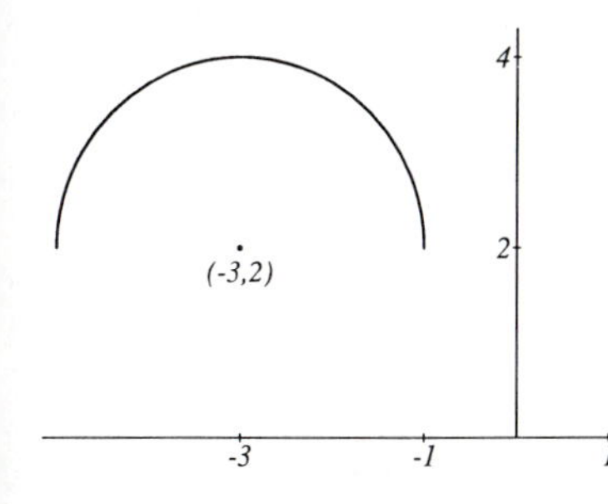

FIGURE 1.10
$y = 2 + \sqrt{-5 - 6x - x^2}$

The equation to be graphed takes the form $y = 2 + \sqrt{4 - (x + 3)^2}$. We eliminate the square root symbol by isolating it on one side of the equation and then squaring both sides. This transforms the original equation into

$$(y - 2)^2 = 4 - (x + 3)^2 \quad \text{or} \quad (y - 2)^2 + (x + 3)^2 = 4.$$

This is the equation of a circle with center $(-3, 2)$ and radius 2. There is a trap here that should be recognized. When we squared both sides of the equation, we enlarged the set of points that satisfied the equation. In the original equation, only the nonnegative square root is used, and so $y \geq 2$ for every point (x, y) on the graph of the equation. It follows that the graph of the original equation is only the upper half of the circle. ■

The Parabola

The *parabola* is the set of all points whose distance from a given point, the *focus*, equals the distance from a given line, the *directrix*. It is assumed that the focus does not lie on the directrix.

We will not derive the algebraic equation for the most general parabola because the algebra is rather complicated. Instead we consider only the case where the directrix is parallel to the x-axis. The case with the directrix parallel to the y-axis is treated in some examples. The most general case of a directrix having arbitrary slope can be treated following a discussion of rotation of axes.

Suppose the directrix is a line $y = q$ and the focus is the point $F(s, p)$ for some constants q, s and p. A point $P(x, y)$ is equally distant from the focus and the directrix if $|PF| = |y - q|$; that is

$$\sqrt{(x - s)^2 + (y - p)^2} = |y - q|.$$

Square both sides and expand to obtain the equivalent equation

$$(x - s)^2 + y^2 - 2yp + p^2 = y^2 - 2yq + q^2.$$

This reduces to the equation

$$2(p - q)y = (x - s)^2 + p^2 - q^2. \tag{1.6}$$

The condition that the focus does not lie on the directrix is equivalent to the condition $p \neq q$. Both sides of the equation can be divided by the coefficient of y to obtain an equation in the form

$$y = \frac{(x - s)^2 + p^2 - q^2}{2(p - q)} = Ax^2 + Bx + C$$

for constants A, B and C with $A \neq 0$. Conversely, every equation of the form $y = Ax^2 + Bx + C$ with $A \neq 0$ is the equation of a parabola. The focus and directrix can be determined in terms of A, B and C. We state the result in a theorem.

THEOREM 1.5

If A, B and C are constants with $A \neq 0$ then the graph of the equation $y = Ax^2 + Bx + C$ is a parabola with focus (s, p) and directrix $y = q$ where

$$s = -\frac{B}{2A} \qquad p = \frac{4AC - B^2 + 1}{4A} \qquad q = \frac{4AC - B^2 - 1}{4A}. \tag{1.7}$$

Proof

Start with the equation $y = Ax^2 + Bx + C$, divide by A, and complete the square to get the equivalent equation

$$\frac{1}{A}y = \left(x + \frac{B}{2A}\right)^2 + \frac{4AC - B^2}{4A^2}.$$

Because we expect to produce constants s, p, and q as in equation (1.6), we match the coefficients in the two equations and set

$$s = -\frac{B}{2A}, \quad 2(p - q) = \frac{1}{A}, \quad p^2 - q^2 = \frac{4AC - B^2}{4A^2}.$$

Because $p^2 - q^2 = (p + q)(p - q) = (p + q)/(2A)$, it is not too difficult to solve for p and q to get the values as stated in the theorem. So we now have produced some constants s, p and q in terms of A, B and C. Now write the equation of the parabola that has focus (s, p) and directrix $y = q$ using the values in terms of A, B, C; the equation turns out to be exactly $y = Ax^2 + Bx + C$. □

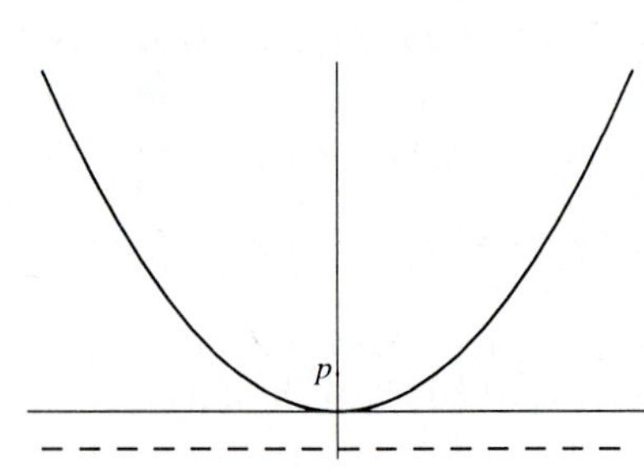

FIGURE 1.11
$4py = x^2$

In the symmetric case with $q = -p$ and $s = 0$, equation (1.6) reduces to

$$4py = x^2.$$

The line through the focus and perpendicular to the directrix is called the *axis* of the parabola. The parabola is symmetric around this line. For the general equation of a parabola given just above, the axis is the line $x = s = -B/2A$.

The point on the parabola nearest the directrix is called the *vertex* of the parabola. The vertex is point of intersection of the graph and the axis of the parabola. If the focus is (s, p) and the directrix is $y = q$ then the vertex has coordinates $(s, (p + q)/2)$. Using the constants in Theorem 1.5, we find the vertex of the parabola having equation $y = Ax^2 + Bx + C$ is

$$\left(\frac{-B}{2A}, \frac{4AC - B^2}{4A}\right).$$

EXAMPLE 4 Sketch the graph of the equations and for each, give the focus, directrix, vertex and axis:

$$y = -2x^2 \qquad y = -2x^2 + 6x.$$

Solution: The focus is at the point (s, p) with the values of s and p computed from the coefficients of the equation of the graph using Theorem 1.5. In case $y = -2x^2 = Ax^2 + Bx + C$ we have from equation (1.7) that $s = 0$ and $p = -1/8$; the focus is $(0, -1/8)$. The directrix is the line $y = q$ where we find $q = 1/8$. The vertex is at the origin as the formula just above indicates.

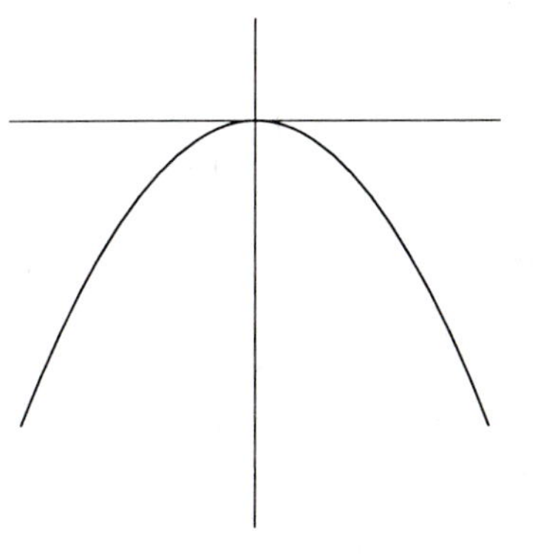

FIGURE 1.12
$y = -2x^2$

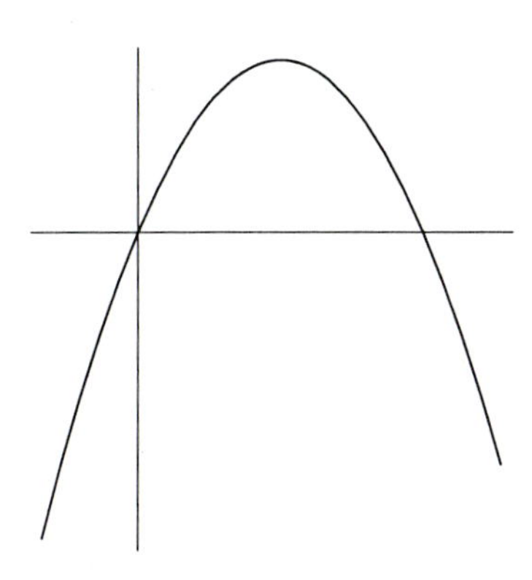

FIGURE 1.13
$y = -2x^2 + 6x$

For the case $y = -2x^2 + 6x = Ax^2 + Bx + C$ we find from equation (1.7) that $(s, p) = (3/2, 35/8)$ and the directrix is $y = q = 37/8$. By inspection of the graph, or by the formula just above, the vertex is found to be $(3/2, 9/2)$. ■

Suppose we now consider a parabola having focus (p, s) and directrix $x = q$. Then we have exactly the previous situation except that the roles of x and y have been reversed. The equation of the parabola must be an equation of the form $x = Ay^2 + By + C$ for some constants $A \neq 0$, B and C. One could work through the relation between A, B, C and p, q, s from the beginning or argue by analogy with the case already completed and find that equations (1.5) give the values.

EXAMPLE 5 Sketch the graph of the equation $x = 2y^2$.

Solution: We note that the graph is a parabola; the values of x are always nonnegative and the parabola passes through the origin. This is enough information to obtain the rough sketch indicated in figure 1.14. ■

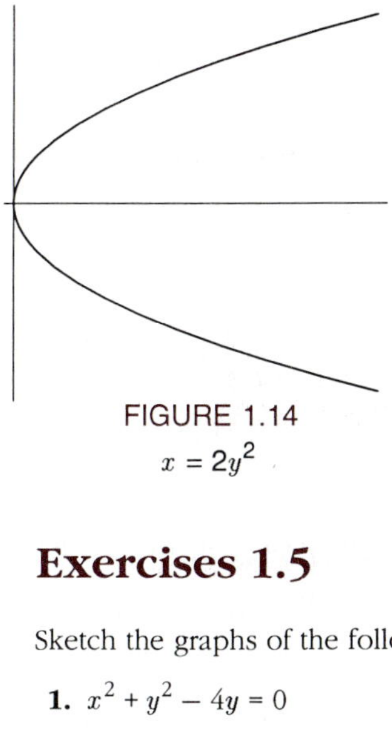

FIGURE 1.14
$x = 2y^2$

Exercises 1.5

Sketch the graphs of the following equations.

1. $x^2 + y^2 - 4y = 0$

2. $4x^2 + 4y^2 = x + y$

3. $12x^2 + 8x = 6y - 12y^2$

4. $x^2 + 6x + y^2 + 25 = 10y$

5. Determine if the origin $(0, 0)$ lies inside or outside of the circles in the previous four exercises.

6. Given the equation of a circle in the form $(x-h)^2+(y-k)^2 = r^2$ and given a point (p, q) describe a foolproof test by which one can quickly determine if the point is inside or outside of the circle.

Sketch the graph of the following equations. Each is a parabola; label the vertex and focus and sketch the axis of symmetry.

7. $y = 4x^2$

8. $y = -2x^2$

9. $y = 8x^2 - 2x$

10. $y = 8x^2 + 2x$

11. $y = -2x^2 + x + 1$

12. $y^2 = x$

13. $y^2 + 4y = x$

14. $x^2 - 8x + y = 0$

Sketch the graphs of the following equations.

15. $y = \sqrt{16 - x^2}$

16. $y = \sqrt{1 + 6x - x^2}$

17. $y = 1 + 6x - x^2$

18. $y = |x|(x + 1)$

19. Let C be a circle of radius r and center at $(0, a)$ on the

y-axis; let P be the parabola with equation $y = x^2$. How many points are common to C and P? The answer will depend on a and r, of course. Consider the problem geometrically and algebraically.

20. Give a geometric reason why there are at most four distinct points (x, y) that satisfy two equations

$$x^2 + ax + y^2 + by = c \quad \text{and} \quad y^2 = px + q$$

for constants a, b, c, p and q.

21. Let A successively take the values $1/2$, $1/4$, $1/8$, $\ldots, 1/2^n, \ldots$. Indicate on a single coordinate axis the shapes of the various graphs of the equation $y = Ax^2 + 2x - 1$ and make a guess at the limiting position of the graphs.

1.6 The ellipse and hyperbola

In this section we consider the remaining two conic curves, the ellipse and the hyperbola. They are defined by using distances from two reference points called the foci.

The Ellipse

An *ellipse* is the set of all points in the plane with the property that the sum of the distances to two fixed points is constant.

Here is a construction that can be used to sketch an ellipse. Take a long string with its two ends fastened at the foci; with a pencil restrained by the string, trace the curve so the pencil keeps the string taut. The resulting curve is, in theory at least, an ellipse because the sum of the distances from the pencil mark to the two foci equals the length of the string.

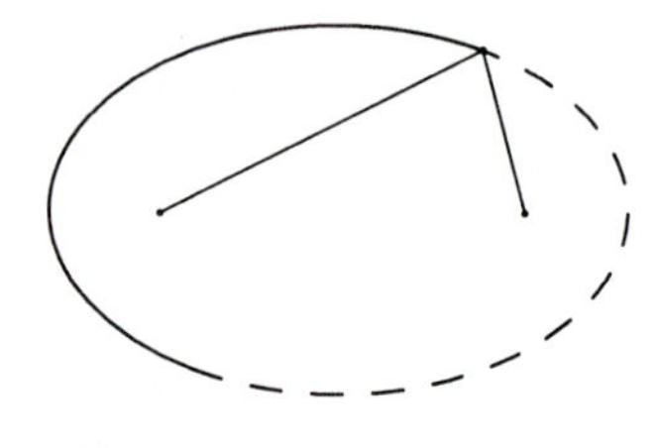

FIGURE 1.15
Drawing an ellipse

We will derive an algebraic equation for the coordinates of the points on the ellipse in the special case that the two foci are symmetrically placed on one of the coordinate axes. To specify a particular ellipse, the two foci must be given and the constant distance must be given. If $2a$ denotes the constant distance then there will be no points on the ellipse unless $2a$ is greater than the distance between the foci. (The constant is written as $2a$ instead of a to simplify the final equation.)

Let (x, y) be a point on the ellipse having foci $(-p, 0)$ and $(p, 0)$, and let the constant distance $2a$ be a number greater than $2p$. Then we have

$$\sqrt{(x + p)^2 + y^2} + \sqrt{(x - p)^2 + y^2} = 2a.$$

Transpose one of the square root terms to the right side of the equation and then square both sides to get

$$(x + p)^2 + y^2 = 4a^2 - 4a\sqrt{(x - p)^2 + y^2} + (x - p)^2 + y^2.$$

Rearrange the terms so that the radical stands alone on one side of the equal sign, collect terms and square again to get the equation

$$\frac{x^2}{a^2} + \frac{y^2}{a^2 - p^2} = 1.$$

Because we assumed $a > p$ at the start, we know there is a real number b such that $b^2 = a^2 - p^2$. Thus the equation satisfied by all points of the ellipse having foci at $(\pm p, 0)$ and constant distance $2a$ is

$$\frac{x^2}{a^2} + \frac{y^2}{b^2} = 1, \tag{1.8}$$

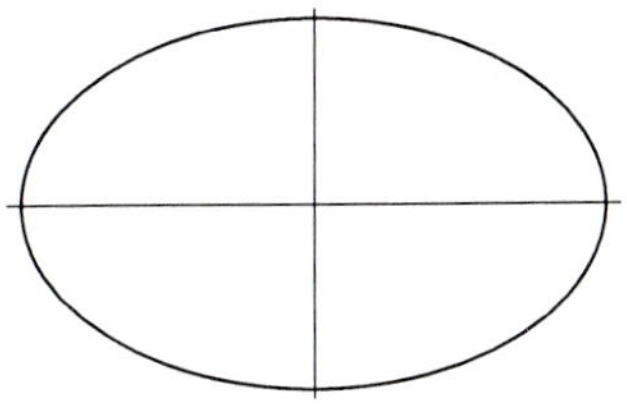

where $a^2 = b^2 + p^2$.

Notice that we must have $a^2 > b^2$ in this case (with the foci on the x-axis). Suppose we change the problem and look for the equation of an ellipse having foci at $(0, -p)$ and $(0, p)$ and constant distance $2a$. Then this is the same problem as the one just completed except the roles of x and y have been exchanged. In this case

FIGURE 1.16
$\frac{x^2}{a^2} + \frac{y^2}{b^2} = 1$

$$\frac{x^2}{b^2} + \frac{y^2}{a^2} = 1$$

is the equation of the ellipse having foci at $(0, \pm p)$ and constant distance $2a$ with $a^2 = b^2 + p^2$. Notice that the equation of the ellipse having foci on the y-axis has $a^2 \geq b^2$. The foci are on the "longer" axis of the ellipse. The longer axis is called the *major axis* and the shorter axis is called the *minor axis.*

EXAMPLE 1 Sketch the graph of the equation $y^2 = 16(36 - x^2)$; give the foci and the length of the major and minor axes.

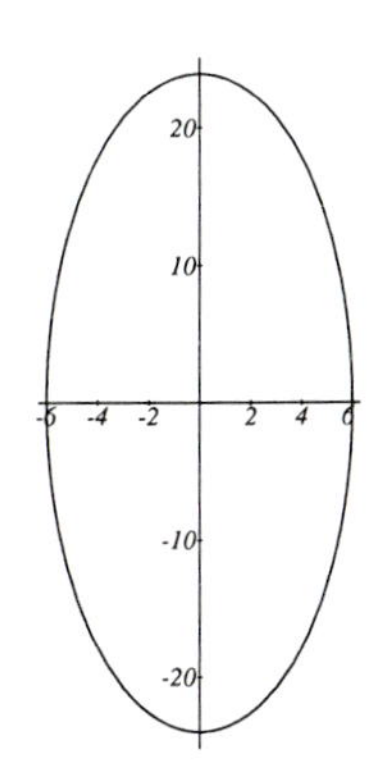

FIGURE 1.17
$(\frac{y}{24})^2 + (\frac{x}{6})^2 = 1$

Solution: Rewrite the equation in the form

$$\left(\frac{y}{6 \cdot 4}\right)^2 + \frac{x^2}{6^2} = 1.$$

This is the equation of an ellipse with major axis of length 24 along the y axis and minor axis of length 6 along the x-axis. The foci lie on the major axis at $(0, \pm p)$ where p is determined from the equation $a^2 = b^2 + p^2$. In this example, we have $24^2 = 6^2 + p^2$ or $p = \pm 6\sqrt{15}$. The graph of the equation is shown in figure 1.17.

Symmetry

The equation of the ellipse makes it obvious that the graph has symmetry in the sense that if (x, y) is a point on the ellipse, then the points $(-x, y)$, $(x, -y)$ and $(-x, -y)$ also lie on the ellipse.

We say the graph of an equation is *symmetric* about the y-axis if whenever (x, y) is on the graph, then $(-x, y)$ is also on the graph. The graph is symmetric about the x-axis if whenever (x, y) is on the graph, then $(x, -y)$ is also on the graph.

Thus the ellipse is symmetric about the x-axis and about the y-axis.

EXAMPLE 2 Sketch the graph of the equation $y = 3\sqrt{16 - x^2}$.

Solution: Starting from the equation $y = 3\sqrt{16 - x^2}$, eliminate the radical to get

$$\left(\frac{y}{3}\right)^2 = 4^2 - x^2.$$

This shows that every point of the graph is a point of the ellipse

$$\frac{y^2}{(3 \cdot 4)^2} + \frac{x^2}{4^2} = 1.$$

However not every point of the ellipse is on the graph of the function. The formula $y = 3\sqrt{16 - x^2}$ shows that y is never negative. Hence the graph of $f(x)$ contains only those points on the ellipse that have nonnegative y coordinate; namely, the upper half of the ellipse.

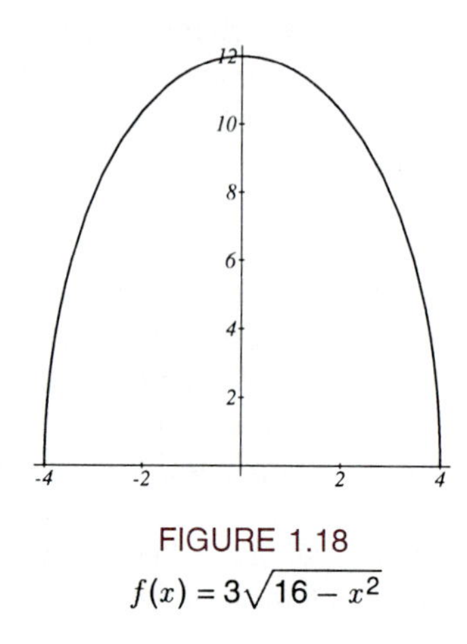

FIGURE 1.18
$f(x) = 3\sqrt{16 - x^2}$

■

Hyperbolas

The *hyperbola* is the set of all points whose distances to two fixed points have a constant difference.

The two fixed points are called the foci of the hyperbola. Suppose we consider the case with the foci given as $F_1(-p, 0)$ and $F_2(p, 0)$ and the constant difference is some number $2a$. Then $P(x, y)$ is on the hyperbola if either $|PF_1| - |PF_2| = 2a$ or $|PF_2| - |PF_1| = 2a$. These lengths involve radicals that are eliminated by squaring; because one difference is the negative of the other, the squaring transforms the two equations into one. We have

$$\sqrt{(x+p)^2 + y^2} - \sqrt{(x-p)^2 + y} = 2a.$$

The radicals can be eliminated by first transposing one radical to the right side, squaring, isolating the remaining radical and squaring again. When the terms are collected, we obtain the equation

$$\frac{x^2}{a^2} - \frac{y^2}{p^2 - a^2} = 1.$$

The constant difference $2a$ is nonzero so there are no points with x coordinate 0 on the curve; it follows that $p^2 - a^2 > 0$. In particular then there is a real number b such that $b^2 = p^2 - a^2$. To summarize this computation we find

$$\frac{x^2}{a^2} - \frac{y^2}{b^2} = 1 \tag{1.9}$$

is the equation of the hyperbola with foci at $(\pm p, 0)$, constant difference $2a$ and with $b^2 = p^2 - a^2$.

Some general observations about the graph of this equation can be made at once. If (x, y) is a point on the hyperbola, then

$$\frac{x^2}{a^2} = \frac{y^2}{b^2} + 1 \geq 1$$

and so $x^2 \geq a^2$. It is easy to see that for any number x satisfying this inequality, there are points on the graph having x as the first coordinate. As $|x|$ increases, so does $|y|$ so that points on the graph get farther and farther away from the origin. There is also symmetry of the graph; if (x, y) is a point on the graph, then the points $(\pm x, \pm y)$ lie on the graph for any choice of signs. Thus this hyperbola is symmetric about each of the coordinate axes.

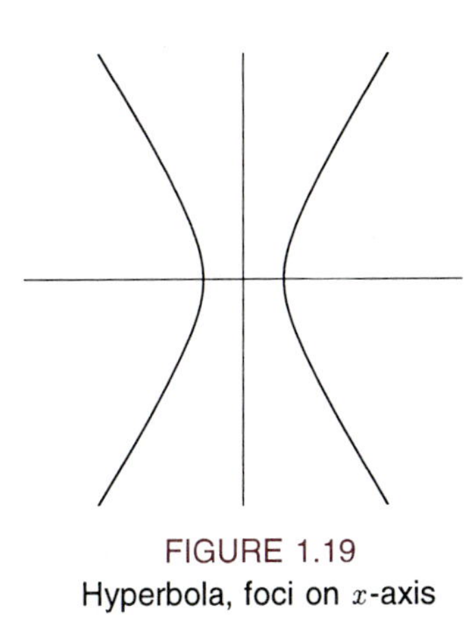

FIGURE 1.19
Hyperbola, foci on x-axis

Branches and Asymptotes

The graph of the hyperbola is not connected; that is it may be necessary to move off the curve when trying to draw a path from one point on the curve to another point on the curve. We refer to the two connected parts of the hyperbola as branches. If we walk along one branch of the graph of the hyperbola given by equation (1.9), the path approaches a straight line called an *asymptote.* Let us prove that this is the case and determine the equation of the lines in question. We can see from the symmetry of the curve that if there are asymptotes, then there must be two of them, one passing through quadrants I and III, the other through quadrants II and IV.

Write equation (1.9) in the form

$$\left(\frac{x}{a} - \frac{y}{b}\right)\left(\frac{x}{a} + \frac{y}{b}\right) = 1$$

and so

$$\left(\frac{x}{a} - \frac{y}{b}\right) = \frac{1}{\left(\frac{x}{a} + \frac{y}{b}\right)}.$$

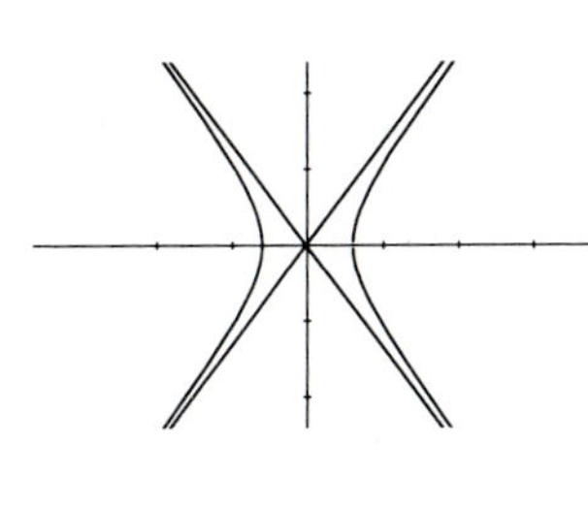

FIGURE 1.20
Hyperbola and asymptotes

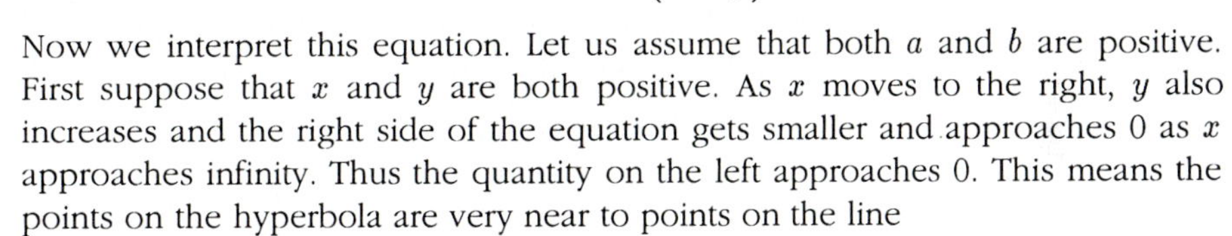

Now we interpret this equation. Let us assume that both a and b are positive. First suppose that x and y are both positive. As x moves to the right, y also increases and the right side of the equation gets smaller and approaches 0 as x approaches infinity. Thus the quantity on the left approaches 0. This means the points on the hyperbola are very near to points on the line

$$\left(\frac{x}{a} - \frac{y}{b}\right) = 0 \quad \text{or} \quad y = \frac{b}{a}x.$$

The same analysis applies if both x and y are negative. When x and y have opposite signs, we consider the equation of the hyperbola rewritten in the form

$$\left(\frac{x}{a} + \frac{y}{b}\right) = \frac{1}{\left(\frac{x}{a} - \frac{y}{b}\right)}$$

and find that the points on the curve are very near to the points on the line $y = -(b/a)x$. Thus we conclude that the two lines $y = \pm(b/a)x$ are the asymptotes to the hyperbola having equation (1.9). The equations of the asymptotes are easily remembered by replacing the "1" in equation (1.9) with "0" to get

$$\frac{x^2}{a^2} - \frac{y^2}{b^2} = \left(\frac{x}{a} - \frac{y}{b}\right)\left(\frac{x}{a} + \frac{y}{b}\right) = 0.$$

The graph of this equation consists of the points on the two lines $y = \pm(b/a)x$.

If the two foci lie on the y-axis rather than the x-axis, the equation of the hyperbola will have the same form as equation (1.9) except that the roles of x

and y will be interchanged. One can verify that

$$\frac{y^2}{a^2} - \frac{x^2}{b^2} = 1 \qquad \textbf{(1.10)}$$

is the equation of the hyperbola having foci $(0, \pm p)$ where $b^2 = p^2 - a^2$ and constant difference $\pm 2a$. The asymptotes of this hyperbola are the two lines $y = \pm(a/b)x$.

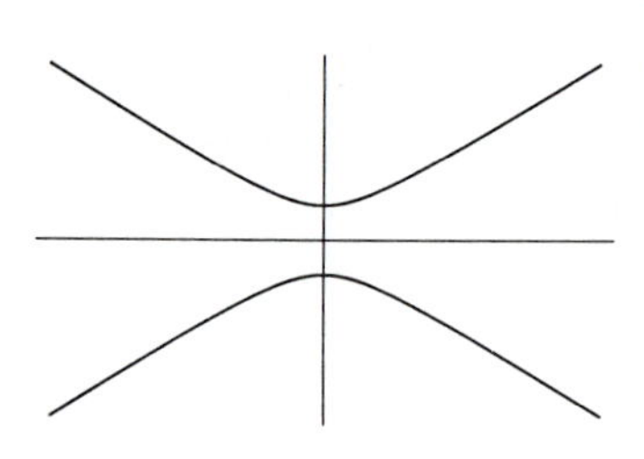

FIGURE 1.21
Hyperbola, foci on y-axis

Translation of Coordinates

We can enlarge the scope of the previous discussion by considering graphs of equations in a translated coordinate system. Here is what we mean by this.

In the standard xy coordinate system, the origin is the center of the coordinate system. A point (a, b) is specified by giving its distance, a, from the vertical line through the origin and its distance b from the horizontal line through the origin; one then assigns the point coordinates (a, b). Suppose (for whatever reason) we assign new coordinates to a point by measuring the horizontal and vertical distance from some point other than the origin. For a specific example, let us use the point $(2, 5)$ as our new "center". For a given point, let u be the horizontal distance from the vertical line through $(2, 5)$ to the point and let v the vertical distance from the horizontal line through $(2, 5)$ to the point. We assign the new coordinates (u, v) to the point. We can form the image of two coordinate systems, one superimposed upon the other. There is the usual xy coordinate system with center $O(0, 0)$ and the new uv coordinate system; the u-axis is the horizontal line through $(2, 5)$ and the v-axis is the vertical line through $(2, 5)$. If a point P has xy coordinates (a, b) and uv coordinates (p, q), then there is a simple relation between these numbers; namely $a = p + 2$ and $b = q + 5$. More simply we can just write

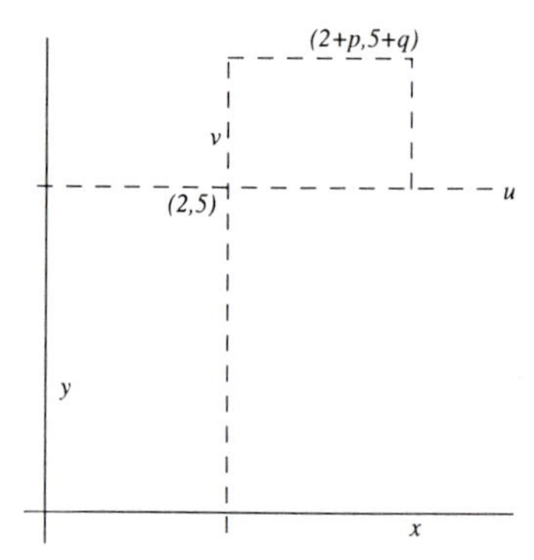

FIGURE 1.22
xy coordinates of point

$$x = u + 2$$
$$y = v + 5.$$

This change in the "names" of the points is useful in the analysis of the graph of some equations.

Suppose we consider the set of all points whose uv coordinates satisfy the equation

$$u^2 + 4v^2 = 16. \qquad \textbf{(1.11)}$$

In the uv system, this set of points is an ellipse crossing the u-axis at ± 4 and crossing the v-axis at ± 2.

To describe this set of points in terms of the original xy coordinates, we simply rewrite equation (1.11) in terms of x and y. Namely, we substitute $x - 2 = u$ and $y - 5 = v$ to get

$$(x - 2)^2 + 4(y - 5)^2 = 16. \qquad \textbf{(1.12)}$$

This is the equation of the ellipse with center at $(2, 5)$. Thus the set of points whose xy coordinates satisfy equation (1.12) is precisely the same as the set of points whose uv coordinates satisfy equation (1.11).

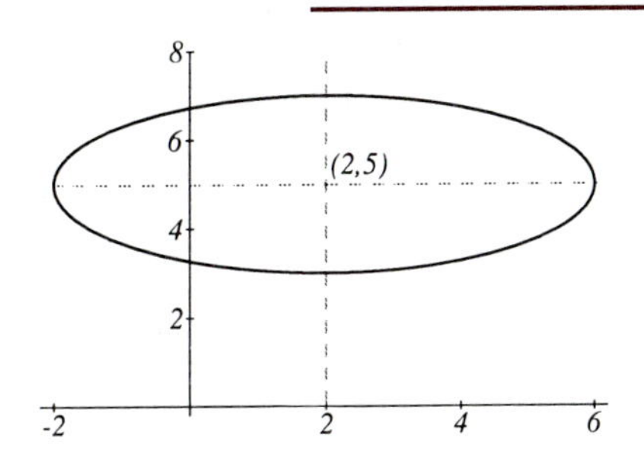

FIGURE 1.23

More generally, an equation of the form

$$\frac{(x-h)^2}{a^2} + \frac{(y-k)^2}{b^2} = 1 \tag{1.13}$$

is the equation of an ellipse with center at (h, k). If we consider the new coordinates as

$$u = x - h$$
$$v = y - k$$

then the u-axis is the horizontal line through (h, k) and the v-axis is the vertical line through (h, k). In this uv coordinate system, equation (1.13) takes on the standard form:

$$\frac{u^2}{a^2} + \frac{v^2}{b^2} = 1.$$

EXAMPLE 3 Describe the graph of the equation $x^2 - 4x + 4y^2 + 24y = 60$.

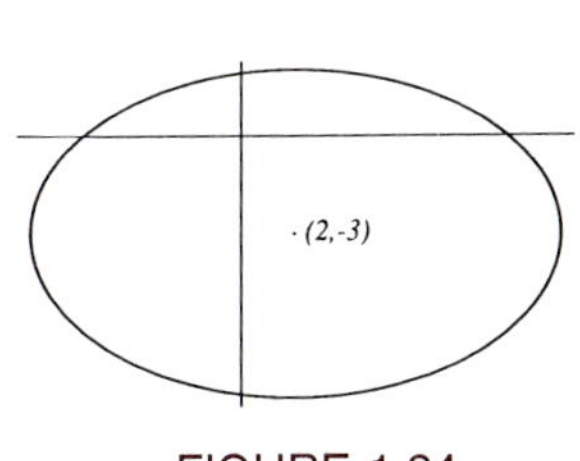

FIGURE 1.24
$x^2 - 4x + 4y^2 + 24y = 60$

Solution: We complete the square on the x and y terms to obtain

$$(x^2 - 4x + 4) + 4(y^2 + 6y + 9) = 60 + 4 + 4 \cdot 9 = 100.$$

Divide this by 100 to get the standard form

$$\frac{(x-2)^2}{10^2} + \frac{(y+3)^2}{5^2} = 1.$$

This is an ellipse with center $(2, -3)$; the graph appears in figure 1.24. ■

EXAMPLE 4 Apply the idea of a change of center to describe the graph of the equation $x^2 - 4x - 4y^2 - 24y = 60$.

Solution: The only difference between this equation and the one in Example 3 is some sign changes. We still begin by completing the square to obtain

$$(x^2 - 4x + 4) - 4(y^2 + 6y + 9) = 60 + 4 - 4 \cdot 9 = 28,$$

and

$$\frac{(x-2)^2}{28} - \frac{(y+3)^2}{7} = 1.$$

If we let $u = x - 2$ and $v = y + 3$, then this is the equation of a hyperbola,

$$\frac{u^2}{28} - \frac{v^2}{7} = 1,$$

which can be sketched using the new coordinate system having center $(2, -3)$.

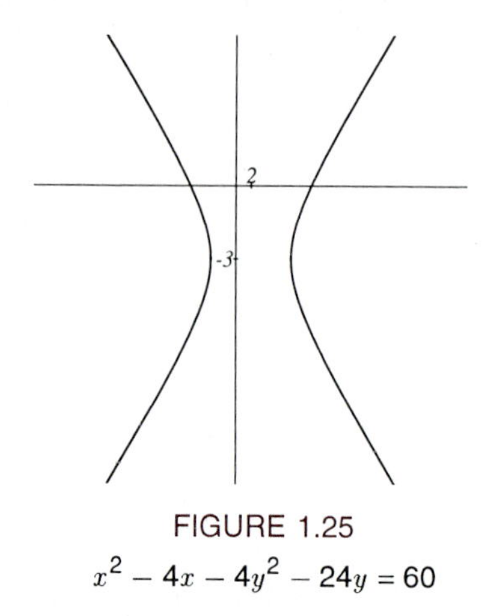

FIGURE 1.25
$x^2 - 4x - 4y^2 - 24y = 60$

■

Exercises 1.6

Sketch the graph of the following equations. Label the points where the graph crosses an axis. If the graph is a hyperbola, sketch the asymptotes.

1. $16x^2 + 64y^2 = 64$

2. $x^2 + 4y^2 = 1$

3. $3x^2 + 12y^2 = 75$

4. $12x^2 + 3y^2 = 75$

5. $3x^2 - 12y^2 = 75$

6. $12x^2 - 3y^2 = 75$

7. $x^2 - 4y^2 = 1$

8. $4x^2 - y^2 = 1$

Use the method of change of center and completing the square to sketch the graphs of the following equations.

9. $x^2 - 2x + 3y^2 - 6y = 32$

10. $4x^2 - 24x + 5y^2 + 5y = 0$

11. $5x^2 - 4y^2 - 8y = 32$

12. $4y^2 - 3x^2 + 12(x + y) - 12 = 0$

13. $4x^2 - 24x + 4y^2 - 14y = 0$

14. $4x^2 - 24x - 4y^2 - 14y = 0$

15. For each ellipse in the first eight exercises, determine if the point $(0.8, 0.3)$ is inside or outside of the ellipse.

16. Give conditions on the numbers p, q that provide a foolproof test to determine if the point (p, q) is inside or outside the ellipse with equation

$$\frac{(x-h)^2}{a^2} + \frac{(y-k)^2}{b^2} = 1.$$

17. Consider the hyperbola with equation $b^2x^2 - a^2y^2 = a^2b^2$ and the line $y = mx$ through the origin with slope m. Find the values of m with the property that the line has some points in common with the hyperbola. Relate these values of m to the slopes of the asymptotes of the hyperbola.

For a point (p, q) on the hyperbola with equation $b^2x^2 - a^2y^2 = a^2b^2$ we will prove later that the line tangent to the hyperbola at (p, q) has slope b^2p/a^2q if $q \neq 0$. Use this information to solve the following problems.

18. Find the equation of the line tangent to the graph of $x^2 - y^2 = 1$ at the point $(2, \sqrt{3})$ and then find where the line crosses the coordinate axes. Make a sketch of the hyperbola and the line.

19. Find the equation of the line tangent to the graph of $x^2/4 - y^2/9 = 1$ at the point $(4, 3\sqrt{3})$ and then find where the line crosses the coordinate axes. Make a sketch of the hyperbola and the line.

20. A tangent line to the graph of $x^2 - 4y^2 = 16$ crosses the x-axis at $(12, 0)$. Find the point on the graph through which the line passes.

1.7 Polar coordinates

There are ways to describe the position of points in the plane other than the rectangular system of xy coordinates. The system of *polar coordinates* is used to locate points. Let P be a point with rectangular coordinates (x, y). The distance from the P to the origin is denoted by r, and we have $r = \sqrt{x^2 + y^2}$. If P is not the origin, the angle from the positive x-axis to the line OP is denoted by θ. Note that r is always nonnegative and we could restrict θ to the interval $0 \leq \theta < 2\pi$ because every point can be located with such r and θ. However, in general, we do not insist that θ be so restricted.

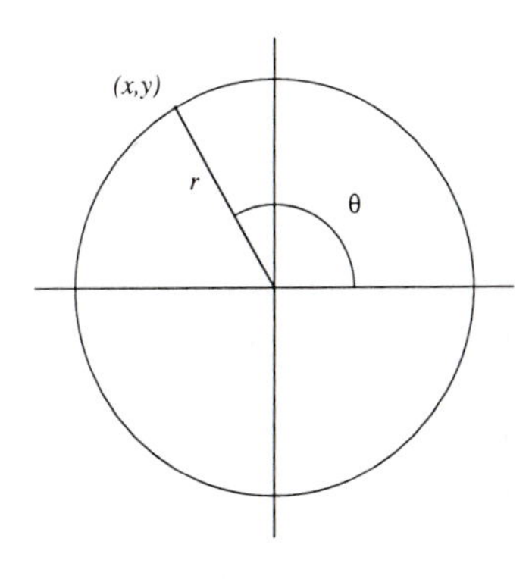

FIGURE 1.26
Polar coordinates

The polar coordinates of a point are denoted by (r, θ) and are related to the rectangular coordinates by the rules

$$x = r\cos\theta \quad \text{and} \quad y = r\sin\theta. \qquad \textbf{(1.14)}$$

Unlike the rectangular coordinate system, the polar coordinate system permits the same point to have more than one set of polar cordinates. For example, the points with polar coordinates $(1, \pi/2)$ and $(1, \pi/2 + 2\pi)$ are exactly the same, namely the point with rectangular coordinates $(1, 1)$. More generally, (r, θ) and $(r, \theta + 2\pi)$ are the polar coordinates of exactly the same point.

The polar coordinates of the origin show an even more degenerate behavior. The distance r from the the origin to itself is $r = 0$; the angle θ is not defined by the procedure mentioned above, but clearly for any θ the point with polar coordinates, $(0, \theta)$ is the origin.

Curves in Polar Coordinates

We use polar coordinates to describe a few curves. The simplest is the equation of a circle with center at the origin that has equation $r = a$. The set of points having polar coordinates (r, θ) that satisfy $r = a$ is the set of points at a distance a from the origin, namely the circle of radius a and center at $(0, 0)$. We can transform from polar coordinates to rectangular coordinates by using equation (1.14); in particular $r^2 = x^2 + y^2$ so the equation $r = a$ implies $a^2 = r^2 = x^2 + y^2$, which confirms the description of the graph as the circle of radius a.

A more general curve might be described by an equation $r = f(\theta)$ where f is a nonnegative function. To "plot" the graph of the equation $r = f(\theta)$ we select an angle θ and move a distance $r = f(\theta)$ along the line through the origin and making an angle θ with the positive x-axis to locate a point on the curve. We give just a few examples where the curve can be recognized by a transformation to rectangular coordinates.

EXAMPLE 1 Describe the graph of the polar equation

$$r = \frac{1}{3 + \cos\theta}.$$

Solution: Clear the fractions to rewrite the equation as

$$3r + r\cos\theta = 1.$$

Substitute $x = r\cos\theta$ and $3r = 3\sqrt{x^2 + y^2}$; after some manipulation we have the rectangular equation

$$9(x^2 + y^2) = (1 - x)^2,$$

which is easily seen to be an ellipse. ■

It can happen that a polar equation given in the form $r = f(\theta)$ does not meet the requirement $r \geq 0$ because f takes on negative values. One could adopt the rigid view that this is not allowed but that would rule out many interesting cases. Instead we adopt the convention that the graph of a polar equation $r = f(\theta)$ consists of all points (r, θ) for which $r = f(\theta) \geq 0$ and the points $(|r|, \pi + \theta)$ if $r = f(\theta) \leq 0$. The effect of this condition is to treat a negative value of r to mean that distance is measured in the "opposite direction" along the line making an angle θ with the positive x-axis; in other words the point is reached by moving a distance $|r|$ from the origin along the line making an angle $\pi + \theta$. The effect of this convention is to make (r, θ) and $(-r, \pi + \theta)$ polar coordinates of the same point. Note that we are merely adopting a convention that allows the description of a wide class of curves by polar equations.

EXAMPLE 2 Describe the graph of the polar equation $r = 4\sin\theta$.

Solution: Multiply the equation by r to get the form $r^2 = 4r\sin\theta$. Using the relation between the polar and rectangular coordinates, we tranform this into the equation $x^2 + y^2 = 4y$. After completing the square, we find the equation is equivalent to $x^2 + (y - 2)^2 = 4$. Hence the graph of the polar equation lies on a circle of radius 4 with center on the y-axis at $(0, 2)$. Notice that $\sin\theta \geq 0$ for $0 \leq \theta \leq \pi$ and $\sin\theta \leq 0$ for $\pi \leq \theta \leq 2\pi$. When θ is in this range, we can write $\theta = \pi + \theta'$ with $0 \leq \theta' \leq \pi$. The point on the graph corresponding to θ is

$$(4|\sin\theta|, \theta + \pi) = (4|\sin(\pi + \theta')|, \theta' + 2\pi).$$

This is the same point as the point with polar coordinates $(4\sin\theta', \theta')$. Thus we get the entire graph of the original equation by restricting θ to the interval $[0, \pi)$. ■

Exercises 1.7

Sketch the curves given by the polar equations. Transform the equations to rectangular coordinates.

1. $r = 2\cos\theta$

2. $r = 6\sin\theta$

3. $r = a \sin 2\theta$

4. $r = a \cos 2\theta$

5. $\theta = 0$

6. $r = 4$

7. $r = b$ (constant)

8. $\theta = a$ (constant)

9. $r(2 - \cos\theta) = 1$

10. $r(3 + \cos\theta) = 1$

11. $r(2 + \sin\theta) = 1$

12. $r(2 - \sin\theta) = 1$

13. If a curve is given by the polar equation $r = f(\theta)$ and f satisfies the condition $f(\theta + \pi) = f(\theta)$ for all θ, then the curve is symmetric about the origin. Illustrate this statement by describing the curve with polar equation $r = \sin 2\theta$.

14. The graph of the equation $r = 1 + \sin\theta$ is an example of a *cardioid* (so named because the curve has a "heart" shape). Sketch the curve.

1.8 Language

In this section we give a very brief introduction into the language used frequently in mathematical statements in this book. In all statements of theorems or definitions it is necessary to pay close attention to the precise meaning of the words and, in particular, to the modifiers and qualifiers.

If-Then Statements

The most common form of statement in a theorem is the if-then statement. In its generic form it reads "If A is true then B is true" where A and B are some conditions or mathematical statements. As an example we cite the following familiar fact.

Theorem. *If x is a real number, then x^2 is a nonnegative real number.*

To prove the correctness of this statement it is not sufficient to consider a specific instance, but rather some argument must be made that applies no matter what number is selected for x. Thus the statement cannot be proved by simply making the observation that $(-7)^2 = 49 > 0$ and $6^2 = 36 > 0$. These are only two specific instances of the theorem. If, however, we were to attempt to show the theorem is false, it would be sufficient to find one instance where it fails. Consider the following example.

Statement. *If x is a real number, then $x^2 + x \geq 0$.*

This statement makes an assertion about every real number but the statement is false and to prove that it is false it is only necessary to find one instance in which it fails. Consider $x = -0.5$. Then $x^2 + x = (-0.5)^2 - 0.5 = -0.25 \not\geq 0$. Thus we have found one instance where the statement fails, and so it cannot be universally true as the statement asserted.

For Every--There Exists Statements

A type of statement used frequently is "For every A there exists B such that C is true". Here A, B and C are some mathematical expressions.

The following is an example of the type of statement indicated by the heading.

Theorem. *For every nonvertical straight line L in the xy plane there exist two numbers m and b such that (x, y) is a point of L if and only $y = mx + b$.*

We have already seen that this is a true theorem. Let us analyze the ingredients. The "for every" part refers to any line in the set of nonvertical lines in the plane. The "there exists" part refers to a pair of constants m and b and the "such that" gives the condition relating the constants to the line. The theorem states that if we select any nonvertical line in the plane it will be possible to describe all the points on that line by selecting two appropriate constants and using them in a certain way to relate the coordinates of points on the line.

Under what conditions would this theorem be false? Because the theorem makes an assertion about every nonvertical line, the theorem would be false if there were at least one line for which the conclusion fails. The conclusion is that "there exists constants m and b such that a certain relation holds between the coordinates of points of the line". Obviously we cannot give an illustration of a failure of this theorem because we have already proved it to be true. But let us change it slightly so that it becomes false.

Statement X. *For every straight line L in the xy plane there exist two numbers m and b such that (x, y) is a point of L if and only $y = mx + b$.*

This statement is false. It makes an assertion about *every* line in the plane; to prove it is false we need only find one example of a line for which it fails. We take the line consisting of all points $(1, y)$, with y any real number. This is a vertical line passing through the x-axis at $x = 1$. How do we show the conclusion is false? Well, we suppose that m and b are two constants that meet the conditions. Then for any points (x, y) on the line we have the relation $y = mx + b$. Let us just try this with two points $(1, 2)$ and $(1, 6)$ both of which are on the line. If the m and b have the required property, then $2 = m(1) + b$ and also $6 = m(1) + b$. Clearly we cannot have both $m + b = 2$ and $m + b = 6$. Hence no such constants exist.

Let us consider a somewhat more complicated statement that is in the same form.

Statement A. *For any polynomial $p = a_n x^n + a_{n-1} + x^{n-a} + \cdots + a_1 x + a_0$ in which every a_i is a real number, $a_n \neq 0$ and n is an integer greater than or equal to 1, there is a real number t such that*

$$a_n t^n + a_{n-1} t^{n-1} + \cdots + a_1 t + a_0 = 0. \qquad \textbf{(1.15)}$$

This statement is false. How does one prove it is false? This statement asserts that a certain property is possessed by every polynomial of positive degree. The property is the existence of a real number t that satisfies equation (1.15). To prove the statement is false, one only needs to find a single polynomial that satisfies the assumptions (real coefficients, positive degree) for which no such t exists. There are many examples of such polynomials; the simplest one is $p = x^2 + 1$. There is no real number t such that $t^2 + 1 = 0$ because for any real number t, we have

$$t^2 \geq 0, \qquad t^2 + 1 \geq 1.$$

Thus it is not possible for $t^2 + 1$ to be 0 when t is a real number.

Now consider the following statement, which is very close to the previous one.

Statement B. *For any polynomial $p = a_n x^n + a_{n-1} + x^{n-a} + \cdots + a_1 x + a_0$ in which every a_i is a real number, $a_n \neq 0$, and n is an odd integer greater than or equal to 1, there is a real number t such that*

$$a_n t^n + a_{n-1} t^{n-1} + \cdots + a_1 t + a_0 = 0.$$

This statement is true; however this is neither obvious nor is it trivial to verify. Notice that the polynomial $x^2 + 1$ which was used to disprove statement A cannot be used to disprove statement B because statement B requires that the degree of the polynomial be odd and $x^2 + 1$ has even degree 2. How does one prove that B is true? We will not prove it here. But when faced with a statement like this one, it may be helpful to examine some particular cases of the general statement just to gain some insight. Let us consider the case in which the polynomial has degree 1. Then the polynomial has the form $a_1x + a_0$ and by assumption, $a_1 \neq 0$. Thus the real number $t = -a_0/a_1$ satisfies the equation

$$a_1t + a_0 = 0.$$

Thus we know B is true in the case $n = 1$. Of course this is only an indication of the truth of B for every odd natural number n. To prove B for every odd-degree polynomial, it will be necessary to study properties of the graph of the equation

$$y = a_nx^n + a_{n-1}x^{n-1} + \cdots + a_1x + a_0.$$

After we make this study we will be able to give a complete proof.

When statements are made in reference to a set of numbers, care must be taken to understand the difference between a statement that asserts a property shared by all members of the set or a property that is possessed by (at least) one member of the set. For example suppose that S is a set of numbers. The statement "For every x in S we have $x^2 + x - 1 > 4$" asserts that *every* element in S has a certain property. The statement "For some x in S we have $x^2 + x - 1 > 4$" asserts that at least one element in the set S has a certain property. Here are some specific examples.

EXAMPLE 1 Let $S = \{x : \text{ either } x < -2 \text{ or } x > 10\}$. Which of the following statements about S are true?

1. For every $t \in S$, $t^2 > 3$.
2. For every $z \in S$, $|z| \geq 10$.
3. There exists $w \in S$ such that $(2w + 3)^2 = 16$.
4. There exists $y \in S$ such that $|6y - 11| < 1$.

Solution: Statement 1 makes an assertion about every element of S. We check the validity of the statement by considering two cases. First suppose $t < -2$. Then $2 < -t$ and $-t$ is positive so $4 < (-t)^2 = t^2$. Thus the statement $t^2 > 3$ is true for the elements of S that are less than -2. Now suppose $t > 10$. Then clearly $t^2 > 100 > 3$ so we have checked that for any element of S the inequality $t^2 > 3$ is true. Thus statement 1 is true.

Next consider statement 2. This statement asserts that every element of S is at least 10 in absolute value. Many elements of S have this property but not all. The number -3 is in S but $|-3| \not\geq 10$. Thus statement 2 is false.

Statement 3 asserts there is (at least) one element w in S such that $(2w + 3)^2 = 16$. This is an equation that can be solved with no difficulty; take the square root of each side to obtain $(2w + 3) = \pm 4$. Thus there are two numbers that make the equation true, $w = 1/2$ and $w = -7/2$. The first choice is not an element of S but the second choice is in S. Thus statement 3 is true.

To determine if statement 4 is true or false, we rewrite the inequality as $-1 < 6y - 11 < 1$ or equivalently $10/6 < y < 2$. No element on this interval lies in S so statement 4 is false. ■

Exercises 1.8

Give the negation of each statement. The negation should be stated in the form of a positive statement.

1. Every car on Interstate 94 has exactly one driver.

2. Every dog in New York City has been registered.

3. If x is any real number, then $x^2 > 0$.

4. If x is any real number, there is some real number y with $x = y^5$.

5. For every car purchased in Wisconsin, there is state sales tax certificate.

6. For every $x > 1$, there is a positive number y such that $y^3 = 1/(x - 1)$.

Determine which of the following statements are true. We use S to denote the set of all numbers that lie between 1 and 2 or between -4 and -3 (including the endpoints).

7. For every $x \in S$, we have $x^2 - x \geq 1$.

8. There is some $x \in S$ with $x^2 - x < 1$.

9. There is some $x \in S$ for which $x - \frac{1}{x} < -3$.

10. For every $x \in S$, the equation $y^2 + 2y = x$ has a solution for y some real number.

2 Functions and Limits

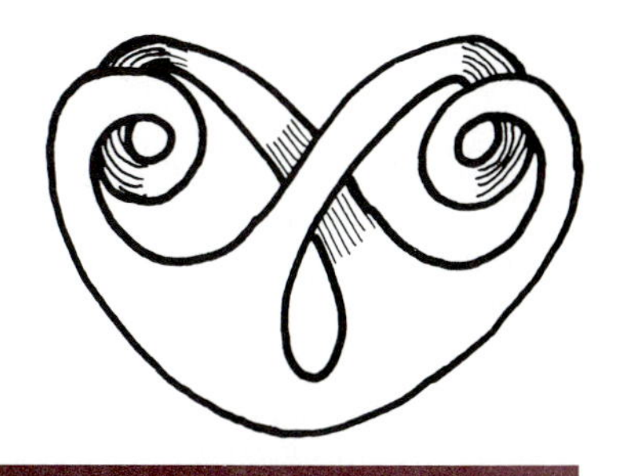

The most basic concept studied in calculus is that of a function. Functions describe relationships between varying quantities, such as speed and applied force, or between the output of a process subjected to an input, such as the depth of liquid in a container subjected to change in volume. They define relationships between data, such as the population of the earth in each year of this century, or represent mathematical quantities, such as the area of a square in terms of the length of one side. Functions appear in every discipline in which numbers play a significant role: computer scientists, economists, engineers and practitioners of many other disciplines routinely use functions.

Although functions can be described in many different ways, here we mainly study examples of functions described by formulas because the methods of calculus are most easily applied to their study. With each function we associate its graph. The graph of a function is one of the most important tools used to study functions. The graph brings geometric ideas to the study of properties of functions and serves to aid the intuition. The graph provides a geometric visualization of the important concept of the limit of a function. In addition to geometric examples, we also present analytic and algebraic examples to illustrate the concept of the limit of a function at a point.

2.1 Examples of functions

The word *function* denotes a correspondence. We say the area of a square is a "function of the length of its side". If A is the area of the square, and if L is the length of one side of a square, then $A = L^2$: A is a function of L. In this case A is determined by the single quantity L and to each L there corresponds an area $A = L^2$.

The distance you travel in your car depends on the amount of time you travel and the speed at which you travel. If v represents the average velocity over the entire trip, t represents the total time, and d is the distance of the trip, then these three variables are related by the equation $d = vt$. We say that d is a function of both v and t. In this case d is determined by the two variables v, t. We can also say that v is a function of d and t by rewriting the relation as $v = d/t$. This equation gives the definition of *average velocity* – namely the distance divided by the time.

Commonly, the word *variable* means a symbol that represents a numerical value that is not fixed but can take on different values. In these two examples, L, v and t denote variables. Some of the examples above presented functions involving several variables. Our interest, for now, is in functions of one variable. We adopt notation for functions that indicates the variable it depends on.

Functional Notation

We use the functional notation $f(x) = y$; f is the function, x is a variable, and y is the result of f acting on x. If $f(x) = x^2$ and if $x \geq 0$ is the length of the side of a square, then $f(x)$ is the area of the square. Other letters can be used for functions and for variables. We might write $A = g(L)$ to denote the area A given as a function g of a variable L. The formula defining the function g is $g(L) = L^2$.

Polynomial Functions

The first large class of functions that we consider is the class of *polynomial functions.* A polynomial in the variable x is a function of the form

$$f(x) = a_n x^n + a_{n-1}x^{n-1} + \cdots + a_1 x + a_0$$

where the a_i are real numbers and n is a nonnegative integer. The largest integer j with $a_j \neq 0$ is called the *degree* of the polynomial.

Here are a few specific examples of polynomial functions of the variables x, t and z:

1. $f(x) = 3x^2 - 5x + 2$ has degree 2,
2. $g(t) = t^3 - t^2 + t - 1$ has degree 3,
3. $p(z) = z^{10} - 2z^5 + 1$ has degree 10.

Each of the functions f, g and p is a polynomial function of one variable. Values of polynomial functions are computed quite easily; we illustrate this by computing some values of the polynomial $f(x)$:

$$f(x) = 3x^2 - 5x + 2$$

$$f(2) = 3 \cdot 2^2 - 5 \cdot 2 + 2 = 12 - 10 + 2 = 4$$

$$f(-1) = 3(-1)^2 - 5(-1) + 2 = 3 + 5 + 2 = 10$$

$$f(w) = 3w^2 - 5w + 2$$

$$f(4z) = 3(4z)^2 - 5(4z) + 2 = 3 \cdot 16z^2 - 5 \cdot 4z + 2 = 48z^2 - 20z + 2$$

$$f(1-t) = 3(1-t)^2 - 5(1-t) + 2 = 3(1 - 2t + t^2) - 5 + 5t + 2 = 3t^2 - t.$$

Rational Functions

Another type of function closely related to polynomial functions is the *rational function.* A rational function is a quotient of two polynomials; for example

$$f(x) = \frac{3x^5 - 7x + 2}{5x^3 - 2x^2 - 3}$$

is a rational function. Its values can be computed for any real number x for which the denominator is not equal to 0. We can compute the value at $x = -1$:

$$f(-1) = \frac{3(-1)^5 - 7(-1) + 2}{5(-1)^3 - 2(-1)^2 - 3} = \frac{6}{-10}.$$

However $f(1)$ cannot be computed because the evaluation would require a division by $5(1)^2 - 2(1) - 3 = 0$ and division by 0 is not allowed. We say the function $f(x)$ is *not defined* at $x = 1$.

Let us consider another type of function that is not a polynomial function but is still quite familiar. The function that assigns to a number x the square root of the number is denoted sqrt(x); the symbolic notation for this function is sqrt(x) $= \sqrt{x}$. To be more precise, the value $u =$ sqrt(x) is the unique nonnegative real number whose square equals x; that is $u^2 = x$. However, not every real number has a square root; there is no u such that $u^2 = -1$. Because the square of a real number is always nonnegative, the square root function sqrt(x) is defined only for nonnegative x. When x is positive, there are two real numbers u whose square is x; if $u^2 = x$ then also $(-u)^2 = x$. Because a function can assign only *one* number to each nonnegative x, we simply declare that sqrt(x) is the *nonnegative* number u that satisfies $u^2 = x$. We collect these remarks into a formal definition.

For every nonnegative real number x, sqrt(x) is the unique nonnegative number whose square equals x.

Domain and Range

The set of numbers for which a function is defined is called the *domain of the function.* The domain of sqrt(x) is the set of nonnegative real numbers. The set of values a function takes is called the *range of the function.* The range of sqrt(x) is the set of nonnegative real numbers; that is if w is a real number and $w \geq 0$, then $w =$ sqrt(w^2) so that w is a value of the square root function.

EXAMPLE 1 Give the domain and range of the rational function $h(x) = x/(x - 2)$.

Solution: The domain of h is the set of all real numbers not equal to 2. The number 2 must be excluded because division by 0 is not permitted. When $x = 2$, the expression $x/(x - 2)$ involves division by 0, but for all other numbers x, the expression $x/(x - 2)$ is a real number.

The range of $h(x) = x/(x - 2)$ is the set of all numbers w for which the equation

$$\frac{x}{x - 2} = w$$

has a solution x. If the equation is solved for x in terms of w, we find

$$x(w - 1) = 2w \quad \text{or} \quad x = \frac{2w}{w - 1}.$$

Of course division by zero is not allowed, so the equation can be solved for x in terms of w whenever $w \neq 1$. Thus the range of this function $h(x)$ is the set of all real numbers w excluding $w = 1$. ■

Now that we have seen several examples of functions, we can give the formal definition of a function. The domain and range of a function are part of its definition.

DEFINITION 2.1

Function

Let D and R be sets of real numbers. A function f with domain D and range contained in R is a rule that assigns to each real number x in D, a single real number y in R. We write $y = f(x)$ to indicate that y is the number assigned by f to x.

Perhaps the simplest function is a constant function. A function $g(x)$ is called a *constant function* if it has the same value at every point; that is for some fixed number c, $g(x) = c$ for every x. For example if $g(x) = -7$ for every x then $g(x)$ is a constant function. The domain of a constant function is the set of all real numbers; the range of a constant function consists of a single number.

Polynomials of degree 0 are constant functions and polynomials of degree greater than 0 are nonconstant functions. Every polynomial has the set of all real numbers as its domain whereas the range of a polynomial can vary depending on the polynomial. The polynomial $f(x) = x$ has all real numbers as its range; the polynomial $s(x) = x^2$ has all nonnegative numbers as its range; the polynomial $p(x) = 1 - 3x^2$ has range equal to the set of all numbers less than or equal to 1. This can be seen as follows. The range consists of all numbers y for which there is some x with $y = 1 - 3x^2$. To find the x in terms of y we solve and find that

$$x = \sqrt{\frac{1-y}{3}}.$$

For x to be a real number, it is necessary that the number under the radical sign be nonnegative. In particular then it is necessary that $1 \geq y$, and whenever this holds, there is an x with $p(x) = y$.

Most functions, but not all, are given by a formula that describes the value of the function at a point x of its domain. When a function is given by a formula, the domain will be the set of all x for which the formula makes sense, unless we specify otherwise. Most restrictions are made to prevent division by 0 or to ensure that a number under a square root sign is nonnegative.

Functions of a variable x can be combined to produce other functions of the same variable. For example, we could start with two functions $f(x)$, $g(x)$ and define new functions:

$$H(x) = f(x) + g(x),$$

$$u(x) = 2f(x) + g(1 - x),$$

$$s(t) = f(t)^2 - 3g(t) + g(t + 1).$$

Suppose we use the functions $f(x) = x^2 - x$ and $g(x) = 2x + 3$. Then the functions defined by the rules above yield the following formulas for $H(x)$, $u(x)$ and $s(t)$:

$$H(x) = (x^2 - x) + (2x + 3) = x^2 + x + 3,$$

$$u(x) = 2(x^2 - x) + \big(2(1 - x) + 3\big) = 2x^2 - 2x + 2 - 2x + 3 = 2x^2 - 4x + 3,$$

$$s(t) = (t^2 - t)^2 - 3(2t + 3) + \big(2(t + 1) + 3\big)$$

$$= t^4 - 2t^3 + t^2 - 6t - 9 + 2t + 2 + 3 = t^4 - 2t^3 + t^2 - 4t - 4.$$

Exercises 2.1

Compute the indicated values using $f(x) = 2x - 3x^2$ and $g(x) = \mathrm{sqrt}(x)$.

1. Compute $f(0)$, $f(2)$ and $f(-2)$.

2. Compute $g(0)$, $g(16)$ and $g(100)$.

3. Compute $g(s)g(t)$, $g(w^2)$ and $\dfrac{g(4z)}{g(z)}$.

4. Compute $f(t + 1)$, $f(s + t)$ and $f(4s)$.

5. Compute $f(x + 2) - f(x)$.

6. Compute $\dfrac{f(x + h) - f(x)}{h}$.

7. If $h(x) = f(x)g(x)^3$, compute $h(0)$, $h(4)$ and $h(16)$.

8. With $h(x)$ as in exercise 7, compute $h(t)$, $h(3x - 1)$ and $h(1/x)$.

9. Find a formula for a nonconstant function $f(x)$ which has domain equal to the set of all real numbers and all of whose values are greater than or equal to 10.

10. Find a nonconstant function $h(t)$ such that for all $t > 0$, we have $1 \le h(t) \le 2$.

11. Find a polynomial $f(x)$ such that $f(0) = 0$, $f(1) = 10$ and $f(x) \ge 0$ for every number x.

12. Find a polynomial $g(x)$ such that $g(0) > 0$, $g(1) < 0$ and $g(2) > 0$.

13. Find a rational function $r(x)$ such that the equation $r(x) = c$ has a solution x for every number c, $r(0) = 1$ and $r(2)$ is undefined.

14. Find a rational function $r(x)$ for which the equation $r(x) = c$ has a solution if and only if $c \ge 0$.

For each rational function below, give the largest possible domain for the function and determine the range of the function.

15. $r(x) = 4x^2 + 8$

16. $r(x) = 9 - x^2$

17. $r(x) = 4/x^2$

18. $r(x) = \dfrac{x}{x(x - 3)}$

19. $r(x) = \dfrac{x + 1}{x - 1}$

20. $r(x) = \dfrac{2x - 1}{x + 2}$

21. $r(x) = \dfrac{3x + 1}{3x - 1}$

22. $r(x) = \dfrac{x^2 + 2x + 1}{(x - 1)}$

23. Give an example of a rational function $r(x)$ that is undefined at $x = 1, 2, 3$ and has positive values at all other numbers x.

24. Let $a_1, \ldots, a_n$ be any n distinct real numbers. Find a rational function $r(x)$ that is undefined at $a_1, \ldots, a_n$ but is defined for all other values of x and $r(x) < 0$ for every x where the function is defined.

2.2 Graphs of functions

Graphs of functions enable us to study functions by representing them in a geometric way. They help create a picture in our minds of the function and permit visualization of abstract properties. If $f(x)$ is a function with domain D, then the graph of $f(x)$ is the set of points

$$\{(x,\ f(x))\ :\ x \in D\}$$

in the coordinate plane. The graph of the function $f(x)$ is the same as the graph of the equation $y = f(x)$.

EXAMPLE 1 Sketch the graph of the function $f(x) = 2x + 1$.

Solution: The graph of the function f is the set of all points of the form $(x, 2x + 1)$ where x can be any real number. The graph is just the graph of the equation $y = 2x + 1$ (figure 2.1). ■

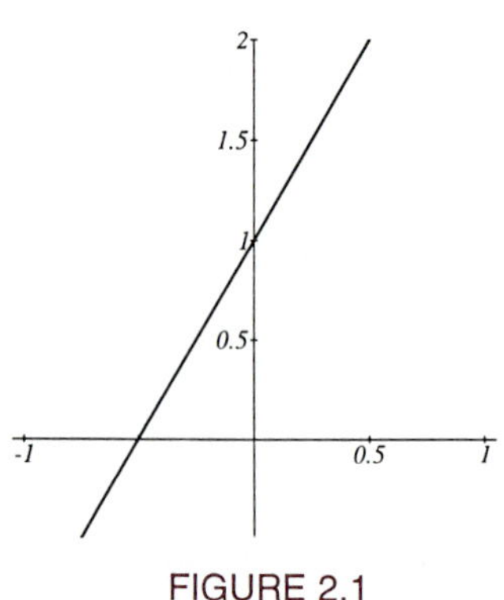

FIGURE 2.1
$f(x) = 2x + 1$

EXAMPLE 2 Suppose the graph of $f(x)$ is as shown in figure 2.2. What are some properties of $f(x)$ that can be inferred from its graph?

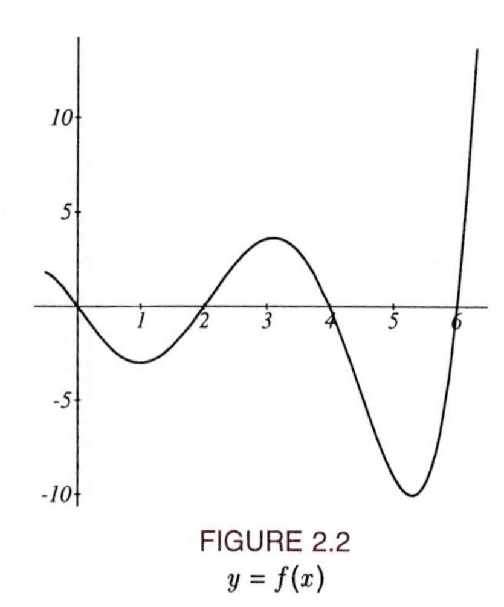

FIGURE 2.2
$y = f(x)$

Solution: The graph reveals some information about the function. For example we can assert that $f(x) \leq 0$ when $0 \leq x \leq 2$ and $4 \leq x \leq 6$. The solutions of the equation $f(x) = 0$ are $x = 0,\ 2,\ 4,\ 6$. We can infer that the function is defined for all x on the interval $[0, 6]$ because for every c on this interval, there is a point on the graph having c as first coordinate.

A property of the function that can be inferred from the graph is whether it is *increasing* or *decreasing* on a particular interval. A function is *increasing* on an interval $[a, b]$ if $a \leq p < q \leq b$ implies $f(p) < f(q)$. Similarly the function is *decreasing* on $[a, b]$ if $a \leq p < q \leq b$ implies $f(p) > f(q)$. In this example, the values of the function are decreasing as x moves from 0 to (about) 1, then the values of $f(x)$ increase until x is just a bit larger than 3, the function values decrease until about $x = 5$ and then increase for as far as the graph is drawn. ■

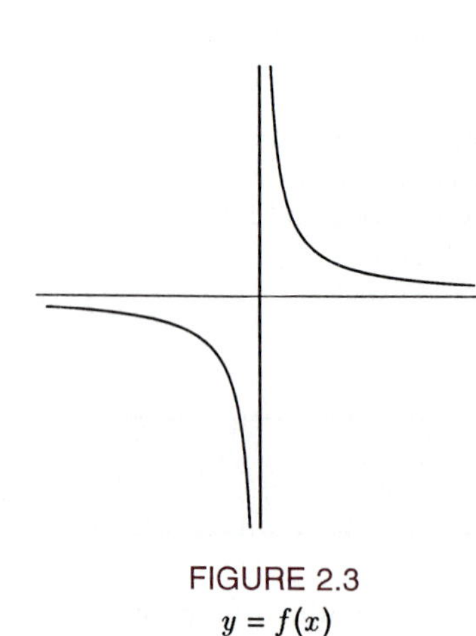

FIGURE 2.3
$y = f(x)$

EXAMPLE 3 State some properties of the function $f(x)$ whose graph is shown in figure 2.3.

Solution: From the graph of this function we see that $f(x)$ is undefined at $x = 0$. The function is defined and is positive for $x > 0$ and is defined and is negative for $x < 0$. There is no solution to the equation $f(x) = 0$ but for any nonzero number c, there is a solution to the equation $f(x) = c$. This can be inferred from the graph because the horizontal line $y = c$ crosses the graph at some point, provided $c \neq 0$. This means that for some t, the point (t, c) is on the line $y = c$ and on the graph of $f(x)$. The point on this graph having first coordinate t has second coordinate $f(t)$ and so $f(t) = c$.

We can also infer from the graph that $f(x)$ is decreasing both on the positive x-axis and on the negative x-axis. ■

EXAMPLE 4 Sketch the graph of the function $a(x) = |x|$.

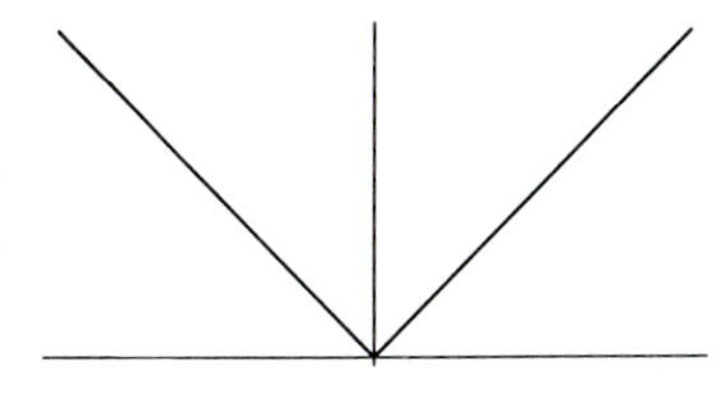

FIGURE 2.4
$y = |x|$

Solution: We remove the absolute value signs by considering two cases.

a. If $x \geq 0$ then $|x| = x$ so this part of the graph coincides with the part of the line $y = x$ for $x \geq 0$.
b. If $x \leq 0$ then $|x| = -x$ so this part of the graph coincides with the part of the line $y = -x$ having $x \leq 0$.

The graph is shown in figure 2.4. ■

The graph of the function $|x|$ illustrates the idea of a function defined by piecing together half-lines and/or segments of straight lines. Such a function is sometimes called *piecewise-linear*. Here is a more complicated example of a piecewise-linear function. Let

$$f(x) = \begin{cases} -1 & \text{if } x < -3 \\ 2x + 5 & \text{if } -3 \leq x \leq 1 \\ -x + 8 & \text{if } 1 < x. \end{cases}$$

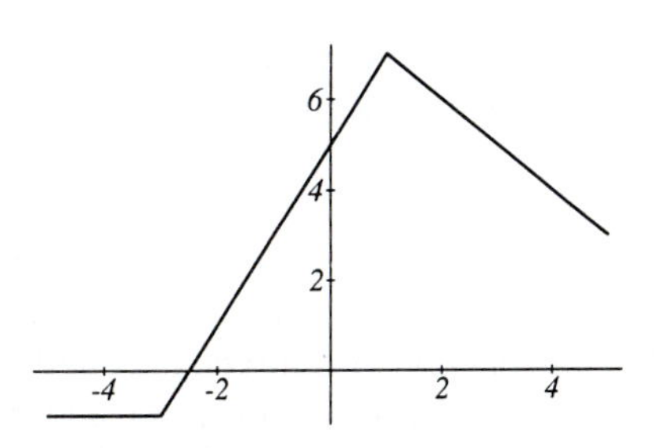

FIGURE 2.5
Piecewise-linear function

The graph of this function is composed of segments of the graphs of three straight lines: $y = -1$, $y = 2x + 5$ and $y = -x + 8$.

EXAMPLE 5 Sketch the graph of the function $\text{sqrt}(x) = \sqrt{x}$.

Solution: We make a rough sketch of this function by first computing a number of points on the curve. Table 2.1 gives some approximate values.

TABLE 2.1
Approximate values

x	$\sqrt{x}$
0	0
1	1
2	1.414
3	1.736
4	2
5	2.236
6	2.449

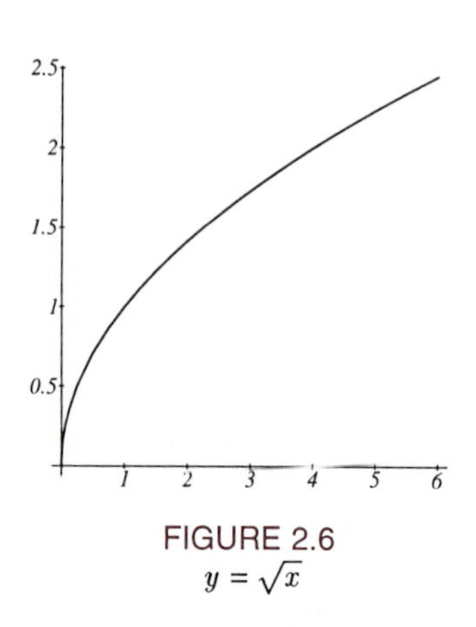

FIGURE 2.6
$y = \sqrt{x}$

Given this rather small number of points, how are we to know what the graph looks like between these points and at the points beyond the limits of the table? One fact that helps us obtain a fairly reasonable sketch is that the function $\sqrt{x}$ grows larger as x increases. That is if $a < b$, then $\sqrt{a} < \sqrt{b}$. This means that between two points determined by the table of values, the graph cannot sharply rise to a peak and then fall back down. Using the points determined by the table of values and this increasing property, one would expect that the graph of the function is reasonably approximated by the figure.

Knowing that the function is increasing and passes through the points determined by the table is still not enough to give a precise rendering of the graph. For example we are unable to state with certainty how the graph bends. The graphs of the two functions $f(x) = \sqrt{x}$ and $g(x) = x^2$ both pass through the two points $(0, 0)$ and $(1, 1)$, both are increasing functions but the "bend" in the two curves is quite different. ■

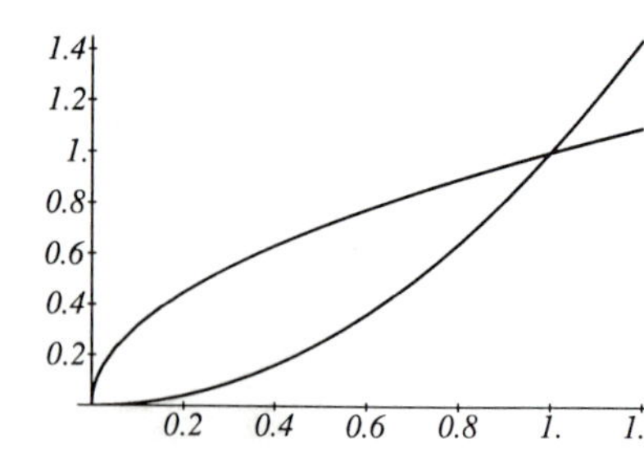

FIGURE 2.7
$y = \sqrt{x}$ and $y = x^2$

Gathering information to analyze this kind of problem will be done with the help of the derivative of the function, a concept introduced in chapter 3. The properties of functions to be studied in later chapters also help us to very accurately sketch the shapes of the graphs of many functions.

Of course, the most interesting and fun way to render accurate graphs is to use one of the many software packages on a personal computer or even at a workstation or mainframe computer. Most of the graphs in this text were rendered with the aid of the *Mathematica* program but other programs also produce excellent results. Several brands of hand-held calculators produce acceptable graphs with only a modest amount of effort. If any of these tools are available to the reader, significant rewards will be returned for the few hours of effort required to gain facility with their use.

Exercises 2.2

Make sketches that could be the graphs of functions having the following properties.

1. The function $f(x)$ is positive when x is negative, is negative if x is positive and is decreasing over the entire x-axis.

2. The function $f(x)$ is positive when $x < 2$, is negative if $2 < x < 4$ and is constant for $4 \leq x$.

3. The function $f(x)$ is defined for all $x \geq 0$, is positive for $x < 5$ and negative for $x > 5$, and is decreasing for all $x > 0$.

4. The function $h(x)$ is undefined at $x = 1$ but is defined and positive for all other values of x and is decreasing for all $x > 5$.

Sketch the graphs of the following functions and label points where the graph crosses either axis.

5. $f(x) = 3x - 2$

6. $g(x) = 4 - x$

7. $h(x) = (3x - 2)(4 - x)$

8. $k(x) = x^2 - 4$

9. $p(x) = x(x - 1)(x - 2)$

10. $q(x) = (x + 2)(x - 2) + 1$

11. Sketch the graphs of $f(x) = x^2$ and $g(x) = x + 1$ on the same coordinate system. Find the points common to both graphs.

12. Sketch the graphs of $f(x) = \sqrt{x}$ and $g(x) = x + 1$ on the same coordinate system. Find the points common to both graphs.

13. Let $f(x) = x^2$. On the same coordinate axes, sketch the graphs of $f(x)$ and $f(x - 2)$ and find the points where the two graphs intersect.

14. Let $g(x) = 3x - 2$. On the same coordinate axes, sketch the graphs of $g(x)$, $g(x + 4)$ and $g(x - 4)$.

15. Let $g(x) = |x|$. Sketch the graphs of $g(x)+3$ and $g(x+3)$ on the same coordinate system.

16. Suppose $f(x)$ is a function defined for all x on the interval $-1 \leq x \leq 1$ and it satisfies the condition $0 \leq f(x) \leq 2$. What restriction does this imply about the graph of f? [Hint: One way to answer this question is to indicate a region of the xy plane in which the graph must lie.]

A function f is increasing if $f(a) < f(b)$ whenever $a < b$. The next problems deal with increasing functions.

17. If f is an increasing function, what does this imply about the graph of f?

18. If f and g are increasing functions, prove that $f + g$ is an increasing function. [Note that the sum, $f + g$, is the function whose value at x is $f(x) + g(x)$.]

2.3 Composite functions

In addition to sums, products and quotients, there is one more important operation that produces new functions from given functions. Given two functions $f(x)$ and $g(x)$, we can form the *composite function* $h(x) = f(g(x))$. For example if $f(x) = 3x^2 - x$ and $g(x) = 2/x$ then the composite is

$$f(g(x)) = 3(g(x))^2 - g(x) = 3\left(\frac{2}{x}\right)^2 - \frac{2}{x}.$$

One must use the terminology carefully here. It is not absolutely precise to refer to *the* composite of f and g because there are really two possible composites of f and g; one could form $f(g(x))$ or $g(f(x))$ and these are usually not the same functions. With f and g as in the example just given, we have

$$g(f(x)) = \frac{2}{f(x)} = \frac{2}{3x^2 - x} \neq f(g(x)).$$

To overcome this ambiguity, we always pay close attention to the order in which the functions are mentioned; if we say "the composite of $f(x)$ and $g(x)$", we mean $f(g(x))$. If we say "the composite of $g(x)$ and $f(x)$", we mean $g(f(x))$.

There is another point to consider, that of the domain of the composite function. The composite $f(g(x))$ is defined at a certain point x if, first of all, $g(x)$ is defined; that is x is in the domain of g. Second, the number $g(x)$ must be in the domain of f because $f(g(x))$ is just f evaluated at $g(x)$. The composite of f and g will have the same domain as g if the range of g is contained in the domain of f.

EXAMPLE 1 Suppose

$$f(x) = \frac{1}{(x-1)(x-4)},$$

and $g(x)$ is some function. What is the domain of the composite $f(g(x))$?

Solution: The composite $h(x) = f(g(x))$ is defined at x if x is in the domain of g and $g(x)$ is in the domain of f. The domain of f consists of all numbers except 1 and 4. Thus the domain of h is the set of all numbers x that lie in the domain of g for which $g(x) \neq 1$ and $g(x) \neq 4$. ■

Formation of composite functions is often a way to describe complicated functions in terms of simpler functions.

EXAMPLE 2 Express $h(x) = \dfrac{\sqrt{x+1}}{\sqrt{x}}$ as a composite of two functions.

Solution: The function h has domain $x > 0$ and can be expressed as a square root of the rational function $r(x) = (x+1)/x$ when r is restricted to the domain $x > 0$. Thus $f(x) = \sqrt{r(x)}$ and so $f(x)$ is the composite of the square root function, $\text{sqrt}(x) = \sqrt{x}$ and $r(x)$. Many other correct solutions are possible. ■

Example 3 shows how a function can be expressed as composites in several ways.

EXAMPLE 3 Express $h(x) = x^4 + 6x^2 + 9$ as a composite in at least two ways.

Solution: We notice that only even powers of x appear in this polynomial so that with $f(x) = x^2 + 6x + 9$ we have

$$h(x) = (x^2)^2 + 6x^2 + 9 = f(x^2).$$

Thus $h(x) = f(g(x))$ with $g(x) = x^2$.

Notice that $h(x) = (x^2+3)^2$ so we also have $h(x) = f(g(x))$ with $g(x) = x^2+3$ and $f(x) = x^2$. ■

Many other solutions are possible in example 3. Here is a way to think about this type of problem. Think of a function as an action performed on a given number to produce another number. The function $h(x) = (x^2+3)^2$ starts with a number x, squares it, adds 3 to that result, and then squares that result. The composite $f(g(x))$ is a process that first applies the action g to x and then applies f to the result. To express h as a composite, let g be the first part of the action that produces h and f the remaining part. For example, g could be the function that squares x and f the function that adds 3 and then squares the result; namely $f(x) = (x+3)^2$. Then $h(x) = f(g(x)) = (x^2+3)^2$ as in our first solution. Alternatively, g could be the function that squares then adds 3, $g(x) = x^2 + 3$, and f the function that squares, $f(x) = x^2$. Other solutions are possible; a simple one is obtained by using $f(x) = x + 9$ and $g(x) = x^4 + 6x^2$. It is important to keep the correct ordering of the steps when viewing functions as composite functions.

Exercises 2.3

Let $f(x) = 3x^2 - 2x$ and $g(x) = 1/(x+1)$.

1. If $h(x) = f(g(x))$, compute $h(0)$, $h(4)$, $h(1-t)$.

2. If $v(x) = g(f(x))$, compute $v(0)$, $v(4)$, $v(1-t)$.

3. If $h(x) = f(g(x))$, compute $h(x)$, $h(t+3)$ and $h(3w+1)$.

4. If $v(x) = g(f(x))$, compute $v(x)$, $v(t+3)$ and $v(3w+1)$.

5. If $h(x) = f(g(x))$, what is the domain of h?

6. If $v(x) = g(f(x))$, what is the domain of v?

In each exercise give a formula for $f(g(x))$ and $g(f(x))$.

7. $f(x) = 3x^2$, $g(x) = 3x - 7$

8. $f(x) = \sqrt{x}$, $g(x) = 2x^2 + 1$

9. $f(x) = \dfrac{1}{x+1}$, $g(x) = \dfrac{x}{x-1}$

10. $f(x) = \dfrac{x+a}{x-a}$, $g(x) = \dfrac{x-a}{x+a}$

11. $f(x) = \begin{cases} \sqrt{1-x} & x \le 1 \\ \sqrt{x-1} & x \ge 1 \end{cases}$, $\quad g(x) = 1 - x^2$

12. $f(x) = \dfrac{x+a}{x-a}$, $g(x) = \dfrac{x-b}{x+b}$

13. Show that $f(g(x))$ is a polynomial of degree 6 if f is a polynomial of degree 3 and g is a polynomial of degree 2.

14. Give an example of two functions f and g, with at least one not a polynomial, but the composite $f(g(x))$ is a polynomial.

15. If $f(x)$ is a polynomial of degree n and $g(x)$ is a polynomial of degree m, prove that $f(g(x))$ is a polynomial of degree nm.

16. If $f(x)$ and $g(x)$ are rational functions show that the composite $f(g(x))$ is also a rational function.

2.4 Bounds and limits

In this section we introduce the idea of a bound for a function and then use this idea to discuss limits of functions. When we study functions, one of the central questions concerns the size of the values of the function — how big does $f(x)$ get when x lies on a certain interval? Bounds for the function deal with this question. The notion of the limit of a function deals with the behavior of a function near a given point. The idea of limit is one of the central ideas of calculus.

A *bound* for a function $f(x)$ is a constant B such that

$$|f(x)| \le B. \tag{2.1}$$

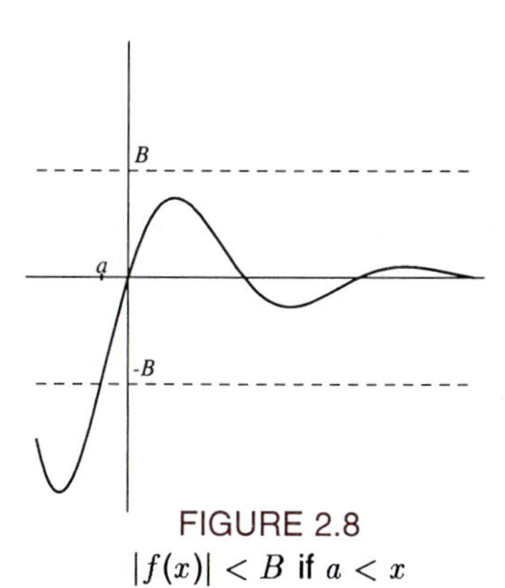

FIGURE 2.8
$|f(x)| < B$ if $a < x$

It is important to be clear about the values of x for which the inequality holds. One might intend inequality (2.1) to hold for all x in the domain of the function $f(x)$. If that is the case then the function is bounded everywhere by B. Most often inequality (2.1) will not hold for all x in the domain but only on a restricted part of the domain. If I is a subset of the real numbers then we say B is a bound for $f(x)$ on I if $|f(x)| \le B$ holds for every x in I.

For example, the function $f(x) = x^2 + 1$ has the set of all real numbers for its domain but there is no bound for the function. To see this, suppose some positive number B is given as a possible candidate for a bound; we can take any large value of x, say $x > 1 + B$ and find that $|f(x)| = x^2 + 1 > B$. Thus B cannot be a bound for $f(x)$ everywhere.

However if we restrict the values of x to lie on some interval, say $0 \le x \le 2$, then B can be found so that inequality (2.1) holds. For example

$$|x^2 + 1| = x^2 + 1 \le 5 \quad \text{for} \quad 0 \le x \le 2.$$

Of course there are other bounds for the function. In fact any number B that is at least 5 serves as a bound for $x^2 + 1$ on the interval $[0, 2]$. No number smaller than 5 can be a bound on this interval because at $x = 2$ the function value is $2^2 + 1 = 5$. In this case we say that the bound $B = 5$ is the *best possible* bound for $x^2 + 1$ on the interval $[0, 2]$.

An equivalent way of writing inequality (2.1) is

$$-B \leq f(x) \leq B.$$

In our example with the function $f(x) = x^2 + 1$ and x on $[0, 2]$ we have seen that

$$-5 \leq x^2 + 1 \leq 5.$$

But now we can give even more precise restrictions. Because the squared term cannot be negative for any value of x, we know that $1 \leq x^2 + 1$. We conclude that when x satisfies $0 \leq x \leq 2$, we have

$$1 \leq x^2 + 1 \leq 5.$$

This is better information than the bounds given before. Sometimes we express bounds in the form

$$A \leq f(x) \leq C \qquad \textbf{(2.2)}$$

for two constants A and C because bounds of this form may give more precise information by being more restrictive than a bound of the form in inequality (2.1). Any function $f(x)$ that is bounded as in inequality (2.2) is also bounded as in inequality (2.1) for some constant B (see exercise 14 following this section).

Bounds for Sums

One of the most useful tools for finding bounds for functions is the *Triangle Inequality*, which follows from Theorem 1.4 of chapter 1:

$$|f(x) + g(x)| \leq |f(x)| + |g(x)|.$$

In fact the repeated use of the triangle inequality for sums of two functions shows the following:

Triangle Inequality for Finite Sums

$$|f_1(x) + f_2(x) + \cdots + f_n(x)| \leq |f_1(x)| + |f_2(x)| + \cdots + |f_n(x)|$$

The triangle inequality allows us to get a bound for the sum of functions by finding a bound for each summand.

EXAMPLE 1 Find a bound B for the function $2x^3 - 7x + 3$ on the interval $[-1, 3]$.

Solution: First apply the triangle inequality to get

$$|2x^3 - 7x + 3| \leq |2x^3| + |-7x| + |3| = 2|x|^3 + 7|x| + 3.$$

Next we estimate the right-hand side. Because $-1 \leq x \leq 3$, it follows that $0 \leq |x| \leq 3$. The largest possible value of $|x|^3$ is $3^3 = 27$ and the largest possible value of $|x|$ is 3; we obtain

$$|2x^3 - 7x + 3| \leq 2 \cdot 27 + 7 \cdot 3 + 3 = 78.$$

Thus 78 is one possible bound for the function on $[-1, 3]$. After we study derivatives of functions and some applications of derivatives in chapter 3, we will be able to show that the bound 78 is not the best possible; in fact $|2x^3-7x+3| \leq 36$ on the interval $[-1,3]$. Even better information will be obtained; namely $-2.05 < 2x^3 - 7x + 3 \leq 36$ for x on $[-1,3]$. But at this point we can only make a fairly accurate sketch of the graph of the function $2x^3 - 7x + 3$ which indicates at least that the earlier bounds are reasonable. ■

Bounds for Products and Quotients

Dealing with products is easier than dealing with sums because we have

$$|f(x)g(x)| = |f(x)||g(x)|.$$

Thus if $|f(x)| \leq B$ and $|g(x)| \leq C$, then

$$|f(x)g(x)| = |f(x)||g(x)| \leq BC.$$

Finding bounds for quotients of the form $f(x)/g(x)$ is not as easy as finding bounds for products. Consider the simple example $h(x) = 1/x$. If x lies on the interval $0 < x < 1$ then we can divide both sides of the inequality $x < 1$ by x to get $1 < 1/x$. This gives a bound for $1/x$ on one side. However there is no constant B with the property that $1/x < B$ for all x on the interval. By taking x sufficiently close to 0, but still positive (for example, $x = 10^{-k}$), it is possible to make $1/x$ larger than any given number (because $1/x = 10^k$).

This shows that even if $g(x)$ is bounded, it may still happen that $1/g(x)$ is not bounded. The problem occurs when $g(x) = 0$ for some x on the interval. To find a bound for the quotient $f(x)/g(x)$ it is necessary to find a bound for $f(x)$ and a bound for g that shows that $|g(x)|$ does not get too near zero. Here is the statement that allows bounds for quotients.

THEOREM 2.1

Bounds for Quotients

Suppose for every x on the set I we have

$$|f(x)| \leq B \quad \text{and} \quad 0 < A \leq |g(x)|.$$

Then

$$\left|\frac{f(x)}{g(x)}\right| \leq \frac{B}{A}.$$

for every x in I.

Proof

The inequality $0 < A \leq |g(x)|$ ensures that the product $A|g(x)|$ is positive for every x on I so we can divide each term in the inequality by the product to get

$$\frac{1}{|g(x)|} \leq \frac{1}{A}.$$

Multiply this by $|f(x)|$ and use the inequality $|f(x)| \leq B$ to get

$$\left|\frac{f(x)}{g(x)}\right| = |f(x)|\frac{1}{|g(x)|} \leq B\frac{1}{A}$$

which proves the statement in the theorem. □

EXAMPLE 2 Find a bound for the function $H(x) = (x^3 - x)/(3x^2 + 2)$ for x on the interval $[-2, 2]$.

Solution: We first find a bound for the numerator. The restriction that x be on the interval $[-2, 2]$ is equivalent to the inequality $|x| \leq 2$. We apply this and the triangle inequality to the numerator to get

$$|x^3 - x| \leq |x|^3 + |-x| \leq 2^3 + 2 = 10.$$

For the denominator, it is necessary to show that $3x^2 + 2$ does not get too close to 0; we have

$$2 \leq 3x^2 + 2 \quad \text{and} \quad \frac{1}{3x^2 + 2} \leq \frac{1}{2}.$$

Putting these two statements together gives us the bound

$$\left|\frac{x^3 - x}{3x^2 + 2}\right| = |x^3 - x|\frac{1}{|3x^2 + 2|} \leq 10 \cdot \frac{1}{2} = 5.$$

Thus we have found the bound $|H(x)| \leq 5$ for x on $[-2, 2]$. ■

In this example, it was comparatively easy to find a bound that kept the denominator away from zero because the denominator was a polynomial involving only even powers of x with positive coefficients. Other functions in the denominator can be more difficult to estimate; we will deal with them as the need arises.

Exercises 2.4

Find a bound of the form $|f(x)| \leq B$ for the function over the given interval. Then repeat each exercise and try to find bounds for the function in the form $A \leq f(x) \leq C$. Look for values of A and C that give more precise information than the bound obtained in the form $|f(x)| \leq B$.

1. $f(x) = 7 - 4x$, $[-3, 10]$

2. $f(x) = 7 - 4x$, $[-10, 3]$

3. $f(x) = 15(x-3)$, $[-3, 3]$

4. $f(x) = -5(x+2)$, $[-3, 8]$

5. $f(x) = x^2 + x + 4$, $[-2, 2]$

6. $f(x) = x^2 + x + 4$, $[-4, 4]$

7. $f(x) = \dfrac{1}{4x - 3}$, $[1, 10]$

8. $f(x) = \dfrac{1}{4x + 3}$, $[1, 10]$

9. $f(x) = \dfrac{x + 2}{x^2 + 2}$, $[-1, 1]$

10. $f(x) = \dfrac{x^2 - 1}{x^2 + 1}$, $[-1, 1]$

11. $f(x) = \dfrac{-3x^2 + 1}{4x^3 + 1}$, $[1, 4]$

12. $f(x) = \dfrac{x - b}{x + b}$, $0 < b < x$

13. For each of the first six exercises, make a fairly accurate sketch of the function over the given interval and use this to help estimate how close your bound is, in each case, to the best possible.

14. If $A \leq f(x) \leq C$ for all x on an interval I, show $|f(x)| \leq B$ for all x on I, where $B = |A| + |C|$. Can you find a smaller value for B that will serve as a bound for $f(x)$ on I?

2.5 Limits of functions

We now turn to the discussion of limits. This concept is a precise formulation of the intuitive idea of "close". What does it mean for a function to be "close" to a value over some interval? Let us begin with an illustration of the idea.

Discussion of Closeness

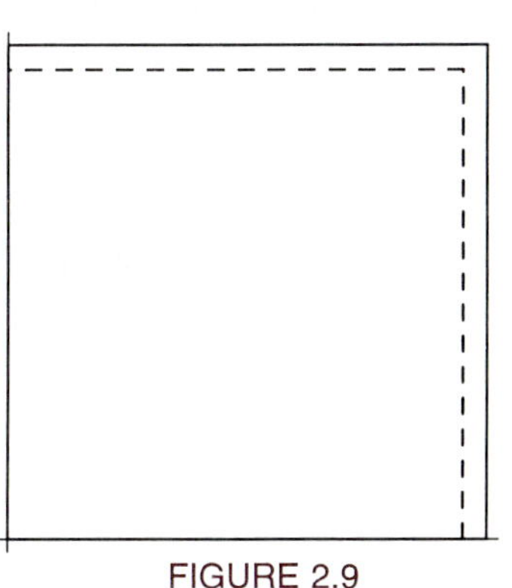

FIGURE 2.9

Consider the problem of physically constructing a fence so that it encloses a square plot of ground having area 400 ft^2. This is a trivial problem in theory; simply make each side exactly 20 ft long and make the corners right angles and the job is done. However, tape measures, cutting tools, wire fence and fence builders have physical limitations; measurements are not absolutely precise. As a practical matter, the chance of successfully completing the job would be much greater if the fence builder were required to construct a fence that encloses an area between 399 ft^2 and 401 ft^2.

We assume for simplicity that the only issue of concern is the problem of cutting the fence material to the required length. The problem for the fence builder is to determine how close to 20 ft she must make the length of the fence to meet this area requirement. Let us translate this into algebraic terms. Let x denote the variable length of the fence for one side. The area of the square is then $f(x) = x^2$. It is necessary that x be restricted so that

$$399 \leq x^2 \leq 401. \qquad \textbf{(2.3)}$$

The problem then is to find an interval $a \leq x \leq b$ so that inequality (2.3) holds for all such x. Let us rewrite this in a more symmetric form. We introduce the permissible area-tolerance ϵ, where $\epsilon = 1$ in our case, and a permissible length-error δ, that is the distance from 20 that x may vary and still keep the area within the tolerance limit of 400 ± 1. The problem is then formulated as follows: Find a positive number δ such that

$$|400 - x^2| < \epsilon \quad \text{whenever} \quad |20 - x| < \delta. \qquad \textbf{(2.4)}$$

Notice that we do not ask for the largest possible δ (that is, the largest possible error) but only for some choice of a positive δ that will make statement (2.4) true.

We examine the inequality that must hold:

$$|400 - x^2| = |(20 - x)(20 + x)| = |20 - x| \cdot |20 + x| < 1.$$

The term $|20 - x|$ is precisely the term bounded by δ; the term $|20 + x|$ can be estimated crudely as follows. Because $19^2 = 361$ and $21^2 = 441$, we can be sure that $19 < x < 21$ (so that $\delta < 1$). This implies

$$39 = 19 + 20 < x + 20 < 21 + 20 = 41.$$

After making some replacements, we find

$$|400 - x^2| = |20 - x| \cdot |20 + x| < \delta \cdot 41.$$

We want to force $|400 - x^2|$ to be less than $\epsilon = 1$; this is accomplished by the requirement $41\delta < \epsilon$ or $\delta < \epsilon/41 = 1/41$.

To check this calculation we compute the area assuming the two extremes, $x = 20 - 1/41$ and $x = 20 + 1/41$ and find

$$\left(20 - \frac{1}{41}\right)^2 = 399.0249853\ldots$$

$$\left(20 + \frac{1}{41}\right)^2 = 400.9762045\ldots.$$

This proves that the area of a square having sides within 1/41 ft of 20 ft will be within 1 ft^2 of 400 ft^2. The computation shows there is still a small amount of freedom remaining in the choice of x, but the value of δ is near to the maximum error allowable. You may have guessed that the source of this room for additional error enters this problem because we used the crude estimate for $|20 + x|$.

If the tolerance ϵ were decreased in this example, the same calculations would show how to select the value of δ so that the inequality (2.4) holds. Given any small positive number ϵ there is a positive δ such that

$$|400 - x^2| < \epsilon \quad \text{whenever} \quad |20 - x| < \delta.$$

In fact the above computation shows that this inequality holds if $\delta = \epsilon/41$ (provided we assume $\epsilon \leq 1$).

The discussion of the fence-length involves a rather complicated way to make a statement about the size of the function x^2 for values of x near $x = 20$. It may be viewed as saying that x^2 is close to 400 if x is close to 20 and it answers the question "How close?". Namely it shows that x^2 will be within ϵ of 400 provided x is within $\delta = \epsilon/41$ of 20.

Let us examine the essentials of this situation. We had a function $f(x) = x^2$, a specific number $L = 400$ and a value $a = 20$. We were given a tolerance ϵ and were asked to find an interval around the point a such that the values of the function were within the tolerance of L. The general problem is formulated as follows:

Given a function $f(x)$, a number L, a point a, and a tolerance $\epsilon > 0$, find a positive number δ such that

$$|f(x) - L| < \epsilon \quad \text{whenever} \quad |x - a| < \delta.$$

Using the terminology of bounds, this says that the function $f(x) - L$ is bounded by ϵ on some interval of positive length having a at its center.

Of course in any given instance, this problem may or may not be solvable. In order that this problem have a solution, there must be special relationships between the objects $f(x)$, L and a.

Let us examine a specific example where f is simple enough so that all the computations with L, a, ϵ and δ can be done explicitly.

EXAMPLE 1 Given the function $f(x) = 3x - 2$, the value $L = 4$ and the point $a = 2$, show that for any positive number ϵ there is a positive number δ such that $|f(x) - L| < \epsilon$ whenever $|x - a| < \delta$.

Solution: Make the substitutions to see that the inequality we want is

$$|(3x - 2) - 4| < \epsilon \tag{2.5}$$

and this should hold whenever $|x - 2| < \delta$. The problem is to find the δ. Rewrite inequality (2.5) in the form

$$|(3x - 2) - 4| = |3x - 6| = 3|x - 2| < \epsilon.$$

How should we select δ so that $|x - 2| < \delta$ implies $3|x - 2| < \epsilon$? This is accomplished by taking δ equal to $\epsilon/3$. Then we have the desired result. That is whenever

$$|x - 2| < \delta = \epsilon/3$$

then

$$|f(x) - L| = |(3x - 2) - 4| < \epsilon.$$

■

Using the language about to be introduced, this example illustrates that the limit as x approaches 2 of the function $3x - 2$ is 4.

Limit at a Point

Here is the definition of the limit of a function at a point.

DEFINITION 2.2

Limit of a Function at a Point

The limit as x approaches a of the function $f(x)$ is L if for any positive number ϵ there exists a positive number δ such that

$$|f(x) - L| < \epsilon \quad \text{whenever} \quad 0 < |x - a| < \delta.$$

When these conditions hold, we write

$$\lim_{x \to a} f(x) = L. \tag{2.6}$$

If there is no number L such that equation (2.6) holds, we say that $\lim_{x \to a} f(x)$ does not exist.

This definition says that the limit as x approaches a of $f(x)$ equals L if the values of $f(x)$ are close to L (within ϵ) whenever x is close enough to a (within δ) and it is possible to find such a δ for every choice of a positive ϵ. The definition makes no assertion about the value of $f(x)$ at $x = a$. As far as the definition is concerned, $f(x)$ need not even be defined at $x = a$. Only the values of the function near $x = a$ are of interest. In particular it is always assumed that the function is defined in some open interval containing a except possibly at the point a itself.

Graphical Interpretation of Limit

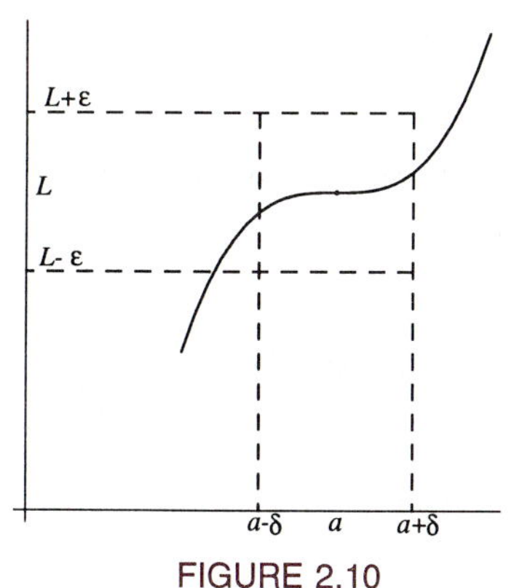

FIGURE 2.10

Consider the graph of the function $y = f(x)$ for values of x near a and suppose the conditions in the definition of limit are satisfied. The definition requires that

$$-\epsilon < f(x) - L < \epsilon,$$

and so

$$L - \epsilon < f(x) < L + \epsilon$$

whenever $0 < |x - a| < \delta$; that is whenever x lies within δ of a excluding $x = a$. This means there is an interval, namely $(a - \delta, a + \delta)$ such that the graph of the function above this interval lies entirely between the two horizontal lines $y = L - \epsilon$ and $y = L + \epsilon$ except possibly for the point where $x = a$. In particular this says that the function $f(x)$ is bounded on some interval having a as midpoint.

Some Basic Limits

EXAMPLE 2 Let $f(x)$ be the constant function with value k. Use the definition of limit to prove

$$\lim_{x \to 3} k = k.$$

Solution: Let ϵ be any positive number. The definition requires that we find a positive number δ such that

$$|f(x) - k| < \epsilon \quad \text{whenever} \quad 0 < |x - 3| < \delta.$$

Because $f(x) = k$, this condition reads

$$|k - k| < \epsilon \quad \text{whenever} \quad 0 < |x - 3| < \delta.$$

The inequality $0 = |k - k| < \epsilon$ is true for every value of x because ϵ is a positive number. Thus any choice of positive number will do for δ.

The role of the number 3 in this limit could be replaced by any number a. The same reasoning proves

$$\lim_{x \to a} k = k \tag{2.7}$$

for any constant k and any point a. ■

EXAMPLE 3 Prove

$$\lim_{x \to a} x = a.$$

Solution: Let ϵ be any positive number. It is necessary to find a positive number δ such that

$$|x - a| < \epsilon \quad \text{whenever} \quad 0 < |x - a| < \delta.$$

The choice $\delta = \epsilon$ has the required property. ■

EXAMPLE 4 Prove

$$\lim_{x \to 2} \frac{x^2 - 4}{x - 2} = 4.$$

Solution: Let $f(x) = (x^2 - 4)/(x - 2)$. The function $f(x)$ is not defined at $x = 2$ because the denominator has the value 0 and division by 0 is not defined. However in the definition of the limit as x approaches 2, we are concerned with values of x *near* 2, but the value $x = 2$ is excluded. Thus

$$f(x) = \frac{x^2 - 4}{x - 2} = \frac{(x + 2)(x - 2)}{x - 2} = x + 2 \quad \text{for} \quad x \neq 2.$$

Now we can deal with the simpler function $x + 2$. Given any positive number ϵ it is required to force the inequality

$$|f(x) - 4| = |(x + 2) - 4| = |x - 2| < \epsilon$$

whenever $0 < |x - 2| < \delta$. Clearly if we select $\delta = \epsilon$ the condition is met so the proof is complete. ■

EXAMPLE 5 Use the definition to prove that $\lim_{x \to 4} 1/x = 1/4$.

Solution: Given any positive number ϵ, it is required that we find some positive δ such that $0 < |x - 4| < \delta$ implies

$$\left| \frac{1}{x} - \frac{1}{4} \right| < \epsilon.$$

We look ahead and note that the δ we are looking for is a bound for $|x - 4|$. Thus we try to express this last inequality in terms of $|x - 4|$. First multiply both sides by the quantity $4x$. We make sure that this is positive by insisting that we only consider values of x near 4; to be specific, let us assume for a first approximation, that $3 < x < 5$ so that $|x - 4| < 1$. (The δ we want should be less than 1 for this assumption to hold.) Then the inequality reads

$$4x\left|\frac{1}{x} - \frac{1}{4}\right| = |4 - x| < 4x\epsilon.$$

The smallest possible value of the right-hand side is 12ϵ because we have restricted x to lie within 1 of 4. So we set $\delta = 12\epsilon$, provided $12\epsilon < 1$, else we take δ as any positive number less than 1. Now assume $0 < |x - 4| < \delta$. Then we have the chain of statements

$$0 < |x - 4| < \delta = 12\epsilon < 4x\epsilon,$$

$$\frac{1}{4x}|x - 4| = \left|\frac{1}{4} - \frac{1}{x}\right| < \epsilon.$$

This proves that the required δ exists and that the limit is correct. ■

The reasoning used in example 5 can also be used to prove

$$\lim_{x \to a} \frac{1}{x} = \frac{1}{a} \qquad \text{for} \quad a \neq 0.$$

Next we show how to prove that a limit does not exist.

EXAMPLE 6 Prove that $\lim_{x \to 0} 1/x$ does not exist.

Solution: Assume that there is a number L such that $\lim_{x \to 0} 1/x = L$. We first observe that $L \neq 0$ because $1/x > 1$ for all x on the interval $(0, 1)$. Next we invoke the definition of limit. Select $\epsilon = 1$ and let δ be the positive number such that

$$\left|\frac{1}{x} - L\right| < 1 \quad \text{whenever} \quad 0 < |x - 0| < \delta. \tag{2.8}$$

We will show that this pair of inequalities cannot hold if x is a sufficiently large multiple of L. We now select such a multiple. Select a positive integer n such that $1 < n(L\delta)$ so that $1/nL < \delta$. Note that any integer larger than $1/L\delta$ will do. If necessary, we can select an even larger value for n to ensure $1 < nL$. Once an n is selected to meet these conditions, any larger value of n will also meet the conditions; in particular $1/(n + 1)L < \delta$. Take $x = 1/(n + 1)L$. Then this is a choice of x for which inequality (2.8) holds and so

$$\left|\frac{1}{x} - L\right| = |(n + 1)L - L| = n|L| < 1.$$

This conflicts with the properties of n selected above. So we have reached an impossible situation arising from the assumption that L met the conditions in the definition of limit. Thus no such L can exist and the limit does not exist. ■

The basic idea of example 6 can be put in a more general context to prove the following.

THEOREM 2.2

If $\lim_{x \to a} f(x) = L$ then there is a positive number δ such that $f(x)$ is bounded on the interval $(a - \delta, a + \delta)$.

Proof

Select $\epsilon = 1$ and let δ be the positive number for which

$$|f(x) - L| < \epsilon \quad \text{whenever} \quad 0 < |x - a| < \delta.$$

For every x on the interval $(a - \delta, a + \delta)$ we have

$$|f(x)| = |f(x) - L + L| < |f(x) - L| + |L| < \epsilon + |L|,$$

which shows that $f(x)$ is bounded by $\epsilon + |L|$ except possibly at the point $x = a$. So we take B to be any number greater than or equal to each of the numbers $\epsilon + |L|$ and $|f(a)|$ (assuming that $f(a)$ is defined). Then B is a bound for $f(x)$ on the interval and the theorem is proved. □

Evaluating the Limit

In certain cases, we may know or believe that $\lim_{x\to a} f(x)$ exists; the process of finding the correct value of L to make the statement $\lim_{x\to a} f(x) = L$ true is called *evaluating the limit.*

EXAMPLE 7 Evaluate the limit $\lim_{x\to 2} 4x - 3$.

Solution: When x is near 2, the function $4x - 3$ is near $4 \cdot 2 - 3 = 5$. It is necessary to prove that this is indeed the correct value. The proof requires that, for any given positive ϵ, we find a positive δ such that

$$|4x - 3 - 5| < \epsilon \quad \text{whenever} \quad 0 < |x - 2| < \delta.$$

Because

$$|4x - 3 - 5| = 4|x - 2|,$$

we see that $\delta = \epsilon/4$ has the required property. ■

Exercises 2.5

Use the definition of limit to prove the following statements are correct.

1. $\lim_{x\to 4} x^2 - 3x = 4$
2. $\lim_{x\to 1} \frac{x}{x+1} = 0.5$
3. $\lim_{x\to 1} 2x^2 + 7 = 9$
4. $\lim_{x\to 0} 3 - 4x^2 = 3$
5. $\lim_{x\to 0} 2x - 3 \neq -2$
6. $\lim_{x\to 3} x^2 - 3 \neq 3$
7. Find the correct L and then prove $\lim_{x\to 1} x^2 + 2 = L$.
8. Find the correct L and then prove $\lim_{x\to 1/2} 3 - 2x^2 = L$.

Determine which of the following limits exist, evaluate the ones that do exist and give reasons for your answers.

9. $\lim_{x\to 1} \frac{x^2 - 1}{x^2}$
10. $\lim_{x\to 0} \frac{x^2 - 1}{x^2}$
11. $\lim_{x\to 1} \frac{1}{x+1}$
12. $\lim_{x\to -1} \frac{1}{x+1}$

In each exercise, find bounds for the function in the form $|f(x) - f(a)| < B$ when x ranges over the indicated set of values and a is the indicated number.

13. $f(x) = x^2 + 4,\ |x - 1| < 1,\ a = 1$
14. $f(x) = x(x - 2),\ |x - 1| < 1,\ a = 1$
15. $f(x) = \frac{x+1}{2x},\ |x - 2| < 1,\ a = 2$
16. $f(x) = -x^3 - 2x,\ |x - 2| < 3,\ a = 2$
17. $f(x) = x + x^{-1},\ |x + 3| < 2,\ a = -3$
18. $f(x) = \frac{x|x| + 1}{x^2 + 1},\ |x| < 4,\ a = 0$
19. Suppose L and M are unequal numbers. Find an $\epsilon > 0$ such that the strip bounded by the two lines $y = L \pm \epsilon$ has no points in common with the strip bounded by the two lines $y = M \pm \epsilon$. Make a sketch to help illustrate the statement. [Hint: First try this with two specific numbers, like $L = 2$ and $M = 4$, to help get the general idea of how to select ϵ.]
20. A rectangular region has sides of length a and b with $a > b$. Let ϵ be a positive number no larger than $a - b$.

A new rectangle is formed by decreasing the length of one side to $a - \epsilon$ and increasing the length of the other side to $b + \epsilon$. Which rectangle has the larger area? Determine the difference in the areas as a function of ϵ.

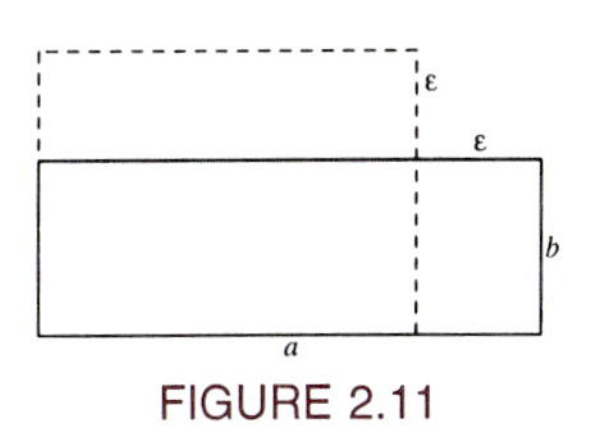

FIGURE 2.11

21. Use Exercise 20 to answer the question: Of all rectangles with fixed perimeter P, what are the dimensions of the one having the largest area?

22. A triangle is formed by the x-axis, the line $x = 2$ and the line $y = mx$ with $0 < m$. Let $A = A(m)$ denote the area of this triangle. By how much does the area $A(m)$ change if m is allowed to increase or decrease by an amount δ?

2.6 Limit theorems

Verifying the definition of the limit of a function by producing the δ for each ϵ is time consuming and sometimes quite difficult. In this section some of this work is avoided by proving fairly general theorems about limits. These general theorems may require the use of ϵ/δ ideas but then that work is done and the general theorems can be applied without using the definition of limit.

The first result shows that a function can have at most one limit at a point.

THEOREM 2.3

Uniqueness of the Limit

If L and M are numbers such that

$$\lim_{x \to a} f(x) = L \quad \text{and} \quad \lim_{x \to a} f(x) = M$$

then $L = M$.

Proof

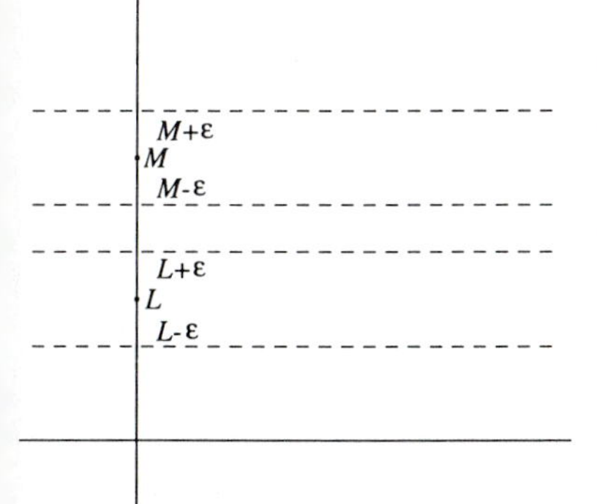

FIGURE 2.12

We give a geometric argument based on the graph of the function. In view of the hypothesis that $\lim_{x \to a} f(x) = L$ we know that for any positive ϵ there is an interval $I_{\delta_1} = (a - \delta_1, a + \delta_1)$ of length $2\delta_1$ such that the graph of $f(x)$ over I_{δ_1} lies between the horizontal lines $y = L - \epsilon$ and $y = L + \epsilon$ except possibly for the point with $x = a$. For the same ϵ, the hypothesis $\lim_{x \to a} f(x) = M$ implies there is an interval $I_{\delta_2} = (a - \delta_2, a + \delta_2)$ such that the graph of $f(x)$ over I_{δ_2} lies between the lines $y = M \pm \epsilon$ except possibly at $x = a$. Now let δ be the smaller of the two numbers δ_1, δ_2. Then for all x satisfying $0 < |x - a| < \delta$, $f(x)$ lies in the strip bounded by the two lines $y = L \pm \epsilon$ and also in the strip bounded by $y = M \pm \epsilon$.

Let us now make the assumption that $L \neq M$ and show that we are then forced into an impossible situation. In view of the assumption $L \neq M$ we can select an ϵ so small that these two strips have no points in common;

for example take $\epsilon = |L - M|/4$. Because $L \neq M$, this ϵ is a positive number so it is a legitimate choice. The points of the graph of $f(x)$ lie in both strips yet the two strips have no points in common. This is an impossible situation so the assumption that $L \neq M$ cannot be valid and the uniqueness of the limit is proved. □

Take note that Theorem 2.3 says a function can have only one limit at a point. It does not say that a function has a limit at any point, but if the function does have a limit, then the limit is unique.

It may happen that there is no number L that meets the conditions of the definition of $\lim_{x \to a} f(x) = L$. We have already seen that the function $f(x) = 1/x$ is unbounded in the interval $(0, 1)$ and so there is no number L such that $\lim_{x \to 0} 1/x = L$.

THEOREM 2.4

General Properties of Limits

Suppose

$$\lim_{x \to a} f(x) = L \quad \text{and} \quad \lim_{x \to a} g(x) = M.$$

Then

i. $\lim_{x \to a} (f(x) + g(x)) = L + M$;

ii. $\lim_{x \to a} (rf(x)) = rL$, for any constant r;

iii. $\lim_{x \to a} f(x)g(x) = LM$.

Proof

Let ϵ be a positive number. To prove property i it is necessary to show that there is a positive δ such that

$$|f(x) + g(x) - (L + M)| < \epsilon \quad \text{whenever} \quad 0 < |x - a| < \delta.$$

Notice that

$$\begin{aligned} |f(x) + g(x) - (L + M)| &= |[f(x) - L] + [g(x) - M]| \\ &\leq |f(x) - L| + |g(x) - M| \end{aligned} \tag{2.9}$$

by the triangle inequality. Thus it is sufficient to show that there is a δ for which $0 < |x - a| < \delta$ implies

$$|f(x) - L| < \epsilon/2 \quad \text{and} \quad |g(x) - M| < \epsilon/2,$$

because these inequalities imply the validity of inequality (2.9).

Because $\epsilon/2$ is a positive number, the hypothesis implies there exist positive numbers δ_1 and δ_2 such that

$$\begin{aligned} 0 < |x - a| < \delta_1 &\quad \text{implies} \quad |f(x) - L| < \epsilon/2 \\ 0 < |x - a| < \delta_2 &\quad \text{implies} \quad |g(x) - M| < \epsilon/2. \end{aligned} \tag{2.10}$$

We take δ equal to the smaller of the numbers δ_1, δ_2; then both conditions in statement (2.10) hold and so this δ has the property we want.

The proof of property ii is left as an exercise. We give a proof of property iii since it involves some subtle manipulation. Given a positive ϵ, we want to show there is a positive δ such that $0 < |x - a| < \delta$ implies

$$|f(x)g(x) - LM| < \epsilon. \tag{2.11}$$

We use the device of adding and subtracting the quantity $Mf(x)$ and the triangle inequality to rearrange the left term as follows:

$$\begin{aligned} |f(x)g(x) - LM| &= |[f(x)g(x) - Mf(x)] + [Mf(x) - LM]| \\ &\leq |f(x)g(x) - Mf(x)| + |Mf(x) - LM| \\ &= |f(x)||g(x) - M| + |M||f(x) - L|. \end{aligned}$$

The problem is now reduced to finding a $\delta > 0$ such that $0 < |x - a| < \delta$ implies both the inequalities

$$|f(x)||g(x) - M| < \epsilon/2 \quad \text{and} \quad |M||f(x) - L| < \epsilon/2.$$

Let us first treat the case $M \neq 0$. The hypothesis that $\lim f(x) = L$ and $\lim g(x) = M$ ensures that for suitable δ the terms $|f(x) - L|$ and $|g(x) - M|$ can be made as small as we like by restricting x close enough to a. The sticky point is the factor $|f(x)|$ in front of $|g(x) - M|$. How large does $|f(x)|$ get? Because we know that $|f(x) - L|$ is small for x near a, we write

$$|f(x)| = |[f(x) - L] + L| \leq |f(x) - L| + |L|.$$

This is used to select a δ to make the right side at most $1 + |L|$. So select numbers δ_1, δ_2, δ_3 such that

$$\begin{aligned} 0 < |x - a| < \delta_1 &\quad \text{implies } |f(x) - L| < 1, \\ 0 < |x - a| < \delta_2 &\quad \text{implies } |f(x) - L| < \frac{\epsilon}{2|M|}, \\ 0 < |x - a| < \delta_3 &\quad \text{implies } |g(x) - M| < \frac{\epsilon}{2(1 + |L|)}. \end{aligned}$$

Finally take δ to be the smallest of δ_1, δ_2, δ_3. Then for all x that satisfy $0 < |x - a| < \delta$, all three of these last inequalities hold; in particular

$$|f(x)| < 1 + |L| \quad |f(x)||g(x) - M| < \epsilon/2, \quad \text{and} \quad |M||f(x) - L| < \epsilon/2.$$

It follows that inequality (2.11) holds for $0 < |x - a| < \delta$ and the proof is complete in the case $M \neq 0$.

For the case $M = 0$ the argument given above does not apply (because we divided by $|M|$) but statement iii is slightly simpler. Let δ_1 be a positive number such that $|x - a| < \delta_1$ implies $|f(x) - L| < 1$ and let δ_2 be a positive number such that $|x - a| < \delta_2$ implies $|g(x)| < \epsilon/(L + 1)$. If δ is the smaller of δ_1 and δ_2 then both inequalities hold and we have

$$|f(x)g(x)| < (L + 1)|g(x)| < (L + 1)\frac{\epsilon}{L + 1} = \epsilon,$$

which is the statement we needed to prove. □

EXAMPLE 1 Evaluate the limits

$$\lim_{x \to 2} 4x - 3 + \frac{1}{x} \quad \text{and} \quad \lim_{x \to 2} \frac{4x - 3}{x}.$$

Solution: We have seen in earlier examples that

$$\lim_{x \to 2} 4x - 3 = 5 \quad \text{and} \quad \lim_{x \to 2} \frac{1}{x} = \frac{1}{2}.$$

Now apply the rule for limits of sums and products to obtain

$$\lim_{x\to 2} 4x - 3 + \frac{1}{x} = \lim_{x\to 2} 4x - 3 + \lim_{x\to 2} \frac{1}{x} = 5 + \frac{1}{2}$$

$$\lim_{x\to 2} \frac{4x-3}{x} = \left(\lim_{x\to 2}(4x-3)\right)\left(\lim_{x\to 2} \frac{1}{x}\right) = 5\left(\frac{1}{2}\right),$$

giving the values of both limits. ∎

Finite Sums and Products of Functions

Analogues of statements i and iii of the theorem are valid for sums and products of more than two functions. If we have

$$\lim_{x\to a} f_1(x) = L_1, \quad \lim_{x\to a} f_2(x) = L_2, \cdots, \lim_{x\to a} f_n(x) = L_n,$$

then

$$\lim_{x\to a} [f_1(x) + f_2(x) + \cdots + f_n(x)] = L_1 + L_2 + \cdots + L_n$$

$$\lim_{x\to a} f_1(x) f_2(x) \cdots f_n(x) = L_1 L_2 \cdots L_n.$$

Proofs of these more general statements can be given using mathematical induction.

Limits of Quotients

Next we deal with quotients $f(x)/g(x)$. In this case, it is necessary to pay attention to the possibility that $g(x)$ may have a limit 0. In some cases, the quotient may still have a limit, in other cases it will not.

THEOREM 2.5

Suppose

$$\lim_{x\to a} f(x) = L \quad \text{and} \quad \lim_{x\to a} g(x) = M.$$

i. If $M \neq 0$, then

$$\lim_{x\to a} \frac{f(x)}{g(x)} = \frac{L}{M}.$$

ii. If $M = 0$ and $L \neq 0$ then $\lim_{x\to a} f(x)/g(x)$ does not exist.

iii. If $M = 0$ and $L = 0$, then $\lim_{x\to a} f(x)/g(x)$ may or may not exist depending on the particular functions $f(x)$ and $g(x)$.

Proof

Using the condition that $M \neq 0$, we first prove

$$\lim_{x\to a} \frac{1}{g(x)} = \frac{1}{M}.$$

It is necessary to prove that for any given $\epsilon > 0$, there is a $\delta > 0$ such that

$$\left|\frac{1}{g(x)} - \frac{1}{M}\right| < \epsilon \quad \text{whenever} \quad 0 < |x - a| < \delta.$$

For a given ϵ, we show how to produce the δ. In view of the assumption that

$\lim g(x) = M \neq 0$, there is a positive δ_1 such that whenever $0 < |x - a| < \delta_1$ we have

$$|g(x) - M| < |M|/2. \tag{2.12}$$

(Here we apply the definition of limit using $\epsilon = |M|/2$.) The statements in the following argument assume that x satisfies $0 < |x - a| < \delta_1$. The strip between the two lines $y = M \pm |M|/2$ contains the graph of $g(x)$ so we have

$$M - |M|/2 < g(x) < M + |M|/2. \tag{2.13}$$

We want to show that $g(x) \neq 0$. If $M > 0$, then

$$0 < M - |M|/2 = M - M/2 = M/2.$$

By combining this with inequality (2.13), we obtain

$$0 < M - |M|/2 < g(x).$$

In an analogous way, we obtain

$$g(x) < M + |M|/2 < 0$$

in the case $M < 0$. So in either case, $g(x)$ does not equal 0.

Now take any positive number ϵ_2 and select δ_2 small enough so that for $0 < |x - a| < \delta_2$ we have both

$$|g(x) - M| < \epsilon_2 \quad \text{and} \quad |g(x) - M| < |M|/2.$$

When we divide both sides of the first of these by $|M||g(x)|$, we obtain

$$\left|\frac{1}{g(x)} - \frac{1}{M}\right| < \frac{\epsilon_2}{|M||g(x)|} < \frac{2\epsilon_2}{|M|^2}.$$

By selecting ϵ_2 to satisfy $2\epsilon_2/M^2 = \epsilon$, we then have the required δ by setting $\delta = \delta_2$. This proves

$$\lim_{x \to a} \frac{1}{g(x)} = \frac{1}{M}.$$

Now we apply the rule for products to obtain statement i:

$$\lim_{x \to a} \frac{f(x)}{g(x)} = \lim_{x \to a} f(x) \cdot \lim_{x \to a} \frac{1}{g(x)} = \frac{L}{M}.$$

The step is valid because the limits as x approaches a of $f(x)$ and of $1/g(x)$ exist.

Now for statement ii we assume that $M = 0$ and $L \neq 0$. We prove that $\lim_{x \to a} f(x)/g(x)$ does not exist by showing that $f(x)/g(x)$ is unbounded in every interval of the form $(a - \delta, a + \delta)$ with positive δ. Let K be any positive number, which we think of as a large number. By the hypothesis that $\lim_{x \to a} g(x) = 0$, there is a positive δ_1 such that $0 < |x - a| < \delta_1$ implies $|g(x) - 0| = |g(x)| < 1/K$. Thus $K < 1/|g(x)|$. There is a δ_2 such that $0 < |x - a| < \delta_2$ implies

$$|f(x) - L| < |L|/2 \quad \text{and so} \quad |f(x)| > |L|/2.$$

For those x satisfying $0 < |x - a| < \delta$, with δ equal to the smaller of δ_1, δ_2, we have

$$\left|\frac{f(x)}{g(x)}\right| = |f(x)|\left|\frac{1}{g(x)}\right| > \frac{|L|}{2}K.$$

Because K is arbitrarily selected, we see that the quotient $|f(x)|/|g(x)|$ is larger than any given number for every x satisfying $0 < |x - a| < \delta$ provided the positive δ is suitably selected.

There is nothing to be proved in statement iii except to give some examples to show that different results may occur. We have already seen the example in which $f(x) = x^2 - 4$ and $g(x) = x - 2$. In this case both functions have a limit 0 as x approaches 2 and for the quotient, we have

$$\lim_{x \to 2} \frac{f(x)}{g(x)} = \lim_{x \to 2} \frac{x^2 - 4}{x - 2} = \lim_{x \to 2} x + 2 = 4.$$

This gives an example where the numerator and denominator each have a limit 0 and the limit of the quotient of the two functions exists.

To illustrate that the limit need not exist, consider a slightly different example in which $f(x) = x^2 - 4$ and $g(x) = (x - 2)^2$. Then

$$\lim_{x \to 2} \frac{f(x)}{g(x)} = \lim_{x \to 2} \frac{x^2 - 4}{(x - 2)^2} = \lim_{x \to 2} \frac{x + 2}{x - 2};$$

this limit does not exist by Theorem 2.5(ii) because the numerator has a nonzero limit but the denominator has zero limit as $x \to 2$. □

Limits of Polynomial and Rational Functions

We use the general results about limits to determine limits of polynomials and rational functions. We begin with powers of x.

THEOREM 2.6

For any integer n and any real number a, we have

$$\lim_{x \to a} x^n = a^n$$

except in the case $n < 0$ and $a = 0$; in that case the limit does not exist.

Proof

First assume n is a positive integer. Then we have already seen that $\lim_{x \to a} x = a$. We apply the rule for products to conclude

$$\lim_{x \to a} x^n = (\lim_{x \to a} x)(\lim_{x \to a} x) \cdots (\lim_{x \to a} x) = aa \cdots a = a^n.$$

Now assume $n = -k$ with k a positive integer and assume that $a \neq 0$. If $g(x) = x^k$, then $\lim_{x \to a} g(x) = \lim_{x \to a} x^k = a^k$ by the case just proved. Because $a^k \neq 0$, the rule for quotients can be applied to give

$$\lim_{x \to a} x^n = \lim_{x \to a} x^{-k} = \lim_{x \to a} \frac{1}{x^k} = \frac{1}{a^k} = a^n$$

as we were required to prove. □

THEOREM 2.7

Limits of Polynomials

Let $f(x) = b_n x^n + b_{n-1} x^{n-1} + \cdots + b_1 x + b_0$ be any polynomial and let a be any real number. Then

$$\lim_{x \to a} f(x) = f(a).$$

Proof

The polynomial is the sum of the functions $f_i(x) = b_i x^i$ for $0 \leq i \leq n$. By the limit result for sums of functions, the result just proved and the result about multiplication by a constant, we have

$$\begin{aligned}\lim_{x\to a} f(x) &= (\lim_{x\to a} b_n x^n) + (\lim_{x\to a} b_{n-1}x^{n-1}) + \cdots + (\lim_{x\to a} b_0)\\ &= b_n a^n + b_{n-1}a^{n-1} + \cdots + b_0\\ &= f(a).\end{aligned}$$

□

This result about polynomials can now be combined with the result about limits of quotients to evaluate the limits of rational functions.

Corollary 2.1

Limits of Rational Functions

Let $p(x)$ and $q(x)$ be polynomials and suppose $q(a) \neq 0$. Then

$$\lim_{x\to a} \frac{p(x)}{q(x)} = \frac{p(a)}{q(a)}.$$

The Squeeze Theorem

The Squeeze Theorem can be applied to evaluate $\lim f(x)$ provided suitable functions can be found to trap $f(x)$.

THEOREM 2.8

The Squeeze Theorem

Let $f(x)$, $g(x)$ and $h(x)$ be functions that satisfy

$$g(x) \leq f(x) \leq h(x)$$

for all x on some open interval containing the point a except possibly at a itself. If

$$\lim_{x\to a} g(x) = L \quad \text{and} \quad \lim_{x\to a} h(x) = L,$$

then

$$\lim_{x\to a} f(x) = L. \qquad \textbf{(2.14)}$$

Proof

We give a proof based on the graphical interpretation of limits. Let ϵ be a given positive number. The statement $\lim_{x\to a} g(x) = L$ implies there is a positive number δ_1 such that all points $(x, g(x))$ lie in the strip between the lines $y = L \pm \epsilon$ provided $0 < |x - a| < \delta_1$. Similarly there is a positive number δ_2 such that all the points $(x, h(x))$ lie in the same strip when $0 < |x - a| < \delta_2$. Now if δ is the smaller of the two numbers δ_1, δ_2, then for all x that satisfy $0 < |x - a| < \delta$, both points $(x, g(x))$ and $(x, h(x))$ lie in the strip bounded by the lines $y = L \pm \epsilon$. For these values of x we have

$$L - \epsilon < g(x) \leq f(x) \leq h(x) < L + \epsilon.$$

This means the points $(x, f(x))$ also lie in the strip and that $|f(x) - L| < \epsilon$ whenever $0 < |x - a| < \delta$. This proves the validity of equation (2.14). □

EXAMPLE 2 Evaluate the limit $\lim_{x \to 0} \sqrt{1 + x^2}$.

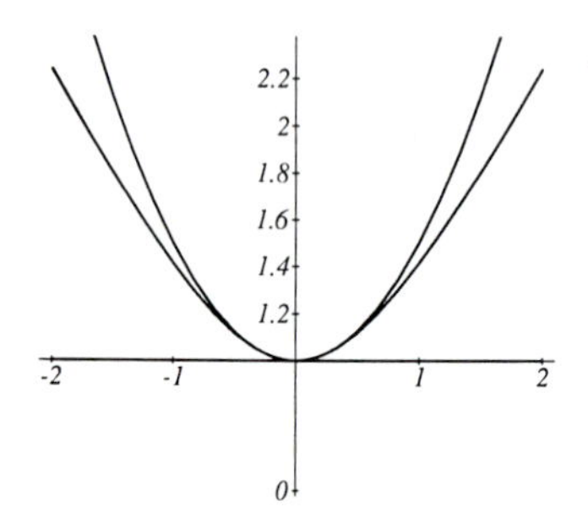

FIGURE 2.13
Graphs of the functions
$1 \leq \sqrt{1 + x^2} \leq 1 + x^2/2$

Solution: We set up a situation where the Squeeze Theorem can be used. For any x, the terms x^2 and x^4 are nonnegative and so we have

$$1 \leq 1 + x^2 \leq 1 + x^2 + \frac{1}{4}x^4 = \left(1 + \frac{1}{2}x^2\right)^2.$$

After taking square roots we obtain

$$1 \leq \sqrt{1 + x^2} \leq 1 + \frac{1}{2}x^2.$$

The limit as x approaches 0 of the constant 1, is 1. The limit of the rightmost term, $1 + x^2/2$, is also 1. Hence $\lim_{x \to 0} \sqrt{1 + x^2} = 1$ by the Squeeze Theorem. ■

One-Sided Limits

The discussion of certain limits is simplified by considering left or right limits. The one-sided version of the limit $\lim_{x \to a} f(x)$ means that x approaches a from *one side only*; that is, x gets close to a through values always larger than a (the right limit) or through values always smaller than a (the left limit). Here are the precise definitions.

DEFINITION 2.3

Right Limit

The limit of the function $f(x)$ as x approaches a from the right is L if for any positive number ϵ there exists a positive number δ such that

$$|f(x) - L| < \epsilon \quad \text{whenever} \quad 0 < x - a < \delta.$$

When these conditions hold, we write

$$\lim_{x \to a^+} f(x) = L. \qquad \textbf{(2.15)}$$

DEFINITION 2.4

Left Limit

The limit of the function $f(x)$ as x approaches a from the left is L if for any positive number ϵ there exists a positive number δ such that

$$|f(x) - L| < \epsilon \quad \text{whenever} \quad 0 < a - x < \delta.$$

When these conditions hold, we write

$$\lim_{x \to a^-} f(x) = L. \qquad \textbf{(2.16)}$$

Take careful notice of the slight differences in the definitions of *limit, right limit*, and *left limit.* In the definition of limit, the variable x is restricted by the condition $0 < |x - a| < \delta$. The absolute value signs permit x to lie on either side of a. In the definition of right limit, x is restricted by $0 < x - a < \delta$. This is equivalent to $a < x < a + \delta$ so that x must be to the right of a. Similarly in the definition of left limit, x is restricted by $0 < a - x < \delta$. This is equivalent to $a - \delta < x < a$ so that x lies to the left of a.

There are several reasons for using one-sided limits. Consider the function $\text{sqrt}(x) = \sqrt{x}$. This function is defined only for $0 \leq x$ and, in particular, it is not defined in any open interval containing 0. Thus the limit as x approaches 0 of sqrt(x) is not meaningful. However the right limit does make sense, and in fact we have

$$\lim_{x \to a^+} \sqrt{x} = 0.$$

Let us give a proof of this fact. For any positive number ϵ we must show there exists a positive number δ such that

$$|\sqrt{x} - 0| = |\sqrt{x}| < \epsilon \quad \text{whenever} \quad 0 < x < \delta.$$

Because the sqrt function is never negative, we have $|\sqrt{x}| = \sqrt{x}$. The condition $0 < \sqrt{x} < \epsilon$ holds if and only if $0 < x < \epsilon^2$. Thus the choice $\delta = \epsilon^2$ has the properties we want.

This example illustrates the idea of computing the limit of a function at an endpoint of its domain. If we encounter a function $f(x)$ whose domain is an interval $[a, b]$ then we can ask for the limits

$$\lim_{x \to a^+} f(x) \qquad \lim_{x \to b^-} f(x)$$

whereas the two-sided limit $\lim_{x \to c} f(x)$ is not meaningful for either $c = a$ or $c = b$ because $f(x)$ is not defined on both sides of c.

Another use of one-sided limits is provided by a function of the following type.

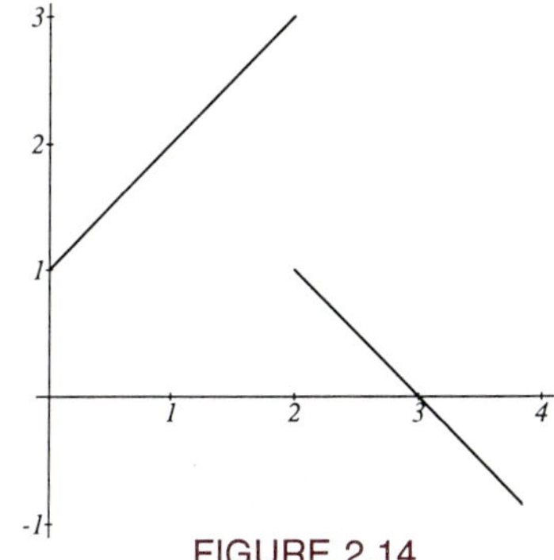

FIGURE 2.14
$y = f(x)$

EXAMPLE 3 Let

$$f(x) = \begin{cases} x + 1 & \text{if } 0 \leq x < 2 \\ 3 - x & \text{if } 2 \leq x \end{cases}$$

Discuss $\lim_{x \to 2} f(x)$, $\lim_{x \to 2^+} f(x)$ and $\lim_{x \to 2^-} f(x)$.

Solution: Notice that the graph of the function has a "jump" at $x = 2$. When x is very close to 2, but to the left of 2, $f(x)$ is very close to $2 + 1 = 3$; however when x is close to 2, but on the right of 2, $f(x)$ is very close to $3 - 2 = 1$. It seems clear that the limit of $f(x)$ as x approaches 2 does not exist. To be a little more precise we note that as x ranges over all values satisfying $0 < |x - 2| < \delta$, for any positive δ, $f(x)$ has some values near 3 and some values near 1; thus there is no single number L close to all the values of $f(x)$. However both the one-sided limits exist:

$$\lim_{x \to 2^+} f(x) = 1 \qquad \lim_{x \to 2^-} f(x) = 3.$$

We prove only the first of these one-sided limits. When we want to determine the right-hand limit, $\lim_{x \to 2^+} f(x)$, the only values of x that are relevant are those satisfying $2 < x$. For these values $f(x) = 3 - x$. For any positive ϵ, we must find a δ such that the inequality

$$|(3 - x) - 1| = |2 - x| < \epsilon$$

holds for all x satisfying $0 < x - 2 < \delta$. Clearly taking $\delta = \epsilon$ satisfies the requirement. A similar discussion can be given to prove the correctness of the left-hand limit. ■

The justification for introducing the one-sided limits is contained in the next result.

THEOREM 2.9

The limit as x approaches a of $f(x)$ exists if and only if both the left and right limits exist and they are equal. That is,

$$\lim_{x \to a^-} f(x) = L \quad \text{and} \quad \lim_{x \to a^+} f(x) = L \tag{2.17}$$

if and only if

$$\lim_{x \to a} f(x) = L.$$

Proof

If $\lim_{x \to a} f(x) = L$, then the definition of limit implies that both the left and right limits exist and are equal to L. The importance of the theorem is the implication that the limit exists when both one-sided limits exist and are equal. So we suppose that equation (2.17) is true and that we are given a positive number ϵ. The definition of the left limit implies that there is a positive δ_1 such that

$$|f(x) - L| < \epsilon \quad \text{whenever} \quad 0 < a - x < \delta_1. \tag{2.18}$$

The definition of right limit implies there is a positive δ_2 such that

$$|f(x) - L| < \epsilon \quad \text{whenever} \quad 0 < x - a < \delta_2. \tag{2.19}$$

Now let δ be the smaller of the two number δ_1 and δ_2. Suppose x satisfies the condition $0 < |x - a| < \delta$. We want to argue that this implies $|f(x) - L| < \epsilon$. Consider two cases. First suppose $0 < x - a$. Then $|x - a| = x - a$ so the supposition implies $0 < x - a < \delta$. By inequality (2.18), we have $|f(x) - L| < \epsilon$. The remaining case is $0 > x - a$. In this case, $|x - a| = a - x$, so we have $0 < a - x < \delta$ and this implies $|f(x) - L| < \epsilon$ by inequality (2.19). Hence we have produced the required δ for the given ϵ and the conditions in the definition of limit are satisfied. □

Exercises 2.6

Evaluate each limit or prove the limit does not exist.

1. $\lim_{x \to 2} x^3 - 4x + 3$

2. $\lim_{x \to -3} x^3 + 8x + 30$

3. $\lim_{x \to 1} \dfrac{x^2 - x + 4}{x^4 - x + 3}$

4. $\lim_{x \to -2} \dfrac{x^2 + 3x + 4}{x^3 + 2x + 3}$

5. $\lim_{t \to 0} \frac{1}{t} - \dfrac{4t^2 + 1}{t}$

6. $\lim_{t \to 1} \dfrac{4}{t - 1} - \dfrac{4t^2 - 8t}{t - 1}$

7. $\lim_{u \to -1} \dfrac{u^2 - 1}{u + 1}$

8. $\lim_{w \to 9} \dfrac{w^4 - 9}{(w + 3)(w - 3)}$

9. $\lim_{x \to 3^+} \sqrt{(x + 2)(x - 3)}$

10. $\lim_{x \to 1^-} \sqrt{4 - x - 3x^2}$

11. Prove $\lim_{x \to 0} x \sin x = 0$.

12. Let $f(x)$ be a function which satisfies $|f(x)| \leq B$ for all x and some constant B. Prove $\lim_{x \to 0} x f(x) = 0$.

13. Let $g(x)$ satisfy $1 \le g(x) \le 2$ for all numbers x. For any a, show that either the limit $\lim_{x\to a} g(x)$ does not exist or $\lim_{x\to a} g(x) = L$ and $1 \le L \le 2$.

14. Let $f(x) = \frac{x}{|x|}$ for $x \ne 0$. Evaluate $\lim_{x\to 0^+} f(x)$ and $\lim_{x\to 0^-} f(x)$ and determine if $\lim_{x\to 0} f(x)$ exists.

15. Let $h(x) = \frac{x - |x|}{2}$. Evaluate $\lim_{x\to a} h(x)$ for every a. [Hint: Consider cases depending on a positive, negative or zero.]

Evaluate the one-sided limits.

16. $\lim_{x\to 2^+} \sqrt{x-2}$

17. $\lim_{x\to 1^-} \sqrt{x - x^2}$

18. $\lim_{x\to 3^+} \frac{|x-3|}{x-3}$

19. $\lim_{x\to 3^-} \frac{|x-3|}{x-3}$

20. Give an example of a piecewise-linear function $f(x)$ that satisfies

$$\lim_{x\to 3^+} f(x) = -5 \quad \text{and} \quad \lim_{x\to 3^-} f(x) = 5.$$

21. Give an example of a piecewise-linear function $f(x)$ that satisfies

$$\lim_{x\to 3^+} f(x) = -5 \quad \text{and} \quad \lim_{x\to 3^-} f(x) = 5$$

$$\lim_{x\to 5^+} f(x) = -3 \quad \text{and} \quad \lim_{x\to 5^-} f(x) = 3.$$

22. Give an example of a piecewise-linear function $f(x)$ that satisfies

$$\lim_{x\to -4^+} f(x) = 3 \quad \text{and} \quad \lim_{x\to 4^-} f(x) = 3.$$

23. Suppose f is an increasing function and ϵ is a given positive number. We wish to verify $\lim_{x\to a} f(x) = L$ by finding the largest positive number δ for which the statement

$$|f(x) - L| < \epsilon \quad \text{whenever} \quad |x - a| < \delta$$

is true. Prove that δ is obtained as follows: Let h and k be numbers such that $f(a+h) = L + \epsilon$ and $f(a-k) = L - \epsilon$. Then δ is the smaller of h and k. [In this problem we make an assumption that such h and k exist.]

2.7 Limits at infinity

In the previous sections, we have discussed limits of functions as the variable approaches a point. We now consider a limit in which the variable increases without bound (or as we say, the variable approaches infinity). To motivate this idea we consider an intuitive example.

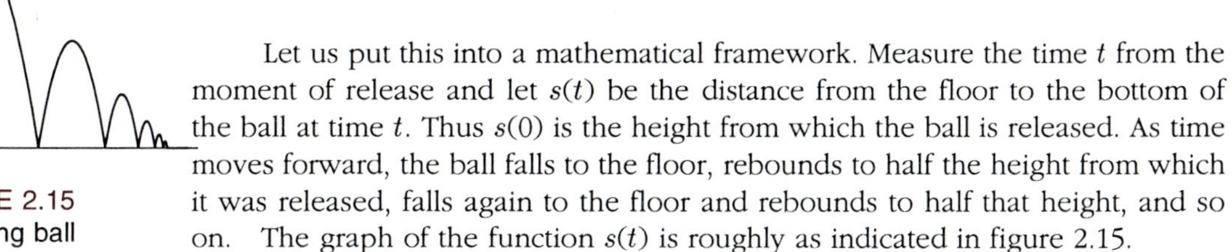

When a ball is dropped onto a solid floor, it rebounds to a height one-half of the height from which it fell. If the ball is released from a height of 3 ft above the floor, what is the ultimate position of the ball?

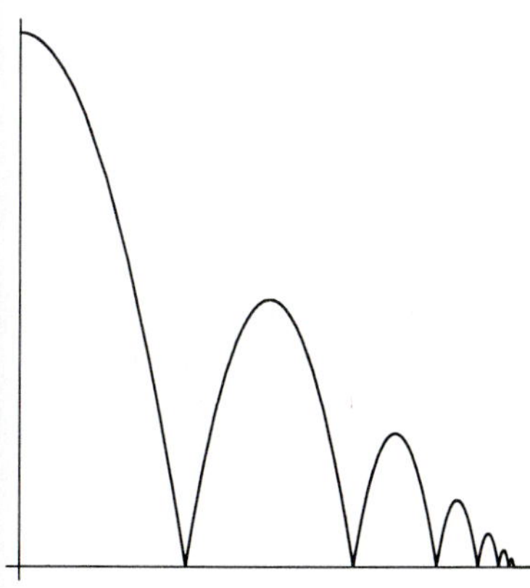

FIGURE 2.15
Bouncing ball

Let us put this into a mathematical framework. Measure the time t from the moment of release and let $s(t)$ be the distance from the floor to the bottom of the ball at time t. Thus $s(0)$ is the height from which the ball is released. As time moves forward, the ball falls to the floor, rebounds to half the height from which it was released, falls again to the floor and rebounds to half that height, and so on. The graph of the function $s(t)$ is roughly as indicated in figure 2.15.

The question posed in this problem asks for the "ultimate" position of the ball. How should this question be interpreted? The word "ultimate" should be interpreted as "after a sufficiently long time" or "for large values of t". Intuitively one would guess that the ultimate position of the ball is "on the floor"—the ball is at height 0. The mathematical formulation of this statement is

$$\lim_{t\to\infty} s(t) = 0.$$

The symbol ∞ is read "infinity" and the symbol $\lim_{t\to\infty}$ is read "the limit as t approaches infinity". (After we study geometric series, we prove that the ball comes to rest after a finite amount of time T, and so $s(t) = 0$ for all $t \geq T$.)

Here is the precise definition of the concept of a limit at infinity.

DEFINITION 2.5

Limit at Infinity

The limit of $f(x)$ as x approaches infinity is L if for any positive ϵ there exists a number N such that

$$|f(x) - L| < \epsilon \quad \text{whenever} \quad x > N.$$

When this is satisfied, we write

$$\lim_{x\to\infty} f(x) = L.$$

The condition $\lim_{x\to\infty} f(x) = L$ can be interpreted in terms of the graph of the function. For a given positive number ϵ, look at the strip between the two lines $y = L + \epsilon$ and $y = L - \epsilon$. There is some number N with the property that every point of the graph of $f(x)$ lies in the strip so long as x lies to the right of N; and moreover, such an N exists for each positive ϵ, no matter how small.

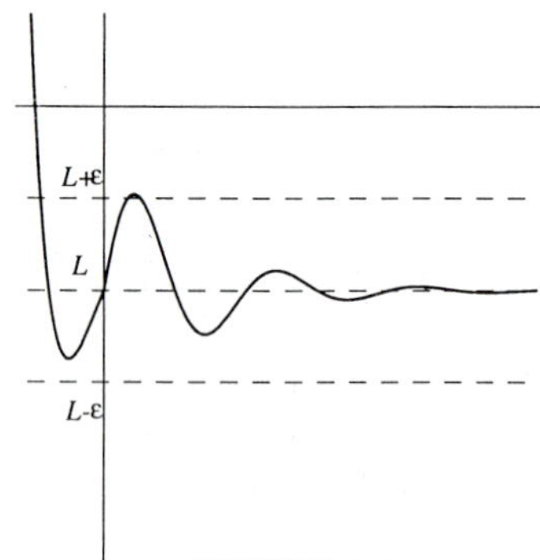

FIGURE 2.16
$\lim_{x\to\infty} f(x) = L$

EXAMPLE 1 Prove $\lim_{x\to\infty} \frac{1}{x} = 0$.

Solution: Select any positive number ϵ and let $N = 1/\epsilon$. If $N < x$ then x is positive and we have $1/x < 1/N = \epsilon$. This is precisely what is required of the definition; namely

$$\left|\frac{1}{x}\right| = \left|\frac{1}{x} - 0\right| < \epsilon \quad \text{whenever} \quad N < x.$$ ■

The Limit Theorems (2.4) and (2.5) that were proved for $\lim_{x\to a}$ are valid in the case where a is replaced by ∞. We do not prove these here because the previous proofs can be adapted to this case with only very slight modifications. We apply some of these results in the next example.

EXAMPLE 2 Evaluate the limit

$$\lim_{x\to\infty} \frac{x}{x+b}.$$

Solution: Observe that we cannot evaluate this limit simply by applying the rule for quotients; because the two limits $\lim_{x\to\infty} x$ and $\lim_{x\to\infty}(x + b)$ do not exist, we cannot evaluate the limit of the quotient $x/(x + b)$ by taking the limit of the numerator divided by the limit of the denominator.

Instead, we use the device of expressing the function in terms of $1/x$, which, in this case, is accomplished by dividing the numerator and denominator by x. Then

$$\begin{aligned}
\lim_{x\to\infty} \frac{x}{x+b} &= \lim_{x\to\infty} \frac{1}{1+\frac{b}{x}} \\
&= \frac{\lim_{x\to\infty} 1}{\lim_{x\to\infty} \left(1+\frac{b}{x}\right)} \\
&= \frac{\lim_{x\to\infty} 1}{\lim_{x\to\infty} 1 + b \lim_{x\to\infty} \frac{1}{x}} \\
&= \frac{1}{1+b\cdot 0} \\
&= 1
\end{aligned}$$

Think about this limit—it ought to seem reasonable to you; x is permitted to get larger and larger while b is a fixed constant. How large is the ratio of x to $x + b$? If x is enormously larger than $|b|$, then $x + b$ is relatively close to x and the ratio is close to 1. ■

Limits at Negative Infinity

The limit as x approaches infinity should be viewed as a process by which the variable x is moved farther and farther to the right along the real line. It is equally meaningful to let x move farther and farther to the left along the negative x-axis. The definition requires only a slight change.

DEFINITION 2.6

Limit at Negative Infinity

The limit of $f(x)$ as x approaches negative infinity is L if for any positive ϵ there exists a number N such that

$$|f(x) - L| < \epsilon \quad \text{whenever} \quad x < N.$$

When this is satisfied, we write

$$\lim_{x\to-\infty} f(x) = L.$$

For any constant a, $\lim_{x\to-\infty} \frac{a}{x} = 0$. Because the proof is almost the same as the corresponding result given above using $x \to \infty$ in place of $x \to -\infty$ here, we omit the details.

EXAMPLE 3 Evaluate $\lim_{x\to-\infty} \dfrac{2+3x}{5+|x|}$.

***Solution*:** Begin by dividing the numerator and denominator by x to produce terms of the shape a/x for constants a. For the second step, we deal with the absolute value sign. Because we are interested in values of x moving to the left on the axis, we can assume at the start that $x < 0$. Then $|x| = -x$ and we have

$$\begin{aligned}
\lim_{x\to-\infty} \frac{2+3x}{5+|x|} &= \lim_{x\to-\infty} \frac{\frac{2}{x}+3}{\frac{5}{x}-1} \\
&= \frac{0+3}{0-1} = -3.
\end{aligned}$$

■

Relations between Infinite and Finite Limits

In principle, one could avoid limits as x approaches either $\pm\infty$ with the following change of variable. Let $x = 1/t$ so that $t = 1/x$. As x approaches ∞, t is positive and approaches 0. As x approaches $-\infty$, t is negative and approaches 0. The reader can now prove the following result.

THEOREM 2.10

For any function $f(x)$ we have

$$\lim_{x\to\infty} f(x) = \lim_{t\to 0^+} f\left(\frac{1}{t}\right) \quad \text{and} \quad \lim_{x\to-\infty} f(x) = \lim_{t\to 0^-} f\left(\frac{1}{t}\right)$$

provided the limits exist.

We give one example to illustrate this idea.

EXAMPLE 4 Let

$$f(x) = \frac{6x^3 - 7x + 2}{2x^3 + x + 1}.$$

Evaluate the two limits $\lim_{x\to\infty} f(x)$ and $\lim_{x\to-\infty} f(x)$.

Solution: We make the change of variable $x = 1/t$ so that

$$f\left(\frac{1}{t}\right) = \frac{6(\frac{1}{t})^3 - 7(\frac{1}{t}) + 2}{2(\frac{1}{t})^3 + (\frac{1}{t}) + 1} = \frac{6 - 7t^2 + 2t^3}{2 + t^2 + t^3}.$$

Now apply the limit theorems to obtain

$$\lim_{t\to 0^+} \frac{6 - 7t^2 + 2t^3}{2 + t^2 + t^3} = \frac{6 - 0 + 0}{2 + 0 + 0} = 3$$

and

$$\lim_{t\to 0^-} \frac{6 - 7t^2 + 2t^3}{2 + t^2 + t^3} = \frac{6 - 0 + 0}{2 + 0 + 0} = 3.$$

It follows that $\lim_{x\to\infty} f(x) = \lim_{x\to-\infty} f(x) = 3$. ■

Functions Having Infinite Limits

There is another way in which the symbol ∞ is used. We have defined limit of $f(x)$ as x approaches ∞; now we assign a meaning to the phrase $f(x)$ approaches ∞.

DEFINITION 2.7

Infinity as a Limit

Let $f(x)$ be a function defined for all x on some interval (a, b). Then the statement

$$\lim_{x\to a^+} f(x) = \infty$$

is true if for any positive number U, there is some number $\delta > 0$ such that

$$f(x) > U \qquad \text{whenever} \qquad 0 < x - a < \delta.$$

Similarly, if $f(x)$ is defined for all x on some interval (c, a), then

$$\lim_{x \to a^-} f(x) = \infty$$

is true if for any positive number V, there is some $\delta > 0$ such that

$$f(x) > V \qquad \text{whenever} \qquad 0 < a - x < \delta.$$

If $f(x)$ is defined for all x on $(a - \delta, a + \delta)$ for some positive number δ and if the two one-sided limits satisfy

$$\lim_{x \to a^-} f(x) = \infty, \qquad \lim_{x \to a^+} f(x) = \infty$$

then we write

$$\lim_{x \to a} f(x) = \infty.$$

The first of these limits says that for any given number U, all the values $f(x)$ are larger than U if x is close enough to a but larger than a. The second says that all the values $f(x)$ are greater than any given V if x is close enough to a but smaller than a. This notion is introduced as a matter of convenience to abbreviate some statements. Take note that the statement

$$\lim_{x \to a} f(x) = \infty$$

does not mean that the limit of $f(x)$ as $x \to a$ exists. Quite the contrary, the limit does not exist because the symbol ∞ is not a number. The limit in the definition is a brief way of stating the property that f has all its values larger than any given number if x is sufficiently close to a.

EXAMPLE 5 Prove the following statement is correct:

$$\lim_{x \to 2^+} \frac{x^2 + 1}{x - 2} = \infty.$$

Solution: When x is near 2, $x^2 + 1$ is near 5 and the denominator is near 0. If also $x > 2$, then the fraction is positive and very large. So intuitively, the statement looks correct. Now we give a formal proof that the condition of the definition is satisfied. Write $f(x) = (x^2 + 1)/(x - 2)$.

Let U be any given number. If $U < 0$ then $f(x) > U$ for all $x > 2$ because $f(x) > 0$. In this case, any choice of δ satisfies the conditions in the definition.

Thus we assume that $U > 0$. Select δ as a positive number smaller than $1/U$ and also smaller than 1. The inequality $0 < x - 2 < \delta$ implies x lies on $(2, 3)$ so that $5 < x^2 + 1$; we also have $1/\delta < 1/(x - 2)$. Use these two inequalities to get

$$f(x) = \frac{x^2 + 1}{x - 2} > \frac{5}{x - 2} > \frac{5}{\delta} > 5U > U.$$

This proves the statement. ■

Similar definitions can be given for the limits

$$\lim_{x \to a^+} f(x) = -\infty, \qquad \lim_{x \to a^-} f(x) = -\infty.$$

The concept in these cases is that the values of f are less than any given number if x is sufficiently close to a and on the correct side of a. Rather than write new definitions for these two limits, we can say more simply

$$\lim_{x \to a^+} f(x) = -\infty \quad \text{is true if and only if} \quad \lim_{x \to a^+} -f(x) = \infty,$$

$$\lim_{x \to a^-} f(x) = -\infty \quad \text{is true if and only if} \quad \lim_{x \to a^-} -f(x) = \infty.$$

Exercises 2.7

Evaluate the limits.

1. $\displaystyle\lim_{x \to \infty} \frac{x+1}{x^2-3}$

2. $\displaystyle\lim_{x \to \infty} \frac{7x - 3x^2 - 4}{7x+2}$

3. $\displaystyle\lim_{x \to -\infty} \frac{5x-3}{2x^2+7x-1}$

4. $\displaystyle\lim_{x \to -\infty} \frac{3x^2 - x}{4x+3}$

5. $\displaystyle\lim_{x \to 0^+} \frac{1}{x}$

6. $\displaystyle\lim_{x \to 0^-} \frac{1}{x}$

7. $\displaystyle\lim_{x \to 1^+} \frac{4-x^2}{1-x}$

8. $\displaystyle\lim_{x \to 1^-} \frac{4-x^2}{1-x}$

9. $\displaystyle\lim_{x \to -3^+} \frac{x+2}{x^2-9}$

10. $\displaystyle\lim_{x \to -3^-} \frac{x+2}{x^2-9}$

11. $\displaystyle\lim_{x \to \infty} \frac{\sqrt{x^2+2x}}{\sqrt{4x^2-3x+1}}$

12. $\displaystyle\lim_{x \to -\infty} \frac{\sqrt{2-x^3}}{\sqrt{(2-x)(x^2+1)}}$

13. Prove a general result giving the value of the limit

$$\lim_{x \to \infty} \frac{a_n x^n + a_{n-1}x^{n-1} + \cdots + a_0}{b_m x^m + b_{m-1}x^{m-1} + \cdots + b_0},$$

where $a_n b_m \neq 0$. [Hint: Consider the three cases $n > m$, $n = m$, $n < m$.]

14. State and prove a version of the Squeeze Theorem for limits as x approaches infinity.

15. Use the version of the Squeeze Theorem for infinite limits to prove the following result: If $|f(x)| < B$ for all x greater than some fixed number, then $\lim_{x \to \infty} \frac{f(x)}{x} = 0$.

A line with equation $y = mx + b$ is called an asymptote for the graph of $y = f(x)$ if either $\lim_{x \to \infty} f(x) - (mx + b) = 0$ or $\lim_{x \to -\infty} f(x) - (mx + b) = 0$. The following problems deal with asymptotes of certain functions.

16. Show that the line $y = x$ is an asymptote for the function $y = \sqrt{x^2 - 1}$.

17. Show that the line $y = 2x - 2$ is an asymptote for $y = 2x^2/(x + 1)$.

18. Find an asymptote $y = mx + b$ for the function $y = 5x^3/(x^2 - 4)$.

19. Find an asymptote $y = mx + b$ for the function $y = (Ax^2 + Bx + C)/(x + D)$.

2.8 Pi and the circle

In this section we use limits to give a careful definition of π and prove the familiar properties relating the circumference of a circle and its radius. The important concept of similar triangles from plane geometry plays an essential role in the discussion.

Similarity

Recall that two triangles are similar if there is a correspondence between the angles of one triangle and the angles of the second triangle such that the corresponding angles are equal. In other words, the vertices can be labeled as ABC for one triangle and $A'B'C'$ for the second triangle so that $\angle A = \angle A'$,

$\angle B = \angle B'$, and $\angle C = \angle C'$. (The notation $\angle A$ means the angle at A.)

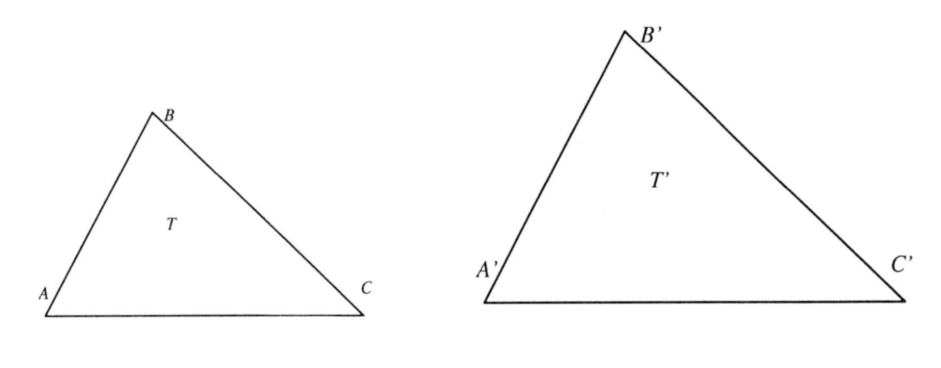

FIGURE 2.17

FIGURE 2.18

The following theorem recalls some facts from geometry that give information about similar triangles. Let $|AB|$ denote the length of the side AB.

THEOREM 2.11

Properties of Similar Triangles

The two triangles T and T' are similar if

$$\angle A = \angle A' \quad \text{and} \quad \frac{|AB|}{|AC|} = \frac{|A'B'|}{|A'C'|}.$$

If the triangles T and T' are similar, then there is a constant t such that

i. $t|AB| = |A'B'|, \quad t|AC| = |A'C'|, \quad t|BC| = |B'C'|$;
ii. $t \cdot$ (perimeter of T) = perimeter of T';
iii. $t^2 \cdot$ (area of T) = area of T'.

Arc Length

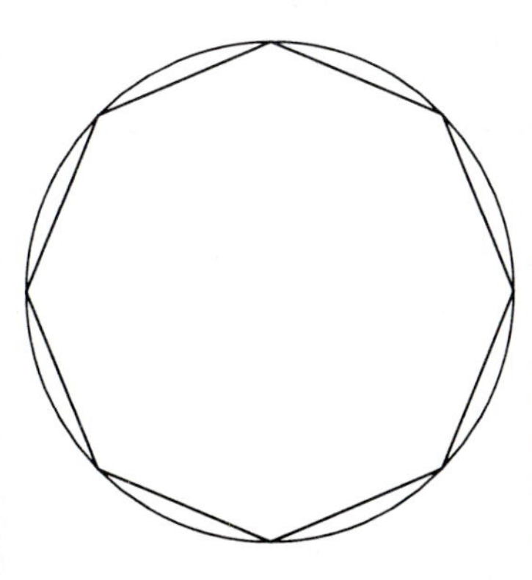

FIGURE 2.19
8-gon

To compute the circumference of a circle, we must have a working definition of the "length" of a curved arc. The length of a straight line is determined by superimposing the line on the x-axis and then taking the difference of the endpoints. We will use a limiting process to compute the circumference of the circle. Begin by inscribing a regular polygon inside a circle of radius R. The length of the chord connecting two consecutive vertices of the polygon is less than the length of the arc connecting them. Thus the perimeter of the polygon is less than the circumference of the circle. Let the perimeter of the regular polygon having n sides inscribed in the circle of radius R be denoted by $P_n(R)$ and let $C(R)$ denote the circumference of the circle of radius R. So far we see that $P_n(R) < C(R)$ for any n, but our goal is to determine the number $C(R)$.

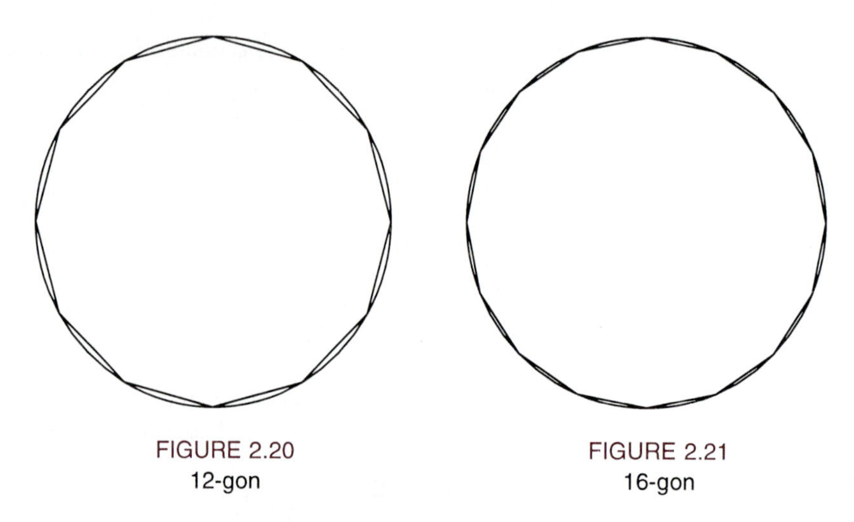

FIGURE 2.20
12-gon

FIGURE 2.21
16-gon

If the number of sides n of the polygon increases, the perimeter $P_n(R)$ increases and gets closer to $C(R)$. This is the idea for our computation of $C(R)$.

THEOREM 2.12

Circumference as a Limit

The circumference $C(R)$ of the circle of radius R is the limit of the perimeters of the regular inscribed n-sided polygons:

$$\lim_{n \to \infty} P_n(R) = C(R). \tag{2.20}$$

There is a technical point here. We can prove that this limit exists, and we could do that formally by using the completeness axiom to be discussed in section 5.4. The geometry of this situation should be sufficient to convince most readers that the limit does indeed exist.

The view taken here is that the circle has some circumference and the theorem gives a means for computing it. Another view that is perfectly reasonable is that the length of the curved arc is *defined* as a limit of the lengths of the inscribed chords. In the latter view, one proves that the limit (2.20) exists and that number is the circumference by definition.

Theorem 2.12 will now be used to prove some facts about circles. The first result shows that the circumference is a constant multiple of the radius.

THEOREM 2.13

Circumference of a Circle

There is a constant π with the property that the circumference of a circle of radius R equals $2\pi R$.

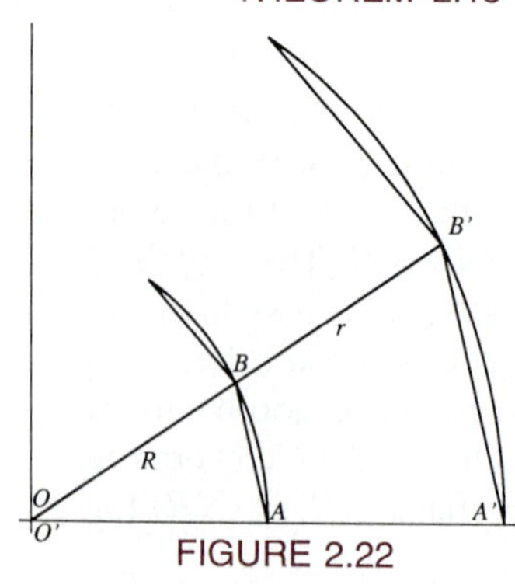

FIGURE 2.22

Proof

The important point in this theorem is that the constant is the same for all circles. Thus, we consider two circles, one of radius R, the other of radius r and inside each inscribe a regular polygon of n sides.

Refer to figure 2.22 and observe that the triangles OAB and $O'A'B'$ are similar ($\angle O = \angle O'$ and $|OA| = |OB| = R$, $|O'A'| = |O'B'| = r$). Thus there is a constant t such that $t|AB| = |A'B'|$ and $t|OA| = |O'A'|$. This implies that $tR = r$ or $t = r/R$. The perimeters of the n-gons are $P_n(R) = n|AB|$ and $P_n(r) = n|A'B'| = nt|AB|$. Thus we have

$$\frac{C(R)}{C(r)} = \lim_{n\to\infty} \frac{P_n(R)}{P_n(r)} = \lim_{n\to\infty} \frac{n|AB|}{nt|AB|} = \frac{1}{t} = \frac{R}{r}.$$

This is equivalent to the equality

$$\frac{C(R)}{R} = \frac{C(r)}{r}.$$

The equality implies that the quotient of the circumference by the radius is the same for any two circles. We use the symbol π to denote one-half this constant so that we write

$$\frac{C(R)}{R} = 2\pi \quad \text{or} \quad C(R) = 2\pi R.$$

This proves the theorem. (The reason for using π to denote one-half the constant rather than the constant itself is that we conform to the tradition that π stands for the ratio of the circumference to the *diameter* and not the ratio of the circumference to the radius.) □

The theorem gives no indication of the numerical value of π which is approximately 3.1415926536. . . . In later chapters we will find methods to approximate the value of π. Approximations are necessary because the exact decimal value cannot be written; it is a nonrepeating infinite decimal.

For later use we isolate one point in this computation.

Corollary 2.2

If ℓ_n is the length of one side of a regular n-sided polygon inscribed in a circle of radius 1, then

$$\lim_{n\to\infty} n\ell_n = 2\pi.$$

Proof

Because $P_n(1) = n\ell_n$, this is just the computation of the circumference of the circle of radius 1. □

Radian Measure

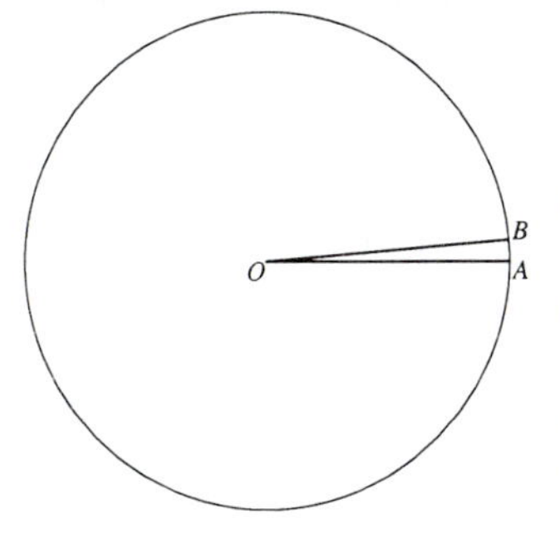

FIGURE 2.23
AOB about 5°

How are angles measured? One begins by placing angles in a standard position and then selecting a unit to represent the angle swept out in one complete revolution. If "degrees" is the unit of measurement, then 360 is the measure of one complete revolution. The measure of the angle AOB in figure 2.23 is 360 times the fraction representing the ratio of the sector AOB to the whole circle. If the sector is $1/360$ of the whole circle, then angle AOB is 1 degree.

The degree is not the most useful unit for measuring angles in the study of trigonometric functions. The use of *radian* measure is more suitable; the radian measure of an angle is closely related to the length of the circular arc subtended by the angle. Given an angle of measure t in degrees, consider a circular arc with t as the central angle in a circle of radius R. The angle swept out by a full circle is 360 degrees so this sector represents the fraction $t/360$ of the entire circular angle. The arc from A to B must have length equal to $t/360$ times the circumference of the circle. Thus

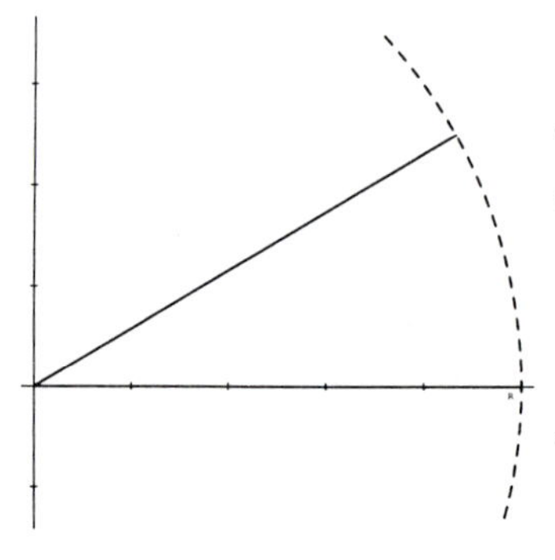

FIGURE 2.24

$$\widehat{AB} = \frac{t}{360} 2\pi R.$$

The radian measure of the angle AOB is the quotient of the length of the arc $\widehat{AB}$ divided by the radius; this gives relations:

$$\text{measure of } \angle AOB \text{ in degrees} = t$$

$$\text{measure of } \angle AOB \text{ in radians} = \frac{2\pi t}{360}.$$

Thus the relationship between degrees t and radians r is

$$\frac{r}{2\pi} = \frac{t}{360}.$$

Note that the radian measure of an angle is a real number without units. Moreover every real number is the radian measure of some angle. A given number r is the radian measure of an angle having measure in degrees equal to $360r/2\pi$.

Exercises 2.8

1. Compute the perimeter of a square inscribed in the circle of radius 1.
2. Compute the perimeter of a square inscribed in the circle of radius R.
3. Compute the perimeter of a regular hexagon inscribed in the circle of radius R.
4. Compute the perimeter of the square that is circumscribed around the circle of radius R.
5. Compute the perimeter of the regular hexagon that is circumscribed around the circle of radius R.
6. Compute the ratio $P_{2n}(R)/P_n(R)$ of the perimeter of the regular polygon with $2n$ sides to the perimeter of the regular polygon of n sides inscribed in the circle of radius R. Express the ratio in terms of the central angle determined by one side of the $2n$-gon.

2.9 Trigonometric functions

The six trigonometric (sine, cosine, tangent, cotangent, secant and cosecant) functions are defined in elementary trigonometry as functions of angles, measured in degrees; the values of the trigonometric functions are ratios of sides of right triangles. For example, if the right triangle ABC has angle A measuring t in degrees, then

$$\sin t = \frac{|BC|}{|AC|} \quad \text{and} \quad \cos t = \frac{|AB|}{|AC|}.$$

We will not use these as the definitions of the trigonometric functions but rather these will be seen as properties of a more general definition that uses radian measure and the points of a circle rather than degrees and triangles. The sine and cosine functions are defined as the coordinates of a point on the unit circle and then the remaining four trigonometric functions are defined using the sine and cosine functions. We recall that an angle having radian measure t is in standard position if it has vertex at the origin, has the positive x-axis as one side and obtains its remaining side by rotating the positive x-axis through t radians; when $t > 0$ the rotation is counterclockwise; when $t < 0$ the rotation is clockwise. When referring to the *unit circle*, we mean the circle of radius 1 having center at the origin.

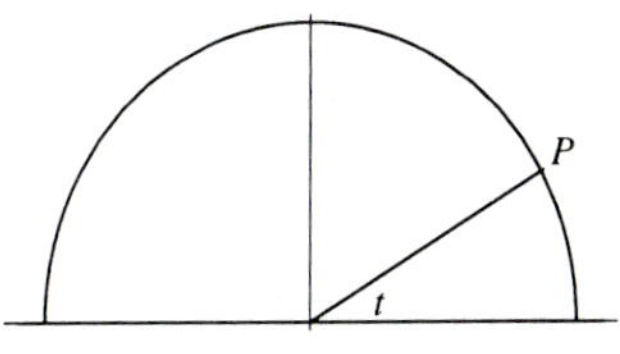

FIGURE 2.25
Definition: $P = (\cos t, \sin t)$

DEFINITION 2.8

Sine and Cosine Functions

For any real number t the two functions $\sin t$ and $\cos t$ are defined by the condition that $(\cos t, \sin t)$ is the point on the unit circle determined by an angle in standard position of radian measure t.

The following theorem contains some of the familiar properties of the trigonometric functions. Each statement follows from the definition and some elementary geometry.

THEOREM 2.14

Elementary Properties of Sine and Cosine

The functions $\cos t$ and $\sin t$ have the following properties, which hold for every number t:

i. $\cos^2 t + \sin^2 t = 1$,
ii. $-1 \leq \cos t \leq 1$ and $-1 \leq \sin t \leq 1$,
iii. $\cos(-t) = \cos t$ and $\sin(-t) = -\sin t$,
iv. $\cos(t + 2\pi) = \cos t$ and $\sin(t + 2\pi) = \sin t$.

Statement iii implies that the cosine function is an *even function* and that the sine function is an *odd function.* (More generally, a function $f(x)$ is called an *even function* if $f(-x) = f(x)$ and an *odd function* if $f(-x) = -f(x)$.)

The statement iv implies that the functions sine and cosine are periodic functions with 2π as a period. (More generally, $f(x)$ is periodic with T as a period if $f(x + T) = f(x)$ for all x.)

Graphs of Sine and Cosine

The definition of the function $f(x) = \cos x$ can be used to obtain a reasonable sketch of the graph. Think of the point $(\cos x, \sin x)$ as moving along the circumference of the unit circle; at $x = 0$, the point is at $(1, 0)$; as x increases away from 0, the point moves up and to the left along the circumference. When x reaches π, the point has moved as far to the left as possible and it begins to

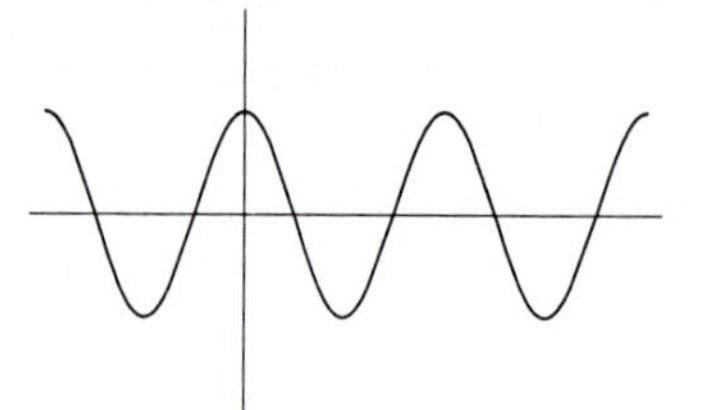

FIGURE 2.26
$y = \cos x$

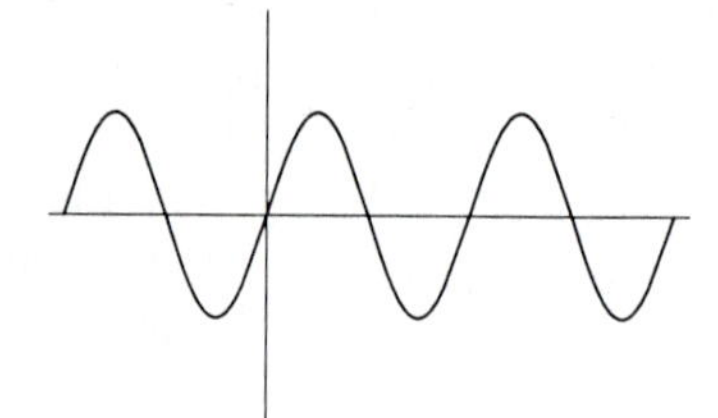

FIGURE 2.27
$y = \sin x$

move down and to the right along the lower half of the circle; when $x = 2\pi$, the point has returned to its starting position. What happens to the function $\cos x$ as x varies in this way? This is asking what happens to the x coordinate of the moving point. The x coordinate begins at 1, decreases in value until reaching 0 when $x = \pi/2$; the value is -1 when $x = \pi$ and then x increases until it returns to its starting value 1 when $x = 2\pi$. This cycle is repeated as x increases. Figure 2.26 shows the graph of the cosine function. Note that the graph crosses the x-axis at $x = \pi/2$, at $x = 3\pi/2$, and more generally at $x = \pi/2 + n\pi$, for any integer n.

The periodic property implies that the graph looks exactly the same over the interval $[a, b]$ as it does over the interval $[a + 2\pi, b + 2\pi]$ for any a, b. So once we know the graph over the interval $[0, 2\pi]$, we can extend the graph in both directions to obtain the graph for all x.

A similar analysis shows that the sine curve has the shape in figure 2.27. We see that $\sin x = 0$ when $x = n\pi$, for any integer n.

Limits Involving Sine and Cosine

We give geometric proofs of the following two limit results.

THEOREM 2.15

For any number a the following two limits hold:

i. $\lim_{x \to a} \sin x = \sin a$,

ii. $\lim_{x \to a} \cos x = \cos a$.

Proof

The definition of limit requires that we produce a positive δ for each given positive ϵ so that certain conditions are met. Let ϵ be any positive number and consider the three horizontal lines $y = \epsilon + \sin a$, $y = \sin a$ and $y = -\epsilon + \sin a$. We assume ϵ is small enough that at least one of the two lines $y = \pm\epsilon + \sin a$ intersects the unit circle. Let δ and γ be positive numbers such that $(\cos(a - \gamma), \sin a + \epsilon)$ and $(\cos(a + \delta), \sin a - \epsilon)$ are the intersections of the lines with the circle near $P = (\cos a, \sin a)$ (see figure 2.28). Measure the length of the arc of the unit circle from the point P to these two intersections and suppose $(\cos(a + \delta), \sin a - \epsilon)$ is the endpoint of the smaller of these lengths. If x is a number such that $0 < |x - a| < \delta$, then the point

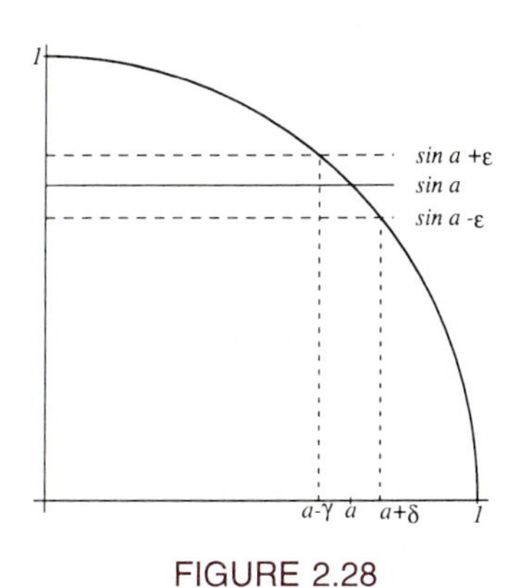

FIGURE 2.28

$Q = (\cos x, \sin x)$ lies on the arc of the circle with distance from P less than δ; thus Q lies in the strip between the lines $y = \pm\epsilon + \sin a$. This means that the difference in the y coordinates of P and Q is at most ϵ; that is

$$|\sin a - \sin x| < \epsilon \quad \text{whenever} \quad 0 < |x - a| < \delta.$$

Thus $\lim_{x \to a} \sin x = \sin a$.

The analogous argument using the vertical lines $x = \pm\epsilon + \cos a$ and $y = \cos a$ will prove the statement $\lim_{x \to a} \cos x = \cos a$. We leave this proof for the reader. □

The concept of continuity of functions is introduced in chapter 3. Theorem 2.15 states that the functions $\sin x$ and $\cos x$ are continuous at every point.

Definitions of the other Trigonometric Functions

The remaining four trigonometric functions are the tangent, secant, cotangent and cosecant functions. These are defined in terms of the sine and cosine functions.

DEFINITION 2.9

Trigonometric Functions

The functions $\tan x$, $\sec x$, cot and $\csc x$ are defined by the equations

$$\tan x = \frac{\sin x}{\cos x}, \quad \cot x = \frac{\cos x}{\sin x}, \quad \sec x = \frac{1}{\cos x}, \quad \csc x = \frac{1}{\sin x}.$$

The domain of each of these functions is the set of all numbers for which the denominator is nonzero.

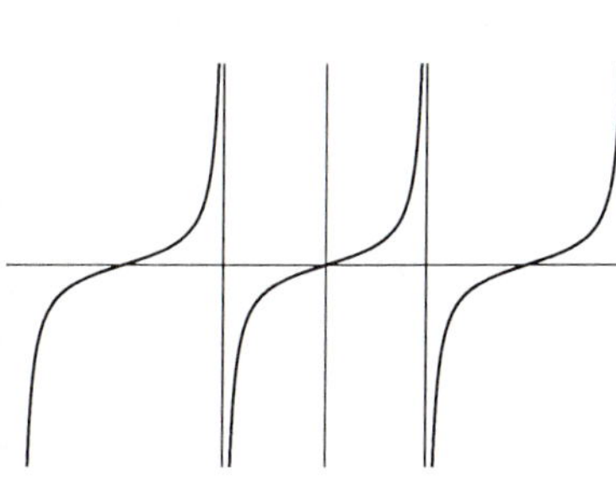
FIGURE 2.29
$y = \tan x$
Asymptotes at $x = \pm\frac{\pi}{2}$

Let us examine the tangent function closely enough to get a sketch of its graph. Both the sine and cosine functions are periodic with 2π as a period, so it follows that $\tan(x + 2\pi) = \tan x$ for all x. Thus the graph of the tangent function is completely known if we know it for any interval of length 2π. Let us consider the interval $I = (-\frac{\pi}{2}, \frac{3\pi}{2})$. The function $\tan x$ is not defined at $x = \pi/2$ or $3\pi/2$ because the cosine function equals 0 at these points; these are the only points on I where $\tan x$ is not defined. Consider the behavior of $\tan x$ as x begins just to the right of $-\pi/2$ and moves farther to the right. Then $\cos x$ is near 0 and positive and $\sin x$ is near 1 and negative; thus $\tan x$ is negative but large in absolute value. The $\sin x$ term is getting smaller, the $\cos x$ term is getting larger, so the combined effect is that $\tan x$ gets smaller in absolute value but stays negative until $x = 0$ where the tangent function is also 0. As x moves from 0 to $\pi/2$, $\tan x$ continues to increase without bound. This is not a precise analysis but is sufficient to produce the sketch shown in figure 2.29.

Two Important Limits

Before finishing this section, we compute two more limits that will arise in later problems. The limits of $\sin x/x$ and $(1 - \cos x)/x$ as x approaches 0 are needed for the computation of the derivative of $\sin x$ and $\cos x$.

THEOREM 2.16

Two Trigonometric Limits

i. $$\lim_{x\to 0}\frac{\sin x}{x}=1$$

ii. $$\lim_{x\to 0}\frac{1-\cos x}{x}=0.$$

Proof

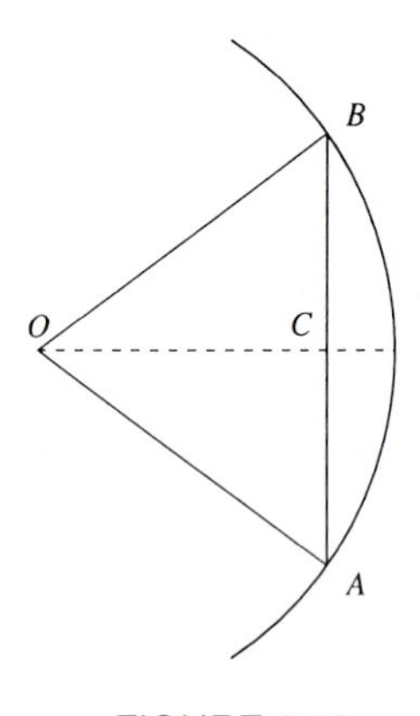

FIGURE 2.30
One side of n-gon

We begin with a regular n-sided polygon inscribed in a circle of radius 1. The central angle subtended by one side of the polygon has radian measure $2\pi/n$. By an inspection of the figure, we determine $\angle BOC=\pi/n$ and that one side, AB, of the n-gon has length

$$\ell_n=2\sin\left(\frac{\pi}{n}\right).$$

By Corollary 2.2 we obtain

$$\lim_{n\to\infty} n\ell_n=\lim_{n\to\infty}2n\sin\left(\frac{\pi}{n}\right)=2\pi.$$

This can be rewritten in the form

$$\lim_{n\to\infty}\frac{\sin\frac{\pi}{n}}{\frac{\pi}{n}}=1. \quad \textbf{(2.21)}$$

This is quite close to statement i; the difference lies in the fact that the limit in the theorem is taken over all real x approaching 0, whereas the limit (2.21) is taken over a special sequence of numbers $x=\pi/n$ with integer values of n. The limit (2.21) will be used to prove i. We are interested in very small values of x so we assume that $0<x<1$. (Similar arguments can be made for negative values of x.) Then there is an integer n that satisfies

$$\frac{\pi}{n+1}<x\le\frac{\pi}{n}. \quad \textbf{(2.22)}$$

In fact, n is the integer that satisfies

$$n\le\frac{\pi}{x}<n+1.$$

Notice that n depends on the choice of x; as x gets smaller, n gets larger and as x approaches 0, n approaches ∞.

For the values we are considering (all near 0) the sine function is increasing. This means that for $u<v$ and $|u|,|v|$ small enough, we have $\sin u<\sin v$. This is easily seen by using the definition of the sine function or by an inspection of the graph. Apply this to the three numbers in inequality (2.22) to get

$$\sin\frac{\pi}{n+1}<\sin x\le\sin\frac{\pi}{n}. \quad \textbf{(2.23)}$$

Combine inequalities (2.22) and (2.23) to get

$$\frac{\sin\frac{\pi}{n+1}}{\frac{\pi}{n}}<\frac{\sin x}{x}\le\frac{\sin\frac{\pi}{n}}{\frac{\pi}{n+1}}.$$

We will show that the terms on the left and right of this inequality both approach 1 as x approaches 0 and n approaches ∞. That is enough to prove statement i in the theorem.

We use the limit (2.21) on the rightmost term as follows:

$$\begin{aligned}\lim_{n\to\infty}\frac{\sin\frac{\pi}{n}}{\frac{\pi}{n+1}} &= \lim_{n\to\infty}\frac{\sin\frac{\pi}{n}}{\frac{\pi}{n}}\,\frac{n+1}{n}\\ &= \lim_{n\to\infty}\frac{\sin\frac{\pi}{n}}{\frac{\pi}{n}}\lim_{n\to\infty}\frac{n+1}{n}\\ &= 1\cdot 1 = 1.\end{aligned}$$

The limit of the left term is proved to be 1 by a similar argument and so statement i is proved.

To show statement ii, we make use of statement i and the fact that $\lim_{x\to 0}(1+\cos x) = 2$;

$$\begin{aligned}\lim_{x\to 0}\frac{1-\cos x}{x} &= \frac{1}{2}\lim_{x\to 0}(1+\cos x)\lim_{x\to 0}\frac{1-\cos x}{x}\\ &= \frac{1}{2}\lim_{x\to 0}\frac{1-\cos^2 x}{x}\\ &= \frac{1}{2}\lim_{x\to 0}\frac{\sin^2 x}{x}\\ &= \frac{1}{2}\lim_{x\to 0}\sin x\lim_{x\to 0}\frac{\sin x}{x}\\ &= \frac{1}{2}\cdot 0\cdot 1 = 0.\end{aligned}$$

This completes the proof of both parts of the theorem. □

The two limits in Theorem 2.16 have some interesting properties that should be noted. One of the limit theorems tells us that

$$\lim_{x\to 0}\frac{f(x)}{g(x)} = \frac{\lim_{x\to 0} f(x)}{\lim_{x\to 0} g(x)}$$

provided all limits exist and $\lim g(x) \neq 0$. The limits in the theorem have the form $f(x)/g(x)$ but, in both cases, $\lim f(x) = 0$ and $\lim g(x) = 0$. The quotients in statements i and ii have different limits in spite of this similar behavior. The quotient of two functions, each having limit 0, is called an *indeterminant form*; limits of indeterminant forms are studied in chapter 4 in the section on L'Hôpital's rule.

Exercises 2.9

Sketch the graph of the functions. Use a plotting calculator or a computer to make accurate sketches of the graphs.

1. $f(x) = 2\sin 4x$

2. $f(x) = 3\cos(4x-\pi)$

3. $f(x) = \cos 2x + \sin 2x$

4. $f(x) = \sin x\cos x$

5. $f(x) = \sin(x^2)$

6. $f(x) = \sin(1/x)$

7. $f(x) = \tan 3x$

8. $f(x) = \sec(x - \frac{\pi}{2})$

9. $f(x) = \csc 3x$

10. $f(x) = \sin\left(\frac{x+1}{x}\right)$

Evaluate the limits that exist.

11. $\lim_{x\to\pi}\sin^2 x\cos 2x$

12. $\lim_{x\to\frac{\pi}{2}}\sin^2 x + \cos^2 x$

13. $\lim_{x\to 0}\frac{\sin 7x}{x}$

14. $\lim_{t\to 0}\frac{\sin 4t}{\sin 3t}$

15. $\lim_{x\to 0}\frac{1-\cos x}{\sin x}$

16. $\lim_{x\to 0}\frac{\sec x}{x} - \frac{1}{x}$

17. $\lim_{x\to 0}\frac{\tan 2x}{x}$

18. $\lim_{t\to\pi}\frac{\sec^2 t - \tan^2 t}{\cos 5t}$

19. $\lim_{x\to\infty}\frac{\sin x}{x}$

20. $\lim_{w\to\infty}\sin\left(\frac{w+1}{w^2}\right)$

21. $\lim\limits_{w \to \infty} \dfrac{w - \cos w^2}{w^2 + 4}$

22. $\lim\limits_{x \to \infty} \dfrac{(x^2 - 1)\tan x}{x^2 \sin x}$

23. Use the central angle determined by one side of the regular n-gon and the trigonometric functions to express the perimeter of the regular n-gon inscribed in the circle of radius R.

24. Use the central angle determined by one side of the regular n-gon and the trigonometric functions to express the perimeter of the regular n-gon circumscribed around the circle of radius R.

3 Continuity and the Derivative

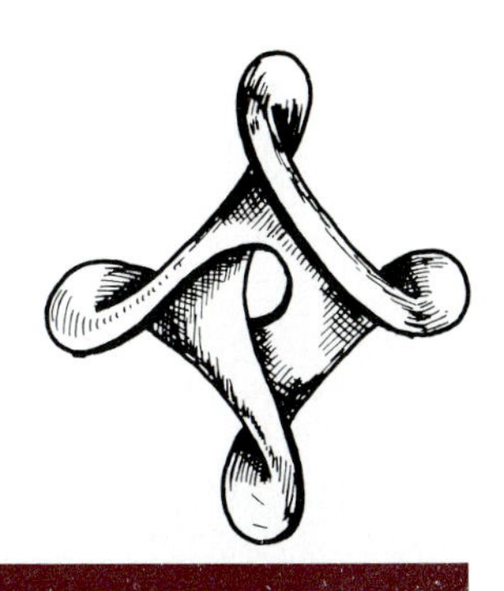

When we describe a physical or mathematical phenomenon using a function to evaluate some related quantity, the derivative of the function describes how the quantity is changing. For example consider a function of time that describes the position of a moving particle. The rate of change of position is velocity; the rate of change of velocity is acceleration.

The derivative of functions are used to help discuss problems about the maximum and minimum values and the growth of functions. It has a very concrete geometric interpretation reflected in the graph of the function.

The notion of continuity is studied in preparation for a study of the derivative. Roughly speaking, continuous functions are those whose graphs have no gaps or breaks; the important property of a continuous function is that its value at a point is "close" to its values at nearby points.

3.1 Continuous functions

In the definition of $\lim_{x\to a} f(x)$, we are concerned with values of x in an interval containing a but with $x \neq a$. The notion of continuity considers the case in which the function is defined at a; the value $f(a)$ plays a critical role. The limit of $f(x)$ as x approaches a may or may not be $f(a)$. When the limit does equal $f(a)$, there is a connection between the values of $f(x)$ for x near a with the value $f(a)$. Here is the precise definition.

DEFINITION 3.1

Continuity of $f(x)$ at a Point

The function $f(x)$ is continuous at a if

i. $f(x)$ is defined at $x = a$,
ii. $\lim_{x\to a} f(x)$ exists,
iii. $\lim_{x\to a} f(x) = f(a)$.

The three conditions in the definition are stated separately to emphasize these essential points. Here is a shorter, but equivalent, statement of continuity at a point.

Alternative Statement

The function $f(x)$ is continuous at $x = a$ if and only if

$$\lim_{x \to a} f(x) = f(a).$$

One can think of continuity as a property that says the value $f(a)$ is close to the value $f(c)$ if a is close to c.

Here are some properties of continuous functions.

THEOREM 3.1

Properties of Continuous Functions

Suppose that $f(x)$ and $g(x)$ are continuous at a. Then

i. $f(x) + g(x)$ is continuous at a,

ii. $f(x)g(x)$ is continuous at a,

iii. $\dfrac{f(x)}{g(x)}$ is continuous at a provided $g(a) \neq 0$.

Proof of these statements follows from corresponding statements about limits as x approaches a.

We have already proved that polynomials are continuous functions. The content of Theorem 2.7 is precisely the Alternative Statement of the definition of continuity for polynomials. Similarly, by Corollary 2.1 on limits of quotients, rational functions are continuous at every point where they are defined. For example the rational function

$$f(x) = \frac{4x^3 - 7x + 1}{(x - 2)(x + 1)}$$

is continuous at every point except at $x = 2$ and $x = -1$; at these two points the denominator is 0 and $f(x)$ is not defined.

Composite Functions

In addition to sums, products and quotients, there is one more general construction of functions for which continuity is preserved, namely composite functions. Suppose we are given two functions $f(x)$ and $g(x)$, and we form the composite function $h(x) = f(g(x))$. To draw a conclusion about the continuity of h at a, we need information about the continuity of g at a and the continuity of f at $g(a)$. Here is the complete statement.

THEOREM 3.2

Continuity of Composite Functions

Suppose $g(x)$ is continuous at $x = a$ and $f(x)$ is continuous at $x = g(a)$. Then the composite function $h(x) = f(g(x))$ is continuous at $x = a$.

Proof

Select any positive number ϵ. Because $f(x)$ is continuous at $g(a)$, there is a positive number δ_1 such that

$$|h(x) - h(a)| = |f(g(x)) - f(g(a))| < \epsilon \quad \text{whenever} \quad 0 < |g(x) - g(a)| < \delta_1.$$

Now by the continuity of $g(x)$ at a (and with δ_1 in place of ϵ in the definition) there is a positive δ such that

$$|g(x) - g(a)| < \delta_1 \quad \text{whenever} \quad 0 < |x - a| < \delta.$$

Thus $0 < |x - a| < \delta$ implies $|h(x) - h(a)| < \epsilon$ as required for the continuity of h at a. □

EXAMPLE 1 Show that the function

$$h(x) = \sqrt{\frac{x^4 + 3x^2 + 1}{2x^6 + 5}}$$

is continuous at every point.

Solution: To see this first observe that h is a composite of

$$f(x) = \sqrt{x} \quad \text{and} \quad g(x) = \frac{x^4 + 3x^2 + 1}{2x^6 + 5}.$$

The function $g(a)$ is continuous and positive at every a and so for every a, $f(a)$ is continuous at $g(a)$. Thus h is continuous at every point. ■

Continuity on an Interval

We say $f(x)$ is continuous on an interval (b, c) if it is continuous at every point of the interval. To be continuous at a point a, the function must be defined on some open interval containing a. This requirement causes some slight difficulty if we want to define continuity of $f(x)$ over the closed interval $[b, c]$. The problem is that the function might not be defined on any open interval containing b (or c). We modify this requirement in the following way. We say $f(x)$ is continuous on $[b, c]$ if $f(x)$ is defined on $[b, c]$, is continuous on (b, c), and if

$$\lim_{x \to b^+} f(x) = f(b) \quad \text{and} \quad \lim_{x \to c^-} f(x) = f(c).$$

For example, the function $f(x) = \sqrt{x}$ is continuous on the interval $[0, 2]$ because it is continuous on $(0, 2)$ and also at $x = 2$ (because it is continuous at every point $a > 0$), and at 0 we have

$$\lim_{x \to 0^+} \sqrt{x} = 0 = f(0).$$

Note that in this example, the function is not defined in any open interval containing 0.

The Graph of a Continuous Function

Consider the graph of some function $f(x)$ shown in figure 3.1. There are several obvious points where the function fails to be continuous.

The function is not continuous at 2 because the graph suggests that

$$\lim_{x \to 2^-} f(x) = 3 \quad \text{and} \quad \lim_{x \to 2^+} f(x) = 5.$$

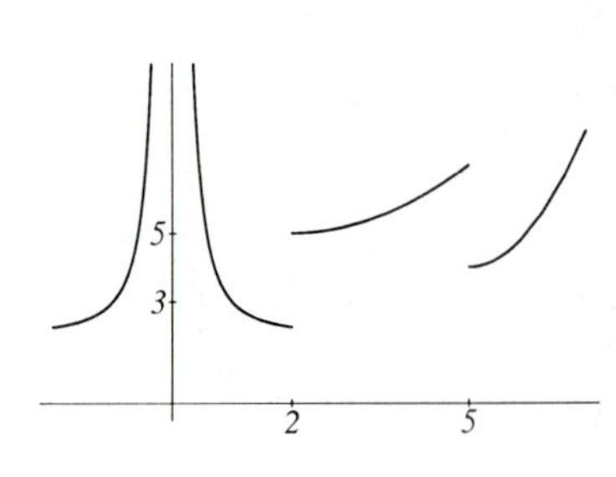

FIGURE 3.1
A discontinuous function

Thus $\lim_{x \to 2} f(x)$ does not exist and f is not continuous at 2. Similarly f is not continuous at 5. The function is not continuous at 0 because it is not even defined at 0. The graph suggests that f is continuous at all other points.

The graph of a continuous function can be "drawn" in a continuous stroke of the pen; that is the graph has no breaks or jumps that would require the pen to be lifted from the paper. Using this nonrigorous and intuitive description of continuity, some nontrivial consequences, which will not be proved now, seem quite plausible.

Intermediate Value Property

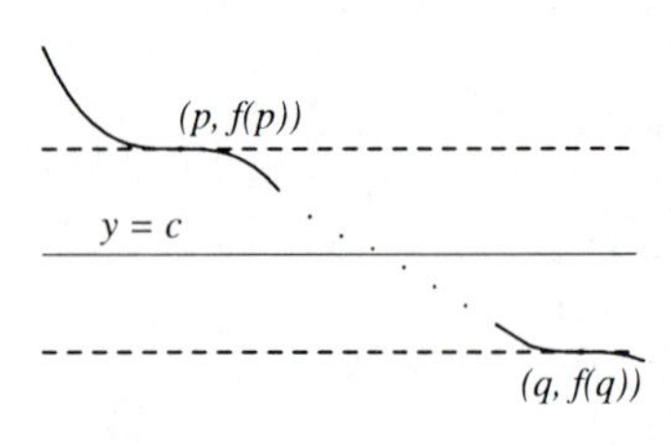

FIGURE 3.2
Graph crosses $y = c$

We give a brief introduction to the consequences of the intermediate value property. Let us restrict our attention to the graph of a continuous function $f(x)$ for x on a closed interval $[a, b]$. We draw the graph by placing our pen at the point $(a, f(a))$ and tracing out the curve in one continuous stroke of the pen until we stop at the point $(b, f(b))$. Let c be a number between the two values $f(a)$ and $f(b)$. Then the horizontal line $y = c$ lies between the horizontal lines through $(a, f(a))$ and $(b, f(b))$. Hence the graph of f crosses the line $y = c$ at some point $(u, c) = (u, f(u))$. Thus there is a number u between a and b that satisfies $f(u) = c$. The conclusion to be drawn from this reasoning is that every number c that lies between two values of the function on the interval $[a, b]$ is itself equal to a value of the function at some point on the interval. Here is a general version of this property.

THEOREM 3.3

Intermediate Value Theorem

Let $f(x)$ be continuous on the interval $[a, b]$ and let p and q be any two points on this interval. If c is a number that satisfies $f(p) \leq c \leq f(q)$, then there is a point r between p and q such that $c = f(r)$.

The intuitive idea of tracing the curve, as described just before the theorem, makes this result plausible but because it makes use of an informal principle about drawing graphs, it is not a rigorous proof.

Here is an illustration of the power of this theorem.

EXAMPLE 2 Let $f(x) = x^7 - 32x^4 + 19$. Prove that there exists a solution to the equation $f(x) = 0$ that lies on the interval $(0, 1)$.

Solution: Because $f(x)$ is a polynomial, it is continuous on every interval. We observe that $f(0) = 19$ and $f(1) = -12$. Because 0 lies between 19 and -12, the Intermediate Value Theorem implies there is a number s on the interval $[0, 1]$ with $f(s) = 0$. The graph of $f(x)$ is shown in the two figures in different scales.

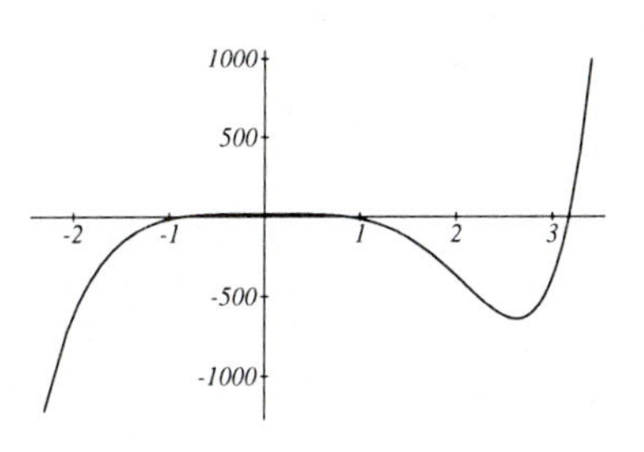

FIGURE 3.3

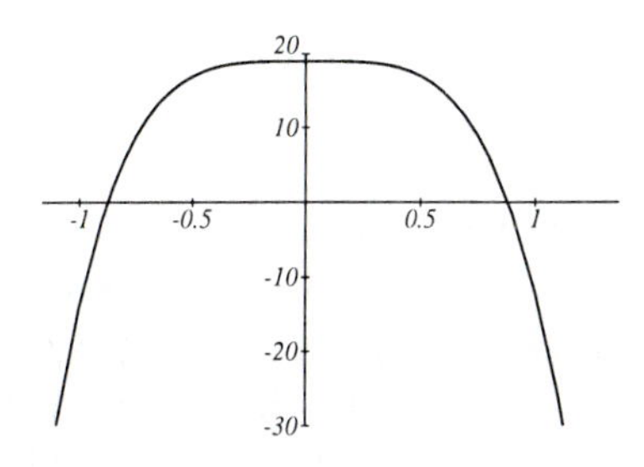

FIGURE 3.4

In this example, an accurate rendering of the graph of the polynomial is very helpful for finding the solutions of $f(x) = 0$. Accurate renderings done by hand may require a great deal of computation so that a sufficiently large number of points can be plotted. Calculators or computers with good graphics output can be very helpful for this.

The Intermediate Value Theorem has theoretical and practical applications. It is the basis for many results about solutions of equations $f(x) = 0$. Some of these ideas are treated more fully in chapter 11.

Exercises 3.1

Find the largest set of numbers on which the function is continuous.

1. $\dfrac{x}{x-3}$ **2.** $\sqrt{5-x}$

3. $\dfrac{2x+3}{x^2-4}$ **4.** $\sin\left(\dfrac{x+1}{x-1}\right)$

5. $\sqrt{x^3+5}$ **6.** $\dfrac{4-6x+x^2}{9-4x-x^2}$

7. $\dfrac{|x-1|}{x-1}$ **8.** $\dfrac{x|x|-x^2}{x+|x|+1}$

Sketch the graphs of the functions and determine the points where the function fails to be continuous.

9. $g(x) = \begin{cases} -1 & x \le -1 \\ x & -1 < x < 1 \\ 0 & 1 \le x \end{cases}$

10. $g(x) = \begin{cases} 2x+4 & x \le -2 \\ x^2-4 & -2 < x < 2 \\ 2x-4 & 2 \le x \end{cases}$

11. $g(x) = \begin{cases} \sqrt{-x} & x \le 0 \\ \sqrt{x} & 0 < x \end{cases}$

12. $g(x) = \begin{cases} x+4 & x \le -4 \\ 3x & -4 < x < 0 \\ 1-x & 0 \le x \end{cases}$

Prove there is at least one solution to each of the following equations; locate some interval $[a, b]$ that contains a solution.

13. $x^4 - 3x + 1 = 0$ **14.** $x^4 + 6x^2 - 3x - 2 = 0$

15. $(x^2+4)(3x^4+2) = 10$ **16.** $\dfrac{x+4}{x^3-8} = 9$

Each of the next equations has exactly three solutions. Find the intervals of the form $[n, n+1]$, with n an integer, that contain a solution.

17. $x^3 - 6x^2 + 2 = 0$ **18.** $x^3 - 4x + 1 = 0$

19. $x^3 - 6x^2 + 5x + 1 = 0$ **20.** $x^3 + 3x^2 - x - 1 = 0$

A function f may be continuous on the intervals (a, c) and (c, b) but fail to be continuous at c for different reasons. If both one-sided limits,

$$\lim_{x\to c^-} f(x), \quad \lim_{x\to c^+} f(x)$$

are finite numbers but they are not equal, f is said to have a *jump discontinuity* at c. If one or both of the one-sided limits fail to exist because the function approaches infinity (plus or minus), f has an *infinite* discontinuity at c. If the one-sided limits exist and are equal, but the function is undefined at c, the discontinuity is *removable.* In this case the function can be defined at c to make it continuous. Some other types of discontinuities are possible also. In the following exercises, classify the discontinuity at c as one of these three types. Illustrate with a graph of the function near c.

21. $\frac{x^2+1}{x-1}$, $c = 1$ **22.** $\frac{x^2-1}{x-1}$, $c = 1$

23. $\frac{1}{x}$, $c = 0$ **24.** $\frac{x}{|x|}$, $c = 0$

25. $x\left(3+\frac{2}{x}\right)$, $c=0$

26. $x\left(3+\frac{2}{x^2}\right)$, $c=0$

27. $\frac{1}{x}+\frac{1}{|x|}$, $c=0$

28. $\frac{x^2+3x+2}{x+1}$, $c=-1$

29. Let f be continuous at $x=0$ and suppose that $f(0)>0$. Prove that there is a positive number s such that $f(x)>0$ for every x on the interval $[-s,s]$.

30. Prove there is a *nonzero* solution to the equation $\sin x = x/3$. [Hint: Apply the Intermediate Value Theorem to $f(x)=\sin x - x/3$ over a suitable interval.]

31. This problem is a generalization of the previous exercise. Let a be a positive number. Find the number of solutions of the equation $\sin x = x/a$. Express the answer as a function of a. [Hint: Sketch the graphs of $y=\sin x$ and $y=x/a$ for some different choices of a, like $a=5, 10, 15$, to get some idea how the number of solutions depends on a.]

32. Let $s(t)$ be the speed of a car at time t. Assume $s(t)$ is a continuous function of t. Suppose that the car is at rest at time $t=0$ (so $s(0)=0$) and accelerates to a speed of 60 mph at time $t=30$ seconds (so $s(30)=60$). The Intermediate Value Theorem implies that for any number u with $0\le u\le 60$ there is a time t with $0\le t\le 30$ and $s(t)=u$. Prove the slightly more complicated fact that there is some time t when the speed of the car was exactly $1+t$ mph.

3.2 Definition of derivative

We consider two different problems—finding the velocity from the position function and finding the tangent line to the graph of a function—and discover there is a common ingredient that is defined below as the derivative.

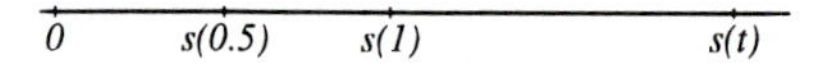

FIGURE 3.5

Suppose an object moves along a straight line and its distance from a fixed point O at time t is given by the function $s(t)$. From a knowledge of this function, how can we determine the velocity at time t? This can be determined by first considering the average velocity. After moving for $t=0.5$ hours, suppose the position is $s(0.5)=28$ miles and after moving for $t=1$ hour the position is $s(1)=48$ miles; the average velocity over the interval $[0.5,1]$ is

$$\frac{s(1)-s(0.5)}{1-0.5}=\frac{48-28}{0.5}=40.$$

If h is a positive number, the position at time $t+h$ is $s(t+h)$; the distance traveled during the time from t to $t+h$ is $s(t+h)-s(t)$. The average velocity over the time interval $[t,t+h]$ is

$$\frac{s(t+h)-s(t)}{h}.$$

Consider what happens when h gets smaller and smaller. The ratio gives the average velocity over smaller and smaller intervals. In the limiting case $h\to 0$, the ratio gives the velocity at time t. We call this the *instantaneous velocity.*

Let us consider the specific example of a falling object subject only to the force of gravity. Its motion is along a straight line. Suppose the position of a freely falling apple is $s(t)$ measured along the line from the point of release. Its position $s(t)$ is given by the formula $s(t) = 16t^2$. (This is a fairly close approximation to the position function of a freely falling object under the influence of gravity with air resistance ignored.) The average velocity over the interval $[2, 2.5]$ is

$$\frac{s(2.5) - s(2)}{2.5 - 2} = \frac{16(2.5)^2 - 16(2)^2}{0.5} = 72.$$

The average velocity over the interval $[2, 2.1]$ is

$$\frac{s(2.1) - s(2)}{2.1 - 2} = \frac{16(2.1)^2 - 16(2)^2}{0.1} = 65.6.$$

The average velocity over the interval $[2, 2 + h]$ is

$$\begin{aligned}\frac{s(2 + h) - s(2)}{2 + h - 2} &= \frac{16(2 + h)^2 - 16(2)^2}{h} \\ &= \frac{16(4 + 4h + h^2) - 16(4)}{h} \\ &= 16(4 + h).\end{aligned}$$

The instantaneous velocity at time $t = 2$ is the limit as h approaches 0 of the average velocity; that is

$$v = \lim_{h \to 0} 16(4 + h) = 64.$$

The instantaneous velocity of the apple at any time t is given by

$$\begin{aligned}v(t) &= \lim_{h \to 0} \frac{s(t + h) - s(t)}{h} = \lim_{h \to 0} \frac{16(t + h)^2 - 16t^2}{h} \\ &= \lim_{h \to 0} \frac{16(t^2 + 2th + h^2) - 16t^2}{h} = \lim_{h \to 0} 16(2t + h) \\ &= 16 \cdot 2t.\end{aligned}$$

The function

$$v(t) = \lim_{h \to 0} \frac{s(t + h) - s(t)}{h}$$

is the *derivative* of the position function $s(t)$ and equals the velocity function of the moving particle.

Tangent Line to a Graph

Suppose we are given the function $y = f(x)$ and we want to find the line tangent to its graph at the point $P = (a, f(a))$. To describe the line, it is necessary to have two pieces of information. One piece is given; namely that the line passes through the point $P(a, f(a))$. For the other piece of information either we need to know another point on the line or we need the slope of the line. How do we find the needed information?

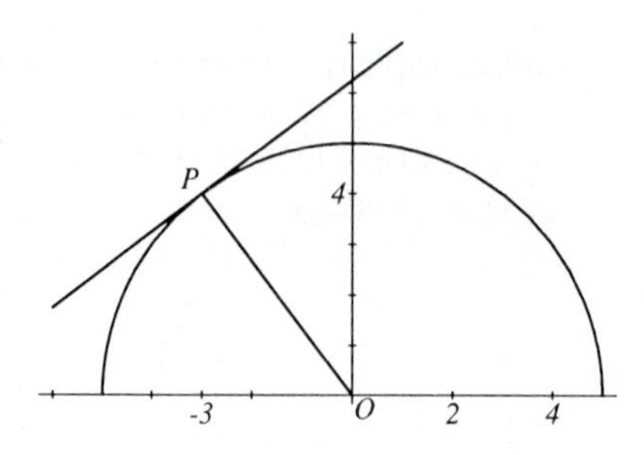

FIGURE 3.6
Tangent at P

Consider the specific case $y = \sqrt{25 - x^2}$ and the point $P = (-3, 4)$ on its graph. In this particular situation, the line tangent to the circle at P is perpendicular to the radius OP, a fact from Euclidean geometry. The radius OP has slope $-4/3$ and so the tangent line has slope $3/4$. (Recall that the product of the slopes of two perpendicular lines equals -1.) We conclude the tangent line to the circle at P has equation

$$y - 4 = \frac{3}{4}(x + 3).$$

In making this calculation, we were aided by a fact from geometry that enabled us to determine the slope of the tangent line to the circle. Unfortunately such useful facts will not be available when we treat graphs of more general functions.

So let us try another way to find the information needed to describe the tangent line by using a method that will be available in other cases.

Let h represent any small number, say $|h| \leq 0.5$, and consider a point on the circle

$$Q = (-3 + h, \sqrt{25 - (-3 + h)^2}).$$

If $h \neq 0$, then P and Q are two different points and there is a unique line passing through them. In particular, the slope of the line through P and Q is

$$m(h) = \frac{\sqrt{25 - (-3 + h)^2} - 4}{(-3 + h) - (-3)} = \frac{\sqrt{16 + 6h - h^2} - 4}{h}.$$

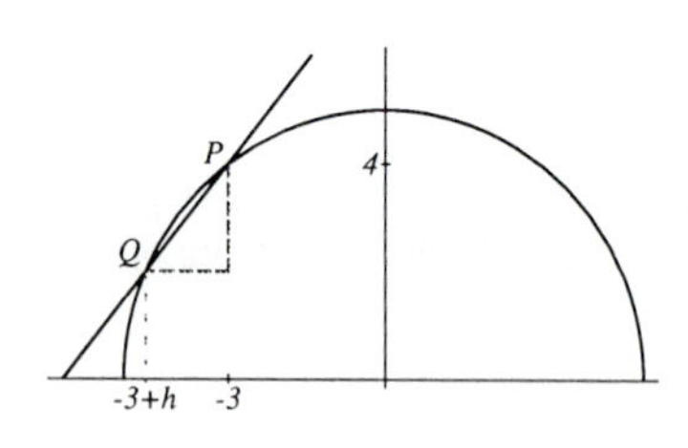

FIGURE 3.7
Chord through P

Now we allow the number h to become very small; as h approaches 0, the point Q moves very near P and the line through P and Q approaches the tangent line to the curve at P. Because $m(h)$ is the slope of the line through P and Q, the slope of the tangent line is

$$\lim_{h \to 0} m(h) = \lim_{h \to 0} \frac{\sqrt{16 + 6h - h^2} - 4}{h} = \lim_{h \to 0} \frac{(16 + 6h - h^2) - 16}{h(\sqrt{16 + 6h - h^2} + 4)} = \frac{3}{4}.$$

Note that we computed the slope of the tangent line by two different methods and reached the same result.

We have now seen two examples where the limiting value of the average change in a function has a practical interpretation. We now give this a formal definition.

DEFINITION 3.2

The Derivative of a Function

Let $f(x)$ be a function defined for all x on an open interval containing the point $x = c$. The derivative of $f(x)$ at c is the number defined by

$$\lim_{h \to 0} \frac{f(c+h) - f(c)}{h},$$

provided this limit exists. The function $f'(x)$ defined by

$$f'(x) = \lim_{h \to 0} \frac{f(x+h) - f(x)}{h}$$

is called the derivative of f.

The definition of the derivative carries a provision; namely that a particular limit must exist. We will use the term *differentiable* to mean that this limit exists. More precisely, we say that $f(x)$ is differentiable at a point c if the derivative of $f(x)$ at c exists; $f(x)$ is *differentiable on an interval* I if the derivative $f'(c)$ exists for every number c on I.

The definition of derivative is made for open intervals. We extend this definition to closed intervals by using one-sided limits at the endpoints. Thus we say f is differentiable on the closed interval $[a, b]$ if the derivative $f'(c)$ exists for all c on (a, b) and if the two limits

$$\lim_{h \to 0^+} \frac{f(a+h) - f(a)}{h} \quad \text{and} \quad \lim_{h \to 0^-} \frac{f(b+h) - f(b)}{h}$$

exist. When attention is restricted to the closed interval, we write $f'(a)$ and $f'(b)$ for the value of the one-sided limits.

Notations for the Derivative

We use several different symbols for the derivative. When we discuss a function $f(x)$, its derivative is written as $f'(x)$. If we are given an equation $y = f(x)$, the derivative may be written as dy/dx or y'. The first has the advantage of reminding us of the variable, x. For example, if we deal with a time variable t and have a function $s = s(t)$, the derivative is written as ds/dt. We write s' only when the variable t is understood so no confusion arises. In some cases we may also write $df(x)/dx$. If it is necessary to emphasize the name of the variable, we say "the derivative of $f(x)$ with respect to x" or "the derivative of $s(t)$ with respect to t". On some occasions we use the notation $D_x f(x)$ for the derivative of f with respect to x. Thus we have the equalities

$$f'(x) = \frac{dy}{dx} = y' = \frac{d}{dx} f(x) = D_x f(x).$$

Computation of Some Derivatives

Let us compute the derivative of some relatively uncomplicated functions.

THEOREM 3.4

Let n be a nonnegative integer and let b be any real number. If $f(x) = bx^n$, then the derivative of $f(x)$ is

$$f'(x) = nbx^{n-1}.$$

Proof

Suppose $n = 0$. Then $x^n = x^0 = 1$ and so $f(x)$ has the value b for every x (including $x + h$ because this is just another choice of the variable).

Then

$$f'(x) = \lim_{h \to 0} \frac{f(x+h) - f(x)}{h} = \lim_{h \to 0} \frac{b - b}{h} = 0$$

for every x. Thus the formula is correct for $n = 0$. Now assume $n > 0$. Then

$$f'(x) = \lim_{h \to 0} \frac{b(x+h)^n - bx^n}{h}.$$

This is expanded using the binomial theorem as

$$\begin{aligned}(x+h)^n - x^n &= x^n + nx^{n-1}h \\ &\quad + \frac{1}{2}n(n-1)x^{n-2}h^2 + \cdots + nxh^{n-1} + h^n - x^n \\ &= nx^{n-1}h + h^2\left[\frac{1}{2}n(n-1)x^{n-2} + \cdots + h^{n-2}\right].\end{aligned}$$

This implies

$$\frac{b(x+h)^n - bx^n}{h} = nbx^{n-1} + hb\left[\frac{1}{2}n(n-1)x^{n-2} + \cdots + h^{n-2}\right]$$

and so

$$f'(x) = \lim_{h \to 0} \frac{b(x+h)^n - bx^n}{h} = nbx^{n-1}$$

as was required by the formula. □

It is worth mentioning one special case of the preceding formula.

Corollary 3.1

If $f(x) = c$ is a constant function on some open interval, then the derivative $f'(x)$ equals 0 on the interval.

EXAMPLE 1 Compute the derivative of the functions

a. $f(x) = 4x^3$. b. $g(x) = -6x^{12}$. c. $s(t) = 30t^{10}$.

Solution:

a. $f'(x) = \dfrac{d}{dx}4x^3 = 3 \cdot 4x^2 = 12x^2$.

b. $g'(x) = \dfrac{d}{dx}(-6x^{12}) = 12(-6)x^{11} = -72x^{11}$.

c. $s'(t) = \dfrac{d}{dt}(30t^{10}) = 10 \cdot 30t^9 = 300t^9$. ■

Additive Property of the Derivative

The additive property states that if $f(x)$ and $g(x)$ are two functions and $F(x) = f(x) + g(x)$ then the derivative of F is the sum of the derivatives of f and g. We write this as

$$\big(f(x) + g(x)\big)' = f'(x) + g'(x).$$

This is easily proved by using the definition of the derivative of $F(x)$ and using the additive property of limits from section 2.6.

EXAMPLE 2 Find the derivative of $F(x) = 3x^5 - 9x^{11}$.

Solution: The polynomial $F(x) = 3x^5 - 9x^{11}$ is the sum of two functions, $f(x) = 3x^5$ and $g(x) = -9x^{11}$. The derivative of $F(x)$ is found using Theorem 3.4 and the additive property as

$$F'(x) = \frac{d}{dx}(3x^5 - 9x^{11}) = \frac{d}{dx}(15x^4) + \frac{d}{dx}(-9x^{11}) = 15x^4 - 99x^{10}. \quad \blacksquare$$

This additive property permits the computation of the derivative of any polynomial function. If

$$f(x) = a_0 + a_1x + a_2x^2 + \cdots + a_nx^n$$

then

$$f'(x) = a_1 + 2a_2x + 3a_3x^2 + \cdots + na_nx^{n-1}.$$

Now we formally state the connection of the derivative with tangent lines.

DEFINITION 3.3

Tangent line

Let $f(x)$ be defined on an open interval containing the point $x = c$ and assume that the derivative $f'(c)$ exists. Then the *tangent line* to the graph of $y = f(x)$ at the point $(c, f(c))$ is the line with slope $f'(c)$. The equation of the tangent line to the graph of $y = f(x)$ at the point $(c, f(c))$ is

$$y - f(c) = f'(c)(x - c).$$

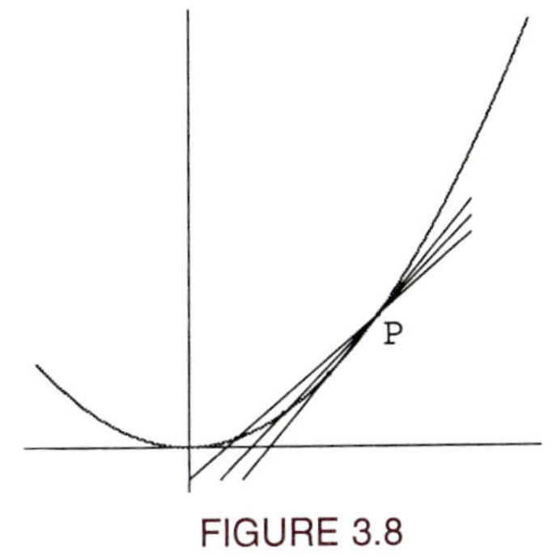

FIGURE 3.8
Lines through P and nearby Q

The motivation for the definition has already been illustrated but we repeat the outline of the idea. For $h \neq 0$ but h small enough so that $c+h$ is still in the interval where $f(x)$ is defined, the points $P(c, f(c))$ and $Q(c+h, f(c+h))$ are distinct and the slope of the line through them is

$$m(h) = \frac{f(c+h) - f(c)}{h}.$$

As h moves toward 0, the point Q moves toward P and the line through these two points approaches the line tangent to the curve at P. Thus the slope of the tangent line is

$$\lim_{h \to 0} m(h) = \lim_{h \to 0} \frac{f(c+h) - f(c)}{h} = f'(c).$$

Linear Approximation

There is an alternative way of defining the tangent line to the graph of a function $y = f(x)$. The tangent line at the point $(a, f(a))$ is the graph of the linear function $g(x) = mx + b$ that is the best approximation to f near a. What does this mean? First we must have $g(a) = f(a)$, so that g and f agree at a. At points near a we want the difference $g(a+h) - f(a+h)$ to be small in comparison with the distance h of the point $a + h$ from a. This condition is implied by the requirement that

$$\lim_{h \to 0} \frac{g(a+h) - f(a+h)}{h} = 0.$$

By adding and subtracting $g(a)$, which equals $f(a)$, in the numerator, we see that this condition implies that $g'(a) = f'(a)$; that is,

$$0 = \lim_{h\to 0} \frac{g(a+h) - f(a+h)}{h} = \lim_{h\to 0} \frac{g(a+h) - g(a)}{h} - \frac{f(a+h) - f(a)}{h} = g'(a) - f'(a).$$

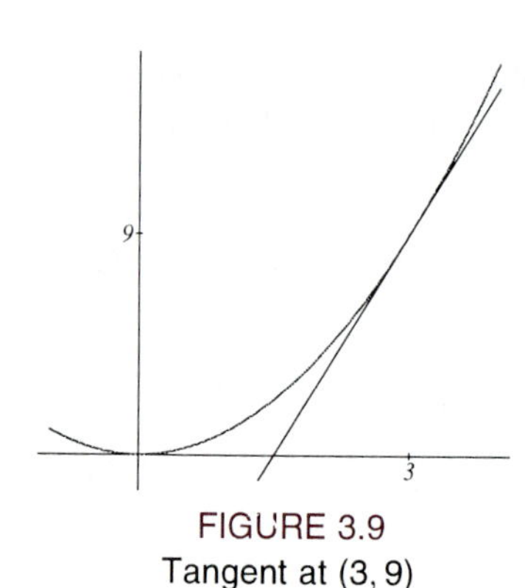

FIGURE 3.9
Tangent at (3, 9)

Hence we see that the tangent to the graph of $y = f(x)$ at $(a, f(a))$ is the line through $(a, f(a))$ with slope $f'(a)$.

EXAMPLE 3 Find the equation of the line tangent to the parabola $y = x^2$ at the point $(3, 9)$.

Solution: The point $(3, 9)$ is on the graph; the slope of the tangent line at this point is the derivative of the function evaluated at $x = 3$. At any point x, we have

$$\frac{dy}{dx} = \frac{d}{dx}x^2 = 2x.$$

The slope of the tangent line at $x = 3$ is $2 \cdot 3 = 6$; the equation of the tangent line is $y - 9 = 6(x - 3)$ or $y = 6x - 9$. ■

EXAMPLE 4 Show that the tangent lines to the two curves $y = x^2$ and $y = x^2 + 2x + 1$ at the point of intersection are perpendicular.

Solution: We first find the point of intersection of the two curves. The x coordinate of the point of intersection is the solution of the equation

$$x^2 = x^2 + 2x + 1, \quad \text{or} \quad x = -\frac{1}{2}.$$

The slope of the tangent line to the curve $y = x^2$ at the point where $x = -1/2$ is the value of the derivative $y' = 2x$; namely the slope is $2(-1/2) = -1$. The slope of the tangent to the second curve is the value of the derivative $y' = 2x + 2$ at $x = -1/2$; namely $-1 + 2 = 1$. Thus the slopes are negative reciprocals of each other and the two tangent lines are perpendicular. ■

Higher Order Derivatives

If the function $f(x)$ is differentiable, we can form the new function $f'(x)$. That function may also be differentiable so we can compute the derivative of $f'(x)$. This derivative is denoted by

$$\frac{d}{dx}f'(x) = f''(x)$$

and is called the *second derivative* of $f(x)$. We can write the second derivative using alternative notation such as

$$\frac{d}{dx}f'(x) = \frac{d^2}{dx^2}f(x) = \frac{d^2y}{dx^2} = y''$$

where $y = f(x)$ in the last two cases.

EXAMPLE 5 If $y = 4x^5 - 3x$ compute y''.

Solution: The first derivative is $y' = 20x^4 - 3$. The derivative of this is the second derivative of y and we find

$$y'' = \frac{d}{dx}(20x^4 - 3) = 80x^3.$$

If the second derivative of $f(x)$ is differentiable, we can compute the third derivative $f'''(x)$. The fourth derivative is usually written as $f^{(4)}(x)$; more generally we write the nth derivative of $f(x)$ as $f^{(n)}(x)$. ■

EXAMPLE 6 Compute all the higher order derivatives of $f(x) = x^4$.

Solution: We have

$$f'(x) = 4x^3, \qquad f'''(x) = 2 \cdot 12x = 24x, \quad f^{(5)}(x) = 0,$$
$$f''(x) = 3 \cdot 4x^2 = 12x^2, \quad f^{(4)}(x) = 24, \qquad f^{(n)}(x) = 0, \quad n \geq 5.$$

■

Exercises 3.2

Compute the derivatives indicated.

1. $\dfrac{d}{dx}(5x^3 - 7x + 2)$

2. $\dfrac{d}{dx}(1 - x + x^2 - x^3)$

3. $\dfrac{d}{dt}(16t^2 - 48t + 19)$

4. $\dfrac{d}{dz}(z^2 + 1)(2z - 3)$

5. $\dfrac{d}{dw}w(w^4 - 2w^2 + 5)$

6. $\dfrac{d}{dx}x^{12}(x^8 + 12x^3)$

7. $\dfrac{d}{dx}(x^2 + 1)(x^2 - 4)$

8. $\dfrac{d}{dt}(t^2+3t+1)(t-1)(t+1)$

Find the equation of the line tangent to the graph of the function $f(x)$ at the indicated points.

9. $f(x) = x^2$ at $(0, 0)$, $(2, 4)$ and $(-2, 4)$

10. $f(x) = x^2 - x$ at the points where $x = 0$, 0.5 and 1

11. $f(x) = -2x^2 + 4$ at $(0, 4)$, $(1, 2)$ and $(\sqrt{2}, 0)$

12. $f(x) = x(x + 1)$ at $(0, 0)$, $(2, 6)$ and $(-2, 2)$

13. $f(x) = 2|x| + 3$ at $(-2, 7)$ and $(2, 7)$

14. $f(x) = 5 - 3x$ at $(a, f(a))$

Sketch the graphs of the following equations and determine all the points on each graph where the tangent line is parallel to the x-axis.

15. $y = x^3 - x$

16. $y = (x - 2)x^2(x + 3)$

17. $y = x^3 - 9x + 15$

18. $y = (x - 1)(x + 1)(x - 3)$

Find the indicated higher order derivatives.

19. If $f(x) = 4x^6 - 3x + 2$, find $f''(x)$.

20. If $f(x) = -3x^4 - 3x^3 + 2x^2$, find $f''(x)$.

21. If $g(x) = x^2(2x^3 - 4x)$, find $g^{(4)}(x)$.

22. If $g(x) = x^6 + 5x^3 - 7x^2 + 4$, find $g^{(4)}(x)$.

23. Find the smallest n such that $g^{(n)}(x) = 0$ for all x when $g(x) = x^7 - x^8$.

24. What is the smallest n such that $g^{(n)}(x) = 0$ for all x when $g(x) = x^m$, m a positive integer?

25. Find the points where the line $y = 2x - 1$ is tangent to the graph of $f(x) = x^2$.

26. Show that the tangents to the graph of $y = x^2$ at two different points are not parallel.

27. Find all pairs of points where the tangents to the graph of $y = x^3$ are parallel.

28. The graphs of $f(x) = x^2$ and $g(x) = x^{1/2}$ intersect at $(1, 1)$. Find the tangent line to each graph at this point and determine if the tangents meet at right angles.

29. The graphs of $f(x) = x^{m/4}$ and $g(x) = x^{-n}$, for positive integers m and n, intersect at $(1, 1)$. Find all pairs of positive integers m and n such that the tangents to the graphs at $(1, 1)$ intersect at right angles.

If $p(x)$ is a polynomial then $p(x+h) - p(x) = hq(x, h)$, where $q(x, h)$ is a polynomial of two variables, namely, a sum of terms of the form $c_{ij}h^i x^j$ for nonnegative integers i, j. Then the derivative $p'(x)$ can be computed by substituting 0 for h in q; that is $p'(x) = q(x, 0)$. Verify this statement by computing $q(x, h)$ and $p'(x)$ for the following polynomials.

30. $p(x) = 3x^2 - 2x$

31. $p(x) = 4x^3 - 6x^2 + 2$

32. $p(x) = x^4 - 1$

33. $p(x) = x^5$

3.3 Product and quotient rules

In this section we see how to compute the derivative of certain combinations of functions in terms of the derivatives of the functions that make up the combination. The first of these is the *product rule*. Before giving the statement and proof of this rule, we will prove an important property of differentiable functions.

THEOREM 3.5

Differentiable Functions Are Continuous

If $f(x)$ is differentiable at a point $x = c$, then $f(x)$ is continuous at c. In particular,

$$\lim_{h\to 0} f(c+h) = f(c).$$

Proof

To prove that $f(x)$ is continuous at c, we must prove that $\lim_{x\to c} f(x) = f(c)$. We will prove the equivalent statement that

$$\lim_{x\to c}[f(x) - f(c)] = 0.$$

Define h by the equation $x = c + h$ so that forcing x to approach c is the same as forcing h to approach 0.

We have

$$\begin{aligned}\lim_{x\to c}[f(x) - f(c)] &= \lim_{x\to c} \frac{[f(x) - f(c)]}{x - c}(x - c)\\ &= \lim_{h\to 0} \frac{[f(c+h) - f(c)]}{(c+h) - c}[(c+h) - c]\\ &= \lim_{h\to 0} \frac{[f(c+h) - f(c)]}{h} \lim_{h\to 0}(h)\\ &= f'(c) \lim_{h\to 0} h\\ &= f'(c)\cdot 0 = 0.\end{aligned}$$

Notice that the assumption of differentiability was used to replace one of the limits with its value $f'(c)$. Without differentiability at c this step would not be valid. □

Now we can prove the product rule.

THEOREM 3.6

Product Rule

Let $f(x)$ and $g(x)$ be differentiable on an open interval I. Then the product $f(x)g(x)$ is differentiable on I and the derivative is

$$\frac{d}{dx} f(x)g(x) = f(x)g'(x) + f'(x)g(x).$$

Proof

We apply the definition of the derivative along with some algebraic manipulation to prove the product rule. We have

$$\begin{aligned}
\frac{d}{dx}[f(x)g(x)] &= \lim_{h\to 0}\frac{f(x+h)g(x+h)-f(x)g(x)}{h} \\
&= \lim_{h\to 0}\frac{f(x+h)g(x+h)-f(x+h)g(x)+f(x+h)g(x)-f(x)g(x)}{h} \\
&= \lim_{h\to 0} f(x+h)\left[\frac{g(x+h)-g(x)}{h}\right] + \lim_{h\to 0}\left[\frac{f(x+h)-f(x)}{h}\right]g(x) \\
&= \lim_{h\to 0} f(x+h)\lim_{h\to 0}\left[\frac{g(x+h)-g(x)}{h}\right] + f'(x)g(x) \\
&= f(x)g'(x)+f'(x)g(x),
\end{aligned}$$

and this proves the product rule. Note that the continuity of $f(x)$ was used to obtain the equality $\lim_{h\to 0} f(x+h) = f(x)$. □

Here are a few examples of the use of the product rule.

EXAMPLE 1 Compute the derivative of $y = (x^2+3)x^5$ using the product rule.

Solution: We have

$$\begin{aligned}
\frac{dy}{dx} &= \left[\frac{d}{dx}(x^2+3)\right]x^5 + (x^2+3)\frac{d}{dx}x^5 \\
&= (2x)x^5 + (x^2+3)5x^4.
\end{aligned}$$

■

EXAMPLE 2 Compute the derivative of $y = (x^3 - 4x + 1)(x^4 + 2x^3)$.

Solution: We treat it as a product and apply the product rule:

$$\begin{aligned}
\frac{dy}{dx} &= (x^3-4x+1)\frac{d}{dx}(x^4+2x^3) + \left[\frac{d}{dx}(x^3-4x+1)\right](x^4+2x^3) \\
&= (x^3-4x+1)(4x^3+6x^2) + (3x^2-4)(x^4+2x^3).
\end{aligned}$$

■

We suggest that the reader multiply out the functions in each of the last two examples, find the derivative using the rule for derivatives of polynomials and then verify that the result is the same as that obtained with the product rule.

Next we use the product rule to evaluate the derivative of the quotient of two differentiable functions.

THEOREM 3.7

Quotient Rule

Let $f(x)$ and $g(x)$ be differentiable functions on an open interval. At every point of the interval where $g(x) \neq 0$, the quotient $f(x)/g(x)$ is differentiable and we have

$$\frac{d}{dx}\frac{f(x)}{g(x)} = \frac{g(x)f'(x) - g'(x)f(x)}{g(x)^2}.$$

Proof

A proof along the lines of the proof of the product rule could be given but is left as an exercise. Instead, we give a proof that is easily remembered and may help the reader reconstruct the quotient rule from the product rule. It has the drawback of assuming the differentiability of the quotient rather than proving this fact. The proof is carried out by applying the product rule to $g(x)$ times the quotient $f(x)/g(x)$, which, of course, equals $f(x)$. If the derivative of $f(x)/g(x)$ exists then

$$\begin{aligned} f'(x) &= \frac{d}{dx}\left[g(x)\left(\frac{f(x)}{g(x)}\right)\right] \\ &= g'(x)\left(\frac{f(x)}{g(x)}\right) + g(x)\frac{d}{dx}\left(\frac{f(x)}{g(x)}\right). \end{aligned}$$

Next multiply this equation by $g(x)$ and rearrange terms to get

$$g(x)f'(x) - g'(x)f(x) = g(x)^2\frac{d}{dx}\left(\frac{f(x)}{g(x)}\right).$$

This is equivalent to the rule stated in the theorem. □

EXAMPLE 3 If $y = x^3/(4x-1)$, find dy/dx.

Solution: By the quotient rule we obtain

$$\frac{dy}{dx} = \frac{(4x-1)3x^2 - x^3(4)}{(4x-1)^2}.$$

■

The following rule is a special case of the quotient rule but, inasmuch as it occurs frequently, it is worth special mention.

THEOREM 3.8

If $g(x)$ is differentiable on an open interval, then the reciprocal function $1/g(x)$ is differentiable at every point of the interval where $g(x) \neq 0$ and its derivative is

$$\frac{d}{dx}\frac{1}{g(x)} = -\frac{g'(x)}{g(x)^2}.$$

Proof

The theorem follows at once from the quotient rule using $f(x) = 1$. □

The quotient rule allows us to extend the computation of the derivative of x^n to cover the case in which n is a negative integer. We have the derivative formula from Theorem 3.4, which was valid only for nonnegative n.

THEOREM 3.9

Let n be any integer. Then

$$\frac{d}{dx}x^n = nx^{n-1}, \tag{3.1}$$

provided $x \neq 0$ when $n < 0$.

Proof

We have already proved this rule in the case where $n \geq 0$. Suppose that $n < 0$; let $k = -n$ so that $k > 0$. Then

$$\begin{aligned}\frac{d}{dx}x^n = \frac{d}{dx}x^{-k} &= \frac{d}{dx}\frac{1}{x^k} \\ &= -\frac{kx^{k-1}}{x^{2k}} = -kx^{-k-1} \\ &= nx^{n-1}.\end{aligned}$$

□

EXAMPLE 4 Find $\frac{dy}{dx}$ if $y = 7x^{-13}$.

Solution: Directly apply equation (3.1) to obtain

$$\frac{dy}{dx} = 7 \cdot \frac{d}{dx}x^{-13} = 7(-13)x^{-14} = -91x^{-14}.$$

■

Exercises 3.3

For each function, compute the derivative with respect to the obvious variable.

1. $(x^2 - 4x)(x + 3)$
2. $x(x^2 - x - 12)$
3. $x + \frac{1}{x}$
4. $4 - x^2 + \frac{2}{x + 1}$
5. $(x^2 + x - 1)(2x^3 + 7x^2 - 1)$
6. $\frac{6x^7 - 3x^2 + 4}{-3x^2 + x + 1}$
7. $\frac{u^2 - 1}{u^2 + 1}$
8. $\frac{u^2 + u - 1}{u^2 - u + 1}$
9. $\frac{u^4 + 3u^2 - 1}{u^3 - 4u^2 + u + 1}$
10. $\frac{u - 1}{(u + 1)(u + 2)(u + 3)}$
11. $\frac{2}{x^2} + \frac{x - 1}{x} - \frac{x + 4}{x^3}$
12. $\left(\frac{x + 1}{x - 1}\right)\left(\frac{x + 2}{5}\right)$
13. Evaluate $f'(3)$ given that $f(x) = x^2 g(x)$ and $g(3) = 1$ and $g'(3) = 0$.
14. Evaluate $f'(2)$ given that $f(x) = \frac{x^2}{g(x)}$ and $g(2) = 4$ and $g'(2) = -10$.
15. Evaluate $f'(-1)$ given that $g(x) = x^3 f(x) + 4x$, $f(-1) = 1$ and $g'(-1) = 0$.
16. Evaluate $f'(10)$ given that $g(x) = x^2/f(x) - xf(x)$, $f(10) = -3$ and $g'(10) = 0$.

Compute the indicated higher order derivatives.

17. Find $f^{(2)}(x)$ if $f(x) = \frac{x}{x + 1}$.
18. Find $f^{(2)}(x)$ if $f(x) = \frac{x^2}{x - 2}$.
19. Find $f^{(3)}(x)$ if $f(x) = \frac{x^2 + 4}{x}$.
20. Find $f^{(3)}(t)$ if $f(t) = \frac{4t^3 - 1}{t^2}$.
21. Find $f^{(4)}(x)$ if $f(x) = \frac{1}{x + 1}$.
22. Find $f^{(n)}(x)$ if $f(x) = \frac{1}{x}$.
23. Find $f^{(n)}(x)$ if $f(x) = \frac{1}{1 - x}$.
24. Find $f^{(n)}(x)$ if $f(x) = \frac{1}{x^2}$.
25. Find $f^{(n)}(x)$ if $f(x) = \frac{2x}{x^2 - 1}$. [Hint: Express $2x/(x^2 - 1)$ in terms of $1/(x - 1)$ and $1/(x + 1)$.]

Find the intervals on which $f'(x) \geq 0$.

26. $f(x) = x^3 - 3x$
27. $f(x) = 4x^5 + 2x$
28. $f(x) = 3x^4 - 4x^3$
29. $f(x) = 6x^5 - 30x^3$
30. Let $u = g(x)$ and $v = h(x)$ be two times differentiable functions of x. Prove the formula $(uv)'' = uv'' + 2u'v' + u''v$.
31. Let u and v be three times differentiable functions of x. Prove the formula $(uv)''' = u'''v + 3u''v' + 3u'v'' + uv'''$.
32. Let $f(x) = u(x)v(x)$. Show that

$$\frac{f'(x)}{f(x)} = \frac{u'(x)}{u(x)} + \frac{v'(x)}{v(x)}.$$

33. Let $f(x) = g_1(x)g_2(x)\cdots g_n(x)$. Show that

$$\frac{f'(x)}{f(x)} = \frac{g_1'(x)}{g_1(x)} + \frac{g_2'(x)}{g_2(x)} + \cdots + \frac{g_n'(x)}{g_n(x)}.$$

[A formal proof can be given using mathematical induction. Give a plausible argument, at least, using Exercise 32.]

34. Let $f(x) = 2x - 2$ for $x \leq 1$ and $f(x) = x^2$ for $1 < x$. Prove that $f'(x)$ does not exist at $x = 1$. [Hint: See Theorem 3.8.]

35. Give an example of a function g that has a derivative $g'(x)$ at every point x but g'' fails to exist at some point. [Hint: Select g so that g' behaves like the function in Exercise 34.]

3.4 Derivatives of trigonometric functions

We begin by reviewing two trigonometric identities.

THEOREM 3.10

Identities Involving Sine and Cosine

For all real numbers u and v we have the identities

1. $\cos(u + v) = \cos u \cos v - \sin u \sin v$,
2. $\sin(u + v) = \sin u \cos v + \cos u \sin v$.

These identities are used in proving the following rules for the derivatives of the sine and cosine functions.

THEOREM 3.11

Derivatives of Sine and Cosine

a. $\dfrac{d}{dx} \sin x = \cos x$ b. $\dfrac{d}{dx} \cos x = -\sin x$

Proof

For the derivative of $\sin x$ we use the definition of the derivative, identity 2 of Theorem 3.10, and the two limits from Theorem 2.16:

$$\begin{aligned}
\frac{d}{dx} \sin x &= \lim_{h\to 0} \frac{\sin(x + h) - \sin x}{h} \\
&= \lim_{h\to 0} \frac{\sin x \cos h + \sin h \cos x - \sin x}{h} \\
&= \lim_{h\to 0} \sin x \frac{\cos h - 1}{h} + \cos x \lim_{h\to 0} \frac{\sin h}{h} \\
&= \sin x \cdot 0 + \cos x \cdot 1 \\
&= \cos x.
\end{aligned}$$

For the proof of b we repeat the analogue of the preceding argument with the $\cos x$ in place of $\sin x$ and using Theorem 3.10, identity 1:

$$\begin{aligned}\frac{d}{dx}\cos x &= \lim_{h\to 0}\frac{\cos(x+h)-\cos x}{h}\\ &= \lim_{h\to 0}\frac{\cos x\cos h - \sin x\sin h - \cos x}{h}\\ &= \lim_{h\to 0}\cos x\left(\frac{\cos h - 1}{h}\right) - \lim_{h\to 0}\sin x\left(\frac{\sin h}{h}\right)\\ &= \cos x\cdot(0) - \sin x\cdot(1)\\ &= -\sin x.\end{aligned}$$

□

EXAMPLE 1 Compute dy/dx for (a) $y = x\sin x$ and (b) $y = \sin x\cos x$.

***Solution*:** Both require the product rule. For (a) we have

$$\frac{d}{dx}(x\sin x) = \left(\frac{d}{dx}x\right)\sin x + x\frac{d}{dx}\sin x = \sin x + x\cos x.$$

For (b) we compute

$$\begin{aligned}\frac{d}{dx}(\sin x\cos x) &= \left(\frac{d}{dx}\sin x\right)\cos x + \sin x\frac{d}{dx}\cos x\\ &= (\cos x)\cos x + \sin x(-\sin x)\\ &= \cos^2 x - \sin^2 x.\end{aligned}$$

■

Derivatives of Tangent, Secant, Cotangent and Cosecant

The derivatives of the remaining four trigonometric functions are now computed by expressing each in terms of the sine and cosine and then using the quotient rules and the rules for derivatives of the sine and cosine. The computations are as follows:

$$\begin{aligned}\frac{d}{dx}\tan x &= \frac{d}{dx}\frac{\sin x}{\cos x}\\ &= \frac{\cos x(\cos x) - \sin x(-\sin x)}{\cos^2 x}\\ &= \frac{\cos^2 x + \sin^2 x}{\cos^2 x} = \frac{1}{\cos^2 x}\\ &= \sec^2 x,\end{aligned}$$

$$\begin{aligned}\frac{d}{dx}\sec x &= \frac{d}{dx}\frac{1}{\cos x}\\ &= \frac{\sin x}{\cos^2 x} = \frac{\sin x}{\cos x}\frac{1}{\cos x}\\ &= \tan x\sec x.\end{aligned}$$

The derivatives of the cotangent and cosecant functions are found in a similar way. The results are summarized here.

$$\frac{d}{dx}\sin x = \cos x \qquad \frac{d}{dx}\cos x = -\sin x$$

$$\frac{d}{dx}\tan x = \sec^2 x \qquad \frac{d}{dx}\cot x = -\csc^2 x$$

$$\frac{d}{dx}\sec x = \sec x \tan x \qquad \frac{d}{dx}\csc x = -\csc x \cot x$$

EXAMPLE 2 Compute the derivatives with respect to x of the functions $g(x) = \tan^2 x$, $h(x) = \sec x \tan x$ and $k(x) = \sin^2 x / \tan x$.

Solution: These are product and quotient rule problems:

$$\begin{aligned} g'(x) &= \frac{d}{dx}\tan^2 x = \frac{d}{dx}(\tan x \tan x) = \left(\frac{d}{dx}\tan x\right)\tan x + \tan x \frac{d}{dx}\tan x \\ &= 2\tan x \sec^2 x, \end{aligned}$$

$$\begin{aligned} h'(x) &= \frac{d}{dx}\sec x \tan x = \left(\frac{d}{dx}\sec x\right)\tan x + \sec x\left(\frac{d}{dx}\tan x\right) \\ &= \sec x \tan x \tan x + \sec x \sec^2 x, \end{aligned}$$

$$\begin{aligned} k'(x) &= \frac{d}{dx}\frac{\sin^2 x}{\tan x} = \frac{\tan x D_x \sin^2 x - \sin^2 x D_x \tan x}{\tan^2 x} \\ &= \frac{\tan x \cdot 2\sin x \cos x - \sin^2 x \cdot \sec^2 x}{\tan^2 x}. \end{aligned}$$

■

Note that in this last computation we used the product rule to evaluate

$$\frac{d}{dx}\sin^2 x = \frac{d}{dx}(\sin x \sin x) = \cos x \sin x + \sin x \cos x = 2\sin x \cos x.$$

In the next sections we will have a better way to find the derivatives of functions of the form $\sin^n x$ using the chain rule (section 3.5). Trigonometric identities can be used to transform functions before differentiating. The example involving $k(x)$, for instance, could have been done by first rewriting the function $k(x)$:

$$k(x) = \frac{\sin^2 x}{\tan x} = \frac{\sin^2 x \cos x}{\sin x} = \sin x \cos x.$$

Thus $k'(x)$ can be computed by differentiating this last expression using the product rule.

Exercises 3.4

Find the derivative $f'(x)$ of $f(x)$.

1. $f(x) = 4\sin x - 3\cos x$

2. $f(x) = \sin x \cos x$

3. $f(x) = 3\sin x + 5\tan x$

4. $f(x) = \tan^2 x$

5. $f(x) = \cos^2 x + \sin^2 x$

6. $f(x) = \cos^2 x \tan x$

7. $f(x) = 4\csc x - 3\sec x$

8. $f(x) = \dfrac{\tan x + \sec x}{\sin x + \cos x}$

9. $f(x) = \dfrac{\sin x}{1 - \cos x}$

10. $f(x) = x \tan x$

11. $f(x) = \dfrac{x \sin x + \cos x}{x \cos x - \sin x}$

12. $f(x) = (x^2 + 1) \cos x$

13. $f(x) = \dfrac{x^3 - \sec x}{x \sin x}$

14. $f(x) = x \sec x - \csc x$

Compute the second derivatives y'' or fourth derivatives $y^{(4)}$ with respect to x as indicated.

15. y'' if $y = \sin x$

16. y'' if $y = \cos x$

17. y'' if $y = \sin^2 x$

18. y'' if $y = 2 \sin x \cos x$

19. y'' if $y = \cos^2 x$

20. y'' if $y = 3 \sin x + 7 \cos x$

21. y'' if $y = \csc x$

22. y'' if $y = \tan x$

23. y'' if $y = \cot x$

24. y'' if $y = \sec x$

25. $y^{(4)}$ if $y = 3 \sin x + 7 \cos x$

26. $y^{(4)}$ if $y = 6 \sin x - 2 \cos x$

3.5 The chain rule

The rules for computing derivatives given in the previous sections are very useful, but they are not sufficiently general to cover many common situations. For example, a formula has been given for the derivative of a polynomial but it is of relatively little help in computing the derivative of the polynomial $f(x) = (x^3 + 5x + 8)^{97}$. The difficulty here is that the polynomial $f(x)$ is not given as a sum of powers of x. In principle, one could carry out the multiplication and express this polynomial as a sum of powers of x with certain real number coefficients and then compute the derivative; that multiplication would be so tedious that one would almost surely look for another method. The *chain rule* provides an easy method for evaluating this derivative. The idea of the chain rule is to consider $f(x)$ as the composite of two functions; namely let $h(x) = x^3 + 5x + 8$ and $g(x) = x^{97}$. Then $f(x) = g(h(x))$. The rule states how the derivative of $f(x)$ can be computed from the derivatives of $g(x)$ and $h(x)$. Its precise statement is

$$\frac{d}{dx} g(h(x)) = g'(h(x))h'(x).$$

When the chain rule is applied to the particular $f(x)$ above, we are able to do the computations quite easily. We find

$$g'(x) = 97x^{96}, \qquad g'(h(x)) = 97(x^3 + 5x + 8)^{96}, \qquad h'(x) = 3x^2 + 5,$$
$$\frac{d}{dx} f(x) = \frac{d}{dx}(x^3 + 5x + 8)^{97} = 97(x^3 + 5x + 8)^{96}(3x^2 + 5).$$

Here is the statement and proof of the chain rule.

THEOREM 3.12

The Chain Rule

Let $h(x)$ be differentiable at a point c and let $g(x)$ be differentiable at the point $h(c)$. Then the function $f(x) = g(h(x))$ is differentiable at c and its derivative is

$$f'(c) = g'(h(c))h'(c). \qquad \textbf{(3.2)}$$

Proof

The derivative of $f(x)$ at c is, by definition,

$$\lim_{k\to 0} \frac{f(c+k)-f(c)}{k}.$$

We will evaluate this in two cases; we first assume that $h'(c) \neq 0$ so that for k sufficiently close to 0, we have $h(c+k)-h(c) \neq 0$. Substitute in terms of g and h to see that the derivative of $f(x)$ at c equals

$$\begin{aligned}\lim_{k\to 0} \frac{g(h(c+k))-g(h(c))}{k} &= \lim_{k\to 0} \frac{g(h(c+k))-g(h(c))}{h(c+k)-h(c)} \frac{h(c+k)-h(c)}{k} \\ &= \lim_{k\to 0} \frac{g(h(c+k))-g(h(c))}{h(c+k)-h(c)} \lim_{k\to 0} \frac{h(c+k)-h(c)}{k} \\ &= \lim_{k\to 0} \frac{g(h(c+k))-g(h(c))}{h(c+k)-h(c)} h'(c).\end{aligned}$$

There is one remaining limit to be evaluated. We introduce a new variable $v = h(c+k)-h(c)$. Note that v is a function of k and when k approaches 0, v also approaches 0 because $h(x)$ is differentiable, hence continuous, at c. Then

$$\begin{aligned}\lim_{k\to 0} \frac{g(h(c+k))-g(h(c))}{h(c+k)-h(c)} &= \lim_{k\to 0} \frac{g(v+h(c))-g(h(c))}{v} \\ &= \lim_{v\to 0} \frac{g(v+h(c))-g(h(c))}{v} \\ &= g'(h(c)).\end{aligned}$$

When this is combined with the first computation in the proof we have obtained formula (3.2) in the case $h'(c) \neq 0$.

Now suppose $h'(c) = 0$. For k near 0, the difference $h(c+k)-h(c)$ may or may not be 0. It is necessary to show that the quantity

$$\frac{f(c+k)-f(c)}{k}$$

is within ϵ of the quantity $g'(h(c))h'(c)$ for all k within δ of 0, where ϵ is any given positive number and δ is to be determined. We have assumed $h'(c) = 0$; our goal is to show that for any positive ϵ there is a positive δ such that

$$\left| \frac{g(h(c+k)-g(h(c))}{k} \right| < \epsilon \quad \text{whenever} \quad |k| < \delta. \tag{3.3}$$

For those choices of k with $h(c+k)-h(c) = 0$ we have

$$\frac{g(h(c+k))-g(h(c))}{k} = 0$$

and so equation (3.3) holds for any choice of ϵ and δ. From now on we can restrict our attention to those values of k for which $h(c+k)-h(c) \neq 0$. Assume a positive ϵ is given. The quotient we are trying to bound can be written as

$$\frac{g(h(c+k)-g(h(c))}{k} = \frac{g(h(c+k)-g(h(c))}{h(c+k)-h(c)} \frac{h(c+k)-h(c)}{k}.$$

In rather imprecise terms we see that the last quotient is near $h'(c) = 0$ and can be made near 0 for sufficiently small k. The other quotient on the right looks as though it should be close to $g'(h(c))$ which is some finite number. The product is no larger that this number multiplied by something very small and hence can be made less than any given ϵ. Now we make this intuitive argument more precise.

The differentiability of g at $h(c)$ implies that there is a positive δ_1 such that

$$\left|\frac{g(h(c)+w)-g(h(c))}{w}-g'(h(c))\right|<1, \quad \text{whenever} \quad |w|<\delta_1.$$

This produces a bound for one of the quotients in equation (3.3) when we set $w = h(c+k) - h(c)$. There is a $\delta_2 > 0$ such that $|k| < \delta_2$ implies $|w| < \delta_1$. We now have

$$\left|\frac{g(h(c+k))-g(h(c))}{h(c+k)-h(c)}\right|<1+|g'(h(c))|, \quad \text{whenever} \quad |k|<\delta_2.$$

Now finally use $\epsilon_3 = \epsilon/(1+|g'(h(c))|)$ to produce a $\delta_3 > 0$ such that

$$\left|\frac{h(c+k)-h(c)}{k}\right|<\epsilon_3, \quad \text{whenever} \quad |k|<\delta_3.$$

By taking δ to be the smaller of the positive numbers δ_2 and δ_3 we obtain the estimate equation (3.3) that is required to complete the proof of the chain rule. □

Let us put the chain rule into a slightly different form. If $u = h(x)$ then the chain rule reads

$$\frac{d}{dx}g(h(x)) = \frac{d}{dx}g(u) = g'(u)\frac{du}{dx}.$$

Note that the term $g'(u)$ can be written as $d\big(g(u)\big)/du$ so the rule has the following form:

$$\frac{d}{dx}g(u) = \frac{d}{du}g(u)\frac{du}{dx}.$$

This form is perhaps the one most easily remembered and most frequently used.

EXAMPLE 1 Compute the derivative of $f(x) = (2x^3 + 3x - 1)^5$.

Solution: Let $u = 2x^3 + 3x - 1$ so that $f(x) = u^5$. Then

$$\begin{aligned}\frac{d}{dx}(2x^3+3x-1)^5 &= \frac{d}{dx}u^5 = \left(\frac{d}{du}u^5\right)\frac{du}{dx}\\ &= 5u^4(6x^2+3)\\ &= 5(2x^3+3x-1)^4(6x^2+3).\end{aligned}$$

■

More generally we have the formula

$$\frac{d}{dx}u^n = \frac{d}{du}u^n\frac{du}{dx} = nu^{n-1}\frac{du}{dx}$$

for any differentiable function $u = h(x)$ and any integer n.

EXAMPLE 2 Compute the derivative of $f(x) = \cos^4 x$.

Solution: Let $u = \cos x$; then we want the derivative with respect to x of u^4:

$$\frac{d}{dx}\cos^4 x = \frac{d}{du}u^4\frac{du}{dx} = 4u^3\frac{d}{dx}\cos x$$
$$= 4\cos^3 x(-\sin x).$$

■

EXAMPLE 3 Compute $\frac{dy}{dx}$ for (a) $y = \sin(5x)$ and (b) $y = x\cos(x^2 + 1)$.

Solution: Both are chain rule problems and, in addition, the second requires the product rule. For (a), let $u = 5x$ and then

$$\frac{d}{dx}\sin(5x) = \frac{d}{du}\sin u\frac{du}{dx} = \cos u\left(\frac{d}{dx}5x\right) = 5\cos 5x.$$

For (b), let $u = x^2 + 1$ and then

$$\frac{d}{dx}x\cos(x^2+1) = x\frac{d}{dx}\cos(x^2+1) + 1\cdot\cos(x^2+1)$$
$$= x\frac{d}{du}\cos u\frac{du}{dx} + \cos(x^2+1)$$
$$= -x\sin u\frac{d}{dx}(x^2+1) + \cos(x^2+1)$$
$$= -(2x)x\sin(x^2+1) + \cos(x^2+1).$$

■

EXAMPLE 4 Suppose $f(x)$ is differentiable at every point. Compute the derivative of $f(x^3)$ and of $f(x)^3$.

Solution: For the derivative of $f(x^3)$, we let $u = x^3$ and apply the chain rule to get

$$\frac{d}{dx}f(x^3) = \frac{d}{du}f(u)\frac{du}{dx} = f'(x^3)(3x^2).$$

For the derivative of the function $f(x)^3$, we let $u = f(x)$ so

$$\frac{d}{dx}f(x)^3 = \frac{d}{du}u^3\frac{du}{dx} = 3u^2\frac{du}{dx}$$
$$= 3f(x)^2f'(x).$$

■

EXAMPLE 5 Suppose $g(x)$ is a function that satisfies $g(7) = 12$ and $g'(7) = -3$. Let $f(x) = g(x^2 + 3)^{-5}$. Compute $f'(2)$.

Solution: This requires two applications of the chain rule:

$$f'(x) = \frac{d}{dx}g(x^2+3)^{-5} = (-5)g(x^2+3)^{-6}\frac{d}{dx}g(x^2+3)$$
$$= (-5)g(x^2+3)^{-6}g'(x^2+3)(2x)$$

Now that we have the derivative of $f(x)$, we can evaluate at $x = 2$ and substitute the values of $g(7)$ and $g'(7)$ (inasmuch as $7 = 2^2 + 3$) to get

$$f'(2) = (-5)g(7)^{-6}g'(7)(2\cdot 2) = (-5)(12)^{-6} - 3\cdot 4.$$

■

Exercises 3.5

In the following y is given as a function of u and u is given as a function of x. Find dy/dx.

1. $y = u^2 - u^{-2}, \quad u = (3x - 7)^4$

2. $y = \dfrac{u}{u^2 + 1}, \quad u = \tan 4x$

3. $y = u^2 + 4u + u^{1/2}, \quad u = \dfrac{x}{x + 1}$

4. $y = \left(\dfrac{u + 4}{u - 4}\right), \quad u = \left(\dfrac{x + 4}{x - 4}\right)$

5. $y = 2u^3 - 4u^2 + 2, \quad u = \sin(4x + 1)$

6. $y = \dfrac{1}{u + 4}, \quad u = \dfrac{1}{2x} - 4$

Find the derivative of the following functions.

7. $f(x) = (2x - 5)^4$

8. $f(x) = (3 + x)^{-2}$

9. $f(x) = (4 - x^2)^2$

10. $f(x) = (x^3 + x + 1)^{-2}$

11. $y = (7x - 3)^5(3 - 7x)^{-5}$

12. $y = (5x^2 + 11)^8(5x^2 - 11)^8$

13. $y = (x - 5)^2(x + 3)^{-2}$

14. $y = (11x + 3)\sqrt{x^2 + 5}$

15. $y = \sin(x^2 + \pi)$

16. $y = (x^3 + 7\cos x + 11)^8$

17. $f(t) = \cos 5t$

18. $y = 8\cos 3x + 3\sin 2x$

19. $f(x) = (x + 1)\sin 2x$

20. $f(t) = \cos^2 5t$

21. $f(x) = \tan^4(1 - 3x)$

22. $f(t) = \left(\dfrac{t + 1}{t - 1}\right)^3$

23. $f(x) = \left(\dfrac{x + \sin x}{1 + x\cos x}\right)^3$

24. $f(t) = \left(\dfrac{t^2 + 1}{t^2 - 1}\right)^3$

The following values for the functions f and g are given:

$$f(0) = 3, \quad f(3) = 3, \quad f'(0) = -A, \quad f'(3) = B$$

$$g(0) = 3, \quad g(3) = 0, \quad g'(0) = -1, \quad g'(3) = 0$$

Let $h(x) = f(g(x))$ and $k(x) = g(f(x))$. Compute the following values.

25. $h'(0)$ and $h'(3)$

26. $k'(0)$ and $k'(3)$

27. If $H(x) = f(g(x^2))$, find $H'(0)$.

28. If $G(x) = f(g(x)^2)$, find $G'(3)$.

Apply the chain rule to evaluate the higher order derivatives.

29. $\dfrac{d^4}{dx^4}(\sin ax)$

30. $\dfrac{d^4}{dx^4}(\cos bx)$

31. y'' where $y = \sin u^3$ and $u = 4x^2 + x$

32. y'' where $y = \dfrac{u^2 + 1}{u^2 - 1}$ and $u = \tan x$

In the study of differential equations, one looks for a function $y = f(x)$ that satisfies an equation involving x, y and derivatives of y. For example the differential equation $y'' = -y$ has a solution $y = \cos x$. (There are infinitely many other solutions.) Find some solutions involving the trigonometric functions, other than the solution $y = 0$, to the following differential equations.

33. $y^{(4)} = y$

34. $y^{(4)} = 16y$

35. $y'' = -9y$

36. $y'' + a^2 y = 0$

37. Let a, b and c be constants. If $y = f(x)$ and $y = g(x)$ are solutions to the differential equation $ay'' + by' + cy = 0$, show that $y = pf(x) + qg(x)$ is also a solution for any constants p, q.

38. For which constants a, b and c will $y = 4\sin 3x + 7\cos 3x$ be a solution to the differential equation $ay'' + by' + cy = 0$?

39. Find formulas analogous to the chain rule $f(g(x))' = f'(g(x))g'(x)$ for the higher order derivatives $f(g(x))''$ and $f(g(x))'''$.

3.6 Derivatives of implicit functions

It sometimes happens that there is a function $y = f(x)$ that is not explicitly defined, but instead is given by some relation. For example suppose that y is the unique real number that satisfies the condition

$$y^3 + y = x^2 + 1 \tag{3.4}$$

for a given real number x. You would be quite right to ask if there really is such a number. It happens to be true that for any given number x there is one and only one number y with the property $y^3 + y = x^2 + 1$. We leave the proof of this to the exercises. Thus for each real number x we have defined a real number y and we denote this number as $y = f(x)$. Because we do not have an explicit formula for $f(x)$ one might think that the derivative of $f(x)$ cannot be computed. This is not the case!

The process of *implicit differentiation* permits evaluation of the derivative $f'(x)$ provided we know the two numbers x and $f(x)$. We proceed on the assumption that $f(x)$ is differentiable at the point x and then apply the chain rule to differentiate both sides of equation (3.4). We obtain

$$\frac{d}{dx}(y^3 + y) = \frac{d}{dx}(x^2 + 1)$$

$$3y^2\frac{dy}{dx} + \frac{dy}{dx} = (3y^2 + 1)\frac{dy}{dx} = 2x.$$

Thus

$$\frac{dy}{dx} = \frac{2x}{3y^2 + 1}.$$

If we need to evaluate $f'(3)$ then we must find the value of y that satisfies equation (3.4) when $x = 3$. By inspection, $y = 2$ and so $f(3) = 2$ and

$$f'(3) = \left.\frac{dy}{dx}\right|_{(x,y)=(3,2)} = \frac{2 \cdot 3}{3 \cdot 2^2 + 1} = \frac{6}{13}.$$

The symbol

$$\left.\frac{dy}{dx}\right|_{(x,y)=(a,b)}$$

means the expression for dy/dx in terms of x and y is evaluated by substituting a for x and b for y.

This example shows that the derivative of an implicitly given function can be computed and values of that derivative can be found at points on the graph of the function, at least at those points where both the x and y coordinates are given.

EXAMPLE 1 A differentiable function $y = f(x)$ is known to satisfy the equation

$$x^2 + y^2 = r^2,$$

with r a constant. Compute the derivative of $f(x)$.

Solution: We differentiate the given equation to get

$$2x + 2y\frac{dy}{dx} = 0 \quad \text{or} \quad \frac{dy}{dx} = -\frac{x}{y}. \tag{3.5}$$

We can give explicitly two differentiable functions that satisfy the equation:

$$y = g(x) = \sqrt{r^2 - x^2} \quad \text{and} \quad y = h(x) = -\sqrt{r^2 - x^2}.$$

In the two cases $y = g(x)$ or $y = h(x)$, the derivative $f'(x)$ is found by substituting the appropriate formula into equations (3.5):

$$g'(x) = -\frac{x}{\sqrt{r^2 - x^2}} \quad \text{if} \quad y = \sqrt{r^2 - x^2}$$

$$h'(x) = \frac{x}{\sqrt{r^2 - x^2}} \quad \text{if} \quad y = -\sqrt{r^2 - x^2}.$$

■

We use this example to point out that it is necessary to assume that there is a differentiable function $y = f(x)$ whose derivative is to be computed implicitly. Just because a function satisfies an equation such as $x^2+y^2 = 1$, it does not follow that it is differentiable. For example let $p(x)$ be defined so that its graph is the upper circle in the first quadrant and the lower circle in the fourth quadrant; that is

$$p(x) = \begin{cases} \sqrt{1-x^2} & \text{if } 0 \le x \le 1 \\ -\sqrt{1-x^2} & \text{if } -1 \le x < 0. \end{cases}$$

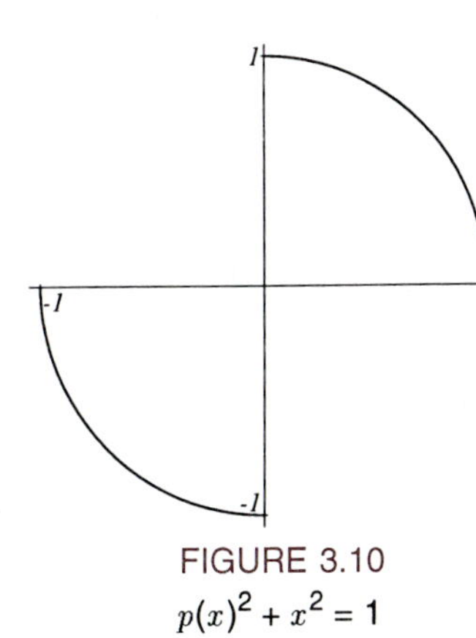

FIGURE 3.10
$p(x)^2 + x^2 = 1$

Then for every x on $[-1, 1]$, $p(x)$ satisfies the equation $x^2 + p(x)^2 = 1$ but the derivative of $p(x)$ at $x = 0$ does not exist everywhere because the function is not continuous at 0 and so not differentiable at 0.

EXAMPLE 2 Let $y = f(x)$ be a differentiable function that satisfies the equation $y^2 + 3xy - x = 10$ for all x on some interval. Compute the second derivative $y'' = d^2y/dx^2$.

Solution: We begin by differentiating the equation with the help of the chain rule and the product rule:

$$2y\frac{dy}{dx} + 3x\frac{dy}{dx} + 3y - 1 = 0.$$

Solve this to obtain

$$y' = \frac{dy}{dx} = \frac{1-3y}{2y+3x}. \tag{3.6}$$

Differentiate this expression using the quotient rule on the right-hand side to obtain

$$y'' = \frac{(2y+3x)(-3y') - (1-3y)(2y'+3)}{(2y+3x)^2}.$$

This can be expressed in terms of x and y only by using expression (3.6) to eliminate the y' terms. ■

Fractional Exponents

We review the definition of x^r when $r = n/m$, with n, m nonzero integers and then show how to compute the derivative of this function.

Assume x is a positive number and that m is a positive integer. We ask the question: Is there a number t such that $t^m = x$? The answer is yes for the following reason. The function $p(t) = t^m$ is continuous for all t and so the Intermediate Value Theorem holds for p. Moreover $p(0) = 0$ and for a suitable choice of the number s, we have

$$0 = p(0) < x \le p(s).$$

By the Intermediate Value Theorem, there is a number t such that $p(t) = x = t^m$. In fact if we insist that $0 \le t$, then there is only one such number t; thus t is uniquely determined by x. This unique number t is denoted by

$$t = x^{1/m}. \tag{3.7}$$

More formally, we say the mth root of the positive number x is the unique positive number t that satisfies $t^m = x$ and this t is denoted by equation (3.7).

This defines exponents of the form $1/m$; the more general rational exponent is defined as follows: For a positive real number x and two positive integers n,

m the fractional power $x^{n/m}$ is defined as the nth power of $x^{1/m}$; that is,

$$x^{n/m} = \left(x^{1/m}\right)^n.$$

Exactly as in the case with integral exponents, the negative rational exponents are defined using reciprocals. If r is a negative rational number then $-r$ is positive, x^{-r} is defined and x^r is its reciprocal; that is,

$$x^r = \frac{1}{x^{-r}}.$$

For any rational number $r = n/m$, we now compute the derivative of the function $y = x^r$.

THEOREM 3.13

If $r = n/m$ is a positive or negative rational number and $y = x^r$, then

$$\frac{dy}{dx} = rx^{r-1}.$$

Proof

Assume $m > 0$ (and n may be positive or negative). If $y = x^{n/m}$, then $y^m = x^n$ for all positive x. Differentiate this equation implicitly to obtain

$$my^{m-1}y' = nx^{n-1}.$$

This computation involves derivatives of functions having integer exponents so these formulas are correct by previous results.

Multiply both sides by y and replace y^m with x^n, then solve for y' to obtain

$$y' = \frac{nx^{n-1}y}{mx^n} = \frac{n}{m}x^{-1}y = rx^{-1}x^r = rx^{r-1}. \qquad \square$$

EXAMPLE 3 Here are some derivatives computed using this rule.

$$\frac{d}{dx}x^{4/5} = \frac{4}{5}x^{-1/5},$$

$$\frac{d}{dx}(x^2+1)^{5/3} = \frac{5}{3}(x^2+1)^{2/3}\frac{d}{dx}(x^2+1)$$

$$= \frac{5}{3}(x^2+1)^{2/3}(2x).$$

■

Exercises 3.6

Assume that $y = f(x)$ is a differentiable function satisfying the given equation. Use implicit differentiation to compute dy/dx.

1. $y^3 + 4x = 3$

2. $y^2 + xy - 3x + y = 1$

3. $x = (y^2 - x^2 + 1)^{1/2}$

4. $\sin y = xy$

5. $\sqrt{2(x+y)} = \sqrt{x} + \sqrt{y}$

6. $\tan y = x$

7. $y^5 = (x^{1/3} + 1)^3$

8. $y^{3/2} + x^{3/2} = 1$

9. $(4y)^{3/2} = (x^2 + 4)^{1/2}$

10. $G(y) = F(x)$

Find the slope of the line tangent to the graph of $y = f(x)$ at the given point when y is implicitly defined by the given equation.

11. $y^2 - x^2 = 9$, $(4, 5)$

12. $2y^2 + xy - 3x^2 = 3$, $(-1, 2)$

13. $\dfrac{x^2 + x - 2}{y^2 + 1} = 5$, $(3, 1)$

14. $\dfrac{x^2 + x - 2}{y^2 + 1} = 5$, $(3, -1)$

If $y = f(x)$ is a twice differentiable function that satisfies the following relation, express y'' in terms of x and y.

15. $x^2 + 3y^2 = 4$

16. $x^2 - xy + y^2 = 16$

17. $2y - y^3 = x$

18. $\sin x \cos y = 1$

Compute y'' at the indicated point for the implicit functions.

19. $3x^2 - 2y^2 = 1$ at $(1, 1)$

20. $2x^2 - 3xy + y^2 = 2$ at $(1, 0)$

21. $\sin^2 x + \cos^2 y = 1$ at $(\pi/4, \pi/4)$

A family of curves can be defined by introducing an arbitrary constant into an equation; for example, the equation $x^2 + y^2 = r^2$ defines the family of circles having center at the origin and radius r for positive r. In each of the following exercises, show that any member of the first family intersects a member of the second family at right angles; that is, the tangents to the respective curves at the point of intersection are perpendicular. Keep in mind that if (x, y) is a point of intersection, then x and y satisfy both equations. So find the slope of the tangent line at a point (x, y) on each curve and compare the results.

22. $y^2 + x^2 = r^2$, $\quad y = mx$

23. $xy = a$, $\quad y^2 - x^2 = b$

24. $x^2 + 2y^2 = c$, $\quad y = bx^2$

25. $y^2 + x^2 = py$, $\quad (x - h)^2 + y^2 = h^2$

26. Let $g(x) = x^3 + x$. Prove that $g(x)$ is an increasing function on the entire real line ($a < b$ implies $g(a) < g(b)$). Use this fact and the fact that $g(x)$ is unbounded on both the positive axis and on the negative axis to prove the statement: For any number c there is one and only one number y such that $y^3 + y = c$.

27. Let $-1 < a_1 < a_2 < \cdots < a_n < 1$ be any n points on the interval $(-1, 1)$. Show there is a function $q(x)$ that satisfies $x^2 + q(x)^2 = 1$ for all x on $[-1, 1]$ but the derivative of $q(x)$ does not exist at the points $x = a_i$, $1 \leq i \leq n$.

28. If $y = f(x)$ is known to satisfy $y' = x + y$ for all x and $f(0) = 1$, determine $f^{(n)}(0)$ for $1 \leq n \leq 4$.

29. If $y = f(x)$ is known to satisfy $y' = xy$ for all x and $f(0) = 2$, compute $f^{(k)}(0)$ for $1 \leq k \leq 5$.

30. If $y = f(x)$ is known to satisfy $y' = 2y$ for all x and $f(0) = 1$, determine $f^{(n)}(0)$ for all positive integers n.

4 Applications of the Derivative

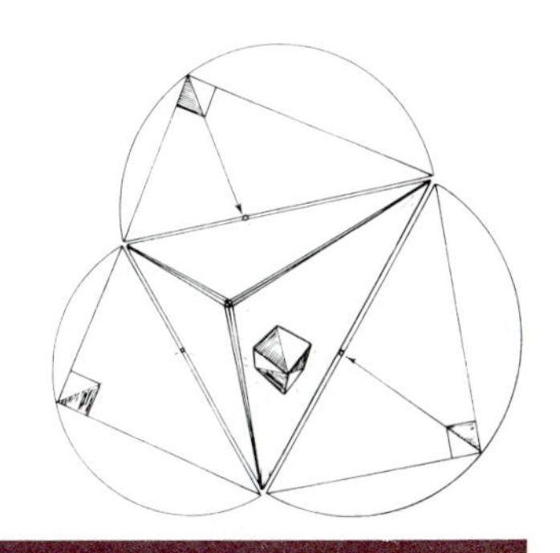

We have already seen in chapter 3 that for a given function f, its derivative f' is a function whose value $f'(a)$ at the point $x = a$ gives the slope of the line tangent to the curve $y = f(x)$ at the point $(a, f(a))$. The simple geometric fact that the graph has a horizontal tangent at a point where the function takes a minimum or maximum value is exploited in this chapter in a variety of situations to study of the behavior of functions. For example, it is used to help determine the points where the maximum and minimum values of a function occur; it is used to study the geometric properties of a function's graph, such as the direction of the bend in a curve at a point; and it is used to study the velocity and acceleration of a moving object. We have studied a number of rather abstract properties of functions in the preceding chapters. Now we see a number of applications of the derivative to solve specific problems.

4.1 Extrema of functions

Many problems to which the methods of calculus are applied involve finding an extreme value of some function. By an *extreme value* we mean either a maximum or a minimum value of a function. For example, if the profit $f(k)$ of the Edsel Motor Company is a function of the number k of television ads run per month, it is of interest (to the company) to determine the maximum value of the function. If $g(t)$ is the distance between two moving objects at time t, the minimum value of the function $g(t)$ occurs at the time of nearest approach. We will see many examples of problems like these in later sections. In this section we lay the theoretical foundation for discussion of extrema of functions; the Mean Value Theorem and some other important concepts are described and applied to solve specific problmes.

We begin this section by stating a theorem that asserts a property of continuous functions; namely that on any closed interval, a continuous function takes on a least and a greatest value. Proving this result requires some subtle properties of the real number system that carry us away from the subject at hand and so we postpone the proof until later. [A proof requires the notion of uniform continuity that is introduced in chapter 5. A proof of the following theorem is given in the exercises following section 5.4.]

THEOREM 4.1

Maximum and Minimum Attained

If the function $f(x)$ is continuous on the closed interval $[a, b]$, then there exist numbers u, v on $[a, b]$ such that the inequality

$$f(u) \leq f(x) \leq f(v)$$

holds for every x on $[a, b]$. In other words, $f(u)$ is the minimum value and $f(v)$ is the maximum value of f on the interval.

The values $f(u)$ and $f(v)$ are the minimum and maximum values of the function on the interval $[a, b]$. We refer to either of these values as *extreme values* of $f(x)$ on $[a, b]$.

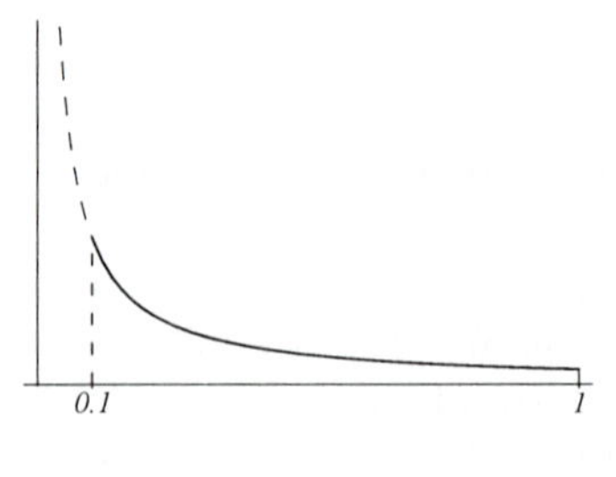

FIGURE 4.1
$f(x) = 1/x$

The conditions that the function be continuous and continuous on the *closed* interval are necessary for the validity of the conclusion. To see this consider the following examples: Let $f(x) = 1/x$. Then $f(x)$ does not have a maximum value on the closed interval $[-1, 1]$. By taking $x = 10^{-n}$ for any positive integer n, we see $f(x) = 10^n$. Thus $f(x)$ takes on values larger than any given number on the interval $[-1, 1]$. This does not conflict with the theorem. The theorem does not apply to this function on this interval because $f(x)$ is not continuous at 0. If we exclude 0 by considering only the interval $(0, 1]$ we are not much better off. The function $1/x$ does not have a maximum value on the interval $(0, 1]$. Even though the function $1/x$ is continuous on the interval, the theorem does not apply in this instance because the interval $(0, 1]$ is not closed. If we finally select a closed interval, such as $[0.1, 1]$ on which $1/x$ is continuous, the theorem can be applied and $1/x$ has its maximum value at $x = 0.1$ and minimum value at $x = 1$.

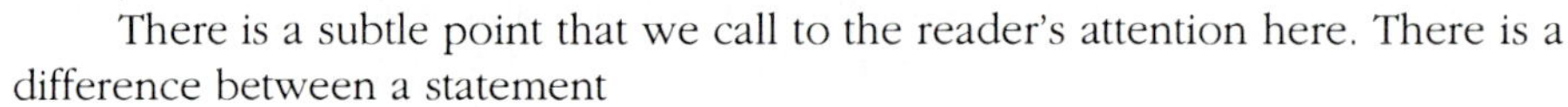

There is a subtle point that we call to the reader's attention here. There is a difference between a statement

$$B \leq f(x) \leq C$$

for some constants B and C and the statement

$$f(u) \leq f(x) \leq f(v).$$

In the first statement, we are assured only that there are some constants that bound the function. The graph of the function lies entirely between the horizontal lines $y = B$ and $y = C$, but the graph might lie in a much narrower strip. In the second statement we are told more: namely that there are constants that bound the function and these constants are actually values of the function. This means that the constants are the best possible in the sense that no larger constant will be a lower bound and no smaller constant will be an upper bound. The graph of the function lies in the strip bounded by the lines $y = f(u)$ and $y = f(v)$ and the graph actually touches both of these lines. (See figures 4.2 and 4.3.)

It is not always true that best possible bounds are actually values of the function. We can give an example over an infinite interval to show this. Consider the function $f(x) = 1/x$ on the interval I consisting of all numbers greater than or equal 1. Then $f(x)$ is continuous on this interval and we have

$$0 \leq f(x) = \frac{1}{x} \leq 1;$$

the number 0 is the best possible lower bound but that lower bound is not a value of the function because there is no solution to the equation $f(x) = 0$. Why is it necessary to use an infinite interval to give this example?

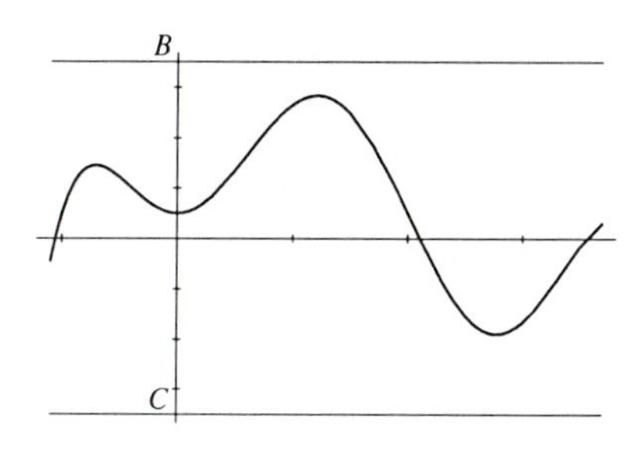

FIGURE 4.2
Bounds for $f(x)$

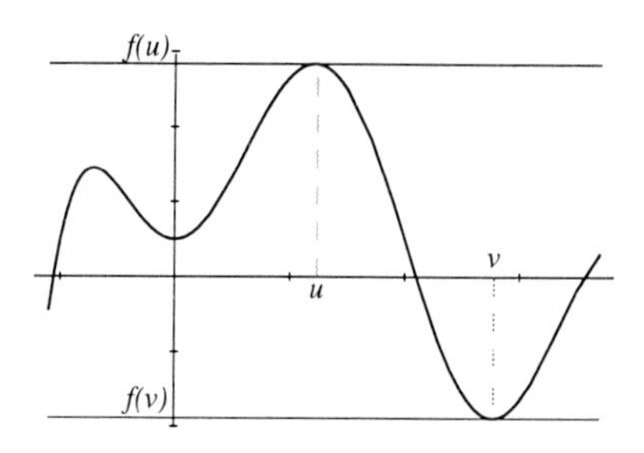

FIGURE 4.3
Sharp bounds for $f(x)$

Location of Extrema

The phrase *extreme value of* $f(x)$ refers to either a minimum value of f or a maximum value of f.

Here is the problem we set out to solve: Given the function $f(x)$ and the interval $[a, b]$, find all the points u for which $f(u)$ is an extreme value of $f(x)$. We will consider only continuous functions because we are then assured that there do exist points where the extreme values occur. The solution is stated in a negative way; it tells which points of the interval cannot be the points where $f(x)$ has extreme values. It remains for us to test the other points of the interval to determine where the function takes on its extreme values.

THEOREM 4.2

Let $f(x)$ be continuous on $[a, b]$ and let c be a point on (a, b) for which $f'(c)$ exists and $f'(c) \neq 0$. Then $f(c)$ is not an extreme value of f on $[a, b]$.

Proof

The proof we are about to give shows that the graph of $f(x)$ stays "close enough" to the tangent line over some interval with c at the center so that on one side of c, $f(x)$ has values less than $f(c)$, and on the other side of c, the values of $f(x)$ are greater than $f(c)$. This shows that $f(c)$ is not an extreme value because there are values of f both less than $f(c)$ and greater than $f(c)$.

We fix the number c and let $f'(c) = m$ so that $y = T(x) = m(x-c)+f(c)$ is the equation of the tangent line to the graph of $f(x)$ at the point $(c, f(c))$. Because the tangent line has nonzero slope, the point $(c, f(c))$ is neither a maximum nor minimum value of the function $T(x)$.

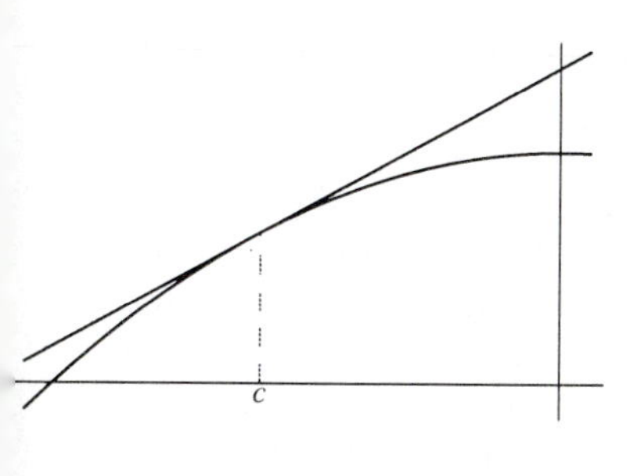

FIGURE 4.4
$f(x)$ close to tangent

By the definition of the derivative, for any positive number ϵ there is a positive number δ such that

$$\left|\frac{f(c+h)-f(c)}{h} - m\right| < \epsilon, \quad \text{whenever} \quad 0 < |h| < \delta. \qquad \textbf{(4.1)}$$

We can assume, no matter what ϵ is selected, that δ is small enough so that the entire interval $(c - \delta, c + \delta)$ lies inside (a, b). (This is possible because c is a point on the open interval.)

Because the assumption is that $m \neq 0$, we have two cases: either $m > 0$ or $m < 0$. We will treat the case $m > 0$ and leave the analogous statement in the other case for the reader. Multiply inequality (4.1) by $|h|$ and use the property of absolute values to obtain

$$-|h|\epsilon + mh < f(c+h) - f(c) < |h|\epsilon + mh, \quad \text{when } 0 < |h| < \delta. \tag{4.2}$$

Let us select the number $\epsilon = m/2$. Because $m > 0$, this is a permissible choice. When $h > 0$, the last inequality becomes

$$h\left(-\frac{m}{2} + m\right) = \frac{hm}{2} < f(c+h) - f(c) < h\left(\frac{m}{2} + m\right).$$

Thus for all h with $0 < h < \delta$ we have (from the left part of the inequality only)

$$f(c) < f(c) + \frac{hm}{2} < f(c+h).$$

Thus $f(c+h)$ is a value of f larger than $f(c)$.

Now suppose we take $-\delta < h < 0$ and we use the right half of inequality (4.2) (and $|h| = -h$) which reads

$$f(c+h) - f(c) < (-h\epsilon + hm) = -h\frac{m}{2} + hm = \frac{hm}{2} < 0.$$

This yields $f(c+h) < f(c)$ and so $f(c+h)$ is a value of f less than $f(c)$. This proves $f(c)$ is not an extreme value of f on $[a, b]$. The case with $m < 0$ is treated in a similar way except that we select $\epsilon = -m/2 > 0$. □

This result tells us that $f(c)$ is an extreme value of f on $[a, b]$, then possibly c is one of the endpoints a or b, possibly c is a point where $f'(c) = 0$ or possibly c is a point where the derivative of f does not exist. Moreover, these are the only possibilities. Consideration of these conditions occurs frequently enough that we give a name to such points.

DEFINITION 4.1

Critical Point

A point c is called a critical point of $f(x)$ if either $f'(c) = 0$ or the derivative of $f(x)$ does not exist at c.

For emphasis, we summarize using this terminology.

Corollary 4.1

Extreme Values

If $f(c)$ is an extreme value of the function $f(x)$ on the interval $[a, b]$, then either c is a critical point of $f(x)$ or c is one of the endpoints, $c = a$ or $c = b$.

This result helps in the solution of the problem: What are the extreme values of the function $f(x)$ on an interval $[a, b]$? To solve this problem we first find the critical points; it is then necessary to determine if the function has extreme values at these points. It is not always easy to decide if the function has an extreme value at a critical point. We study this problem in more detail using the second derivative in a later section. For now we illustrate a method that can be applied in many cases.

EXAMPLE 1 Find the extreme values of the function $f(x) = x(1 - x)$ over the interval $[0, 2]$.

Solution: The extreme values of $f(x)$ occur at either an endpoint of the interval or at a critical point. We first look for the critical points. The derivative of $f(x)$ is

$$f'(x) = 1 - 2x.$$

The derivative exists throughout the interval $[0, 2]$ and equals 0 at $x = 1/2$. Thus $x = 1/2$ is the only critical point of f and the extreme values of $f(x)$ on the interval $[0, 2]$ must occur at one of the points 0, 1/2, or 2. We compute the values to get $f(0) = 0$, $f(1/2) = 1/4$, $f(2) = -1$. Thus the maximum value of f on $[0, 2]$ is $f(1/2) = 1/4$ and the minimum value is $f(2) = -1$. ■

Rolle's Theorem

The Theorem 4.2 and Corollary 4.1 give excellent computational tools for determining the extreme values of a function. These results are now used to prove some very important facts about differentiable functions. We apply Corollary 4.1 to prove Rolle's Theorem. The last two results as well as Rolle's Theorem are tools used to prove the Mean Value Theorem (given later). Although the proof of the Mean Value Theorem is our main goal, we delay proving it directly because the proof would involve some version of all these preliminary results anyway; they have been given the status of theorems to keep the main ideas separate.

THEOREM 4.3

Rolle's Theorem

Let $f(x)$ be continuous on the interval $[a, b]$ (with $a < b$) and assume $f(a) = f(b)$. If the derivative $f'(x)$ exists for every x on the open interval (a, b), then there is a point c on (a, b) with $f'(c) = 0$.

Proof

If $f(x)$ is constant on $[a, b]$, then $f'(x) = 0$ for every x on (a, b) and the theorem is true in this case. Suppose that $f(x)$ is not constant. Then at least one of the extreme values of the function is different from $f(a)$; thus there is a point c for which $f(c)$ is an extreme value not equal to $f(a)$. Because $f(a) = f(b) \neq f(c)$, c cannot be either a or b; in particular c is on the open interval (a, b). The derivative $f'(c)$ exists by assumption and by the previous corollary we must have $f'(c) = 0$. (If $f'(c) \neq 0$ then $f(c)$ is not an extreme value.) □

This theorem has a very nice geometric interpretation. Note that the condition $f'(c) = 0$ means that the tangent to the graph of $f(x)$ at the point $(c, f(c))$ has slope 0; that is, the tangent is horizontal. If the graph of the continuous function $f(x)$ crosses the x-axis at the point $P(a, 0)$ and also crosses the x-axis at $Q(b, 0)$, then there must be a critical point c on the open interval (a, b). If the derivative $f'(x)$ exists at every point of the interval (a, b), then in fact there must be a point c where $f'(c) = 0$, and so there is a point where the graph has a horizontal tangent.

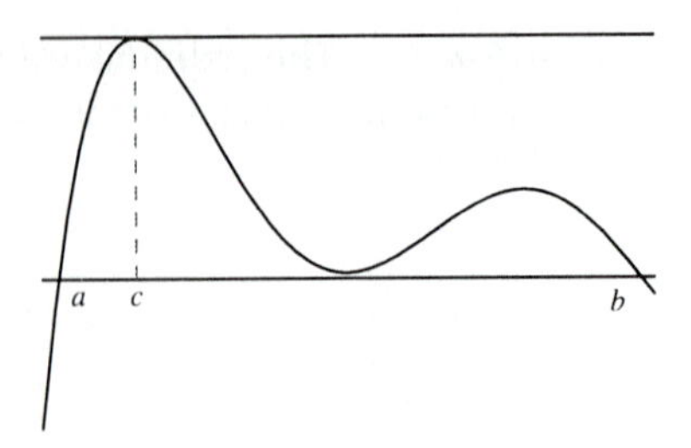

FIGURE 4.5
Two of three horizontal tangents

With just a very simple device, we can extend the application of Rolle's Theorem to a situation involving two functions. The next theorem says that two differentiable functions having equal values at a and at b must have parallel tangents at some point between a and b.

THEOREM 4.4

Extended Rolle's Theorem

Let the two functions $h(x)$ and $g(x)$ be continuous on the interval $[a, b]$ and differentiable on the open interval (a, b). Suppose

$$h(a) = g(a) \qquad \text{and} \qquad h(b) = g(b).$$

Then there is a point c on the interval (a, b) such that $h'(c) = g'(c)$.

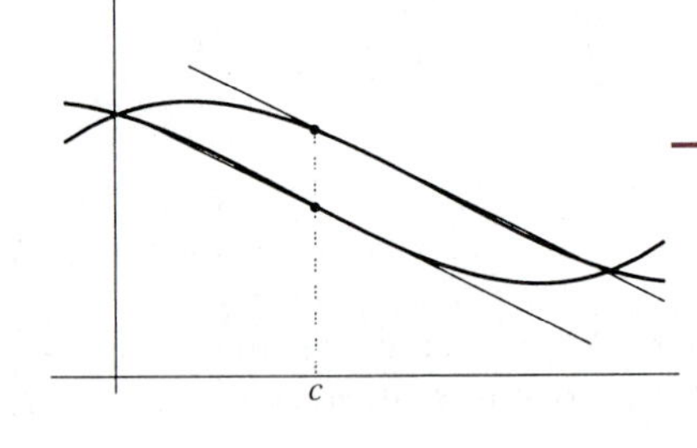

FIGURE 4.6
Intersecting curves have parallel tangents

Proof

Let $f(x) = h(x) - g(x)$. Then the hypothesis of Rolle's Theorem holds for $f(x)$ and the conclusion of Rolle's Theorem implies there is a point c on the interval such that

$$0 = f'(c) = h'(c) - g'(c).$$

Thus $h'(c) = g'(c)$ and the theorem is proved. □

This result has a rather startling interpretation when $g(x)$ and $h(x)$ are position functions and $x = t$ is a time variable. Suppose two moving objects start from the same point at the same time and move along possibly different paths at variable rates of speed so that after time t the distances from the starting point are $g(t)$ and $h(t)$. If both arrive at the same destination point at the same time, then there must have been some instant of time when both were moving at exactly the same velocity. Because the velocity functions are $g'(t)$ and $h'(t)$, the Extended Rolle's Theorem implies that there is some time $t = t_1$ between the starting time and the finishing time where $g'(t_1) = h'(t_1)$. (Of course, it is necessary that the two functions $g(t)$ and $h(t)$ be differentiable for this argument to hold.)

Exercises 4.1

Find the critical points of the functions on the indicated interval and determine if the function has a maximum or minimum or neither at each critical point. To test each critical point for an extreme value, an accurate sketch of the curve over an interval containing the critical point can be helpful.

1. $x^2(x-2)$ on $[-4, 4]$

2. $x^3(x-2)^2$ on $[-4, 4]$

3. $\dfrac{x+1}{x^2+4}$ on $(-4, 4)$

4. $\dfrac{x^2+1}{x^2-4}$ on $(-4, 4)$

5. $\sqrt{\dfrac{x+1}{x-1}}$ on $(1, 10)$

6. $\sqrt{\dfrac{x^2+1}{x+1}}$ on $(-1, 5]$

7. $1-\cos x$ on $[-\pi, \pi]$

8. $3+2\sin x$ on $[0, 2\pi]$

9. $2x-\cos 2x$ on $[-2\pi, 2\pi]$

10. $x+\sin(x/2)$ on $[0, 4\pi]$

For each function and interval, verify the hypothesis of Rolle's Theorem and then locate the point that is guaranteed by Rolle's Theorem to have certain properties.

11. x^2-x on $[0, 1]$

12. x^3-x^2 on $[0, 1]$

13. $\dfrac{x^2-1}{x+2}$ on $[-1, 1]$

14. $\dfrac{x^2-4}{2x+11}$ on $[-2, 2]$

For each pair of functions and the given interval, verify the hypothesis of the Extended Rolle's Theorem and then find the x coordinate where the graphs of the two functions have parallel tangents.

15. $h(x) = x^2$, $g(x) = 2x^2 - x$ over the interval $[0, 1]$

16. $h(x) = x^2+4x^3$, $g(x) = 3x^3-2x^2$ over the interval $[0, 1]$

17. $h(x) = \dfrac{x^2}{x+2}$, $g(x) = \dfrac{1}{x+2}$ over the interval $[-1, 1]$

18. $h(x) = \cos x$, $g(x) = 1 - 2x/\pi$ over the interval $[0, \pi]$ [Find an approximate solution accurate to one decimal.]

The next few exercises deal with the roots of polynomials and the application of Rolle's Theorem to obtain proofs of these results.

19. Let $p(x) = ax^2+bx+c$, with a, b, c constants. Use Rolle's Theorem to prove that there are at most two solutions to the equation $p(x) = 0$. [Of course this can also be done using the quadratic formula.]

20. Use Rolle's Theorem to prove that the equation $x^3 + 4x + c = 0$ has only one solution no matter what the value of the constant c.

21. If $p(x)$ is a polynomial and there are exactly n different solutions to the equation $p(x) = 0$, prove that there are at least $n-1$ solutions to the equation $p'(x) = 0$. [Hint: Apply Rolle's Theorem over each interval $[a, b]$ where $p(a) = p(b) = 0$.]

22. If $p(x)$ is a polynomial of degree n, prove there are no more than n solutions to the equation $p(x) = 0$. [Hint: Use Exercise 21 and mathematical induction; i.e., prove the statement is true for $n = 1$ and then, based on the assumption that the statement is true for a positive integer $n = k$, prove it is true for $n = k+1$.]

4.2 The mean value theorem

The Mean Value Theorem (MVT) gives a connection between values of a function at two points and a value of the derivative of the function at a point between the two points. This theorem is used many times throughout the remainder of this text; it is fair to say that this theorem enters in some way into the discussion of most of the theoretical concepts to be introduced in the next few chapters.

THEOREM 4.5

The Mean Value Theorem

Let f be continuous on the closed interval $[a, b]$ and assume the derivative $f'(x)$ exists at every point of the open interval (a, b). Then there is a point c on the open interval (a, b) that satisfies

$$f'(c) = \frac{f(b) - f(a)}{b - a}. \tag{4.3}$$

The quotient on the right side of equation (4.3) is the slope of the line connecting the two points $P(a, f(a))$ and $Q(b, f(b))$. The derivative $f'(c)$ is the slope of the line tangent to the graph at $(c, f(c))$. The MVT states that there is some point c between a and b where the tangent line at $(c, f(c))$ is parallel to the line PQ.

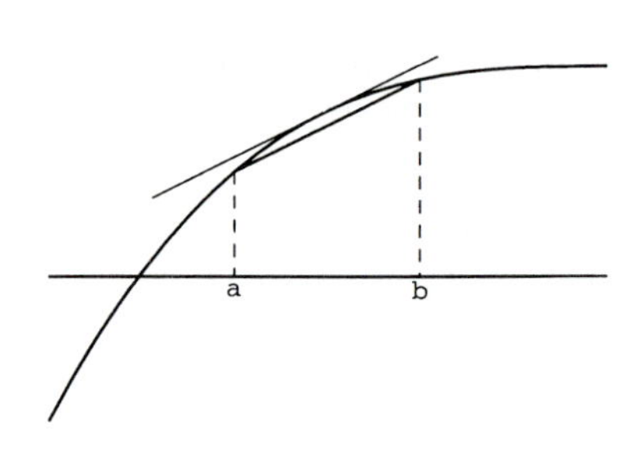

FIGURE 4.7
Tangent parallel to chord

Proof

The equation of the line through $P(a, f(a))$ and $Q(b, f(b))$ is

$$g(x) = f(a) + \frac{f(b) - f(a)}{b - a}(x - a).$$

We intend to apply the Extended Rolle's Theorem to g and f. First we verify that the hypothesis of the theorem holds. By construction, the graphs of both f and g pass through P and Q; that is $f(a) = g(a)$ and $f(b) = g(b)$. By hypothesis, f is differentiable; also g is differentiable because it is a polynomial in x and its derivative is the constant

$$g'(x) = \frac{f(b) - f(a)}{b - a}.$$

The Extended Rolle's Theorem implies there is a point c on (a, b) with $f'(c) = g'(c)$; equivalently this says

$$f'(c) = \frac{f(b) - f(a)}{b - a},$$

which completes the proof. □

The difference $f(b) - f(a)$ is the change in the function over the interval $[a, b]$ and the quotient

$$\frac{f(b) - f(a)}{b - a}$$

is the average value or *mean value* of the change in the function over the interval. (Think of this quotient as the change in the function per unit distance.) The derivative $f'(c)$ is the "instantaneous" rate of change of the function at c. The Mean Value Theorem says that the "mean value of the change in f over an interval" equals the "instantaneous change in f" at some point of the interval.

EXAMPLE 1 Verify the statement of the Mean Value Theorem for the case of $f(x) = x + 1/x$ on the interval $[1, 2]$.

Solution: The function f is continuous and its derivative f' exists on the interval $[1, 2]$ so the Mean Value Theorem can be applied. The chord connecting $(1, f(1)) = (1, 2)$ and $(2, f(2)) = (2, 5/2)$ has slope

$$\frac{f(2) - f(1)}{2 - 1} = \frac{5/2 - 2}{2 - 1} = \frac{1}{2}.$$

The Mean Value Theorem says that there is some point c with $1 < c < 2$ such that $f'(c) = 1/2$. Compute the derivative of $f(x)$ to get the equation

$$1 - \frac{1}{x^2} = \frac{1}{2}.$$

This equation has the solutions $x = \pm\sqrt{2}$ and the solution $x = \sqrt{2}$ does lie on the interval $[1, 2]$. ■

Exercises 4.2

The following exercises can be done with the help of the Mean Value Theorem.

1. Suppose f is differentiable on the interval $[-1, 1]$ and $f'(x) < 1$ on this interval. If $f(-1) = 3$, prove $f(1) < 5$.

2. Suppose f is differentiable on the interval $[1, 3]$ and $|f'(x)| \leq 4$ for all x on the interval. If $f(1) = 9$ prove $1 \leq f(3) \leq 17$.

3. Suppose f is differentiable on the interval $[-2, 2]$ and $f'(x) \geq 6$ on this interval. If $f(2) = 199$ prove $f(-2) \leq 175$.

4. Suppose f is differentiable on the interval $[a, b]$ and $f'(x) < C$ on this interval. Prove $f(b) < f(a)+(b-a)C$.

5. If $|f'(x)| \leq 1$ for all x prove $|f(u) - f(v)| \leq |u - v|$.

6. Prove that $|\cos u - \cos v| \leq |u - v|$ for any u and v.

7. Prove that $|\sin u - \sin v| \leq |u - v|$ for any u and v.

When f satisfies the hypothesis of the Mean Value Theorem on $[a, b]$, there is a number c between a and b such that $f(b) - f(a) = f'(c)(b - a)$. Find a number c that satisfies this condition for the following functions and intervals.

8. $x^3 + 3x - 7$, $[0, 2]$

9. $x^3 - x + 2$, $[-2, 2]$

10. $x^3 + x^2 - 1$, $[-1, 2]$

11. $x^4 - 5x^2 + 3x$, $[-1, 2]$

12. $x^4 - 10x^2 + 2x$, $[1, 3]$

13. $x^4 + 4x$, $[-2, 2]$

14. We have proved that if $f'(x) > 0$ for all x on $[a, b]$ then $a \leq u < v \leq b$ implies $f(u) < f(v)$. Modify the proof of this result to prove: If $f'(x) \geq 0$ for all x on $[a, b]$ then $a \leq u \leq v \leq b$ implies $f(u) \leq f(v)$.

15. Suppose $f'(x)$ exists and is positive on $[a, b]$. Prove that the equation $f(x) = 0$ has at most one solution on $[a, b]$. Draw the same conclusion if $f'(x) < 0$ on $[a, b]$.

16. Suppose that $f''(x) > 0$ on $[a, b]$. Prove that the equation $f(x) = 0$ has at most two solutions on $[a, b]$.

17. Let f be a function whose derivative is given by $f'(x) = 1/x$ for $x > 0$. [This function is studied in chapter 7.] Suppose that $f(1) = 0$. Prove that

$$(x - 1)/x < f(x) < x - 1$$

for all $x > 0$. [Hint: Do the cases $x > 1$ and $0 < x \leq 1$ separately.]

18. Suppose f is a function whose derivative is given by $f'(x) = 1/(1 + x^2)$ for all x. Assume $f(0) = 0$. Prove

$$x/(1 + x^2) < f(x) < x$$

for $x > 0$. Find a similar inequality valid for $x < 0$.

4.3 Applications of the mean value theorem

We have seen that the derivative of a constant function equals 0. Consider the converse question: If $f'(x) = 0$ for all x, does it follow that f is a constant function? The Mean Value Theorem is used to give an affirmative answer.

THEOREM 4.6

A Function with 0 Derivative Is Constant

Let f be a function whose derivative is zero at every point of the interval $[a, b]$. Then f is constant on $[a, b]$.

Proof

By hypothesis, $f'(x)$ exists and is 0 at each point of $[a, b]$ so the assumptions in the Mean Value Theorem are valid for this $f(x)$ and any interval contained in $[a, b]$. Let $p < q$ be any two points on $[a, b]$. We can apply the MVT to f on the interval $[p, q]$ to obtain a point c between p and q such that $f(q) - f(p) = f'(c)(q - p)$. By assumption $f'(c) = 0$, because c is on $[a, b]$, so it follows that $f(q) - f(p) = 0$; that is $f(p) = f(q)$ for any points p and q on $[a, b]$; f is constant on the interval. □

This theorem tells us that the only functions having derivative equal to 0 are the constant functions. This is a special instance of a more general question: If we know the derivative of a function, what is the function? We can apply the Mean Value Theorem to answer this question. The answer to this question will be important when we take up the indefinite integral later in chapter 5.

THEOREM 4.7

Functions with Equal Derivatives Differ by a Constant

Let $f(x)$ and $g(x)$ be two functions for which $f'(x) = g'(x)$ for all x on some interval $[a, b]$. Then there is a constant C such that

$$f(x) = g(x) + C$$

for all x on $[a, b]$.

Proof

The hypothesis implies that the function $h(x) = f(x) - g(x)$ has derivative zero on $[a, b]$ and so it is a constant function. Thus $h(x) = f(x) - g(x) = C$ and $f(x) = g(x) + C$ for some constant C. □

EXAMPLE 1 Let $f(x) = \sin^2 x$ and $g(x) = -\cos^2 x$. Then the derivatives are

$$f'(x) = 2\sin x\cos x, \qquad g'(x) = -2\cos x(-\sin x),$$

and it follows that $f'(x) = g'(x)$ for all x. Thus the functions differ by a constant. The equation $\sin^2 x - (-\cos^2 x) = C$ should come as no surprise; in fact the constant is $C = 1$. ■

EXAMPLE 2 Let $f(x)$ be a function such that $f'(x) = 3$ for all x. Prove that $f(x) = 3x + C$ for some constant C.

Solution: We know that the function $g(x) = 3x$ has derivative $g'(x) = 3$ for all x and so the derivatives of the two functions $f(x)$ and $g(x)$ are equal for every x. It follows from the theorem that $f(x) = g(x) + C$ for some constant C; thus $f(x) = 3x + C$. ■

The two theorems just proved give information about solutions of the problem of finding all functions $f(x)$ that satisfy the condition $f'(x) = g(x)$ for a given function $g(x)$. For certain choices of $g(x)$, we might be able to guess one function having g as derivative. In this favorable situation, all such functions are then obtained by adding an arbitrary constant to the solution. This is the situation illustrated in the last example.

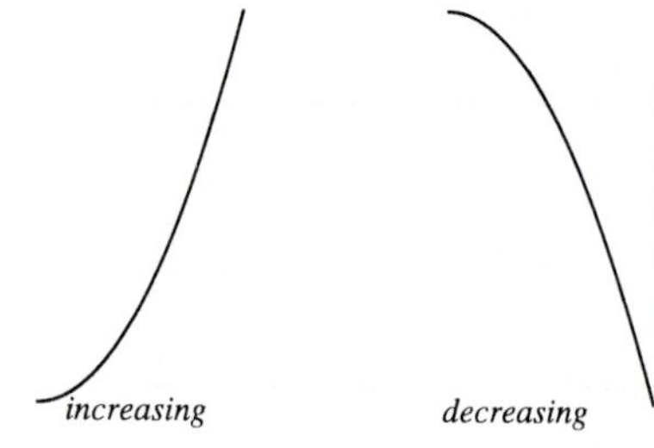

FIGURE 4.8

Increasing and Decreasing Functions

A function $f(x)$ is *increasing* on the interval I if for any numbers b and c on I with $b < c$, it follows that $f(b) < f(c)$. In other words, as the variable gets larger, the function values get larger. Analogously, $f(x)$ is *decreasing* on I if for two numbers b and c on I with $b < c$, we have $f(b) > f(c)$. The Mean Value

Theorem provides a convenient test for determining if a differentiable function is either increasing or decreasing on an interval.

THEOREM 4.8

Test for Increasing or Decreasing Functions

Let $f(x)$ have a derivative at every point of the interval $I = [u, v]$.
  a. If $f'(x) > 0$ for every x on I, then $f(x)$ is increasing on I.
  b. If $f'(x) < 0$ for every x on I, then $f(x)$ is decreasing on I.

Proof

Let $b < c$ be two points on I. Then the Mean Value Theorem applies to $f(x)$ on the interval $[b, c]$. There is a number p between b and c such that

$$f(c) - f(b) = f'(p)(c - b).$$

The term $(c - b)$ is positive; if we have case (a) that $f'(x) > 0$ for all x then, in particular, $f'(p) > 0$ and so $f(c) - f(b) > 0$ or $f(c) > f(b)$. Thus the function is increasing. Similarly if the derivative is negative on the interval, then $f(c) - f(b) < 0$, or $f(b) > f(c)$ and the function is decreasing. □

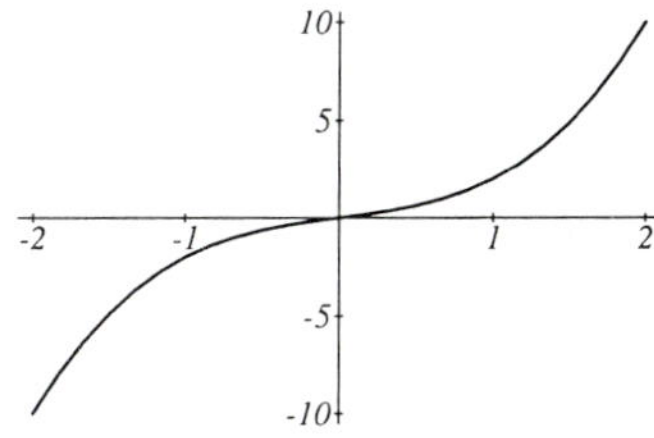

FIGURE 4.9
$x^3 + x$ is increasing

EXAMPLE 3 Let $f(x) = x^3 + x$. Show that f is increasing over the entire x-axis. Sketch the graph of f and confirm this for x on $[-2, 2]$.

Solution: We first compute the derivative of f and find $f'(x) = 3x^2 + 1 > 0$ for every number x. Thus $f(x)$ is increasing over the entire real number axis. The graph of f is shown in figure 4.9. ■

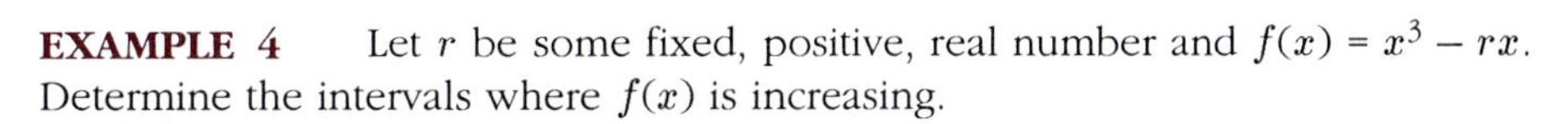

EXAMPLE 4 Let r be some fixed, positive, real number and $f(x) = x^3 - rx$. Determine the intervals where $f(x)$ is increasing.

Solution: The function is increasing on the intervals where $f'(x)$ is positive. The derivative is $f'(x) = 3x^2 - r$. The graph of $f'(x)$ is a parabola opening upward and crossing the x-axis at the two points $\pm\sqrt{r/3}$. Thus $f'(x) < 0$ for all x on the interval $(-\sqrt{r/3}, \sqrt{r/3})$ and the original function $x^3 - rx$ is decreasing on this interval. It is increasing on every interval that lies entirely to the left of $-\sqrt{r/3}$ or entirely to the right of $\sqrt{r/3}$. ■

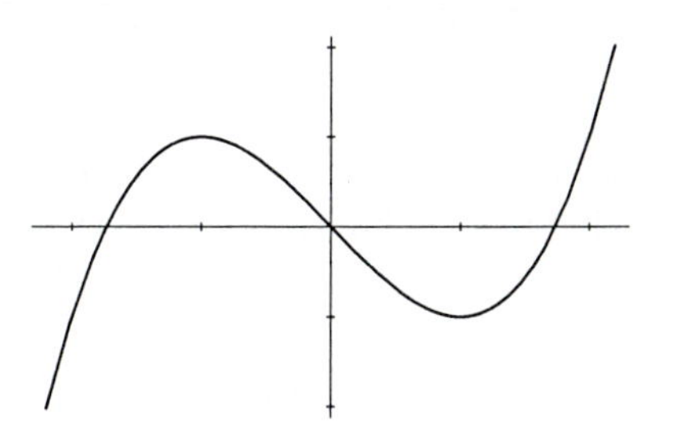

FIGURE 4.10
$y = f(x) = x^3 - rx$

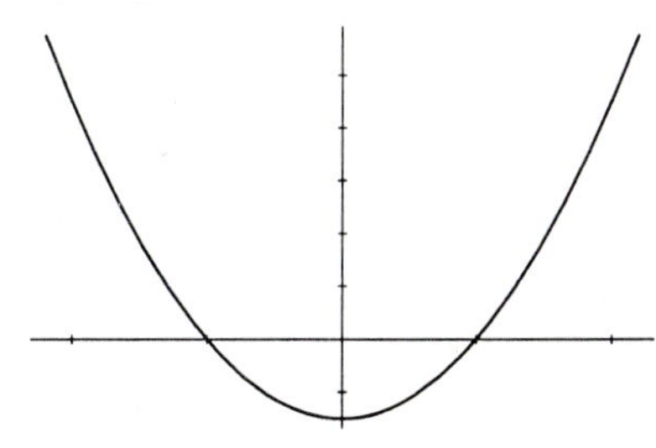

FIGURE 4.11
$y = f'(x) = 3x^2 - r$

EXAMPLE 5 Determine the extrema of the function $f(x) = \frac{3x^2+4}{x^2+10}$.

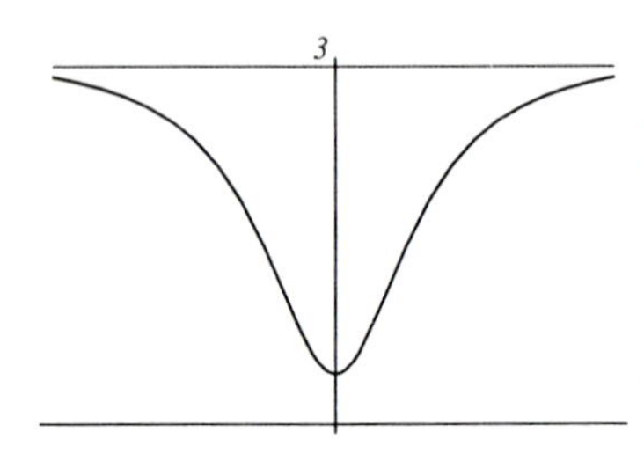

FIGURE 4.12
One critical point, $x = 0$

Solution: The function is defined and differentiable for all numbers x. Begin by locating the critical points:

$$f'(x) = \frac{(x^2+10)6x - (3x^2+4)2x}{(x^2+10)^2} = \frac{52x}{(x^2+10)^2}.$$

Thus there is only one critical point, $x = 0$. Moreover we have

$$\begin{cases} f'(x) > 0 & \text{if } x > 0, \\ f'(x) < 0 & \text{if } x < 0. \end{cases}$$

Thus $f(x)$ is a decreasing function on the negative x-axis and an increasing function on the positive x-axis. Thus the absolute minimum value is $f(0) = 4/10 = 0.4$. There is no absolute or local maximum even though the function is bounded. Note that

$$\lim_{x\to\infty} \frac{3x^2+4}{x^2+10} = \lim_{x\to-\infty} \frac{3x^2+4}{x^2+10} = 3.$$

There is no solution to the equation $f(x) = 3$. This implies $0.4 \leq f(x) < 3$ for all x and these bounds are best possible. ■

A Method of Proving Inequalities

The theory of increasing functions can sometimes be used to prove inequalities of the sort $f(x) \geq g(x)$ for all x on some interval $I = [a, b]$. The procedure is to show that $f(a) \geq g(a)$ and the difference $f(x) - g(x)$ is an increasing function. In other words, $f(x) - g(x)$ is nonnegative at $x = a$ and the difference gets larger as x moves to the right. Here is a precise statement.

THEOREM 4.9

Let $f(x)$ and $g(x)$ be differentiable on an interval $[a, b]$ (where we allow $b = \infty$) and suppose

$$f(a) \geq g(a) \quad \text{and} \quad f'(x) \geq g'(x) \quad \text{for all } x \text{ on } [a, b].$$

Then $f(x) \geq g(x)$ for all x on $[a, b]$.

Proof

The hypothesis on the derivatives of f and g implies that

$$f'(x) - g'(x) \geq 0$$

for all x on $[a, b]$. Thus $f(x) - g(x)$ is an increasing function on $[a, b]$. Because the smallest value of this difference is $f(a) - g(a)$ and this is nonnegative by assumption, it follows that $f(x) - g(x) \geq 0$ for all x on $[a, b]$. □

Here are some applications of this theorem to two specific examples.

EXAMPLE 6 Prove $(1 + x^2)^{1/2} \leq 1 + x$ for all $x \geq 0$.

Solution: Let $f(x) = (1 + x^2)^{1/2}$ and $g(x) = 1 + x$. We will prove $f(x) \leq g(x)$

by showing $g(x) - f(x)$ is an increasing function and equals 0 at $x = 0$. Compute the derivative:

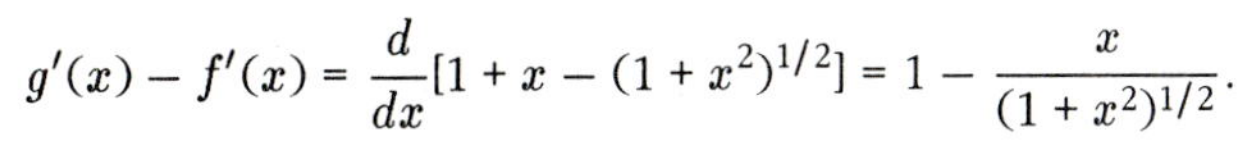

$$g'(x) - f'(x) = \frac{d}{dx}[1 + x - (1 + x^2)^{1/2}] = 1 - \frac{x}{(1 + x^2)^{1/2}}.$$

We argue that this derivative is positive for all $x \geq 0$ as follows:

$$x^2 < 1 + x^2 \quad \text{implies} \quad x = \sqrt{x^2} < \sqrt{1 + x^2},$$

and so

$$\frac{x}{\sqrt{1 + x^2}} < 1.$$

This proves that $g'(x) - f'(x)$ is positive and thus $g(x) - f(x)$ is an increasing function for all $x \geq 0$. The smallest value of the difference occurs at $x = 0$. The smallest value is

$$g(0) - f(0) = 1 - 1 = 0,$$

and so $g(x) - f(x) \geq g(0) - f(0) = 0$. Thus $g(x) \geq f(x)$ for all nonnegative x and

$$1 + x \geq \sqrt{1 + x^2}.$$

■

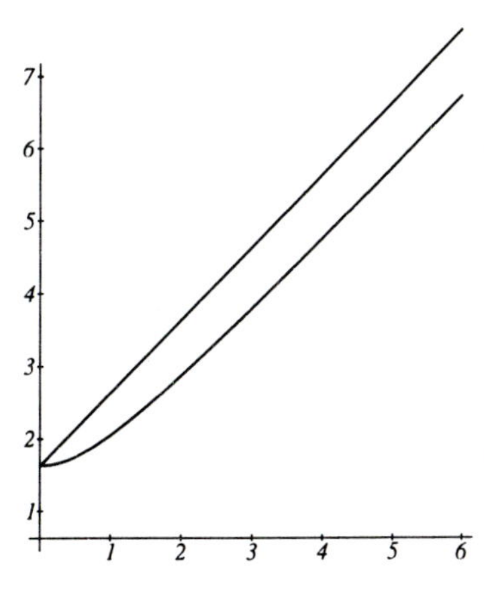

FIGURE 4.13
$(1 + x^2)^{1/2} \leq 1 + x$

EXAMPLE 7 Prove $\sin x \leq x$ for all $x \geq 0$.

***Solution*:** Let $f(x) = x - \sin x$ so that $f'(x) = 1 - \cos x$. We know from the properties of the trigonometric functions that the difference $1 - \cos x$ is nonnegative for every x; thus $f(x)$ is a nondecreasing function. Because $f(0) = 0$, it follows that $f(x) \geq 0$ for $x \geq 0$ and

$$x - \sin x > 0 \quad \text{for} \quad 0 < x.$$

Because $x = \sin x$ at $x = 0$ we have $x \geq \sin x$ for all $x \geq 0$.

Even though $f'(x) \geq 0$ for all x (positive or negative), why does this proof not show that $\sin x \leq x$ for all numbers x? ■

Exercises 4.3

The following problems ask for information about increasing or decreasing of certain functions. Answers to questions of this kind can be inferred from *accurate* renderings of the graphs of the functions. However, calculator or computer graphics used to produce accurate renderings might not always display precise results because of difficulties with the resolution of the viewing screen or the printing device. In such cases, it is worthwhile to have an analytic method to verify the information provided by the computer graphic. For each function, determine the intervals on which the function is increasing and the intervals on which it is decreasing. If possible, use an accurate rendering of the graph to predict the results and then verify the predictions by computation.

1. $f(x) = x^3 - 6x + 8$ **2.** $f(x) = x^3 - 6x^2 + 8$

3. $f(x) = x^3 + 6x + 8$ **4.** $f(x) = (x + 1)^3$

5. $f(x) = x + \frac{1}{x}$ **6.** $f(x) = x - \frac{1}{x}$

7. $f(x) = x^4 + 10x^2 - 100$ **8.** $f(x) = 400 - 80x + x^5$

9. $f(x) = x^2(x + 1)^2$ **10.** $f(x) = \frac{x - 4}{4x - 1}$

11. $f(x) = 200x^3 - 75x^2 + x + 12$

12. $f(x) = 26 - 78x + 306x^2 - 400x^3$

13. Find all functions $f(x)$ that satisfy $f'(x) = 2x$.

14. Find all functions $f(x)$ that satisfy $f'(x) = 4x + 3$.

15. Find all functions $f(x)$ that satisfy $f'(x) = x^2 + x$.

16. Find all functions $f(x)$ that satisfy $f'(x) = \sin x$.

17. A student learning calculus by working problems observed that the derivative of $\sin^2 x$ is $2 \sin x \cos x$ and the derivative of $-\cos^2 x$ is also $2 \sin x \cos x$ came to the conclusion that $\sin^2 x = -\cos^2 x$. How would you explain the error?

18. Suppose $f(x)$ is differentiable and increasing on (a, b). If $g(x)$ is differentiable on (a, b), then $g(x)$ and $f(g(x))$ have the same critical points on (a, b). [This is a generalization of the idea used earlier to say that $\sqrt{g(x)}$ and $g(x)$ have the same critical points on any interval where $g(x) \geq 0$.]

Verify the following inequalities by showing that an appropriate function is increasing (or decreasing) or by applying other results from this chapter.

19. $x \cos x \leq \sin x$ for $0 \leq x \leq \pi$

20. $(1+x)^{1/3} \leq 1 + \frac{1}{3}x$ for $x \geq 0$

21. $|\tan x| \geq |x|$ for $|x| < \pi/2$

22. The graph of the polynomial $p(x) = x^3 + ax^2 + bx + c$ might have the general shape of the cubic curve, $y = x^3$ (which is increasing on the entire real axis) or it might have the shape similar to the curve $y = x^3 - x$ (which has a "wiggle" because the graph crosses the x-axis at -1, 0 and 1). Show that the graph of $p(x)$ is more like the curve $y = x^3$ than $y = x^3 - x$ precisely when $a^2 \geq 3b$; when $a^2 < 3b$, the graph of $p(x)$ has a "wiggle". Explain why the condition does not involve the constant c.

23. Suppose $f(x)$ is a function that satisfies $f(0) = 0$ and $f'(x) = 1/(1+x^2)$. Prove the inequality $u/(1+u^2) \leq f(u) \leq u$ for all $u \geq 0$.

4.4 Indeterminate forms

We have seen that Rolle's Theorem about one function has an extension to a statement involving two functions. The same type of extension can be made for the Mean Value Theorem. This result is also known as the Cauchy Mean Value Theorem (named after the French mathematician Auguste Henri Cauchy).

THEOREM 4.10

The Extended Mean Value Theorem

Let $g(x)$ and $f(x)$ be continuous on the interval $[a, b]$ and differentiable on the open interval (a, b). Then there is a point c on the open interval (a, b) such that

$$g'(c)\big(f(b) - f(a)\big) = f'(c)\big(g(b) - g(a)\big). \qquad (4.4)$$

REMARK:

a. In the case $g(a) \neq g(b)$, the conclusion has the more symmetric form

$$\frac{f(b) - f(a)}{g(b) - g(a)} = \frac{f'(c)}{g'(c)}. \qquad (4.5)$$

b. If the Mean Value Theorem is applied separately to the two functions, we are assured that there exist points c_1 and c_2 on (a, b) such that

$$\frac{f(b) - f(a)}{b - a} = f'(c_1) \quad \text{and} \quad \frac{g(b) - g(a)}{b - a} = g'(c_2);$$

thus

$$f(b) - f(a) = \frac{f'(c_1)}{g'(c_2)}\big(g(b) - g(a)\big).$$

The Extended Mean Value Theorem says that we can take the two numbers c_1 and c_2 to be equal in the last equation.

Proof

First consider the case in which $g(a) - g(b) = 0$. Then by Rolle's Theorem there is a point c such that $g'(c) = 0$ and the conclusion holds for this choice of c inasmuch as both sides of equation (4.4) equal 0; the function $f(x)$ plays no particular role in this case.

Now assume $g(a) \neq g(b)$. We will produce a situation where the Extended Rolle's Theorem can be applied. We must have two functions that agree at a and at b. Let $G(x) = ug(x) + v$ for two constants u and v to be determined. We want to have

$$f(a) = G(a) = ug(a) + v \quad \text{and} \quad f(b) = G(b) = ug(b) + v.$$

The two equations can be solved to find

$$u = \frac{f(b) - f(a)}{g(b) - g(a)} \quad \text{and} \quad v = f(b) - ug(b).$$

Notice that it was necessary to divide by $g(b) - g(a)$ to solve for u and v; this step is valid because $g(a) \neq g(b)$. The Extended Rolle's Theorem applied to $f(x)$ and $G(x)$ on $[a, b]$ implies there is a point c on (a, b) such that

$$f'(c) = G'(c) = ug'(c) = \frac{f(b) - f(a)}{g(b) - g(a)} g'(c);$$

this is equivalent to the form we want so the theorem is proved. □

Indeterminate Forms

The Extended Mean Value Theorem can be applied to evaluate certain limits. We have considered the limit of a quotient

$$\lim_{x \to a} \frac{f(x)}{g(x)}$$

and have proved that the limit exists if the limits of the numerator and denominator both exist and $\lim g(x) \neq 0$. Moreover the limit might still exist even if $\lim g(x) = 0$ so long as $\lim f(x) = 0$ as well. The following rule, called *L'Hôpital's rule,* gives certain conditions under which such a limit might be evaluated. We present a general form that applies to one-sided limits.

THEOREM 4.11

L'Hôpital's Rule

A. Let $f(x)$ and $g(x)$ be continuous on $[a, q]$ and differentiable on (a, q). Suppose

$$\lim_{x \to a^+} f(x) = \lim_{x \to a^+} g(x) = 0. \tag{4.6}$$

Then

$$\lim_{x \to a^+} \frac{f(x)}{g(x)} = \lim_{x \to a^+} \frac{f'(x)}{g'(x)},$$

provided this limit exists.

B. Let $f(x)$ and $g(x)$ be continuous on $[p, b]$ and differentiable on (p, b). Suppose

$$\lim_{x \to b^-} f(x) = \lim_{x \to b^-} g(x) = 0. \tag{4.7}$$

Then

$$\lim_{x \to b^-} \frac{f(x)}{g(x)} = \lim_{x \to b^-} \frac{f'(x)}{g'(x)},$$

provided this limit exists.

Proof

We will prove part A. The proof of part B can be carried out in an analogous manner. For any x on (a, q) we consider the interval $I = (a, x)$. The Extended Mean Value Theorem applies to $f(x)$ and $g(x)$ on I so there is a point z between x and a such that

$$\frac{f(x) - f(a)}{g(x) - g(a)} = \frac{f'(z)}{g'(z)}.$$

Note that $f(a) = g(a) = 0$ because f and g are continuous at a and equation (4.6) holds. Thus we have

$$\frac{f(x)}{g(x)} = \frac{f'(z)}{g'(z)}.$$

The number z changes as we change the choice of x but z always remains between x and a. As x approaches a, z also approaches a and so

$$\lim_{x \to a^+} \frac{f(x)}{g(x)} = \lim_{z \to a^+} \frac{f'(z)}{g'(z)}.$$

If we now change the name of z to x in the rightmost term, we have the conclusion of the theorem. (Do not be confused by this change of z to x. The symbol x in $\lim f(x)/g(x)$ is a "dummy" variable in that for any function $h(x)$ we have $\lim_{z \to c} h(z) = \lim_{x \to c} h(x)$.) □

EXAMPLE 1 Evaluate $\lim_{x \to 0} \dfrac{\sin x}{x}$ with the aid of L'Hôpital's rule.

Solution: Both $f(x) = \sin x$ and $g(x) = x$ are differentiable over an interval containing 0 and both functions have limit 0 as x approaches 0. thus L'Hôpital's rule can be applied:

$$\lim_{x \to 0} \frac{\sin x}{x} = \lim_{x \to 0} \frac{f'(x)}{g'(x)} = \lim_{x \to 0} \frac{\cos x}{1} = \cos 0 = 1.$$ ■

REMARK:

a. Take careful notice of the statement of L'Hôpital's rule. It does *not* say that $\lim f(x)/g(x)$ can be evaluated by taking the derivative of $f(x)/g(x)$. The rule uses the derivative of the numerator divided by the derivative of the denominator, and this rule can be applied only when certain conditions are satisfied by the two functions.

b. L'Hôpital's rule can be used with limits in which x goes to infinity; the rule takes the form

$$\lim_{x \to \infty} \frac{f(x)}{g(x)} = \lim_{x \to \infty} \frac{f'(x)}{g'(x)},$$

provided $\lim_{x \to \infty} f(x) = \lim_{x \to \infty} g(x) = 0$ and the limit of $f'(x)/g'(x)$ exists. We will not carry out the proof of this case here; a sketch of the method is provided in the exercises.

Exercises 4.4

Evaluate the limits.

1. $\displaystyle \lim_{x \to 2} \frac{x^2 - 4}{x - 2}$

2. $\displaystyle \lim_{x \to 1} \frac{x^2 - 3x - 2}{2x - 2}$

3. $\displaystyle \lim_{x \to 0} \frac{\sin 3x}{x}$

4. $\displaystyle \lim_{x \to 0} \frac{\sin 5x}{\sin 2x}$

5. $\lim_{x\to 0} \dfrac{\cos x}{\sin x}$

6. $\lim_{x\to 0} \dfrac{\cos 3x - 1}{x^2}$

7. $\lim_{x\to 3} \dfrac{x^3 - 3x^2 + x - 3}{x^2 - 9}$

8. $\lim_{x\to 4} \dfrac{\sqrt{x} - 2}{x - 4}$

9. $\lim_{x\to 2} \dfrac{f(x) - f(2)}{x - 2}$

10. $\lim_{x\to a} \dfrac{g(x) - g(a)}{x - a}$

11. $\lim_{x\to 2} \dfrac{x^{1/3} - 2^{1/3}}{x - 2}$

12. $\lim_{x\to 0} \dfrac{\sin ax}{\sin bx}$

13. $\lim_{x\to 0} \dfrac{\sqrt{1+x} - \sqrt{1-x}}{x}$

14. $\lim_{h\to 0} \dfrac{f(u+h) - f(u-h)}{2h}$

In each of the next exercises, find a constant k so that the limit exists and is *not zero*, and determine the limit.

15. $\lim_{x\to 0} \dfrac{\sin 2x - kx}{x^3}$

16. $\lim_{x\to 0} \dfrac{\cos x - 1 - kx^2}{x^4}$

17. $\lim_{x\to 0} \dfrac{\tan x - kx}{x^3}$

18. $\lim_{x\to 0} \dfrac{\sin x - x - kx^3}{x^5}$

The next exercise indicates how to prove L'Hôpital's rule in the case of limits at $\pm\infty$.

19. Let $x = 1/t$. For any function $h(x)$ show

$$\lim_{x\to\infty} h(x) = \lim_{t\to 0^+} h(1/t) \quad \text{and}$$

$$\lim_{x\to -\infty} h(x) = \lim_{t\to 0^-} h(1/t).$$

Use this and L'Hôpital's rule for limits at 0 to prove a version of Theorem 4.11 in which a and b are replaced by ∞ and $-\infty$.

Rewrite each of the following limits in a form to which L'Hôpital's rule can be applied and then evaluate the limits that exist.

20. $\lim_{x\to\infty} \sqrt{x^2 - 1} - x$

21. $\lim_{x\to 0} \left(\dfrac{1}{\sin x} - \dfrac{1}{x}\right)$

22. $\lim_{x\to\infty} x \sin(1/x)$

23. $\lim_{x\to\infty} x \sin(1/x^2)$

24. $\lim_{x\to\infty} x^2 \sin(1/x)$

25. $\lim_{x\to\infty} \sqrt{x^2 + x} - x$

26. $\lim_{x\to\infty} x\sqrt{x^2 + 1} - x^2$

27. $\lim_{x\to\infty} (x+1)^{1/3} - x^{1/3}$

28. $\lim_{x\to 0^+} \left(\dfrac{1}{\sqrt{x+1}} - \dfrac{1}{\sqrt{x}}\right)$

4.5 Maximum and minimum function values

A question frequently asked about functions concerns the range of values of the function; in particular, what are the smallest and largest values of the function? If we wish to examine a particular function $f(x)$, there are several forms that a question about its extreme values might take. If we are interested in the range of values of $f(x)$ when x is restricted to lie near a certain point, we refer to this as a *local* situation and we use the following terms when referring to the maximum and minimum values of f at a point u.

DEFINITION 4.2

Local Maximum and Local Minimum

The function $f(x)$ has a *local maximum* at the point u if there is some open interval (p, q) containing u such that

$$f(u) \geq f(w) \quad \text{for all} \quad w \in (p, q).$$

The function $f(x)$ has a *local minimum* at the point v if there is an open interval (p, q) containing v such that

$$f(v) \leq f(w) \quad \text{for all} \quad w \in (p, q).$$

In cases where we are interested in the variation of f over its entire domain, we refer to the *absolute* maximum and minimum of f.

DEFINITION 4.3

Absolute Maximum and Absolute Minimum

The function $f(x)$ has an absolute maximum at the point u if

$$f(u) \geq f(x)$$

for every number x in the domain of f.

The function $f(x)$ has an absolute minimum at the point v if

$$f(v) \leq f(x)$$

for every number x in the domain of f.

The case of local extrema of differentiable functions has been discussed in connection with Rolle's Theorem. The result Corollary 4.1 tells us that an extreme value of the differentiable function $f(x)$ over the interval $[a, b]$ must occur either at a critical point or at one of the endpoints of the interval. A local maximum or a local minimum is an extreme value of the function over a suitable interval.

EXAMPLE 1 Find the local extreme values of $f(x) = x^3 - 3x + 1$ over the interval $[-3, 2]$.

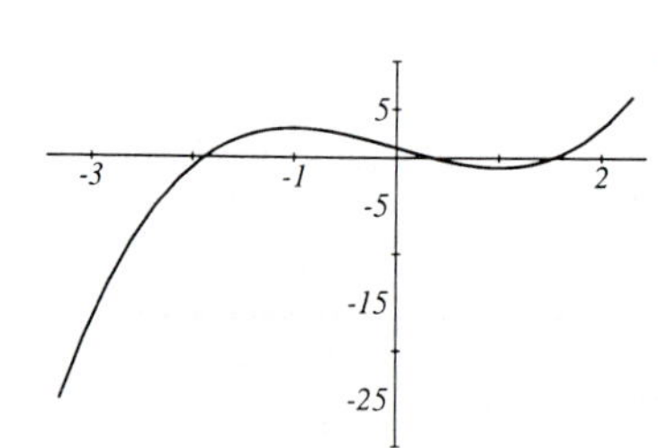

FIGURE 4.14
$f(x) = x^3 - 3x + 1$

Solution: The polynomial function is differentiable at every point so the extreme values of $f(x) = x^3 - 3x + 1$ must occur at points where $f'(x) = 0$ (the critical points) or at the endpoints of the interval. We begin by locating the critical points:

$$f'(x) = 3x^2 - 3 = 0 \quad \text{at} \quad x = \pm 1.$$

The points at which a local maximum or local minimum might occur are the two critical points $x = \pm 1$ and the two end points $x = -3$ or 2. An inspection of the graph suggests that there is a local maximum at (or near) $x = -1$ and a local minimum at (or near) $x = 1$. We carry out the following analysis without using the graph.

Let us test $x = 1$. We wish to determine if there is an open interval containing 1 such that either $f(x) \geq f(1)$ for all x on the interval or $f(x) \leq f(1)$ for all x on the interval. Let $x = 1 + h$ so that as h ranges over an interval $(-\delta, \delta)$, x ranges over an interval $(1 - \delta, 1 + \delta)$. Then the difference in the values of f at 1 and $1 + h$ is

$$\begin{aligned} f(1 + h) - f(1) &= (1 + h)^3 - 3(1 + h) + 1 - 1 + 3 - 1 \\ &= h^3 + 3h^2 = h^2(h + 3). \end{aligned}$$

For sufficiently small $|h|$, namely $|h| < 3$, this difference is positive. Thus $f(x) > f(1)$ for all x on $(1 - 3, 1 + 3) = (-2, 4)$ and $f(1)$ is a local minimum of f. A similar analysis shows that $f(x)$ has a local maximum at $x = -1$.

Now we compare the four values $f(-3) = -17$, $f(-1) = 3$, $f(1) = -1$ and $f(2) = 3$. The minimum value of $f(x)$ on $[-3, 2]$ is -17 at the endpoint $x = -3$. The maximum value of $f(x)$ on the interval $[-3, 2]$ is 3 at the two points $x = -1$ and $x = 2$. ■

Absolute Extrema

The technique of studying the derivative of a function does not provide a direct method of finding an absolute maximum or absolute minimum of a function. We are often able to determine local extrema by testing critical points. The problem of determining the absolute maximum or the absolute minimum of a function can sometimes be reduced to consideration of an appropriate interval and treating the problem as a local one.

EXAMPLE 2 Find the maximum and minimum values of the function

$$f(x) = \frac{x}{x^2 + 1}.$$

Solution: We first notice that $f(x)$ is defined and differentiable at every point and we consider an interval $[-N, N]$ (with N very large). Then the extreme values of $f(x)$ over the interval must occur either at an endpoint or at a critical point. First find the critical points;

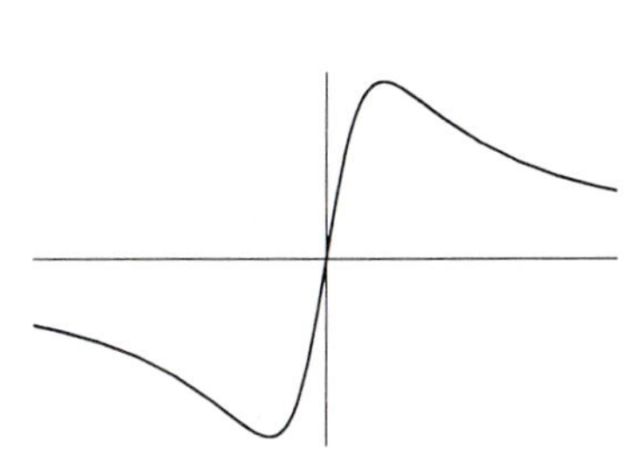

FIGURE 4.15
Absolute extrema at $x = \pm 1$

$$f'(x) = \frac{(x^2+1) - x(2x)}{(x^2+1)^2} = \frac{1 - x^2}{(x^2+1)^2}.$$

The critical points are $x = \pm 1$ and we find $f(-1) = -1/2$ and $f(1) = 1/2$. The extrema of $f(x)$ on $[-N, N]$ occur at one of the points ± 1, $\pm N$. When we take N very large, the values $f(\pm N)$ are very close to 0 because

$$\lim_{x \to \pm\infty} f(x) = \lim_{x \to \pm\infty} \frac{x}{x^2+1} = 0.$$

In particular, we can select N so large that $|f(x)| < 0.1$ for all x that satisfy $|x| \geq N$. Then the extreme values of $f(x)$ over $[-N, N]$ are $\pm\frac{1}{2}$ at $x = \pm 1$ and all values of $f(x)$ outside this interval are smaller in absolute value than 1/2. Hence the absolute extrema of $f(x)$ occur at $x = \pm 1$. Once again this analysis confirms the information obtained from the graph. ■

There are some situations where absolute extrema can be predicted to occur at a critical point. In the next theorem we state such a result but we postpone the proof until after we study a bit more about derivatives, especially second derivatives.

THEOREM 4.12

Assume the second derivative of $f(x)$ exists and $f''(x) > 0$ for every x on the interval $I = [a, b]$. If there is a point c on I such that $f'(c) = 0$, then there is only one such point and $f(c)$ is the minimum of $f(x)$ on I.

Similarly if $f''(x) < 0$ for every x on I, and if there is a critical point c on I, then there is only one critical point and $f(c)$ is the maximum of $f(x)$ on I.

EXAMPLE 3 Determine the absolute extrema of $f(x) = x^2 + x^{-2}$.

Solution: We first notice that the function is defined and differentiable for all numbers x except $x = 0$. Let I denote either the set of all positive numbers or the set of all negative numbers. Then $f(x)$ is defined and differentiable on I. Let us compute both the first and second derivatives:

$$f'(x) = 2x - 2x^{-3} = \frac{2x^4 - 2}{x^3},$$

$$f''(x) = 2 + 6x^{-4} = \frac{2x^4 + 6}{x^4}.$$

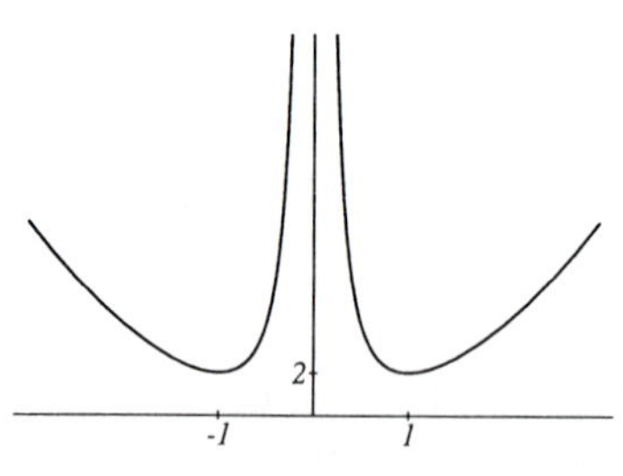

FIGURE 4.16
$f(x) = x^2 + x^{-2}$

Thus the second derivative is positive for every x on I. The critical points are the solutions of $f'(x) = 0$ on I; the equation $2x^4 - 2 = 0$ has exactly one solution on I. If I consists of the positive numbers, then $x = 1$ is the only critical point. If I consists of the negative numbers, then $x = -1$ is the only critical point. Thus f has an absolute minimum at this critical point.

There could be some confusion here because it seems that we have found two points giving the extreme values whereas the theorem says there should be just one extreme value. The explanation is this: The theorem asserting that there is just one critical point applies only to a function that is twice differentiable on the interval. Thus we can apply this theorem to the function $f(x) = x^2 + x^{-2}$ over the interval of positive real numbers on which it is twice differentiable. The conclusion of the theorem then implies that $x = 1$ gives the absolute minimum of this function. We can also apply the theorem to $f(x) = x^2 + x^{-2}$ over the interval of all negative numbers. In this case again, there is an absolute minimum at $x = -1$.

It just happens to be a coincidence that $f(1) = f(-1)$ in this case. ■

Absolute Extrema of Polynomials

Of course, any given function might not have an absolute maximum or an absolute minimum. Polynomial functions exhibit this behavior. The following theorem gives the full story about absolute extrema of polynomial functions.

THEOREM 4.13

Absolute Extrema of Polynomials

Let

$$f(x) = a_0 + a_1x + \cdots + a_nx^n$$

be a polynomial with leading coefficient $a_n \neq 0$ and degree $n > 0$.

a. If n is odd, then $f(x)$ has no absolute minimum and no absolute maximum.
b. If n is even and $a_n > 0$, then $f(x)$ has no absolute maximum but $f(x)$ does have an absolute minimum.
c. If n is even and $a_n < 0$, then $f(x)$ has no absolute minimum but $f(x)$ does have an absolute maximum.

Proof

We can write the polynomial as the product of two functions:

$$f(x) = a_nx^n\left(1 + \frac{a_{n-1}}{a_nx} + \cdots + \frac{a_1}{a_nx^{n-1}} + \frac{a_0}{a_nx^n}\right).$$

The second factor has a limit 1 as x approaches either plus or minus infinity. We can interpret this to mean that $f(x)$ is close to the function a_nx^n for large values of $|x|$; in particular,

$$\lim_{x\to\pm\infty} f(x) = \lim_{x\to\pm\infty} a_nx^n.$$

It is clear that this last limit is not a finite number; what is at issue here is whether the function becomes positively infinite or negatively infinite as x approaches either $\pm\infty$.

Suppose n is odd. Then $\lim_{x\to\infty} x^n = \infty$ and $\lim_{x\to-\infty} x^n = -\infty$.

Regardless of the sign of a_n, the values of $f(x)$ grow arbitrarily large or arbitrarily small (negatively large). Thus $f(x)$ has neither an absolute minimum nor an absolute maximum.

Now suppose n is even and $a_n > 0$. Then

$$\lim_{x\to\infty} f(x) = \lim_{x\to\infty} a_n x^n = +\infty,$$
$$\lim_{x\to-\infty} f(x) = \lim_{x\to-\infty} a_n x^n = +\infty.$$

These limits at infinity imply that $f(x)$ has no maximum value. Also the first limit implies there is some number N_1 such that $f(x) > a_0$ for all $x > N_1$. Similarly, the second limit implies there is some N_2 such that $f(x) > a_0$ for all $x < N_2$. Note that $f(0) = a_0$ so it follows that $0 \leq N_1$ and $0 \geq N_2$. The function $f(x)$ has a minimum value, say $m = f(c)$, with c on the closed interval $[N_2, N_1]$. This minimum value of $f(x)$ on $[N_2, N_1]$ is the absolute minimum of $f(x)$, for the following reason. Inasmuch as 0 is on $[N_2, N_1]$ and $f(0) = a_0$, it follows that $c \leq a_0$. But for any point x not on the interval $[N_2, N_1]$, we have $f(x) \geq a_0$ and so $f(x) \geq m$; thus m is the least value of f. This proves statement b; similar arguments can be applied to prove statement c. □

EXAMPLE 4 Find the absolute extrema of the polynomial

$$f(x) = 7 + 4x^3 - 3x^4.$$

Solution: Because the polynomial has even degree 4, and the coefficient of the x^4 term is negative, the polynomial behaves like $-3x^4$ for $|x|$ large enough. There is no absolute minimum; there is an absolute maximum that must occur at a critical point of the function. We first find the critical points:

$$f'(x) = 12x^2 - 12x^3 = 12x^2(1 - x).$$

The critical points are $x = 0$ and $x = 1$. Compute the values $f(0) = 7$ and $f(1) = 8$. Because we know there is an absolute maximum and it must occur at 0 or 1, we conclude that the maximum value of $f(x)$ is 8 at $x = 1$.

Figures 4.17 and 4.18 show graphs of the polynomial $f(x) = 7 + 4x^3 - 3x^4$ and of the polynomial $g(x) = -3x^4$ in two scales. When x lies on a fairly short interval and the function values are not too large, the graphs look quite different. However when the scale is changed so that x ranges over a large interval where the values of the functions are quite large (in absolute value), the difference between the two functions become negligible in comparison to the size of the function values.

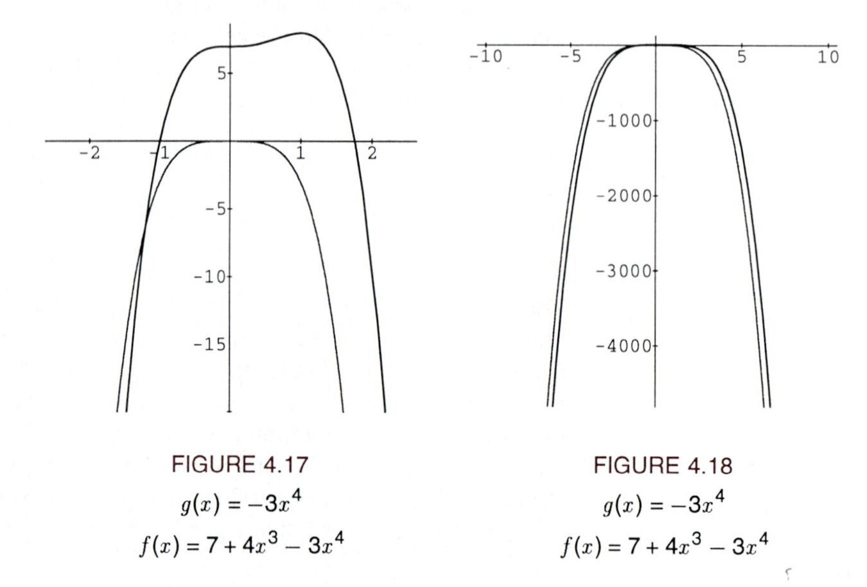

FIGURE 4.17
$g(x) = -3x^4$
$f(x) = 7 + 4x^3 - 3x^4$

FIGURE 4.18
$g(x) = -3x^4$
$f(x) = 7 + 4x^3 - 3x^4$

Just for further illustration of this point, consider the polynomial

$$f(x) = 2x^5 - 30(x + 1)(x - 1)(x + 2).$$

This differs from the polynomial $g(x) = 2x^5$ by a polynomial of degree 3 that crosses the x-axis at -2, -1, and 1 (see figure 4.19). Thus for values of x not too far from the origin, f and g behave quite differently. For values of x very large, the two functions are relatively close (see figure 4.20); for example, $f(100)$ and $g(100)$ are close to $2 \cdot 10^{10}$, but

$$g(100) - f(100) = 30 \cdot 101 \cdot 99 \cdot 102 \approx 3 \cdot 10^7,$$

that is roughly $1/1000$ of the value of the functions.

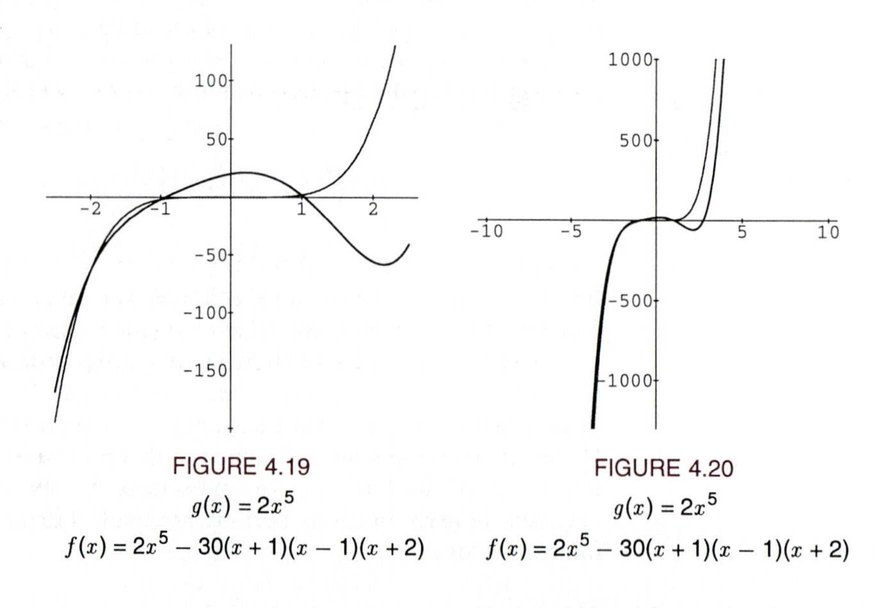

FIGURE 4.19
$g(x) = 2x^5$
$f(x) = 2x^5 - 30(x + 1)(x - 1)(x + 2)$

FIGURE 4.20
$g(x) = 2x^5$
$f(x) = 2x^5 - 30(x + 1)(x - 1)(x + 2)$

Exercises 4.5

Find the absolute extrema and the local extrema of the following polynomials. Examine accurate graphs of the functions to confirm the analytical computations.

1. $f(x) = 4 + 8x^2 + 4x^4$

2. $f(x) = 15x - 10x^3$

3. $f(x) = 16x^6 - 4x^4$

4. $f(x) = 30 - 5x^4 - 20x^5$

5. $f(x) = x(x-1)(x-2)$

6. $f(x) = x^3(x-2)^2$

7. $f(x) = -x^4(x-1)(x-2)$

8. $f(x) = -x(x-1)^4(x-2)$

Determine the local and absolute extrema of the following functions.

9. $f(x) = x^5(x-a)^4$

10. $f(x) = x^m(x-a)^n$

11. $f(x) = \sqrt{4-x^2}$

12. $f(x) = x\sqrt{4-x^2}$

13. $f(x) = x^2\sqrt{4-x^2}$

14. $f(x) = x^n\sqrt{4-x^2}$

15. $f(x) = \dfrac{(x-b)^m}{(x-a)^m}$

16. $f(x) = \sin x + \cos x$

Give examples of polynomials $f(x)$ having the indicated properties.

17. $f(x) \geq 0$ for all x and f has a local minimum at $x = 0$.

18. $f(x)$ has an absolute minimum at $x = 2$.

19. $f(x)$ has an local minimum at $x = 2$ and a local maximum at $x = 4$.

20. $f(x)$ has local extrema at $x = 1$, 2 and 3.

21. Let a and b be positive constants and let n and m be positive integers with $n > m$. Then $ax^n > bx^m$ for all x larger than some fixed number N. Find N in terms of a, b, n and m.

22. Let $f(x) = 0.007x^3 + 0.0005x - 199886$ and let $g(x) = 48966x^2 + 10^{40}x + 10^{38}$. Consider the two inequalities, $f(x) > g(x)$ and $f(x) < g(x)$. Find a value of N so that one of these inequalities holds for all $x > N$. [There is no need to find the smallest such N. Crude exaggeration might be the quickest road to a solution.]

23. Let $f(x)$ be a function defined and differentiable at all numbers x. Prove that an absolute maximum or an absolute minimum of $f(x)$ occurs at a point c where $f'(c) = 0$.

4.6 More maximum and minimum problems

We consider some problems whose solutions depend on finding the maximum or minimum values of some function.

EXAMPLE 1 Find the point on the line $y = 4x - 1$ nearest the point $(-1, 3)$.

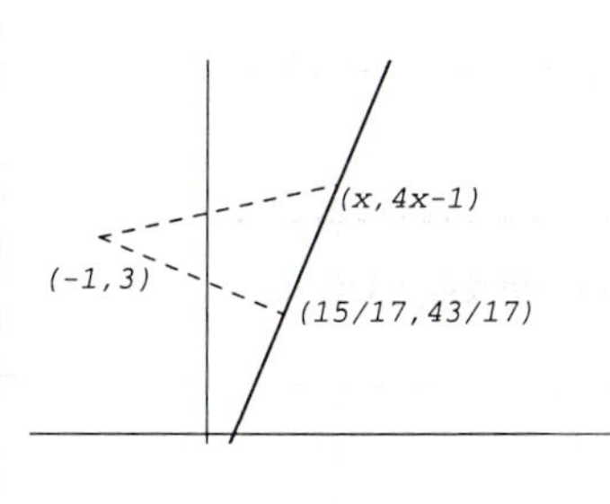

FIGURE 4.21
Shortest distance is along the perpendicular

Solution: This problem asks for the minimum value of the function that measures the distance from the point $(-1, 3)$ to a point $(x, 4x-1)$ on the given line. The distance is

$$D(x) = \sqrt{(x+1)^2 + (4x-1-3)^2}.$$

The points at which $D(x)$ takes on a minimum are exactly the same as the points at which

$$H(x) = D(x)^2 = (x+1)^2 + (4x-4)^2$$

takes on its minimum. The function $H(x)$ is somewhat easier to work with because the square root sign has been removed. We observe that $H(x)$ is defined and differentiable for all x; the absolute minimum occurs at a critical point. First collect the terms in H to obtain $H(x) = 17x^2 - 30x + 17$ and $H'(x) = 34x - 30$. There is a unique critical point, namely $x = 30/34 = 15/17$. The point on the graph nearest to $(-1, 3)$ is $(15/17, 43/17)$. ■

It is very likely that the reader could solve this problem, without using any calculus, by recognizing that the shortest distance from a point to a line is measured along a perpendicular to the line (see figure 4.21). Let us verify that the solution has this property. Two lines are perpendicular if the product of their slopes equals -1. The slope of the given line $y = 4x - 1$ is 4. The line from $(-1, 3)$ to $(15/17, 43/17)$ has slope

$$\frac{\frac{43}{17} - 3}{\frac{15}{17} + 1} = \frac{-8}{32} = -\frac{1}{4}.$$

Thus the shortest segment from $(-1, 3)$ to the line $y = 4x - 1$ is along a perpendicular to the line.

A more general version of this example can be found in the exercises at the end of this section. There you are asked to show that the shortest distance from a point to a graph of a differentiable function is always measured along a perpendicular to the curve (a line that is perpendicular to the line tangent to the curve).

EXAMPLE 2 What is the largest area of a rectangle having two vertices on a semicircle of radius 10 and one side along the diameter (see figure 4.22)?

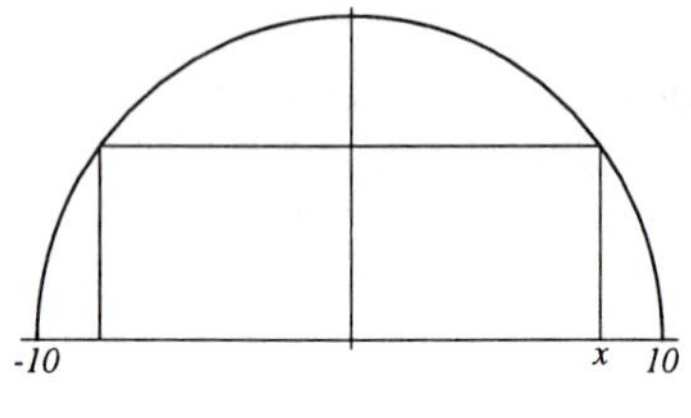

FIGURE 4.22

Solution: We introduce coordinates into the problem by assuming the circle has center at the origin; the equation of the semicircle is $y = \sqrt{100 - x^2}$. Let one vertex of the rectangle be at $(x, 0)$ with $0 \leq x \leq 10$. Then the dimensions of the rectangle are $2x$ and $\sqrt{100 - x^2}$ and its area is $A(x) = 2x\sqrt{100 - x^2}$. Our problem is to find the maximum value of $A(x)$ for x on the interval $[0, 10]$. At the two endpoints we find $A(0) = A(10) = 0$ which is clearly the minimum value of the function. Thus the maximum values occur at critical points. We compute the derivative:

$$A'(x) = 2(100 - x^2)^{1/2} + 2x(100 - x^2)^{-1/2}(-x) = \frac{2(100 - 2x^2)}{\sqrt{100 - x^2}}.$$

The critical point for the function is the solution of $100 - 2x^2 = 0$; namely $x = \sqrt{50}$. The maximum area is

$$A(\sqrt{50}) = 2\sqrt{50}\sqrt{100 - 50} = 100.$$

The semicircle has area $50\pi \approx 157.08$. Thus the maximum rectangle has area roughly 2/3 the area of the semicircle. ■

EXAMPLE 3 An open top box is to be made from a square piece of cardboard by cutting a square piece from each corner and folding up the sides. If the piece of cardboard has sides of length b, what is the largest possible volume of the constructed box?

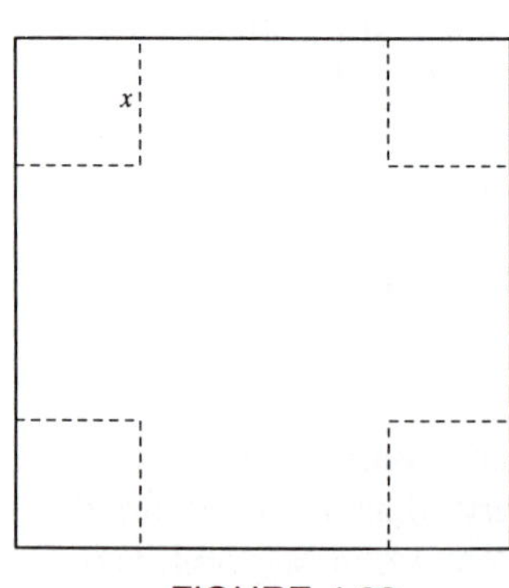

FIGURE 4.23
Box construction

Solution: Let x denote the side of the small square to be cut. After folding the sides, the dimensions of the box are $b - 2x$, $b - 2x$, and x. The volume V is then easily expressed in terms of the variable x as

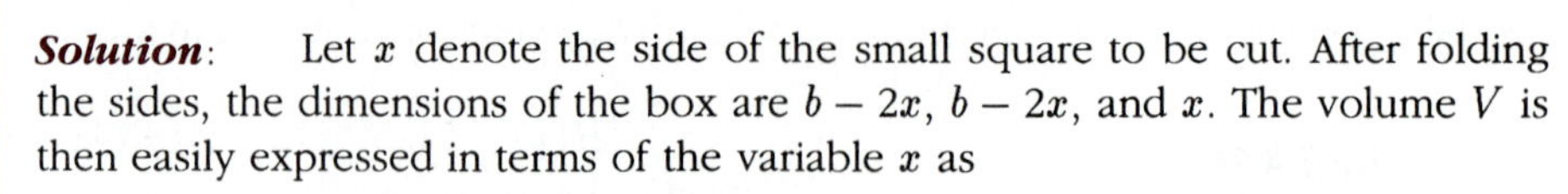

$$V = V(x) = x(b - 2x)(b - 2x) = b^2x - 4bx^2 + 4x^3.$$

Our task is to find the value of x that makes V as large as possible; that is we want to find the maximum value of V. There are restrictions placed on x by the physical constraints of the problem. First of all, x cannot be negative because that does not correspond to a possible way of cutting the cardboard. Second, we cannot cut away more than half of the side from each corner because the length of one side after removing the cut-outs is $b - 2x$; thus we must restrict x to satisfy $0 \leq x \leq b/2$.

The precise formulation of our problem is: Find the maximum value of $V(x)$ for x on the interval $[0, b/2]$.

Observe that $V(0) = V(b/2) = 0$. This is surely the minimum value of the volume function because no volume can be negative; thus the maximum value must occur at one of the critical points of the function $V(x)$. These are computed by setting the derivative equal to 0;

$$V'(x) = b^2 - 8bx + 12x^2 = (b - 6x)(b - 2x) = 0.$$

The critical points are $b/2$ and $b/6$. The volume at $b/2$ is 0, as was seen before, and $V(b/6) = 2b^3/27$ is the maximum volume. The box with maximum volume has dimensions $b/6$, $2b/3$, $2b/3$.

There is some lack of symmetry in this solution that makes it somewhat difficult to guess the result before doing the analytical computation. A somewhat impulsive, but not unreasonable, guess would have been that the maximum volume is attained by forming a cube; that is by making all sides equal. This would occur with $x = b/3$, in which case the volume would be $V(b/3) = b^3/27$. However, the actual solution shows it is possible to construct a box with twice the volume of this cube. ■

EXAMPLE 4 A homeowner wants to enclose parı of the backyard with a fence to make a rectangular shaped garden. There is an existing fence 20 feet long that is to be a part of the boundary of the garden. What are the dimensions of the garden having maximum area if 40 feet of new fence material is available? What are the dimensions if 70 feet of new fence material is available? Bear in mind that the existing fence might be only part of one side.

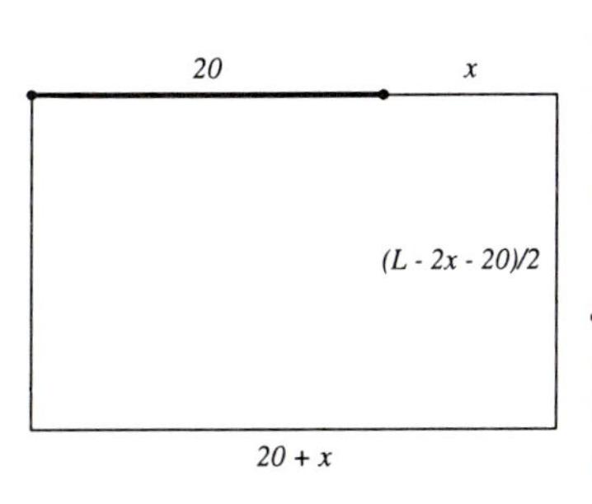

FIGURE 4.24

Solution: Rather than work the problem twice with two different lengths, we introduce a variable for the length and work the problem just once. Let L be the length of available new fence material, so that either $L = 40$ or $L = 70$ for the two cases. Let x be the length of the new fence added to the existing 20-foot side. Then two parallel sides have lengths $x + 20$ each. Because the total length of new fence is L, the two remaining sides must have lengths $\frac{1}{2}(L - 2x - 20)$. The area of the rectangle is

$$A(x) = (x + 20)\frac{1}{2}(L - 2x - 20) = \frac{1}{2}\left(- 2x^2 + (L - 60)x + 20L - 400\right).$$

The problem requires that we find the maximum value of function $A(x)$. In order that the problem be meaningful, x must be nonnegative and $L - 2x - 20$ must be nonnegative. Thus $A(x)$ is defined over the interval $[0, \frac{1}{2}L - 10]$. The maximum of A occurs either at a critical point or at an endpoint of the interval. We solve for the critical points:

$$A'(x) = \frac{1}{2}(-4x + L - 60) = 0 \quad \text{at} \quad x = \frac{L - 60}{4}.$$

For the case $L = 40$, we find $x = -5$. This point is not on the interval where the area function is defined. Hence $A(x)$ has no critical points on the interval and the maximum occurs at an endpoint. Because $A = 0$ at the right-hand endpoint, the maximum occurs at the left-hand endpoint, $x = 0$. The dimensions of the garden with maximum area are 20×10.

In the case $L = 70$, the critical point is $x = (70 - 60)/4 = 2.5$. For this value of x the area is

$$A = A(2.5) = (22.5)\frac{1}{2}(70 - 5 - 20) = (22.5)(22.5) = 506.25.$$

The area at the endpoint, $x = 0$, is

$$A = A(0) = (20)\frac{1}{2}(50) = 500.$$

This corresponds to the dimensions 20×25. Thus the garden with maximum area has dimensions 22.5×22.5. ■

EXAMPLE 5 An offshore oil drilling rig is 300 feet from the nearest point P of the straight shoreline. The holding tank for the oil is 1000 feet down the shoreline from P. It cost \$50 per foot to construct a pipeline underwater and \$30 per foot to construct a pipeline overland. How far from P should the pipeline from the rig to the tank reach shore to minimize the cost of the construction?

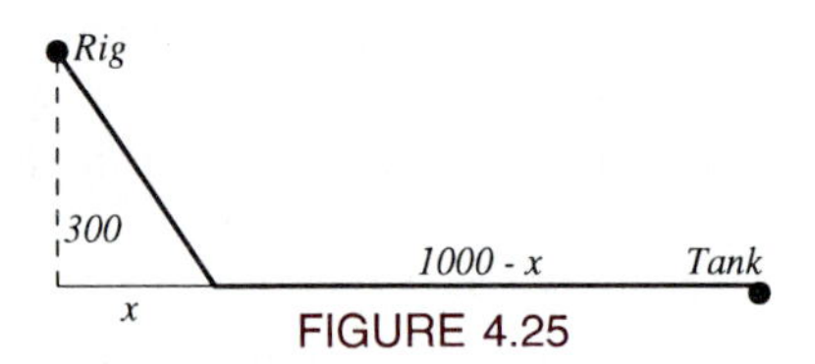

FIGURE 4.25

Solution: Let x denote the distance from P to the point where the pipeline comes ashore. The cost of the underwater pipeline is $50\sqrt{300^2 + x^2}$ and the cost of the overland pipeline is $30(1000 - x)$. The total cost of the pipeline is

$$C(x) = 50\sqrt{300^2 + x^2} + 30(1000 - x).$$

Common sense imposes the restriction $0 \leq x \leq 1000$. We want to find the minimum of $C(x)$ when x is restricted to this interval. We first check the two endpoints:

$$C(0) = 50(300) + 30(1000) = 45,000$$

$$C(1000) = 50\sqrt{300^2 + 1000^2} \approx 52,201.5.$$

Next we find the critical points by setting the derivative equal to 0 to get

$$C'(x) = 50(300^2 + x^2)^{-1/2}x - 30 = 0.$$

After some manipulation we get

$$50x = 30\sqrt{300^2 + x^2}.$$

After squaring both sides, collecting terms and solving for x we get $x = 225$. Compute the cost of construction by evaluating the cost function at $x = 225$ to get $C(225) = \$42,000$. Because this amount is less than the amounts at the two endpoints, the minimum occurs at $x = 225$. Figure 4.25 is drawn approximately to scale to show the path of minimum cost. ■

Line of Best Fit To a Set of Points

Suppose we are given n points $P_1(x_1, y_1)$, $P_2(x_2, y_2), \ldots, P_n(x_n, y_n)$ in the plane and we are given the problem of finding a line $y = mx + b$ that is the best fit to the given points. The words *best fit* must be made precise before the computation can be made. A line that is best fit means that some quantity determined by the points and the line is made as small as possible for all possible lines. This type of problem occurs, for example, in a laboratory where some process is known to give an output y that is a polynomial of degree one depending on a variable input x; that is $y = mx + b$ for some unknown m and b. To find the coefficients m and b, the process is subjected to inputs x_i and the outputs y_i are measured. Laboratory experiments are subject to error so the points $P_i(x_i, y_i)$ might not all lie on a straight line. One tries to give an estimate of the coefficients m and b from the knowledge of the (possibly inaccurate) data points $P_i(x_i, y_i)$. If $P_i(x_i, y_i)$ is on the line $y = mx + b$, then $y_i = mx_i + b$; if P_i is not on the line, then $y_i - mx_i - b$ is the vertical distance from the line to the point. The sign of this difference depends on whether the point P_i is above or below the line. One way to define the "line of best fit" is to select m and b to minimize the sum of these differences. However, the sum of the differences is not a very good measure of closeness because a large error caused by a point far above the line might be corrected by a large error caused by a point below the line.

A somewhat better quantity to minimize would be the sum of the absolute values $|y_i - mx_i - b|$. It turns out that dealing with the absolute values is awkward because $|x|$ is not a differentiable function of x (at $x = 0$). The process of finding the minimum of a function that has absolute values as part of its definition may require analysis of many cases.

Instead of minimizing the sum of the terms $|y_i - mx_i - b|$, it is more convenient to sum the square of the differences $(y_i - mx_i - b)^2$. The definition of "line of best fit" that seems easiest to work with using the methods of calculus is the following:

The line $y = mx + b$ is called a line of best fit to the points $P_1(x_1, y_1)$, $P_2(x_2, y_2), \ldots, P_n(x_n, y_n)$ if the constants m and b are selected to make the sum

$$D(m, b) = (y_1 - mx_1 - b)^2 + (y_2 - mx_2 - b)^2 + \cdots + (y_n - mx_n - b)^2 \quad \textbf{(4.8)}$$

as small as possible.

The line of best fit according to this definition is referred to as the line of best fit *in the least-squares sense* when it is necessary to emphasize which definition of best fit is used.

Thus the line of best fit is determined by finding the minimum value of the function $D(m, b)$. This problem is different from the other examples in that this is a function depending on two independent variables m and b. Up to now we have dealt only with functions of one variable. Later we will make a careful study of the extreme values of functions of two or more variables. However, the function $D(m, b)$ is sufficiently simple that our knowledge of one-variable techniques is adequate for determining its extreme values. We begin by assuming that m is held fixed and we treat b as a variable. We write $f(b) = D(m, b)$ so f is a function of one variable b. Multiply the terms in $D(m, b)$ and collect the coefficients to obtain

$$f(b) = U - 2Vb + nb^2,$$

where

$$\begin{aligned} U &= (y_1 - mx_1)^2 + (y_2 - mx_2)^2 + \cdots + (y_n - mx_n)^2, \\ V &= (y_1 - mx_1) + (y_2 - mx_2) + \cdots + (y_n - mx_n) \\ &= (y_1 + y_2 + \cdots + y_n) - (x_1 + x_2 + \cdots + x_n)m. \end{aligned}$$

The graph of the function $f(b)$ is a parabola that has a unique absolute minimum at the point where $f'(b) = -2V + 2nb = 0$; that is $b = V/n$ gives the minimum value of $f(b)$ for the given m; this minimum is

$$f(V/n) = U - 2V(V/n) + n(V/n)^2 = U - V^2/n.$$

We pause here to consider what we have just computed. By fixing a value of m, we are considering all lines $y = mx + b$ with slope m and variable y intercept b. From all such lines, we have determined the unique one that makes the sum of squares in equation (4.8) as small as possible. The number $f(V/n)$ is still a function of the particular m.

The next step is to allow m to vary and then determine the value that makes $f(V/n)$ as small as possible for all choices of m. Let $g(m) = U - V^2/n$. As a function of m, U is a quadratic polynomial and V is a polynomial of degree 1. Thus $g(m)$ is just a quadratic polynomial:

$$g(m) = A + 2Bm + Cm^2$$

where

$$A = (y_1^2 + \cdots + y_n^2) - \frac{1}{n}(y_1 + \cdots + y_n)^2,$$

$$B = -(x_1y_1 + \cdots + x_ny_n) + \frac{1}{n}(y_1 + \cdots + y_n)(x_1 + \cdots + x_n),$$

$$C = (x_1^2 + \cdots + x_n^2) - \frac{1}{n}(x_1 + \cdots + x_n)^2.$$

It is possible to prove that the coefficient C is nonnegative regardless of the choice of the numbers x_i and so $g(m)$ is a parabola (in the variable m) that has an absolute minimum at the solution of $g'(m) = 0$, i.e., at $m = -B/C$. This gives the value of m from which we can then obtain the value of $b = V/n$ and thereby determine the line giving the best fit.

EXAMPLE 6 Find the line of best fit to the points $(-3, 3)$, $(0, 5)$, $(1, 2)$, and $(2, 1)$.

Solution: We have selected an example in which the sum $x_1 + \cdots + x_n$ equals 0. This simplifies the formulas in the above development. We have

$$V = (3 + 5 + 2 + 1) = 11$$
$$B = -(-3 \cdot 3 + 0 \cdot 5 + 1 \cdot 2 + 2 \cdot 1) = 5$$
$$C = \left((-3)^2 + 0^2 + 1^2 + 2^2\right) = 14.$$

Thus $m = -B/C = -5/14$ and $b = V/n = 11/4$ and the line of best fit is $y = -5x/14 + 11/4$.

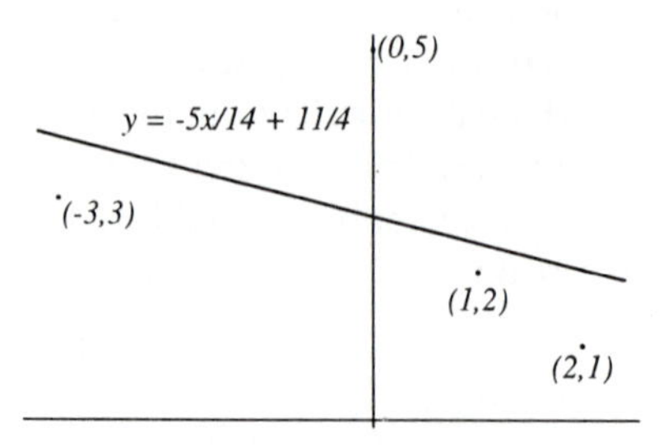

FIGURE 4.26
Line of best fit

Exercises 4.6

In these exercises, the phrase *normal line to the curve* refers to a line perpendicular to the tangent line to the curve.

1. Find the normal line to $y = 6x + 5$ passing through $(0, 1)$; prove that the shortest distance from $(0, 1)$ to a point on the graph of $y = 6x + 5$ is along a normal to the line.

2. Do the general case of Exercise 1. Prove that the shortest distance from a point (p, q) to a point on the line $y = mx + b$ is

$$D = \frac{|q - mp - b|}{\sqrt{m^2 + 1}}$$

and this distance occurs along the normal to the line.

3. Find the normal line to $y = x^2$ passing through $(2, -1)$. Prove that the shortest line from $(2, -1)$ to a point of the graph of $y = x^2$ is a normal line. [Hint: It is not necessary to find the shortest distance from point to curve explicitly.]

4. What are the dimensions of the rectangle of largest area inscribed in a circle of radius 1?

5. What are the dimensions of the isosceles triangle of largest area inscribed in a circle of radius R?

6. What is the maximum area of an isosceles triangle of perimeter P?

7. Let $f(x)$ be twice differentiable and let (a, b) be a point not on the graph of $f(x)$. Prove that the shortest line from (a, b) to a point on the graph of $f(x)$ is a line normal to the graph.

8. What is the largest possible product of two positive numbers whose sum is 10?

9. If two numbers have sum 4, what is the smallest possible sum of their squares?

10. Find the largest area of a rectangle inscribed in the ellipse $a^2y^2 + b^2x^2 = a^2b^2$. [You may assume without proof that the sides of the rectangle are parallel to the coordinate axes.]

11. Find the shortest distance from a point on the ellipse $a^2y^2 + b^2x^2 = a^2b^2$ to the point $(p, 0)$. Assume $0 < b < a < p$; do this by finding the critical points of the distance function and considering the values of x where the problem is defined.

12. A box with a square bottom and volume V is to be made out of material that costs \$1.00 per square foot for the sides and top and \$4.00 per square foot for the bottom. What are the dimensions (in terms of V) of the box that has the least material cost?

13. What is the largest volume of a right circular cylinder inscribed in a sphere of radius R?

14. What is the largest area of a rectangle having perimeter 4?

15. Do the problem given in example 5 with all data exactly the same except that the cost of constructing the pipeline overland is a dollars per foot. Interpret your results for some extreme cases such as $a = \$49$ or $a = \$1$.

16. The strength of a wooden beam having a rectangular cross section is proportional to the area of the cross section. What are the cross-sectional dimensions of the strongest beam that can be cut from a log with circular cross section of radius 1?

17. What point on the circle of radius 1 and having center at the origin maximizes the sum of the distances from $(1, 0)$ and $(0, 1)$? What point makes the sum of the distances a minimum?

18. What is the maximum volume of a box having the sum of its length and girth (= cross-sectional circumference) equal to 100 inches? Assume the cross section is a square. [Remark: This is the form of the restriction on size imposed by the U.S. Post Office for mailing certain types of parcels.]

19. A piece of wire of length L is cut into two pieces. One piece is made into a square frame and the other is bent into a circle. (a) How long should each piece be if the sum of the areas of the square and the circle is to be a maximum? (b) How long should each piece be if the sum of the areas of the square and the circle is to be a minimum?

20. If the sum of two numbers is 4, what is the smallest and largest possible sum of their cube roots?

21. Two corridors of unequal width meet at right angles. What is the longest stick that can be carried horizontally around the corner? Assume the corridors have widths a and b; express the result in terms of a and b. [Hint: There are several ways to do this problem; they all require considerable algebraic manipulation. One way is to set up the coordinate system so that the origin is at the inside corner and then consider the length of the segment of the line $y = mx$ having its endpoints on the outside walls of the corridors. Find the m to make this length as small as possible. The smallest such length equals the length of the longest stick that can be carried around the corner without getting stuck.]

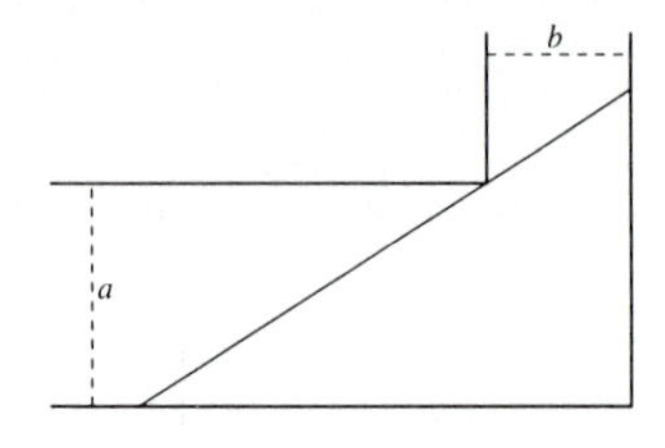

FIGURE 4.27

22. Find the line of best fit to the points $(-2, 0)$, $(-1, 8)$ and $(2, 0)$.

23. Find the line of best fit to the points $(-2, 4)$, $(-1, 2)$ and $(2, 6)$.

24. Find the line of best fit to the points $(-3, 4)$, $(-1, 2)$ and $(3, 6)$.

25. Find the line of best fit to the points $(-3, 4)$, $(1, 2)$ and $(2, 6)$.

26. Find the line of best fit to the points $(-3, 4)$, $(-2, 4)$, $(-1, 2)$, $(1, 2)$ and $(2, 6)$.

27. Find the line of best fit to the points $(0, 0)$, (a, b) and (c, d).

28. Let n be any positive integer and $x_1, x_2, \ldots, x_n$ any n real numbers. Let

$$S_n = x_1^2 + x_2^2 + \cdots + x_n^2 - \frac{1}{n}(x_1 + x_2 + \cdots + x_n)^2.$$

Prove $S_n \geq 0$. [One way: Regard S_n as a function of x_n with the other variables held fixed and then consider the minimum value of the function and prove

$S_n \geq S_{n-1}$. Thus $S_n \geq S_{n-1} \geq \cdots S_2 \geq S_1 = 0$.]

4.7 The second derivative

In this section we show how the second derivative can be used to give information about the graph of the function and the extreme values of the function.

The Second Derivative Test for Extrema

Suppose we are given the problem of determining the minimum and maximum values of $f(x)$ for x on the interval $[a, b]$. If f is differentiable on the interval then the first step is to find the solutions of the equation $f'(x) = 0$. There are two possibilities that might occur.

1. There might be no solutions to the equation $f'(x) = 0$; in this case (assuming the continuity of $f'(x)$) the function $f'(x)$ is either always positive or always negative. Thus $f(x)$ is either always increasing or always decreasing. In this case the extrema occur at endpoints.
2. There might be solutions (perhaps many solutions) of the equation $f'(x) = 0$ and it is then necessary to determine, for each solution, whether the function has a maximum value, a minimum value, or neither of these.

Keep in mind that the critical points are only the candidates for the locations of the local maximum or minimum; there is the possibility that the function has no local extremum at some particular critical point.

We introduce the second derivative test to provide a method for deciding if a function has a local extremum at a critical point. Even with this test, it is still necessary to examine the values of the function at the endpoints. If the derivative $f'(x)$ is not continuous at certain points, then it is necessary to also examine the behavior of $f(x)$ at these points.

Suppose the function $y = f(x)$ has a local maximum value at $x = a$; figure

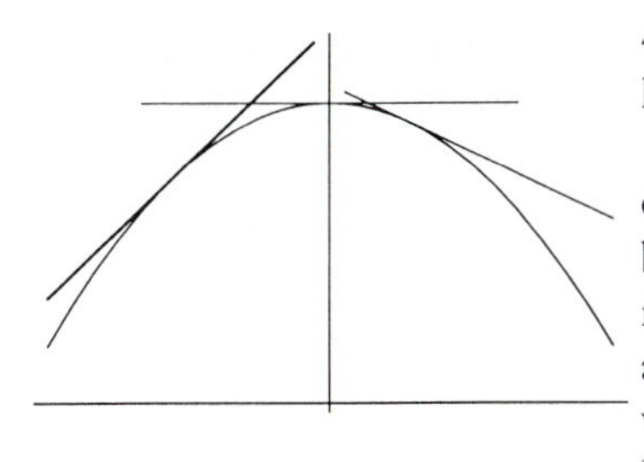

FIGURE 4.28
Tangents near a maximum

4.28 indicates the approximate shape of the graph of $f(x)$. Consider the tangent lines to the curve at points near $x = a$.

As x moves from the left of a, reaches $x = a$, then moves on to the right of a, the tangent line has slope that is positive, decreases toward zero, then becomes negative. Thus the slope of the tangent line is decreasing on some open interval with a at its center. In other words, the function $f'(x)$ is decreasing on an interval containing a. Because a decreasing function has a negative derivative, we conclude the derivative of $f'(x)$ is negative (or possibly zero) on this interval. In particular then $f''(a) \leq 0$. A similar line of reasoning leads to the conclusion $f''(a) \geq 0$ when $f(x)$ has a local minimum at $x = a$. Notice that the condition $f''(a) = 0$ can occur in either case.

This gives a geometric motivation for the following test.

THEOREM 4.14

Second Derivative Test

Let $f(x)$ be a function such that $f''(x)$ exists for each x on the interval $[b, c]$ and suppose a is a number with $b < a < c$ and $f'(a) = 0$.

1. If $f''(a) < 0$, then $f(a)$ is a local maximum.
2. If $f''(a) > 0$, then $f(a)$ is a local minimum.
3. If $f''(a) = 0$, then $f(a)$ could be a local maximum or minimum or neither of these.

Proof

Assume $f''(a) < 0$. Use the limit definition of the derivative to get

$$\lim_{h \to 0} \frac{f'(a+h) - f'(a)}{h} = \lim_{h \to 0} \frac{f'(a+h)}{h} = f''(a) < 0.$$

The definition of limit implies there is a $\delta > 0$ such that $f'(a+h)/h < 0$ for $0 < |h| < \delta$. In particular this means $f'(a+h)$ and h have opposite signs; when h is negative (and so $a+h$ is to the left of a), $f'(a+h)$ is positive and f is increasing on the interval to the left of a; when h is positive (and so $a+h$ is to the right of a), $f'(a+h)$ is negative and f is decreasing. Hence the shape of the graph shows that $f(x)$ has a local maximum at $x = a$. This proves statement 1.

If you are not convinced that this proves the graph must have the indicated shape, here is a line of reasoning that does not refer to the graph. Let h be a positive number satisfying $0 < h < \delta$. We can apply the Mean Value Theorem to the function $f(x)$ on the interval $[a, a+h]$. The conclusion is that there is some number c with $a < c < a+h$ such that

$$f(a+h) - f(a) = hf'(c).$$

Because c is to the right of a, it follows from the paragraph above that $f'(c) < 0$ and so

$$f(a+h) \leq f(a).$$

The same line of reasoning shows $f(a+h) \leq f(a)$ when h satisfies $-\delta < h < 0$. Thus $f(a) \geq f(x)$ for all x on the interval $(a-\delta, a+\delta)$. Thus $f(a)$ is a local maximum.

Analogous reasoning is used to prove statement 2; only some signs and the direction of a few inequalities have to be changed. We leave this for the reader.

We give some examples to show that no conclusion can be inferred in the case $f''(a) = 0$. The function $f(x) = x^4$ has an absolute minimum at the critical point $x = 0$; $f''(0) = 12 \cdot 0 = 0$ serves to illustrate that a local minimum can occur at a point where both $f'(x)$ and $f''(x)$ equal 0.

The function $f(x) = -x^4$ has an absolute maximum at $x = 0$ and $f'(0) = f''(0) = 0$ so a local maximum can occur at a point where both $f'(x)$ and $f''(x)$ equal 0.

The function $f(x) = x^3$ has a critical point at $x = 0$ because $f'(0) = 0$. Moreover $f''(0) = 0$ and $f(x)$ has neither a local maximum nor a local minimum at $x = 0$. Thus the function might have neither a maximum nor a minimum at a critical point where the second derivative is zero. □

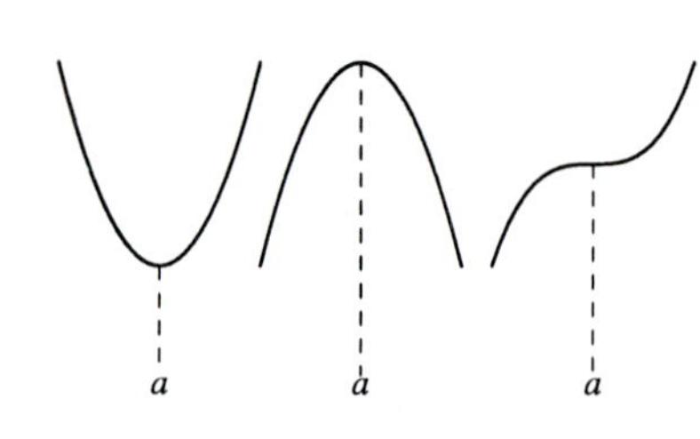

FIGURE 4.29
$f'(a) = f''(a) = 0$

Here are some examples where the second derivative test helps to determine the behavior of the function at a critical point.

EXAMPLE 1 Of all triangles having base b and altitude h, which one has the smallest perimeter?

Solution: We first describe the triangles in terms of coordinates in the plane. For the base, we take the interval $[0, b]$ so that $O(0, 0)$ and $R(b, 0)$ are two of the vertices; the altitude is given as h so for the third vertex of the triangle, we take a point $Q(x, h)$ on the line $y = h$. There is one unknown quantity, x, which is allowed to vary over all real numbers because for any choice of x, a triangle ORQ meeting the given conditions is obtained. For this triangle, let P denote its perimeter. The perimeter depends on x so $P = P(x)$ is the function we want to minimize. We evaluate $P(x)$ as

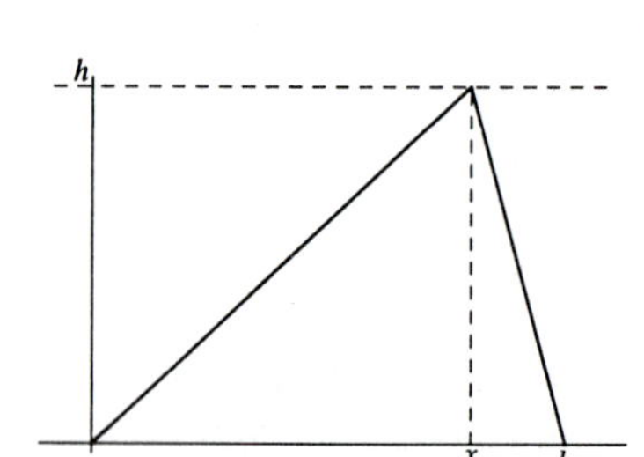

FIGURE 4.30
Base b, area $bh/2$

$$P(x) = b + \sqrt{x^2 + h^2} + \sqrt{(x - b)^2 + h^2}.$$

The critical points are found by first evaluating the derivative:

$$P'(x) = \frac{x}{\sqrt{x^2 + h^2}} + \frac{(x - b)}{\sqrt{(x - b)^2 + h^2}}.$$

Next we solve $P'(x) = 0$; transpose one term, square to eliminate the radicals, and do some cancellation to get the equation $x^2 = (x - b)^2$. Extract the square roots to finally get $x = \pm(x - b)$. Only the minus sign leads to a meaningful solution, $x = b/2$. So we have found the only critical point. Is it a local maximum, a local minimum, or neither of these? The second derivative test helps here:

$$P''(x) = \frac{h^2}{\sqrt{(x^2 + h^2)^3}} + \frac{h^2}{\sqrt{((x - b)^2 + h^2)^3}}.$$

Clearly the terms h^2 are nonnegative and the square root terms are nonnegative. Thus $P''(x) \geq 0$ for all x. More to the point, we have

$$P''(\frac{b}{2}) = \frac{16h^2}{\sqrt{(b^2 + 4h^2)^3}} > 0.$$

Thus the second derivative test ensures that $P(x)$ has a minimum at $x = b/2$. ■

The result of this example can be rephrased as follows:

Of all triangles with a given base and a given area, an isosceles triangle has the smallest perimeter.

Here is another application of the second derivative test to prove a result that was stated earlier.

THEOREM 4.15

Assume the second derivative of $f(x)$ exists and $f''(x) > 0$ for every x on the interval $I = [a, b]$. If there is a point c on I such that $f'(c) = 0$, then there is only one such point and $f(c)$ is a minimum of $f(x)$ on I.

Similarly, if $f''(x) < 0$ for every x on I, and if there is a critical point c on I, then there is only one critical point and $f(c)$ is the maximum of $f(x)$ on I.

Proof

Assume $f''(x) > 0$ for every x on the interval I. Then the function $f'(x)$ is increasing on I because its derivative is positive. If $c_1 < c_2$ are two points in I then $f'(c_1) < f'(c_2)$ by the definition of increasing functions. In particular, this means that the equation $f'(x) = 0$ has no more than one solution on I. Because it is assumed that there is at least one solution, we conclude there is exactly one point c satisfying $f'(c) = 0$. We apply the second derivative test to this critical point; we have $f''(c) > 0$ and so f has a local minimum at $x = c$. It is necessary to show that $f(c)$ is an absolute minimum.

Let u be any point on I other than c. We want to show $f(u) \geq f(c)$. We consider two cases depending on the size of u compared to c.

First assume $u > c$. We apply the Mean Value Theorem to f on the interval $[c, u]$ and find there is some number z with $c < z < u$ and

$$\frac{f(u) - f(c)}{u - c} = f'(z).$$

We know that $f'(x)$ is an increasing function so $c < z$ implies

$$0 = f'(c) < f'(z).$$

Thus the two quantities $f'(z)$ and $u - c$ are positive; it follows that $f(u) - f(c)$ is positive or, equivalently, that $f(u) > f(c)$.

In the second case with $u < c$, we make the same line of reasoning to conclude $f'(z)$ and $u - c$ are negative so $f(u) - f(c)$ is positive again.

The second statement of the theorem in which the second derivative is assumed negative, is proved in exactly the same way; the only difference is that some of the inequalities are reversed. □

EXAMPLE 2 Let $f(x)$ and $f''(x)$ be positive on the interval $[a, b]$ and assume that f''' exists on the interval. For any number w with $a < w < b$, let $A(w)$ be the area of the trapezoid formed by the lines $x = a$, $x = b$, $y = 0$ and the tangent line to the graph of $f(x)$ at $(w, f(w))$. For which value of w is the area $A(w)$ a maximum?

Solution: We use the second derivative test to prove a rather surprising geometric fact.

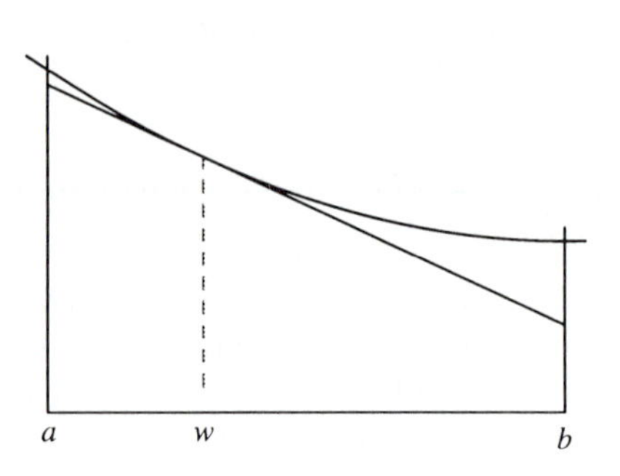

FIGURE 4.31
Find maximum trapezoid

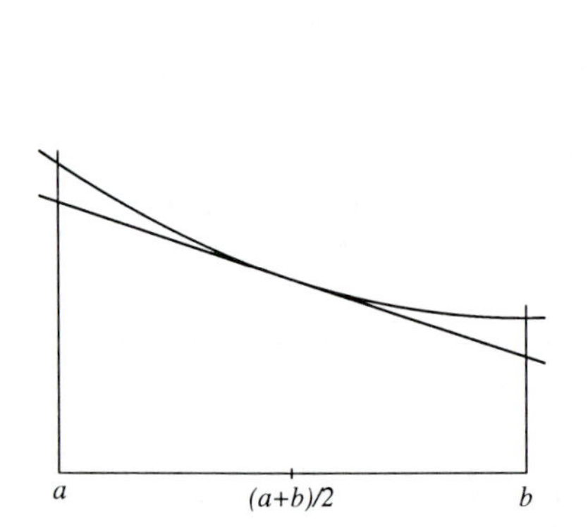

FIGURE 4.32
Trapezoid has area $A_{\max}$

The tangent line at the point $(w, f(w))$ has the equation

$$y = f'(w)(x - w) + f(w).$$

The area of the trapezoid having parallel sides of lengths u and v at a perpendicular distance h apart is $A = h(u + v)/2$. Using this formula, we find that the trapezoid formed in the figure has area

$$\begin{aligned} A(w) &= \frac{1}{2}(b - a)[f'(w)(a - w) + f(w) + f'(w)(b - w) + f(w)] \\ &= \frac{1}{2}(b - a)[2f(w) + f'(w)(a + b - 2w)]. \end{aligned}$$

We compute the derivative of $A(w)$ with respect to w:

$$\begin{aligned} A'(w) &= \frac{1}{2}(b - a)[2f'(w) + f''(w)(a + b - 2w) - 2f'(w)] \\ &= \frac{1}{2}(b - a)[f''(w)(a + b - 2w)]. \end{aligned}$$

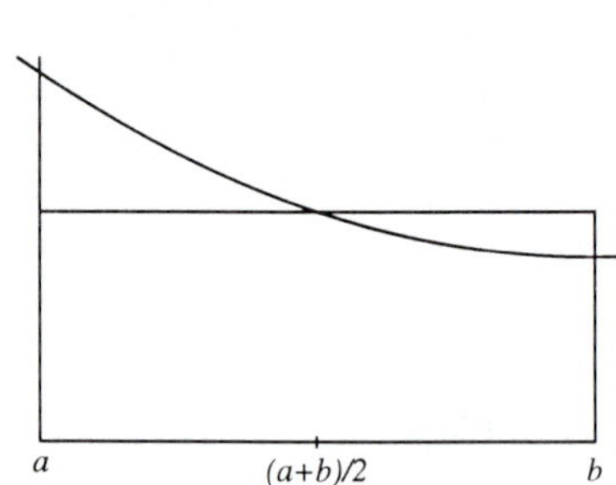

FIGURE 4.33
Rectangle has area $A_{\max}$

By assumption, $f''(w) > 0$, so the only critical point is $w = (a + b)/2$. We apply the second derivative test to determine if we have a local extremum:

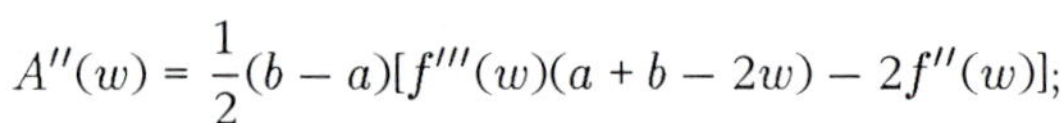

$$A''(w) = \frac{1}{2}(b - a)[f'''(w)(a + b - 2w) - 2f''(w)];$$

in particular

$$A''\left(\frac{a + b}{2}\right) = \frac{1}{2}(b - a)\left[-2f''\left(\frac{a + b}{2}\right)\right] < 0$$

because the second derivative of f is positive. The second derivative test implies $A(w)$ has a local maximum at the midpoint of the interval. Moreover, because the second derivative of $A(w)$ is negative and there is only one critical point, we conclude from the last theorem that the local maximum is an absolute maximum.

This result is surprising because we get the same answer, namely the midpoint of the interval, for all functions having a positive second derivative.

It is worth observing that the maximum value of the area function $A(w)$ is

$$A_{\max} = A\left(\frac{a + b}{2}\right) = (b - a)f\left(\frac{a + b}{2}\right);$$

this is the area of the rectangle having as one side the horizontal line through the point on the graph at the midpoint of the interval. ■

EXAMPLE 3 A container is to be made in the shape of a cylinder with top and bottom. The side of the cylinder is made from a material that costs s cents per unit of area, whereas the circular top and bottom are made from a material that costs t cents per unit of area. If the container is to hold 1 unit of volume, find the dimensions of the container that costs the least.

Solution: Let the radius of the circular base be r and the height of the container be h. Then the volume V is the area of the base times the height; namely

$$V = 1 = \pi r^2 h.$$

The area of the top and bottom combined is $2\pi r^2$ and the area of the side is the circumference of the base times the height, $2\pi rh$. The cost of the material is

$$C = 2\pi r^2 t + 2\pi rhs.$$

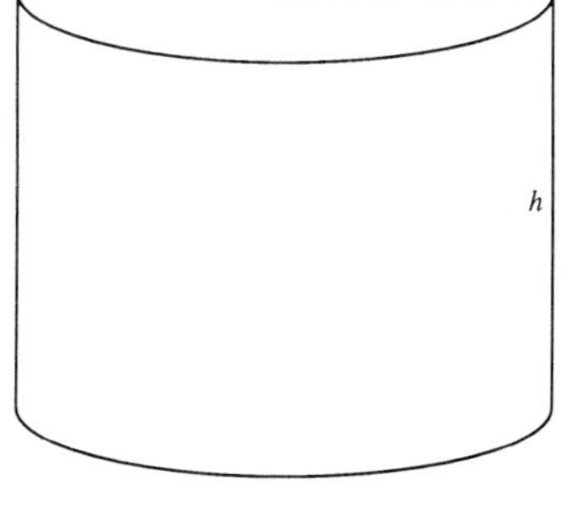

FIGURE 4.34
Cylinder

To determine the minimum of C, it is first necessary to use the volume condition to express C as a function of only one variable. We substitute $h = 1/\pi r^2$ to get

$$C = C(r) = 2\pi r^2 t + 2\pi rs\left(\frac{1}{\pi r^2}\right) = 2\pi r^2 t + \frac{2s}{r}.$$

The cost function $C(r)$ is defined for all positive r. To determine the minimum cost, we begin by locating the critical points of $C(r)$:

$$C'(r) = 4\pi rt - \frac{2s}{r^2} = 0 \quad \text{if} \quad r^3 = \frac{s}{2\pi t}.$$

Thus there is a unique critical point $r_0 = (s/2\pi t)^{1/3}$; it is necessary to determine if the cost is a minimum or a maximum at the critical point. We use the second derivative test:

$$C''(r) = 4\pi t + \frac{4s}{r^3},$$

$$C''(r_0) = 4\pi t + \frac{4s}{r_0^3} = 4\pi t + \frac{4s \cdot 2\pi}{s} > 0.$$

Thus C has a local minimum at r_0. The ratio of diameter to height that produces the container of least cost is

$$\frac{2r}{h} = \frac{2r_0}{\frac{1}{\pi r_0^2}} = 2r_0^3\pi = \frac{s}{t}.$$

This solution has a nice interpretation. The cost of the container is a minimum if the ratio (diameter/height) equals the inverse of the ratio (unit cost of top material/unit cost of side material).

We interpret this for a few cases. If the top and bottom material cost the same as the material for the sides, so that $s = t$, the least expensive container has $2r/h = 1$ or height equals twice the radius (i.e., the height equals the diameter). If the top and bottom material cost twice as much as the side material, so $2s = t$, then $4r = h$. This should seem reasonable; if the top is expensive, make it small. ■

Fermat's Principle of Least Time

In this subsection, we apply the second derivative test to find the path taken by a beam of light. Fermat's principle asserts that a beam of light travels from point A to point B along a path that requires the least amount of time. This does not always turn out to be the straight line path from A to B. For example, light has a different velocity in water than it does in air. If a light beam passes through a water tank and then through the air, the path need not be a straight line. There is a change of direction on the boundary between the two media. This is the concept of *refraction*. Of course, if the velocity of light is constant throughout the region containing A and B, then the path taken by a beam of light from A to B is the straight line path. We consider a fairly simple situation in the following example.

EXAMPLE 4 A beam of light travels in the xy plane so that its velocity through the upper half plane is v_1 and its velocity through the lower half plane is v_2. Find the path taken by a light beam originating at $A(0, a)$ and terminating at $B(b, c)$ (with $c < 0$).

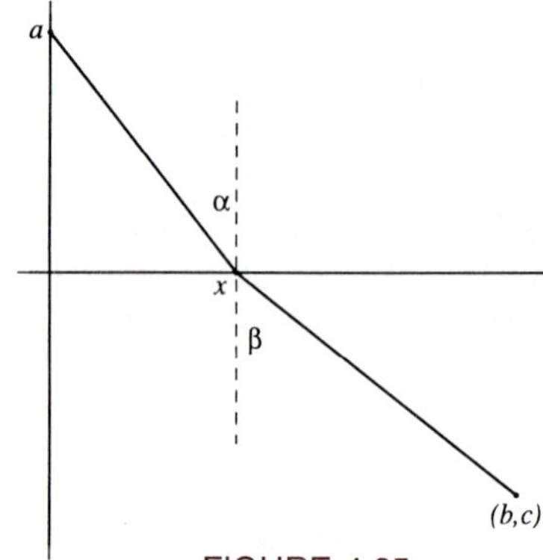

FIGURE 4.35
Refraction and Snell's law

Solution: The light follows a straight line from $(0, a)$ to a point $(x, 0)$ on the x-axis and then a straight line to (b, c). The restriction on the choice of the coordinate x is that the total time taken for the light to pass from A to B must be a minimum. We do not determine an explicit formula for x in terms of a, b, c (because that would require the solution of a fourth degree polynomial equation) and instead we derive a geometric condition that must hold. If the angles α and β are as indicated in figure 4.35, we prove *Snell's law* which asserts

$$\frac{\sin\alpha}{\sin\beta} = \frac{v_1}{v_2}.$$

To see this we allow x to be a variable point and compute the time required for light to follow the given path. The time along the leg in each half plane is the distance divided by the velocity; the total time is

$$T = T(x) = \frac{1}{v_1}\sqrt{x^2 + a^2} + \frac{1}{v_2}\sqrt{(b - x)^2 + c^2}.$$

This function is defined for all values of x so the minimum occurs at a critical point. We compute both the first and second derivatives:

$$T'(x) = \frac{1}{v_1}\frac{x}{\sqrt{x^2 + a^2}} - \frac{1}{v_2}\frac{b - x}{\sqrt{(b - x)^2 + c^2}},$$

$$T''(x) = \frac{1}{v_1}\frac{a^2}{\sqrt{(x^2 + a^2)^3}} + \frac{1}{v_2}\frac{c^2}{\sqrt{[(b - x)^2 + c^2]^3}}.$$

From this we see that $T''(x) > 0$ for every x; thus the first derivative $T'(x)$ is an increasing function; there can be only one solution to the equation $T'(x) = 0$ and at this critical point, the function T has a minimum. Assume x is the solution of this equation. We see that

$$\sin\alpha = \frac{x}{\sqrt{x^2 + a^2}} \quad \text{and} \quad \sin\beta = \frac{b - x}{\sqrt{(b - x)^2 + c^2}}.$$

The equation $T'(x) = 0$ can be rewritten as

$$\frac{\sin\alpha}{v_1} = \frac{\sin\beta}{v_2}$$

which yields a proof of Snell's law.

Note that this law implies $\alpha = \beta$ when $v_1 = v_2$, a fact that was predictable. ■

Exercises 4.7

Use the second derivative test to make certain that the solution you obtain is at the correct extreme.

1. A water cup is made by cutting a sector from a circular disc of radius 6 inches and joining the two edges of the sector to make a cone. What is the maximum volume of such a cup?

2. A rectangle of perimeter P is rotated the about one of its sides to produce a cylinder. What are the dimensions of the rectangle that produce the cylinder having the largest possible volume?

3. A notebook page is to have a total of 93.5 square inches and must have 1-inch margins on the sides and bottom and a 1.5-inch margin at the top. What are the dimensions of the page that gives the maximum area of printed material?

4. Prove that the rectangle inscribed in a triangle has area at most one-half the area of the triangle.

5. A jogger moves north along a straight road toward an intersection 1 mile ahead with another road crossing the first at right angles. He is able to run at 6 miles per hour on the road but could cut across the field and move at 4.5 miles per hour. If his destination is a point 2 miles west of the intersection, how far west of the intersection should he return to the road to reach the destination in the least time?

6. Redo exercise 5 with the jogger's velocity along the road equal to v and through the field equal to vk, $0 < k < 1$. Interpret the result for k near 0 and near 1. At what ratio k does it become better to stay on the road for the entire distance?

7. A beam of light passes through air and strikes the surface of a liquid at an angle of $45°$ and "bends" downward by an additional $3°$. Use a calculator or trigonometric tables to estimate the ratio of the velocity of light through air to the velocity of light in the liquid. If a beam of light passes through the air and strikes the surface of the same liquid at $30°$ from the vertical, what angle from the vertical does the beam make inside the liquid?

8. Use Snell's law to derive the reflection principle for the path of light reflecting off a surface: The angle of incidence equals the angle of reflection.

9. Use the principle of least time to derive the reflection principle: If a light beam from $(0, a)$ is reflected off the x-axis to a point (b, c) with $b > 0$ and $c > 0$, then the two angles made by the path with the x-axis are equal.

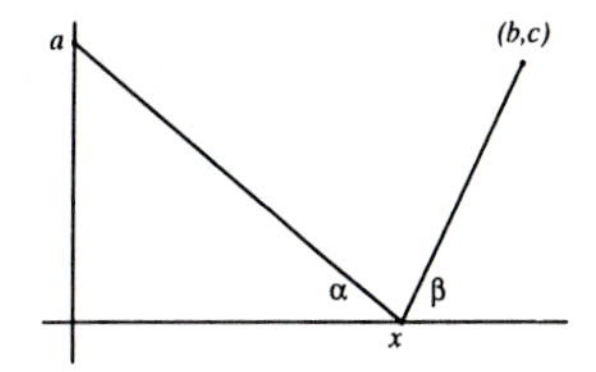

FIGURE 4.36

4.8 Concavity

The term *concavity* refers to the direction of the "bend" of the graph of a function. We have seen earlier how the two functions $f(x) = x^2$ and $g(x) = \sqrt{x}$ are both increasing on the interval $[0, 1]$ and both pass through $(0, 0)$ and $(1, 1)$. Yet the graphs look quite different because the graph of $y = x^2$ bends upward, whereas the graph of $y = \sqrt{x}$ bends downward.

The direction of bend is described as *concave* or *convex*. We can distinguish between the two directions of bending in the curve by introducing a chord; if the graph bends up, it lies below the chord connecting the endpoints; if the graph bends down, it lies above the chord connecting the endpoints (see figure 4.38).

Here is the precise definition of these concepts.

DEFINITION 4.4

Geometric Definition of Convex and Concave Functions

A function $f(x)$ is *convex* over the interval $[a, b]$ if for every pair of numbers

u, v satisfying $a < u < v < b$, the graph of $f(x)$ for $u < x < v$ lies below the chord connecting $(u, f(u))$ and $(v, f(v))$.

A function $f(x)$ is *concave* over the interval $[a, b]$ if for every pair of numbers u, v satisfying $a < u < v < b$, the graph of $f(x)$ for $u < x < v$ lies above the chord connecting $(u, f(u))$ and $(v, f(v))$.

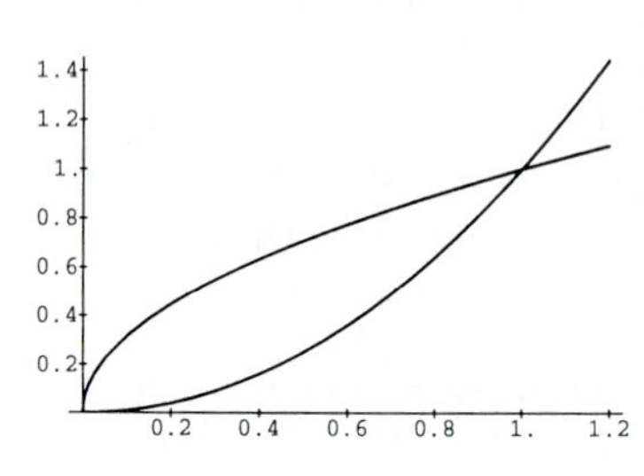

FIGURE 4.37
$y = \sqrt{x}$ and $y = x^2$

The definitions of concave and convex are given in terms of a geometric condition satisfied by the graph of the function. The geometric terms "above" and "below" can be translated into inequalities to transform these definitions into more analytical terms.

Given the points $a < u < v < b$, the line through the points $P(u, f(u))$ and $Q(v, f(v))$ is the graph of the equation

$$T(x) = \frac{f(v) - f(u)}{v - u}(x - u) + f(u).$$

The statement that the graph of f lies above the chord PQ is equivalent to the statement $f(x) > T(x)$ for all x on (u, v). The statement that the graph of f lies below the chord is equivalent to $f(x) < T(x)$. We rewrite these using the definition of T to obtain the following statement:

DEFINITION 4.5

Analytic Definition of Concave and Convex Functions

The function $f(x)$ is concave on the interval $[a, b]$ if for every triple of points u, x, v satisfying $a \leq u < x < v \leq b$ we have

$$\frac{f(x) - f(u)}{x - u} > \frac{f(v) - f(u)}{v - u}.$$

The function $f(x)$ is convex on the interval $[a, b]$ if for every triple of points u, x, v satisfying $a \leq u < x < v \leq b$ we have

$$\frac{f(x) - f(u)}{x - u} < \frac{f(v) - f(u)}{v - u}.$$

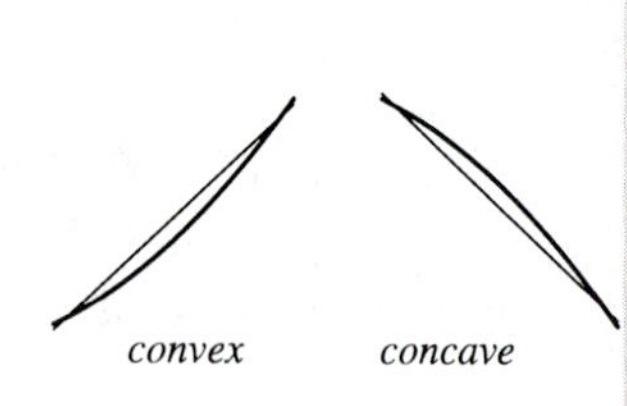

FIGURE 4.38

The following result provides a test for concavity for functions having second derivatives.

THEOREM 4.16

Second Derivative Test for Concavity and Convexity

Let $f(x)$ be a function for which the second derivative $f''(x)$ exists for every x on the interval $[a, b]$. Then,

1. If $f''(x) > 0$ for each x on (a, b), then $f(x)$ is convex on $[a, b]$.
2. If $f''(x) < 0$ for each x on (a, b), then $f(x)$ is concave on $[a, b]$.

Proof

Let u and v be two numbers satisfying $a \leq u < v \leq b$. The line through the points $A(u, f(u))$ and $B(v, f(v))$ has equation

$$y = \frac{f(v) - f(u)}{v - u}(x - v) + f(v).$$

In the following, we restrict x to lie on the interval $[u, v]$. The vertical

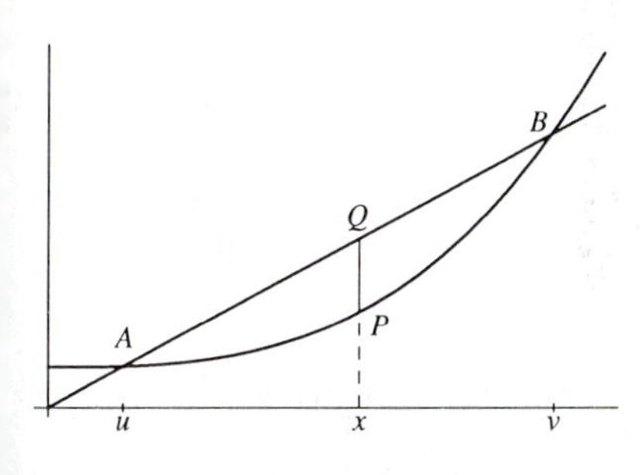

FIGURE 4.39
Convex function $f(x)$ and chord

distance from the point $P(x, f(x))$ on the graph to the point $Q(x, y)$ on the chord is

$$F(x) = \frac{f(v) - f(u)}{v - u}(x - v) + f(v) - f(x);$$

when Q is above P, $F(x)$ is positive; when Q is below P, $F(x)$ is negative. We also note the following:

$$F(u) = F(v) = 0,$$

$$F'(x) = \frac{f(v) - f(u)}{v - u} - f'(x),$$

$$F''(x) = -f''(x).$$

With this preparation, let us begin the proof of the theorem. Suppose that $f''(x) > 0$ for all x. Then $F''(x) < 0$ and so the function $F'(x)$ has negative first derivative and therefore is decreasing on the interval $[u, v]$. Because a decreasing function can cross the x-axis at most once, there is at most one point c with $F'(c) = 0$. In view of the equations $F(u) = F(v) = 0$, Rolle's Theorem implies that there is at least one point c on (u, v) with $F'(c) = 0$. This implies that the function $F(x)$ has exactly one critical point, namely $x = c$, on $[u, v]$. Moreover, the statement $F''(c) < 0$ implies that $F(x)$ has a local maximum at c. We know $F(x)$ is continuous and therefore has both a maximum and a minimum on $[u, v]$ and these must occur either at a critical point or at an endpoint. We have just seen there is only one critical point and the function has a maximum value there. Hence the minimum must occur at the endpoint; thus $F(u) = F(v) = 0$ is the minimum value of F. In particular, this means $F(x) \geq 0$ for all x on $[u, v]$. In view of the interpretation mentioned at the beginning of the proof of the sign of $F(x)$, we conclude that the graph lies below the chord; $f(x)$ is convex over $[a, b]$.

In a similar way, the assumption $f''(x) < 0$ for all x leads to the conclusion that $f(x)$ is concave on $[a, b]$. □

FIGURE 4.40
Inflection point

Inflection Points

The last theorem makes no assertion about points where the second derivative is zero. There are several possible situations at such a point. If $f''(c) = 0$ then, in a small interval centered at c, $f''(x)$ might be greater than or equal to 0, $f''(x)$ might be less than or equal to 0, or $f''(x)$ might have one sign on half of the interval and the opposite sign on the other half. If $f''(x)$ is positive on the left of c and also on the right of c, then $f(x)$ is concave to the left and to the right of c. Similarly, $f(x)$ is convex on each side of c if $f''(x)$ is negative on an interval to the left and right of c. The remaining case is that $f''(x)$ is positive on one side of c and negative on the other side of c. In this case we call c an *inflection point* of $f(x)$ (see figure 4.40). If c is an inflection point, then the graph of $f(x)$ is convex on one side of c and concave on the other side. Thus the concavity changes at an inflection point.

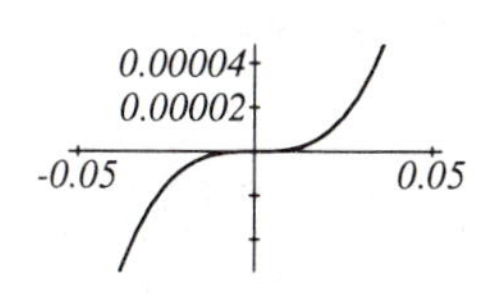

FIGURE 4.41
x^3 over $[-0.05, 0.05]$

EXAMPLE 1 The function $f(x) = x^3$ has an inflection point at $x = 0$.

Solution: Because $f''(x) = 6x$ we see that $f''(x)$ is negative on the left of 0, equals 0 at 0, is positive on the right of 0. The graph is concave to the left of 0, convex to the right of 0. Figure 4.41 shows the graph in a very small interval around 0. ■

EXAMPLE 2 Prove that the equation

$$f(x) = x^4 - 7x + 2 = 0$$

has exactly two real number solutions.

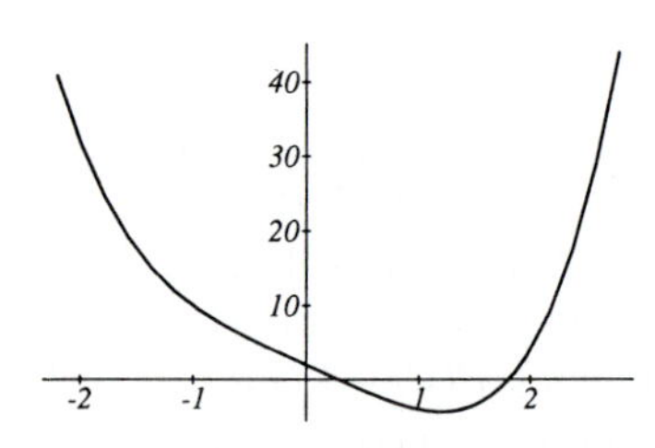

FIGURE 4.42
$x^4 - 7x + 2 = 0$ has two solutions

Solution: The idea behind this solution is that $f''(x) = 12x^2 \geq 0$ for all x and so the graph of the function is convex. Thus a chord with endpoints on the graph cannot touch the graph at a third point. The conclusion is reached by selecting part of the x-axis connecting two solutions as the chord; this chord cannot touch the graph at any point between the two solutions. Here are the details.

Let us first show that there must be at least two solutions. Observe that $f(0) = 2 > 0$ and $f(1) = -4 < 0$; by the Intermediate Value Theorem, there is a point r with $0 < r < 1$ and $f(r) = 0$. Also $f(2) = 11 > 0$ so there is a point s with $1 < s < 2$ and $f(s) = 0$. Thus there are at least two solutions of the equation, one on the interval (0,1) and one on the interval (1,2). Suppose there were a third solution t. Rename these three points so that a is the smallest b is the largest, and c is the remaining one.

Then $a < c < b$ and these points satisfy $f(a) = f(c) = f(b) = 0$ and the three points $P(a, 0)$, $R(c, 0)$ and $Q(b, 0)$ lie on the curve $y = f(x)$ and also lie on the chord PQ. Notice that $f''(x) = 12x^2 > 0$ for all x on $[a, b]$.

The theorem just proved showed that the graph of a function having a positive second derivative on $[a, b]$ cannot touch the chord PQ between P and Q. Hence there cannot be three such points and the original equation has exactly two solutions.

One could solve this problem in others ways but this method illustrates the use of concavity. An inspection of the graph of $f(x) = x^4 - 7x + 2$ might also be convincing. ■

Exercises 4.8

Find the intervals over which the functions are concave or convex and determine the inflection points.

1. $f(x) = x^3 - 3x + 1$

2. $f(x) = x^2 - \frac{1}{x}$

3. $f(x) = x^4 - 4x$

4. $f(x) = x^4 + 8x$

5. $f(x) = 8x^2 - x^4$

6. $f(x) = \frac{x+1}{x}$

7. $f(x) = \sin x$

8. $f(x) = \tan x$

9. $f(x) = \cos x \sin x$

10. $f(x) = \sin^2 x$

11. Show that the graph of a cubic polynomial $f(x) = ax^3 + bx^2 + cx + d$ has exactly one inflection point and it lies midway between the points where f has a local maximum and local minimum.

12. Let $f(x)$ have positive second derivative on the interval $[a, b]$. Prove that the graph of $f(x)$ for $a \leq x \leq b$ lies above the tangent line at $(c, f(c))$ for any c on $[a, b]$. [Hint: The function $g(x) = f'(c)(x - c) + f(c) - f(x)$ is positive at any point x where $(x, f(x))$ is below the tangent line to the graph of $f(x)$ at $(c, f(c))$. Show $g(x) \geq 0$ for x on $[a, b]$ by checking where $g'(x)$ is decreasing and observing that $g(c) = 0$.]

13. Let $f(x)$ have negative second derivative on the interval $[a, b]$. Prove that the graph of $f(x)$ for $a \leq x \leq b$ lies below the tangent line at $(c, f(c))$ for any c on $[a, b]$.

14. Let $f(x)$ and $g(x)$ be concave and differentiable functions on an interval I. Prove that $f(x) + g(x)$ is a concave function on I.

15. Give an example of two differentiable and concave functions $f(x)$ and $g(x)$ such that the product $f(x)g(x)$ is not a concave function.

4.9 Velocity and acceleration

Consider an object that moves in a straight line. Its position is measured from some fixed reference point O with one direction selected as the positive direction and some unit of length given. Let $s(t)$ denote its position at time t. Thus at time $t = a$, the object is $s(a)$ units from O; at some time b its position is $s(b)$ and so the change in distance during the time interval $a \leq t \leq b$ is $s(b) - s(a)$.

$O \qquad s(a) \qquad s(b)$

The *average velocity* of the object during the time interval from $t = a$ to $t = b$ is the distance divided by the time; namely

$$v_{\text{ave}} = \frac{s(b) - s(a)}{b - a}.$$

This definition of average velocity is used to define the velocity at a particular instant of time, the so-called *instantaneous velocity.* The velocity at time t is defined by a limiting process. Consider a small change in the time from t to $t + h$, determine the average velocity over this time interval and let h approach 0. Thus the velocity at time t is

$$v(t) = \lim_{h \to 0} \frac{s(t+h) - s(t)}{h} = \frac{d}{dt}s(t) = s'(t).$$

We refer to this relation by saying "the instantaneous rate of change of the position function is the velocity function".

The acceleration is treated in the analogous way. The average rate of change of the velocity is the acceleration. The instantaneous acceleration is given as

$$a(t) = \frac{d}{dt}v(t) = v'(t).$$

EXAMPLE 1 Suppose an apple is thrown from the top of the Washington Monument so that it moves in a vertical line and its position t seconds after release is $s(t) = 16t^2 - 1.8t$ for the time interval $0 \leq t \leq T$, where T is the time the apple hits the ground. Here we take $s(0) = 0$ as the point of release and the positive direction is downward. The units are feet and seconds.

1. In what direction was the apple thrown?
2. If the point of release was 555 feet above the ground, at what time does the apple reach the ground?
3. At what velocity does the apple strike the ground?
4. What is the acceleration function for the apple?

***Solution*:** The velocity at time t is

$$v(t) = \frac{d}{dt}s(t) = \frac{d}{dt}(16t^2 - 1.8t) = 32t - 1.8.$$

Thus the velocity at time $t = 0$ is $v(0) = -1.8$—and we discover that the apple was thrown upward (the negative direction). The time T required for the apple to strike the ground satisfies the condition $s(T) = 555$ or

$$16T^2 - 1.8T = 555.$$

The solutions of this equation are found by applying the quadratic formula:

$$T = \frac{1.8 \pm \sqrt{1.8^2 + 4 \cdot 555 \cdot 16}}{32} \approx \frac{1.8 \pm 188.4761}{32} \approx 5.946 \text{ or } -5.834.$$

The negative time has no meaning in this problem so the solution is $T = 5.946$ sec. The velocity at impact is

$$v(T) = v(5.946) = 32(5.946) - 1.8 = 188.472 \text{ ft/sec}.$$

(Just for comparison of units, this is 128.5 miles/hour.) The acceleration function is the derivative of the velocity;

$$a(t) = \frac{d}{dt}v(t) = \frac{d}{dt}(32t - 1.8) = 32.$$

Thus the acceleration is a constant, $a(t) = 32\,\text{ft/sec}^2$. ■

A reasonable question to ask after examining this example is How does one obtain a formula for $s(t)$? For an answer to this, it is necessary to apply a basic principle of physics, namely Newton's law, which asserts that an object of mass m, that is subjected to a force F, experiences an acceleration a, and these three quantities are related by the equation $F = ma$ when the units are suitably selected.

In the example of the falling apple, the only force acting on the apple during its fall is the force of gravity attracting the apple toward the ground. (There are other forces, such as resistance caused by the air, but these are ignored here for simplicity.) The acceleration due to gravitational force is constant (or nearly so as long as the objects are fairly close to the surface of the earth). This constant is traditionally denoted by the letter g and has a numerical value of approximately $32\,\text{ft/sec}^2$. In numerical examples we assume that $g = 32\,\text{ft/sec}^2$. An object of mass m is attracted toward the earth's surface with a force gm. Thus for any object acted on only by the force of gravity, the acceleration is constant: $a(t) = 32\,\text{ft/sec}^2$. The constant g is called *the acceleration due to gravity.*

The velocity $v(t)$ is a function whose derivative is the constant 32. Thus

$$v(t) = 32t + v_0,$$

where v_0 is a constant.

The position function, $s(t)$ has $v(t) = 32t + v_0$ as its derivative. Thus

$$s(t) = 16t^2 + v_0 t + s_0$$

where s_0 is a constant. The two constants can be interpreted as certain properties at time $t = 0$. We see $v(0) = v_0$ so v_0 is the velocity at time $t = 0$. Similarly, $s(0) = s_0$ so that s_0 is the position of the object at time $t = 0$.

We summarize the discussion to this point. If an object falls in a straight line subject only to the force of gravity then its position function $s(t)$, velocity function $v(t)$, and acceleration function $a(t)$ satisfy:

$$a(t) = g \qquad (\approx 32 \text{ ft/sec}^2),$$
$$v(t) = gt + v_0, \qquad v_0 = \text{velocity at time } t = 0,$$
$$s(t) = \tfrac{1}{2}gt^2 + v_0 t + s_0, \qquad s_0 = \text{position at time } t = 0.$$

Exercises 4.9

For each position function $s(t)$, find the velocity at the times when $s(t) = 0$ and then find the position s at the times when the velocity is 0. Sketch a graph of the height as a function of time.

1. $s(t) = 64 - 16t^2$

2. $s(t) = -16t^2 + 30t + 25$

3. $s(t) = 32t^2 + 64t - 4$

4. $s(t) = -16t^2 + 84t - 80$

A ball is thrown vertically upward and $s(t)$ denotes its height above some fixed point for the time t the ball is in the air. Find the maximum height reached by the ball and the length of time required to reach that height when $s(t)$ is given.

5. $s(t) = -16t^2 + 160t$

6. $s(t) = -16t^2 + 96t + 50$

7. $s(t) = -16t^2 + 0.5t + 500$

8. $s(t) = -16t^2 + 500$

In all the following problems involving the acceleration due to gravity use $g = 32$ ft/sec^2.

9. A ball is released from the top of a 100-foot building with initial velocity $v_0 = 0$. How much time passes before the ball hits the ground?

10. Repeat exercise 9 under the assumption that the initial velocity is 2 feet/sec.

11. Redo exercise 9 with the assumption that the initial velocity is v_0 feet/sec and obtain the falling time as a function of v_0. Use this information to determine the truth or falsity of the statement: If the ball is thrown downward twice as fast, it hits the ground in half the time.

12. A tennis ball is hit directly upward and returns to the ground in 16 seconds. What was the initial velocity and what height was attained by the ball?

13. A tennis ball is hit directly upward and reaches a maximum height M above the ground. Each height less than M is attained by the ball two times, once going up and once going down. Show that the absolute value of the velocity of the ball is the same when it passes through any given height.

14. A racing car is able to generate enough power to maintain a constant acceleration of 65 ft/sec^2. How long does it take to travel 1/4 mile if it starts from rest? What is its velocity as it crosses the 1/4-mile mark?

15. An automobile maker advertises that its car can go from 0 to 60 (miles per hour) in 12.8 seconds. Find the acceleration assuming that the acceleration is constant. (Take care to use compatible units.) How long does it take to reach 90 miles/hour at this acceleration?

4.10 Motion in the plane

In this section we consider the motion of a particle in a plane rather than along a straight line. The position of the particle at time t is given by specifying its two coordinates, both of which are functions of t. Suppose that $P(t) = (x(t), y(t))$ is the position at time t. Notice that $x(t)$ and $y(t)$ are real-valued functions of the time variable t, but the values of $P(t)$ are points in the plane. This is a different use of the idea of a function than has been seen previously inasmuch as the values of $P(t)$ are points in place of real numbers.

Motion of a Projectile

We consider the problem of describing the motion of a particle that moves in a (vertical) plane subject only to the (downward) force of gravity. We suppose that ground level is along the x-axis and upward vertical distance is measured along the positive y-axis.

To help clarify the ideas, we consider an example of a baseball thrown from ground level at some initial velocity and at some angle of elevation. (In spring training camp, a mechanical device known as Iron Mike can be preset to throw the baseball at a given initial velocity and angle of elevation.)

EXAMPLE 1 Suppose Iron Mike throws a baseball with an initial velocity of v_0 at an angle α above the horizontal. Describe the path of the baseball.

Solution: Measure time t from the instant the ball is thrown and assume the point of release is at the origin $(0, 0)$. Let $P(t) = (x(t), y(t))$ be the position of the particle at time t so that $P(0) = (0, 0)$.

We consider the motion of the ball broken into two separate directions: the function $y(t)$ measures motion in the vertical direction and $x(t)$ measures motion in the horizontal direction. We treat these separately according to the theory of motion along a straight line. First consider the vertical motion.

The only force acting on the ball is the force of gravity acting vertically downward. The vertical acceleration is $y''(t)$ at time t so from Newton's law, we have the equation

$$y''(t)m = -gm \qquad m = \text{mass of projectile.}$$

The minus sign is used because the upward direction is positive but the force is acting downward. The description of the vertical motion is then

$$y(t) = -\frac{1}{2}gt^2 + y'(0)t + y(0).$$

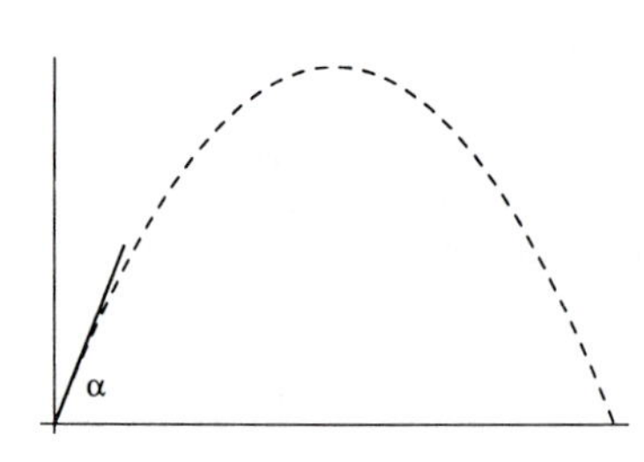

FIGURE 4.43

We know $y(0) = 0$ from the given data at time $t = 0$. The term $y'(0)$ is the initial velocity in the vertical direction. This number has not been given explicitly. The initial velocity is v_0 in a direction along a line making an angle α with the positive x-axis.

Regard this velocity as composed of two velocities, one vertical and one horizontal. The magnitudes of these are determined from the right triangle having hypotenuse v_0 and angle α. We find the vertical velocity $y'(0)$ and the horizontal velocity, $x'(0)$ are given by

$$y'(0) = v_0 \sin\alpha \qquad x'(0) = v_0 \cos\alpha.$$

Thus

$$y(t) = -\frac{1}{2}gt^2 + (v_0 \sin\alpha)t.$$

The horizontal motion is somewhat easier to describe because there is no force acting in the horizontal direction. Thus the horizontal acceleration is 0; $x''(t) = 0$. This means that $x'(t)$ is constant and the constant must be the same as the initial velocity. Thus $x'(t) = v_0 \cos\alpha$. It follows that

$$x(t) = (v_0 \cos\alpha)t + x(0) = (v_0 \cos\alpha)t$$

because $x(0) = 0$. We now have the position at any time t given by

$$P(t) = ((v_0 \cos\alpha)t, -\frac{1}{2}gt^2 + (v_0 \sin\alpha)t).$$ ■

EXAMPLE 2 An athlete throws a javelin with an initial velocity of 120 feet/sec at an angle of 30 degrees ($= \pi/6$ radians). Assuming level ground, what is the maximum height of the javelin above the ground and how far away does the javelin go before it strikes the ground?

Solution: Assume the distances are measured from the point of release so at time $t = 0$ the javelin is at the origin. Then at time t its position is

$$\begin{aligned} P(t) &= ((v_0 \cos(\pi/6))t, -\frac{1}{2}gt^2 + (v_0 \sin(\pi/6))t) \\ &= (60\sqrt{3}t, -16t^2 + 60t). \end{aligned}$$

We use units of feet and seconds so $g = 32$. The highest point reached is the maximum value of $-16t^2 + 60t$ for $t > 0$. This occurs at $t = 60/32$ and

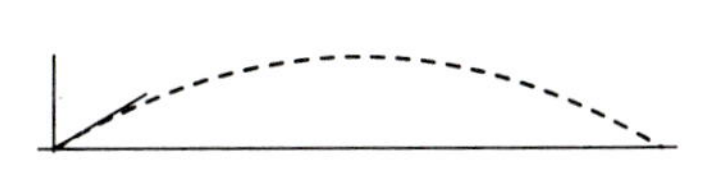

$$y(60/32) = -16(60/32)^2 + 60(60/32) = 60^2/64 = 56.25.$$

The particle strikes the ground at the time when $y(t) = 0$. The equation $-16t^2 + 60t = 0$ has two solutions: $t = 0$ and $t = 60/16$. The particle is initially on the ground at $t = 0$ and returns to the ground at $t = 60/16$. The distance from the origin where the projectile hits the ground is

FIGURE 4.44
30° angle of elevation

$$x(60/16) = (60\sqrt{3})(60/16) \approx 389.7.$$

Note that the particle reaches its highest point in exactly half the time it takes to return to the ground. ■

Exercises 4.10

Let $P(t) = (x(t), y(t))$ be the position of a moving particle at time t. We assume level ground with the origin at ground level. For each example, (a) At what time $t > 0$ does the object return to the ground? (b) What is the cosine of the initial angle of elevation of the motion?

1. $P(t) = (3t, -16t^2 + 4t)$ **2.** $P(t) = (12t, -16t^2 + 5t)$

In all these problems use $g = 32\,\text{ft/sec}^2$.

3. An archer releases an arrow at an initial velocity of v_0 ft/sec in the horizontal direction aimed directly at the center of a circular target 12 inches in diameter. If the target is 90 feet from the point of release, how large must v_0 be if the arrow hits the target?

4. A baseball player hits a ball at home plate and gives the ball an initial velocity of 180 feet/sec at an angle of elevation of 13°. Will the ball get over a fence 12 ft higher than the point of contact (of bat and ball) located 350 feet from home plate? [This requires a calculator. Watch the units!]

5. A professional football kicker is able to impart an initial velocity to a kicked ball of 70 mi/hr. At what initial angle should he kick the ball to send it downfield the maximum distance? At what initial angle should he kick the ball so that it remains in the air the longest possible time?

6. If an athlete is able to throw the 16-pound shot a distance of 60 feet, what initial velocity does he give to the shot assuming the optimal angle of release?

7. A road up the side of a hill is a straight line making an angle $\beta < \pi/2$ (in radians) with the horizontal. A projectile is fired from the base of the hill at an angle α above the horizontal with some given initial velocity. Prove that distance along the road where the projectile lands will be a maximum distance from the point at the base of the hill if $\alpha = \frac{1}{2}\beta + \frac{\pi}{4}$.

4.11 Related rates

The phrase *related rates* refers to a situation in which there are two quantities related to each other by some equation; both quantities change with time and one asks for the rate of change of one of the quantities when information is given about the rate of change of the other.

EXAMPLE 1 A six-foot-tall man is walking directly away from a 10-foot-high street light. The man's shadow is observed get longer and so one concludes that the tip of the shadow is moving faster than the man. What is the relation between the velocity of the walker and the velocity of the end of the shadow?

Solution: The variable quantities in the problem are the position of the walker and the position of the tip of the shadow; both of these change with time. The problem asks us to find the relation between the velocity of the walker and the velocity of the shadow. We must find expressions for each of these to relate them.

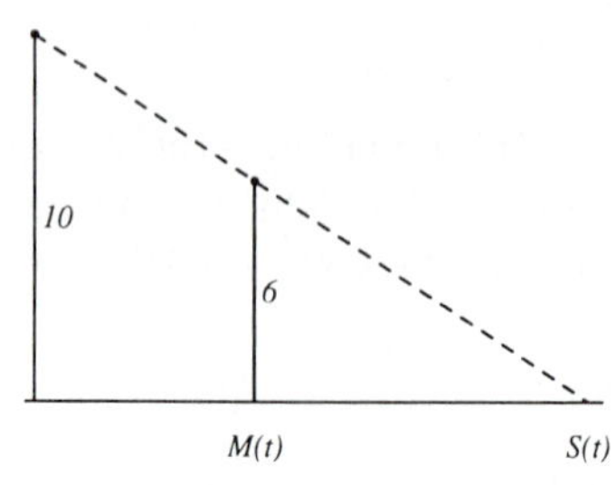

FIGURE 4.45

We begin by arranging a coordinate system to enable us to describe the objects in terms of functions. Assume the man walks along the x-axis, that the street light is along the y-axis, and the variable t measures time from some convenient starting point; for example, the instant the man passes the light. The distance of the man from the light at time t is denoted by $M(t)$ and the distance of the tip of the shadow from the light is $S(t)$.

The relation between $M(t)$ and $S(t)$ is found from the similar triangles:

$$\frac{S(t)}{10} = \frac{S(t) - M(t)}{6} \quad \text{or} \quad M(t) = 0.4S(t).$$

To get a relation between the velocities of the walker and the shadow, we differentiate this last equation to get

$$M'(t) = 0.4S'(t).$$

At any given instant of time, the velocity of the walker is 0.4 times the velocity of the shadow.

If a brisk walker maintains a constant velocity of 5 miles/hour then the relation between the velocities yields $0.4S'(t) = 5$ or $S'(t) = 12.5$ miles/hour. ■

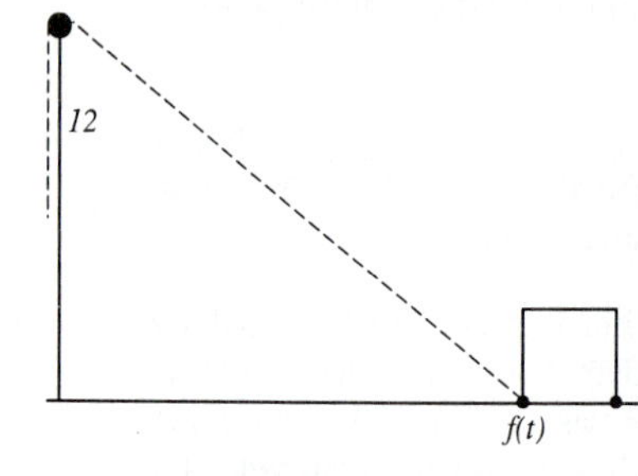

FIGURE 4.46

EXAMPLE 2 A rope passes over a pulley 12 feet above the level ground and is attached to a box that can slide along the ground. What is the relation between the rate at which the rope is pulled and the rate at which the box moves along the ground?

Solution: Place the pulley at the point $(0, 12)$ and assume the box slides along the x-axis so that it is at the point $(f(t), 0)$ at time t. Let $h(t)$ be the length of the rope from the pulley to the box at time t. The rate at which the rope is pulled is the rate of change of $h(t)$ with respect to t, namely $h'(t)$. The problem asks for the relation between this and $f'(t)$, the velocity of the box along the ground.

From the distance formula we obtain

$$h(t)^2 = 12^2 + f(t)^2 \quad \text{or} \quad h(t) = \sqrt{12^2 + f(t)^2},$$

and by taking derivatives with respect to t in the first equation, we find

$$2h(t)h'(t) = 2f(t)f'(t) \quad \text{or} \quad f'(t) = \frac{h(t)h'(t)}{f(t)}.$$

If the rope is being pulled at a constant rate of 0.5 ft/sec $= h'(t)$ then the box is moving at the rate $(0.5)h(t)/f(t)$. This depends on the position of the box $f(t)$. For example, we compute the velocity of the box when it is at the point $(50, 0)$; the value of $h(t)$ is then

$$h = \sqrt{(12^2 + 50^2)} = \sqrt{2644} \approx 51.4,$$

and

$$f'(t) = \frac{h(t)h'(t)}{f(t)} = \frac{(51.4)(0.5)}{50} = 0.514 \text{ ft/sec}.$$

This shows that the box moves along the floor slightly faster than a point on the rope moves toward the pulley. The reader may wish to prove that this occurs at every instant of time for any rate of pulling the rope. ■

EXAMPLE 3 A triangular frame has two fixed sides meeting at right angles. The third side has fixed length l and is constrained so each endpoint remains on one of the fixed sides but the endpoints are allowed to slide. Describe the motion of one endpoint in terms of the motion of the other endpoint and in particular, determine the velocity of one endpoint in terms of the velocity of the other endpoint.

Test your intuition by making some guesses before reading the solution. Do you think the two velocities have the same magnitude? Always? Sometimes? Never? What do you think happens to the velocity as one endpoint nears the origin (that is when the other endpoint gets as far away from the origin as possible)?

Solution: Take one of the fixed sides to be the positive y-axis and the other fixed side to be the positive x-axis and l = the length of the third side. At time t let the end points be at $(0, P(t))$ and $(Q(t), 0)$. The condition in the problem requires the distance between the endpoints to equal the constant l; that is

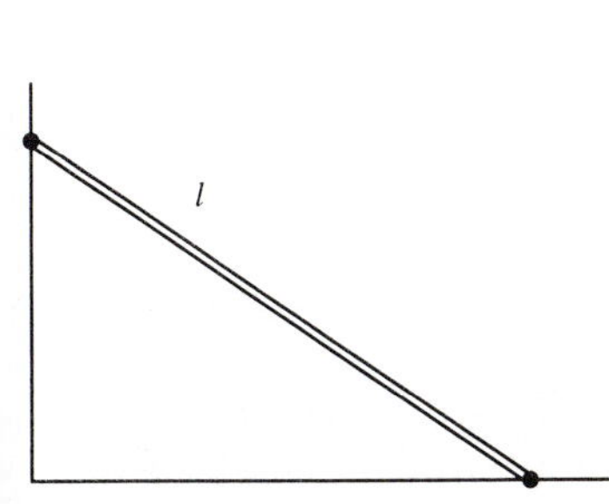

FIGURE 4.47

$$P(t)^2 + Q(t)^2 = l^2$$

for all time t. This relation between the two functions can be differentiated with respect to t to obtain a relation between the velocities of the two points:

$$2P(t)P'(t) + 2Q(t)Q'(t) = 0.$$

If we consider times when the positions and velocities are not 0, we can then rearrange this equation to yield

$$\frac{Q(t)}{P(t)} = -\frac{P'(t)}{Q'(t)}.$$

Thus at the instant of time when the triangle is isosceles, that is $P(t) = Q(t)$, we have $P'(t) = -Q'(t)$; the velocities have equal magnitudes but opposite signs. When the legs of the triangle are unequal, the velocities are not equal (in magnitude). In fact the endpoint of the longer leg always has the smaller velocity.

Let us take a specific example. Assume the endpoint on the x-axis moves with constant velocity $P'(t) = v > 0$. Then the velocity of the point on the y-axis is

$$Q'(t) = -\frac{vP(t)}{Q(t)} = -\frac{vx}{\sqrt{l^2 - x^2}},$$

where we have written $x = P(t)$ and used $\sqrt{l^2 - P(t)^2} = Q(t)$. If x is close to 0, then $Q'(t)$ is close to 0; when x is close to l, then $Q'(t)$ is very large in magnitude. ■

Did your intuition agree with these results?

Exercises 4.11

1. A walker of height h is moving directly away at velocity v from a light at height $H > h$. Express the rate of change of the length of the walker's shadow in terms of the given constants and the length of the shadow at any given time.

2. A point moves along the line $y = 2x - 3$. Find the relation between the rate of change of the x coordinate and the rate of change of the y coordinate. If the velocity in the x direction is 10 ft/sec what is the velocity in the y direction?

3. An equilateral triangle expands so its area changes at the rate of 2 square miles per year. What is the rate of change of one side expressed in terms of its area at a given time?

4. A point moves along the graph of the differentiable

function $y = f(x)$. What is the relation between the rates of change of the x coordinate and the y coordinate at a general point (x, y) on the graph?

5. A rectangle is shrinking in such a way that its length is always four times its width. If the perimeter decreases at the constant rate of 1 inch per year, how fast is the area of the rectangle changing when the perimeter is 100 miles?

6. A sphere of radius R has surface area $4\pi R^2$ and volume $4\pi R^3/3$. If the sphere is a balloon whose radius is changing at a constant rate δ per unit time, what is the rate of change of the surface area and the volume? If the volume is changing at a rate of γ per unit time, what is the rate of change of the surface area in terms of γ?

7. A cube is inscribed in a sphere whose radius increases at the rate of 1 inch per hour. Express the rate of change of the surface area of the cube in terms of the radius at a given time.

8. A 12-foot ladder leans against a vertical building. If the base of the ladder is pulled away from the wall at 1 ft/sec what is the velocity of the top of the ladder at the times when the base of the ladder is 3 ft, 6 ft, and 9 ft from the base of the wall?

9. Person Y walks along the y-axis on the interval $[2, 10]$ while person X walks along the x-axis in such a way that a tree at $(1, 1)$ is always directly between the two walkers. Find the relation between the velocities of the two walkers. What is the largest and smallest velocity of X when Y has constant velocity 2 ft/sec and where is X at these times?

10. A water container has the shape of an inverted cone with circular base of radius 10 and height 20. If water flows out of the bottom of the container at a constant rate of 1 unit of volume per unit time, how fast is the depth of the water changing inside the container? [Hint: The volume of a cone is one-third the product of the area of the base times the height.]

11. A lighthouse is 1 mile from a straight shoreline and the light revolves at the rate of four revolutions per minute. What is the function describing the motion of the point on the shoreline where the light beam is seen? What is the velocity of the point of light at the nearest point to the lighthouse and at the point one mile along the shore?

12. A row boat is attached to a dock 10 feet above the water level (see figure 4.48). If someone on the dock pulls the rope at 2 ft/sec, how fast is the boat moving toward the dock at the times when the boat is 40, 30, 20, or 10 feet away from the dock?

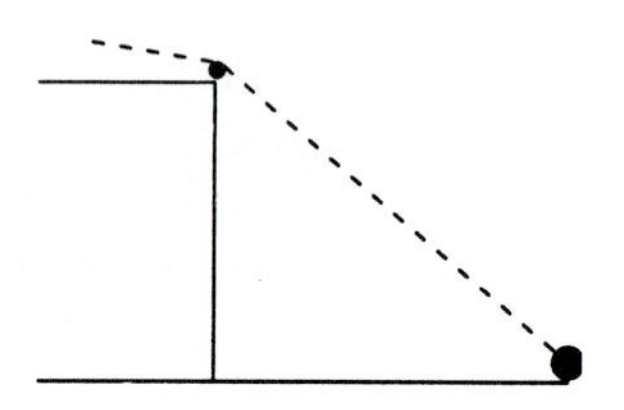

FIGURE 4.48

13. A hot air balloon rises vertically at a rate of 2 feet per second. An observer 1000 feet from the point of release watches the balloon rise. What is the rate of change of the angle of inclination above the horizontal of the observers line of sight?

14. One end of a straight shaft of length L is attached to the rim of a rotating wheel of radius 1 $(1 < L)$; the other end of the shaft slides along a straight line that passes through the center of the wheel. If the wheel rotates 1 revolution per second, describe the velocity of the sliding end of the shaft. At which points does the sliding end have the maximum and minimum velocities? Assume the point on the rim has position $(\cos t, \sin t)$ at time t and the other end of the shaft is at $(x(t), 0)$. [Comment: This is an extremely difficult problem that should be approached as a group project for discussion. It is suggested that a computer be used to generate numerical data from which reasonable guesses can be made. Attempts to verify the accuracy of the guesses might require substantial computation. Do not expect a closed-form solution.]

5 The Definite Integral

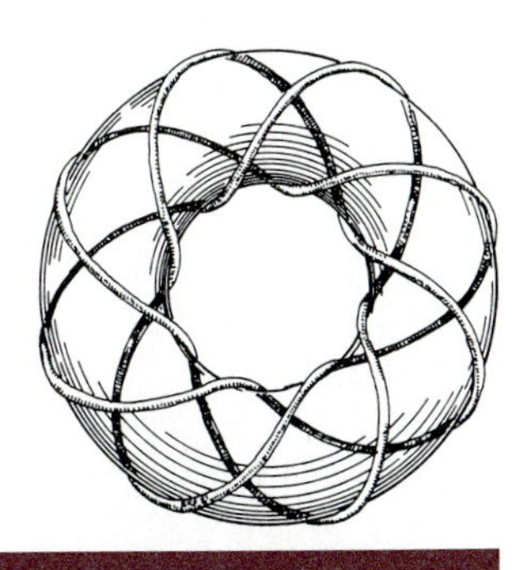

Up to this point we have studied topics dealing with differential calculus. The next major topic is the study of *integral* calculus. The use of integrals as a computational tool goes back to Isaac Newton (1642–1727), who is generally given credit for most of the original ideas in the study of calculus. Although the ideas of the derivative and integral can be found in works before Newton's, he was the first to realize the great computational power of these methods.

The first application of the definite integral that we study is the computation of areas of regions in the plane bounded by curves. Everyone in high school learns the formula πr^2 for the area of a circle of radius r; this formula, and many others like it, are derived using integral calculus. In addition to computing areas, we use the definite integral to solve a variety of problems such as computing volumes, finding the work done by variable force, computing the length of curves, computing the area of a surface in three dimensions, and even evaluating certain infinite sums. In this chapter we derive some of the basic properties of the integral.

5.1 Summation notation

Before beginning the study of the integral, we must introduce some notation that is used frequently to simplify writing sums. As the reader will soon see, summation of long strings of terms is an important tool in the discussion of the integral.

The greek letter sigma (Σ) is used as symbol for writing sums. If $f(k)$ is a function of the integer variable k, then the sum of the values $f(1)$, $f(2)$, $f(3)$ and $f(4)$ is denoted by

$$\sum_{k=1}^{4} f(k).$$

More generally,

$$\sum_{k=1}^{n} f(k) = f(1) + f(2) + f(3) + \cdots + f(n).$$

This statement is written in some computer programming languages by some descriptive phrase such as $Sum[f(k), \{k, 1, n\}]$ that means "evaluate the sum of the terms $f(k)$ as k ranges through the integers from 1 to n".

Here are a few examples of sums written using the sigma notation:

$$\sum_{k=1}^{5} k^2 = 1^2 + 2^2 + 3^2 + 4^2 + 5^2 = 55,$$

$$\sum_{j=0}^{10} (j-3) = -3 - 2 - 1 - 0 + 1 + 2 + 3 + 4 + 5 + 6 + 7 = 22,$$

$$\sum_{k=1}^{5} \frac{k+1}{k+2} = \frac{2}{3} + \frac{3}{4} + \frac{4}{5} + \frac{5}{6} + \frac{6}{7} = \frac{547}{140},$$

$$\sum_{i=3}^{n} (-1)^i i^2 = -3^2 + 4^2 - 5^2 + 6^2 - \cdots + (-1)^n n^2,$$

$$\sum_{i=1}^{n} c = c + c + c + \cdots + c = nc,$$

$$\sum_{j=-1}^{n+1} jx^j = -x^{-1} + 0 + x + 2x^2 + 3x^3 + \cdots + (n+1)x^{n+1}.$$

A *sequence* is a set of numbers indexed by the positive integers (or by the set of nonnegative integers in some cases). We use the subscript notation a_1, a_2, ..., or the more compact form a_k to denote a generic sequence.

If a_k is a sequence of numbers then the sum

$$\sum_{k=1}^{n} a_k$$

is the sum of the numbers $a_1, a_2, \ldots, a_n$. The index k plays no special role and is called a "dummy variable" because it can be replaced with any other variable without changing the meaning of the expression; i.e.,

$$\sum_{k=1}^{n} a_k = \sum_{i=1}^{n} a_i = \sum_{m=1}^{n} a_m.$$

The sum of a certain set of numbers can be expressed in many different ways using the sigma notation. The sum

$$\frac{1}{4^3} + \frac{1}{5^3} + \frac{1}{6^3} + \frac{1}{7^3} + \cdots + \frac{1}{20^3}$$

of the reciprocals of the cubes of the integers from 4 to 20 can be expressed in the forms

$$\sum_{k=4}^{20} \frac{1}{k^3} = \sum_{k=1}^{17} \frac{1}{(k+3)^3} = \sum_{j=0}^{16} \frac{1}{(j+4)^3}.$$

Many other expressions represent the same sum.

Algebraic Manipulation of Sums

If a_k and b_k are two sequences of numbers, then

$$\sum_{j=1}^{n} a_j + \sum_{j=1}^{n} b_j = \sum_{j=1}^{n} (a_j + b_j),$$

because each side of the equality stands for the sum of the $2n$ numbers a_j and b_j for $1 \leq j \leq n$. Note that the range of the dummy variable j is the same in both summations on the right side of the equation.

If we want to combine the sum

$$\sum_{k=1}^{12} k^3 + \sum_{k=3}^{12} \frac{1}{k^2}$$

into a single summation, it is necessary to account for the two extra terms in the first summation. One way to combine these is

$$\sum_{k=1}^{12} k^3 + \sum_{k=3}^{12} k^2 = 1^3 + 2^3 + \sum_{k=3}^{12} \left(k^3 + \frac{1}{k^2} \right).$$

EXAMPLE 1 Let $f(x)$ be the polynomial of degree 5 whose coefficients are alternately +1 and -1 as the powers of x increase. Write the possible $f(x)$ using the sigma notation.

Solution: If the constant term of f is +1, then

$$f(x) = 1 - x + x^2 - x^3 + x^4 - x^5 = \sum_{i=0}^{5} (-1)^i x^i.$$

If the constant term is -1, then

$$f(x) = -1 + x - x^2 + x^3 - x^4 + x^5 = \sum_{i=0}^{5} (-1)^{i+1} x^i.$$

There are many other correct solutions to this problem that can be found using some of the variations in the dummy variable or the range of the summation, for example the last sum could be written $\sum_{m=4}^{9} (-1)^m x^{m-4}$. ■

Closed Form and Telescoping Sums

The notation $\sum_{m=1}^{100} m$ is a short way to write the sum of the first 100 integers but it is just a notation for the sum and does not tell its numerical value. To *evaluate* the sum means to give its numerical value or a formula that is easily evaluated for the numerical values. Here is an example of a formula that can be used to evaluate the sum of the first 100 integers or the sum of the first n integers:

$$1 + 2 + 3 + \cdots + n = \sum_{m=1}^{n} m = \frac{n(n+1)}{2}. \qquad \textbf{(5.1)}$$

There are several ways to prove this formula; the most usual way is by mathematical induction. Here is an alternative proof in which the induction is hidden.

Let $S = 1 + 2 + 3 + \cdots + n$ so that

$$\begin{aligned} 2S &= [1 + 2 + 3 + \cdots + n] \\ &\quad + [n + (n-1) + (n-2) + \cdots + 1] \\ &= (1+n) + (2+n-1) + (3+n-2) + \cdots + (n+1) \\ &= (n+1) + (n+1) + (n+1) + \cdots + (n+1) \qquad (n \text{ times}) \\ &= n(n+1). \end{aligned}$$

Equation (5.1) follows.

A formula like Equation (5.1) is called a closed form for the summation. For most sequences a_n, a closed form for $\sum_{k=1}^{m} a_k$ is difficult, if not impossible, to find. There are some exceptions to this statement. There is a class of sequences for which the closed form is particularly easy to find. We describe these now.

If $f(k)$ is a function of the integer variable k then a sum of the form

$$\sum_{k=p}^{q} \big(f(k) - f(k+1)\big) = T$$

is called a *telescoping sum.** It is easily evaluated because most of the terms cancel leaving the result $T = f(p) - f(q+1)$. The pattern becomes fairly clear after writing down the sum from $p = 1$ to $q = 5$ representing T. Here is an evaluation using the manipulations with the sigma notation. We have

$$\begin{aligned} T &= \sum_{k=p}^{q} f(k) - \sum_{k=p}^{q} f(k+1) \\ &= \sum_{k=p}^{q} f(k) - \sum_{m=p+1}^{q+1} f(m) \\ &= \Big(f(p) + \sum_{k=p+1}^{q} f(k)\Big) - \Big(f(q+1) + \sum_{k=p+1}^{q} f(k)\Big) \\ &= f(p) - f(q+1). \end{aligned}$$

EXAMPLE 2 Evaluate the sum $\displaystyle\sum_{k=1}^{100} \frac{1}{k(k+1)}$.

Solution: The identity

$$\frac{1}{k(k+1)} = \frac{1}{k} - \frac{1}{k+1}$$

shows that the sum is a telescoping sum with $f(k) = 1/k$. Thus

$$\sum_{k=1}^{100} \frac{1}{(k+1)k} = \sum_{k=1}^{100} \left(\frac{1}{k} - \frac{1}{k+1}\right) = 1 - \frac{1}{101}.$$ ■

Here is an application of a telescoping series to find a formula for the sum of the squares of the first n integers.

*The name comes from the image of the captain of the pirate ship peering through his four-foot-long telescope then pressing the ends into the palms of his hands and collapsing the telescope to a six-inch length.

EXAMPLE 3 Derive the formula

$$\sum_{k=1}^{n} k^2 = \frac{n(n+1)(2n+1)}{6}. \qquad \textbf{(5.2)}$$

Solution: We first use a telescoping series with $f(k) = k^3$ to get

$$\sum_{k=1}^{n} \left((k+1)^3 - k^3\right) = (n+1)^3 - 1.$$

Next, expand the general term of the summation to get

$$(k+1)^3 - k^3 = k^3 + 3k^2 + 3k + 1 - k^3 = 3k^2 + 3k + 1.$$

Then we have

$$\begin{aligned}
\sum_{k=1}^{n} \left((k+1)^3 - k^3\right) &= \sum_{k=1}^{n} (3k^2 + 3k + 1) \\
&= 3\sum_{k=1}^{n} k^2 + 3\sum_{k=1}^{n} k + \sum_{k=1}^{n} 1 \\
&= 3\sum_{k=1}^{n} k^2 + 3\frac{n(n+1)}{2} + n.
\end{aligned}$$

Now combine the two expressions for the sum to get

$$3\sum_{k=1}^{n} k^2 = (n+1)^3 - 1 - \frac{3n(n+1)}{2} - n.$$

After some rearrangement of the terms over a common denominator, the formula (5.2) is obtained. ■

Geometric Sum

A sum of the form

$$S = a + ar + ar^2 + \cdots + ar^n = \sum_{k=0}^{n} ar^k$$

is called a *geometric sum.* A closed form for S can be found in a straightforward manner. Notice that if $r = 1$, then $S = a(n+1)$ because there are $n+1$ terms being summed and each equals a. From now on, we assume that $r \neq 1$. Then we have

$$\begin{aligned}
S - rS &= \sum_{k=0}^{n} ar^k - \sum_{k=0}^{n} ar^{n+1} \\
&= \left(a + \sum_{k=1}^{n} ar^k\right) - \left(\sum_{k=1}^{n} ar^k + ar^{n+1}\right) \\
&= a = ar^{n+1}.
\end{aligned}$$

Because $r \neq 1$ we obtain

$$S = \frac{a(1 - r^{n+1})}{1 - r}.$$

For example if we take $r = 1/10$ and $a = 3$, we find

$$3 + \frac{3}{10} + \frac{3}{10^2} + \cdots + \frac{3}{10^n} = \frac{3(1 - 10^{-n-1})}{1 - 10^{-1}} = \frac{10^{n+1} - 1}{3 \cdot 10^n}.$$

Exercises 5.1

Write the following sums using the sigma notation.

1. $1 - 2 + 3 - 4 + 5 - 6 + 7 - 8 + 9 - 10$

2. $1 + 3 + 5 + 7 + 9 + 11 + 13 + 15 + 17 + 19$

3. $\frac{1}{3} - \frac{2}{4} + \frac{3}{5} - \frac{4}{6} + \frac{5}{7} - \frac{6}{8} + \frac{7}{9} - \frac{8}{10} + \frac{9}{11}$

4. $x + \frac{x^2}{2} + \frac{x^3}{3} + \frac{x^4}{4} + \frac{x^5}{5} + \frac{x^6}{6} + \frac{x^7}{7}$

Evaluate the sums.

5. $\sum_{k=0}^{7}(k^2 - k)$

6. $\sum_{k=0}^{10}(2k - 6)$

7. $\sum_{k=3}^{8}\frac{1}{k-2}$

8. $\sum_{k=1}^{4}2^k$

Use some computing aid to evaluate the sums.

9. $\sum_{k=1}^{30}\frac{1}{k}$

10. $\sum_{k=1}^{30}(-1)^{k+1}\frac{1}{k}$

11. $\sum_{k=1}^{100}\frac{1}{k^2}$

12. $\sum_{k=1}^{100}(-1)^{k+1}\frac{1}{k^2}$

13. Use summation manipulations to show
$\sum_{k=p}^{q}\left(f(k+2)-f(k)\right) = f(q+2)+f(q+1)-f(p)-f(p+1)$

Use Exercise 13 to evaluate the sums.

14. $\sum_{k=1}^{20}\frac{1}{k2^k} - \frac{1}{4(k+2)2^k}$

15. $\sum_{k=1}^{100}\frac{1}{k(k+2)}$

16. Use a telescoping series $\sum(f(k+1) - f(k))$ with $f(k) = k^4$ to find a formula for the sum of the cubes of the first n integers. [Hint: This requires the formulas for sum of first and second powers.]

17. Use a telescoping series with $f(k) = k^5$ to find a formula for the sum of the 4th powers of the first n integers. [Hint: This requires the formulas for sum of the first, second and third powers.]

Use formulas in the text along with formulas derived in Exercises 16 and 17 to evaluate the sums.

18. $\sum_{k=1}^{n}(2k^2 + 3k - 4)$

19. $\sum_{k=1}^{20}(k^3 - 3k^2)$

20. $\sum_{k=1}^{25}(k^4 - 8k^3 + k^2 - 8k)$

21. $\sum_{k=1}^{50}(k + 3k^2)^2$

Find a closed form expression for the sums.

22. $\sum_{k=0}^{14}3 \cdot 5^k$

23. $\sum_{k=0}^{20}\left(\frac{1}{2}\right)^k$

24. $\sum_{k=0}^{50}\left(-\frac{2}{3}\right)^k$

25. $\sum_{k=10}^{50}\left(-\frac{2}{3}\right)^k$

26. $\sum_{k=18}^{30}(2^k - (0.2)^k)$

27. $\sum_{k=10}^{100}(1 - (0.5)^k)$

5.2 Lower and upper sums

In this section we are led to the definition of the definite integral by a natural attempt to compute the area of a region of the plane having curved boundaries.

Let us consider the problem of finding the area of the region A, bounded at the left and right by the lines $x = a$ and $x = b$, at the bottom by the x-axis, and at the top by the graph of a continuous, nonnegative function $y = f(x)$.

This problem is made difficult by the curved boundary. If the region had the horizontal line $y = c$ as its upper boundary, then A would be the area of a rectangle with sides of lengths c and $(b - a)$; thus $A = c(b - a)$. Because we know the area of a rectangle, we can use this information to *approximate* the area A of a nonrectangular region.

The function $f(x)$ is continuous so it has a minimum value $m = f(p)$ and a maximum value $M = f(q)$ on the interval $[a, b]$. The curve $y = f(x)$ then lies between the two lines $y = m$ and $y = M$. The smaller rectangle of height m is completely within the region A and the region A is completely within the larger rectangle of height M. Thus we obtain a first estimate of A:

$$f(p)(b-a) \leq A \leq f(q)(b-a), \quad \begin{array}{l} f(p) = m = \text{min of } f \text{ on } [a,b], \\ f(q) = M = \text{max of } f \text{ on } [a,b]. \end{array} \tag{5.3}$$

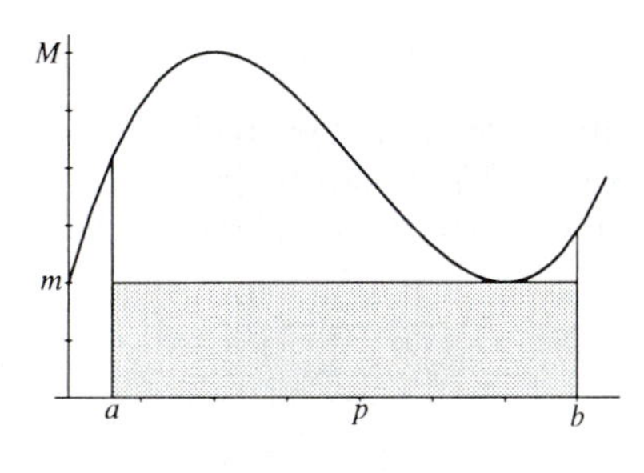

FIGURE 5.1

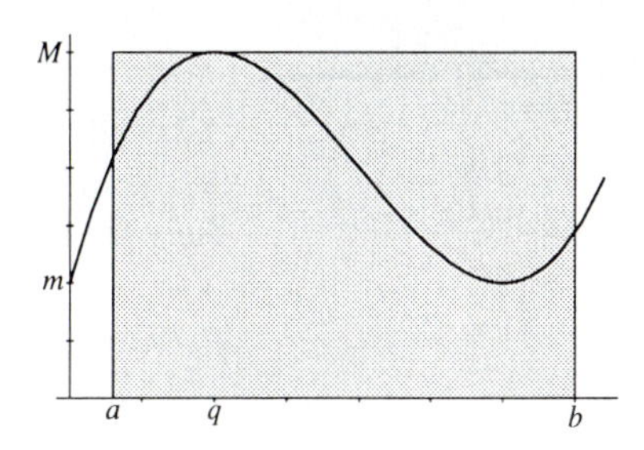

FIGURE 5.2

This is only a crude estimate of the area of the region; but we can also estimate how far off the approximation is from the truth. The error involved in this approximation is no worse than the distance between the two extremes in Equation (5.3); that is the number we wish to compute, the area A, is known to be a point on the interval $[f(p)(b-a), f(q)(b-a)]$ of length $(f(q)-f(p))(b-a)$. The difference $f(q)-f(p)$ is the difference between the maximum and minimum values of f on $[a,b]$. The simple method just described can be summarized as follows:

- The area A of the region bounded by the lines $x = a$, $x = b$ and $y = 0$ and the graph of the nonnegative, continuous function $y = f(x)$ lies between the numbers $m(b-a)$ and $M(b-a)$, where m = minimum value of $f(x)$ on the interval $[a,b]$ and M = maximum value of $f(x)$ on the interval $[a,b]$.
- The approximation $A \approx m(b-a)$ underestimates A with an error at most $(M-m)(b-a)$.
- The approximation $A \approx M(b-a)$ overestimates A with an error at most $(M-m)(b-a)$.

In the case where the minimum and maximum values of $f(x)$ are very close to each other, $(M-m)$ is very small and the approximations to A by using either $m(b-a)$ or $M(b-a)$ might be very good. We exploit a property of continuous functions that can be described roughly as follows: Over a very short interval, the maximum and minimum values of a continuous function differ by a very small amount.

Here is how this idea can be used to improve the first estimate. Instead of approximating the region with a rectangle of width $b-a$ and height $f(p)$ inside the region, we break the interval into two subintervals $[a,c]$ and $[c,b]$ with $a < c < b$. We then apply the approximation method to each of the subintervals. It might happen that the minimum value of $f(x)$ on one of the subintervals is larger than the minimum of $f(x)$ on all of $[a,b]$; thus one of the two lower rectangles contributes a bit more of the actual area than the original single rectangle of width $b-a$. As a result the lower approximation might be closer to the actual area, and certainly will be no worse than the approximation with a single rectangle. Similarly, the area enclosed by the two upper rectangles is be a bit less than the single rectangle but is larger than the actual area of the region under the curve.

If an improved estimate was obtained by partitioning the interval $[a, b]$ into two subintervals $[a, c]$ and $[c, b]$, then we could improve again by further dividing each of the two subintervals into two subintervals. Of course there is no particular need to divide into two intervals each time. To develop this idea systematically, we begin with the notion of a *partition* of the interval. A partition of the interval $[a, b]$ is a sequence

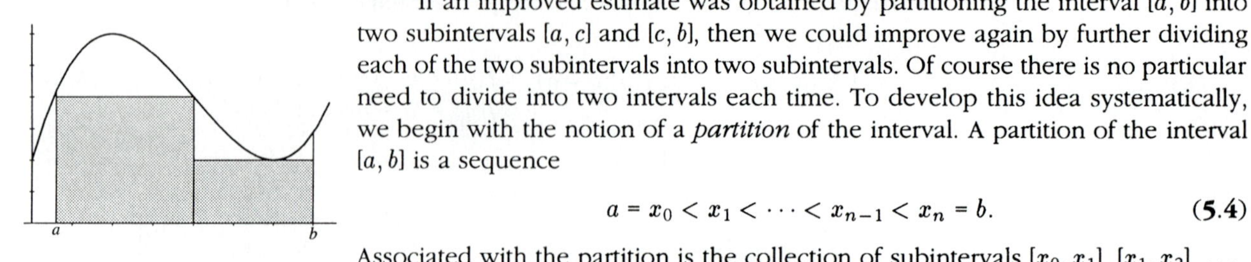

FIGURE 5.3

$$a = x_0 < x_1 < \cdots < x_{n-1} < x_n = b. \qquad \textbf{(5.4)}$$

Associated with the partition is the collection of subintervals $[x_0, x_1]$, $[x_1, x_2]$, ..., $[x_{n-1}, x_n]$ of $[a, b]$. If all the intervals $[x_{i-1}, x_i]$ $(1 \leq i \leq n)$ have the same length, the partition is called a *regular partition*; in this case we have

$$a = x_0 \quad x_1 \quad x_2 \quad \cdots \quad x_{n-1} \quad x_n = b$$

$$x_i - x_{i-1} = \frac{b-a}{n} \quad \text{for } i = 1, 2, \ldots, n.$$

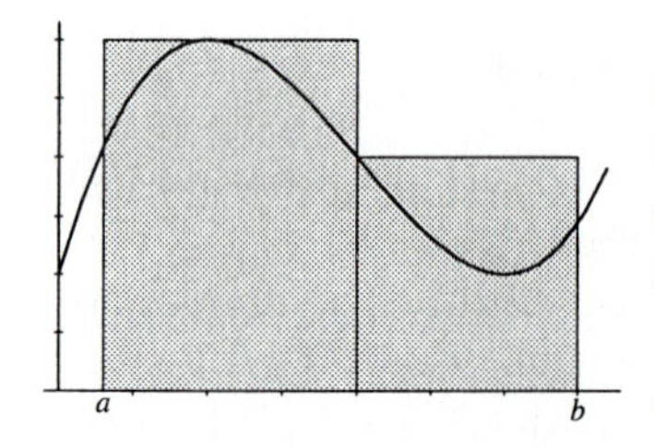

FIGURE 5.4

Let A_i denote the area under the curve $y = f(x)$ and above the ith interval $[x_{i-1}, x_i]$; using the idea summarized above, let m_i and M_i be the minimum and maximum values of $f(x)$ over the interval $[x_{i-1}, x_i]$. Then A_i satisfies the inequality

$$m_i(x_i - x_{i-1}) \leq A_i \leq M_i(x_i - x_{i-1}).$$

The area A is the sum of the regions over the intervals of the partition, so

$$A = A_1 + A_2 + \cdots + A_n = \sum_{i=1}^{n} A_i.$$

Now add the inequalities to obtain

$$\sum_{i=1}^{n} m_i(x_i - x_{i-1}) \leq A \leq \sum M_i(x_i - x_{i-1}).$$

This becomes somewhat simpler when the partition is regular.

When the partition is regular, each difference $x_i - x_{i-1}$ has the value $(b-a)/n$ and so the above estimate of the area becomes

$$(m_1 + m_2 + \cdots + m_n)\left(\frac{b-a}{n}\right) \leq A \leq (M_1 + M_2 + \cdots + M_n)\left(\frac{b-a}{n}\right). \qquad \textbf{(5.5)}$$

Now we give an interpretation of this estimate. When the number n of subintervals in the regular partition is very large, the intervals $[x_{i-1}, x_i]$ are very short; the difference $(M_i - m_i)$ between the maximum and minimum values of $f(x)$ on this interval are very small. The difference between the two extreme sides of inequality (5.5) is just the average of the small terms, $(M_i - m_i)(b-a)$. If we let

$$\frac{(M_1 - m_1) + (M_2 - m_2) + \cdots + (M_n - m_n)}{n} = E,$$

then the approximation of A by either side of inequality (5.5) involves an error of at most

$$\frac{(M_1 - m_1) + (M_2 - m_2) + \cdots + (M_n - m_n)}{n}(b-a) = (b-a)E.$$

But the term $(b-a)$ is constant whereas it seems that E can be made as small as we like by selecting a partition with large enough n.

Thus it should seem reasonable that as $n \to \infty$, the extreme sides of the inequality (5.5) approach the same limit and this limit is the area A under the curve. We have not *proved* that the two limits are equal, of course, but this should indicate that it is reasonable. (It will be proved later that E approaches 0 as n gets large.)

Using this idea as motivation we now give the definition of the *lower sums* and *upper sums* of $f(x)$ over $[a, b]$. Although the motivation required the function to take on only nonnegative values, we can make the definitions for more general functions; *we allow functions $f(x)$ that take on positive and/or negative values.*

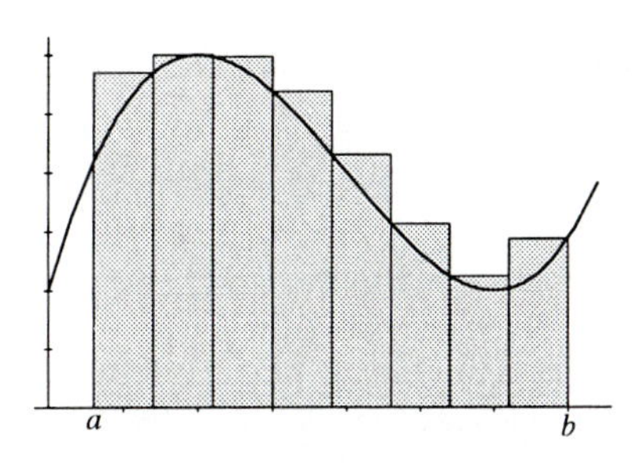

FIGURE 5.5
Upper sum for P_8

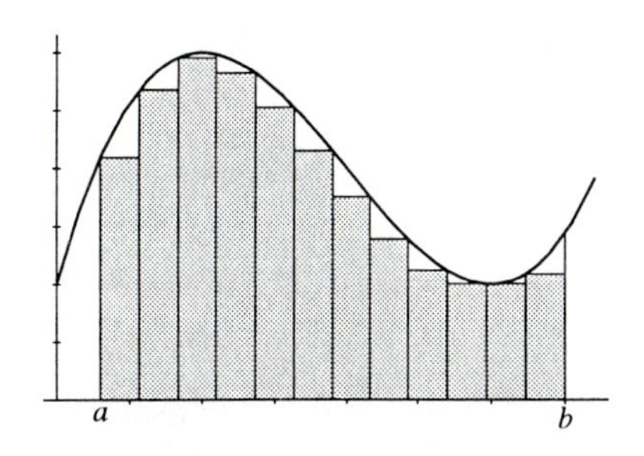

FIGURE 5.6
Lower sum for P_{12}

Let P_n denote the partition of $[a, b]$ into n subintervals:

$$P_n : a = x_0 < x_1 < \cdots < x_{n-1} < x_n = b, \tag{5.6}$$

and let points p_i, q_i be selected on the interval $[x_{i-1}, x_i]$ such that $f(p_i)$ is the minimum value of f on $[x_{i-1}, x_i]$ and $f(q_i)$ is the maximum value of f on $[x_{i-1}, x_i]$. Now form the sums

$$L(P_n) = f(p_1)(x_1 - x_0) + f(p_2)(x_2 - x_1) + \cdots + f(p_n)(x_n - x_{n-1}), \tag{5.7}$$

$$U(P_n) = f(q_1)(x_1 - x_0) + f(q_2)(x_2 - x_1) + \cdots + f(q_n)(x_n - x_{n-1}). \tag{5.8}$$

Then $L(P_n)$ is called the *lower sum* of $f(x)$ over $[a, b]$ for the partition P_n. Similarly $U(P_n)$ is called the *upper sum* of $f(x)$ for P_n.

In the case that $f(x)$ is nonnegative on $[a, b]$, the area A below the curve $y = f(x)$ and above the interval $[a, b]$ satisfies the inequalities $L(P_n) \leq A$ and $A \leq U(P_m)$ for any two partitions, P_n and P_m.

For any continuous function $f(x)$ that might take on negative and positive values, we have

$$L(P_n) \leq U(P_n). \tag{5.9}$$

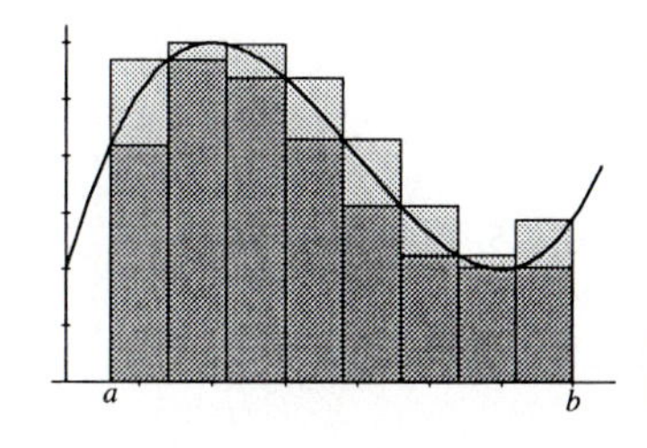

FIGURE 5.7
Upper and lower sums

This is easily proved as follows. For every i, $f(p_i) \leq f(q_i)$ because the minimum value of f is no larger than the maximum value of f on the interval $[x_{i-1}, x_i]$. It follows that

$$f(p_i)(x_i - x_{i-1}) \leq f(q_i)(x_i - x_{i-1})$$

for each $i = 1, 2, \ldots, n$. By adding these inequalities, we obtain inequality (5.9).

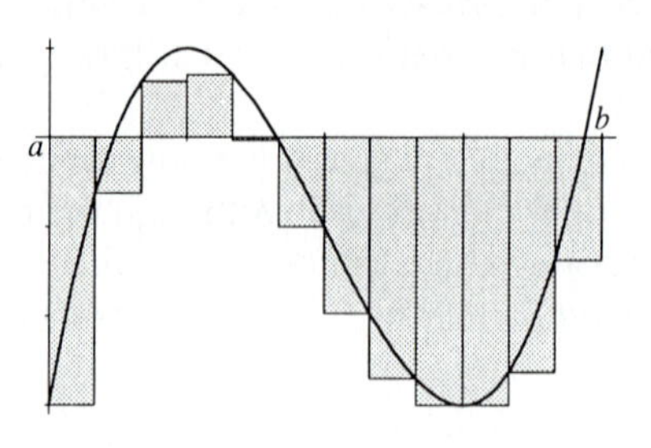

FIGURE 5.8
Lower sum for a general $f(x)$

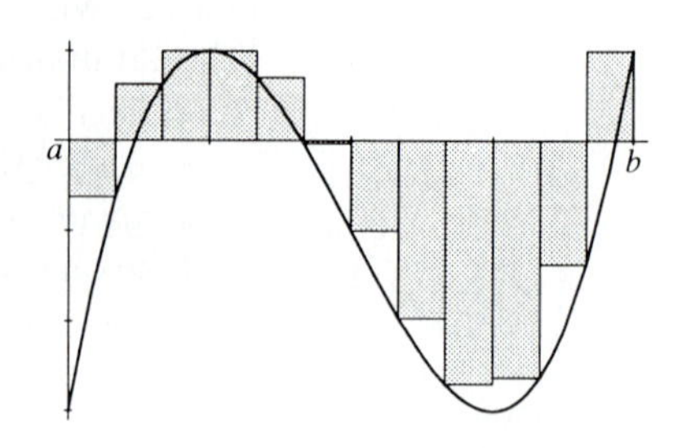

FIGURE 5.9
Upper sum for general $f(x)$

EXAMPLE 1 Compute some lower and upper sums for the function $f(x) = 3x^2 - 1$ over $[1, 2]$ and thereby estimate the area of the region bounded by the lines $x = 1$, $x = 2$, $y = 0$ and the graph of $3x^2 - 1$.

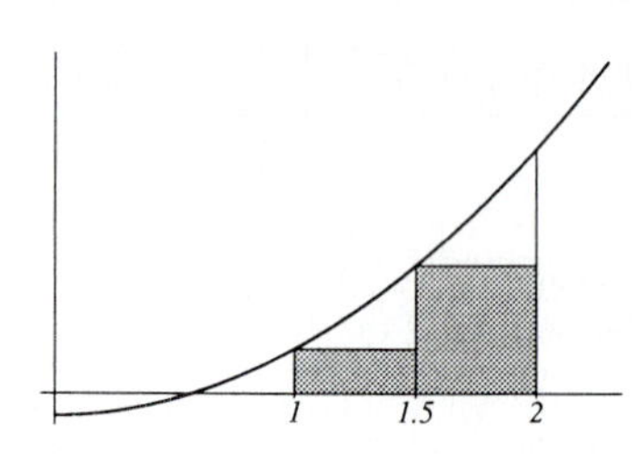

FIGURE 5.10
$L(P_2)$ for $3x^2 - 1$

Solution: We compute $L(P_n)$ and $U(P_n)$ for two partitions P_n of $[1, 2]$. In doing so, it is necessary to find the minimum and maximum of $3x^2 - 1$ over each subinterval of the partition. The function $3x^2 - 1$ is increasing (its derivative is positive over the entire interval) and so the minimum of $3x^2 - 1$ over $[x_i, x_{i+1}]$ occurs at the left endpoint x_i and the maximum occurs at the right endpoint, x_{i+1}.

We begin with a partition into two subintervals of length 1/2:

$$P_2 : 1 < \frac{3}{2} < 2.$$

The lower and upper sums are

$$\begin{aligned} L(P_2) &= \frac{1}{2}f(1) + \frac{1}{2}f\left(\frac{3}{2}\right) \\ &= \frac{1}{2}[3(1)^2 - 1 + 3\left(\frac{3}{2}\right)^2 - 1] = \frac{31}{8} = 3.875, \end{aligned}$$

$$\begin{aligned} U(P_2) &= \frac{1}{2}f\left(\frac{3}{2}\right) + \frac{1}{2}f(2) \\ &= \frac{1}{2}\left[3\left(\frac{3}{2}\right)^2 - 1 + 3(2)^2 - 1\right] = \frac{67}{8} = 8.375. \end{aligned}$$

If A denotes the area of the region, then $3.875 < A < 8.375$. Of course, this is not very precise information about the value of A. We can get better information by using a finer partition. Next, we use subintervals of length 1/4:

$$P_4 : 1 < \frac{5}{4} < \frac{6}{4} < \frac{7}{4} < 2.$$

The lower and upper sums for P_4 are

$$L(P_4) = \frac{1}{4}f(1) + \frac{1}{4}f\left(\frac{5}{4}\right) + \frac{1}{4}f\left(\frac{6}{4}\right) + \frac{1}{4}f\left(\frac{7}{4}\right)$$
$$= \frac{1}{4}\left[3(1)^2 - 1 + 3\left(\frac{5}{4}\right)^2 - 1 + 3\left(\frac{6}{4}\right)^2 - 1 + 3\left(\frac{7}{4}\right)^2 - 1\right]$$
$$= \frac{125}{32} \approx 3.906,$$

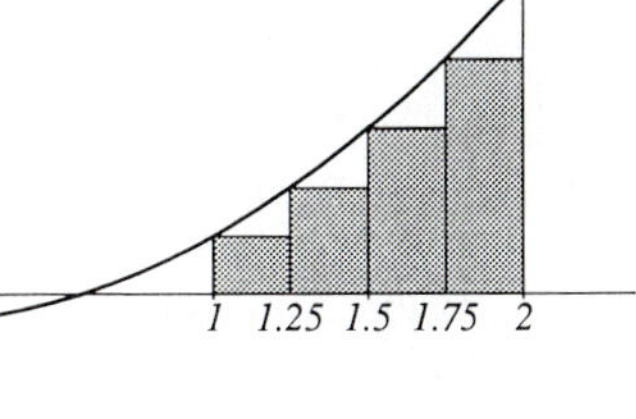

FIGURE 5.11
$L(P_4)$ for $3x^2 - 1$

$$U(P_4) = \frac{1}{4}f\left(\frac{5}{4}\right) + \frac{1}{4}f\left(\frac{6}{4}\right) + \frac{1}{4}f\left(\frac{7}{4}\right) + \frac{1}{4}f(2)$$
$$= \frac{1}{4}\left[3\left(\frac{5}{4}\right)^2 - 1 + 3\left(\frac{6}{4}\right)^2 - 1 + 3\left(\frac{7}{4}\right)^2 - 1 + 3(2)^2 - 1\right]$$
$$= \frac{229}{32} \approx 7.156.$$

This computation slightly improves the approximation of A in that we now know $3.906 < A < 7.156$.

As the number of subintervals gets larger, the amount of computation grows rapidly. A calculator or computer can be used to relieve the tedium. Some computer-generated lower and upper sums for regular partitions are given in table 5.1.

TABLE 5.1
Lower and Upper Sums for $3x^2 - 1$

n	$L(P_n)$	$U(P_n)$
2	3.87500	8.37500
4	4.90625	7.15625
8	5.44531	6.57031
12	5.62847	6.37847
16	5.72070	6.28320
20	5.77625	6.22625
50	5.91020	6.09020
100	5.95505	6.04505
200	5.97751	6.02251
300	5.98501	6.01501
1000	5.99550	6.00450

The values in the table suggest that as n increases, the lower sums increase toward 6 and the upper sums decrease toward 6. By applying the Fundamental Theorem of Calculus given in the next section, we prove that the area A is exactly 6. ■

Exercises 5.2

Compute the lower and upper sums for $f(x)$ and the given partition. Verify in each case that $L(P) \leq U(P)$.

1. $f(x) = 2x + 4,\ P : -2 < 0 < 2$

2. $f(x) = 2x + 4,\ P : -2 < -1 < 0 < 1 < 2$

3. $f(x) = x^2,\ P : -1 < -0.5 < 0 < 0.5 < 1$

4. $f(x) = x^2,\ P : -1 < -0.75 < -0.5 < 0 < 0.5 < 0.75 < 1$

5. $f(x) = \frac{1}{x},\ P : 1 < 1.5 < 2 < 2.5 < 3$

6. $f(x) = \frac{1}{x}$, $P : 1 < 1.25 < 1.5 < 1.75 < 2 < 2.25 < 2.5 < 2.75 < 3$

Each function below is decreasing on the given interval and so the maximum and minimum values of the function on any subinterval occur at the left endpoint and the right endpoint, respectively. Use a regular partition with n subintervals and compute the lower and upper sums and then compute the difference between the upper and lower sums for each choice of n.

7. $\frac{1}{x^2}$; $[1, 2]$; $n = 4, 8$

8. $\sin \pi x$; $[0.5, 1]$; $n = 1, 2$

9. $\cos \pi x$; $[0, 0.5]$; $n = 1, 2$

10. $\frac{1}{1 + x^2}$; $[0, 1]$; $n = 2, 4$

Using a calculator or computer, make a table of lower and upper sums for the given functions using regular partitions with n subintervals of the given interval.

11. $f(x) = x^2$; $[0, 1]$; $n = 4, 8, 16$

12. $f(x) = x^2$; $[-1, 1]$; $n = 4, 8, 16$

13. $f(x) = x^3$; $[0, 1]$; $n = 4, 8, 16$

14. $f(x) = x^3$; $[-1, 1]$; $n = 4, 8, 16$

In the next four exercises, $L(P)$ and $U(P)$ are supposed to be the lower and upper sums for some unknown function f over the interval $[0, 2]$. In each case there is something wrong and no such f exists. Explain why such numbers cannot be the lower or upper sums for any continuous function. Draw pictures to help your intuition.

15. $L(P) = -1.95$, $U(P) = -1.98$, $P : 0 < 1.1 < 2$

16. $L(P) = 11.43$, $U(P) = 11.34$, $P : 0 < 0.7 < 1.2 < 2$

17. $L(P) = 7$, $P : 0 < 1 < 2$, and $L(P') = 5.75$, $P' : 0 < 0.5 < 1 < 1.5 < 2$

18. $U(P) = 13.475$, $P : 0 < 0.6 < 1.4 < 2$, and $U(P') = 13.75$, $P' : 0 < 0.2 < 0.6 < 1.4 < 1.8 < 2$

19. In Exercises 1 and 2 of this section compute the exact value of the area under the graph of the function and the indicated interval of the x-axis. Then show that the area equals the average $(L(P) + U(P))/2$ of the lower and upper sums. Draw a picture to help explain why this happens and why it will always happen if the graph is a straight line.

20. Suppose $f(x)$ is an increasing function on $[a, b]$ and P_n is a regular partition of $[a, b]$ into n subintervals. Write an expression for the upper and lower sums for f for this partition.

21. Let f be a continuous function on $I = [a, b]$ and suppose there is a partition P of I such that the lower and upper sums of f for P are equal; that is $U(P) = L(P)$. Prove that f is constant on I.

5.3 Definition of the definite integral

For any function $f(x)$ that is continuous on an interval $[a, b]$, we can define the lower and upper sums. The next theorem shows that there is exactly one real number that is greater than or equal to every lower sum and is also less than or equal to every upper sum. This number is called the definite integral of the function over the interval.

THEOREM 5.1

Existence of the Definite Integral

Let $f(x)$ be a continuous function over the interval $[a, b]$. There is a unique real number that is greater than or equal to every lower sum for $f(x)$ over $[a, b]$ and is also less than or equal to every upper sum for $f(x)$ over $[a, b]$.

The proof of this theorem is given in section 5.4.

DEFINITION 5.1

The Definite Integral

If $f(x)$ is continuous on $[a, b]$, the *definite integral of* $f(x)$ over $[a, b]$ is the unique real number that is greater than or equal to every lower sum and less than or equal to every upper sum of $f(x)$ over $[a, b]$. This number is denoted by the symbol

$$\int_a^b f(x)\,dx,$$

which is read "the integral from a to b of $f(x)\,dx$".

Using the notation for definite integral of $f(x)$ just introduced, we restate the property in the previous theorem as follows:

$$L(P_n) \leq \int_a^b f(x)\,dx \leq U(P_n)$$

for every partition P_n of $[a, b]$.

In the definition of $\int_a^b f(x)\,dx$, it is assumed that $a < b$. It will be helpful to have the integral defined even when $a \geq b$. If $a > b$, then $\int_b^a f(x)\,dx$ is defined as

$$\int_a^b f(x)\,dx = -\int_b^a f(x)\,dx.$$

In case the upper and lower limits are equal to a, the original definition allows us to compute the integral. On the interval $[a, a]$ there is only one lower sum and only one upper sum; namely $f(a)(a - a) = 0$. Thus

$$\int_a^a f(x)\,dx = 0.$$

The next property of the integral is more subtle and requires proof.

THEOREM 5.2

Additive Property of the Integral

If $f(x)$ is continuous on an interval containing the three numbers a, b and c, then

$$\int_a^b f(x)\,dx + \int_b^c f(x)\,dx = \int_a^c f(x)\,dx.$$

Proof

The proof of this is carried out only in the case $a < b < c$. The proof in the other possible cases can be derived from this one. For simplicity of notation, let

$$S = \int_a^b f(x)\,dx + \int_b^c f(x)\,dx.$$

By definition, $\int_a^c f(x)\,dx$ is the unique number greater than or equal to every lower sum of $f(x)$ over $[a, c]$ and also less than or equal to every upper sum of $f(x)$ over $[a, c]$. We prove that the sum

$$S = \int_a^b f(x)\,dx + \int_b^c f(x)\,dx$$

has this same property and so it must equal $\int_a^c f(x)\,dx$.

Let $P : a = x_0, x_1, \ldots, x_n = c$ be a partition of $[a, c]$. Suppose for the moment that one of the points in the partition equals b, say $x_k = b$. Then the lower sum for $f(x)$ on the interval is

$$L(P) = f(p_1)(x_1 - x_0) + f(p_2)(x_2 - x_1) + \cdots + f(p_n)(x_n - x_{n-1}).$$

Because $x_k = b$, this is a lower sum for $f(x)$ over $[a, b]$ plus a lower sum over $[b, c]$. Each of these is bounded above by the integral of $f(x)$ over the appropriate interval; thus we obtain

$$\begin{aligned} L(P) &= f(p_1)(x_1 - x_0) + \cdots + f(p_k)(b - x_{k-1}) \\ &\qquad + f(p_{k+1})(x_{k+1} - b) + \cdots + f(p_n)(x_n - x_{n-1}) \\ &\le \int_a^b f(x)\,dx + \int_b^c f(x)\,dx. \end{aligned} \tag{5.10}$$

This proves that S is greater than or equal to every lower sum for a partition containing b as one of its points.

Now suppose that P is a partition not containing b as an endpoint of one of its intervals. Create a new partition by inserting b:

$$P' : a = x_0 < x_1 < \cdots < x_k < b < x_{k+1} < \cdots < x_n = c.$$

Then P' is a finer partition than P and the lower sum corresponding to P is no greater than the lower sum corresponding to P'; that is

$$L(P) \le L(P').$$

Now P' is a partition of $[a, c]$ containing b as an endpoint of one of the subintervals so we conclude $L(P') \le S$ by Equation (5.10). By combining this and the previous inequality, we obtain

$$L(P) \le S = \int_a^b f(x)\,dx + \int_b^c f(x)\,dx$$

for every partition P of $[a, c]$.

By applying the analogous reasoning to a consideration of upper sums, we conclude

$$S = \int_a^b f(x)\,dx + \int_b^c f(x)\,dx \le U(P)$$

for every upper sum $U(P)$ of $f(x)$ over $[a, c]$. Thus S is the unique number greater than or equal to every lower sum and less than or equal to every upper sum; that is,

$$\int_a^c f(x)\,dx = S = \int_a^b f(x)\,dx + \int_b^c f(x)\,dx. \qquad \square$$

If $f(x)$ is positive on $[a, b]$, then the area A of the region bounded by the graph of f, the x-axis, and the lines $x = a$ and $x = b$ satisfies the inequalities

$$L(P) \le A \quad \text{and} \quad A \le U(P),$$

for any partition P of $[a, b]$. Thus the area equals the definite integral of f over $[a, b]$. The definition of the definite integral does not give an efficient way of computing this number. It does little good to say that

$$A = \int_1^2 3x^2 - 1\,dx$$

is the area under the curve $y = 3x^2 - 1$ and above the interval $[1, 2]$ unless we have some means of evaluating that number. Of course we could appeal to the definition that gives this number in terms of lower and upper sums, but the direct application of the definition involves the evaluation of infinitely many lower and upper sums. One can hope to evaluate integrals of only the simplest functions using the definition. Fortunately there is an alternate and very practical method for solving this problem. The solution is the content of the Fundamental Theorem of Calculus (Theorem 5.4), which makes a remarkable connection between the integral $\int_a^b f(x)\,dx$ and the functions having $f(x)$ as derivative. The connection is the antiderivative.

DEFINITION 5.2

Antiderivative

Let $f(x)$ be a function defined on an interval $[a, b]$. A function $F(x)$ is an antiderivative of $f(x)$ over $[a, b]$ if F is continuous on $[a, b]$ and $F'(x) = f(x)$ for every x on (a, b).

It is possible to find an antiderivative of some functions by inspecting differentiation rules. For example,

$$\frac{d}{dx}x^3 = 3x^2$$

implies that x^3 is an antiderivative of $3x^2$ over any interval. More generally,

$$\frac{d}{dx}x^{n+1} = (n+1)x^n$$

for any number n. If $n + 1 \neq 0$, then we see that

$$\frac{x^{n+1}}{n+1} \quad \text{is an antiderivative of } x^n.$$

If $F(x)$ is an antiderivative for $f(x)$, then $F(x) + c$ is also an antiderivative of $f(x)$ for any constant c. Theorem 4.7 from section 4.3 implies that every antiderivative of $f(x)$ must have this form. Thus if one antiderivative of $f(x)$ is known, then all others are obtained by adding a constant to the one already found.

The definition of antiderivative requires a function f and an interval. We frequently omit reference to the interval if we mean an antiderivative of f on any interval where f is continuous.

The next theorem summarizes a few properties of antiderivatives.

THEOREM 5.3

1. If $F(x)$ is an antiderivative of $f(x)$ and a is any constant, then $aF(x)$ is an antiderivative of $af(x)$.
2. If $F(x)$ is an antiderivative of $f(x)$ and $G(x)$ is an antiderivative of $g(x)$, then $F(x) + G(x)$ is an antiderivative of $f(x) + g(x)$.

Proof

If $F(x)$ is an antiderivative of $f(x)$, and if a is any constant, then

$$\big(aF(x)\big)' = aF'(x) = af(x).$$

If $F(x)$ is an antiderivative of $f(x)$ and $G(x)$ is an antiderivative of $g(x)$, then

$$\big(F(x) + G(x)\big)' = F'(x) + G'(x) = f(x) + g(x). \qquad \square$$

EXAMPLE 1 Find an antiderivative of $f(x) = 3x^5 - 2x + 8x^{-3}$.

Solution: An antiderivative for the sum of the three terms is found by taking the sum of antiderivatives for each individual term. The individual terms of the form cx^n have antiderivative $cx^{n+1}/(n+1)$ for $n + 1 \neq 0$. The solution is

$$F(x) = 3\frac{x^6}{6} - 2\frac{x^2}{2} + 8\frac{x^{-2}}{-2} = \frac{x^6}{2} - x^2 - 4x^{-2}.$$

We check the result by computing the derivative of $F(x)$:

$$F'(x) = \frac{6x^5}{2} - 2x + (-2)(-4)x^{-3} = 3x^5 - 2x + 8x^{-3} = f(x). \qquad \blacksquare$$

Now we can state the theorem that relates the definite integral and antiderivatives. This theorem is called the *Fundamental Theorem* because it describes a relation between the two basic concepts of calculus: the derivative and the integral.

THEOREM 5.4

Fundamental Theorem of Calculus

Let $f(x)$ be continuous on the interval $[a, b]$ and let $F(x)$ be any antiderivative of $f(x)$. Then

$$\int_a^b f(x)\,dx = F(b) - F(a).$$

Notice the power of this result. We are able to evaluate the area A mentioned at the beginning of this section by simply finding an antiderivative for $f(x) = 3x^2 - 1$. By inspection we find $F(x) = x^3 - x$ is an antiderivative for $3x^2 - 1$ and so

$$\int_1^2 3x^2 - 1\,dx = F(2) - F(1) = (2^3 - 2) - (1^3 - 1) = 6.$$

The effort required for this exact computation should be compared with the effort required to compute the lower and upper sum estimates given in the previous section.

The proof and further discussion of the Fundamental Theorem are given in section 5.6.

Exercises 5.3

Find all antiderivatives of the following functions.

1. $f(x) = 4x^3 + 2x$

2. $f(x) = 12x^{-7} - \dfrac{4}{x^5}$

3. $f(x) = \sin x + \cos 2x$

4. $f(x) = \dfrac{x^2 + 2x - 4}{x^4}$

5. $f(t) = 2t^4 - \cos t$

6. $f(w) = (w + 1)^3$

7. $f(z) = (z - 4)^4$

8. $f(t) = 3(t + 2)^2$

9. $f(x) = \sqrt{4z}$

10. $f(w) = \dfrac{5}{\sqrt{1 + w}}$

11. $f(x) = 5x^{2/3} - 3x^{3/2}$

12. $f(t) = \dfrac{t^{1/2} + t^{5/2}}{t^2 + 1}$

Apply the Fundamental Theorem to evaluate the definite integrals.

13. $\displaystyle\int_1^2 4x^3 + 2x\,dx$

14. $\displaystyle\int_1^3 12x^{-7} - \frac{4}{x^5}\,dx$

15. $\displaystyle\int_0^\pi \sin x\,dx$

16. $\displaystyle\int_{-2}^{-1} \frac{x^2 + 2x - 4}{x^4}\,dx$

17. $\displaystyle\int_0^{\frac{\pi}{2}} t^4 - \cos t\,dt$

18. $\displaystyle\int_1^2 (w + 1)^3\,dw$

19. $\displaystyle\int_4^5 (z - 4)^4\,dz$

20. $\displaystyle\int_0^2 t^4 + 2t^2 - 4\,dt$

21. $\displaystyle\int_1^3 \sqrt{4z}\,dz$

22. $\displaystyle\int_3^8 \frac{5\,dw}{\sqrt{1 + w}}$

23. $\displaystyle\int_8^{16} 5x^{2/3} - 3x^{3/2}\,dx$

24. $\displaystyle\int_0^4 \frac{t^{1/2} + t^{5/2}}{t^2 + 1}$

Find a closed form expression for the sums.

25. $\displaystyle\sum_{k=1}^{n} \int_k^{k+1} x\,dx$

26. $\displaystyle\sum_{k=1}^{n} \int_k^{k+1} f(x)\,dx$

Use a calculator or computer to estimate the following integrals by taking a regular partition P with some number of subintervals and trapping the integral between $L(P)$ and $U(P)$ so that $U(P) - L(P) \leq 0.05$. Can you replace the tolerance 0.05 with a smaller number (like 10^{-5}) and still do the estimate in a reasonable time?

27. $\displaystyle\int_1^2 \frac{dx}{x}$

28. $\displaystyle\int_0^1 \frac{dx}{1 + x^2}$

29. Return to the exercises at the end of section 5.2 and apply the Fundamental Theorem of Calculus to determine the exact value of the areas that were approximated by lower and upper sums.

30. Evaluate $\displaystyle\int_{-1}^{1} |x|\,dx$. [Hint: $\displaystyle\int_{-1}^{1} = \int_{-1}^{0} + \int_{0}^{1}$.]

31. Evaluate $\displaystyle\int_{-10}^{3} (x + |x|)\,dx$.

5.4 Proof of the existence of the definite integral

The proof of Theorem 5.1 is rather long and requires the introduction of two new ideas, the completeness property of the real numbers and the concept of uniform continuity. We introduce the first of these immediately; the second is introduced during the course of the proof.

Completeness Property of the Real Numbers

Let S be a nonempty set of real numbers. If there is a number b such that $x \leq b$ for every x in S, then there is a smallest number B with the property $x \leq B$ for every x in S. In other words, if the set S has an upper bound, then it has a least upper bound.

Similarly, if there is a number m with $m \leq x$ for every x in S, then there is a largest number M with the property $M \leq x$ for every x in S. In other words, if the set S has a lower bound, then it has a greatest lower bound.

Now we turn to the proof of Theorem 5.1. Let M be the maximum value of $f(x)$ on $[a, b]$; let P_n be any partition of $[a, b]$. We keep the notation given in the equations (5.7) and (5.8). Then $L(P_n) \leq M(b-a)$. This is easily seen because $f(p_i) \leq M$ for each i and so the definition of $L(P_n)$, equation (5.7), implies this inequality. Thus the set of numbers consisting of all lower sums of $f(x)$ over $[a, b]$ is bounded above by $M(b-a)$. Hence, by the Completeness Property, the set of lower sums has a least upper bound that we denote by $I_{\text{lower}}(f(x))$. This can be restated as

$$L(P_n) \leq I_{\text{lower}}(f(x)) \tag{5.11}$$

for every partition, P_n. Moreover, $I_{\text{lower}}(f(x))$ is the smallest number with this property.

Similarly, if m is the minimum value of $f(x)$ on $[a, b]$, then

$$m(b-a) \leq U(P_m) \tag{5.12}$$

for every partition, P_m. Thus there is a real number $I_{\text{upper}}(f(x))$ that satisfies

$$I_{\text{upper}}(f(x)) \leq U(P_m) \tag{5.13}$$

for every partition, P_m and $I_{\text{upper}}(f(x))$ is the largest number with this property.

The important point in Theorem 5.1 is that the same number I satisfies both inequalities (5.11) and (5.13). We eventually prove $I_{\text{lower}} = I_{\text{upper}}$ and this common value is the number I.

The proof is broken into several intermediate results; first we show that $I_{\text{lower}}(f(x)) \leq I_{\text{upper}}(f(x))$ and then finally show the equality of these numbers. An important tool is the idea of a *refinement* of a partition. Let

$$P : a = x_0 < x_1 < \cdots < x_k = b$$
$$Q : a = u_0 < u_1 < \cdots < u_r = b$$

be two partitions of $[a, b]$. Then Q is a *refinement* of P if each point x_i of P is one of the points of Q. It may be helpful to think of the refinement Q as obtained from P by the addition of more points. For example, the partition

$$Q : 0 < 0.5 < 1 < 1.25 < 1.5 < 2$$

is a refinement of the partition

$$P : 0 < 1 < 2,$$

obtained by the addition of three points.

We now begin the proof of Theorem 5.1. There are four steps, the first three of which give properties of partitions.

Step 1. If the partition Q is a refinement of the partition P, then

$$L(P) \leq L(Q) \qquad \text{and} \qquad U(Q) \leq U(P).$$

In other words, by the addition of points to a partition, the lower sum increases (or stays the same) and the upper sum decreases (or stays the same).

To prove this, let us start with the partition $P : a = x_0 < x_1 < \cdots < x_n = b$ and refine it by the addition of one point u, say with $x_t < u < x_{t+1}$ for some t. Denote the new partition by Q. Then the lower sums for P and Q are exactly the same except for the terms

$$f(p_t)(x_{t+1} - x_t) \quad \text{in} \quad L(P),$$
$$f(p_t')(u - x_t) + f(p_t'')(x_{t+1} - u) \quad \text{in} \quad L(Q).$$

In these expressions, $f(p_t)$ is the minimum of f on $[x_t, x_{t+1}]$, $f(p_t')$ is the minimum of f on $[x_t, u]$, and $f(p''_t)$ is the minimum of f on $[u, x_{t+1}]$. It follows that

$$f(p_t) \leq f(p'_t) \quad \text{and} \quad f(p_t) \leq f(p''_t),$$

inasmuch as the minimum over the entire interval is no larger than the minimum over any part of the interval. We conclude

$$f(p_t)(x_{t+1} - x_t) = f(p_t)(u - x_t) + f(p_t)(x_{t+1} - u) \leq f(p'_t)(u - x_t) + f(p''_t)(x_{t+1} - u),$$

and it follows that $L(P) \leq L(Q)$.

Now suppose Q is a refinement of P obtained by the addition of more than one point. Consider a chain of partitions

$$P,\ Q_1,\ Q_2, \ldots, Q_h,\ Q,$$

with each partition Q_i obtained from the partition before it by the addition of one point. By suitably selecting the points added, we eventually reach the partition Q. We then have

$$L(P) \leq L(Q_1) \leq L(Q_2) \leq \cdots \leq L(Q),$$

which proves the inequality we want for lower sums.

A similar line of reasoning is used to prove that the upper sums do not increase; namely $U(Q) \leq U(P)$ for any refinement Q of P. We encourage the reader to carry out this case as an exercise.

Step 2. If P and R are any two partitions of $[a, b]$ then

$$L(P) \leq U(R). \tag{5.14}$$

We know that $L(Q) \leq U(Q)$ for any partition Q, but it is necessary to prove the stronger statement of inequality (5.14) that a lower sum for any partition is no greater than an upper sum for any other partition. Let Q be the partition obtained by taking all of the points of P and all of the points of R. Then Q is a refinement of P so $L(P) \leq L(Q)$. But Q is also a refinement of R so $U(Q) \leq U(R)$. These last two inequalities can be combined to give us

$$L(P) \leq L(Q) \leq U(Q) \leq U(R),$$

which proves our assertion.

Next we prove relations between I_{lower}, I_{upper} and the lower and upper sums.

Step 3. For any two partitions P and Q of $[a, b]$ we have

$$L(P) \leq I_{\text{lower}}(f(x)) \leq I_{\text{upper}}(f(x)) \leq U(Q). \tag{5.15}$$

Recall the definition of $I_{\text{lower}}(f(x))$; it is the *smallest* number that is greater than or equal to every lower sum. We have just seen in step 2 that $U(R)$ is greater than or equal to every lower sum. Thus $I_{\text{lower}}(f(x)) \leq U(R)$, for *every* partition R. But $I_{\text{upper}}(f(x))$ is the *largest* number that is less than or equal to every upper sum; so we conclude $I_{\text{lower}}(f(x)) \leq I_{\text{upper}}(f(x))$.

The remaining part of the proof is to show $I_{\text{lower}}(f(x)) = I_{\text{upper}}(f(x))$. We accomplish this in step 4 but first it is necessary to introduce a variation on the idea of continuity. The notion of *uniform continuity* plays an important role in the discussion of the definite integral.

DEFINITION 5.3

Uniform Continuity

A function f is uniformly continuous on a set S of real numbers if for any positive number ϵ there exists a positive number δ with the property that whenever u and v are points in S with $|u - v| < \delta$ then $|f(u) - f(v)| < \epsilon$.

Let us compare the definitions of continuity and uniform continuity. The definition of continuity at a point c implies that for any given $\lambda > 0$, there is an $\eta > 0$ such that

$$|x - c| < \eta \quad \text{implies} \quad |f(x) - f(c)| < \lambda.$$

For any u on the interval $(c - \eta, c + \eta)$ around c, the function values $f(u)$ and $f(c)$ lie between $f(c) - \lambda$ and $f(c) + \lambda$. It follows that $|f(u) - f(c)| < \lambda$. This is a property just like that in the definition of uniform continuity except that it holds only for points on the small interval of width 2η centered at c. To satisfy the definition of continuity, we first select the point c and any positive number λ and then produce the number η with certain properties. If we keep the same λ but move to another point c', there will be another η' such that the same property holds on the interval of length $2\eta'$ centered at c'. There is a possibility that the η that was selected for the point c does not satisfy the required conditions for the point c'. The importance of uniform continuity is that we can select a single η that suffices for every choice of c on the interval $[a, b]$. To satisfy the definition of uniform continuity, we are first given the positive number ϵ and then produce the positive δ that has certain properties for every pair of points closer than δ. The difference between function values is small if the points are within δ anywhere on the interval $[a, b]$.

It appears on the surface that uniform continuity is more restrictive than continuity. However, there is no difference when we are talking about continuity on a closed interval as we now prove.

THEOREM 5.5

Continuity Implies Uniform Continuity

If $f(x)$ is continuous on the interval $[a, b]$, then $f(x)$ is uniformly continuous on $[a, b]$.

Proof

The proof is given in an argument by contradiction. We suppose the theorem is false, derive some logically correct statements from this assumption and reach a statement that we know to be false. Hence the assumption from which we began could not have been correct. That assumption was that the theorem was false; hence the theorem must be true after all.

Suppose the theorem is false. Then there is a positive number ϵ for which there is no positive number δ with the property:

$$\mathcal{P}(\epsilon): \quad |f(u) - f(v)| < \epsilon \quad \text{whenever} \quad |u - v| < \delta$$

for all points u and v on $[a, b]$. We fix this number ϵ and we now assume that there is no positive δ for which $\mathcal{P}(\epsilon)$ is true for all u and v on $[a, b]$. Break the interval $[a, b]$ into two subintervals $[a, c]$ and $[c, b]$ with $c = (b+a)/2$. We claim that $\mathcal{P}(\epsilon)$ must fail on at least one of the two subintervals. Why is this? Just suppose that statement $\mathcal{P}(\epsilon)$ holds on both intervals $[a, c]$ and $[c, b]$ for possibly different choices of δ. By selecting δ' to be the smaller of the two possible choices of δ, we can assume that the same δ' works for each of the two intervals. Now to show that the property holds on the full interval $[a, b]$ for the given ϵ it is necessary to find δ that makes $\mathcal{P}(\epsilon)$ true for every pair u, v on $[a, b]$. When u and v lie in the same subinterval $[a, c]$ or $[c, b]$ and are within δ' of each other, then the function values at u and v are within ϵ. Thus we have only to consider points u and v that are not both in the same subinterval. For such points, we invoke the continuity of f at c and use the idea that the continuity at c restricts the values of f at points close to c. For any positive number ϵ' there is some number $\lambda > 0$ such that $|f(x) - f(c)| < \epsilon'$ whenever $|x - c| < \lambda$. Select $\epsilon' = \epsilon/2$. If u and v satisfy

$$c - \lambda < u < c < v < c + \lambda$$

then

$$\begin{aligned} |f(u) - f(v)| &= |(f(u) - f(c)) + (f(c) - f(v))| \\ &\leq |f(u) - f(c)| + |f(v) - f(c)| \\ &< 2\epsilon' = \epsilon. \end{aligned}$$

By taking the mesh δ smaller than the minimum of the two numbers δ' and λ, we find that $\mathcal{P}(\epsilon)$ holds on the full interval $[a, b]$. However this conclusion is reached on the basis that $\mathcal{P}(\epsilon)$ holds on each subinterval. Because we are supposing that $\mathcal{P}(\epsilon)$ does not hold on $[a, b]$, it follows that $\mathcal{P}(\epsilon)$ must fail on one of the subintervals $[a, c]$ or $[c, b]$. We select $[a_1, b_1]$ to be the half interval on which $\mathcal{P}(\epsilon)$ fails. Thus either $[a_1, b_1] = [a, c]$ or $[a_1, b_1] = [c, b]$.

Now we repeat this same argument with the interval $[a_1, b_1]$ replacing $[a, b]$ to obtain a subinterval $[a_2, b_2]$ of $[a_1, b_1]$ of half the length and on which $\mathcal{P}(\epsilon)$ fails. Continuing this way we obtain an infinite decreasing collection of intervals

$$[a, b] \supset [a_1, b_1] \supset [a_2, b_2] \supset \cdots$$

with the length of each subinterval half the length of the preceding one and $\mathcal{P}(\epsilon)$ failing on each subinterval $[a_i, b_i]$.

Let S be the set of all the left endpoints

$$S = \{a_1, a_2, a_3, \ldots\}$$

of the subintervals obtained. The set S is bounded because every a_i is on $[a, b]$ and so $a_i < b$. By the completeness property of the real numbers, there is a least upper bound A of S. The number A satisfies $a_i \leq A$ for every i and if A' is any number less than A, there is some a_k greater than A'. Because b is an upper bound for S, it follows that $A \leq b$ and so A is on $[a, b]$. Now invoke the continuity of f at A using the number $\epsilon' = \epsilon/2$ (again). There is some η such that $|f(A) - f(u)| < \epsilon'$ whenever $|A - u| < \eta$. Next we argue that the interval $(A - \eta, A + \eta)$ contains one of the intervals $[a_k, b_k]$. Because A is the least upper bound of the a_i, there is some L such that $A - \eta \leq a_L < A$ and hence all of the a_n for $n > L$ are in the interval $(A - \eta, A)$. The interval $[a_n, b_n]$ has length

$$b_n - a_n = \frac{b - a}{2^n},$$

and so there is some $n \geq L$ such that $(b - a)/n < \eta$ and $[a_n, b_n]$ is contained in the interval $(A - \eta, A + \eta)$.

Now for any u and v on $[a_n, b_n]$ we have

$$\begin{aligned} |f(u) - f(v)| &= |(f(u) - f(A)) + (f(A) - f(v))| \\ &< |f(u) - f(A)| + |f(v) - f(A)| \\ &< 2\epsilon' = \epsilon. \end{aligned}$$

This implies that the statement $\mathcal{P}(\epsilon)$ is true for $[a_n, b_n]$ and $\delta = b_n - a_n$. However we had arranged, by our assumption, that no such δ existed for this given ϵ. We have reached an impossible situation based on the assumption that f was not uniformly continuous on $[a, b]$. This assumption cannot hold and the theorem is true. □

REMARK: This proof is rather complicated partly because the statement is quite delicate and depends on the fact that f is continuous on a *closed* interval. If the interval in question is not closed, that is if we consider (a, b), $(a, b]$ or $[a, b)$, the theorem might fail. A function f might be continuous on $[a, b)$ (or any of the nonclosed intervals) but not be uniformly continuous on $[a, b)$. If we were to apply the proof just given to a case in which we assumed only that f is continuous on $[a, b)$ the proof would break down at the point where we used the assumption that f is continuous at A. The sequence of inequalities $a_1 \leq a_2 \leq \cdots < b$ implies that $A \leq b$. It could happen that $A = b$. If b is not on the interval where f is assumed continuous, then we cannot complete the argument as given. An example is given in the exercises following this section to show that a continuous function need not be uniformly continuous over a nonclosed interval.

The relevance for us and for the proof of Theorem 5.1 is the following statement:

Corollary 5.1

If $f(x)$ is continuous on $[a, b]$ and if ϵ is any positive number, then there is a partition $P : a = x_0 < x_1 < \cdots < x_n = b$ such that for any two points u and v on the same subinterval $[x_{i-1}, x_i]$ we have

$$|f(u) - f(v)| < \epsilon.$$

Proof

Because f is continuous, it is uniformly continuous. We apply the definition of uniform continuity. For the given ϵ there is a δ such that $|f(u) - f(v)| < \epsilon$ whenever $|u - v| < \delta$. Thus we select any partition P with the property $|x_i - x_{i-1}| < \delta$ for every i. Then two points on the same subinterval must be less than δ apart and the function values differ by at most ϵ. □

We can state the property in somewhat less precise terms as follows: Given any positive number ϵ, we can select a partition with the property that the difference between the maximum and minimum values of the function on any subinterval is less than ϵ.

Now we give the last step needed to finish the proof of Theorem 5.1.

Step 4. For any given positive number ϵ, there is a positive number η with the property that for any partition P of $[a, b]$ into subintervals of length at most η we have $U(P) - L(P) < \epsilon$.

We are given the positive number ϵ and must produce a number η with certain properties. By the uniform continuity property [using $\epsilon/(b - a)$ in place of the ϵ in the definition], there is a number $\eta > 0$ with the property that for any two points u, v on $[a, b]$ with $|u - v| < \eta$ we have

$$|f(u) - f(v)| < \frac{\epsilon}{b - a}.$$

We have selected the η and now we show that a partition into subintervals of length less than η has upper and lower sums differing by no more than ϵ.

Let $P : a = x_0 < x_1 < \cdots < x_n = b$ be any partition for which every subinterval has length less than η. Thus

$$x_i - x_{i-1} < \eta \qquad \text{for} \quad 1 \leq i \leq n.$$

Let p_i and q_i be the points on $[x_i, x_{i+1}]$ for which $f(p_i)$ is the minimum and $f(q_i)$ is the maximum value of f on the subinterval. Because p_i and q_i lie on an interval of length less than η, the difference $|f(p_i) - f(q_i)|$ is less than $\epsilon/(b-a)$. Then, applying the definition of lower and upper sums, we have

$$\begin{aligned} U(P) - L(P) &= \big(f(q_1) - f(p_1)\big)(x_1 - x_0) + \cdots + \big(f(q_n) - f(p_n)\big)(x_n - x_{n-1}) \\ &\leq \frac{\epsilon}{b-a}(x_1 - x_0) + \cdots + \frac{\epsilon}{b-a}(x_n - x_{n-1}) \\ &= \frac{\epsilon}{b-a}\big((x_1 - x_0) + (x_2 - x_1) + \cdots + (x_n - x_{n-1})\big) \\ &= \frac{\epsilon}{b-a}\big(x_n - x_0\big) = \frac{\epsilon}{b-a}[b-a] \\ &= \epsilon. \end{aligned}$$

This proves step 4. Now we have the information needed to prove

$$I_{\text{upper}}(f(x)) - I_{\text{lower}}(f(x)) = 0.$$

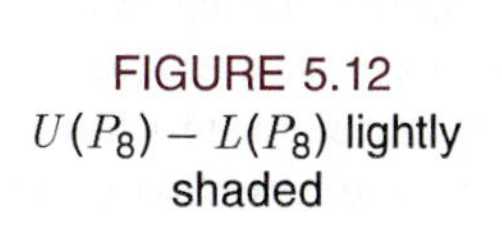

FIGURE 5.12
$U(P_8) - L(P_8)$ lightly shaded

Let us write the difference of these numbers as

$$\delta = I_{\text{upper}}(f(x)) - I_{\text{lower}}(f(x)) \tag{5.16}$$

so that by step 3, $\delta \geq 0$. We also know from the definitions that for any partition P of $[a, b]$, we have

$$\begin{aligned} L(P) &\leq I_{\text{lower}}(f(x)), \\ U(P) &\geq I_{\text{upper}}(f(x)). \end{aligned}$$

Thus we conclude

$$U(P) - L(P) \geq I_{\text{upper}}(f(x)) - I_{\text{lower}}(f(x)) = \delta \geq 0,$$

for every partition P. If $\delta \neq 0$, then $\delta > 0$ and so in step 4 the number $\epsilon = \delta/2$ is positive and there is a partition P with the property

$$\epsilon = \delta/2 > U(P) - L(P).$$

But the last two inequalities are not consistent; it is impossible for both to be true. The last inequality was proved on the assumption that $\delta \neq 0$. Hence this assumption must be false; $\delta = 0$ and $I_{\text{upper}}(f(x)) = I_{\text{lower}}(f(x))$.

Now if I is a number that is greater than or equal to every lower sum, then $I_{\text{lower}}(f(x)) \leq I$ because $I_{\text{lower}}(f(x))$ is the smallest number with that property. If I also is less than or equal to every upper sum, then $I \leq I_{\text{upper}}(f(x))$ because $I_{\text{upper}}(f(x))$ is the largest number with that property. It now follows that $I = I_{\text{lower}}(f(x)) = I_{\text{upper}}(f(x))$; there is only one number that lies between every lower and upper sum. This completes the proof of Theorem 5.1.

Exercises 5.4

1. The function $f(x) = 3x^2$ is uniformly continuous on $[0, 5]$. Find a number $\delta > 0$ such that the inequality $|f(u) - f(v)| < 0.2$ holds whenever u and v are on the interval $[0, 5]$ and $|u - v| < \delta$.

2. The function $f(x) = x^3 - 3x$ is uniformly continuous on $[-2, 2]$. Find a number $\delta > 0$ such that $|f(u) - f(v)| < 0.01$ whenever u and v are on the interval $[-2, 2]$ and $|u - v| < \delta$.

3. Show that the function $f(x) = 1/x$ is not uniformly continuous on the interval $[-1, 0)$ even though it is continuous on the interval. [Hint: Consider numbers $u = -10^{-n}$ and $v = -10^{-(n+r)}$ and the differences $|f(u) - f(v)|$ with $|u - v| < 10^{-n}$.]

The following exercises make use of uniform continuity to prove a basic theorem that was quoted without proof in chapter 4.

4. Prove that a function f that is continuous on $[a, b]$ is bounded; that is $|f(x)| < M$ for some M and all x on $[a, b]$. [Hint: Use the uniform continuity of f, with $\epsilon = 1$, to conclude there is some $\delta > 0$ such that $|f(u) - f(v)| < 1$ if $|u - v| < \delta$. Then take a partition of the interval $a = x_0 < x_1 < \cdots < x_n = b$ with every subinterval of length at most δ. Then the inequality $-1 + f(x_i) \le f(x) \le f(x_i) + 1$ holds for every x on $[x_{i-1}, x_i]$. This shows f is bounded on $[x_{i-1}, x_i]$. Conclude f is bounded on $[a, b]$.]

5. Prove that for a continuous function f on $[a, b]$ there are numbers u and v on $[a, b]$ such that $f(u)$ is the minimum of f and $f(v)$ is the maximum of f on the interval. [Hint: We show how to prove that f takes on its maximum value. The case for the minimum value is done similarly. By Exercise 4, there is an upper bound for the set of values of f on $[a, b]$ and so there is a least upper bound M. Thus $f(x) \le M$ for all x on $[a, b]$. If there is a number v on the interval such that $f(v) = M$, then we are done. Suppose then $f(x) \ne M$ for x on $[a, b]$. Consider the function

$$g(x) = \frac{1}{M - f(x)}, \qquad x \text{ on } [a, b].$$

Then g is continuous, because the denominator is never 0, and so g is bounded. Let $g(x) \le M'$ for all x on $[a, b]$. Now show this implies that M is not the smallest upper bound for f. This conflicts with the choice of M as the least upper bound so the assumption $f(x) \ne M$ cannot be true for every x.]

6. The function f is known to be uniformly continuous on $[0, 1]$ and the inequality $|f(u) - f(v)| < 0.15$ holds for every pair of numbers u and v on $[0, 1]$ satisfying $|u - v| < 0.05$. If we also know $f(0) = 3$, find numbers m and M such that $m \le f(x) \le M$ for every x on $[0, 1]$.

5.5 Riemann sums

We introduce a slightly different theoretical way of computing the definite integral of a continuous function than the method of the previous sections that used the upper and lower sums. The method is more general and has applications later in this text and also in practical applications where the definite integral is used to compute physical quantities.

Let f be continuous on the interval $[a, b]$ and let $P : a = x_0 < x_1 < \cdots < x_n = b$ be a partition of $[a, b]$. On each subinterval select a number u_i so $x_{i-1} \le u_i \le x_i$. The the sum

$$R = \sum_{i=1}^{n} f(u_i)(x_i - x_{i-1}) \tag{5.17}$$

is called a *Riemann sum* for f over $[a, b]$. The value of R depends on many variables: Clearly R depends on f and $[a, b]$ and also on the partition P and the choice of the numbers u_i on the subintervals. We regard f and $[a, b]$ as fixed for this discussion, so different Riemann sums arise by taking different partitions and then for each partition by taking different choices for the u_i. If we select u_i to be the point on the subinterval where f has its minimum value, then R is the lower sum $L(P)$. Similarly, by selecting $f(u_i)$ to be the maximum of f on the ith subinterval R the upper sum $U(P)$. Note that for the general Riemann sum with arbitrary choices of the u_i on each subinterval, we have

$$L(P) \le R \le U(P).$$

This is easily proved because $f(u_i)$ lies between the minimum and maximum values of f on each subinterval.

Our goal in this section is to show that the Riemann sums corresponding to a sequence of partitions approach the definite integral, regardless of the choices of the u_i, provided only that the partitions have their longest subinterval approaching length zero. We make this a bit more precise now.

Norm of a Partition

For a partition $P : a = x_0 < x_1 < \cdots < x_n = b$ of $[a, b]$, the *norm of* P is the longest of the lengths $x_i - x_{i-1}$, for $1 \leq i \leq n$. We denote the norm of P by $\| P \|$ so that

$$\| P \| = \text{maximum number in } \{x_i - x_{i-1} : 1 \leq i \leq n\}.$$

If the partition P is regular, then all subintervals have the same length and $\| P \| = (b - a)/n$.

THEOREM 5.6

The Definite Integral as a Riemann Sum

Let f be a continuous function on the interval $[a, b]$ and let $P_1, P_2, \ldots, P_n, \ldots$ be any sequence of partitions of $[a, b]$ such that the norms satisfy $\lim_{n \to \infty} \| P_n \| = 0$. Let

$$R(P_n) = \sum_{i=1}^{n} f(u_i)(x_i - x_{i-1})$$

be any Riemann sum of f corresponding to the partition P_n. Then

$$\lim_{n \to \infty} R(P_n) = \int_a^b f(x)\, dx. \tag{5.18}$$

Proof

Let ϵ be any positive number. We have to prove that there is some integer N such that

$$\left| \int_a^b f(x)\, dx - R(P_n) \right| < \epsilon \quad \text{for all} \quad n > N.$$

By step 4 in section 5.4, there is a positive number δ such that any partition Q of $[a, b]$ with $\| Q \| < \delta$ satisfies

$$U(Q) - L(Q) < \epsilon.$$

For the sequence of partitions P_k that are given in the theorem, we have $\lim_{n \to \infty} \| P_n \| = 0$. Hence, by the definition of limit, there is some N such that $\| P_n \| < \eta$ for all $n > N$. It follows then that $U(P_n) - L(P_n) < \epsilon$ and so the length of the interval $[L(P_n), U(P_n)]$ is at most ϵ. This interval contains the number $R(P_n)$, because the Riemann sum lies between the lower and upper sum, and the number $\int_a^b f(x)\, dx$ because the definite integral lies between the lower and upper sums. It follows that

$$\left| \int_a^b f(x)\, dx - R(P_n) \right| < \epsilon$$

for all $n > N$. This proves that the limit (5.18) is true. □

One fairly obvious choice of partitions of $[a, b]$ is the sequence of regular partitions. Let P_n be the regular partition into subintervals of length $(b - a)/n$. Then the norm of P_n is $(b - a)/n$ so the condition $\lim_{n\to\infty} \| P_n \| = 0$ holds.

We give a restatement of the previous theorem for the case in which regular partitions are used and the selection of the point u_i on $[x_i, x_{i+1}]$ is made either as the left endpoint $u_i = x_i$ or the right endpoint $u_i = x_{i+1}$.

THEOREM 5.7

Let f be a continuous function on $[a, b]$. Then the definite integral of f equals each of the following limits in which $h_n = (b - a)/n$:

$$\int_a^b f(x)\,dx = \lim_{n\to\infty} \left(\sum_{i=1}^{n} f\big(a + (i-1)h_n\big)\right)\left(\frac{b-a}{n}\right),$$

$$\int_a^b f(x)\,dx = \lim_{n\to\infty} \left(\sum_{i=1}^{n} f(a + ih_n)\right)\left(\frac{b-a}{n}\right).$$

The proof amounts to observing that the points $a + ih_n = a + (b - a)i/n$ are the endpoints of the subintervals in a regular partition of $[a, b]$ and then applying the previous theorem.

EXAMPLE 1 Express the definite integral of $f(x) = x^2$ over the interval $[0, 1]$ as a limit of Riemann sums of regular partitions and then evaluate the limit.

Solution: The regular partition into n subintervals is given by

$$P_n : 0 < \frac{1}{n} < \frac{2}{n} < \cdots < \frac{n-1}{n} < \frac{n}{n} = 1,$$

so $x_i = i/n$. Then using the right endpoint of each subinterval we have

$$\int_0^1 x^2\,dx = \lim_{n\to\infty} \sum_{i=1}^{n} \left(\frac{i}{n}\right)^2 \frac{1}{n}$$
$$= \lim_{n\to\infty} \frac{1^2 + 2^2 + 3^2 + \cdots + n^2}{n^3}.$$

(A similar limit is obtained using the right endpoint of each subinterval.) The limit can be evaluated by first using the formula developed in section 5.1 for the sum of the squares of the first n integers:

$$\sum_{k=1}^{n} k^2 = \frac{n(n+1)(2n+1)}{6}.$$

We see that

$$\lim_{n\to\infty} \frac{1^2 + 2^2 + \cdots + n^2}{n^3} = \lim_{n\to\infty} \frac{n(n+1)(2n+1)}{6n^3} = \frac{1}{3}.$$

This is one of the rather rare instances where the limit of Riemann sums can be evaluated explicitly. ■

Exercises 5.5

1. Let $f(x) = x^3+1$. Sketch the graph of f over the interval $[-1, 1]$ and draw the region whose area is computed by the Riemann sum corresponding to the partition $P : -1 < -0.5 < 0 < 0.5 < 1$ and using the midpoint of each subinterval to evaluate the function. Use the Riemann sum to obtain an approximation of the area under the curve.

2. Let $f(x) = 3x + 1$. Sketch the graph of f over the interval $[0, 2]$ and make three drawings showing the region whose area is computed by a Riemann sum corresponding to the partition $P : 0 < 0.5 < 1 < 1.5 < 2$ and using (a) the left-hand endpoint, (b) the midpoint and (c) the right-hand endpoint of each subinterval to evaluate the function. Use the Riemann sums as approximations of the area under the curve. Which approximation seems to be nearest the actual area under the graph?

Evaluate the definite integral of the given function over the indicated interval by computing the limit of Riemann sums taken over regular partitions.

3. $3x + 2$ over $[0, 1]$

4. $x - 4$ over $[-1, 1]$

5. $2x^2$ over $[0, 2]$

6. $3x^2 + 1$ over $[1, 3]$

7. $x^2 + x + 2$ over $[0, 2]$

8. $ax^2 + bx + c$ over $[0, 1]$

Find some function and some interval such that each of the following sums is a Riemann sum for the function over the interval.

9. $\displaystyle\sum_{i=1}^{n}\left(2+\frac{i}{n}\right)^3\frac{1}{n}$

10. $\displaystyle\sum_{i=0}^{n-1}\left(\frac{4n+i}{n^2}\right)$

Evaluate the limits as the value of some integral.

11. $\displaystyle\lim_{k\to\infty}\frac{1}{n^4}\sum_{k=1}^{n}k^3$

12. $\displaystyle\lim_{n\to\infty}\sum_{i=1}^{n}\frac{32i^5}{n^6}$

13. $\displaystyle\lim_{k\to\infty}\frac{1}{n^{a+1}}\sum_{k=1}^{n}k^a;\quad a\neq -1$

14. A Riemann sum for the integral $\int_1^4 dx/x$ is formed by taking a regular partition P_n with subintervals $I_j = [1+3(j-1)/n, 1+3j/n]$ and point u_j as the midpoint of I_j. Prove that the Riemann sum so obtained is less than the value of the integral. [Hint: Example 2 of section 4.7 may be useful here.]

5.6 Proof of the fundamental theorem of calculus

Let $f(x)$ be a continuous function on $[a, b]$ and let $F(x)$ be any antiderivative of $f(x)$. The Fundamental Theorem of Calculus (Theorem 5.4) states that

$$\int_a^b f(x)\,dx = F(b) - F(a).$$

We give a proof of this theorem in this section after developing the following idea.

Functions Defined by Integrals

The idea of a function defined by a definite integral with a variable upper limit provides the key tool in the proof of the Fundamental Theorem. Assume $a < b$ and let t be any number that satisfies $a \le t \le b$; let $A(t)$ be the function defined as

$$A(t) = \int_a^t f(x)\,dx.$$

We intend to compute the derivative of $A(t)$ with respect to t and prove

$$A'(t) = f(t),$$

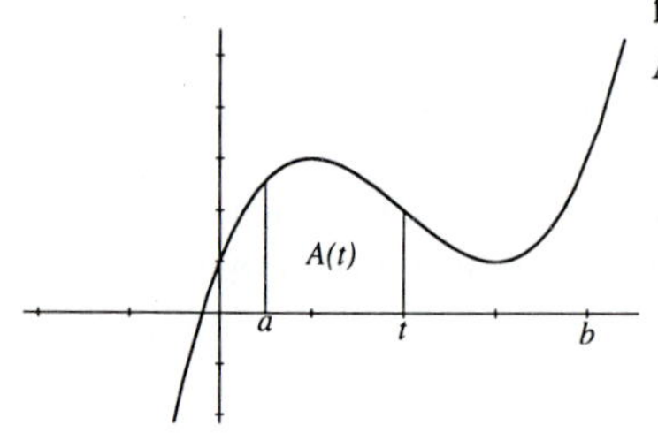

FIGURE 5.13
$A(t)$ = integral from a to t

for all t on $[a, b]$. In other words, we show that $A(x)$ is an antiderivative of $f(x)$. According to the definition of the derivative, we have

$$\begin{aligned} A'(t) &= \lim_{h\to 0} \frac{1}{h}\{A(t+h) - A(t)\} \\ &= \lim_{h\to 0} \frac{1}{h}\left\{\int_a^{t+h} f(x)\,dx - \int_a^t f(x)\,dx\right\} \\ &= \lim_{h\to 0} \frac{1}{h}\int_t^{t+h} f(x)\,dx. \end{aligned} \tag{5.19}$$

The next step is to evaluate the limit that remains here. We treat the case in which $h \geq 0$, the case with negative h can be done similarly. Let u be a point on the interval $[t, t+h]$ such that $f(u)$ is the minimum value of f on $[t, t+h]$. Similarly, let v be a point where $f(v)$ is a maximum value of f on the interval. Using the lower and upper sum for the partition of $[t, t+h]$ consisting of only one subinterval, we find

$$f(u)(t+h-t) = f(u)h \leq \int_t^{t+h} f(x)\,dx \leq f(v)h.$$

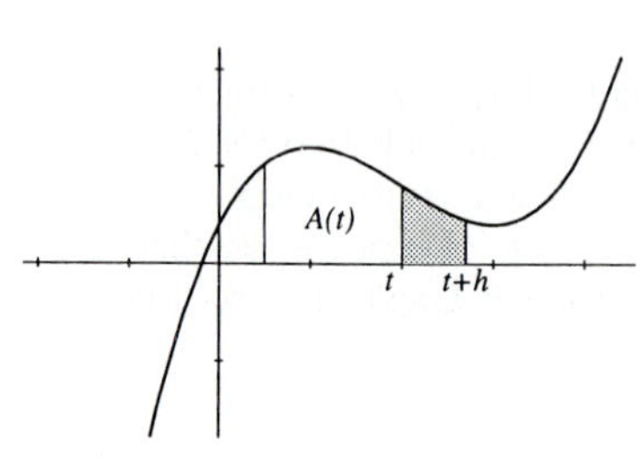

FIGURE 5.14
$A(t+h) - A(t)$ =
shaded region

We rewrite this last inequality as

$$f(u) \leq \frac{1}{h}\int_t^{t+h} f(x)\,dx \leq f(v).$$

Of course the points u and v depend on the particular value of h; when h changes to a smaller value, a new choice of u and v on $[t, t+h]$ must be made. As h approaches 0, u and v each approach t because, for all h, we have

$$t \leq u \leq t+h \quad \text{and} \quad t \leq v \leq t+h.$$

Using the continuity of f and the limits

$$\lim_{h\to 0} u = t \quad \text{and} \quad \lim_{h\to 0} v = t,$$

we obtain

$$f(t) = \lim_{h\to 0} f(u) \leq \lim_{h\to 0} \frac{1}{h}\int_t^{t+h} f(x)\,dx \leq \lim_{h\to 0} f(v) = f(t).$$

In view of equation (5.19), this proves

$$A'(t) = f(t).$$

If we change the name of the variable from t to x, we have $A'(x) = f(x)$ and $A(x)$ is one antiderivative of $f(x)$. Moreover note that $A(a) = 0$ and so

$$\int_a^b f(x)\,dx = A(b) = A(b) - A(a),$$

which is precisely the conclusion of the Fundamental Theorem for the particular antiderivative $A(x)$ of $f(x)$. However, the conclusion is stated for *any* antiderivative of $f(x)$. We have already seen that if $F(x)$ is another antiderivative of $f(x)$, then $F(x) = A(x) + c$ for some constant c. Then

$$F(b) - F(a) = \{A(b) + c\} - \{A(a) + c\} = A(b) - A(a) = \int_a^b f(x)\,dx.$$

This completes the proof of the Fundamental Theorem of Calculus.

EXAMPLE 1 Evaluate the definite integral

$$\int_0^5 2x - 3x^5\,dx.$$

Solution: One antiderivative of $2x - 3x^5$ is

$$F(x) = x^2 - 3x^6/6 = x^2 - x^6/2.$$

Thus

$$\int_0^5 2x - 3x^5\,dx = F(5) - F(0) = 5^2 - 5^6/2 - 0 = -7787.5.$$ ■

It is be necessary to write expressions like $F(b) - F(a)$ rather often so we adopt a short notation for this:

$$F(x)\Big|_a^b = F(b) - F(a).$$

Using this notation, the last integral can be expressed as

$$\int_0^5 2x - 3x^5\,dx = x^2 - x^6/2\Big|_0^5 = 5^2 - 5^6/2 - 0 = -7787.5.$$

The above proof of the Fundamental Theorem emphasizes the derivative of a function defined as an integral. This idea is significant enough to warrant repeating it as a theorem.

THEOREM 5.8

The Derivative of the Integral

Let $f(x)$ be continuous over an interval $[a, b]$. Define the function $F(x)$ by

$$F(x) = \int_a^x f(u)\,du,$$

for any number x on $[a, b]$. Then $\frac{d}{dx}F(x) = f(x)$.

Do not be confused by the variable u in the integral. The symbols

$$\int_a^b f(x)\,dx \quad \text{and} \quad \int_a^b f(u)\,du$$

represent the same number, namely the unique number that lies between all lower sums for f of the form (5.7) and upper sums for f of the form (5.8). It is a bit confusing to write

$$\int_a^x f(x)\,dx$$

because the various appearances of x in this symbol do not all refer to the same thing. The limits x and a stand for the endpoints of an interval over which the integral is taken; the x in $f(x)\,dx$ is simply a generic symbol indicating the name of the variable. The variable x in the part $f(x)\,dx$ of the integral symbol is often called a *dummy variable*. It can be replaced with any other symbol and the value of the integral remains the same. We avoid this ambiguity by writing the integral as in the last theorem.

Linearity of the Integral

The Fundamental Theorem can be used to give an easy proof of the following property of the integral.

THEOREM 5.9

Linearity of the Definite Integral

Let $f(x)$ and $g(x)$ be continuous functions on $[a, b]$. Then

$$\int_a^b f(x) + g(x)\,dx = \int_a^b f(x)\,dx + \int_a^b g(x)\,dx.$$

Proof

Let $F(x)$ be an antiderivative of $f(x)$ and $G(x)$ an antiderivative of $g(x)$. Then

$$\frac{d}{dx}[F(x) + G(x)] = F'(x) + G'(x) = f(x) + g(x).$$

This shows that $F(x) + G(x)$ is an antiderivative of $f(x) + g(x)$. We can now apply the Fundamental Theorem to obtain

$$\begin{aligned}\int_a^b f(x) + g(x)\,dx &= [F(x) + G(x)]\Big|_a^b \\ &= F(b) - F(a) + G(b) - G(a) \\ &= \int_a^b f(x)\,dx + \int_a^b g(x)\,dx.\end{aligned}$$

□

Exercises 5.6

Given $\int_1^3 f(x)\,dx = 10$ and $\int_1^3 g(x)\,dx = -4$, evaluate the following integrals.

1. $\int_1^3 4f(x) - g(x)\,dx$ **2.** $\int_1^3 3g(x) - 2f(x)\,dx$

3. $\int_3^1 af(x) + bg(x)\,dx$ **4.** $\int_3^3 4f(x) - 2g(x)\,dx$

Evaluate the integral $\int_a^b f(x)\,dx$ and sketch the graph of $f(x)$ and the region whose area is given by the integral.

5. $\int_1^5 2x - 1\,dx$ **6.** $\int_{-2}^3 4 - x^2\,dx$

7. $\int_{-2}^{-1} 3 - x + x^2\,dx$ **8.** $\int_{-2}^2 (x - 2)(x + 2)\,dx$

9. $\int_0^1 x(1 - x)\,dx$ **10.** $\int_0^1 x^2(1 - x)\,dx$

11. $\int_0^1 x^3(1 - x)\,dx$ **12.** $\int_0^1 x^4(1 - x)\,dx$

Evaluate the derivative of $F(x)$ for each function defined by an integral.

13. $F(x) = \int_0^x 2u - 3\,du$ **14.** $F(x) = \int_4^x 3u^2 + 4u\,du$

15. $F(x) = \int_x^0 \frac{t-1}{t+1}\,dt$ **16.** $F(x) = \int_x^1 \frac{z^2+z}{z+1}\,dz$

17. $F(x) = \int_0^{x^3} \frac{u\,du}{u+2}$ [Hint: Use the chain rule.]

18. $F(x) = \int_{u(x)}^{v(x)} H(t)\,dt$ [Hint: $\int_u^v = \int_0^v + \int_u^0$.]

19. $F(x) = \int_1^{x^2+x} \sin u\,du$ **20.** $F(x) = \int_{x^2}^{x^3} (w + 1)^2\,dw$

Evaluate the derivatives of $G(x)$ in two ways; first by using the theorem about the derivative of the integral and second by evaluating the integral and then differentiating the result.

21. $G(x) = \int_1^{x+1} z^2 + z\,dz$ **22.** $G(x) = \int_{x+1}^1 z^3 - 2z\,dz$

23. $G(x) = \int_{x-1}^{x+1} z^2\, dz$ **24.** $G(x) = \int_{-2x}^{2x} 4t + 3\, dt$

25. The *Mean Value Theorem for Integrals* states that if f is a continuous function on $[a, b]$ then there is a number c on (a, b) such that

$$\int_a^b f(t)\, dt = (b - a)f(c).$$

Show this is true by applying the Mean Value Theorem to the function

$$g(x) = \int_a^x f(t)\, dt, \qquad a \le x \le b.$$

26. The *average value* of a continuous function f over an interval $[a, b]$ is the number

$$\text{AVE} = \frac{1}{b-a}\int_a^b f(x)\, dx.$$

Show there is a point c on (a, b) where AVE = $f(c)$.

27. What is the average value of $\sin x$ on $[0, \pi]$? on $[0, \pi/2]$?

28. Let $g(x)$ be a continuous function on $[a, b]$ and let $F(x) = \int_a^x g(t)\, dt$. If $F(x)$ is an increasing function on $[a, b]$, prove that $g(t) \ge 0$ for all t on $[a, b]$. Give an interpretation of this result in terms of areas but give a proof independent of geometric considerations.

29. Let $g(x)$ be a continuous function on $[a, b]$. Suppose that $g(x) \ge 0$ for all x and $g(c) > 0$ for some c on $[a, b]$. Prove

$$\int_a^b g(x)\, dx > 0.$$

[Hint: Use continuity to show there is an interval of positive length ℓ with center at c over which $g(x) > g(c)/2 > 0$. Then show there is some partition P such that the lower sum $L(P)$ for $g(x)$ over $[a, b]$ is at least $\ell g(c)/2$.]

30. Let $f(x)$ and $g(x)$ be continuous on $[a, b]$. Suppose $f(x) \ge g(x)$ for all x and that $f(c) > g(c)$ for some c on $[a, b]$. Use Exercise 29 to prove

$$\int_a^b f(x)\, dx > \int_a^b g(x)\, dx.$$

5.7 Computation of areas

Let $f(x)$ be a continuous and nonnegative function over the interval $[a, b]$. The area A of the region bounded by the x-axis, the lines $x = a$ and $x = b$ and the graph of $f(x)$ is a number that is greater than or equal to any lower sum and less than or equal to any upper sum of $f(x)$ over $[a, b]$; that is,

$$A = \int_a^b f(x)\, dx.$$

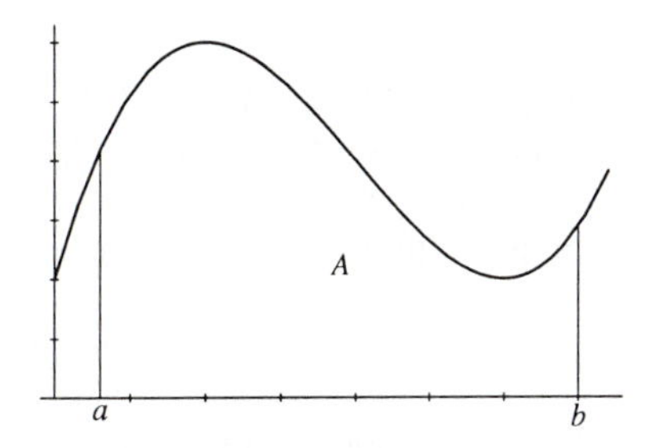

FIGURE 5.15
Area under a positive function

The Fundamental Theorem provides a means for computing this number. Here are some examples of such computations.

EXAMPLE 1 Find the area below the graph of the function $f(x) = x^3$ over the interval $[0, 4]$.

Solution: The area in question is indicated in figure 5.16 and is equal to the integral of x^3 over the interval $[0, 4]$. To apply the Fundamental Theorem, we first find an antiderivative of x^3; $F(x) = x^4/4$ is one antiderivative of x^3. Thus

$$A = \int_0^4 x^3\,dx = \left.\frac{x^4}{4}\right|_0^4 = \frac{4^4}{4} - 0 = 4^3.$$ ■

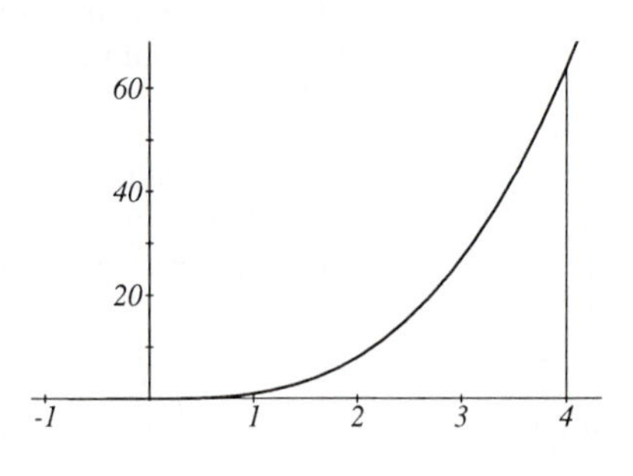

FIGURE 5.16
$y = x^3$

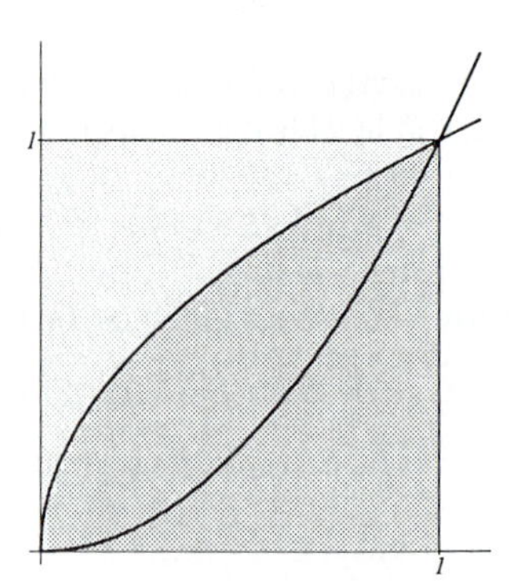

FIGURE 5.17

EXAMPLE 2 Let S be the square with sides on the x- and y-axes and diagonal corners at $(0, 0)$ and $(1, 1)$. The curve $y = \sqrt{x},\ 0 \le x \le 1,$ separates S into two regions; let A_1 be the area of the larger region. The curve $y = x^2$ $0 \le x \le 1,$ also separates S into two regions; let A_2 be the area of the larger region. Prove $A_1 = A_2$.

Solution: Let A denote the area of the region bounded by the x-axis, the curve $y = \sqrt{x}$, and the line $x = 1$; this is the dark region in figure 5.17. To compute A by the Fundamental Theorem, we need to know an antiderivative of $x^{1/2}$. By inspection, one is found to be $2x^{3/2}/3$; thus

$$A = \int_0^1 x^{1/2}\,dx = 2x^{3/2}/3\Big|_0^1 = \frac{2}{3} - 0 = \frac{2}{3}.$$

The unit square S has area 1, so the larger region has area $A_1 = A = 2/3$ and the smaller region has area $1 - 2/3 = 1/3$.

Next we deal with the other curve. An antiderivative of x^2 is $x^3/3$ so the area bounded by $y = x^2$, the x-axis and the line $x = 1$ is

$$\int_0^1 x^2\,dx = \left.\frac{x^3}{3}\right|_0^1 = \frac{1}{3} - 0 = \frac{1}{3}.$$

The larger part of the square S has area $A_2 = 1 - 1/3 = 2/3$, which is equal to A_1, as we wished to prove. ■

Functions That Change Sign

If $f(x)$ is a positive function over the interval $[a, b]$, then the integral of f over this interval is an area. How can we interpret the integral in case the function takes on negative values?

Suppose f takes on only negative values so that $f(x) \leq 0$ for all x on $[a, b]$. Then $-f(x)$ is a nonnegative function and the area below the graph of $-f(x)$ over $[a, b]$ is

$$\int_a^b -f(x)\,dx = -\int_a^b f(x)\,dx.$$

Thus the integral $\int_a^b f(x)\,dx$ is the negative of the area bounded by the x-axis and the graph of $f(x)$ over the interval $[a, b]$. We could have reasoned that every upper sum for $f(x)$ is the negative of a lower sum for $|f(x)|$ (see figures 5.18 and 5.19) and reached the same conclusion.

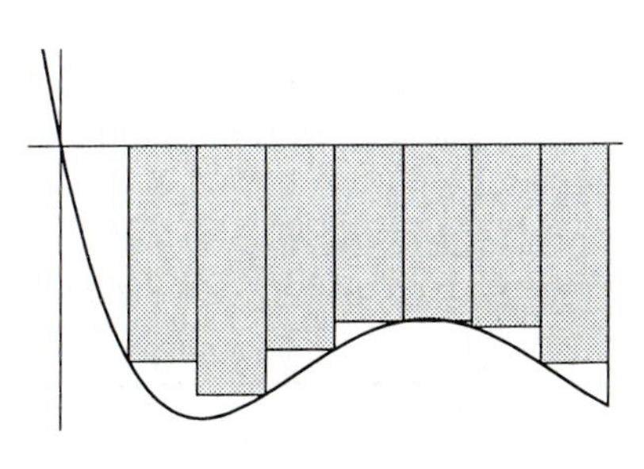

FIGURE 5.18
An upper sum for $f(x)$

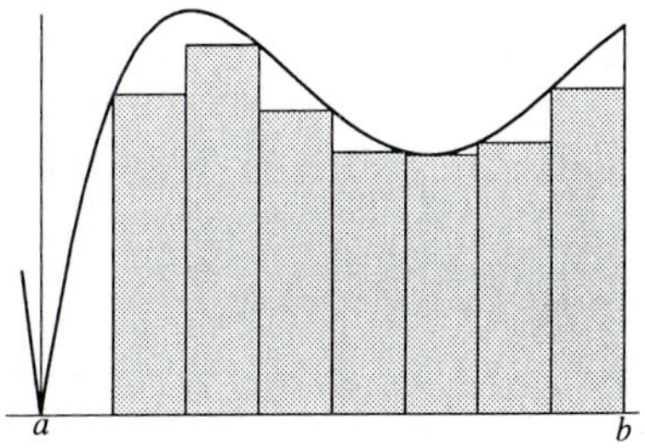

FIGURE 5.19
A lower sum for $|f(x)|$

If $f(x)$ takes on both positive and negative values on $[a, b]$, we interpret the integral as a *signed sum* of areas in the following manner. For illustration, suppose

$$f(x) \geq 0 \quad \text{for} \quad a \leq x \leq c,$$
$$f(x) \leq 0 \quad \text{for} \quad c \leq x \leq b.$$

Then

$$\int_a^b f(x)\,dx = \int_a^c f(x)\,dx + \int_c^b f(x)\,dx = A - B,$$

where A is the area of the region bounded by the graph of f above the x-axis over $[a, c]$ and B is the area of the region below the x-axis over $[c, b]$ bounded by the graph of f.

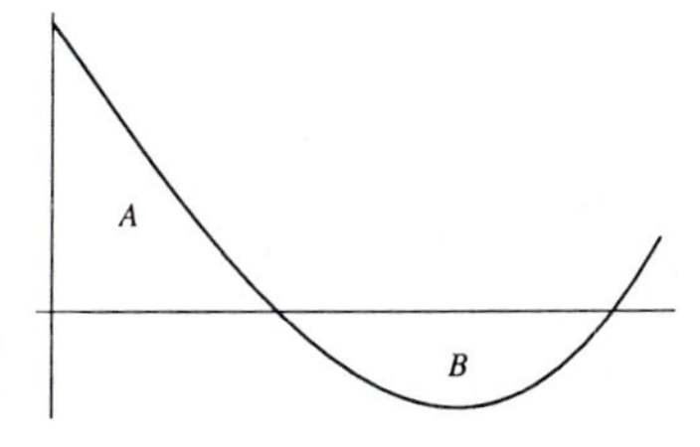

FIGURE 5.20

Regions Bounded by Two Curves

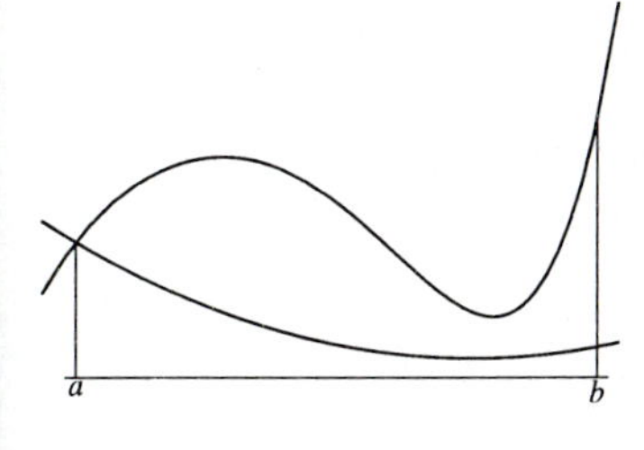

FIGURE 5.21

Let us consider the computation of the area of regions having a somewhat more complicated boundary than the regions considered above. Suppose that $f(x)$ and $g(x)$ are continuous functions over the interval $[a, b]$ and let us suppose further that $f(x) \geq g(x)$ so that the graph of $f(x)$ lies above the graph of $g(x)$. Let A be the area of the region enclosed by the two graphs and the lines $x = a$ and $x = b$. The region might look like one of those in figures 5.21, 5.22 or 5.23. The area is computed by applying the Fundamental Theorem to a suitable function.

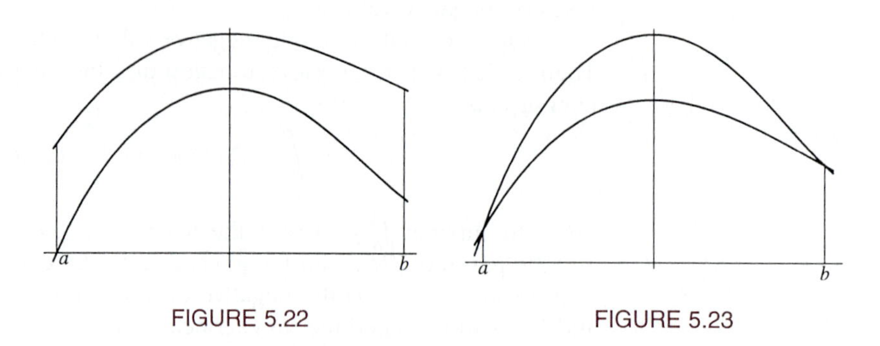

FIGURE 5.22

FIGURE 5.23

THEOREM 5.10

Let $f(x)$ and $g(x)$ be continuous on $[a, b]$ and assume $f(x) \geq g(x)$ for all x on this interval. Then the area of the region enclosed by the graphs of $f(x)$ and $g(x)$ and the lines $x = a$ and $x = b$ is

$$A = \int_a^b f(x) - g(x)\, dx.$$

Proof

The functions $f(x)$ and $g(x)$ are not assumed to be positive so that some or all of the region in question might lie below the x-axis. We have previously discussed the areas of regions bounded above by positive functions, so we now make some small changes to ensure that the functions we deal with are positive. To avoid repetition, let us agree to consider only choices of x on $[a, b]$.

Let M be some number such that $g(x) > M$ for all x. For example we could take M to be the minimum value of $g(x)$. We translate our picture by "raising" everything a distance $|M|$ so that the functions are above the x-axis. Let $G(x) = g(x) - M$ and $F(x) = f(x) - M$. Then $G(x) = g(x) - M$ is positive for all x and also $F(x) \geq G(x)$. The region bounded by $F(x)$ and $G(x)$ and the lines $x = a$ and $x = b$ has the same area as the region bounded by $f(x)$ and $g(x)$ and $x = a$ and $x = b$ inasmuch as the regions are congruent. This area is computed by first finding the area between the x-axis and the upper curve $y = F(x)$ and then subtracting the area between the x-axis and the lower curve $y = G(x)$; so the area A is

$$A = \int_a^b F(x)\, dx - \int_a^b G(x)\, dx.$$

Because the two integrals are taken over the same interval $[a, b]$, the linearity property permits us to combine this difference into a single integral. In view of the equality

$$F(x) - G(x) = [f(x) - M] - [g(x) - M] = f(x) - g(x),$$

we have

$$A = \int_a^b F(x) - G(x)\, dx = \int_a^b f(x) - g(x)\, dx,$$

as we wanted to prove. □

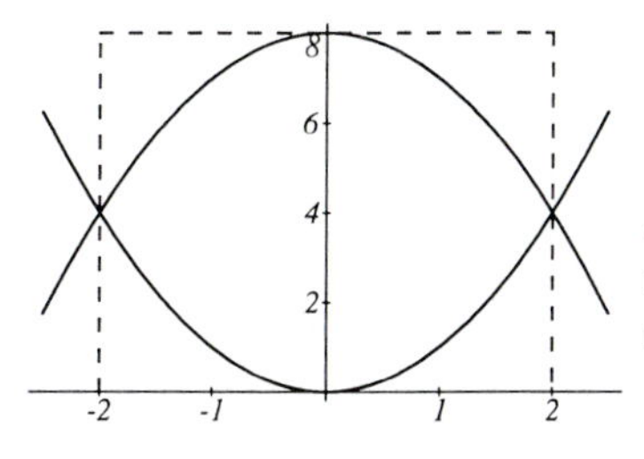

FIGURE 5.24
Area bounded by
x^2 and $8 - x^2$
is inside the
4×8 rectangle

EXAMPLE 3 Find the area enclosed by the two curves $y = x^2$ and $y = 8 - x^2$.

Solution: We first make a sketch of the two graphs to get some idea of the region. The curves intersect at points where $x^2 = 8 - x^2$: at $x = \pm 2$. Over the interval $[-2, 2]$ we have $8 - x^2 \geq x^2$ so the enclosed area is

$$\int_{-2}^{2} (8 - x^2) - x^2 \, dx = \int_{-2}^{2} 8 - 2x^2 \, dx = 8x - \frac{2x^3}{3}\Big|_{-2}^{2}$$

$$= \left(16 - \frac{16}{3}\right) - \left(-16 + \frac{16}{3}\right) = \frac{64}{3}.$$

■

EXAMPLE 4 Find the area that lies inside the square with vertices at $(0, 0)$, $(0, 1)$, $(1, 0)$ and $(1, 1)$ and lies above the curve $y = (x - 1/2)(4x + 1)$.

Solution: We begin by examining a sketch of the graph of the curve and its relation to the unit square.

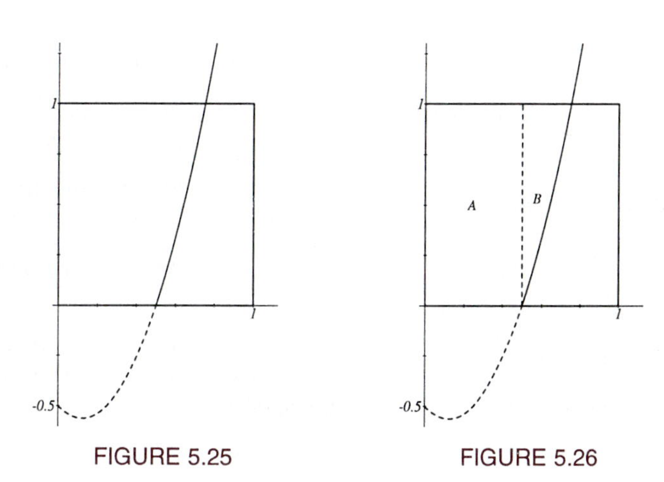

FIGURE 5.25 FIGURE 5.26

The area we want to compute is not quite in the form described above; the lower boundary consists of the line $y = 0$ for $0 \leq x \leq 0.5$ and the graph of the given function for $0.5 \leq x \leq 0.75$. To compute the area, we break it into two smaller regions where the method of integration can be applied. The part of the square above the interval $[0, 0.5]$ is in the region and contributes an area equal to $A = 1 \times 0.5 = 0.5$. The remaining region, over the interval $[0.5, 0.75]$ is bounded above by the line $y = 1$ and bounded below by the graph of $y = (x - 1/2)(4x + 1)$.

Hence this part of the square has area

$$\begin{aligned}
B &= \int_{0.5}^{0.75} 1-(x-1/2)(4x+1)\,dx \\
&= \int_{0.5}^{0.75} 1.5+x-4x^2\,dx \\
&= \left(1.5x-\frac{4x^3}{3}+\frac{x^2}{2}\right)\Big|_{0.5}^{0.75} \\
&= 1.5(0.75)-\frac{4(0.75)^3}{3}+\frac{(0.75)^2}{2}-1.5(0.5)+\frac{4(0.5)^3}{3}-\frac{(0.5)^2}{2} \\
&= \frac{13}{96} \approx 0.13541667.
\end{aligned}$$

The area within the square above the curve is $A+B = 61/96 \approx 0.63541667$. ■

Exercises 5.7

Sketch the region bounded by the indicated graphs and lines and then compute the area of the region.

1. $y = 0,\ x = 2,\ x = 4,\ f(x) = 3x - 1$

2. $y = 0,\ x = 1,\ x = a > 1,\ f(x) = 1/x^2$

3. $y = -1,\ x = 2,\ x = 6,\ f(x) = 4x + 1$

4. $x = 0,\ x = \pi,\ f(x) = \sin x$

5. $y = \sin x,\ y = x,\ x = \pi/2$

6. $y = 4x + 2,\ y = 3x - 1,\ x = 0$

7. $f(x) = \sin x,\ g(x) = \cos x,\ x = \frac{\pi}{4},\ x = \frac{3\pi}{4}$

8. $f(x) = x^2,\ g(x) = 2x + 3$

9. $f(x) = x^2,\ g(x) = mx + 3$

10. $f(x) = \sqrt{x},\ y = 2,\ x = 16$

11. $f(x) = x^3,\ g(x) = x$ [Be careful here.]

12. $f(x) = \sqrt{x},\ 4y = x$

13. Find the area of the region that lies above the graph of $f(x) = (3 - 4x)/2$ and inside the square with corners at $(0,0)$, $(0,1)$, $(1,1)$ and $(1,0)$.

14. Find the area of the region that lies in the strip $-1 \le y \le 1$ and below the graph of $f(x) = 2x - x^2$.

15. The area of the region bounded by the two curves $y^2 = 4 - x$ and $3y = x$ can be found using the techniques of this section by breaking the region into two parts, the first with $-12 \le x \le 3$ and the second with $3 \le x \le 4$. If we change our point of view and regard y as the independent variable and x as a function of y then a more direct application of the ideas of this section can be used. Show that the area equals $\int_{-4}^{1}(4-y^2)-3y\,dy$.

16. Use the idea in Exercise 15 to find the area bounded by the graphs of $y^2 = 12 - 3x$ and $3y = x + 1$.

17. Find the area of the region bounded on the left by $y^2 = x$ and on the right by the vertical line $x = 4$.

18. Find the region bounded by the graphs of $y = x^{1/3}$ and $y = x^3$.

19. The area of a circle of radius 1 is

$$\pi = 2\int_{-1}^{1} \sqrt{1 - x^2}\,dx.$$

Use this value and substitution to find the area bounded by the ellipse with equation $b^2x^2 + a^2y^2 = a^2b^2$.

20. Let $A(N)$ denote the area under the graph of $y = x^{-2}$ over the interval $[1, N]$. Show that $\lim_{N\to\infty} A(N)$ exists, and thus an unbounded region can have finite area.

21. Find the area of the set of points (x, y) that satisfy the two conditions $-2 \le x \le 2$ and the shortest distance from (x, y) to the line $y = -x$ is at most 1.

22. Find the area of the set of points (x, y) that satisfy the two conditions $-3 \le x \le 3$ and the shortest distance from (x, y) to the line $y = 2x$ is at most 2.

5.8 Indefinite integrals

If $F(x)$ is an antiderivative of $f(x)$ on an interval $[a,b]$, then

$$\int_a^b f(x)\,dx = F(x)\Big|_a^b = F(b) - F(a).$$

The computation of the left-hand side depends on finding the antiderivative $F(x)$ of $f(x)$. In doing so, the interval $[a,b]$ that appears in the definite integral plays no significant role. To focus our attention on the main issue of finding antiderivatives, we introduce the *indefinite integral* of $f(x)$. The symbol

$$\int f(x)\,dx$$

is called the *indefinite integral of* $f(x)$ and it equals any function whose derivative is $f(x)$; i.e., the indefinite integral of $f(x)$ is the antiderivative of $f(x)$. The function f in the symbol $\int f(x)\,dx$ is called the *integrand.* The first important point to notice is that the antiderivative of a function is not unique. The function x^3 is an antiderivative of the function $3x^2$ because $\frac{d}{dx}x^3 = 3x^2$. However the function $x^3 + 7$ is also an antiderivative of $3x^2$; in fact, for any constant c, $x^3 + c$ is an antiderivative of $3x^2$. More generally we have already proved the following result in Theorem 4.7 of chapter 4.

THEOREM 5.11

Let $f(x)$ be continuous on an interval I and let $F(x)$ be an antiderivative for $f(x)$. Then every antiderivative of $f(x)$ has the form $F(x) + c$ for some constant c.

The Constant of Integration

The proper use of the indefinite integral symbol is to write

$$\int f(x)\,dx = F(x) + c.$$

The constant c is called *the constant of integration.* In some problems, additional information might be available to determine a particular constant of integration.

EXAMPLE 1 Find the antiderivative of $3x^2 - 2$ whose graph passes through the point $(1, 2)$.

Solution: By inspection we find

$$\int 3x^2 - 2\,dx = x^3 - 2x + c.$$

Every antiderivative of $3x^2 - 2$ has the form $F(x) = x^3 - 2x + c$ for some constant c. The graph of $F(x)$ passes through $(1, 2)$ if $F(1) = 2$. This requires

$$2 = F(1) = 1^3 - 2(1) + c = -1 + c.$$

Thus $c = 3$ and $x^3 - 2x + 3$ is the solution to this problem. ■

Here are some examples of indefinite integrals.

$$\int x^n\,dx = \frac{1}{n+1}x^{n+1} + c, \qquad n \neq -1,$$

$$\int \cos x\,dx = \sin x + c,$$

$$\int \sin x\,dx = -\cos x + c,$$

$$\int (t+1)^2\,dt = \frac{1}{3}(t+1)^3 + c,$$

$$\int (aw+b)\,dw = \frac{1}{2}aw^2 + bw + c, \qquad a, b \text{ are constants.}$$

Each formula is verified by computing the derivative of the right-hand side with respect to the variable and verifying that it is indeed an antiderivative of the indicated function. The process of evaluating $\int f(x)\,dx$, for a given $f(x)$, is called *integration of* $f(x)$. In the above examples, the integral is evaluated by inspection; that is we rely on our knowledge of certain functions and their derivatives to guess the antiderivative and simply verify that the guess is correct. Some more systematic techniques for evaluating an indefinite integral are discussed in later sections.

Differential Equations

We are all familiar with equations of the type

$$3x^2 - 21x - 98 = 0$$

in which x stands for some unknown number. By a solution of the equation we mean any number that satisfies the equation. This quadratic equation happens to have two solutions. A *differential equation* is an equation involving an unknown function $y = f(x)$, its derivatives and functions of x. For example,

$$y' + 2x = 0 \qquad y'' + y = 0 \qquad y' - \sin y = 0$$

are differential equations. A solution of a differential equation is a function that satisfies the equation. Some differential equations, such as $y' + 2x = 0$, can be solved by a simple integration. If $y' + 2x = 0$, then $y' = -2x$, and so the unknown function y is an antiderivative of $-2x$; that is

$$y = \int -2x\,dx = -x^2 + c.$$

The second example above is not so easily solved, partly because it involves a second derivative and partly because of the two terms involving the unknown function y. Some solutions of the equation $y'' + y = 0$ can be guessed from our knowledge of the trigonometric functions; $y = \sin x$ and $y = \cos x$ are two of the infinitely many solutions. The third equation above is very complicated and no function studied to this point is a solution. These examples show that some differential equations can be solved quite easily using indefinite integrals, whereas others require more advanced techniques. We consider only a few examples of the simplest kind.

Separation of Variables

A differential equation is solvable by the method of *separation of variables* if it is possible to rearrange the terms of the equation so that one side of the equal sign is expressed in terms of the unknown function y and the other side involves only the variable x.

EXAMPLE 2 Solve the differential equation $2y'x^2 = 6 + 4x^3$.

Solution: We assume $x \neq 0$; then the equation can be rewritten as

$$y' = \frac{6}{2x^2} + \frac{4x^3}{2x^2} = 3x^{-2} + 2x.$$

The solution is

$$y = \int 3x^{-2} + 2x\,dx = -3x^{-1} + x^2 + c.$$

■

EXAMPLE 3 Find the solution $y = f(x)$ of the differential equation

$$y'' = x + 3$$

that satisfies

$$f(6) = 1 \quad \text{and} \quad f'(6) = 0.$$

Solution: We integrate the equation once to obtain y':

$$y' = \int x + 3\,dx = \frac{x^2}{2} + 3x + c.$$

The condition $f'(6) = 0$ translates into the condition

$$f'(6) = \frac{6^2}{2} + 3(6) + c = 36 + c = 0.$$

Thus $c = -36$ and $y' = x^2/2 + 3x - 36$. Another integration yields

$$y = \int \frac{x^2}{2} + 3x - 36\,dx = \frac{x^3}{6} + \frac{3x^2}{2} - 36x + C.$$

The constant C is found by using the condition $f(6) = 1$. We have

$$1 = f(6) = \frac{6^3}{6} + \frac{3(6^2)}{2} - 36(6) + C = -126 + C.$$

Thus $C = 127$ and $y = x^3/6 + 3x^2/2 - 36x + 127$ is the solution. ■

In the next example we deal with the second derivative in a slightly different manner.

EXAMPLE 4 Solve the differential equation $y'' = 14x^6 + x^3$.

Solution: This differential equation involves an unknown function $y = f(x)$ and its second derivative. Write $u = y' = f'(x)$ so that $u' = y''$. The original equation can be written in terms of u and x as

$$u' = 14x^6 + x^3.$$

The solution is

$$u = \int 14x^6 + x^3\,dx = 2x^7 + \frac{1}{4}x^4 + c,$$

for some constant c. To find y we use this information about u to get

$$y' = u = 2x^7 + \frac{1}{4}x^4 + c,$$

$$y = \int 2x^7 + \frac{1}{4}x^4 + c\,dx = \frac{2x^8}{8} + \frac{1}{4}\frac{x^5}{5} + cx + c_1.$$

This process required two integrations so there are two constants in the solution, c and c_1. The correctness of the solution can be verified by computing the second derivative of y. ■

EXAMPLE 5 An oscillator applies a force $\cos 4t$ directed along the x-axis to a particle of mass 1 at time t. If the particle is at rest at the origin at time $t = 0$, find its position at all future time, t on the assumption that the particle is constrained to move along the x-axis.

Solution: Let $(s(t), 0)$ denote the position of the particle at time t. The force F and acceleration a are related by

$$F = ma = a$$

because the mass equals 1. The acceleration is $a = s''(t)$ so the position function satisfies the differential equation

$$s''(t) = \cos 4t.$$

The velocity function is $v(t) = s'(t)$ so we have

$$v'(t) = \cos 4t.$$

An integration of this equation yields

$$v(t) = \frac{1}{4}\sin 4t + v_0.$$

The constant v_0 satisfies $v(0) = (1/4)\sin 0 + v_0 = v_0$. We are told that $v(0) = 0$, so $v_0 = 0$ and

$$v(t) = s'(t) = \frac{1}{4}\sin 4t.$$

Integrate this again to obtain

$$s(t) = -\frac{1}{16}\cos 4t + s_0,$$

where the constant s_0 satisfies

$$s(0) = -\frac{1}{16}\cos 0 + s_0 = -\frac{1}{16} + s_0.$$

From the equation $s(0) = 0$ follows $s_0 = 1/16$ and

$$s(t) = \frac{1 - \cos 4t}{16}.$$

■

Exercises 5.8

Evaluate the indefinite integrals.

1. $\int x^2 + x^{-2}\,dx$

2. $\int (x-1)(x+2)\,dx$

3. $\int \frac{1}{\sqrt{x}}\,dx$

4. $\int \cos x\,dx$

5. $\int \sqrt{x}(x^2 + 3x - 1)\,dx$

6. $\int t^2(t+1)(t-1)\,dt$

7. $\int \frac{w^4 - 3w^2 + 2}{w^2}\,dw$

8. $\int (z-2)^4\,dz$

9. $\int (x^{1/2}-2)(x^{1/2}+2)\,dx$

10. $\int \frac{1+2t+t^2}{(t+1)^4}\,dt$

Find all solutions to the differential equations.

11. $y' = 4 - x^2$

12. $y'(x+2)^2 = 1$

13. $y'' = x + 1$

14. $y'' = \sin x$

15. $s'' = 4\cos 2t$

16. $s'' = 3\sin(2t+1)$

Find a solution $y = f(x)$ of the differential equation that satisfies the given conditions.

17. $y' = 1 - x^3$, $f(1) = 0$

18. $y'(x+2)^2 = 1$, $f(1) = 1$

19. $y' = 2x - 4\cos 2x$, $f(0) = 4$

20. $y'' = x + 1$, $f(0) = 0$, $f'(0) = 4$

21. $y'' = x + 1$, $f(0) = 4$, $f'(0) = 0$

22. $y'' = \sin x$, $f(\pi) = 1$, $f'(\pi) = 2$

23. Show that for any constants c_1 and c_2, the function $f(x) = c_1 \cos x + c_2 \sin x$ is a solution of the differential equation $y'' + y = 0$.

24. Let $y = f(x)$ be a solution to the differential equation $y'' + y = 0$. Suppose that $f(0) = f'(0) = 0$. Prove that $f^{(n)}(0) = 0$ for every positive integer n.

25. It can be proved that the only solution of the equation $y'' + y = 0$ that satisfies $f(0) = f'(0) = 0$ is the zero-function $f(x) = 0$ for all x. Use this fact to show that every solution of $y'' + y = 0$ has the form $f(x) = c_1 \cos x + c_2 \sin x$ for some constants c_1 and c_2. [Hint: Let $g(x)$ be a solution and let $f(x) = g(0)\cos x + g'(0)\sin x$. Show $g(x) - f(x)$ is a solution with some special properties.]

5.9 Integration by substitution

In this section we introduce a technique that is useful for finding indefinite integrals of derivatives of composite functions. The derivative of the composite function $f(g(x))$ is given by the chain rule as

$$\frac{d}{dx} f(g(x)) = f'(g(x))g'(x).$$

It follows that

$$\int f'(g(x))g'(x) = f(g(x)) + c.$$

To apply this formula to evaluate an integral $\int h(x)\,dx$, it is necessary to find the functions $f(x)$ and $g(x)$ that satisfy $h(x) = f'(g(x))g'(x)$. We regard this as a *change of variable* problem in the following sense.

Introduce a new variable u related to x by the rule $u = g(x)$. Then we compare two integrals:

$$\int f'(u)\,du = f(u) + c \qquad \int f'(g(x))g'(x)\,dx = f(g(x)) + c.$$

The right side of the two equations are equal, $f(u) = f(g(x))$. The expressions $f'(u)\,du$ and $f'(g(x))g'(x)\,dx$ are equal if we agree on the following conventions:

$$u = g(x) \quad \text{implies} \quad du = g'(x)\,dx.$$

This convention is easily remembered by regarding the numerator and denominator of du/dx as separate quantities; that is $u = g(x)$ implies

$$\frac{du}{dx} = g'(x) \quad \text{and} \quad du = g'(x)\,dx.$$

Here are some examples that use this idea.

EXAMPLE 1 Evaluate

$$\int 2x(x^2 + 4)^{30}\, dx.$$

Solution: Let $u = x^2 + 4$ so that $du = 2x\, dx$. Then

$$\int 2x(x^2 + 4)^{30}\, dx = \int u^{30}\, du = \frac{u^{31}}{31} + c = \frac{(x^2 + 4)^{31}}{31} + c.$$

It is a good habit to check the integration by computing the derivative of the result. ■

EXAMPLE 2 Evaluate

$$\int x \sin x^2\, dx.$$

Solution: Let

$$u = x^2 \quad \text{so that} \quad du = 2x\, dx.$$

The term $2x\, dx$ must be divided by 2 to make it agree with the corresponding part of the given integral. Thus

$$\int x \sin x^2\, dx = \int (\sin u)\frac{1}{2}\, du = -\frac{1}{2}\cos u + c = -\frac{1}{2}\cos x^2 + c.$$ ■

The next example illustrates a slightly different way to handle the adjustment of a constant factor.

EXAMPLE 3 Evaluate

$$\int x\sqrt{1 - x^2}\, dx.$$

Solution: Let

$$u = 1 - x^2 \quad \text{and then} \quad du = -2x\, dx.$$

Then

$$\begin{aligned}\int x\sqrt{1 - x^2}\, dx &= -\frac{1}{2}\int \sqrt{1 - x^2}(-2x\, dx)\\ &= -\frac{1}{2}\int u^{1/2}\, du\\ &= -\frac{1}{2}\frac{u^{3/2}}{3/2} + c\\ &= -\frac{1}{3}(1 - x^2)^{3/2} + c.\end{aligned}$$

In this computation, we multiplied the term following the integral sign by (-2) to agree with the du term and then compensated by dividing the entire integral by (-2) so the value did not change. To ensure that the constants have been correctly used, check the result by computing the derivative of the answer. ■

Change of Variable in Definite Integrals

When the substitution method is used to evaluate a definite integral, the following formula might eliminate one step of the work.

THEOREM 5.12

Change of Variable Formula

Let $f(x)$ and $g(x)$ be functions such that $g(x)$ is differentiable with $g'(x)$ and $f(g(x))$ continuous on $[a, b]$. Then

$$\int_a^b f(g(x))g'(x)\,dx = \int_{g(a)}^{g(b)} f(u)\,du.$$

Proof

By assumption, $f(g(x))$ is continuous on $[a, b]$ so $f(u)$ is continuous on $[g(a), g(b)]$ (or on $[g(b), g(a)]$ if $g(b) < g(a)$). Let $F(x)$ be an antiderivative of $f(x)$ that interval. Then

$$\frac{d}{dx}F(g(x)) = F'(g(x))g'(x) = f(g(x))g'(x)$$

so $F(g(x))$ is an antiderivative of $f(g(x))g'(x)$. The Fundamental Theorem implies

$$\int_a^b f(g(x))g'(x)\,dx = F(g(x))\Big|_a^b = F(g(b)) - F(g(a)).$$

Because $F(u)$ is an antiderivative of $f(u)$, the Fundamental Theorem implies

$$\int_{g(a)}^{g(b)} f(u)\,du = F(u)\Big|_{g(a)}^{g(b)} = F(g(b)) - F(g(a)).$$

The two integrals in the theorem have the same value, $F(g(b)) - F(g(a))$ and so the theorem is proved. □

EXAMPLE 4 Evaluate the definite integral

$$\int_0^1 \frac{2x}{(1 + x^2)^3}\,dx.$$

Solution: Let $u = 1 + x^2$ so that $du = 2x\,dx$. In the change of variable formula, the role of $g(x)$ is now played by $u = 1 + x^2$ and the interval $[a, b]$ is $[0, 1]$. In particular, when $x = 0$ then $u = g(0) = 1$ and when $x = 1$ then $u = g(1) = 2$. Thus

$$\int_0^1 \frac{2x}{(1 + x^2)^3}\,dx = \int_1^2 \frac{du}{u^3} = \frac{u^{-2}}{-2}\Big|_1^2 = -\frac{1}{2}\left[\frac{1}{4} - 1\right] = \frac{3}{8}.$$ ■

EXAMPLE 5 Evaluate the integral

$$\int_0^{\pi/4} \cos 2t\,dt.$$

Solution: We know an antiderivative for $\cos x$, so make a change of variable $u = 2t$ so that $du = 2dt$; then $u = 0$ when $t = 0$ and $u = \pi/2$ when $t = \pi/4$. We have

$$\int_0^{\pi/4} \cos 2t\,dt = \int_0^{\pi/2} \frac{1}{2}\cos u\,du = \frac{1}{2}\sin u\Big|_0^{\pi/2} = \frac{1}{2}.$$ ■

When indefinite integrals are evaluated, the result can be checked by differentiating the result to verify that the antiderivative has been correctly found. It is not as easy to check the evaluation of a definite integral. It is possible to make some rather crude approximations to show that the answer is reasonable. We illustrate this with the integral in example 4.

The function

$$f(x) = \frac{2x}{(1+x^2)^3}$$

is nonnegative over the interval $[0, 1]$ and so the definite integral represents the area bounded by the graph of $f(x)$ and the x-axis over that interval. In particular, the result should be positive. We can be somewhat more precise. If m and M are the minimum and maximum values of $f(x)$ over $[0, 1]$, then

$$m(1-0) = m \leq \int_0^1 \frac{2x}{(1+x^2)^3}\,dx \leq M(1-0) = M$$

as we see by considering the lower and upper sum for the partition consisting of just one subinterval. If we are satisfied with a crude estimate, it is not necessary to compute the precise values of m and M; clearly $m = 0$. From the conditions $0 \leq x \leq 1$ and $1 \leq (1+x^2)^3$, we see that

$$f(x) = 2\frac{x}{(1+x^2)^3} \leq 2.$$

Thus $M \leq 2$ and the value of the integral should lie between 0 and 2. Thus the value $3/8$ is on the correct interval at least. In chapter 11 we study methods to find numerical estimates of definite integrals much more precisely than illustrated by this rather crude estimate.

Exercises 5.9

Use an appropriate substitution to evaluate the integrals.

1. $\int (4x+5)^3\,dx$

2. $\int (6-2x)^7\,dx$

3. $\int \sqrt{20-3x}\,dx$

4. $\int \frac{x\,dx}{(x^2+1)^2}$

5. $\int \frac{x^2\,dx}{(x^3-3)^7}$

6. $\int \frac{w\,dw}{\sqrt{3w^2+7}}$

7. $\int x(1+x^2)^n\,dx,\ n \neq -1$

8. $\int 4t(1+2t^2)^4\,dt$

9. $\int \frac{z^3\,dz}{(z^4-3)^6}$

10. $\int x^2\sqrt{3+x^3}\,dx$

11. $\int x^m\sqrt{a+x^{m+1}}\,dx$

12. $\int 3x\sin x^2\,dx$

13. $\int (x+1)\cos(x^2+2x)\,dx$

14. $\int t\sin t^2 + t\cos t^2\,dt$

15. $\int \sin 4x\,dx$

16. $\int 6\cos(3t+2)\,dt$

17. $\int \cos(6-\pi x)\,dx$

18. $\int \sin 3w - \cos 5w\,dw$

Evaluate the following definite integrals by making an appropriate substitution.

19. $\int_0^1 (4x+1)^2\,dx$

20. $\int_{-1}^4 \sqrt{20-3x}\,dx$

21. $\int_0^1 \frac{x\,dx}{(x^2+1)^2}$

22. $\int_{-1}^1 \frac{x\,dx}{(x^2+1)^2}$

23. $\int_2^3 \frac{t^2\,dt}{(t^3-3)^7}$

24. $\int_{-3}^{-1} \frac{w\,dw}{\sqrt{3w^2+7}}$

25. $\int_0^2 x(1+x^2)^n\,dx,\quad n \neq -1$

Let $f(x)$ be a function defined on an interval I of the form $[-a, a]$ for some positive number a. We call $f(x)$ an *even function* if $f(-x) = f(x)$ for all x on I. We call $f(x)$ an *odd function* if $f(-x) = -f(x)$ for all x on I. These names are appropriate because sums of even powers of x are even functions; $f(x) = x^2$ and $7 + 5x^2 - 3x^4 + 9x^8$ are examples of even functions. Sums of odd powers of x are odd functions; $f(x) = x^3$ and $-4x + 7x^5 - 3x^9$ are examples of odd functions. Polynomials are not the only even or odd functions; for example $f(x) = \sin x$ is an odd function and $f(x) = \cos x$ is an even function. The following exercises ask you to prove facts about even and odd functions.

Determine if the following functions are even, odd or neither.

26. $f(x) = \dfrac{x^2 + 8}{x^4 - 3x^2}$

27. $f(x) = x^3 \sin x$

28. $f(x) = \sin(x^3) + x \cos x$

29. $f(x) = \dfrac{x^3 + 1}{x^3 - 1}$

30. $f(x) = (x^2 - 1)\cos x + \dfrac{x \sin x}{\cos x}$

31. $f(x) = g(x) + g(-x)$ for any function $g(x)$ defined on $[-a, a]$

32. $f(x) = g(x) - g(-x)$ for any function $g(x)$ defined on $[-a, a]$

33. Prove that any function $g(x)$ defined on $[-a, a]$ can be expressed as a sum of an even function and an odd function. [Hint: Look at Exercises 31 and 32.]

34. If $f(x)$ is an odd function on the interval $[-a, a]$, show

$$\int_{-a}^{a} f(x)\,dx = 0.$$

[Hint: Separate the integral as a sum over the intervals $[-a, 0]$ and $[0, a]$ and then make a change of variable $x = -t$ in one of the integrals.]

35. If $f(x)$ is an even function on the interval $[-a, a]$, show

$$\int_{-a}^{a} f(x)\,dx = 2\int_{0}^{a} f(x)\,dx.$$

[See hint for Exercise 34.]

36. Let $p(x)$ and $q(x)$ be polynomials. Prove that $p(x^2)/q(x^2)$ is an even function and $xp(x^2)/q(x^2)$ is an odd function. [If you are very good with polynomials you may wish to try to prove that any even, rational function equals $p(x^2)/q(x^2)$ for some polynomials p and q and that every odd, rational function has the form $xp(x^2)/q(x^2)$ for some polynomials p and q.]

5.10 Areas in polar coordinates

Some regions in the xy-plane are more conveniently described using *polar coordinates* (r, θ) in place of the usual rectangular coordinates (x, y).

To review briefly, the polar coordinates of a point having rectangular coordinates (x, y) are the numbers (r, θ), where r is the nonnegative distance from the origin to the point and θ is the angle (in radians) from the positive x-axis to the radial line from the origin to the point. If the point is the origin, then $r = 0$ and θ is undefined (or θ has any value you like). The algebraic relations between the rectangular and polar coordinates are

$$x = r\cos\theta, \qquad r = \sqrt{x^2 + y^2},$$

$$y = r\sin\theta, \qquad \tan\theta = \frac{y}{x} \quad x \neq 0.$$

Special cases arise for the last equation when $x = 0$. If the point has rectangular coordinates $(0, y)$, then the polar coordinates are (r, θ) with

$$r = |y|, \qquad \theta = \begin{cases} \pi/2 & \text{if } y > 0 \\ -\pi/2 & \text{if } y < 0 \\ \text{undefined} & \text{if } y = 0. \end{cases}$$

Regions in Polar Coordinates

If $f(\theta)$ is a nonnegative function, then the region consisting of all points with polar coordinates satisfying $\alpha \le \theta \le \beta$ and $0 \le r \le f(\theta)$ has an area that can be computed as an integral

$$A = \frac{1}{2}\int_{\alpha}^{\beta} f(\theta)^2\,d\theta.$$

We must assume some condition on α and β to ensure that no region of positive area is covered more than once; for example $0 \le \alpha < \beta \le 2\pi$ or $-\pi \le \alpha < \beta \le \pi$ is often adequate. This formula is used here to illustrate some integral computations but the proof of the area formula is postponed until section 15.9.

EXAMPLE 1 Compute the area of the circle of radius a.

Solution: The region inside the circle of radius a with center at the origin consists of the points having polar coordinates that satisfy $0 \le r \le a$ and $0 \le \theta \le 2\pi$. Using $r = f(\theta) = a$ in the formula for the area, we obtain

$$A = \frac{1}{2}\int_0^{2\pi} a^2\, d\theta = \frac{1}{2}a^2(2\pi) = \pi a^2.$$

Thus we obtain the familiar formula. ■

EXAMPLE 2 Compute the area of the region enclosed by the graph of the equation $r = a(1 - \cos\theta)$ for $-\pi \le \theta \le \pi$ and a a positive constant.

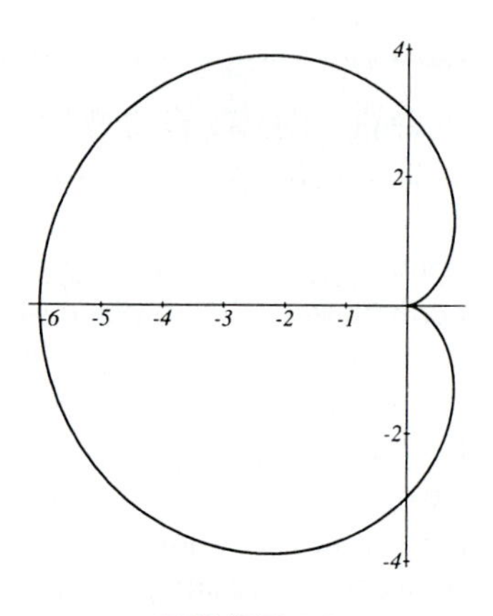

FIGURE 5.27
$r = 3(1 - \cos\theta)$

Solution: We note that the function $a(1 - \cos\theta)$ is nonnegative. As θ varies over the interval $[-\pi, \pi]$, the curve traced is called a *cardioid* (because it is roughly heart shaped). Using the formula for the enclosed area, we find

$$\begin{aligned} A &= \frac{1}{2}\int_{-\pi}^{\pi} a^2(1 - \cos\theta)^2\, d\theta \\ &= \frac{a^2}{2}\int_{-\pi}^{\pi} 1 - 2\cos\theta + \cos^2\theta\, d\theta \\ &= \frac{a^2}{2}\int_{-\pi}^{\pi} 1 - 2\cos\theta + \frac{1}{2}\left(1 + \cos 2\theta\right) d\theta \\ &= \frac{a^2}{2}\left(2\pi + 0 + \frac{1}{2}(2\pi) + 0\right) = \frac{3}{2}\pi a^2. \end{aligned}$$

■

Some computations of areas in polar coordinates require integration of combinations of trigonometric functions. The following identities help to evaluate integrals involving $\sin^2 x$ or $\cos^2 x$:

Double Angle Formulas

$$\sin^2 x = \frac{1 - \cos 2x}{2} \qquad \cos^2 x = \frac{1 + \cos 2x}{2}.$$

Exercises 5.10

Find the area of the regions R described using polar coordinates. Make a sketch of each region before doing the computation of the area.

1. R is the set of points with polar coordinates (r, θ) satisfying $0 \le r \le 3$, $0 \le \theta \le \alpha$ with α a constant between 0 and 2π.

2. R is the set of points with polar coordinates (r, θ) satisfying $r = \theta$, $0 \le \theta \le \pi$.

3. R is the set of points with polar coordinates (r, θ) satisfying $0 \le r \le \sqrt{\theta}$, $0 \le \theta \le \pi$.

4. R is the set of points with polar coordinates (r, θ) satisfying $0 \le r \le 4\cos\theta$, $-\pi/2 \le \theta \le \pi/2$.

Each of the following polar curves $r = f(\theta)$ encloses a region of the plane. Make a sketch of the region and then find its area.

5. $r = 2\sin\theta$

6. $r = 6$

7. $r = 3(1 + \cos\theta)$

8. $r = 4(1 + \sin\theta)$

9. $r = \sin 2\theta$

10. $r = \cos 2\theta$

11. Suppose g is a continuous, positive function defined on $I = [-1, 1]$. The graph of the polar equation $r = g(\cos\theta)$ is a closed curve enclosing a region of area A. Let m and M be the minimum and maximum values of $g(x)$ for x on I. Prove that A satisfies the inequality

$$\pi m^2 \le A \le \pi M^2.$$

Give an analytical proof using properties of the integral and also a geometric proof by interpreting the terms of the inequality as areas of certain regions.

6 Computations Using the Definite Integral

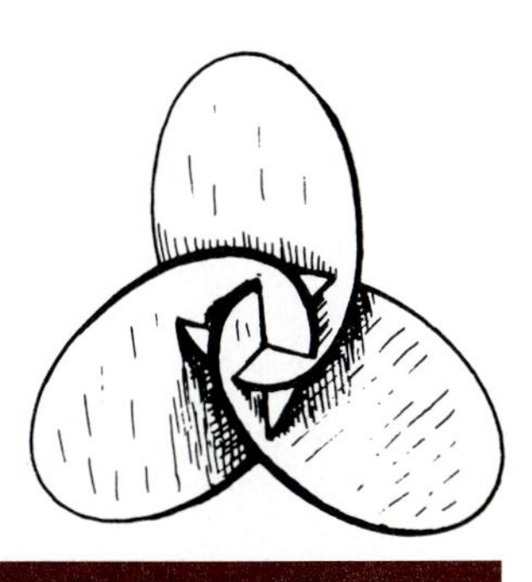

The definite integral is an important computational tool in mathematical and physical problems. Several applications are illustrated in this chapter—the volume of a sphere, the length of the graph of a function, area of the surface of a sphere, the work done by a variable force and determination of the center of mass are examples considered.

6.1 Volumes of certain solids

The first application of the definite integral is to the computation of the volume of certain solid bodies. We begin with a very simple solid constructed as follows: Suppose a region R in the xy plane has area A. A solid is formed by erecting a vertical "fence" of constant height h along the boundary of R. The region enclosed by R at the bottom, a "slice" congruent to R at the top, and the fence is called the *cylinder of height h over R*.

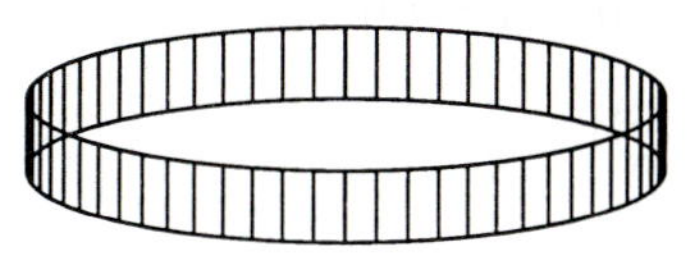

FIGURE 6.1
Cylinder over region R

Every cross section of this solid parallel to the base is a region congruent to R. The *volume* of the cylinder is hA, the height times the area of the base. If the base R is a square with side a, and if the fence has height a, then the cylinder is a cube with volume a^3. If the base R is a rectangle with sides a and b, then $A = ab$ and the volume is hab, the familiar formula for the volume of a rectangular parallelopiped. We make use of this basic volume computation to determine the volume of certain more complicated solids.

Volumes from Known Cross Sections

Suppose a solid S is given. Establish reference points by taking the x-axis as a line through S. For a point x on the axis, the cross section at x is the slice of the solid, perpendicular to the axis, that meets the axis at x.

Let the area of this slice be $A(x)$. We assume that the boundary of the solid consists of "smooth" surfaces (this term is not precisely defined here) to ensure that $A(x)$ is a continuous function on $[a, b]$, the interval of the x-axis cut off by S. Then the volume of the solid is

$$\mathrm{vol}(S) = \int_a^b A(x)\,dx. \qquad \textbf{(6.1)}$$

Let us indicate why this integral formula is correct. We prove that for any partition P of $[a, b]$, the volume satisfies

$$L(P) \leq \text{vol}\,(S) \leq U(P)$$

where $L(P)$ and $U(P)$ are the lower sum and upper sum for $A(x)$ over the partition P.

Let $P : a = x_0 < x_1 < \cdots < x_n = b$ be a partition of $[a, b]$ and consider the slices of the solid between the cross sections at the x_i. The ith piece of the solid between x_{i-1} and x_i, is not quite a cylinder, as described above, because not all the cross sections are congruent. We can estimate the volume of the ith slice by comparing it to two cylinders. The volume, V_i of the piece of the solid between x_{i-1} and x_i satisfies

$$A(p_i)(x_i - x_{i-1}) \leq V_i \leq A(q_i)(x_i - x_{i-1}),$$

where $A(p_i)$ and $A(q_i)$ are the minimum and maximum values of A on $[x_{i-1}, x_i]$. To see this, consider the cylinder of height $h = x_i - x_{i-1}$ over the slice of area $A(p_i)$ at p_i. Because $A(p_i)$ is the smallest value of A, this cylinder has volume no larger than the volume of the ith piece of S. Similarly the cylinder of height h with base of area $A(q_i)$ has volume no smaller than the volume of the ith piece of the solid.

Just as we have done in other applications, we sum these inequalities and get

$$L(P) \leq V \leq U(P),$$

where $L(P)$ and $U(P)$ are the lower and upper sums for the function $A(x)$ corresponding to the partition P. It follows that equation (6.1) the correct volume.

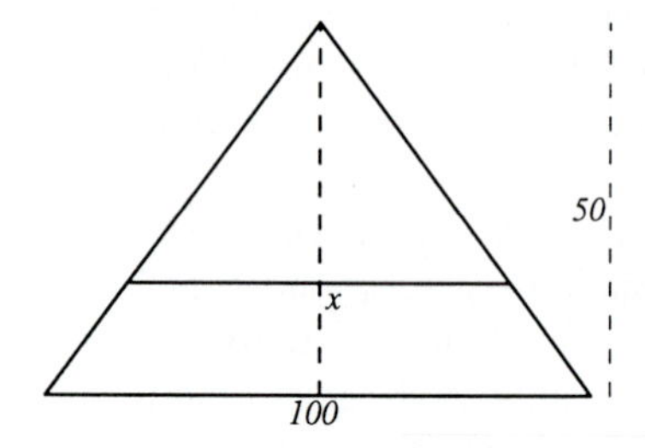

FIGURE 6.2
Cross section of pyramid

EXAMPLE 1 The vertex of a pyramid is 50 feet above the center of the base, which is square with side 100 feet. Determine its volume.

Solution: Figure 6.2 shows the vertical cross section containing the vertex. We select the x-axis starting at the vertex and running perpendicular to the base. At the point x, the horizontal slice of the pyramid is a square with side $2x$, as an argument with similar triangles shows. The cross section at x has area $A(x) = (2x)^2$ and the volume of the pyramid is

$$V = \int_0^{50} (2x)^2\, dx = 4\frac{x^3}{3}\bigg|_0^{50} = \frac{4}{3} \cdot 50^3.$$ ■

If a solid has constant cross-sectional area $A(x) = k$, then the integral formula gives its volume as kh, the cross-sectional area times the height. For example, a cylinder having all cross sections equal to a circle of radius R has volume $\pi R^2 h$ even if the cylinder is skewed.

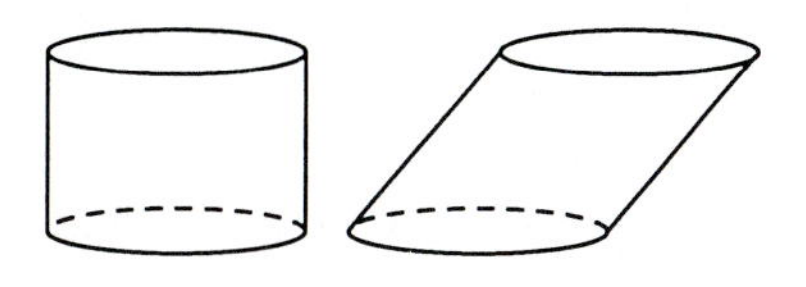

FIGURE 6.3
Equal cross sections and volumes

Volumes of Solids of Revolution

Let $f(x)$ be a continuous and nonnegative function over the interval $[a, b]$ and let R be the region between the interval and the graph of f.

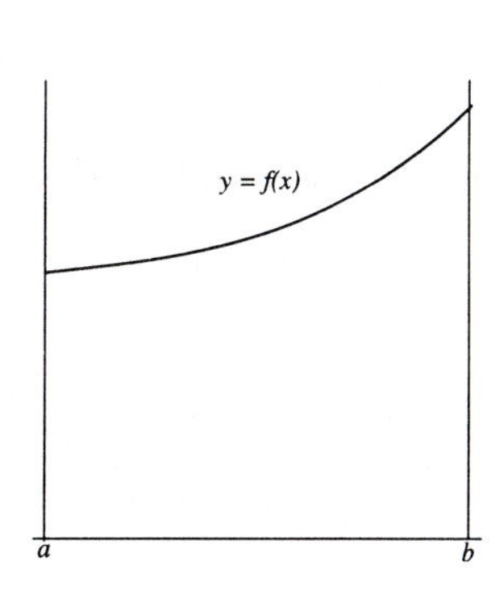

FIGURE 6.4
Graph to be rotated

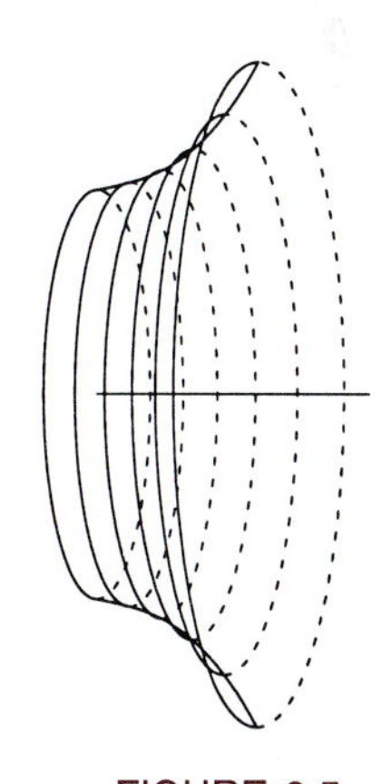

FIGURE 6.5
Solid of revolution

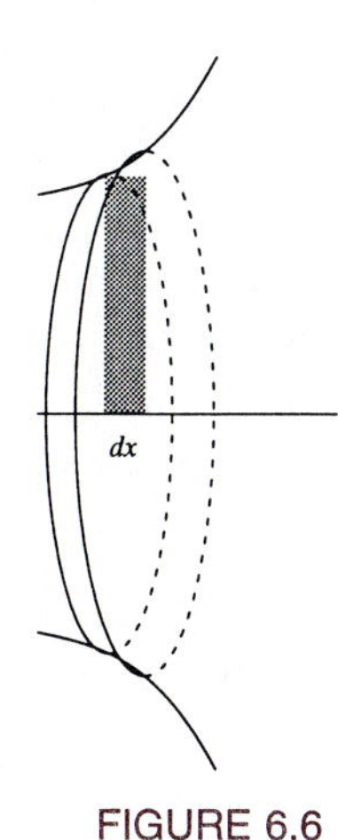

FIGURE 6.6

A solid S is formed by revolving the region R around the x-axis. We compute the volume S using the method of cross sections just described. The cross section through x is a circle of radius $f(x)$ and area $A(x) = \pi f(x)^2$. Thus we have the formula

$$\text{Volume of the solid of revolution} = \int_a^b \pi f(x)^2 \, dx. \qquad \textbf{(6.2)}$$

There is a convenient way to remember this formula. Think of a very thin piece of the solid constructed by revolving the rectangle with base at x having width dx and height $f(x)$. It contributes a volume $\pi f(x)^2 \, dx$ to the total solid; the integral of this quantity performs an addition of the volumes of all these slices as x varies from a to b.

EXAMPLE 2 Find the volume of a sphere of radius a.

Solution: The equation of a circle of radius a is $x^2 + y^2 = a^2$ and the upper half of the circle is the graph of the function

$$f(x) = \sqrt{a^2 - x^2}$$

over the interval $[-a, a]$. When the upper half of the circle is revolved around the x-axis, the solid traced out is the sphere of radius a. The formula for the volume of this solid is

$$\pi \int_{-a}^{a} (\sqrt{a^2 - x^2}\,)^2\,dx = \pi \int_{-a}^{a} a^2 - x^2\,dx = \pi \left(a^2 x - \frac{x^3}{3}\right)\Bigg|_{-a}^{a} = \frac{4}{3}\pi a^3. \quad \blacksquare$$

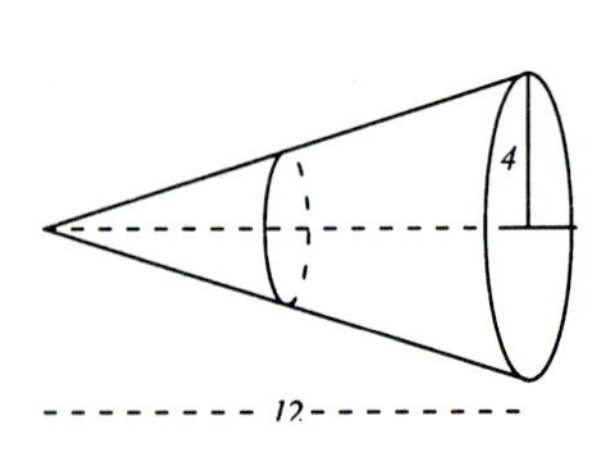

FIGURE 6.7
Circular cone

EXAMPLE 3 Compute the volume of the right circular cone having height 12 and base of radius 4.

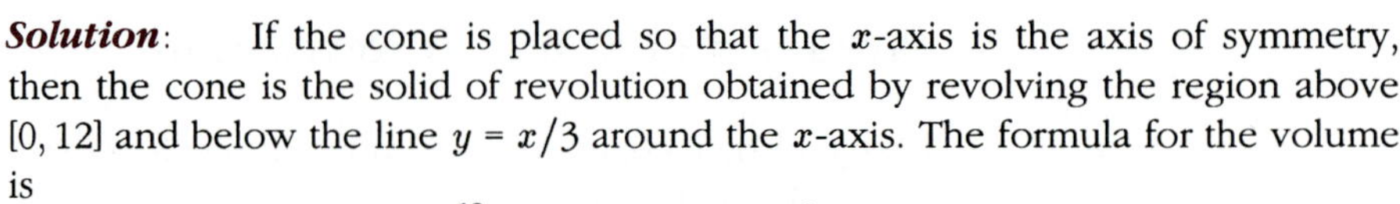

Solution: If the cone is placed so that the x-axis is the axis of symmetry, then the cone is the solid of revolution obtained by revolving the region above $[0, 12]$ and below the line $y = x/3$ around the x-axis. The formula for the volume is

$$\int_0^{12} \pi (x/3)^2\,dx = \pi \frac{1}{9}\frac{x^3}{3}\Bigg|_0^{12} = 64\pi. \quad \blacksquare$$

Solids with Holes

Suppose $f(x)$ and $g(x)$ are continuous on $[a, b]$ and

$$f(x) \geq g(x) \geq 0$$

for all x on the interval. The region between the two curves can be revolved around the x-axis to obtain a solid with a hole. The volume of the solid equals

$$\pi \int_a^b f(x)^2 - g(x)^2\,dx. \qquad \textbf{(6.3)}$$

This can be seen by first revolving the region bounded by $[a, b]$ and the graph of f to obtain a solid with volume given by equation (6.2). From this volume, subtract the volume of the hole, which is given by equation (6.2) using g in place of f.

EXAMPLE 4 Find the volume of the "doughnut" shape obtained by revolving the circle $(y-4)^2 + x^2 = 9$ around the x-axis. (This solid is an example of a *torus*.)

Solution: Let

$$f(x) = 4 + \sqrt{9 - x^2},$$
$$g(x) = 4 - \sqrt{9 - x^2},$$

so that the graph of f is the upper half of the circle and the graph of g is the lower half of the circle. The volume of the torus is

$$\begin{aligned} \pi \int_{-3}^{3} f(x)^2 - g(x)^2\,dx &= \pi \int_{-3}^{3} (4 + \sqrt{9 - x^2})^2 - (4 - \sqrt{9 - x^2})^2\,dx \\ &= \pi \int_{-3}^{3} 16\sqrt{9 - x^2}\,dx. \end{aligned}$$

We have not found an antiderivative for the function $\sqrt{9 - x^2}$ in any examples to this point. The antiderivative of this function is evaluated later by the method of trigonometric substitutions. Nevertheless, the numerical value of the definite integral can be found by using the formula for the area of a circle. The integral

$$\int_{-3}^{3} \sqrt{9 - x^2}\, dx$$

is the area of the region under the curve $y = \sqrt{9 - x^2}$ and over the interval $[-3, 3]$; namely, half of the area of a circle of radius 3, $9\pi/2$. Thus the volume of the given torus is $16\pi(9\pi/2) = 72\pi^2$. ■

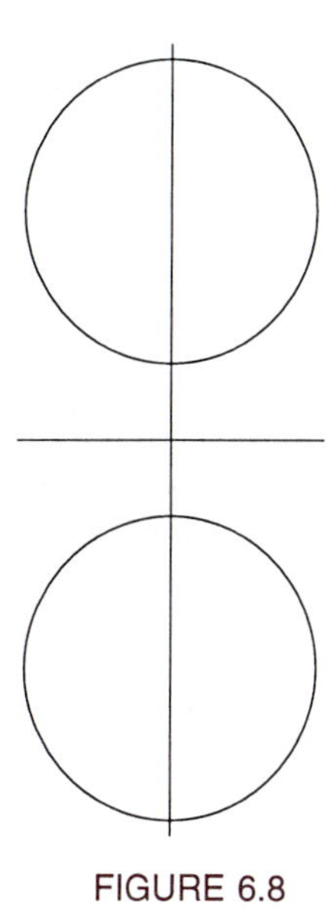

FIGURE 6.8

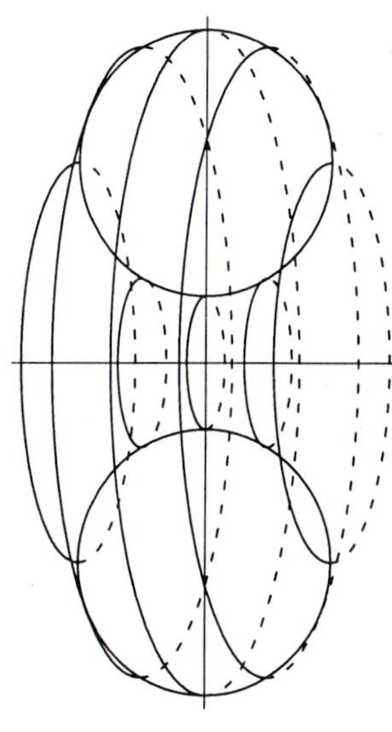

FIGURE 6.9

Exercises 6.1

Sketch the region in the xy plane bounded by the graphs indicated and then find the volume of the solid obtained by rotating the region around the x-axis.

1. $y = 2x + 1,\ y = 0,\ x = 0,\ x = 3$

2. $y = 3x - 2,\ y = 0,\ x = 1,\ x = 3$

3. $y = x^2,\ y = 0,\ x = 2$

4. $y = 3x^2 + 1,\ y = 0,\ x = 0,\ x = 3$

5. $y = x^2,\ y = 4x - 3$

6. $y = 4x^2 + 1,\ y = 5$

7. $y = \sqrt{1 - x^2},\ y = 0$

8. $y = \sqrt{x},\ y = x/2$

9. $y = |x|,\ y = 0,\ x = \pm 3$

10. $y = |x|,\ y = 1 - |x|$

11. The region of the xy plane bounded by the x-axis and the graph of $y = 4 - x^2$ is the base of a solid all of whose vertical cross sections parallel to the y-axis are squares. What is the volume of this solid?

12. Repeat Exercise 11 if the vertical cross sections are equilateral triangles in place of squares.

13. A swimming pool is 10 yards wide and 25 yards long. The bottom of the pool is an inclined plane so that the depth of the pool along the 10-yard side is 4 feet at one end and 8 feet at the other end. What is the volume of water in the pool?

14. A swimming pool is 10 yards wide and 25 yards long. The bottom of the pool is an inclined plane so that the depth of the pool along the 25-yard side is 4 feet at one end and 8 feet at the other end. What is the volume of water in the pool?

15. The region bounded by $y = x^2$ and $x = 1$ is revolved around the y-axis. Show that the horizontal cross section of the solid at height y has area $A(y) = \pi - \pi(\sqrt{y})^2$ and the volume of the solid is found by integrating this function with respect to y.

16. Use the method of Exercise 15 to determine the volume of the solid obtained by rotating around the y-axis the region bounded by $y = x^3$, $y = x$ and $0 \leq x \leq 1$.

17. The graph of $y = 1/x$ over the interval $[1, a]$, $a > 1$, is revolved around the x-axis. What is the volume of the solid and what is the limiting value of the volume as a increases without bound?

18. A circular hole of radius A is drilled through a sphere of radius $2A$ with the center of the hole following a diameter of the sphere. What is the volume of the remaining part of the sphere?

19. A hemispherical bowl of radius 10 inches is filled with fruit punch to a depth of 8 inches. What is the volume of punch in the bowl?

20. The general torus is obtained by revolving the circle with equation $(y-h)^2 + x^2 = R^2$ (with $R \leq h$) around the x-axis. Find the volume of this torus as a function of R and h.

21. Find the volume of a pyramid with square base of side A and having vertex of height H above the base. Explain why it does not matter in this problem if the vertex is directly over the center of the base or over any other point of the base.

22. A storage shed is constructed with a rectangular base, 8 feet wide across the front and 5 feet deep, front to back. The roof is flat (a plane) attached 6 feet from the ground at the front and 8 feet high at the back. What volume is enclosed by the shed?

23. A (poorly constructed) storage shed has a rectangular base with dimensions $a \times b$, but the four corner posts that hold up the flat roof are not of equal length. Two diagonally opposite posts have lengths h_1 and h_2 whereas the other two diagonally opposite corner posts have lengths h_3 and h_4. What is the volume enclosed by the shed? [Remark: Because the roof is a plane, it follows that $h_1+h_2 = h_3+h_4$.][Hint: Use similar triangles to determine the area of a vertical cross section and integrate.]

24. A doughnut maker wants her product to have maximum volume and still be able to stack doughnuts flat in a tall, narrow box having having a 5 in. × 5 in. cross section. (The doughnuts have the shape of a general torus described in Exercise 20.) How high must the box be to contain 12 of these plump doughnuts?

25. (Continuation of Exercise 24.) After making some doughnuts of maximum volume and total diameter 5 in., the baker realizes that too much dough is used. She decides to make the doughnuts only 90% of the maximum volume but still keeping the total diameter of 5 in. In this case, what are the dimensions of the doughnut and how high must the container be? [This is a difficult problem involving a cubic equation for which an approximate solution is necessary. A computational or graphing device might help.]

6.2 Work

Work is a concept defined in physics as force times distance. (Work is also what is required to learn calculus; we do not use that definition in this section.) The usual units of work are foot-pounds, dyne-centimeter (= erg), and newton-meter (= joule). For example, the work required to carry a 4-pound calculus book up 30 feet to the fourth floor of the mathematics building is $4 \cdot 30 = 120$ foot-pounds. More generally, if a *constant* force F is exerted to move an object a distance D, the work done is

$$W = F \cdot D.$$

If the force is not constant, an integral is used to compute the work. We indicate briefly how this is derived.

Suppose an object is moved from $x = a$ to $x = b$ along the x-axis and the force acting on the object at point x is $F(x)$. We assume that $F(x)$ is a continuous function. Partition the interval $[a, b]$ into n subintervals, $a = x_0 < x_1 < \cdots < x_n = b$. On the ith subinterval let $F(p_i)$ and $F(q_i)$ be the minimum and maximum values of F. If the work done moving the object over $[x_{i-1}, x_i]$ is W_i, then

$$F(p_i)(x_i - x_{i-1}) \leq W_i \leq F(q_i)(x_i - x_{i-1}).$$

Add these inequalities for $i = 1, 2, \ldots, n$ to get

$$L(P) \leq W_1 + \cdots + W_n = W \leq U(P),$$

where $L(P)$ and $U(P)$ are the lower and upper sums for the function $F(x)$ over

$[a, b]$ corresponding to the partition P. Thus we obtain the integral formula for the work done when a force $F(x)$ is applied along a straight line:

$$W = \int_a^b F(x)\,dx.$$

A rather standard configuration where work can be measured involves expansion and compression of a spring, as we consider in the next example.

EXAMPLE 1 Hooke's law describes the behavior of a spring; it states that the amount of force required to displace a spring x units from its equilibrium position is directly proportional to x. Suppose a spring is hanging at equilibrium and a 3-pound weight displaces the spring 1 inch downward. How much work is done displacing the spring 3 inches downward?

Solution: If $F(x)$ is the force required to displace the spring x units, then, according to Hooke's law, $F(x) = kx$ for some constant k. In this problem we are told that a force of 3 pounds produces a displacement of 1 inch (= 1/12 foot); i.e., $F(1/12) = 3$. Thus

$$F(1/12) = k\frac{1}{12} = 3, \qquad k = 36.$$

Therefore the force is $F(x) = 36x$. The work done displacing the spring from 0 to 3 inches (= 0.25 feet) is

$$W = \int_0^{0.25} F(x)\,dx = \int_0^{0.25} 36x\,dx = 18x^2\Big|_0^{0.25} = 1.125 \text{ ft-lb.}$$ ■

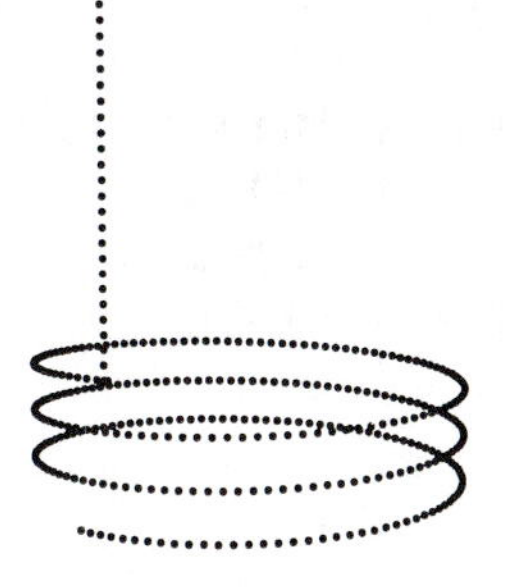

FIGURE 6.10

In the next example we follow tradition and use the Greek letter ρ (rho) to denote density.

EXAMPLE 2 A heavy cable with density ρ (in pounds per foot) is coiled on the ground. One end of the cable is attached to a crane that lifts the cable until the entire cable of length 50 feet is off the ground. How much work is done lifting the cable?

Solution: Establish a coordinate system so that distance x is measured along the length of the cable; the end attached to the crane is $x = 0$, the other end is $x = 50$. The force required to lift the cable is not constant because greater force is required as more of the cable comes off the ground and is held by the crane. The point $x = 0$ is lifted 50 ft; the point $x = 10$ is lifted 40 ft; the general point x is lifted $50 - x$ feet. To approximate the work done, partition the interval [0, 50] into n subintervals $[x_{i-1}, x_i]$. The weight of the ith segment of cable is $\rho(x_i - x_{i-1})$; the work W_i done in lifting this segment of cable satisfies the inequality

$$(50 - x_i)\rho(x_i - x_{i-1}) \leq W_i \leq (50 - x_{i-1})\rho(x_i - x_{i-1}),$$

inasmuch as each part of the segment is lifted at least $50 - x_i$ units but no more than $50 - x_{i-1}$ units. The total work is the sum of the W_i; the sum of the left- and right-hand sides of the inequality are lower and upper sums for the function $(50 - x)\rho$. Thus the work done is

$$W = \int_0^{50} (50 - x)\rho\,dx = \rho(50x - \frac{x^2}{2})\Big|_0^{50} = 1250\rho.$$

Note that this equals the work done lifitng the entire coil to a height of 25 feet. ■

Exercises 6.2

1. A hanging spring is extended 4 inches from its equilibrium position by a weight of 20 pounds. How much work is done in extending the spring 2 inches from equilibrium?

2. A coiled car spring is compressed 1 inch when a force of 400 pounds is applied to it. How much work is required to compress the spring 4 inches from its equilibrium position?

3. If a spring is compressed 1 inch from its equilibrium position by a force of 8 pounds, how much work is done in stretching the spring 4 inches beyond its equilibrium position?

4. A bucket hangs at the end of a 50-foot rope into a well. The bucket initially holds 50 pounds of water but has a leak that uniformly releases b pounds of water each foot the bucket is raised ($0 \leq b \leq 1$). The empty bucket weights 10 pounds. How much work is done lifting the bucket 50 feet?

5. A 180-pound fireman has a very long coiled fire hose on the ground. If the hose weighs 1/2 pound per foot, how much work is done when the fireman picks up one end of the hose and allows it to uncoil as he climbs to the top of a 30-foot ladder?

6. If two particles have positive charges of q and Q, respectively, and are separated by a distance x, then the repelling force exerted by one particle on the other is kqQ/x^2 for some constant k. If the particle with charge Q is at the origin, how much work is required to move the particle of charge q from a to b along the x-axis?

7. A cylindrical tank of radius R and height h is filled with fluid of density ρ. The bottom of the tank is a piston that forces the water up and over the top of the tank. How much work is done emptying the tank? [Hint: A thin horizontal slice at a depth x from the top of the tank with thickness Δx has mass $\rho R^2 \Delta x$ and the work done lifting it out of the tank is $\rho R^2 x \Delta x$. Now set up an integral for the problem.]

8. A conical tank (vertex down) has top of radius r and height h. If it is filled with liquid of density ρ, how much work is done in pumping all the liquid out over the top of the tank?

9. A holding tank has the shape of a right circular cylinder with base of diameter 20 feet and height 12 feet. Water is pumped out of the tank by a suction device that floats on the surface of the water and forces the water over the top of the tank. If the water is initially 8 feet deep, how much work is done pumping all of the water out? [Useful fact: Water weighs about 62.5 lb/ft^3.]

10. A particle of mass m moves along the x-axis under the action of a variable force F. Its velocity v is a function of time. Assume the motion is such that v is also a function of position, $v = v(x)$. The kinetic energy is $k(x) = mv^2/2$. Show that the work done in moving the particle from $x = a$ to $x = b$ is the change in kinetic energy $k(b) - k(a)$. [Hint: By Newton's law

$$F = m\frac{dv}{dt} = m\frac{dv}{dx}\frac{dx}{dt} = mv\frac{dv}{dx}.$$

Use this to change variables in the integral that gives the work.]

11. Newton's law of gravitation asserts that two bodies of masses m and M exert an attractive force on each other of magnitude $F = GMm/r^2$, where r is the distance between the bodies and G is a universal constant. How much work is done if the mass M remains fixed and m moves from $r = a$ to $r = b$? If M and m are masses of planets and m moves off to another galaxy (infinitely far away), how much work is done?

6.3 Arc length

Let $f(x)$ be continuous on the interval $I = [a, b]$. The graph of $y = f(x)$ for x on I is called a *curve* or an *arc*. A definite integral is used to compute the length of the arc. Before we set out to compute this length, it is necessary to agree on what we mean by the "length" of a curved line. We have already agreed to the meaning of the length of a straight line, so we use straight lines to approximate the length of the curve.

We take the intuitive idea that a curve from point A to point B that deviates only a little from a straight line should have length only a little larger than the

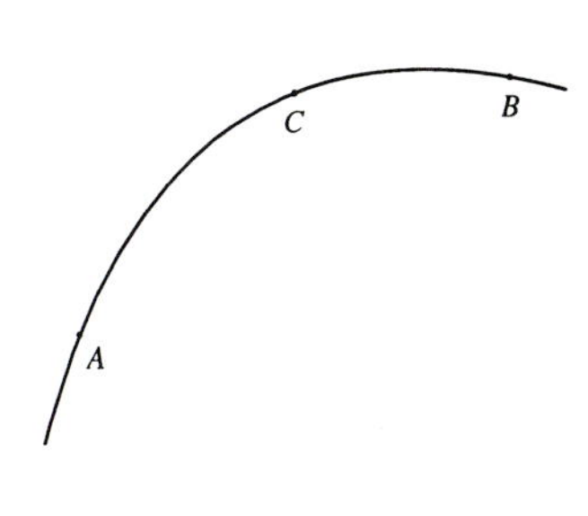

FIGURE 6.11

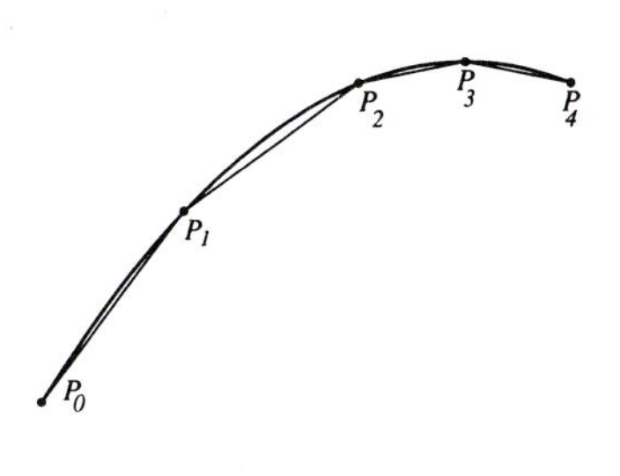

FIGURE 6.12

length of the straight line from A to B. We take another intuitive idea: the length of a curve from point A to point B should equal the sum of the lengths of the curves from A to C and from C to B when C is between A and B.

Let $P : a = x_0 < x_1 < \cdots < x_{n-1} < x_n = b$ be a partition of I and let $P_i(x_i, f(x_i))$ denote the corresponding point on the curve. The line C_i connecting P_i and P_{i+1} has length $|C_i|$, which is an approximation to the length of the arc of the curve from P_i to P_{i+1}. The polygonal path $P_0P_1 \cdots P_{n-1}P_n$ made up of the segments C_0, $C_1, \cdots$, C_{n-1} is called the *chord approximation* to the curve determined by the partition P. The sum of the lengths

$$L_C(P) = |C_0| + |C_1| + \cdots + |C_{n-1}|$$

is called the *length* of the chord approximation to the curve. As P ranges over all partitions of I we obtain a set of lengths $\{L_C(P)\}$. If this set of numbers has an upper bound, then it has a least upper bound, and this least upper bound is called the *length of the curve.*

It is not obvious that the set of lengths of the chord approximations to the curve has an upper bound; in fact it is possible for some continuous function $f(x)$ to behave so badly that the chord approximations have arbitrarily large lengths and, therefore, no upper bound. We do not discuss these pathological functions here; instead, we consider a restricted class of functions for which the length of the curve exists and can be computed as a definite integral. The restriction we impose on the function is that $f'(x)$ exists and is continuous on the interval I.

DEFINITION 6.1

Smooth Curve

The graph of the function $f(x)$ defined for x on an interval I is called a *smooth curve* if $f'(x)$ exists and is continuous on I.

We also say the function $f(x)$ is a *smooth function* on $[a, b]$ if both $f(x)$ and $f'(x)$ are continuous on $[a, b]$.

We begin by showing how the assumption that the curve is smooth is used to prove that the length of the curve is a finite number. The first consequence of smoothness is that $f'(x)^2$ has a maximum and minimum value on I because it is continuous.

THEOREM 6.1

A Smooth Curve Has Bounded Chord Approximations

Let $y = f(x)$ define a smooth curve over the interval $I = [a, b]$, let p be a point on I at which $f'(x)^2$ takes its minimum value, and let q be a point on I at which $f'(x)^2$ takes its maximum value. Then for any partition P of I, the length of the chord approximation satisfies

$$(b - a)\sqrt{1 + f'(p)^2} \leq L_C(P) \leq (b - a)\sqrt{1 + f'(q)^2}.$$

Proof

Let $P : a = x_0 < x_1 < \cdots < x_{n-1} < x_n = b$ be a partition of I. The length of the chord C_i is

$$|C_i| = \sqrt{(x_{i+1} - x_i)^2 + (f(x_{i+1}) - f(x_i))^2}$$
$$= (x_{i+1} - x_i)\sqrt{1 + \left(\frac{f(x_{i+1}) - f(x_i)}{x_{i+1} - x_i}\right)^2}$$
$$= (x_{i+1} - x_i)\sqrt{1 + f'(z_i)^2},$$

where z_i is a point with $x_i < z_i < x_{i+1}$ satisfying

$$f(z_i) = \frac{f(x_{i+1}) - f(x_i)}{x_{i+1} - x_i}$$

whose existence is ensured by the Mean Value Theorem. Because the maximum value of $f'(x)^2$ on all of $[a, b]$ is $f'(q)^2$, we have

$$|C_0| + \cdots + |C_{n-1}|$$
$$= (x_1 - x_0)\sqrt{1 + f'(z_0)^2} + \cdots + (x_n - x_{n-1})\sqrt{1 + f'(z_{n-1})^2}$$
$$\leq (x_1 - x_0)\sqrt{1 + f'(q)^2} + \cdots + (x_n - x_{n-1})\sqrt{1 + f'(q)^2}$$
$$= [(x_1 - x_0) + \cdots + (x_n - x_{n-1})]\sqrt{1 + f'(q)^2}$$
$$= (b - a)\sqrt{1 + f'(q)^2}.$$

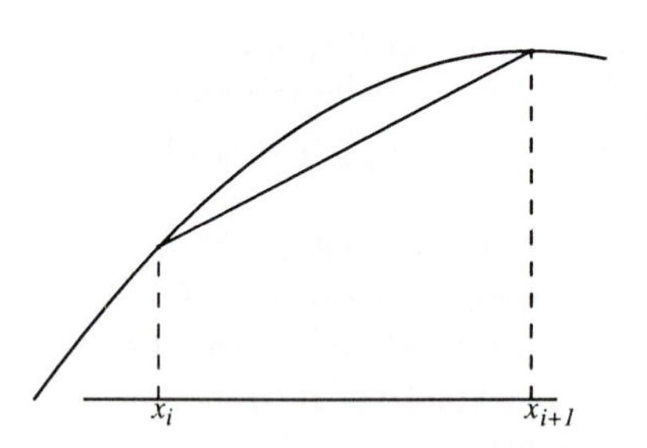

FIGURE 6.13

Thus $L_C(P) \leq (b-a)\sqrt{1 + f'(q)^2}$, proving one of the two inequalities stated in the conclusion of the theorem. The other inequality is proved in the same way by using $f'(p)^2 \leq f'(z_i)^2$ in place of $f'(z_i)^2 \leq f'(q)^2$ in the proof given. □

The proof gives an estimate of the length of a curve. We state it as the next theorem.

THEOREM 6.2

A Bound for the Arc Length

Let $y = f(x)$ define a smooth curve over the interval $I = [a, b]$ and let $f'(q)^2$ be the maximum value of $f'(x)$ for x on I. Then the length of the graph of $f(x)$ over I is at most $(b - a)\sqrt{1 + f'(q)^2}$.

Proof

In the previous proof we saw that the number $u = (b - a)\sqrt{1 + f'(q)^2}$ is an upper bound for every chord approximation, $L_C(P)$. The length of the graph is the smallest number that is an upper bound for all the chord approximations. Thus the length of the graph is less than $u = (b - a)\sqrt{1 + f'(q)^2}$. □

Now that we know the curve has a length, we compute that length.

THEOREM 6.3

Arc-Length Formula

If $y = f(x)$ is a smooth curve defined on an interval $I = [a, b]$, then the length of the curve is

$$\ell = \int_a^b \sqrt{1 + f'(x)^2}\, dx. \qquad \textbf{(6.4)}$$

Proof

The proof is carried out by showing that the length of the arc lies between each lower sum and upper sum for the function $\sqrt{1 + f'(x)^2}$ over the interval I and so equals the definite integral in equation (6.4).

Let $P : a = x_0 < x_1 < \cdots < x_n = b$ be a partition of $[a, b]$ and let p_i and q_i be points on $[x_{i-1}, x_i]$ at which $f'(x)^2$ takes on its minimum and maximum values. Let ℓ_i be the length of the curve $y = f(x)$ over $[x_{i-1}, x_i]$. By Theorem 6.1 we have

$$(x_i - x_{i-1})\sqrt{1 + f'(p_i)^2} \le \ell_i \le (x_i - x_{i-1})\sqrt{1 + f'(q_i)^2}.$$

If we add these inequalities for $i = 1, 2, \ldots, n$, the left-hand sides sum to the lower sum $L(P)$ for the function $h(x) = \sqrt{1 + f'(x)^2}$, the middle terms sum to the length ℓ of the curve, and the right-hand sides sum to the upper sum $U(P)$ for $h(x)$. Thus the length ℓ must be the unique number that lies between every lower sum and upper sum, namely

$$\ell = \int_a^b \sqrt{1 + f'(x)^2}\, dx. \qquad \square$$

The hypothesis of the theorem restricts attention to smooth curves. This involves some fairly serious restrictions inasmuch as it eliminates curves having a vertical tangent like a semicircle. We take up the study of parametric curves in section 13.5 and are able to consider a wider class of curves (including those having vertical tangents) and discover that an analogue of equation (6.4) can be used to determine their lengths.

EXAMPLE 1 Let us first apply the arc-length formula to the simplest possible case. Find the length of the line $y = 5x + 2$ between $x = 2$ and $x = 6$.

Solution: We apply the arc-length formula with $f(x) = 5x + 2$. Then

$$1 + f'(x)^2 = 1 + 5^2 = 26,$$

and

$$\int_2^6 \sqrt{1 + f'(x)^2}\, dx = \int_2^6 \sqrt{26}\, dx = \sqrt{26}x \Big|_2^6 = 4\sqrt{26}$$

is the length. Of course this could have been found by applying the distance formula to find the length of the segment connecting $(2, f(2)) = (2, 12)$ and $(6, f(6)) = (6, 32)$. ■

EXAMPLE 2 Let

$$f(x) = \frac{3x^4 + 4}{12x}.$$

Find the length of the graph of f over the interval $[1, 4]$.

Solution: To apply the arc-length formula we first compute the term

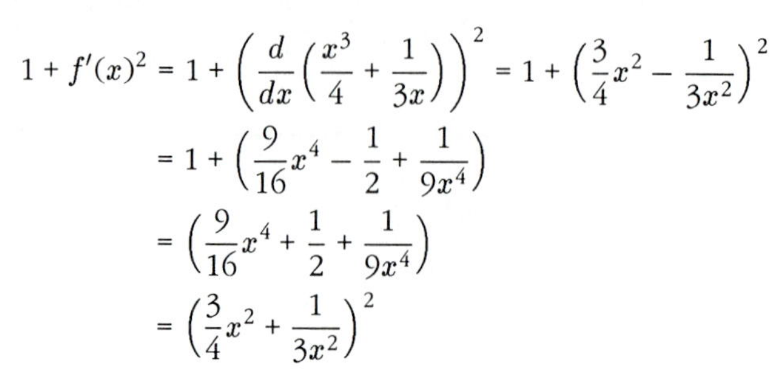

$$
\begin{aligned}
1 + f'(x)^2 &= 1 + \left(\frac{d}{dx}\left(\frac{x^3}{4} + \frac{1}{3x}\right)\right)^2 = 1 + \left(\frac{3}{4}x^2 - \frac{1}{3x^2}\right)^2 \\
&= 1 + \left(\frac{9}{16}x^4 - \frac{1}{2} + \frac{1}{9x^4}\right) \\
&= \left(\frac{9}{16}x^4 + \frac{1}{2} + \frac{1}{9x^4}\right) \\
&= \left(\frac{3}{4}x^2 + \frac{1}{3x^2}\right)^2
\end{aligned}
$$

The length of the curve is

$$\int_1^4 \frac{3}{4}x^2 + \frac{1}{3x^2}\,dx = \frac{1}{4}x^3 - \frac{1}{3x}\Big|_1^4 = 16.$$

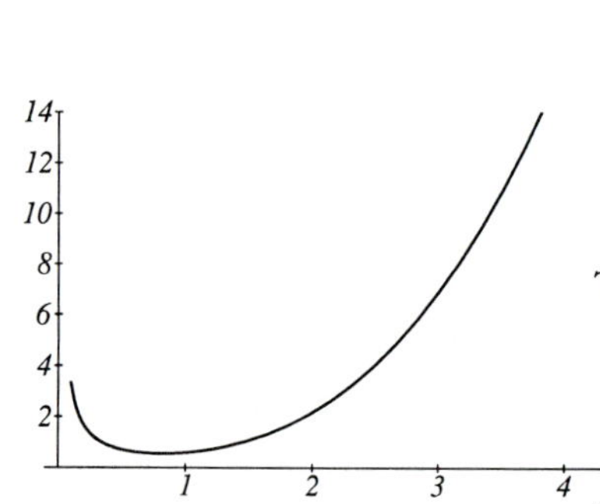

FIGURE 6.14

Let us make a crude estimate of the length just to see that the answer is reasonable. The length of the curve is greater than the straight line distance from $(1, f(1)) = (1, 7/12)$ to $(4, f(4) = (4, 193/12)$. The distance between endpoints is

$$\sqrt{(4-1)^2 + \left(\frac{193}{12} - \frac{7}{12}\right)^2} = \frac{\sqrt{997}}{2} \approx 15.787.$$

The straight line distance is only slightly smaller than the arc length; this gives some confidence in the definite integral computation. (If the straight line length had been 25, we would know there was an error; if the straight line length had been 5, we would be very suspicious.) Because the curve length differs only slightly from the straight line length, we expect the graph to be quite flat over $[1, 4]$ (see figure 6.14). ■

Exercises 6.3

Find the length of the graph of $y = f(x)$ over the indicated interval.

1. $y = 3x - 1$, $[1, 4]$

2. $y = mx + c$, $[a, b]$

3. $y = \frac{1}{3}(x^2 + 2)^{3/2}$, $[1, 3]$

4. $y = (x^4 + 3)/6x$, $[1, 2]$

5. $y = (2x^6 + 1)/8x^2$, $[1, 2]$

6. $y = 1/3x + x^3/4$, $[1, 2]$

7. $y = (15x^{4/5} - 40x^{6/5})/48$, $[1, 64]$

8. A canonball is fired with initial velocity 400 ft/sec at an angle of elevation $\pi/4$. Write a definite integral whose value equals the distance the canonball travels before it reaches its highest point. [Hint: Use the formulas in section 4.10 to describe the path and then find a function $y = f(x)$ by eliminating t whose graph also describes the path. Then use the formula for arc length.]

9. The graph of the equation $y^{2/3} + x^{2/3} = a^{2/3}$ is called an *asteroid.* Sketch this curve and find its length. [Hint: If (x, y) is on the curve then so are the four points $(\pm x, \pm y)$ so, by symmetry, the total length is four times the length of the part in the first quadrant.]

10. Let both $f(x)$ and $g(x)$ be smooth functions on $[0, 1]$ with the following properties:
(1) Both $f(x)$ and $g(x)$ are increasing functions.
(2) For every t on $[0, 1]$, the length of the graph of $f(x)$ from $(0, f(0))$ to $(t, f(t))$ equals the length of the graph of $g(x)$ from $(0, g(0))$ to $(t, g(t))$.
Prove that $g(x) = f(x) + c$ for some constant c and every x on $[0, 1]$.

11. The graph of f is a smooth curve passing through $(0, 0)$ and $(1, 1)$. The function has the property

$$|f(x + s) - f(x)| \leq |s|$$

for all x and all s with $|s| < 0.5$. Give a plausible argument using arc length to prove that the graph of f is a straight line between $(0, 0)$ and $(1, 1)$.

6.4 Surface area of solids of revolution

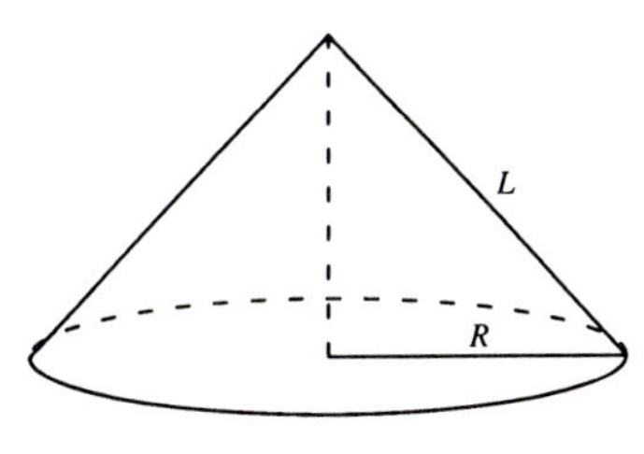

FIGURE 6.15
Cone with slant height L

We have seen how the area of a region in the xy plane can be computed with the aid of the definite integral. The area of a curved surface is not so easily computed. We make this computation for the area of certain surfaces of revolution. Let us begin with a familiar example often studied in elementary geometry: the right circular cone.

Let the cone have a circular base of radius R, height h and slant height L. Then $L^2 = R^2 + h^2$ so $L > R$. If the cone is cut by a straight line from vertex to base and then spread flat, the region covered is a sector of a circle of radius L. The arc of the circle included by the sector has length $2\pi R$, the circumference of the base of the cone. The area of the sector is the fraction $2\pi R/2\pi L = R/L$ of the area of the full circle, πL^2. The area of the surface of the cone is

$$S = \pi RL.$$

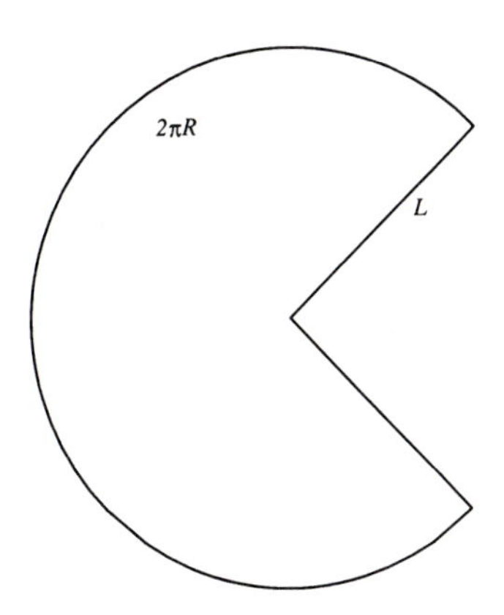

FIGURE 6.16

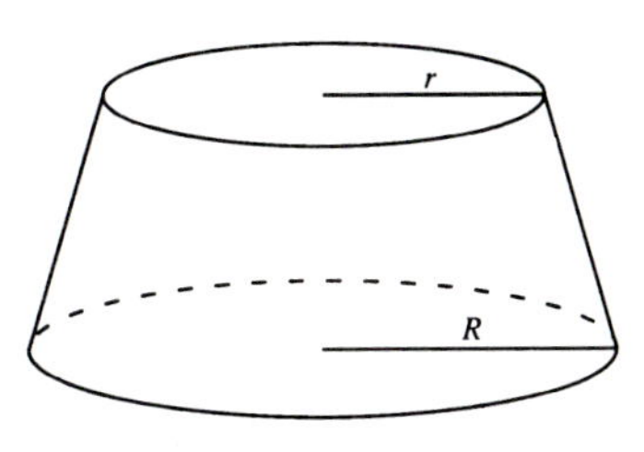

FIGURE 6.17
Frustum of a cone

A *frustum* is the part of the cone that remains after the cone is cut by a plane parallel to the base and the top is removed. If the top of the frustum has radius r, the base has radius R and the slant height is ℓ, then the surface area of the frustum is $\pi(r + R)\ell$. (The top and bottom of the frustum are not included in this area computation.) The formula is derived by subtracting the lateral area of the small upper cone that is removed from the lateral area of the large cone.

The lateral area of the frustum is used to determine the area of a surface of revolution in the same way that the length of a chord was used to determine the length of an arc.

Let $f(x)$ and $f'(x)$ be continuous on the interval $[a, b]$. If the graph of f is revolved around the x-axis, it traces out a surface that we refer to as a *surface of revolution*. Our goal is to compute the area of this surface. If the chord from $(a, f(a))$ to $(b, f(b))$ is used as an approximation to the curve $y = f(x)$, then the chord traces out the surface of a frustum of a cone. The radii of the two bases of the frustum are $f(a)$ and $f(b)$ so the surface area of the frustum is

$$S = \pi\big(f(b) + f(a)\big)\sqrt{(b - a)^2 + (f(b) - f(a))^2}.$$

The chord is shorter than the curve so the area of the frustum is less than the area of the surface. This gives a first approximation to the surface area. To get better approximations, we partition the interval and use the same idea on each of the small intervals. However some complications arise here that were not present in either the arc-length or solid of revolution computations. Here are the details.

Let

$$P : a = x_0 < x_1 < \cdots < x_n = b$$

be a partition of $[a, b]$. Over each subinterval $[x_{i-1}, x_i]$ we revolve the chord C_i from $\big(x_{i-1}, f(x_{i-1})\big)$ to $\big(x_i, f(x_i)\big)$ around the x-axis to generate a frustum with surface area

$$S_i = \pi\big(f(x_i) + f(x_{i-1})\big)\sqrt{(x_i - x_{i-1})^2 + (f(x_i) - f(x_{i-1}))^2}. \tag{6.5}$$

The sum $S_1 + S_2 + \cdots + S_n$ is called the *frustum approximation* to the surface area determined by the partition P for the function f. If the set of frustum approximations taken over all partitions of $[a, b]$ has an upper bound, then the least upper bound of this set of numbers is the area of the surface. It is necessary to prove that there is an upper bound for this set of numbers.

THEOREM 6.4

Frustum Approximations are Bounded

Let $f(x)$ be a positive function such that $f'(x)$ is continuous on $[a, b]$. Let p, p_0, q and q_0 be numbers on $[a, b]$ such that

$$\begin{gathered} f(p) \le f(x) \le f(q) \\ f'(p_0)^2 \le f'(x)^2 \le f'(q_0)^2 \end{gathered}$$

for all x on $[a, b]$. Then the frustum approximation for the partition P satisfies

$$\begin{aligned} 2\pi f(p)\sqrt{1 + f'(p_0)^2}(b - a) &\le S_1 + S_2 + \cdots + S_n \\ &\le 2\pi f(q)\sqrt{1 + f'(q_0)^2}(b - a). \end{aligned}$$

In particular, the set of all frustum approximations to the area of the surface of revolution has an upper bound.

Proof

The number $\big(f(x_i) + f(x_{i-1})\big)/2$ is the midpoint of the interval with endpoints $f(x_{i-1})$ and $f(x_i)$. The Intermediate Value Theorem and the continuity of f imply there is a point c between x_{i-1} and x_i with

$$f(c) = \frac{f(x_i) + f(x_{i-1})}{2}.$$

In particular this number is at most $f(q)$. By the Mean Value Theorem, there is a number c_0 such that

$$f'(c_0) = \frac{f(x_i) - f(x_{i-1})}{x_i - x_{i-1}}$$

and so

$$\begin{aligned}
&\sqrt{(x_i - x_{i-1})^2 + (f(x_i) - f(x_{i-1}))^2} \\
&\quad = \sqrt{1 + \left(\frac{f(x_i) - f(x_{i-1})}{x_i - x_{i-1}}\right)^2}(x_i - x_{i-1}) \\
&\quad = \sqrt{1 + f'(c_0)^2}(x_i - x_{i-1}) \\
&\quad \leq \sqrt{1 + f'(q_0)^2}(x_i - x_{i-1})
\end{aligned}$$

The frustum approximation determined by the partition P is the sum of the S_i given by equation (6.5). The inequalities just written imply

$$S_i \leq 2\pi f(q)\sqrt{1 + f'(q_0)^2}(x_i - x_{i-1}).$$

By taking the sum of these for $i = 1, 2, \ldots, n$ we obtain the bound stated in the theorem. □

We now know the area of the surface of revolution is determined as the least upper bound of a certain set of numbers; we next give a practical way to compute this number as an integral.

THEOREM 6.5

Area of a Surface of Revolution

If f is a positive function for which $f'(x)$ is continuous on $[a, b]$, then the area S of the surface obtained by revolving the graph of f around the x-axis is

$$S = \int_a^b 2\pi f(x)\sqrt{1 + f'(x)^2}\, dx. \qquad \textbf{(6.6)}$$

Proof

Let $F(x) = 2\pi f(x)\sqrt{1 + f'(x)^2}$. The expected proof would involve lower and upper sums for the function $F(x)$ along with an inequality involving the frustum approximation. Unfortunately there are some difficulties in constructing such a natural proof. The problem can be traced to the fact that the bound in Theorem 6.4 involves evaluating $2\pi f(x)$ and $\sqrt{1 + f'(x)^2}$ at two different points rather than evaluating their product $F(x)$ at one point. Here is a proof that takes these difficulties into account.

Let $P : a = x_0 < x_1 < \cdots < x_n = b$ be a partition of $[a, b]$ and let p_i, p_i', q_i and q_i' be points on $[x_{i-1}, x_i]$ such that

$$\begin{aligned}
f(p_i) &\leq f(x) \leq f(q_i), \\
f'(p_i')^2 &\leq f'(x)^2 \leq f'(q_i')^2,
\end{aligned}$$

for all x on $[x_{i-1}, x_i]$. Then the contribution S_i of the ith piece to the frustum approximation of the surface area satisfies

$$2\pi f(p_i)\sqrt{1 + f'(p_i')^2}(x_i - x_{i-1}) \leq S_i \leq 2\pi f(q_i)\sqrt{1 + f'(q_i')^2}(x_i - x_{i-1}). \qquad \textbf{(6.7)}$$

Let $L'(P)$ be the sum of the left-hand sides of these inequalities (for $i = 1, 2, \ldots, n$) and $U'(P)$ the sum of the right-hand sides. The sum $L'(P)$ is not a lower sum for the function F because the terms involve function values at two points, p_i and p_i', in place of a minimum value of F. Similarly, $U'(P)$ is not an upper sum. The frustum approximation satisfies

$$L'(P) \leq S_1 + \cdots + S_n \leq U'(P). \tag{6.8}$$

Next we show that the number S, given by the definite integral of inequality (6.6), also is trapped between these two extremes.

Let s_i and t_i be points on $[x_{i-1}, x_i]$ for which $F(s_i) \leq F(x) \leq F(t_i)$ for all x on the ith subinterval. Then

$$2\pi f(p_i)\sqrt{1 + f'(p_i')^2} \leq 2\pi f(s_i)\sqrt{1 + f'(s_i)^2} = F(s_i)$$

and

$$F(t_i) = 2\pi f(t_i)\sqrt{1 + f'(t_i)^2} \leq 2\pi f(q_i)\sqrt{1 + f'(q_i')^2}.$$

Multiply this inequality by the positive number $x_i - x_{i-1}$ and add the inequalities; the terms involving F have sums equal to the lower sum $L(P)$ and upper sum $U(P)$. Using the fact that the definite integral S lies between every lower and upper sum for F, we obtain

$$L'(P) \leq L(P) \leq S \leq U(P) \leq U'(P).$$

At this point of the proof we have proved that, for any partition P, the interval

$$I(P) = [L'(P), U'(P)]$$

contains the frustum approximation for the partition P and also the number S, that does not depend on P. The continuity of $F(x)$ is used to show that for any given number ϵ, there is a partition P such that the length of the interval $I(P)$ is at most ϵ. Once this is established, it follows that the distance between the frustum approximation and the integral S can be made as small as desired by selecting a suitable partition; the theorem follows.

The concept of uniform continuity was discussed in connection with the existence of the definite integral. We use it here. Given the continuous function $h(x)$ over an interval $[a, b]$, and given any positive number ϵ, there is a number δ such that whenever two points u, v of $[a, b]$ satisfy $|u - v| < \delta$, then $|h(u) - h(v)| < \epsilon$. When this is applied to the two continuous functions $f(x)$ and $\sqrt{1 + f'(x)^2}$, we draw the conclusion that

$$\left| 2\pi f(q_i)\sqrt{1 + f'(q_i')^2} - 2\pi f(p_i)\sqrt{1 + f'(p_i')^2} \right| \leq \epsilon'$$

for any given positive number ϵ' provided the partition P has maximum subinterval length sufficiently small. It follows that

$$|U'(P) - L'(P)| < \epsilon'(b - a)$$

and so the length of the interval $I(P)$ can be made as small as we want by the selection of a suitably fine partition. The proof is complete. □

EXAMPLE 1 Find the surface area of a sphere of radius a.

Solution: Let $f(x) = \sqrt{a^2 - x^2}$. Then the upper semicircle of radius a is the graph of $y = f(x)$ and the sphere is obtained by revolving this graph around the x-axis. The formula in the theorem gives the surface area as

$$S = 2\pi \int_{-a}^{a} y\sqrt{1 + (y')^2}\, dx.$$

Because y satisfies the equation $y^2 + x^2 = a^2$, we obtain by implicit differentiation

$$y' = -\frac{x}{y}$$

$$y\sqrt{1 + (y')^2} = y\sqrt{1 + \left(\frac{x}{y}\right)^2} = y\frac{\sqrt{y^2 + x^2}}{y} = a.$$

The surface area is

$$S = 2\pi \int_{-a}^{a} a\,dx = 2\pi a x\Big|_{-a}^{a} = 4\pi a^2. \quad ■$$

Exercises 6.4

Find the area of the surface obtained by revolving the graph of $y = f(x)$ over the given interval about the x-axis.

1. $y = 3x$, $[0, h]$

2. $y = 3x + 2$, $[0, h]$

3. $y = (x - 1)^3$, $[1, 2]$

4. $y = x^3/3$, $[0, 1]$

5. $y = (x + 1)^3$, $[0, a]$

6. $y = \sqrt{x}$, $[0, 4]$

7. $y = 3\sqrt{1 - x}$, $[0, 1]$

8. $y = \sqrt{3x}$, $[4, 16]$

9. $y = \dfrac{3\sqrt{x} - x\sqrt{x}}{3}$, $[0, 3]$

10. $y = 2\sqrt{x}$, $[0, 8]$

11. Use integration to verify the formula $2\pi r h$ for the lateral area of a right circular cylinder having base of radius r and height h.

12. Use integration to derive the formula for the lateral area of a right circular cone having base of radius r and height h.

13. The curve $4y = x^2$ between $(0, 0)$ and $(2, 1)$ is revolved around the y-axis. Find the area of the surface. [Hint: Treat y as the independent variable in place of x and use the discussion in the text with the roles of y and x exchanged.]

14. Find the surface area of the torus obtained by rotating the circle $(x - a)^2 + y^2 = b^2$ $(a > b > 0)$ around the y-axis.

15. Find the area of the surface obtained by rotating the first-quadrant arch of the asteroid, $x^{2/3} + y^{2/3} = a^{2/3}$ around the x-axis.

16. Two parallel planes separated by a distance d intersect a sphere of radius R $(2R > d)$. Show that the area of the surface of the sphere between the two planes depends only on d and not on the location of the section. [Method: Compute the area of the surface of revolution generated by the curve $y = \sqrt{R^2 - x^2}$ over the interval $[c, c + d]$ for $-R \leq c \leq c + d \leq R$ and show the result does not depend on c.] [This proves that if you slice an orange into pieces of equal width, all slices have the same amount of peel.]

6.5 Moments and center of mass

In this section, we discuss the concept of *moment* and show how it is used to determine the center of mass of a discrete set of points or a plane homogeneous region.

Discrete Points on a Line

We motivate the idea of moment with the familiar seesaw used by children. We know from experience that a 10-foot-long board supported in the middle will remain horizontal if children of equal weights are seated at the ends of the board. Suppose the symmetric situation is changed just slightly. Assume the board is supported by a fulcrum placed 5.1 feet from one end and a 104-pound child sits on the short end while a 100-pound child sits at the longer end. After the horizontal board pivots at the fulcrum, which child hits the ground?

The answer is determined by computing the moment of each child about the fulcrum; the moment is the product of the mass and the distance from the fulcrum. In our example, the child weighing 104 pounds is at distance 4.9 feet and has moment

$$(104)(4.9) = 509.6\,\text{ft-lb},$$

while the 100-pound child is at distance 5.1 feet and has moment

FIGURE 6.18

$$(100)(5.1) = 510\,\text{ft-lb}.$$

The rotation slowly carries the 100-pound child to the ground because his is the slightly greater moment.

The idea of moment can be applied to a system of many masses placed at points along the x-axis. Suppose n weights having masses $m_1, m_2, \ldots, m_n$ are placed at the points $x_1, x_2, \ldots, x_n$, respectively. Then the *moment about the origin* of the system of n points is

$$M_0 = x_1 m_1 + x_2 m_2 + \cdots + x_n m_n.$$

FIGURE 6.19

The *center of mass* of the system is the point $\overline{x}$ at which all the mass of the system could be concentrated without changing the moment. The total mass of the system is

$$M = m_1 + m_2 + \cdots + m_n$$

so the center of mass $\overline{x}$ satisfies the equation $M\overline{x} = M_0$, or

$$\overline{x} = \frac{M_0}{M}.$$

In the example mentioned at the beginning of the section, we place the origin at the fulcrum, the 104-pound boy at $x_1 = -4.9$ and the 100-pound boy at $x_2 = 5.1$. The mass and moment are

$$\begin{aligned} M &= 104 + 100 = 204, \\ M_0 &= 104x_1 + 100x_2 = (104)(-4.9) + (100)(5.1) = -509.6 + 510 = 0.4. \end{aligned}$$

The center of mass is

$$\overline{x} = \frac{0.4}{204} = 0.001961.$$

The moment of the system would be the same if all the mass were concentrated at the point $\overline{x}$ very slightly to the right of the origin.

A system of masses is in *equilibrium* if its moment equals zero. No rotation of the seesaw occurs if the system is in equilibrium.

EXAMPLE 1 Four children having weights 45, 52, 55 and 61 are placed on a seesaw at points (measured from the fulcrum) -6, 6, 5.5 and -5.3 respectively (everything is measured in some convenient units). Find the moment of this system and its center of mass.

FIGURE 6.20

Solution: The moment of the system is

$$M_0 = 45(-6) + 52(6) + 55(5.5) + 61(-5.3) = 21.2.$$

The total mass is

$$M = 45 + 52 + 55 + 61 = 213,$$

so the center of mass is $\overline{x}$ given by

$$\overline{x} = \frac{21.2}{213} = 0.09953.$$

If the board is moved slightly so that the fulcrum is at the point 0.09953, then the system will be in equilibrium (and all the children get impatient). ■

Discrete Points in the Plane

We extend the situation from a set of points on a line to a set of points in the plane. Instead of calculating moments about a point, we now compute moments about a line.

If a mass m is located at point (x_0, y_0), then its moment about the x-axis M_x is the product of the mass and the distance from the x-axis:

$$M_x = my_0.$$

(Take care to notice that the moment about the x-axis is the product of the mass and the y coordinate of the point.)

The moment about the y-axis M_y is

$$M_y = mx_0.$$

If we are given a system of n masses $m_1, m_2, \ldots, m_n$, located at points $(x_1, y_1), (x_2, y_2), \ldots, (x_n, y_n)$ respectively then the moments M_x and M_y of the system about the x-axis and the y-axis are

$$M_x = m_1y_1 + m_2y_2 + \cdots + m_ny_n,$$
$$M_y = m_1x_1 + m_2x_2 + \cdots + m_nx_n.$$

The *center of mass* is the point $(\overline{x}, \overline{y})$ where all the mass of the system could be placed so that moments about the x- and y-axes would remain the same as the moments of the original system. If the total mass is

$$M = m_1 + m_2 + \cdots + m_n,$$

then

$$\overline{x} = \frac{M_y}{M} = \frac{m_1x_1 + m_2x_2 + \cdots + m_nx_n}{m_1 + m_2 + \cdots + m_n},$$
$$\overline{y} = \frac{M_x}{M} = \frac{m_1y_1 + m_2y_2 + \cdots + m_ny_n}{m_1 + m_2 + \cdots + m_n}.$$

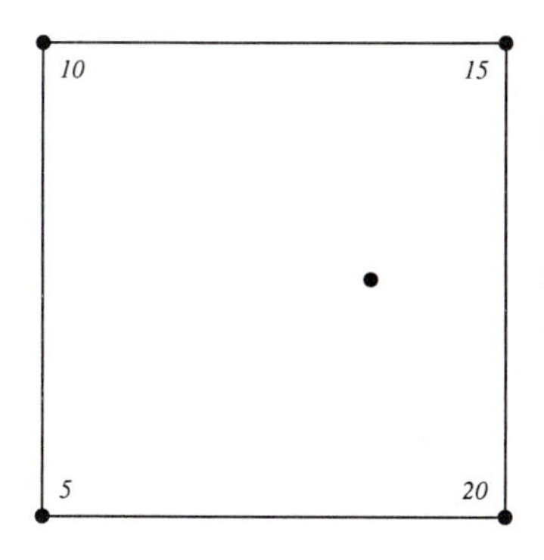

FIGURE 6.21

EXAMPLE 2 Four weights of 5, 10, 15 and 20 pounds are placed cyclically around the vertices of a square. Locate the center of mass.

Solution: Let the coordinates be selected so that $(0, 0)$ has mass 5, $(0, 1)$ has mass 10, $(1, 1)$ has mass 15, and $(1, 0)$ has mass 20. Then

$$M = 5 + 10 + 15 + 20 = 50,$$
$$M_x = 5(0) + 10(1) + 15(1) + 20(0) = 25,$$
$$M_y = 5(0) + 10(0) + 15(1) + 20(1) = 35,$$
$$\overline{x} = \frac{M_y}{M} = \frac{35}{50} = 0.7,$$
$$\overline{y} = \frac{M_x}{M} = \frac{25}{50} = 0.5.$$

The nonsymmetric sizes of the weights at the vertices results in the center of mass at (0.7, 0.5) rather than the center of the square. This ought to seem reasonable because the two upper vertices have the same total mass as the lower two vertices, implying that $\overline{y} = 0.5$ looks right. The total mass on the left side is less than on the right so that $\overline{x}$ ought to be shifted to the right of center. ■

Moments of Continuous Regions

Up to now we have considered a discrete set of masses on a line or in a plane and have seen how to determine the center of mass. Next we take up the same problem for a continuous distribution of points. We do not treat the most general case but only the case of a plane *lamina*. By a lamina we mean a thin, homogeneous material covering a region of the plane bounded by graphs of continuous functions. Homogeneous means that the material has constant density ρ. The effect of this assumption is that the mass of any piece of the material is ρ times the area. The word *thin* means the material is three dimensional but its thickness is so small compared to other dimensions that we treat the material as if it were two dimensional.

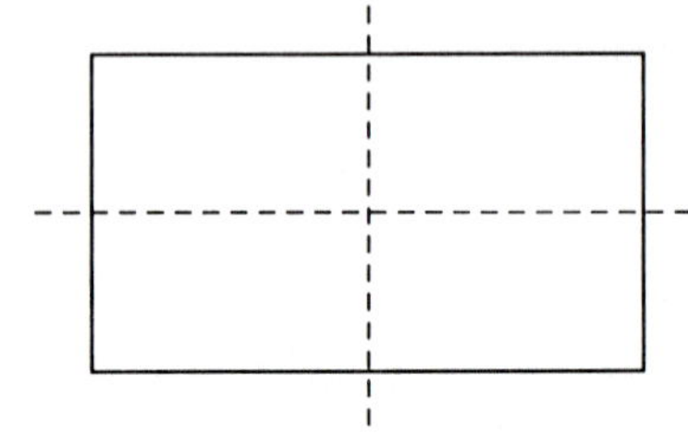

FIGURE 6.22

The center of mass of a lamina is the "balance point"; the lamina will be in equilibrium if it remains horizontal when supported by a pin at the center of mass. If the lamina has an axis of symmetry, the the center of mass lies on that axis. If there is more than one axis of symmetry, the center of mass lies on each axis, hence, must be at the point of intersection of the axes of symmetry. For example, a lamina in the shape of a square has center of mass at the center of the square because each diagonal line is an axis of symmetry; the center of mass lies on each of the diagonals. A rectangle has two axes of symmetry so its center of mass is at their intersection, the center of the rectangle.

We begin by computing the moment of a lamina bounded by the graph of a positive function, the x-axis and two vertical lines.

THEOREM 6.6

Moments of a Lamina

Assume that a lamina of constant density ρ covers a region R bounded above by the graph of a positive, continuous function $f(x)$, below by the x-axis and by the vertical lines $x = a$ and $x = b$. Then the moments M_x and M_y of the lamina about the x-axis and the y-axis are given by the integrals

$$M_x = \frac{\rho}{2}\int_a^b f(x)^2\,dx, \qquad M_y = \rho\int_a^b x f(x)\,dx.$$

The mass of the lamina is

$$M = \rho\int_a^b f(x)\,dx.$$

The center of mass of the lamina is the point $(\overline{x}, \overline{y}) = (M_y/M, M_x/M)$.

Proof

Begin with any partition $P : a = x_0 < x_1 < \cdots < x_n = b$, and let p_i and q_i be points on $[x_{i-1}, x_i]$ where f takes its minimum and maximum values. Let R_i denote the strip of the region bounded by the graph of $f(x)$ above the interval $[x_{i-1}, x_i]$. Let $M_x(R_i)$ denote the moment about the x-axis of the strip R_i. We obtain an estimate of $M_x(R_i)$ by considering a rectangle contained within R_i and another rectangle containing R_i. The rectangle with base $[x_{i-1}, x_i]$ and height $f(p_i)$ is contained within R_i. The moment of the rectangle is easily determined; it equals the same moment as the one-point system having all the mass of the rectangle concentrated at its center of mass. The mass of the rectangle is $\rho f(p_i)(x_i - x_{i-1})$, the density times area. The center of mass of the rectangle is the center of the rectangle $\big((x_{i-1} + x_i)/2, f(p_i)/2\big)$ and so the moment of the rectangle is

$$\rho f(p_i)(x_i - x_{i-1})\frac{f(p_i)}{2} = \frac{\rho}{2} f(p_i)^2(x_i - x_{i-1}).$$

The moment of R_i is slightly greater than this because there is a bit more area in R_i than there is in the rectangle. Thus we have

$$\frac{\rho}{2} f(p_i)^2(x_i - x_{i-1}) \le M_x(R_i).$$

Similarly, by considering the rectangle of height $f(q_i)$ containing R_i, we obtain

$$M_x(R_i) \le \frac{\rho}{2} f(q_i)^2(x_i - x_{i-1}).$$

When we add these inequalities for $i = 1, 2, \ldots, n$, the sum of the terms $(\rho/2)f(p_i)^2(x_i - x_{i-1})$ is a lower sum, $L(P)$ for the function $(\rho/2)f(x)^2$; the sum of the terms $M_x(R_i)$ is the moment $M_x(R)$ of the entire region; the sum of the terms $\frac{\rho}{2} f(q_i)^2(x_i - x_{i-1})$ is an upper sum $U(P)$ for the function $(\rho/2)f(x)^2$. The inequality

$$L(P) \le M_x(R) \le U(P)$$

holds for every partition, and so the term $M_x(R)$ lies between all lower sums and all upper sums. It follows that

$$M_x(R) = \frac{\rho}{2} \int_a^b f(x)^2\, dx.$$

The computation of the moment about the y-axis is slightly more complicated and we replace the now familiar reasoning with lower and upper sums by an argument that uses Theorem 5.5 that permits the use of regular partitions and a limit expression for the integral. In particular, if $g(x)$ is continuous over $[a, b]$ and if P_n is the regular partition of $[a, b]$ having n subintervals, then

$$\lim_{n\to\infty} L(P_n) = \lim_{n\to\infty} U(P_n) = \int_a^b g(x)\, dx.$$

The reason for this will become apparent during the course of the argument. Let $g(x) = xf(x)$ and let P_n be the regular partition of $[a, b]$ given by

$$P_n : a = x_0 < x_1 < \cdots < x_n = b, \quad \text{where} \quad x_i = a + i\frac{b-a}{n}.$$

Let $f(p_i)$ be the minimum value of $f(x)$ on $[x_{i-1}, x_i]$ with $x_{i-1} \le p_i \le x_i$. The region R_i below the graph of f and above the interval $[x_{i-1}, x_i]$ contains the rectangle with the same base and height $f(p_i)$. The center of mass of the rectangle is $\big((x_{i-1} + x_i)/2, f(p_i)/2\big)$ and its mass is $\rho f(p_i)/n$. (Remember $x_i - x_{i-1} = 1/n$.) Let m_i denote the moment of the rectangle about the y-axis so that

$$m_i = \rho f(p_i)\frac{1}{n}\frac{x_{i-1} + x_i}{2}.$$

Let s_i be the point on $[x_{i-1}, x_i]$ where $g(x)$ [$= xf(x)$] takes its minimum value. If we had $m_i = \rho g(s_i)/n$ then we could complete the proof in the usual way by taking the sum of the moments for each subinterval and thereby get the lower sum for $g(x)$ less than the moment about the y-axis. Unfortunately we do not know the moment m_i satisfies this equality. So we now estimate how far apart the two terms m_i and $\rho g(s_i)$ can be. We have

$$\begin{aligned} m_i - \rho g(s_i)/n & \\ &= \rho f(p_i)\frac{x_{i-1}+x_i}{2}\frac{1}{n} - \rho f(s_i)s_i\frac{1}{n} \\ &= \frac{\rho}{n}\left[\big(f(p_i)-f(s_i)\big)\left(\frac{x_{i-1}+x_i}{2}\right) + f(s_i)\left(\frac{x_{i-1}+x_i}{2} - s_i\right)\right]. \end{aligned}$$

We must estimate the size of this difference. The term $f(p_i) - f(s_i)$ is the difference of two values of f at two points on $[x_{i-1}, x_i]$. This can be made as small as we like by the uniform continuity property. More precisely, let ϵ be any positive number. Then there is an integer N such that whenever u and v are points on $[a, b]$ with $|u - v| < 1/N$ then $|f(u) - f(v)| < \epsilon$. We now assume that the number n is greater than N. To bound $x_{i-1} + x_i$, we let c be the larger of $|a|$ and $|b|$ so that $|x_{i-1} + x_i| \leq 2c$. Then

$$\left|\big(f(p_i) - f(s_i)\big)\left(\frac{x_{i-1}+x_i}{2}\right)\right| < c\epsilon.$$

This takes care of one of the two terms; the other term is bounded by letting u denote the maximum value of f on $[a, b]$ and observing that $(x_{i-1} + x_i)/2$ and s_i both lie on the interval $[x_{i-1}, x_i]$ so the difference between them is at most the length of the interval, $1/n$. Then $|f(s_i)((x_{i-1} + x_i)/2 - s_i)| < u/n$. So we finally obtain

$$\begin{aligned} |m_i - \frac{\rho}{n}g(s_i)| & \\ &= \frac{\rho}{n}\left|\big(f(p_i)-f(s_i)\big)\left(\frac{x_{i-1}+x_i}{2}\right) + f(s_i)\left(\frac{x_{i-1}+x_i}{2} - s_i\right)\right| \\ &\leq \frac{\rho}{n}\left|\big(f(p_i)-f(s_i)\big)\left(\frac{x_{i-1}+x_i}{2}\right)\right| + \left|f(s_i)\left(\frac{x_{i-1}+x_i}{2} - s_i\right)\right| \\ &\leq \frac{\rho}{n}\left(\epsilon c + \frac{u}{n}\right). \end{aligned}$$

It is important to observe that this last quantity does not depend on the particular interval inasmuch as it is independent of i. This estimate implies

$$\frac{\rho}{n}g(s_i) - \frac{\rho}{n}\left(c\epsilon + \frac{u}{n}\right) \leq m_i \leq M_y(R_i), \tag{6.9}$$

where $M_y(R_i)$ is the moment of the region under the graph of f over the interval $[x_{i-1}, x_i]$. Now let t_i be the point on the interval $[x_{i-1}, x_i]$ such that $g(t_i)$ is the maximum value of g. By the analogous reasoning, we obtain the upper estimate

$$M_y(R_i) \leq \frac{\rho}{n}g(t_i) + \frac{\rho}{n}\left(c\epsilon + \frac{u}{n}\right). \tag{6.10}$$

Next we add inequalities (6.9) for $i = 1, 2, \ldots, n$. The sum of the terms $\rho g(s_i)/n$ is just the lower sum $L(P_n)$ for the partition P_n for the function $\rho g(x) = \rho x f(x)$; the sum of the terms $\rho(c\epsilon + u/n)/n$ is $n\big(\rho(c\epsilon + u/n)/n\big) = \rho(c\epsilon + u/n)$; the sum of the terms $M_y(R_i)$ is the moment of the region R about the y-axis. Treating inequality (6.10) in the same manner, we obtain

$$L(P_n) - \rho(c\epsilon + u/n) \leq M_y(R) \leq U(P_n) + \rho(c\epsilon + u/n).$$

If we take the limit as n goes to ∞, we have

$$\lim_{n\to\infty} L(P_n) = \lim_{n\to\infty} U(P_n) = \rho \int_a^b x f(x)\,dx,$$

and

$$\lim_{n\to\infty} \rho(c\epsilon + u/n) = \rho c\epsilon.$$

It follows that

$$\rho \int_a^b x f(x)\,dx - \rho c\epsilon \le M_y(R) \le \rho \int_a^b x f(x)\,dx + \rho c\epsilon.$$

But ϵ was an arbitrary positive number; the only way this last inequality can hold for every possible positive ϵ is for

$$M_y(R) = \rho \int_a^b x f(x)\,dx.$$

Once the moments about the x- and y-axes have been determined, the center of mass is computed by the ratios in the last statement of the theorem. □

EXAMPLE 3 Find the center of mass of the lamina in the shape of a quarter circle of radius 10.

Solution: The lamina is bounded above by the graph of

$$f(x) = \sqrt{100 - x^2}$$

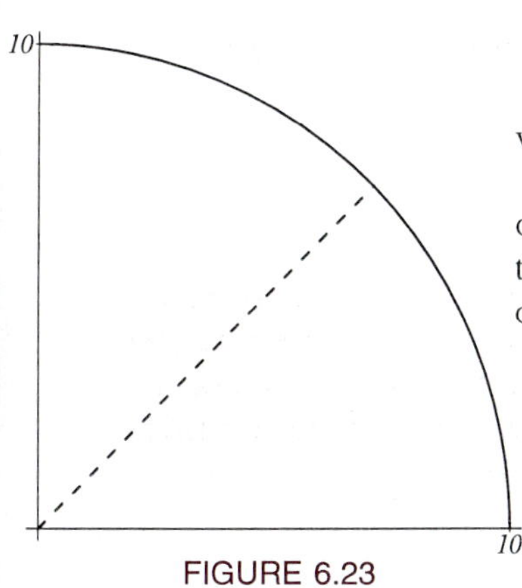

FIGURE 6.23

with $0 \le x \le 10$, by the x-axis below, and the y-axis on the left.

There is an axis of symmetry, the line $y = x$. Thus the center of mass lies on this line and has the form (c, c). We can solve the problem by computing the moment about either the x-axis or the y-axis. We do both as a check and to observe the differences in the level of difficulty. We have

$$\begin{aligned} M_x &= \frac{\rho}{2} \int_a^b f(x)^2\,dx = \frac{\rho}{2} \int_0^{10} (\sqrt{100 - x^2})^2\,dx \\ &= \frac{\rho}{2} \int_0^{10} 100 - x^2\,dx = \frac{\rho}{2}\left(100x - \frac{x^3}{3}\right)\Big|_0^{10} = \frac{1000\rho}{3}, \end{aligned}$$

$$M = \frac{1}{4}\pi 10^2 \rho = 25\pi\rho,$$

$$\begin{aligned} c = \overline{y} &= \frac{M_x}{M} = \frac{1000\rho}{3} \frac{1}{25\pi\rho}, \\ &= \frac{40}{3\pi} \approx 4.24413. \end{aligned}$$

If we compute the moment about the y-axis, the integral is slightly more complicated:

$$M_y = \rho \int_a^b x f(x)\, dx = \rho \int_0^{10} x\sqrt{100 - x^2}\, dx$$

$$= \frac{-\rho}{2} \int_{100}^{0} u^{1/2}\, du \qquad (u = 100 - x^2, \quad du = -2x\, dx)$$

$$= \frac{-\rho}{2} \frac{2}{3} u^{3/2} \Big|_{100}^{0} = \frac{1000\rho}{3}.$$

We get the same result as in the computation of M_x; thus $\overline{x} = \overline{y}$ is seen as a result of the direct computation.

As a check on the answer, note that the center of mass of the square lamina with side 10 is $(5, 5)$. Because some of the square in the upper right corner is removed to produce the quarter circle, the center of mass ought to be displaced slightly to the left and below $(5, 5)$, just as the computation shows. ■

The moments and center of mass of a region can be computed from a knowledge of the moments and centers of mass of individual parts of the region. We illustrate with an example.

EXAMPLE 4 A square lamina with side 2 rests atop a rectangular lamina of height 4 and width 6 with the leftmost side of the square aligned with the leftmost side of the rectangle. Find the center of mass of the combined lamina.

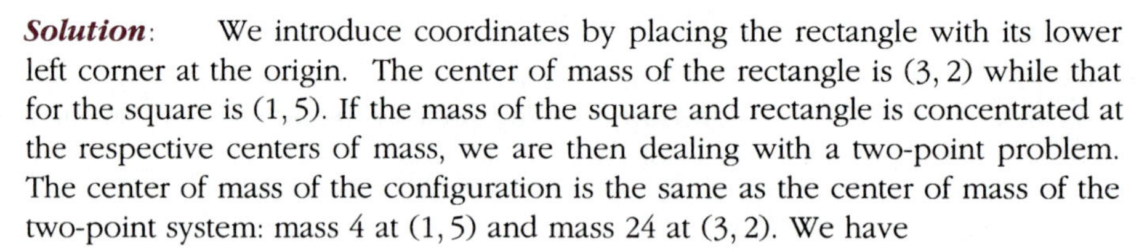

FIGURE 6.24

***Solution*:** We introduce coordinates by placing the rectangle with its lower left corner at the origin. The center of mass of the rectangle is $(3, 2)$ while that for the square is $(1, 5)$. If the mass of the square and rectangle is concentrated at the respective centers of mass, we are then dealing with a two-point problem. The center of mass of the configuration is the same as the center of mass of the two-point system: mass 4 at $(1, 5)$ and mass 24 at $(3, 2)$. We have

$$M_x = 4 \cdot 5 + 24 \cdot 2 = 68,$$
$$M_y = 4 \cdot 1 + 24 \cdot 3 = 76.$$

Because the total mass is $4 + 24 = 28$, we have

$$\overline{x} = \frac{M_y}{M} = \frac{76}{28} = 2.714,$$
$$\overline{y} = \frac{M_x}{M} = \frac{68}{28} = 2.429.$$

■

The technique illustrated in this example can be applied to more complicated regions if they can be decomposed into simpler regions for which the center of mass is easily determined.

Lamina Bounded by Two Curves

Let $f(x)$ and $g(x)$ be continuous functions and assume $f(x) \geq g(x)$ for all x on $[a, b]$. We compute the center of mass of the lamina R bounded by the lines $x = a$, $x = b$ and the graphs of $f(x)$ and $g(x)$. Let the constant density be denoted by ρ so that the mass of the lamina is

$$M = \rho \int_a^b f(x) - g(x)\, dx.$$

The moments $M_x(R)$ and $M_y(R)$ of the lamina about the x- and y-axes are expressed as integrals.

Let A be the lamina bounded above by the graph of $f(x)$ above $[a, b]$, and let B be the lamina bounded above by the graph of $g(x)$ above $[a, b]$. Then symbolically we have $A = B + R$; that is the lamina A is composed of two smaller laminas, R and B. The moments of the regions about the x- or y-axis satisfy

$$M_x(A) = M_x(B) + M_x(R),$$
$$M_y(A) = M_y(B) + M_y(R).$$

The moments for A and B have been determined in the previous theorem. This permits the computation of the moments for R as the difference of the moment for A and the moment for B. We state the final result in the next theorem.

THEOREM 6.7

Moments and Center of Mass of a Lamina

Let $f(x)$ and $g(x)$ be continuous functions with $f(x) \geq g(x)$ for all x on $[a, b]$. The lamina with density ρ bounded above and below by the graphs of f and g and by the lines $x = a$ and $x = b$ has moments

$$M_x = \frac{\rho}{2} \int_a^b f(x)^2 - g(x)^2\, dx,$$
$$M_y = \rho \int_a^b x\big(f(x) - g(x)\big)\, dx.$$

The mass of the lamina is

$$M = \rho \int_a^b f(x) - g(x)\, dx.$$

The center of mass of the lamina is the point $(M_y/M, M_x/M)$.

As an aid to memory, these formulas can be set up as follows: Consider the thin strip with x as the midpoint of the base, the rectangular approximation in the lamina of height $f(x) - g(x)$, and width dx. The center of this rectangle is its center of mass, so the moment of the rectangle about the x-axis is $(\rho/2)\big(f(x) + g(x)\big)\big(f(x) - g(x)\big)\, dx$. The integration forms the sum of these approximations as x ranges from a to b. Similarly the moment about the y-axis of the approximating rectangle is $\rho x(f(x) - g(x))\, dx$ and the integration gives the sum of all the approximating rectangles.

Exercises 6.5

1. Find the moment about the origin of the system of point masses on the x-axis: 10 at $(3, 0)$, 20 at $(6, 0)$, 7 at $(-12, 0)$, 4 at $(-10, 0)$ and 5 at $(-5, 0)$.

2. A 12-foot seesaw rests on the fulcrum at the 6-foot mark. Four of five children are sitting on it arranged as follows: an 80-pound child is placed at the 6-foot mark with a 75-pound child 1 foot in front of him. At the other end there is a 45-pound child at the 6-foot mark and a 60-pound child at the 5-foot mark. What is the weight of a child sitting at the 4-foot mark if the seesaw is at equilibrium?

3. Compute the center of mass of the system consisting of three point masses at the given positions: 100 at $(3, 0)$, 50 at $(-4, 0)$ and 25 at $(-5, 0)$.

4. Compute the center of mass of the system consisting of three point masses at the given positions: 100 at $(3, -1)$, 50 at $(-4, 3)$ and 25 at $(-5, 1)$.

5. Compute the center of mass of the system consisting of five point masses at the given positions: 10 at $(-4, 0)$, 20 at $(-3, 0)$, 15 at $(-1, 0)$, 40 at $(0.5, 0)$ and 50 at $(2, 0)$.

6. Compute the center of mass of the system consisting of five point masses at the given positions: 10 at $(-4, 1)$, 20 at $(-3, 2)$, 15 at $(-1, 3)$, 40 at $(0.5, -2)$ and 50 at $(2, -1)$.

7. A system of four point masses has center of mass at the origin. Three of the masses are: 15 at $(1, -1)$, 20 at $(2, 0)$ and 25 at $(-1, 1)$. The remaining mass m is located at $(3, y)$. Find m and y.

8. Find the center of mass of the lamina bounded by the x-axis and the graph of $f(x) = x(4 - x)$.

9. Find the center of mass of the semicircle bounded by the x-axis and the graph of $y = \sqrt{R^2 - x^2}$.

10. Find the center of mass of the lamina enclosed by the graphs of the two functions $f(x) = x^2$ and $g(x) = 4x + 5$.

11. Find the center of mass of the lamina consisting of a square with side 2 and a semicircle of diameter 2 placed atop the square so the diameter of the semicircle coincides with the upper side of the square.

12. Find the center of mass of the lamina enclosed by the x-axis, the vertical lines $x = 1$ and $x = 3$, and the semicircle $y = 3 + \sqrt{1 - (x - 2)^2}$.

13. Find the center of mass of a circular lamina having boundary $x^2 + y^2 = 4^2$ from which a hole having boundary $(x - 2)^2 + y^2 = 1$ has been cut. [Hint: You might think of the hole as a disk with negative mass added to the original solid circular disk.]

14. Find the center of mass of the upper half of the lamina of Exercise 13.

15. Prove the First Theorem of Pappus: If a plane region of area A bounded by the graphs of the two functions $f(x) \geq g(x) \geq 0$ is revolved about the x-axis, the volume of the solid of revolution equals the area A times the distance traveled by the center of mass of the plane region.

16. A thin wire of uniform density ρ per unit length has the shape of the graph of $y = f(x)$ for $a \leq x \leq b$. Give an argument based on chord approximations to show that the mass M, moment around the x-axis M_x and the moment around the y-axis M_y are given by the integrals

$$M = \int_a^b \sqrt{1 + f'(x)^2}\, dx,$$

$$M_x = \int_a^b f(x)\sqrt{1 + f'(x)^2}\, dx,$$

$$M_y = \int_a^b x\sqrt{1 + f'(x)^2}\, dx.$$

17. Prove the Second Theorem of Pappus: If the graph of $y = f(x)$ is an arc in the first quadrant and is rotated about the x-axis, the area of the surface generated equals the product of the length of the arc and the distance traveled by the center of mass of the arc.

18. Use the First Theorem of Pappus to derive the formula for the volume of a cone having height H and circular base of radius R.

19. Use the known formulas for the volume of a sphere and area of a circle along with the First Theorem of Pappus to determine the center of mass of a semicircle.

20. Use the Second Theorem of Pappus to derive the formula $\pi R\ell$ for the surface area of a cone of height h, circular base of radius R and slant height $\ell = \sqrt{h^2 + R^2}$.

21. Use the known formulas for the surface area of a sphere and circumference of a circle along with the Second Theorem of Pappus to determine the center of mass of a semicircle of radius R.

7 Transcendental Functions

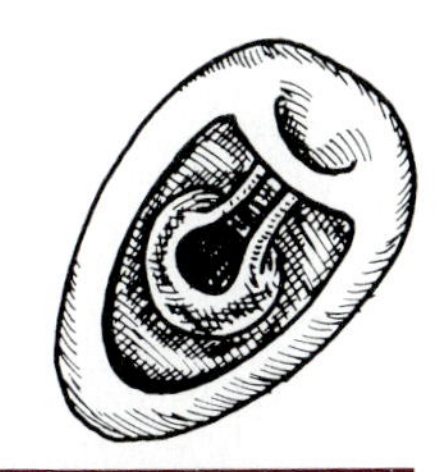

In this chapter we study some functions that arise naturally in the solution of many mathematical and physical problems. We begin with the study of inverse functions in general and then define the natural logarithm and its inverse, the exponential function. The logarithm and exponential functions are used to describe physical phenomena such as the rate at which the temperature of a cold drink left in a warm room increases—Newton's law of cooling—or the rate of growth of the streptococcus bacteria in a culture. We also use these functions to complete the definition of real exponents (up to now expressions like 2^x have been defined only for rational x). Another application of the natural logarithm function is to simplify differentiation of complicated expressions and the computation of certain limits with variable exponents. Next, the inverse trigonometric functions are defined and studied. Their derivatives are computed and seen to be very important for the evaluation of integrals involving square roots of quadratic polynomials and other functions that occur frequently in mathematical and physical problems. In this chapter we finish defining the collection of basic functions that are studied in the traditional calculus course.

7.1 Inverse functions

We begin with the definition of the inverse of a function.

DEFINITION 7.1

Inverse Function

Let g be a function with domain D and range R. The inverse of g is a function f with domain R such that for every x in the domain of g,

$$f(g(x)) = x.$$

We prove below that when f is the inverse of g, then g is the inverse of f so that there really is symmetry in the definition.

EXAMPLE 1 Find the inverse of the function $g(x) = x^3$.

Solution: We want an f such that

$$f(x^3) = x$$

for all x. We see at once that the function $f(x) = x^{1/3}$ has this property. The domain and range of both functions is the set of all numbers. ■

Not every function has an inverse function; for a given $g(x)$, there might not be any function that satisfies $f(g(x)) = x$ for every x. We find conditions on g that ensure the existence of an inverse function but first, here is a computation to show why the inverse function is useful.

Suppose c is a value of g and we want to find the solutions x to the equation

$$g(x) = c. \qquad (7.1)$$

If it is known that $f(x)$ is the inverse of $g(x)$, then we can express the solution to the Equation (7.1) by applying f to both sides:

$$f(c) = f(g(x)) = x.$$

Thus the solution to Equation (7.1) is the value of f at c. Inasmuch as f is a function, there is only one value $f(c)$ and so the equation $g(x) = c$ must have only one solution. For any given number c, there is at most one x that satisfies $g(x) = c$. In particular, if both $g(u) = c$ and $g(v) = c$ then $u = v$. Thus we have discovered a condition that must be satisfied if g has an inverse. Namely,

If $g(x)$ is a function that has an inverse, then $g(u) = g(v)$ implies $u = v$.

More directly, if f is the inverse of g and $g(u) = g(v)$, then

$$u = f(g(u)) = f(g(v)) = v.$$

In the next theorem we prove that this condition is sufficient to ensure the existence of the inverse function. First we introduce some formalities.

DEFINITION 7.2

One-to-One Function

A function $g(x)$ is one to one on the set D if the equality $g(u) = g(v)$, with u and v in D, implies $u = v$.

The one-to-one property of a function can be interpreted in terms of the graph of the function. Namely a function is one to one if and only if each horizontal line crosses the graph of the function at most once. The line $y = c$ crosses the graph of $g(x)$ at the points (u, c) where $g(u) = c$. When g is one to one, there is at most one such u for any given c.

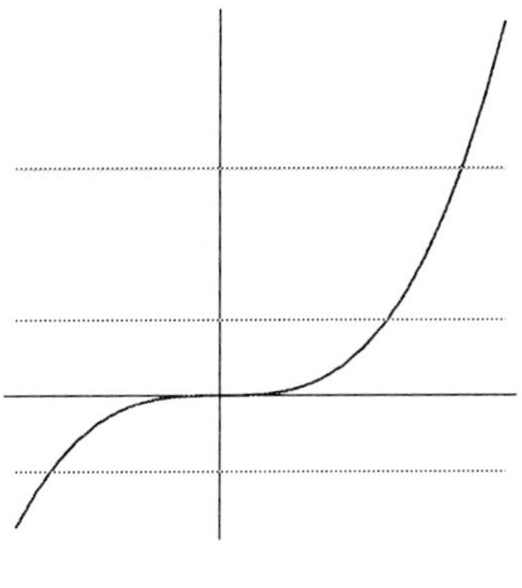

FIGURE 7.1
A one-to-one function

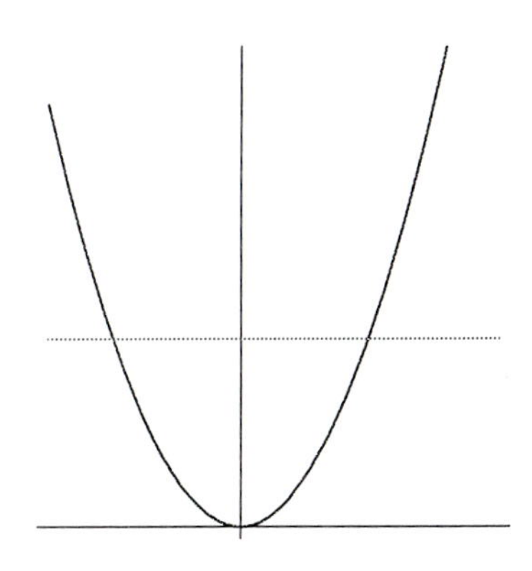

FIGURE 7.2
Not one to one

Now we show that the one-to-one condition is all that is necessary to ensure that a function have an inverse.

THEOREM 7.1

Existence of the Inverse Function

Let $g(x)$ be a function with domain D. Then there exists an inverse function $f(x)$ that satisfies $f(g(x)) = x$ for all x in D if and only if g is one to one on D.

Proof

If f is the inverse function, then $g(u) = g(v)$ implies $f(g(u)) = f(g(v))$. Because $f(g(x)) = x$ for all x in D, we obtain $u = f(g(u)) = f(g(v)) = v$. Thus g is one to one.

Now suppose g is one to one on D. Let R be the range of g. Thus c is in R if and only if there is some u in D with $g(u) = c$. The one-to-one property implies that there is only one u in D that satisfies this condition. For each c in R, we define $f(c) = u$. Thus f is a function defined on the range of g and f satisfies

$$f(c) = u \quad \text{if and only if} \quad g(u) = c.$$

Substitute for c to see that $f(g(u)) = u$ for every u in D. Thus f is the inverse of g.

□

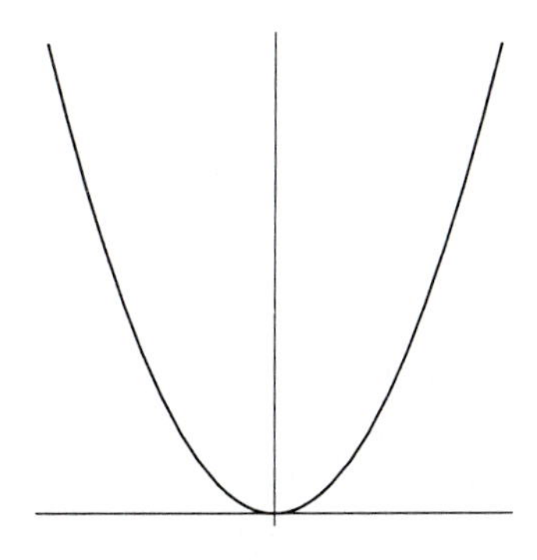

FIGURE 7.3
One to one on the positive x-axis or the negative x-axis

EXAMPLE 2 Suppose $g(x) = x^2$. Does g have an inverse? Can the domain of g be restricted so that an inverse exists on the restricted domain?

Solution: The domain of g is the set of all real numbers. The inverse function does not exist because g is not one to one on its domain; for example $g(-2) = g(2) = 4$ but $-2 \neq 2$. Even though g has no inverse, we can define a closely related function that does have an inverse by restricting the domain of g. Let $G(x) = x^2$ for all $x \geq 0$; G is not defined for negative x. On the set D of nonnegative numbers, G is one to one and the inverse function is $f(x) = \sqrt{x}$. Of course the square root symbol denotes the nonnegative square root of x.

We could also define $H(x) = x^2$ for all $x \leq 0$. Then H is one to one on the set of nonpositive numbers and its inverse is $F(x) = -\sqrt{x}$. ■

The Inverse of the Inverse

Suppose $f(x)$ is the inverse of $g(x)$. Does f have an inverse? Does your intuition tell you that f has an inverse? To prove that it does, we need only prove that f is one to one on its domain. Let u and v be numbers in the domain of f such that

$$f(u) = f(v).$$

Because the domain of f is the range of g, there are numbers a and b in the domain of g such that $g(a) = u$ and $g(b) = v$. Substituting into the equation involving f, we find

$$f(u) = f(g(a)) = a \quad \text{and} \quad f(v) = f(g(b)) = b.$$

Thus $f(u) = f(v)$ implies $a = b$ and $u = g(a) = g(b) = v$; hence f is one to one. So your intuition is correct—the inverse of a function has an inverse and moreover taking the inverse twice returns us to the original function as we now show.

THEOREM 7.2

If $f(x)$ is the inverse of $g(x)$ on the domain D, then $g(x)$ is the inverse of $f(x)$ on the domain of f.

Proof

Let u be any number in the domain of g and let $g(u) = c$. Then c is in the domain of f and $f(c) = u$. The inverse of f is a function $G(x)$ that is defined on the domain of g and that satisfies $G(f(c)) = G(u) = c$. Clearly the functions g and G are the same; that is, $G(u) = g(u) = c$ for every u in the domain of g. □

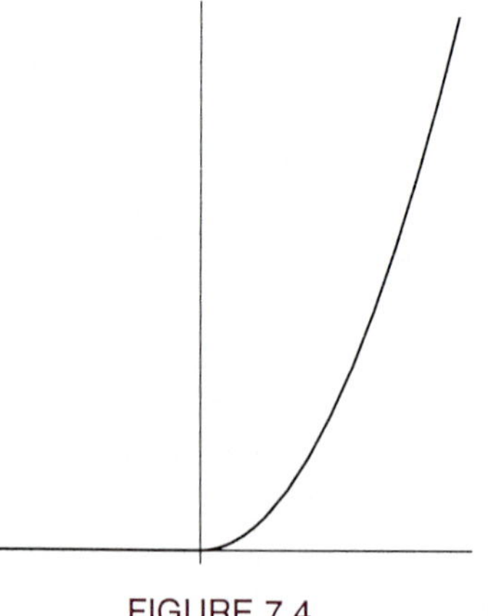

FIGURE 7.4
One to one on $x \geq 0$

If g has inverse f, then we have both equations

$$f(g(x)) = x, \quad x \text{ in the domain of } g,$$
$$g(f(x)) = x, \quad x \text{ in the domain of } f.$$

Remark: It is important that the domain of the inverse function equal the range of the given function or else false conclusions might be drawn. For example the function $g(x) = \sqrt{x}$ has range consisting of all nonnegative numbers, so the domain of the inverse function $f(x) = x^2$ is the set of all nonnegative numbers. If one incorrectly took the domain of f as all numbers, then g would not be the inverse of f.

Computation of the Inverse Function

If it is known that a function has an inverse, how does one compute it? Finding an explicit formula might not be easy, or even possible. One procedure to try is the following. Suppose we are given the function $g(x)$ that is one to one on its domain D. Start from the equation

$$g(x) = c,$$

with c an unspecified constant and, if possible, "solve" for x in terms of c; that is, obtain an expression

$$x = f(c).$$

The function $f(c)$ is the inverse function with c as the independent variable. Replace c by x to put f in the usual form of a function of x.

This method has an additional bonus. If $g(x)$ is given but it is not known that $g(x)$ is one to one, and if it is possible to solve for x in terms of c in an unambiguous way, then we discover that the function g is one to one.

EXAMPLE 3 Let $g(x) = x/(x-4)$ with domain D consisting of all numbers except $x = 4$. Determine if there is an inverse function and find it if possible.

***Solution*:** Consider the equation

$$\frac{x}{x-4} = c,$$

and solve for x in terms of c:

$$x = cx - 4c,$$
$$x(1-c) = -4c,$$
$$x = \frac{-4c}{1-c} = \frac{4c}{c-1}.$$

We have solved for x in terms of c; namely,

$$x = f(c) = \frac{4c}{c-1}.$$

After replacing c by x, we see the inverse of g is the function $f(x) = 4x/(x-1)$. Let us examine some details in this computation. In solving for x, we divided by $c - 1$ so it is necessary to assume $c \neq 1$. This causes no problem because 1 is not a value of g. (We know this because our calculations show that there is no value of x that satisfies $g(x) = 1$.) Thus the domain of $f(x)$ is all numbers except $x = 1$. The domain of $f(x)$ is precisely the set of values of $g(x)$. ■

Here is an example in which we show directly that a function is one to one.

EXAMPLE 4 Show the function $f(x) = x|x| + 10$ is one to one and find its inverse.

***Solution*:** The domain of f is the set of all real numbers. Let u and v be two numbers with $f(u) = f(v)$. Then $u|u| + 10 = v|v| + 10$ and so $u|u| = v|v|$. Suppose $u > 0$. Then $|u| = u$ and $u|u| = u^2 > 0$. Because $u^2 = v|v|$, it follows that $v > 0$ also; we have $u^2 = v^2$ and both u and v are positive; so we have $u = v$. Similarly if $u < 0$, we have $u|u| = -u^2$ and it follows that $v < 0$ and again $u^2 = v^2$, that implies $u = v$. Finally if $u = 0$, then $v = 0$. So in all cases, $f(u) = f(v)$ implies $u = v$. Thus f is one to one and the inverse function exists. How do we find it?

Let c be any number and we try to solve $f(x) = c$ for x in terms of c. The equation is $x|x| + 10 = c$ or $x|x| = c - 10$. We consider cases to remove the absolute value sign. Assume $c \geq 10$ so that $c - 10 \geq 0$ and it follows that $x \geq 0$ and $x|x| = x^2$. Thus $x = \sqrt{c-10}$.

If $c - 10 < 0$ and $x < 0$, then $x|x| = -x^2 = c - 10$. Because $10 - c > 0$ we have

$$10 - c = x^2,$$
$$\sqrt{10-c} = \sqrt{x^2} = |x| = -x.$$

We now have the function

$$h(x) = \begin{cases} \sqrt{x-10} & \text{if } x \geq 10 \\ -\sqrt{10-x} & \text{if } x < 10. \end{cases}$$

and h is the inverse of f. ■

Notation for the Inverse Function

We use the notation $g^{-1}(x)$ to denote the inverse function of g. The previous equations can be written as

$$g(g^{-1}(x)) = x, \quad \text{and} \quad g^{-1}(g(x)) = x.$$

The reader should take care to distinguish this use of the exponent -1 from the common use of designating the reciprocal. If it is necessary to denote $1/g(x)$ with an exponent, we write $g(x)^{-1}$ or $(g(x))^{-1}$ so that no confusion with the inverse function arises.

The Graph of the Inverse Function

Suppose $g(x)$ has domain D and the inverse $g^{-1}(x)$ exists. We want to compare the graphs of these two functions. If (a, b) is a point on the graph of g, then $b = g(a)$. This is equivalent to $a = g^{-1}(b)$. This means that the point (b, a) is on the graph of g^{-1}. Conversely, the same reasoning shows that if (u, v) is a point on the graph of g^{-1}, then (v, u) is on the graph of g. Thus the graph of g^{-1} is obtained from the graph of g by reversing the coordinates of all the points. This operation has a nice geometric interpretation. The point (a, b) is the "mirror image" of the point (b, a) if the mirror is placed along the line $y = x$.

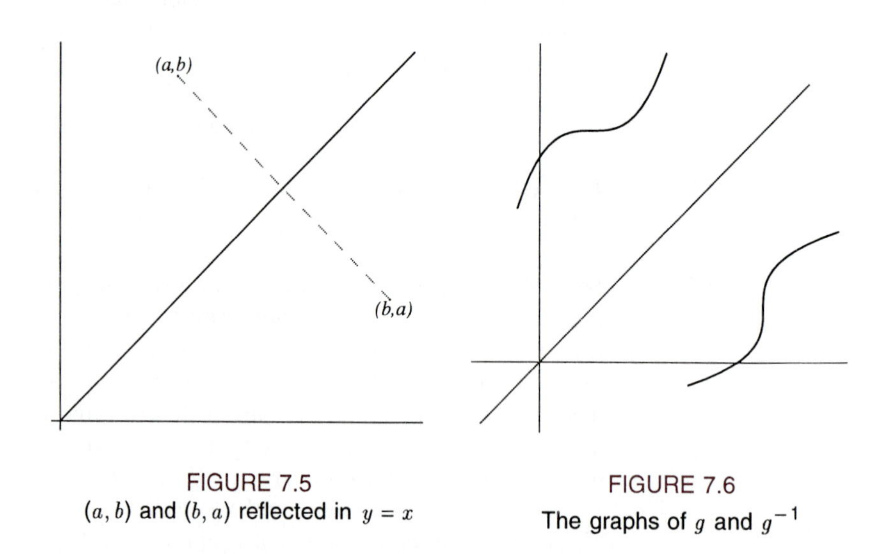

FIGURE 7.5
(a, b) and (b, a) reflected in $y = x$

FIGURE 7.6
The graphs of g and g^{-1}

Thus the graph of $g^{-1}(x)$ is the reflection in the line $y = x$ of the graph of $g(x)$.

Exercises 7.1

Prove that each function is one to one on the domain of all numbers, and then find the inverse function.

1. $f(x) = 2x - 3$

2. $f(x) = ax + b,\ a \neq 0$

3. $f(x) = x^3 + 5$

4. $f(x) = x^5$

5. $f(x) = 2x + |x|$

6. $f(x) = x|x| + x$

Find a suitable domain on which each function is one to one and then find the inverse function.

7. $f(x) = \dfrac{1}{x}$

8. $f(x) = \dfrac{x}{x+1}$

9. $f(x) = \dfrac{x+a}{x-a},\ a \neq 0$

10. $f(x) = x^2 + 2x$

11. If $f(x)$ and $g(x)$ are one-to-one functions each with domain equal to the set of all numbers, prove that the composite $h(x) = f(g(x))$ is also one to one on its domain.

12. If f, g and h are as in exercise 11, then $h(x) = f(g(x))$ expresses $h(x)$ in terms of f and g. Express the inverse of h in terms of the inverses of f and g.

13. Let f be a function with the property $f(x) = f^{-1}(x)$ for all x in the domain of f. State a property involving symmetry that the graph of f must have. Give a nontrivial example of a function that satisfies these conditions.

14. Suppose that $f(x)$ is a polynomial for which the inverse function f^{-1} exists and is also a polynomial. Prove that $f(x) = ax + b$ for some constants a and b.

15. Suppose f is a continuous, one-to-one function defined on [0, 1] and $f(0) < f(1)$. Show that f is increasing on [0, 1]. [Hint: The continuity is used in the form of the Intermediate Value Theorem. Suppose $0 \leq u < v \leq 1$. Show first $f(0) \leq f(u)$, then show $f(u) < f(v)$. Both cases use the same idea: If $f(u) < f(0)$ then $f(u) < f(0) < f(1)$ so there is some point c between u and 1 where $f(c) = f(0)$. This violates the one-to-one property.]

16. Give an example of a one-to-one function on [0, 1] that is neither increasing nor decreasing. [Hint: The function cannot be continuous in view of exercise 15.]

17. If f is one to one and differentiable on $[a, b]$ show that either $f'(x) \geq 0$ for every x or $f'(x) \leq 0$ for every x. Also state and prove the converse result.

18. Let $g(x) > 0$ for all $x \geq a$ and $F(x) = \int_a^x g(t)\,dt$. Prove that F is one to one on the interval $[a, \infty)$.

7.2 The natural logarithm function

By applying the rules for differentiation, the antiderivatives of the function $f(x) = x^n$ are easily determined to be

$$\frac{x^{n+1}}{n+1} + C,$$

provided $n + 1 \neq 0$. In case $n = -1$, no antiderivative of $x^n = 1/x$ has the form kx^m for any constants k and m. The antiderivatives of $1/x$ are not any of the familiar functions studied up to this point. Even though no familiar function is an antiderivative of $1/x$, the properties of a definite integral with a variable upper limit allow us to write an expression for a function having $1/x$ as derivative. We define a function $\ln x$ to be the particular antiderivative of $1/x$ given by the formula

$$\ln x = \int_1^x \frac{dt}{t}, \qquad x > 0. \tag{7.2}$$

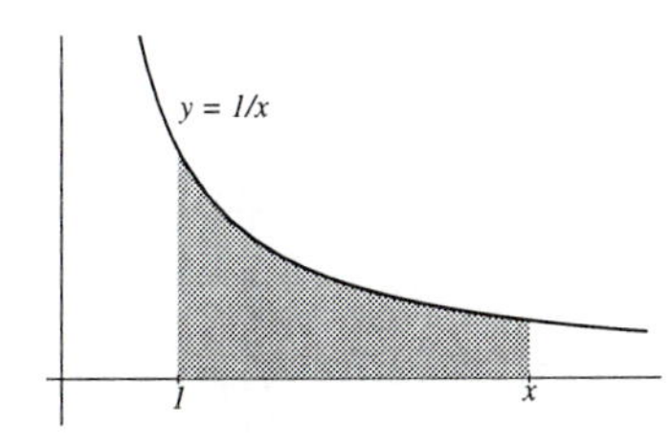

FIGURE 7.7
Shaded area = $\ln x$

Note that the restriction $x > 0$ is imposed because the function $1/t$ is not continuous at $t = 0$ and the integral can be evaluated only on an interval where the function is continuous. We are following the conventional notation $\ln x$ for the function rather than the expected $\ln(x)$. No particular significance should be attached to the omission of the parentheses; it is a typographical convenience.

The function $\ln x$ is called the *natural logarithm*. It is very closely related to the logarithm function $\log_{10} x$ studied in elementary mathematics.

The function might seem difficult to study inasmuch as its definition is certainly different from the definition of any of the functions studied to this point. To feel that a function is "familiar" to us, we need some information about it. For example, the function $\sin x$ is considered a familiar function because we know some geometric properties ($\sin x$ is the y coordinate of a certain point on the unit circle), we can easily compute some of its values [$\sin(\pm k\pi/2)$ for any integer k], we know some relations it satisfies [like formulas involving $\sin(u+v)$] and we know the shape of its graph with reasonable accuracy. We develop similar information about $\ln x$. Here are some basic properties of the natural logarithm function.

THEOREM 7.3

General Properties of the Natural Logarithm

a. $\dfrac{d}{dx}\ln x = \dfrac{1}{x}$;

b. $\ln x$ is defined and continuous on $(0, \infty)$;

c. $\ln x < 0$ for $0 < x < 1$,
   $\ln 1 = 0$,
   $\ln x > 0$, for $x > 1$;

d. $\ln x$ is an increasing function over its domain.

Proof

Because the function $1/t$ is continuous on $(0, \infty)$, the integral in Equation (7.2) is defined for all positive x and the derivative of $\ln x$ is given by the Fundamental Theorem of Calculus as $1/x$. Thus property a holds. Property b follows because a differentiable function is continuous.

To prove property c, observe that for $x > 1$, $\ln x$ is the area of the region bounded by the graph of the positive function $1/t$ above the interval $[1, x]$ (on the t-axis). Thus $\ln x$ is positive. For $0 < x < 1$, the number

$$\int_x^1 \frac{dt}{t} = -\int_1^x \frac{dt}{t} = -\ln x$$

is positive because it is the area under the graph of $1/t$ above $[x, 1]$; thus $-\ln x$ is positive and $\ln x$ is negative. The equation $\ln 1 = 0$ follows immediately from the definition of $\ln x$ because

$$\ln 1 = \int_1^1 \frac{dt}{t} = 0.$$

The derivative of $\ln x$ is $1/x$, which is positive for $x > 0$. A function with a positive derivative is increasing, so property d is proved. □

The above general properties of the natural logarithm are shared by many functions. The following are more special properties of the natural logarithm function.

THEOREM 7.4

Special Properties of the Natural Logarithm

a. $\ln uv = \ln u + \ln v$ for any positive numbers u, v.

b. $\ln x^r = r \ln x$ for any rational number r and positive number x.

Proof

Let u be a fixed positive number and let x be a variable positive number; let

$$f(x) = \ln ux.$$

Compute the derivative of $f(x)$ using the chain rule:

$$f'(x) = \frac{1}{ux}\frac{d(ux)}{dx} = \frac{1}{ux}u = \frac{1}{x}.$$

Thus $f(x)$ and $\ln x$ have the same domain and the same derivative; it follows that $f(x)$ and $\ln x$ differ by a constant:

$$\ln ux = f(x) = \ln x + C.$$

Set $x = 1$ and use the fact that $\ln 1 = 0$ to obtain $\ln u = C$. So

$$\ln ux = \ln x + \ln u$$

for all positive x; this proves property a.

The proof of property b uses the same idea as the proof of property a. The derivative of $\ln x^r$ can be computed by the chain rule:

$$\frac{d}{dx}\ln x^r = \frac{1}{x^r}\frac{d}{dx}x^r = \frac{rx^{r-1}}{x^r} = \frac{r}{x}.$$

The last term r/x equals the derivative of $r \ln x$ so $\ln x^r$ and $r \ln x$ have the same derivative at every point. It follows that

$$\ln x^r = r \ln x + C$$

for some constant C. To find C, substitute $x = 1$ to obtain

$$\ln 1^r = \ln 1 + C.$$

From $\ln 1^r = \ln 1 = 0$, we conclude $C = 0$ and property b is proved. □

Note that it was necessary to restrict r in property b to be rational because x^r is defined only for rational r. In section 7.3 we use property b to define x^r for all real r. For now we prove some information about the range of $\ln x$.

THEOREM 7.5

Range of ln x

For every real number z there is a unique positive number x such that $\ln x = z$. In particular, the range of the natural logarithm function is the set of all real numbers.

Proof

Because $\ln x$ is increasing and $\ln 1 = 0$, it follows that $\ln 2 > 0$. For any positive real number z, there is a nonnegative integer n such that

$$n \ln 2 \leq z < (n + 1) \ln 2.$$

Using property b of Theorem 7.4, we conclude z lies between $\ln 2^n$ and $\ln 2^{n+1}$. The natural logarithm is a continuous function and so the Intermediate Value Theorem applies; there is some x between 2^n and 2^{n+1} that satisfies $\ln x = z$. Thus every positive number is in the range of ln. If z

is negative, then $-z$ is positive and there is some x with $\ln x = -z$. Then

$$z = -\ln x = \ln x^{-1},$$

which shows that every negative number is also a value of ln. Because $0 = \ln 1$, we see that every real number is in the range of ln. The theorem asserts slightly more; namely that there is a *unique* solution to $\ln x = z$ for every z. This follows from the increasing property of ln; if $u < v$, then $\ln u < \ln v$ so it is not possible for two different choices of x to satisfy $\ln x = z$ for a given z. □

EXAMPLE 1 Prove

$$\lim_{x\to\infty} \ln x = \infty.$$

Solution: It is necessary to prove that for any number k, there is some N such that $\ln x > k$ whenever $x > N$. For any given number k, the theorem just proved implies there is some number N with $\ln N = k$. We know that $\ln x$ is an increasing function so it follows that

$$\ln x > k \quad \text{whenever} \quad x > N.$$ ■

Some Values of ln x

The values of the logarithm function are easily interpreted as areas and hence some values can be approximated by using lower and upper sums.

EXAMPLE 2 Give numerical approximations for the values ln 2 and ln 3.

Solution: The number ln 2 is the area under the graph of $1/x$ and above the interval [1, 2] (figure 7.7) and so it lies between each lower sum and each upper sum for a partition of the interval. Suppose we use a regular partition with n subintervals $P_n : 1 = x_0 < x_1 < \cdots < x_n = 2$. Then each subinterval has length $1/n$, and we easily determine that

$$x_i = 1 + \frac{i}{n} = \frac{n+i}{n}, \qquad 0 \le i \le n.$$

Table 7.1
Lower and Upper Sums Approximating ln 2

n	$L(P_n)$	$U(P_n)$
2	0.583333	0.833333
3	0.616667	0.783333
4	0.634524	0.759524
5	0.645635	0.745635
10	0.668771	0.718771
15	0.676758	0.710091
20	0.680803	0.705803
30	0.684883	0.70155
40	0.686936	0.699436
50	0.688172	0.698172
60	0.688998	0.697331
80	0.690032	0.696282
100	0.690653	0.695653

Table 7.2
Lower and Upper Sums Approximating ln 3

n	$L(P_n)$	$U(P_n)$
2	0.83333	1.50000
3	0.90793	1.35238
4	0.950000	1.28333
5	0.976934	1.24360
10	1.034900	1.16823
15	1.055480	1.14437
20	1.06602	1.13269
30	1.07672	1.12116
40	1.08213	1.11546
50	1.08540	1.11206
60	1.08758	1.10981
80	1.09033	1.10699
100	1.09198	1.10531

The function $1/x$ is decreasing so the minimum value of $1/x$ on $[x_{i-1}, x_i]$ occurs at the right-hand endpoint and the maximum value occurs at the left-hand endpoint. The lower and upper sums for this partition are

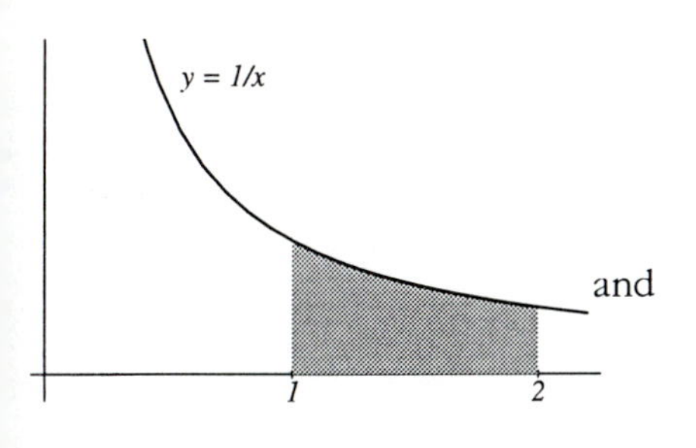

FIGURE 7.8
ln 2

$$L(P_n) = \left[\frac{1}{x_1} + \frac{1}{x_2} + \cdots + \frac{1}{x_n}\right]\frac{1}{n}$$
$$= \left[\left(\frac{n}{n+1}\right) + \left(\frac{n}{n+2}\right) + \cdots + \left(\frac{n}{n+n}\right)\right]\frac{1}{n},$$

and

$$U(P_n) = \left[\frac{1}{x_0} + \frac{1}{x_1} + \cdots + \frac{1}{x_{n-1}}\right]\frac{1}{n}$$
$$= \left[\left(\frac{n}{n+0}\right) + \left(\frac{n}{n+1}\right) + \cdots + \left(\frac{n}{n+(n-1)}\right)\right]\frac{1}{n}.$$

For small values of n, these sums can be evaluated by hand; for large values 0.25 of n a computer can do these sums very efficiently. The results of the calculations are shown in table 7.1. For any choice of n we have

$$L(P_n) < \ln 2 = \int_1^2 \frac{dt}{t} \leq U(P_n)$$

and in particular using $n = 100$, we see from table 7.1 that

$$0.690653 < \ln 2 < 0.695653. \qquad \textbf{(7.3)}$$

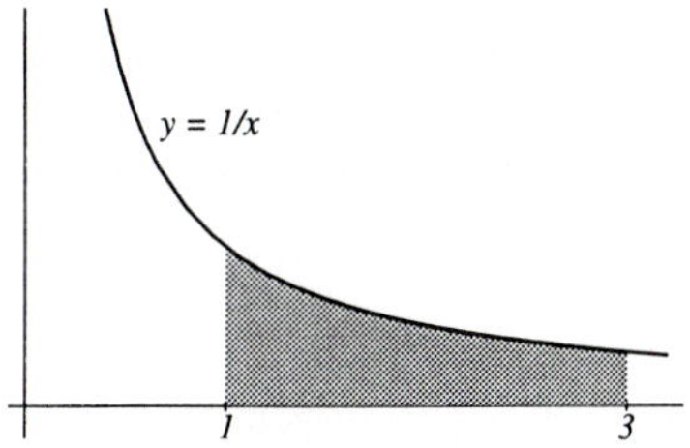

FIGURE 7.9
ln 3

To estimate ln 3, we select regular partitions P_n of $[1, 3]$:

$$P_n \ : \ 1 = x_0 < \cdots < x_n = 3, \qquad x_i = 1 + \frac{2i}{n}.$$

The lower sum $L(P_n)$ for $[1, 3]$ is a sum of terms $2/(n + 2i)$ for $1 \leq i \leq n$ and the upper sum $U(P_n)$ is a sum of terms $2/(n + 2i)$ for $0 \leq i \leq n - 1$. For every n we have

$$L(P_n) < \ln 3 < U(P_n),$$

and for $n = 100$ we have

$$1.09198 < \ln 3 < 1.10531. \qquad \textbf{(7.4)}$$

Some values of the lower and upper sums are shown in table 7.2. ■

EXAMPLE 3 Use estimates of ln 2 and ln 3 to give estimates of ln 12 and ln(4/9).

Solution: From the equation $12 = 2^2 \cdot 3$ and properties a and b of Theorem 7.4, we find

$$\ln 12 = \ln 2^2 3 = \ln 2^2 + \ln 3 = 2 \ln 2 + \ln 3.$$

Now use the inequalities (7.3) and (7.4) to get

$$2(0.69065) + 1.09198 = 2.47328 < \ln 12$$
$$< 2(0.69565) + 1.10531 = 2.49661.$$

The estimate of ln(4/9) is found as follows:

$$\ln(4/9) = \ln(2/3)^2 = 2 \ln(2/3)$$
$$= 2(\ln 2 - \ln 3) < 2(0.69565 - 1.09198) = -0.79266.$$

Similarly we also have the inequality

$$2(0.69065 - 1.10531) = -0.82932 < 2(\ln 2 - \ln 3) = \ln(2/3)^2.$$

We conclude

$$-0.82932 < \ln(4/9) < -0.79266.$$

When we take up the study of series approximations, we will be able to obtain more accurate estimates such as $\ln 12 \approx 2.484906$ and $\ln(2/3) \approx -0.405465$ with the errors in each case at most 10^{-6}. ■

The Number e

Because every real number is in the range of the natural logarithm function, there is a number e that satisfies

$$\ln e = 1 \quad \text{or} \quad \int_1^e \frac{dt}{t} = 1.$$

We have seen that $\ln 2 < 1$ and $1 < \ln 3$, so e must lie between 2 and 3 by the Intermediate Value Theorem. A slightly better approximation can be made using the approximate values $\ln 2 \approx 0.692$ and $\ln 3 \approx 1.10$. From these estimates we conclude

$$3\ln 2 - \ln 3 \approx 3(0.692) - 1.10 = 0.976 \approx 1.$$

From the equation

$$3\ln 2 - \ln 3 = \ln 2^3 - \ln 3 = \ln(8/3)$$

we conclude that $\ln(8/3)$ is close to 1 so e is close to $8/3 \approx 2.666$. Later we obtain the much better estimate $e \approx 2.7182818$ by different methods involving series.

The number e is an important constant that appears frequently in applications. In section 7.4 e plays a fundamental role in the discussion of exponential functions.

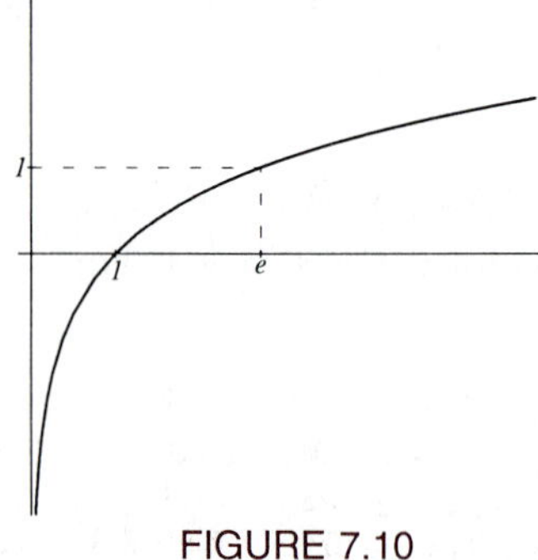

FIGURE 7.10
$y = \ln x$

The Graph of $\ln x$

There is sufficient information about the natural logarithm function given in this section to obtain a fairly good sketch of its graph. For example we know $\ln x$ is defined and increasing for $x > 0$, it has negative second derivative $-1/x^2$ and so its graph is concave, it has every number in its range and satisfies $\ln 1 = 0$. A rendering of the graph is given in figure 7.10.

Exercises 7.2

Use the approximations $\ln 2 \approx 0.693$ and $\ln 3 \approx 1.099$ to estimate each of the following numbers.

1. $\ln 144$

2. $\ln 216$

3. $\ln \sqrt{3}$

4. $\ln 6^{1/4}$

5. $\ln 0.66666667$

6. $\ln 4/81$

Give the range and the largest possible domain of each function.

7. $f(x) = \ln(x^2 + 1)$

8. $f(x) = \ln(1 - x^2)$

9. $f(x) = \ln\left(\frac{x-1}{x+1}\right)$

10. $f(x) = x \ln|x - 1|$

11. Obtain an approximation for

$$\ln 5 = \int_1^5 \frac{dt}{t}$$

by computing the lower and upper sums for $f(x) = 1/x$ using the partition $P : 4 < 17/4 < 18/4 < 19/4 < 5$

of the interval [4, 5] and then using the approximation $2\ln 2 \approx 2 \cdot 0.693$ to obtain the area over the interval $[1, 4]$.

12. In the text, we described how $8/3$ was an approximation to e. Use the same idea to show that $2^{11}/3^6$ gives a reasonable approximation to e and then look closely at the properties of $\ln x$ to show that $8/3 < e < 2^{11}/3^6$.

13. For each partition in tables 7.1 and 7.2, let

$$A(P) = \left(L(P) + U(P)\right)/2$$

be the average of the lower and upper sums. Show that $L(P) \leq \ln 2 \leq A(P)$ (and similarly for $\ln 3$) and thereby obtain a better estimate for $\ln 2$ and $\ln 3$.

14. Apply Example 2 of section 4.7 to each subinterval of a regular partition of [1, 2] to get another lower bound for $\ln 2$ of the form $2\sum_{i=1}^{n}(2(n+i)+1)^{-1} \leq \ln 2$.

15. Sketch a region under the curve $y = 1/x$ and above some interval of the x-axis whose area is $\ln\left(1 + \frac{1}{x}\right)$. [Hint: Use the relation

$$\ln\left(1 + \frac{1}{x}\right) = \ln\left(\frac{x+1}{x}\right) = \ln(x+1) - \ln x.]$$

16. Sketch the graphs of $\ln x^2$ and $(\ln x)^2$ on the same coordinate axes.

7.3 Derivatives and integrals involving ln x

The derivative formula

$$\frac{d}{dx}\ln x = \frac{1}{x}$$

and the chain rule for derivatives yields the formula

$$\frac{d}{dx}\ln g(x) = \frac{g'(x)}{g(x)}$$

for any positive differentiable function $g(x)$.

EXAMPLE 1 For x on the interval $(-\pi/2, \pi/2)$, evaluate the derivative of $\ln\cos x$.

Solution: First notice that $\ln\cos x$ is defined only for those x that satisfy $\cos x > 0$. This condition is satisfied by x on $(-\pi/2, \pi/2)$ so $\ln\cos x$ is defined for these x. It is differentiable because $\ln\cos x$ is the composite of two differentiable functions, $\cos x$ and $\ln x$. The derivative is

$$\frac{d}{dx}\ln\cos x = \frac{1}{\cos x}\frac{d}{dx}\cos x = -\frac{\sin x}{\cos x}.$$

■

EXAMPLE 2 Find the domain and derivative of $\ln|x|$.

Solution: Because $|x| > 0$ for every $x \neq 0$, the function $\ln|x|$ is defined for all nonzero x. The absolute value function is not differentiable at $x = 0$ but is differentiable at all other points. Because

$$|x| = \begin{cases} x & \text{for } x > 0 \\ -x & \text{for } x < 0 \end{cases},$$

we have

$$\frac{d}{dx}|x| = \begin{cases} 1 & \text{for } x > 0 \\ -1 & \text{for } x < 0 \end{cases}.$$

An application of the chain rule gives us

$$\frac{d}{dx}\ln|x| = \frac{1}{|x|}\frac{d}{dx}|x| = \begin{cases} 1/|x| & \text{for } x > 0 \\ -1/|x| & \text{for } x < 0. \end{cases}$$

In case $x > 0$ we have $|x| = x$ and $1/|x| = 1/x$; in case $x < 0$ we have $|x| = -x$ and so $-1/|x| = 1/x$. Thus we have the unified formula

$$\frac{d}{dx}\ln|x| = \frac{1}{x} \quad \text{for} \quad x \neq 0.$$ ■

The result of the computation in example 2 yields a general formula for the derivative of $\ln|u|$ where u is a differentiable function of x:

$$\frac{d}{dx}\ln|u| = \frac{1}{u}\frac{du}{dx} \quad \text{if} \quad u \neq 0.$$

EXAMPLE 3 Compute the derivative of $\ln|\cos x|$ at points where $\cos x \neq 0$.

Solution: An application of the chain rule yields

$$\frac{d}{dx}\ln|\cos x| = \frac{1}{\cos x}\frac{d}{dx}\cos x = -\frac{\sin x}{\cos x} = -\tan x.$$ ■

The computation provides an antiderivative for $\tan x$:

$$\int \tan x\, dx = -\ln|\cos x| + C.$$

EXAMPLE 4 Find the local extrema, the range and the behavior at 0 and ∞ of

$$f(x) = \frac{\ln x}{x}$$

and then sketch the graph of the function.

Solution: The function $f(x)$ is defined for all $x > 0$ and its derivative is

$$f'(x) = \frac{1 - \ln x}{x^2}.$$

Because the denominator is always positive, the sign of $f'(x)$ is the same as the sign of $1 - \ln x$. The solution of $1 - \ln x = 0$ is $x = e$ and we have seen that $2 < e < 3$. Thus $f'(x) > 0$ when $0 < x < e$ and $f'(x) < 0$ when $e < x$. In other words,

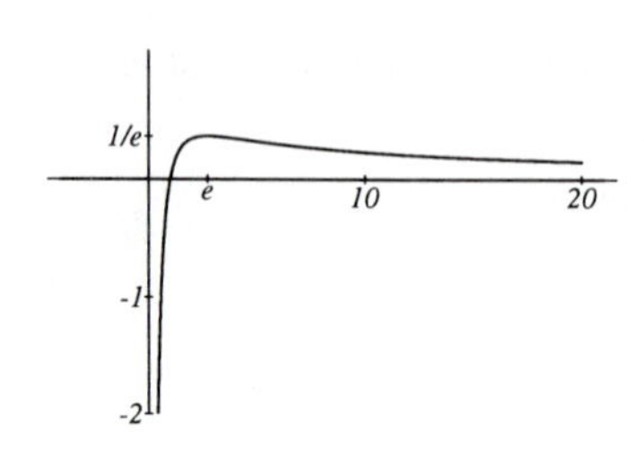

FIGURE 7.11
$y = (\ln x)/x$

$$\frac{\ln x}{x} \quad \text{is increasing if} \quad 0 < x < e,$$

$$\frac{\ln x}{x} \quad \text{is decreasing if} \quad e < x,$$

$$\frac{\ln x}{x} \quad \text{has an absolute maximum at} \quad x = e.$$

The behavior at ∞ is found by taking the limit with the help of L'Hôpital's rule:

$$\lim_{x\to\infty}\frac{\ln x}{x} = \lim_{x\to\infty}\frac{1/x}{1} = 0.$$

The behavior at the right of 0 is straightforward; because $\ln x$ approaches $-\infty$, we have

$$\lim_{x\to 0^+}\frac{\ln x}{x} = -\infty.$$

A maximum value occurs at $x = e$ and the function is decreasing for larger values of x and increasing for smaller values of x, it follows that the absolute maximum value of the function is $f(e) = 1/e$ and the range is $(-\infty, 1/e]$. ■

Logarithmic Differentiation

The special properties of the logarithm function can be exploited to simplify more complicated expressions and sometimes make the computation of derivatives less troublesome.

EXAMPLE 5 Compute the derivative of

$$f(x) = \ln \frac{(x^2-1)^3}{(x+4)(x-7)}.$$

Solution: Direct computation of $f'(x)$ by the chain rule would require an application of the quotient and product rules to differentiate the inside function. The properties in Theorem 7.4 permit simplification before taking the derivative, as follows:

$$\begin{aligned} f(x) &= \ln \frac{(x^2-1)^3}{(x+4)(x-7)} \\ &= \ln(x^2-1)^3 - \ln(x+4)(x-7) \\ &= 3\ln(x^2-1) - \ln(x+4) - \ln(x-7). \end{aligned}$$

Now the derivative is computed as

$$f'(x) = 3\frac{1}{x^2-1}(2x) - \frac{1}{x+4} - \frac{1}{x-7}.$$ ■

The idea of example 5 can be applied in cases where we want to differentiate a complicated product or quotient even when the logarithm function is not present.

EXAMPLE 6 Find the derivative of $g(x) = (x+1)^3(x-4)^2(x+10)$.

Solution: Begin by taking the logarithm of the absolute value of the function:

$$\begin{aligned} \ln|g(x)| &= \ln|(x+1)^3(x-4)^2(x+10)| = \ln|(x+1)|^3|(x-4)|^2|(x+10)| \\ &= 3\ln|x+1| + 2\ln|x-4| + \ln|x+10|. \end{aligned}$$

Now take the derivative of each side to obtain

$$\begin{aligned} \frac{g'(x)}{g(x)} &= \frac{3}{x+1} + \frac{2}{x-4} + \frac{1}{x+10}, \\ g'(x) &= g(x)\left[\frac{3}{x+1} + \frac{2}{x-4} + \frac{1}{x+10}\right]. \end{aligned}$$ ■

The idea of taking the logarithm of the absolute value of a product before differentiating is called *logarithmic differentiation*; it yields the following formula:

Extended Product Rule

If $g(x) = h_1(x)h_2(x)\cdots h_k(x)$ with each $h_i(x)$ a differentiable function, then

$$g'(x) = g(x)\left[\frac{h_1'(x)}{h_1(x)} + \frac{h_2'(x)}{h_2(x)} + \cdots + \frac{h_k'(x)}{h_k(x)}\right].$$

This formula is obtained by differentiating the logarithm of $|g(x)|$. Because $\ln|g(x)|$ is defined only at points where $g(x) \neq 0$, the formula is not valid at the points where $g(x) = 0$.

EXAMPLE 7 Use logarithmic differentiation to compute the derivative of

$$g(x) = x\sqrt{1 - x^2}\sin x.$$

Solution: Rewrite $\sqrt{1 - x^2}$ as $(1 - x^2)^{1/2}$ and then take the logarithm of the absolute value to get

$$\begin{aligned}\ln|g(x)| &= \ln|x|\,|1 - x^2|^{1/2}|\sin x| \\ &= \ln|x| + \frac{1}{2}\ln|1 - x^2| + \ln|\sin x|.\end{aligned}$$

Take the derivative of both sides then multiply by $g(x)$ to get

$$\begin{aligned}g'(x) &= g(x)\left[\frac{1}{x} - \frac{x}{1 - x^2} + \frac{\cos x}{\sin x}\right] \\ &= \sqrt{1 - x^2}\sin x - x(1 - x^2)^{-1/2}\sin x + x\sqrt{1 - x^2}\cos x.\end{aligned}$$

■

Integration Using ln *u*

If $g(x)$ is a differentiable function of x, then

$$\frac{d}{dx}\ln|g(x)| = \frac{g'(x)}{g(x)}.$$

From the derivative formula we obtain an antiderivative rule:

$$\int \frac{g'(x)}{g(x)}\,dx = \ln|g(x)| + C.$$

Here are several examples of indefinite integrals that can be evaluated by a substitution of the form

$$u = g(x) \quad du = g'(x)\,dx$$

$$\int \frac{g'(x)}{g(x)}\,dx = \int \frac{du}{u} = \ln|u| + C = \ln|g(x)| + C.$$

EXAMPLE 8 Evaluate the integrals by using a suitable substitution.

$$\text{(a)} \int \frac{4x}{x^2 + 7}\,dx \qquad \text{(b)} \int \frac{1}{x\ln x}\,dx.$$

Solution: (a) Let $u = x^2 + 7$ so $du = 2x\,dx$. Then

$$\int \frac{4x}{x^2+7}\,dx = \int \frac{2du}{u} = 2\ln|u| + C = 2\ln(x^2+7) + C.$$

(b) Let $u = \ln x$ so $du = dx/x$. Then

$$\int \frac{1}{x\ln x}\,dx = \int \frac{du}{u} = \ln|u| + C = \ln|\ln x| + C.$$

■

Exercises 7.3

Evaluate the derivative $f'(x)$.

1. $f(x) = \ln(x^2 - 4)$
2. $f(x) = \ln(4x^2 - 2x + 2)$
3. $f(x) = (x-1)\ln(x-1)$
4. $f(x) = \cos x \ln(\sin^2 x)$
5. $f(x) = \ln\left(\dfrac{x-1}{x+1}\right)$
6. $f(x) = (\ln x)^4 - \ln(x^4)$

Use the special properties of the logarithm function to simplify the functions and then evaluate the derivative $f'(x)$.

7. $f(x) = \ln\left(\dfrac{(x^2+4)^2(x-1)}{(x^2+9)(x+1)^3}\right)$
8. $f(x) = \ln\left(\dfrac{1+\cos^2 x}{2+\sin x}\right)$
9. $f(x) = (x-1)\ln\left(\dfrac{(x^2-1)^3}{(x^2+4)^2}\right)$
10. $f(x) = \ln\left(\dfrac{(x-1)(x-2)(x-3)}{(6-x)(7-x)(8-x)}\right)$
11. $f(x) = \dfrac{(x^3+x+1)\sin(x^2-1)}{(x^2-4x+12)\cos x}$
12. $f(x) = (x+1)(x+2)(x+3)(x+4)$

Evaluate the integrals.

13. $\displaystyle\int \frac{dx}{x+4}$
14. $\displaystyle\int \frac{dx}{7x-2}$
15. $\displaystyle\int \frac{4x^3}{1+x^4}\,dx$
16. $\displaystyle\int \frac{(2t+1)}{t^2+t+4}\,dt$
17. $\displaystyle\int \frac{\sin x}{3+\cos x}\,dx$
18. $\displaystyle\int \frac{t^2+4t+1}{1+t^2}\,dt$
19. $\displaystyle\int_1^{2e} \frac{dx}{x}$
20. $\displaystyle\int_4^{12} \frac{dt}{t+4}$
21. $\displaystyle\int_1^{e} \frac{\ln x}{x}\,dx$
22. $\displaystyle\int \tan 4x\,dx$
23. $\displaystyle\int \cot x\,dx$

Evaluate the limits that exist using L'Hôpital's rule whenever appropriate.

24. $\displaystyle\lim_{x\to 0^+} \ln x$
25. $\displaystyle\lim_{x\to 0^+} x\ln x$
26. $\displaystyle\lim_{x\to\infty} \frac{\ln(x^2+1)}{x}$
27. $\displaystyle\lim_{x\to\infty} \ln\left(\frac{x}{x+1}\right)$
28. $\displaystyle\lim_{x\to\infty} (x+1)^{-3}(\ln x+1)^2$
29. $\displaystyle\lim_{x\to 1^+} \frac{1-x+\ln x}{(x-1)^2}$
30. Which number is larger, π^e or e^π? [Hint: Determine what numbers satisfy $e^x < x^e$ by taking logarithms and using the properties of $\ln x/x$ given in example 4.]
31. Let $p(x)$ be any nonzero polynomial. Prove
$$\lim_{x\to\infty} \frac{\ln|p(x)|}{x} = 0.$$
32. Let m be any positive number. Prove
$$\lim_{x\to\infty} \frac{\ln x}{x^m} = 0.$$
[This can be interpreted as saying that $\ln x$ grows more slowly than any positive power of x.]
33. Let $p(x)$ be polynomial such that $p(0) = 0$. Prove $\lim_{x\to 0^+} p(x)\ln x = 0$. [Hint: $p(x) = x^m q(x)$ for some integer $m \geq 1$ and some polynomial $q(x)$ with $q(0) \neq 0$.]

7.4 The exponential function

For a positive number a and an integer n, the nth power a^n has been defined as the product of a with itself n times (for $n > 0$) or the product of $1/a$ with itself $|n|$ times (for $n < 0$). The nth root of a has been defined as the unique number y such that $y^n = a$ and we write $y = a^{1/n}$. This permits the definition of rational number exponents: $a^{n/m} = (a^n)^{1/m}$. At this point of the discussion of exponents, no meaning has been given for the expression $3^{\sqrt{2}}$ or similar terms involving nonrational exponents. We now use the logarithm function to cure this deficiency; we use it to define a^x for any real number x and positive number a.

DEFINITION 7.3

Real Exponents

Let a be a positive number and x any real number. Then a^x is the unique real number that satisfies $\ln a^x = x \ln a$.

If we set $z = x \ln a$, then Theorem 7.5 says there is a unique solution y to the equation $\ln y = z = x \ln a$. That solution is denoted by $y = a^x$. Because the domain of the logarithm function is the set of positive reals, the solution $y = a^x$ must be a positive real number. Thus

$$a^x > 0 \quad \text{for any} \quad a > 0 \quad \text{and every} \quad x.$$

Inasmuch as a^x has already been defined when x is rational, it is necessary to check that the new definition and the old definition give the same value in case both can be applied. That they do agree is the content of statement b of Theorem (7.4), which says $\ln x^r = r \ln x$ for rational r.

The definition of a^x must seem somewhat peculiar to the reader inasmuch as it involves the seemingly extraneous logarithm function. To get acquainted with the definition, we use it to prove some familiar properties of exponents.

THEOREM 7.6

Properties of Exponents

Let a be a positive number and let x and y be any real numbers. Then

a. $a^{x+y} = a^x a^y$,

b. $a^{-x} = 1/a^x$,

c. $\left(a^x\right)^y = a^{xy}$.

Proof

The proofs use the one-to-one property of the logarithm that can be stated as $\ln u = \ln v$ implies $u = v$. We prove property a by applying the definitions of a^{x+y} and a^x, a^y:

$$\ln a^{x+y} = (x + y) \ln a = x \ln a + y \ln a = \ln a^x + \ln a^y = \ln a^x a^y.$$

Thus $a^{x+y} = a^x a^y$.

To prove property b we have

$$\ln a^{-x} = -x \ln a = -1(x \ln a) = -1(\ln a^x) = \ln(a^x)^{-1}.$$

Thus $a^{-x} = (a^x)^{-1} = 1/a^x$.

To prove property c, we apply the definitions several times:

$$\ln\left(a^x\right)^y = y \ln a^x = y(x \ln a) = (xy) \ln a = \ln a^{xy}.$$

The one-to-one property of the logarithm implies

$$(a^x)^y = a^{xy}. \qquad \square$$

For a positive number a, the function $f(x) = a^x$ is called the *exponential function with base* a. When the number e, defined by the equation $\ln e = 1$, is the base, the exponential function has some special properties we now examine.

THEOREM 7.7

Properties of e^x

For every number x, the following hold:

a. $\ln e^x = x$,

b. $e^{\ln x} = x$ if $x > 0$.

Proof

By definition, $\ln e^x = x \ln e$; but $\ln e = 1$ so property a holds. To prove property b, write the definition again; for any number u, e^u is the unique number logarithm is u; that is $\ln e^u = u$. Substitute $u = \ln x$ to obtain

$$\ln e^{\ln x} = \ln x.$$

The one-to-one property of the logarithm function, $\ln z = \ln w$ implies $z = w$, now gives

$$e^{\ln x} = x. \qquad \square$$

Other special properties of e will be seen shortly when the derivatives of e^x and of a^x are computed.

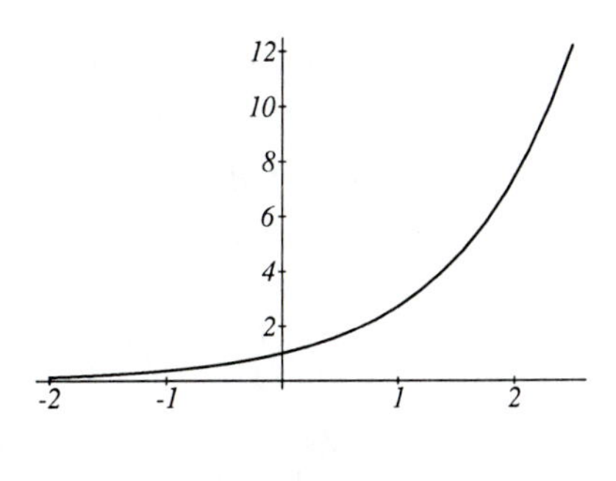

FIGURE 7.12
$y = e^x$

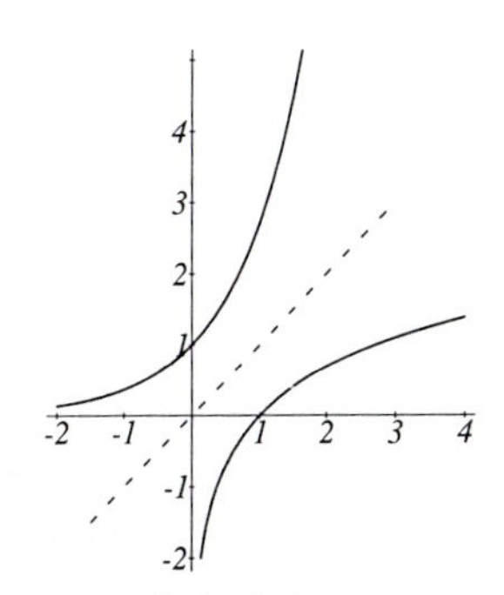

FIGURE 7.13
e^x is $\ln x$ reflected

It follows from the theorem that $L(x) = \ln x$ and $E(x) = e^x$ are inverse functions; that is $L(E(x)) = x$ and $E(L(x)) = x$. From a knowledge of the graph of $y = \ln x$ we obtain the graph of $y = e^x$. A point (p, q) is on the graph of one if and only if (q, p) is on the graph of the other. The figures 7.12 and 7.13 show the graph of e^x in different scales and its reflection in the line $y = x$.

Exercises 7.4

Simplify the following expressions so no logarithms remain.

1. $e^{\ln 3}$

2. $e^{2\ln x}$

3. $\ln e^{2/3}$

4. $\ln e^{-7x+3}$

5. $e^{-\ln 3x+1}$

6. $e^{\ln x - \ln 5}$

7. $\ln e^{\sqrt{x}}$

8. $\ln \sqrt{e^x}$

9. $\ln(\ln e^2) - \ln 2$

10. $e^{-7x+3\ln x}$

11. Use the definitions given in this section to prove the properties: (a) $1^x = 1$ for all numbers x; (b) $a^0 = 1$ for all positive numbers a.

12. Use the definitions of the exponential $E(x) = e^x$ and the properties of the logarithm function to prove that $E(x)$ is an increasing function.

13. Use the definitions of the exponential $y = a^x$ and the properties of the logarithm function to prove that a^x is an increasing function when $a > 1$ and a^x is a decreasing function when $0 < a < 1$.

14. Sketch the graphs of the three functions 2^x, e^x and 3^x on the same coordinate system to indicate their relative positions.

15. Sketch the graph of e^{-x}.

16. Sketch the graphs of 2^{-x}, e^{-x} and 3^{-x} on the same coordinate system.

17. Show that $f(x) = e^x + e^{-x}$ is an even function and sketch its graph.

18. Show that $g(x) = e^x - e^{-x}$ is an odd function and sketch its graph.

Find all solutions x of the following equations.

19. $2^{2x} = 3^{-x}$

20. $e^{3x^2} = 2^{4x}$

21. $e^{2x} - 2e^x + 1 = 0$

22. $e^{2x} + 2e^x - 3 = 0$

23. $3^{2x} - 3^{x+1} + 2 = 0$

24. $4^x - 3 \cdot 2^x - 4 = 0$

25. The equation $e^{-x} = x$ cannot be solved by purely algebraic manipulations. Use the graphs of $y = e^{-x}$ and $y = x$ to argue that there is exactly one solution to the equation and that solution lies on the interval $[0, 1]$. Use some calculator or computer calculations to argue more precisely that the unique solution lies on the interval $[0.5, 0.6]$.

7.5 The derivative of the exponential function

The derivative of the exponential function $E(x) = e^x$ can be determined by differentiating the equation $x = \ln e^x$ with the use of the chain rule. We get

$$1 = \frac{d}{dx}x = \frac{d}{dx}\ln e^x = \frac{1}{e^x}\frac{d}{dx}e^x.$$

Thus

$$\frac{d}{dx}e^x = e^x.$$

The function e^x is equal to its own derivative at every point. If $g(x)$ is a differentiable function, then the chain rule gives the formula

$$\frac{d}{dx}e^{g(x)} = e^{g(x)}g'(x). \tag{7.5}$$

EXAMPLE 1 Compute the derivative of the functions

(a) $f(x) = e^{3x}$ (b) $g(x) = e^{-4x^2}$.

Solution: Both problems require the application of Equation (7.5).

$$\text{(a)} \quad \frac{d}{dx}e^{3x} = e^{3x}\frac{d}{dx}(3x) = 3e^{3x}.$$

$$\text{(b)} \quad \frac{d}{dx}e^{-4x^2} = e^{-4x^2}\frac{d}{dx}(-4x^2) = -8xe^{-4x^2}.$$

■

EXAMPLE 2 Find the points where the function $f(x) = e^{\sin x}$ has local minima and local maxima and then sketch the graph.

Solution: It is necessary to find the solutions to $f'(x) = 0$ and then determine the sign of the second derivative at these points. The first derivative is

$$f'(x) = e^{\sin x}\frac{d}{dx}\sin x = e^{\sin x}\cos x.$$

Because $e^x > 0$ for all x, $f'(x) = 0$ precisely when $\cos x = 0$. Let c be any number for which $\cos c = 0$. At this point, $\sin c$ is either $+1$ or -1 (because $\sin^2 c + \cos^2 c = 1$). The second derivative of f is

$$f''(x) = -e^{\sin x}\sin x + e^{\sin x}\cos^2 x,$$

and

$$f''(c) = -e^{\sin c}\sin c$$

because the $\cos c$ term is 0. The exponential term $e^{\sin c}$ is positive so the second derivative at c has sign opposite to that of $\sin c$. By the second derivative test we conclude

$$f(c) \quad \text{is a local maximum if} \quad \sin c = 1, \quad c = \frac{\pi}{2} + 2k\pi,$$

$$f(c) \quad \text{is a local minimum if} \quad \sin c = -1, \quad c = -\frac{\pi}{2} + 2k\pi,$$

where k is any integer.

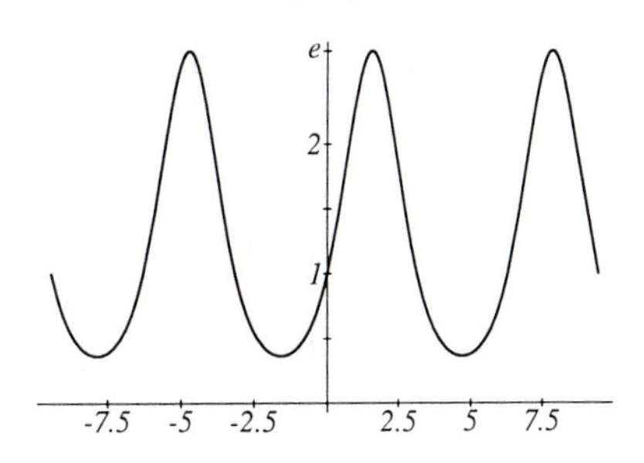

FIGURE 7.14

$y = e^{\sin x}$

The graph is shown in the figure. ■

Derivative of the Exponential with General Base

The exponential function $f(x) = a^x$ to a base a not equal to e has a more complicated derivative than the exponential e^x. We can use the definition of a^x to compute the derivative. Differentiate the equation

$$\ln a^x = x \ln a$$

to get

$$\frac{d}{dx} \ln a^x = \frac{1}{a^x} \frac{d}{dx} a^x = \frac{d}{dx}(x \ln a) = \ln a.$$

Thus

$$\frac{d}{dx} a^x = a^x \ln a.$$

When $a = e$, the constant $\ln a$ on the right equals 1; this simplification is the main reason why the base e is commonly used rather than some other base such as base 10.

For example

$$\frac{d}{dx} 2^x = 2^x \ln 2 \approx (0.69) 2^x,$$
$$\frac{d}{dx} 10^x = 10^x \ln 10 \approx (2.30) 10^x.$$

Integrals Involving Exponentials

The derivative formula

$$\frac{d}{dx} e^x = e^x$$

implies the integral formula

$$\int e^x \, dx = e^x + C.$$

The chain rule formula, Equation (7.5), implies a corresponding integral formula. The substitution $u = g(x)$, $du = g'(x)\,dx$ is used to make the evaluation

$$\int e^{g(x)} g'(x) \, dx = \int e^u \, du = e^u + C = e^{g(x)} + C.$$

EXAMPLE 3 Evaluate

$$\int e^{-7x} \, dx.$$

Solution: Let $u = -7x$ and $du = -7\,dx$. Then

$$\int e^{-7x} \, dx = -\frac{1}{7} \int e^u \, du = -\frac{1}{7} e^u + C = -\frac{1}{7} e^{-7x} + C.$$ ■

EXAMPLE 4 Evaluate the integral

$$\int \frac{e^{-x}}{(1 + e^{-x})^2} \, dx.$$

Solution: Make the substitution $u = 1 + e^{-x}$, $du = -e^{-x}\,dx$ and get

$$\int \frac{e^{-x}}{(1 + e^{-x})^2} \, dx = \int \frac{-du}{u^2} = u^{-1} + C$$
$$= \frac{1}{1 + e^{-x}} + C.$$ ■

Exercises 7.5

Compute the derivative $f'(x)$ of the indicated functions.

1. $f(x) = e^{4x}$
2. $f(x) = e^{-3x} + e^{3x}$
3. $f(x) = xe^{2x} + e$
4. $f(x) = (2x + 1)e^{x^2-2x}$
5. $f(x) = e^{\sqrt{x}} - \sqrt{e^x}$
6. $f(x) = (e^{2x} - 2e^x + 1)^{1/2}$
7. $f(x) = (4e^{-3x} - x^3)^{1/3}$
8. $f(x) = \dfrac{e^{2x} - 1}{e^{4x}}$
9. $f(x) = \cos e^x$
10. $f(x) = e^{\cos x}$
11. $f(x) = e^{\ln x}$
12. $f(x) = \ln e^x$
13. Use the definitions to prove $a^x = e^{x \ln a}$ for any positive a and any x.

Compute the derivatives of the indicated functions. Use the equation in exercise 13 to rewrite the terms having variable exponents as a power of e.

14. $f(x) = 3^{-2x}$
15. $f(x) = 10^x + 10^{-x}$
16. $f(x) = (1/2)^{x^2}$
17. $f(x) = (3x - 1)3^{2x+3}$

Evaluate the following integrals.

18. $\int e^{-3x}\,dx$
19. $\int xe^{x^2}\,dx$
20. $\int e^{ax+b}\,dx$
21. $\int \dfrac{e^x}{1+e^x}\,dx$
22. $\int \sqrt{e^x}\,dx$
23. $\int \dfrac{1}{1+e^{-x}}\,dx$
24. $\int \dfrac{1}{1+e^x}\,dx$
25. $\int e^x \sin e^x\,dx$
26. $\int \dfrac{e^{1/x}}{x^2}\,dx$
27. $\int \dfrac{e^x}{\sqrt{1+e^x}}\,dx$

Use logarithmic differentiation to find $f'(x)$.

28. $f(x) = x^x$, $x > 0$
29. $f(x) = (x^2 + 1)^{1/x}$
30. $f(x) = x^{\cos x}$, $x > 0$
31. $f(x) = (e^x + 1)^{\ln x}$
32. Find the point on the graph of $f(x) = e^{ax}$ where the tangent line passes through the origin.
33. Find the volume of the solid obtained by rotating around the x-axis the region below the graph of $y = e^{-x}$ and above the interval $[0, \ln a]$.
34. Find the volume of the solid obtained by rotating around the x-axis the region below the graph of $y = 1/\sqrt{x}$ and above the interval $[4, 16]$.

7.6 Applications of the exponential function

We pose and solve a number of practical problems in which the exponential function plays an important role. We begin with a section on exponential growth and decay. Here we study several physical phenomena having a common feature; there is a quantity varying with time at a rate of change proportional to the quantity itself. In other words, the quantity is represented by a function of time whose derivative is proportional to the function at each instant of time.

Growth of Bacteria

Many physical phenomena are approximated by continuous functions even though the quantities measured are discrete. For example the number of bacteria cells in a growing culture is an integer $N(t)$ at time t; unless $N(t)$ is constant, it is not a continuous function because the graph has jumps as the values change from one integer to another. In laboratory experiments, one hopes to gather data over a fairly short time and use that data to make predictions about future behavior. The bacteriologist makes some assumptions about the function $N(t)$ from some observable values at $t = t_1, t_2, \ldots, t_n$ and then attempts to predict function values $N(t)$ for all t.

A commonly used, but somewhat oversimplified, technique is to assume that $N(t)$ is a continuous and differentiable function. One gains some information about the rate of change of $N(t)$ from observable data and then expresses the rate of change as the derivative $N'(t)$ to obtain a relation between the unknown function and its derivative.

In the case of the growing bacteria culture, the data suggests that the rate of change of $N(t)$ is proportional to $N(t)$ when the conditions within the culture are favorable. It says roughly that when there are more bacteria present in the culture, the rate of change of $N(t)$ is greater. The rate of change of N from time t to $t + h$ is the familiar quotient

$$\frac{N(t+h) - N(t)}{h}$$

and the instantaneous rate of change is $N'(t)$. The equation

$$N'(t) = kN(t) \tag{7.6}$$

is the mathematical formulation of the statement that the rate of change of the number of cells is a constant times the number of cells present. The k in the equation is a constant of proportionality that must be determined by observing the particular bacterial culture. Equation (7.6) is an example of a differential equation; there is an unknown function and an equation relating the function and its derivative.

The unknown function satisfying Equation (7.6) is a constant multiple of its derivative. We already saw functions with this property when we discussed the derivatives of the exponential function. The function $g(t) = e^{kt}$ (with k constant) satisfies

$$g'(t) = ke^{kt} = kg(t).$$

Thus $N(t) = e^{kt}$ is a solution to Equation (7.6). We have found this solution by guessing; we relied on our memory to produce the function e^{kt}. Are there other solutions to (7.6)? Yes there are, and here is how they can be found.

If $N(t)$ satisfies Equation (7.6), then

$$\frac{N'(t)}{N(t)} = k,$$

and so

$$\int \frac{N'(t)}{N(t)}\,dt = \int k\,dt = kt + C.$$

The first integral can be evaluated by the substitution $u = N(t)$, $du = N'(t)\,dt$ to obtain

$$\int \frac{N'(t)}{N(t)}\,dt = \int \frac{du}{u} = \ln|u| + c = \ln|N(t)| + c.$$

We have

$$\ln|N(t)| + c = kt + C,$$

where c and C are constants. There is no need to carry both c and C in the computation inasmuch as $a = C - c$ is just another constant. Also in this problem, $N(t) > 0$ because it represents the number of cells in a culture and so the absolute value signs can be removed. We have

$$\ln N(t) = kt + a.$$

To get an expression for $N(t)$ we raise e to powers:

$$e^{\ln N(t)} = e^{kt+a} = e^{kt}e^{a}.$$

Use the relation $e^{\ln u} = u$ and the fact that $e^a = A$ is a constant to conclude

$$N(t) = Ae^{kt}, \qquad A = \text{constant}.$$

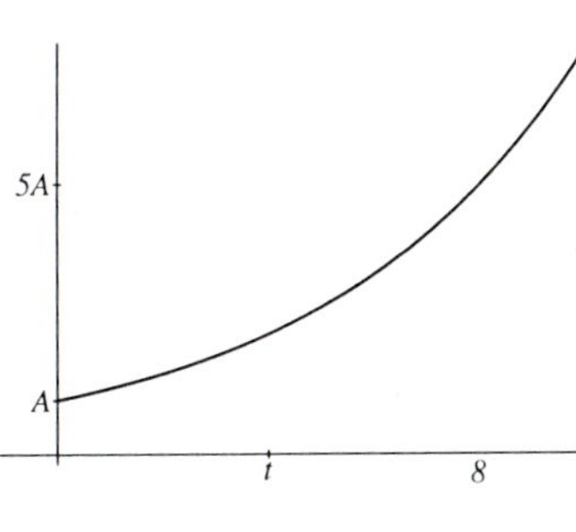

FIGURE 7.15
$N(t) = Ae^{0.2t}$

So we have found the function $N(t)$ subject to determining two constants, A and k. These constants must be determined by data from the experiment. The constant A is the number of cells present at time $t = 0$ because the formula gives $N(0) = Ae^0 = A$. We see that a bacteria culture has *exponential growth*.

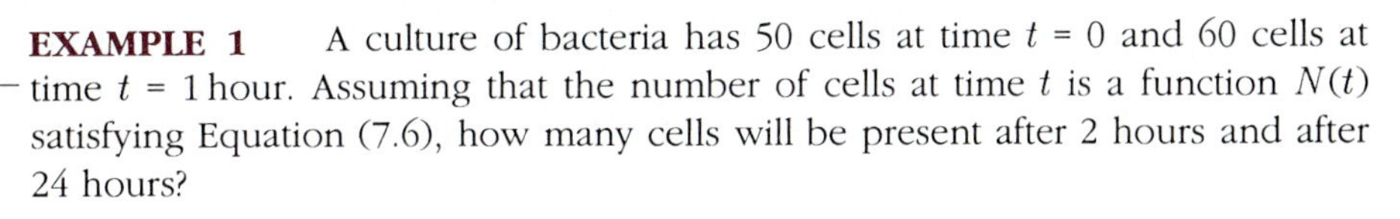

EXAMPLE 1 A culture of bacteria has 50 cells at time $t = 0$ and 60 cells at time $t = 1$ hour. Assuming that the number of cells at time t is a function $N(t)$ satisfying Equation (7.6), how many cells will be present after 2 hours and after 24 hours?

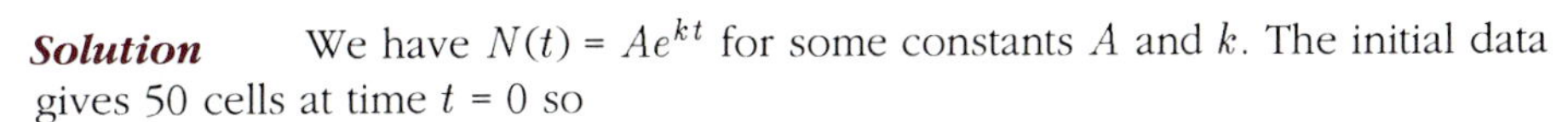

Solution We have $N(t) = Ae^{kt}$ for some constants A and k. The initial data gives 50 cells at time $t = 0$ so

$$50 = N(0) = Ae^0 = A.$$

So far we know $N(t) = 50e^{kt}$. Measure time in hours, so after 1 hour we have

$$60 = N(1) = 50e^k \qquad \text{or} \qquad e^k = \frac{60}{50} = 1.2.$$

There is no need to solve explicitly for k in this problem, if we use the relation

$$e^{kt} = \left(e^k\right)^t = (1.2)^t.$$

Thus

$$N(t) = 50(1.2)^t.$$

After 2 hours, the number of cells is

$$N(2) = 50(1.2)^2 = 72.$$

After 24 hours, the number of cells is

$$N(24) = 50(1.2)^{24} \approx 50(79.4968) = 3974.84.$$

The number of cells must be an integer but $N(t)$ is a continuous function approximating the actual number of cells. We are safe in saying that if the number of cells increases by a rule described by Equation (7.6), then after 24 hours there will be about 4000 cells. ■

Radioactive Decay

Certain substances, such as isotopes of radium, plutonium, and uranium, decay at a rate proportional to the quantity of material present. The process of decay, called radioactive decay, is measured in terms of the *half-life* of the substance. The half-life of a substance is the time required for one half of the substance to be lost by decay. Radium has a half-life of 1600 years; this means that the 1 milligram of radium in your glow-in-the-dark wrist watch will reduce to 0.5 milligrams of radium 1600 years from now. (This should not significantly affect the resale value of your watch.) Some man-made isotopes of carbon have a half-life measured in minutes whereas other substances have half-lives measured in fractions of seconds.

If $G(t)$ is the quantity of a radioactive substance at time t, then the rate of change of G is proportional to G; that is,

$$G'(t) = kG(t).$$

The function G satisfies the same equation as the number of bacteria in a culture. The solution to this differential equation is

$$G(t) = Ae^{kt}.$$

The constant A is the quantity at time $t = 0$ because $G(0) = Ae^0 = A$. The constant k (or e^k) is determined by the particular material in question.

EXAMPLE 2 A certain isotope of plutonium has a half-life of 13 years. How much of a 5-gram sample of this isotope will remain after 1 year?

Solution: If $G(t)$ is the number of grams of the sample after t years, then $G(0) = 5$ and

$$G(t) = 5e^{kt}$$

for some constant k. After 13 years, half the material has decayed and so

$$2.5 = G(13) = 5e^{13k}.$$

Divide by 5 and then take the $1/13$ power of each side to obtain

$$(1/2)^{1/13} = e^k.$$

Replace $1/2$ by 2^{-1} and substitute for e^k in the formula for $G(t)$ to get

$$G(t) = 5\left(e^k\right)^t = 5\left(2^{-1/13}\right)^t = 5(2)^{-t/13}.$$

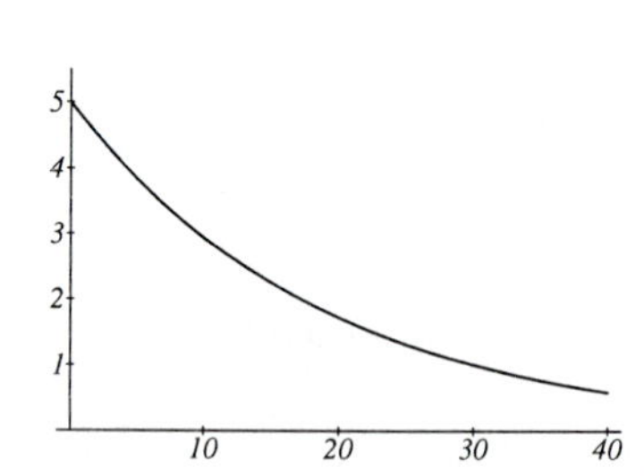

FIGURE 7.16
$G(t) = 5(2)^{-t/13}$

The quantity of material remaining after 1 year is

$$G(1) = 5(2)^{-1/13}.$$

To put the answer in a more meaningful form, it is necessary to find some approximation of the power of 2 that remains. My hand calculator gives

$$G(1) = 5(0.948) = 4.740.$$

The approximation of numbers like $2^{-1/13}$ is usually done with the help of a calculator or computer. Other approximations that must be done by hand are discussed after the theory of series is developed. The graph of $G(t)$ shown in figure 7.16 is produced by machine; approximate values of G can be read directly from the graph. ■

Newton's Law of Cooling

If a container of boiling water is removed from the kitchen stove and allowed to cool in surroundings at room temperature 72°F, how does the temperature of the water change as a function of time? Newton's law of cooling says that the rate of change of the temperature of the water is proportional to the difference of the temperature between the water and its surroundings. If $T(t)$ is the temperature of the water at time t and if T_s is the temperature of the surrounding air, then

$$T'(t) = k(T(t) - T_s). \tag{7.7}$$

If the container of boiling water is so large that its presence changes the temperature of the air in the room, then T_s changes with time and Equation (7.7) is somewhat more complicated to solve. We treat only the simpler case in which T_s is constant. Then Equation (7.7) can be solved by direct integration. Divide the equation by $(T(t) - T_s)$ to get

$$k = \frac{T'(t)}{T(t) - T_s},$$

which implies

$$\int k\,dt = \int \frac{T'(t)}{T(t) - T_s}\,dt.$$

Perform the integration by using the substitution $u = T(t) - T_s$, $du = T'(t)\,dt$ and get

$$kt + C = \ln|T(t) - T_s|.$$

In our example we have $T(t) - T_s > 0$ so the absolute value signs can be removed; after raising to powers we obtain

$$e^{kt+C} = e^{\ln(T(t)-T_s)} = T(t) - T_s,$$

and finally

$$T(t) = T_s + Ae^{kt} \qquad \textbf{(7.8)}$$

for some constant $A = e^C$. The description of $T(t)$ contains three constants: T_s the constant temperature of the surroundings, A and k. In our example we are given $T_s = 72$ and the initial temperature of the water $T(0) = 212$, the boiling point of water. With this information we can determine A: $212 = 72 + Ae^0 = 72 + A$ or $A = 140$. One additional piece of information is needed to determine k. Suppose the temperature of the water is found to be 180°F after 10 minutes. Then we have

$$180 = 72 + 140\,e^{10k} \quad \text{and so} \quad \frac{108}{140} = e^{10k}.$$

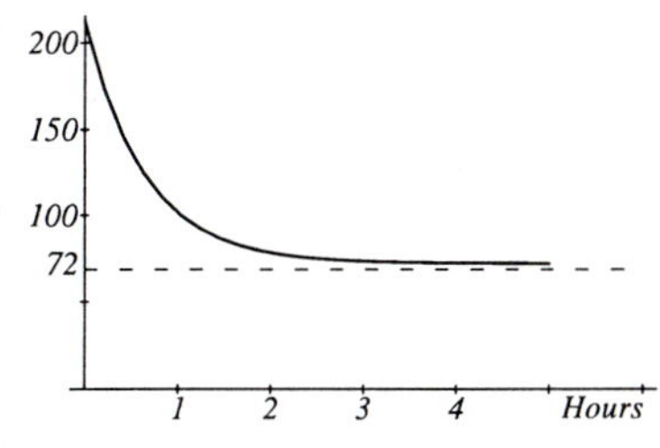

FIGURE 7.17
$$T(t) = 72 + 140\left(\frac{108}{140}\right)^{t/10}$$

Raise both sides of the last equation to the power 1/10 and substitute into the equation for $T(t)$ to get the final form:

$$T(t) = 72 + 140\left(\frac{108}{140}\right)^{t/10}.$$

After an hour the temperature of the water will be $T(60) \approx 101.5$ and after 2 hours it will be $T(120) \approx 78.2$. The graph of $T(t)$ shown in figure 7.17 shows that the temperature falls rather quickly toward the limiting temperature of 72°F.

The solution to equation (7.7) defined in equation (7.8) can be applied to describe the temperature function of an object cooling (or heating) in a surrounding medium having constant temperature. We summarize the general situation that has been illustrated in this example.

THEOREM 7.8

Newton's Law of Cooling

If material with temperature $T(t)$ at time t is placed in a surrounding medium having constant temperature T_s, then $T(t)$ satisfies the equation

$$T'(t) = k(T(t) - T_s),$$

for some constant k. The solution of this equation is

$$T(t) = T_s + Ae^{kt}$$

for some constant A.

Exercises 7.6

Find the solution of the differential equation that satisfies the given initial condition. Sketch the graph of the solution.

1. $\dfrac{dT(t)}{dt} = -3T(t),\ T(0) = 10$
2. $\dfrac{dT(t)}{dt} = 0.05T(t),\ T(0) = 100$
3. $\dfrac{dT(t)}{dt} = 2T(t) - 4,\ T(0) = 6$
4. $\dfrac{dT(t)}{dt} = 2T(t) + 4,\ T(0) = 6$
5. $\dfrac{dG}{dt} = -0.1G,\ G(0) = 10$
6. $\dfrac{dG}{dt} = -0.1G + 40,\ G(0) = 0$
7. $\dfrac{dG}{dt} = -0.1G - 40,\ G(0) = 0$
8. $\dfrac{dG}{dt} = 0.1G + 40,\ G(0) = 0$
9. $\dfrac{dG}{dt} = 0.1G - 40,\ G(0) = 0$
10. A bacteria culture contains 2000 cells at time $t = 0$ and 3000 cells at time $t = 2$ hours. How many cells are present after 4 hours?
11. A bacteria culture contains 2000 cells at time $t = 1$ and 3000 cells at time $t = 3$ hours. How many cells are present after 5 hours?
12. A bacteria culture contains 2000 cells at time $t = 0$ and 3000 cells at time $t = 6$ hours. How many cells were present after 4 hours?
13. The half-life of radium-226 is 1620 years. How long does it take for a sample to lose one tenth of its size?
14. A man-made element loses one fifth of its size in 2.7 hours. What is its half-life?
15. One bacteria culture has 100 cells at time 0 and 150 cells 1 day later. A second culture has 200 cells at time 0 and 220 cells 1 day later. In how many days will the number of cells in the first culture exceed the number in the second culture?
16. An oversimplified description of the population growth of the United States asserts that the rate of increase of the population is proportional to the population size. Assuming this is correct predict the population 25 years forward if the present population is 225 million and five years ago it was 209 million.
17. A Las Vegas gambler loses at a rate proportional to the amount of money he has at any time. If he loses at the rate of one quarter of his present cash per hour, how long will it take for him to lose half of his original cash?
18. On a cold winter's night, the outside temperature is -15°F and the temperature inside your house is 72°F when the power fails and the furnace shuts down. One hour later the inside temperature has dropped to 64°F. Assuming that the outside temperature remains constant and the furnace remains off, what will the temperature be after 24 hours and how long will it take for the inside temperature to reach 32°F?
19. If the temperature function $T(t)$ satisfies the differential equation $T' = kT - T_0$, what is the limiting value of T (as $t \to \infty$)? For what values of t will the temperature $T(t)$ differ from the limiting value by no more than 1% of the limiting value?

7.7 Indeterminate forms

Suppose we consider the behavior of the function $h(x) = (1 + x)^{1/x}$ as x approaches 0 from the right. The exponent $1/x$ gets very large, so one might be inclined to guess that $h(x)$ grows without bound. But the term $(1 + x)$ gets close to 1 and 1 to any positive power equals 1; thus one might also guess that the limit of $h(x)$ is 1. In fact, neither of these guesses is correct; we see below that $\lim_{x\to 0^+}(1 + x)^{1/x} = e$.

More generally let $f(x)$ be a positive function and $g(x)$ an arbitrary function. We wish to investigate limits of the form

$$\lim_{x\to p^+} f(x)^{g(x)}.$$

Two-sided limits and left-sided limits as well as limits as x approaches $\pm\infty$ can also be handled in the same way as this case.

This limit certainly exists if

$$\lim_{x\to p^+} g(x) = A \qquad \text{and} \qquad \lim_{x\to p^+} f(x) = B \neq 0,$$

for some real numbers A and B, because then we have

$$\lim_{x\to p^+} f(x)^{g(x)} = B^A.$$

(The proof is given below.) There are other cases in which the limit exists even though the separate functions do not have finite limits. In some cases, the limit can be evaluated using a variation on L'Hôpital's rule. We describe some of these cases.

The limit

$$\lim_{x\to p^+} f(x)^{g(x)} \tag{7.9}$$

is called an *indeterminate form* of the type 0^0 if

$$\lim_{x\to p^+} f(x) = \lim_{x\to p^+} g(x) = 0.$$

This is called an indeterminate form because the limit (7.9) cannot be found by using the equation

$$\lim_{x\to p^+} f(x)^{g(x)} = \left(\lim_{x\to p^+} f(x)\right)^{\lim_{x\to p^+} g(x)}. \tag{7.10}$$

Equation (7.10) is valid under suitable conditions, which we describe after proving the next theorem.

There are two other indeterminate forms, namely 1^∞ and ∞^0 that we consider. All three cases are based on the following technique.

THEOREM 7.9

Exponential Limits

Let $f(x)$ be a positive function in an interval (p, q) and suppose $g(x)$ is another function such that the limit

$$\lim_{x\to p^+} g(x)\ln f(x)$$

exists and is equal to L. Then

$$\lim_{x\to p^+} f(x)^{g(x)} = e^L.$$

The analogous statement is true for left-sided limits, two-sided limits and for limits as x approaches $\pm\infty$.

Proof

The natural logarithm function $\ln u$ is continuous for $u > 0$. When $u = u(x)$ is a function of x, the continuity of $\ln x$ implies

$$\lim_{x\to p^+} \ln u(x) = \ln\left(\lim_{x\to p^+} u(x)\right).$$

Take the case

$$u(x) = f(x)^{g(x)}$$

to get

$$L = \lim_{x \to p^+} g(x) \ln f(x) = \lim_{x \to p^+} \ln f(x)^{g(x)} = \ln \lim_{x \to p^+} f(x)^{g(x)}.$$

It follows from the properties of the exponential function that

$$e^L = e^{\ln \lim_{x \to p^+} f(x)^{g(x)}} = \lim_{x \to p^+} f(x)^{g(x)}. \qquad \square$$

In order that this be a useful theorem, it is necessary to evaluate the limit

$$\lim_{x \to p^+} g(x) \ln f(x). \qquad \textbf{(7.11)}$$

In the case mentioned above, namely,

$$\lim_{x \to p^+} g(x) = A \qquad \text{and} \qquad \lim_{x \to p^+} f(x) = B \neq 0,$$

we have

$$\lim_{x \to p^+} g(x) \ln f(x) = A \ln B$$

and so

$$\lim_{x \to p^+} f(x)^{g(x)} = e^{A \ln B} = B^A.$$

In the three cases, 0^0, 1^∞ and ∞^0, it is possible that the limit can be evaluated by a variation on L'Hôpital's rule. We illustrate the three possibilities.

EXAMPLE 1 Evaluate the limit of x^x as x approaches 0 from the right.

***Solution*:** This is a 0^0 form. Let $y = x^x$ so that $\ln y = x \ln x$. Then

$$\begin{aligned} \lim_{x \to 0^+} \ln y = \lim_{x \to 0^+} x \ln x &= \lim_{x \to 0^+} \frac{\ln x}{1/x} \\ &= \lim_{x \to 0^+} \frac{1/x}{-1/x^2} = \lim_{x \to 0^+} (-x) \\ &= 0. \end{aligned}$$

Thus

$$\lim_{x \to 0^+} y = \lim_{x \to 0^+} x^x = e^0 = 1. \qquad \blacksquare$$

EXAMPLE 2 Evaluate the limit

$$\lim_{x \to \infty} (1 + kx)^{1/x},$$

for any constant k.

***Solution*:** This is a ∞^0 form. Set $y = (1 + kx)^{1/x}$ so that

$$\ln y = \frac{\ln(1 + kx)}{x}.$$

We can apply L'Hôpital's rule to obtain

$$\begin{aligned} \lim_{x \to \infty} y = \lim_{x \to \infty} \frac{\ln(1 + kx)}{x} \\ = \lim_{x \to \infty} \frac{\frac{k}{1 + kx}}{1} \\ = 0. \end{aligned}$$

Thus

$$\lim_{x \to \infty} y = \lim_{x \to \infty} (1 + kx)^{1/x} = e^0 = 1. \qquad \blacksquare$$

EXAMPLE 3 Evaluate the limit

$$\lim_{x \to 0^+} (1 + kx)^{1/x},$$

for any constant k.

***Solution*:** This is a 1^∞ form. We use the same notation and follow the same reasoning as in example 2 so that $y = (1 + kx)^{1/x}$ and

$$\begin{aligned}\lim_{x \to 0^+} \ln y &= \lim_{x \to 0^+} \frac{\ln(1 + kx)}{x} \\ &= \lim_{x \to 0^+} \frac{\frac{k}{1 + kx}}{1} \\ &= k.\end{aligned}$$

Thus

$$\lim_{x \to 0^+} y = \lim_{x \to 0^+} (1 + kx)^{1/x} = e^k. \qquad \blacksquare$$

Notice that by taking different values for the constant k, any positive number can be obtained as the limit of an indeterminate form of exponential type.

We give the usual warning that the hypothesis of L'Hôpital's theorem must be verified before applying the rule; that is, the expression must really be an indeterminate form before applying the derivative procedure.

An Exponential Limit and Compound Interest

The evaluation of limits by first taking logarithms is especially suited for the evaluation of

$$\lim_{t \to \infty} \left(1 + \frac{p}{t}\right)^t,$$

where p is a constant. This limit occurs in a number of contexts, one of which is illustrated below.

Let

$$u = \left(1 + \frac{p}{t}\right)^t$$

so that

$$\lim_{t \to \infty} \ln u = \lim_{t \to \infty} \ln\left(1 + \frac{p}{t}\right)^t = \lim_{t \to \infty} t \ln\left(1 + \frac{p}{t}\right).$$

We evaluate this with the aid of L'Hôpital's rule:

$$\begin{aligned}\lim_{t \to \infty} t \ln\left(1 + \frac{p}{t}\right) &= \lim_{t \to \infty} \frac{\ln(1 + p/t)}{1/t} \\ &= \lim_{t \to \infty} \frac{(1 + p/t)^{-1}(-p/t^2)}{(-1/t^2)} \\ &= \lim_{t \to \infty} \frac{p}{1 + (p/t)} \\ &= p.\end{aligned}$$

Thus we have

$$\lim_{t\to\infty}\left(1+\frac{p}{t}\right)^t = e^p. \tag{7.12}$$

As a special case with $p = 1$ we have

$$\lim_{n\to\infty}\left(1+\frac{1}{n}\right)^n = e.$$

If n is restricted to be a positive integer that is not too large, this expression can be used to obtain an approximation of the number e. Table 7.3 shows a few values of

$$\left(1+\frac{1}{n}\right)^n \approx e.$$

With $n = 100$, the approximation has an error less than 0.1. (We admit that this is not a particularly good method of approximation.)

TABLE 7.3
Approximations to e

n	8	12	24	36	60	84	100
$(1+1/n)^n$	2.5657	2.6130	2.6637	2.6814	2.6959	2.7022	2.7048

We can deduce that the approximations are all less than e by showing that the function

$$f(x) = \left(1+\frac{1}{x}\right)^x$$

is increasing for all $x > 1$. We can show $f(x)$ is increasing by proving that the derivative $f'(x)$ is positive for all $x > 1$. We first compute

$$\ln f(x) = x\ln\left(1+\frac{1}{x}\right)$$

and then differentiate each side to get

$$\begin{aligned}\frac{f'(x)}{f(x)} &= x\left(\frac{1}{1+1/x}\right)\left(\frac{-1}{x^2}\right)+\ln\left(1+\frac{1}{x}\right)\\ &= x\frac{x}{1+x}\left(\frac{-1}{x^2}\right)+\ln\frac{x+1}{x}\\ &= \ln x + 1 - \ln x - \frac{1}{x+1}.\end{aligned}$$

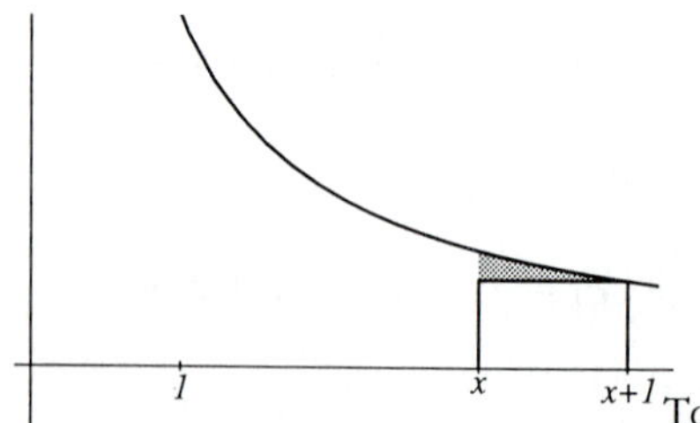

FIGURE 7.18
$\ln(1+1/x) - 1/(x+1) > 0$

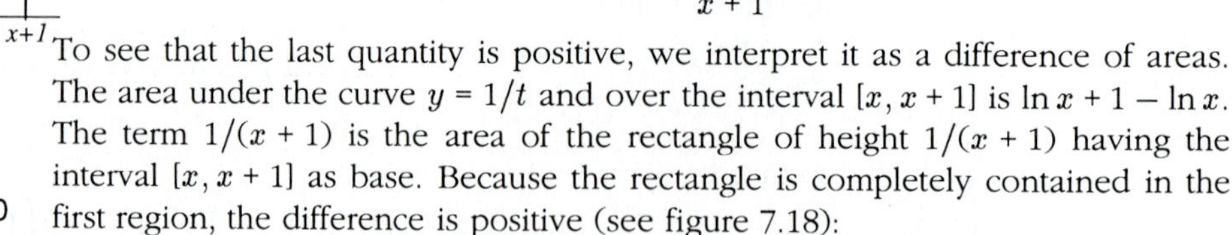

To see that the last quantity is positive, we interpret it as a difference of areas. The area under the curve $y = 1/t$ and over the interval $[x, x+1]$ is $\ln x + 1 - \ln x$. The term $1/(x+1)$ is the area of the rectangle of height $1/(x+1)$ having the interval $[x, x+1]$ as base. Because the rectangle is completely contained in the first region, the difference is positive (see figure 7.18):

$$\ln(x+1) - \ln x - \frac{1}{x+1} = \ln\left(1+\frac{1}{x}\right) - \frac{1}{x+1} > 0.$$

From $f(x) > 0$, it follows that $f'(x) > 0$ for all $x > 1$. Thus $f(x)$ is increasing for $x > 1$.

Compound Interest

If you deposit \$100 in an account at FNB (First National Bank) that pays interest at the rate of 8% per year, then after one year your account contains the initial amount plus the interest earned; namely

$$100 + 100(0.08) = 100(1 + 0.08) = 108.$$

If FNB wants to attract more deposits, they offer to compound the interest at intervals more frequently than once a year. If the interest is compounded every 6 months then after 6 months your account is worth

$$100 + 100(0.08/2) = 100(1 + 0.04) = 104;$$

at the end of the second 6 months it is worth

$$104 + 104(0.08/2) = 104(1 + 0.04) = 100(1 + 0.04)^2 = 108.16.$$

More generally if you deposit an amount A in an account that pays interest at the rate of p per year (with p as a decimal, not a percentage) then after one year your account is worth $A(1 + p)$. If the bank generously agrees to compound the interest n times per year in equally spaced intervals, then at the end of the year the account is worth

$$A\left(1 + \frac{p}{n}\right)^n.$$

If the bank compounds every day, then $n = 365$. If the bank agrees to compound every hour, then $n = 24 \cdot 365 = 8760$.

What happens to your account if the bank agrees to compound the interest every second? Will you become wealthy after making only a modest deposit? In the limiting case, the number of compounding periods n approaches infinity. The value of your account at the end of one year with compounding every instant is

$$\lim_{n\to\infty} A\left(1 + \frac{p}{n}\right)^n = Ae^p.$$

TABLE 7.4

Value of \$1.00 after 1 year at 100p% compounded n times

	4%	5%	6%	7%	8%	9%	10%
$n = 6$	1.0407	1.0511	1.0615	1.0721	1.0827	1.0934	1.1043
$n = 12$	1.0407	1.0512	1.0617	1.0723	1.0830	1.0938	1.1047
$n = 18$	1.0408	1.0512	1.0617	1.0724	1.0831	1.0939	1.1049
$n = 24$	1.0408	1.0512	1.0618	1.0724	1.0831	1.0940	1.1049
$n = 30$	1.0408	1.0512	1.0618	1.0724	1.0832	1.0940	1.1050
$n = \infty$	1.0408	1.0513	1.0618	1.0725	1.0833	1.0942	1.1052

Table 7.4 shows some values of the multiplier for various compounding intervals n and rates of interest p. If \$1 is placed in an account on which the interest is paid at the rate of 10% per year compounded at six equal intervals, the table shows the account will be worth \$1.1043 at the end of the year. The table row marked $n = \infty$ shows the limit as n approaches infinity of the account value. Even though the table values are rounded, it is clear that increasing the number of compounding periods does not produce a dramatic increase in the account value.

Exercises 7.7

Evaluate the limits that exist.

1. $\lim_{x\to\infty}\left(\frac{x+1}{x}\right)^x$

2. $\lim_{x\to\infty}\left(\frac{x+a}{x}\right)^x$

3. $\lim_{x \to \infty} (1 + x)^{1/x^2}$

4. $\lim_{x \to 0^+} \left(\frac{x+1}{x}\right)^x$

5. $\lim_{x \to 0^+} \left(\frac{x+a}{x}\right)^x$

6. $\lim_{x \to 0^+} (1 + x)^{1/x^2}$

7. $\lim_{x \to \infty} \left(1 + \frac{1}{4x+3}\right)^{3x}$

8. $\lim_{x \to \infty} \left(1 + \frac{1}{ax+b}\right)^{cx}$

9. Prove the inequality

$$\frac{1}{t+1} < \ln\left(1 + \frac{1}{t}\right) = \ln t + 1 - \ln t < \frac{1}{t}$$

for all $t > 0$ by interpreting the three terms as areas related to the curve $y = 1/x$ above the interval $[t, t+1]$.

10. For $x > 0$, let

$$f(x) = \left(1 + \frac{1}{x}\right)^x \quad \text{and} \quad g(x) = \left(1 + \frac{1}{x}\right)^{x+1}.$$

Use logarithmic differentiation and the exercise 9 to prove that $f(x)$ is an increasing function and $g(x)$ is a decreasing function.

11. With $f(x)$ and $g(x)$ as in the exercise 10, prove

$$\lim_{x \to \infty} f(x) = \lim_{x \to \infty} g(x) = e$$

and then conclude $f(x) < e < g(x)$ for all $x > 0$.

12. With the aid of a computer or hand calculator, use exercise 11 to estimate the number e by evaluating $f(n)$ and $g(n)$ for $n = 5,\ 10,\ 15,\ 20$ and some larger values of n that your device can handle.

13. Given the opportunity to place for 1 year your new found wealth of \$1000 either in a bank that pays 6.75% per year compounded monthly or in a savings and loan institution that pays 7% compounded twice per year, which would you select assuming that your only consideration is to make the most money from your deposit?

14. What interest rate having one compounding period per year produces the same interest as an annual interest rate of 8% compounded four times per year?

15. Show that e^x grows faster than any power of x by proving $\lim_{x \to \infty} e^x/x^m = \infty$ for any number m.

16. For any polynomial $p(x)$, prove $\lim_{x \to \infty} p(x)e^{-x} = 0$.

7.8 Inverse trigonometric functions

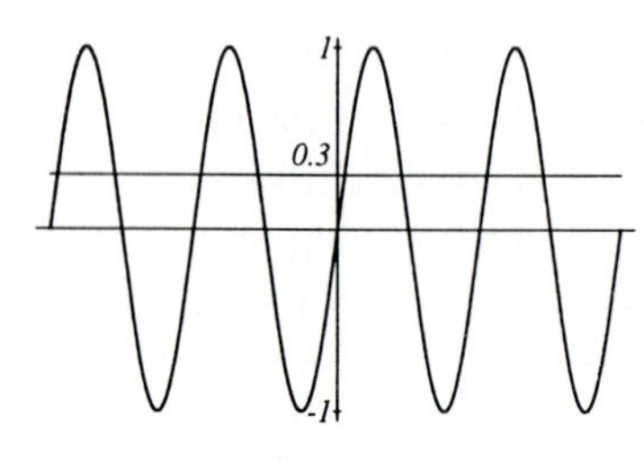

FIGURE 7.19
$\sin x = 0.3$
has many solutions

If we are required to solve the equation $\sin x = 0.3$ for x, an inspection of the graph of $\sin x$ reveals that there are infinitely many solutions. In particular the results in section 2.3 imply that the function $\sin x$ does not have an inverse on its domain. If we restrict our attention to an interval $[a, b]$ of length π on which $\sin x$ is increasing, then $\sin a = -1$ and $\sin b = 1$ and each number between -1 and $+1$ is a value of $\sin x$ for exactly one x on the interval. Thus $\sin x$ is one to one on such an interval. When the domain of $\sin x$ is restricted to any interval on which $\sin x$ is increasing, an inverse function exists. The same conclusion applies over an interval on which $\sin x$ is decreasing.

A convenient, but somewhat arbitrary, interval is the symmetric one with zero at the midpoint.

Inverse Sine

If the domain of $\sin x$ is restricted to the interval $[-\pi/2, \pi/2]$, then $\sin x$ has an inverse function that is denoted by $\arcsin x$.

The domain of $\arcsin x$ is $[-1, 1]$; the range of $\arcsin x$ is $[-\pi/2, \pi/2]$.

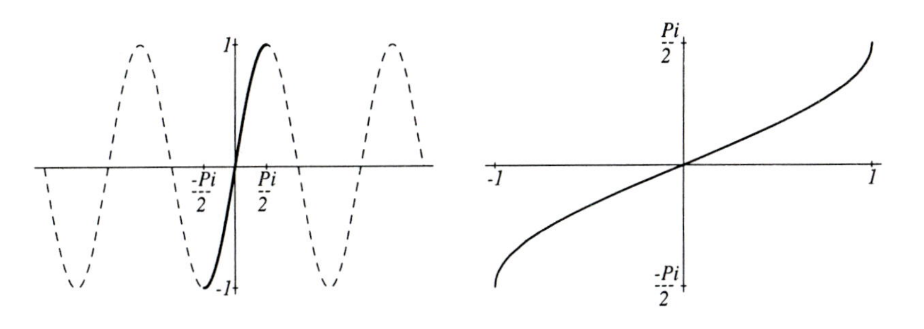

FIGURE 7.20
$\sin x$ 1-to-1 on $[-\pi/2, \pi/2]$

FIGURE 7.21
$y = \arcsin x$

The following relations are consequences of the defining properties of inverse functions:

$$\arcsin(\sin x) = x \quad \text{if} \quad -\frac{\pi}{2} \leq x \leq \frac{\pi}{2},$$

$$\sin(\arcsin x) = x \quad \text{if} \quad -1 \leq x \leq 1,$$

$$\sin u = v \quad \text{if and only if} \quad \arcsin v = u, \quad |u| \leq \frac{\pi}{2}, \; |v| \leq 1.$$

From the equation $\sin(-x) = -\sin x$ we conclude

$$\arcsin(-\sin x) = \arcsin\big(\sin(-x)\big) = -x. \tag{7.13}$$

Any number u on $[-1, 1]$ can be expressed as $u = \sin x$ for some x on $[-\pi/2, \pi, 2]$ and so $\arcsin u = x$ and after substituting in Equation (7.13) we get

$$\arcsin(-u) = -\arcsin u.$$

The graph of $y = \arcsin x$, figure 7.21, can be sketched by reflecting the graph of $y = \sin x$ in the line $y = x$; if (p, q) is on the graph of $\sin x$, then (q, p) is on the graph of $\arcsin x$.

Inverse Cosine Function

The function $\cos x$ is decreasing on $[0, \pi]$ and takes every value between -1 and $+1$ exactly once over this interval. We define the inverse function of the cosine function having domain restricted to $[0, \pi]$.

Inverse Cosine

If the domain of $\cos x$ is restricted to the interval $[0, \pi]$, then $\cos x$ has an inverse function that is denoted by $\arccos x$.

The domain of $\arccos x$ is $[-1, 1]$; the range of $\arccos x$ is $[0, \pi]$.

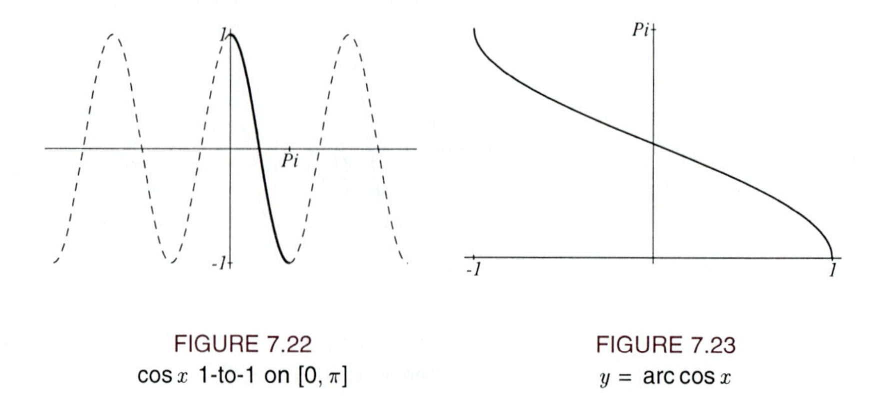

FIGURE 7.22
$\cos x$ 1-to-1 on $[0, \pi]$

FIGURE 7.23
$y = \arccos x$

The following relations are consequences of the defining properties of inverse functions:

$$\arccos(\cos x) = x \quad \text{if} \quad 0 \leq x \leq \pi,$$
$$\cos(\arccos x) = x \quad \text{if} \quad -1 \leq x \leq 1,$$
$$\cos u = v \quad \text{if and only if} \quad \arccos v = u, \quad 0 \leq u \leq \pi, \ |v| \leq 1.$$

The equation $\cos(-x) = \cos x$ cannot be used to prove the analogous formula in the same way as we derived $\arcsin(-x) = -\arcsin x$. The problem is that we cannot have both x and $-x$ in the restricted domain $[0, \pi]$ of cos (unless $x = 0$). The correct starting point for such a formula is the equation

$$\cos(\pi - x) = -\cos x \quad \text{or} \quad \pi - x = \arccos(-\cos x).$$

If we let $u = \cos x$ and $x = \arccos u$ then we obtain

$$\pi - \arccos u = \arccos(-u) \quad \text{for} \quad |u| \leq 1.$$

The graph of $y = \arccos x$ is shown in figure 7.23.

Inverse Tangent Function

The tangent function is increasing on the interval $(-\pi/2, \pi/2)$ and its range is the set of all real numbers. The tangent with restricted domain has an inverse function.

Inverse Tangent

If the domain of $\tan x$ is restricted to the interval $(-\pi/2, \pi/2)$, then $\tan x$ has an inverse function that is denoted by $\arctan x$.

The domain of $\arctan x$ is all real numbers; the range of $\arctan x$ is $(-\pi/2, \pi/2)$.

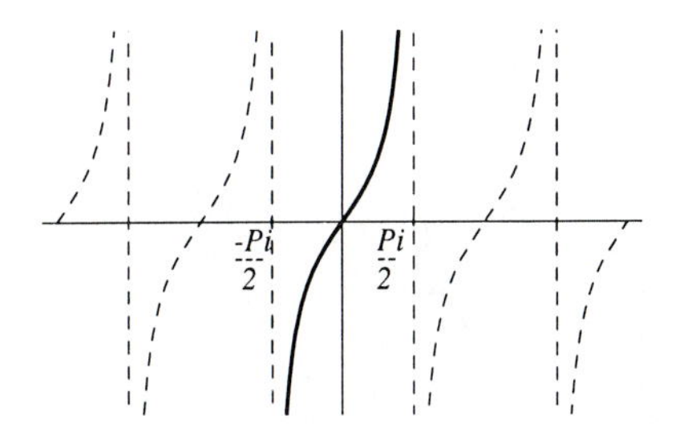

FIGURE 7.24
$\tan x$ 1-to-1 on $(-\pi/2, \pi/2)$

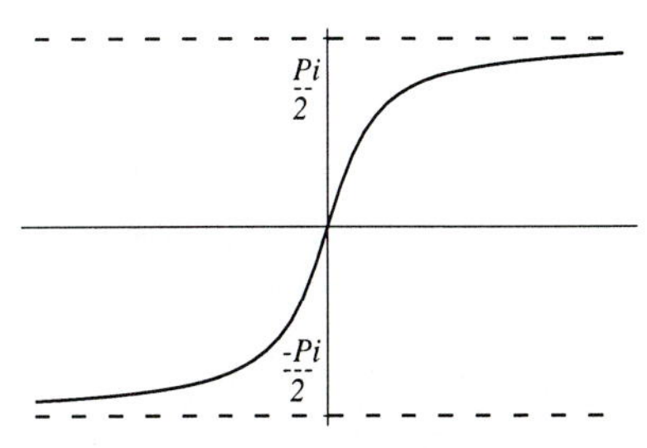

FIGURE 7.25
$y = \arctan x$

The following relations are consequences of the defining properties of inverse functions:

$$\arctan(\tan x) = x \quad \text{if} \quad -\frac{\pi}{2} < x < \frac{\pi}{2},$$

$$\tan(\arctan x) = x \quad \text{for any number } x,$$

$$\tan u = v \quad \text{if and only if} \quad \arctan v = u, \quad |u| < \frac{\pi}{2}.$$

From the equation $\tan(-x) = -\tan x$ we conclude

$$\arctan(-x) = -\arctan x$$

by the same reasoning that was used to prove the corresponding equation for $\arcsin x$. The graph of $y = \arctan x$ is shown in figure 7.25.

We can read from the graph the following useful limits:

$$\lim_{x\to\infty} \arctan x = \frac{\pi}{2} \qquad \lim_{x\to-\infty} \arctan x = -\frac{\pi}{2}. \tag{7.14}$$

Inverse of Secant, Cosecant and Cotangent

The remaining three trigonometric functions can be treated similarly; because they are used less often, we are very brief.

The range of the secant function is the set of all numbers y with $|y| \geq 1$; the domain is the set of points where $\cos x \neq 0$. The selection of the restricted domain on which $\sec x$ is one to one is not as clear cut as in the three previous cases. We select the domain most often used.

Inverse Secant

If the domain of $\sec x$ is restricted to the interval $[0, \pi]$ except that $x \neq \pi/2$ then $\sec x$ has an inverse function denoted by $\operatorname{arcsec} x$.

The domain of $\operatorname{arcsec} x$ is the set of all x satisfying $|x| \geq 1$; the range of $\operatorname{arcsec} x$ is the union of the two half-open intervals $[0, \pi/2)$ and $(\pi/2, \pi]$.

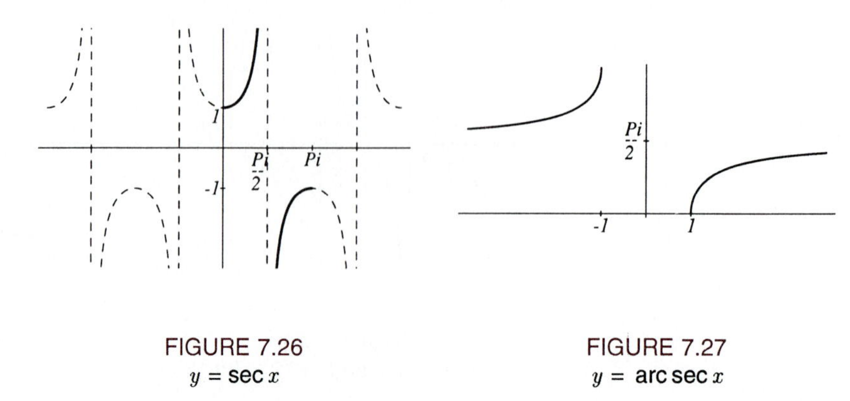

FIGURE 7.26
$y = \sec x$

FIGURE 7.27
$y = \text{arc sec } x$

The cosecant function is undefined at points where the sine function equals zero and has the same range as the secant function.

Inverse Cosecant

If the domain of $\csc x$ is restricted to $[-\pi/2, \pi/2]$ except that $x \neq 0$, then $\csc x$ has an inverse function denoted by $\text{arc csc } x$.

The domain of $\text{arc csc } x$ is all x for which $|x| \geq 1$; the range of $\text{arc csc } x$ is the union of the two half-open intervals $[-\pi/2, 0)$ and $(0, \pi/2]$.

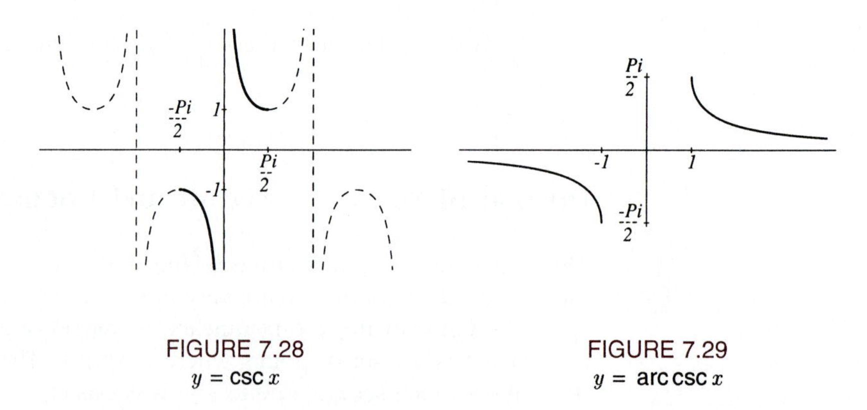

FIGURE 7.28
$y = \csc x$

FIGURE 7.29
$y = \text{arc csc } x$

The domain of the cotangent function is the set of points where $\sin x \neq 0$; the range is the set of all real numbers. The cotangent function is one to one on the interval $(0, \pi)$.

Inverse Cotangent

If the domain of $\cot x$ is restricted to the interval $(0, \pi)$ then $\cot x$ has an inverse function denoted by $\text{arc cot } x$.

The domain of $\text{arc cot } x$ is all real numbers and the range is $(0, \pi)$.

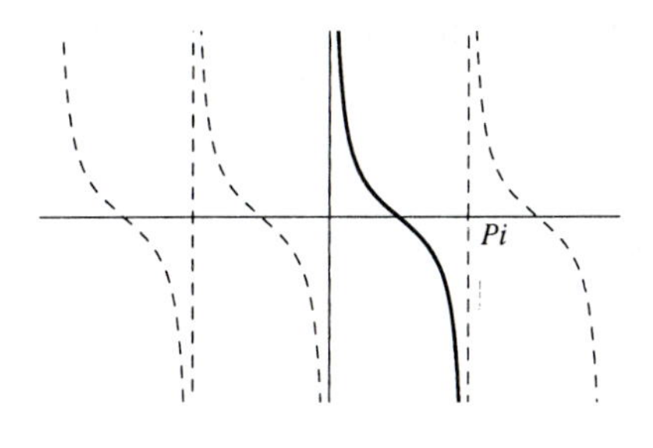

FIGURE 7.30
$y = \cot x$

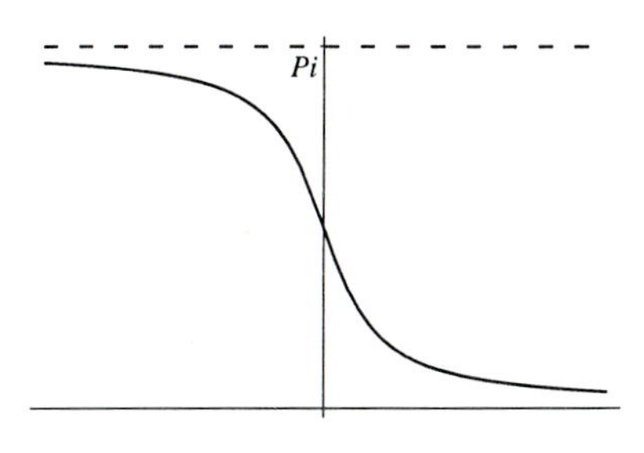

FIGURE 7.31
$y = \text{arc}\cot x$

Numerical Computations

EXAMPLE 1 Evaluate the following inverse functions:

a. $\arcsin(\sqrt{2}/2)$,
b. $\arccos -1/2$,
c. $\arctan -1$,
d. $\arcsin(\sin(7\pi/2))$.

Solution:

a. Because $\sin(\pi/4) = \sqrt{2}/2$ and $\pi/4$ is on the interval $[-\pi/2, \pi/2]$, the definition of the inverse sine function implies $\arcsin(\sqrt{2}/2) = \pi/4$.
b. Because $\cos(2\pi/3) = -1/2$ and $2\pi/3$ is on the interval $[0, \pi]$, the definition of the inverse cose function implies $\arccos(-1/2) = 2\pi/3$.
c. Because $\tan(-\pi/4) = -1$ and $-\pi/4$ is on the interval $[-\pi/2, \pi/2]$, the definition of inverse tangent implies $\arctan(-1) = -\pi/4$.
d. If x is on the interval $[-\pi/2, \pi/2]$, then $\arcsin(\sin x) = x$. But $x = 7\pi/2$ is not on the correct interval so we use the relation

$$\sin(7\pi/2) = \sin(7\pi/2 - 4\pi) = \sin(-\pi/2)$$

to conclude

$$\arcsin(\sin(7\pi/2)) = \arcsin(\sin(-\pi/2)) = -\pi/2.$$

■

Algebraic Relations

For any x in the domain of $\arcsin x$, the relation

$$\sin(\arcsin x) = x$$

holds. We want to evaluate slightly more complicated expressions such as $\cos(\arcsin x)$. The method is to express the cosine function in terms of the sine function so that we eventually have only to evaluate $\sin(\arcsin x)$. We begin with the relation

$$\sin^2 u + \cos^2 u = 1$$

and conclude that

$$\cos u = \pm\sqrt{1 - \sin^2 u}.$$

Now substitute $u = \arcsin x$ into this equation to get

$$\cos(\arcsin x) = \pm\sqrt{1 - \sin^2(\arcsin x)} = \pm\sqrt{1 - x^2}.$$

We deal with the ambiguous sign as follows. The range of arc sin is the interval $[-\pi/2, \pi/2]$ and the cosine function is nonnegative on this interval. Thus all values of $\cos(\arcsin x)$ are nonnegative and we obtain

$$\cos(\arcsin x) = \sqrt{1 - x^2} \qquad |x| \leq 1. \tag{7.15}$$

In the same way one can prove

$$\sin(\arccos x) = \sqrt{1 - x^2} \qquad |x| \leq 1. \tag{7.16}$$

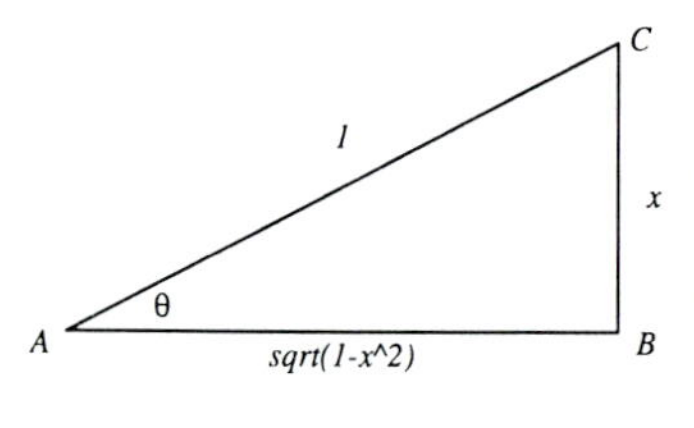

FIGURE 7.32
$\theta = \arcsin x$

These relations are used to compute the derivatives of arc sin and arc cos in the next section.

A quick way to compute compound expressions like $\sin(\arccos x)$ for $0 < x < 1$ is to form a right triangle with one side of length x, one side of length 1 and the other side determined by the Pythagorean theorem. Label the parts of the triangle so that one acute angle is the inverse trigonometric function to be evaluated and then determine the other trigonometric functions of that angle from the usual triangle relations. We illustrate this in figure 7.32 with the case of trigonometric functions of $\arcsin x$.

EXAMPLE 2 If the right triangle ABC has hypotenuse $AC = 1$, side $BC = x$ and side $AB = \sqrt{1 - x^2}$, then the angle A equals $\arcsin x$. We can then read from the triangle the following formulas:

$$\cos(\arcsin x) = \sqrt{1 - x^2},$$
$$\tan(\arcsin x) = \frac{x}{\sqrt{1 - x^2}},$$
$$\sec(\arcsin x) = \frac{1}{\sqrt{1 - x^2}}.$$

■

Exercises 7.8

Determine the following values, giving exact values in the first eight exercises, and using a calculator or tables to approximate the others. Look at an accurate graph to help determine if the approximation seems reasonable.

1. $\arcsin(\sqrt{3}/2)$

2. $\arccos(1/2)$

3. $\arctan 0$

4. $\arccos(-1/\sqrt{2})$

5. $\arcsin(-1/\sqrt{2})$

6. $\arctan 1$

7. $\arccos(-\sqrt{3}/2)$

8. $\arcsin(-\sqrt{3}/2)$

9. $\arccos(0.1)$

10. $\arccos(-0.1)$

11. $\arctan(100)$

12. $\arcsin 1$

Assume $x > 0$ in the next few exercises and then evaluate the expressions by referring to a suitable right triangle.

13. $\sin(\arccos x)$

14. $\tan(\arccos x)$

15. $\sec(\arccos x)$

16. $\sin(\arctan x)$

17. $\cos(\arctan x)$

18. $\sec(\arctan x)$

7.9 Derivatives of the inverse trigonometric functions

The derivatives of the inverse trigonometric functions can be found by starting with the general formula for the derivative of the inverse of a function. If $f(x)$ and $g(x)$ are inverse functions so that

$$f(g(x)) = x,$$

we differentiate both sides to obtain

$$f'(g(x))g'(x) = 1 \quad \text{or} \quad g'(x) = \frac{1}{f'(g(x))}.$$

Here is the statement of the derivative formulas.

THEOREM 7.10

Derivatives of the Inverse Trigonometric Functions

For all x in the respective domains of the inverse trigonometric functions, we have:

$$\frac{d}{dx}\arcsin x = \frac{1}{\sqrt{1-x^2}},$$

$$\frac{d}{dx}\arccos x = \frac{-1}{\sqrt{1-x^2}},$$

$$\frac{d}{dx}\arctan x = \frac{1}{1+x^2},$$

$$\frac{d}{dx}\operatorname{arccot} x = \frac{-1}{1+x^2},$$

$$\frac{d}{dx}\operatorname{arcsec} x = \frac{1}{|x|\sqrt{x^2-1}},$$

$$\frac{d}{dx}\operatorname{arccsc} x = \frac{-1}{|x|\sqrt{x^2-1}}.$$

Proof

To compute the derivative of the arcsine function, we differentiate both sides of the equation

$$\sin(\arcsin x) = x$$

to obtain

$$\cos(\arcsin x)\frac{d}{dx}\arcsin x = 1.$$

The value $\cos(\arcsin x) = \sqrt{1-x^2}$ has been proved in Equation (7.15); substitute this into the derivative formula to conclude

$$\frac{d}{dx}\arcsin x = \frac{1}{\sqrt{1-x^2}}.$$

We also compute the derivative of the arctangent function and then leave the remaining ones for the reader to verify.

We differentiate both sides of the equation

$$\tan(\arctan x) = x$$

to get

$$\sec^2(\arctan x)\frac{d}{dx}(\arctan x) = 1.$$

Because $\sec^2 u = 1 + \tan^2 u$ for all u, we have

$$\sec^2(\operatorname{arc}\tan x) = 1 + \tan^2(\operatorname{arc}\tan x) = 1 + x^2.$$

Thus we have the derivative formula

$$\frac{d}{dx}\operatorname{arc}\tan x = \frac{1}{1+x^2}.$$

Differentiate both sides of the equation

$$\sec(\operatorname{arc}\sec x) = x$$

to get

$$\sec(\operatorname{arc}\sec x)\tan(\operatorname{arc}\sec x)\frac{d}{dx}(\operatorname{arc}\sec x) = 1.$$

Because $\sec^2 u - 1 = \tan^2 u$, we have $\tan^2(\operatorname{arc}\sec x) = x^2 - 1$. If $x \geq 1$ then $\operatorname{arc}\sec x \geq 0$ and $\tan(\operatorname{arc}\sec x) = \sqrt{x^2-1}$. If $x \leq -1$ then $\operatorname{arc}\sec x \leq 0$ and $\tan(\operatorname{arc}\sec x) = -\sqrt{x^2-1}$. In both cases an inspection of the signs implies the derivative of $\operatorname{arc}\sec x$ is as indicated in the theorem using absolute value signs to unify the cases. □

EXAMPLE 1 Compute the derivative of $\operatorname{arc}\sin\left(\frac{x}{a}\right)$ with a constant.

***Solution*:** We use the formula for the derivative of the arcsine function and the chain rule to get

$$\frac{d}{dx}\operatorname{arc}\sin\left(\frac{x}{a}\right) = \frac{1}{\sqrt{1-(x/a)^2}}\frac{1}{a} = \frac{1}{\sqrt{a^2-x^2}}.$$ ■

EXAMPLE 2 Compute the derivative of $\operatorname{arc}\tan\left(\frac{x}{a}\right)$ with a constant.

***Solution*:** Apply the derivative formula and the chain rule to get

$$\frac{d}{dx}\operatorname{arc}\tan\left(\frac{x}{a}\right) = \frac{1}{1+(x/a)^2}\frac{1}{a} = \frac{a}{a^2+x^2}.$$ ■

Each derivative formula gives rise to the corresponding integral formula. Here are three integrals that arise in later computations. We omit the analogous formulas involving the three remaining inverse functions.

Integrals Involving Inverse Trigonometric Functions

$$\int \frac{dx}{\sqrt{a^2-x^2}} = \operatorname{arc}\sin\left(\frac{x}{a}\right) + c,$$

$$\int \frac{dx}{a^2+x^2} = \frac{1}{a}\operatorname{arc}\tan\left(\frac{x}{a}\right) + c,$$

$$\int \frac{dx}{x\sqrt{x^2-a^2}} = \frac{1}{a}\operatorname{arc}\sec\left(\frac{x}{a}\right) + c \quad x > 0.$$

The first two formulas have been verified in the examples worked above; the third formula can be verified by differentiation.

EXAMPLE 3 Evaluate the integral

$$\int \frac{2x+1}{4+x^2}\,dx.$$

Solution: The integral can be written as the sum of two integrals, one of which is evaluated by a substitution, the other of which involves the arctangent function:

$$\int \frac{2x+1}{4+x^2}\,dx = \int \frac{2x}{4+x^2}\,dx + \int \frac{1}{4+x^2}\,dx.$$

To evaluate the first integral on the right, make a substitution $u = 4 + x^2$, so $du = 2x\,dx$ and obtain

$$\int \frac{2x}{4+x^2}\,dx = \int \frac{du}{u} = \ln|u| + c = \ln|4+x^2| + c.$$

The value of the second integral is a standard one involving the arctangent function whose value is given by the formula above. The final result is

$$\int \frac{2x+1}{4+x^2}\,dx = \ln|4+x^2| + \frac{1}{2}\arctan\left(\frac{x}{2}\right) + C.$$ ■

Exercises 7.9

Find $f'(x)$.

1. $f(x) = \arcsin(2x+1)$

2. $f(x) = \arctan(x^2)$

3. $f(x) = \operatorname{arcsec}(1-x)$

4. $f(x) = x \arcsin(\sqrt{x})$

5. $f(x) = \arctan(a/x)$

6. $f(x) = \operatorname{arcsec}(x^2+1)$

7. $f(x) = \arcsin(x/a)$

8. $f(x) = \ln|\arctan 5x|$

9. $f(x) = e^{\arcsin x}$

10. $f(x) = x \arccos(1/x)$

11. $f(x) = \operatorname{arccsc}(5x - x^2)$

12. $f(x) = \operatorname{arccot}(a/x)$

Evaluate the integrals.

13. $\displaystyle\int_0^1 \frac{dx}{1+x^2}$

14. $\displaystyle\int_0^{3/2} \frac{dx}{\sqrt{9-x^2}}$

15. $\displaystyle\int \frac{dx}{\sqrt{5^2-x^2}}$

16. $\displaystyle\int \frac{dt}{3^2+t^2}$

17. $\displaystyle\int \frac{dx}{x\sqrt{x^2-4}}$

18. $\displaystyle\int \frac{dx}{16+9x^2}$

19. $\displaystyle\int \frac{dt}{1+(t-3)^2}$

20. $\displaystyle\int \frac{dx}{\sqrt{1-(x+2)^2}}$

21. $\displaystyle\int \frac{dx}{\sqrt{10-6x-x^2}}$

22. $\displaystyle\int \frac{t+1}{\sqrt{1-t^2}}\,dt$

23. $\displaystyle\int \frac{x^2}{1+x^2}\,dx$

24. $\displaystyle\int \frac{2t+5}{1+t^2}\,dt$

25. $\displaystyle\int_2^{2\sqrt{2}} \frac{dt}{t\sqrt{6t^2-24}}$

26. $\displaystyle\int_3^{3+1/\sqrt{2}} \frac{dt}{\sqrt{6t-t^2-8}}$

27. From the formulas for the derivatives of $\operatorname{arcsec} x$ and $\operatorname{arccsc} x$ we find that, at each point of the domain common to the two functions, $(\operatorname{arcsec} x + \operatorname{arccsc} x)' = 0$ and it follows that $\operatorname{arcsec} x + \operatorname{arccsc} x = c$, a constant. Find the constant c.

7.10 Improper integrals

In this section we extend the definition given in chapter 5 of the definite integral over a bounded interval to define the integral over an unbounded interval. The result might or might not be a real number. Later in the section we consider the integration of functions that are not continuous.

If $f(x)$ is defined and continuous on $[a, \infty)$, then for any $b > a$ the integral of f from a to b has already been defined. The symbol

$$\int_a^\infty f(x)\,dx \tag{7.17}$$

is called an *improper integral* and is defined by the equation

$$\int_a^\infty f(x)\,dx = \lim_{b\to\infty} \int_a^b f(x)\,dx$$

provided the limit exists. When the limit exists we say the improper integral *converges*; if the limit does not exist, the improper integral *diverges*.

Similarly for a function $g(x)$ that is continuous on $(-\infty, a)$ we define

$$\int_{-\infty}^a g(x)\,dx = \lim_{b\to-\infty} \int_b^a g(x)\,dx$$

provided the limit exists. For a function $h(x)$ defined and continuous on $(-\infty, \infty)$ we define

$$\int_{-\infty}^\infty h(x)\,dx = \int_{-\infty}^c h(x)\,dx + \int_c^\infty h(x)\,dx$$

for any fixed number c, provided both integrals converge. Note that the additivity property of the integral implies that the sum does not depend on the choice of c in case the integrals converge.

It should be mentioned that the improper integrals do not enjoy all of the properties proved for definite integrals over finite intervals in chapter 5. We do not dwell on the pathology here and only give some illustration of their use.

EXAMPLE 1 Verify that

$$\text{i.}\quad \int_0^\infty e^{-x}\,dx = 1, \qquad \text{ii.}\quad \int_{-\infty}^{-1} \frac{dx}{x} \quad \text{diverges.}$$

Solution: i. Apply the definition directly to get

$$\begin{aligned}\int_0^\infty e^{-x}\,dx &= \lim_{b\to\infty} \int_0^b e^{-x}\,dx \\ &= \lim_{b\to\infty} \left(-e^{-b} + 1\right) \\ &= 1.\end{aligned}$$

Thus the integral converges to 1.

ii. Apply the definition to get

$$\begin{aligned}\int_{-\infty}^{-1} \frac{dx}{x} &= \lim_{b\to-\infty} \int_b^{-1} \frac{dx}{x} \\ &= \lim_{b\to-\infty} \left(\ln|-1| - \ln|b|\right) \\ &= -\infty.\end{aligned}$$

The integral diverges. ■

The integral in next example also diverges but for a different reason.

EXAMPLE 2 Show the divergence of the improper integral

$$\int_0^\infty \sin x\,dx.$$

Solution: To determine if the integral converges, we must consider $\lim_{b\to\infty}(1 - \cos b)$. This limit does not exist because the values of $\cos b$ range from -1 to $+1$ as b ranges over any interval of length 2π. In particular, when b is very large, the values of $\cos b$ do not stay close (within ϵ) to any specific limit. Hence the integral diverges. ■

This example can be used to explain a technical point in the definition of the improper integral of the type

$$\int_{-\infty}^{\infty} f(x)\,dx. \tag{7.18}$$

One might reasonably ask why we did not define the improper integral as

$$\lim_{a\to\infty}\int_{-a}^{a} f(x)\,dx.$$

For the case $f(x) = \sin x$, we conclude from the actual definition and the example just given that

$$\int_{-\infty}^{\infty}\sin x\,dx = \int_{-\infty}^{0}\sin x\,dx + \int_{0}^{\infty}\sin x\,dx$$

does not exist. However, the otherwise plausible definition yields

$$\int_{-a}^{a}\sin x\,dx = \cos a - \cos(-a) = \cos a - \cos a = 0$$

showing that the two definitions do not produce the same results. It can be shown that when the improper integral (7.18) exists, then its value is equal to $\lim_{a\to\infty}\int_{-a}^{a} f(x)\,dx$.

Improper integrals can be interpreted as areas, at least in the case where the graph of the function lies above the x-axis. Let p be any positive number and

$$f(x) = \frac{1}{x^p} \qquad x > 0.$$

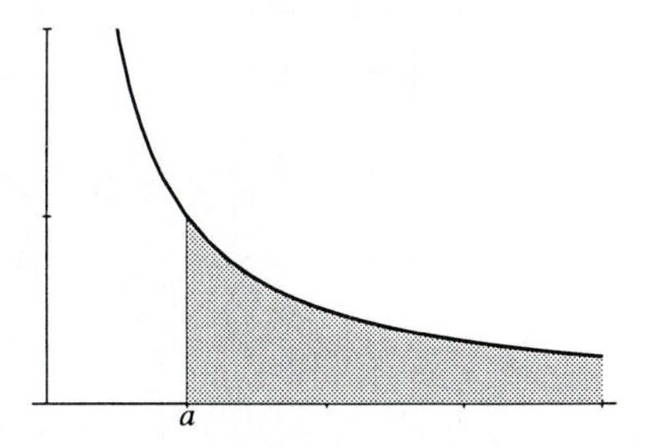

FIGURE 7.33
$y = 1/x^p$

The unbounded region below the graph of f, above the x-axis and to the right of $x = a$ has area

$$\int_{a}^{\infty}\frac{dx}{x^p}.$$

Let us determine the values of p for which the integral converges and thus for which p, the area under the curve, is finite. For $p \neq 1$ we have

$$\int_a^\infty \frac{dx}{x^p} = \lim_{b\to\infty} \frac{1}{1-p}\left(\frac{1}{b^{p-1}} - \frac{1}{a^{p-1}}\right).$$

If $p < 1$ then $1/b^{p-1} = b^{1-p}$ goes to ∞ as b increases so the integral does not converge. If $p > 1$ then $1/b^{p-1}$ has limit 0 and the integral converges. In the remaining case $p = 1$ we see that the integral does not converge because

$$\int_a^\infty \frac{dx}{x} = \lim_{b\to\infty} \int_a^b \frac{dx}{x} = \lim_{b\to\infty} (\ln|b| - \ln|a|) = \infty.$$

To summarize these computations we have for any $a > 0$

$$\int_a^\infty \frac{dx}{x^p} = \begin{cases} \dfrac{1}{(p-1)a^{p-1}} & \text{if } p > 1; \\ \text{diverges} & \text{if } p \leq 1. \end{cases}$$

Comparison Tests

The evaluation of the integral (7.18) amounts to the evaluation of a limit if an antiderivative of $f(x)$ is known. If the antiderivative of f is either unknown or is difficult to compute it might be possible to determine if the improper integral converges without finding its exact value. The next theorem gives one very useful test.

THEOREM 7.11

Comparison Test for Improper Integrals

Suppose $f(x)$ and $g(x)$ are continuous functions and

$$0 \leq f(x) \leq g(x)$$

for all x with $a \leq x$.

i. If $\int_a^\infty g(x)\,dx$ converges then $\int_a^\infty f(x)\,dx$ converges.

ii. If $\int_a^\infty f(x)\,dx$ diverges then $\int_a^\infty g(x)\,dx$ diverges.

Proof

Because f and g are nonnegative functions, the improper integrals can be regarded as areas. The region V over $[a, \infty)$ and under the graph of f is

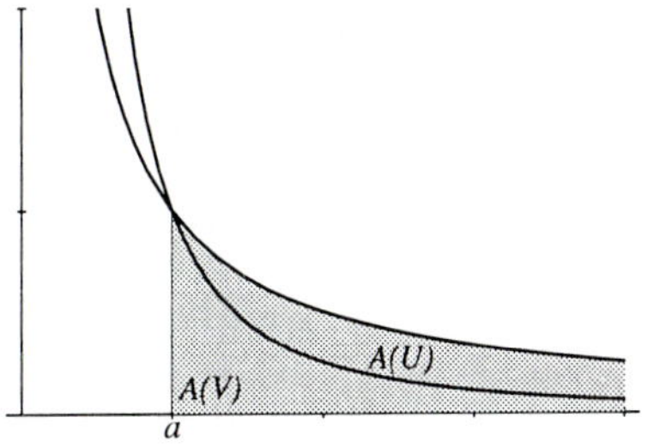

A(U)
A(V)
a

FIGURE 7.34

completely contained in the region U over $[a, \infty)$ and under the graph of g. Thus if U has finite area $A(U)$ then V must also have finite area $A(V)$ and $A(V) \leq A(U)$. If the smaller region V has infinite area then the larger region also has infinite area.

□

The analogous statements about improper integrals of positive functions from $-\infty$ to a are also true.

EXAMPLE 3 By finding a suitable function for comparison, determine the convergence or divergence of the integrals

$$\text{i.} \quad \int_0^\infty \frac{dx}{1 + x^4}, \qquad \text{ii.} \quad \int_{10}^\infty \frac{dx}{\sqrt{x - 5}}.$$

Solution: As x gets large, $x^4 + 1$ is approximately equal to x^4 and we have

$$0 < \frac{1}{1 + x^4} < \frac{1}{x^4}$$

for all x. Because $\int_0^\infty \frac{dx}{x^4}$ converges, the integral $\int_0^\infty \frac{dx}{1 + x^4}$ of the smaller function also converges.

For the second integral, we have $\sqrt{x - 5}$ is very close to $\sqrt{x}$ for large x and $\sqrt{x - 5} < \sqrt{x}$ implies

$$\frac{1}{\sqrt{x}} < \frac{1}{\sqrt{x - 5}}.$$

Because the integral $\int_{10}^\infty \frac{dx}{\sqrt{x}}$ of the smaller function diverges, the integral $\int_{10}^\infty \frac{dx}{\sqrt{x - 5}}$ also diverges.

■

EXAMPLE 4 Show that $\int_0^\infty e^{-x^2}\, dx$ converges.

Solution: We do not know an antiderivative for the function e^{-x^2} in any explicit and usable form (even though an antiderivative does exist) so we try the comparison test. The function xe^{-x^2} has an antiderivative $-e^{-x^2}/2$, and for

$x \geq 1$ we have $e^{-x^2} \leq xe^{-x^2}$. Thus

$$\begin{aligned}\int_0^{\infty} e^{-x^2}\,dx &= \int_0^1 e^{-x^2}\,dx + \int_1^{\infty} e^{-x^2}\,dx \\ &\leq \int_0^1 e^{-x^2}\,dx + \int_1^{\infty} xe^{-x^2}\,dx \\ &= \int_0^1 e^{-x^2}\,dx + \frac{1}{2e}.\end{aligned}$$

It follows that the original integral converges. ■

Example 4 illustrates a slight variation on the comparison test. If $b > a$ and if $0 \leq f(x) \leq g(x)$ for all $x \geq b$ and if

$$\int_b^{\infty} g(x)\,dx$$

converges, then $\int_a^{\infty} f(x)\,dx$ converges. This follows because the integral from a to b is some finite number.

Unbounded Functions on Bounded Intervals

The definite integral

$$\int_0^1 \frac{dx}{1-x^2} \tag{7.19}$$

lies outside the scope of the discussion of the definite integral given earlier because the function $f(x) = 1/(1-x^2)$ fails to be continuous on $[0, 1]$. However, $f(x)$ is continuous on $[0, c]$ for every c with $0 \leq c < 1$. A reasonable interpretation of the integral (7.19) would be

$$\int_0^1 \frac{dx}{1-x^2} = \lim_{c\to 1^-} \int_0^c \frac{dx}{1-x^2},$$

provided the limit exists. More generally suppose $f(x)$ is continuous on $[a, b)$ but not necessarily continuous at b. We make the definition

$$\int_a^b f(x)\,dx = \lim_{c\to b^-} \int_a^c f(x)\,dx,$$

provided the limit exists.

The integral (7.19) does not exist because for $0 < c < 1$ we have

$$\int_0^c \frac{dx}{1-x^2} = \ln\left|\frac{1+c}{1-c}\right|,$$

and this becomes infinite as c approaches 1.

The integral $\int_0^1 \frac{dx}{\sqrt{1-x}}$ does exist because

$$\begin{aligned}\int_0^1 \frac{dx}{\sqrt{1-x}} &= \lim_{c\to 1^-} \int_0^c (1-x)^{-1/2}\,dx \\ &= \lim_{c\to 1^-} -2(1-x)^{1/2}\Big|_0^c \\ &= \lim_{c\to 1^-} -2(1-c)^{1/2} + 2 \\ &= 2.\end{aligned}$$

Exercises 7.10

Evaluate the improper integrals that converge.

1. $\int_0^\infty e^{-2x}\,dx$
2. $\int_1^\infty \frac{dx}{x^{3/2}}$
3. $\int_0^\infty xe^{-x}\,dx$
4. $\int_0^\infty \frac{dx}{1+x^2}$
5. $\int_1^\infty \frac{dx}{x(x+1)}$
6. $\int_e^\infty \frac{\ln x}{x}\,dx$
7. $\int_e^\infty \frac{dx}{x\ln x}$
8. $\int_4^\infty \frac{dx}{x^2-2}$
9. $\int_0^1 \frac{dx}{(x^2-1)}$
10. $\int_{-1}^0 \frac{2x+1}{x^2(x-1)}\,dx$
11. $\int_0^1 \frac{dx}{\sqrt{1-x}}$
12. $\int_0^1 \frac{dx}{(1-x)^{2/3}}$
13. $\int_0^1 \frac{dx}{(1-x)^{3/2}}$
14. $\int_0^1 \frac{dx}{(1-x)^p}$
15. $\int_0^a \frac{dx}{\sqrt{a^2-x^2}}$
16. $\int_a^\infty \frac{dt}{t\sqrt{t^2-a^2}}$

Find a suitable function to use for comparison and determine if the integrals converge or diverge.

17. $\int_2^\infty \frac{e^{-x}}{x^2-2}\,dx$
18. $\int_4^\infty \frac{e^x}{x-1}\,dx$
19. $\int_2^\infty \frac{dx}{x^3-x^2+x+1}$
20. $\int_1^\infty \frac{|\sin x|}{x^3}\,dx$

The Limit Comparison Test states that if $f(x)$ and $g(x)$ are positive functions such that

$$\lim_{x\to\infty}\frac{f(x)}{g(x)} = L$$

for a *positive* number L, then $\int_a^\infty f(x)\,dx$ and $\int_a^\infty g(x)\,dx$ either both converge or both diverge. Use this test to determine the convergence or divergence of the following integrals by comparing each with the integral of some suitable function. The suitable function should be one that is easy to integrate, while at the same time meeting the limit condition with a positive L.

21. $\int_0^\infty \frac{dx}{\sqrt{x+5}}$
22. $\int_0^\infty \frac{x\,dx}{\sqrt{x^3+8}}\,dx$
23. $\int_0^\infty \frac{dx}{e^x+3x}$
24. $\int_1^\infty \frac{xe^{-x}}{5x-1}\,dx$
25. $\int_2^\infty \frac{x-3}{x^3-4}\,dx$
26. $\int_0^\infty \frac{x^2+1}{x^5+3x+4}\,dx$

27. The graph of $y = 1/(1+x^2)$ extends infinitely far to the left and right yet the area between it and the x-axis is finite. Determine this area.

28. Suppose $p(x)$ is a polynomial of degree n and $q(x)$ a polynomial of degree m. Use the Limit Comparison Test to give conditions on n and m that ensure the convergence of

$$\int_a^\infty \frac{p(x)}{q(x)}\,dx$$

for any a larger than each solution of $q(x) = 0$.

8 Methods of Integration

In the previous chapters we have seen many examples of problems whose solution depends on the integration of some function. The evaluation of a definite integral by the Fundamental Theorem of Calculus requires having an antiderivative of a function. So far, methods for finding antiderivatives have been limited to simple substitutions. The collection of functions for which an antiderivative can be found is expanded considerably in this chapter, thereby enlarging the applicability of the integral to the solution of practical problems.

When the reader is required to evaluate an indefinite integral, one of the options is to look it up in a table. If you have access to extensive tables of integrals and, if you are lucky, the integral you want is found in the table, then your task is complete. The use of tables is introduced here only to help the reader become familiar with some types of integrals that can be evaluated. Another possibility is to use one of the many software packages capable of symbolic integration to carry out the solution of an integration problem. If that is not available to you, or if your computer package is not successful with your particular integral, you might have to compute the integral by hand. Often there is more to a problem than just the evaluation of an integral. A specific problem might require the interpretation of the numerical results and an analysis of the methods used to reach the numerical results. The methods of evaluating integrals discussed in this chapter might give additional insight to the problems to which they are applied. Knowledge of the basic techniques of integration is still valuable in spite of alternative means for reaching an answer. In many cases it is faster to work out an integral by hand than to look it up or boot up your computer to have it evaluate the integral. The main goal of this chapter is to provide the reader with methods for evaluating many indefinite integrals by hand.

We give a detailed treatment of the method of partial fractions, which is used to find the antiderivative of rational functions. We then apply the theory of partial fractions to show that several large classes of functions have antiderivatives that can be computed exactly. The classes of functions include rational functions of sines and cosines and rational functions of certain algebraic functions.

8.1 Integral formulas and integral tables

Before we take up some by-hand methods of integration, we discuss the use of tables. A rather short table of indefinite integrals is provided in the end pages. The correctness of these tables can be verified by differentiation. Most of the entries in the table are created by the methods to be discussed in the remaining sections of this chapter. It usually happens that the integral you want to evaluate is not exactly like one in the table. Sometimes a simple substitution can bring your integral into the form of one in the table. We concentrate on that idea in this section.

Completing the Square

The tables contain many integrals involving quadratic polynomials such as

$$\text{(T)} \quad \int \frac{du}{u^2 + a^2} = \frac{1}{a} \arctan\left(\frac{u}{a}\right) + c,$$

$$\text{(L)} \quad \int \frac{du}{u^2 - a^2} = \frac{1}{2a} \ln\left|\frac{u - a}{u + a}\right| + c.$$

The a appearing in the integrals is a constant and $a \neq 0$. The constant is written as a^2 rather than just a because its sign is important; a^2 is always positive and usually a is taken positive also. The integral

$$\int \frac{dx}{x^2 + m}$$

is evaluated by using (T) if $m > 0$ or using (L) if $m < 0$. If $m > 0$, then $m = a^2$ with $a = \sqrt{m}$ and $x^2 + m = x^2 + a^2$. If $m < 0$, then $-m > 0$ and $-m = a^2$ with $a = \sqrt{|m|}$ and $x^2 + m = x^2 - a^2$.

Quadratic polynomials can be written as a constant plus a constant multiple of a squared term by applying the familiar process of completing the square. An integral of the form

$$\int \frac{dx}{x^2 + px + q}$$

can be put into a form so that (T) or (L) can be applied. Begin by completing the square in the denominator:

$$x^2 + px + q = \left(x^2 + px + \frac{p^2}{4}\right) + q - \frac{p^2}{4} = \left(x + \frac{p}{2}\right)^2 + \frac{4q - p^2}{4}.$$

Now make the change of variable

$$u = x + \frac{p}{2} \qquad du = dx$$

and let $m = (4q - p^2)/4$ so that the integral becomes

$$\int \frac{dx}{x^2 + px + q} = \int \frac{du}{u^2 + m}.$$

This is evaluated using either (T) or (L) depending on the sign of $m = (4q - p^2)/4$. Of course if $4q - p^2 = 0$, then neither formula is used because the integral reduces to $\int u^{-2}\,du = -u^{-1} + c$.

EXAMPLE 1 Put the integral into a form that permits evaluation using either

(T) or (L):

$$\int \frac{dx}{4x^2 + 4x + 5}$$

***Solution*:** Begin by completing the square:

$$4x^2 + 4x + 5 = 4\left(x^2 + x + \frac{5}{4}\right) = 4\left(\left(x^2 + x + \frac{1}{4}\right) - \frac{1}{4} + \frac{5}{4}\right)$$
$$= 4\left(x + \frac{1}{2}\right)^2 + 4 = (2x + 1)^2 + 4.$$

Next make the substitution $u = 2x + 1$, $du = 2dx$ and we have

$$\int \frac{dx}{4x^2 + 4x + 5} = \int \frac{dx}{(2x + 1)^2 + 4} = \frac{1}{2} \int \frac{du}{u^2 + 4}.$$

Now apply (T) to finish the problem. ■

Here are two integrals found in the tables:

$$\text{(A)} \quad \int \frac{du}{\sqrt{a^2 - u^2}} = \arcsin\left(\frac{u}{a}\right) + c,$$

$$\text{(B)} \quad \int \frac{du}{\sqrt{u^2 - a^2}} = \ln\left|u + \sqrt{u^2 - a^2}\right| + c.$$

EXAMPLE 2 Find substitutions to make evaluation of the integrals possible with the use of formula (A) or (B):

$$\text{(i)} \quad \int \frac{dx}{\sqrt{2x^2 - 4x - 6}}, \qquad \text{(ii)} \quad \int \frac{dx}{\sqrt{4x - x^2}}.$$

***Solution*:** (i) Begin by completing the square of the term under the square root sign:

$$2x^2 - 4x - 6 = 2(x^2 - 2x - 3) = 2[(x^2 - 2x + 1) - 4] = 2[(x - 1)^2 - 2^2].$$

The integral (i) has the form

$$\int \frac{dx}{\sqrt{2x^2 - 4x + 6}} = \frac{1}{\sqrt{2}} \int \frac{dx}{\sqrt{(x - 1)^2 - 2^2}}.$$

This can be evaluated using integral (B) after making a substitution $u = x - 1$.

(ii) Once again complete the square of the term under the square root sign:

$$4x - x^2 = -[(x^2 - 4x + 4) - 4] = -[(x - 2)^2 - 4] = 4 - (x - 2)^2.$$

The integral (ii) has the form

$$\int \frac{dx}{\sqrt{4x - x^2}} = \int \frac{dx}{\sqrt{4 - (x - 2)^2}},$$

which can be evaluated using (A) after making the substitution $u = x - 2$. ■

The reader could profit by completing the integrations in this example and then verifying the correctness of the results by differentiating.

Here is a very short table of integrals that is used, along with the other integral formulas (A), (B), (T) and (L) above, in the exercises. The correctness of these integrals should be verified by differentiation. A method for evaluating these integrals is discussed in the later sections of this chapter.

$$\int \sqrt{u^2+m}\,du = \frac{1}{2}\left[u\sqrt{u^2+m} + m\ln\left|u+\sqrt{u^2+m}\right|\right] + c$$

$$\int \sqrt{a^2-u^2}\,du = \frac{1}{2}\left[u\sqrt{a^2-u^2} + a^2\arcsin\left(\frac{u}{a}\right)\right] + c$$

$$\int \frac{\sqrt{a^2-u^2}}{u}\,du = \sqrt{a^2-u^2} - a\ln\left|\frac{a+\sqrt{a^2-u^2}}{u}\right| + c$$

$$\int \frac{\sqrt{u^2-a^2}}{u}\,du = \sqrt{u^2-a^2} - a\,\mathrm{arcsec}\left(\frac{u}{a}\right) + c$$

Exercises 8.1

Use any appropriate substitution and/or any of the integral formulas given in this section to evaluate the following integrals.

1. $\int \frac{dx}{16-x^2}$

2. $\int \frac{dx}{x^2+16}$

3. $\int \frac{dx}{x^2+6x-7}$

4. $\int \frac{dx}{6x-x^2}$

5. $\int \sqrt{x(x+2)}\,dx$

6. $\int \sqrt{1-x-x^2}\,dx$

7. $\int \frac{dx}{x^2+6x}$

8. $\int \frac{dx}{8x^2+16x+12}$

9. $\int \sqrt{9-x^2}\,dx$

10. $\int \sqrt{x^2-9}\,dx$

11. $\int \sqrt{x^2+9}\,dx$

12. $\int \sqrt{4x^2+12}\,dx$

13. $\int \sqrt{12-4x^2}\,dx$

14. $\int_0^1 \sqrt{1-x^2}\,dx$

15. $\int_{-4}^4 \sqrt{16+x^2}\,dx$

16. $\int_{-4}^4 \sqrt{16-x^2}\,dx$

17. $\int \frac{\sqrt{2x-x^2}}{x-1}\,dx$

18. $\int \frac{\sqrt{x^2+2x}}{x+1}\,dx$

19. $\int \frac{\sqrt{x^2-8x}}{2x-8}\,dx$

20. $\int \frac{\sqrt{6x-x^2}}{x-3}\,dx$

21. $\int \frac{dx}{x\sqrt{1-x^2}}$ [Hint: Try $x = 1/u$.]

22. Find the area of the circle of radius a by integration.

23. A segment of length 1 has its endpoints on a circle of radius 1 and separates the interior of the circle into two regions. Find the area of the smaller region.

24. Find the length of the arc of the parabola $y = x^2$ between $(0,0)$ and $(1,1)$.

25. The interior of the ellipse $2x^2 + y^2 = 1$ is cut into two regions by the vertical line $x = a$ so that the larger region has area twice as large as the smaller region. Find an equation involving trigonometric functions that must be satisfied by a.

8.2 Integration by parts

Let $u = u(x)$ and $v = v(x)$ be differentiable functions of x. The product rule for derivatives,

$$\frac{d}{dx}(uv) = uv' + vu',$$

implies a formula for antiderivatives:

$$uv = \int uv'\,dx + \int vu'\,dx.$$

We rewrite this to isolate one of the integrals so that

$$\int uv'\,dx = uv - \int vu'\,dx. \tag{8.1}$$

Use of this equation to evaluate the left-hand side is called *integration by parts.* It is surprisingly useful for evaluating certain integrals or at least for transforming certain integrals into other integrals of a simpler form. Here are some examples of its use.

EXAMPLE 1 Evaluate the integral

$$\int xe^x\,dx.$$

Solution: We set

$$\int uv'\,dx = \int xe^x\,dx$$

with the idea of using Equation (8.1) to transform the integral into an easier integral. The first step is to decide what part of the expression $xe^x\,dx$ is selected as u and what part is selected as $v'\,dx$. There are two fairly obvious choices. Let us try both and compare the results. Let

$$u = x \qquad \text{and} \qquad v'\,dx = e^x\,dx.$$

Then $v = e^x$ and $u'\,dx = dx$ and the integration by parts equation becomes

$$\int xe^x\,dx = uv - \int vu'\,dx = xe^x - \int e^x\,dx = xe^x - e^x + c.$$

This choice of u and v was very successful. The other obvious choice of u and v is

$$u = e^x \qquad \text{and} \qquad v'\,dx = x\,dx.$$

Then $u'\,dx = e^x\,dx$ and $v = x^2/2$. The integration by parts formula is

$$\int xe^x\,dx = \frac{1}{2}x^2e^x - \frac{1}{2}\int x^2e^x\,dx.$$

Although this formula is correct, it does not help much in finding the value of the integral on the left because the integral on the right side is more complicated than the integral we started with; thus the first choice is best. ■

The use of the integration by parts formula always presents a choice—what part should we select for u and what part for $v'\,dx$? One observation might be helpful. After making the choice of $v'\,dx$ we must integrate to find v. We had better select $v'\,dx$ so that it is possible to do the integration. A rule of thumb (that sometimes leads to success) says that $v'\,dx$ should be the most complicated part of the integrand that can be integrated easily. Whatever is selected for u must be differentiated because $u'\,dx$ appears on the right.

Before moving on to other examples, we introduce a bit of shorthand for the integration by parts formula. When we introduced substitutions in chapter 4, the notation $u = g(x)$ and $du = g'(x)\,dx$ was very convenient. We use the same notation here. We write $dv = v'\,dx$ and $du = u'\,dx$ so that the integration by parts formula reads

$$\int u\,dv = uv - \int v\,du.$$

This form is more easily remembered and we use it from now on.

EXAMPLE 2 Use integration by parts to evaluate

$$\int x^2 \cos x \, dx.$$

Solution: Let $u = x^2$ and $dv = \cos x$. Then $du = 2x \, dx$ and $v = \sin x$. Then the formula gives

$$\int x^2 \cos x \, dx = x^2 \sin x - \int 2x \sin x \, dx.$$

We have made some progress because the x^2 term in the first integral has been replaced by an x but there is still more work to be done. We can evaluate the remaining integral using the parts formula once again. We evaluate $\int 2x \sin x \, dx$ by parts using $u = 2x$ and $dv = \sin x \, dx$. Then $du = 2dx$ and $v = -\cos x$ and

$$\int 2x \sin x \, dx = -2x \cos x - \int (-\cos x) 2dx = -2x \cos x + 2 \sin x + c.$$

This value can be substituted above to yield

$$\int x^2 \cos x \, dx = x^2 \sin x + 2x \cos x - 2 \sin x + C.$$ ■

Examination of this example ought to convince you that for any polynomial $p(x)$, repeated use of integration by parts leads to evaluation of the integrals

$$\int p(x) \sin x \, dx \quad \text{and} \quad \int p(x) \cos x \, dx.$$

EXAMPLE 3 Find the antiderivatives of $\ln x$.

Solution: At first it might not appear that this is an integration by parts problem because there does not seem to be a product in $\int \ln x \, dx$. But there is: let $u = \ln x$ and $dv = dx$ so that $du = dx/x$ and $v = x$. Then

$$\int \ln x \, dx = x \ln x - \int x \frac{dx}{x} = x \ln x - \int dx = x \ln x - x + c.$$ ■

Integration by Parts in Definite Integrals

Integration by parts works just as well for definite integrals. Again starting from the product rule for derivatives for two differentiable functions $u = u(x)$ and $v = v(x)$, we have

$$u \frac{dv}{dx} = \frac{d}{dx}(uv) - v \frac{du}{dx}.$$

This yields a definite integral equation

$$\int_a^b u \frac{dv}{dx} \, dx = \int_a^b \frac{d}{dx}(uv) \, dx - \int_a^b v \frac{du}{dx} \, dx.$$

This can be written more succinctly as

$$\begin{aligned} \int_a^b u \, dv &= uv\Big|_a^b - \int_a^b v \, du \\ &= u(b)v(b) - u(a)v(a) - \int_a^b v \, du. \end{aligned}$$

EXAMPLE 4 Evaluate

$$\int_0^1 \arctan x \, dx.$$

Solution: We use integration by parts and set $u = \arctan x$, $dv = dx$ so that $du = 1/(x^2+1)^2$ and $v = x$. Then

$$\begin{aligned}\int_0^1 \arctan x \, dx &= x \arctan x \Big|_0^1 - \int_0^1 \frac{x\,dx}{x^2+1} \\ &= \frac{\pi}{4} - 0 - \frac{1}{2} \ln |x^2+1| \Big|_0^1 \\ &= \frac{\pi}{4} - \frac{1}{2} \ln 2.\end{aligned}$$

■

The Reappearing Integral

There are a few situations where integration by parts is successful even when it looks like we are about to go in circles. The following example illustrates a curious phenomenon.

EXAMPLE 5 Evaluate the integral

$$\int e^{-2x} \sin x \, dx.$$

Solution: Use integration by parts by setting $u = e^{-2x}$ and $dv = \sin x$. Then $du = -2e^{-2x}\,dx$ and $v = -\cos x$ and we have

$$\int e^{-2x} \sin x \, dx = -e^{-2x} \cos x - \int 2e^{-2x} \cos x \, dx. \tag{8.2}$$

We are now faced with a second integral only slightly different from the original one. We press on and use integration by parts on this integral also. Once again let $u = e^{-2x}$ and $dv = \cos x \, dx$ so that $du = -2e^{-2x}$ and $v = \sin x$. We have then

$$\int e^{-2x} \cos x \, dx = e^{-2x} \sin x - \int (-2e^{-2x}) \sin x \, dx.$$

It looks as though we are back to where we started because the integral we started with has reappeared on the right (multiplied by a constant). But something nice happens when we substitute this value back into Equation (8.2):

$$\begin{aligned}\int e^{-2x} \sin x \, dx &= -e^{-2x} \cos x - 2\left(e^{-2x} \sin x - \int (-2e^{-2x}) \sin x \, dx\right) \\ &= -e^{-2x} \cos x - 2e^{-2x} \sin x - 4 \int e^{-2x} \sin x \, dx.\end{aligned}$$

The original integral on the left has reappeared on the right. Now transpose the integral on the right to the left side of the equal sign and we have the integral we want standing alone:

$$5 \int e^{-2x} \sin x \, dx = -e^{-2x} \cos x - 2e^{-2x} \sin x.$$

Of course, a constant of integration must be added so the final result is

$$\int e^{-2x} \sin x\, dx = \frac{1}{5}\left(-e^{-2x} \cos x - 2e^{-2x} \sin x\right) + c. \qquad \blacksquare$$

The idea of the reappearing integral can be applied to evaluate integrals of the form

$$\int e^{ax} \sin bx\, dx \quad \text{and} \quad \int e^{ax} \cos bx\, dx$$

as well as some other examples. See the exercises following this section for other ways to evaluate these integrals.

The Constant of Integration

When the integration by parts formula is used, an integration is used to determine v from dv. In all the examples, we have not added a constant of integration at this step. Here is the reason for omitting the constant. Given the integral $\int u\, dv$ to evaluate, the function v is determined up to a constant; let k be any constant so that $\int dv = v + k$. The parts formula then becomes

$$\int u\, dv = u(v + k) - \int (v + k)\, du.$$

The constant k disappears because $\int k\, du = ku$. In some very rare instances, there is an advantage to retaining the constant k because, for some particular choice of k, the integral $\int (v + k)\, du$ might be easier to evaluate than the integral $\int v\, du$.

EXAMPLE 6 Use integration by parts to evaluate the integral

$$\int x \arctan x\, dx.$$

Solution: Let $u = \arctan x$ and $dv = x\, dx$. Then

$$du = \frac{dx}{x^2 + 1} \quad \text{and} \quad v = \int x\, dx = \frac{1}{2}x^2 + k$$

and

$$\int x \arctan x\, dx = \left(\frac{1}{2}x^2 + k\right) \arctan x - \int \frac{1}{2}\frac{x^2 + 2k}{x^2 + 1}\, dx.$$

The integral that remains on the right becomes trivial to evaluate if we select $k = 1/2$. With this choice, the final result is

$$\int x \arctan x\, dx = \left(\frac{1}{2}x^2 + \frac{1}{2}\right) \arctan x - \frac{x}{2} + c. \qquad \blacksquare$$

Exercises 8.2

Evaluate the integrals using integration by parts. Check your work by differentiating the answer.

1. $\int x \sin 2x\, dx$
2. $\int xe^{-x}\, dx$
3. $\int x \ln x\, dx$
4. $\int x^2 e^{3x}\, dx$
5. $\int (\ln x)^2\, dx$
6. $\int e^{2x} \cos 3x\, dx$

7. $\displaystyle\int \frac{x^2}{\sqrt{x+2}}\,dx$

8. $\displaystyle\int x^3 \cos x\,dx$

9. $\displaystyle\int \frac{\ln w}{w^2}\,dw$

10. $\displaystyle\int t \sec t\,dt$

11. $\displaystyle\int \sin(\ln x)\,dx$

12. $\displaystyle\int \frac{\ln x}{x^{3/2}}\,dx$

13. $\displaystyle\int x(x+1)^{100}\,dx$

14. $\displaystyle\int_0^1 x(1-x)^{20}\,dx$

15. $\displaystyle\int t^2(3t-4)^{12}\,dt$

16. $\displaystyle\int_0^1 x^2(1-x)^{20}\,dx$

17. $\displaystyle\int (x+1)^2 \ln(x+1)\,dx$

18. $\displaystyle\int \sin^2 x\,dx$

19. $\displaystyle\int \cos^2 x\,dx$

20. $\displaystyle\int_1^e \sqrt{x}\ln x\,dx$

21. $\displaystyle\int \frac{x^7\,dx}{(1+x^4)^{5/2}}$

22. $\displaystyle\int t^3 \cos t\,dt$

23. $\displaystyle\int_0^\pi e^x \sin x\,dx$

24. $\displaystyle\int (x^2-3x)e^{-4x}\,dx$

25. Let $p(x)$ be a polynomial of degree n. Prove

$$\int p(x)e^x\,dx = \left(p(x) - p'(x) + \cdots + (-1)^n p^{(n)}(x)\right)e^x + c$$
$$= \sum_{i=0}^{n} (-1)^i p^{(i)}(x)e^x.$$

26. Let $p(x)$ be a polynomial of degree n. Prove

$$\int p(x)\sin x\,dx = \left(-p(x) + p''(x) + \cdots\right)\cos x$$
$$+ \left(p'(x) - p'''(x) + \cdots\right)\sin x + c,$$

where the coefficient of $\cos x$ is the alternating sum of the even derivatives of $p(x)$ and the coefficient of $\sin x$ is the alternating sum of the odd derivatives of $p(x)$.

27. Let $p(x)$ be a polynomial. State and prove a formula for $\int p(x)\cos x\,dx$ analogous to the formula for $\int p(x)\sin x\,dx$ given in exercise 26.

28. Let $f(x)$ and $g(x)$ be twice differentiable functions that satisfy

$$f''(x) = kf(x), \qquad g''(x) = hg(x),$$

for some constants h and k with $h \neq k$. Prove

$$\int f(x)g(x)\,dx = \frac{1}{h-k}\left(f(x)g'(x) - f'(x)g(x)\right) + c.$$

[Hint: The equation $g''(x) = hg(x)$ implies $h\int g(x)\,dx = g'(x) + c$. Use integration by parts twice.]

Use exercise 28 to evaluate the following integrals.

29. $\displaystyle\int e^{ax}\sin bx\,dx$

30. $\displaystyle\int e^{ax}\cos bx\,dx$

31. $\displaystyle\int \sin mx \cos nx\,dx,\ m^2 \neq n^2$

32. $\displaystyle\int \cos mx \cos nx\,dx,\ m^2 \neq n^2$

33. For a continuous function g, let $g_{(n)}(x)$ denote an nth antiderivative of g; that is, $g_{(1)}(x) = \int g(x)\,dx$, $g_{(2)}(x) = \int g_{(1)}(x)\,dx$, and so on. Prove the generalized integration by parts formula

$$\int f(x)g(x)\,dx = f(x)g_{(1)}(x) - f'(x)g_{(2)}(x) + \cdots$$
$$+ (-1)^{n-1}f^{(n-1)}(x)g_{(n)}(x)$$
$$+ (-1)^n \int f^{(n)}(x)g_{(n)}(x)\,dx.$$

8.3 Reduction formulas

The integration of x^3e^x can be carried out by repeatedly using integration by parts. After one application of the integration by parts process, there remains an integral involving x^2e^x. Another application further reduces the computation to an integral involving xe^x, and then finally the integral is evaluated with one additional integration by parts. At each step there is an integral involving x^ne^x that is converted into a simpler integral involving $x^{n-1}e^x$. Even for modest values of n, such as $n = 6$, the computation becomes very repetitive and somewhat tedious. It is much more economical to look for a formula for the integral of x^ne^x that can be applied. Here is the formula in a slightly more general form:

$$\int x^n e^{kx}\, dx = \frac{1}{k} x^n e^{kx} - \frac{n}{k} \int x^{n-1} e^{kx}\, dx, \quad n \geq 1.$$

This is called a *reduction formula* because some parameter in the original integral (in this case the exponent n) has been reduced to a smaller value. It is derived using integration by parts by selecting u and dv as follows:

$$u = x^n \quad \text{so} \quad du = nx^{n-1}\, dx$$

$$dv = e^{kx}\, dx \quad \text{so} \quad v = \frac{1}{k} e^{kx}.$$

If n is a positive integer, then repeated application of the reduction formula reduces the exponent on the x term to 0, so all that remains to integrate is a constant times e^{kx}.

EXAMPLE 1 Evaluate the indefinite integral of $x^2 e^{3x}$.

Solution: We apply the reduction formula twice with $k = 3$ and $n = 2$ in the first application and $n = 1$ in the second application:

$$\begin{aligned}\int x^2 e^{3x} /, dx &= \frac{1}{3} x^2 e^{3x} - \frac{2}{3} \int x\, e^{3x}\, dx \\ &= \frac{1}{3} x^2 e^{3x} - \frac{2}{3} \left(\frac{1}{3} x\, e^{3x} - \frac{1}{3} \int e^{3x}\, dx \right) \\ &= \frac{1}{3} x^2 e^{3x} - \frac{2}{9} x e^{3x} + \frac{2}{27} e^{3x} + c.\end{aligned}$$

■

In the next two sections we take up the problem of integrating rational functions. One important step in the solution of that general problem is the evaluation of the integral

$$R_n = \int \frac{dx}{(x^2 + a^2)^n}, \tag{8.3}$$

where the constant $a \neq 0$ and $n \geq 1$. In the case $n = 1$, this integral is the standard one involving the inverse tangent function:

$$R_1 = \int \frac{dx}{(x^2 + a^2)} = \frac{1}{a} \arctan\left(\frac{x}{a}\right) + c.$$

The derivation of a reduction formula for general n has a novel twist at the end. We apply integration by parts using

$$u = \frac{1}{(x^2 + a^2)^n} \quad \text{and so} \quad du = \frac{-2nx\, dx}{(x^2 + a^2)^{n+1}},$$

$$dv = dx \quad \text{and so} \quad v = x.$$

Then

$$\begin{aligned}R_n &= \frac{x}{(x^2 + a^2)^n} + 2n \int \frac{x^2\, dx}{(x^2 + a^2)^{n+1}} \\ &= \frac{x}{(x^2 + a^2)^n} + 2n \int \frac{x^2 + a^2\, dx}{(x^2 + a^2)^{n+1}} - 2n \int \frac{a^2\, dx}{(x^2 + a^2)^{n+1}} \\ &= \frac{x}{(x^2 + a^2)^n} + 2n \int \frac{dx}{(x^2 + a^2)^n} - 2na^2 \int \frac{dx}{(x^2 + a^2)^{n+1}} \\ &= \frac{x}{(x^2 + a^2)^n} + 2nR_n - 2na^2 R_{n+1}.\end{aligned}$$

Something seems to have gone wrong because the exponent n has increased to $n + 1$. Don't despair. We take advantage of the equation just derived and solve for R_{n+1} in terms of R_n to get the reduction formula

$$\int \frac{dx}{(x^2 + a^2)^{n+1}} = \frac{1}{2na^2}\left[\frac{x}{(x^2 + a^2)^n} + (2n - 1)\int \frac{dx}{(x^2 + a^2)^n}\right].$$

EXAMPLE 2 Evaluate the indefinite integral

$$\int \frac{dx}{(x^2 + 1)^2}.$$

Solution: We apply the reduction formula with $n + 1 = 2$ (so $n = 1$) to get

$$\begin{aligned}\int \frac{dx}{(x^2 + 1)^2} &= \frac{x}{2(x^2 + 1)} + \frac{1}{2}\int \frac{dx}{(x^2 + 1)} \\ &= \frac{x}{2(x^2 + 1)} + \frac{1}{2}\arctan x + c.\end{aligned}$$

■

Integrals of Powers of the Sine and Cosine

Integrals of the type

$$S_n = \int \sin^n x\, dx \qquad \text{and} \qquad C_n = \int \cos^n x\, dx$$

can be evaluated by a reduction formula along with the use of simple trigonometric identities. To evaluate S_n, we use integration by parts with

$$\begin{aligned} u &= \sin^{n-1} x \quad \text{so that} \quad du = (n - 1)\sin^{n-2} x \cos x \\ dv &= \sin x\, dx \quad \text{so that} \quad v = -\cos x. \end{aligned}$$

Then we have

$$\begin{aligned} S_n &= -\cos x \sin^{n-1} x + (n - 1)\int \sin^{n-2} x \cos^2 x\, dx \\ &= -\cos x \sin^{n-1} x + (n - 1)\int \sin^{n-2} x(1 - \sin^2 x)\, dx \\ &= -\cos x \sin^{n-1} x + (n - 1)\int \sin^{n-2} x\, dx - (n - 1)\int \sin^n x\, dx \\ &= -\cos x \sin^{n-1} x + (n - 1)S_{n-2} - (n - 1)S_n. \end{aligned}$$

Solve this for S_n in terms of S_{n-2} to get the reduction formula

$$S_n = \int \sin^n x\, dx = -\frac{1}{n}\cos x \sin^{n-1} x + \frac{n - 1}{n}\int \sin^{n-2} x\, dx. \qquad \textbf{(8.4)}$$

This formula reduces the exponent n by 2. If n is an odd positive integer, repeated application of the reduction formula eventually reduces the power to 1. If n is an even positive integer, repeated application reduces the power to 0.

EXAMPLE 3 Evaluate the two integrals

$$\int \sin^2 x\, dx \qquad \text{and} \qquad \int \sin^3 x\, dx.$$

Solution: The reduction formula with $n = 2$ gives

$$\int \sin^2 x\,dx = -\frac{1}{2}\cos x \sin x + \frac{1}{2}\int \sin^0 x\,dx$$
$$= -\frac{1}{2}\cos x \sin x + \frac{x}{2} + c.$$

The reduction formula with $n = 3$ gives

$$\int \sin^3 x\,dx = -\frac{1}{3}\cos x \sin^2 x + \frac{2}{3}\int \sin^{3-2} x\,dx$$
$$= -\frac{1}{3}\cos x \sin^2 x + \frac{2}{3}(-\cos x) + c.$$

■

The idea used to obtain the reduction formula (8.4) can be applied to obtain the following formula for reduction of the powers of the cosine:

$$\int \cos^n x\,dx = \frac{1}{n}\sin x \cos^{n-1} x + \frac{n-1}{n}\int \cos^{n-2} x\,dx. \qquad \textbf{(8.5)}$$

Integrals of Products of Sines and Cosines

Integrals of the form

$$\int \sin^m x \cos^n x\,dx \qquad \textbf{(8.6)}$$

can also be evaluated by using a suitable reduction formula but in some cases a much simpler method is available. For example, the formula for $m \neq -1$

$$\int \sin^m x \cos x\,dx = \frac{\sin^{m+1} x}{m+1} + c$$

can be found with a simple substitution $u = \sin x$, $dx = \cos x\,dx$. The exponent m on the sine term need not be positive or even an integer.

The analogous substitution, $u = \cos x$, can be used if $m = 1$ and n is any exponent.

More generally, if $n = 2k + 1$ is an odd integer, the integral (8.6) can be evaluated by using the identity

$$\cos^{2k} x = (\cos^2 x)^k = (1 - \sin^2 x)^k$$

to replace all but one power of the cosine term. Similarly, if $m = 2k + 1$ is an odd integer, the identity

$$\sin^{2k} x = (1 - \cos^2 x)^k$$

is used to replace all but one power of of the sine term.

EXAMPLE 4 Evaluate the integral

$$\int \sin^m x \cos^5 x\,dx.$$

Solution: Use the idea just mentioned to replace $\cos^4 x$ with $(1 - \sin^2 x)^2$ and then use the substitution $u = \sin x$ to obtain

$$\begin{aligned}\int \sin^m x \cos^5 x\,dx &= \int \sin^m x(1-\sin^2 x)^2 \cos x\,dx\\ &= \int u^m(1-u^2)^2\,du\\ &= \frac{u^{m+1}}{m+1} - \frac{2u^{m+3}}{m+3} + \frac{u^{m+5}}{m+5} + c\\ &= \frac{\sin^{m+1} x}{m+1} - \frac{2\sin^{m+3} x}{m+3} + \frac{\sin^{m+5} x}{m+5} + c.\end{aligned}$$

Notice that the computation in this example does not require m to be a positive integer but the answer must be modified if $m = -1, -3$, or -5. ■

The integral (8.6) in which both m and n are even integers cannot be handled by a simple substitution. One could look for a reduction formula but there is an alternate way that is usually faster. Recall the *double-angle formulas* from trigonometry:

$$\sin^2 x = \frac{1}{2}[1-\cos 2x] \qquad \cos^2 x = \frac{1}{2}[1+\cos 2x].$$

These formulas can be used to reduce the exponents of the sine and cosine terms at the expense of changing the angle from x to $2x$ and introducing some constants.

EXAMPLE 5 Evaluate the integral

$$\int \sin^4 x \cos^2 x\,dx.$$

Solution: Because the exponents on both the sine and cosine terms are even integers, we use the double-angle formulas and get

$$\begin{aligned}\int \sin^4 x\cos^2 x\,dx &= \frac{1}{8}\int [1-\cos 2x]^2[1+\cos 2x]\,dx\\ &= \frac{1}{8}\int 1 - \cos 2x - \cos^2 2x + \cos^3 2x\,dx\\ &= \frac{1}{16}\int 1-\cos u - \cos^2 u + \cos^3 u\,du \qquad (u = 2x)\\ &= \frac{1}{16}\left(\frac{\sin u\cos^2 u}{3} - \frac{\sin u}{3} - \frac{\sin u\cos u}{2} + \frac{u}{2}\right) + c\\ &= \frac{1}{16}\left(\frac{\sin 2x\cos^2 2x}{3} - \frac{\sin 2x}{3} - \frac{\sin 2x\cos 2x}{2} + x\right) + c.\end{aligned}$$

The integration of the powers of the cosine was done using the reduction formula (8.5) several times. ■

Integrals of other trigonometric functions are discussed in later sections.

Exercises 8.3

Evaluate the integrals.

1. $\int x^3 e^{-2x}\,dx$
2. $\int \frac{dx}{(x^2+1)^3}$
3. $\int (x^4 - 2x^2 + 1)e^x\,dx$
4. $\int \frac{x+2}{(x^2+4)^2}\,dx$
5. $\int (1+x)^2 e^{4x}\,dx$
6. $\int \frac{x\,dx}{(1+x)^{12}}$

7. $\displaystyle\int \frac{dx}{(x^2+2x+5)^2}$

8. $\displaystyle\int \cos^4 x \sin x\, dx$

9. $\displaystyle\int \sin^5 x\, dx$

10. $\displaystyle\int \cos^4 x\, dx$

11. $\displaystyle\int \sin^3 x \cos^5 x\, dx$

12. $\displaystyle\int \sin^4 x \cos^4 x\, dx$

13. $\displaystyle\int \sin^3 3x \cos^2 3x\, dx$

14. $\displaystyle\int \sin^2 4x \cos^5 4x\, dx$

15. $\displaystyle\int \tan^3 x\, dx$ [Hint: Write $\tan x$ as $\sin x/\cos x$.]

16. Derive a reduction formula for $\int x^m (\ln x)^n\, dx$ in which the power of $\ln x$ is reduced. Assume m and n are positive integers.

17. Derive a reduction formula for $\int x^m e^{-x^2}\, dx$, and use it to evaluate $\int x^5 e^{-x^2}\, dx$.

18. Use a reduction formula to prove the formula

$$\int_0^{\pi/2} \sin^{2n} x\, dx = \frac{(2n-1)(2n-3)\cdots(3)(1)}{2n(2n-2)\cdots(4)(2)}\frac{\pi}{2}.$$

19. Use a reduction formula to find an expression similar to the formula in exercise 18 for the value of

$$\int_0^{\pi/2} \sin^{2n+1} x\, dx.$$

20. Develop a reduction formula to evaluate the definite integral

$$\int_0^1 x^n(1-x)^m\, dx$$

for positive integers n and m.

8.4 Partial fraction decomposition of rational functions

The goal of this and the next section is to describe a method for evaluating the integral of any rational function. This technique involves algebraic computations more complicated than in any of the previous methods, but this should expected because we describe a method to integrate a very large class of functions. This section is devoted to the necessary algebra of polynomials and rational functions. The next section deals with the evaluation of integrals of rational functions. The method shown here is useful for evaluating some other types of functions as well as rational functions.

Here is an illustration of the ideas we study. If you are asked to give the integral of the function

$$f(x) = \frac{1}{x-1} + \frac{1}{x-2} + \frac{1}{x-3}$$

you would immediately answer by inspection:

$$\int f(x)\, dx = \ln|x-1| + \ln|x-2| + \ln|x-3| + c.$$

If you were asked to give the integral of the function

$$g(x) = \frac{3x^2-12x+11}{(x-1)(x-2)(x-3)},$$

you would not do it by inspection (if at all). But wait a minute. It happens that $f(x)$ and $g(x)$ are the same function! This is easily seen by combining the three terms in f over a common denominator. The task of integrating $g(x)$ is made much easier when g is written as the sum of the three simple fractions defining f. The *method of partial fractions* is the process of writing the rational function $g(x)$ in the form $f(x)$. Of course more complicated choices for g force us to consider more complicated choices for the terms in f. Before we can state the precise theorem, some properties of polynomials are developed.

Taylor's Theorem for Polynomials

Taylor's theorem provides an expression for the polynomial $p(x)$ in terms of the values of the derivatives of p evaluated at a given point. We recall that a polynomial $p(x)$ has degree n if x^n is the highest power of x having a nonzero coefficient in $p(x)$. In other words, if $p(x)$ has degree n, then

$$p(x) = c_0 + c_1 x + c_2 x^2 + \cdots + c_n x^n, \quad \text{where} \quad c_n \neq 0.$$

In the next theorem we use the factorial notation $k!$ (read k factorial) defined for nonnegative integers k by the rules

$$0! = 1 \quad \text{and} \quad k! = 1 \cdot 2 \cdot 3 \cdots (k-1) \cdot k \quad \text{for} \quad k > 0.$$

Recall that $p^{(k)}(x)$ means the kth derivative of $p(x)$.

THEOREM 8.1

Taylor's Theorem for Polynomials

Let $p(x)$ be a polynomial of degree $n \geq 1$ and let a be any number. Then

$$p(x) = p(a) + p'(a)(x-a) + \frac{p''(a)}{2}(x-a)^2 + \cdots + \frac{p^{(k)}(a)}{k!}(x-a)^k + \cdots + \frac{p^{(n)}(a)}{n!}(x-a)^n.$$

Proof

Let us first treat the case in which $a = 0$. We have

$$p(x) = c_0 + c_1 x + c_2 x^2 + \cdots + c_n x^n \qquad \textbf{(8.7)}$$

and we must prove

$$c_k = p^{(k)}(0)/k!, \quad \text{for} \quad k = 0, 1, \cdots, n.$$

For the constant term c_0, we just evaluate both sides of Equation (8.7) at $x = 0$ to get $p(0) = c_0$. Next differentiate $p(x)$ to get

$$p'(x) = c_1 + 2c_2 x + \cdots + kx^{k-1} + \cdots + nx^{n-1}.$$

Evaluate both sides of this equation at $x = 0$ to get $p'(0) = c_1$. Differentiate again:

$$p''(x) = 2c_2 + 3 \cdot 2c_3 x + \cdots + k(k-1)x^{k-2} + \cdots + n(n-1)x^{n-2}$$

and $p''(0) = 2c_2$. This gives the value of c_2 as $p''(0)/2$ as the formula predicts. We do one more step in the same manner. The third derivative of p is

$$p^{(3)}(x) = 3 \cdot 2 \cdot 1c_3 + 4 \cdot 3 \cdot 2c_4 x + \cdots + n(n-1)(n-2)c_n c^{n-3},$$

and evaluation at $x = 0$ gives the value of c_3 as

$$c_3 = \frac{p^{(3)}(0)}{3 \cdot 2 \cdot 1}.$$

The pattern should be clear by now; the formula for the coefficients could be proved by a formal mathematical induction argument but we omit it.

Now we turn to the case of a general value of a. If we replace x by $x + a$, then

$$p(x + a) = c_0 + c_1(x + a) + c_2(x + a)^2 + \cdots + c_n(x + a)^n,$$

can be expanded and written in the standard form as a sum of constants times powers of x. Let $g(x) = p(x + a)$ so that $g(x)$ is a polynomial and

$$g(x) = p(x + a) = b_0 + b_1 x + b_2 x^2 + \cdots + b_n x^n$$

for some coefficients b_i. We apply the computation of the coefficients proved in the first part of this theorem for the case $a = 0$, namely

$$b_k = \frac{g^{(k)}(0)}{k!} \quad \text{for} \quad k = 0, 1, \ldots, n.$$

Observe that $g^{(k)}(x) = p^{(k)}(x + a)$ and $g^{(k)}(0) = p^{(k)}(a)$ so that

$$b_k = \frac{p^{(k)}(a)}{k!}.$$

Finally we get back to $p(x)$ by observing that $g(x - a) = p(x + a - a) = p(x)$ so

$$p(x) = b_0 + b_1(x - a) + b_2(x - a)^2 + \cdots + b_n(x - a)^n.$$

This proves the theorem. □

For a polynomial $p(x)$ and any number a, there is a polynomial $q(x)$ such that

$$p(x) = q(x - a).$$

Taylor's theorem shows how to compute the polynomial $q(x)$:

$$q(x) = p(a) + p'(a)x + \frac{p''(a)}{2!}x^2 + \cdots + \frac{p^{(n)}(a)}{n!}x^n,$$

where n is the degree of $p(x)$. We say $q(x - a)$ is *a polynomial in* $(x - a)$ or that $p(x)$ is expressed as a polynomial in $(x - a)$ when written as the sum of constants times powers of $(x - a)$. Of course when $p(x)$ is written in the standard form, we say it is a polynomial in x.

EXAMPLE 1 Let $p(x) = 2 - 4x - 3x^2 + x^3$. Express $p(x)$ as a polynomial in $x - 2$; i.e., write $p(x)$ as a sum of constants times powers of $(x - 2)$.

Solution: To apply Taylor's theorem, we first compute the derivatives of p:

$$\begin{aligned} p(x) &= 2 - 4x - 3x^2 + x^3, \\ p'(x) &= -4 - 6x + 3x^2, \\ p^{(2)}(x) &= -6 + 6x, \\ p^{(3)}(x) &= 6. \end{aligned}$$

Next we compute the coefficients given in Taylor's theorem:

$$p(2) = -10,$$
$$\frac{p'(2)}{1!} = \frac{-4}{1} = -4,$$
$$\frac{p^{(2)}(2)}{2!} = \frac{6}{2} = 3,$$
$$\frac{p^{(3)}(2)}{3!} = \frac{6}{6} = 1.$$

Thus

$$p(x) = -10 - 4(x - 2) + 3(x - 2)^2 + (x - 2)^3.$$ ■

Notice the "flavor" of the information in Taylor's theorem. It tells us that a knowledge of all the derivatives $p^{(k)}(a)$ of $p(x)$ at one point a, determines all the coefficients of p and hence determines all the values of p at all points x.

Some properties of polynomials can be easily deduced from Taylor's theorem. We give an example in the next theorem.

THEOREM 8.2

Properties of Polynomials

a. Let $p(x)$ be a polynomial and a a number. Then

$$p(x) = p(a) + q(x)(x - a)$$

for some polynomial $q(x)$.

b. If $p(x)$ and $r(x)$ are polynomials and a is some number such that

$$p(a) = r(a),$$

then there is some polynomial $q(x)$ such that

$$p(x) = r(x) + q(x)(x - a).$$

Proof

Taylor's theorem shows that $p(x) - p(a)$ is a polynomial that has $(x - a)$ as a factor so $p(x) - p(a) = q(x)(x - a)$ where $q(x)$ is a polynomial that can be written explicitly in terms of the derivatives of p evaluated at $x = a$. A particular case of property a states that $p(a) = 0$ implies $p(x) = (x - a)q(x)$ for some polynomial $q(x)$.

For the proof of property b we observe that $p(x) - r(x)$ is a polynomial whose value at $x = a$ is zero. Thus by property a applied to $p(x) - r(x)$ we have $p(x) - r(x) = (x - a)q(x)$ for some polynomial q. □

We now give the first step in the decomposition of rational functions.

THEOREM 8.3

Let $g(x)$ be a polynomial with $g(0) \neq 0$ and $h(x)$ any polynomial. For any positive integer n, there are constants $b_0, b_1, \ldots, b_{n-1}$ and a polynomial

$q(x)$ such that

$$\frac{h(x)}{x^n g(x)} = \frac{b_0}{x} + \frac{b_1}{x^2} + \cdots + \frac{b_{n-1}}{x^n} + \frac{q(x)}{g(x)}.$$

Proof

The proof is given for the case $n = 1$ first and then that case is used to proceed to the cases $n = 2, 3, \ldots$. Let $c_0 = h(0)/g(0)$. This is a real number because of the assumption $g(0) \neq 0$. Then the two polynomials $h(x)$ and $c_0 g(x)$ have the same value $h(0)$ at $x = 0$. By Theorem 8.2, there is a polynomial $q_0(x)$ such that

$$h(x) = c_0 g(x) + x q_0(x).$$

Divide both sides of this equation by $xg(x)$ to obtain

$$\frac{h(x)}{xg(x)} = \frac{c_0}{x} + \frac{q_0(x)}{g(x)}.$$

This is the conclusion of the theorem in the case $n = 1$. For the case $n = 2$, divide the last equation by x to obtain

$$\frac{h(x)}{x^2 g(x)} = \frac{c_0}{x^2} + \frac{q_0(x)}{xg(x)}.$$

Now treat the fraction $q_0(x)/\big(xg(x)\big)$ just as in the case $n = 1$ of the theorem. There is a constant c_1 and a polynomial $q_1(x)$ such that

$$\frac{q_0(x)}{xg(x)} = \frac{c_1}{x} + \frac{q_1(x)}{g(x)}$$

and so

$$\frac{h(x)}{x^2 g(x)} = \frac{c_0}{x^2} + \frac{c_1}{x} + \frac{q_1(x)}{g(x)}$$

This is the form stated in the conclusion of the theorem (except for the numbering of the constants) for the case $n = 2$. For $n = 3$, we divide the last equation by x to obtain

$$\frac{h(x)}{x^3 g(x)} = \frac{c_0}{x^3} + \frac{c_1}{x^2} + \frac{q_1(x)}{xg(x)}$$
$$= \frac{c_0}{x^3} + \frac{c_1}{x^2} + \frac{c_2}{x} + \frac{q_2(x)}{g(x)}.$$

The pattern is now established to show how the proof is completed for any positive integer n. □

In this theorem, the powers of x are singled out for special attention. This was done only as a matter of convenience. We could have just as easily singled out powers of $(x - a)$ for any number a. Here is the statement in the more general case, the proof of which can be carried out in the same manner as the proof of Taylor's theorem by replacing x in the previous theorem with $x - a$.

THEOREM 8.4

Let a be any number and $g(x)$ a polynomial with $g(a) \neq 0$. For any positive integer n, there are constants $b_0, b_1, \ldots, b_{n-1}$ and a polynomial $q(x)$ such that

$$\frac{h(x)}{(x-a)^n g(x)} = \frac{b_0}{x-a} + \frac{b_1}{(x-a)^2} + \cdots + \frac{b_{n-1}}{(x-a)^n} + \frac{q(x)}{g(x)}.$$

With the idea suggested by the preceding theorem, we can treat one important case of the partial fraction decomposition problem, namely the case in which the denominator of the rational function is a product of polynomials of the first degree.

THEOREM 8.5

Partial Fraction Decomposition: First Case

Let $h(x)/f(x)$ be a rational function for which the denominator $f(x)$ has the factorization

$$f(x) = (x-a_1)^{n_1}(x-a_2)^{n_2}\cdots(x-a_k)^{n_k}$$

with numbers $a_1, a_2, \ldots, a_k$ no two of which are equal and some positive exponents $n_1, n_2, \ldots, n_k$. Then

$$\frac{h(x)}{f(x)} = u(x) + L_1 + L_2 + \cdots + L_k \tag{8.8}$$

where each L_i has the form

$$L_i = \frac{c_1}{(x-a_i)} + \frac{c_2}{(x-a_i)^2} + \cdots + \frac{c_{n_i}}{(x-a_i)^{n_i}},$$

for some constants c_j (that also depend on i) and where $u(x)$ is a polynomial. If the degree of $h(x)$ is less than the degree of $f(x)$, then $u(x) = 0$.

Proof

Write

$$f(x) = (x-a_1)^{n_1} g(x)$$

where $g(x)$ is the product of all the remaining terms. Because the a_j are all different, $g(a_1) \neq 0$. By Theorem 8.4 we have

$$\frac{h(x)}{f(x)} = \frac{h(x)}{(x-a_1)^{n_1} g(x)} = L_1 + \frac{q(x)}{g(x)},$$

where $q(x)$ is some polynomial. This shows how to reduce the number of distinct factors in the denominator of the rational function because we are now faced with the term $q(x)/g(x)$ with $k-1$ distinct factors in place of the original $h(x)/f(x)$ that had k distinct factors in its denominator. When this argument is repeated k times, we have $L_1 + \cdots + L_k$ plus a term having only a 1 in the denominator. That term is the polynomial $u(x)$.

Now suppose that the degree of $h(x)$ is less than the degree of $f(x)$. We show that the polynomial $u(x)$ is zero. Multiply both sides of Equation (8.8) by $f(x)$ to clear the denominators. We get the equation

$$h(x) = f(x)u(x) + f(x)L_1 + \cdots + f(x)L_k.$$

Notice that each term $f(x)L_i$ is a polynomial of degree less than the degree of $f(x)$ because the denominator of L_i cancels some factor in $f(x)$. If the polynomial $u(x)$ is not 0, then the degree of $f(x)u(x)$ is at least as great as the degree of $f(x)$. But the degree of $h(x)$ is less than the degree of $f(x)$, so the only possible case is $u(x) = 0$. □

EXAMPLE 2 Find the partial fraction decomposition of

$$\frac{x^2 - 6}{(x-1)(x-2)(x-3)}.$$

Solution: We have the case in which the degree of the numerator, 2, is less than the degree of the denominator, 3, so Theorem 8.5 implies there are constants a, b and c such that

$$\frac{x^2 - 6}{(x-1)(x-2)(x-3)} = \frac{a}{x-1} + \frac{b}{x-2} + \frac{c}{x-3}.$$

The theorem does not tell us how to compute the constants. There are several good methods to find them. We illustrate a method that works very efficiently in this example because all the factors in the denominator have exponent 1. Clear the denominators to obtain the equation

$$x^2 - 6 = a(x-2)(x-3) + b(x-1)(x-3) + c(x-1)(x-2). \qquad \textbf{(8.9)}$$

Because this equation is true for all numbers x, we select some specific choices of x. The choice $x = 1$ is good because two of the terms on the right become zero. Substitute $x = 1$ and get

$$1 - 6 = a(1-2)(1-3) + b \cdot 0 + c \cdot 0 = 2a \quad \text{and} \quad a = \frac{-5}{2}.$$

To find b, let $x = 2$ in Equation (8.9) to obtain

$$4 - 6 = a \cdot 0 + b(2-1)(2-3) + c \cdot 0 = -b \quad \text{and} \quad b = 2.$$

Finally, to find c substitute $x = 3$ in Equation (8.9) to obtain

$$9 - 6 = a \cdot 0 + b \cdot 0 + c(3-1)(3-2) \quad \text{and} \quad c = \frac{3}{2}.$$

We have the equation

$$\frac{x^2 - 6}{(x-1)(x-2)(x-3)} = \frac{-5}{2(x-1)} + \frac{2}{x-2} + \frac{3}{2(x-3)}.$$ ■

The polynomial $u(x)$ in Equation (8.8) can be computed before finding any of the terms L_i by the long division method. If the rational function $h(x)/f(x)$ has a numerator with degree larger than or equal to the degree of the denominator, one may divide $f(x)$ into $h(x)$ and obtain a quotient with a remainder having degree less than the degree of $f(x)$. In other words,

$$h(x) = f(x)u(x) + r(x),$$

and

$$\frac{h(x)}{f(x)} = u(x) + \frac{r(x)}{f(x)},$$

where the degree of $r(x)$ is less than the degree of $f(x)$.

EXAMPLE 3 Find the partial fraction decomposition of

$$\frac{x^4 + 3x^3 - x^2 - 1}{x^2(x+1)}.$$

Solution: By dividing $x^3 + x^2$ into the numerator, we find

$$x^4 + 3x^3 - x^2 - 1 = (x^3 + x^2)(x + 2) - 3x^2 - 1$$

and

$$\frac{x^4 + 3x^3 - x^2 - 1}{x^2(x + 1)} = x + 2 + \frac{-3x^2 - 1}{x^2(x + 1)}.$$

The general theorem can be applied to the remaining quotient to say

$$\frac{-3x^2 - 1}{x^2(x + 1)} = \frac{a}{x} + \frac{b}{x^2} + \frac{c}{x + 1}.$$

To find the constants a, b and c, clear of fractions to get the equation

$$-3x^2 - 1 = ax(x + 1) + b(x + 1) + cx^2,$$

which holds for all x. Substitute $x = 0$ to get $-1 = b$. Substitute $x = -1$ to get $-4 = c$. There is no value that can be substituted for x to make both the coefficients of b and c equal 0 as was possible in the earlier example. This difficulty always happens when the denominator has a factor raised to a power higher than the first. If we rewrite the equation using the values found for b and c some cancellation is possible that allows an easy finish:

$$-3x^2 - 1 = ax(x + 1) - 1(x + 1) - 4x^2,$$
$$x^2 + x = ax(x + 1).$$

Thus $a = 1$ and we have the final form

$$\frac{x^4 + 3x^3 - x^2 - 1}{x^2(x + 1)} = x + 2 + \frac{1}{x} + \frac{-1}{x^2} + \frac{-4}{x + 1}.$$

■

Factorization of Polynomials

The first case of the decomposition into partial fractions does not cover all possibilities because not every polynomial is a product of factors of the first degree. The polynomial $x^2 + 1$ cannot be written as a product of two factors of degree one having real coefficients. Of course if one permits the use of complex numbers as coefficients, then further factorization is possible. However we avoid complex numbers in our discussinon so it is necessary to deal with quadratic factors.

A quadratic polynomial $ax^2 + bx + c$ (with $a \neq 0$) is called *irreducible* if it cannot be written as a product of two polynomials of degree one having real number coefficients. We always restrict polynomials to have real numbers as coefficients so this assumption will not be mentioned each time it is used. One might recognize an irreducible quadratic polynomial by its discriminant, $D = b^2 - 4ac$. The quadratic is irreducible if and only if $D < 0$. This is true for the following reason. By the quadratic formula we know the roots of the equation

$$ax^2 + bx + c = 0$$

are

$$r_+,\ r_- = \frac{-b \pm \sqrt{D}}{2a}$$

and so

$$ax^2 + bx + c = a(x - r_+)(x - r_-).$$

If $D \geq 0$ then r_+ and r_- are real numbers and we have a factorization as a product of polynomials with real coefficients. In case $D < 0$, the roots of the polynomial are nonreal. No factorization as a product of degree one factors with real coefficients is possible because any such factorization produces a real root of the polynomial.

Irreducible quadratic polynomials and polynomials of degree 1 are the factors from which all polynomials are obtained, as stated in following the fundamental theorem about factorization of polynomials with real coefficients.

THEOREM 8.6

Factorization of Polynomials

Every nonzero polynomial $f(x)$ can be written as a product

$$f(x) = c(x - a_1)^{n_1} \cdots (x - a_k)^{n_k}(x^2 + p_1 x + q_1)^{m_1} \cdots (x^2 + p_t x + q_t)^{m_t} \quad \textbf{(8.10)}$$

where c is the leading coefficient of f, no two of the a_i are equal, no two of the pairs (p_j, q_j) are equal, the exponents n_i and m_j are positive integers and the quadratic polynomials $x^2 + p_i x + q_i$ are irreducible.

The proof of this theorem requires tools that would carry us quite far from the subject of integration so we do not give it. As a consequence of this theorem, we see that the study of rational functions that we made in Theorem 8.5 covers a significant part of the general case. All that remains to study is the case of a rational function $h(x)/f(x)$ where f might have quadratic factors.

It is appropriate to mention here that the success of the method of partial fractions depends on the ability to write the denominator of a rational function as a product of factors of degree 1 and irreducible quadratic factors. This is usually not an easy task. The degree 1 factors can be found by finding the real roots of the polynomial. The irreducible quadratic factors can be found by finding the complex roots of the polynomial. The problem of finding the exact roots of a polynomial is not one that we try to solve here. A discussion of numerical methods for finding the approximate roots of a polynomial is given in chapter 11. The point of view that we follow in the discussion of partial fractions is that the reader has available some method for determining the factorization of a polynomial.

Partial Fractions in the General Case

We have all the necessary preparations to give the general theorem about the decomposition of rational functions into partial fractions.

THEOREM 8.7

Partial Fraction Decomposition: General Case

Let $h(x)/f(x)$ be a rational function for which the denominator $f(x)$ has the factorization given in Equation (8.10). Then

$$\frac{h(x)}{f(x)} = u(x) + L_1 + \cdots + L_k + Q_1 + \cdots + Q_t,$$

where

$$L_i = \frac{c_1}{(x - a_i)} + \frac{c_2}{(x - a_i)^2} + \cdots + \frac{c_{n_i}}{(x - a_i)^{n_i}},$$

for come constants $c_1, \ldots, c_{n_i}$;

$$Q_j = \frac{b_1 + s_1 x}{x^2 + p_j x + q_j} + \frac{b_2 + s_2 x}{(x^2 + p_j x + q_j)^2} + \cdots + \frac{b_{m_j} + s_{m_j} x}{(x^2 + p_j x + q_j)^{m_j}}$$

for some constants b_k, s_k, and where $u(x)$ is a polynomial. If the degree of $h(x)$ is less than the degree of $f(x)$, then $u(x) = 0$.

The proof of this theorem follows roughly the same lines as the proof of the decomposition theorem in the case where the denominator had only factors of the first degree. There are some technical complications with the quadratic terms, so we do not present the proof here. An outline of an argument to prove the theorem is given in the exercises following this section.

This is a very complicated theorem to state in its most general form. In relatively small examples, the theorem is not so complicated. Here is a specific example.

EXAMPLE 4 Give the form of the decomposition into partial fractions for the function

$$r(x) = \frac{x^3 + x - 2}{(x-1)^2(x+3)(x^2 - x + 2)^2(x^2 + 2x + 7)}.$$

Solution: The degree of the numerator is less than the degree of the denominator so the polynomial $u(x)$ in the theorem is 0 for this example. The denominator has two factors of degree 1 and two factors of degree 2, both irreducible. There are four parts in the decomposition, one corresponding to each of the factors in the denominator.

Corresponding to the factor $(x-1)^2$ having exponent 2, there is a sum

$$L_1 = \frac{c_1}{x-1} + \frac{c_2}{(x-1)^2}.$$

Corresponding to the factor $x + 3$ with exponent 1, there is the single term

$$L_2 = \frac{c_3}{x+3}.$$

Corresponding to the quadratic term $(x^2 - x + 2)^2$ having exponent 2, there is a sum

$$Q_1 = \frac{b_1 + s_1 x}{x^2 - x + 2} + \frac{b_2 + s_2 x}{(x^2 - x + 2)^2}.$$

Corresponding to the quadratic term $(x^2 + 2x + 7)$ having exponent 1, there is a single term

$$Q_2 = \frac{b_3 + s_3 x}{x^2 + 2x + 7}.$$

The conclusion of the theorem is that

$$\frac{x^3 + x - 2}{(x-1)^2(x+3)(x^2 - x + 2)^2(x^2 + 2x + 7)} = L_1 + L_2 + Q_1 + Q_2. \qquad \blacksquare$$

The theorem states that there exist constants c_i, b_j and s_j such that this equation holds. The theorem does not tell how to determine these constants. The determination of the constants might be a very time consuming part of this process. There are many computer software packages that can do this computation very efficiently (with limitations on the degrees of the polynomials in the numerator and denominator). We indicate one method for doing these computations by hand in example 5. There are many special devices that can be used to determine the constants in the partial fraction decomposition, but most ultimately rely on solving systems of linear equations. The trick is to make the system as easy to solve as possible. We give examples in which the general method of solving large systems of linear equations is avoided as much as possible.

EXAMPLE 5 Find the partial fraction decomposition of the rational function

$$\frac{2x^2 + 2x - 3}{x(x^2 + 1)}.$$

Solution: The general theorem implies there exist three constants A, B and C such that

$$\frac{2x^2 + 2x - 3}{x(x^2 + 1)} = \frac{A}{x} + \frac{B + Cx}{x^2 + 1}.$$

First clear the fractions by multiplying both sides by $x(x^2 + 1)$ to get

$$2x^2 + 2x - 3 = A(x^2 + 1) + (B + Cx)x.$$

Set $x = 0$ and the equation becomes $-3 = A$. Put the value of A into the equation and collect the known terms on the left to get

$$\begin{aligned} 2x^2 + 2x - 3 - (-3)(x^2 + 1) &= (B + Cx)x, \\ 5x^2 + 2x &= Bx + Cx^2. \end{aligned}$$

The coefficients of equal powers of x on each side must be equal. Thus $B = 2$ and $C = 5$; the partial fraction decomposition is

$$\frac{2x^2 + 2x - 3}{x(x^2 + 1)} = \frac{-3}{x} + \frac{2 + 5x}{x^2 + 1}.$$ ■

The method used to determine the unknown coefficients might be described roughly by saying: after the denominators have been cleared, evaluate both sides at the points where the degree 1 factors of the original denominator are 0. This determines some of the coefficients; insert the newly found values for these coefficients into the equation and transpose the terms with known coefficients to the left side of the equal sign. Some cancellation may be possible. If the equation is simple enough, compare coefficients of equal powers of x to determine the remaining constants. We illustrate this method with one more example that is a bit more complicated.

EXAMPLE 6 Find the partial fraction decomposition of the rational function

$$r(x) = \frac{45 + 73x + 31x^2 + 40x^3 + 5x^4 + 6x^5}{(x - 1)(x + 3)(x^2 + 4)^2}.$$

Solution: According to the decomposition theorem, there exist six constants A, B, C, D, E and F such that

$$r(x) = \frac{A}{x - 1} + \frac{B}{x + 3} + \frac{C + Dx}{x^2 + 4} + \frac{E + Fx}{(x^2 + 4)^2}.$$

To find these constants, first clear the denominators to get the equation

$$\begin{aligned}45+73x + 31x^2 + 40x^3 + 5x^4 + 6x^5 &= \\ A(x+3)(x^2+4)^2 &+ B(x-1)(x^2+4)^2 \\ &+ (C+Dx)(x^2+4)(x-1)(x+3) + (E+Fx)(x-1)(x+3).\end{aligned} \tag{8.11}$$

This equation holds for all x so we could multiply out the right-hand side and collect the coefficients of x. The polynomial on the left must be the same, term by term, as the polynomial on the right. After equating the coefficients of like powers of x, we get a system of six equations in six unknowns to solve. In this example, that involves a lot of computation so we try to simplify that process by selecting some values for x that enable us to easily compute some of the constants. We select $x = 1$ and $x = -3$ because these choices force all but one term on the right to equal 0. Setting $x = 1$ gives the equation

$$45+73 + 31 + 40 + 5 + 6 = A(1+3)(1+4)^2$$

$$200 = A \cdot 4 \cdot 25 = 100A \qquad \text{and} \qquad A = 2.$$

Setting $x = -3$ gives the equation

$$45 + 73(-3) + 31(-3)^2 + 40(-3)^3 + 5(-3)^4 + 6(-3)^5 = B\big((-3)-1\big)\big((-3)^2+4\big)^2,$$

$$-2028 = B(-4)(169) = -676B \qquad \text{and} \qquad B = 3.$$

We have found the constants associated with the degree 1 terms. The remaining four constants cannot be found immediately by substituting some value for x that makes all but one term drop out of Equation (8.11), because there is no real number x for which $x^2 + 4 = 0$. For now, substitute the values just found for A and B and transpose these two terms to the left-hand side of Equation (8.11). We get

$$-3 - 7x + 7x^2 + 2x^4 + x^5 = (C+Dx)(x^2+4)(x-1)(x+3) + (E+Fx)(x-1)(x+3).$$

The right side has a factor $(x-1)(x+3)$ so the left side must have the same factor. Do the long division by $(x-1)(x+3)$ and get

$$\begin{aligned}1 + 3x + x^3 &= (C+Dx)(x^2+4) + (E+Fx) \\ &= (4C+E) + (4D+F)x + Cx^2 + Dx^3.\end{aligned}$$

Now compare the coefficients on each side; the coefficient of x^3 gives $D = 1$ and the coefficient of x^2 gives $C = 0$. Comparing the coefficients of 1 and x gives

$$1 = 4C + E = 0 + E, \quad \text{and} \quad 3 = (4D+F) = 4 + F.$$

So finally $E = 1$ and $F = -1$ and the partial fraction decomposition of $r(x)$ is

$$r(x) = \frac{2}{x-1} + \frac{3}{x+3} + \frac{x}{x^2+4} + \frac{1-x}{(x^2+4)^2}.$$ ■

Exercises 8.4

Find the partial fraction decomposition of each of the rational functions.

1. $\dfrac{1}{x(x-4)}$

2. $\dfrac{x}{(x-1)(x+2)}$

3. $\dfrac{x^2+1}{x(x-1)(x+1)}$

4. $\dfrac{1}{x(x^2+4)}$

5. $\dfrac{2x+x^2}{(x-3)(x^2+4)}$

6. $\dfrac{1}{x^3(x^2+4)}$

7. $\dfrac{x - x^2}{(x-2)^2(x^2+1)}$

8. $\dfrac{x + x^2}{(x-2)(x^2+4)^2}$

9. $\dfrac{2 + x^2}{(x^2-2x+5)(x^2+4)}$

10. $\dfrac{1 + 2x^2}{(x^2-4x+5)(x^2+1)}$

The next two exercises show a systematic method of finding the coefficients corresponding to the degree 1 factors in the partial fraction decomposition of a rational function.

11. If $h(x)$ and $g(x)$ are polynomials and $g(a) \neq 0$ and if

$$r(x) = \frac{h(x)}{(x-a)^n g(x)} = \frac{c_n}{(x-a)^n} + \cdots + \frac{c_1}{x-a} + \frac{q(x)}{g(x)}$$

is the first step of the partial fraction decomposition of $r(x)$, show that

$$c_n = \lim_{x\to a}(x-a)^n\, r(x) = \lim_{x\to a}\frac{h(x)}{g(x)} = \frac{h(a)}{g(a)}.$$

12. Keeping all the notation of exercise 11, and assuming $n > 1$, show that

$$r_2(x) = r(x) - \frac{c_n}{(x-a)^n} = \frac{h_2(x)}{(x-a)^{n-1}g(x)}$$

for some polynomial $h_2(x)$. Thus the next coefficient c_{n-1} can be found by the method indicated in exercise 11 applied to $r_2(x)$ in place of $r(x)$. [Comment: Repetition of this idea permits computation of all the coefficients of the terms $c_i/(x-a)^i$ in the partial fraction decomposition of $r(x)$.]

Apply the method described in exercises 11 and 12 to obtain the partial fraction decomposition of the rational functions.

13. $r(x) = 1/[(x-2)^2(x-1)^2]$

14. $r(x) = 1/[x^3(x+1)^2]$

15. $r(x) = (x+3)/[(x+1)^2(x-1)^2(x^2+9)]$

16. $r(x) = (x^2-4x+7)/[(x-2)^3(x^2+x+1)^2]$

17. There is exactly one polynomial $p(x)$ of degree 3 that satisfies

$$p(1) = 1, \quad p'(1) = -2, \quad p''(1) = 0, \quad p'''(1) = 12.$$

Find $p(x)$.

18. If $p(x)$ is a polynomial of degree 2 such that $p(1) = 1$, $p'(1) = 2$ and $p''(1) = 6$, find $p(0)$, $p'(0)$ and $p''(0)$.

19. Let a and $s_0, s_1, \ldots, s_n$ be any given numbers. Prove there is one and only one polynomial $p(x)$ with degree at most n that satisfies

$$p^{(k)}(a) = s_k \quad \text{for} \quad k = 0, 1, \ldots, n.$$

The next set of exercises outlines a proof of the partial fraction decomposition theorem for the general case of quadratic factors in the denominator. Here is the statement that must be proved: If $u(x) = x^2 + bx + c$ is an irreducible quadratic and $g(x)$ is a polynomial not having $u(x)$ as a factor, then

$$\frac{h(x)}{u(x)^n g(x)} = \frac{A_1 + B_1 x}{u(x)} + \cdots + \frac{A_n + B_n x}{u(x)^n} + \frac{q(x)}{g(x)}, \qquad \textbf{(8.12)}$$

for some constants A_i and B_i and some polynomial $q(x)$.

The idea of the proof is to reduce to the case of factors of degree 1, that has already been proved in detail in the text. The reduction to the degree 1 case is accomplished by the use of even polynomials, i.e., polynomials $k(x)$ such that $k(x) = k(-x)$. Every even polynomial is a sum of constants times even powers of x.

20. If $g(x)$ is a polynomial then $g(x)g(-x)$ is an even function and there is a polynomial $G(x)$ such that $G(x^2) = g(x)g(-x)$.

21. Let $u(x) = x^2 + c$ be an irreducible quadratic and $g(x)$ a polynomial that does not have $u(x)$ as a factor. Then $u(x)$ is not a factor of $g(-x)$ either.

22. Let $h(x)$ and $g(x)$ be any polynomials. Then there exist polynomials $P(x)$, $Q(x)$ such that $h(x)g(-x) = P(x^2) + xQ(x^2)$. [Hint: the polynomial $h(x)g(-x)$ is a sum of all the terms involving only even powers of x plus the sum of the terms involving only odd powers of x.]

23. Keep all the notation of exercises 20, 21 and 22. Make the observation

$$\frac{h(x)}{(x^2+c)^n g(x)} = \frac{h(x)g(-x)}{(x^2+c)^n G(x^2)} = \frac{P(x^2)}{(x^2+c)^n G(x^2)} + x\frac{Q(x^2)}{(x^2+c)^n G(x^2)}.$$

Apply the first case of the partial fraction decomposition theorem to the rational functions of t

$$\frac{P(t)}{(t+c)^n G(t)} \quad \text{and} \quad \frac{Q(t)}{(t+c)^n G(t)}$$

to obtain (after substituting $t = x^2$)

$$\frac{h(x)}{(x^2+c)^n g(x)} = \left[\frac{A_1}{(x^2+c)} + \cdots + \frac{A_n}{(x^2+c)^n} + \frac{r_1(x^2)}{G(x^2)}\right] + x\left[\frac{B_1}{(x^2+c)} + \cdots + \frac{B_n}{(x^2+c)^n} + \frac{r_2(x^2)}{G(x^2)}\right] \qquad \textbf{(8.13)}$$

for some polynomials r_1, r_2.

24. Show that the term

$$\frac{r_1(x^2) + x r_2(x^2)}{G(x^2)}$$

appearing in exercise 23 is equal to

$$\frac{q(x)}{g(x)}$$

for some polynomial $q(x)$. [Hint: Multiply both sides of Equation (8.13) by $(x^2+c)^n g(x)$ and use the fact that

$g(-x)$ does not have $x^2 + c$ as a factor.]

25. Complete the proof of the general statement, equation (8.12), by using the results of exercise 24 after a substitution of $(x + m)^2 + s = x^2 + bx + c = u(x)$ in place of $x^2 + c$.

8.5 Integration of rational functions

By the partial fraction decomposition theorem, evaluating the integral

$$\int \frac{p(x)}{q(x)}\,dx$$

for polynomials $p(x)$ and $q(x)$ is reduced to the problem of evaluating integrals of the form

$$\int \frac{A}{(x+a)^n}\,dx \quad \text{and} \quad \int \frac{Bx + C}{(x^2 + bx + c)^n}\,dx,$$

where A, B, C, a, b and c are constants.

In the first case there is no difficulty; make the substitution $u = x + a$ so that $du = dx$ and we have

$$\int \frac{A}{(x+a)^n}\,dx = \int \frac{A}{u^n}\,du.$$

We have the evaluation

$$\int \frac{A}{(x+a)^n}\,dx = \begin{cases} \dfrac{A}{(1-n)(x+a)^{n-1}} + c & \text{if } n \neq 1 \\ A \ln|x+a| + c & \text{if } n = 1. \end{cases}$$

The evaluation of the integrals with the a quadratic term in the denominator is somewhat more complicated. For a first step, a substitution can be made to simplify the denominator. Complete the square of the quadratic term $x^2 + bx + c = (x + b/2)^2 + (c - b^2/4)$; let $t = x + b/2$; then $Ax + B = At + B'$ with $B' = B + Ab/2$, a constant. Then $m = c - b^2/4$ is a positive constant when the original quadratic is irreducible and we have

$$\int \frac{Ax + B}{(x^2 + bx + c)^n}\,dx = \int \frac{At + B'}{(t^2 + m)^n}\,dt.$$

The evaluation of this integral is carried out by treating the integrand as the sum of two integrals,

$$\int \frac{At + B'}{(t^2+m)^n}\,dt = \int \frac{At}{(t^2+m)^n}\,dt + \int \frac{B'}{(t^2+m)^n}\,dt,$$

and dealing with each separately.

The integral

$$\int \frac{At}{(t^2+m)^n}\,dt$$

can be evaluated by a substitution $u = t^2 + m$, $du = 2t\,dt$:

$$\int \frac{At}{(t^2+m)^n}\,dt = \frac{1}{2}\int \frac{A}{u^n}\,du.$$

It now follows easily that

$$\int \frac{At}{(t^2+m)^n}\,dt = \begin{cases} \dfrac{A}{2(1-n)(t^2+m)^{n-1}} & \text{if } n \neq 1 \\ \frac{1}{2} A \ln(t^2+m) & \text{if } n = 1. \end{cases}$$

Evaluation of the integral

$$I_n = \int \frac{dt}{(t^2+m)^n}$$

is accomplished using a reduction formula from section 8.2 which reads

$$I_{n+1} = \frac{1}{2nm}\left[\frac{t}{(t^2+m)^n} + (2n-1)I_n\right] \quad n > 0. \tag{8.14}$$

By repeated application of the reduction formula, the integral I_k, for any positive integer k, can be expressed in terms of rational functions and I_1. However, I_1 has already been evaluated in section 7.9. Recall that the quadratic $x^2 + bx + c = t^2 + m$ was assumed irreducible. The irreducibility implies that $m > 0$ and so we can write the constant m as a square of another constant: $m = a^2$. Then

$$I_1 = \int \frac{dt}{t^2+a^2} = \frac{1}{a}\arctan\left(\frac{t}{a}\right) + c.$$

Here is a summary of the steps used to evaluate $\int \frac{Ax+B}{(x^2+bx+c)^n}\,dx$ when the quadratic term is irreducible:

1. Complete the square of the quadratic term, make a change of variable $t = x - b/2$ to reduce to the form $\int \frac{At+B'}{(t^2+m)^n}\,dt$.
2. Break this into the sum of two integrals

$$\int \frac{At+B'}{(t^2+m)^n}\,dt = \int \frac{At\,dt}{(t^2+m)^n} + \int \frac{B'\,dt}{(t^2+m)^n}.$$

The first integral on the right can be evaluated by a substitution $u = t^2 + m$; the second integral can be reduced to a sum of rational functions and an arctangent function by the reduction formula (8.14).

EXAMPLE 1 Evaluate the integral

$$\int \frac{x^4 - 4x^3 + 4x^2 - 16x + 9}{(x^2+4)(x^2+1)^2}\,dx.$$

Solution: We carry out the partial fraction decomposition and obtain

$$\frac{x^4 - 4x^3 + 4x^2 - 16x + 9}{(x^2+4)(x^2+1)^2} = \frac{1}{x^2+4} + \frac{2-4x}{(x^2+1)^2}.$$

The problem requires that we evaluate the integrals

$$\int \frac{dx}{x^2+4}, \qquad \int \frac{2-4x}{(x^2+1)^2}\,dx.$$

The first is straightforward;

$$\int \frac{1}{x^2+4}\,dx = \frac{1}{2}\arctan\left(\frac{x}{2}\right) + c.$$

The integral of the second term is broken into the sum of two integrals, one of which is evaluated by a substitution, the other by a reduction formula:

$$\int \frac{-4x\,dx}{(x^2+1)^2} = \int \frac{-2\,du}{u^2} \qquad (u = x^2+1)$$
$$= \frac{2}{u} + c = \frac{2}{x^2+1} + c,$$

$$\int \frac{2\,dx}{(x^2+1)^2} = \frac{x}{(x^2+1)} + \int \frac{2\,dx}{x^2+1}$$
$$= \frac{x}{(x^2+1)} + \arctan x + c.$$

Now we combine all the parts to record the final result:

$$\int \frac{x^4 - 4x^3 + 4x^2 - 16x + 9}{(x^2+4)(x^2+1)^2}\,dx = \frac{1}{2}\arctan\left(\frac{x}{2}\right) + \frac{2+x}{x^2+1} + \arctan x + c.$$

A problem as complicated as this should be checked by differentiating the answer to verify that it is correct. ■

Exercises 8.5

Apply the partial fraction decomposition found in the exercises of section 8.4 to evaluate the integrals.

1. $\displaystyle\int \frac{dx}{x(x-4)}$

2. $\displaystyle\int \frac{x\,dx}{(x-1)(x+2)}$

3. $\displaystyle\int \frac{(x^2+1)\,dx}{x(x-1)(x+1)}$

4. $\displaystyle\int \frac{dx}{x(x^2+4)}$

5. $\displaystyle\int \frac{(2x+x^2)\,dx}{(x-3)(x^2+4)}$

6. $\displaystyle\int \frac{dx}{x^3(x^2+4)}$

7. $\displaystyle\int \frac{(x-x^2)\,dx}{(x-2)^2(x^2+1)}$

8. $\displaystyle\int \frac{(x+x^2)\,dx}{(x-2)(x^2+4)^2}$

9. $\displaystyle\int \frac{(2+x^2)\,dx}{(x^2-2x+5)(x^2+4)}$

10. $\displaystyle\int \frac{(1+2x^2)\,dx}{(x^2-4x+5)(x^2+1)}$

Carry out the partial fraction decomposition of the rational functions and then evaluate the integrals.

11. $\displaystyle\int \frac{dx}{(x-1)(x+1)}$

12. $\displaystyle\int \frac{x\,dx}{(x-1)^2(x+1)}$

13. $\displaystyle\int \frac{(x-3)\,dx}{x(x-1)(x+1)}$

14. $\displaystyle\int \frac{(1+x)\,dx}{x(x+1)^3}$

15. $\displaystyle\int \frac{dx}{(x-1)(x^2+1)}$

16. $\displaystyle\int \frac{dx}{(x-1)(x^2+1)^2}$

17. $\displaystyle\int \frac{dx}{(x-1)^2(x^2+1)^2}$

18. $\displaystyle\int \frac{x-3\,dx}{(x^2+x+1)(x^2+1)}$

19. $\displaystyle\int \frac{dx}{(x^2+1)^2(x^2+x+1)^2}$

20. $\displaystyle\int_1^2 \frac{dx}{x^2+2x}$

21. $\displaystyle\int_0^1 \frac{dx}{(x+1)^2(x+3)}$

22. $\displaystyle\int_0^1 \frac{x^2\,dx}{16-x^4}$

23. $\displaystyle\int_0^2 \frac{dx}{x^3+1}$

24. $\displaystyle\int_0^1 \frac{1}{x^5+1}$

25. State and prove general conditions on $p(x)$ and m that ensure that the integral of $p(x)/(x-a)^m$ is a rational function. Here $p(x)$ is a polynomial and m is a positive integer.

26. Let $p(x)$ be a quadratic polynomial such that $p(0) = 1$ and

$$\int \frac{p(x)}{x^2(x-1)^2}\,dx$$

is a rational function. Find p.

In principle, the integral of a rational function can be evaluated, but the computational difficulties might be serious. For example, the problem of factoring the denominator as a product of irreducible factors is nontrivial. Rather than trying to find the exact value of the following definite integrals, obtain an approximate value by using a Riemann sum with a regular partition having n subintervals. Use a calculator or computer to do the arithmetic with a greater number of subintervals if possible.

27. $\displaystyle\int_0^1 \frac{dx}{x^4-3x^2-3},\quad n = 4, 8, 16$

28. $\displaystyle\int_0^2 \frac{(x-3)\,dx}{x^3+5x^2-16x-80},\quad n = 4, 8$

29. $\displaystyle\int_{-1}^0 \frac{x^2\,dx}{x^4-6x^3+13x^2-12x+4},\quad n = 4, 8$

30. $\displaystyle\int_0^2 \frac{(x^2+2)\,dx}{x^5+5x+1},\quad n = 4, 8$

8.6 Integrals of algebraic functions

The main goal of this section is to show a method to evaluate integrals of certain algebraic functions. The most practical case is that in which integrals are evaluated in terms of familiar functions. We know in general that if $f(x)$ is any continuous function, then the Fundamental Theorem of Calculus provides an antiderivative of $f(x)$, namely,

$$F(x) = \int_a^x f(t)\,dt.$$

Because $F'(x) = f(x)$, we have found an antiderivative of $f(x)$. However this expression as an integral is not a satisfactory form for the antiderivative of $f(x)$ except in theory. In practice, we want to find an antiderivative of $f(x)$ by giving an explicit function that is expressed in terms of familiar functions. For the purpose of this discussion, the familiar functions are rational functions, nth root functions, trigonometric functions and their inverses, the exponential and logarithmic functions and any composites of these functions, as well as sums, products and quotients of these functions. To appreciate some of the statements to follow, the reader should be aware that the integral of a familiar function might not be a familiar function. For example, the function

$$E(x) = e^{-x^2}$$

is a familiar function because it is a composite of the exponential function e^x and the rational function, $-x^2$. However the antiderivative

$$\int e^{-x^2}\,dx$$

is not a familiar function. It is not possible to write down an explicit function (using only a finite number of familiar functions) whose derivative equals $E(x)$. The proof of this fact is difficult and is not given. Our goal is the reverse problem. We want to show that certain integrals can be explicitly evaluated in terms of familiar functions. We begin with some terminology necessary to describe the functions we will study.

Algebraic Functions

We expand our earlier definition of polynomial to permit two independent variables. A polynomial in x and w is a function of the form

$$P(x, w) = a_{00} + a_{10}x + a_{01}w + a_{11}xw + \cdots + a_{ij}x^iw^j + \cdots + a_{nm}x^nw^m,$$

where the coefficients a_{ij} are constants. We say $P(x, w)$ is a function of two variables x and w. The polynomial $P(x, w)$ is nonzero if at least one of the coefficients a_{ij} is nonzero. A rational function of x and w is a quotient

$$R(x, w) = \frac{P(x, w)}{Q(x, w)}$$

of two polynomials $P(x, w)$ and $Q(x, w)$. Thus the function

$$R(x, w) = \frac{x^2 - xw + x^3}{x + 4w^2}$$

is an example of a rational function of x and w.

A function $f(x)$ is called an *algebraic function* if there is a nonzero polynomial function $P(x, w)$ such that

$$P(x, f(x)) = 0$$

for every x in the domain of f.

EXAMPLE 1 Show that each of the functions

a. $f(x) = x^{1/n}$, n = a positive integer,

b. $b(x) = \sqrt{x^2 + m}$,

c. $h(x) = \sqrt{a^2 - x^2}$,

d. $k(x) = \sqrt{1 + x^3}$,

is an algebraic function.

Solution: For a we observe $f(x)^n = x$ so that the polynomial $P(x, w) = w^n - x$ has the property $P(x, f(x)) = 0$ for all x in the domain of f. Thus $f(x)$ is an algebraic function.

The function b in b satisfies $b(x)^2 = x^2 + m$ so $(x, b(x))$ satisfies the polynomial $P(x, w) = w^2 - x^2 - m$.

Similarly, $(x, h(x))$ satisfies the polynomial $P(x, w) = w^2 + x^2 - a^2$ and $(x, k(x))$ satisfies the polynomial $P(x, w) = w^2 - x^3 - 1$. Thus the four functions are algebraic. ■

The functions e^x, $\ln x$, $\sin x$ and $\cos x$ are not algebraic functions. A function that is not algebraic is called *transcendental.* To prove that the four functions in the example just above are algebraic functions, we simply produced the nonzero polynomial satisfied by each function. It is usually more difficult to prove that a function, like e^x for example, is transcendental because we must show there is no nonzero polynomial $P(x, w)$ that satisfies $P(x, e^x) = 0$. We will not prove that any particular function is transcendental because it will not be needed in the development to follow but see the exercises following this section for an indication of how to prove e^x is a transcendental function.

The integral of an algebraic function cannot always be evaluated in terms of familiar functions. The function $k(x) = \sqrt{1 + x^3}$ is algebraic but the integral $\int k(x)\,dx$ cannot be evaluated using only familiar functions. How can we tell the good guys from the bad guys in the collection of algebraic functions? Rather than attempting a complete answer to this question, we will show that any rational function of x and $\sqrt{ax^2 + bx + c}$ can be integrated. There are other algebraic functions that can be integrated but the description of large classes is more complicated and we leave that topic for the reader to investigate. (There is a practical problem that will arise when evaluating such an integral—it will be necessary to factor the denominators of some rational functions to apply the method of partial fractions. We will not give methods for factoring but will assume that any polynomial that arises can be factored by hand or machine.) The method we use can be applied to many other algebraic functions as well.

Simple Algebraic Functions

For any positive integer n, the function $x^{1/n}$ is algebraic. Any rational function $R(x^{1/n})$ of $x^{1/n}$ can be made into a rational function of t by setting $x = t^n$. Then

$dx = nt^{n-1}\,dt$. It follows that the integral

$$\int R(x^{1/n})\,dx$$

of a rational function of $x^{1/n}$ is transformed into the integral of a rational function of t; namely

$$\int R(x^{1/n})\,dx = n\int R(t)t^{n-1}\,dt,$$

and so the integral can be evaluated.

If a rational function involves two roots, say $x^{1/n}$ and $x^{1/m}$, it can be viewed as a rational function of $x^{1/mn}$ because $x^{1/n} = \left(x^{1/mn}\right)^m$ and $x^{1/m} = \left(x^{1/mn}\right)^n$.

EXAMPLE 2 Evaluate the integral

$$\int \frac{x^{1/2}}{1+x^{1/3}}\,dx.$$

Solution: Because $x^{1/3} = \left(x^{1/6}\right)^2$ and $x^{1/2} = \left(x^{1/6}\right)^3$, it follows that the integrand is a rational function of $x^{1/6}$. We make the substitution

$$x = t^6, \qquad dx = 6t^5\,dt,$$

and obtain

$$\int \frac{x^{1/2}}{1+x^{1/3}}\,dx = \int \frac{t^3}{1+t^2}\,6t^5\,dt.$$

We first carry out the long division

$$\frac{6t^8}{1+t^2} = 6\left(t^6 - t^4 + t^2 - 1 + \frac{1}{1+t^2}\right).$$

The integral equals

$$6\left(\frac{t^7}{7} - \frac{t^5}{5} + \frac{t^3}{3} - t + \arctan t\right) + c$$

$$= \frac{6}{7}x^{7/6} - \frac{6}{5}x^{5/6} + 2x^{1/2} - 6x^{1/6} + 6\arctan x^{1/6} + c.$$

■

Rational functions of $\sqrt{ax^2+bx+c}$

The general rational function of x and $\sqrt{ax^2+bx+c}$ can be written as

$$R(x, \sqrt{ax^2+bx+c}) = \frac{p(x) + r(x)\sqrt{ax^2+bx+c}}{q(x) + s(x)\sqrt{ax^2+bx+c}},$$

for some polynomials $p(x)$, $q(x)$, $r(x)$ and $s(x)$. If we want to integrate this function, we begin by completing the square of the term under the radical sign and factoring out a constant so that the radical has one of the forms

$$f(x) = \sqrt{a^2 - x^2} \qquad \text{or} \qquad f(x) = \sqrt{x^2 + m}$$

where the constant m can be positive or negative and a is positive. We will prove that integrals of functions like

$$\frac{1}{\sqrt{x^2+4}}, \qquad \frac{x^2}{\sqrt{x^2-1}}, \qquad \frac{3x - x^3\sqrt{9-x^2}}{1+\sqrt{9-x^2}}$$

can be evaluated explicitly in terms of familiar functions.

The method for proving that an antiderivative of $R(x, f(x))$ can be explicitly computed involves finding a rational function $x = x(t)$ of one variable t with the property that $f(x(t))$ is also a rational function. Because $f(x)$ is the square root of a polynomial, it is necessary to find $x(t)$ so that either $x(t)^2 + m$ or $a^2 - x(t)^2$ is the square of a rational function. There are many possibilities for the choice of $x(t)$; one choice is described in Theorem 8.8.

THEOREM 8.8

Integration of Algebraic Functions

Let $R(x, w)$ be a rational function of x and w.

a. The integral

$$\int R(x, \sqrt{a^2 - x^2})\, dx$$

is transformed into the integral of a rational function of t by the substitution

$$x(t) = \frac{a(1 - t^2)}{1 + t^2}, \quad dx = \frac{-4at}{(1 + t^2)^2}\, dt, \quad t = \frac{\sqrt{a^2 - x^2}}{a + x}.$$

b. The integral

$$\int R(x, \sqrt{x^2 + m})\, dx$$

is transformed into the integral of a rational function of t by the substitution

$$x(t) = \frac{t^2 - m}{2t}, \quad dx = \frac{t^2 + m}{2t^2}\, dt, \quad t = x + \sqrt{x^2 + m}.$$

Proof

a. If we let

$$x(t) = \frac{a(1 - t^2)}{1 + t^2},$$

then

$$\begin{aligned} a^2 - x(t)^2 &= a^2 - \frac{a^2(1 - t^2)^2}{(1 + t^2)^2} \\ &= a^2\left[\frac{(1 + t^2)^2 - (1 - t^2)^2}{(1 + t^2)^2}\right] \\ &= a^2\left[\frac{4t^2}{(1 + t^2)^2}\right]. \end{aligned}$$

Thus we obtain

$$\sqrt{a^2 - x^2} = \frac{2at}{1 + t^2}$$

and this is a rational function of t. The derivative of $x(t)$ is easily computed from the definition of $x(t)$ so we make the change of variable $x = x(t)$ and we have

$$\int R(x, \sqrt{a^2 - x^2})\, dx = \int R\left(\frac{a(1 - t^2)}{1 + t^2}, \frac{2at}{1 + t^2}\right) \frac{-4at}{(1 + t^2)^2}\, dt.$$

Because the function $R(x, w)$ is a rational function of x and w, it follows that the integrand on the right is a rational function of t. Because an antiderivative can be found for every rational function, this integral can be evaluated. To return to the original variable, it is necessary to solve the equation

$$x = x(t) = \frac{a(1-t^2)}{1+t^2}$$

for t in terms of x. After a bit of manipulation, one finds

$$t = \frac{\sqrt{a^2 - x^2}}{a + x}.$$

b. We set

$$x = x(t) = \frac{t^2 - m}{2t}.$$

After some algebra is carried out, we find

$$\sqrt{x^2 + m} = \frac{t^2 + m}{2t}.$$

Compute dx/dt and make the substitution to get

$$\int R(x, \sqrt{x^2 + m})\, dx = \int R\left(\frac{t^2 - m}{2t}, \frac{t^2 + m}{2t}\right) \frac{t^2 + m}{2t^2}\, dt.$$

Just as in the previous case, the remaining integrand is a rational function and this integral can be evaluated. To return to the original variables we solve for t in terms of x and find $t = x + \sqrt{x^2 + m}$. □

We illustrate this theorem with two examples.

EXAMPLE 3 Evaluate the integral

$$\int \frac{dx}{\sqrt{x^2 + 5}}.$$

Solution: The substitution indicated in Theorem 8.8 is

$$x = x(t) = \frac{t^2 - 5}{2t}, \quad dx = \frac{t^2 + 5}{2t^2}\, dt, \quad t = x + \sqrt{x^2 + 5}.$$

Using these we have

$$\begin{aligned} \int \frac{dx}{\sqrt{x^2 + 5}} &= \int \frac{2t}{t^2 + 5} \frac{t^2 + 5}{2t^2}\, dt \\ &= \int \frac{dt}{t} = \ln|t| + c \\ &= \ln|x + \sqrt{x^2 + 5}| + c. \end{aligned}$$

■

EXAMPLE 4 Evaluate the integral

$$\int \frac{dx}{x\sqrt{1 - x^2}}.$$

Solution: We make the substitution

$$x(t) = \frac{1 - t^2}{1 + t^2}, \quad dx = \frac{-4t}{(1 + t^2)^2}\, dt, \quad t = \frac{\sqrt{1 - x^2}}{1 + x}$$

so that

$$\begin{aligned}
\int \frac{dx}{x\sqrt{1-x^2}} &= \int \frac{1+t^2}{1-t^2}\frac{1+t^2}{2t}\frac{-4t}{(1+t^2)^2}\,dt \\
&= 2\int \frac{dt}{t^2-1} = \int \frac{1}{t-1} - \frac{1}{t+1}\,dt \\
&= \ln\left|\frac{t-1}{t+1}\right| + c \\
&= \ln\left|\frac{\sqrt{1-x^2}-(x+1)}{\sqrt{1-x^2}+(x+1)}\right| + c \\
&= \ln\left|\frac{\sqrt{1-x^2}-1}{x}\right| + c.
\end{aligned}$$

The algebraic manipulations needed to obtain the last two equalities are left for the reader to supply. ■

The reader might wonder how the substitutions in the theorem are found. They are based on the identity

$$(s^2+b)^2 = (s^2-b)^2 + (2sb)^2.$$

If we set $b = 1$ and both sides of this equation are divided by $(s^2+1)^2$, we obtain a sum of the squares of two rational functions of s equal to 1. If we divide both sides by $(2s)^2$ we obtain b^2 as the difference of the squares of two rational functions. These substitutions will come up again in the paragraph on rational functions of sines and cosines below.

The integration performed by reduction to rational functions usually involves considerable computation as was seen in the two worked examples. The theorem assures us that certain integrals can be evaluated; in practice, there may be alternative methods that permit evaluation with less computation. The method of trigonometric substitutions, to be studied in section 8.7, is sometimes a less complicated method for evaluating integrals involving $\sqrt{a^2-x^2}$ or $\sqrt{x^2+m}$. Why do we bother with this method if there are simpler methods available?

One reason for providing this technique is that the method of partial fractions can be done by computer. Symbolic integration packages usually use substitutions and the partial fraction method for integrals of the type considered in this section. An understanding of this technique is vital to a programmer wanting to write symbolic integration packages.

Another reason, perhaps more practical, is that definite integrals are often evaluated by numerical methods rather than by actually finding antiderivatives. We have discussed upper and lower sums for approximating integrals. Other methods are discussed in chapter 11. Any method of numerically approximating a definite integral of a function $f(x)$ will require that (many) values of $f(x)$ be computed. When $f(x)$ is a rational function, its values are computed very easily, whereas the values of more complicated algebraic functions may be more difficult to compute. Thus numerical methods can be carried out more easily for the integral of a rational function. Although we do not discuss it here, there is a very nice theory of approximation of functions by rational functions. Thus a definite integral of a complicated function can be approximated by the integral of a rational function and the methods described here are valuable.

Rational Functions of Sine and Cosine

The idea of transforming an algebraic function into a rational function that was employed in Theorem 8.8 to evaluate $\int R(x, \sqrt{ax^2 + bx + c})\,dx$ can also be applied to integrate rational functions $R(\cos\theta, \sin\theta)$ of $\sin\theta$ and $\cos\theta$. We show that integrals like

$$\int \frac{\sin\theta\,d\theta}{4 + 3\cos\theta} \qquad \text{and} \qquad \int \frac{4\sin\theta + \sin^2\theta\cos\theta}{4 + \sin\theta - 2\cos\theta}\,d\theta$$

can be evaluated by transforming them into integrals of rational functions.

THEOREM 8.9

Integration of $R(\cos\theta, \sin\theta)$

Let $R(u, v)$ be a rational function of two variables. Then the integral

$$\int R(\cos\theta, \sin\theta)\,d\theta$$

is transformed into the integral of a rational function of t by the substitution

$$\cos\theta = \frac{1 - t^2}{1 + t^2}, \quad \sin\theta = \frac{2t}{1 + t^2}, \quad d\theta = \frac{2\,dt}{1 + t^2}. \tag{8.15}$$

The transformation back to the original variables is made using the equation

$$t = \frac{\sin\theta}{1 + \cos\theta}.$$

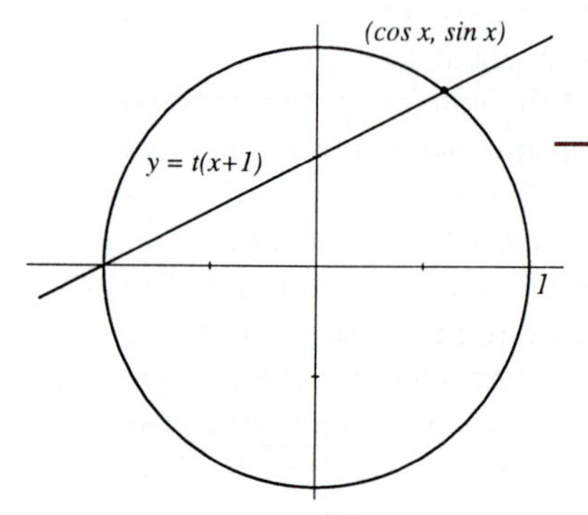

FIGURE 8.1

Proof

The point $P = (\cos\theta, \sin\theta)$ lies on the graph of the equation $x^2 + y^2 = 1$. The line through P and the point $(-1, 0)$ has equation

$$y = t(x + 1), \qquad \text{where} \qquad t = \frac{\sin\theta}{1 + \cos\theta}.$$

(See figure 8.1.) Solve the two equations $x^2 + y^2 = 1$ and $y = t(x + 1)$ simultaneously by eliminating the y term to obtain the equation

$$(t^2 + 1)x^2 + 2tx - (1 - t^2) = (x + 1)\big((t^2 + 1)x - (1 - t^2)\big) = 0.$$

Thus we obtain $x = -1$ or $x = (1 - t^2)/(1 + t^2)$. Use the equation $y = t(x + 1)$ to determine y from these values of x. We find that the line and circle intersect at the points

$$(-1, 0) \qquad \text{and} \qquad \left(\frac{1 - t^2}{1 + t^2}, \frac{2t}{1 + t^2}\right).$$

Because we had started by assuming the points of intersection were $(-1, 0)$ and $(\cos\theta, \sin\theta)$, we have obtained the expressions in Equations (8.15) for $\cos\theta$ and $\sin\theta$. To obtain $d\theta$ in terms of t, we differentiate the equation

$$\sin\theta = \frac{2t}{1 + t^2}$$

with respect to t to get

$$\begin{aligned}\frac{d\theta}{dt} &= \frac{1}{\cos\theta}\frac{2(1 - t^2)}{(1 + t^2)^2} \\ &= \frac{1 + t^2}{1 - t^2}\frac{2(1 - t^2)}{(1 + t^2)^2} \\ &= \frac{2}{1 + t^2}.\end{aligned}$$

The function $R(u, v)$ is a rational function so that

$$\int R\left(\frac{1-t^2}{1+t^2}, \frac{2t}{1+t^2}\right)\frac{2}{1+t^2}\,dt$$

is the integral of a rational function of t that equals the original integral $\int R(\cos\theta, \sin\theta)\,d\theta$. □

This theorem can be applied to evaluate the integral of the secant function.

EXAMPLE 5 Evaluate the integral

$$\int \sec\theta\,d\theta.$$

Solution: From $\sec\theta = 1/\cos\theta$, we see this is a rational function of $\cos\theta$. Make the substitutions given in Equations (8.15) to obtain

$$\begin{aligned}\int \sec\theta\,d\theta &= \int \frac{d\theta}{\cos\theta} = \int \frac{1+t^2}{1-t^2}\frac{2}{1+t^2}\,dt\\ &= \int \frac{2}{1-t^2}\,dt = \int \frac{1}{t+1} + \frac{1}{1-t}\,dt\\ &= \ln|1+t| - \ln|1-t| + c = \ln\left|\frac{1+t}{1-t}\right| + c.\end{aligned}$$

The translation back to the variable θ is made a bit easier if we already know the answer we want to reach. We observe that

$$\tan\theta = \frac{\sin\theta}{\cos\theta} = \frac{2t}{1-t^2}.$$

Then

$$\begin{aligned}\frac{1+t}{1-t} &= \frac{(1+t)^2}{1-t^2} = \frac{1+t^2}{1-t^2} + \frac{2t}{1-t^2}\\ &= \frac{1}{\cos\theta} + \tan\theta = \sec\theta + \tan\theta.\end{aligned}$$

So finally we have the value of the integral in terms of θ:

$$\int \sec\theta\,d\theta = \ln|\sec\theta + \tan\theta| + c.$$

This is one of the basic integrals that is used frequently. ■

Exercises 8.6

Transform each of the integrals into the integral of a rational function of t by making a suitable substitution. No need to carry out the integration.

1. $\displaystyle\int \frac{1}{x+\sqrt{x^2-1}}\,dx$

2. $\displaystyle\int \frac{1}{\sqrt{1-x^2}}\,dx$

3. $\displaystyle\int \frac{x^2}{\sqrt{x^2+4}}\,dx$

4. $\displaystyle\int \frac{x^2}{\sqrt{4-x^2}}\,dx$

5. $\displaystyle\int \frac{x^n}{x+\sqrt{1-x^2}}\,dx$

6. $\displaystyle\int \frac{1+\sqrt{x^2-1}}{x\sqrt{x^2-1}}\,dx$

7. $\displaystyle\int \frac{\cos\theta}{3-2\sin\theta}\,d\theta$

8. $\displaystyle\int \frac{\cos\theta}{2-3\sin\theta}\,d\theta$

9. $\displaystyle\int \frac{\cos\theta}{\cos\theta-2\sin\theta}\,d\theta$

10. $\displaystyle\int \frac{1}{a+b\sin\theta}\,d\theta$

11. $\displaystyle\int_0^{\pi} \frac{\cos\theta}{4-2\cos\theta}\,d\theta$

12. $\displaystyle\int_0^{2\pi} \frac{\sin\theta}{1+\sin\theta+\sin^2\theta}\,d\theta$

13. If $R(x, w)$ is a rational function of x and w and $p(x)$ is any polynomial, show that $R(x, \sqrt{p(x)})$ can be expressed in the form

$$R(x, \sqrt{p(x)}) = u(x) + v(x)\sqrt{p(x)}$$

for some rational functions $u(x)$ and $v(x)$ of x.

14. Show that an integral of the form

$$\int \left(\frac{x+a}{x+b}\right)^{1/n} dx$$

with $a \neq b$ and n a positive integer can be transformed into the integral of a rational function of t by a substitution $t^n = (x+a)/(x+b)$.

15. Use the idea in exercise 14 to evaluate the integral

$$\int \left(\frac{x-1}{x+1}\right)^{1/2} dx.$$

16. Prove e^x is a transcendental function. [Hint: Assume there is some polynomial

$$P(x,y) = a_0(x) + a_1(x)y + \cdots + a_n(x)y^n, \quad a_n(x) \neq 0$$

for which $P(x, e^x) = 0$ for all x. From all such possible polynomials, assume this one has n as small as possible and from all those having y degree n, assume the degree m of $a_n(x)$ is as small as possible. Now show that

$$P(x, e^x) - \frac{1}{n}\frac{d}{dx}P(x, e^x) = 0$$

and the left-hand side can be written as $Q(x, e^x)$ where $Q(x, y)$ is a polynomial having either y degree less than n or y degree equal to n but the coefficient of y^n is a polynomial of degree less than m. This produces a conflict with the minimal choices for P thus showing that no such P exists.]

Convert the following definite integrals into definite integrals of rational functions and then estimate the value, without finding an explicit antiderivative, by using Riemann sums for regular partitions with 2, 4 and 8 subintervals. This computation will require a calculator or computer to do the arithmetic. The location of the minimum and maximum value of the function on each subinterval is made easier by first checking where the function is increasing or decreasing. If you have a computer available, use finer partitions containing 16, 32 and 64 subintervals.

17. $\displaystyle\int_0^1 \frac{x^2}{\sqrt{4-x^2}}\,dx$

18. $\displaystyle\int_1^3 \frac{1}{x^2\sqrt{x^2+16}}\,dx$

8.7 Trigonometric and other substitutions

When simple substitutions were introduced in chapter 5, the motivation came from the chain rule from which we observed

$$\int f(g(x))g'(x)\,dx = \int f(u)\,du,$$

where $u = g(x)$. We now consider a variation of this theme that allows us to transform integrals of certain complicated functions into integrals that have been considered already. Starting from an integral

$$\int f(x)\,dx,$$

we select some function $h(u)$ and set

$$x = h(u) \quad \text{and} \quad dx = h'(u)\,du$$

to transform the given integral into a new integral

$$\int f(x)\,dx = \int f(h(u))h'(u)\,du.$$

Of course, if this is to be useful, the integral on the right should be simpler to evaluate than the integral on the left. We consider a number of examples.

Trigonometric Substitutions

The trigonometric identity

$$(a \sin \theta)^2 + (a \cos \theta)^2 = a^2$$

can be used to evaluate some integrals involving the radical $\sqrt{a^2 - x^2}$. We make a substitution

$$x = a \sin \theta, \qquad \sqrt{a^2 - x^2} = \sqrt{a^2 - a^2 \sin^2 \theta} = |a \cos \theta|.$$

The absolute value signs can be removed if we restrict θ to an interval where $a \cos \theta \geq 0$.

EXAMPLE 1 Evaluate the integral

$$\int \sqrt{4 - x^2}\, dx.$$

Solution: Let $x = 2 \sin \theta$ so that

$$\sqrt{4 - x^2} = 2 \cos \theta, \qquad dx = 2 \cos \theta \, d\theta.$$

Note that θ must be restricted so that $\cos \theta \geq 0$ for this substitution to be valid. Then we have

$$\int \sqrt{4 - x^2}\, dx = \int 2 \cos \theta (2 \cos \theta)\, dt = 4 \int \cos^2 \theta \, d\theta$$
$$= 2 \cos \theta \sin \theta + 2\theta + c.$$

To return to the original variable, we have

$$2 \cos \theta = \sqrt{4 - x^2}, \quad \sin \theta = \frac{x}{2}, \quad \theta = \arcsin\left(\frac{x}{2}\right),$$

and

$$\int \sqrt{4 - x^2}\, dx = \frac{x}{2}\sqrt{4 - x^2} + 2 \arcsin\left(\frac{x}{2}\right) + c.$$ ■

The trigonometric identity

$$a^2 + (a \tan \theta)^2 = (a \sec \theta)^2$$

can be used to evaluate some integrals involving $\sqrt{x^2 \pm a^2}$. If an integral involves $\sqrt{x^2 + a^2}$ we try the substitution

$$x = a \tan \theta, \quad dx = a \sec^2 \theta,$$
$$\sqrt{x^2 + a^2} = \sqrt{a^2 \tan^2 \theta + a^2} = |a \sec \theta|.$$

If θ is suitably restricted, the absolute value signs can be removed.

EXAMPLE 2 Evaluate the integral

$$\int \frac{dx}{\sqrt{x^2 + 9}}.$$

Solution: We make the substitution

$$x = 3 \tan \theta, \quad \sqrt{x^2 + 9} = 3 \sec \theta, \quad dx = 3 \sec^2 \theta \, d\theta$$

and we have

$$\int \frac{dx}{\sqrt{x^2+9}} = \int \frac{3\sec^2\theta\, d\theta}{3\sec\theta} = \int \sec\theta\, d\theta$$
$$= \ln|\sec\theta + \tan\theta| + c = \ln|x + \sqrt{x^2+9}| + c.$$

Note that the denominator 3 has been absorbed in the constant. ■

For integrals involving $\sqrt{x^2 - a^2}$ we try the substitution

$$x = a\sec\theta, \quad dx = a\sec\theta\tan\theta\, d\theta$$
$$\sqrt{x^2 - a^2} = \sqrt{a^2\sec^2\theta - a^2} = |a\tan\theta|.$$

EXAMPLE 3 Evaluate the integral

$$\int \frac{\sqrt{x^2-1}}{x}\, dx.$$

Solution: We make the substitution

$$x = \sec\theta, \quad dx = \sec\theta\tan\theta\, d\theta,$$
$$\sqrt{x^2-1} = \sqrt{\sec^2\theta - 1} = |\tan\theta|.$$

Then we have

$$\int \frac{\sqrt{x^2-1}}{x}\, dx = \int \frac{\tan\theta}{\sec\theta}\sec\theta\tan\theta\, d\theta$$
$$= \int \tan^2\theta\, d\theta = \int (\sec^2\theta - 1)\, d\theta = \tan\theta - \theta + c$$
$$= \sqrt{x^2-1} - \operatorname{arc\,sec} x + c.$$

■

Integrals of Powers of Secant and Tangent

The examples just above show that integrals of algebraic functions can lead to the evaluation of

$$\int \sec^m x \tan^n x\, dx.$$

In some cases, it may be possible to express this integral in terms of powers of $\sin x$ and $\cos x$ and then use the reduction formulas from section 8.3. We can also deal with the integral as it stands. There are two easy cases that we take care of first.

We plan to let $u = \tan x$ and $du = \sec^2 x\, dx$. If we can express everything in terms of the tangent, except for a $\sec^2 x$ term this will work well. Suppose the power of the secant is an even integer, $m = 2k$. We split the secant term as a product and use the identity $\sec^2 x = 1 + \tan^2 x$ to get

$$\int \sec^{2k} x \tan^n x\, dx = \int (\sec^2 x)^{k-1}\tan^n x \sec^2 x\, dx$$
$$= \int (1+\tan^2 x)^{k-1}\tan^n x\sec^2 x\, dx$$
$$= \int (1+u^2)^{k-1} u^n\, du.$$

Multiply out the integrand and perform the integration.

In the other easy case, we plan to let $u = \sec x$ and $du = \sec x \tan x$. After splitting off a factor $\sec x \tan x$, it must be possible to express the remaining factor in terms of the secant function. Suppose the power of the tangent is an odd integer $n = 2k + 1$. We use the identity $\tan^2 x = \sec^2 x - 1$ to get

$$\begin{aligned}\int \sec^m x \tan^{2k+1} x\, dx &= \int \sec^{m-1} x(\tan^2 x)^k \sec x \tan x\, dx\\ &= \int \sec^{m-1} x(\sec^2 x - 1)^k \sec x \tan x\, dx\\ &= \int u^{m-1}(u^2 - 1)^k\, du.\end{aligned}$$

Multiply out the integrand and perform the integration.

The case not covered by these two easy cases has an even power of the tangent times an odd power of the secant. The even power of the tangent is replaced by a power of $(\sec^2 x - 1)$ and, after expanding, it is necessary to integrate powers of the secant function. This is done by a reduction formula:

$$\int \sec x\, dx = \ln|\sec x + \tan x| + c,$$

$$\int \sec^2 x\, dx = \tan x + c,$$

$$\int \sec^m x\, dx = \frac{1}{m-1}\left[\tan x \sec^{m-2} x + (m-2)\int \sec^{m-2} x\, dx\right],$$

for $m \neq 1, 2$. This reduction formula is obtained using integration by parts with $u = \sec^{m-2}$ and $dv = \sec^2 x\, dx$. Although this reduction formula is correct for all m, it can be avoided when m is an even integer by the method described above for dealing with even powers of the secant.

Exercises 8.7

Use some trigonometric substitution to evaluate the integrals.

1. $\int \sqrt{4 - x^2}\, dx$

2. $\int \sqrt{16 + x^2}\, dx$

3. $\int \frac{x^2}{\sqrt{25 - x^2}}\, dx$

4. $\int \frac{x^2}{\sqrt{1 + 4x^2}}\, dx$

5. $\int \frac{1}{\sqrt{(1 - 9x^2)^3}}\, dx$

6. $\int \frac{x^2}{\sqrt{(9 - x^2)^3}}\, dx$

7. $\int \sqrt{a^2 - x^2}\, dx$

8. $\int (a^2 - x^2)^{3/2}\, dx$

9. $\int \frac{dx}{\sqrt{9 - x^2}}$

10. $\int \frac{x^2\, dx}{\sqrt{4 - x^2}}$

Evaluate the trigonometric integrals.

11. $\int \sec^3 \theta\, d\theta$

12. $\int \sec^2 \theta \tan^2 \theta\, d\theta$

13. $\int \sec^3 \theta \tan^3 \theta\, d\theta$

14. $\int \sec^4 \theta \tan^3 \theta\, d\theta$

15. $\int \sec^3 \theta \tan^2 \theta\, d\theta$

16. $\int_0^{\pi/4} \sec^3 \theta \tan \theta\, d\theta$

9 Taylor Polynomials and Sequences

The functions $\sin x, \ln x$, and e^x are quite familiar but how does one compute numerical values like $\sin 0.2$, $\ln 3.7$, and $e^{2.1}$? You might answer that the values are computed by pressing the appropriate keys on a calculator but that is adequate only for a very limited selection of functions. Furthermore, how did the calculator's developer know how to compute the values? We give the answer in this chapter. We study techniques to numerically evaluate or approximate function values and determine the accuracy of estimations. We begin with the idea that polynomial functions are easy to evaluate at given real numbers; we will show that most familiar functions can be closely approximated by polynomials. It is proved, for example, that the sine function can be approximated by polynomials with any desired degree of accuracy over any range of the variable. The polynomial $x - x^3/6$ serves as an excellent approximation to the values of $\sin x$ so long as $|x|$ is small. Because phrases like "excellent approximation" and "$|x|$ is small" are rather vague, we must make these statements precise.

The theory of approximation of functions by Taylor polynomials leads us to the study of power series in chapter 10. As a preparation for this study, we introduce sequences and their limits. This topic contains aspects of interest in its own right but the main reason for introducing sequences is that they are very useful in the study of infinite series to be taken up in chapter 10.

9.1 Taylor polynomials

We define the Taylor polynomials associated with a differentiable function and show how they can be used to approximate function values. Suppose $f(x)$ is n times differentiable at $x = 0$. The nth *Taylor polynomial* centered at 0 for $f(x)$ is

$$P_n(x) = f(0) + f'(0)x + \cdots + \frac{f^{(k)}(0)}{k!}x^k + \cdots + \frac{f^{(n)}(0)}{n!}x^n.$$

The factorial notation is used here (read $m!$ as "m factorial"): $0! = 1$ and $m! = 1 \cdot 2 \cdot 3 \cdots m$ for any positive integer m. We also use the convention that $f^{(0)}(x) = f(x)$. The phrase "centered at 0" is used to emphasize that the derivatives are evaluated at 0 and the polynomial is given in powers of x.

If a is any number and $f(x)$ is n times differentiable at $x = a$, then the Taylor polynomials centered at a are defined later in this section. We concentrate on the Taylor polynomials centered at 0 for now since the theory of the polynomials centered at a can be obtained by performing a substitution $x + a$ for x in the function.

For $k \le n$, the coefficient of x^k in the nth Taylor polynomial is $f^{(k)}(0)/k!$. Note that this is correct even for $k = 0$ because of the convention that $f^{(0)}(x) = f(x)$.

EXAMPLE 1 Find the third, fourth and fifth Taylor polynomials centered at $x = 0$ for $f(x) = e^x$.

Solution: All derivatives of e^x equal e^x, so we have

$$\frac{f^{(k)}(0)}{k!} = \frac{e^0}{k!} = \frac{1}{k!}.$$

In the nth Taylor polynomial for e^x, the coefficient of x^k is $1/k!$. Thus the first few Taylor polynomials are

$$\begin{aligned}
P_0(x) &= 1,\\
P_1(x) &= 1 + x,\\
P_2(x) &= 1 + x + \frac{x^2}{2!},\\
P_3(x) &= 1 + x + \frac{x^2}{2!} + \frac{x^3}{3!},\\
P_4(x) &= 1 + x + \frac{x^2}{2!} + \frac{x^3}{3!} + \frac{x^4}{4!},\\
P_5(x) &= 1 + x + \frac{x^2}{2!} + \frac{x^3}{3!} + \frac{x^4}{4!} + \frac{x^5}{5!}.
\end{aligned}$$

■

EXAMPLE 2 Find the mth Taylor polynomial (for any positive m) centered at $x = 0$ for $f(x) = \sin x$.

Solution: We first compute the derivatives up to the fifth at $x = 0$:

$$\begin{aligned}
f(x) &= \sin x, & f^{(3)}(x) &= -\cos x,\\
f'(x) &= \cos x, & f^{(4)}(x) &= \sin x,\\
f^{(2)}(x) &= -\sin x, & f^{(5)}(x) &= \cos x.
\end{aligned}$$

There is an obvious pattern that enables us to write all the derivatives:

$$\begin{aligned}
f(x) = f^{(4)}(x) = f^{(8)}(x) = \cdots &= f^{(4k)}(x) = \sin x,\\
f'(x) = f^{(5)}(x) = f^{(9)}(x) = \cdots &= f^{(4k+1)}(x) = \cos x,\\
f^{(2)}(x) = f^{(6)}(x) = f^{(10)}(x) = \cdots &= f^{(4k+2)}(x) = -\sin x,\\
f^{(3)}(x) = f^{(7)}(x) = f^{(11)}(x) = \cdots &= f^{(4k+3)}(x) = -\cos x.
\end{aligned}$$

Evaluate the derivatives at $x = 0$. We have for any nonnegative integer k

$$\begin{aligned} f^{(4k)}(0) &= 0, \\ f^{(4k+1)}(0) &= 1, \\ f^{(4k+2)}(0) &= 0, \\ f^{(4k+3)}(0) &= -1. \end{aligned}$$

This information determines the coefficients of all the Taylor polynomials:

$$\begin{aligned} P_3(x) &= x - \frac{x^3}{6}, \\ P_4(x) &= x - \frac{x^3}{6}, \\ P_5(x) &= x - \frac{x^3}{6} + \frac{x^5}{120}, \\ P_6(x) &= x - \frac{x^3}{6} + \frac{x^5}{120}, \\ &\vdots \\ P_{2n-1}(x) &= x - \frac{x^3}{6} + \frac{x^5}{120} + \cdots - \frac{(-1)^n}{(2n-1)!}x^{2n-1}, \\ P_{2n}(x) &= P_{2n-1}(x). \end{aligned}$$

Note that $P_{2n-1}(x) = P_{2n}(x)$ because $f^{(2n)}(0) = 0$; this makes the coefficient of x^{2n} in $P_{2n}(x)$ equal to zero. ■

Here is a list of Taylor polynomials for a few familiar functions. The reader can verify these by computing the derivatives and evaluating at 0.

Function	Taylor Polynomial
$f(x) = e^x$	$P_n(x) = 1 + x + \frac{x^2}{2!} + \frac{x^3}{3!} + \cdots + \frac{x^n}{n!}$,
$f(x) = \sin x$	$P_{2n+1}(x) = x - \frac{x^3}{3!} + \frac{x^5}{5!} + \cdots + (-1)^n \frac{x^{2n+1}}{(2n+1)!}$,
$f(x) = \cos x$	$P_{2n}(x) = 1 - \frac{x^2}{2!} + \frac{x^4}{4!} + \cdots + (-1)^n \frac{x^{2n}}{2n!}$,
$f(x) = \ln(x + 1)$	$P_n(x) = x - \frac{x^2}{2} + \frac{x^3}{3} - \cdots + (-1)^{n+1} \frac{x^n}{n}$.

For the first three functions, there is no restriction on x because each function is n times differentiable at every point. The function $\ln x$ is not defined at 0, so we cannot define the Taylor polynomials of $\ln x$ centered at 0. We consider the function $\ln(x + 1)$ instead. The function $f(x) = \ln(x + 1)$ is not defined if $x \leq -1$ but it is defined and n times differentiable on the interval $(-1, b]$ for any $b > -1$ and any positive n.

Approximations by Taylor Polynomials

We estimate how closely the values of the Taylor polynomials approximate the values of a function $f(x)$. The nth Taylor polynomial of $f(x)$ differs from $f(x)$ by an expression that depends on the size of $f^{(n+1)}(x)$. Before giving the statement of the precise result we need some properties of the Taylor polynomials.

The derivatives of the Taylor polynomial centered at 0 can be easily evaluated at $x = 0$. Directly from the definition of $P_n(x)$ we find $P_n(0) = f(0)$. After differentiation, we find $P_n'(0) = f'(0)$. More generally, we have

$$P_n^{(k)}(0) = f^{(k)}(0), \qquad 0 \le k \le n. \tag{9.1}$$

In other words, the derivatives up to the nth derivative of $f(x)$ and $P_n(x)$ have the same value at $x = 0$.

Thus the graphs of $P_1(x)$ and $f(x)$ pass through the point $(0, f(0))$ and the curves have the same slope at this point. The Taylor polynomial $P_1(x)$ is a linear approximation to $f(x)$, and we might expect the approximation to be fairly good when x is very close to 0. $P_2(x)$ might be a better approximation to $f(x)$ because the graph of $P_2(x)$ passes through $(0, f(0))$ with the same slope as the graph of f and also bends in the same direction as the graph of f because the second derivatives of P_2 and f are equal at 0. We prove that the Taylor polynomials $P_n(x)$ give better and better approximations to f as n increases provided that f satisfies some mild restrictions.

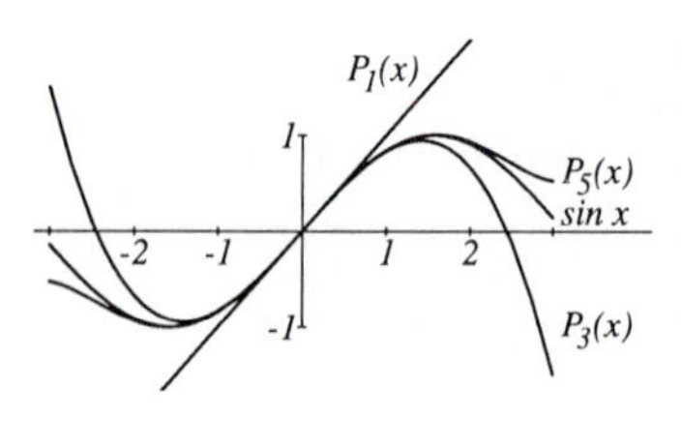

FIGURE 9.1
Taylor polynomials

EXAMPLE 3 Sketch the graphs of $\sin x$ and its Taylor polynomials P_1, P_3 and P_5.

The graphs of the three Taylor polynomials, shown in figure 9.1, are very close to the graph of $\sin x$ for values of x close to 0. As x moves away from 0, the graphs get farther and farther apart. ■

In the course of the next proof we use the following property of the definite integral, which follows directly from properties of lower and upper sums.

THEOREM 9.1

Integral Inequalities

A. If $k(x)$ is continuous and $0 \le k(x)$ for all x satisfying $a \le x \le b$, then

$$0 \le \int_a^b k(t)\,dt.$$

B. If $h(x)$ and $k(x)$ are continuous and $h(x) \le k(x)$ for all x satisfying $a \le x \le b$, then

$$\int_a^b h(t)\,dt \le \int_a^b k(t)\,dt.$$

Proof

A. Let m be the minimum value of $k(x)$ on $[a, b]$. Then $m \ge 0$ and $m(b - a)$ is a lower sum. Thus

$$0 \le m(b - a) \le \int_a^b k(t)\,dt.$$

Statement B follows from statement A because $0 \le k(x) - h(x)$ implies

$$0 \le \int_a^b k(t) - h(t)\,dt = \int_a^b k(t)\,dt - \int_a^b h(t)\,dt. \qquad \square$$

To see how close $P_n(x)$ is to $f(x)$, we consider the difference. Let

$$R(x) = f(x) - P_n(x).$$

Then

$$R(0) = R'(0) = \cdots = R^{(n)}(0) = 0$$

as can be seen from Equation (9.1). We make use of this property in the form

$$\int_0^x R^{(k+1)}(t)\,dt = R^{(k)}(x) - R^{(k)}(0) = R^{(k)}(x), \quad 0 \le k \le n. \tag{9.2}$$

With these preparations out of the way, we get to the main theorem about approximations by Taylor polynomials. It gives an estimate of the difference between the function $f(x)$ and its nth Taylor polynomial $P_n(x)$ on the assumption that the $(n+1)$st derivative of f is continuous. The continuity of $f^{(n+1)}(x)$ ensures that $f^{(n+1)}(x)$ has a minimum and a maximum value on any closed interval.

THEOREM 9.2

Taylor's Theorem: Polynomial Approximations

Assume that the first $n+1$ derivatives of $f(x)$ are continuous on a closed interval I containing 0 and that m is the minimum value of $f^{(n+1)}(x)$ and M is the maximum value of $f^{(n+1)}(x)$ on I. If $0 \le x$ for any n or if $x \le 0$ and n is odd, then

$$\frac{m\,x^{n+1}}{(n+1)!} \le f(x) - P_n(x) \le \frac{M\,x^{n+1}}{(n+1)!}.$$

If $x \le 0$ and n is even, then

$$\frac{-m\,x^{n+1}}{(n+1)!} \le P_n(x) - f(x) \le \frac{-M\,x^{n+1}}{(n+1)!}.$$

Proof

Let $R(x) = f(x) - P_n(x)$. Then

$$\begin{aligned} R'(x) &= f'(x) - P_n'(x), \\ R''(x) &= f''(x) - P_n''(x), \\ &\vdots \\ R^{(n)}(x) &= f^{(n)}(x) - P_n^{(n)}(x), \\ R^{(n+1)}(x) &= f^{(n+1)}(x). \end{aligned}$$

The last equality holds because $P_n(x)$ is a polynomial of degree at most n, so it follows that $P_n^{(n+1)}(x) = 0$.

By definition of m and M we have

$$m \le R^{(n+1)}(x) \le M \tag{9.3}$$

for all x on I. We intend to apply Theorem 9.1 to these two inequalities over the interval $[0, x]$, when $x \ge 0$ and over the interval $[x, 0]$ when $x \le 0$.

Assume $0 \le x$. Then

$$\int_0^x m\,dt \le \int_0^x R^{(n+1)}(t)\,dt \le \int_0^x M\,dt$$

$$mx \le R^{(n)}(x) \le Mx.$$

Because this inequality holds for all nonnegative x on I, we may integrate again:

$$\int_0^x mt\,dt \le \int_0^x R^{(n)}(t)\,dt \le \int_0^x Mt\,dt,$$

$$\frac{mx^2}{2} \le R^{(n-1)}(x) \le \frac{Mx^2}{2}.$$

Integrate yet again:

$$\int_0^x \frac{mt^2}{2}\,dt \le \int_0^x R^{(n-1)}(t)\,dt \le \int_0^x \frac{Mt^2}{2}\,dt,$$

$$\frac{mx^3}{2\cdot 3} \le R^{(n-2)}(x) \le \frac{Mx^3}{2\cdot 3}.$$

The integration can be repeated until we reach the integral of $R'(x)$:

$$\int_0^x \frac{mt^n}{n!}\,dt \le \int_0^x R'(t)\,dt \le \int_0^x \frac{Mt^n}{n!}\,dt,$$

$$\frac{mx^{n+1}}{(n+1)!} \le R(x) = f(x) - P_n(x) \le \frac{Mx^{n+1}}{(n+1)!}.$$

This is the the inequality we wanted to prove for the case $0 \le x$.

For the case $x \le 0$ it is necessary to integrate from x to 0; some minus signs appear that did not occur in the previous case. Observe first that

$$\int_x^0 R^{(k+1)}(t)\,dt = R^{(k)}(0) - R^{(k)}(x) = -R^{(k)}(x),$$

and similarly

$$\int_x^0 \frac{mt^k}{k!}\,dt = \left.\frac{mt^{k+1}}{(k+1)!}\right|_x^0 = -\frac{mx^{k+1}}{(k+1)!}.$$

If we start with inequality (9.3) and integrate $n+1$ times from x to 0 we obtain

$$(-1)^{n+1}\frac{mx^{n+1}}{(n+1)!} \le (-1)^{n+1}R(x)$$

$$= (-1)^{n+1}(f(x) - P_n(x)) \le (-1)^{n+1}\frac{Mx^{n+1}}{(n+1)!}.$$

When n is odd, $n+1$ is even and $(-1)^{n+1} = 1$; the inequality is the same as that reached in the first case. When n is even, then $n+1$ is odd and $(-1)^{n+1} = -1$ and we have the inequality stated in the theorem. □

We change the notation slightly and we write

$$f(x) = P_n(x) + R_n(x).$$

Then $R_n(x)$ is called the nth *remainder term*. The previous theorem gives inequalities bounding the remainder term. Here is an alternate description of the remainder term that does not require separation into cases. This form is somewhat easier to remember and is often applied when making numerical computations.

THEOREM 9.3

Alternative Form of the Remainder

Assume the first $n+1$ derivatives of $f(x)$ are continuous on a closed interval I containing 0. For each x on this interval there is a number z on the interval such that the remainder term $R_n(x) = f(x) - P_n(x)$ has the form

$$R_n(x) = \frac{f^{(n+1)}(z)}{(n+1)!}x^{n+1}. \qquad \textbf{(9.4)}$$

Proof

Let x be a given number. If $x = 0$, then any number z makes Equation (9.4) correct because

$$R_n(0) = f(0) - P_n(0) = 0 \quad \text{and} \quad \frac{f^{(n+1)}(z)}{(n+1)!}0^{n+1} = 0.$$

So assume that $x \neq 0$. Then, by the inequalities in Theorem 9.2, the number

$$\frac{R_n(x)(n+1)!}{x^{n+1}} = \frac{(f(x) - P_n(x))(n+1)!}{x^{n+1}}$$

lies between m, the minimum of $f^{(n+1)}(x)$, and M, the maximum of $f^{(n+1)}(x)$ on the interval I. Because $f^{(n+1)}(x)$ is continuous, every number between the minimum and maximum is a value of $f^{(n+1)}(x)$ by the Intermediate Value Theorem. Hence there is some number z on I that satisfies

$$f^{(n+1)}(z) = \frac{R_n(x)(n+1)!}{x^{n+1}}.$$

It follows that

$$R_n(x) = \frac{f^{(n+1)}(z)}{(n+1)!}x^{n+1},$$

as we wanted to prove. □

Some writers call this theorem the Extended Mean Value Theorem. The designation is reasonable because, in the case $n = 0$, the theorem is just the Mean Value Theorem on the interval $[0, x]$ or $[x, 0]$.

Here is one more useful form for the estimate of the remainder term, the proof of which follows directly from the previous theorem.

THEOREM 9.4

Absolute Value of the Remainder Term

Assume the first $n+1$ derivatives of $f(x)$ are continuous on a closed interval I containing 0. Let W_{n+1} be the maximum value of $|f^{(n+1)}(x)|$ on I. Then

$$|f(x) - P_n(x)| = |R_n(x)| \leq \frac{W_{n+1}|x|^{n+1}}{(n+1)!} \quad \text{for } x \text{ on } I,$$

where $P_n(x)$ is the nth Taylor polynomial of $f(x)$ centered at 0.

Let us apply this approximation method to a few specific cases.

EXAMPLE 4 Compute the numerical value of e with an error at most 10^{-4}.

Solution: If $f(x) = e^x$ then the number we want is $f(1)$. We have already seen that $2 < e < 3$ when we studied the natural logarithm function in chapter 7. We use this crude approximation to get a better one. Because the exponential is a positive and increasing function, we have

$$0 < f^{(n+1)}(x) = e^x < 3$$

for all x on $[0, 1]$.

Then Theorem 9.2 implies

$$0 < e^x - P_n(x) \leq \frac{3x^{n+1}}{(n+1)!}.$$

We evaluate this at $x = 1$ and obtain

$$0 < e - P_n(1) \leq \frac{3}{(n+1)!}.$$

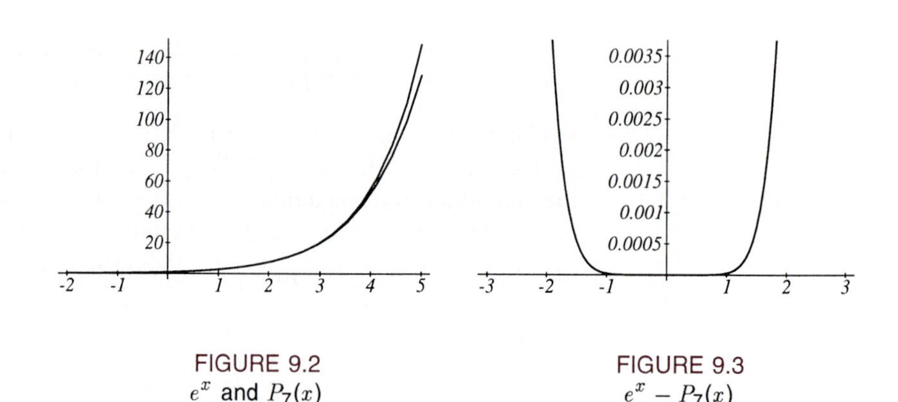

FIGURE 9.2
e^x and $P_7(x)$

FIGURE 9.3
$e^x - P_7(x)$

We must select n so that

$$\frac{3}{(n+1)!} < 10^{-4} \qquad \text{or} \qquad 3 \times 10^4 < (n+1)!.$$

A computation of factorials gives $8! = 40,320 > 4 \times 10^4$ so $n + 1 = 8$ and $n = 7$ is the right choice:

$$\begin{aligned} P_7(1) &= 1 + \frac{1}{1!} + \frac{1}{2!} + \frac{1}{3!} + \frac{1}{4!} + \frac{1}{5!} + \frac{1}{6!} + \frac{1}{7!} \\ &= 1 + 1 + \frac{1}{2} + \frac{1}{6} + \frac{1}{24} + \frac{1}{120} + \frac{1}{720} + \frac{1}{5040} \\ &= \frac{685}{252} \approx 2.71825. \end{aligned}$$

We conclude

$$0 < e - 2.71825 < 10^{-4} \quad \text{and} \quad 2.71825 < e < 2.71835. \qquad \blacksquare$$

EXAMPLE 5 Estimate the value $\sin 1$ with an error at most 10^{-3}.

Solution: For any nonnegative integer n, the minimum and maximum values of the nth derivative of $\sin x$ are $m = -1$ and $M = 1$. Thus

$$\frac{-x^{n+1}}{(n+1)!} \leq \sin x - P_n(x) \leq \frac{x^{n+1}}{(n+1)!}.$$

Substituting $x = 1$ and rewriting this using absolute values signs yields

$$|\sin 1 - P_n(1)| \leq \frac{1}{(n+1)!}.$$

We must select a value of n so that

$$\frac{1}{(n+1)!} \leq 10^{-3} \qquad \text{or} \qquad 10^3 \leq (n+1)!.$$

If we compute the factorial of the first few values of n we find that $7! = 5040$ is the first that is larger than 10^3. So we take $n + 1 = 7$ and $n = 6$. Thus

$$|\sin 1 - P_6(1)| \leq \frac{1}{(6+1)!} = \frac{1}{5040} < 10^{-3}.$$

We have

$$P_6(x) = x - \frac{x^3}{3!} + \frac{x^5}{5!}$$

and

$$P_6(1) = 1 - \frac{1}{6} + \frac{1}{120} = \frac{101}{120} \approx 0.841667.$$

Thus $\sin 1 \approx 101/120$ with an error no greater than $1/5040 \approx 0.0002$. ∎

EXAMPLE 6 Estimate the greatest difference between $\sin x$ and its third Taylor polynomial $P_3(x)$ when x is restricted to lie on the interval $[-0.5, 0.5]$.

Solution: For any nonnegative integer n, the minimum and maximum values of the nth derivative of $\sin x$ are $m = -1$ and $M = 1$. Thus we can use $W = 1$ in the absolute value form of the remainder and conclude

$$|\sin x - P_3(x)| \leq \frac{|x|^4}{4!}.$$

If x is restricted to lie on $[-0.5, 0.5]$, then $|x|^4 < 0.0625$ and so

$$|\sin x - P_3(x)| = \left|\sin x - \left(x - \frac{x^3}{6}\right)\right| \leq 0.0625/4! \approx 0.00260.$$

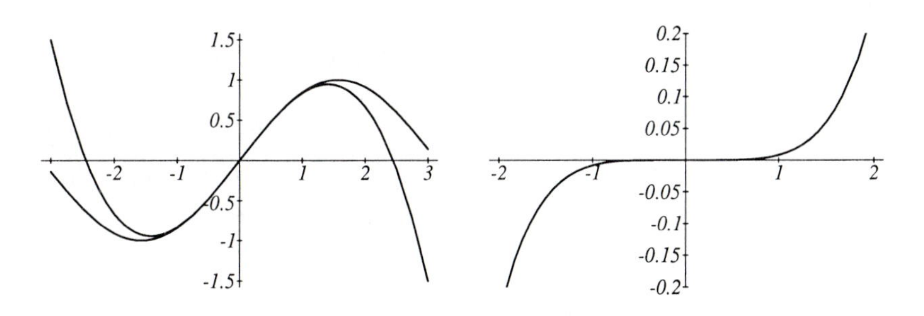

FIGURE 9.4
$\sin x$ and $P_3(x)$

FIGURE 9.5
$\sin x - P_3(x)$

In fact the greatest difference is much less than this. The third and fourth Taylor polynomials for $\sin x$ are equal, so we can use $n = 4$ in the estimate of the remainder. Namely

$$|\sin x - P_3(x)| = |\sin x - P_4(x)| \leq \frac{|x|^5}{5!}.$$

With $|x| \leq 0.5$ we get the estimate

$$|\sin x - P_3(x)| \leq 0.000260.$$

Thus $P_3(x) = x - x^3/6$ is a good approximation to $\sin x$ when x stays close to 0.

Figures 9.4 and 9.5 show the graphs of $\sin x$ and $P_3(x)$ and the difference between the two functions. ■

EXAMPLE 7 Find an interval $I = [-a, a]$ with the property that for every x on I, $\cos x$ differs from its second Taylor polynomial $P_2(x)$ by no more than 10^{-2}.

Solution: We use the alternative form of the remainder term given in Theorem 9.4 and the fact that the third derivative of $\cos x$ is $\sin x$ to get

$$|\cos x - P_2(x)| = \left|\sin z \frac{x^3}{6}\right| < \frac{|x|^3}{6}$$

for some z depending on x. The difference between $\cos x$ and $P_2(x)$ is less than 10^{-2} if

$$\frac{|x|^3}{6} < 10^{-2} \quad \text{or} \quad |x| < (6 \times 10^{-2})^{1/3} \approx 0.39149.$$

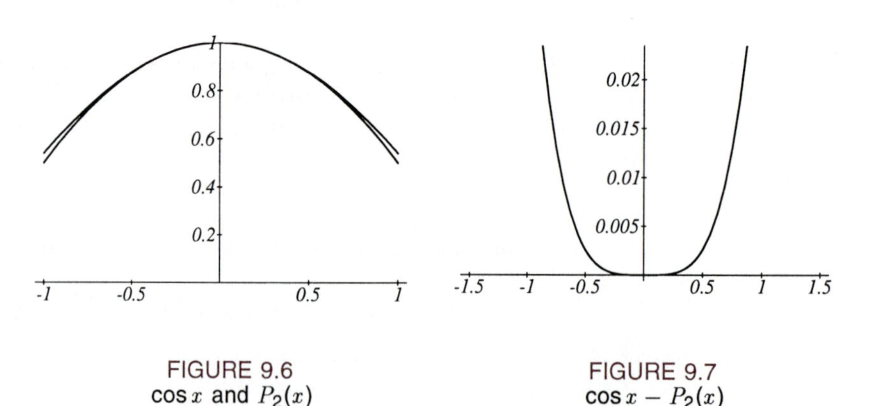

FIGURE 9.6
$\cos x$ and $P_2(x)$

FIGURE 9.7
$\cos x - P_2(x)$

If $a = (6 \times 10^{-2})^{1/3}$ then for all x on $[-a, a]$ we have

$$\left|\cos x - \left(1 - \frac{x^2}{2}\right)\right| < 10^{-2}.$$

The figures show $\cos x$ and $P_2(x)$ and the difference between them. As x grows, the difference increases very fast. ■

Taylor Polynomials centered at *a*

Some of the functions we deal with do not have derivatives at 0 and so the Taylor polynomials centered at 0 are not defined. In many cases the Taylor polynomials centered at some other point are useful. We give the definition and state the estimates that hold for the Taylor polynomials centered at a.

DEFINITION 9.1

Taylor Polynomials

Let $f(x)$ be a function whose nth derivative exists at $x = a$. The *nth Taylor polynomial* centered at a for $f(x)$ is

$$P_n(x) = f(a) + f'(a)(x - a) + \cdots + \frac{f^{(k)}(a)}{k!}(x - a)^k + \cdots + \frac{f^{(n)}(a)}{n!}(x - a)^n.$$

If g is defined by the equation $g(x) = f(x + a)$, then $f(a) = g(0)$ and $f^{(k)}(a) = g^{(k)}(0)$. The results proved about Taylor polynomials of g centered at 0 translate into results about Taylor polynomials of f centered at a. We state the theorems without any further proof.

THEOREM 9.5

Taylor's Theorem: Polynomial Approximations

Assume that the first $n + 1$ derivatives of $f(x)$ are continuous for all x satisfying $|x - a| \le r$ and that m is the minimum value of $f^{(n+1)}(x)$ and M is the maximum value of $f^{(n+1)}(x)$ on $[a - r, a + r]$. Let $P_n(x)$ be the nth Taylor polynomial of $f(x)$ centered at a.

A. If $0 \le x - a \le r$ or if $-r \le x - a \le 0$ and n is odd then

$$\frac{m(x - a)^{n+1}}{(n + 1)!} \le f(x) - P_n(x) \le \frac{M(x - a)^{n+1}}{(n + 1)!}.$$

If $-r \le x - a \le 0$ and n is even, then the correct inequality is

$$\frac{-m(x - a)^{n+1}}{(n + 1)!} \le P_n(x) - f(x) \le \frac{-M(x - a)^{n+1}}{(n + 1)!}.$$

B. For each x on the interval $I = [a - r, a + r]$ there is a number z also on I such that the remainder term $R_n(x) = f(x) - P_n(x)$ has the form

$$R_n(x) = \frac{f^{(n+1)}(z)}{(n + 1)!}(x - a)^{n+1}. \qquad \mathbf{(9.5)}$$

C. Let W be the maximum value of $|f^{(n+1)}(x)|$ on $I = [a - r, a + r]$. Then

$$|f(x) - P_n(x)| \le \frac{W|x - a|^{n+1}}{(n + 1)!} \quad \text{for} \quad |x - a| \le r.$$

EXAMPLE 8 Let $f(x) = x^3$. Show that for any a and $n \ge 3$, the nth Taylor polynomials of f centered at a equal x^3. Write the third Taylor polynomial centered at a.

***Solution*:** The nth derivative of f exists and is continuous for all x so the nth Taylor polynomials are defined with any point as center. For any point x and any center a we have

$$x^3 = f(x) = P_n(x) + \frac{f^{(n+1)}(z)}{(n + 1)!}(x - a)^{n+1}$$

for some number z. The fourth derivative of $f(x) = x^3$ is 0. If $n \ge 3$ then

$f^{(n+1)}(z) = 0$ for every number z and so $x^3 = f(x) = P_n(x)$. For any a, we have

$$f'(a) = 3a^2, \qquad \frac{f''(a)}{2!} = 3a, \qquad \frac{f^{(3)}(a)}{3!} = 1, \qquad f^{(4)}(a) = 0.$$

Thus

$$x^3 = P_3(x) = a^3 + 3a^2(x-a) + 3a(x-a)^2 + (x-a)^3.$$

This equality can also be verified by multiplying out the right-hand side and collecting terms. If x is replaced by $y + a$ we obtain the familiar binomial expansion

$$(y+a)^3 = a^3 + 3a^2y + 3ay^2 + y^3.$$

■

EXAMPLE 9 Compute the fourth Taylor polynomial centered at 1 for $f(x) = \ln x$ and then estimate how close $P_4(1.2)$ is to $\ln 1.2$.

Solution: Let $f(x) = \ln x$. The first five derivatives of f must be evaluated at 1:

$$\begin{aligned}
f(x) &= \ln x & f(1) &= 0,\\
f'(x) &= \frac{1}{x} & f'(1) &= 1,\\
f''(x) &= -\frac{1}{x^2} & \frac{f''(1)}{2!} &= -\frac{1}{2},\\
f^{(3)}(x) &= \frac{2}{x^3} & \frac{f^{(3)}(1)}{3!} &= \frac{1}{3},\\
f^{(4)}(x) &= -\frac{2\cdot 3}{x^4} & \frac{f^{(4)}(1)}{4!} &= -\frac{1}{4},\\
f^{(5)}(x) &= \frac{2\cdot 3\cdot 4}{x^5} & \frac{f^{(5)}(1)}{5!} &= \frac{1}{5}.
\end{aligned}$$

From the definition of P_4 we obtain

$$P_4(x) = (x-1) - \frac{1}{2}(x-1)^2 + \frac{1}{3}(x-1)^3 - \frac{1}{4}(x-1)^4.$$

To estimate how close this polynomial is to $\ln x$ we apply part A of Theorem 9.5. First we must specify an interval. We are interested in $x = 1.2$, so we select the interval $[0.8, 1.2]$ having 1 at its center. (There is considerable flexibility in selecting the interval; other choices of the interval lead to slightly different estimates.) The maximum value of $f^{(5)}(x)$ on this interval is at the left endpoint and the minimum is at the right endpoint because it is a decreasing function. For $1 \le x \le 1.2$ we have

$$\frac{f^{(5)}(1.2)(x-1)^5}{5!} \le \ln x - P_4(x) \le \frac{f^{(5)}(0.8)(x-1)^5}{5!}.$$

Evaluate at $x = 1.2$ to get

$$\frac{f^{(5)}(1.2)(0.2)^5}{5!} \le \ln 1.2 - P_4(1.2) \le \frac{f^{(5)}(0.8)(0.2)^5}{5!},$$

$$0.000026 \le \ln 1.2 - P_4(1.2) \le 0.000195.$$

The inequality tells how far apart the values $\ln 1.2$ and $P_4(1.2)$ can be. We evaluate $P_4(1.2)$ to get

$$\begin{aligned} P_4(1.2) &= 0.2 - \frac{1}{2}(0.2)^2 + \frac{1}{3}(0.2)^3 - \frac{1}{4}(0.2)^4 \\ &= 0.182267. \end{aligned}$$

This can be used to give an estimate of $\ln 1.2$ even better than the value of $P_4(1.2)$. We have

$$\begin{aligned} 0.000026 + P_4(1.2) \leq \ln 1.2 \leq 0.000195 + P_4(1.2), \\ 0.182293 \leq \ln 1.2 \leq 0.182462. \end{aligned}$$

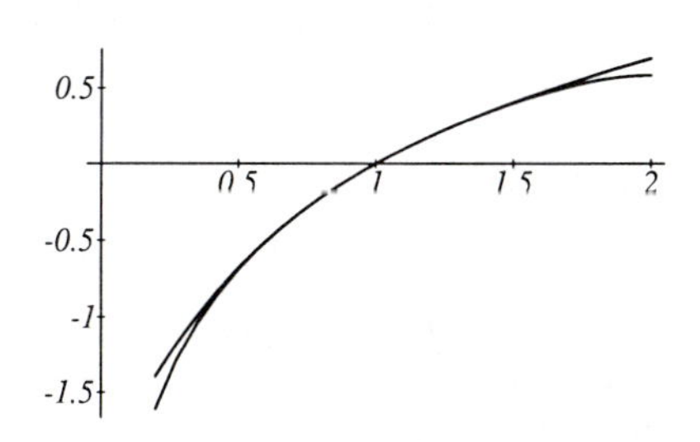

FIGURE 9.8
$\ln x$ and $P_4(x)$

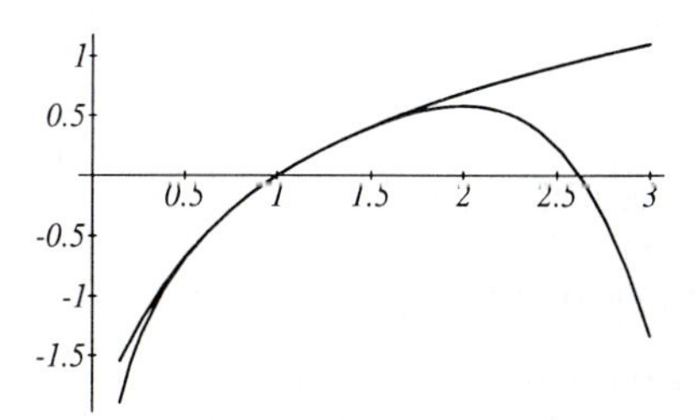

FIGURE 9.9
$\ln x$ and $P_4(x)$

Figure 9.8 showing $\ln x$ and $P_4(x)$ taken over a short interval centered at 1 might lead you to think the two curves are very close. When viewed over a slightly longer interval as in figure 9.9, it becomes apparent that the polynomial moves quite far away from $\ln x$. ■

Exercises 9.1

Find the Taylor polynomial for the given function $f(x)$ and center a.

1. $P_5(x)$ for $f(x) = \dfrac{1}{2+x}$, $a = 0$

2. $P_5(x)$ for $f(x) = \dfrac{1}{2+x}$, $a = 1$

3. $P_7(x)$ for $f(x) = \cos 2x$, $a = 0$

4. $P_7(x)$ for $f(x) = \sin\left(x + \dfrac{\pi}{3}\right)$, $a = 0$

5. $P_7(x)$ for $f(x) = \sin x$, $a = \frac{\pi}{3}$

6. $P_7(x)$ for $f(x) = (x-1)^3(x+1)^4$, $a = 0$

7. $P_7(x)$ for $f(x) = (x-1)^3(x+1)^4$, $a = 1$

8. $P_7(x)$ for $f(x) = (x-1)^3(x+1)^4$, $a = -1$

9. $P_4(x)$ for $f(x) = \sqrt{4+x}$, $a = 0$

10. $P_4(x)$ for $f(x) = (16+x)^{1/4}$, $a = 0$

11. $P_4(x)$ for $f(x) = \ln(1 + \sin x)$, $a = 0$

12. $P_5(x)$ for $f(x) = \ln(1 + e^x)$, $a = 0$

13. $P_3(x)$ for $f(x) = \arcsin x$, $a = 0$

14. $P_3(x)$ for $f(x) = \arctan x$, $a = 0$

15. $P_{10}(x)$ for $f(x) = \dfrac{x}{x^2 - 1}$, $a = 0$ [Hint: Partial fractions will help.]

Use Taylor polynomials for the indicated function f and the estimates given in this section to approximate the number N with an error no greater than E.

16. $N = e^{-1}$, $f(x) = e^{-x}$, $E = 10^{-3}$

17. $N = e^2$, $f(x) = e^x$, $E = 10^{-2}$

18. $N = \cos 0.2$, $f(x) = \cos x$, $E = 10^{-4}$

19. $N = \sin(22/7)$, $f(x) = \sin x$, $E = 10^{-3}$

20. $N = \ln 2$, $f(x) = \ln(1+x)$, $E = 10^{-4}$

21. $N = \ln 0.95$, $f(x) = \ln(1+x)$, $E = 10^{-3}$

22. $N = \sqrt{5}$, $f(x) = \sqrt{4+x}$, $E = 10^{-3}$

23. $N = (17)^{1/4}$, $f(x) = (16 + x)^{1/4}$, $E = 10^{-2}$

24. Estimate the largest difference between e^x and its fifth Taylor polynomial centered at 0 as x varies over the interval $[-1, 4]$.

25. Estimate the largest difference between e^{-x} and its third Taylor polynomial centered at 0 as x varies over the interval $[0, 4]$.

26. Estimate the largest difference between $\ln(x + 1)$ and its fifth Taylor polynomial centered at 0 as x varies over the interval $[0, 3]$.

27. Estimate the largest difference between $\cos 2x$ and its fourth Taylor polynomial centered at 0 as x varies over the interval $[0, \pi/4]$.

28. If $f(x) = x^3 + x + 1$ and $P_1(x)$ is the first Taylor polynomial of f centered at 0, find the number z such that $f(x) - P_1(x) = f''(z)x^2/2$ and verify that z lies between 0 and x.

29. If $f(x)$ is n times differentiable on $[-r, r]$ and $f(x)$ is an odd function $[f(-x) = -f(x)]$, then Taylor polynomial $P_n(x)$ centered at 0 is an odd function. [Hint: One solution uses two facts. If $g(x)$ is an odd function then $g(0) = 0$ because $g(-0) = -g(0)$ and $g(-0) = g(0)$. Also if $g(x)$ is differentiable and odd, then $g'(x)$ is even; if $g(x)$ is differentiable and even, then $g'(x)$ is odd.]

30. If $f(x)$ is n times differentiable on $[-r, r]$ and $f(x)$ is an even function $(f(-x) = f(x))$, then Taylor polynomial $P_n(x)$ centered at 0 is an even function. [See the hints in exercise 29.]

31. If $P_n(x)$ is the nth Taylor polynomial centered at 0 for $f(x)$, then $P_n(-x)$ is the nth Taylor polynomial centered at 0 for $f(-x)$.

32. Using the method of Taylor polynomials, derive the binomial theorem

$$(x+a)^m = x^m + max^{m-1} + \cdots + \frac{m!}{k!(m-k)!}a^{m-k}x^k + \cdots + a^m$$

for any positive integer m. [Hint: Let $f(x) = x^m$. Show

$$f^{(k)}(x) = m(m-1)\cdots(m-k+1)x^{m-k} = \frac{m!}{(m-k)!}x^{m-k}$$

and then compute the mth Taylor polynomial of f centered at a. Finally replace x by $x + a$.]

9.2 Sequences

A *sequence* is a function whose domain is the set of all integers greater than some fixed integer. The most usual domain of a sequence is the set of nonnegative integers or the set of positive integers. If a sequence is mentioned without specifying its domain, it can be assumed that the domain is the largest set of nonnegative integers for which the sequence is defined. It is common to use a subscript notation a_n for the value of the sequence a at the integer n rather than the function notation $a(n)$. Some examples of sequences are

$$\begin{aligned} a_n &= \frac{1}{n}, \quad n \geq 1, \\ b_n &= \frac{3+n}{n!}, \quad n \geq 0, \\ c_k &= (-1)^k \frac{k+1}{k-3}, \quad k \geq 4, \\ d_i &= 3i^2, \quad i \geq -7, \\ u_n &= \frac{f^{(n)}(0)}{n!}, \quad n \geq 0. \end{aligned}$$

The individual values $a_1, a_2, a_3, \ldots$ are called the *terms* of the sequence a_n. The dots "..." indicate that the listing of terms is to continue in the same manner as suggested by the first few values. Sometimes a sequence is described by giving the first few terms and implying that the remaining terms are determined by an obvious rule. For example

$$a_1 = 1^2, \quad a_2 = 2^2, \quad a_3 = 3^2, \quad a_4 = 4^2, \ldots$$

suggests that the sequence is $a_n = n^2$ for $n \geq 1$. The brief notation must be used with care and it is usually better to indicate the sequence precisely if there is any possibility of ambiguity. When we write $a_n = n^2$, we have specified *the general term* of the sequence, i.e., rule or formula for computing every term of the sequence.

Bounded Sequences

A sequence is a function so the basic properties of functions can be directly applied to sequences. For example, a sequence a_n is *bounded* if there is some number B such that

$$|a_n| \leq B \quad \text{for all} \quad n.$$

Bounds can also be given in the alternative form

$$A \leq a_n \leq C$$

for some constants A and C.

EXAMPLE 1 Show that each of the sequences

$$a_n = \frac{10}{n}, \qquad b_k = (-1)^k \frac{k+1}{k+4}, \qquad c_m = \frac{2^m}{m!}$$

is bounded.

***Solution*:** For $n \geq 1$ we have $10/n \leq 10$; thus $|a_n| \leq 10$.

For any $k \geq 0$ we have

$$k + 1 < k + 4 \quad \text{and so} \quad \frac{k+1}{k+4} < 1.$$

Thus

$$|b_k| = \left|(-1)^k \frac{k+1}{k+4}\right| = \frac{k+1}{k+4} < 1$$

and b_k is bounded.

The first few values of the sequence c_m are

$$c_1 = 2, \quad c_2 = 2, \quad c_3 = \frac{4}{3}, \quad c_4 = \frac{2}{3}$$

and all values after these are smaller because of the following reasoning. For $m \geq 2$ we have $2/(m+1) < 1$ and so

$$c_{m+1} = \frac{2^{m+1}}{(m+1)!} = \frac{2^m}{m!}\frac{2}{m+1} < \frac{2^m}{m!} = c_m.$$

Thus we have

$$c_1 = c_2 > c_3 > c_4 > c_5 > \cdots.$$

It follows that $0 < c_m \leq 2$ for every m and the sequence c_m is bounded. ■

Increasing and Decreasing Sequences

When we study limits of sequences, the increasing and decreasing sequences play an important role.

DEFINITION 9.2

Increasing and Decreasing Sequences

Let a_n be a sequence. We say

i. a_n is increasing if $a_k < a_{k+1}$ for all k,
ii. a_n is nondecreasing if $a_k \leq a_{k+1}$ for all k,
iii. a_n is decreasing if $a_k > a_{k+1}$ for all k,
iv. a_n is nonincreasing if $a_k \geq a_{k+1}$ for all k.

The sequence a_n is called *monotone* if it has any of these properties.

Here are some examples of monotone sequences.

EXAMPLE 2 The sequence $a_n = 2n$ is increasing because

$$a_n = 2n < 2(n+1) = a_{n+1}$$

for every n. The sequence $b_k = 1 - k^2$ is decreasing because

$$b_k = 1 - k^2 > 1 - (k+1)^2 = b_{k+1}$$

for every k. ■

EXAMPLE 3 Determine if a_n is a bounded or monotone sequence when

$$\text{i.}\quad a_n = \frac{n}{n+1} \qquad \text{ii.}\quad a_n = \frac{3^n + 4^n}{5^n} \qquad \text{iii.}\quad a_n = \frac{2^n}{n!}.$$

Solution: i. The first few terms of the sequence are $1/2, 2/3, 3/4, \ldots$ so we are inclined to try to prove the sequence is increasing. We have $a_n < a_{n+1}$ if and only if $a_n/a_{n+1} < 1$. We compute this ratio and find

$$\frac{a_n}{a_{n+1}} = \left(\frac{n}{n+1}\right) \div \left(\frac{n+1}{n+2}\right) = \frac{n(n+2)}{(n+1)^2}$$

$$= \frac{n^2 + 2n}{n^2 + 2n + 1} < 1.$$

Hence $a_n < a_{n+1}$ for all n and the sequence is monotone increasing. The sequence is also bounded because

$$a_1 = \frac{1}{2} < a_n < 1$$

for all n.

ii. The idea used in part i, namely comparing ratios of consecutive terms, is a bit messy for this sequence because the quotient remains somewhat complicated. However we can write a_n as the sum of two sequences that are visibly decreasing. Namely,

$$a_n = \frac{3^n + 4^n}{5^n} = \left(\frac{3}{5}\right)^n + \left(\frac{4}{5}\right)^n.$$

We have

$$\left(\frac{3}{5}\right)^{n+1} = \left(\frac{3}{5}\right)^n \frac{3}{5} < \left(\frac{3}{5}\right)^n,$$

$$\left(\frac{4}{5}\right)^{n+1} = \left(\frac{4}{5}\right)^n \frac{4}{5} < \left(\frac{4}{5}\right)^n,$$

and so

$$a_{n+1} = \left(\frac{3}{5}\right)^{n+1} + \left(\frac{4}{5}\right)^{n+1} < \left(\frac{3}{5}\right)^n + \left(\frac{4}{5}\right)^n = a_n.$$

Thus a_n is monotone decreasing. It is also bounded because all the terms are positive yet less than the first term;

$$0 < a_n < a_1.$$

iii. This sequence appeared in Example 1 of section 9.1. The first few terms are

$$c_1 = 2 = c_2 > c_3 = \frac{4}{3} > c_4 = \frac{2}{3} > \ldots$$

and for $n \geq 2$ we have

$$c_{n+1} = c_n \frac{2}{n+1} < c_n.$$

The sequence is monotone and nonincreasing. It is bounded because $0 < c_n < c_1 = 2$ for all n. ■

Exercises 9.2

For each sequence determine if it is monotone and/or bounded.

1. $a_n = 3^n$

2. $a_n = 3^{-n}$

3. $a_n = (-3)^{-n}$

4. $(-3)^n$

5. $b_k = 1 - k^{-1}$

6. $b_k = k^{-1}(1 - k^{-1})$

7. $b_k = k - \frac{1}{k}$

8. $c_j = \frac{3^j}{j!}$

9. $c_j = \frac{1000^j}{j!}$

10. $c_j = \frac{j^j}{j!}$

11. $d_i = \frac{1 \cdot 3 \cdot 5 \cdots (2i-1)}{(2i)!}$

12. $d_i = \frac{(i!)^2}{(2i)!}$

13. $f_n = \frac{n-1}{n+1}$

14. $g_n = \frac{n(n+1)}{(n+2)(n+3)}$

15. $a_n = \frac{n^2 - 2n + 1}{n^2 + 2n + 1}$

16. $a_n = \frac{(n+1)(n+4)}{(n+2)(n+3)}$

17. If c_m is a bounded sequence show that $a_n = (-1)^n c_n$ is also a bounded sequence.

18. If c_m is a bounded and monotone sequence with all terms positive show that $a_n = c_n^2$ is also a bounded and monotone sequence.

19. Give an example of a bounded sequence a_n such that $b_k = a_k^k$ is not a bounded sequence.

20. The *arithmetic mean* of n numbers $a_1, \ldots, a_n$ is

$$m_n = \frac{a_1 + \cdots + a_n}{n}.$$

Prove that m_k is an increasing sequence if a_k is an increasing sequence.

21. The *geometric mean* of n positive numbers $a_1, \ldots, a_n$ is

$$g_n = (a_1 a_2 \cdots a_n)^{1/n}.$$

Prove that g_n is a decreasing sequence if a_n is a decreasing sequence.

22. Let $f(x)$ be a function defined on $[0, \infty)$ and let a_n be the sequence defined by $a_n = f(n)$ for $n \geq 0$. Show that the following statement is true: If $f(x)$ is a bounded function on $[0, \infty)$, then a_n is a bounded sequence. Show by example that the following statement is false: If a_n is a bounded sequence, then $f(x)$ is a bounded function on $[0, \infty)$.

9.3 Limits of sequences

The definition of $\lim_{n\to\infty} a_n$ is the same for the sequence a_n as for any function but we repeat it here inasmuch as the context is a bit different from what was considered in earlier chapters.

DEFINITION 9.3

Limit of a Sequence

If a_n is a sequence, then

$$\lim_{n\to\infty} a_n = L$$

means that for any given positive number ϵ there is some number N such that

$$|a_n - L| < \epsilon$$

whenever $n > N$.

If no such L exists for the sequence a_n, then $\lim_{n\to\infty} a_n$ does not exist.

The properties of limits of functions carry over to limits of sequences. Some of the relevant theorems are summarized here. We abbreviate $\lim_{n\to\infty}$ by lim since no other limits are used here.

THEOREM 9.6

Properties of Limits of Sequences

If $\lim a_n$ and $\lim b_n$ exist, then

i. $\lim(a_n + b_n) = \lim a_n + \lim b_n$,

ii. $\lim a_n b_n = (\lim a_n)(\lim b_n)$,

iii. $\lim a_n/b_n = (\lim a_n)/(\lim b_n)$ provided $\lim b_n \neq 0$.

The Squeeze Theorem for sequences is used quite often.

THEOREM 9.7

Squeeze Theorem

If a_n, b_n, c_n are sequences such that

$$a_n \le b_n \le c_n \quad \text{for all} \quad n$$

and if $\lim a_n = \lim c_n = L$, then $\lim b_n = L$.

We give a few examples of limits of sequences.

EXAMPLE 1 Evaluate the limit $\lim_{n\to\infty} a_n$ for the cases

$$\text{i.} \quad a_n = \frac{1}{n^p}, \quad p > 0;$$

$$\text{ii.} \quad b_n = (-1)^n \left(\frac{1}{2}\right)^n;$$

$$\text{iii.} \quad c_n = \frac{\sin(n\pi/2)}{n}.$$

Solution: i. Because $p > 0$, $a_n = 1/n^p$ is a decreasing function of n. We can force a_n to be within 10^{-r} of 0 by selecting $n > 10^{r/p}$. This is good enough to satisfy the definition of the limit, and so

$$\lim_{n\to\infty} \frac{1}{n^p} = 0.$$

ii. The limit of the sequence b_n is 0 because, for any given $\epsilon > 0$, we have

$$|b_n - 0| = \left|(-1)^n \frac{1}{2^n}\right| = \frac{1}{2^n} < \epsilon$$

provided we take n large enough to satisfy

$$\frac{1}{\epsilon} < 2^n \quad \text{or} \quad n > \frac{\ln \epsilon^{-1}}{\ln 2}.$$

iii. For an integer m, the possible values of $\sin(m\pi/2)$ are 0,1 and -1. We replace the sine term with its extreme values to obtain

$$\frac{-1}{n} \leq c_n = \frac{\sin(n\pi/2)}{n} \leq \frac{1}{n}.$$

Because $\lim(-1/n) = \lim(1/n) = 0$, the Squeeze Theorem implies $\lim c_n = 0$. ■

EXAMPLE 2 Show that the limit $\lim t_n$ does not exist if

$$\text{i. } t_n = (-1)^n \frac{n}{n+1} \quad \text{or ii. } t_n = \frac{2^n}{n^2}.$$

Solution: i. We show the limit does not exist by proving that there cannot be one number L with the property that all terms t_k of the sequence are within 0.1 of L for every k larger than some fixed number.

For $n \geq 10$, $|t_n|$ lies between 0.9 and 1. For even n greater than 10, t_n is positive and lies between 0.9 and 1. For odd n greater than 10, t_n is negative and lies between -1 and -0.9. Thus half the terms of the sequence are near +1 and half are near -1. Thus it is impossible for all the terms to be near any single value. Note that this sequence is bounded but still has no limit.

ii. We show that t_n increases without bound so in fact $\lim t_n = \infty$. There are several methods to show this; we select a method using the natural logarithm function and the idea that $\lim(\ln t_n) = \infty$ implies $\lim t_n = \infty$. We have

$$\begin{aligned}\ln t_n &= \ln\left(\frac{2^n}{n^2}\right) = n \ln 2 - 2 \ln n \\ &= n\left(\ln 2 - \frac{2 \ln n}{n}\right).\end{aligned}$$

The term $\ln n/n$ approaches 0 as $n \to \infty$. This can be seen easily by applying L'Hôpital's rule to the function $\ln x/x$. It follows that $\ln t_n$ is close to $n \ln 2$ for large n and, because $\ln 2 > 0$, $\ln t_n$ approaches ∞. Thus t_n is an unbounded sequence approaching positive infinity. ■

Next we prove a very general theorem about the existence of the limit of bounded monotone sequences. The proof makes use of the Completeness Property of the real numbers stated in section 5.2.

THEOREM 9.8

Bounded Monotone Sequences Have a Limit

Let a_n be a bounded and monotone sequence. Then the limit $\lim a_n$ exists. If a_n is increasing or nondecreasing, the limit is the smallest number L that satisfies $a_n \leq L$ for all n. If a_n is decreasing or nonincreasing, the limit is the largest number L that satisfies $L \le a_n$ for all n.

Proof

We give the proof for the case that a_n is increasing. The proofs in the remaining cases follow this one very closely and are left as an exercise.

Assume that a_n is increasing and bounded. There is some number B satisfying $a_n \le B$ for all n. The Completeness Property of the real numbers states that there is a smallest number, call it L, that satisfies

$$a_n \le L \quad \text{for all} \quad n.$$

We show

$$\lim_{n\to\infty} a_n = L$$

by verifying the conditions in the definition of limit. Let ϵ be a positive number. Then $L - \epsilon < L$, and so $L - \epsilon$ cannot be greater than or equal to every term of the sequence a_n (L was the smallest number larger than or equal to every a_n). There must be at least one number N such that

$$L - \epsilon < a_N.$$

The increasing property of the sequence and the definition of L imply

$$L - \epsilon < a_N < a_{N+1} < \cdots < a_n < \cdots \le L$$

for every $n \ge N$. Thus

$$L - \epsilon < a_n \le L < L + \epsilon \quad \text{or} \quad |a_n - L| < \epsilon$$

for every $n > N$. This is exactly what is required by the definition of the limit so we have $\lim a_n = L$. □

This is an important theorem that is used many times in this and the next chapter. It is one of the only theorems that asserts the existence of a limit based on properties of the given sequence alone. We will see other tests that assert the existence of the limit of a sequence based on comparison with a second sequence.

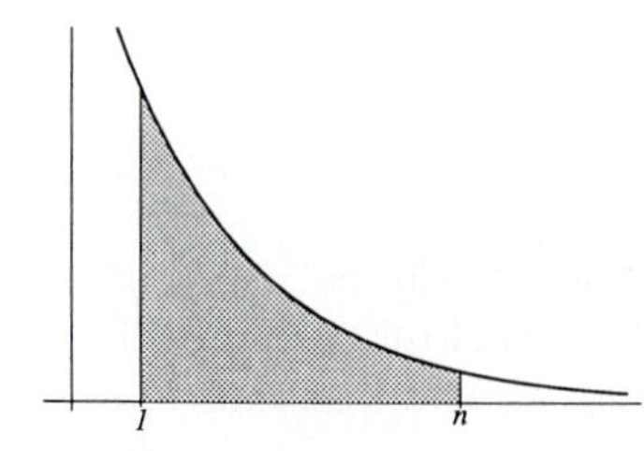

FIGURE 9.10
$\int_1^n e^{-x}/x\,dx$

EXAMPLE 3 The sequence

$$a_n = \int_1^n \frac{e^{-x}}{x}\,dx$$

is monotone and bounded and so $\lim a_n$ exists.

Solution: We have $a_1 = 0$ and for $n > 1$, a_n is the area above the interval $[1, n]$ under the graph of the positive function $f(x) = e^{-x}/x$. Because a_{n+1} is the area of a larger region it follows that $a_n < a_{n+1}$. Thus the sequence is monotone.

To show the sequence is bounded we first note that for $x \ge 1$ we have

$$\frac{e^{-x}}{x} < e^{-x}$$

and so

$$a_n = \int_1^n \frac{e^{-x}}{x}\,dx < \int_1^n e^{-x}\,dx = e - e^{-n} < e.$$

This shows $0 < a_n < e$ so the sequence is bounded and monotone and hence has a limit L. Of course this computation does not tell us what the limit is but it does show $L \le e$. ■

Next we consider a sequence defined by *recursion* instead of by a formula. Recursion means that the first term (or first few terms) is specified and the nth term of the sequence is given as a function of $a_1, a_2, \ldots, a_{n-1}$.

EXAMPLE 4 Let the sequence a_n be recursively defined by the relations

$$a_1 = 1, \qquad a_{n+1} = \sqrt{a_n + 1} \quad \text{for} \quad n \ge 1.$$

Show that a_n is a bounded monotone sequence and find its limit.

Solution: The first few terms after a_1 are

$$a_2 = \sqrt{2} \approx 1.41421,$$

$$a_3 = \sqrt{1 + \sqrt{2}} \approx 1.55377,$$

$$a_4 = \sqrt{1 + \sqrt{1 + \sqrt{2}}} \approx 1.59805,$$

$$a_5 = \sqrt{1 + \sqrt{1 + \sqrt{1 + \sqrt{2}}}} \approx 1.61185.$$

Let us first show that the sequence is increasing. From the computations we see that

$$a_1 < a_2 < a_3 < a_4 < a_5.$$

To make progress we observe

$$a_5^2 = 1 + a_4 < 1 + a_5 = a_6^2.$$

Because all terms are positive, the inequality $a_5^2 < a_6^2$ implies $a_5 < a_6$. We have extended the string of inequalities to include one more term. More generally, if we have shown, for some k, that

$$a_{k-1} < a_k$$

then we have

$$a_k^2 = 1 + a_{k-1} < 1 + a_k = a_{k+1}^2.$$

It follows that $a_k < a_{k+1}$. This inductive reasoning proves the sequence is increasing.

To show the sequence is bounded, we pull a rabbit out of the hat. Let u be the positive root of the equation $x^2 - x - 1 = 0$ so that

$$u^2 = u + 1.$$

By the quadratic formula we have

$$u = \frac{1 + \sqrt{5}}{2} \approx 1.61803$$

and so we know

$$0 < a_n < u \tag{9.6}$$

for $n = 1, 2, 3, 4, 5$ by the computations above. We show inequality (9.6) is true for every n. Suppose k is some index for which $0 < a_k < u$. Then

$$a_{k+1}^2 = 1 + a_k < 1 + u = u^2$$

and it follows that $0 < a_{k+1} < u$. Thus the sequence is bounded, monotone increasing and has a limit. Let $L = \lim a_n$. At this point we know only that $L \leq u$. The exact value of L can be found using the defining relation for the sequence and the observation that

$$\lim_{n\to\infty} a_n = \lim_{n\to\infty} a_{n+1} = L.$$

We have

$$L = \lim_{n\to\infty} a_{n+1} = \lim_{n\to\infty} \sqrt{1+a_n} = \sqrt{1+L}. \tag{9.7}$$

It follows that $L^2 = L + 1$ and so L is one of the two roots of the quadratic equation $x^2 - x - 1 = 0$. Only one solution of this equation is positive, namely $u = (1+\sqrt{5})/2$, so it follows that $L = u$.

The reader might wonder why we bother with the computation showing a_n is bounded and monotone. Why not start with Equation (9.7) to get the equation $L^2 = 1 + L$ and determine L from it? The answer is that the computation in Equation (9.7) is not valid unless we know that the limit of the sequence exists. The limit is known to exist because the sequence was shown to be bounded and monotone. ∎

Exercises 9.3

Find the limit of the sequences that converge.

1. $\dfrac{2n+1}{1-3n}$

2. $(-1)^n \dfrac{2}{n+5}$

3. $3^{1/n}$

4. $\left(\dfrac{1}{n}\right)^{2n}$

5. $\dfrac{(k^3-1)(4k^2+1)}{8k^2(3k^2+4)}$

6. $\dfrac{(11k^2-4)(k^2+4)}{9k^2(k^2+13)}$

7. $\dfrac{\ln(n+1)}{n}$

8. $\dfrac{2^{k-1}}{5^{k+1}}$

9. $\dfrac{k(k-1)}{k^2+1}$

10. $\dfrac{k(k^2-1)}{k^2+2k}$

11. $\ln\left(\dfrac{k}{k+1}\right)$

12. $\ln\left(\dfrac{k+1}{k}\right)$

13. $\left(1+\dfrac{1}{n}\right)^n$

14. $\left(1+\dfrac{3}{k}\right)^k$

15. $\dfrac{\sin(n\pi/2)}{n}$

16. $\dfrac{\cos n\pi}{n^2}$

17. $\dfrac{n!}{n^{20}}$

18. $\dfrac{(n!)^2}{3^{2n}}$

19. Let a_n be recursively defined by $a_1 = p$ and $a_{n+1} = \sqrt{1+a_n}$ for $n \geq 1$. In the example worked in this section, p was equal to 1 and the sequence was shown to be bounded and increasing with limit $L = (1+\sqrt{5})/2$. Show that precisely the same conclusion holds when $0 < p < (1+\sqrt{5})/2$.

20. Let a_n be recursively defined by $a_1 = p$ and $a_{n+1} = \sqrt{1+a_n}$ for $n \geq 1$. If $p > (1+\sqrt{5})/2$ show the sequence is bounded and decreasing and determine its limit.

21. Let a_n be recursively defined by $a_1 = 2$ and $a_{n+1} = \sqrt{3+a_n}$. Write the first four terms of the sequence, show that it is bounded and monotone, and then determine the limit of the sequence.

22. Let $f(x)$ be a continuous function on $I = [0, c]$ for some $c > 0$ and suppose that f satisfies the condition $x \leq f(x) \leq c$ for every x on I. Use f and some positive constant $b < c$ to define a sequence recursively as $a_1 = b$, $a_2 = f(a_1) = f(b)$, $a_3 = f(a_2) = f(f(b))$, ..., $a_{n+1} = f(a_n) = f(f(f\cdots(f(b))\cdots))$ (n f s). Show that $\lim_{n\to\infty} a_n = L$ exists and that L is a solution to the equation $f(x) = x$.

23. Let f be a decreasing function on $[0, \infty)$ and let a_n be the sequence defined by $a_n = f(n)$. Prove that

$$\lim_{n\to\infty} a_n = \lim_{x\to\infty} f(x)$$

in the sense that if either limit exists, then both limits exist and they are equal. [Comment: Compare this with the last exercise in section 9.2.]

24. Let C and C' be the two circles with equations $(x-1)^2 + (y-1)^2 = 1$ and $(x+1)^2 + (y-1)^2 = 1$. These are tangent to the x-axis. Construct a sequence C_n of circles as follows: C_1 is the small circle tangent to the x-axis and to C and C'; C_2 is the small circle tangent to C_1 and C and C'. Continuing this way, C_{n+1} is the small circle tangent to C_n and to C and C'. Let A_n be the sum of the diameters of the circles $C_1, C_2, \ldots, C_n$. Show that the sequence A_n is monotone and bounded. What do you think its limit is? [Hint: A picture is worth a thousand computations.]

25. A sequence T_n of geometric figures is constructed as follows: T_1 is an equilateral triangle with sides of length one. T_2 is obtained by replacing the middle third of each side of T_1 with an outward pointing equilateral triangle with sides of length $1/3$. Thus T_2 has 12 sides. T_3 is obtained from T_2 by the analogous construction: replace the middle third of each side with an outward pointing equilateral triangle with sides of length $1/9$. Continue this way, obtaining T_n by replacing the middle third of each side of T_{n-1} with an equilateral triangle with sides of length $1/3^{n-1}$. Find the perimeter p_n of the region enclosed by T_n and determine the limit of p_n as n increases without bound. [This limit curve is called the "snowflake" curve.] Determine the area of T_n. [It might be easiest to first determine what area is added when passing from T_{n-1} to T_n.] What is the limit of the areas as n increases without bound?

26. The curve $y = e^{-x} \sin x$, for $x \geq 0$, is revolved around the x-axis to produce a solid that looks like a necklace with infinitely many smaller and smaller beads. Determine the volume v_n of the nth bead (the one over the interval $[(n-1)\pi, n\pi]$). Show that $v_{n+1} = v_n R$ for a constant R independent of n and conclude that $v_n = v_1 R^{n-1}$. Use a geometric sequence to compute the total volume in terms of v_1 without evaluating any improper integrals.

10 Power Series

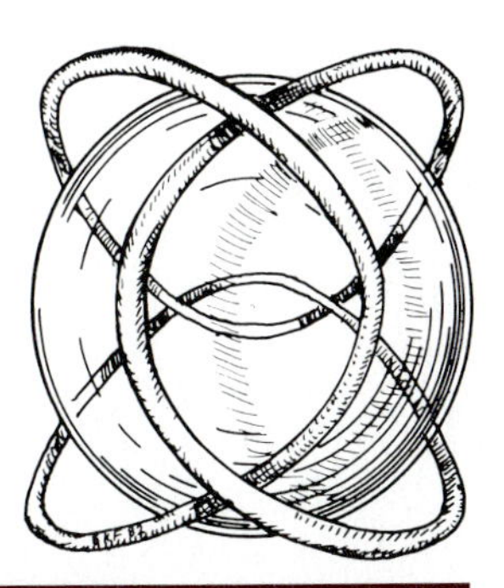

In this chapter we introduce a new way of looking at many of the functions that have been studied up to this point—functions are represented as power series. Power series are introduced as limits of Taylor polynomials of infinitely differentiable functions. We then consider power series that are not determined by the Taylor polynomials of some differentiable function and discover that convergent power series actually define new functions that are infinitely differentiable.

Power series are important for several different reasons. They provide a convenient computational tool for doing certain numerical problems. The theory of power series also provides a way of defining functions, for example, those that arise as solutions of certain differential equations. They also provide a language for expressing the solution of problems that otherwise could not be solved such as finding the antiderivative of certain functions. After developing the theory, we illustrate the application of power series to problems of evaluating definite integrals, evaluating limits of some indeterminate forms and finding solutions of differential equations.

10.1 Taylor series

Let $f(x)$ be a function for which the nth derivative $f^{(n)}(x)$ exists for $|x| \leq a$ and for every positive integer n. Then the nth Taylor polynomial (centered at 0) $P_n(x)$ is defined for every n and we have seen that the values $f(x)$ can be approximated by values $P_n(x)$. Moreover, for some functions that we have considered such as e^x and $\sin x$, the approximation gets better as n gets larger. It is reasonable to ask what happens if we let n increase without bound. In particular, is it true that

$$\lim_{n\to\infty} P_n(x) = f(x)? \qquad \textbf{(10.1)}$$

Of course this limit is correct when $x = 0$ because $P_n(0) = f(0)$ for every n. We are interested in knowing if the limit is correct for any other values of x. Our first theorem about this problem translates Equation (10.1) into a condition on the bounds for the derivatives of f.

THEOREM 10.1

Limits of Taylor Polynomials

Let f be a function whose nth derivative exists for every n and every x on an interval I containing 0 and let $P_n(x)$ be the nth Taylor polynomial centered at 0 for $f(x)$. Let W_n be a constant satisfying

$$|f^{(n)}(x)| \leq W_n \qquad \text{for all } x \text{ on } I.$$

If t is a point on I such that

$$\lim_{n\to\infty} \frac{W_{n+1}\, t^{n+1}}{(n+1)!} = 0, \tag{10.2}$$

then

$$\lim_{n\to\infty} P_n(t) = f(t).$$

Proof

By Theorem 9.4 we have the inequality

$$0 \leq |f(x) - P_n(x)| \leq \frac{W_{n+1}\, x^{n+1}}{(n+1)!}$$

for all x on I. If t is a number satisfying Equation (10.2) then the Squeeze Theorem implies

$$0 = \lim_{n\to\infty} |f(t) - P_n(t)|,$$

which is equivalent to the conclusion of the theorem. □

Now it is time to look at some functions where the assumptions in this theorem can be verified. The following limit is important.

For any number x,

$$\lim_{n\to\infty} \frac{x^n}{n!} = 0. \tag{10.3}$$

One might get a feeling for why this limit is 0 by considering how the sequence changes as n increases. To get $x^{n+1}/(n+1)!$ from $x^n/n!$ we multiply $x^n/n!$ by $x/(n+1)$. When n is much larger than x, we are multiplying by a very small number so the numbers $x^{n+1}/(n+1)!$ are getting smaller and smaller (in absolute value) when x is fixed and $n > x$. This intuitive argument is made more precise in the exercises.

EXAMPLE 1 If $f(x)$ is $\sin x$, $\cos x$, or e^x and $P_n(x)$ is the nth Taylor polynomial of $f(x)$ centered at 0 then

$$f(x) = \lim_{n\to\infty} P_n(x) \tag{10.4}$$

for every number x.

Solution: If f is either $\sin x$ or $\cos x$ then the nth derivative of f is one of the four functions $\pm \sin x$, $\pm \cos x$ and so

$$|f^{(n)}(x)| \leq 1$$

for any number x and any nonnegative integer n. Take $W_n = 1$. The interval I

can be any interval whatever in these two cases. Then

$$\lim_{n\to\infty} \frac{W_{n+1}\, x^{n+1}}{(n+1)!} = \lim_{n\to\infty} \frac{x^{n+1}}{(n+1)!} = 0$$

by Equation (10.3). Thus Equation (10.4) holds for any x.

If $f(x) = e^x$ then $f^{(n)}(x) = e^x$ for every n. If $I = [a, b]$, then

$$|f^{(n)}(x)| = e^x \leq e^b$$

for all x on I and every n. Take $W_n = e^b$ and then

$$\lim_{n\to\infty} \frac{W_{n+1}\, x^{n+1}}{(n+1)!} = e^b \lim_{n\to\infty} \frac{x^{n+1}}{(n+1)!} = 0.$$

Once again Equation (10.4) holds for any x. ■

Power Series

Let $P_n(x)$ be the Taylor polynomial centered at 0 for e^x. Then

$$P_n(x) = 1 + x + \frac{x^2}{2!} + \frac{x^3}{3!} + \cdots + \frac{x^n}{n!}.$$

For each n, $P_{n+1}(x)$ is obtained from $P_n(x)$ by adding one term:

$$P_{n+1}(x) = P_n(x) + \frac{x^{n+1}}{(n+1)!}.$$

The limit as n approaches infinity of the $P_n(x)$ can be regarded as an infinite sum:

$$\begin{aligned} e^x &= \lim_{n\to\infty} P_n(x) \\ &= 1 + x + \frac{x^2}{2!} + \cdots + \frac{x^n}{n!} + \cdots \\ &= \sum_{k=0}^{\infty} \frac{x^k}{k!}. \end{aligned}$$

The infinite sum is a suggestive way to write the limit of the $P_n(x)$. If a_k is a sequence of constants (independent of x) we call a sum of the form

$$\sum_{k=0}^{\infty} a_k x^k$$

a *power series* in x. We can add a hundred or a million terms by the ordinary rules of addition. The sum of an infinite number of terms requires a special definition because it is not possible to simply add infinitely many terms by the usual rules. We give the definition of the sum of the terms in a general sequence.

DEFINITION 10.1

Convergence and Divergence

Let u_k be a sequence. The sum $\sum_{k=0}^{\infty} u_k$ is called the *infinite series* corresponding to the sequence. For each nonnegative integer n, the sum

$$s_n = u_0 + u_1 + \cdots + u_n = \sum_{k=0}^{n} u_k$$

is called the nth *partial sum* of the infinite series. The series converges to

S if

$$\lim_{n\to\infty} s_n = S.$$

If $\lim_{n\to\infty} s_n$ does not exist, the series diverges.

If f is a function for which $f^{(n)}(0)$ exists for all n the power series

$$\lim_{n\to\infty} P_n(x) = \sum_{k=0}^{\infty} \frac{f^{(k)}(0)}{k!} x^k \quad \textbf{(10.5)}$$

is called the *Taylor series* of f in powers of x. If the limit is a (finite) number S for some particular x, then we say the infinite series *converges* to S at x.

We restate Theorem 10.1 above using these terms.

THEOREM 10.2

Convergence of Taylor Series

Let f be a function whose nth derivative exists for every n and every x on an interval I containing 0 and let $P_n(x)$ be the nth Taylor polynomial centered at 0 for $f(x)$. Let W_n be a constant satisfying

$$|f^{(n)}(x)| \leq W_n \qquad \text{for all } x \text{ on } I.$$

If t is a point on I such that

$$\lim_{n\to\infty} \frac{W_{n+1} t^{n+1}}{(n+1)!} = 0 \quad \textbf{(10.6)}$$

then the Taylor series of f converges at $x = t$ to $f(t)$; i.e.,

$$f(t) = \sum_{k=0}^{\infty} \frac{f^{(k)}(0)}{k!} t^k.$$

The example with the three functions $\sin x$, $\cos x$ and e^x worked above implies that the Taylor series of each function converges to the function value for every x. We have for all x

$$e^x = 1 + x + \frac{x^2}{2!} + \frac{x^3}{3!} + \cdots = \sum_{k=0}^{\infty} \frac{x^k}{k!},$$

$$\sin x = x - \frac{x^3}{3!} + \frac{x^5}{5!} - \cdots = \sum_{k=0}^{\infty} (-1)^k \frac{x^{2k+1}}{(2k+1)!},$$

$$\cos x = 1 - \frac{x^2}{2!} + \frac{x^4}{4!} - \cdots = \sum_{k=0}^{\infty} (-1)^k \frac{x^{2k}}{(2k)!}.$$

Next we look at an example where the Taylor series converges to the function for only certain x in the domain of the function.

Geometric Series

The Taylor series centered at 0 of

$$g(x) = \frac{1}{1-x}$$

plays a special role in the theory of infinite series. Other series are compared to this one when we discuss convergence of general series. Let us compute the coefficients and discuss the convergence. We have the derivatives

$$\begin{aligned} g'(x) &= \frac{1}{(1-x)^2}, \\ g''(x) &= \frac{2}{(1-x)^3}, \\ g^{(3)}(x) &= \frac{2 \cdot 3}{(1-x)^4}, \\ &\vdots \\ g^{(n)}(x) &= \frac{n!}{(1-x)^{n+1}}. \end{aligned}$$

The coefficient of x^n in the Taylor series for g is

$$\frac{g^{(n)}(0)}{n!} = \frac{n!}{1}\frac{1}{n!} = 1$$

and so the Taylor series is

$$1 + x + x^2 + x^3 + x^4 + \cdots = \sum_{k=0}^{\infty} x^k.$$

This series is called a *geometric series.* If we want to apply Theorem 10.2 to prove convergence, we must find a bound W_n for $g^{(n)}(x)$ on a suitable interval I containing 0. We must avoid $x = 1$ because the derivatives of g are undefined at 1. Take any number r with $0 < r < 1$ and let $I = [-r, r]$. Then the derivatives of g are continuous and $|g^{(n)}(x)|$ is an increasing function on I. The maximum occurs at the right endpoint $x = r$ so the bound we take for $|g^{(n)}(x)|$ is

$$W_n = g^{(n)}(r) = \frac{n!}{(1-r)^{n+1}}.$$

The Taylor series converges at $x = t$ if

$$\begin{aligned} 0 &= \lim_{n\to\infty} \frac{W_{n+1}|t|^{n+1}}{(n+1)!} \\ &= \lim_{n\to\infty} \frac{(n+1)!}{(1-r)^{n+2}} \frac{|t|^{n+1}}{(n+1)!} \\ &= \frac{1}{1-r} \lim_{n\to\infty} \left(\frac{|t|}{1-r}\right)^{n+1}. \end{aligned}$$

This limit is correct if $|t| < 1 - r$. However t must also lie on $[-r, r]$, so the choice of r giving the longest interval of choices for t is $r = 1/2$. The argument proves that the geometric series converges to $g(x)$ for every x on $(-0.5, 0.5)$.

This is not the best possible result because the geometric series actually converges on a larger interval. The estimate given by Taylor's theorem is not the only way to prove a convergence theorem. Here is an alternative treatment that works for this special series.

THEOREM 10.3

Convergence of the Geometric Series

For all x on the open interval $(-1, 1)$ the geometric series

$$\sum_{k=0}^{\infty} x^k = 1 + x + x^2 + x^3 + \cdots$$

converges to $1/(1 - x)$.

Proof

For any positive integer n let

$$s_n = 1 + x + x^2 + \cdots + x^n.$$

Multiply by $1 - x$ to get

$$(1 - x)s_n = (1 + x + x^2 + \cdots + x^n) - x(1 + x + x^2 + \cdots + x^n),$$
$$(1 - x)s_n = 1 - x^{n+1},$$
$$s_n = \frac{1 - x^{n+1}}{1 - x} = \frac{1}{1 - x} - \frac{x^{n+1}}{1 - x}.$$

Thus

$$\frac{1}{1 - x} - (1 + x + x^2 + \cdots + x^n) = \frac{x^{n+1}}{1 - x}.$$

When $|x| < 1$, we have $\lim_{n \to \infty} x^{n+1} = 0$ and so

$$\frac{1}{1 - x} = 1 + x + x^2 + \cdots + x^n + \cdots \quad \text{for} \quad |x| < 1. \qquad \square$$

Decimals and Geometric Series

If a is a constant, the series

$$a + ax + ax^2 + ax^3 + \cdots = \sum_{k=0}^{\infty} ax^k = \frac{a}{1 - x} \qquad \textbf{(10.7)}$$

is also called a geometric series; it converges for all x on $(-1, 1)$. We give an example of a very common occurrence of geometric series.

Every real number can be represented as a finite or infinite decimal of a special type. The number 12.357 is a finite decimal that can be written as

$$12.357 = 12 + \frac{3}{10} + \frac{5}{10^2} + \frac{7}{10^3}.$$

The real number $\pi \approx 3.14159\cdots$ can be written as an infinite sum

$$\pi = 3 + \frac{1}{10} + \frac{4}{10^2} + \frac{1}{10^3} + \cdots + \frac{d_k}{10^k} + \cdots,$$

where each number d_k is an integer satisfying $0 \leq d_k \leq 9$. The digits d_k do not "repeat" in a cyclical pattern and there is no practical formula to determine d_k in terms of k. (This statement is not obvious but a proof would require a major digression from the subject at hand.)

The number $0.3333333\cdots$ is an infinite and repeating decimal that can be written as the infinite sum

$$0.333333\cdots = \frac{3}{10} + \frac{3}{10^2} + \frac{3}{10^3} + \cdots + \frac{3}{10^n} + \cdots.$$

The sum is a geometric series that can be evaluated using Equation (10.7). We have

$$\begin{aligned}\frac{3}{10} + \frac{3}{10^2} + \frac{3}{10^3} + \cdots + \frac{3}{10^n} + \cdots &= \frac{3}{10}\left(1 + \frac{1}{10} + \frac{1}{10^2} + \frac{1}{10^3} + \cdots\right) \\ &= \frac{3}{10}\left(\frac{1}{1 - 1/10}\right) = \frac{3}{10 - 1} \\ &= \frac{1}{3}.\end{aligned}$$

Of course this sum comes as no surprise but the underlying idea can be used to show that every decimal that is eventually repeating is a rational number. Here is a slightly more complicated example.

Consider the number $7.976353535\cdots$ where the digits 35 repeat indefinitely. This is a rational number and to express it as the quotient of two integers we first write it as a series. We have

$$\begin{aligned}7.976353535\cdots &= 7.976 + 0.00035 + 0.0000035 + \cdots \\ &= \frac{7976}{10^3} + \frac{35}{10^5} + \frac{35}{10^7} + \cdots \\ &= \frac{7976}{10^3} + \frac{35}{10^5}\left(1 + \frac{1}{10^2} + \frac{1}{10^4} + \cdots\right) \\ &= \frac{7976}{10^3} + \frac{35}{10^5}\left(\frac{1}{1 - 1/10^2}\right) \\ &= \frac{7976}{10^3} + \frac{35}{10^3}\left(\frac{1}{99}\right) \\ &= \frac{789,659}{99,000}.\end{aligned}$$

Taylor Series of ln(*x*+*1*)

Let $f(x) = \ln(x+1)$ and $P_n(x)$ equal the nth Taylor polynomial centered at 0 of $f(x)$. We want to apply Theorem 10.2 to determine the values of t that satisfy

$$\begin{aligned}\ln(t+1) &= \lim_{n\to\infty} P_n(t) \\ &= t - \frac{t^2}{2} + \frac{t^3}{3} - \frac{t^4}{4} + \cdots \qquad \textbf{(10.8)} \\ &= \sum_{k=0}^{\infty} (-1)^k \frac{t^{2k+1}}{2k+1}.\end{aligned}$$

We prove the Taylor series converges to $\ln(x+1)$ for all x on the interval $(-0.5, 1]$. After studying the integration of power series, we can prove that the convergence is valid for x on $(-1, 1]$.

The function $\ln(x+1)$ and all its derivatives are defined for all x greater than -1. The derivatives are seen to be

$$f^{(n)}(x) = (-1)^{n+1}\frac{(n-1)!}{(x+1)^n} \qquad (n \geq 1).$$

We next select an interval I over which to estimate the derivatives. Take $I = [a, b]$ with $-1 < a \leq 0 \leq b$. All the derivatives of $f(x)$ exist on I. For any positive n,

$|f^{(n)}(x)|$ is a decreasing function on I so its maximum occurs at $x = a$. For a bound W_{n+1} satisfying $|f^{(n+1)}(x)| \leq W_{n+1}$, we take

$$W_{n+1} = |f^{(n+1)}(a)| = \frac{n!}{(a+1)^{n+1}}.$$

The Taylor series of f converges at $x = t$ if

$$\lim_{n\to\infty} \frac{W_{n+1}t^{n+1}}{(n+1)!} = 0. \qquad \textbf{(10.9)}$$

We must determine for which t this limit holds. Substitute the value of W_{n+1} and use the fact that $(n+1)! = n!(n+1)$ to see that the condition (10.9) on t reduces to

$$0 = \lim_{n\to\infty} \frac{W_{n+1}t^{n+1}}{(n+1)!} = \lim_{n\to\infty} \left(\frac{t}{a+1}\right)^{n+1} \left(\frac{1}{n+1}\right). \qquad \textbf{(10.10)}$$

If c is a number satisfying $-1 < c < 1$, then

$$\lim_{n\to\infty} c^n = 0.$$

If the fraction $t/(a+1)$ is between -1 and 1, then Equation (10.10) holds. If $t = \pm(a+1)$ then Equation (10.10) still holds because of the factor $1/(n+1)$ and so the Taylor series converges at t whenever $-(a+1) \leq t \leq a+1$. In addition t was restricted to lie on $[a,b]$ so that $a \leq t$. We conclude Equation (10.8) is valid whenever $a \leq t$ and $-(a+1) \leq t \leq a+1$. Take $a = -1/2$ and conclude that Equation (10.8) holds for all t satisfying $-1/2 \leq t \leq 1/2$. Then take $a = 0$ to conclude that Equation (10.8) holds for t satisfying $0 \leq t \leq 1$. Thus the Taylor series of $\ln(x+1)$ converges to the function value at any x on the interval $(-0.5, 1]$:

$$\ln(x+1) = x - \frac{x^2}{2} + \frac{x^3}{3} - \cdots = \sum_{k=1}^{\infty} (-1)^{k+1}\frac{x^k}{k}, \quad -\frac{1}{2} \leq x \leq 1. \qquad \textbf{(10.11)}$$

By taking $x = 1$ we obtain an infinite sum converging to $\ln 2$:

$$\ln 2 = 1 - \frac{1}{2} + \frac{1}{3} - \frac{1}{4} + \frac{1}{5} - \cdots.$$

The Taylor series for $\ln(x+1)$ presents a phenomenon that can be seen in other examples. The function $\ln(x+1)$ on the left side of Equation (10.11) is defined for all $x > -1$ but it is equal to the right-hand side of the equation only for those values where the right side converges, namely for x on $(-1, 1]$. For a value outside this range, such as $x = 3$, the limit of $P_n(3)$ as n approaches infinity does not exist, as we see later when we study conditions for convergence of general power series.

Exercises 10.1

Find the Taylor series centered at 0 for the functions $f(x)$ and find some interval over which the series converges to f. Do this by finding a bound for the nth derivative and applying Theorem 10.2. [More efficient methods are developed in later sections for determining precisely where the series converge.]

1. $f(x) = \dfrac{1}{1+x}$

2. $f(x) = \dfrac{1}{1+3x}$

3. $f(x) = \sin 2x$

4. $f(x) = \ln(1+3x)$

5. $f(x) = e^{x+1}$

6. $f(x) = 5\cos 3x$

7. $f(x) = 4e^{5x-3}$

8. $f(x) = \cos \pi x$

Find the Taylor series centered at 0 for the rational functions by finding a formula for the nth derivatives. The use of partial fractions might simplify the computation of the derivatives of the functions having products in the denominator.

9. $f(x) = \dfrac{1}{2+x}$

10. $f(x) = \dfrac{1}{1-x}$

11. $f(x) = \dfrac{1}{1-3x}$

12. $f(x) = \dfrac{1}{2+5x}$

13. $f(x) = \dfrac{1}{(1+x)(2+x)}$

14. $f(x) = \dfrac{1}{(2+x)(1-x)}$

15. $f(x) = \dfrac{x-2}{(x+1)(x+3)}$

16. $f(x) = \dfrac{1-3x}{(5x+2)(x-4)}$

17. Let c be any number satisfying $-1 < c < 1$. Prove $\lim_{n\to\infty} c^n = 0$. [Hint: For $0 < c < 1$ show $\lim_{n\to\infty} \ln c^n = -\infty$ and so $\lim c^n = 0$. Use the case $c > 0$ to prove the same result if $c < 0$ by considering even and odd powers of c.]

18. For any real number x, show

$$\lim_{n\to\infty} \frac{x^n}{n!} = 0.$$

[Hint: First assume that $x > 0$. Select an integer $N > x$. Because we are interested in the limit as n gets large, assume $n > N$. Then

$$\frac{x^n}{n!} = \frac{x^N}{N!}\frac{x}{N+1}\frac{x}{N+2}\cdots\frac{x}{n}.$$

Now use the inequality $x/(N+k) < x/N$ to conclude

$$\begin{aligned}\frac{x^n}{n!} &\le \frac{x^N}{N!}\frac{x}{N}\frac{x}{N}\cdots\frac{x}{N}\\ &\le \frac{x^N}{N!}\left(\frac{x}{N}\right)^{n-N}.\end{aligned}$$

Because N is fixed, $x^N/N!$ is constant; because $0 < x/N < 1$, the limit as n goes to infinity of $(x/N)^n$ is zero. Argue in a similar manner for the case $x < 0$.]

19. Prove that the number e is not rational. [Hint: Suppose $e = a/b$ with a and b positive integers. Let

$$m = b!\left(e - \sum_{k=0}^{b} \frac{1}{k!}\right).$$

First prove that m is a nonzero integer. Next show that

$$m = \sum_{k=b+1}^{\infty} \frac{b!}{k!}$$

and that $b!/k! < (1/k)^{k-b}$ for $b < k$. Compare the infinite series for m with a geometric series to conclude $0 < m < 1/b$. Argue that this conflicts with the assertion that m is an integer and so the supposition that e is rational must be false.]

20. According to Equation (10.11), the infinite series $\sum_{k=1}^{\infty}(-1)^{k+1}/k$ converges to $\ln 2$. Evaluate the sum $\sum_{k=1}^{20}(-1)^{k+1}/k$ and compare it to the approximation of ln 2 given in section 7.2. [One concludes that this series is not very efficient for computing ln 2.]

21. Use the bounds in the text to determine some n for which the nth Taylor polynomial $P_n(x)$ for $\ln 1 + x$ satisfies $|\ln 2 - P_n(1)| < 10^{-4}$.

22. Define f by $f(x) = e^{-1/x^2}$ for $x \neq 0$ and $f(0) = 0$. With some effort it can be proved that f is infinitely differentiable at every point and $f^{(n)}(0) = 0$ for all n. What is the Taylor series of f centered at 0? Does the Taylor series at x converge to $f(x)$? [Hint: To compute $f^{(n)}(0)$, guess a general form for $f^{(n)}(x)$ at $x \neq 0$ and take the limit as x approaches 0.]

Express the following infinite repeating decimals as the quotient of two integers.

23. $0.777777777\cdots$

24. $0.999999999\cdots$

25. $0.52525252\cdots$

26. $0.123451234512345\cdots$

27. $0.782222222\cdots$

28. $2.049494949\cdots$

29. $3.141614161416\cdots$

30. $12.704587587587\cdots$

Use long division to express each rational number as a finite or infinite repeating decimal. Write the decimal as a series and verify that the sum is the correct rational number.

31. $\frac{1}{6}$

32. $\frac{4}{7}$

33. $\frac{18}{7}$

34. $\frac{17}{8}$

35. A ball rebounds to a height ρh if dropped from a height h where the constant ρ satisfies $0 < \rho < 1$ and depends on the composition of the ball. If the ball is dropped from a height of 10 feet and $\rho = 1/2$, how far does the ball travel if allowed to continue bouncing until it comes to rest. Rework the problem with the general constant ρ in place of $1/2$.

10.2 Convergence of infinite series

In the first section of this chapter we discussed the Taylor series

$$\sum_{n=0}^{\infty} a_n x^n \tag{10.12}$$

obtained from a given infinitely differentiable function $f(x)$ and, in Taylor's theorem, found some conditions on x and f that ensure convergence of the series to $f(x)$. Those conditions were stated in terms of the maximum of the derivatives of f over an interval and hence depended on knowledge of f. Now we take a different starting point. We consider the series (10.12) where we are given the sequence a_n of coefficients but we are not given any function f that determines the coefficients. The first problem is to find the x for which the power series converges. After we describe several tests and methods for determining the convergence of series, we eventually prove that the values of x for which the series converges are either all numbers, an interval $(-r, r)$ for some number r along with 0, 1 or 2 of its endpoints, or the series converges only for $x = 0$. Before getting to this result, we develop some important facts about convergence of power series. There is a second problem that we only touch on. Once we know a series converges, it would be logical to ask for a formula for the sum of the series at x. In the case of Taylor series, we started with a function $f(x)$ and showed the convergence of the series to $f(x)$ for certain x. If we start with only the series, there is no f except the series itself. In most cases it is not possible to give a closed form formula for the convergent series. Although this may seem a disadvantage, it serves to show that the familiar functions that we have dealt with and given names to are but a small number of the possible functions that can be defined by convergent power series.

We begin with some formalities about the manipulation of convergent power series. For brevity we often write u_k in place of $a_k x^k$ in the discussion of series where the particular x plays no special role. The definition (10.1) of convergence and divergence should be in the foreground of the readers mind.

We use the abbreviated notation $\sum u_k$ to indicate the sum of the terms u_k as k ranges from some starting point (usually $k = 0$ or $k = 1$) to ∞. If more than one summation appears in a statement, it is assumed that each begins at the same starting value. The reason for this casual approach is that the convergence of a series is not affected by the starting point. If $\sum_{k=0}^{\infty} u_k$ converges then so does $\sum_{k=7}^{\infty} u_k$ because the two differ only by the sum of a finite number of terms. Of course, the actual sum of the series is affected by the starting point but not the convergence or divergence.

THEOREM 10.4

Addition Theorem

Let $\sum u_k$ and $\sum v_k$ be convergent series and let b be any constant. Then $\sum bu_k$ and $\sum(u_k \pm v_k)$ are convergent series and

$$\sum bu_k = b\sum u_k,$$
$$\sum(u_k \pm v_k) = \sum u_k \pm \sum v_k.$$

Proof

Let s_n be the nth partial sum of $\sum u_k$ and t_n the nth partial sum of $\sum v_k$. Then

$$bs_n = bu_0 + bu_1 + \cdots + bu_n$$

is the nth partial sum of the series $\sum bu_k$ and so

$$\sum bu_k = \lim_{n\to\infty} bs_n = b \lim_{n\to\infty} s_n = b\sum u_k.$$

Similarly for the sum or difference of two convergent series we have

$$\begin{aligned}\sum (u_k \pm v_k) &= \lim_{n\to\infty} (s_n \pm t_n) \\ &= \lim_{n\to\infty} s_n \pm \lim_{n\to\infty} t_n \\ &= \sum u_k \pm \sum v_k .\end{aligned}$$

Note that the second equality is valid because both $\lim s_n$ and $\lim t_n$ are known to exist. □

These natural looking properties of infinite series are used frequently without mention. It is important to note that all the series involved must be convergent before the addition rules can be used.

EXAMPLE 1 Find the sum of the series

$$\sum_{k=0}^{\infty} 4\left(\frac{1}{3}\right)^k \quad \text{and} \quad \sum_{k=0}^{\infty} \frac{3^k + 4^k}{5^k}$$

using the geometric series $\sum_{k=0}^{\infty} x^k = 1/(1-x)$ for $|x| < 1$.

Solution: We have

$$\sum_{k=0}^{\infty} \left(\frac{1}{3}\right)^k = \frac{1}{1-\frac{1}{3}} = \frac{3}{2}$$

and so

$$\sum_{k=0}^{\infty} 4\left(\frac{1}{3}\right)^k = 4\sum_{k=0}^{\infty} \left(\frac{1}{3}\right)^k = 4\left(\frac{3}{2}\right) = 6.$$

For the second series we have

$$\sum_{k=0}^{\infty} \left(\frac{3}{5}\right)^k = \frac{5}{2},$$

$$\sum_{k=0}^{\infty} \left(\frac{4}{5}\right)^k = 5,$$

and so

$$\sum_{k=0}^{\infty} \frac{3^k + 4^k}{5^k} = \sum_{k=0}^{\infty} \left(\frac{3^k}{5^k} + \frac{4^k}{5^k}\right) = \sum_{k=0}^{\infty} \left(\frac{3}{5}\right)^k + \sum_{k=0}^{\infty} \left(\frac{4}{5}\right)^k = \frac{5}{2} + 5 = \frac{15}{2}.$$ ■

We use the Addition Theorem for a critical step in the proof of the next theorem.

THEOREM 10.5

Absolute Value Test

If u_k is a sequence of numbers for which $\sum |u_k|$ converges, then $\sum u_k$ also converges. Moreover we have the estimate

$$\left|\sum u_k\right| \le \sum |u_k|.$$

Proof

Let $w_k = u_k + |u_k|$ and $y_k = -u_k + |u_k|$. We prove the two series $\sum w_k$ and $\sum y_k$ converge by showing the sequence of partial sums for each is a bounded, monotonic sequence and thus has a limit. We have

$$0 \leq w_k = u_k + |u_k| \leq 2|u_k| \quad \text{and} \quad 0 \leq y_k = -u_k + |u_k| \leq 2|u_k|.$$

These are verified by considering the two cases $u_k \geq 0$ and $u_k \leq 0$. Now let

$$s_n = w_1 + w_2 + \cdots + w_n$$

be the nth partial sum for the series $\sum w_k$. Then

$$s_{n+1} = s_n + w_{n+1} \geq s_n$$

because $w_{n+1} \geq 0$. Thus the sequence s_n is monotonic. To see that it is also bounded we observe

$$s_n = w_1 + w_2 + \cdots + w_n \leq 2|u_1| + 2|u_2| + \cdots + 2|u_n| \leq \sum_{k=1}^{\infty} 2|u_k|.$$

By assumption, $\sum |u_k|$ converges to a limit U and so $s_n \leq 2U$ for all n. By the Bounded Monotonic Sequence Theorem,

$$\sum (u_k + |u_k|) = \sum w_k = \lim s_n$$

exists. By the same reasoning, $\sum(-u_k + |u_k|)$ also converges. Now we apply the Addition Theorem using the fact

$$2u_k = (u_k + |u_k|) - (-u_k + |u_k|)$$

to get

$$\sum u_k = \frac{1}{2}\left(\sum (u_k + |u_k|) - \sum(-u_k + |u_k|)\right)$$

is convergent.

To prove the estimate we observe that $u_k + |u_k| \geq 0$ for all k and so

$$0 \leq \sum (u_k + |u_k|) = \sum u_k + \sum |u_k|$$

and

$$-\sum u_k \leq \sum |u_k|.$$

Similarly the inequality $0 \leq -u_k + |u_k|$ implies

$$\sum u_k \leq \sum |u_k|.$$

Thus

$$\left|\sum u_k\right| \leq \sum |u_k|. \qquad \square$$

The proof requires some steps that seem very mysterious. The introduction of w_k and y_k was necessary to produce a series of nonnegative terms; a series of nonnegative terms always has a monotonic sequence of partial sums. That was a necessary ingredient to permit the use of the only available theorem that asserts the existence of a limit — namely the Bounded Monotonic Sequence Theorem.

The point of this theorem is that convergence of $\sum u_k$ is sometimes easier to establish when all the u_k are known to be nonnegative. We might be able to establish convergence of $\sum a_n x^n$ by proving the convergence of $\sum |a_n||x|^n$.

Absolute Convergence

We say a given series $\sum u_k$ *converges absolutely* if $\sum |u_k|$ converges. By the theorem just proved, if $\sum u_k$ converges absolutely then it converges. Later we see examples of series that converge but do not converge absolutely. One such example is the alternating harmonic series

$$\sum_{k=1}^{\infty}(-1)^{k+1}\frac{1}{k} = 1 - \frac{1}{2} + \frac{1}{3} - \frac{1}{4} + \cdots .$$

Bounded Terms Test for Convergence

Here is the first theorem giving a test for the convergence of a general power series.

THEOREM 10.6

Bounded Terms Test

Let $\sum a_n x^n$ be a given power series. Suppose r is a positive number such that the sequence of terms $|a_n|r^n$ is bounded, i.e.,

$$0 \leq |a_n r^n| \leq B$$

for all n and some constant B. Then

$$\sum a_n x^n \qquad \text{converges absolutely for} \qquad |x| < r.$$

Proof

We prove $\sum |a_n x^n|$ converges for $|x| < r$; this implies the convergence of $\sum a_n x^n$ as well. Let x be a fixed number with $|x| < r$. In particular we emphasize

$$\left|\frac{x}{r}\right| < 1.$$

Because $|a_{n+1}x^{n+1}| \geq 0$, we have

$$s_n = |a_0| + |a_1 x| + |a_2 x^2| + \cdots + |a_n x^n| \leq s_n + |a_{n+1}x^{n+1}| = s_{n+1}.$$

Thus the sequence s_n of partial sums of the series $\sum |a_n x^n|$ is a monotonic sequence. Next we prove it is also bounded. For any n,

$$\begin{aligned} s_n &= \sum_{k=0}^{n} |a_k x^k| = \sum_{k=0}^{n} |a_k r^k| \left|\frac{x}{r}\right|^k \\ &\leq \sum_{k=0}^{n} B\left|\frac{x}{r}\right|^k \leq \sum_{k=0}^{\infty} B\left|\frac{x}{r}\right|^k \\ &= \frac{B}{1 - \left|\frac{x}{r}\right|}. \end{aligned}$$

The final term is independent of n, so it is a bound for all the partial sums. The sequence s_n of partial sums is bounded and monotonic and therefore has a limit. This proves the series $\sum |a_n x^n|$ converges and hence the original series $\sum a_n x^n$ converges for $|x| < r$. □

EXAMPLE 2 Show that the power series

$$x - \frac{x^2}{2} + \frac{x^3}{3} - \cdots = \sum_{k=1}^{\infty} (-1)^{k+1} \frac{x^k}{k}$$

converges for all x on $(-1, 1)$.

Solution: We apply Theorem 10.6 using $r = 1$. It is necessary to verify the hypothesis about the boundedness of the individual terms $|a_n r^n|$. In this case $a_0 = 0$ and for $n \geq 1$ we have

$$|a_n r^n| = \left|(-1)^{n+1} \frac{1^n}{n}\right| = \frac{1}{n} \leq 1.$$

The terms are bounded so the series converges for $|x| < 1$. ■

EXAMPLE 3 Use the Bounded Terms Test to show the series

$$\sum_{k=0}^{\infty} \frac{x^{2k}}{k!} \tag{10.13}$$

converges for every x.

Solution: Let x be any number and let r be any number with $r > |x|$. We show that the sequence $q_k = r^{2k}/k!$ is bounded. Let N be an integer larger than r^2 and let B be the largest of the numbers in the finite list $q_0, q_1, \ldots, q_N$. For $k > N$ we have

$$q_k = \frac{r^{2k}}{k!} = \frac{r^{2N}}{N!} \frac{r^2}{N+1} \frac{r^2}{N+2} \cdots \frac{r^2}{k}.$$

Because each fraction $r^2/(N + j)$ is less than 1, it follows that

$$q_k < q_N \leq B.$$

Thus B is an upper bound for the sequence $r^{2k}/k!$ of positive terms. The Bounded Terms Test implies that the series (10.13) converges at x because $|x| < r$. ■

Now that we have at least one test to prove convergence, we also need some test to prove divergence. Several tests are discussed in later sections, but we can give the following rather simple test now.

THEOREM 10.7

nth Term Divergence Test

If x is a number such that $\lim_{k\to\infty} a_k x^k \neq 0$ then $\sum a_k x^k$ diverges.

Proof

We prove the contrapositive form of the statement; namely, if the series $\sum a_k x^k$ converges then $\lim_{k\to\infty} a_k x^k = 0$. Let $s_n = \sum_{k=0}^{n} a_k x^k$ be the nth partial sum of the series. By the convergence assumption we know $\lim_{n\to\infty} s_n$ exists and equals some number S. Now we have

$$s_{n-1} + a_n x^n = s_n$$

and

$$\lim_{n\to\infty} a_n x^n = \lim_{n\to\infty} s_n - s_{n-1} = S - S = 0.$$ □

This theorem can be used to assert the divergence of a series $\sum a_k x^k$ if it can be shown that $\lim a_k x^k \neq 0$, but it can never be used to assert the convergence of a series.

EXAMPLE 4 Show that the Taylor series

$$x - \frac{x^2}{2} + \frac{x^3}{3} - \cdots = \sum_{k=1}^{\infty} (-1)^{k+1} \frac{x^k}{k}$$

of $\ln(x + 1)$ does not converge if $x = 3$. More generally, determine all positive c for which $\lim_{n\to\infty}(-1)^{n+1}c^n/n$ does not equal 0 and thereby determine some points where the series diverges.

Solution: The general term of the series evaluated at $x = 3$ is $t_n = (-1)^{n+1}3^n/n$. We temporarily ignore the alternating signs and observe that

$$|t_n| = \frac{3^n}{n} < \frac{3^{n+1}}{n+1} = |t_{n+1}|.$$

Thus $|t_n|$ is an increasing sequence so all the terms are greater than $t_1 = 3$. Thus the terms t_n lie either to the right of 3 or to the left of -3 and so the limit of the t_n is not 0. Thus the nth term test implies the series does not converge at $x = 3$.

It should be clear that one could show that the series fails to converge at $x = c$ provided $c > 0$ and the sequence c^n/n is increasing for n larger than some fixed number. We have

$$\frac{c^n}{n} < \frac{c^{n+1}}{n+1}$$

if and only if

$$\frac{n+1}{n} < c. \tag{10.14}$$

If c is greater than 1, then there is some N such that

$$\frac{N+1}{N} = 1 + \frac{1}{N} < c$$

and for every $n \geq N$, inequality (10.14) holds. It follows that for any $c > 1$, the limit of the sequence $(-1)^{n+1}c^n/n$ is not 0 and hence the Taylor series for $\ln(1 + x)$ does not converge at $x = c$. ■

In the last example we found all the points x where the nth term of the series did not have a 0 limit and concluded the series did not converge at these points. We could not conclude that we had found all the points where the series diverged because a series could diverge for some reason other than the that given by the nth term test. When dealing with a general series $\sum a_k x^k$, the nth term test can be applied to conclude the divergence of a series but it can never be used to conclude the convergence of the series. Even though $\lim a_k x^k = 0$, we still cannot conclude that $\sum a_k x^k$ converges.

Exercises 10.2

Apply the Bounded Terms Test to find an interval of nonzero length where the series converges absolutely and then apply the nth term test to find a point where the series does not converge.

1. $\displaystyle\sum_{k=1}^{\infty} \frac{x^k}{k}$

2. $\displaystyle\sum_{k=1}^{\infty} \frac{(2x)^k}{k^2}$

3. $\sum_{k=0}^{\infty} \frac{x^{2k}}{k+4}$

4. $\sum_{k=0}^{\infty} (-1)^k \frac{x^k}{2k+1}$

5. $\sum_{k=0}^{\infty} (-1)^k \frac{(3x)^k}{k!}$

6. $\sum_{k=0}^{\infty} \frac{(2x)^{3k}}{k(k+1)}$

7. $\sum_{k=0}^{\infty} (-1)^k \frac{x^{2k}}{5^k(k^2+1)}$

8. $\sum_{k=1}^{\infty} \frac{(4x)^{3k}}{10^k k(k+1)}$

9. $\sum_{k=0}^{\infty} \frac{x^k}{2^k+k}$

10. $\sum_{k=0}^{\infty} (-1)^k (2^{-k}+k)(2x)^k$

11. $\sum_{k=1}^{\infty} (-1)^k \frac{x^k}{(200)^k k(k+1)(k+2)}$

12. By applying only the nth term test, what conclusion, if any, can be drawn about the convergence or divergence of the following series ? (a) $\sum (1.03)^k$, (b) $\sum k/(k+1)$, (c) $\sum (2k+1)/k^2$, (d) $\sum k/(k^3+k+1)$.

13. If a_n is a sequence such that $\sum_{n=0}^{\infty} a_n$ converges then $\sum_{n=0}^{\infty} a_n x^n$ converges for $|x| < 1$.

14. Given a sequence a_n of positive numbers such that $\sum_{k=0}^{\infty} a_k$ converges, explain why the series $\sum_{k=0}^{\infty} \epsilon_n a_n x^n$ converges for $|x| < 1$ when ϵ_n is a sign, ± 1, for all n.

15. A series of constants $\sum_{k=1}^{\infty} a_k$ has nth partial sum given by

$$s_n = (2n+1)/(n+1).$$

Find the sum of the series. What is the general term a_k?

16. A series of constants $\sum_{k=1}^{\infty} a_k$ has nth partial sum given by

$$s_n = (3n + (-1)^n)/(3n+1).$$

Find the sum of the series. What is the general term a_k?

17. Let a_k be a sequence of positive constants such that $\sum a_k$ converges. Prove $\sum a_k^2$ converges. [Hint: Show $a_k^2 \le a_k$ for all but a finite number of k.]

18. If the series $\sum a_k x^k$ converges at $x = t$ show that the sequences $a_n t^n$ and $|a_n t^n|$ are bounded.

19. If $\sum a_k x^k$ converges at $x = t$, show that $\sum |a_k| x^k$ converges for $|x| < t$. [Hint: The sequence $a_k t^k$ is bounded by the previous exercise so the Bounded Terms Test can be applied to the series $\sum |a_k| x^k$.]

10.3 The interval of convergence

The first theorem in this section states that the set of points where a power series converges is a single point, an interval or the whole line. The other theorems in this section help us determine which case applies and what the interval is in that case.

THEOREM 10.8

Let $\sum a_k x^k$ be a given power series. One of the following statements is true.

i. The series converges only for $x = 0$.
ii. The series converges for all numbers x.
iii. There is a positive number R such that the series converges if $|x| < R$ and diverges if $|x| > R$. Either convergence or divergence of the series at $x = \pm R$ is possible and depends on the particular series.

Proof

Suppose that neither statement i nor statement ii is true for the given series. We must show that statement iii is true.

Let C be the set of all numbers x such that $\sum a_k x^k$ converges. Statement i does not hold, so there is some number $t \ne 0$ in C and, because statement ii does not hold, there is some number y that is not in C. We prove that $|y|$ is an upper bound for C.

Let c be any element of C so that $\sum a_k c^k$ converges to some value S. Let

$$s_n = \sum_{k=0}^{n} a_k c^k$$

be the nth partial sum. Then $s_{n+1} = s_n + a_{n+1}c^{n+1}$ and

$$\lim_{n\to\infty} s_{n+1} = \lim_{n\to\infty} s_n = S.$$

Thus

$$\lim_{n\to\infty} a_{n+1}c^{n+1} = \lim_{n\to\infty} s_{n+1} - s_n = S - S = 0.$$

We now show that this implies the sequence $a_n c^n$ is bounded. Take $\epsilon = 1$ in the definition of limit to conclude there is some number N such that

$$|a_n c^n - 0| = |a_n c^n| < 1 \quad \text{for} \quad n > N$$

Now take B to be the largest of the numbers

$$|a_0|, |a_1 c^1|, |a_2 c^2|, \ldots, |a_N c^N|, 1$$

so that $|a_n c^n| \leq B$ for all n. Now we are in a position to apply the Bounded Term Test. We conclude that the series converges for all x satisfying $|x| < |c|$. This computation proves that if c is an element of C, then C also contains every number x such that $|x| < |c|$. In particular, because statement i does not hold by assumption, C contains some nonzero positive number.

Now let y be any number where the series diverges and let c be any number where the series converges. We cannot have $|y| < |c|$ because we have just seen that this inequality would imply the convergence of the series at y. Hence $c \leq |c| \leq |y|$. Thus $|y|$ is an upper bound for the set C. The completeness property of the real numbers implies there is a smallest number R such that $c \leq R$ for every c in C. For this choice of R, we show that statement iii holds. Because C contains a nonzero positive number, it follows that $R > 0$. Take a number x with $|x| < R$. If x is not in C, then the series diverges at x so $|x|$ is an upper bound for C. But R is the least upper bound for S so $|x| \geq R$, an impossible situation. Thus the series converges at every number x with $|x| < R$. Most of statement iii has been proved. If y is a number with $|y| > R$ then y is not in C and so the series diverges at y. There remains the assertion that some series converge and other series diverge at x if $|x| = R$. Such examples are given later when we discuss convergence at the endpoints. □

Statement iii implies that the set of points where the series converges contains the interval $(-R, R)$ and might also contain one or both of the endpoints, $\pm R$. The decision about convergence at the endpoints must be made on a case-by-case basis.

Radius and Interval of Convergence

The positive number R in statement iii of Theorem 10.8 is called the *radius of convergence* of the series. The interval $(-R, R)$ is called the *interval of convergence.* If statement i holds in the theorem, we set $R = 0$; if statement ii holds, we set $R = \infty$ so the radius of convergence is defined for every series.

EXAMPLE 1 Find the radius and interval of convergence of the geometric series $\sum_{k=0}^{\infty} x^k$.

Solution: We saw already that

$$\frac{1}{1-x} = \sum_{k=0}^{\infty} x^k \qquad \text{if} \qquad |x| < 1.$$

We show that the series does not converge at $x = 1$. For $x = 1$ the nth partial sum is

$$s_n = \sum_{k=0}^{n} 1^k = n + 1.$$

Thus $\lim s_n = \lim n + 1 = \infty$ and the series diverges. As we have seen in the proof of the last theorem, $y = 1$ is an upper bound for the set of points where the series converges and clearly 1 is the least upper bound. It follows that the radius of convergence is $R = 1$ and the interval of convergence is $(-1, 1)$. We have just shown that the series diverges at the right endpoint. For completeness, let us show that it diverges at the left endpoint also.

For $x = -1$ the nth partial sum of the geometric series is

$$s_n = \sum_{k=0}^{n} (-1)^k = 1 - 1 + 1 - 1 + \cdots + (-1)^n = \begin{cases} 0 & \text{if } n \text{ odd,} \\ 1 & \text{if } n \text{ even.} \end{cases}$$

The terms s_n oscillate between the two values 0 and 1; the limit $\lim s_n$ does not exist and the series diverges at $x = -1$. ■

EXAMPLE 2 Find the radius of convergence of the series $\sum_{k=1}^{\infty} k! x^k$.

Solution: We show the radius of convergence is 0. Suppose the series converges for some number $x \neq 0$. If $x \geq 1$ then the nth Term Divergence test implies $\sum k! x^k$ diverges because $k! x^k$ does not approach 0. If $0 < x < 1$ select an integer n such that $nx \geq 2$ and let $k = n + N$ for an index $k > n$. We have

$$k! x^k = n! x^n (n+1)x \cdot (n+2)x \cdots (n+N)x > n! x^n 2 \cdot 2 \cdots 2 = n! x^n 2^N.$$

As k grows, n remains fixed and N grows; clearly $\lim_{k \to \infty} k! x^k \neq 0$, so the series diverges for $x \neq 0$. ■

The Bounded Terms Test for power series permits us to conclude $f(x) = \sum a_k x^k$ converges absolutely for $|x| < R$ if the sequence $a_k R^k$ is bounded. The largest R for which this holds is the radius of convergence of the power series. We give two tests, the Root Test and the Ratio Test, that are very useful for determining the radius of convergence. Before describing these tests, we make an important reduction that is essentially a restatement of the Bounded Terms Test applied to a series of constants.

THEOREM 10.9

The series $\sum_{k=0}^{\infty} a_k$ converges absolutely if there is a positive number ϵ such that the sequence $|a_k|(1 + \epsilon)^k$ is bounded.

Proof

Let

$$f(x) = \sum_{k=0}^{\infty} a_k x^k.$$

Assume the sequence $|a_k|(1+\epsilon)^k$ is bounded. The Bounded Terms Test implies $f(x)$ converges absolutely for $|x| < 1+\epsilon$. In particular $1 < 1+\epsilon$ so

$$f(1) = \sum_{k=0}^{\infty} a_k$$

converges absolutely. □

The next two theorems give conditions on the sequence a_k that permit us to invoke this theorem.

THEOREM 10.10

The Root Test

Let a_k be a sequence such that

$$\lim_{k\to\infty} |a_k|^{1/k} = L.$$

If $L < 1$ then $\sum a_k$ converges absolutely. If $L > 1$ then both $\sum a_k$ and $\sum |a_k|$ diverge. If $L = 1$ then the series can either converge or diverge.

Proof

For any positive number ϵ there is some integer N such that

$$\left||a_k|^{1/k} - L\right| < \epsilon \qquad \text{for all} \quad k \geq N.$$

This can be rewritten as

$$L - \epsilon < |a_k|^{1/k} < L + \epsilon.$$

We are interested only in small numbers ϵ, so let us assume that $\epsilon < L$ in the case that $L > 0$. In the case $L = 0$, the terms $L - \epsilon$ in the following argument should be replaced by 0. We rewrite the last inequality as

$$(L-\epsilon)^k < |a_k| < (L+\epsilon)^k \quad \text{for} \quad k \geq N. \tag{10.15}$$

Note that the series $\sum_{k=0}^{\infty} a_k$ converges if and only if the series $\sum_{k=N}^{\infty} a_k$ converges inasmuch as the two series differ only by the sum of a finite number of terms. We show convergence or divergence of $\sum_{k=N}^{\infty} a_k$.

Suppose $L > 1$. Then select ϵ so that $L - \epsilon > 1$. Then the left inequality of (10.15) implies

$$\infty = \lim_{k\to\infty} (L-\epsilon)^k \leq \lim_{k\to\infty} |a_k|$$

and so $\sum |a_k|$ diverges by the n term divergence test. We also see that the sequence a_k is unbounded and so $\sum a_k$ diverges by the same test.

Suppose $L < 1$. Select a positive number ϵ that satisfies

$$(L+\epsilon)(1+\epsilon) < 1.$$

Such an ϵ exists because

$$\lim_{t \to 0^+} (L + t)(1 + t) = L < 1.$$

Use the second half of inequality (10.15) to conclude

$$|a_k|(1+\epsilon)^k < (L+\epsilon)^k(1+\epsilon)^k = [(L+\epsilon)(1+\epsilon)]^k < 1.$$

Now apply Theorem 10.9 to conclude that $\sum |a_k|$ converges. □

Application of the Root Test

EXAMPLE 3 Determine the convergence or divergence of the two series

$$\sum_{k=1}^{\infty} \frac{1}{k^k} \quad \text{and} \quad \sum_{k=1}^{\infty} \frac{2^k}{k^2}.$$

Solution: For the first series, the Root Test yields

$$\lim_{k\to\infty} |a_k|^{1/k} = \lim_{k\to\infty} \left(\frac{1}{k^k}\right)^{1/k} = \lim_{k\to\infty} \frac{1}{k} = 0.$$

The limit is less than 1 so the series converges by the Root Test.

For the second series, the Root Test yields

$$\lim_{k\to\infty} |a_k|^{1/k} = \lim_{k\to\infty} \left(\frac{2^k}{k^2}\right)^{1/k} = \lim_{k\to\infty} 2\left(\frac{1}{k^2}\right)^{1/k}$$

To evaluate the limit that remains here, take the logarithm to get

$$\ln\left(\frac{1}{k^2}\right)^{1/k} = -\frac{2}{k}\ln k;$$

this has limit 0 and so $(1/k^2)^{1/k}$ has the limit $e^0 = 1$; thus

$$\lim_{k\to\infty} \left(\frac{2^k}{k^2}\right)^{1/k} = 2 \cdot 1 = 2.$$

The limit is greater than 1 and so the series diverges. ■

The Ratio Test

The next test is probably the most commonly used test for convergence because it is often quite easy to apply.

THEOREM 10.11

The Ratio Test

Let a_k be a sequence such that the limit

$$\lim_{k\to\infty} \left|\frac{a_{k+1}}{a_k}\right| = L.$$

If $L < 1$ then $\sum a_k$ converges absolutely. If $L > 1$ then both $\sum a_k$ and $\sum |a_k|$ diverge. If $L = 1$ then the series can either converge or diverge.

Proof

For any positive number ϵ, there is some integer N such that

$$\left|\frac{|a_{k+1}|}{|a_k|} - L\right| < \epsilon \quad \text{for} \quad k \geq N.$$

This can be rewritten as

$$L - \epsilon < \frac{|a_{k+1}|}{|a_k|} < L + \epsilon,$$

which implies

$$(L - \epsilon)|a_k| < |a_{k+1}| < (L + \epsilon)|a_k| \quad \text{for} \quad k \geq N. \tag{10.16}$$

For $k > N$, we use this inequality repeatedly to obtain

$$|a_k| < |a_{k-1}|(L + \epsilon) < |a_{k-2}|(L + \epsilon)^2 < \cdots < |a_N|(L + \epsilon)^{k-N}.$$

Assume $L < 1$. Select $\epsilon > 0$ so that $(L + \epsilon)(1 + \epsilon) < 1$ (as in the proof of the Root Test) so that we obtain

$$\begin{aligned} |a_k|(1 + \epsilon)^k &< |a_N|(L + \epsilon)^{k-N}(1 + \epsilon)^k \\ &= \frac{|a_N|}{(L + \epsilon)^N}[(L + \epsilon)(1 + \epsilon)]^k \\ &< \frac{|a_N|}{(L + \epsilon)^N}. \end{aligned}$$

Thus the sequence $|a_k|(1 + \epsilon)^k$ is bounded and the two series $\sum a_k$ and $\sum |a_k|$ converge by Theorem 10.9.

If $L > 1$, take $\epsilon > 0$ so that $L - \epsilon > 1$. By the same iterative argument given in the previous case we obtain

$$(L - \epsilon)^{k-N}|a_N| < |a_k|.$$

Because $L - \epsilon > 1$, the terms $(L - \epsilon)^{k-N}$ are unbounded and hence $|a_k|$ and a_k are unbounded sequences; both sums $\sum |a_k|$ and $\sum a_k$ diverge by the nth term test. □

Application of the Ratio Test

EXAMPLE 4 Apply the Ratio Test to determine the convergence or divergence of the two series

$$\sum_{k=1}^{\infty} \frac{2^k}{k!} \quad \text{and} \quad \sum_{k=0}^{\infty} \frac{2^k}{k(k+2)}.$$

Solution: With $a_k = 2^k/k!$ we have $|a_k| = a_k$ and

$$\frac{a_{k+1}}{a_k} = \frac{2^{k+1}}{(k+1)!} \cdot \frac{k!}{2^k} = \frac{2}{k+1}.$$

The limit of this ratio is $L = 0$ so the ratio test implies convergence of the first series.

Note that this series converges to $e^2 - 1$ as can be seen by examination of the Taylor series for e^x evaluated at $x = 2$.

Now consider the second series. If $a_k = 2^k/k(k+2)$ then $|a_k| = a_k$ and

$$\frac{a_{k+1}}{a_k} = \frac{2^{k+1}}{(k+1)(k+3)} \cdot \frac{k(k+2)}{2^k} = \frac{2k(k+2)}{(k+1)(k+3)}.$$

As k approaches ∞, the ratio approaches 2. The Ratio Test implies the series diverges. ■

The Root Test and the Ratio Test are quite effective when limit is less than 1 and the conclusion gives convergence of the series. When the limit is greater than 1, the conclusion gives the divergence of the series of absolute values, rather than the divergence of the original series. Of course if the terms of the series are all nonnegative then the original series and the series of absolute values coincide and a conclusion is reached about the original series.

Computing the Radius of Convergence

The Ratio Test and the Root Test can be used to determine the radius of convergence as we demonstrate in the following theorem.

THEOREM 10.12

Radius of Convergence

Let $f(x) = \sum a_k x^k$ be a power series with radius of convergence R. If

$$\lim_{k\to\infty} |a_k|^{1/k} = L,$$

then $R = 1/L$. If

$$\lim_{k\to\infty} \left|\frac{a_{k+1}}{a_k}\right| = L,$$

then $R = 1/L$.

Proof

Let x be any real number. Apply the Root Test to the series $\sum a_k x^k$ to get

$$\lim_{k\to\infty} |a_k x^k|^{1/k} = |x| \lim_{k\to\infty} |a_k|^{1/k} = |x|L.$$

The series converges if $|x|L < 1$ and diverges if $|x|L > 1$. This is equivalent to saying that the series converges if $|x| < 1/L$ and diverges if $|x| > 1/L$. It follows from the definition of radius of convergence that $R = 1/L$.

The proof of the analogous statement for the Ratio Test is similar to the one for the Root Test and is left to the reader to carry out. □

EXAMPLE 5 Find the radius of convergence of the series $\sum a_k x^k$ if

$$\text{i.}\quad a_k = \frac{k}{3^k} \qquad \text{and} \qquad \text{ii.}\quad a_k = \frac{2\cdot 4\cdots(2k-2)(2k)}{1\cdot 3\cdots(2k-1)(2k+1)}.$$

Solution: i. Apply the Root Test to the coefficient sequence to get

$$\lim_{k\to\infty} \left|\frac{k}{3^k}\right|^{1/k} = \frac{1}{3}\lim_{k\to\infty} k^{1/k} = \frac{1}{3}.$$

Thus $1/L = R = 3$ is the radius of convergence.

ii. The Ratio Test applies here and we have

$$\begin{aligned}\lim_{k\to\infty}\left|\frac{a_{k+1}}{a_k}\right| &= \lim_{k\to\infty}\frac{2\cdot 4\cdots(2k)(2k+2)}{1\cdot 3\cdots(2k+1)(2k+3)}\,\frac{1\cdot 3\cdots(2k-1)(2k+1)}{2\cdot 4\cdots(2k-2)(2k)}\\ &= \lim_{k\to\infty}\frac{2k+2}{2k+3}\\ &= 1.\end{aligned}$$

The radius of convergence in this case is $R = 1$. ■

Exercises 10.3

Use any of the tests described in this section to find the radius of convergence of each series.

1. $\displaystyle\sum_{k=1}^{\infty}\frac{x^k}{k}$

2. $\displaystyle\sum_{k=1}^{\infty}\frac{(2x)^k}{k^2}$

3. $\displaystyle\sum_{k=0}^{\infty}\frac{x^{2k}}{k+4}$

4. $\displaystyle\sum_{k=0}^{\infty}(-1)^k\frac{x^k}{2k+1}$

5. $\displaystyle\sum_{k=0}^{\infty}(-1)^k\frac{(3x)^k}{k!}$

6. $\displaystyle\sum_{k=0}^{\infty}\frac{(-x)^k}{(k+1)^k}$

7. $\displaystyle\sum_{k=0}^{\infty}\frac{(2k+5)(4x)^k}{k^3(k^2+1)}$

8. $\displaystyle\sum_{k=1}^{\infty}(-1)^k\frac{x^k}{(20)^k k(k+1)(k+2)}$

9. Suppose the series $\sum a_k x^k$ has $(-1,1)$ as its interval of convergence. What is the interval of convergence of $\sum a_k(x/2)^k$? What is the interval of convergence of $\sum a_k(cx)^k$ for a constant c? What is the interval of convergence of $\sum(-1)^k a_k(cx)^k$ for a constant c?

10. Let $p(k) = c_0 + c_1 k + \cdots + c_m k^m$ be a polynomial in k. Find the radius of convergence of the series

$$\sum_{k=N}^{\infty}\frac{x^k}{p(k)}$$

where $p(k) \neq 0$ for $k \geq N$.

11. Let $p(k)$ and $q(k)$ be polynomials. Find the radius of convergence of the power series having general term $q(k)x^k/p(k)$.

12. Revisit the first few exercises of this section in view of exercises 10 and 11.

13. Let a_n be a sequence such that $\lim_{n\to\infty} a_n = A$ is some finite number. Show that the power series $\sum a_n x^n$ converges for $|x| < 1$. [Hint: Show that the sequence a_n is bounded and then use the Bounded Terms Test.]

14. Let a_n be a sequence such that $\lim_{n\to\infty} a_n = A$ is some finite number. If the power series $\sum a_n x^n$ converges for $x = 1$ or $x = -1$, show that $A = 0$.

15. Use exercise 14 to show that series of the form $\sum p(k)x^k/q(k)$ converge for $|x| < 1$ but fail to converge at either $x = 1$ or $x = -1$ when $p(k)$ and $q(k)$ are polynomials of the same degree.

10.4 Differentiation and integration of series

The formula

$$f(x) = \sum_{k=0}^{\infty} a_k x^k \qquad |x| < R$$

defines the function f at every x on the interval of convergence of the series. We show that the derivative of f exists and can be expressed as a power series obtained by differentiating the series for f term by term. That is

$$f'(x) = \sum_{k=1}^{\infty} k a_k x^{k-1} \qquad |x| < R.$$

Similarly the definite integral of f can be computed as the series obtained by integrating the series for f term by term:

$$\int_0^x f(t)\,dt = \sum_{k=0}^{\infty} \int_0^x a_k t^k\,dt = \sum_{k=0}^{\infty} \frac{a_k}{k+1} x^{k+1} \qquad |x| < R.$$

Even though these are rather natural looking formulas, some care is need to carry out the proof. We begin by showing the differentiated series converges.

THEOREM 10.13

Convergence of the Derivative Series

Let the power series $\sum_{k=0}^{\infty} a_k x^k$ have nonzero radius of convergence R. Then the series

$$\sum_{k=1}^{\infty} k a_k x^{k-1} \tag{10.17}$$

converges for all x with $|x| < R$.

Proof

Let u be any number satisfying $0 < u < R$. We first show that the sequence $|n a_n u^{n-1}|$ is bounded. The proof uses the fact

$$\lim_{n \to \infty} n c^n = 0 \qquad \text{if} \qquad 0 < c < 1.$$

A proof of this fact is sketched in the exercises following this section. This limit implies that the sequence nc^n is bounded when $0 < c < 1$.

Let t be a number satisfying $u < t < R$ so that $c = u/t < 1$. Then

$$\begin{aligned} |n a_n u^{n-1}| &= |n a_n t^n| \left|\frac{u}{t}\right|^n \frac{1}{u} \\ &= |a_n t^n| \frac{1}{u} \cdot n \left|\frac{u}{t}\right|^n. \end{aligned}$$

The sequence $|a_n t^n|$ is bounded because the series $\sum a_k x^k$ converges at $x = t$. The term $1/u$ is constant and the terms $n(u/t)^n$ are bounded because of the remark above and the fact that $0 < u/t < 1$. Thus the sequence $|n a_n u^n|$ is bounded. By the Bounded Terms Test, the series (10.17) converges for every x satisfying $|x| < u$. For any x satisfying $|x| < R$, there is some number u with $|x| < u < R$ and so the series (10.17) converges at x. □

REMARK: If the series $\sum a_k x^k$ converges at $x = R$, it does not follow that the derivative series $\sum k a_k x^k$ converges at $x = R$. For example it has been shown that the Taylor series for $\ln(1 + x)$ has radius of convergence $R = 1$ and the Taylor series converges at $x = 1$. The derivative series is a geometric series that does not converge at $x = 1$.

The argument used in Theorem 10.13 can be applied to prove the analogous theorem about the series obtained using term-by-term integration.

THEOREM 10.14

Convergence of the Integrated Series

Let the power series $\sum_{k=0}^{\infty} a_k x^k$ have nonzero radius of convergence R. Then the series

$$\sum_{k=0}^{\infty} \frac{a_k x^{k+1}}{k+1} \qquad \textbf{(10.18)}$$

converges for all x with $|x| < R$.

Proof

Let u be any number with $0 < u < R$. Then the terms $|a_n u^n|$ are bounded because the series converges at u. Then

$$\left|\frac{a_n u^{n+1}}{n+1}\right| = |a_n u^n| \frac{u}{n+1}.$$

Because u is fixed, the terms $u/(n+1)$ are bounded and so the terms $|a_n u^{n+1}/(n+1)|$ are bounded. It follows that the series (10.18) converges for all x with $|x| < u$. For any x satisfying $|x| < R$, there is a u with $|x| < u < R$ so the series (10.18) converges for all x with $|x| < R$. □

Next we establish the continuity of a convergent power series. We need to know the series is a continuous function so we can assert its integral exists.

THEOREM 10.15

Continuity of Power Series

If a power series $f(x) = \sum a_k x^k$ has radius of convergence $R > 0$, then f is continuous at every point x with $|x| < R$.

In addition, if R is finite and if the series converges at $x = R$ (or at $x = -R$) then f is continuous at R (or at $-R$).

REMARK: The statement about continuity at the endpoints, R or $-R$, is known as *Abel's Theorem.*

Proof

We treat the case of continuity at a nonendpoint first. Let x and u be points on $(-R, R)$. For any given $\epsilon > 0$ it is necessary to show there is a $\delta > 0$ such that $|x - u| < \delta$ implies $|f(x) - f(u)| < \epsilon$. In more general terms, we want an upper estimate of $|f(x) - f(u)|$ when we know a bound for $|x - u|$. We have

$$|f(x) - f(u)| = \left|\sum a_k (x^k - u^k)\right| \le \sum |a_k||(x^k - u^k)|.$$

To put this in a more usable form, we apply the Mean Value Theorem to the function $g_n(x) = x^n$. For each n, there is a point z_n between x and u (and depending on n) such that

$$g_n(x) - g_n(u) = x^n - u^n = g'(z_n)(x - u) = n z_n^{n-1}(x - u).$$

We have the inequality

$$|f(x) - f(u)| \leq \sum |a_k||(x^k - u^k)| = \sum |a_k| \cdot k \cdot |z_k|^{k-1}|x - u|.$$

Because all of the numbers z_n are between x and u, there is some number w that satisfies $|z_n| < w < R$ for all n. The series $\sum k a_k x^{k-1}$ converges if $|x| < R$ by Theorem 10.13 so $S = \sum k|a_k|w^{k-1}$ is some finite positive number. We have

$$|f(x) - f(u)| \leq S|x - u|$$

so a δ that works for the given ϵ is $\delta = \epsilon/S$. Hence the conditions in the definition of continuity at u are satisfied by f.

This proof breaks down if $u = R$ (or if $u = -R$) because the derivative series might not converge at R even if the original series converges at R. The proof of the endpoint case requires a slightly more delicate argument.

The change of variable $x = tR$ transforms the series into one with the variable t and having radius of convergence equal to 1, so we assume at the start that $R = 1$ and that the series converges at $x = R = 1$. The case for convergence at $x = -R$ is treated in a similar way and is left to the reader.

Let x be any point where the series converges and let ϵ be any positive number. We must show there is a positive number δ such that

$$|f(x) - f(1)| < \epsilon \quad \text{whenever} \quad 1 - x < \delta. \qquad \textbf{(10.19)}$$

For each positive integer n let

$$b_n = \sum_{k=n}^{\infty} a_k,$$

so b_n is the "remainder" of the series after the $n - 1$ partial sum is removed. Then

$$\lim_{n\to\infty} b_n = \lim_{n\to\infty} \Big(f(1) - \sum_{k=0}^{n-1} a_k\Big) = 0$$

because the series converges to $f(1)$ by assumption. Also

$$a_n = b_n - b_{n+1}.$$

We have

$$\begin{aligned} f(x) - f(1) &= \sum_{k=0}^{\infty} a_k(x^k - 1) \\ &= \sum_{k=0}^{\infty} (b_k - b_{k+1})(x^k - 1) \\ &= b_0(x^0 - 1) + \sum_{k=1}^{\infty} b_k\big((x^k - 1) - (x^{k-1} - 1)\big) \\ &= 0 + \sum_{k=1}^{\infty} b_k(x - 1)x^{k-1}. \end{aligned}$$

Because the b_n approach 0, there is some N such that $|b_j| < \epsilon/2$ for $j > N$. We assume that $0 < x < 1$ because we are interested in values of x near 1.

Then we have

$$\begin{aligned}
|f(x) - f(1)| &= |x-1|\left|\sum_{k=1}^{N} b_k x^{k-1} + \sum_{k=N+1}^{\infty} b_k x^{k-1}\right| \\
&\le (1-x)\left[\sum_{k=1}^{N} |b_k| x^{k-1} + \sum_{k=N+1}^{\infty} |b_k| x^{k-1}\right] \\
&\le (1-x)\sum_{k=1}^{N} |b_k| x^k + \frac{\epsilon}{2}(1-x)\sum_{k=N+1}^{\infty} x^{k-1} \\
&\le (1-x)\sum_{k=1}^{N} |b_k| + \frac{\epsilon}{2}.
\end{aligned}$$

In the last equality we used $(1-x)\sum_{k=N+1}^{\infty} x^{k-1} = x^N < 1$. Now finally we take

$$\delta = \frac{\epsilon}{2\sum_{k=1}^{N} |b_k|},$$

so that whenever $1 - x < \delta$, $|f(x) - f(1)| < \epsilon$. Thus f is continuous at $x = 1$. □

Now that we know a convergent power series is a continuous function, we can integrate the power series. We show the integral can be found by integrating term by term.

THEOREM 10.16

Term-By-Term Integration of Series

Let

$$f(x) = \sum_{k=0}^{\infty} a_k x^k$$

be a power series with radius of convergence $R > 0$. Then

$$\int_0^x f(t)\,dt = \sum_{k=0}^{\infty} \int_0^x a_k t^k\,dt = \sum_{k=0}^{\infty} a_k \frac{x^{k+1}}{k+1}$$

for all x with $|x| < R$.

Proof

For convenience, let us write

$$F(x) = \int_0^x f(t)\,dt$$

$$G(x) = \sum_{k=0}^{\infty} a_k \frac{x^{k+1}}{k+1}.$$

For any positive integer n, we write the series $f(x)$ as the sum of its first n terms plus the remaining terms:

$$f(x) = \sum_{k=0}^{n} a_k x^k + \sum_{k=n+1}^{\infty} a_k x^k.$$

Then

$$F(x) - \sum_{k=0}^{n} \frac{a_k x^{k+1}}{k+1} = \int_0^x \Big(f(t) - \sum_{k=0}^{n} a_k t^k\Big)\, dt = \int_0^x \sum_{k=n+1}^{\infty} a_k t^k \, dt. \qquad \textbf{(10.20)}$$

We estimate the difference and determine the limit as n grows to infinity. First we fix x and keep in mind that $0 \le |t| \le |x|$ in the integral. Then

$$\begin{aligned}
\Big|\int_0^x \sum_{k=n+1}^{\infty} a_k t^k \, dt\Big| &\le \int_0^x \sum_{k=n+1}^{\infty} |a_k||t|^k \, dt \\
&\le \int_0^x \sum_{k=n+1}^{\infty} |a_k||x|^k \, dt \\
&\le \Big(\sum_{k=n+1}^{\infty} |a_k||x|^k\Big)\Big|\int_0^x dt\Big| \\
&\le \Big(\sum_{k=n+1}^{\infty} |a_k||x|^k\Big)|x|.
\end{aligned}$$

Using Equation (10.20), we obtain the estimate

$$\Big|F(x) - \sum_{k=0}^{n} \frac{a_k x^{k+1}}{k+1}\Big| \le \Big(\sum_{k=n+1}^{\infty} |a_k||x|^k\Big)|x|.$$

Now we prove that the right side approaches zero as n grows to infinity. Because x is inside the interval of convergence of f, it follows that $\sum_{k=0}^{\infty} |a_k||x|^k$ converges (by the Bounded Terms Test) to some number $A(x)$. Then

$$A(x) = \lim_{n\to\infty} \sum_{k=0}^{n} |a_k x^k|$$

implies

$$0 = \lim_{n\to\infty} \Big(A(x) - \sum_{k=0}^{n} |a_k x^k|\Big) = \lim_{n\to\infty} \sum_{k=n+1}^{\infty} |a_k x^k|.$$

This gives the limit we set out to prove, namely

$$0 \le \lim_{n\to\infty} \Big|F(x) - \sum_{k=0}^{n} \frac{a_k x^{k+1}}{k+1}\Big| \le \lim_{n\to\infty} \Big(\sum_{k=n+1}^{\infty} |a_k||x|^k\Big)|x| = 0.$$

It follows that

$$F(x) = \int_0^x \sum_{k=0}^{\infty} a_k t^k \, dt = \sum_{k=0}^{\infty} \frac{a_k x^{k+1}}{k+1}.$$

□

EXAMPLE 1 Integrate the geometric series to obtain a series converging to $\ln(1 - x)$.

Solution: The geometric series

$$\frac{1}{1-x} = 1 + x + x^2 + x^3 + \cdots + x^n + \cdots$$

has radius of convergence $R = 1$. Apply the theorem on term-by-term integration to obtain, for $|x| < 1$,

$$\int_0^x \frac{dt}{1-t} = -\ln(1-t)\Big|_0^x = -\ln(1-x)$$

$$\int_0^x (1 + t + t^2 + t^3 + \cdots + t^n + \cdots)\, dt = x + \frac{x^2}{2} + \frac{x^3}{3} + \cdots + \frac{x^n}{n} + \cdots.$$

Thus we have the series

$$\ln(1-x) = -x - \frac{x^2}{2} - \frac{x^3}{3} - \cdots - \frac{x^n}{n} - \cdots$$

that converges for $|x| < 1$.

By replacing x with $-x$, the series becomes the Taylor series of $\ln(1+x)$,

$$\ln(1+x) = x - \frac{x^2}{2} + \frac{x^3}{3} - \cdots + (-1)^{n+1}\frac{x^n}{n} - \cdots.$$

This series converges for $|x| < 1$.

After we study alternating series, it will be seen that the series

$$1 - \frac{1}{2} + \frac{1}{3} - \frac{1}{4} + \cdots$$

converges. Thus the power series

$$f(x) = x - \frac{x^2}{2} + \frac{x^3}{3} - \cdots + (-1)^{n+1}\frac{x^n}{n} - \cdots$$

converges for all x on $(-1, 1]$. We have just seen that $f(x) = \ln(1+x)$ for $|x| < 1$ so we ask about the equality at $x = 1$. By Abel's Theorem about convergence at an endpoint we know $f(x)$ is continuous at $x = 1$. Of course $\ln(1+x)$ is also continuous at $x = 1$. We conclude $f(1) = \ln 2$ by the following reasoning. From $f(x) = \ln(1+x)$ for $|x| < 1$, it follows that $\lim_{x\to 1} f(x) = \lim_{x\to 1} \ln(1+x)$. By the continuity of each function we get $f(1) = \ln(1+1) = \ln 2$. Notice that this improves on an earlier result in which we had only shown equality of the series and $\ln(1+x)$ for x on the interval $[-1/2, 1]$. ■

Now we can prove that the derivative of a convergent power series is found by differentiating the series term by term.

THEOREM 10.17

Term-By-Term Differentiation of Series

Let

$$f(x) = \sum_{k=0}^{\infty} a_k x^k$$

be a power series with radius of convergence R. Then $f'(x)$ exists for every x on $(-R, R)$ and moreover

$$f'(x) = \sum_{k=1}^{\infty} k a_k x^{k-1} \qquad \text{for} \qquad |x| < R.$$

Proof

Let $g(x) = \sum_{k=1}^{\infty} k a_k x^{k-1}$. We have shown that $g(x)$ converges for every x

on $(-R, R)$. We can integrate $g(x)$ term by term to get

$$\int_0^x g(t)\,dt = \sum_{k=1}^{\infty} \int_0^x k a_k t^{k-1}\,dt = \sum_{k=1}^{\infty} a_k x^k.$$

This series is the series for $f(x)$ except for the constant term so we have

$$f(x) = a_0 + \int_0^x g(t)\,dt.$$

Now apply the Fundamental Theorem of Calculus to obtain

$$f'(x) = g(x),$$

as we wished to prove. □

The derivative f' of the convergent power series is again a convergent power series and so it too has a derivative f''. By iterating this argument, we conclude that a convergent power series is differentiable infinitely many times at each point of its interval of convergence. Recall that the discussion of Taylor series began with a function that was infinitely differentiable. We now close the circle of ideas and return to the Taylor series of an infinitely differentiable function.

THEOREM 10.18

Convergent Power Series Are Taylor Series

Let $f(x) = \sum_{k=0}^{\infty} a_k x^k$ be a power series with radius of convergence $R > 0$. Then $f^{(n)}(x)$ exists for every n and every x with $|x| < R$. The coefficients of the power series are given by

$$a_k = \frac{f^{(k)}(0)}{k!},$$

and so the power series is the Taylor series centered at zero of $f(x)$.

Proof

The series for f

$$f(x) = a_0 + a_1 x + a_2 x^2 + \cdots + a_n x^n + \cdots$$

yields a series for $f'(x)$ by differentiating term by term to get

$$f'(x) = a_1 + 2a_2 x + 3a_3 x^2 + \cdots + n a_n x^{n-1} + \cdots.$$

The series for f' converges if $|x| < R$, and so its derivative can be found using term-by-term differentiation with the result

$$f''(x) = 2a_2 + 2 \cdot 3x + 3 \cdot 4x^2 + \cdots + n(n-1)a_n x^{n-2} + \cdots.$$

After differentiating n times, we get

$$f^{(n)}(x) = n(n-1)\cdots 3 \cdot 2 \cdot 1\, a_n + (n+1)n \cdots 3 \cdot 2a_{n+1}x + \cdots$$

and

$$\frac{f^{(n)}(0)}{n!} = a_n.$$

The function f is infinitely differentiable at each x, $|x| < R$, and the coefficients a_k are given exactly as in Taylor's formula. Thus the convergent power series is the Taylor series centered at zero of f. □

EXAMPLE 2 The Taylor series for $\sin x$ is given by

$$\sin x = x - \frac{x^3}{3!} + \frac{x^5}{5!} + \cdots = \sum_{k=0}^{\infty}(-1)^k \frac{x^{2k+1}}{(2k+1)!}.$$

Use it to find the Taylor series for $\cos x$.

***Solution*:** Because $\cos x$ is the derivative of $\sin x$, we differentiate the series term by term to get

$$\cos x = (\sin x)' = \sum_{k=0}^{\infty}(-1)^k(2k+1)\frac{x^{2k}}{(2k+1)!}$$

$$= \sum_{k=0}^{\infty}(-1)^k \frac{x^{2k}}{(2k)!} = 1 - \frac{x^2}{2!} + \frac{x^4}{4!} - \cdots.$$

■

EXAMPLE 3 The Taylor series of $1/(1-x)^m$ for positive m can be found by differentiating the geometric series. For $m = 1, 2, 3$ and $|x| < 1$ we have

$$\frac{1}{1-x} = 1 + x + x^2 + x^3 + x^4 + \cdots + x^n + \cdots,$$

$$\frac{1}{(1-x)^2} = 1 + 2x + 3x^2 + 4x^3 + \cdots + nx^{n-1} + \cdots,$$

$$\frac{2}{(1-x)^3} = 2 + 6x + 12x^2 + \cdots + n(n-1)x^{n-2} + \cdots.$$

■

Exercises 10.4

For each $f(x)$, guess the form of the nth term, then find $f'(x)$ and $\int_0^x f(t)\,dt$ for x a point on the interval of convergence of f and express the answer as a summation.

1. $f(x) = 1 - x + \frac{x^2}{2!} - \frac{x^3}{3!} + \frac{x^4}{4!} - \cdots$

2. $f(x) = 2x - \frac{(2x)^3}{3!} + \frac{(2x)^5}{5!} - \frac{(2x)^7}{7!} + \cdots$

3. $f(x) = \frac{x}{5} - \frac{x^2}{2 \cdot 5^2} + \frac{x^3}{3 \cdot 5^3} - \frac{x^4}{4 \cdot 5^4} + \cdots$

Each of the following series is the derivative or integral of a series closely related to the geometric series. Find a closed form sum of each.

4. $\sum_{k=3}^{\infty} kx^{k-1}$

5. $\sum_{k=1}^{\infty} kx^k$

6. $\sum_{k=0}^{\infty} \frac{x^{k+2}}{k+1}$

7. $\sum_{k=1}^{\infty} kx^{k+1}$

8. $\sum_{k=2}^{\infty} k(k+1)x^{k+3}$

9. $\sum_{k=2}^{\infty} k(x^k - x^{k-1})$

Find power series centered at 0 for $F(x)$ and give the interval of convergence.

10. $F(x) = \int_0^x \frac{\sin t}{t}\,dt$

11. $F(x) = \int_0^x \frac{1-\cos t}{t^2}\,dt$

12. $F(x) = \int_0^x \frac{1-e^{-t}}{t}\,dt$

13. $F(x) = \int_0^x \frac{\ln(1+t)}{t}\,dt$

14. The Bessel function of order 0 is given by the series

$$J_0(x) = \sum_{k=0}^{\infty} \frac{(-1)^n}{(n!)^2}\left(\frac{x}{2}\right)^{2n}.$$

Find the radius of convergence of the series and then use term-by-term differentiation to show that $y = J_0(x)$ is a solution of the differential equation $xy''+y'+xy = 0$.

15. Let $P_{2n}(x)$ be the sum of the first n nonzero terms in the series for $J_0(x)$ given in Exercise 14. Give an accurate sketch of $P_2(x)$, $P_4(x)$, and $P_6(x)$ to get some indication of the shape of the graph of J_0 near the origin.

16. Let a_n be a sequence such that $\lim_{n\to\infty} a_n = A$ and let

$$f(x) = \sum_{k=0}^{\infty} a_k x^k.$$

Then $f(x)$ converges for $|x| < 1$ (by the Bounded Terms Test). Show that $\lim_{x\to 1^-}(1-x)f(x) = A$.

17. Let $f(x) = \sum_{k=0}^{\infty} a_k x^k$ and suppose $f(x) = 0$ for every x satisfying $|x| < r$ for some positive r. Show that $a_k = 0$ for every k. [Hint: Evaluate f and its derivatives at 0.]

18. Use Exercise 17 to prove that

$$\sum_{k=0}^{\infty} b_k x^k = \sum_{k=0}^{\infty} c_k x^k$$

for all x on $(-r, r)$ for some $r > 0$, then $b_k = c_k$ for every k.

19. If f is an even function (so that $f(-x) = f(x)$ for all x) and if $f(x) = \sum_{k=0}^{\infty} a_k x^k$ for $|x| < 1$, then $a_k = 0$ for every odd k.

20. If f is an odd function (so that $f(-x) = -f(x)$ for all x) and if $f(x) = \sum_{k=0}^{\infty} a_k x^k$ for $|x| < 1$ then $a_k = 0$ for every even k.

Express the functions $f(x)$ as a power series and then use the theorem on the continuity of convergent power series to determine a constant c that makes $f(x)$ continuous for all x.

21. $f(x) = \dfrac{\cos x - 1}{x^2}$ for $x \neq 0$, $f(0) = c$

22. $f(x) = \dfrac{\sin 2x - 2x}{x^3}$ for $x \neq 0$, $f(0) = c$

23. $f(x) = \dfrac{e^{2x} - 1 - 2x}{x^2}$ for $x \neq 0$, $f(0) = c$

24. $f(x) = \dfrac{e^x - 1 - x}{x}$ for $x \neq 0$, $f(0) = c$

10.5 Computation of Taylor series

In this section we show how the use of substitutions, derivatives, integrals and other manipulations can give relatively simple constructions for the Taylor series of certain functions.

We begin by enlarging the class of functions for which Taylor series are considered. Up to now we have discussed only Taylor series centered at 0 of a function $g(x)$:

$$g(x) = \sum_{k=0}^{\infty} a_k x^k \qquad |x| < R$$

where

$$a_n = \frac{g^{(n)}(0)}{n!}. \tag{10.21}$$

For any number a, let

$$f(x) = g(x-a).$$

Then $f(x)$ has a power series representation obtained from the series for g by substituting $x - a$ in place of x; the series converges whenever $|x - a| < R$. We have

$$f(x) = \sum_{k=0}^{\infty} a_k (x-a)^k \qquad |x-a| < R.$$

This is the Taylor series of f centered at a. The derivatives of f satisfy $f^{(n)}(x) = g^{(n)}(x-a)$ and $f^{(n)}(a) = g^{(n)}(0)$. The coefficients are given by Equation (10.21) and satisfy

$$a_n = \frac{f^{(n)}(a)}{n!}.$$

If $P_n(x)$ is the Taylor polynomial centered at a of $f(x)$, then

$$\lim_{n\to\infty} P_n(x) = \sum_{k=0}^{\infty} a_k(x-a)^k \qquad |x-a| < R.$$

The theorems dealing with differentiation, integration and convergence of Taylor series centered at 0 carry over to the more general setting of Taylor series centered at a.

Here are a few examples showing Taylor series centered at a nonzero point.

EXAMPLE 1 For any number a, find the Taylor series of e^x centered at a and discuss the convergence of the series.

Solution: For any nonnegative integer n, the nth derivative of $f(x) = e^x$ is e^x and so the nth coefficient of the Taylor series centered at a for e^x is

$$\frac{f^{(n)}(a)}{n!} = \frac{e^a}{n!}.$$

The Taylor series is

$$e^x = \sum_{k=0}^{\infty} \frac{e^a}{k!}(x-a)^k.$$

For any number $r > 0$, we have

$$\lim_{k\to\infty} \frac{e^a}{k!}|r-a|^k = e^a \lim_{k\to\infty} \frac{|r-a|^k}{k!} = 0.$$

It follows that the sequence $e^a(r-a)^k/k!$ is bounded. By the Bounded Terms Test, the series converges for all x with $|x-a| < |r-a|$. Inasmuch as r was an arbitrary number, the series converges for all x. ■

The series for e^x centered at a could have been found using the series for e^x and substituting $x - a$ for x to get

$$e^{x-a} = \sum_{k=0}^{\infty} \frac{(x-a)^k}{k!},$$

which converges for all x. After multiplication by e^a, we get the same series as in the example.

Here is an example with somewhat more complicated calculations required.

EXAMPLE 2 Find the Taylor series for $\sin x$ centered at $a = \pi/4$.

Solution: Let $f(x) = \sin x$. Compute the derivatives at $\pi/4$:

$$f^{(4k)}(x) = \sin x, \qquad f^{(4k)}\left(\tfrac{\pi}{4}\right) = \tfrac{1}{\sqrt{2}},$$

$$f^{(4k+1)}(x) = \cos x, \qquad f^{(4k+1)}\left(\tfrac{\pi}{4}\right) = \tfrac{1}{\sqrt{2}},$$

$$f^{(4k+2)}(x) = -\sin x, \qquad f^{(4k+2)}\left(\tfrac{\pi}{4}\right) = -\tfrac{1}{\sqrt{2}},$$

$$f^{(4k+3)}(x) = -\cos x, \qquad f^{(4k+3)}\left(\tfrac{\pi}{4}\right) = -\tfrac{1}{\sqrt{2}}.$$

The series starts off as

$$\sin x = \frac{1}{\sqrt{2}} + \frac{1}{\sqrt{2}}\left(x - \frac{\pi}{4}\right) - \frac{1}{\sqrt{2}}\frac{1}{2!}\left(x - \frac{\pi}{4}\right)^2 - \frac{1}{\sqrt{2}}\frac{1}{3!}\left(x - \frac{\pi}{4}\right)^3 + \frac{1}{\sqrt{2}}\frac{1}{4!}\left(x - \frac{\pi}{4}\right)^4 + \cdots.$$

We can write the general term of the series more easily if we group all the even powers and then all the odd powers. We find

$$\sin x = \frac{1}{\sqrt{2}}\sum_{n=0}^{\infty}(-1)^n\frac{\left(x - \frac{\pi}{4}\right)^{2n}}{(2n)!} + \frac{1}{\sqrt{2}}\sum_{n=0}^{\infty}(-1)^n\frac{\left(x - \frac{\pi}{4}\right)^{2n+1}}{(2n+1)!}.$$

The series for $\sin x$ and $\cos x$ centered at 0 and evaluated at $x - \pi/4$ are visible in this equation. We can read the identity

$$\sin x = \frac{1}{\sqrt{2}}\sin\left(x - \frac{\pi}{4}\right) + \frac{1}{\sqrt{2}}\cos\left(x - \frac{\pi}{4}\right)$$

from the series obtained for $\sin x$. ■

Simple Substitutions

We next illustrate how a Taylor series can be found from a known series by making a change of variable. If we start with the geometric series

$$\frac{1}{1-x} = 1 + x + x^2 + x^3 + \cdots = \sum_{k=0}^{\infty} x^k \qquad |x| < 1$$

and replace x with $-x$ we obtain the series

$$\frac{1}{1+x} = 1 - x + x^2 - x^3 + \cdots = \sum_{k=0}^{\infty}(-1)^k x^k \qquad |x| < 1.$$

In each of these series we replace x by x^2 to get new series

$$\frac{1}{1-x^2} = 1 + x^2 + x^4 + x^6 + \cdots = \sum_{k=0}^{\infty} x^{2k} \qquad |x| < 1,$$

$$\frac{1}{1+x^2} = 1 - x^2 + x^4 - x^6 + \cdots = \sum_{k=0}^{\infty}(-1)^k x^{2k} \qquad |x| < 1.$$

Some care should be taken to calculate the correct radius of convergence of a transformed series. If

$$f(x) = \sum_{k=0}^{\infty} a_k x^k$$

has radius of convergence R and if we make a substitution $x = cu^m$ for some constant c and positive integer m, then we obtain a new series

$$g(u) = f(cu^m) = \sum_{k=0}^{\infty} a_k c^k u^{mk}$$

in the variable u. This series converges when $|x| = |cu^m| < R$. The restriction on u is $|u| < (R/|c|)^{1/m}$. If we rewrite the series for g using the variable x in place of u, we have

$$g(x) = \sum_{k=0}^{\infty} a_k c^k x^{mk} \qquad |x| < \left|\frac{R}{c}\right|^{1/m}.$$

Here is an illustration of this type of transformation.

EXAMPLE 3 Find the Taylor series centered at 0 for the function

$$f(x) = \frac{1}{16 + x^4}$$

and determine its radius of convergence.

Solution: The function can be written as

$$f(x) = \frac{1}{16}\frac{1}{1 + \frac{x^4}{16}} = \frac{1}{16}\frac{1}{1 + \left(\frac{x}{2}\right)^4}.$$

If we start with the series

$$\frac{1}{1+x} = \sum_{k=0}^{\infty}(-1)^k x^k \qquad |x| < 1$$

and make the replacement of $(x/2)^4$ for x, we get

$$\frac{1}{1 + \left(\frac{x}{2}\right)^4} = \sum_{k=0}^{\infty}(-1)^k \left(\frac{x}{2}\right)^{4k} \qquad \left|\frac{x}{2}\right|^4 < 1.$$

The condition $|x/2|^4 < 1$ is equivalent to $|x| < 2$. We obtain

$$f(x) = \frac{1}{16 + x^4} = \frac{1}{16}\left(\sum_{k=0}^{\infty}(-1)^k 2^{-4k} x^{4k}\right)$$

$$= \sum_{k=0}^{\infty}(-1)^k 16^{-k-1} x^{4k}.$$

■

The computation of the series in example 3 was surely much easier than the method that requires computation of the derivatives $f^{(n)}(0)$. In fact now that the series is known, we can give the values of these derivatives.

EXAMPLE 4 If

$$f(x) = \frac{1}{16 + x^4}$$

compute $f^{(20)}(0)$ and $f^{(21)}(0)$.

Solution: It would certainly be possible to compute $f^{(20)}(x)$ by hand if you had nothing else to do this afternoon. If you do have something else to do, then the Taylor series for f found in example 3 makes the computation trivial. The coefficient of x^n in the Taylor series for f is

$$\frac{f^{(n)}(0)}{n!} = \begin{cases} (-1)^k 16^{-k-1} & \text{if } n = 4k, \\ 0 & \text{otherwise.} \end{cases}$$

For $n = 20 = 4 \cdot 5$, we have $k = 5$ and

$$f^{(20)}(0) = -\frac{(20)!}{16^6} \approx -1.45012 \times 10^{11}$$

and $f^{(21)}(0) = 0$ because 21 is not an integer multiple of 4. ■

Series for Arctangent

The Taylor series for the function $f(x) = \arctan x$ can be obtained by integrating the series for $1/(1 + x^2)$. We have

$$\int_0^x \frac{dt}{1+t^2} = \arctan x - \arctan 0 = \arctan x$$

and

$$\begin{aligned}\int_0^x \frac{dt}{1+t^2} &= \int_0^x \left(1 - t^2 + t^4 - \cdots + (-1)^n t^{2n} + \cdots\right) dt \\ &= x - \frac{x^3}{3} + \frac{x^5}{5} - \cdots + (-1)^n \frac{x^{2n+1}}{2n+1} + \cdots \\ &= \sum_{k=0}^{\infty} (-1)^k \frac{x^{2k+1}}{2k+1}.\end{aligned}$$

Because the series for $1/(1 + x^2)$ has radius of convergence $R = 1$, it follows that

$$\arctan x = \sum_{k=0}^{\infty} (-1)^k \frac{x^{2k+1}}{2k+1} \qquad \text{for} \qquad |x| < 1.$$

After studying alternating series, we can show that the series for $\arctan x$ converges at $x = \pm 1$. The series at $x = 1$ can be used to estimate π. The equation $\tan(\pi/4) = 1$ implies $\arctan 1 = \pi/4$ and so

$$\pi = 4 \arctan 1 = 4\left(1 - \frac{1}{3} + \frac{1}{5} - \frac{1}{7} + \cdots\right).$$

If you use a computer to sum a large number of terms of this series, you discover that the convergence to the limit is very slow. The sum of the first n nonzero terms of the series is shown in table 10.1 for various values of n. Even for 1000 terms, the sum is not very close to $\pi \approx 3.14159$.

TABLE 10.1
Approximation of π

n	$\sum_{k=0}^{n} \frac{(-1)^k}{2k+1}$
50	3.16120
100	3.15149
150	3.14822
200	3.14657
250	3.14558
500	3.14359
1000	3.14259

Series that converge much faster, that is, get closer to the limit with a smaller number of terms, must be used for better approximations of π.

A Series for Logarithms

We have already derived a series for $\ln(1 + x)$ that converges for $|x| < 1$. We give an improved series here that is easily derived and converges more rapidly on some intervals.

Start with the two geometric series

$$\frac{1}{1-x} = 1 + x + x^2 + x^3 + \cdots = \sum_{k=0}^{\infty} x^k,$$

$$\frac{1}{1+x} = 1 - x + x^2 - x^3 + \cdots = \sum_{k=0}^{\infty} (-1)^k x^k,$$

both having radius of convergence $R = 1$. We form the sum of the two series and then integrate to get a series involving the natural logarithm:

$$\frac{1}{1-x} + \frac{1}{1+x} = 2 + 2x^2 + 2x^4 + \cdots = 2\sum_{k=0}^{\infty} x^{2k},$$

$$\int_0^x \frac{dt}{1-t} + \int_0^x \frac{dt}{1+t} = 2\sum_{k=0}^{\infty}\int_0^x t^{2k}\,dt,$$

$$-\ln(1-x) + \ln(1+x) = 2\sum_{k=0}^{\infty}\frac{x^{2k+1}}{2k+1}$$

$$\ln\left(\frac{1+x}{1-x}\right) = 2\left(x + \frac{x^3}{3} + \frac{x^5}{5} + \cdots\right).$$

This series converges for all x with $|x| < 1$. It is very useful because any positive number u can be expressed as

$$u = \frac{1+x}{1-x}$$

by taking

$$x = \frac{u-1}{u+1}.$$

Note that $u > 0$ implies $|x| < 1$. If we want to approximate $\ln 2$ we set $u = 2$ and find $x = 1/3$. The series gives

$$\ln 2 = 2\left(\frac{1}{3} + \frac{1}{3}\left(\frac{1}{3}\right)^3 + \frac{1}{5}\left(\frac{1}{3}\right)^5 + \frac{1}{7}\left(\frac{1}{3}\right)^7 + \cdots\right).$$

We use the computer to give some partial sums s_n of this numerical series. The convergence is quite fast at $x = 1/3$, so we carry the computations to 15 decimals. The results are shown in table 10.2.

TABLE 10.2
Approximation of ln 2

n	s_n
1	0.69135 80246 91358
2	0.69300 41152 26337
3	0.69313 47573 32288
4	0.69314 60473 90827
5	0.69314 70737 59785
6	0.69314 71702 56012
7	0.69314 71795 48241
8	0.69314 71804 59244
9	0.69314 71805 49812
10	0.69314 71805 58916

Multiplication of Series

If two power series

$$f(x) = \sum_{k=0}^{\infty} a_k x^k \quad \text{and} \quad g(x) = \sum_{k=0}^{\infty} b_k x^k$$

converge for $|x| < A$ then the product $f(x)g(x)$ has a power series representation

$$f(x)g(x) = \sum_{k=0}^{\infty} c_k x^k,$$

where that converges for $|x| < A$ and the coefficients are given by

$$\begin{aligned} c_0 &= a_0 b_0, \\ c_1 &= a_0 b_1 + a_1 b_0, \\ &\vdots \\ c_n &= a_0 b_n + a_1 b_{n-1} + \cdots + a_n b_0. \end{aligned}$$

Then the c_n are determined as if we were multiplying two polynomials. We do not prove this convergence result because, in fact, this is a seldom used formula because of the difficulty finding a closed formula for the general term c_n.

The multiplication is very useful when one of the power series is a finite series, that is a polynomial. In this case the proof of the multiplication formula can be given quite easily using the limit of partial sums. We do not carry this out here.

EXAMPLE 5 Find the Taylor series centered at 0 for $(1 - x^2)\arctan x$.

Solution: We have the series for $\arctan x$ so we need only multiply it by $(1 - x^2)$. We find

$$\begin{aligned} (1 - x^2)\arctan x &= (1 - x^2)\left(\sum_{k=0}^{\infty}(-1)^k \frac{x^{2k+1}}{2k+1}\right) \\ &= \sum_{k=0}^{\infty}(-1)^k \frac{x^{2k+1}}{2k+1} - \sum_{k=0}^{\infty}(-1)^k \frac{x^{2k+3}}{2k+1} \\ &= \left(x - \frac{x^3}{3} + \frac{x^5}{5} - \cdots\right) - \left(x^3 - \frac{x^5}{3} + \frac{x^7}{5} - \cdots\right) \\ &= x - \left(\frac{1}{3} + 1\right)x^3 + \left(\frac{1}{5} + \frac{1}{3}\right)x^5 + \cdots \\ &= x + \sum_{n=1}^{\infty}(-1)^n \frac{4n}{4n^2 - 1} x^{2n+1}. \end{aligned}$$

■

Division of Power Series

One power series can be divided by another and, if the denominator has nonzero constant term, the quotient is a power series. It is seldom possible to find more than a few terms of the quotient by this method but for some problems a few terms might suffice.

EXAMPLE 6 Find the terms up to x^7 in the Taylor series for $\tan x$ by dividing the series for $\sin x$ by the series for $\cos x$.

Solution: Because all terms after x^7 are to be ignored, we carry out the long division using only the first part of the $\sin x$ series and the first part of the $\cos x$ series:

$$\frac{x - \dfrac{x^3}{6} + \dfrac{x^5}{120} - \dfrac{x^7}{5040}}{1 - \dfrac{x^2}{2} + \dfrac{x^4}{24} - \dfrac{x^6}{720} + \dfrac{x^8}{8!}}$$

to get the result

$$\tan x = x + \frac{x^3}{3} + \frac{2x^5}{15} + \frac{17x^7}{315} + \text{terms with higher powers of } x. \qquad \blacksquare$$

If a power series

$$f(x) = a_0 + a_1 x + a_2 x^2 + \cdots$$

has a nonzero constant term $a_0 \neq 0$ then the function $1/f(x)$ has a Taylor series centered at 0. The first few terms of the power series can be found by the following device that we illustrate using $f(x) = \cos x$.

The series for the cosine function is

$$\cos x = 1 - \frac{x^2}{2!} + \frac{x^4}{4!} - \cdots = 1 - T$$

where $-T$ is the sum of the terms involving the nonzero powers of x. Then

$$\begin{aligned}\frac{1}{\cos x} &= \frac{1}{1-T} = 1 + T + T^2 + T^3 + \cdots \\ &= 1 + \left(\frac{x^2}{2!} - \frac{x^4}{4!} + \cdots\right) + \left(\frac{x^2}{2!} - \frac{x^4}{4!} + \cdots\right)^2 \\ &= 1 + \frac{x^2}{2} + \frac{5x^4}{24} + \cdots .\end{aligned}$$

The drawback to this technique is the difficulty in carrying out the computations. If only a few terms of the series are needed, this might prove satisfactory. With a bit of additional work we find the first few terms for the Taylor series of $\sec x = 1/\cos x$ are

$$\sec x = 1 + \frac{x^2}{2} + \frac{5x^4}{24} + \frac{61x^6}{72} + \frac{277x^8}{8064} + \cdots .$$

Taylor Series of Rational Functions

The Taylor series of some rational functions can be found by making simple substitutions in the geometric series. To find the Taylor series of $1/(1-x)$ centered at some point $a \neq 1$ we use the identity

$$\frac{1}{1-x} = \frac{1}{(1-a)-(x-a)} = \frac{1}{1-a} \frac{1}{\left(1 - \dfrac{x-a}{1-a}\right)}.$$

The first fraction on the right is a constant and the second has the form

$$\frac{1}{1-t} = 1 + t + t^2 + t^3 + \cdots$$

with

$$t = \frac{x-a}{1-a}.$$

We have

$$\frac{1}{1-x} = \left(\frac{1}{1-a}\right)\left(1 + \left(\frac{x-a}{1-a}\right) + \left(\frac{x-a}{1-a}\right)^2 + \left(\frac{x-a}{1-a}\right)^3 + \cdots\right).$$

Note that it was necessary to impose the restriction $a \neq 1$ because the function $1/(1-x)$ does not have a Taylor series centered at 1 because it is not even continuous at $x = 1$.

For the Taylor series of the rational function

$$\frac{1}{c+bx},$$

we make the substitution $t = bx/c$ so the function has the form

$$\frac{1}{c}\frac{1}{\left(1+\frac{bx}{c}\right)} = \frac{1}{c}\frac{1}{1+t}.$$

The Taylor series centered at any point except $t = -1$ can be written down almost immediately.

For more complicated rational functions, the method of partial fractions can be useful for simplifying the function. Terms with quadratic denominators can sometimes be reduced to the geometric series case by completing the square.

EXAMPLE 7 Find the Taylor series centered at 1 for the function

$$\frac{-2x+5}{x(x^2-2x+5)}.$$

Solution: Apply the method of partial fractions to get

$$\frac{-2x+5}{x(x^2-2x+5)} = \frac{1}{x} - \frac{x}{x^2-2x+5} = \frac{1}{x} - \frac{x}{(x-1)^2+4}.$$

The Taylor series of the first fraction is

$$\frac{1}{x} = \frac{1}{1+(x-1)} = 1-(x-1)+(x-1)^2-(x-1)^3+\cdots.$$

For the second fraction we compute

$$\frac{x}{x^2-2x+5} = \frac{x}{(x-1)^2+4} = \frac{x-1}{(x-1)^2+4} + \frac{1}{(x-1)^2+4}$$

and finally get

$$\frac{x-1}{(x-1)^2+4} = \frac{x-1}{4}\left(1-\left(\frac{x-1}{2}\right)^2+\left(\frac{x-1}{2}\right)^4-\left(\frac{x-1}{2}\right)^6+\cdots\right),$$

$$\frac{1}{(x-1)^2+4} = \frac{1}{4}\left(1-\left(\frac{x-1}{2}\right)^2+\left(\frac{x-1}{2}\right)^4-\left(\frac{x-1}{2}\right)^6+\cdots\right).$$

Now finally combine the three series to get

$$\frac{-2x+5}{x(x^2-2x+5)} = \sum_{k=0}^{\infty}(-1)^k(x-1)^k + \frac{1}{4}[1+(x-1)]\sum_{k=0}^{\infty}(-1)^k\left(\frac{x-1}{2}\right)^{2k}. \blacksquare$$

Exercises 10.5

Find the Taylor series centered at a for the given rational functions. Use partial fractions, substitutions and the geometric series to simplify the computations.

1. $\dfrac{1}{(x+1)(x-1)},\ a = 0$

2. $\dfrac{1}{(x^3+1)(x^3-1)},\ a = 0$

3. $\dfrac{x+1}{(x+2)(x-1)},\ a = 0$

4. $\dfrac{x+1}{(x+2)(x-1)},\ a = 2$

5. $\dfrac{x-3}{x(x+1)(x-1)},\ a = 3$

6. $\dfrac{(x+1)^2}{(x+3)^2(x-1)},\ a = -1$

7. $\dfrac{x^2}{(x^2+1)^3(x^2-1)^2},\ a = 0$ [Hint: $t = x^2$]

Use known series for familiar functions (e^x, $\sin x$, $\ln x$) to find the Taylor series centered at a for the given functions.

8. e^{x^2+2x}, $a = -1$

9. $\sin(x^2 + \pi)$, $a = \pi$

10. $\sin(x^2 + \pi)$, $a = 0$

11. $\ln(x + 4)$, $a = 1$

12. Use the Taylor series for $\sin x$ and long division to find the first four nonzero terms of the Taylor series for $x/\sin x$.

13. Use the multiplication of power series to find the first five nonzero terms of a power series for $e^x \cos x$. With almost no additional computation, do the same for $e^{-x} \cos x$.

14. Use the computation in exercise 13 to estimate $\int_0^{0.1} e^{-x} \cos x \, dx$ and give a bound for the size of the error.

15. Use the first n nonzero terms of the series for $\ln(1 + x)$ and for $\ln[(1 + x)/(1 - x)]$ to approximate $\ln(0.3)$ with $n = 4, 6, 8, 10, 12$. Which series seems to give more accurate results at least insofar as these computations show?

16. Find an interval $(-h, h)$ such that the difference

$$\left| \frac{\sin x}{x} - \cos x \right|$$

is at most 0.005. [Hint: The difference in the Taylor series of $\sin x/x$ and $\cos x$ is an alternating series whose partial sums differ from the function value by at most the first term omitted.]

17. Use the binomial theorem to write the first four nonzero terms of a power series for $(\cos x)^p = (1 - x^2/2 + \cdots)^p$ for a positive constant p.

18. Use exercise 17 to find a value of p such that

$$\frac{\sin x}{x} - (\cos x)^p = x^3 \sum_{k=0}^{\infty} a_k x^k$$

for some sequence $\{a_k\}$ of constants. For the value of p so determined, make accurate plots of the functions $\sin x/x$ and $\cos^p x$ over intervals $(-0.5, 0.5)$ and $(-\pi, \pi)$.

19. If $f(x) = \sum_{k=0}^{\infty} a_k x^k$ converges for $|x| < R$, use multiplication of series to show that

$$\frac{f(x)}{1 - x} = \sum_{k=0}^{\infty} \left(a_0 + a_1 + \cdots + a_k\right) x^k.$$

Comment on the interval for this series.

20. If $f(x) = \sum_{k=0}^{\infty} a_k x^k$ converges for $|x| < R$, use multiplication of series to find an expression similar to that in exercise 19 for $f(x)/(1 + x)$.

Use exercises 19 and 20 to write down the first six nonzero terms of the following series.

21. $\dfrac{\sin x}{1 - x}$

22. $\dfrac{\ln(1 + x)}{1 - x}$

23. $\dfrac{\cos x}{1 + x}$

24. $\dfrac{e^{-x}}{1 + x}$

25. If $f(x) = \sum_{k=0}^{\infty} a_k x^k$ converges for $|x| < R$, use multiplication of series and partial fractions to find a power series for $f(x)/(1 - x^2)$.

26. Assume $f(x)$ equals its convergent Taylor series for $|x| < R$. Show that $(f(x) + f(-x))/2$ equals the sum of the terms with even exponents in the series for f and that $(f(x) - f(-x))/2$ is the sum of the terms with odd exponents.

Use the idea of exercise 26 and knowledge of familiar series to find a closed form sum of the following series.

27. $\displaystyle\sum_{k=0}^{\infty} \frac{x^{2k}}{(2k!)}$

28. $\displaystyle\sum_{k=1}^{\infty} \frac{x^{2k-1}}{2k - 1}$

29. $\displaystyle\sum_{k=1}^{\infty} \frac{x^{2k-1}}{(2k)!}$

30. $\displaystyle\sum_{k=0}^{\infty} \frac{x^{2k}}{(2k + 1)!}$

31. Let $B(x) = x/(e^x - 1)$. Use the division of power series to find the terms up to x^7 of the Taylor series for $B(x)$ centered at 0.

32. Let $B(x)$ be the function defined in exercise 31. Show that the Taylor series of $B(x)$ has a nonzero coefficient of x but all other odd powers of x have zero coefficient. [Hint: Show $B(x) - 1 - x/2$ is an even function.]

33. Suppose $f(x) = \sum_{k=0}^{\infty} a_k x^k$ converges for $|x| < 1$ and $\lim_{n\to\infty} a_n = A$. Prove $(1 - x)f(x)$ is continuous on $[0, 1]$ and that $\lim_{x\to 1^-}(1 - x)f(x) = A$. [Hint: Use the theorem on the continuity of power series at the endpoints and a telescoping series.]

10.6 Applications of power series

In this section we solve a number of different problems using power series. Some of the problems have been solved in other ways in earlier sections by other methods.

Indeterminate Forms

Some of the limits that were computed by L'Hôpital's rule can be computed quite easily using Taylor series.

EXAMPLE 1 Evaluate

$$\lim_{x\to 0} \frac{\sin x}{x}$$

by using the Taylor series for $\sin x$.

Solution: We have

$$\frac{\sin x}{x} = \frac{x - \frac{x^3}{3!} + \frac{x^5}{5!} - \cdots}{x} = 1 - \frac{x^2}{3!} + \frac{x^4}{5!} - \cdots,$$

from which we conclude

$$\lim_{x\to 0} \frac{\sin x}{x} = 1$$

by simply substituting 0 for x in the series. ■

The main idea behind this example is that $\sin 0 = 0$ so the Taylor series for $\sin x$ has 0 as constant term and the series can be written as x times another power series. The same is true of the functions $e^x - 1 - x$ and $\cos x - 1$ in example 2 except that both have a factor x^2 instead of just x.

EXAMPLE 2 Evaluate the limit

$$\lim_{x\to 0} \frac{e^x - 1 - x}{\cos x - 1}.$$

Solution: Use the Taylor series for each function to get

$$e^x - 1 - x = \frac{x^2}{2!} + \frac{x^3}{3!} + \cdots = x^2\left(\frac{1}{2!} + \frac{x}{3!} + \cdots\right),$$

$$\cos x - 1 = \frac{x^2}{2!} - \frac{x^4}{4!} + \cdots = x^2\left(\frac{1}{2!} - \frac{x^2}{4!} + \cdots\right).$$

Thus

$$\frac{e^x - 1 - x}{\cos x - 1} = \frac{\frac{1}{2!} + \frac{x}{3!} + \cdots}{\frac{1}{2!} - \frac{x^2}{4!} + \cdots},$$

and the limit as x approaches 0 is found by substituting 0 for x. The limit is 1. ■

The explanation of L'Hôpital's rule is easily understood if the two functions involved have Taylor series representations. The rule amounts to cancelling a power of $x - a$ as we see in the following formulation of L'Hôpital's rule.

THEOREM 10.19

L'Hôpital's Rule by Taylor Series

Suppose $f(x)$ and $g(x)$ are functions that equal their convergent Taylor series centered at a point a. Suppose further that

$$f(a) = f'(a) = \cdots = f^{(k)}(a) = 0,$$
$$g(a) = g'(a) = \cdots = g^{(k)}(a) = 0, \qquad g^{(k+1)}(a) \neq 0,$$

for some positive integer k. Then

$$\lim_{x \to a} \frac{f(x)}{g(x)} = \frac{f^{(k+1)}(a)}{g^{(k+1)}(a)}.$$

Proof

The Taylor series centered at a for f is the sum of terms

$$\frac{f^{(n)}(a)}{n!}(x-a)^n$$

for $n \geq 0$. By assumption, the first k of these terms equal 0 so that

$$\begin{aligned} f(x) &= \frac{f^{(k+1)}(a)}{(k+1)!}(x-a)^{k+1} + \frac{f^{(k+2)}(a)}{(k+2)!}(x-a)^{k+2} + \cdots \\ &= (x-a)^{k+1}\left(\frac{f^{(k+1)}(a)}{(k+1)!} + \frac{f^{(k+2)}(a)}{(k+2)!}(x-a) + \cdots\right). \end{aligned}$$

Similarly we have

$$g(x) = (x-a)^{k+1}\left(\frac{g^{(k+1)}(a)}{(k+1)!} + \frac{g^{(k+2)}(a)}{(k+2)!}(x-a) + \cdots\right)$$

with the additional assumption that $g^{(k+1)}(a) \neq 0$. The quotient of these two functions can be simplified by cancelling a common factor $(x-a)^k$ and can be expressed as

$$\frac{f(x)}{g(x)} = \frac{\dfrac{f^{(k+1)}(a)}{(k+1)!} + \dfrac{f^{(k+2)}(a)}{(k+2)!}(x-a) + \cdots}{\dfrac{g^{(k+1)}(a)}{(k+1)!} + \dfrac{g^{(k+2)}(a)}{(k+2)!}(x-a) + \cdots}.$$

The limit as x approaches a can be found by substituting $x = a$ in this expression. All terms are 0 except for the first term in the denominator and possibly the first term in the numerator. The result stated in the theorem follows. □

Series Solutions of Differential Equations

Using series to solve differential equations is based on the uniqueness of the coefficients of a convergent series. If two power series are equal

$$\sum a_k x^k = \sum b_k x^k$$

for all x satisfying $|x| < A$ for some positive A, then $a_k = b_k$ for every k. This follows because a convergent power series can be differentiated term by term and the coefficients computed using the value of the derivatives at 0. We do not prove any general theorems about series solutions but simply give some examples that illustrate the ideas.

EXAMPLE 3 Assume that $y = f(x)$ is a function that is equal to its Taylor series for $|x| < R$ for some positive R and that y satisfies the differential equation

$$y' = 2y.$$

Find y.

Solution: This is an easy differential equation to solve without the use of series but we use series anyway just to illustrate the technique.

Let

$$y = \sum_{k=0}^{\infty} a_k x^k.$$

Then

$$y' = \sum_{k=1}^{\infty} k a_k x^{k-1} = a_1 + 2a_2 x + 3a_3 x^2 + \cdots + n a_n x^{n-1} + \cdots,$$

$$2y = 2\sum_{k=0}^{\infty} a_k x^k = 2a_0 + 2a_1 x + 2a_2 x^2 + \cdots + 2a_j x^j + \cdots.$$

Because the two series are equal at every x on the interval of convergence, the coefficients of equal powers of x must be the same in the two series. We have

$$a_1 = 2a_0, \qquad 2a_2 = 2a_1, \qquad 3a_3 = 2a_2$$

and more generally, the coefficients of x^m satisfy

$$(m+1)a_{m+1} = 2a_m$$

for all $m \geq 0$. This is a recursion formula in which each term a_{m+1} is expressed in terms of the preceding coefficient by the rule

$$a_{m+1} = \frac{2}{m+1} a_m.$$

But then a_m can be expressed in terms of the coefficient before it by

$$a_m = \frac{2}{m} a_{m-1},$$

which yields

$$a_{m+1} = \frac{2}{m+1}\frac{2}{m} a_{m-1}.$$

We can continue backing up until we reach a_0. Thus

$$\begin{aligned} a_{m+1} &= \frac{2}{m+1}\frac{2}{m}\frac{2}{m-1}\cdots\frac{2}{2}\frac{2}{1}a_0 \\ &= \frac{2^{m+1}}{(m+1)!}a_0. \end{aligned}$$

It follows that

$$y = \sum_{k=0}^{\infty} a_k x^k = a_0 \sum_{k=0}^{\infty} \frac{2^k}{k!} x^k.$$

The general term of this sum can be written as $(2x)^k/k!$, which is the general term of the series for e^{2x}. Thus the solution of the differential equation is

$$y = a_0 e^{2x}$$

with a_0 any constant. It is usually not possible to identify the series solution in closed form as a familiar function even though the examples in this section might suggest the contrary.

The "easy way" to solve the differential equation $y' = 2y$ is to integrate the equation in the form $y'/y = 2$ with the result $\ln y = 2x + c$ and $y = Ae^{2x}$ for some A. Of course the two techniques give the same result. ■

We give an additional example that is slightly more complicated.

EXAMPLE 4 Assume that $y = f(x)$ is a function that is equal to its Taylor series for $|x| < R$ for some positive R and that y satisfies the differential equation

$$y' = xy + 1.$$

Find the Taylor series of y.

Solution: Let

$$y = \sum_{k=0}^{\infty} a_k x^k,$$

so that

$$y' = \sum_{k=1}^{\infty} k a_k x^{k-1} = a_1 + 2a_2 x + 3a_3 x^2 + \cdots + k a_k x^{k-1} + \cdots,$$

$$xy + 1 = 1 + a_0 x + a_1 x^2 + \cdots + a_j x^{j+1} + \cdots.$$

These two power series converge and are equal for $|x| < R$; the coefficient of x^n in one series must be the same as the coefficient of x^n in the other series. We have

$$a_1 = 1, \qquad 2a_2 = a_0, \qquad 3a_3 = a_1,$$

for the first few terms and, more generally, for $n > 0$ we compare the coefficient of x^n to conclude

$$(n+1)a_{n+1} = a_{n-1}.$$

The correct way to view this recursion formula is to see that each coefficient can be computed in terms of the coefficient two steps earlier and then that coefficient is expressed in terms of the coefficient two steps before it. Moving back two steps at a time, we eventually reach either a_1 or a_0. In either case,

$$a_{n+1} = \frac{1}{n+1} a_{n-1} = \frac{1}{n+1}\frac{1}{n-1} a_{n-3} = \cdots.$$

For an even index $n = 2m$, we have

$$a_{2m} = \frac{1}{(2m)(2m-2)(2m-4)\cdots 4 \cdot 2} a_0.$$

For an odd index $n = 2m+1$, we have

$$a_{2m+1} = \frac{1}{(2m+1)(2m-1)(2m-3)\cdots 5 \cdot 3} a_1.$$

We had seen $a_1 = 1$ earlier, so the coefficients of odd powers of x are uniquely determined. The coefficients of the even powers of x all contain a factor a_0. We write the series for y as a sum of the series of odd powers plus the series of even powers in the form

$$y = 1 + a_0 \sum_{m=0}^{\infty} \frac{1}{(2m)(2m-2)(2m-4)\cdots 4 \cdot 2} x^{2m}$$

$$+ \sum_{m=0}^{\infty} \frac{1}{(2m+1)(2m-1)(2m-3)\cdots 5 \cdot 3} x^{2m+1}.$$

With a bit of manipulation and the Bounded Terms Test, one can show each of the series converges for all x. The a_0 is an arbitrary constant; for any choice of a_0, the function y satisfies the given differential equation. ■

Antiderivatives

There are functions for which no antiderivative can be expressed using only a finite number of familiar functions. The functions e^{x^2} and $(\sin x)/x$ are two examples. Both of these functions equal their Taylor series so that the antiderivatives can be expressed as convergent power series. Definite integrals of these functions can be expressed as infinite series of constants.

EXAMPLE 5 Find an antiderivative of the function $\sin x/x$.

Solution: We express the function as a convergent power series

$$\frac{\sin x}{x} = 1 - \frac{x^2}{3!} + \frac{x^4}{5!} - \cdots + (-1)^n \frac{x^{2n}}{(2n+1)!} + \cdots$$

and find the antiderivative using term-by-term integration

$$\int_0^x \frac{\sin t}{t}\, dt = \int_0^x 1 - \frac{t^2}{3!} + \frac{t^4}{5!} - \cdots + (-1)^n \frac{t^{2n}}{(2n+1)!} + \cdots\, dt$$

$$= x - \frac{x^3}{3!\,3} + \frac{x^5}{5!\,5} - \cdots + (-1)^n \frac{x^{2n+1}}{(2n+1)!\,(2n+1)} + \cdots.$$

■

The integral in example 6 is important in the study of statistics and probability because of its relation to the normal distribution or "bell curve".

EXAMPLE 6 Express the integral

$$\int_0^c e^{-x^2}\, dx$$

for a constant c as a convergent infinite series.

Solution: If we start with the Taylor series for e^x and replace x with $-x^2$, we obtain the Taylor series

$$e^{-x^2} = 1 - x^2 + \frac{x^4}{2!} - \cdots + (-1)^n \frac{x^{2n}}{n!} + \cdots$$

for e^{-x^2}, and this converges for all x. Using term-by-term integration we find

$$\int_0^c e^{-x^2}\, dx = \sum_{k=0}^{\infty} (-1)^k \int_0^c \frac{x^{2k}}{k!}\, dx$$

$$= \sum_{k=0}^{\infty} (-1)^k \frac{c^{2k+1}}{(2k+1)k!}.$$

Later we study alternating series and see that, for $c > 0$, this number can be approximated by taking the sum of the first n terms with an error no greater than the absolute value of the $(n+1)$st term. For example, the error involved in approximating this integral with the value

$$1 - \frac{c^3}{3} + \frac{c^5}{5 \cdot 2!}$$

is no greater than

$$\left|\frac{c^7}{7 \cdot 3!}\right|.$$

For a specific example we have

$$\int_0^{1/2} e^{-x^2}\,dx \approx 1 - \frac{1}{3 \cdot 2^3} + \frac{1}{5 \cdot 2 \cdot 2^5} = 0.9615$$

with an error no greater than $1/(2^7 \cdot 6) \approx 0.0013$. ■

Exercises 10.6

Use series to evaluate the limits.

1. $\lim_{x\to 0} \dfrac{e^x + e^{-x} - 2\cos x}{x^2}$

2. $\lim_{x\to 0} \dfrac{\tan 3x}{e^{-2x} + e^{2x}}$

3. $\lim_{t\to 0} \dfrac{e^{-t} - \cos\sqrt{t}}{\ln(1+t)}$

4. $\lim_{x\to 0} \dfrac{e^{-x}\sin x}{\ln(1+x)}$

5. $\lim_{x\to\infty} x\sin(2/x)$

6. $\lim_{t\to\infty} t^3\left(\arctan(2/t) - (2/t)\right)$

7. $\lim_{x\to 0} \dfrac{f(x) - f(0) - f'(0)x}{x^2}$ for f twice differentiable.

8. $\lim_{x\to 0} \dfrac{f(x) - P_n(x)}{x^n}$ where $P_n(x)$ is the nth Taylor polynomial of f centered at 0, where $f^{(n+1)}$ exists at 0.

9. The differential equation $xy'' + (1-x)y' + ny = 0$ is called *Laguerre's equation.* It is known that when n is a positive integer, the equation has a polynomial solution $L_n(x)$ of degree n satisfying $L_n(0) = 1$. Treat the polynomial solution as a series solution with all but a finite number of coefficients equal to 0 and find $L_1(x)$, $L_2(x)$, $L_3(x)$, and $L_4(x)$.

10. Find a series solution, centered at 0, of the equation $(x+1)y' + y = 0$.

11. Find a series solution, centered at 0, of the equation $(x-1)y' + y = 0$. [Hint: You might want to use exercise 10 and a substitution of $-x$ for x.]

12. Find a series solution in powers of x for the differential equation
$$x^2y'' - xy = \sin x.$$
[Hint: Replace the $\sin x$ term with its power series centered at 0 and assume y has a power series representation centered at 0.]

13. By looking for series solutions centered at 0 for the differential equation $xy'' + 2y' + xy = 0$, show how to discover the solution $y = x^{-1}\sin x$.

14. Attempt to solve the differential equation $x^2y' - y = -(x+1)$ by assuming $y = \sum a_k x^k$ and then following the power series method discussed in this section. Show that the series obtained converges only at $x = 0$. Conclude that this particular equation has no solution that can be expressed as a power series centered at 0.

10.7 Additional convergence tests

The tests for convergence that we have studied up to this point do not give conclusive results for every possible series. For example, when the limit in either the Ratio or the Root Test is 1, some other test must be applied. Now we give several more tests that can be applied to determine the convergence or divergence of a series of nonnegative terms.

Series of Nonnegative Terms

THEOREM 10.20

Nonnegative Series Test

If $a_k \geq 0$ for all k and if the partial sums are bounded by some number B,

$$s_n = \sum_{k=0}^{n} a_k < B,$$

then the series $\sum a_k$ converges.

Proof

The sequence of partial sums is monotone because the a_k are nonnegative, and so

$$s_{n+1} = s_n + a_{n+1} \geq s_n .$$

Now the conclusion follows from the principle that a bounded monotone sequence has a limit. □

The Integral Test

The integral test can be applied if there is a suitable continuous function $f(x)$ whose values at integers k give the sequence $a_k = f(k)$. When the integral test can be applied, it relates the convergence or divergence of the infinite series $\sum a_k$ to that of an improper integral of $f(x)$.

THEOREM 10.21

The Integral Test

Let $f(x)$ be continuous, nonnegative and decreasing on $[1, \infty)$. Then the infinite series $\sum_{k=1}^{\infty} f(k)$ and the improper integral $\int_1^{\infty} f(x)\,dx$ either both converge or both diverge.

Proof

Let n be a positive integer. Take the partition $P : 1 < 2 < 3 < \cdots < n$ of the interval $[1, n]$ and form the lower and upper sums for $f(x)$ relative to this partition. Inasmuch as f is a decreasing function, the maximum of f over $[i, i+1]$ is $f(i)$ and the minimum over the interval is $f(i+1)$. The interval lengths are equal to 1 so

$$L(P) = f(2) + \cdots + f(n)$$
$$U(P) = f(1) + \cdots + f(n-1)$$

are lower and upper sums for f over $[1, n]$. It follows that

$$f(2) + \cdots + f(n) < \int_1^n f(x)\,dx < f(1) + \cdots + f(n-1). \qquad \textbf{(10.22)}$$

Suppose the improper integral converges to L. Then every partial sum of the infinite series is bounded by $f(1) + L$ because

$$f(1) + f(2) + \cdots + f(n) < f(1) + \int_1^n f(x)\,dx < f(1) + \int_1^{\infty} f(x)\,dx.$$

A bounded monotone sequence has a limit, so it follows that

$$\sum_{k=1}^{\infty} f(k) = \lim_{n\to\infty} \sum_{k=1}^{n} f(k)$$

exists and is less than $f(1) + L$. Thus the convergence of the improper integral implies the convergence of the series.

Now suppose the series converges to a sum S. Let

$$q_n = \int_1^n f(x)\,dx.$$

By inequality (10.22) we have for every n

$$q_n < f(1) + \cdots + f(n-1) < \sum_{k=1}^{\infty} f(k) = S$$

so q_n is a bounded sequence. It is also monotone because

$$q_{n+1} = \int_1^{n+1} f(x)\,dx = q_n + \int_n^{n+1} f(x)\,dx \geq q_n.$$

Thus the sequence q_n has a limit and the limit is the improper integral. Thus the convergence of the series implies the convergence of the improper integral. □

The proof shows the integral is not too far from the sum of the series in case both converge. We have the inequalities

$$\int_1^{\infty} f(x)\,dx < \sum_{k=1}^{\infty} f(k) < f(1) + \int_1^{\infty} f(x)\,dx. \qquad \textbf{(10.23)}$$

EXAMPLE 1 Use the integral test to estimate the sum of the series

$$\sum_{k=1}^{\infty} \frac{1}{k^2}.$$

Solution: If we let $f(x) = 1/x^2$, then inequality (10.23) gives

$$\int_1^{\infty} \frac{dx}{x^2} = 1 < \sum_{k=1}^{\infty} \frac{1}{k^2} < 1 + \int_1^{\infty} \frac{dx}{x^2} = 2.$$

Clearly this is not a very good estimate of the sum. Inequality (10.23) can be used in a slightly different way to get a better estimate. Let s_n be the sum of the first n terms of the series. We ask how close an estimate is s_n to the sum of the series. The difference

$$\sum_{k=1}^{\infty} \frac{1}{k^2} - s_n = \sum_{k=n+1}^{\infty} \frac{1}{k^2}$$

can be estimated by applying the integral estimate to the remainder. In particular, we have

$$\int_{n+1}^{\infty} \frac{dx}{x^2} < \sum_{k=n+1}^{\infty} \frac{1}{k^2} < \frac{1}{(n+1)^2} + \int_{n+1}^{\infty} \frac{dx}{x^2}.$$

After evaluating the integrals we find

$$\frac{1}{n+1} < \sum_{k=1}^{\infty} \frac{1}{k^2} - s_n < \frac{1}{(n+1)^2} + \frac{1}{n+1},$$

and so we have the estimate

$$s_n + \frac{1}{n+1} < \sum_{k=1}^{\infty} \frac{1}{k^2} < s_n + \frac{n+2}{(n+1)^2}.$$

With some sample choices for n, we obtain estimates of the infinite series. For example, with $n = 9$ we find $s_9 \approx 1.54$ and so

$$1.54 + \frac{1}{10} = 1.64 < \sum_{k=1}^{\infty} \frac{1}{k^2} < 1.54 + \frac{11}{100} = 1.65.$$

Of course, there is a slight problem with this estimate because we have not computed the sum s_9 exactly. The exact value of s_9 is $9,778,141/6,350,400$ which should be used in place of 1.54. ■

Here is another application of the integral test.

p-Series

The series

$$\sum_{k=1}^{\infty} \frac{1}{k^p} = \frac{1}{1^p} + \frac{1}{2^p} + \frac{1}{3^p} + \cdots$$

is called a p-series; its convergence or divergence is determined by the integral test using the function $f(x) = 1/x^p$. Since we have already evaluated the improper integral, we just repeat the conclusion now:

$$\int_1^{\infty} \frac{dx}{x^p} = \begin{cases} \dfrac{1}{p-1} & \text{if } p > 1, \\ \text{diverges} & \text{if } p \leq 1. \end{cases}$$

We conclude

$$\sum_{k=1}^{\infty} \frac{1}{k^p} \quad \begin{cases} \text{converges} & \text{if } p > 1, \\ \text{diverges} & \text{if } p \leq 1. \end{cases}$$

In the special case $p = 1$ the p-series is called the *harmonic series.* The harmonic series

$$1 + \frac{1}{2} + \frac{1}{3} + \cdots = \sum_{k=1}^{\infty} \frac{1}{k}$$

diverges. This fact is frequently used in making comparisons with other series.

Exercises 10.7

Determine if the series converge. Indicate what test you use to arrive at an answer. Do not overlook the Ratio and Root Tests from earlier sections.

1. $\displaystyle\sum_{k=2}^{\infty} \frac{1}{k(k^2-1)}$

2. $\displaystyle\sum_{k=2}^{\infty} \frac{2^k}{k(k^2-1)}$

3. $\displaystyle\sum_{k=1}^{\infty} \frac{(1.1)^{-k^2}}{k^k}$

4. $\displaystyle\sum_{k=1}^{\infty} \frac{5^{k^2}}{k^k}$

5. $\displaystyle\sum_{k=1}^{\infty} \frac{1}{3k+2}$

6. $\displaystyle\sum_{k=1}^{\infty} \frac{k+1}{k^3+1}$

7. $\sum_{k=1}^{\infty} \frac{k}{10k^2+1}$

8. $\sum_{k=1}^{\infty} \frac{\sqrt{k}}{10k^{5/2}+1}$

9. Determine for which values of the variable c the series $\sum_{k=0}^{\infty} c^{k^2}$ converges.

10. Prove the following error estimate based on the integral test. If $f(x)$ is a continuous, positive and decreasing function and if

$$S = \sum_{k=1}^{\infty} f(k)$$

is a convergent series having nth partial sum s_n, then

$$\int_{n+1}^{\infty} f(x)\,dx < S - s_n < \int_{n}^{\infty} f(x)\,dx.$$

[Hint: Draw a picture to represent s_n and the integrals.]

11. Based on the estimate given in exercise 10, how many terms of the series

$$\sum_{k=1}^{\infty} \frac{1}{k^4}$$

must be taken to estimate the infinite sum with an error no greater than 9×10^{-6}?

10.8 Alternating series and conditional convergence

In this section we consider series of constants that might include both positive and negative terms. We have seen that the series $\sum a_k$ converges if the sum of the absolute values $\sum |a_k|$ converges; the series converges if it converges absolutely. The most subtle series to deal with are the series having both positive and negative terms and not converging absolutely. We give such a series a special designation.

DEFINITION 10.2

Conditionally Convergent Series

The series $\sum a_k$ is conditionally convergent if $\sum |a_k|$ diverges but $\sum a_k$ converges.

As an example of a conditionally convergent series we cite the alternating harmonic series

$$1 - \frac{1}{2} + \frac{1}{3} - \cdots + (-1)^n \frac{1}{n} + \cdots.$$

We have already seen in the discussion of the Taylor series for $\ln(x+1)$ that this series converges to $\ln 2$. The series of absolute values

$$1 + \frac{1}{2} + \frac{1}{3} + \cdots + \frac{1}{n} + \cdots$$

is the harmonic series that diverges.

The alternating harmonic series is conditionally convergent. Any conditionally convergent series must have both positive and negative terms. The series with terms of both sign most easily dealt with is an alternating series according to the following definition.

Alternating Series

If a_k is a sequence of positive terms then the terms in

$$\sum_{k=0}^{\infty}(-1)^k a_k$$

alternate in sign. Such a series is called an *alternating series*.

The next theorem gives a criterion for the convergence of alternating series satisfying an additional condition as well as an estimate of its sum.

THEOREM 10.22

Alternating Series Test

Let a_k be a monotone decreasing secuence of positive numbers having limit 0. Then the alternating series

$$\sum_{k=0}^{\infty}(-1)^k a_k$$

converges to a sum S. For any $n \geq 0$,

$$\left|S - \sum_{k=0}^{n-1}(-1)^k a_k\right| \leq a_n,$$

so the error in approximating S by the nth partial sum is no greater than the absolute value of the first term omitted.

Proof

We begin with the observation that $a_n - a_{n+1}$ is nonnegative because the sequence a_k is decreasing. Thus partial sums of the form

$$s_{2n+1} = (a_0 - a_1) + (a_2 - a_3) + \cdots + (a_{2n} - a_{2n+1})$$

are sums of positive terms and form an increasing sequence inasmuch as

$$s_{2(n+1)+1} = s_{2n+1} + (a_{2(n+1)} - a_{2(n+1)+1}) > s_{2n+1}.$$

We show these partial sums are bounded by rearranging the terms so that

$$s_{2n+1} = a_0 - (a_1 - a_2) - (a_3 - a_4) - \cdots - (a_{2n-1} - a_{2n}) - a_{2n+1}.$$

This shows that $s_{2n+1} < a_0$ because all the terms enclosed in parenthesis are positive. Hence the sequence $q_n = s_{2n+1}$ is a bounded increasing sequence and has a limit. Let

$$S = \lim_{n\to\infty} s_{2n+1}.$$

So far we have considered only half of the partial sums. Now we consider the other half, namely

$$s_{2n} = a_0 - a_1 + a_2 - a_3 + \cdots + a_{2n} = s_{2n-1} + a_{2n}.$$

Now we use the assumption that $\lim a_k = 0$ to conclude

$$\lim_{n\to\infty} s_{2n} = \lim_{n\to\infty} s_{2n-1} + 0 = S.$$

It follows from this and the definition of limit that

$$\lim_{k\to\infty} s_k = S.$$

Now we consider the estimation of the sum by the partial sums. We have already seen that the sequence $q_n = s_{2n+1}$ is increasing and has limit S. In particular s_{2n+1} is less than the limit; *i.e.* $s_{2n+1} \leq S$ for all n. The sequence of partial sums s_{2n} is decreasing because

$$s_{2(n+1)} = s_{2n} - (a_{2n+1} - a_{2(n+1)}) < s_{2n}.$$

Inasmuch as S is the limit of this decreasing sequence, every term is greater than or equal to the limit. For every n we have

$$s_{2n-1} < S < s_{2n} = s_{2n-1} + a_{2n}$$

and

$$0 < S - s_{2n-1} < a_{2n}.$$

Similarly,

$$s_{2n+1} = s_{2n} - a_{2n+1} < S < s_{2n},$$

and so

$$-a_{2n+1} < S - s_{2n+1} < 0.$$

The estimates for even sums and for odd sums can be unified into the statement

$$|S - s_k| < a_{k+1}$$

for any k.

Note that the approximation by a partial sum overestimates the sum if the first omitted term is negative and underestimates the sum if the first omitted term is positive. In other words, the remainder term has the same sign as the first term omitted. □

EXAMPLE 1 Estimate the sum of the alternating series

$$\sum_{k=1}^{\infty} (-1)^{k+1} \frac{1}{k^2}.$$

Solution: First observe that the series converges to a sum S because the sequence $1/k^2$ is a decreasing positive sequence with a limit 0. The sum of the first 49 terms

$$\sum_{k=1}^{49} (-1)^{k+1} \frac{1}{k^2} = \frac{790\ 09958\ 03121\ 18871\ 07106\ 06072\ 18267\ 71424\ 62911}{960\ 40768\ 39297\ 31549\ 80025\ 57630\ 75964\ 78009\ 60000} \approx 0.82267$$

differs from S by no more than $1/50^2 = 0.0004$. Because the first term omitted is negative, the approximation overestimates the sum and we have

$$0.82267 - 0.0004 = 0.82227 < S < 0.82267.$$ ■

The Alternating Series Test can be used with Taylor series as in the next example.

EXAMPLE 2 Treat the Taylor series centered at 0 for $\sin x$ as an alternating series and estimate $\sin 0.5$ with an error no greater than 10^{-3}.

Solution: The Taylor series for $\sin x$ is

$$\sin x = x - \frac{x^3}{3!} + \frac{x^5}{5!} - \cdots + (-1)^n \frac{x^{2n+1}}{(2n+1)!} + \cdots.$$

For $x = 0.5$, this is an alternating series and the terms $x^{2n+1}/(2n+1)!$ form a decreasing sequence. The Taylor polynomial $P_{2n-1}(x)$ is the sum of the first n terms and the alternating series estimate gives

$$|\sin x - P_{2n-1}(x)| < \frac{x^{2n+1}}{(2n+1)!}.$$

Substituting $x = 1/2$ and $n = 2$ we have

$$|\sin 0.5 - P_3(0.5)| < \frac{1}{2^5\, 5!} \approx 0.00026.$$

Thus $P_3(0.5) = 0.5 - (0.5)^3/6 \approx 0.4979$ is the approximation. ■

Example of Endpoint Convergence

The following example is instructive because it shows that the convergence at endpoints of the interval of convergence might be quite unpredictable.

Let

$$f(x) = \sum_{k=1}^{\infty} \frac{x^k}{k^2}.$$

The Bounded Terms Test shows that the series converges for $|x| < 1$. At $x = 1$, the series converges because it is a p-series with $p = 2$. At $x = -1$ the series converges by several tests. It is absolutely convergent and therefore convergent. It is also an alternating series with terms decreasing in absolute value. The derivative of the convergent series can be computed term by term and will converge for $|x| < 1$. We have

$$f'(x) = \sum_{k=1}^{\infty} \frac{x^{k-1}}{k}.$$

At $x = 1$, this is the harmonic series and does not converge. At $x = -1$ this is an alternating series that does converge. Thus $f(x)$ converges on $[-1, 1]$ whereas $f'(x)$ converges on $[-1, 1)$. The second derivative of f is

$$f''(x) = \sum_{k=2}^{\infty} \frac{k-1}{k} x^{k-2}.$$

This series does not converge at either ± 1 by the nth Term Test but does converge, of course, for $|x| < 1$.

Rearrangements and Conditionally Convergent Series

Suppose the series $\sum_{k=0}^{\infty} a_k$ converges to a sum S and we form a new sequence b_k whose terms are the same as the terms in the sequence a_k but not necessarily in the same order. Does the series $\sum_{k=0}^{\infty} b_k$ converge, and if yes, does it converge to S? Our knowledge that all rearrangements of finite sums are equal might lead us to guess that the answer is yes. In many cases that is correct but, surprisingly, it is not always correct. Let us first state the positive result.

THEOREM 10.23

Rearrangement of Absolutely Convergent Series

If the series $\sum_{k=0}^{\infty} a_k$ converges absolutely to S and if the sequence b_k is a rearrangement of the sequence a_k then $\sum_{k=0}^{\infty} b_k$ converges to S.

The proof is omitted but a proof in the case $a_k \geq 0$ is sketched in the exercises after this section.

The rearrangement of a series that is convergent but not absolutely convergent can produce a series with a different sum. We illustrate with the following example. The alternating harmonic series converges to $\ln 2$ as we have seen but the exact sum is not important for this example. We need only know that the series converges to some limit A. We write

$$A = 1 - \frac{1}{2} + \frac{1}{3} - \frac{1}{4} + \frac{1}{5} - \cdots . \tag{10.24}$$

Multiply the series by $1/2$ and insert some 0 terms to get

$$\frac{A}{2} = 0 + \frac{1}{2} + 0 - \frac{1}{4} + 0 + \frac{1}{6} + 0 - \frac{1}{8} + \cdots .$$

Add this to the original series to get

$$\frac{3A}{2} = 1 + \frac{1}{3} - \frac{1}{2} + \frac{1}{5} + \frac{1}{7} - \frac{1}{4} + \cdots .$$

The pattern for this series is to take the terms from the original series in a rearranged order; take the first two terms with odd denominators, then one with an even denominator, then two with odd denominators and one with an even denominator. It is not too difficult to write a formula for the general term of the rearranged series. The important point is that the rearranged series has sum $3A/2 \neq A$. Thus it is possible to rearrange the terms of a conditionally convergent series to obtain a series that converges to a different sum.

This is only an indication of what can happen when the terms of a conditionally convergent series are rearranged. Riemann showed in 1867 that any conditionally convergent series can be rearranged so that the sum is any number that you like. We will not give the details of this but we illustrate the proof using the alternating harmonic series.

Suppose we want to rearrange the terms of the series to produce a new series whose sum is some given number, let us take 3 just to be specific. We select the series b_k as follows. Let $b_1, b_2, \ldots, b_N$ be the first positive terms of the series (10.24) with N selected so that

$$\sum_{k=1}^{N} b_k > 3 \geq \sum_{k=1}^{N-1} b_k .$$

Then take $b_{N+1}, \ldots, b_{N+K}$ to be the first unused negative terms of series (10.24) with K selected so that

$$\sum_{k=1}^{N+K} b_k < 3 \leq \sum_{k=1}^{N+K-1} b_k .$$

Continue selecting unused positive terms, then unused negative terms, so that the partial sums move just beyond 3 then back just below 3. It can be shown that it is possible to continue this procedure indefinitely and that the resulting infinite sequence actually converges to 3.

The point of this digression is that one must take care manipulating infinite series. There is a difference in the behavior of absolutely convergent series and conditionally convergent series. An absolutely convergent series converges because the terms in the sum are very small; a conditionally convergent series converges because of the cancellation of positive and negative terms. Rearranging the series affects the cancellation.

Exercises 10.8

Determine if the following series converge absolutely, converge conditionally, or diverge. For each convergent series give an estimate of the sum with an error less than 10^{-1}.

1. $\sum_{k=1}^{\infty}(-1)^k\frac{k+1}{k^2}$

2. $\sum_{k=1}^{\infty}(-1)^k\frac{2k-3}{k^2}$

3. $\sum_{k=1}^{\infty}(-1)^k\frac{k^2-1}{k^2}$

4. $\sum_{k=1}^{\infty}(-2)^k$

5. $\sum_{k=1}^{\infty}\frac{\cos k\pi}{k^2+1}$

6. $\sum_{k=1}^{\infty}\frac{k\cos k\pi}{k^2+k+1}$

Find all the values of x for which the series converge. Be sure to determine the behavior at the endpoints. Describe the tests used to make the determination.

7. $\sum_{k=0}^{\infty}\frac{x^k}{5^k}$

8. $\sum_{k=0}^{\infty}(-1)^k\frac{x^k}{2^k}$

9. $\sum_{k=0}^{\infty}\frac{x^{3k}}{5^k}$

10. $\sum_{k=0}^{\infty}(-1)^k\frac{x^{2k}}{2^k}$

11. $\sum_{k=0}^{\infty}(-1)^k\frac{k+1}{k+2}x^k$

12. $\sum_{k=0}^{\infty}\frac{(x-2)^k}{2k^k}$

13. $\sum_{k=0}^{\infty}(-1)^k\frac{k}{k^2+1}(x-1)^{4k}$

14. $\sum_{k=0}^{\infty}(-1)^k\frac{(x+3)^k}{(1.5)^k}$

15. $\sum_{k=0}^{\infty}(-1)^k\frac{(2k+1)(2k-1)\cdots 3}{(2k+2)!}x^k$

16. $\sum_{k=1}^{\infty}(-1)^k\frac{k}{k(k+1)(k+2)}x^k$

17. Let a_k be a sequence of positive terms and let b_k be a rearrangement of the sequence a_k. If the series $\sum a_k$ converges to a sum S, then $\sum b_k$ also converges to S. [Hint: Let s_n be the nth partial sum of the series $\sum a_k$ and let t_n be the nth partial sum of the series $\sum b_k$. Then s_k is monotone increasing and bounded and has limit S. For any positive integer m, show there is a positive integer m' such that $t_m \le s_{m'} \le S$. Thus the sequence t_k is bounded and has limit $T \le S$. By reversing the roles of a_k and b_k show $S \le T$.]

18. Use the integral test to estimate how many of the first positive terms of the alternating harmonic series are needed to have the sum greater than 3.

19. The series

$$1-1+1-1+1-1+\cdots$$

diverges because the partial sums are $s_{2n+1} = 1$ and $s_{2n} = 0$, and so the limit of the partial sums does not exist. If the terms are grouped in pairs to produce a new series

$$(1-1)+(1-1)+(1-1)+\cdots,$$

the series converges because every partial sum is 0. Thus grouping terms might produce a convergent series even if the original series is divergent. Suppose that a_k is a sequence with $\lim a_k = 0$ and b_n is the series obtained by grouping the terms in $\sum_{k=0}^{\infty} a_k$ in pairs; that is, $b_n = a_{2n} + a_{2n+1}$. Show that $\sum b_k$ is convergent if and only if $\sum a_k$ is convergent. [Notice that the series above with $a_k = \pm 1$ does not satisfy the condition $\lim a_k = 0$.]

10.9 The hyperbolic and binomial series

The use of convergent power series to define functions is illustrated here; the hyperbolic functions could be defined using the exponential function but we select the power series method just to make the point that one can work with certain series.

We define two functions as power series obtained from the series for the sine and cosine functions by removing the minus signs. We define the hyperbolic sine function, $\sinh x$, and the hyperbolic cosine function, $\cosh x$ as follows:

$$\sinh x = x + \frac{x^3}{3!} + \frac{x^5}{5!} + \cdots = \sum_{k=0}^{\infty} \frac{x^{2k+1}}{(2k+1)!},$$

$$\cosh x = 1 + \frac{x^2}{2!} + \frac{x^4}{4!} + \cdots = \sum_{k=0}^{\infty} \frac{x^{2k}}{(2k)!}.$$

We know that the Taylor series for $\sin x$ converges absolutely for every x. Because the series for $\sinh x$ is obtained from the series for $\sin x$ by replacing the coefficients with their absolute values, it follows that $\sinh x$ converges for all x. Analogously the $\cosh x$ series converges for all x.

By using term-by-term differentiation, we get the formulas

$$\begin{aligned} \frac{d}{dx} \sinh x &= \cosh x, \\ \frac{d}{dx} \cosh x &= \sinh x. \end{aligned} \qquad \textbf{(10.25)}$$

These hyperbolic functions can be expressed in terms of the exponential function. We use the series for e^x, replace x by $-x$ to get the series for e^{-x} with the result

$$e^x = 1 + x + \frac{x^2}{2!} + \frac{x^3}{3!} + \cdots = \sum_{k=0}^{\infty} \frac{x^k}{k!},$$

$$e^{-x} = 1 - x + \frac{x^2}{2!} - \frac{x^3}{3!} + \cdots = \sum_{k=0}^{\infty} (-1)^k \frac{x^k}{k!}.$$

It follows that

$$\begin{aligned} \sinh x &= \frac{e^x - e^{-x}}{2}, \\ \cosh x &= \frac{e^x + e^{-x}}{2}. \end{aligned} \qquad \textbf{(10.26)}$$

The analogy with the trigonometric functions carries over to some identities that are easily verified using Equations (10.26). We have

$$\begin{aligned} &\cosh^2 x - \sinh^2 x = 1, \\ &\sinh 2x = 2 \sinh x \cosh x, \\ &\sinh u \cosh v = \frac{1}{2}\Big[\sinh(u+v) + \sinh(u-v)\Big], \\ &\sinh(-x) = -\sinh x, \\ &\cosh(-x) = \cosh x. \end{aligned}$$

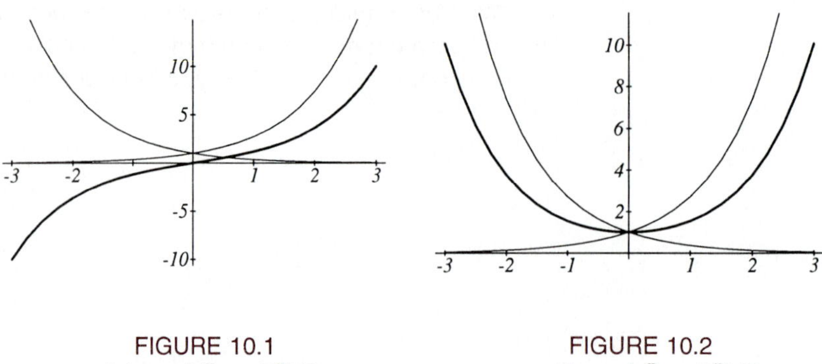

FIGURE 10.1
$\sinh x = (e^x - e^{-x})/2$

FIGURE 10.2
$\cosh x = (e^x + e^{-x})/2$

Many of the familiar trigonometric identities involving $\sin x$ and $\cos x$ have counterpart identities involving $\sinh x$ and $\cosh x$.

As is evident from the power series representations, $\sinh x$ is an odd function and $\cosh x$ is an even function. The graphs can be sketched quite easily by first sketching the two exponentials e^x and e^{-x} and then adding or subtracting y coordinates.

The ordinary trigonometric functions ($\sin x$ and $\cos x$) are sometimes called *circular functions* because for any number t, the point $(\cos t, \sin t)$ lies on the unit circle. The names of the hyperbolic functions derive from the fact that for any number t, the point $(\cosh t, \sinh t)$ lies on the unit hyperbola with equation $x^2 - y^2 = 1$.

Just as the circular trigonometric functions were used to evaluate some integrals involving $\sqrt{a^2 - x^2}$, the hyperbolic functions can be used to integrate some functions involving $\sqrt{a^2 + x^2}$. This is illustrated in the exercises following this section.

The Binomial Series

If m is any real number and $1 + x \geq 0$, then

$$f(x) = (1 + x)^m$$

is defined. We show how to find the Taylor series of $f(x)$ and determine where it converges to f.

We first introduce some notation to describe the coefficients. For any number m and any positive integer k, we write

$$\binom{m}{k} = \frac{m(m-1)(m-2)\cdots(m-k+1)}{k!}. \qquad \textbf{(10.27)}$$

The symbol $\binom{m}{k}$ is read "m choose k". Notice that the numerator is the product of k terms starting with m and decreasing by 1. The denominator is the product of the k terms 1, 2, ..., k. For the case $k = 0$, we define

$$\binom{m}{0} = 1.$$

If m and k are positive integers, then

$$\binom{m}{k} = \begin{cases} \text{a positive integer} & \text{if } 0 \leq k \leq m, \\ 0 & \text{if } k > m. \end{cases}$$

It is not obvious that $\binom{m}{k}$ is an integer when $m \geq k$ are integers. One can show that this number equals the number of ways of choosing k objects from a set of m objects and thus it is an integer. For example, $\binom{m}{2}$ is the number of pairs of socks that can be selected from a drawer containing m socks. This is the reason that the symbol $\binom{m}{k}$ is read "m choose k". When m is not an integer then the numbers $\binom{m}{k}$ are nonzero for all k. The numbers defined in Equation (10.27) are called the *binomial coefficients* because they appear in the binomial series as we now show.

THEOREM 10.24

The Binomial Series

If m is a nonnegative integer we have the binomial formula

$$(1+x)^m = \sum_{k=0}^{m} \binom{m}{k} x^k$$

for all numbers x.

If m is any real number other than a nonnegative integer then

$$(1+x)^m = \sum_{k=0}^{\infty} \binom{m}{k} x^k \qquad \text{for} \qquad |x| < 1.$$

Proof

Let

$$f(x) = (1+x)^m.$$

Then the kth derivative of f is easily found to be

$$f^{(k)}(x) = m(m-1)(m-2)\cdots(m-k+1)(1+x)^{m-k},$$

and the coefficient of x^k in the Taylor series for f is

$$\frac{f^{(k)}(0)}{k!} = \frac{m(m-1)(m-2)\cdots(m-k+1)}{k!} = \binom{m}{k}.$$

If m is a nonnegative integer, then the coefficients are all 0 after the mth. Moreover the $(m+1)$st derivative of f is identically 0 so $f(x)$ equals its mth Taylor polynomial. The binomial formula is just a restatement of this fact.

From now on we assume that m is not a nonnegative integer. Let

$$g(x) = \sum_{k=0}^{\infty} \binom{m}{k} x^k.$$

We will prove that g converges for $|x| < 1$ using the Ratio Test. If we set

$$a_k = \binom{m}{k},$$

then

$$\lim_{k\to\infty} \left|\frac{a_{k+1}}{a_k}\right| =$$

$$\lim_{k\to\infty} \left| \frac{m(m-1)\cdots(m-k+1)(m-k)}{(k+1)!} \frac{k!}{m(m-1)\cdots(m-k+1)} \right|$$

$$= \lim_{k\to\infty} \left|\frac{m-k}{k+1}\right| = 1.$$

Thus the radius of convergence is $R = 1$ and the series $g(x)$ converges absolutely for all x, $|x| < 1$, by the Ratio Test.

Now we are left with the task of proving that $g(x) = (1 + x)^m$. The expected line of proof would be to use the estimate of the remainder term in Taylor's theorem to show that the nth partial sum of the series $g(x)$ differs from $(1 + x)^m$ by some term that approaches 0 as n approaches ∞. If the reader carries out this plan, it will be found that convergence of g to $(1 + x)^m$ will be valid for $-0.5 < x < 1$ rather than for the interval $-1 < x < 1$ that we are trying to prove. This same type of difficulty arose in the discussion of the geometric series, where convergence was proved only for $|x| < 0.5$ when an alternate proof showed convergence for $|x| < 1$. We give an argument not encountered before this point to prove g converges to the expected value for all x with $|x| < 1$.

We first show

$$(1 + x)g'(x) = mg(x). \qquad \textbf{(10.28)}$$

Because $g(x)$ is a convergent power series, it can be differentiated term by term to give

$$\begin{aligned}(1 + x)g'(x) &= (1 + x)\sum_{k=1}^{\infty} k\binom{m}{k}x^{k-1} \\ &= \sum_{k=1}^{\infty} k\binom{m}{k}x^{k-1} + \sum_{k=1}^{\infty} k\binom{m}{k}x^{k}.\end{aligned}$$

Next we collect the coefficients of equal powers of x. The first sum has a constant term m whereas the second sum has 0 constant term. For $n > 0$ the coefficient of x^n is

$$\begin{aligned}&(n + 1)\binom{m}{n + 1} + n\binom{m}{n} \\ &= \frac{m(m - 1)\cdots(m - n + 1)(m - n)}{n!} + \frac{m(m - 1)\cdots(m - n + 1)}{(n - 1)!} \\ &= \frac{m(m - 1)\cdots(m - n + 1)}{n!}[(m - n) + n] \\ &= m\binom{m}{n}.\end{aligned}$$

So we have

$$(1 + x)g'(x) = m + m\sum_{n=1}^{\infty}\binom{m}{n}x^n = mg(x).$$

Next we set

$$h(x) = \frac{g(x)}{(1 + x)^m}$$

and compute the derivative of h. We use the quotient rule to get

$$\begin{aligned}h'(x) &= \frac{(1 + x)^m g'(x) - mg(x)(1 + x)^{m-1}}{(1 + x)^{2m}} \\ &= \frac{(1 + x)g'(x) - mg(x)}{(1 + x)^{m+1}} \\ &= 0.\end{aligned}$$

Because the derivative of h is 0, h is a constant function and so

$$g(x) = c(1 + x)^m \qquad \textbf{(10.29)}$$

for some constant c. From the definition of g we see that $g(0) = 1$ and from Equation (10.29) we find $g(0) = c$. Thus $c = 1$ and $g(x) = (1 + x)^m$ for all x with $|x| < 1$. □

Exercises 10.9

1. Make an accurate plot of the function $\sinh x$ and its Taylor polynomials $P_3(x)$, $P_5(x)$, $P_7(x)$ on the same coordinate system.

2. Make an accurate plot of the function $\cosh x$ and its Taylor polynomials $P_2(x)$, $P_4(x)$, $P_6(x)$ on the same coordinate system.

3. Find the Taylor polynomial that approximates the values of $\sinh x$ over the interval $[-0.3, 0.3]$ with an error no greater than 10^{-3}.

4. Find the Taylor polynomial that approximates the values of $\cosh x$ over the interval $[-0.6, 0.6]$ with an error no greater than 10^{-2}.

5. Verify the identity $2 \sinh x \cosh x = \sinh 2x$ by using the exponential functions that define the hyperbolic functions and then by multiplying the Taylor series for $\sinh x$ and $\cosh x$.

6. Find the Taylor series for $\arcsin x$ by using the binomial series with exponent $-1/2$ and the integral of $1/\sqrt{1 - x^2}$. Give the interval of convergence of the series.

7. The derivative $y = \sinh x$ is a positive function so $\sinh x$ is increasing and has an inverse. Show that the inverse is $\ln(x + \sqrt{x^2 + 1})$. [Hint: Transform the equation $2y = e^x - e^{-x}$ into a quadratic equation in e^x, solve for e^x and then for x.]

Use a substitution of the form $x = a \sinh t$, the identity involving $\sinh t$ and $\cosh t$ and the inverse of the hyperbolic sine from exercise 7 to evaluate the following integrals.

8. $\displaystyle\int \frac{dx}{\sqrt{x^2 + a^2}}$

9. $\displaystyle\int \sqrt{x^2 + a^2}\, dx$

10. Use the binomial series to approximate $(1.2)^{1/4}$ with an error no greater than 10^{-3}.

11. Use the binomial series to approximate $(0.9)^{1/6}$ with an error no greater than 10^{-4}.

12. Use the binomial series for $(1 + x)^m$ to obtain a series for $(a + x)^m$ where $a > 0$. [Hint: $(a + x)^m = a^m(1 + y)^m$, where $y = x/a$.]

13. Use the binomial series to approximate $17^{1/4}$ with an error no greater than 10^{-3}. [Hint: Use exercise 12 with $a = 2^4$ and $17 = 2^4 + 1$.]

14. Use the binomial series to approximate $7^{1/3}$ with an error no greater than 10^{-3}. [Hint: Use exercise 12 and $7 = 2^3 - 1$.]

11 Numerical Computations

The great value of calculus is its computational power. By this point of the text, the reader has seen many problems for which the derivative or integral of some function is essential to the solution. When dealing with real-world problems, it is often necessary to evaluate integrals or solve equations involving complicated functions but it may not always be necessary to obtain an exact solution to obtain useful information. So in this chapter we consider some techniques for making approximations with a high degree of accuracy. We consider the basic problem of finding approximate solutions of an equation $f(x) = 0$. We begin with several methods that rely on the assumption that $f(x)$ is continuous or even differentiable. The method used can be described in rough terms by saying that we guess a solution and then try to improve the guess to a better guess. If all goes well, the process converges toward an exact solution. This is the idea of *iteration*, that is key in Newton's method.

Another computation, frequently encountered, is the evaluation of a definite integral. In practice, functions arise for which it is difficult or even impossible to find an antiderivative. When this happens, the Fundamental Theorem of Calculus cannot be applied to evaluate the integral and some other method must be used. The approximation of definite integrals has been done using lower and upper sums in chapter 5. However the difference between the upper and lower sums can be relatively large and might not be close enough to the actual value if the function oscillates rapidly over the interval of integration. In this chapter we present several other methods for approximating the numerical value of a definite integral with particular emphasis on estimating the size of the error. These techniques usually give much better results with less calculation than the lower or upper sum method.

Any numerical method necessarily involves arithmetic calculation. The amount of computation increases as the desired accuracy increases. Long computations are generally best done on a hand calculator or on a computer for which a suitable program has been written. We do not dwell on the machine aspect of computation but rather on the underlying principles. We assume that the reader has some means available to carry out the computations.

11.1 Solution of equations

Suppose that we want to solve an equation $f(x) = 0$ where $f(x)$ is a continuous function. If we can find two numbers a and b such that $f(a)$ and $f(b)$ have opposite signs, for example, $f(a) > 0$ and $f(b) < 0$, then the Intermediate Value Theorem tells us that there is some point c between a and b at which $f(c) = 0$. This can be interpreted as a (rather crude) approximation of a solution c; that is the equation $f(x) = 0$ has a solution on the interval (a, b) if $a < b$ or on the interval (b, a) if $b < a$. This would be very good information if the length of the interval were very small; for example, if $|a - b| = 10^{-6}$, then we know that a solution c must satisfy $|a - c| \leq 10^{-6}$. Thus we could take $x = a$ as an approximation to a solution with an error of at most 10^{-6}. A slightly better guess might be the midpoint $x = (a + b)/2$ so that the distance from x to the exact solution is at most half the interval length, 0.5×10^{-6}.

Of course it is unlikely that one would be able to make an initial guess of a and b this close together. Here is a method that can be used to reach this state starting from any guess.

Interval Bisection

Given the function $f(x)$ and an interval (a_1, b_1) such that

$$f(a_1) > 0 \quad f(b_1) < 0. \tag{11.1}$$

Of course this same idea with trivial modifications works if $f(a_1) < 0$ and $f(b_1) > 0$. Let $m = (a_1 + b_1)/2$ be the midpoint of the interval. Compute $f(m)$; if $f(m) = 0$ then we are very lucky and have found a solution. If $f(m) \neq 0$ then either $f(m) > 0$ or $f(m) < 0$. In these cases, we replace (a_1, b_1) with a new choice of interval (a_2, b_2) as follows:

$$\begin{aligned} f(m) > 0 &\quad \text{implies} \quad (a_2, b_2) = (m, b_1), \\ f(m) < 0 &\quad \text{implies} \quad (a_2, b_2) = (a_1, m). \end{aligned}$$

In either case we now have an interval (a_2, b_2) of length $b_2 - a_2 = (b_1 - a_1)/2$ such that $f(a_2)$ and $f(b_2)$ have opposite signs.

Now we repeat the procedure of taking the midpoint, examining the sign of the value of $f(x)$ at this midpoint, and replacing the interval with either the left or right half interval. If this is repeated n times, we reach an interval (a_n, b_n) such that

1. There is a solution to $f(x) = 0$ on the interval (a_n, b_n);
2. $b_n - a_n = (b - a)/2^{n-1}$.

If we stop the computations at the interval (a_n, b_n), we take, as the approximate solution, the midpoint of the last interval, namely

$$c_n = \frac{a_n + b_n}{2}.$$

Then the distance from c_n to a true solution is at most half the length of (a_n, b_n); that is, the error is at most $(b_n - a_n)/2 = (b - a)/2^n$.

As the number of repetitions increases, the interval length approaches zero so that one knows, for example with $n = 10$, the solution is approximated by $c_{10} = (a_{10} + b_{10})/2$ with an error at most $(b_{10} - a_{10})/2 = (b - a)/2^{10}$, which is roughly $(b - a) \times 10^{-3}$. We refer to this method of approximating a solution as the *interval bisection method*.

EXAMPLE 1 Let $f(x) = x^4 - 7x + 2$. It was shown in chapter 4.4 that there are exactly two solutions of the equation $f(x) = 0$. By trial and error, or by inspection, we see that

$$f(0) = 2 > 0, \quad f(1) = -4 < 0, \quad f(2) = 4 > 0,$$

so there is one solution on $(0, 1)$ and one solution on $(1, 2)$. Use the interval-bisection method to approximate the two solutions with an error at most 10^{-2}.

Solution: Each interval $(0, 1)$ and $(1, 2)$ has length 1, so the possible error in using the midpoint of the nth subinterval is $1/2^n$. By taking $n = 7$ we get a maximum error less than 10^{-2}. We begin with the interval $J_1 = (0, 1)$. Note that f is positive at the left endpoint and negative at the right endpoint. At the midpoint, we have $f(0.5) = -1.43\cdots$, so the new interval is $J_2 = (0, 0.05)$. Evaluate f at the midpoint of J_2 and find $f(0.25) > 0$ so $J_3 = (0.25, 0.5)$. Repeat this several more times; the steps in general are: If $J_i = (a, b)$ and if $c = (a + b)/2$ then

$$J_{i+1} = \begin{cases} (c, b) & \text{if } f(a) \text{ and } f((a + b)/2) \text{ have the same sign} \\ (a, c) & \text{otherwise.} \end{cases}$$

The result of the interval bisection is shown in the table for both intervals $(0, 1)$ and $(1, 2)$.

TABLE 11.1
Intervals containing a solution of *f(x) = 0*

J_2	J_3	J_4	J_5	J_6	J_7
(0, 0.5)	(0.25, 0.5)	(0.25, 0.375)	(0.25, 0.2875)	(0.2688, 0.2875)	(0.2781, 0.2875)
(1.5, 2)	(1.75, 2)	(1.75, 1.875)	(1.75, 1.8125)	(1.7813, 1.8125)	(1.7969, 1.8125)

The midpoints of the seventh intervals are approximately 0.2828 and 1.8047. In each case the interval has length 1/64 (approximately, because of round-off error in my calculator) so the estimated root differs from the true root by no more than $1/128 \approx 0.0078$. As a check on the method we evaluate f at the approximate roots to find $f(0.2828) \approx 0.0268$ and $f(1.8047) \approx -0.025$. ■

REMARK: The approximations in example 1 are not particularly good. Many iterations are usually required to reach a small enough interval to give good accuracy. However this method has the advantage that it predicts the maximum error after n steps as $(b - a)/2^{n+1}$ no matter what equation is to be solved. It has a disadvantage in that it does not take into account any values of the function except for their sign. One might think that in case the values $f(a_j)$ are very small, a good method should take this into account and converge faster toward the true solution. The next method is superior in this regard.

Even though the interval bisection method has some drawbacks, it is worth noticing that almost no calculus is required to understand the principle. Only the Intermediate Value Theorem was needed. The next method uses more sophisticated calculus, the derivative in particular, and the method gives better results.

Newton's Method

Next we present a method for solving $f(x) = 0$ for a function having a derivative $f'(x)$ at every point. The motivation for Newton's method can be drawn from the geometry of curves and their tangents or from the theory of Taylor polynomials. We present both motivations. We must say at the outset that Newton's method has some drawbacks in that there are cases where the method fails to converge to a solution. This is relatively rare, however, and for most cases it is satisfactory.

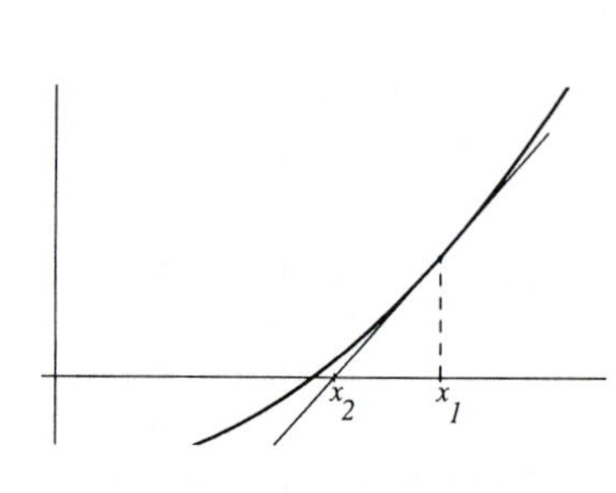

FIGURE 11.1
Newton approximates $f(x) = 0$

Consider the graph of the curve $y = f(x)$ and suppose that it crosses the x-axis somewhere; in other words we suppose that there is a solution of the equation $f(x) = 0$. Make some guess at the solution, $x = x_1$, and consider figure 11.1 showing the curve and the point x_1.

If our guess x_1 is not too far from the true solution, then the tangent line to the curve at the point $(x_1, f(x_1))$ might intersect the x-axis at a point x_2, which is nearer than x_1 to the true solution. The equation of the tangent line is

$$y - f(x_1) = f'(x_1)(x - x_1).$$

The graph crosses the x-axis at a point x_2, which is found by setting $y = 0$; that is $x_2 = x_1 - f(x_1)/f'(x_1)$. We then replace the first guess x_1 with x_2 and repeat the entire computation to produce x_3. The general step produces a sequence x_1, x_2, x_3, x_4, ..., where

$$x_{n+1} = x_n - \frac{f(x_n)}{f'(x_n)}. \qquad \textbf{(11.2)}$$

The sequence produced this way depends on $f(x)$ of course, and on the initial guess x_1. We call the sequence x_n the *Newton approximations* to the solution of $f(x) = 0$ with initial guess x_1. If the sequence x_n converges, then it converges to a solution of the equation $f(x) = 0$. We state this formally.

THEOREM 11.1

If $f(x)$ and $f'(x)$ are continuous on an interval containing x_1 and all the numbers x_n defined by Equation (11.2) and if the $\lim_{n\to\infty} x_n = c$ exists, then c satisfies $f(c) = 0$.

Proof

The definition of the sequence x_n implies

$$f(x_n) = f'(x_n)(x_n - x_{n+1}).$$

The assumption that the sequence x_n converges to c and the continuity of $f(x)$ and $f'(x)$ imply that

$$\begin{aligned} f(c) &= \lim_{n\to\infty} f(x_n) \\ &= \lim_{n\to\infty} f'(x_n)(x_n - x_{n+1}) \\ &= f'(c)(c - c) \\ &= 0. \end{aligned}$$

In these equations we used the fact that

$$\lim_{n\to\infty} x_n = \lim_{n\to\infty} x_{n+1} = c. \qquad \square$$

Notice that when $f(x)$ is a polynomial, then $f(x)$ and $f'(x)$ are continuous everywhere so the assumptions in this theorem are satisfied. Newton's method can be applied to give a solution of $f(x) = 0$ whenever the sequence x_n has a limit.

EXAMPLE 2 Let us consider again the function $f(x) = x^4 - 7x + 2$. It was seen earlier in this section that this function has a root on the interval $(0, 1)$ and another root on the interval $(1, 2)$. Find approximate values for these roots.

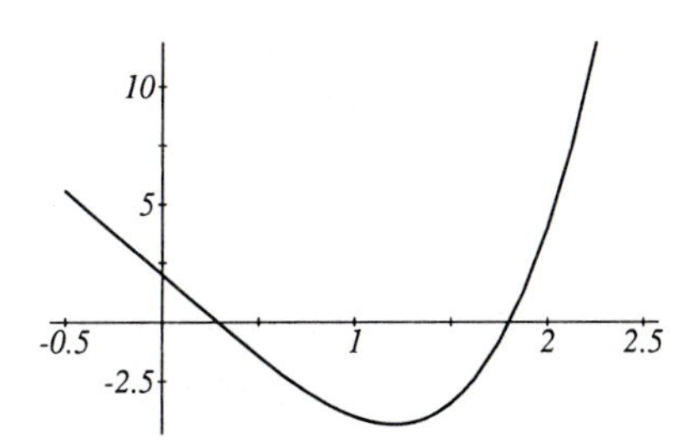

FIGURE 11.2

$x^4 - 7x + 2 = 0$ has two solutions

Solution: The derivative of the polynomial is $f'(x) = 4x^3 - 7$, and we have

$$x_{n+1} = x_n - \frac{x_n^4 - 7x_n + 2}{4x_n^3 - 7} = \frac{3x_n^4 - 2}{4x_n^3 - 7}.$$

As a first guess we try $x_1 = 0$. Using a pocket calculator to do the computations, one finds that

$$\begin{aligned} x_1 &= 0.0, \\ x_2 &= 0.285714286, \\ x_3 &= 0.286679129, \\ x_4 &= 0.286679195, \\ x_5 &= 0.286679195. \end{aligned}$$

Thus the sequence has converged at least as far as the accuracy of the calculator is able to detect.

Let us repeat the computation using $x_1 = 1$ as first guess:

$$\begin{aligned} x_1 &= 1.0, \\ x_2 &= -0.333333333, \\ x_3 &= 0.279611399, \\ x_4 &= 0.286669387, \\ x_5 &= 0.286679195, \\ x_6 &= 0.286679195. \end{aligned}$$

This gives exactly the same final result as the previous start. Finally, we try one more first guess $x_1 = 2$ on the interval containing the other solution:

$$\begin{aligned} x_1 &= 2.0, \\ x_2 &= 1.84, \\ x_3 &= 1.807502688, \\ x_4 &= 1.806227114, \\ x_5 &= 1.806225190, \\ x_6 &= 1.806225190. \end{aligned}$$

We find the second root with good accuracy. ■

There is some potential deception that comes about when calculating with any machine (pocket calculator or computer). Most machines do computations with a limited number of decimal places (known as the precision of the machine). This is usually 8 or 10 decimals with simple hand-held calculators. Thus two numbers that differ in some decimal place beyond the precision of the machine are called equal because the machine cannot distinguish between them. In the last computation above, the hand calculator believes $x_5 = x_6$ because the two numbers agree in all the decimals places the machine can store. In fact, the general formula for x_{n+1} shows that x_6 is not equal to x_5, but differs from it by only a very small amount. In spite of such possible errors, we still say the solutions of the equation

$$x^4 - 7x + 2 = 0$$

are given by the approximate values

$$x = 0.286679195 \quad \text{and} \quad x = 1.806225190.$$

As a check on the validity of the solutions we evaluate the polynomial $f(x) = x^4 - 7x + 2$ at the two solutions and find

$$f(0.286679195) \approx 3.16 \times 10^{-7}$$
$$f(1.806225190) \approx -1.08 \times 10^{-9}.$$

Compare this with the accuracy obtained by the interval bisection method used above.

Nonconvergence of Newton's Method

Here is an example that illustrates one of the difficulties that can occur with Newton's method.

EXAMPLE 3 Let $f(x) = x^3 - 6x^2 - 33x + 82$. Show that there is a a solution of $f(x) = 0$ on the interval $(0, 5)$ but Newton's method does not produce a solution if the first guess is $x_1 = 5$.

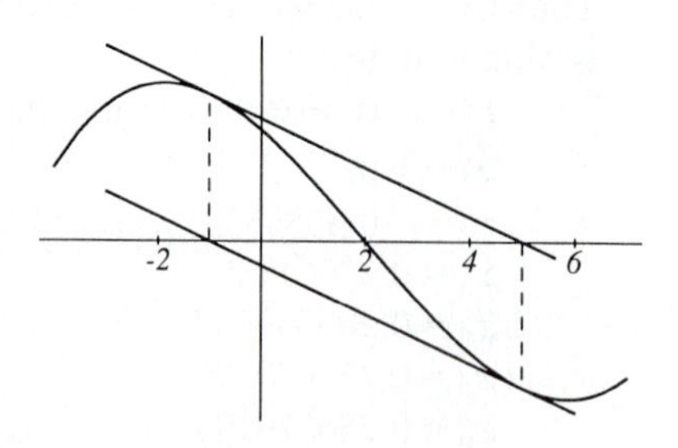

FIGURE 11.3
Newton's method fails

Solution: We observe that $f(0) = 82 > 0$ and $f(5) = -108 < 0$ so there is a solution to $f(x) = 0$ between 0 and 5. If we take $x_1 = 5$ as a first approximation, the sequence x_n oscillates and does not converge to a limit:

$x_1 = 5,$
$x_2 = -1,$

$x_3 = 5,$
$x_4 = -1$, etc.

Newton's method converges to a solution by starting with a different first approximation, for example $x_1 = 4$. ■

Taylor's Theorem Derivation

We promised to give another motivation for Newton's method inspired by the use of Taylor series. Suppose that $f(x)$ has continuous second derivatives in an interval containing all the points we consider. Suppose that x_1 is a first guess at a solution to the equation $f(x) = 0$ and that $x_1 + h$ is the true solution, where h is not known. Apply Taylor's Theorem to write the function as a polynomial plus remainder so that we have

$$f(x) = f(x_1) + f'(x_1)(x - x_1) + \frac{1}{2}f''(z)(x - x_1)^2$$

for some z between x and x_1. Now evaluate this at the point $x_1 + h$ to get

$$0 = f(x_1 + h) = f(x_1) + f'(x_1)h + \frac{1}{2}f''(z)h^2. \qquad \textbf{(11.3)}$$

If h is small, in particular smaller in absolute value than 1, then h^2 is usually much smaller than h. We take the view that the term $f''(z)h^2/2$ is small in comparison with the other terms, and we find an approximation for h; it is a solution of

$$0 = f(x_1) + f'(x_1)h, \quad \text{or } h = -\frac{f(x_1)}{f'(x_1)}.$$

Inasmuch as the true solution was $x_1 + h$, we have recovered precisely the earlier formulation of the next guess, $x_2 = x_1 + h = x_1 - f(x_1)/f'(x_1)$.

Convergence of Newton's Method

The Taylor Equation (11.3) can be used, in the presence of additional hypothesis, to prove a convergence theorem for Newton's method. In fact, we are able to prove that the approximations form a monotone sequence in certain circumstances. We describe the precise situation in the next theorem. We use the phrase "*$g(x)$ has constant sign on* $[a, b]$" to mean that either $g(x) < 0$ for all x on $[a, b]$ or $g(x) > 0$ for all x on $[a, b]$.

THEOREM 11.2

Convergence of Newton's Method

Let $f(x)$ be a function that is continuous on an interval $[a, b]$ and which satisfies the conditions:

i. The values $f(a)$ and $f(b)$ have opposite signs;
ii. Both $f'(x)$ and $f''(x)$ exist and have constant sign on $[a, b]$. Then there is exactly one number c with $a < c < b$ and $f(c) = 0$. For any choice of x_0 on $[a, b]$, let x_n be the sequence of Newton approximations defined by

$$x_{n+1} = x_n - \frac{f(x_n)}{f'(x_n)} \quad \text{for } n \geq 0.$$

Then the sequence x_n converges to c.

More precisely,

a. If $f'(x) > 0$, $f''(x) > 0$, and $x_0 = b$, then the sequence $\{x_n\}$ is decreasing with limit c.
b. If $f'(x) > 0$, $f''(x) < 0$, and $x_0 = a$, then the sequence $\{x_n\}$ is increasing with limit c.
c. If $f'(x) < 0$, $f''(x) > 0$, and $x_0 = a$, then the sequence $\{x_n\}$ is increasing with limit c.
d. If $f'(x) < 0$, $f''(x) < 0$, and $x_0 = b$, then the sequence $\{x_n\}$ is decreasing with limit c.

Proof

Because the first derivative of $f(x)$ has constant sign on the interval, the function $f(x)$ is either increasing on $[a, b]$ or decreasing on $[a, b]$. In either case the graph of f crosses the x-axis at most once so there is at most one value of x on the interval for which $f(x) = 0$. The condition that $f(a)$ and $f(b)$ have opposite signs implies that there is at least one solution to $f(x) = 0$ (by the Intermediate Value Theorem). Hence there is exactly one solution that we denote by c.

Now assume we have case a. Figure 11.1 gives an indication of the shape of the curve and the first approximations in this case. We select $x_0 = b$ so that $x_0 > c$. We prove that for every n, we have $c \leq x_{n+1} \leq x_n$. Suppose that we know

$$c \leq x_k \leq x_{k-1} \leq \cdots \leq x_0$$

for some integer $k \geq 0$. Based on this, we prove now that $c \leq x_{k+1} \leq x_k$. By Taylor's Theorem, we have the analogue of Equation (11.3); namely

$$f(x) = f(x_k) + f'(x_k)(x - x_k) + \frac{1}{2}f''(z)(x - x_k)^2$$

for some z between x and x_k. Evaluate this at c and use $f(c) = 0$ to conclude

$$f(x_k) + f'(x_k)(c - x_k) = -\frac{1}{2}f''(z)(c - x_k)^2 \leq 0. \qquad \textbf{(11.4)}$$

The inequality holds because the second derivative is positive by assumption. Also the first derivative is positive by assumption, so we obtain the inequality

$$c - x_k \leq -\frac{f(x_k)}{f'(x_k)}.$$

It follows that

$$c \leq x_k - \frac{f(x_k)}{f'(x_k)} = x_{k+1}. \qquad \textbf{(11.5)}$$

This is one of the two inequalities we wanted to show. To show that x_{k+1} is less than x_k we recall that $f(x)$ is increasing on $[a, b]$ and that $f(c) = 0$. From $c \leq x_k$, we conclude that $f(x_k) \geq 0$. We also know $f'(x_k) \geq 0$ so Equation (11.5) implies $x_{k+1} \leq x_k$. This proves that the sequence of Newton approximations is decreasing and bounded. Hence the sequence has a limit. The general theory implies that the sequence converges to a solution of $f(x) = 0$ and so the sequence x_n must converge to c because c is the only solution.

The other three cases are done in a similar way using only very slight changes so these are left to the reader. The theorem proves the convergence of Newton's approximations under the assumption that the first and second derivatives do not change sign on the interval. These are fairly restrictive conditions but can often be satisfied by taking a sufficiently small interval. Of course, there are many situations where the conditions about the signs of the derivatives are not satisfied but Newton's approximations still converge to a solution.

Notice that the theorem makes a suggestion about the initial guess. If the first and second derivatives have the same sign on $[a, b]$, use the right-hand endpoint b as the first guess. If the first and second derivatives have opposite signs on $[a, b]$, use the left-hand endpoint a as the first guess. In many cases, the recommended choice of initial guess results in faster convergence to the limit. □

EXAMPLE 4 Find an approximate solution of the equation $\tan x = x$ with $x > 0$.

Solution: The solutions of the equation $\tan x = x$ can be represented geometrically as the x coordinates of the points where the graphs of $y = \tan x$ and $y = x$ cross. There are infinitely many such points one of which is on the interval $\pi < x < 3\pi/2$. To apply Newton's method, we begin by writing the equation in the form

$$f(x) = \sin x - x \cos x = 0.$$

Then f and all derivatives of f are continuous everywhere. Moreover

$$f(\pi) = 0 - \pi(-1) > 0, \qquad f(3\pi/2) = -1 - (3\pi/2)0 < 0.$$

This shows in a nongeometric way that there is a solution between π and $3\pi/2$. For the derivatives we compute

$$f'(x) = x \sin x \qquad \text{and} \qquad f''(x) = \sin x + x \cos x.$$

Using the fact that $\sin x \le 0$ and $\cos x \le 0$ on the interval $[\pi, 3\pi/2]$, we find $f'(x) < 0$ and $f''(x) < 0$ for all x on the interval. So we have a situation in which the derivatives have constant sign and Theorem 11.2 applies. Part d of that theorem implies that the sequence of Newton approximations starting with $x_1 = 3\pi/2$ is monotone decreasing and has a solution of the equation as its limit. The computation of the successive terms of the Newton sequence is somewhat complicated because we have

$$\begin{aligned} g(x) &= x - \frac{f(x)}{f'(x)} = x - \frac{\sin x - x \cos x}{x \sin x} \\ &= x - \frac{1}{x} + \cot x. \end{aligned}$$

The Newton approximations are found with the aid of a calculator to be

$$\begin{aligned} x_1 &= \frac{3\pi}{2} = 4.7123\,8898, \\ x_2 &= g(x_1) = 4.5001\,8239, \\ x_3 &= g(x_2) = 4.4934\,1954, \\ x_4 &= g(x_3) = 4.4934\,0946, \\ x_5 &= g(x_4) = 4.4934\,0946. \end{aligned}$$

As far as the calculator can tell, the sequence has reached its limit. To test this as a solution of the original equation, we compute (to within the accuracy of my calculator)

$$\tan(4.4934\,0946) - 4.4934\,0946 = 0.0000\,004 = 4 \times 10^{-7},$$

which indicates that the approximate solution is quite good. To get some idea how close to the true solution, we randomly select a slightly smaller number and test the value of the function. We reduce the number by 1.46×10^{-6} and evaluate

$$\tan(4.493408) - 4.493408 = -0.000029.$$

Thus we see $\tan x - x$ changes sign on the interval $[4.493408, 4.4934\,0946]$ and the true solution of the equation $\tan x - x = 0$ lies between 4.4934 08 and 4.4934 0946. ■

Rate of Convergence

One may justifiably ask why we use Newton's method to solve an equation $f(x) = 0$ when we have a method like interval bisection that is easier to apply (because it usually involves less computation). The answer is that Newton's method converges quickly to its limit when it converges at all. The rate of convergence is "quick" (more technically the rate is called *quadratic*) in the following sense: there is a constant C, depending on f such that

$$|c - x_{k+1}| \leq C|c - x_k|^2 \tag{11.6}$$

where c is the limit of the sequence x_k. The inequality is interpreted as follows. After the kth step, we have an approximation x_k to the unknown number c and $|c - x_k|$ is the error in the approximation. If we assume this error is smaller than 1, then its square is much smaller. The error in the next step is $|c - x_{k+1}|$, which is at most the constant times the square of the previous error. To derive inequality (11.6) we go back to Equation (11.4) and rearrange the terms to obtain

$$0 = f'(x_k)\left[c + \frac{f(x_k)}{f'(x_k)} - x_k\right] + \frac{1}{2}f''(z)(c - x_k)^2.$$

Use the definition of x_{k+1} to rewrite this equation as

$$|c - x_{k+1}| = \left|\frac{f''(z)}{2f'(x_k)}\right| |c - x_k|^2.$$

To obtain a constant C that was predicted we find constants m and M such that $|f''(x)| \leq M$ and $|2f'(x)| \geq m$ for all x on an interval in which we are working and then take $C = M/m$. More directly, we can simply take C as the maximum value of $|f''(x)/2f'(x)|$ on the interval when this is easily determined.

The rapid convergence of Newton's method is the principal reason for its widespread use.

Iteration

If $g(x)$ is a given function and c is a number, the *iterates* of c are the numbers

$$c_1 = g(c), \quad c_2 = g(g(c)), \quad c_3 = g(g(g(c))), \ldots \tag{11.7}$$

obtained by repeatedly applying the function. More precisely, we let $c_{n+1} = g(c_n)$ for $n \geq 1$ and set $c_1 = g(c)$ to obtain the sequence c_n of iterates. In the study of dynamical systems, one is concerned with the sequence of iterates of numbers (usually complex numbers, but we discuss only iterates of real numbers) and the convergence or nonconvergence of the iterate sequence. This topic is very closely related to Newton's method for finding solutions of $f(x) = 0$. Suppose that

$f(x)$ is a differentiable function and the function $g(x)$ is defined by the equation

$$g(x) = x - \frac{f(x)}{f'(x)}. \tag{11.8}$$

For any number c, the iterates in Equation (11.7) are precisely the Newton approximations to the solutions of $f(x) = 0$ with $x_1 = c$ as the initial guess. Information about the convergence of Newton's approximations gives information about the iterates of c by g. Of course, if one starts with the function $g(x)$, the Newton theory can be applied only if we find the function $f(x)$ satisfying Equation (11.8). How do we find $f(x)$? Equation (11.8) can be rewritten in the form

$$\frac{f'(x)}{f(x)} = \frac{1}{x - g(x)}.$$

The expression $f'(x)/f(x)$ is the derivative of $\ln|f(x)|$ so we find

$$\ln|f(x)| - \ln|f(a)| = \int_a^x \frac{dt}{t - g(t)}, \tag{11.9}$$

where a is some given point such that $1/(t - g(t))$ is continuous on $[a, x]$. In principle the integration can be carried out whenever $g(x)$ is a rational function and in many other cases as well. Note that the interval $[a, x]$ cannot contain any point t satisfying $g(t) = t$ because the integrand is not continuous at such a point. We give an example to illustrate some ideas.

EXAMPLE 5 Let $g(x) = x^2 - x$. As the number c varies, what are the possible limits $\lim_{n\to\infty} c_n$ when c_n is the sequence of iterates given by $c_1 = g(c)$, $c_2 = g(c_1)$, $\ldots$, $c_{n+1} = g(c_n)$, $\ldots$?

Solution: We begin by finding a function $f(x)$ such that the sequence of iterates c_n is the sequence of Newton approximations to the solutions of $f(x) = 0$. Substitute the definition of $g(t)$ into Equation (11.9) to get

$$\begin{aligned}\ln|f(x)| - \ln|f(a)| &= \int_a^x \frac{dt}{t - (t^2 - t)} = \int_a^x \frac{dt}{1 - (t-1)^2} \\ &= \frac{1}{2}\ln\left|\frac{x}{x-2}\right| - \frac{1}{2}\ln\left|\frac{a}{a-2}\right|.\end{aligned}$$

Collect the terms involving a into one constant and raise to powers to get

$$|f(x)| = A\sqrt{\left|\frac{x}{x-2}\right|}, \qquad A = \text{constant}.$$

The absolute value signs can be removed if we consider the intervals on which x lies. The expression $x/(x-2)$ is positive if $x > 2$ or $x < 0$ and is negative if $0 < x < 2$. The constant A plays no significant role because the A cancels in the quotient $f(x)/f'(x)$. So we set $A = 1$ and obtain

$$f(x) = \begin{cases} \sqrt{\dfrac{x}{x-2}} & \text{if } x > 2 \text{ or } x \le 0, \\[2ex] \sqrt{\dfrac{x}{2-x}} & \text{if } 0 \le x < 2. \end{cases}$$

We have left $f(x)$ undefined at $x = 2$. The points $x = 0$ and $x = 2$ are special points in this problem. All the iterates of 0 are equal to 0 because $g(0) = 0^2 - 0 = 0$. Similarly all the iterates of 2 equal 2 because $g(2) = 2^2 - 2 = 2$. (The points 0 and 2 are called *fixed points* of $g(x)$.) However, the fact that f is undefined at 2 makes that point exceptional. At all other points, f is continuous and differentiable and

Theorem 11.2 can be applied. If $c \neq 2$ is some point such that the iterates of c given in Equation (11.7) form a convergent sequence, then the sequence must converge to a point L where $f(L) = 0$. All solutions of $f(x) = 0$ are easily found; the only possibility is $L = 0$. Thus we are able to assert: if the iterates of some point $c \neq 2$ converge to a limit, the limit is 0. ■

One of the most interesting problems in the subject is to determine the numbers c having the property that the sequence of iterates of c by application of a function g converges. This is often illustrated with high-speed computer graphics. If the related function $f(x)$ has many different solutions to $f(x) = 0$, the iterates of a point c under repeated application of g either converges toward one of the solutions, or does not converge at all. Each point on the line is "painted" a color depending on the solution toward which it converges or a different color if the iterates do not converge at all. For the very simple function $g(x) = x^2$, the iterates of a point c are just the powers c^{2^n}. The points 0 and 1 are special because they are the fixed points of g so we could paint 0 blue and 1 black. All points x with $|x| > 1$ are painted green to indicate that their iterates do not converge. All points x with $|x| < 1$ are painted red to indicate that their iterates converge to 0. Note the chaotic behavior near the boundary point 1. A point near 1 but just slightly smaller will have its iterates converge to 0 whereas a point very close to 1 but larger will have its iterates diverging to infinity. The lack of continuity in the behavior on nearby points gives the general problem unusual interest. When this idea is extended to the complex plane, beautiful pictures result that show the complexity of the problem. [For example see: James Gleick, *Chaos—Making a New Science*, Penguin books, 1987; or the article by Ivars Peterson, *Zeroing in on chaos*, Science News, February 28, 1987.]

Exercises 11.1

Use the method of interval bisection to approximate a solution of each equation on the indicated interval. Repeat the bisection enough times so that the final approximation is within $1/64 \approx 0.016$ of the true solution. Check your work by evaluating the function at the approximate solution to see if the number is really "close" to 0.

1. $x^2 - 2 = 0$, $[1, 2]$ **2.** $x^2 - 10 = 0$, $[3, 4]$

3. $x^2 + x - 3 = 0$, $[1, 2]$ **4.** $x^2 - 3x + 1 = 0$, $[0, 1]$

5. $x^3 - 2 = 0$, $[1, 2]$ **6.** $x^3 - 16 = 0$, $[2, 3]$

7. $x^3 - x^2 + x - 2 = 0$, $[1, 2]$ **8.** $x^4 + x^2 - 1 = 0$, $[0, 1]$

Use Newton's method to approximate the solutions of the following equations. In each case, use the initial guess x_0 near a solution and then compute the next four Newton approximations x_1, x_2, x_3 and x_4. Compare these computations with the results of exercises 1–8.

9. $x^2 - 2 = 0$, $x_0 = 2$ **10.** $x^2 - 10 = 0$, $x_0 = 3$

11. $x^2 + x - 3 = 0$, $x_0 = 1$ **12.** $x^2 - 3x + 1 = 0$, $x_0 = 0$

13. $x^3 - 2 = 0$, $x_0 = 1$ **14.** $x^3 - 16 = 0$, $x_0 = 2$

15. $x^3 - x^2 + x - 2 = 0$, $x_0 = 1$ **16.** $x^4 + x^2 - 1 = 0$, $x_0 = 1$

17. Give an accurate sketch of the graph of the function $f(x) = x^3 - 6x^2 - 33x + 82$ over the interval $[-5, 9]$ and draw the tangent lines at the points where $x = 5$ and $x = -1$. [See example 3.] Then draw the tangent line at $x = 4$, locate where it crosses the x-axis and then find the second Newton approximation.

18. Find approximations to all three real solutions of $x^3 - 6x^2 - 37x + 82 = 0$.

19. Give sketches that could be the graphs of four functions that satisfy the four cases considered in Theorem 11.2 and indicate the tangent lines and the second Newton approximations.

20. Difficulties arise if one tries to apply Newtons's method to a polynomial that has no zeros. If $f(x) = x^2 + 1$, then there is no real-number solution to $f(x) = 0$. Thus an application of Newton's method, with any first approximation, must lead to a divergent sequence x_n. Select several numbers for x_1 and compute the terms x_n in Newton's method for $2 \leq n \leq 8$. If you have a machine to do the computations, extend the computations to $n = 20$.

For each $g(x)$, compute the iterates $c_n = g(g(\cdots g(c)\cdots))$ for $1 \leq n \leq 8$ for each of the given points c.

21. $g(x) = x^2 - x$, $c = 1.9, 1.99, 2, 2.01, 2.1$

22. $g(x) = 1 - x^2$, $c = 1, 0.5, -0.5$

23. $g(x) = x^3 - x^2$, $c = -1, 0, 1$

24. $g(x) = x^3 - 2x^2 + 2x - 1$, $c = -0.5, 1, 2$

25. Let $g(x) = x^2$. Determine the possible limits of the sequence of iterates c, $g(c)$, $g(g(c))$, $g(g(g(c)))$, $\ldots$ as c ranges over all real numbers.

26. Let $g(x) = 1 - x^2$. Use Newton's method to determine the possible limits of the sequence of iterates c, $g(c)$, $g(g(c))$, $g(g(g(c)))$, $\ldots$ as c ranges over all real numbers.

11.2 Numerical integration

In this section we consider ways of approximating the value $\int_a^b f(x)\,dx$. Of course, when an antiderivative of $f(x)$ is known, the Fundamental Theorem of Calculus can be applied to obtain the value of the integral (provided of course that the values of the antiderivative can be computed). When the antiderivative is not known, approximation methods must be used to estimate the value of the integral. It can happen in practice that the solution to a problem is given as an integral $\int_a^b f(x)\,dx$ where the function $f(x)$ is not explicitly known but is given, instead, by its numerical values at some set of points. In this case, there is no possibility of evaluating the integral by anything other than numerical methods.

As with any approximation technique, it is important to estimate the accuracy of the computations. We present a number of different estimation rules and determine an upper bound for the error in each case. There is a common theme used in all cases. To estimate the definite integral

$$I = \int_a^b f(x)\,dx$$

the function $f(x)$ is replaced by another function $A(x)$ with the properties that $\int_a^b A(x)\,dx$ is easy to compute and the error term

$$E = \int_a^b f(x) - A(x)\,dx$$

can be estimated and is "small". Then $\int_a^b A(x)\,dx$ is the approximation of I with E as the error involved in the approximation. We use polynomial functions for the choice of $A(x)$ over a single interval. The case in which the integral is approximated by upper or lower sums corresponds to $A(x) =$ constant. For greater accuracy, the interval $[a, b]$ can be partitioned into subintervals and different polynomial functions are used to approximate $f(x)$ over each subinterval. We will treat only the simplest cases; namely those in which $A(x)$ is a polynomial of small degree, usually at most degree two.

We begin with a very general method of approximating the definite integral of a function $f(x)$ by using Taylor polynomials. The estimates are based on the following inequalities that were proved in Theorem 9.5.

THEOREM 11.3

General Approximation Theorem

Let $f(x)$ have $n + 1$ continuous derivatives at each point of an interval $[a, b]$.

Let c be some point on $[a, b]$ and let

$$P_n(x) = f(c) + f'(c)(x - c) + \cdots + \frac{f^{(n)}(c)(x - c)^n}{n!}$$

be the nth Taylor polynomial of f centered at c. Let m_{n+1} and M_{n+1} be the minimum and maximum values of $f^{(n+1)}(x)$ on $[a, b]$.

If $c \leq x \leq a$ and n is arbitrary or if $a \leq x \leq c$ and n is odd then

$$\frac{m_{n+1}(x - c)^{n+1}}{(n+1)!} \leq f(x) - P_n(x) \leq \frac{M_{n+1}(x - c)^{n+1}}{(n+1)!}.$$

If $a \leq x \leq c$ and n is even, then the correct inequality is

$$\frac{-m_{n+1}(x - c)^{n+1}}{(n+1)!} \leq P_n(x) - f(x) \leq \frac{-M_{n+1}(x - c)^{n+1}}{(n+1)!}.$$

We think of $P_n(x)$ as an approximation to $f(x)$ so the integral of $P_n(x)$ is an approximation of the integral of $f(x)$. The error in the approximation of the integral is

$$E_n = \int_a^b f(x)\,dx - \int_a^b P_n(x)\,dx. \qquad \textbf{(11.10)}$$

The size of this error will be estimated by integrating the inequalities in the General Approximation Theorem. Some care must be used when n is even because the second inequality must be used in the integration from a to c, whereas the first inequality is used in the integration from c to b. Writing m and M for the minimum and maximum of $f^{(n+1)}(x)$, we find the result of the integration is as follows:

For n odd,

$$\frac{m\left[(b - c)^{n+2} - (a - c)^{n+2}\right]}{(n+2)!} \leq E_n \leq \frac{M\left[(b - c)^{n+2} - (a - c)^{n+2}\right]}{(n+2)!}. \qquad \textbf{(11.11)}$$

For n even,

$$\frac{m(b - c)^{n+2} - M(a - c)^{n+2}}{(n+2)!} \leq E_n \leq \frac{M(b - c)^{n+2} - m(a - c)^{n+2}}{(n+2)!}. \qquad \textbf{(11.12)}$$

These estimates are rather complicated, but they become manageable by taking small values of n and making good choices of c; we will use c as either an endpoint (so that either $a - c = 0$ or $b - c = 0$) or as the midpoint (so that $(a-c)^2 = (b-c)^2$). An alternative form expresses the error in terms of the value of some derivative of f. We show this now for c equal to one of the two endpoints.

Some One-Point Approximation Rules

For $n = 0$, the Taylor polynomial centered at c is $P_0(x) = f(c)$. The application of inequality (11.12) to the cases $c = a$ and $c = b$ yields the following theorem.

THEOREM 11.4

Let $f'(x)$ be continuous on $[a, b]$. Then there are numbers z_1 and z_2 on $[a, b]$

such that the following hold:

Left Endpoint Rule: $$\int_a^b f(x) = f(a)(b-a) + f'(z_1)\frac{(b-a)^2}{2}$$

Right Endpoint Rule: $$\int_a^b f(x) = f(b)(b-a) - f'(z_2)\frac{(b-a)^2}{2}$$

In the Left or Right Endpoint Rules the number z appears in the equations because every number lying between m_1 and M_1, the minimum and maximum of $f'(x)$, is a value of $f'(x)$ at some point.

The equations given here are to be viewed as approximation formulas. The point z is usually unknown except that $a \leq z \leq b$ and so the term containing z is an "error term" that is usually estimated by replacing $f'(z)$ with its minimum or maximum value. These are called *one-point rules* because the function is evaluated at only one point on the interval. We will consider some two- and three-point rules later.

The error in the approximation using the endpoint rules can be interpreted geometrically.

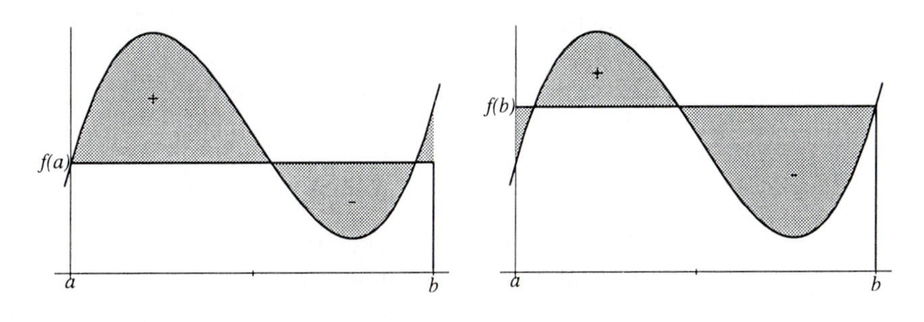

FIGURE 11.4
Left endpoint approximation

FIGURE 11.5
Right endpoint approximation

The area computed by the endpoint rule is bounded by the rectangle; the area bounded by the curve $y = f(x)$ is the value of the integral. The unshaded region inside the rectangle is counted by both the endpoint approximation and the integral. The region (+), along with the little shaded corner, is counted when computing the integral but not with the endpoint rule. The region (−) is counted with the endpoint rule but not by the integral. In both the left and right endpoint cases, there is some balancing of errors so it is difficult to tell from the figures how close either approximation is to the true value or which one is closer to the true value.

Now we use the endpoint estimates in a specific example.

EXAMPLE 1 Approximate

$$I = \int_2^{2.5} \frac{dx}{x}$$

using each of the two methods described in Theorem 11.4 and give an upper bound for the errors.

Solution: Use $f(x) = 1/x$ and $f'(x) = -1/x^2$.

The Left Endpoint approximation gives

$$\int_2^{2.5} \frac{dx}{x} = f(2)(2.5 - 2) + E = 0.25 + E$$

with

$$E = f'(z)\frac{(b-a)^2}{2} = -\frac{1}{z^2}\frac{(0.5)^2}{2} = -\frac{1}{8z^2},$$

where $2 \leq z \leq 2.5$. The largest possible error occurs with $z = 2$, so

$$-\frac{1}{32} \leq E \leq 0.$$

Thus

$$0.25 - \frac{1}{32} = 0.21875 \leq I \leq 0.25.$$

The Right Endpoint approximation gives

$$\int_2^{2.5} \frac{dx}{x} = f(2.5)(2.5 - 2) + E = 0.20 + E$$

with

$$E = -f'(z)\frac{(b-a)^2}{2} = \frac{1}{z^2}\frac{(0.5)^2}{2} = \frac{1}{8z^2}.$$

The largest possible error occurs with $z = 2$, so

$$0 \leq E \leq \frac{1}{32},$$

and we have

$$0.2 \leq I \leq 0.2 + \frac{1}{32} = 0.23125.$$

Clearly the two estimates together provide better information than either one individually; we can now conclude

$$0.21875 \leq I \leq 0.23125.$$

A computer calculation gives $I = \ln 2.5 - \ln 2 \approx 0.223144$. ■

We will show how to improve this estimate by taking a partition and applying an approximation to each subinterval. Before doing this, let us look at two ways of improving the estimate even for just one interval.

The Midpoint and Trapezoidal Rules

The approximation of the integral of $f(x)$ given by the Left or Right Endpoint rules uses a constant function to approximate f. We now improve this by approximating $f(x)$ with its first Taylor polynomial. We will use the midpoint of the interval as the center for the Taylor polynomial. Let $c = (a+b)/2$ so the first Taylor polynomial is

$$P_1(x) = f(c) + f'(c)(x - c),$$

and

$$\int_a^b P_1(x)\,dx = f(c)(b-a) + \frac{1}{2}f'(c)[(b-c)^2 - (a-c)^2]$$
$$= f\left(\frac{a+b}{2}\right)(b-a).$$

The cancellation occurs because $(b-c)^2 = (a-c)^2$ when c is midway between a and b. This fortunate cancellation and inequality (11.11) for $n = 1$ shows

$$\frac{m_2}{3!}\left(\frac{b-a}{2}\right)^3 \leq \int_a^b f(x)\,dx - f\left(\frac{a+b}{2}\right)(b-a) \leq \frac{M_2}{3!}\left(\frac{b-a}{2}\right)^3 \qquad \textbf{(11.13)}$$

If we assume the continuity of $f''(x)$, then every number between m_2 and M_2 can be expressed as $f''(z)$ for some z and we have an alternative form.

THEOREM 11.5

Midpoint Rule

If $f''(x)$ is continuous on $[a, b]$ then there is a number z on $[a, b]$ such that

$$\int_a^b f(x) = f\left(\frac{a+b}{2}\right)(b-a) + f''(z)\left(\frac{b-a}{2}\right)^3.$$

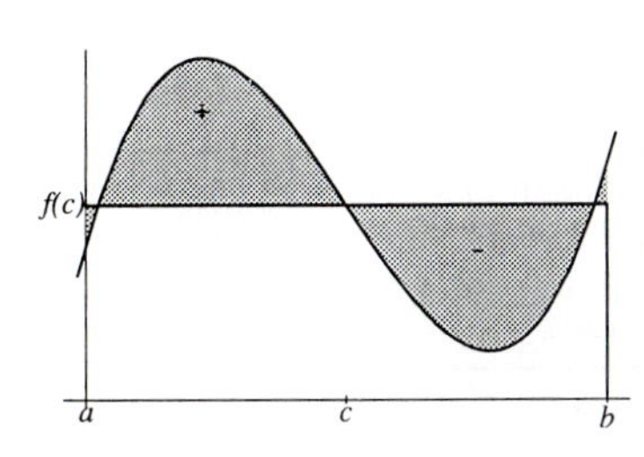

FIGURE 11.6
Rectangle = midpoint approximation

EXAMPLE 2 Approximate

$$I = \int_2^{2.5} \frac{dx}{x}$$

using the Midpoint Rule and give an upper bound for the error.

Solution: The midpoint approximation gives

$$\int_2^{2.5} \frac{dx}{x} = f(2.25)(2.5-2) + E = \frac{2}{9} + E = 0.22222 + E$$

with

$$E = f''(z)\left(\frac{b-a}{2}\right)^3 = \frac{2}{z^3}\frac{1}{64}.$$

Because $2 \leq z \leq 2.5$, the largest possible error occurs with $z = 2$. So we have

$$-\frac{1}{256} \leq E \leq \frac{1}{256}$$

and

$$-\frac{1}{256} + \frac{2}{9} \leq I \leq \frac{1}{256} + \frac{2}{9}.$$

In decimal form we have

$$0.218315\ldots \leq \int_2^{2.5} \frac{dx}{x} \leq 0.226128\ldots.$$

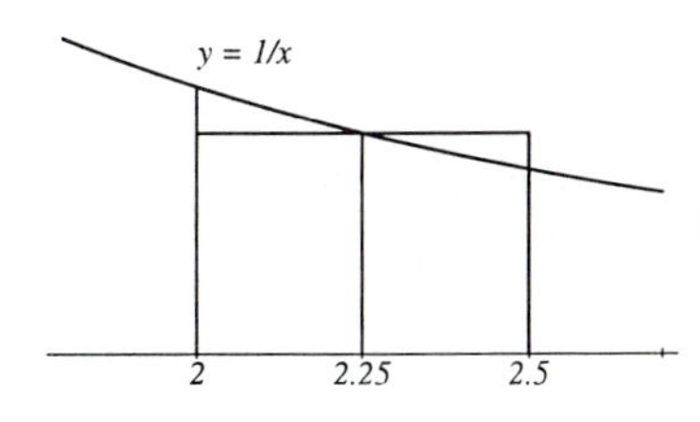

FIGURE 11.7
Midpoint approximation

Figure 11.7 shows the region whose area is computed by the Midpoint Rule; it is the rectangle of height $1/2.25$ and width 0.5. The integral is the area below the curve over the interval [2, 2.5]. The figure reveals some balancing of errors so that a fairly good approximation is to be expected. ■

The Midpoint Rule is motivated by approximating the function $f(x)$ with the horizontal line passing through the point on its graph at the midpoint of the interval. There is another straight line approximation that is reasonable to try. We could use the straight line passing through the two points on the graph at the ends of the interval. The line through $(a, f(a))$ and $(b, f(b))$ has the equation

$$y = \frac{f(b)-f(a)}{b-a}(x-a) + f(a).$$

One could compute the area under this line and then derive an estimate of the difference between it and the area under the graph of f by steps similar to those used for the Midpoint Rule. It is much easier to observe that the area of the trapezoid is just the average of the areas of the two rectangles having base $[a, b]$ and opposite side either the line $y = f(a)$ or $y = f(b)$. In other words, this approximation is just the average of the Left Endpoint and Right Endpoint approximations. We quickly admit that this may not produce the best error term since, in the process of adding two terms with errors, we cannot keep track of the signs of the errors (one may be positive the other negative but we will not know that). This problem shows up by the appearance of the θ in the theorem. Here is the statement of the Trapezoidal Rule.

THEOREM 11.6

Trapezoidal Rule

Assume $f''(x)$ is continuous on $[a, b]$. Then there is a number z on $[a, b]$ and a number θ with $|\theta| \leq 1$ such that

$$\int_a^b f(x)\,dx = \frac{f(a) + f(b)}{2}(b - a) + \frac{f''(z)(b - a)^3\theta}{2}.$$

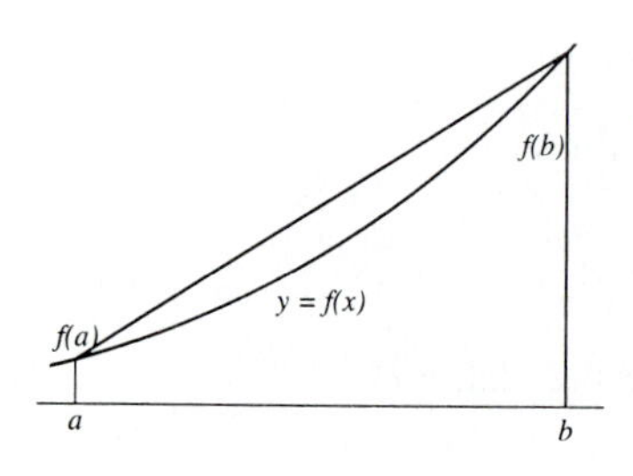

FIGURE 11.8
Trapezoid approximates the integral

Proof

If we just average the left and right endpoint approximations given in Theorem 11.4 we have

$$\int_a^b f(x)\,dx = \frac{f(a) + f(b)}{2}(b - a) + \frac{(f'(z_1) - f'(z_2))(b - a)^2}{2} \qquad \textbf{(11.14)}$$

for some numbers z_1, z_2 on $[a, b]$. By the Mean Value Theorem applied to $f'(x)$ we obtain some number z between z_1 and z_2 such that

$$f'(z_1) - f'(z_2) = f''(z)(z_1 - z_2) = f''(z)(b - a)\theta$$

where

$$\theta = \frac{z_1 - z_2}{b - a}.$$

Because $a \leq z_1, z_2 \leq b$, it follows that $|z_1 - z_2| \leq b - a$ and $|\theta| \leq 1$. After we substitute $f''(z)(b - a)\theta$ in place of the difference $f'(z_1) - f'(z_2)$ in Equation (11.14), the formula stated in the theorem is obtained. □

The Trapezoidal Rule is an example of a *two-point* rule because the function is evaluated at two points on the interval. The extra bit of computation is rewarded by a slightly better error term. The second derivative appears in the error term in place of the first derivative that appeared in the one-point rule. If the second derivative is smaller than the first derivative by any significant amount, then the Trapezoidal Rule is more accurate. Furthermore the length of the interval $(b - a)$ appears with an exponent 3 rather than exponent 2. When the interval length is smaller than 1, the higher power of $(b - a)$ gives a smaller number. Both the Trapezoidal Rule and the Midpoint Rule give more accurate estimates than the other one-point rules when the interval length $(b - a)$ is small (smaller than 1). The improved error term in the Midpoint Rule is a fortunate coincidence that comes from the symmetry of the integration.

Partitioned Intervals

The approximation formulas given above are not very good if the integration is performed over a long interval. To remedy this the interval $[a, b]$ is partitioned into small subintervals and the approximation formula is used on each subinterval and the results added. Surprisingly, the accumulation of errors is small and becomes smaller as the number of subintervals increases.

We illustrate the method using the left endpoint approximation and then state the formulas in the other cases.

THEOREM 11.7

Let $f'(x)$ be continuous on $[a, b]$ and let $P : a = x_0 < x_1 < \cdots < x_n = b$ be a regular partition. There exists a number z on $[a, b]$ such that

$$I = \int_a^b f(x)\,dx = \big(f(x_0) + f(x_1) + \cdots + f(x_{n-1})\big)\left(\frac{b-a}{n}\right) + \frac{f'(z)(b-a)^2}{2n}.$$

Proof

Since the partition is regular, every subinterval has the same length, namely

$$x_i - x_{i-1} = \frac{b-a}{n}.$$

The value I equals the sum of the integrals of f over $[x_{i-1}, x_i]$ as i ranges over $i = 1, 2, \ldots, n$. We apply the Left Endpoint Rule in Theorem 11.4 to get

$$\int_{x_{i-1}}^{x_i} f(x)\,dx = f(x_{i-1})\left(\frac{b-a}{n}\right) + \frac{f'(z_i)(b-a)^2}{2n^2}$$

where z_i is some point on $[x_{i-1}, x_i]$. If we add these terms, we do not get the error term exactly the form stated in the theorem. One more step is required. If m_1 and M_1 are the minimum and maximum values of $f'(x)$ on $[a, b]$, then $m_1 \le f'(z_i) \le M_1$. Add the n inequalities for $i = 1, 2, \ldots, n$ and divide by n to obtain the inequality

$$m_1 \le \frac{f'(z_1) + \cdots + f'(z_n)}{n} \le M_1.$$

Every number between m_1 and M_1 is a value of $f'(x)$ by the Intermediate Value Theorem so there is some z on $[a, b]$ that satisfies

$$f'(z) = \frac{f'(z_1) + \cdots + f'(z_n)}{n}.$$

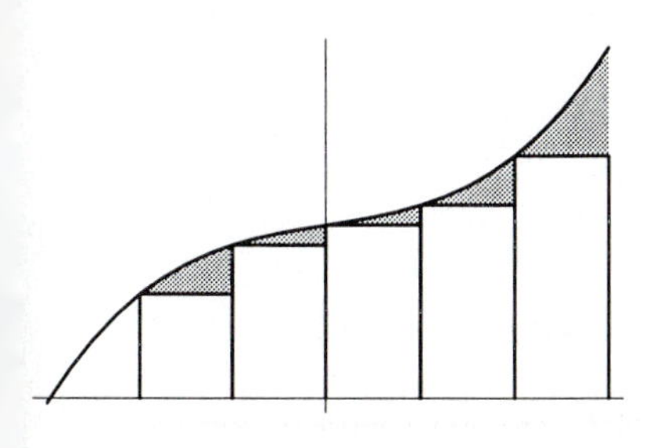

FIGURE 11.9
Shaded area = LEP$_n$ error

Now we have

$$\begin{aligned}\int_a^b f(x)\,dx &= \sum_{i=1}^{n} \int_{x_{i-1}}^{x_i} f(x)\,dx \\ &= \sum_{i=1}^{n} f(x_{i-1})\left(\frac{b-a}{n}\right) + \left(\sum_{i=1}^{n} f'(z_i)\right)\left(\frac{(b-a)^2}{2n^2}\right) \\ &= \sum_{i=1}^{n} f(x_{i-1})\left(\frac{b-a}{n}\right) + \frac{f'(z)(b-a)^2}{2n}.\end{aligned}$$

□

The device of writing the accumulated error terms as a single term makes the formulas look nice and also has some practical value. The same idea is used in the other cases. Here are the one-point and two-point rules; the proofs use the same ideas as in the previous theorem except that in the cases where the mysterious θ appears, absolute values, the triangle inequality and the continuity of $|f''(x)|$ may be needed. In all the formulas, n is a positive integer and

$$x_i = a + \frac{i(b-a)}{n}.$$

The number θ satisfies $|\theta| \leq 1$ and z is some unspecified point on $[a, b]$.

Left Endpoint Rule

$$\int_a^b f(x)\,dx = \mathrm{LEP}_n + \frac{f'(z)(b-a)^2}{2n},$$

$$\mathrm{LEP}_n = \left[f(x_0) + f(x_1) + \cdots + f(x_{n-1})\right]\left(\frac{b-a}{n}\right).$$

Right Endpoint Rule

$$\int_a^b f(x)\,dx = \mathrm{REP}_n - \frac{f'(z)(b-a)^2}{2n},$$

$$\mathrm{REP}_n = \left[f(x_1) + f(x_2) + \cdots + f(x_n)\right]\left(\frac{b-a}{n}\right).$$

Midpoint Rule

$$\int_a^b f(x)\,dx = \mathrm{MP}_n + \frac{f''(z)(b-a)^3}{8n^2},$$

$$\mathrm{MP}_n = \left[f\left(\frac{x_0 + x_1}{2}\right) + \cdots + f\left(\frac{x_{n-1} + x_n}{2}\right)\right]\left(\frac{b-a}{n}\right).$$

Trapezoidal Rule

$$\int_a^b f(x)\,dx = \mathrm{T}_n + \frac{f''(z)(b-a)^3\theta}{2n^2},$$

$$\mathrm{T}_n = \frac{1}{2}\left[f(x_0) + 2f(x_1) + \cdots + 2f(x_{n-1}) + f(x_n)\right]\left(\frac{b-a}{n}\right).$$

EXAMPLE 3 Compare the approximation rules by finding how many subintervals of $[1, 2]$ are required to approximate

$$I = \int_1^2 \frac{dx}{1 + x^6}$$

with an error no greater than 10^{-4}.

Solution: Let $f(x) = (1 + x^6)^{-1}$. Then the derivatives of f are

$$f'(x) = \frac{-6x^5}{(1 + x^6)^2} \qquad f''(x) = \frac{6x^4(7x^6 - 5)}{(1 + x^6)^3}.$$

The first step is to estimate the size of the first and second derivatives over $[1, 2]$. We find $f''(x)$ is positive on $[1, 2]$. This implies that $f'(x)$ is increasing on the interval and satisfies

$$-1.5 = f'(1) \leq f'(x) \leq f'(2) \leq -0.0455.$$

The second derivative is positive but neither increasing nor decreasing on $[1, 2]$. With a bit of computation with the third derivative, one can find the maximum value of $f''(x)$ and show

$$f''(2) \approx 0.1548 \leq f''(x) \leq 3.23.$$

Now $[1, 2]$ is partitioned into n subintervals of equal length, and we find the following error bounds for the integral of $f(x)$ (for some z with $1 \leq z \leq 2$):

$$\begin{aligned}
\text{Left Endpoint:} &\quad -\frac{1.5}{2n} \leq E = \frac{f'(z)(1)^2}{2n} \leq -\frac{0.0455}{2n}, \\
\text{Right Endpoint:} &\quad \frac{0.0455}{2n} \leq E = -\frac{f'(z)(1)^2}{2n} \leq \frac{1.5}{2n}, \\
\text{Midpoint:} &\quad -\frac{3.23}{8n^2} \leq E = -\frac{f''(z)(1)^3}{8n^2} \leq -\frac{0.1540}{8n^2}, \\
\text{Trapezoidal:} &\quad -\frac{3.23}{2n^2} \leq E = \frac{f''(z)(1)^3\theta}{2n^2} \leq \frac{3.23}{2n^2}.
\end{aligned}$$

In these statements we have replaced $f'(z)$ and $f''(z)$ by their extreme values and, in the last case, allowed for the possibility that θ may be as large as 1 or as small as -1.

Next we must select n so that $|E| < 10^{-4}$. In the Left or Right Endpoint cases, this is accomplished if

$$\frac{0.75}{n} < 10^{-4} \quad \text{or} \quad 7500 < n.$$

In the Midpoint case, we require

$$\left|-\frac{0.40375}{n^2}\right| < 10^{-4} \quad \text{or} \quad \sqrt{4037.5} < 64 \leq n.$$

In the Trapezoidal Rule case, we require

$$\frac{1.615}{n^2} < 10^{-4} \quad \text{or} \quad \sqrt{16150} \approx 128 \leq n.$$

The least computation would be needed using the Midpoint Rule.

A computer calculation gives the numerical value of the integral as $I = 0.1372197$.

The reader might wonder why we selected the particular function $f(x) = (1 + x^6)^{-1}$ to illustrate the approximation of the integral when it is possible to find an antiderivative and apply the Fundamental Theorem of Calculus. In this case the exact result can be given but it is not a pretty sight. In fact, the exact value may be of little help in solving a numerical problem involving the integral of this $f(x)$. Just to emphasize the value of numerical computation, we give the exact value of the integral:

$$\int_1^2 \frac{dx}{1+x^6} = -\frac{\pi}{12} + \frac{\arctan 2}{3} - \frac{\arctan(2-\sqrt{3})}{6} + \frac{\arctan(4-\sqrt{3})}{6} - \frac{\arctan(2+\sqrt{3})}{6} + \frac{\arctan(4+\sqrt{3})}{6} - \frac{\sqrt{3}\ln(5-2\sqrt{3})}{12} + \frac{\sqrt{3}\ln(2-\sqrt{3})}{12} - \frac{\sqrt{3}\ln(2+\sqrt{3})}{12} + \frac{\sqrt{3}\ln(5+2\sqrt{3})}{12}.$$

■

Convex and Concave Functions

We can be somewhat more precise when using the approximation formulas if some information is available about the sign for the derivative or the second derivative. For example if $f'(x) \geq 0$ for all x on $[a, b]$ then the Left Endpoint Rule has a nonnegative error term and the Right Endpoint Rule has a nonpositive error term. It follows then

$$\mathrm{LEP}_n \leq \int_a^b f(x)\,dx \leq \mathrm{REP}_n.$$

Similarly if the $f'(x) \leq 0$ for all x on $[a, b]$, the above statement is true if the inequalities are reversed.

Of course both of these inequalities are rather obvious when we consider the geometry of the graph. When $f'(x) \geq 0$, the function is increasing and so the rectangle having base $[x_{i-1}, x_i]$ with height $f(x_{i-1})$ lies below the graph and the rectangle with height $f(x_i)$ lies above the graph.

There is a similar situation for convex functions.

THEOREM 11.8

Assume $f''(x) \geq 0$ for all x on $[a, b]$. Then for any positive n, the Midpoint Rule and the Trapezoidal Rule satisfy the inequality

$$\mathrm{MP}_n \leq \int_a^b f(x)\,dx \leq \mathrm{T}_n.$$

If $f''(x) \leq 0$ for all x on $[a, b]$ then

$$\mathrm{T}_n \leq \int_a^b f(x)\,dx \leq \mathrm{MP}_n.$$

Proof

The function f is convex when $f''(x) \geq 0$ for x on $[a, b]$. Then the error term in the Midpoint Rule is nonnegative and the midpoint approximation is less than the integral:

$$\mathrm{MP}_n \leq \int_a^b f(x)\,dx.$$

The Trapezoidal Rule will overestimate the integral, but the error that we have given does not show this because we do not know the sign of θ. However we can obtain the inequality by remembering that the graph of a convex function lies below a chord (see section 4.7). Thus the trapezoid encloses an area larger than the area under the graph.

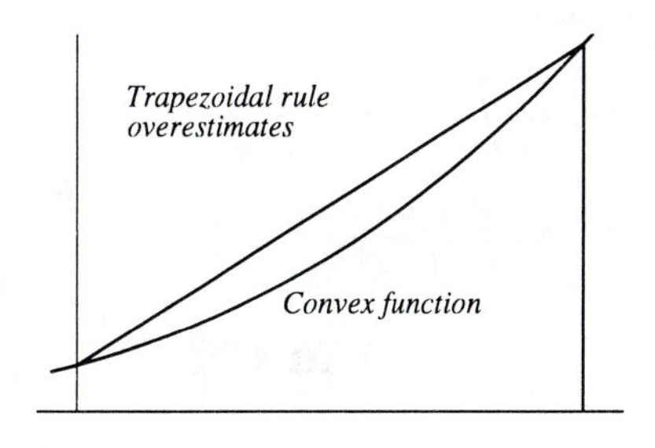

FIGURE 11.10

A similar argument can be given in the case of a concave function to prove the second statement of the theorem. □

The information given in this theorem may be better than the information given by using any one rule and its error term.

EXAMPLE 4 Estimate the integral

$$I = \int_1^2 \frac{dx}{1 + x^6}$$

by applying both the Midpoint and the Trapezoidal Rules.

Solution: Notice that the second derivative of $(1 + x^6)^{-1}$ is positive on $[1, 2]$ so the value of the integral lies between MP_2 and T_2. We use the case $n = 2$ with the partition $1 < 1.5 < 2$ for each rule. We have

$$MP_2 = \frac{1}{2}\left[\frac{1}{1 + (1.25)^6} + \frac{1}{1 + (1.75)^6}\right] \approx 0.12067$$

$$T_2 = \frac{1}{2}\left[\frac{1}{1 + 1} + \frac{2}{1 + (1.5)^6} + \frac{1}{1 + 2^6}\right] \approx 0.2256.$$

The maximum error using MP_2 is

$$\frac{f''(z)(b - a)^3}{8n^2} \leq \frac{3.23}{8 \cdot 4} \leq 0.101.$$

The maximum error using T_2 is

$$\frac{f''(z)(b - a)^3\theta}{2n^2} \leq \frac{3.23}{2 \cdot 4} \leq 0.403.$$

The Midpoint approximation has a bound for the error that is nearly the same size as the approximation. In the Trapezoidal case, the error bound is twice the size of the approximation. Thus these estimates are not very useful. We have already remarked above that $I \approx 0.13722$. ■

Exercises 11.2

Use the Midpoint Rule and the Trapezoidal Rule and partitions into $n = 2$, 4, and 8 subintervals to approximate the integrals and compare the results.

1. $\displaystyle\int_0^1 \frac{dx}{1+x^2}$

2. $\displaystyle\int_0^2 \frac{dx}{1+x^2}$

3. $\displaystyle\int_0^1 \frac{dx}{1+x^4}$

4. $\displaystyle\int_0^2 \frac{dx}{1+x^4}$

5. Approximate $\int_0^{0.5} \cos x\, dx$ by replacing $\cos x$ with a suitable Taylor polynomial and using estimates (11.11) or (11.12) to ensure that the error is at most 10^{-4}.

6. Carry out a detailed analysis of the four approximation methods as done in example 3 but using the integral $\int_1^2 (1+x^3)^{-1}\, dx$ in place of the integral in that example.

7. Estimate the integral $\int_0^{0.5} \frac{\arctan x}{x}\, dx$ to within 10^{-4}. [Hint: Find the Taylor series and note that it is an alternating series so a partial sum approximates the function to within the first missing term (Theorem 10.11).]

11.3 Simpson's rule

The approximations of a definite integral of $f(x)$ given in section 11.2 were based on the approximation of $f(x)$ by a constant or a polynomial of degree 1. We now try the next case in which we use a Taylor polynomial of degree 2 to approximate the function and we derive a three-point rule called *Simpson's Rule.* The three points at which the function is evaluated are the two end points and the midpoint; we first derive the approximation and error estimate for one interval.

THEOREM 11.9

Simpson's Rule: One Interval

Let $f(x)$ be a function with a continuous fourth derivative $f^{(4)}(x)$ on an interval $[a, b]$. Then

$$\int_a^b f(x)\, dx = \left[f(a) + 4f\left(\frac{b+a}{2}\right) + f(b)\right]\left(\frac{b-a}{6}\right) + E$$

where the error term E can be expressed as

$$E = \frac{7f^{(4)}(t)\gamma}{36}\left(\frac{b-a}{2}\right)^5$$

for some number t satisfying $a \le t \le b$ and some number γ satisfying $|\gamma| \le 1$.

Proof

We select the notation to emphasize the symmetry of the situation; let $c = (b+a)/2$ and $h = (b-a)/2$ so that c is the midpoint of the interval $[a, b]$ and h is half the length of the interval. We have $b - c = h$ and $a - c = -h$. We begin with the Taylor polynomials $P_2(x)$ for $f(x)$ centered at c as an approximation to f and determine the error involved when the integral of f is approximated by the integral of P_2. For any x on $[a, b]$ there is some z such that

$$f(x) = f(c) + f'(c)(x - c) + \frac{1}{2}f''(c)(x - c)^2 + \frac{1}{6}f'''(z)(x - c)^3 \quad \textbf{(11.15)}$$

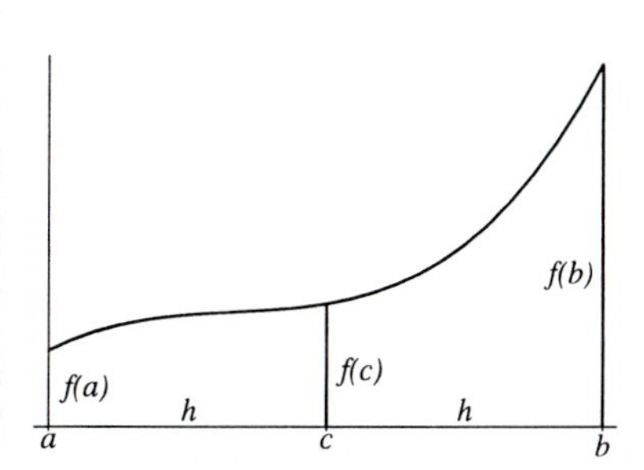

FIGURE 11.11
Data for Simpson's rule

by Taylor's Theorem 9.3. The second Taylor polynomial for f is

$$P_2(x) = f(c) + f'(c)(x - c) + \frac{1}{2}f''(c)(x - c)^2.$$

We use $P_2(x)$ to approximate the integral and get

$$\begin{aligned}\int_a^b f(x)\,dx &= \int_{c-h}^{c+h} f(x)\,dx \\ &= \int_{c-h}^{c+h} P_2(x)\,dx + E_2 \\ &= \int_{c-h}^{c+h} \left[f(c) + f'(c)(x - c) + \frac{1}{2}f''(c)(x - c)^2\right] dx + E_2 \\ &= 2f(c)h + \frac{1}{3}f''(c)h^3 + E_2.\end{aligned} \quad \textbf{(11.16)}$$

To estimate the error term E_2 we let the minimum value of $f'''(x)$ on $[a, b]$ be $m_3 = f'''(u)$ and the maximum value of $f'''(x)$ be $M_3 = f'''(v)$, where $a \le u, v \le b$. By inequality (11.12) we have

$$|E_2| \le \frac{(M_3 - m_3)h^4}{4!}.$$

The term $(M_3 - m_3)$ can be expressed in terms of the fourth derivative by applying the Mean Value Theorem to $f'''(x)$:

$$M_3 - m_3 = f'''(v) - f'''(u) = f^{(4)}(w)(v - u) = f^{(4)}(w)(2h)\theta,$$

where

$$\theta = \frac{v - u}{2h} = \frac{v - u}{b - a}$$

and $|\theta| \le 1$ (since $a \le u, v \le b$). We now have good control over the error E_2 in that

$$E_2 = \frac{f^{(4)}(w)h^5\theta}{12}.$$

There is a small problem with Equation (11.16) because the estimate requires that we evaluate both f and f'' at c. We prefer to evaluate the original function at several points rather than evaluate the function and its derivatives at a single point. So we now try to eliminate the term $f''(c)$ from Equation (11.16). Evaluate the function at $x = a$ and $x = b$ from the expression (11.15) to get

$$f(a) = f(c - h) = f(c) - f'(c)h + \frac{1}{2}f''(c)h^2 - \frac{1}{6}f'''(z_1)h^3,$$

$$f(b) = f(c + h) = f(c) + f'(c)h + \frac{1}{2}f''(c)h^2 + \frac{1}{6}f'''(z_2)h^3,$$

for some z_1, z_2 on $[a.b]$. Add these two equations and solve for the term containing the second derivative to get

$$f''(c)h^2 = f(a) + f(b) - 2f(c) + \frac{\big(f'''(z_1) - f'''(z_2)\big)h^3}{6}.$$

Using the Mean Value Theorem as has been done several times before, we can write

$$f'''(z_1) - f'''(z_2) = f^{(4)}(w_1)(z_1 - z_2) = f^{(4)}(w_1)(2h)\theta_1$$

for some w_1 between z_1 and z_2 and some θ_1 satisfying $|\theta_1| \leq 1$. Putting several equations together we obtain

$$\begin{aligned}\int_a^b f(x)\,dx &= \int_a^b P_2(x)\,dx + E_2 \\ &= 2f(c)h + \frac{1}{3}f''(c)h^3 + E_2 \\ &= 2f(c)h + \frac{h}{3}[f(a) + f(b) - 2f(c)] - \frac{f^{(4)}(w_1)h^5\theta_1}{9} + \frac{f^{(4)}(w)h^5\theta}{12} \\ &= \frac{h}{3}[f(a) + 4f(c) + f(b)] + E_2^*,\end{aligned}$$

where

$$E_2^* = \left[\frac{f^{(4)}(w)\theta}{12} - \frac{f^{(4)}(w_1)\theta_1}{9}\right]h^5.$$

The main term is the one predicted in the statement of the theorem but the error term is not quite in the best form. Just a bit more work is required to simplify this. Let $|f^{(4)}(x)|$ have a maximum value M for $a \leq x \leq b$. Using the triangle inequality, we find

$$\begin{aligned}\left|\frac{f^{(4)}(w)\theta}{12} - \frac{f^{(4)}(w_1)\theta_1}{9}\right| &\leq \left|\frac{f^{(4)}(w)\theta}{12}\right| + \left|\frac{f^{(4)}(w_1)\theta_1}{9}\right| \\ &\leq \frac{M}{12} + \frac{M}{9} = \frac{7M}{36}.\end{aligned}$$

The function $|7f^{(4)}(x)/36|$ takes on the maximum value $7M/36$ and all values smaller to some point (possibly 0 or some larger number). We allow ourselves the flexibility to multiply $|7f^{(4)}(x)/36|$ by a constant γ between 0 and 1 to ensure that any value between 0 and $7M/36$ is attained by $|7f^{(4)}(x)/36|\gamma$. Finally, allowing the multiplier γ to range between -1 and $+1$ ensures that any given number between $-7M/36$ and $7M/36$ equals a value of $7f^{(4)}(x)\gamma/36$. It follows that for some suitable point t and some number γ with $|\gamma| \leq 1$ we have

$$\frac{7f^{(4)}(t)\gamma}{36} = \frac{f^{(4)}(w)\theta}{12} - \frac{f^{(4)}(w_1)\theta_1}{9},$$

and so

$$E_2^* = \frac{7}{36}f^{(4)}(t)\gamma h^5$$

with t on $[a, b]$ and $|\gamma| \leq 1$. Thus the formula stated in the theorem is achieved. □

Integrals of Polynomials of Low Degree

The error in the approximation of a definite integral by Simpson's Rule involves the fourth derivative of the function. The presence of the fourth derivative has an interesting consequence for functions whose fourth derivative equals 0. Suppose $f(x) = x^k$ with $k = 0$, 1, 2 or 3. Then $f^{(4)}(x) = 0$ for all x and the error term in

Simpson's Rule is 0. Stated more directly we have

$$\int_a^b x^k\,dx = \left[a^k + 4\left(\frac{b+a}{2}\right)^k + b^k\right]\left(\frac{b-a}{6}\right) \quad \text{for} \quad k = 0, 1, 2, 3.$$

Now if $f(x) = px^3 + qx^2 + rx + s$ is a polynomial of degree at most 3, then Simpson's Rule gives an exact evaluation of the integral of f; that is,

$$\int_a^b f(x)\,dx = \left[f(a) + 4f\left(\frac{a+b}{2}\right) + f(b)\right]\left(\frac{b-a}{6}\right).$$

Of course, the integration could be checked without any reference to Simpson's Rule by just evaluating and comparing the left and right sides of the equations. However, it is rather unlikely that you would think of these formulas without the reference to Simpson's Rule.

Partitioned Intervals

Now we give Simpson's Rule for an interval that has been partitioned into an even number of subintervals. We restrict our attention to partitions having an even number of subintervals because the formula makes use of the midpoints of intervals. The notation becomes simpler when this restriction is made.

Let $[a, b]$ be partitioned as

$$a = x_0 < x_1 < x_2 < \cdots < x_{2n-1} < x_{2n} = b. \tag{11.17}$$

We will think of this as a partition into n intervals $[x_0, x_2]$, $[x_2, x_4]$, ..., $[x_{2n-2}, x_{2n}]$ with the odd numbered points x_{2i+1} playing the role of the midpoints of the intervals. We then apply Simpson's Rule on each interval $[x_{2i-2}, x_{2i}]$ and apply the theorem just proved to determine the error. For simplicity let us write

$$S_{2n} = [f(x_0) + 4f(x_1) + 2f(x_2) + 4f(x_3) + \cdots + 4f(x_{2n-1}) + f(x_{2n})]\frac{b-a}{6n}$$

and call S_{2n} the Simpson approximation corresponding to the partition of $[a, b]$ into $2n$ equally spaced subintervals. Here is the statement of the approximation and error.

THEOREM 11.10

Simpson's Rule: General Case

Let $f(x)$ be a function having a continuous fourth derivative $f^{(4)}(x)$ at each point of $[a, b]$. Consider a partition of $[a, b]$ given in (11.17) into an even number of subintervals of length $h = (b - a)/2n$. Then

$$\int_a^b f(x)\,dx = S_{2n} + E_n$$

where the error term satisfies

$$E_n = \frac{7f^{(4)}(w)\theta}{36n^4}\left(\frac{b-a}{2}\right)^5$$

for some point w on $[a, b]$ and some number θ with $|\theta| \le 1$.

Proof

The one-interval version of Simpson's Rule applied to the interval $[x_{2i-2}, x_{2i}]$

gives the approximation

$$\int_{x_{2i-2}}^{x_{2i}} f(x)\,dx =$$

$$[f(x_{2i-2}) + 4f(x_{2i-1}) + f(x_{2i})]\left(\frac{b-a}{6n}\right) + \frac{7f^{(4)}(z_i)\gamma_i(b-a)^5}{36 \cdot 2^5 \cdot n^5},$$

where z_i is some point on the interval $[x_{2i-2}, x_{2i}]$ and $|\gamma_i| \leq 1$. We form the sum as i varies from 1 to n of these terms. Use the fact that there is some w on $[a, b]$ and some γ with $|\gamma| \leq 1$ such that

$$nf^{(4)}(w)\gamma = \sum_{i=1}^{n} f^{(4)}(z_i)\gamma_i$$

to get the formula and error as stated in the theorem. □

Let us apply Simpson's Rule to two integrals that were approximated in section 11.2.

EXAMPLE 1 How many subintervals of $[1, 2]$ are required to approximate

$$I = \int_1^2 \frac{dx}{1 + x^6}$$

by Simpson's Rule with an error no greater than 10^{-4}.

Solution: We worked this example with the one- and two-point rules in section 11.2. The error terms required computing the second and third derivatives of $f(x) = (1 + x^6)^{-1}$ and finding an upper bound for the absolute value of the derivatives. In the present case, it is necessary to find a bound for the fourth derivative. That task is the hardest part of the problem. With the aid of a computer we find

$$f^{(4)}(x) = \frac{72x^2(42x^{18} - 245x^{12} + 140x^6 - 5)}{(1 + x^6)^5}.$$

It is a nontrivial problem to find the maximum and minimum values of this function on $[1, 2]$. The computer can plot the graph and we discover by inspection of it that $-180 \leq f^{(4)}(x) \leq 20$.

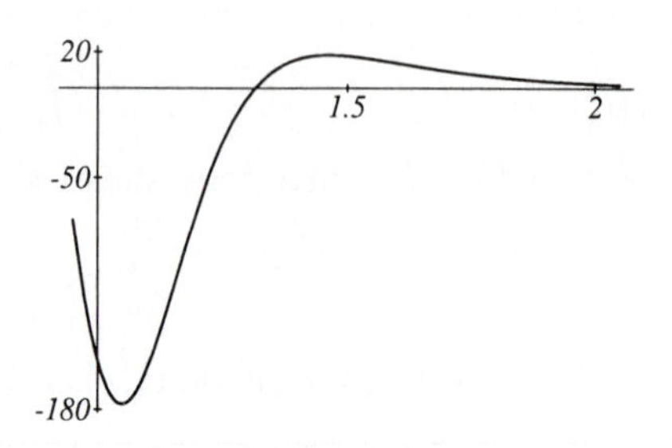

FIGURE 11.12
Plot of $f^{(4)}(x)$

Using Simpson's Rule with $2n$ subintervals involves an error of

$$|SE_n| = \frac{7|f^{(4)}(z)||\gamma|1^5}{36 \cdot 32n^4} \leq \frac{7 \cdot 180 \cdot 1 \cdot 1}{1152n^4}.$$

To ensure that the error is at most 10^{-4} we need only take n to satisfy

$$\frac{7 \cdot 180}{1152n^4} \leq 10^{-4}.$$

This holds if

$$\frac{1260}{1152} \times 10^4 \leq n^4$$

or

$$11 \leq n.$$

So Simpson's Rule with $2n = 22$ subintervals gives the accuracy desired. ■

EXAMPLE 2 Estimate the integral

$$\int_2^{2.5} \frac{dx}{x}$$

using Simpson's Rule with $n = 2$ and determine the maximum error involved in using this approximation.

Solution: The partition of $[2, 2.5]$ with $n = 2$ is $2 < 9/4 < 5/2 = 2.5$ and the Simpson approximation is

$$S_2 = \frac{1}{3}\frac{1}{4}\left[\frac{1}{2} + 4\left(\frac{4}{9}\right) + \frac{2}{5}\right] = \frac{241}{12 \cdot 90} \approx 0.223148.$$

The error in using this approximation

$$E = \frac{7}{1152}\frac{1}{2^4}\frac{1}{2^5}\frac{24}{z^5}$$

for some z between 2 and 2.5. The largest value for $1/z^5$ occurs if $z = 2$. Thus

$$0 \leq E \leq \frac{7}{1152}\frac{1}{2^4}\frac{1}{2^5}\frac{24}{2^5} \approx 8.9 \times 10^{-6}.$$

Simpson's Rule gives a very good approximation with only two subintervals. ■

The geometric interpretation of the Left and Right Endpoint Rules, the Midpoint and Trapezoidal Rules is not difficult. For a given function $f(x)$, the Trapezoidal Rule for the interval $[a, b]$ determines the area of the trapezoid with verticies at $(a, 0)$, $(a, f(a))$, $(b, f(b))$, and $(b, 0)$. There is an interpretation of the area computed by Simpson's Rule but it is more subtle than any of the previous rules. One can show that there is a polynomial $h(x) = px^2 + qx + r$ whose graph passes through the three points $(a, f(a))$, $(c, f(c))$ and $(b, f(b))$ where $c = (a+b)/2$. (In fact there is only one such polynomial.) Thus $h(x) = f(x)$ for $x = a$, c and b. Then Simpson's Rule gives the area under the graph of $h(x)$:

$$\begin{aligned}\int_a^b h(x)\,dx &= \left(\frac{b-a}{6}\right)\left[h(a) + 4h(c) + h(b)\right]\\ &= \left(\frac{b-a}{6}\right)\left[f(a) + 4f(c) + f(b)\right]\\ &= S_2.\end{aligned}$$

Thus Simpson's Rule can be viewed as one that approximates $f(x)$ by pieces of parabolas and then computes the area under the parabolas.

Simpson's Rule is an example of a three-point rule. The theory of four-point, five-point, ... rules is treated in the subject of Numerical Analysis. The more points used in the approximation rule, the smaller the error in general. Accurate integration rules are of great interest because the highly accurate, high-speed computers can be used to take advantage of the otherwise complicated rules.

Just to indicate that detailed error analysis is of practical significance, consider the Hubble Space Telescope launched into space in early 1990. The telescope project cost approximately $1.5 billion and was developed over a 12-year period with countless hours devoted to its construction. Three months after it was put into orbit, the NASA scientists discovered that the reflecting mirrors were constructed to specifications that were slightly inaccurate and consequently, the photos from the Hubble telescope were not significantly better than those obtained from earth-based observatories. It really does pay to keep control over the error estimates.

Having said this, we must admit that some of the error estimates given in this chapter are not the best possible. The error in Simpson's Rule, for example, might be improved by a different analysis to be slightly smaller. However only the constants in the error terms can be improved (the 7/1152 can be replaced by a slightly smaller value), not the power of the interval length or the order of the derivative of the function. A similar disclaimer applies to the errors in the Midpoint Rule and the Trapezoidal Rule. A proof of better error terms would have required different tools than those developed earlier in the text, such as the error estimates by the use of Taylor polynomials.

Exercises 11.3

Use Simpson's Rule with the given number of subintervals to estimate the integrals. In each case give an estimate of the error involved in the approximation. If you have machine computing facilities available, extend the range on n to get more accurate results. If the Fundamental Theorem of Calculus can be easily applied to obtain the exact value, find it and compare it to the Simpson approximation.

1. $\int_0^2 x^4\,dx,\ n = 2, 4, 8, 10$ **2.** $\int_1^3 \frac{dx}{x},\ n = 2, 4, 8$

3. $\int_0^1 x^5\,dx,\ n = 2, 4, 8, 10$ **4.** $\int_1^5 \frac{dx}{x},\ n = 2, 4, 8$

5. $\int_0^1 \frac{dx}{1+x^2},\ n = 2, 4, 8$ **6.** $\int_{-1}^1 \frac{dx}{1+x^2},\ n = 2, 4, 8$

7. $\int_0^{0.5} \frac{dx}{1-x^2},\ n = 2, 4, 8$ **8.** $\int_2^3 \frac{dx}{1-x^2},\ n = 2, 4, 8$

9. The integral $\int_0^1 \frac{dx}{1+x^2}$ has a value $\pi/4$. Apply Simpson's Rule to approximate the value numerically and thereby obtain an approximation to π. Make a table showing the maximum error in the approximation for $2n$ subintervals with $1 \leq n \leq 10$ and do the computations for each value of n. Compare the results with the known value of π.

10. It is not possible to express an antiderivative of $(1 + ax^3)^{1/2}$ (a = positive constant) with a finite number of familiar functions. Use the approximation techniques in this chapter to obtain some lower and upper estimates (in terms of a) for the definite integral

$$F(a) = \int_0^1 \frac{dx}{\sqrt{1+ax^3}}.$$

11. To approximate the integral $\int_0^1 \sqrt{1+x^4}\,dx$ by Simpson's Rule (or any of the integration approximations discussed in this chapter) it is necessary to compute the value of $\sqrt{1+x^4}$ at many points. For most values of x an approximation must be used that will involve some error that is not taken into account by the error estimates for Simpson's Rule. Suppose your calculator can compute square roots with error at most 5×10^{-8}. Make an estimate of the error incurred by using Simpson's Rule with n = 2, 4, 8 (or general n) that takes into account the possible error in computing the square roots. [Hint: Assume the worst case that the maximum error occurs every time a square root is evaluated and that the errors accumulate when taking sums.]

12. The owner of a shopping center wants to beautify the area along one edge of the parking lot by covering it with hand-laid bricks. The bricklayer charges $6.10

per square foot for the work. To determine the area of the irregular region the owner (who knows Simpson's Rule) makes measurements from the edge of the parking lot out to the edge of the region to be covered at intervals of 4 feet along the edge of the parking lot. The distances are shown in figure 11.13. If the budget for this project is $1000, will the job be under or over budget?

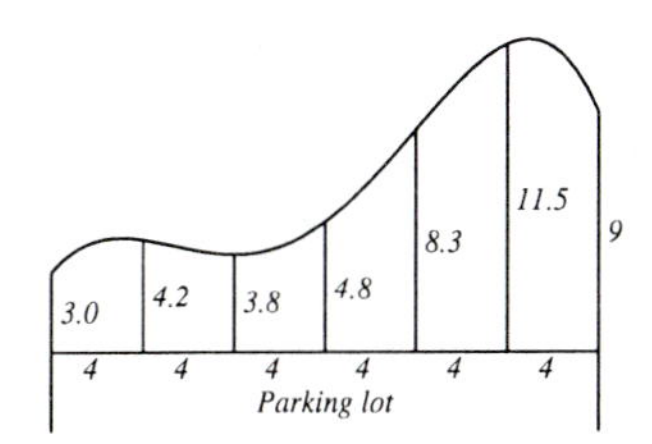

FIGURE 11.13

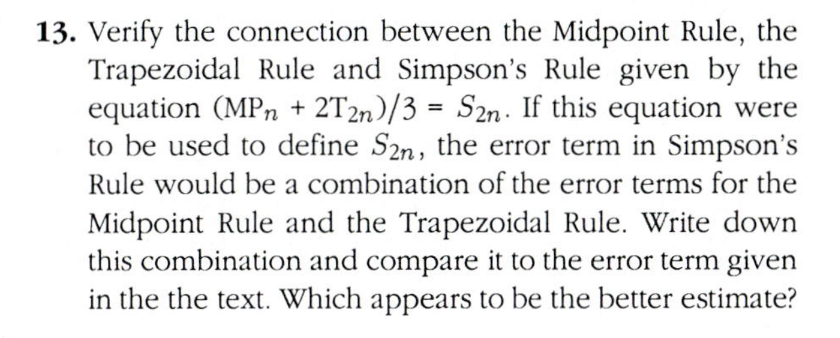

13. Verify the connection between the Midpoint Rule, the Trapezoidal Rule and Simpson's Rule given by the equation $(\mathrm{MP}_n + 2\mathrm{T}_{2n})/3 = S_{2n}$. If this equation were to be used to define S_{2n}, the error term in Simpson's Rule would be a combination of the error terms for the Midpoint Rule and the Trapezoidal Rule. Write down this combination and compare it to the error term given in the the text. Which appears to be the better estimate?

12 Vectors in Two and Three Dimensions

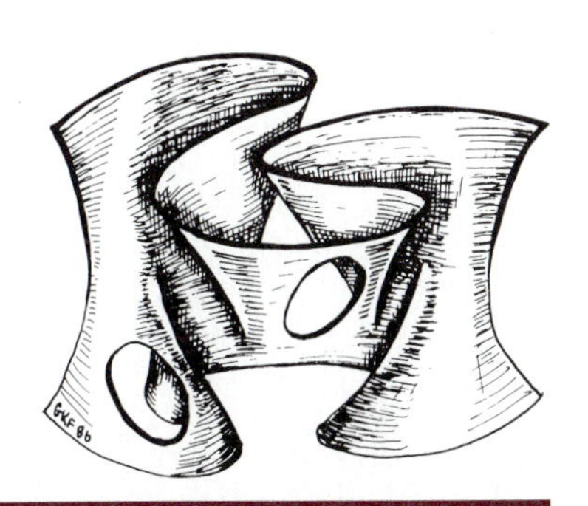

In this chapter we begin the study of calculus in three dimensions. The coordinate system is introduced first and followed by the description of some planes. Vectors are introduced in both two and three dimensions as a device to simplify the language. A vector can be associated with each point in two- or three-dimensional space but a vector contains more information; it has a magnitude and a direction. Vectors are used to state simply and derive easily facts that could be stated and derived in a more complicated way without the use of vectors. The derivation of the equation of a plane is just one example where vector notation is efficient. The advantage of the vector notation is shown in chapter 13 when we discuss space curves, tangent lines, curvature and many other concepts.

12.1 Coordinates in three dimensions

The familiar xy coordinate system gives a very useful way to locate points in a plane. To locate points in space we need a three-dimensional coordinate system. We introduce the xyz coordinates by starting with the usual xy plane and placing the third axis, the z-axis, through the origin and perpendicular to the xy plane. There are two possible ways to introduce coordinates on the z-axis; the positive direction could be above or below the xy plane. We adopt the *right-handed system* in which the positive z-axis is above the xy plane. In order to picture the three-dimensional system on a two-dimensional page, we adopt the common convention that the xy plane is pictured horizontally and the viewer is standing somewhere on the first quadrant roughly on the line $y = x$. The view is as if you are standing in the middle of a classroom and you face the northwest corner, the point where the floor meets the north and west walls is the origin; the west wall on your left and floor meet along the positive x-axis, the north wall on your right and the floor meet along the positive y-axis; the north and west walls meet along the positive z-axis.

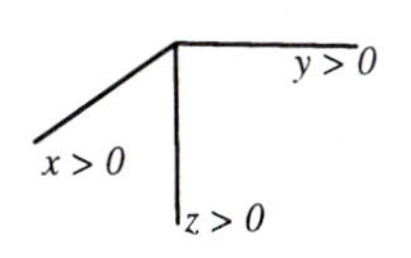

FIGURE 12.1
Left-hand system

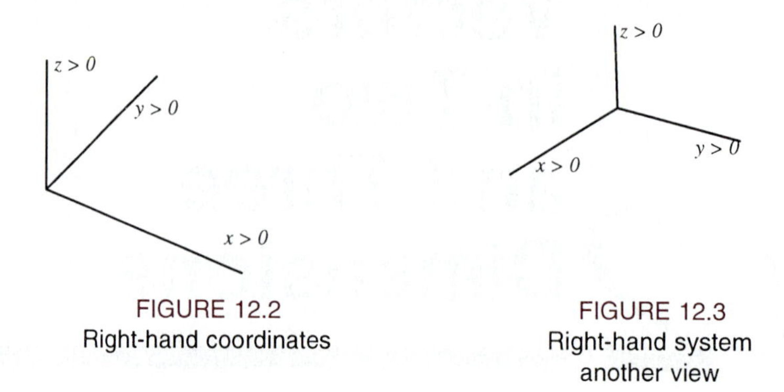

FIGURE 12.2
Right-hand coordinates

FIGURE 12.3
Right-hand system
another view

Coordinates of a Point

Given a point P in space, we assign coordinates (a, b, c) to P by passing a line parallel to the z-axis through P. The point $Q(a, b)$ is the point where the line through P parallel to the z-axis meets the xy plane. The line OQ from the origin to Q is parallel to a line through P that meets the z-axis at some point with coordinate c. Then P is given the coordinates (a, b, c). To help visualize the location of the point, it may be useful to draw in the lines just referred to. For example to locate the point $(1, 2, 3)$ we begin by sketching the xyz coordinate axes, locate the point $Q(1, 2)$ in the xy plane, draw in the line through $(1, 2)$ and parallel to the z-axis, draw OQ and the line parallel to OQ from the point 3 on the z-axis.

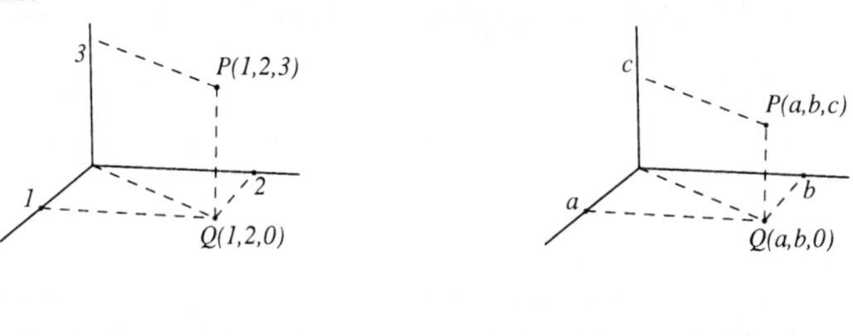

FIGURE 12.4
Locate $P(1, 2, 3)$ in space

FIGURE 12.5
Locate $P(a, b, c)$ in space

EXAMPLE 1 Which of the three points $Q = (-1, 2, 3), (1, -2, 3)$, or $(1, 2, -3)$ lies directly above or below the point $P(1, -2, -3)$?

Solution: We interpret the phrase P lies "directly above or below" Q to mean that the line through P and Q is parallel to the z-axis. If the line through P and Q is parallel to the z-axis, then this line meets the xy plane at some point $(a, b, 0)$ and $(a, b, ?) = P$ and $(a, b, ?) = Q$. Thus P and Q must have the same x and y coordinates 1, -2 in this case. It follows that $Q = (1, -2, 3)$ is directly above $P(1, -2, -3)$. ■

Two unequal but intersecting lines determine a plane; that is, there is one and only one plane containing the two lines. The plane determined by the x-axis and the y-axis is called (naturally) the xy plane; the plane determined by the x-axis and the z-axis is called the xz plane; the plane determined by the y-axis and the z-axis is called the yz plane. We refer to these three planes as the *coordinate planes.* The equation of a coordinate plane is very simple. A point with coordinates $(a, b, 0)$ lies in the xy plane; the set of all such points can be simply described by the equation $z = 0$. In other words the set of all points (x, y, z) satisfying the equation $z = 0$ is precisely the xy plane. Similarly a point with coordinates $(0, u, v)$ lies in the yz plane; the equation of the yz plane is $x = 0$. A point with coordinates $(p, 0, q)$ lies in the xz plane; the equation of the xz plane is $y = 0$.

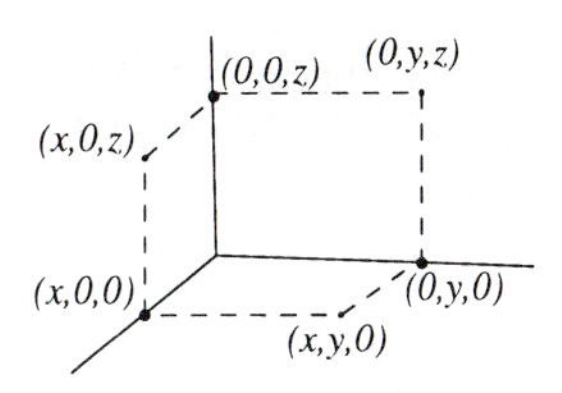

FIGURE 12.6
Coordinate axes and coordinate planes

A point on the x-axis has coordinates $(x, 0, 0)$; a point on the y-axis has coordinates $(0, y, 0)$; a point on the z-axis has coordinates $(0, 0, z)$.

We can describe slightly more general planes that are parallel to the coordinate planes with equations of the form $x = a$, $y = b$, or $z = c$. For example the set of all points (x, y, z) satisfying the equation $y = 2$ is a plane parallel to the xz plane and passing through the point $(0, 2, 0)$ on the y-axis.

The equation $z = 4$ describes all points in the plane parallel to the xy plane passing through $(0, 0, 4)$. Similarly the equation $y = -2$ describes all points in the plane parallel to the xz plane passing through $(0, -2, 0)$.

The equation of a general plane is developed after we have studied some facts about vectors.

The Distance Formula

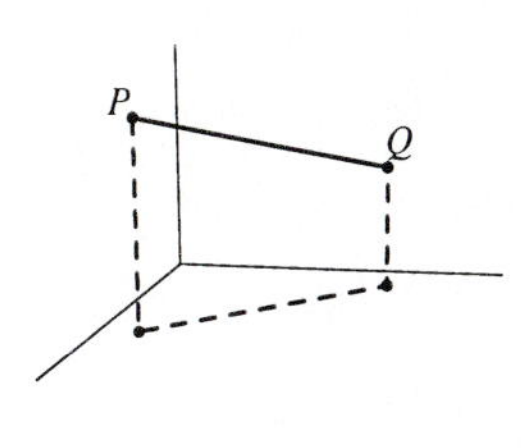

FIGURE 12.7

The formula for the distance between two points in the xy plane has a natural extension to give the distance between $P(x_1, y_1, z_1)$ and $Q(x_2, y_2, z_2)$ as

$$d(P, Q) = \sqrt{(x_1 - x_2)^2 + (y_1 - y_2)^2 + (z_1 - z_2)^2}.$$

Of course if the two points P and Q happen to lie in the xy plane, then $d(P, Q)$ reduces to the usual distance formula in two dimensions.

The distance of $P(x_1, y_1, z_1)$ from the origin $O(0, 0, 0)$ is $\sqrt{x_1^2 + y_1^2 + z_1^2}$. If R is some nonnegative number, the set of all points $P(x, y, z)$ at a distance R from the origin lie on the sphere of radius R and the coordinates satisfy the equation

$$R^2 = x^2 + y^2 + z^2.$$

More generally the sphere with center $C(a, b, c)$ and radius R is the set of all points $P(x, y, z)$ whose distance from C equals R. The equation $d(P, C) = R$ is equivalent to $d(P, C)^2 = R^2$ because the quantities are all nonnegative. Hence the equation of the sphere with center $C(a, b, c)$ and radius R is

Equation of the Sphere

$$(x - a)^2 + (y - b)^2 + (z - c)^2 = R^2.$$

By completing the square one finds that the set of points whose coordinates satisfy an equation of the form

$$x^2 + y^2 + z^2 + px + qy + rz + s = 0$$

is either a sphere, a point or the empty set depending on whether the "radius" is positive, zero or negative.

Exercises 12.1

Sketch the location the points P and Q and determine the distance $d(P, Q)$.

1. $P(1, 1, 1)$, $Q(2, 0, 1)$

2. $P(1, -1, 1)$, $Q(2, 0, -1)$

3. $P(1, -1, -1)$, $Q(-2, 0, 1)$

4. $P(-1, -1, -1)$, $Q(2, 0, 1)$

Give the equation of the plane satisfying the given conditions

5. The plane that passes through $(1, 2, 3)$ and is parallel to xy plane

6. The plane that passes through $(1, 2, 3)$ and is parallel to yz plane

7. The plane that passes through $(2, -2, 4)$ and is parallel to xz plane

8. The plane that passes through $(2, -2, 4)$ and is parallel to yz plane

Decide if the given point P is inside, on or outside the sphere with the given equation.

9. $x^2 + y^2 + z^2 = 9$, $P(-1, -2, 1)$

10. $x^2 + y^2 + z^2 = 4$, $P(1, -1, 1.5)$

11. $(x - 1)^2 + (y + 2)^2 + z^2 = 9$, $P(-1, -2, 2)$

12. $(x - 1)^2 + (y + 2)^2 + z^2 = 16$, $P(-1, -2, 3)$

Find the equation of a sphere satisfying the given conditions. Decide if there is one or more than one correct solution.

13. Center at $(1, 2, 3)$ and radius 3

14. Center at $(-1, 2, -3)$ and radius 9

15. Center on the plane $y = 4$ and passing through the point $(0, 0, 0)$

16. Center on the plane $x = 1$ and on the plane $z = -2$ and passing through the point $(4, 0, -1)$

17. Center on the plane $x = 2$, tangent to the plane $z = 0$

18. Center on the plane $z = 0$, radius $R = 1$ and passing through $(0, 0, 0)$

Without solving simultaneous equations determine if the two spheres with the given equations have any common points by considering the distance between centers and the radius of each sphere.

19. $x^2 + y^2 + z^2 = 4$, $(x - 1)^2 + (y + 2)^2 + (z - 2)^2 = 4$

20. $x^2 + y^2 + z^2 = 4$, $(x + 1)^2 + (y - 1)^2 + (z - 3)^2 = 2$

21. $x^2 + y^2 + z^2 + 2x - 4y = 9$, $(x + 1)^2 + (y + 1)^2 + (z - 1)^2 = 10$

22. $x^2 + y^2 + z^2 + 4x - 2y + 6z = 18$, $(x + 2)^2 + (y + 3)^2 + (z - 1)^2 = 25$

23. $x^2 - 4x + y^2 + 6y + (z + c)^2 = 3 - c^2$, $x^2 + 2x + y^2 - 4y + (z - c)^2 = -1$

12.2 Vectors

Vectors are used as an aid to intuition and as a convenience to simplify notation. Vectors are used in many circumstances: to describe a force acting in a certain direction; to describe the position or velocity of a moving object; to describe geometric properties. We begin with a coordinate-free definition and some elementary properties.

A vector is a line segment in the xy plane or in xyz space with a specified direction. A vector can be defined by specifying two points P and Q, one called the *initial point*, the other called the *terminal point.* A vector with initial point P and terminal point Q is represented by an arrow from P to Q. The symbol $\overrightarrow{PQ}$ is used to denote this vector. The direction of the vector represented this way is the line through P and Q oriented from P moving toward Q. The magnitude of $\overrightarrow{PQ}$ is the length of the line segment PQ. Two vectors are equal if the have the same direction and magnitude. For example, the vector with initial and terminal points P,Q equals the vector having initial and terminal points P',Q' provided the line through PQ is parallel to the line through $P'Q'$, the magnitudes $|PQ|$ and $|P'Q'|$ are equal, and the directions along the two lines are the same. Given any vector $\overrightarrow{PQ}$ and any point U, there is a vector equal to $\overrightarrow{PQ}$ having U as initial point. On the line through U parallel to PQ locate the point V so that the distance from U to V equals the distance from P to Q. Of course there are two possible choices for such a V; select the one so that the direction from U to V is the same as the direction from P to Q. Then $\overrightarrow{PQ} = \overrightarrow{UV}$.

In this text we use boldface letters to denote vectors so, for example, we write $\mathbf{v} = \overrightarrow{PQ}$.

The zero vector $\mathbf{0}$ is exceptional. It has magnitude 0 and no specific direction. It is equally correct to say that the zero vector points in *every* direction.

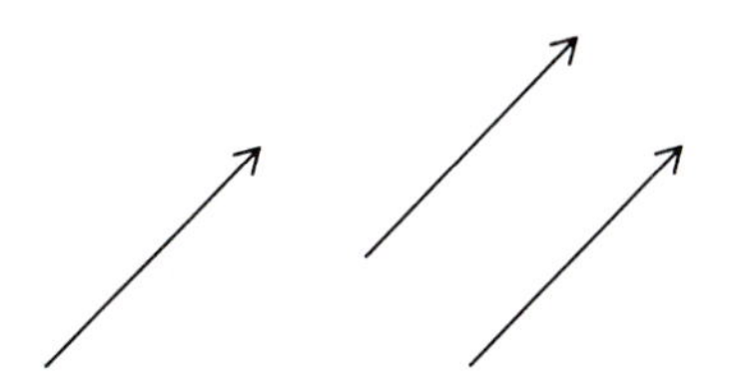

FIGURE 12.8
Equal vectors

Addition of Vectors

Addition of vectors is given by the *parallelogram law.* If we are given two vectors $\overrightarrow{PQ}$ and $\overrightarrow{QR}$ with the terminal point of the first vector equal to the initial point of the second vector, then the sum is

$$\overrightarrow{PQ} + \overrightarrow{QR} = \overrightarrow{PR}.$$

This is called the parallelogram law because the vectors $\overrightarrow{PQ}$ and $\overrightarrow{QR}$ form two adjacent sides of a parallelogram and their sum is the diagonal of the parallelogram. If we want to add two vectors $\overrightarrow{PQ}$ and $\overrightarrow{AB}$ where the terminal point of one is not the initial point of the other, it is necessary to replace one of the vectors by an equal vector to make the terminal and initial points coincide. For example, we would replace the vector $\overrightarrow{AB}$ by an equal vector having initial point Q and then apply the parallelogram law.

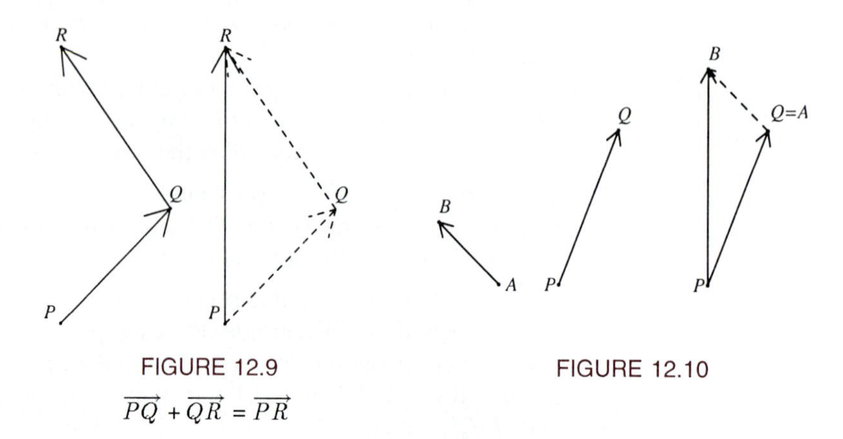

FIGURE 12.9
$\overrightarrow{PQ} + \overrightarrow{QR} = \overrightarrow{PR}$

FIGURE 12.10

Magnitude

The magnitude of a vector $\mathbf{v} = \overrightarrow{PQ}$ is the distance $d(P, Q)$ from its initial point P to its terminal point Q and is denoted by $|\mathbf{v}|$. The absolute value symbol is used for vectors just as for real numbers. The magnitude of a vector is never negative so we have $|\mathbf{v}| \geq 0$ for any vector $\mathbf{v}$. If $\mathbf{v} \neq \mathbf{0}$ then $|\mathbf{v}| > 0$. We make use of the geometric property of triangles that says that the side of any triangle has length no greater than the sum of the lengths of the other two sides. This, along with the parallelogram law, gives the rule

$$|\mathbf{v} + \mathbf{u}| \leq |\mathbf{v}| + |\mathbf{u}|, \tag{12.1}$$

which is called the Triangle Inequality. It is identical in form to the Triangle Inequality for the sum of two real numbers.

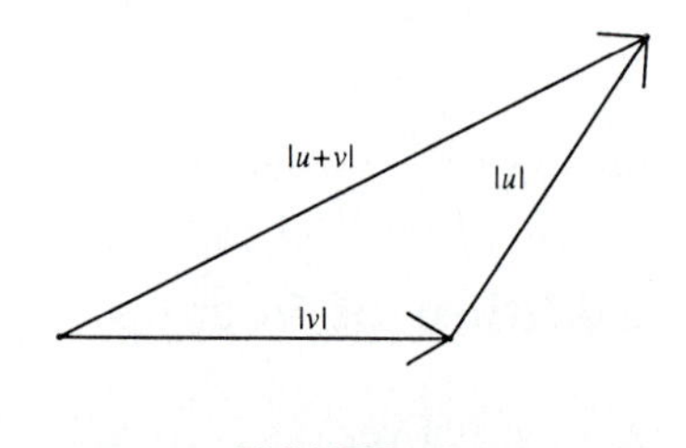

FIGURE 12.11
Triangle inequality

Scalar Multiples

Real numbers are called *scalars*; multiplication of a scalar times a vector is defined as follows: If t is a positive real number and $\mathbf{v}$ is a vector then $t\mathbf{v}$ is the vector having magnitude t times the magnitude of $\mathbf{v}$ and having the same direction as $\mathbf{v}$. If t is a negative real number then $t\mathbf{v}$ has the magnitude $|t|$ times the magnitude of $\mathbf{v}$; the direction of $t\mathbf{v}$ is parallel to $\mathbf{v}$ but in the opposite direction. If $t = 0$ then $t\mathbf{v}$ is the zero vector having magnitude 0 and no specific direction.

Thus for any scalar t we have

$$|t\mathbf{v}| = |t||\mathbf{v}|.$$

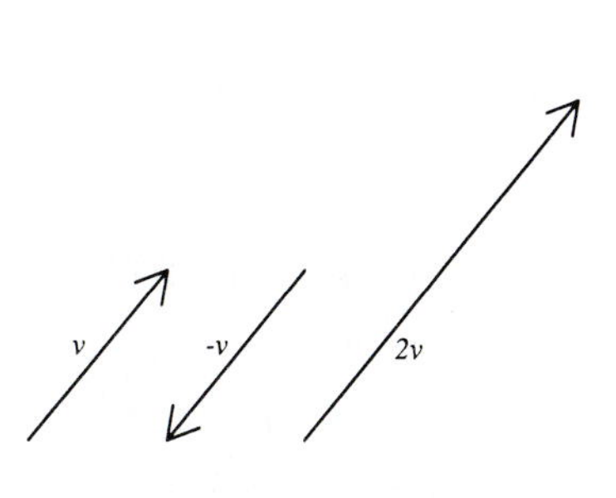

FIGURE 12.12
Scalar multiples

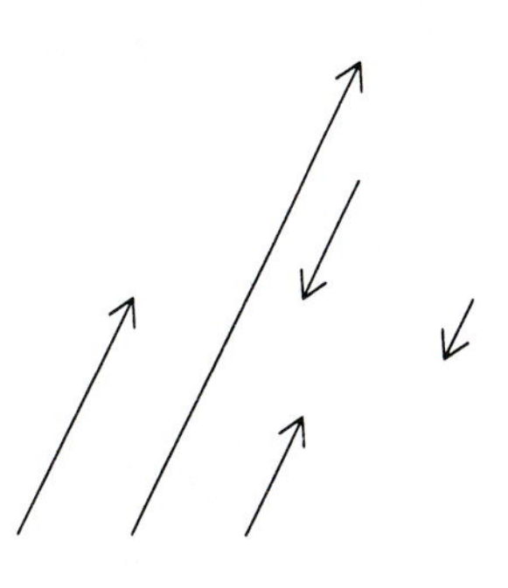

FIGURE 12.13
Parallel vectors

The vector **v** is parallel to $t\mathbf{v}$. If the two vectors have the same initial point, then **v** and $t\mathbf{v}$ lie on the same line and point in the same direction when $t > 0$ and point in opposite directions when $t < 0$. When we say two vectors are *parallel* we mean one is a scalar multiple of the other.

Coordinates for Vectors

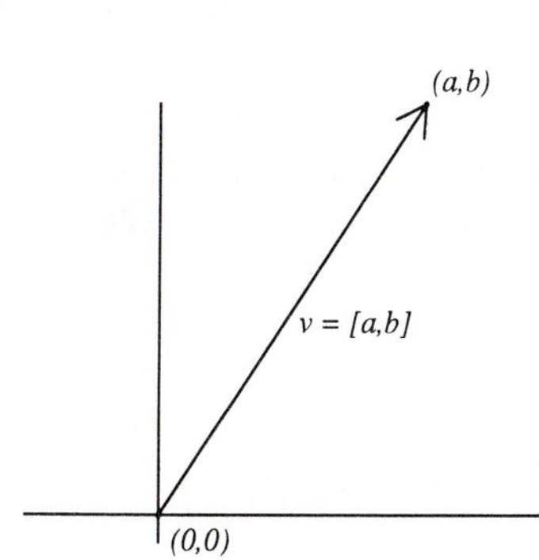

FIGURE 12.14
Coordinates of **v**

Up to this point the information we have given about vectors is equally valid if the vectors are considered as two-dimensional objects in the xy plane or as three-dimensional objects in the xyz space. Now we want to introduce coordinates for vectors. It is necessary to decide where our vectors lie. We begin with vectors in the xy plane and then give a parallel treatment for vectors in xyz space in the next section.

Let **v** be a vector in the xy plane. The coordinates of **v** are defined by finding a point $P = (a, b)$ so that $\mathbf{v} = \overrightarrow{OP}$ where $O = (0, 0)$. Note that there is only one such point P for the given vector. Then we set

$$\mathbf{v} = [a, b]$$

and call $[a, b]$ the *coordinate form* of **v**. Thus $\mathbf{v} = [a, b]$ if $(0, 0)$ is the initial point of **v** and (a, b) is the terminal point of **v**.

Suppose $\mathbf{v} = [a, b]$ and $\mathbf{u} = [c, d]$. How do we find the coordinate form of $\mathbf{v} + \mathbf{u}$? Some elementary analytic geometry can be applied to show that $(0, 0)$, (a, b), (c, d), and $(a + c, b + d)$ are the four vertices of a parallelogram. Two of the sides of the parallelogram are **v** and **u** and so the diagonal must be their sum. The diagonal has initial point $(0, 0)$ and terminal point $(a + c, b + d)$ so we have the addition rule

$$[a, b] + [c, d] = [a + c, b + d].$$

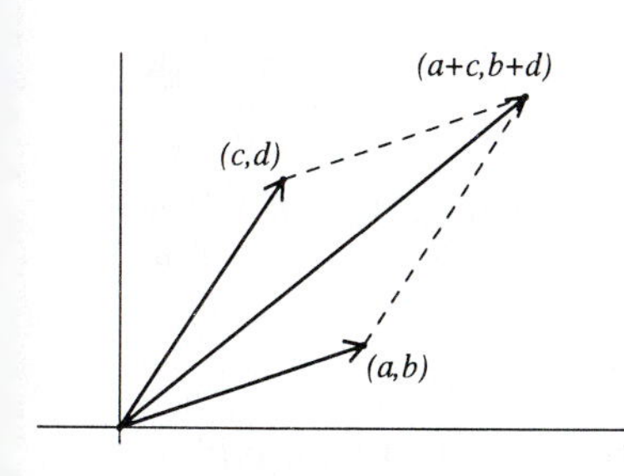

FIGURE 12.15
$[a, b] + [c, d] = [a + c, b + d]$

The magnitude of the vector $\mathbf{v} = [a, b]$ is

$$|\mathbf{v}| = |[a, b]| = \sqrt{a^2 + b^2}.$$

It is easily checked that for any real number t, the magnitude of $[ta, tb]$ is $|t|\sqrt{a^2 + b^2}$. The following equation gives the rule for multiplication by scalars:

$$t[a, b] = [ta, tb].$$

We use the subtraction symbol $-\mathbf{v}$ in place of the slightly more complicated notation $(-1)\mathbf{v}$. Thus if $\mathbf{v} = [a, b]$ then $-\mathbf{v} = [-a, -b]$.

EXAMPLE 1 Let $\mathbf{v} = [2, -3]$ and $\mathbf{w} = [-1, 1]$. Compute the magnitudes of (i)

$|\mathbf{v}+\mathbf{w}|$, (ii) $|\mathbf{v}-\mathbf{w}|$ and (iii) $|2\mathbf{v}-3\mathbf{w}|$.

Solution: Magnitudes are computed by first finding the coordinate form of each vector.

$$\begin{aligned}
\text{i.} \quad \mathbf{v}+\mathbf{w} &= [2,-3]+[-1,1]=[1,-2] \\
|\mathbf{v}+\mathbf{w}| &= \sqrt{1^2+(-2)^2}=\sqrt{5}. \\
\text{ii.} \quad \mathbf{v}-\mathbf{w} &= [2,-3]-[-1,1]=[3,-4] \\
|\mathbf{v}-\mathbf{w}| &= \sqrt{3^2+(-4)^2}=\sqrt{25}=5. \\
\text{iii.} \quad 2\mathbf{v}-3\mathbf{w} &= 2[2,-3]-3[-1,1]=[7,-9] \\
|\mathbf{v}-\mathbf{w}| &= \sqrt{7^2+(-9)^2}=\sqrt{130}.
\end{aligned}$$

■

EXAMPLE 2 Let $\mathbf{v} = [3, 1]$ and $\mathbf{u} = [-1, 5]$. Find the center of the parallelogram having $\mathbf{v}$ and $\mathbf{u}$ as two sides.

Solution: The center of a parallelogram is the midpoint of the diagonal. The diagonal of the parallelogram is represented by the vector

$$\mathbf{v}+\mathbf{u} = [3,1]+[-1,5] = [2,6].$$

The midpoint of the diagonal is the terminal point of a vector half as long as the diagonal vector, namely $[2/2, 6/2] = [1, 3]$. So the center is $(1, 3)$. ■

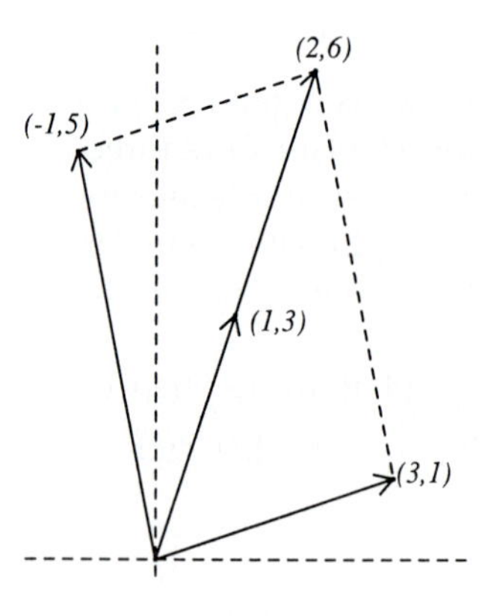

FIGURE 12.16 Center of the parallelogram

EXAMPLE 3 Find the coordinate form of the vector with initial point $P(2, 5)$ and terminal point $Q(4, -3)$.

Solution: Let $\overrightarrow{PQ} = [a, b]$ so that we must find a and b. With O equal to the origin we have

$$\overrightarrow{OP} + \overrightarrow{PQ} = \overrightarrow{OQ}.$$

Write each vector in coordinate form to get

$$[2,5]+[a,b] = [4,-3].$$

We now solve for $[a, b]$ and find

$$[a,b] = [4,-3]-[2,5] = [2,-8].$$

■

The idea used in example 3 can be applied to two general points. If $\mathbf{v}$ has initial point $P(a_1, a_2)$ and terminal point $Q(b_1, b_2)$ then the coordinate form of $\mathbf{v}$ is

$$\mathbf{v} = \overrightarrow{OQ} - \overrightarrow{OP} = [b_1, b_2] - [a_1, a_2] = [b_1 - a_1, b_2 - a_2].$$

One way to help remember this equation is that $\mathbf{v}$ is the vector you must add to $\overrightarrow{OP}$ to reach $\overrightarrow{OQ}$.

Unit Vectors

A vector of magnitude one is called a *unit vector*. Every vector is a scalar multiple of a unit vector as we now show. Let $\mathbf{v} = [a, b]$ be any nonzero vector and let $t = 1/\sqrt{a^2 + b^2}$. Then $t = 1/|\mathbf{v}|$ and $t|\mathbf{v}| = |t\mathbf{v}| = 1$. Thus $t\mathbf{v}$ is a unit vector explicitly given as

$$\frac{1}{|\mathbf{v}|}\mathbf{v} = \left[\frac{a}{\sqrt{a^2 + b^2}}, \frac{b}{\sqrt{a^2 + b^2}}\right],$$

and, of course, having the same direction as $\mathbf{v}$. Clearly then

$$\mathbf{v} = |\mathbf{v}|\mathbf{u}, \qquad \text{where} \qquad \mathbf{u} = \frac{1}{|\mathbf{v}|}\mathbf{v}$$

expresses $\mathbf{v}$ as a scalar multiple of the unit vector $\mathbf{u}$.

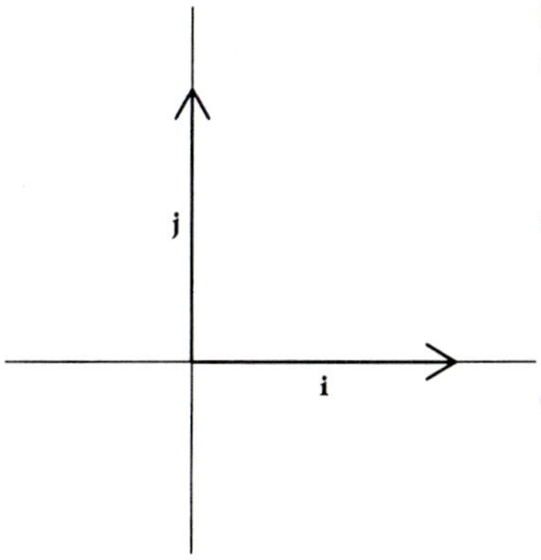

We single out two special unit vectors that are used repeatedly. Let

$$\mathbf{i} = [1, 0] \qquad \text{and} \qquad \mathbf{j} = [0, 1].$$

FIGURE 12.17

Then $\mathbf{i}$ is a unit vector along the positive x-axis and $\mathbf{j}$ is a unit vector along the positive y-axis.

A sum of the form $a\,\mathbf{i} + b\,\mathbf{j}$ is called a *linear combination* of $\mathbf{i}$ and $\mathbf{j}$. Every vector in the xy plane can be expressed as a linear combination of $\mathbf{i}$ and $\mathbf{j}$. To see this let $\mathbf{v} = [a, b]$ be any vector. Then

$$\mathbf{v} = [a, b] = [a, 0] + [0, b] = a[1, 0] + b[0, 1] = a\,\mathbf{i} + b\,\mathbf{j}.$$

Because unit vectors are useful for many computations, we describe all unit vectors in the xy plane. If $\mathbf{u} = [a, b]$ is a unit vector, then the point (a, b) lies on the circle having equation $x^2 + y^2 = 1$. Let θ be the radian measure of the arc from $(1, 0)$ to (a, b) measured in the positive direction, so then $(a, b) = (\cos\theta, \sin\theta)$. It follows that

$$\mathbf{u} = [\cos\theta, \sin\theta] = \cos\theta\;\mathbf{i} + \sin\theta\;\mathbf{j}.$$

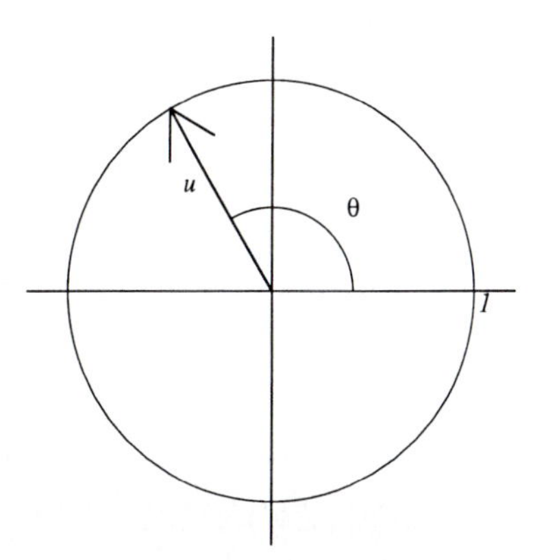

FIGURE 12.18
Unit vector

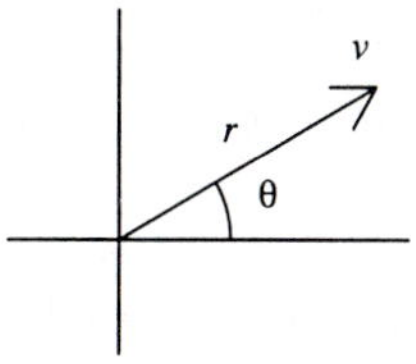

By combining this form of the unit vectors with the fact that every vector is a scalar multiple of a unit vector we obtain the *polar form* of a general vector. Any vector $\mathbf{v}$ can be written in the form

$$\mathbf{v} = r\cos\theta\;\mathbf{i} + r\sin\theta\;\mathbf{j}$$

FIGURE 12.19

where $r = |\mathbf{v}|$ and $0 \le \theta < 2\pi$.

If (x, y) is any point then the vector $[x, y]$ can be written in polar form as $[x, y] = [r\cos\theta, r\sin\theta]$. The numbers (r, θ) are called the *polar coordinates* of (x, y). The equations relating the four quantities are

$$\begin{cases} x = r\cos\theta \\ y = r\sin\theta \end{cases} \qquad \begin{cases} r = \sqrt{x^2 + y^2} \\ \theta = \arctan(y/x) \end{cases}.$$

An exception in the form of θ must be made if $x = 0$. In that case $\theta = \pi/2$ when $y > 0$ and $\theta = 3\pi/2$ when $y < 0$. If $x = y = 0$ then the point (x, y) is the origin and θ is undefined.

Exercises 12.2

Let $\mathbf{v} = 2\mathbf{i} - 3\mathbf{j}$, $\mathbf{w} = -\mathbf{i} + \mathbf{j}$ and $\mathbf{u} = \mathbf{i} + 3\mathbf{j}$. Find the coordinate form and the magnitude of each vector.

1. $\mathbf{v} + \mathbf{w} + \mathbf{u}$

2. $2\mathbf{v} - \mathbf{w} - 2\mathbf{u}$

3. $(-\mathbf{v} + 3\mathbf{w}) + (3\mathbf{v} - \mathbf{w} + \mathbf{u})$

4. $\left(1/|\mathbf{v}|\right)\mathbf{v} + \left(1/|\mathbf{w}|\right)\mathbf{w}$

Let P, Q, R, O be points in the xy plane. Sketch the vectors to illustrate the following equations.

5. $(\overrightarrow{OP} + \overrightarrow{PR}) + \overrightarrow{RQ} = \overrightarrow{OP} + (\overrightarrow{PR} + \overrightarrow{RQ})$

6. $\overrightarrow{OP} + \overrightarrow{PR} + \overrightarrow{RQ} = \overrightarrow{OQ}$

7. $2\overrightarrow{OP} + \overrightarrow{PR} + \overrightarrow{PQ} = \overrightarrow{OR} + \overrightarrow{OQ}$

8. $2\overrightarrow{OP} + 3\overrightarrow{PR} + \overrightarrow{RQ} = 2\overrightarrow{OR} + \overrightarrow{PQ}$

Let $\mathbf{v} = [-3, 2]$, $\mathbf{w} = [-2, -1]$ and $\mathbf{u} = [1, 1]$ be the coordinate form of three vectors. Write each of the following vectors as a linear combination of $\mathbf{i}$, $\mathbf{j}$.

9. $\mathbf{v} - \mathbf{w} - 2\mathbf{u}$

10. $2\mathbf{v} + \mathbf{w} + \mathbf{u}$

11. $3(\mathbf{v} - 4\mathbf{w}) - 2(2\mathbf{v} - 2\mathbf{w} + 3\mathbf{u})$

12. $\left(1/|\mathbf{v}|\right)\mathbf{v} + \left(1/|\mathbf{w}|\right)\mathbf{w}$

Express each of the following vectors in polar form $r(\cos\theta\,\mathbf{i} + \sin\theta\,\mathbf{j})$. Find the exact value of θ for the vectors given in coordinate form and approximate values of θ in the other cases.

13. $\mathbf{v} = [2, 2]$

14. $\mathbf{u} = [1, \sqrt{3}]$

15. $\mathbf{w} = [-\sqrt{3}, 1]$

16. $\mathbf{v} = [-32, 32]$

17. $\mathbf{u} = 4\mathbf{i} + 3\mathbf{j}$

18. $\mathbf{u} = 4\mathbf{i} - 3\mathbf{j}$

19. $\mathbf{u} = 8\mathbf{i} + 6\mathbf{j}$

20. $\mathbf{u} = -16\mathbf{i} + 12\mathbf{j}$

21. Using the computations in the previous exercises, find the polar coordinates of the points $(2, 2)$, $(1, \sqrt{3})$, $(4, 3)$ and $(4, -3)$.

22. As $\mathbf{u}$ ranges over all unit vectors in the xy plane, what is the smallest value of $|\mathbf{u} + \mathbf{i}|$? What is the largest value of $|\mathbf{u} + \mathbf{i}|$? Do you think that every value between the smallest and the largest occurs as a length of the form $|\mathbf{u} + \mathbf{i}|$?

23. If $\mathbf{u}$ and $\mathbf{v}$ range over all unit vectors, what are the possible values of $|\mathbf{u} + \mathbf{v}|$?

24. Let $\mathbf{v}$ be any vector. Show that $\mathbf{v}$ can be expressed as a sum of unit vectors.

12.3 Coordinates for vectors in three dimensions

Next we discuss the three-dimensional version of the ideas developed in the previous section for vectors in the xy plane. We begin with coordinates for vectors. Let $O(0, 0, 0)$ denote the origin of xyz space and let $\mathbf{v}$ be a vector. Let $P(a, b, c)$ be a point such that the initial point of $\mathbf{v}$ is the origin and $P(a, b, c)$ is the terminal point; that is

$$\mathbf{v} = \overrightarrow{OP}.$$

Then we set

$$\mathbf{v} = [a, b, c]$$

and call this the *coordinate form* of $\mathbf{v}$.

The magnitude of $\mathbf{v}$ is

$$|\mathbf{v}| = \sqrt{a^2 + b^2 + c^2}.$$

For any real number t the vector $t\mathbf{v}$ has the coordinate form

$$t\mathbf{v} = [ta, tb, tc].$$

If $\mathbf{v} = [a, b, c]$ and $\mathbf{w} = [p, q, r]$ then the coordinate form of their sum is

$$\mathbf{v} + \mathbf{w} = [a, b, c] + [p, q, r] = [a + p, b + q, c + r].$$

This is called the *parallelogram law* just as in the two-dimensional case. The four points $(0, 0, 0)$, (a, b, c), (p, q, r), and $(a + p, b + q, c + r)$ are the vertices of a parallelogram although the geometry is somewhat more complicated in this three-dimensional case than it was in the two-dimensional case.

EXAMPLE 1 Let $\mathbf{v} = [2, -1, 3]$ and $\mathbf{w} = [-1, 0, 2]$. Compute the magnitude of $\mathbf{v} + 2\mathbf{w}$ and $\mathbf{v} - 2\mathbf{w}$.

Solution: First find the coordinate form for each sum and then compute the length. For the first vector, we have

$$\mathbf{v} + 2\mathbf{w} = [2, -1, 3] + 2[-1, 0, 2] = [0, -1, 7],$$
$$|\mathbf{v} + 2\mathbf{w}| = \sqrt{(-1)^2 + 7^2} = \sqrt{50}.$$

For the difference of the vectors we have

$$\mathbf{v} - 2\mathbf{w} = [2, -1, 3] - 2[-1, 0, 2] = [4, -1, -1],$$
$$|\mathbf{v} - 2\mathbf{w}| = \sqrt{4^2 + (-1)^2 + (-1)^2} = \sqrt{18}.$$

■

Unit Vectors

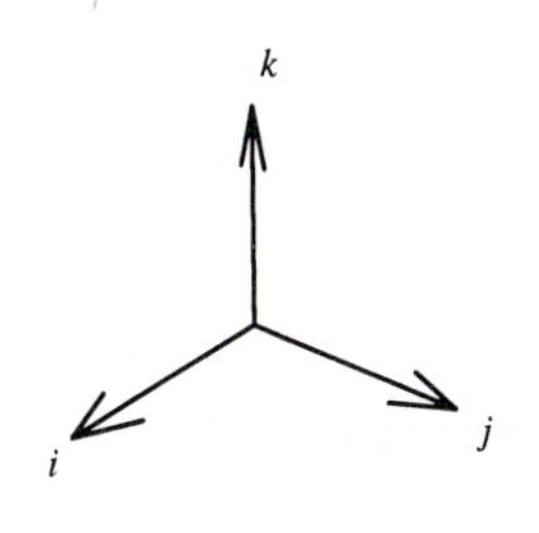

FIGURE 12.20

For any nonzero vector $\mathbf{v}$, the vector

$$\mathbf{u} = \frac{\mathbf{v}}{|\mathbf{v}|}$$

is a unit vector and $\mathbf{v} = |\mathbf{v}|\mathbf{u}$ gives the original vector as a scalar multiple of a unit vector. We single out three particular unit vectors for special use. The unit vectors pointing along the positive coordinate axes are

$$\mathbf{i} = [1, 0, 0], \qquad \mathbf{j} = [0, 1, 0], \qquad \mathbf{k} = [0, 0, 1].$$

Every vector is a linear combination of $\mathbf{i}$, $\mathbf{j}$, $\mathbf{k}$; for an arbitrary $\mathbf{w} = [a, b, c]$ we have, just as in the two-dimensional case,

$$[a, b, c] = a[1, 0, 0] + b[0, 1, 0] + c[0, 0, 1] = a\,\mathbf{i} + b\,\mathbf{j} + c\,\mathbf{k}.$$

The numbers a, b, c are called the components of $\mathbf{w}$. From now on we usually use the $\mathbf{i}$, $\mathbf{j}$, $\mathbf{k}$ and components to describe vectors.

EXAMPLE 2 Let $\mathbf{v} = 2\mathbf{i} + 3\mathbf{j} - 4\mathbf{k}$ and $\mathbf{w} = -\mathbf{i} + 3\mathbf{k}$. Express the vector $\mathbf{u} = -2\mathbf{v} + 3\mathbf{w}$ as a linear combination of $\mathbf{i}$, $\mathbf{j}$, $\mathbf{k}$. Find the magnitude of $\mathbf{u}$ and a unit vector that has the same direction as $\mathbf{u}$.

Solution: First find the sum by the addition rules to be

$$\mathbf{u} = -2\mathbf{v} + 3\mathbf{w} = -2(2\mathbf{i} + 3\mathbf{j} - 4\mathbf{k}) + 3(-\mathbf{i} + 3\mathbf{k})$$
$$= (-4 - 3)\mathbf{i} + (-6 + 0)\mathbf{j} + (8 + 9)\mathbf{k} = -7\mathbf{i} - 6\mathbf{j} + 17\mathbf{k}.$$

The length of **u** is given by

$$|\mathbf{u}| = \sqrt{(-7)^2 + (-6)^2 + 17^2} = \sqrt{374}.$$

The unit vector with the same direction as **u** is

$$\frac{\mathbf{u}}{|\mathbf{u}|} = -\frac{7}{\sqrt{374}}\,\mathbf{i} - \frac{6}{\sqrt{374}}\,\mathbf{j} + \frac{17}{\sqrt{374}}\,\mathbf{k}.$$

■

The unit vectors in three dimensions are a bit more complicated to describe than the simple polar form that was used in the two-dimensional case. It can be shown that every unit vector has the form

$$\mathbf{u} = \sin\phi\cos\theta\ \mathbf{i} + \sin\phi\sin\theta\ \mathbf{j} + \cos\phi\,\mathbf{k}$$

for some θ and ϕ satisfying $0 \le \theta < 2\pi$, $0 \le \phi \le \pi$. It follows that every vector $\mathbf{v} = xbi + y\,\mathbf{j} + z\,\mathbf{k}$ can be expressed in the form

$$\mathbf{v} = r(\sin\phi\cos\theta\ \mathbf{i} + \sin\phi\sin\theta\ \mathbf{j} + \cos\phi\,\mathbf{k})$$

where $r = |\mathbf{v}|$ and θ, ϕ are restricted to the intervals indicated just above. This expression of a general vector is sometimes called the *spherical form* of **v**. We can write

$$\begin{cases} x = r\sin\phi\cos\theta \\ y = r\sin\phi\sin\theta \\ z = r\cos\phi \end{cases}$$

so then (r, θ, ϕ) are called the *spherical coordinates* of the point (x, y, z).

Some illustrations are left for the exercises that follow.

Exercises 12.3

Let $\mathbf{v} = 2\mathbf{i} - 3\mathbf{j} + \mathbf{k}$, $\mathbf{w} = -\mathbf{i} + \mathbf{j} + \mathbf{k}$ and $\mathbf{u} = \mathbf{i} + 3\mathbf{j} - 4\mathbf{k}$. Find the coordinate form and the magnitude of each vector.

1. $\mathbf{v} + \mathbf{w} + \mathbf{u}$

2. $2\mathbf{v} - \mathbf{w} - 2\mathbf{u}$

3. $(-\mathbf{v} + 3\mathbf{w}) + (3\mathbf{v} - \mathbf{w} + \mathbf{u})$

4. $\left(1/|\mathbf{v}|\right)\mathbf{v} + \left(1/|\mathbf{w}|\right)\mathbf{w}$

Let $\mathbf{v} = [-3, 2, 0]$, $\mathbf{w} = [0, -2, -1]$ and $\mathbf{u} = [1, 1, 1]$ be the coordinate form of the vectors. Write each of the following vectors as a linear combination of **i**, **j** and **k**.

5. $\mathbf{v} - \mathbf{w} - 2\mathbf{u}$

6. $2\mathbf{v} + \mathbf{w} + \mathbf{u}$

7. $(\mathbf{v} - 4\mathbf{w}) + (2\mathbf{v} - 2\mathbf{w} + 3\mathbf{u})$

8. $\left(1/|\mathbf{v}|\right)\mathbf{v} + \left(1/|\mathbf{w}|\right)\mathbf{w}$

9. Find a scalar a such that $\mathbf{i} + 2\mathbf{j} - 2\mathbf{k}$ and $-2\mathbf{i} + a\mathbf{j} + 4\mathbf{k}$ are parallel.

10. Show that there is no scalar b such that $\mathbf{i} + 2\mathbf{j} - 2\mathbf{k}$ and $-2\mathbf{i} + b\mathbf{j} - 4\mathbf{k}$ are parallel.

11. Find the coordinate form of the vector with initial point $P(1, 1, 4)$ and terminal point $Q(-1, 1, 3)$. Repeat with $P(0, -2, 3)$ and $Q(4, 0, -2)$.

12. Express as a linear combination of **i**, **j** and **k** the vector having initial point $P(3, -2, 1)$ and terminal point $Q(-1, 2, 3)$. Repeat with $P(1, 2, -2)$ and $Q(4, 2, -3)$.

13. Find the scalar t so that $\mathbf{v} = t(\mathbf{i} - 2\mathbf{j} + 3\mathbf{k})$ is a unit vector.

14. Prove there is no scalar t such that $\mathbf{v} = 2\mathbf{i} - 2t\,\mathbf{j} + 3t\,\mathbf{k}$ is a unit vector.

15. Find all scalars t such that $\mathbf{v} = (1/2)(\mathbf{i} - 2t\,\mathbf{j} + 3t\,\mathbf{k})$ is a unit vector.

16. Show that every unit vector in three dimensions has the form

$$\mathbf{u} = \sin\phi\cos\theta\ \mathbf{i} + \sin\phi\sin\theta\ \mathbf{j} + \cos\phi\,\mathbf{k}$$

for some numbers ϕ, θ satisfying $0 \le \theta < 2\pi$, $0 \le \phi \le \pi$. [Hint: Write $\mathbf{u} = a\,\mathbf{i} + b\,\mathbf{j} + c\,\mathbf{k}$. Show $a\,\mathbf{i} + b\,\mathbf{j} = r(\cos\theta\ \mathbf{i} + \sin\theta\ \mathbf{j})$ for some r with $0 \le r \le 1$ and $r^2 + c^2 = 1$. Then show there is a ϕ with $r = \sin\phi$ and $c = \cos\phi$.]

Find the spherical coordinates (r, θ, ϕ) of the following points given in the usual rectangular coordinate system.

17. $(x, y, z) = (3, 0, 3)$

18. $(x, y, z) = (4, 0, 0)$

19. $(x, y, z) = (1, 1, 0)$

20. $(x, y, z) = (4, 4, -4)$

21. $(x, y, z) = (2, 2, 2)$

22. $(x, y, z) = (1, 1, \sqrt{6})$

23. Describe in words the set of points whose spherical

coordinates have the form

a. $(r, \theta, \phi) = (2, \theta, 0)$,
b. $(r, \theta, \phi) = (1, \theta, \phi)$,
c. $(r, \theta, \phi) = (r, 0, \phi)$.

[The notation here means that the unrestricted variables are allowed to range over all possible values subject to the usual restrictions $r \geq 0$, $0 \leq \theta < 2\pi$ and $0 \leq \phi \leq \pi$.]

12.4 The dot product

We define the dot product of two vectors in this section. The definition is given in terms of the coordinates of the vector and so the definitions are slightly different for the two- and three-dimensional cases. The dot product can be interpreted geometrically as a function of lengths and the angle between the two vectors. The interpretation is the same for two and three dimensions.

Let $\mathbf{u} = [a_1, a_2]$ and $\mathbf{v} = [b_1, b_2]$ be vectors in the xy plane. The *dot product* of $\mathbf{u}$ and $\mathbf{v}$ is

$$\mathbf{u} \cdot \mathbf{v} = a_1 b_1 + a_2 b_2.$$

For two vectors in three dimensions, $\mathbf{u} = [a_1, a_2, a_3]$ and $\mathbf{v} = [b_1, b_2, b_3]$, the dot product is

$$\mathbf{u} \cdot \mathbf{v} = a_1 b_1 + a_2 b_2 + a_3 b_3.$$

Notice that the dot product of two vectors is a scalar, not a vector. Here are a few properties of this product. In most cases, we give the proof only in the three-dimensional case. The proof in the two-dimensional case can be obtained by setting the third coordinates equal to 0. Let us rewrite the definition of the dot product using the $\mathbf{i}$, $\mathbf{j}$, $\mathbf{k}$ notation:

Dot Product

$$(a_1\,\mathbf{i} + a_2\,\mathbf{j} + a_3\,\mathbf{k}) \cdot (b_1\,\mathbf{i} + b_2\,\mathbf{j} + b_3\,\mathbf{k}) = a_1 b_1 + a_2 b_2 + a_3 b_3.$$

The dot product of a vector with itself gives the square of its length:

$$\mathbf{v} \cdot \mathbf{v} = |\mathbf{v}|^2. \qquad \textbf{(12.2)}$$

We prove this by evaluating everything in terms of coordinates. If $\mathbf{u} = [a_1, a_2, a_3]$ then

$$\mathbf{u} \cdot \mathbf{u} = a_1^2 + a_2^2 + a_3^2 = |\mathbf{u}|^2. \qquad \textbf{(12.3)}$$

The dot product of two vectors is the same when taken in either order,

$$\mathbf{v} \cdot \mathbf{u} = \mathbf{u} \cdot \mathbf{v}. \qquad \textbf{(12.4)}$$

This follows because both sides of the equation equal $a_1 b_1 + a_2 b_2 + a_3 b_3$.

Scalars can be factored out of the product. For any real numbers s and t we have

$$(s\mathbf{u}) \cdot (t\mathbf{v}) = (st)\mathbf{u} \cdot \mathbf{v}. \qquad \textbf{(12.5)}$$

To verify this we compute the coordinates of all the objects:

$$\begin{aligned} s\mathbf{u} &= [sa_1, sa_2, sa_3], \\ t\mathbf{v} &= [tb_1, tb_2, tb_3], \\ (s\mathbf{u})\cdot(t\mathbf{v}) &= (sa_1)(tb_1) + (sa_2)(tb_2) + (sa_3)(tb_3) \\ &= (st)(a_1b_1 + a_2b_2 + a_3b_3) \\ &= (st)\mathbf{u}\cdot\mathbf{v}. \end{aligned}$$

The dot product distributes over addition:

$$\mathbf{u}\cdot(\mathbf{v}+\mathbf{w}) = \mathbf{u}\cdot\mathbf{v} + \mathbf{u}\cdot\mathbf{w}. \tag{12.6}$$

Again, this is verified by working out the coordinates and applying the definition. We leave the verification to the reader.

EXAMPLE 1 Let $\mathbf{u} = [1, 2, -3]$ and $\mathbf{v} = [-3, 2, -1]$. Compute the dot products $\mathbf{u}\cdot\mathbf{v}$, $\mathbf{u}\cdot\mathbf{u}$ and $\mathbf{u}\cdot(2\mathbf{v} - 3\mathbf{u})$.

Solution: We apply the definition and properties of the dot product to get

$$\begin{aligned} \mathbf{u}\cdot\mathbf{v} &= [1, 2, -3]\cdot[-3, 2, -1] = 1(-3) + 2(2) - 3(-1) = 4, \\ \mathbf{u}\cdot\mathbf{u} &= [1, 2, -3]\cdot[1, 2, -3] = 1(1) + 2(2) - 3(-3) = 14, \\ \mathbf{u}\cdot(2\mathbf{v} - 3\mathbf{u}) &= 2(\mathbf{u}\cdot\mathbf{v}) - 3(\mathbf{u}\cdot\mathbf{u}) = 2(4) - 3(14) = -34. \end{aligned}$$

■

The properties listed to this point look very much like properties of multiplication of real numbers. There is a significant difference, however, because the dot product of nonzero vectors can be 0. For example

$$[1, 0, 1]\cdot[1, 0, -1] = 1\cdot 1 + 0\cdot 0 + 1\cdot(-1) = 0.$$

We prove that the dot product of two vectors equals 0 precisely when the two vectors are perpendicular to each other. This geometric property is one of the most important properties of the dot product. In fact a more general statement can be proved. For two given vectors $\mathbf{u}$ and $\mathbf{v}$, the angle between them is the angle formed when the two vectors have the same initial point.

THEOREM 12.1

Geometric Dot Product

Let $\mathbf{u}$ and $\mathbf{v}$ be two vectors and let θ be the angle between them with $0 \le \theta \le \pi$. Then

$$\mathbf{u}\cdot\mathbf{v} = |\mathbf{u}|\,|\mathbf{v}|\cos\theta.$$

Proof

We give a proof for vectors in the xy plane and leave the three-dimensional case to the exercises. We write the vectors in polar form as

$$\mathbf{u} = |\mathbf{u}|(\cos\alpha\ \mathbf{i} + \sin\alpha\ \mathbf{j}) \qquad \mathbf{v} = |\mathbf{v}|(\cos\beta\ \mathbf{i} + \sin\beta\ \mathbf{j}).$$

Then

$$\mathbf{u}\cdot\mathbf{v} = |\mathbf{u}|\,|\mathbf{v}|(\cos\alpha\cos\beta + \sin\alpha\sin\beta).$$

Now we use the trigonometric identity that says

$$\cos(\alpha - \beta) = \cos\alpha\cos\beta + \sin\alpha\sin\beta.$$

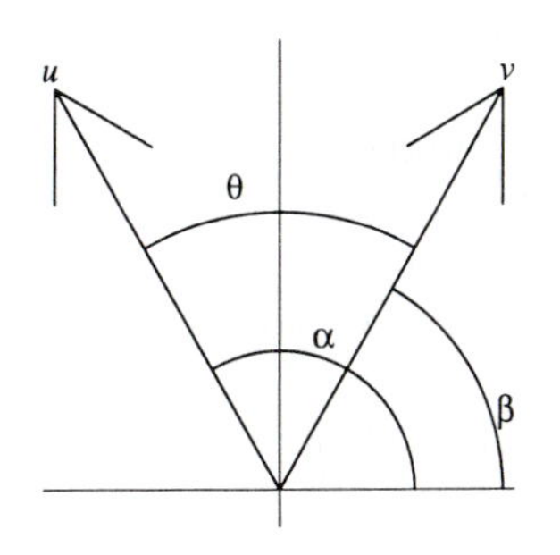

FIGURE 12.21
$\theta = \alpha - \beta$

Thus we have

$$\mathbf{u} \cdot \mathbf{v} = |\mathbf{u}|\,|\mathbf{v}| \cos(\alpha - \beta) = |\mathbf{u}|\,|\mathbf{v}| \cos\theta.$$

The angle between the two vectors is $\theta = \alpha - \beta$ and by selecting the notation properly, we can arrange to have $0 \leq \theta \leq \pi$. □

Two vectors that are perpendicular to each other when placed with the same initial point are called *orthogonal vectors.* Two vectors are orthogonal when the angle between them is $\pm\pi/2$. The zero vector is an exceptional case and we simply decree that the zero vector is perpendicular to every vector. The dot product gives a method to determine when two vectors are orthogonal.

THEOREM 12.2

Orthogonal Vectors

Let **u** and **v** be nonzero vectors. Then **u** and **v** are orthogonal if and only if $\mathbf{u} \cdot \mathbf{v} = 0$.

Proof

If θ is the angle between the two vectors then the dot product is

$$\mathbf{u} \cdot \mathbf{v} = |\mathbf{u}|\,|\mathbf{v}| \cos\theta. \qquad \textbf{(12.7)}$$

Because the vectors are nonzero, the lengths $|\mathbf{u}|$ and $|\mathbf{v}|$ are nonzero so the dot product is zero if and only if $\cos\theta = 0$. This occurs if an only if $\theta = \pm\pi/2 + 2k\pi$ with some integer k. In other words, the dot product is zero if and only if the vectors are orthogonal. □

EXAMPLE 2 Determine if any two of the vectors $\mathbf{u} = 3\mathbf{i} - 2\mathbf{j} + 2\mathbf{k}$, $\mathbf{v} = 2\mathbf{i} + 3\mathbf{j} - \mathbf{k}$ and $\mathbf{w} = 2\mathbf{i} + 4\mathbf{j} + \mathbf{k}$ are orthogonal. Find the cosine of the angle between the vectors that are not orthogonal.

***Solution*:** We begin by computing the dot products of each pair:

$$\mathbf{u} \cdot \mathbf{v} = (3\mathbf{i} - 2\mathbf{j} + 2\mathbf{k}) \cdot (2\mathbf{i} + 3\mathbf{j} - \mathbf{k}) = 3(2) - 2(3) + 2(-1) = -2,$$
$$\mathbf{u} \cdot \mathbf{w} = (3\mathbf{i} - 2\mathbf{j} + 2\mathbf{k}) \cdot (2\mathbf{i} + 4\mathbf{j} + \mathbf{k}) = 3(2) - 2(4) + 2(1) = 0,$$
$$\mathbf{v} \cdot \mathbf{w} = (2\mathbf{i} + 3\mathbf{j} - \mathbf{k}) \cdot (2\mathbf{i} + 4\mathbf{j} + \mathbf{k}) = 2(2) + 3(4) - (1) = 15.$$

Because $\mathbf{u} \cdot \mathbf{w} = 0$ the vectors **u** and **w** are orthogonal. Neither of the other two pairs of vectors are orthogonal because the dot products are nonzero.

To find the cosine of the angle between two vectors we solve for the cosine term in the equation (12.7) to obtain

$$\cos\theta = \frac{\mathbf{u} \cdot \mathbf{v}}{|\mathbf{u}|\,|\mathbf{v}|}.$$

If α is the angle between **u** and **v** and β is the angle between **v** and **w** then

$$\cos\alpha = \frac{\mathbf{u} \cdot \mathbf{v}}{|\mathbf{u}|\,|\mathbf{v}|} = \frac{-2}{\sqrt{17}\sqrt{14}} \approx -0.12962,$$
$$\cos\beta = \frac{\mathbf{w} \cdot \mathbf{v}}{|\mathbf{w}|\,|\mathbf{v}|} = \frac{15}{\sqrt{21}\sqrt{14}} \approx 0.87482.$$

■

In the next example we apply the dot product to prove a theoretical fact: If we are given two nonzero vectors then the first vector is orthogonal to the sum of the second vector and some scalar multiple of the first.

EXAMPLE 3 Is there a scalar t such that $\mathbf{u}$ is orthogonal to $\mathbf{v} + t\mathbf{u}$?

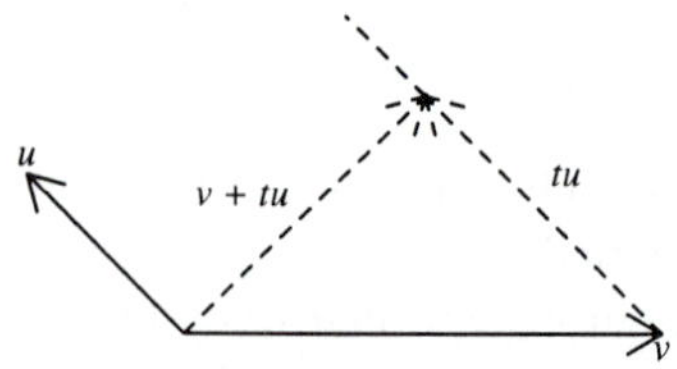

FIGURE 12.22
$\mathbf{u}$ orthogonal to $\mathbf{v} + t\mathbf{u}$

Solution: The vectors $\mathbf{u}$ and $\mathbf{v} + t\mathbf{u}$ are orthogonal if and only if

$$\mathbf{u} \cdot (\mathbf{v} + t\mathbf{u}) = \mathbf{u} \cdot \mathbf{v} + t\mathbf{u} \cdot \mathbf{u} = 0.$$

If $\mathbf{u} \neq \mathbf{0}$ then $\mathbf{u} \cdot \mathbf{u} = |\mathbf{u}|^2 \neq 0$ and the equation can be solved for t. Thus $t = -\mathbf{u} \cdot \mathbf{v}/\mathbf{u} \cdot \mathbf{u}$ gives the solution. Note that there is only one choice of t. We summarize this computation: If $\mathbf{u}$ is nonzero, then $\mathbf{u}$ is orthogonal to

$$\mathbf{v} - \left(\frac{\mathbf{u} \cdot \mathbf{v}}{\mathbf{u} \cdot \mathbf{u}}\right) \mathbf{u}.$$

■

The computation in example 3 becomes a bit easier to remember if the $\mathbf{u}$ has unit length. When $\mathbf{u}$ is a unit vector then $\mathbf{u} \cdot \mathbf{u} = 1$ and so $\mathbf{u}$ and $\mathbf{v} - (\mathbf{v} \cdot \mathbf{u})\mathbf{u}$ are orthogonal. The idea in this computation is used in the discussion of projections.

Projections

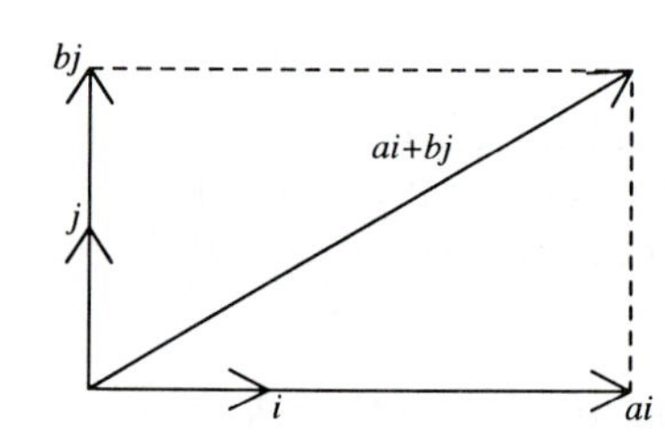

FIGURE 12.23

If $\mathbf{v}$ is any vector in the xy plane then $\mathbf{v} = a\mathbf{i} + b\mathbf{j}$ for some numbers a and b. Then a is the *component* of $\mathbf{v}$ in the direction of $\mathbf{i}$ and b is the *component* of $\mathbf{v}$ in the direction of $\mathbf{j}$. The numbers a and b can be computed from $\mathbf{v}$ as

$$\begin{cases} a = \mathbf{v} \cdot \mathbf{i} \\ b = \mathbf{v} \cdot \mathbf{j}. \end{cases}$$

Here we use $\mathbf{i} \cdot \mathbf{i} = \mathbf{j} \cdot \mathbf{j} = 1$ and $\mathbf{i} \cdot \mathbf{j} = 0$. The sum $\mathbf{v} = a\mathbf{i} + b\mathbf{j}$ expresses $\mathbf{v}$ as a sum of two orthogonal vectors that happen to be parallel to the coordinate axes.

We want to make an analogous computation for components not necessarily parallel to the coordinate axis and for vectors in three dimensions. What does this mean?

If $\mathbf{u}$ is a nonzero vector and $\mathbf{v}$ is any vector then it is possible to write

$$\mathbf{v} = \mathbf{u}' + \mathbf{w}$$

where $\mathbf{u}'$ is parallel to $\mathbf{u}$ and $\mathbf{w}$ is orthogonal to $\mathbf{u}$. This is intuitively clear by looking at some pictures. We give a complete proof that makes use of the notion of *projections*, which we now define. The projection of $\mathbf{v}$ on $\mathbf{u}$ is the vector

$$\mathbf{pr_u}(\mathbf{v}) = \left(\frac{\mathbf{v} \cdot \mathbf{u}}{\mathbf{u} \cdot \mathbf{u}}\right) \mathbf{u}.$$

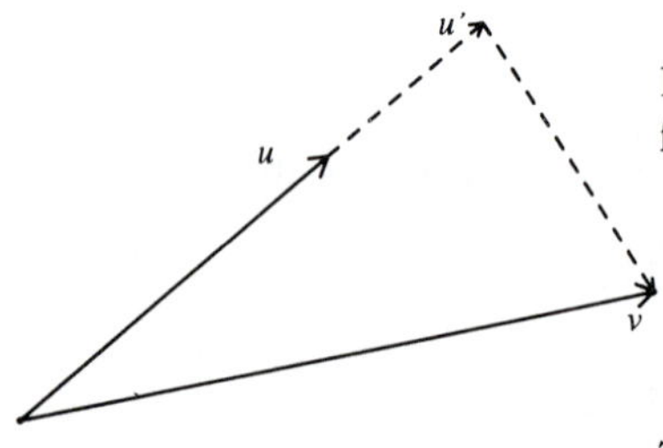

FIGURE 12.24
$\mathbf{u}' = \mathbf{pr_u}(\mathbf{v})$

Note that the length of $\mathbf{u}$ does not play a role in the value of this projection. By this we mean the following. Suppose we replace $\mathbf{u}$ by a parallel vector $t\mathbf{u} = \mathbf{w}$ for a nonzero scalar t. Then

$$\begin{aligned} \mathbf{pr_w}(\mathbf{v}) &= \left(\frac{\mathbf{v} \cdot \mathbf{w}}{\mathbf{w} \cdot \mathbf{w}}\right) \mathbf{w} = \left(\frac{\mathbf{v} \cdot t\mathbf{u}}{t\mathbf{u} \cdot t\mathbf{u}}\right) t\mathbf{u} \\ &= \left(\frac{\mathbf{v} \cdot \mathbf{u}}{\mathbf{u} \cdot \mathbf{u}}\right) \mathbf{u} = \mathbf{pr_u}(\mathbf{v}). \end{aligned}$$

Thus the projection of $\mathbf{v}$ on $\mathbf{u}$ is the same as the projection of $\mathbf{v}$ on $t\mathbf{u}$ for any scalar $t \neq 0$.

The formulas are sometimes easier to remember for unit vectors so we assume that the original vector $\mathbf{u}$ is a unit vector. When $\mathbf{u}$ is a unit vector, the projection of $\mathbf{v}$ in the direction of $\mathbf{u}$ is the vector

$$\mathbf{pr_u}(\mathbf{v}) = (\mathbf{v} \cdot \mathbf{u})\mathbf{u}. \tag{12.8}$$

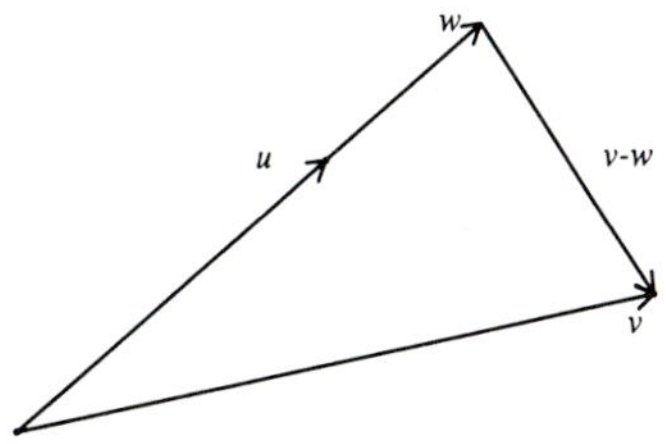

Note that the length of $\mathbf{pr_u}(\mathbf{v})$ is $|\mathbf{v} \cdot \mathbf{u}|$, because $|\mathbf{u}| = 1$, and that $\mathbf{pr_u}(\mathbf{v})$ is parallel to $\mathbf{u}$. The projection has the property that

$$\mathbf{v} - \mathbf{pr_u}(\mathbf{v})$$

is orthogonal to $\mathbf{u}$. We check this by computing the dot product:

$$\begin{aligned}\mathbf{u} \cdot (\mathbf{v} - \mathbf{pr_u}(\mathbf{v})) &= \mathbf{u} \cdot \mathbf{v} - \mathbf{u} \cdot (\mathbf{v} \cdot \mathbf{u})\mathbf{u} \\ &= \mathbf{u} \cdot \mathbf{v} - (\mathbf{v} \cdot \mathbf{u})\mathbf{u} \cdot \mathbf{u} \\ &= \mathbf{u} \cdot \mathbf{v} - (\mathbf{v} \cdot \mathbf{u})1 \\ &= 0.\end{aligned}$$

FIGURE 12.25
$\mathbf{w} = \mathbf{pr_u}(\mathbf{v})$ orthogonal to $\mathbf{v} - \mathbf{w}$

So finally we can write

$$\mathbf{v} = \mathbf{pr_u}(\mathbf{v}) + (\mathbf{v} - \mathbf{pr_u}(\mathbf{v})),$$

which expresses $\mathbf{v}$ as a sum of a vector parallel to $\mathbf{u}$ and a vector orthogonal to $\mathbf{u}$. If we write $\mathbf{w} = \mathbf{v} - \mathbf{pr_u}(\mathbf{v})$ and replace the projection by using Equation (12.8) we obtain

$$\mathbf{v} = (\mathbf{v} \cdot \mathbf{u})\mathbf{u} + \mathbf{w} \quad \text{where} \quad \mathbf{u} \cdot \mathbf{w} = 0.$$

The scalar $\mathbf{v} \cdot \mathbf{u}$ is called the *component of* $\mathbf{v}$ in the direction of $\mathbf{u}$. (Remember that $\mathbf{u}$ is a unit vector.)

EXAMPLE 4 Let $\mathbf{u}$ be the unit vector $(\mathbf{i} + \mathbf{j})/\sqrt{2}$ and let $\mathbf{v} = \mathbf{i} + 3\mathbf{j}$. Find the projection of $\mathbf{v}$ on $\mathbf{u}$ and express $\mathbf{v}$ as a sum of a vector parallel to $\mathbf{u}$ and a vector orthogonal to $\mathbf{u}$.

Solution: First compute the projection of $\mathbf{v}$ on $\mathbf{u}$:

$$\begin{aligned}\mathbf{u}' = \mathbf{pr_u}(\mathbf{v}) = (\mathbf{v} \cdot \mathbf{u})\mathbf{u} &= [(\mathbf{i} + 3\mathbf{j}) \cdot \frac{1}{\sqrt{2}}(\mathbf{i} + \mathbf{j})]\mathbf{u} \\ &= \frac{4}{\sqrt{2}}\mathbf{u} = \frac{4}{\sqrt{2}}\frac{1}{\sqrt{2}}(\mathbf{i} + \mathbf{j}) \\ &= 2(\mathbf{i} + \mathbf{j}).\end{aligned}$$

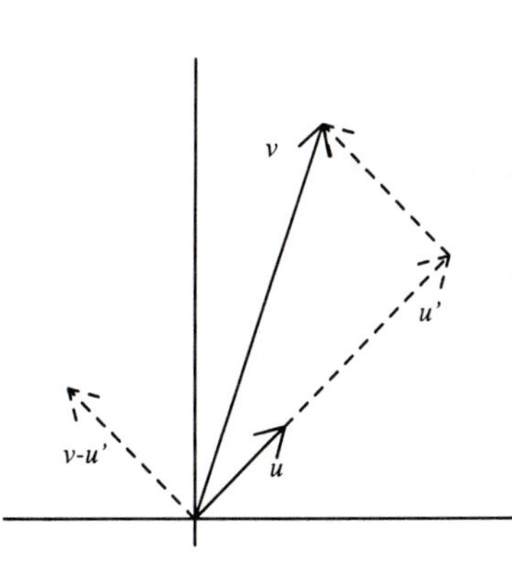

FIGURE 12.26

The vector orthogonal to $\mathbf{u}$ is

$$\mathbf{v} - \mathbf{pr_u}(\mathbf{v}) = (\mathbf{i} + 3\mathbf{j}) - (2\mathbf{i} + 2\mathbf{j}) = -\mathbf{i} + \mathbf{j}.$$

Then we have

$$\mathbf{v} = \mathbf{pr_u}(\mathbf{v}) + \big(\mathbf{v} - \mathbf{pr_u}(\mathbf{v})\big) = 2(\mathbf{i} + \mathbf{j}) + (-\mathbf{i} + \mathbf{j}).$$

As a check, we verify that the two vectors on the right are indeed orthogonal by computing their dot product:

$$2(\mathbf{i} + \mathbf{j}) \cdot (-\mathbf{i} + \mathbf{j}) = -2 + 2 = 0.$$

■

Next we try the same idea in three dimensions.

EXAMPLE 5 Let $\mathbf{v} = 3\mathbf{i} + \mathbf{j} - 2\mathbf{k}$ and let $\mathbf{u}$ be a unit vector parallel to $\mathbf{w} = 3\mathbf{i} - 6\mathbf{j} + 6\mathbf{k}$. Find the projection $\mathbf{pr_w}(\mathbf{v})$ of $\mathbf{v}$ on $\mathbf{w}$ and then express $\mathbf{v}$ as a sum of two vectors, one parallel to $\mathbf{w}$ and one orthogonal to $\mathbf{w}$.

Solution: First we find the unit vector $\mathbf{u}$ by dividing $\mathbf{w}$ by its length:

$$\mathbf{u} = \frac{1}{|\mathbf{w}|}\mathbf{w} = \frac{1}{\sqrt{3^2+6^2+6^2}}(3\mathbf{i}-6\mathbf{j}+6\mathbf{k}) = \frac{1}{3}(\mathbf{i}-2\mathbf{j}+2\mathbf{k}).$$

Next we compute the projection of $\mathbf{v}$ in the direction of $\mathbf{u}$ as

$$\begin{aligned}\mathbf{pr_u}(\mathbf{v}) = (\mathbf{v}\cdot\mathbf{u})\mathbf{u} &= \left[(3\mathbf{i}+\mathbf{j}-2\mathbf{k})\cdot\left(\frac{1}{3}\right)(\mathbf{i}-2\mathbf{j}+2\mathbf{k})\right]\mathbf{u}\\ &= \left(\frac{1}{3}\right)\big(3(1)+(-2)-2(2)\big)\mathbf{u} = -\mathbf{u}\\ &= -\left(\frac{1}{3}\right)(\mathbf{i}-2\mathbf{j}+2\mathbf{k}).\end{aligned}$$

Then subtract this from $\mathbf{v}$ to get the vector

$$\mathbf{v}-\mathbf{pr_u}(\mathbf{v}) = (3\mathbf{i}+\mathbf{j}-2\mathbf{k}) + \left(\frac{1}{3}\right)(\mathbf{i}-2\mathbf{j}+2\mathbf{k}) = \left(\frac{1}{3}\right)(10\mathbf{i}+\mathbf{j}-4\mathbf{k}).$$

Finally, we have

$$\begin{aligned}\mathbf{v} &= \mathbf{pr_u}(\mathbf{v}) + (\mathbf{v}-\mathbf{pr_u}(\mathbf{v}))\\ &= -\left(\frac{1}{3}\right)(\mathbf{i}-2\mathbf{j}+2\mathbf{k}) + \left(\frac{1}{3}\right)(10\mathbf{i}+\mathbf{j}-4\mathbf{k}).\end{aligned}$$

As a check on the arithmetic, let us verify that the two vectors in the sum on the right are really orthogonal. Their dot product is

$$-\left(\frac{1}{3}\right)(\mathbf{i}-2\mathbf{j}+2\mathbf{k})\cdot\left(\frac{1}{3}\right)(10\mathbf{i}+\mathbf{j}-4\mathbf{k}) = \frac{1}{9}\big(1(10)-2(1)+2(-4)\big) = 0,$$

which shows the two vectors are orthogonal. ■

Exercises 12.4

Compute the dot products given that $\mathbf{u} = 2\mathbf{i}-3\mathbf{j}+\mathbf{k}$, $\mathbf{v} = -2\mathbf{i}-\mathbf{j}+3\mathbf{k}$, and $\mathbf{w} = 2\mathbf{j}-6\mathbf{k}$.

1. $\mathbf{u}\cdot\mathbf{v}$

2. $\mathbf{w}\cdot\mathbf{v}$

3. $(\mathbf{u}+\mathbf{v})\cdot(\mathbf{u}-\mathbf{v})$

4. $(\mathbf{w}+6\mathbf{u})\cdot(2\mathbf{v}+\mathbf{w})$

5. $\mathbf{u}\cdot(\mathbf{v}+2\mathbf{w})$

6. $(2\mathbf{w}-4\mathbf{v})\cdot\mathbf{u}$

7. Find a number c so that $2\mathbf{i}+c\mathbf{j}+\mathbf{k}$ is orthogonal to $-\mathbf{i}-\mathbf{j}+2\mathbf{k}$.

8. Show that there is no number c such that $\mathbf{i}+c\mathbf{j}$ is orthogonal to $2\mathbf{i}+c\mathbf{j}-3\mathbf{k}$.

9. Find the cosine of the angle between $\mathbf{i}+\mathbf{j}+2\mathbf{k}$ and $\mathbf{i}+\mathbf{j}-2\mathbf{k}$.

10. A tetrahedron has its four vertices at $(0,0,0)$, $(1,0,0)$, $(0,1,0)$, $(0,0,1)$. It has four faces and four edges and each vertex has three edges emanating from it. Compute the cosine of the angles between the three edges emanating from each vertex.

11. Give an example of three nonzero vectors $\mathbf{u}$, $\mathbf{v}$, $\mathbf{w}$ such that $\mathbf{u}$ is orthogonal to $\mathbf{w}$, $\mathbf{v}$ is orthogonal to $\mathbf{w}$ but $\mathbf{u}$ is not parallel to $\mathbf{v}$.

12. Prove that the vector $a\,\mathbf{i}+b\,\mathbf{j}$ is orthogonal to the line having equation $ax+by=c$. [Hint: Find a vector parallel to the line by using two points on the line.]

For each pair of vectors $\mathbf{u}$ and $\mathbf{v}$, find the two projections $\mathbf{pr_u}(\mathbf{v})$ and $\mathbf{pr_v}(\mathbf{u})$.

13. $\mathbf{u} = \mathbf{i}+\mathbf{j}$, $\mathbf{v} = \mathbf{j}$

14. $\mathbf{u} = 2\mathbf{i}-2\mathbf{j}$, $\mathbf{v} = 2\mathbf{i}+\mathbf{j}$

15. $\mathbf{u} = \mathbf{i}+\mathbf{j}+\mathbf{k}$, $\mathbf{v} = \mathbf{i}-\mathbf{j}-\mathbf{k}$

16. $\mathbf{u} = 2\mathbf{i}-3\mathbf{j}+\mathbf{k}$, $\mathbf{v} = 2\mathbf{i}+\mathbf{j}-\mathbf{k}$

Write the vector $\mathbf{v}$ as a sum of a vector parallel to $\mathbf{u}$ and a vector orthogonal to $\mathbf{u}$. Make sketches showing the vectors.

17. $\mathbf{u} = \mathbf{i}+\mathbf{j}$, $\mathbf{v} = \mathbf{j}$

18. $\mathbf{u} = 2\mathbf{i}-2\mathbf{j}$, $\mathbf{v} = 2\mathbf{i}+\mathbf{j}$

19. $\mathbf{u} = \mathbf{i}+3\mathbf{j}$, $\mathbf{v} = \mathbf{i}-3\mathbf{j}$

20. $\mathbf{u} = 2\mathbf{i}-3\mathbf{j}$, $\mathbf{v} = 2\mathbf{i}$

21. What can you conclude about two nonzero vectors $\mathbf{u}$ and $\mathbf{v}$ that satisfy $(\mathbf{u}\cdot\mathbf{v})\mathbf{u} = (\mathbf{u}\cdot\mathbf{v})\mathbf{v}$?

22. Prove that the relation

$$4(\mathbf{u} \cdot \mathbf{v}) = |\mathbf{u} + \mathbf{v}|^2 - |\mathbf{u} - \mathbf{v}|^2$$

holds for all vectors. [Hint: Express the right side in terms of dot products. This shows that the dot product can be determined from a knowledge of lengths only. Coordinates are not needed to make the definition.]

23. Let $\mathbf{v}$ and $\mathbf{u}$ be vectors in xyz space and θ the angle between them. Prove the formula $\mathbf{v} \cdot \mathbf{u} = |\mathbf{v}|\,|\mathbf{u}| \cos\theta$ for the dot product. [Hint: Apply the law of cosines to the triangle with sides $\mathbf{v}$, $\mathbf{u}$, $\mathbf{v} - \mathbf{u}$ to conclude

$$2|\mathbf{v}||\mathbf{u}| \cos\theta = |\mathbf{v}|^2 + |\mathbf{u}|^2 - |\mathbf{v} - \mathbf{u}|^2.$$

Then use the coordinate form of the vectors to conclude that the right-hand side of this equation equals $2(\mathbf{v} \cdot \mathbf{u})$.]

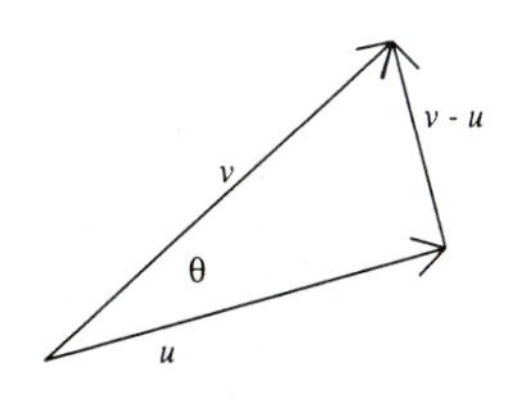

FIGURE 12.27

24. If $\mathbf{u}$ and $\mathbf{v}$ are unit vectors such that $\mathbf{u} + \mathbf{v}$ is also a unit vector, find the angle between $\mathbf{u}$ and $\mathbf{v}$.

25. If $\mathbf{u}$ and $\mathbf{v}$ are unit vectors such that $\mathbf{u} + \mathbf{v}$ is also a unit vector, find the magnitude of $\mathbf{u} - \mathbf{v}$.

26. Prove that $\mathbf{u} = \mathbf{0}$ is the only vector that satisfies the three equations $\mathbf{u} \cdot \mathbf{i} = 0$, $\mathbf{u} \cdot \mathbf{j} = 0$, $\mathbf{u} \cdot \mathbf{k} = 0$.

27. Prove that vectors $\mathbf{v}$ and $\mathbf{w}$ must be equal if they satisfy the three conditions $\mathbf{v} \cdot \mathbf{i} = \mathbf{w} \cdot \mathbf{i}$, $\mathbf{v} \cdot \mathbf{j} = \mathbf{w} \cdot \mathbf{j}$, $\mathbf{v} \cdot \mathbf{k} = \mathbf{w} \cdot \mathbf{k}$. [Hint: Apply exercise 26 with $\mathbf{u} = \mathbf{v} - \mathbf{w}$.]

28. Let $\mathbf{w}$ be a nonzero vector. Show by example that the equation $\mathbf{u} \cdot \mathbf{w} = \mathbf{v} \cdot \mathbf{w}$ does not imply the equality of $\mathbf{u}$ and $\mathbf{v}$.

12.5 The cross product

We introduce the cross product in this section. The definition in terms of coordinates given just below maight seem quite odd so we give a bit of motivation by considering some properties that the cross product should have.

First, the cross product of $\mathbf{u}$ and $\mathbf{v}$, written as $\mathbf{u} \times \mathbf{v}$, should be a vector orthogonal to both $\mathbf{u}$ and $\mathbf{v}$. There are infinitely many vectors orthogonal to two given vectors. For example, any scalar multiple of $\mathbf{k}$ is orthogonal to $\mathbf{i}$ and $\mathbf{j}$. The definition below implies that $\mathbf{i} \times \mathbf{j} = \mathbf{k}$ and that $\mathbf{j} \times \mathbf{i} = -\mathbf{k}$.

The next property we would like the cross product to have is that it should be a geometric property depending only on the position of the two vectors relative to each other and not dependent on the particular location in space of the two vectors. One way of interpreting this condition is to say the cross product is invariant under rotation. This means that if $\mathbf{u} \times \mathbf{v} = \mathbf{w}$ and if the entire system of coordinates is held rigid but rotated about the origin so that $\mathbf{u}$ moves to $\mathbf{u}'$, $\mathbf{v}$ moves to $\mathbf{v}'$ and $\mathbf{w}$ moves to $\mathbf{w}'$, then $\mathbf{u}' \times \mathbf{v}' = \mathbf{w}'$.

We can use the rotation idea to compute $\mathbf{j} \times \mathbf{k}$. Rotate the coordinate system so that $\mathbf{i}$ moves to $\mathbf{j}$ and $\mathbf{j}$ moves to $\mathbf{k}$, then $\mathbf{k}$ moves to $\mathbf{i}$. Thus we have $\mathbf{j} \times \mathbf{k} = \mathbf{i}$. (If you have very flexible hands and fingers, you may be able to visualize this by forming a $\mathbf{i}$, $\mathbf{j}$, $\mathbf{k}$ triple with your first two fingers and thumb and then performing the rotation. As an alternative to prevent sprains I recommend three pencils and two rubber bands.) One further rotation gives rise to $\mathbf{k} \times \mathbf{i} = \mathbf{j}$.

When the order of the two vectors is reversed in the product, the sign is changed. Thus we reach the following rules for the cross products of the basic unit vectors:

$$\begin{aligned} \mathbf{i} \times \mathbf{j} &= \mathbf{k}, & \mathbf{j} \times \mathbf{i} &= -\mathbf{k}, \\ \mathbf{j} \times \mathbf{k} &= \mathbf{i}, & \mathbf{k} \times \mathbf{j} &= -\mathbf{i}, \\ \mathbf{k} \times \mathbf{i} &= \mathbf{j}, & \mathbf{i} \times \mathbf{k} &= -\mathbf{j}. \end{aligned} \tag{12.9}$$

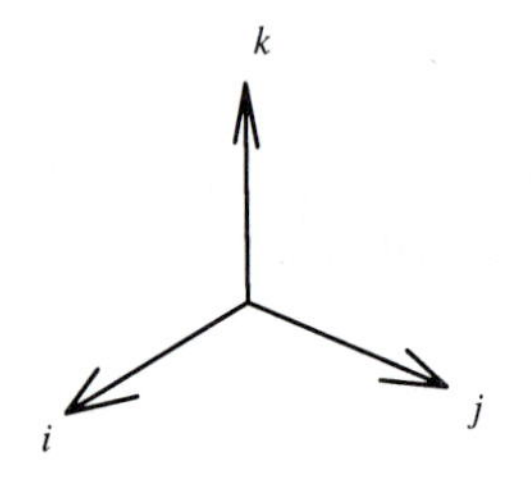

FIGURE 12.28

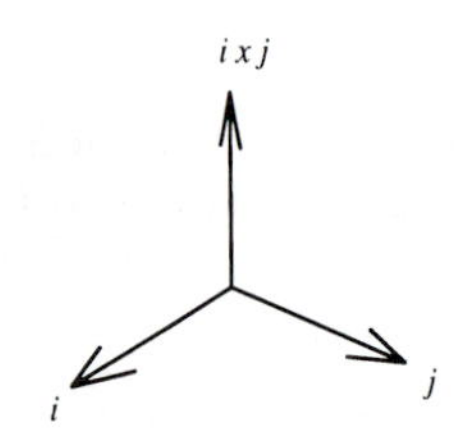

FIGURE 12.29

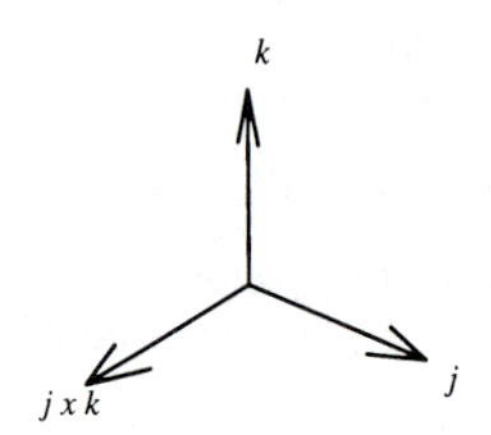

FIGURE 12.30

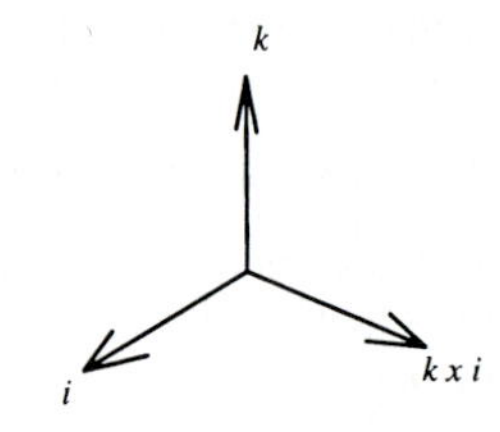

FIGURE 12.31

Another decision must be made about products like $\mathbf{i} \times \mathbf{i}$. This is supposed to be a vector orthogonal to $\mathbf{i}$ and $\mathbf{i}$. This time there is not only an infinite number of vectors orthogonal to $\mathbf{i}$, but they can be selected in an infinite number of directions. One way out is to decide that $\mathbf{u} \times \mathbf{u} = \mathbf{0}$ for any vector $\mathbf{u}$. Since $\mathbf{0}$ is orthogonal to every vector, the orthogonal property of the cross product still holds.

The last property of the cross product that we impose is the *linearity* property. It asserts that

$$(r\mathbf{u} + s\mathbf{v}) \times \mathbf{w} = r(\mathbf{u} \times \mathbf{w}) + s(\mathbf{v} \times \mathbf{w})$$

for all vectors $\mathbf{u}, \mathbf{v}$ and $\mathbf{w}$ and all real numbers r and s. The linearity property, along with the cross products already given for $\mathbf{i}$, $\mathbf{j}$, $\mathbf{k}$, allow the computation of the cross product of any two vectors.

Now let

$$\begin{aligned} \mathbf{u} &= [u_1, u_2, u_3] = u_1\mathbf{i} + u_2\mathbf{j} + u_3\mathbf{k} \\ \mathbf{v} &= [v_1, v_2, v_3] = v_1\mathbf{i} + v_2\mathbf{j} + v_3\mathbf{k}. \end{aligned}$$

We compute

$$\begin{aligned} \mathbf{u} \times \mathbf{v} =& (u_1\mathbf{i} + u_2\mathbf{j} + u_3\mathbf{k}) \times (v_1\mathbf{i} + v_2\mathbf{j} + v_3\mathbf{k}) \\ =& u_1v_1(\mathbf{i} \times \mathbf{i}) + u_1v_2(\mathbf{i} \times \mathbf{j}) + u_1v_3(\mathbf{i} \times \mathbf{k}) \\ &+ u_2v_1(\mathbf{j} \times \mathbf{i}) + u_2v_2(\mathbf{j} \times \mathbf{j}) + u_2v_3(\mathbf{j} \times \mathbf{k}) \\ &+ u_3v_1(\mathbf{k} \times \mathbf{i}) + u_3v_2(\mathbf{k} \times \mathbf{j}) + u_3v_3(\mathbf{k} \times \mathbf{k}) \\ =& u_1v_1(\mathbf{0}) + u_1v_2(\mathbf{k}) + u_1v_3(-\mathbf{j}) \\ &+ u_2v_1(-\mathbf{k}) + u_2v_2(\mathbf{0}) + u_2v_3(\mathbf{i}) \\ &+ u_3v_1(\mathbf{j}) + u_3v_2(-\mathbf{i}) + u_3v_3(\mathbf{0}) \\ =& (u_2v_3 - u_2v_2)\mathbf{i} + (u_3v_1 - u_1v_3)\mathbf{j} + (u_1v_2 - u_2v_1)\mathbf{k}. \end{aligned}$$

The assumptions we have made about multiplication allow us to determine the cross product of any two vectors. Let us make this as a formal definition now.

DEFINITION 12.1

The Cross Product

The *cross product* of vectors $\mathbf{u} = [u_1, u_2, u_3]$ and $\mathbf{v} = [v_1, v_2, v_3]$ is

$$\mathbf{u} \times \mathbf{v} = [u_2v_3 - u_3v_2,\ u_3v_1 - u_1v_3,\ u_1v_2 - u_2v_1].$$

This definition was derived by making some assumptions about the desired outcome. We should now check to see that we have indeed achieved the goal. We show that the cross product of two vectors is orthogonal to each vector.

THEOREM 12.3

Properties of the Cross Product

For any vectors $\mathbf{u}$ and $\mathbf{v}$ we have:

i. $(\mathbf{u} \times \mathbf{v}) \cdot \mathbf{u} = \mathbf{0}$, ii. $(\mathbf{u} \times \mathbf{v}) \cdot \mathbf{v} = \mathbf{0}$,

iii. $\mathbf{u} \times \mathbf{v} = -\mathbf{v} \times \mathbf{u}$, iv. $\mathbf{u} \times \mathbf{u} = \mathbf{0}$.

Proof

The proof of properties i and ii can be carried out by direct computation using the definition of the cross product and of the dot product. Apply the definition of the cross product and the definition of the dot product to get

$$\begin{aligned}(\mathbf{u} \times \mathbf{v}) \cdot \mathbf{u} &= [u_2v_3 - u_3v_2,\ u_3v_1 - u_1v_3,\ u_1v_2 - u_2v_1] \cdot [u_1, u_2, u_3] \\ &= (u_2v_3 - u_3v_2)u_1 + (u_3v_1 - u_1v_3)u_2 + (u_1v_2 - u_2v_1)u_3 = 0.\end{aligned}$$

The algebraic manipulation, the collection of terms and the verification of the second statement is left to the reader.

The result of properties i and ii is that $\mathbf{u} \times \mathbf{v}$ is orthogonal to both $\mathbf{v}$ and $\mathbf{u}$. It is also quite straightforward to verify, using the definition, that

$$\mathbf{u} \times \mathbf{v} = -\mathbf{v} \times \mathbf{u}$$

because the exchange of the u_i for the v_i in the definition of the cross product produces only a sign change in each term. The definition also shows that statement iv holds since each coefficient is 0 when $u_i = v_i$ for each i. This statement can also be proved as a consequence of property iii; it implies that $\mathbf{u} \times \mathbf{u}$ is a vector equal to its negative. However the only vector that is equal to its negative is the zero vector. □

The definition of the cross product is difficult to remember; here is a device that might be of some help. It requires the use of determinants. A determinant is a real number that is associated to a square array of real numbers; we use the 2×2 and the 3×3 arrays only.

The value of a 2×2 determinant is given by the formula

$$\begin{vmatrix} a & b \\ c & d \end{vmatrix} = ad - bc.$$

For example,

$$\begin{vmatrix} 1 & 5 \\ 3 & 8 \end{vmatrix} = 1(8) - 3(5) = -7.$$

The 3×3 determinant is defined in terms of the 2×2 case as follows:

$$\begin{vmatrix} x & y & z \\ a & b & c \\ d & e & f \end{vmatrix} = \begin{vmatrix} b & c \\ e & f \end{vmatrix} x - \begin{vmatrix} a & c \\ d & f \end{vmatrix} y + \begin{vmatrix} a & b \\ d & e \end{vmatrix} z. \tag{12.10}$$

Take careful notice of the signs. There is a minus sign in front of the middle terms that is often forgotten.

As an example, we compute the 3×3 determinant

$$\begin{vmatrix} -2 & 3 & 1 \\ -4 & 0 & 2 \\ -3 & 1 & 9 \end{vmatrix} = \begin{vmatrix} 0 & 2 \\ 1 & 9 \end{vmatrix}(-2) - \begin{vmatrix} -4 & 2 \\ -3 & 9 \end{vmatrix} 3 + \begin{vmatrix} -4 & 0 \\ -3 & 1 \end{vmatrix} 1$$
$$= (-2)(-2) - (-30)(3) + (-4)(1) = 90.$$

Determinants are used in a symbolic way to evaluate the cross product of two vectors. Assume $\mathbf{u} = [u_1, u_2, u_3]$ and $\mathbf{v} = [v_1, v_2, v_3]$. Then

$$\mathbf{u} \times \mathbf{v} = \begin{vmatrix} \mathbf{i} & \mathbf{j} & \mathbf{k} \\ u_1 & u_2 & u_3 \\ v_1 & v_2 & v_3 \end{vmatrix}$$
$$= \begin{vmatrix} u_2 & u_3 \\ v_2 & v_3 \end{vmatrix} \mathbf{i} - \begin{vmatrix} u_1 & u_3 \\ v_1 & v_3 \end{vmatrix} \mathbf{j} + \begin{vmatrix} u_1 & u_2 \\ v_1 & v_2 \end{vmatrix} \mathbf{k}$$
$$= (u_2 v_3 - u_2 v_2)\mathbf{i} - (u_1 v_3 - u_3 v_1)\mathbf{j} + (u_1 v_2 - u_2 v_1)\mathbf{k}.$$

EXAMPLE 1 Compute the cross product $\mathbf{u} \times \mathbf{v}$ if $\mathbf{u} = \mathbf{i} - 2\mathbf{j}$ and $\mathbf{v} = 3\mathbf{i} + \mathbf{j} + 2\mathbf{k}$.

Solution: We use the determinant method to make the computation:

$$\mathbf{u} \times \mathbf{v} = \begin{vmatrix} \mathbf{i} & \mathbf{j} & \mathbf{k} \\ 1 & -2 & 0 \\ 3 & 1 & 2 \end{vmatrix}$$
$$= (-2(2) - 0(1))\mathbf{i} - (1(2) - (0)(3))\mathbf{j} + (1(1) - (-2)(3))\mathbf{k}$$
$$= -4\mathbf{i} - 2\mathbf{j} + 7\mathbf{k}.$$

Now to verify that the cross product of $\mathbf{u}$ and $\mathbf{v}$ is orthogonal to each vector $\mathbf{u}$ and $\mathbf{v}$, we compute the dot products:

$$(\mathbf{u} \times \mathbf{v}) \cdot \mathbf{u} = (-4\mathbf{i} - 2\mathbf{j} + 7\mathbf{k}) \cdot (\mathbf{i} - 2\mathbf{j}) = -4(1) - 2(-2) = 0,$$
$$(\mathbf{u} \times \mathbf{v}) \cdot \mathbf{v} = (-4\mathbf{i} - 2\mathbf{j} + 7\mathbf{k}) \cdot (3\mathbf{i} + \mathbf{j} + 2\mathbf{k}) = -4(3) - 2(1) + 7(2) = 0.$$

The general rule is reinforced by the computations in this case. ■

Direction and Length of the Cross Product

The next task is to determine the direction and to compute the length of the cross product of two vectors. We know that the cross product is orthogonal to each of the two vectors so the direction of the cross product is along the line perpendicular to the plane containing the two vectors. The problem is to decide along which of the two possible directions along the line the cross product lies. To determine the length of the cross product we could write the coordinates of two vectors and then apply the definition of the cross product and write a formula for the length in terms of coordinates; this is not what we have in mind. We want the length of $\mathbf{u} \times \mathbf{v}$ in terms of the lengths of $\mathbf{u}$ and $\mathbf{v}$ and the angle between them.

We compute the length and direction of the product

$$\mathbf{w} = \mathbf{u} \times \mathbf{v}.$$

Let us first do this for vectors having a special position. Let $\mathbf{u} = s\,\mathbf{i}$ and $\mathbf{v} = r(\cos\theta\,\mathbf{i} + \sin\theta\,\mathbf{j})$ with $s = |\mathbf{u}| \geq 0$, $t = |\mathbf{v}| \geq 0$ and θ the angle between $\mathbf{u}$ and $\mathbf{v}$ measured from $\mathbf{u}$. In other words, θ is the angle through which one would rotate $\mathbf{u}$ to make it lie in the same direction as $\mathbf{v}$.

Then $\mathbf{u}$ points along the positive x-axis and $\mathbf{v}$ is in the xy plane and lies in the upper half-plane if $0 \leq \theta \leq \pi$ and in the lower half-plane if $\pi < \theta < 2\pi$. Then the cross product is

$$\begin{aligned}\mathbf{u} \times \mathbf{v} &= s\,\mathbf{i} \times r(\cos\theta\,\mathbf{i} + \sin\theta\,\mathbf{j})\\ &= sr\cos\theta\,\mathbf{i} \times \mathbf{i} + sr\sin\theta\,\mathbf{i} \times \mathbf{j}\\ &= sr\sin\theta\,\mathbf{k}.\end{aligned}$$

Because s and r are both positive, we see that $\mathbf{u} \times \mathbf{v}$ is pointing along the positive z-axis when $\sin\theta > 0$ and along the negative z-axis when $\sin\theta < 0$. In either case the length can be written as

$$|\mathbf{u} \times \mathbf{v}| = |\mathbf{u}|\,|\mathbf{v}|\,|\sin\theta|. \qquad \textbf{(12.11)}$$

Now suppose $\mathbf{u}$ and $\mathbf{v}$ are any two vectors in three-dimensional space. Place the two vectors with initial points at the origin. Rotate the coordinate space so that $\mathbf{u}$ points along the positive x-axis and the other vector, $\mathbf{v}$, is in the xy plane. The rotated vectors have the same lengths as before rotation. The angle between the vectors is the same after rotation as it was before rotation. The cross product of the rotated vectors has the same length as the cross product of the two original vectors but now formula (12.11) can be applied to express the length of the cross product in terms of the lengths of the original vectors and the angle between them. Moreover the direction of the cross product vector is determined by the direction of the cross product of the rotated vectors. It follows that Equation (12.11) holds for all vectors in three-dimensional space.

If the idea of rotating the coordinate system is difficult to visualize, there is a slightly different view that might help. If you want to determine the direction of $\mathbf{u} \times \mathbf{v}$, imagine a right-handed coordinate system with the origin at the common initial point of $\mathbf{u}$ and $\mathbf{v}$ arranged so that $\mathbf{u}$ points along the positive x-axis and $\mathbf{v}$ is in the xy plane. If the angle from $\mathbf{u}$ to $\mathbf{v}$ is between 0 and π, the cross product $\mathbf{u} \times \mathbf{v}$ points along the positive z-axis . If the angle from $\mathbf{u}$ to $\mathbf{v}$ is between π and 2π, the cross product points along the negative z-axis.

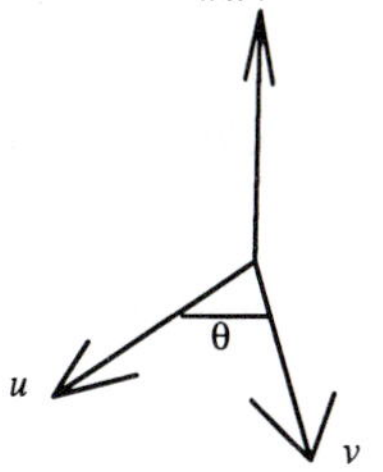

Here is a moving picture to help visualize the cross product. Let $\mathbf{u} = \mathbf{i}$ and $\mathbf{v} = \cos\theta\,\mathbf{i} + \sin\theta\,\mathbf{j}$. The cross product

$$\mathbf{u} \times \mathbf{v} = \sin\theta\,\mathbf{k}$$

is a function of θ. When $\theta = 0$, $\mathbf{u}$ and $\mathbf{v}$ are parallel and their cross product is $\mathbf{0}$. As θ increases, the product points along the positive z-axis and grows longer until $\theta = \pi/2$ and then the product vector grows shorter until $\theta = \pi$ when the product is $\mathbf{0}$ again. As θ continues past π the cross product vector points along the negative z-axis growing from length 0 to length 1 at $\theta = 3\pi/2$ and the decreasing to 0 again at $\theta = 2\pi$.

u x v
θ
u
v

FIGURE 12.32

Area of a Parallelogram

The formula for the length of the cross product has a nice geometric application.

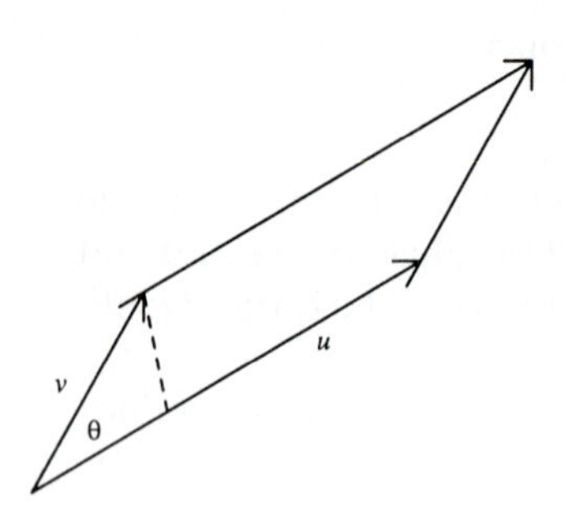

FIGURE 12.33

Consider the parallelogram having $\mathbf{u}$ and $\mathbf{v}$ as adjacent sides and let θ be the angle from $\mathbf{u}$ to $\mathbf{v}$ with $0 \leq \theta \leq \pi$. The perpendicular line from $\mathbf{u}$ to the opposite parallel side has length $|\mathbf{v}| \sin\theta$ and so the area of the parallelogram is $|\mathbf{u}||\mathbf{v}| \sin\theta$ by the formula $area = base \quad times \quad height$. It follows then that the area of the parallelogram with sides $\mathbf{u}$ and $\mathbf{v}$ is the magnitude of the cross product $|\mathbf{u} \times \mathbf{v}|$. For any two vectors where the angle from $\mathbf{u}$ to $\mathbf{v}$ might be between π and 2π, we obtain the same formula, namely,

$$area = |\mathbf{u} \times \mathbf{v}|.$$

EXAMPLE 2 Find the area of the parallelogram in the xy plane having vertices $(0, 0)$, $(2, 3)$, $(4, -1)$ and $(6, 2)$.

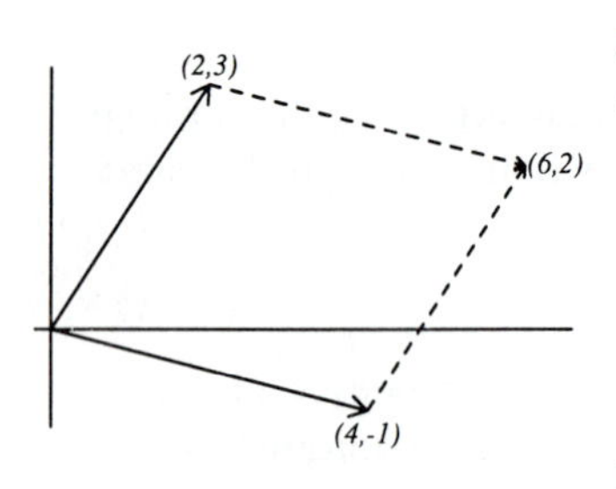

FIGURE 12.34

Solution: Let $\mathbf{u} = [2, 3] = 2\mathbf{i} + 3\mathbf{j}$ and $\mathbf{v} = [4, -1] = 4\mathbf{i} - \mathbf{j}$. Then $\mathbf{u}$ and $\mathbf{v}$ are sides of the parallelogram and the area equals

$$\begin{aligned} |\mathbf{u} \times \mathbf{v}| &= |(2\mathbf{i} + 3\mathbf{j}) \times (4\mathbf{i} - \mathbf{j})| \\ &= |8\mathbf{i} \times \mathbf{i} - 2\mathbf{i} \times \mathbf{j} + 12\mathbf{j} \times \mathbf{i} - 3\mathbf{j} \times \mathbf{j}| \\ &= |-2\mathbf{k} - 12\mathbf{k}| \\ &= 14. \end{aligned}$$

It is worth a remark that the determinant

$$\begin{vmatrix} 2 & 3 \\ 4 & -1 \end{vmatrix} = -14$$

is the negative of the area of the parallelogram. This is an example of a general phenomenon. ■

In general, the parallelogram with vertices $(0, 0)$, (a, b), (c, d) and $(a+c, b+d)$ has area equal to the absolute value of the determinant

$$\begin{vmatrix} a & b \\ c & d \end{vmatrix}.$$

A proof based on example 2 is left to the exercises at the end of the section.

EXAMPLE 3 Find the area of the triangle having the vertices $P(1, 1, 2)$, $Q(0, -1, -1)$ and $R(1, 1, 1)$.

Solution: The area of the triangle is one-half the area of the parallelogram having adjacent sides $\overrightarrow{PQ}$ and $\overrightarrow{PR}$. The area of the parallelogram is the absolute value of the cross product $\overrightarrow{PQ} \times \overrightarrow{PR}$. To compute this we first write these vectors in coordinate form:

$$\overrightarrow{PQ} = \overrightarrow{OQ} - \overrightarrow{OP} = [0, -1, -1] - [1, 1, 2] = [-1 - 2, -3],$$

$$\overrightarrow{PR} = \overrightarrow{OR} - \overrightarrow{OP} = [1, 1, 1] - [1, 1, 2] = [0, 0, -1].$$

The area of the triangle PQR is one half of

$$\begin{aligned} |\overrightarrow{PQ} \times \overrightarrow{PR}| &= |[-1, -2, -3] \times [0, 0, -1]| \\ &= |(-\mathbf{i} - 2\mathbf{j} - 3\mathbf{k}) \times (-\mathbf{k})| \\ &= |(-\mathbf{j} - 2\mathbf{i})| \\ &= \sqrt{5}. \end{aligned}$$

The area of triangle PQR is $\sqrt{5}/2$. Generalizations of this area computation can be found in the exercises. ■

Volume of a Parallelepiped

There is a three-dimensional analogue of the parallelogram area computation. A parallelepiped is a three-dimensional solid bounded by three pairs of parallel planes. Every cross section parallel to the base is a parallelogram congruent to a side. The volume of a parallelepiped is the area of the base times the height, the height being measured along a line perpendicular to the base.

A parallelepiped can be specified by giving vectors representing three sides having a common vertex. Suppose the three sides are **u**, **v** and **w**. We have to select one side of the parallelepiped to be called the base, so let it be the side whose edges are **u** and **v**. Then the area of the base is the absolute value of the cross product $\mathbf{u} \times \mathbf{v}$. The volume of the parallelepiped can be computed if we know the height. The height is the distance from the base up to the terminal point of the vector **w**. This height is the projection of **w** onto a unit vector perpendicular to the base. Fortunately we know a vector perpendicular to the base, namely $\mathbf{u} \times \mathbf{v}$ so the height of the parallelepiped is the dot product of **w** with a unit vector parallel to $\mathbf{u} \times \mathbf{v}$:

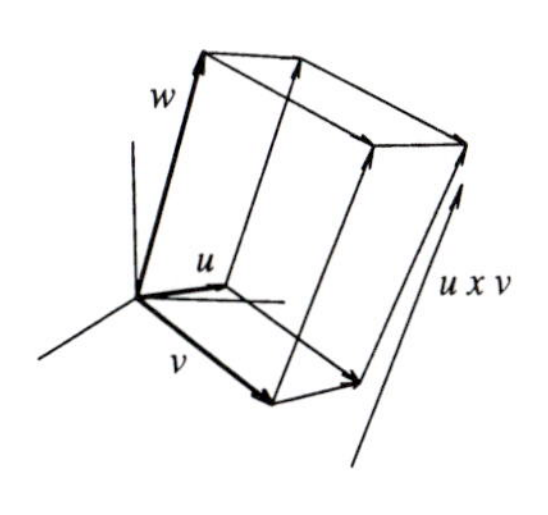

FIGURE 12.35

$$\text{height} = \pm\mathbf{w} \cdot \frac{\mathbf{u} \times \mathbf{v}}{|\mathbf{u} \times \mathbf{v}|}.$$

The ambiguous sign is present because we do not know in general if the cross product points up or down from the base; choose the sign to make the height positive. The volume of the parallelepiped is the height times the area of the base $|\mathbf{u} \times \mathbf{v}|$. So the volume formula is

$$\begin{aligned}\text{volume} &= |\mathbf{u} \times \mathbf{v}| \left(\pm\mathbf{w} \cdot \frac{\mathbf{u} \times \mathbf{v}}{|\mathbf{u} \times \mathbf{v}|}\right) \\ &= \pm\mathbf{w} \cdot (\mathbf{u} \times \mathbf{v}).\end{aligned}$$

The expression $\mathbf{w} \cdot (\mathbf{u} \times \mathbf{v})$ can be written without the parenthesis as $\mathbf{w} \cdot \mathbf{u} \times \mathbf{v}$ without introducing any ambiguity. The product makes sense in only one grouping because $(\mathbf{w} \cdot \mathbf{u}) \times \mathbf{v}$ would be the cross product of a scalar $\mathbf{w} \cdot \mathbf{u}$ and a vector **v**, which is meaningless. The product $\mathbf{w} \cdot \mathbf{u} \times \mathbf{v}$ is called the *scalar triple product* of the three vectors. The value of the product is a scalar quantity.

The geometric interpretation of the scalar triple product as a volume permits us to write down some identities at once. In the volume computation we selected one side of the parallelepiped to be the base. Of course any of the sides could be called the base and so all of the following quantities equal the volume:

$$\begin{aligned}\text{volume} &= |\mathbf{w} \cdot \mathbf{u} \times \mathbf{v}| = |\mathbf{u} \cdot \mathbf{v} \times \mathbf{w}| = |\mathbf{v} \cdot \mathbf{w} \times \mathbf{u}| \\ &= |\mathbf{w} \cdot \mathbf{v} \times \mathbf{u}| = |\mathbf{u} \cdot \mathbf{w} \times \mathbf{v}| = |\mathbf{v} \cdot \mathbf{u} \times \mathbf{w}|.\end{aligned}$$

EXAMPLE 4 Compute the volume of the parallelepiped with vertices $(0, 0, 0)$, $(1, 1, 0)$, $(1, 0, 1)$, $(0, 1, 1)$, $(2, 1, 1)$, $(1, 2, 1)$, $(1, 1, 2)$ and $(2, 2, 2)$.

Solution: Let $\mathbf{u} = [1, 1, 0]$, $\mathbf{v} = [1, 0, 1]$ and $\mathbf{w} = [0, 1, 1]$ so these are three sides of the parallelepiped with terminal points at three of the vertices. The origin is another vertex and the remaining four vertices are the terminal points of $\mathbf{u} + \mathbf{v}$, $\mathbf{w} + \mathbf{u}$, $\mathbf{w} + \mathbf{v}$ and $\mathbf{w} + \mathbf{v} + \mathbf{u}$. The volume of the parallelepiped is the absolute value of the scalar triple product

$$\begin{aligned}\mathbf{u}\cdot\mathbf{v}\times\mathbf{w} &= (\mathbf{i}+\mathbf{j})\cdot(\mathbf{i}+\mathbf{k})\times(\mathbf{j}+\mathbf{k})\\ &= (\mathbf{i}+\mathbf{j})\cdot(\mathbf{k}-\mathbf{i}-\mathbf{j})\\ &= -2.\end{aligned}$$

The volume is 2.

Just to illustrate the identities, we compute the scalar triple product in a different order and get

$$\begin{aligned}\mathbf{w}\cdot\mathbf{v}\times\mathbf{u} &= (\mathbf{j}+\mathbf{k})\cdot(\mathbf{i}+\mathbf{k})\times(\mathbf{i}+\mathbf{j})\\ &= (\mathbf{j}+\mathbf{k})\cdot(\mathbf{j}+\mathbf{k}-\mathbf{i})\\ &= 2.\end{aligned}$$

■

The scalar triple product can be expressed as a 3×3 determinant. Let P be the parallelepiped formed with three sides given by the vectors

$$\begin{aligned}\mathbf{u} &= u_1\,\mathbf{i}+u_2\,\mathbf{j}+u_3\,\mathbf{k},\\ \mathbf{v} &= v_1\,\mathbf{i}+v_2\,\mathbf{j}+v_3\,\mathbf{k},\\ \mathbf{w} &= w_1\,\mathbf{i}+w_2\,\mathbf{j}+w_3\,\mathbf{k}.\end{aligned}$$

Then the volume of P is the absolute value of the determinant

$$\begin{vmatrix} u_1 & u_2 & u_3\\ v_1 & v_2 & v_3\\ w_1 & w_2 & w_3\end{vmatrix}. \qquad \textbf{(12.12)}$$

To see why this holds we evaluate the scalar triple product using the determinant computation for the cross product. We have

$$\begin{aligned}\mathbf{u}\cdot\mathbf{v}\times\mathbf{w} &= (u_1\,\mathbf{i}+u_2\,\mathbf{j}+u_3\,\mathbf{k})\cdot\begin{vmatrix}\mathbf{i} & \mathbf{j} & \mathbf{k}\\ v_1 & v_2 & v_3\\ w_1 & w_2 & w_3\end{vmatrix}\\ &= (u_1\,\mathbf{i}+u_2\,\mathbf{j}+u_3\,\mathbf{k})\cdot\left[\begin{vmatrix} v_2 & v_3\\ w_2 & w_3\end{vmatrix}\mathbf{i}-\begin{vmatrix} v_1 & v_3\\ w_1 & w_3\end{vmatrix}\mathbf{j}+\begin{vmatrix} v_1 & v_2\\ w_1 & w_2\end{vmatrix}\mathbf{k}\right]\\ &= \begin{vmatrix} v_2 & v_3\\ w_2 & w_3\end{vmatrix}u_1-\begin{vmatrix} v_1 & v_3\\ w_1 & w_3\end{vmatrix}u_2+\begin{vmatrix} v_1 & v_2\\ w_1 & w_2\end{vmatrix}u_3.\end{aligned}$$

From the definition of a 3×3 determinant given in Equation (12.10) we see that the last line of the computation is exactly the definition of the determinant (12.12). The volume is the absolute value of the scalar triple product which equals the absolute value of the 3×3 determinant.

The volume of the parallelepiped described in the example 4 was found to be 2. We recompute it as a determinant. The volume is the absolute value of the determinant

$$\begin{aligned}\begin{vmatrix} 1 & 1 & 0\\ 1 & 0 & 1\\ 0 & 1 & 1\end{vmatrix} &= \begin{vmatrix} 0 & 1\\ 1 & 1\end{vmatrix}(1)-\begin{vmatrix} 1 & 1\\ 0 & 1\end{vmatrix}(1)+\begin{vmatrix} 1 & 0\\ 0 & 1\end{vmatrix}(0)\\ &= (-1)(1)-(1)(1) = -2.\end{aligned}$$

EXAMPLE 5 Use the cross product and the dot product to determine if the three vectors $\mathbf{u} = \mathbf{i}+2\mathbf{j}-\mathbf{k}$, $\mathbf{v} = -2\mathbf{j}+3\mathbf{k}$ and $\mathbf{w} = 2\mathbf{i}+2\mathbf{j}+\mathbf{k}$ lie in a plane.

Solution: If three vectors $\mathbf{u}$, $\mathbf{v}$ and $\mathbf{w}$ do not lie in a plane, then they determine a parallelepiped having nonzero volume and so $\mathbf{u} \cdot \mathbf{v} \times \mathbf{w} \neq 0$. If the three vectors do lie in a plane, then the determine a parallelepiped that has 0 volume so the scalar triple product is 0. So we compute the scalar triple product to answer the question:

$$\mathbf{u} \cdot \mathbf{v} \times \mathbf{w} = \begin{vmatrix} 1 & 2 & -1 \\ 0 & -2 & 3 \\ 2 & 2 & 1 \end{vmatrix}$$
$$= (-8)(1) - (-6)(2) + (4)(-1)$$
$$= 0.$$

Thus the vectors do lie in a plane. ∎

There is another way to reason that the scalar triple product is zero if and only if the three vectors lie in a plane. The equation $\mathbf{u} \cdot \mathbf{v} \times \mathbf{w} = 0$ means $\mathbf{u}$ is orthogonal to $\mathbf{v} \times \mathbf{w}$. The cross product $\mathbf{v} \times \mathbf{w}$ is orthogonal to both $\mathbf{v}$ and $\mathbf{w}$ so the three vectors $\mathbf{u}$, $\mathbf{v}$ and $\mathbf{w}$ are all orthogonal to the same vector, $\mathbf{v} \times \mathbf{w}$, and they must lie in a plane. Hence $\mathbf{u}$ is in the plane containing $\mathbf{v}$ and $\mathbf{w}$.

Exercises 12.5

Expand and simplify the expressions.

1. $\mathbf{i} \times (\mathbf{j} - \mathbf{k})$

2. $\mathbf{j} \times (\mathbf{i} + \mathbf{j} + \mathbf{k})$

3. $(\mathbf{i} + 2\mathbf{j}) \times (\mathbf{j} - \mathbf{k})$

4. $(2\mathbf{i} + 2\mathbf{j} + \mathbf{k}) \times (\mathbf{j} + 3\mathbf{k})$

5. $\mathbf{i} \times [(\mathbf{j} + \mathbf{k}) \times (\mathbf{j} - \mathbf{k})]$

6. $(\mathbf{i} + \mathbf{j}) \times (\mathbf{i} - \mathbf{j})$

7. $\mathbf{k} \cdot (2\mathbf{i} + \mathbf{j}) \times (\mathbf{j} - \mathbf{k})$

8. $\mathbf{i} \times [(\mathbf{i} + \mathbf{k}) \times (\mathbf{i} - \mathbf{k})]$

9. $(\mathbf{j} + \mathbf{k}) \cdot (\mathbf{i} - \mathbf{j}) \times (\mathbf{j} - \mathbf{k})$

10. $(\mathbf{i} - \mathbf{k}) \cdot (\mathbf{i} - 3\mathbf{j}) \times (\mathbf{i} - 2\mathbf{j})$

11. $\mathbf{i} \cdot (2\mathbf{i} + 3\mathbf{j} + 3\mathbf{k}) \times (\mathbf{i} - 3\mathbf{j} - 3\mathbf{k})$

12. $(-\mathbf{i} + \mathbf{j} - 2\mathbf{k}) \times (4\mathbf{i} + \mathbf{j} - \mathbf{k})$

13. $\mathbf{j} \times [(\mathbf{i} + 2\mathbf{k}) \times (3\mathbf{j} - 2\mathbf{k})]$

14. $\mathbf{k} \times [(-2\mathbf{i} + \mathbf{k}) \times (\mathbf{i} + \mathbf{k})]$

Use cross products to compute the areas of the figures.

15. The parallelogram with vertices $(0,0)$, $(1,2)$, $(2,-3)$ and $(3,-1)$

16. The parallelogram with vertices $(1,3)$, $(1,2)$, $(2,-3)$ and $(1,-2)$

17. The triangle with vertices $(0,0)$, $(1,2)$ and $(2,-3)$

18. The triangle with vertices $(0,0)$, $(4,2)$ and $(2,4)$

19. The triangle with vertices $(1,6)$, $(1,2)$ and $(2,-3)$

20. The triangle with vertices $(2,-2)$, $(-1,3)$ and $(0,3)$

Find the volume of the parallelepiped with the given vectors representing three adjacent sides.

21. $\mathbf{u} = \mathbf{i} + \mathbf{j}$, $\mathbf{v} = \mathbf{j} + \mathbf{k}$, $\mathbf{w} = \mathbf{i} + \mathbf{k}$

22. $\mathbf{u} = 2\mathbf{i} - \mathbf{j}$, $\mathbf{v} = \mathbf{i} + \mathbf{j} - 2\mathbf{k}$, $\mathbf{w} = -\mathbf{i} + \mathbf{k}$

23. $\mathbf{u} = \mathbf{i} + \mathbf{j} + \mathbf{k}$, $\mathbf{v} = -\mathbf{i} + \mathbf{j} + \mathbf{k}$, $\mathbf{w} = \mathbf{i} + \mathbf{k}$

24. $\mathbf{u} = \mathbf{i} - \mathbf{j} - 2\mathbf{k}$, $\mathbf{v} = 2\mathbf{i} - \mathbf{j} + \mathbf{k}$, $\mathbf{w} = \mathbf{i} - 3\mathbf{j} + \mathbf{k}$

25. List the eight vertices of the parallelepiped that has one vertex at $(1,2,3)$ and three sides emanating from that vertex have ends at $(2,3,8)$, $(3,5,5)$ and $(5,6,7)$. Also find the volume of the parallelepiped.

26. Find a unit vector orthogonal to both $\mathbf{u} = \mathbf{i} + \mathbf{j}$ and $\mathbf{v} = -\mathbf{i} + 2\mathbf{j} + \mathbf{k}$. Then find the perpendicular distance from the point of $(2,-1,2)$ to the plane containing $\mathbf{u}$ and $\mathbf{v}$.

27. A parallelepiped has two sides $\mathbf{u} = \mathbf{i} + \mathbf{j}$ and $\mathbf{v} = 2\mathbf{i} - \mathbf{j} + \mathbf{k}$ emanating from the origin. The third side from the origin is a vector $\mathbf{w}$ of length 2. What is the maximum possible volume of the parallelepiped?

28. Show by example that the cross product is not associative; that is, for some vectors we have $(\mathbf{u} \times \mathbf{v}) \times \mathbf{w} \neq \mathbf{u} \times (\mathbf{v} \times \mathbf{w})$.

29. Let $\mathbf{u}$ be a nonzero vector and $\mathbf{v}$ a vector for which $\mathbf{u} \times \mathbf{v} = \mathbf{0}$. What can you conclude about $\mathbf{v}$?

30. Prove the identity $\mathbf{u} \times (\mathbf{v} \times \mathbf{w}) = (\mathbf{u} \cdot \mathbf{w})\mathbf{v} - (\mathbf{u} \cdot \mathbf{v})\mathbf{w}$. [Hint: This can be done by writing all vectors in coordinate form. The rotation of the coordinate system so that one vector lies along the positive x-axis simplifies some of the computations.]

31. Let $\mathbf{u} \cdot \mathbf{v} = 0$ so that $\mathbf{u}$ is orthogonal to $\mathbf{v}$. Then $\mathbf{u} \times \mathbf{v}$ is orthogonal to both $\mathbf{u}$ and $\mathbf{v}$ and so $\mathbf{v} \times (\mathbf{u} \times \mathbf{v})$ is a scalar multiple of $\mathbf{u}$. Find the scalar.

32. Compute the area A_1 of the parallelogram with two sides given by the vectors $\mathbf{u} = \mathbf{i} + 2\mathbf{j} - \mathbf{k}$ and $\mathbf{w} = 3\mathbf{i} + \mathbf{j} + 2\mathbf{k}$. Then compute the area A_2 of the projection of this parallelogram into the xy plane. The projection is the parallelogram with sides $\mathbf{u}' = \mathbf{i} + 2\mathbf{j}$ and $\mathbf{w}' = 3\mathbf{i} + \mathbf{j}$. Compare the ratio of the areas A_2/A_1 with the cosine of the angle between the xy plane and the plane containing $\mathbf{u}$ and $\mathbf{w}$.

33. Use cross products to prove that the parallelogram with vertices at $(0, 0)$, (a, b), (c, d) and $(a + c, b + d)$ has area equal to the absolute value of the determinant

$$\begin{vmatrix} a & b \\ c & d \end{vmatrix}.$$

34. Find a determinant whose absolute value gives the area of the parallelogram three of whose vertices are (a, b), (c, d) and (e, f).

12.6 Equations of lines

We use vectors to describe lines in three dimensions. By a description of the line we mean a mathematical statement from which we can determine precisely which points lie on the line. We introduce the parametric form of lines that is easily obtained from the vector description. More generally the parametric form of a curve in three dimensions is the set of points (x, y, z) with the coordinates determined by equations

$$\begin{cases} x = f(t), \\ y = g(t), \\ z = h(t), \end{cases}$$

where f, g and h are functions of a variable t called the parameter.

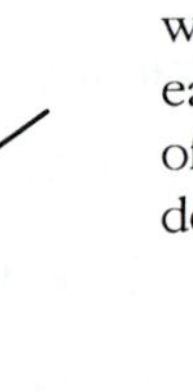

FIGURE 12.36

Consider first the simplest case. Let $P(a, b, c)$ be a point other than the origin $O(0, 0, 0)$. Then there is a unique line L passing through O and P. How can we tell which points are on this line? The vector $\mathbf{v} = \overrightarrow{OP} = [a, b, c]$ having initial point O and terminal point P lies in the line L. For any real number r, the multiple $r\mathbf{v}$ of $\mathbf{v}$ is parallel to $\mathbf{v}$ and if we regard it as having O as initial point, then its terminal point is on L. Conversely if Q is any point on L then there is some real number s such that the length of $s\mathbf{v}$ equals the distance from O to Q. Thus the vector having initial point O and terminal point Q must be $\pm s\mathbf{v}$. We draw the conclusion that the points on L are the terminal points of the vectors $t\mathbf{v}$ as t ranges over all real numbers. A point (x, y, z) lies on L if and only if

$$[x, y, z] = t\mathbf{v} = t[a, b, c] = [ta, tb, tc]$$

for some t. The vector equation $\mathbf{r}(t) = t\mathbf{v} = t[a, b, c]$ represents the line through the origin and (a, b, c).

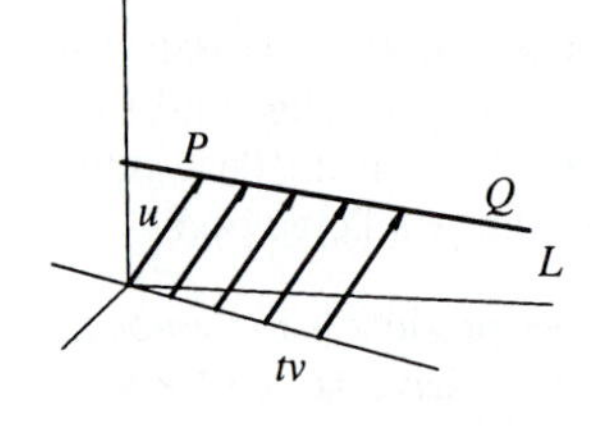

FIGURE 12.37

Now consider a more general line; let $P(a, b, c)$ and $Q(a', b', c')$ be two different points and let L be the line passing through them. The vector $\mathbf{v} = \overrightarrow{PQ}$ is parallel to L. The coordinate form of $\mathbf{v}$ is

$$\mathbf{v} = \overrightarrow{PQ} = \overrightarrow{OQ} - \overrightarrow{OP} = [a', b', c'] - [a, b, c] = [a' - a, b' - b, c' - c].$$

The terminal points of all scalar multiples of $\mathbf{v}$ form a line through the origin parallel to L. To get the line L itself we add the vector $\overrightarrow{OP}$ to each vector $t\mathbf{v}$. The set of terminal points of the vectors $\mathbf{r}(t) = \overrightarrow{OP} + t\mathbf{v}$ form a line parallel to

L and passing through P. Note that $\mathbf{r}(0) = \overrightarrow{OP}$ and

$$\mathbf{r}(1) = \overrightarrow{OP} + 1\mathbf{v} = \overrightarrow{OP} + (\overrightarrow{OQ} - \overrightarrow{OP}) = \overrightarrow{OQ}.$$

Thus the line with vector representation $\mathbf{r}(t) = \overrightarrow{OP} + t\mathbf{v}$ is the line L through P and Q.

Here is a summary of the above discussion.

THEOREM 12.4

Vector Form of a Line

The line passing through two points P and Q consists of the terminal points of all vectors

$$\mathbf{r}(t) = \overrightarrow{OP} + t\overrightarrow{PQ},$$

where t runs through all real numbers.

The line L has many different vector representations. If $\mathbf{u} + t\mathbf{v}$ is one representation of L, then $\mathbf{v}$ can be replaced by any vector parallel to $\mathbf{v}$, like $-3\mathbf{v}$ for example. Thus $\mathbf{u} - 3t\mathbf{v}$ also represents the line L. Moreover $\mathbf{u}$ is a vector from the origin to a point on the line. Thus $\mathbf{u}$ can be replaced with any vector $\mathbf{u}'$ from the origin to a point on L, like $\mathbf{u}' = \mathbf{u} + 4\mathbf{v}$. Then $\mathbf{u}' + t\mathbf{v}$ also is a vector form for the line L.

EXAMPLE 1 Find the line passing through $(0, 1, 2)$ that is parallel to the line through $P(1, 0, 2)$ and $Q(2, 2, 1)$. Give three different representations of the same line.

Solution: The line parallel to the line through P and Q and passing through the origin has the vector form

$$t\mathbf{v} = t([1, 0, 2] - [2, 2, 1]) = [-t, -2t, t].$$

Add the vector $[0, 1, 2]$ to every point of this line to get the parallel line passing through $(0, 1, 2)$. The vector form of the line is

$$\mathbf{r}(t) = [0, 1, 2] + [-t, -2t, t] = [-t, 1 - 2t, 2 + t].$$

To give some alternative forms of the same line we could replace $[0, 1, 2]$ by any point on the line. Let us use $\mathbf{r}(-2) = [2, 5, 0]$ so that $(2, 5, 0)$ is a point on the line. Then

$$\mathbf{r}_1(t) = [2, 5, 0] + t[-1, -2, 1] = [2 - t, 5 - 2t, t]$$

is another equation of the line.

The vector $\mathbf{v}$ is parallel to the line and can be replaced by any other vector parallel to it. We arbitrarily select the parallel vector $\mathbf{v}' = -6\mathbf{v} = [6, 12, -6]$. Using the point $(0, 1, 2)$ on the line and calling the parameter s in place of t (just for variety) gives us the equation

$$\mathbf{r}_2(s) = [0, 1, 2] + s[6, 12, -6] = [6s, 1 + 12s, 2 - 6s]$$

that represents the line. Of course many other possible representations give the same line. ■

Parametric Form of Lines

The parametric equations of a line are scalar equations giving conditions on the coordinates of points so that the points satisfying the conditions are the points on the line. We derive the parametric form of the lines from the vector form.

A line through the origin and the point (a, b, c) has vector form $\mathbf{r}(t) = t[a, b, c]$. Thus a point (x, y, z) is on the line if and only if there is a number t such that $[x, y, z] = t[a, b, c]$. The parametric equations of the line passing through the origin and the point (a, b, c) are

$$\begin{cases} x = ta, \\ y = tb, \\ z = tc, \end{cases}$$

with the parameter t ranging over all real numbers.

Now consider the more general line through two points. The parametric form of the line can be obtained from the vector form.

A point (x, y, z) is on the line through $P(a, b, c)$ and $Q(a', b', c')$ if and only if there is some real number t such that

$$[x, y, z] = [a, b, c] + t[a' - a, b' - b, c' - c] = [a + t(a' - a), b + t(b' - b), c + t(c' - c)].$$

The parametric form of the line through $P(a, b, c)$ and $Q(a', b', c')$ is

$$\begin{cases} x = a + t(a' - a), \\ y = b + t(b' - b), \\ z = c + t(c' - c). \end{cases}$$

We change the notation slightly: let $[p, q, r]$ be any vector parallel to the line; for example let $[p, q, r] = [a' - a, b' - b, c' - c]$. Then the parametric form of this line is given as follows.

THEOREM 12.5

Parametric Form of a Line

The line parallel to the vector $[p, q, r]$ and passing through (a, b, c) is the set of all (x, y, z) given parametrically by

$$\begin{cases} x = a + tp, \\ y = b + tq, \\ z = c + tr. \end{cases}$$

If we wish to eliminate the parameter, the line can be written in *symmetric form* as

$$\frac{x - a}{p} = \frac{y - b}{q} = \frac{z - c}{r}$$

because each fraction equals t. This assumes $pqr \neq 0$. If one or more of these constants is zero, then we must alter the definition of the symmetric form. For example suppose $p = 0$ but $qr \neq 0$. Then the symmetric form of the line would be

$$x = a, \quad \frac{y - b}{q} = \frac{z - c}{r}.$$

Similar adjustments must be made if one of the other constants is zero.

EXAMPLE 2 Find the vector form, the parametric form and the symmetric form of the line passing through the points $(1, 3, -2)$ and $(2, 5, -1)$. Use this information to find the point where the line intersects the yz plane.

***Solution*:** The line through the points is parallel to the vector

$$\mathbf{v} = [1, 3, -2] - [2, 5, -1] = [-1, -2, -1].$$

The line through the origin parallel to the line we are looking for has vector form $t\mathbf{v} = [-t, -2t, -t]$. The line through the two given points has vector form

$$[1, 3, -2] + t[-1, -2, -1] = [1 - t, 3 - 2t, -2 - t].$$

The parametric form of this line is

$$\begin{cases} x = 1 - t, \\ y = 3 - 2t, \\ z = -2 - t. \end{cases}$$

The symmetric form is found by solving each equation for t to obtain

$$1 - x = \frac{3 - y}{2} = -2 - z \quad (= t).$$

To find a point on the line that is in the yz plane we look for the point on the line having x coordinate 0. We first find the value of the parameter t for which $x = 0$. Because $x = 1 - t$, we have $x = 0$ when $t = 1$. With $t = 1$ we find $y = 3 - 2t = 1$, and $z = -2 - t = -3$. Thus $(0, 1, -3)$ is on the line through $(1, 3, -2)$ and $(2, 5, -1)$. ■

Parallel and Intersecting Lines

In the xy plane, two lines are either parallel or they have a point of intersection. However in three dimensional space there is a third possibility: lines might be *skewed*, that is neither intersecting nor parallel. If two lines are represented in vector form as $L_1 : \mathbf{u} + t\mathbf{v}$ and $L_2 : \mathbf{u}' + s\mathbf{v}'$ for parameters t and s, then the lines are parallel only if the vectors $\mathbf{v}$ and $\mathbf{v}'$ are parallel. In other words, L_1 is parallel to L_2 if and only if $r\mathbf{v} = \mathbf{v}'$ for some scalar r.

The two lines have a point of intersection if there are values of t and s such that $\mathbf{u} + t\mathbf{v} = \mathbf{u}' + s\mathbf{v}'$. (There will be such an s and t if and only if the vector $\mathbf{u} - \mathbf{u}'$ lies in the plane containing the nonparallel vectors $\mathbf{v}$ and $\mathbf{v}'$. When $\mathbf{v}$ and $\mathbf{v}'$ are parallel, there is a common point only when the lines are equal.)

If lines are written are written in parametric form we have

$$L_1 : \begin{cases} x = a + tp \\ y = b + tq \\ z = c + tr, \end{cases} \qquad L_2 : \begin{cases} x = a' + sp' \\ y = b' + sq' \\ z = c' + sr'. \end{cases}$$

A common point on the lines L_1 and L_2 is found by finding the values of t and s (if they exist) to satisfy the simultaneous equations

$$\begin{aligned} x &= a + tp = a' + sp', \\ y &= b + tq = b' + sq', \\ z &= c + tr = c' + sr'. \end{aligned}$$

EXAMPLE 3 Determine if the line through $P(0, 1, -2)$ and $Q(2, -1, -3)$ intersects the line through the origin and the point $R(4, -2, a)$ for each number a.

***Solution*:** The line through P and Q has vector form

$$\mathbf{r}(t) = [0, 1, -2] + t[2, -2, -1].$$

The line through the origin and R has the vector form

$$\mathbf{r}_1(s) = s[4, -2, a].$$

If there is a point of intersection of the two lines then it occurs where s and t satisfy

$$[2t, 1 - 2t, -2 - t] = [4s, -2s, as].$$

The equality of the x coordinates implies $2t = 4s$ or $t = 2s$. From the equality of the y coordinates we deduce $1 - 2t = -2s$. Because $t = 2s$ we may substitute and get $1 - 2t = 1 - 4s = -2s$. It follows that $s = 1/2$ and $t = 2s = 1$. Note that we have not yet used the third coordinate. If there is an intersection point of the lines, the z coordinates must match using the values of t and s already found. We have $-2 - t = -2 - 1 = -3$ and $as = a(1/2) = a/2$. Because the third coordinates must be equal at the point of intersection, $-3 = a/2$ or $a = -6$. Thus the line through $(0, 0, 0)$ and $(4, -2, -6)$ meets the line through P and Q at the point $(2, -1, -3)$. For any other choice of a, the lines do not intersect. ■

REMARK: In example 3 and the discussion before it, we considered two lines represented parametrically. It is usually clearer to use different letters to denote the parameters. For example, given a pair of lines $\mathbf{r}_1(t)$ and $\mathbf{r}_2(t)$, using the same parameter can lead to some confusion. To find where the two lines intersect, it is not enough to try and solve $\mathbf{r}_1(t) = \mathbf{r}_2(t)$ for a single choice of t. It is necessary to solve $\mathbf{r}_1(t) = \mathbf{r}_2(s)$ for some t and s.

Distance from a Point to a Line

Here is an example of a problem that shows the difference between vector methods and analytic methods. Suppose we are given a line $\mathbf{r}(t) = \mathbf{u} + t\mathbf{v}$ in vector form and a point $P(a, b, c)$ and we are asked to find the (shortest) distance from P to a point on the line. For the nonvector method, we write the coordinates of the points on the line

$$[x, y, z] = \mathbf{u} + t\mathbf{v} = [u_1 + tv_1, u_2 + tv_2, u_3 + tv_3]$$

and compute the distance (really the square of the distance) from the point to P to (x, y, z):

$$d(t) = (a - u_1 - tv_1)^2 + (b - u_2 - tv_2)^2 + (c - u_3 - tv_3)^2.$$

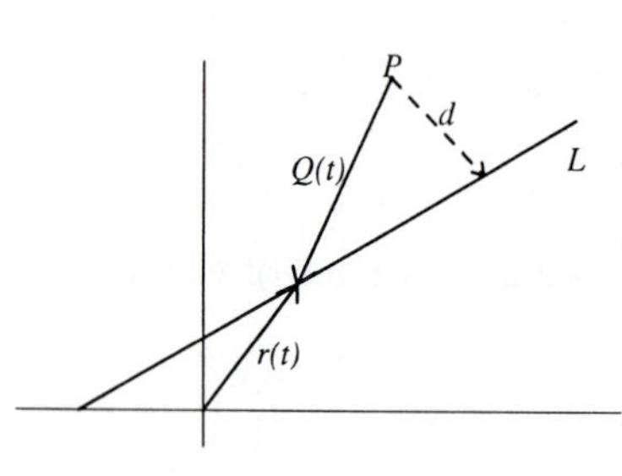

FIGURE 12.38

Then we find the minimum value of the distance function as a function of t. The minimum occurs at a point where the derivative $d'(t) = 0$. The solution is

$$t = \frac{(a - u_1)v_1 + (b - u_2)v_2 + (c - u_3)v_3}{v_1^2 + v_2^2 + v_3^2}. \tag{12.13}$$

We have the value of t giving the nearest point on the line to P. We use it to find the point on the line and then compute its distance from P.

Now we give the vector solution to the same problem. Denote by $\mathbf{Q}(t)$ the vector from P to the general point $\mathbf{r}(t) = \mathbf{u} + t\mathbf{v}$ on the line. Thus $\mathbf{Q}(t) = \mathbf{u} + t\mathbf{v} - [a, b, c]$. The vector $\mathbf{Q}(t)$ has minimum length when it is perpendicular to the line because we know that the shortest distance from a point to a line is measured along a perpendicular to the line. The line is parallel to the vector $\mathbf{v}$ so we require a value of t so that $\mathbf{Q}(t)$ and $\mathbf{v}$ are orthogonal. They are orthogonal if their dot product is 0 so we must solve

$$0 = \mathbf{Q}(t) \cdot \mathbf{v} = \mathbf{u} \cdot \mathbf{v} + t\mathbf{v} \cdot \mathbf{v} - [a, b, c] \cdot \mathbf{v}.$$

The solution is

$$t = \frac{[a, b, c] \cdot \mathbf{v} - \mathbf{u} \cdot \mathbf{v}}{\mathbf{v} \cdot \mathbf{v}}. \tag{12.14}$$

This gives the value of t corresponding to the nearest point to P on the line. If we only want the distance d to the line and not the nearest point on the line, we can make it easier. Select any point on the line, say the terminal point of $\mathbf{u}$ and call it R. Then we have the vector $\mathbf{v}$ parallel to the line, and the vector $\overrightarrow{RP}$ from R to the given point P. These form two sides of a right triangle and the third side is the perpendicular distance from P to the line. The length of that third side is $d = |\overrightarrow{RP}| \sin\theta$ where θ is the angle between $\overrightarrow{RP}$ and $\mathbf{v}$. Thus the distance we want to compute is closely related to the cross product of $\mathbf{v}$ and $\overrightarrow{RP}$ which has length $|\overrightarrow{RP} \times \mathbf{v}| = |\overrightarrow{RP}|\,|\mathbf{v}|| \sin\theta|$. We have the shortest distance from P to a point on the line is

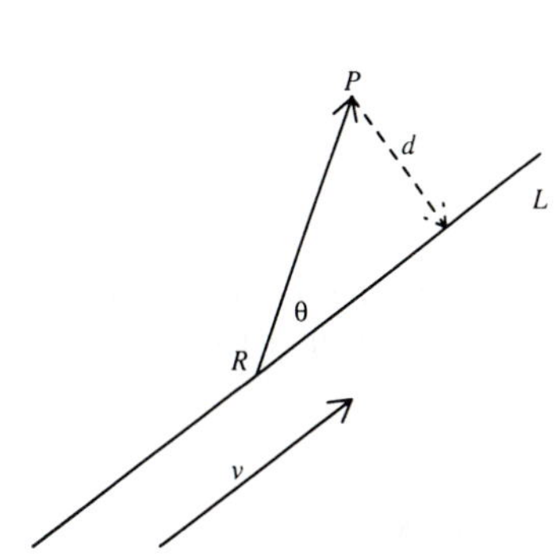

FIGURE 12.39

$$d = \frac{|\mathbf{v} \times \overrightarrow{RP}|}{|\mathbf{v}|}, \tag{12.15}$$

where R is any point on the line. Let us give an example using this compact formula.

EXAMPLE 4 Find the shortest distance from $P(1, 2, 3)$ to the line through the points $(-1, -1, 4)$ and $(-2, 3, 2)$.

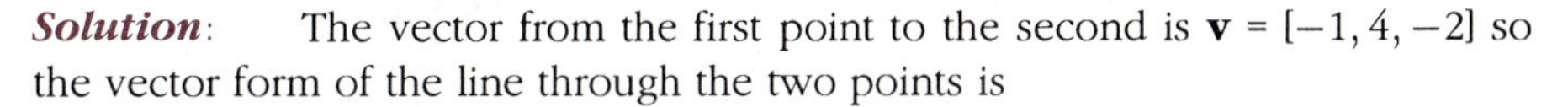

Solution: The vector from the first point to the second is $\mathbf{v} = [-1, 4, -2]$ so the vector form of the line through the two points is

$$\mathbf{r}(t) = [-1, -1, 4] + t[-1, 4, -2].$$

Select a point R on the line, say $R = (-1, -1, 4)$; P is given as $(1, 2, 3)$. The vector $\overrightarrow{RP}$ is $[2, 3, -1]$. Formula (12.15) gives the shortest distance from P to the line as

$$d = \frac{|\mathbf{v} \times \overrightarrow{RP}|}{|\mathbf{v}|} = \frac{|[-1, 4, -2] \times [2, 3, -1]|}{\sqrt{21}}$$
$$= \frac{|[2, -5, -11]|}{\sqrt{21}} = \frac{\sqrt{150}}{\sqrt{21}} = \frac{5\sqrt{2}}{\sqrt{7}}.$$

The task of finding the nearest point is a bit more work. From the formula derived above, the point nearest P on the line corresponds to the value of t given by Equation (12.14) as

$$t = \frac{[1, 2, 3] \cdot [-1, 4, -2] - [-1, -1, 4] \cdot [-1, 4, -2]}{[-1, 4, -2] \cdot [-1, 4, -2]} = \frac{12}{21} = \frac{4}{7}.$$

The nearest point on the line is

$$[-1, -1, 4] + \frac{4}{7}[-1, 4, -2] = \frac{1}{7}[-11, 9, 20].$$ ■

Exercises 12.6

Find the vector form, parametric form and the symmetric form for the equation of the line described.

1. The line through $(0, 0, 0)$ and $(1, 2, 1)$

2. The line through $(1, 1, 1)$ and $(2, 2, 2)$

3. The line through $(-1, -1, 0)$ and $(-2, -3, 1)$

4. The line through $(10, 4, 1)$ and $(1, 2, 0)$

5. The line through $(2, -4, -1)$ parallel to $[2, 4, 6]$

6. The line through $(1, 4, 5)$ parallel to $[-1, 4, 5]$

Find the points where the line intersects the three coordinate planes.

7. $\mathbf{r}(t) = [1, -2, 3] + t[6, 3, 1]$

8. $\dfrac{x-1}{3} = \dfrac{y+4}{6} = \dfrac{z}{2}$

9. $\mathbf{r}(t) = [2, 2, 5] + t[0, 3, -1]$

10. $\dfrac{x+2}{4} = \dfrac{y-4}{8} = \dfrac{z+6}{2}$

Determine if the pair of lines are parallel, intersecting, or skew. If they intersect, find the common point.

11. $\mathbf{r}_1(t) = [1, 2, -3] + t[0, -4, 6]$ and $\mathbf{r}_2(s) = [3, -6, 3] + s[-2, 0, 6]$

12. $\mathbf{r}_1(t) = [2, -2, 1] + t[1, -2, 4]$ and $\mathbf{r}_2(s) = [4, -4, 5] + s[-2, 1, -5]$

13. $\mathbf{r}_1(t) = [1, 0, -3] + t[2, 4, -8]$ and $\mathbf{r}_2(s) = [1, 4, 3] + s[-3, -6, 12]$

14. $\mathbf{r}_1(t) = [1, -2, 0] + t[10, 4, -8]$ and $\mathbf{r}_2(s) = [-3, -6, 4] + s[-5, -4, -4]$

Find the shortest distance from the point P to a point on the line L.

15. $L : \mathbf{r}(t) = [1, -2, 3] + t[6, 3, 1],\ P(1, 1, 1)$

16. $L : x - 2 = \dfrac{y-3}{4} = \dfrac{z-2}{2},\ P(0, 0, 0)$

17. $L : \mathbf{r}(t) = [2, -3, 3] + t[-1, 2, 1],\ P(1, 0, -1)$

18. $L : \dfrac{x-2}{3} = \dfrac{y+2}{6} = \dfrac{z+2}{3},\ P(3, 2, 0)$

19. Find the shortest distance between the two parallel lines $\mathbf{u} + t\mathbf{w}$ and $\mathbf{v} + s\mathbf{w}$. Express the answer in terms of $\mathbf{u}$, $\mathbf{v}$ and $\mathbf{w}$.

20. Find the shortest distance between the two skew lines $[1, 2, 1] + t[2, 2, -1]$ and $[1, 1, 1] + s[1, -1, -4]$. [Hint: The vector $\mathbf{u} = [2, 2, -1] \times [1, -1, -4]$ is orthogonal to both lines. If P is a point on the first line and Q is a point on the second line, the the projection of $\overrightarrow{PQ}$ onto $\mathbf{u}$ is the shortest vector from one line to the other because it is orthogonal to each line and has initial point in one line, terminal point in the other line.]

21. Two walkers set out along straight line paths in the xy plane. At time t walker 1 is at $(2-t, -t)$ and walker 2 is at $(2t-3, -2t-1)$. Where will the walker's paths cross? Will the two walkers meet at the point of intersection?

22. Two walkers set out along straight line paths in the xy plane. At time t walker 1 is at $(x_0 + at, y_0 + bt)$ and walker 2 is at (ct, dt) for some constants x_0, y_0, a, b, c, d. Find choices of constants that illustrate each of the following possibilities:
a. The two paths never cross.
b. The two walkers meet at a point.
c. The two paths cross but the walkers do not meet.

23. Find an equation of the line that is orthogonal to both $[1, 2, 1]$ and $[1, 1, -1]$ and passes through the origin.

24. Find an equation of the line that is orthogonal to both $[2, 2, 0]$ and $[1, 0, 1]$ and passes through the point $(2, 2, 6)$.

25. Use the methods of this section to derive the formula for the distance d from the point (x_0, y_0) to the line $ax + by + c = 0$ in the xy plane:

$$d = \frac{|ax_0 + by_0 + c|}{\sqrt{a^2 + b^2}}.$$

12.7 Equations of planes

We turn our attention to the description of planes. A plane can be uniquely specified in several different ways. One could give three points that do not lie on a straight line. There is only one plane containing these three points. Another way to specify a plane makes use of a *normal* vector which we now define.

A vector $\mathbf{N}$ is *normal* to the plane $\mathcal{P}$ if $\mathbf{N}$ is orthogonal to every vector having initial and terminal points in $\mathcal{P}$. Thus a normal vector to $\mathcal{P}$ is orthogonal to every line in $\mathcal{P}$. We also say that $\mathcal{P}$ is normal to $\mathbf{N}$ if $\mathbf{N}$ is normal to $\mathcal{P}$.

Specifying a normal vector is not enough information to determine a plane uniquely. If $\mathbf{N}$ is normal to $\mathcal{P}$ then $\mathbf{N}$ is also normal to every plane parallel to $\mathcal{P}$.

A plane can be uniquely specified by giving a point on the plane in addition to the normal vector. We use this idea to obtain the equation of a plane.

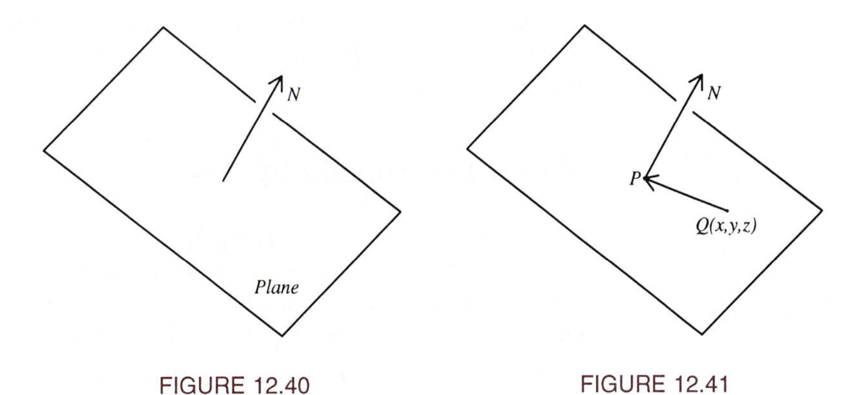

FIGURE 12.40

FIGURE 12.41

Let $\mathbf{N} = [a, b, c]$ be a vector and $P(x_0, y_0, z_0)$ a point. We want to describe the plane passing through P and having $\mathbf{N}$ as a normal vector. Then the point $Q(x, y, z)$ is in the plane normal to $\mathbf{N}$ through P if and only if the vector $\overrightarrow{PQ} = [x - x_0,\ y - y_0,\ z - z_0]$ is orthogonal to $\mathbf{N}$. This statement is equivalent to the equation

$$\mathbf{N} \cdot [x - x_0,\ y - y_0,\ z - z_0] = a(x - x_0) + b(y - y_0) + c(z - z_0) = 0$$

and so we have found an equation that determines the points in the plane. We bring these facts together in the following statement.

THEOREM 12.6

Equation of a Plane

The plane with normal vector $\mathbf{N} = [a, b, c]$ and passing through (x_0, y_0, z_0) is the set of all (x, y, z) satisfying

$$a(x - x_0) + b(y - y_0) + c(z - z_0) = 0.$$

The equation of the plane may also be written in the form

$$ax + by + cz = d \qquad \textbf{(12.16)}$$

where d is the constant $d = ax_0 + by_0 + cz_0$. Every equation of the form of equation (12.16) represents a plane. A normal vector to the plane can be read from the coefficients; $\mathbf{N} = a\,\mathbf{i} + b\,\mathbf{j} + c\,\mathbf{k}$ is normal to the plane. To put the equation in the form given in Theorem 12.6 it is necessary to find only one point on the plane. For example if $a \neq 0$ then the point $[d/a, 0, 0]$ is on the plane and the theorem gives the equivalent form of the equation as $a(x - d/a) + b(y - 0) + c(z - 0) = 0$. Infinitely many other equivalent forms of the equation can be written by using other points in the plane.

EXAMPLE 1 Find the equation of the plane passing through the point $(2, 2, 3)$ and having normal vector $\mathbf{N} = [2, -1, -4]$. Determine if the point $(4, 2, 4)$ is on the plane.

Solution: The point (x, y, z) is on the plane in question if

$$\mathbf{N} \cdot [x - 2, y - 2, z - 3] = 0.$$

The equation of the plane is

$$2(x-2)-(y-2)-4(z-3)=0.$$

Evaluate the left side of the equation at the coordinates of the point $(4, 2, 4)$ to get $2(4-2)-(2-2)-4(4-3)=0$. The coordinates do satisfy the equation of the plane so the point is on the plane. ■

Planes Determined by Three Points

Now suppose $P(a_1, b_1, c_1)$, $Q(a_2, b_2, c_2)$ and $R(a_3, b_3, c_3)$ are three given points and we want the equation of the plane determined by P, Q and R. We assume that the three points do not lie on a line (otherwise there are infinitely many planes containing the three points). The first step is to find a normal vector to the plane. The two vectors $\overrightarrow{PQ}$ and $\overrightarrow{PR}$ are parallel to the plane and they do not have the same direction (because P, Q, R do not lie on a line). The cross product $\mathbf{N} = \overrightarrow{PQ} \times \overrightarrow{PR}$ is orthogonal to both $\overrightarrow{PQ}$ and $\overrightarrow{PR}$ and hence is a normal to the plane. We now apply the analysis of the previous case to obtain the equation of the plane. We write it in vector form only.

THEOREM 12.7

Equation of the Plane Through Three Points

For three points,

$$P(a_1, b_1, c_1),\ Q(a_2, b_2, c_2) \text{ and } R(a_3, b_3, c_3)$$

not on a line, the unique plane passing through P, Q and R is the set of all (x, y, z) satisfying

$$\overrightarrow{PQ} \times \overrightarrow{PR} \cdot [x-a_1,\ y-b_1,\ z-c_1] = 0.$$

EXAMPLE 2 Find the equation of the plane passing through $P(1, -1, 1)$, $Q(1, 2, 0)$ and $R(0, 2, 1)$.

***Solution*:** The two vectors

$$\overrightarrow{PQ} = [0, 3, -1] = 3\mathbf{j} - \mathbf{k}$$

$$\overrightarrow{PR} = [-1, 3, 0] = -\mathbf{i} + 3\mathbf{j}$$

are parallel to the plane so their cross product is normal to the plane:

$$\mathbf{N} = (3\mathbf{j} - \mathbf{k}) \times (-\mathbf{i} + 3\mathbf{j}) = 3\mathbf{k} + \mathbf{j} + 3\mathbf{i}.$$

Select one point on the plane, say $Q(1, 2, 0)$ and apply Theorem 12.6 to get the equation

$$3(x-1) + (y-2) + 3z = 0$$

of the plane passing through P, Q and R. As a check on the arithmetic we can verify that the coordinates of P,Q and R do satisfy the equation. They do! ■

Vector Form of the Plane

There is yet another way of representing the points in a plane that is analogous to

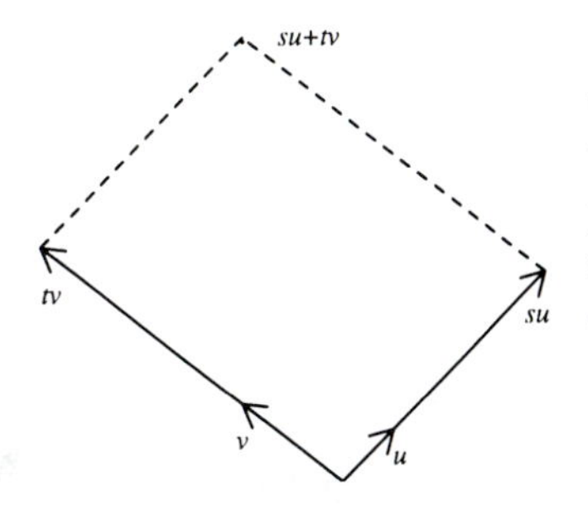

FIGURE 12.42

the vector form of a line given in the previous section. Consider two nonparallel vectors **u** and **v**. For any real numbers s, t the terminal point of $s\mathbf{u} + t\mathbf{v}$ lies in the plane through the origin containing the terminal points of **u** and **v** (both viewed as having initial points at the origin). The vector form of the plane through the origin containing the two vectors **u** and **v** is

$$\mathbf{Q}(s,t) = s\mathbf{u} + t\mathbf{v}.$$

If we take a third vector **w** and add it to every vector in this plane through the origin, the set of terminal points lie in a parallel plane passing through the terminal point of **w**. Hence we have another representation of a plane.

THEOREM 12.8

Vector Form of a Plane

Let **u**, **v** and **w** be three vectors with **u** and **v** not parallel. Then **u** and **v** lie in a unique plane through the origin consisting of the terminal points of all vectors of the form

$$\mathbf{Q}(s,t) = s\mathbf{u} + t\mathbf{v} \qquad s \text{ and } t \text{ real.}$$

The plane that passes through the terminal point of **w** and is parallel to the plane $\mathbf{Q}(s,t)$ consists of all the terminal points of the vectors

$$\mathbf{Q}_1(s,t) = \mathbf{w} + s\mathbf{u} + t\mathbf{v} \qquad s \text{ and } t \text{ real.}$$

EXAMPLE 3 Find the vector form of the plane passing through the points through $P(1,-1,1)$, $Q(1,2,0)$ and $R(0,2,1)$.

***Solution*:** The vectors

$$\mathbf{u} = \overrightarrow{PQ} = [0,3,-1] = 3\mathbf{j} - \mathbf{k}$$

$$\mathbf{v} = \overrightarrow{PR} = [-1,3,0] = -\mathbf{i} + 3\mathbf{j}$$

are nonparallel but are parallel to the plane $\mathcal{P}$ containing P, Q and R. The vectors $\mathbf{Q}(s,t) = s\mathbf{u} + t\mathbf{v}$ determine the plane through the origin parallel to the plane $\mathcal{P}$. The plane $\mathcal{P}$ is found by adding a vector with terminal point in $\mathcal{P}$ to all the vectors $\mathbf{Q}(s,t)$. Select $\mathbf{w} = \overrightarrow{OP} = [1,-1,1]$ and the vector form of $\mathcal{P}$ is

$$\mathbf{Q}_1(s,t) = [1,-1,1] + s\mathbf{u} + t\mathbf{v} = [1-t, -1+3s+3t, 1-s].$$ ■

EXAMPLE 4 Find the point where the line through the origin and $(1,2,3)$ meets the plane of example 3.

***Solution*:** The vector form of the line is $\mathbf{r}(t) = t[1,2,3]$ where t is any number. The vector form of the plane is

$$\mathbf{Q}_1(s,t) = [1,-1,1] + s\mathbf{u} + t\mathbf{v} = [1-t, -1+3s+3t, 1-s].$$

Before going any farther, we have to make a change of notation. The parameter t appears in two equations but the t in $\mathbf{r}(t)$ has nothing to do with the t in $\mathbf{Q}(s,t)$. To avoid confusion, we change the name of the parameter and write the line through the origin and $(1,2,3)$ as

$$\mathbf{r}(h) = h[1,2,3] = [h, 2h, 3h].$$

Now the point of intersection of the line and plane occurs for the values of s, t

and h that satisfy

$$\mathbf{r}(h) = \mathbf{Q}(s,t), \qquad \text{or}$$
$$[h, 2h, 3h] = [1-t, -1+3s+3t, 1-s].$$

We must solve the three equations

$$h = 1-t, \qquad 2h = -1+3s+3t, \qquad 3h = 1-s.$$

Use the first equation to substitute for h in the second and third equations to get

$$2(1-t) = -1+3s+3t, \quad 3(1-t) = 1-s,$$

or

$$3 = 3s+5t, \qquad 2 = -s+3t.$$

Eliminate s from this system of two equations by adding 3 times the second to the first to get $9 = 14t$ or $t = 9/14$. Then $s = -2+3t = -1/14$ and $h = 1-t = 5/14$. The common point is $(5/14)[1, 2, 3]$. ■

If it is required to convert the equation of a plane to vector form one begins by finding three points in the plane P, Q, R; then a vector form is

$$\mathbf{Q}(s,t) = \overrightarrow{OP} + s\overrightarrow{PQ} + t\overrightarrow{PR}.$$

If the vector form $\mathbf{Q}(s,t) = \mathbf{w}+s\mathbf{u}+t\mathbf{v}$ of a plane is given, the equation is obtained quickly by first finding a normal to the plane, $\mathbf{N} = \mathbf{u}\times\mathbf{v}$ and a point in the plane, like the terminal point of $\mathbf{w}$, and then the equation is $(\mathbf{u}\times\mathbf{v})\cdot([x,y,z]-\mathbf{w}) = 0$.

Distance between Parallel Planes

Two planes are parallel if their normal vectors are parallel. Suppose we have two planes with equations

$$\mathcal{P}_1: \quad ax+by+cz = d,$$
$$\mathcal{P}_2: \quad a'x+b'y+c'z = d'.$$

We determine the normal vectors to be $\mathbf{N}_1 = [a,b,c]$ and $\mathbf{N}_2 = [a',b',c']$. Thus $\mathcal{P}_1$ is parallel to $\mathcal{P}_2$ if and only if there is a scalar s with $[a,b,c] = s[a',b',c']$. When these equations relating a, b, c and a', b', c' hold, we can change the equation of $\mathcal{P}_2$ by multiplying it by s to get

$$\mathcal{P}_2: \quad s(a'x+b'y+c'z) = sd' = ax+by+cz.$$

After the multiplication by a constant factor, the equations of two parallel planes have the same coefficients of x, y, z and only the constant terms are different.

Let us now show how the constant terms are used to determine the distance between the planes. The distance between $\mathcal{P}_1$ and $\mathcal{P}_2$ is measured along a line perpendicular to each plane. The length of that line can be computed if we know the coordinates of the ends of the line. Determination of these endpoints may be possible but there is an easier way to find the distance between the planes. Select a point Q in the plane $\mathcal{P}_1$ and a point R in the plane $\mathcal{P}_2$. Then the vector $\overrightarrow{QR}$ has initial point in one plane and terminal point in the other plane. The projection of $\overrightarrow{QR}$ in the direction perpendicular to the two planes is a vector with initial point in one plane, terminal point in the other plane and its direction is perpendicular to both planes. Thus the length of the projection is the distance between the planes. Let us compute this explicitly. The normal to the planes is $\mathbf{N} = [a,b,c]$. Let $Q = (x_0, y_0, z_0)$ be a point in $\mathcal{P}_1$ so that $d = ax_0+by_0+cz_0$. Let

$R = (x_1, y_1, z_1)$ be a point in $\mathcal{P}_2$ so that $sd' = ax_1 + by_1 + cz_1$. Then

$$\overrightarrow{QR} = [x_1 - x_0, y_1 - y_0, z_1 - z_0]$$

and the projection of $\overrightarrow{QR}$ in the direction of $\mathbf{N}_1$ is

$$\mathbf{pr}_{\mathbf{N}_1}(\overrightarrow{QR}) = \overrightarrow{QR} \cdot \frac{1}{|\mathbf{N}_1|}\mathbf{N}_1.$$

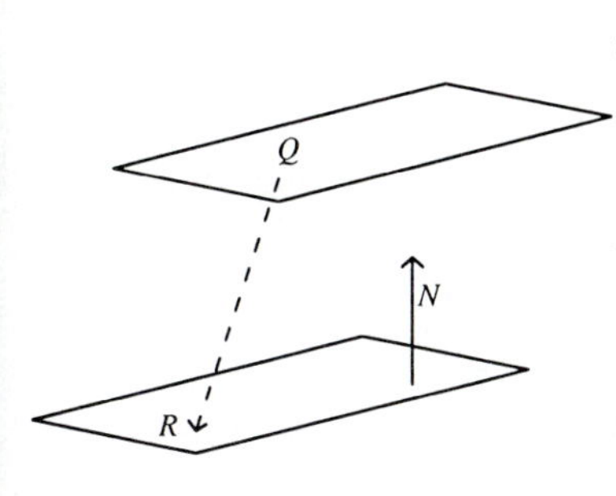

FIGURE 12.43

Now compute the dot product

$$\begin{aligned}\overrightarrow{QR} \cdot \mathbf{N}_1 &= a(x_1 - x_0) + b(y_1 - y_0) + c(z_1 - z_0)\\ &= ax_1 + by_1 + cz_1 - ax_0 - by_0 - cz_0\\ &= sd' - d.\end{aligned}$$

Thus the distance between the planes is the absolute value of this quantity divided by the length of $\mathbf{N}_1$:

$$\text{Distance between } \mathcal{P}_1 \text{ and } \mathcal{P}_2 = \frac{|sd' - d|}{\sqrt{a^2 + b^2 + c^2}}.$$

EXAMPLE 5 The planes with equations $2x - 3y + z = 7$ and $6x - 9y + 3z = 9$ are parallel. What is the distance between them?

Solution: We multiply the first equation by 3 so the equations are the same except for the constant terms, $6x - 9y + 3z = 21$. A common normal vector to these planes is $\mathbf{N} = [6, -9, 3]$ and the formula just derived for the distance between the planes gives us

$$\text{Distance} = \frac{|21 - 9|}{\sqrt{6^2 + (-9)^2 + 3^2}} = \frac{12}{\sqrt{126}} = \frac{4}{\sqrt{14}}.$$ ■

There are other closely related distance problems that can be solved by the same principle used in finding the distance between parallel planes. For example, the distance from a point Q to a plane $\mathcal{P}$ is found by first locating any point R in $\mathcal{P}$ and then computing the length of the projection of $\overrightarrow{QR}$ in the direction of a normal to $\mathcal{P}$.

Similarly, the distance from a plane to a parallel line is found by selecting any point Q on the line and using the method just described to find the distance from Q to the plane.

The distance between two lines $\mathbf{r}_1(t) = \mathbf{w}_1 + t\mathbf{v}_1$ and $\mathbf{r}_2(t) = \mathbf{w}_2 + t\mathbf{v}_2$ is found by first computing a vector orthogonal to both lines (like $\mathbf{N} = \mathbf{v}_1 \times \mathbf{v}_2$), finding a point Q_1 in the first line, a point Q_2 in the second line and then projecting $\overrightarrow{Q_1Q_2}$ in the direction of $\mathbf{N}$.

FIGURE 12.44

Intersection of Two Planes

Two different planes may be parallel, and so have no common points, or they may intersect and have a line in common. We give a fairly typical example to show how to find the line of intersection of two planes.

EXAMPLE 6 Find the line of intersection of the two planes having equations $4x + 3y + 2z = 10$ and $8x - 4y - 6z = 5$.

Solution: If (x, y, z) is a point on both planes, then the coordinates satisfy both equations. There are three variables but only two equations so we should expect to be able to solve for two of the variables in terms of the third. Let us try to express x and y in terms of z. Rewrite the equations of the planes in the form

$$4x + 3y = 10 - 2z,$$
$$8x - 4y = 5 + 6z.$$

Eliminate the x terms by multiplying the first equation by -2 and adding to the second equation to get

$$-10y = -15 + 10z \quad \text{or} \quad y = \frac{-15 + 10z}{-10} = \frac{3 - 2z}{2}.$$

Now go back to one of the original equations and substitute this value in place of y. We use the first equation to get

$$4x + 3\left(\frac{3 - 2z}{2}\right) = 10 - 2z \quad \text{or} \quad x = \frac{11 + 2z}{8}.$$

This is a slightly disguised parametric form of the line we are looking for. We write $z = t$ and obtain

$$\begin{cases} x = (11 + 2t)/8, \\ y = (3 - 2t)/2, \\ z = t, \end{cases}$$

which is the parametric form of the line.

It is worth checking the work to verify that the line really does lie in both planes. Substitute the coordinates of a point on the line into the equations of the planes:

$$4\left(\frac{11 + 2t}{8}\right) + 3\left(\frac{3 - 2t}{2}\right) + 2t = \frac{11}{2} + t + \frac{9}{2} - 3t + 2t = 10$$

and

$$8\left(\frac{11 + 2t}{8}\right) - 4\left(\frac{3 - 2t}{2}\right) - 6t = 11 + 2t - 6 + 4t - 6t = 5.$$

Everything checks so the line is on both planes. ■

If two planes intersect, we define the angle between the planes as the angle between the normal vectors to the two planes (the angle is always taken between 0 and $\pi/2$).

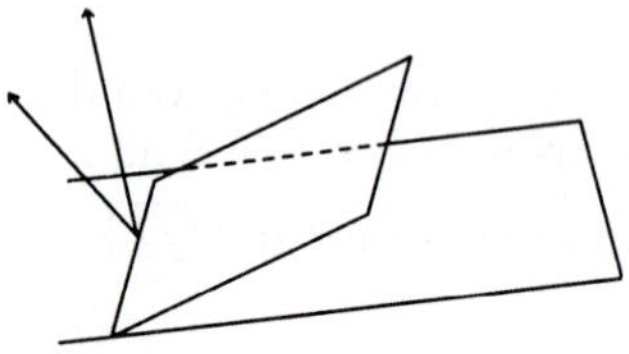

FIGURE 12.45
Angle between planes

EXAMPLE 7 Find the angle between the two planes of example 6.

Solution: The plane with equation $4x + 3y + 2z = 10$ has $\mathbf{N}_1 = 4\mathbf{i} + 3\mathbf{j} + 2\mathbf{k}$ as a normal and the plane with equation $8x - 4y - 6z = 5$ has $\mathbf{N}_2 = 8\mathbf{i} - 4\mathbf{j} - 6\mathbf{k}$ as a normal. The cosine of the angle between the normal vectors is

$$\cos\theta = \frac{\mathbf{N}_1 \cdot \mathbf{N}_2}{|\mathbf{N}_1||\mathbf{N}_2|} = \frac{8}{\sqrt{29}\sqrt{116}} = \frac{8}{58}.$$

Use a calculator to determine the angle as $\theta = \arccos(8/58) \approx 1.43242$ in radians and about 82.072 degrees. ■

Projection onto a Plane

We have defined the projection $\mathbf{pr_u}(\mathbf{v})$ of a vector $\mathbf{v}$ onto a vector $\mathbf{u}$ and saw that $\mathbf{v} = \mathbf{pr_u}(\mathbf{v}) + \mathbf{v}_1$, where $\mathbf{v}_1$ is orthogonal to $\mathbf{u}$ and $\mathbf{pr_u}(\mathbf{v})$ is parallel to $\mathbf{u}$.

We now want to define the projection of a vector into a plane and obtain an analogous decomposition of the vector as a sum of a vector parallel to the plane and a vector orthogonal to the plane.

Let $\mathcal{P}$ be a plane with vector equation $\mathbf{Q}(s,t) = s\mathbf{v} + t\mathbf{w}$ for some nonparallel vectors $\mathbf{v}$ and $\mathbf{w}$. Then $\mathcal{P}$ is a plane through the origin. Recall that the projection of a vector in the direction of $\mathbf{w}$ is the same as the projection of the vector in the direction of any vector parallel to $\mathbf{w}$. The same is true of the projection into a plane; that is, the projection of a vector into $\mathcal{P}$ is the same as the projection of the vector into any plane parallel to $\mathcal{P}$. So there is no harm in discussing the special case of projection into a plane through the origin inasmuch as every plane is parallel to one through the origin.

Let

$$\mathbf{N} = \frac{\mathbf{v} \times \mathbf{w}}{|\mathbf{v} \times \mathbf{w}|}$$

so that $\mathbf{N}$ is a unit vector normal to $\mathcal{P}$.

Now let $\mathbf{u}$ be any vector. Let $\mathbf{u}_1 = \mathbf{u} - (\mathbf{u} \cdot \mathbf{N})\mathbf{N}$. Then

$$\mathbf{u} = (\mathbf{u} \cdot \mathbf{N})\mathbf{N} + \mathbf{u}_1 = \mathbf{pr}_{\mathbf{N}}(\mathbf{u}) + \mathbf{u}_1$$

is the expression of $\mathbf{u}$ as a sum of a vector parallel to $\mathbf{N}$ and a vector $\mathbf{u}_1$ orthogonal to $\mathbf{N}$. The vector orthogonal to $\mathbf{N}$ lies in $\mathcal{P}$ because $\mathcal{P}$ is the plane (through the origin) orthogonal to $\mathbf{N}$. We call $\mathbf{u}_1$ the *projection of* $\mathbf{u}$ *onto the plane* $\mathcal{P}$ and we write

$$\mathbf{pr}_{\mathcal{P}}(\mathbf{u}) = \mathbf{u}_1 = \mathbf{u} - (\mathbf{u} \cdot \mathbf{N})\mathbf{N}$$

for this vector. Thus

$$\mathbf{u} = \mathbf{pr}_{\mathbf{N}}(\mathbf{u}) + \mathbf{pr}_{\mathcal{P}}(\mathbf{u}) \tag{12.17}$$

with $\mathbf{pr}_{\mathcal{P}}(\mathbf{u})$ in $\mathcal{P}$ and $\mathbf{pr}_{\mathbf{N}}(\mathbf{u})$ orthogonal to $\mathcal{P}$.

We summarize these computations.

THEOREM 12.9

Projection onto a Plane

Let $\mathbf{N}$ be a unit normal to the plane $\mathcal{P}$ and let $\mathbf{u}$ be any vector. Then

$$\mathbf{u} = \mathbf{pr}_{\mathbf{N}}(\mathbf{u}) + \mathbf{pr}_{\mathcal{P}}(\mathbf{u}) \qquad \begin{cases} \mathbf{pr}_{\mathbf{N}}(\mathbf{u}) = (\mathbf{u} \cdot \mathbf{N})\mathbf{N} \\ \mathbf{pr}_{\mathcal{P}}(\mathbf{u}) = \mathbf{u} - (\mathbf{u} \cdot \mathbf{N})\mathbf{N} \end{cases}$$

expresses $\mathbf{u}$ as the sum of a vector $\mathbf{pr}_{\mathcal{P}}(\mathbf{u})$ in $\mathcal{P}$ and a vector $\mathbf{pr}_{\mathbf{N}}(\mathbf{u})$ orthogonal to $\mathcal{P}$.

The simplest case of a projection is one that we implicitly use every time we locate a point in three-dimensional space. Suppose we project a vector $\mathbf{u} = a\mathbf{i} + b\mathbf{j} + c\mathbf{k}$ onto the xy plane. Begin with a unit vector normal to the plane, namely $\mathbf{k}$, and compute $\mathbf{pr}_{\mathbf{xy}}(\mathbf{u})$ according to the prescription given in the theorem:

$$\mathbf{pr}_{\mathbf{xy}}(\mathbf{u}) = a\mathbf{i} + b\mathbf{j} + c\mathbf{k} - \big((a\mathbf{i} + b\mathbf{j} + c\mathbf{k}) \cdot \mathbf{k}\big)\,\mathbf{k} = a\mathbf{i} + b\mathbf{j}.$$

The technical definition of projection into the xy plane agrees with the commonly used interpretation of projection. In a similar fashion one could show that the projection of $a\mathbf{i} + b\mathbf{j} + c\mathbf{k}$ into the yz plane is $b\mathbf{j} + c\mathbf{k}$ and the projections of $a\mathbf{i} + b\mathbf{j} + c\mathbf{k}$ into the xz plane is $a\mathbf{i} + c\mathbf{k}$.

EXAMPLE 8 Find the projection of the vector $\mathbf{u} = 2\mathbf{i} + \mathbf{j} - 2\mathbf{k}$ into the plane

$\mathcal{P}$ passing through the three points $(1,0,0)$, $(0,1,0)$ and $(0,0,1)$.

***Solution*:** We first find a normal vector to the plane. The vectors

$$\mathbf{v} = [1,0,0] - [0,1,0] = [1,-1,0] = \mathbf{i} - \mathbf{j},$$
$$\mathbf{w} = [1,0,0] - [0,0,1] = [1,0,-1] = \mathbf{i} - \mathbf{k}$$

lie in the plane $\mathcal{P}$ so $\mathbf{q} = \mathbf{v} \times \mathbf{w} = \mathbf{i} + \mathbf{j} + \mathbf{k}$ is a vector normal to $\mathcal{P}$. The unit vector normal to $\mathcal{P}$ is

$$\mathbf{N} = \frac{\mathbf{q}}{|\mathbf{q}|} = \frac{1}{\sqrt{3}}(\mathbf{i} + \mathbf{j} + \mathbf{k}).$$

Then

$$\begin{aligned}\mathbf{pr}_{\mathcal{P}}(2\mathbf{i} + \mathbf{j} - 2\mathbf{k}) &= 2\mathbf{i} + \mathbf{j} - 2\mathbf{k} - \mathbf{pr}_{\mathbf{N}}(2\mathbf{i} + \mathbf{j} - 2\mathbf{k}) \\ &= 2\mathbf{i} + \mathbf{j} - 2\mathbf{k} - \left[(2\mathbf{i} + \mathbf{j} - 2\mathbf{k}) \cdot \frac{(\mathbf{i} + \mathbf{j} + \mathbf{k})}{\sqrt{3}}\right] \frac{(\mathbf{i} + \mathbf{j} + \mathbf{k})}{\sqrt{3}} \\ &= 2\mathbf{i} + \mathbf{j} - 2\mathbf{k} - \frac{1}{\sqrt{3}}\frac{1}{\sqrt{3}}(\mathbf{i} + \mathbf{j} + \mathbf{k}) \\ &= (5\mathbf{i} + 2\mathbf{j} - 7\mathbf{k})/3.\end{aligned}$$

The decomposition of $\mathbf{u}$ as a sum of a vector in $\mathcal{P}$ and a vector orthogonal to $\mathcal{P}$ is

$$2\mathbf{i} + \mathbf{j} - 2\mathbf{k} = \frac{1}{3}(5\mathbf{i} + 2\mathbf{j} - 7\mathbf{k}) + \frac{1}{3}(\mathbf{i} + \mathbf{j} + \mathbf{k}).$$ ■

Next we consider the projection of a parallelogram into a plane and determine the relation between the area of the original parallelogram and its projection. Let us begin with an example that is not too difficult to calculate. Consider the parallelogram with sides given by the vectors $\mathbf{i} + \mathbf{j}$ and $\mathbf{i} + \mathbf{k}$. The area of this parallelogram is the absolute value of the cross product,

$$|(\mathbf{i} + \mathbf{j}) \times (\mathbf{i} + \mathbf{k})| = |-\mathbf{j} - \mathbf{k} + \mathbf{i}| = \sqrt{3}.$$

Project the parallelogram onto the xy plane. The projection is the parallelogram having sides $\mathbf{i} + \mathbf{j}$ and $\mathbf{i}$ (the projections into the xy plane of $\mathbf{i} + \mathbf{j}$ and $\mathbf{i} + \mathbf{k}$). The area of the projected parallelogram is

$$|(\mathbf{i} + \mathbf{j}) \times \mathbf{i}| = |-\mathbf{k}| = 1.$$

Obviously the areas are different. How can we predict the area of the projected parallelogram? The answer is we take the cross product whose absolute value gave us the area of the original parallelogram and dot that with a unit vector orthogonal to the plane into which the projection is made, to get the area of the projected parallelogram (up to sign). In the example we would have

$$(\mathbf{i} + \mathbf{j}) \times (\mathbf{i} + \mathbf{k}) \cdot \mathbf{k} = (-\mathbf{j} - \mathbf{k} + \mathbf{i}) \cdot \mathbf{k} = -1.$$

The absolute value of this is the area we just computed. Let us state and prove the general result.

THEOREM 12.10

Area of Projections

The area of the projection of the parallelogram with sides $\mathbf{u}$ and $\mathbf{v}$ into any plane having unit vector normal vector $\mathbf{N}$ is the absolute value of $\mathbf{u} \times \mathbf{v} \cdot \mathbf{N}$.

Proof

Let $\mathcal{P}$ be the plane through the origin having $\mathbf{N}$ as normal vector. The vectors $\mathbf{u}$ and $\mathbf{v}$ can be expressed as a sum of a vector parallel to $\mathbf{N}$ and a vector in $\mathcal{P}$ as

$$\mathbf{u} = (\mathbf{u}\cdot\mathbf{N})\mathbf{N} + \mathbf{pr}_{\mathcal{P}}(\mathbf{u}),$$
$$\mathbf{v} = (\mathbf{v}\cdot\mathbf{N})\mathbf{N} + \mathbf{pr}_{\mathcal{P}}(\mathbf{v}).$$

The parallelogram with sides $\mathbf{u}$ and $\mathbf{v}$ projects to the parallelogram with sides $\mathbf{pr}_{\mathcal{P}}(\mathbf{u})$ and $\mathbf{pr}_{\mathcal{P}}(\mathbf{v})$ and so the area of the projection is the absolute value of $\mathbf{pr}_{\mathcal{P}}(\mathbf{u}) \times \mathbf{pr}_{\mathcal{P}}(\mathbf{v})$. Because $\mathbf{N}$ is a unit vector orthogonal to both $\mathbf{pr}_{\mathcal{P}}(\mathbf{u})$ and $\mathbf{pr}_{\mathcal{P}}(\mathbf{v})$, it follows that $\mathbf{N}$ is parallel to $\mathbf{pr}_{\mathcal{P}}(\mathbf{u}) \times \mathbf{pr}_{\mathcal{P}}(\mathbf{v})$. In other words, $\mathbf{pr}_{\mathcal{P}}(\mathbf{u}) \times \mathbf{pr}_{\mathcal{P}}(\mathbf{v}) = s\mathbf{N}$ for some scalar s. Now we observe that

$$|\mathbf{pr}_{\mathcal{P}}(\mathbf{u}) \times \mathbf{pr}_{\mathcal{P}}(\mathbf{v})| = |s\mathbf{N}| = |s|$$

equals the absolute value of

$$\mathbf{pr}_{\mathcal{P}}(\mathbf{u}) \times \mathbf{pr}_{\mathcal{P}}(\mathbf{v}) \cdot \mathbf{N} = s\mathbf{N}\cdot\mathbf{N} = s.$$

The proof of the theorem will be finished when we show

$$\mathbf{pr}_{\mathcal{P}}(\mathbf{u}) \times \mathbf{pr}_{\mathcal{P}}(\mathbf{v}) \cdot \mathbf{N} = \mathbf{u}\times\mathbf{v}\cdot\mathbf{N}. \qquad \textbf{(12.18)}$$

Observe that for any scalar r, $r\mathbf{N}\times\mathbf{w}\cdot\mathbf{N} = 0$ for any vector $\mathbf{w}$ because the scalar triple product is zero whenever the three vectors lie in a plane. Now we know

$$\begin{aligned}\mathbf{u}\times\mathbf{v} &= \big((\mathbf{u}\cdot\mathbf{N})\mathbf{N} + \mathbf{pr}_{\mathcal{P}}(\mathbf{u})\big) \times \big((\mathbf{v}\cdot\mathbf{N})\mathbf{N} + \mathbf{pr}_{\mathcal{P}}(\mathbf{v})\big)\\ &= \mathbf{pr}_{\mathcal{P}}(\mathbf{u}) \times \mathbf{pr}_{\mathcal{P}}(\mathbf{v}) + (\mathbf{u}\cdot\mathbf{N})\mathbf{N}\times\mathbf{pr}_{\mathcal{P}}(\mathbf{v}) + \mathbf{pr}_{\mathcal{P}}(\mathbf{u}) \times \mathbf{pr}_{\mathcal{P}}(\mathbf{u}).\end{aligned}$$

When we form the dot product with $\mathbf{N}$, all the terms of the form $s\mathbf{N}\times\mathbf{w}\cdot\mathbf{N}$ or $\mathbf{w}\times s\mathbf{N}\cdot\mathbf{N}$ equal zero and we are left with

$$\mathbf{u}\times\mathbf{v}\cdot\mathbf{N} = \mathbf{pr}_{\mathcal{P}}(\mathbf{u}) \times \mathbf{pr}_{\mathcal{P}}(\mathbf{u}) \cdot \mathbf{N},$$

as we wished. This completes the proof of the theorem. □

EXAMPLE 9 Compute the area of the projection into the xy plane of the parallelogram having $\mathbf{u} = \mathbf{i} + 2\mathbf{j}$ and $\mathbf{v} = \mathbf{i} + 2\mathbf{k}$ as two sides.

Solution: The vector $\mathbf{k}$ is a unit normal to the xy plane so the area of the projection is the absolute value of

$$\mathbf{u}\times\mathbf{v}\cdot\mathbf{k} = (\mathbf{i}+2\mathbf{j})\times(\mathbf{i}+2\mathbf{k})\cdot k = -2,$$

so the area is 2. The area of the parallelogram before projection is the absolute value of

$$(\mathbf{i}+2\mathbf{j})\times(\mathbf{i}+2\mathbf{k}) = -2\mathbf{j} - 2\mathbf{k} + 2\mathbf{i};$$

that is, $2\sqrt{3}$. ■

Exercises 12.7

Find the equation of the form $ax + by + cz = d$ for the plane described.

1. The plane with normal $\mathbf{N} = [2, 2, -3]$ passing through $(0, 0, 0)$
2. The plane with normal $\mathbf{N} = [1, -4, 3]$ passing through $(1, -2, 5)$
3. The plane passing through $(1, 0, 1)$ parallel to the plane with equation $2x - y - 3z = 5$
4. The plane passing through $(4, 2, 10)$ parallel to the plane with equation $x + y + z = 0$

5. The plane passing through $P(1,-1,0)$, $Q(1,0,-1)$, and $R(0,-1,1)$

6. The plane passing through $P(2,1,1)$, $Q(1,1,-1)$, and $R(3,1,1)$

7. The plane passing through $P(1,0,0)$, $Q(0,1,0)$, and $R(0,0,1)$

8. The plane passing through $P(3,1,1)$, $Q(0,0,1)$, and $R(0,1,0)$

9. The plane having vector equation $\mathbf{Q}(s,t) = [1,2,3] + s[0,1,1] + t[1,2,0]$

10. The plane having vector equation $\mathbf{Q}(s,t) = [1,1,-3] + s[2,-1,1] + t[-1,2,5]$

Find a vector form for the plane having the given equation.

11. $2x - 2y + z = 0$

12. $2x - 2y + z = 6$

13. $4x - 2y + 3z = 10$

14. $-2x + 3y - 5z = 2$

15. Find the distance between the parallel planes having equations $2x - y + z = 3$ and $2x - y + z = 6$.

16. Find the distance between the parallel planes having equations $2x + 4y - 6z = 0$ and $3x - 6y + 9z = 6$.

17. Find the distance between the parallel planes having vector equations $\mathbf{Q}_1(s,t) = [1,1,0] + s[0,2,3] + t[1,0,2]$ and $\mathbf{Q}_2(s,t) = [2,0,1] + s[0,2,3] + t[1,0,2]$.

18. The line with vector form $\mathbf{r}(t) = [1,1,1] + t[2,-1,3]$ is parallel to the plane with vector form $\mathbf{Q}(s,t) = s[2,-1,3] + t[1,-1,0]$. Find the distance between them.

19. Find the line of intersection of the two planes with equations $2x + 4y - 6z = 0$ and $2x - y + z = 0$.

20. Find the line of intersection of the two planes with equations $x + 3y - 2z = 4$ and $x - 2y + z = 5$.

21. Find the line of intersection of the two planes with equations $\mathbf{Q}_1(s,t) = [1,1,0] + s[0,2,3] + t[2,0,4]$ and $\mathbf{Q}_2(s,t) = [0,1,1] + s[1,2,-3] + t[1,0,2]$.

22. Find the line of intersection of the two planes with equations $\mathbf{Q}_1(s,t) = [1,1,1] + s[1,2,3] + t[2,-3,4]$ and $\mathbf{Q}_2(s,t) = [1,1,1] + s[1,2,-3] + t[1,0,2]$.

23. Find the cosine of the angle between the planes with equations $2x + 4y - 6z = 0$ and $2x - y + z = 0$.

24. Find the cosine of the angle between the planes with equations $x - 4y - z = 2$ and $2x - y + 3z = 2$.

25. Find a constant c so that the planes with equations $2x + y + z = 10$ and $x - 3y + cz = 0$ meet at right angles. Is there another choice of c such that the two planes are parallel?

26. Find the area of the projection of the parallelogram with sides $\mathbf{i} - \mathbf{j} + \mathbf{k}$ and $2\mathbf{i} + \mathbf{j} + \mathbf{k}$ into the plane having normal vector $(\mathbf{i} + \mathbf{j})/\sqrt{2}$.

27. Find the area of the projection of the parallelogram with sides $\mathbf{i} + 3\mathbf{k}$ and $3\mathbf{i} + \mathbf{j}$ into the plane having normal vector $(\mathbf{i} + \mathbf{j} - \mathbf{k})/\sqrt{3}$.

28. The parallelogram having sides $\mathbf{u} = \mathbf{i} + \mathbf{j} + \mathbf{k}$ and $\mathbf{v} = 2\mathbf{i} - \mathbf{j} - \mathbf{k}$ is projected into each of the coordinate planes. Find the area of the parallelogram and of the three projections.

29. Derive the following formula for the distance D from a point (x_0, y_0, z_0) to the plane with equation $ax+by+cz = m$:

$$D = \frac{|ax_0 + by_0 + cz_0 - m|}{\sqrt{a^2 + b^2 + c^2}}.$$

30. Let A be the area of a parallelogram in three-dimensional space and let A_1, A_2, A_3 be the areas of the projections of the parallelogram into the xy, xz and yz planes. Show $A^2 = A_1^2 + A_2^2 + A_3^2$. [Hint: For any vector $\mathbf{w}$, show $|\mathbf{w}|^2 = (\mathbf{w} \cdot \mathbf{i})^2 + (\mathbf{w} \cdot \mathbf{j})^2 + (\mathbf{w} \cdot \mathbf{k})^2$.]

31. Let $\mathcal{P}_1$ and $\mathcal{P}_2$ be planes that intersect with θ the angle between them. If a parallelogram of area A in $\mathcal{P}_1$ is projected into $\mathcal{P}_2$, show that the area of the projection is $A\cos\theta$.

13 Vector Functions

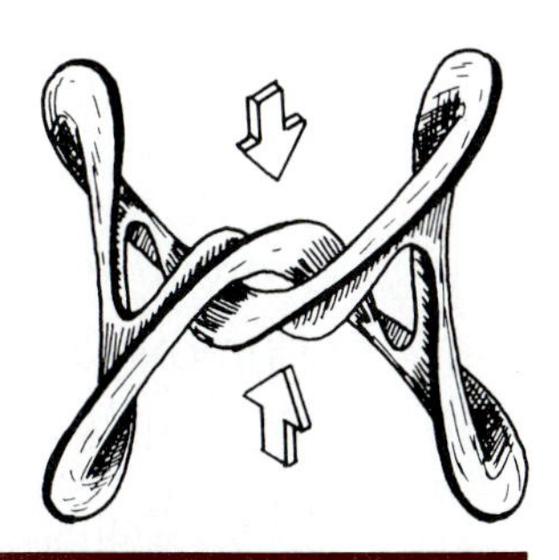

In this chapter we study vector-valued functions of a single variable. This notion provides an efficient way of describing certain geometric and physical concepts. For example, the vector description of curves in the plane, which is equivalent to the parametric description of curves, allows the systematic study of curves that are more general than the curves arising as the graph of a function of one variable.

After establishing basic properties of differentiation and integration of vector functions, we apply them to describe curves in two and three dimensions, determine their tangents and normals and establish an arc-length formula. The concept of curvature is introduced and it is shown that only straight lines and arcs of circles have constant curvature. We study Newton's laws of motion as applied to particles moving along a curved path in three dimensions. The vector formulation permits an easy description of the forces that are felt by a passenger in a car as it moves around a curve. We also apply the theory to study Kepler's laws of planetary motion that show the orbit of a planet around the sun is elliptical.

13.1 Vector functions

To describe lines in three-dimensional space, we considered functions of the parameter t defined by $\mathbf{r}(t) = \mathbf{u} + t\mathbf{v}$ for fixed vectors $\mathbf{u}$ and $\mathbf{v}$. This function provides an example of the following concept. A *vector-valued function* is a rule which assigns a unique vector $\mathbf{r}(t)$ to each real number t in some set D. We refer to $\mathbf{r}$ as a *vector function*, $\mathbf{r}(t)$ is its value at t and D as the *domain* of $\mathbf{r}$. The most direct way to construct examples of vector functions is by starting with three real-valued functions $x(t)$, $y(t)$ and $z(t)$ of the variable t, defined on some common domain D, and forming the vector-valued function

$$\mathbf{r}(t) = x(t)\,\mathbf{i} + y(t)\,\mathbf{j} + z(t)\,\mathbf{k}. \qquad \textbf{(13.1)}$$

The functions $x(t)$, $y(t)$ and $z(t)$ are called the *coordinate functions* of $\mathbf{r}(t)$.

We state and prove some properties of vector functions analogous to the basic properties of real-valued functions of one variable.

Limit of a Vector Function

We first give a definition of limit that does not mention the coordinate functions and closely follows the definition of the limit of a real-valued function.

DEFINITION 13.1

Limit of a Vector Function

Let **r** be a vector-valued function defined on an open interval containing a. The vector **u** is the limit of $\mathbf{r}(t)$ as t approaches a if for every positive number ϵ there is some number δ such that $|\mathbf{r}(t) - \mathbf{u}| < \epsilon$ whenever $0 < |t - a| < \delta$. When this holds we write

$$\lim_{t \to a} \mathbf{r}(t) = \mathbf{u}.$$

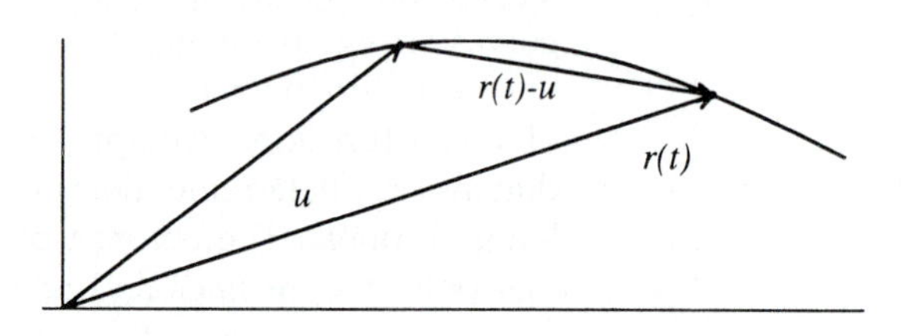

FIGURE 13.1
$\mathbf{r}(t)$ approaches **u**

The intuitive ideas about limits of real-valued functions that the reader may have developed generally carry over to the limits of vector-valued functions. For example, if $\lim_{t \to a} \mathbf{r}(t) = \mathbf{u}$ then the vector $\mathbf{r}(t)$ is "close" to **u** whenever t is "close" to a. The notion of two vectors being close might be somewhat less intuitive than the notion of two numbers being close. Two vectors are close if their difference has small magnitude. The ϵ and δ in the definition give a quantitative precision to small and close.

The definition of a limit of a vector function can be interpreted in terms of the coordinate functions. If $\mathbf{r}(t)$ is given as in Equation (13.1) then the limit of $\mathbf{r}(t)$ is determined by the limits of the coordinate functions as

$$\lim_{t \to a} \mathbf{r}(t) = \left(\lim_{t \to a} x(t)\right) \mathbf{i} + \left(\lim_{t \to a} y(t)\right) \mathbf{j} + \left(\lim_{t \to a} z(t)\right) \mathbf{k}.$$

This can be proved using the definitions of limits of real-valued functions but we leave it to the exercises.

From the definition one can prove, for vector functions $\mathbf{r}(t)$ and $\mathbf{s}(t)$, that

$$\lim_{t \to a}(c_1\mathbf{r}(t) + c_2\mathbf{s}(t)) = c_1 \lim_{t \to a} \mathbf{r}(t) + c_2 \lim_{t \to a} \mathbf{s}(t),$$

for any constants c_1 and c_2.

Continuity of a Vector Function

The vector function $\mathbf{r}(t)$ is continuous at $t = a$ if $\mathbf{r}(a)$ is defined and

$$\lim_{t\to a} \mathbf{r}(t) = \mathbf{r}(a).$$

The vector function $\mathbf{r}(t)$ is continuous at a point if and only if all the coordinate functions of $\mathbf{r}(t)$ are continuous at the point. Thus, for example, the function

$$\mathbf{r}(t) = \left(\frac{t^2+1}{t}\right)\mathbf{i} + \sin t\,\mathbf{j} + \left(\frac{t}{t+1}\right)\mathbf{k}$$

is continuous at every point except $t = 0$ and $t = -1$ because at any other point, all three coordinate functions are continuous. At $t = 0$ or $t = -1$, at least one of the coordinate functions is not continuous, so $\mathbf{r}(t)$ is not continuous either.

Derivative of a Vector Function

Once the definition of the limit is given, the definition of the derivative can be written down at once by imitation of the definition of the derivative of real-valued functions. The derivative of the vector function $\mathbf{r}(t)$ is defined to be

$$\mathbf{r}'(t) = \lim_{h\to 0} \frac{\mathbf{r}(t+h) - \mathbf{r}(t)}{h},$$

provided this limit exists. The derivative of a vector function is also a vector function. The derivative is easily computed in terms of the coordinate functions that define $\mathbf{r}$. If $\mathbf{r}(t)$ is defined as in Equation (13.1) then

$$\begin{aligned}\mathbf{r}'(t) &= \lim_{h\to 0} \frac{\mathbf{r}(t+h) - \mathbf{r}(t)}{h} \\ &= \lim_{t\to a} \frac{x(t+h) - x(t)}{h}\,\mathbf{i} + \lim_{t\to a} \frac{y(t+h) - y(t)}{h}\,\mathbf{j} + \lim_{t\to a} \frac{z(t+h) - z(t)}{h}\,\mathbf{k} \\ &= x'(t)\,\mathbf{i} + y'(t)\,\mathbf{j} + z'(t)\,\mathbf{k}.\end{aligned}$$

Thus the derivative of $\mathbf{r}(t)$ exists whenever the derivatives of the three coordinate functions, $x(t)$, $y(t)$ and $z(t)$, exist.

EXAMPLE 1 Find the derivative of the function $\mathbf{r}(t) = 3t^2\,\mathbf{i} - (1+4t)\,\mathbf{j} + \cos t\,\mathbf{k}$.

Solution: We have

$$\begin{aligned}\mathbf{r}'(t) &= (3t^2)'\,\mathbf{i} - (1+4t)'\,\mathbf{j} + (\cos t)'\,\mathbf{k} \\ &= 6t\,\mathbf{i} - 4\,\mathbf{j} - \sin t\,\mathbf{k}.\end{aligned}$$

■

Product Rules

Next we consider some product rules. If $a(t)$ is a real-valued function and $\mathbf{r}(t)$ is a vector-valued function of t then the product $a(t)\mathbf{r}(t)$ is a vector-valued function. If $\mathbf{s}(t)$ is vector function then $\mathbf{r}(t) \cdot \mathbf{s}(t)$ is a real-valued function and $\mathbf{r}(t) \times \mathbf{s}(t)$ is a vector-valued function. The derivatives of these various products follow the pattern for the derivative of the product of two real-valued functions. We just state the formulas. Each formula can be verified by expressing the vector functions in terms of coordinate functions, expanding the products and computing the derivatives of the functions on each side of the equation.

Product Rules

$$\frac{d}{dt}\big(a(t)\mathbf{r}(t)\big) = a'(t)\mathbf{r}(t) + a(t)\mathbf{r}'(t),$$

$$\frac{d}{dt}\big(\mathbf{s}(t)\cdot\mathbf{r}(t)\big) = \mathbf{s}'(t)\cdot\mathbf{r}(t) + \mathbf{s}(t)\cdot\mathbf{r}'(t),$$

$$\frac{d}{dt}\big(\mathbf{s}(t)\times\mathbf{r}(t)\big) = \mathbf{s}'(t)\times\mathbf{r}(t) + \mathbf{s}(t)\times\mathbf{r}'(t).$$

These rules are valid whenever the derivatives of the respective functions exist. Some of the proofs are requested in the exercises at the end of this section.

The chain rule for derivatives also has its counterpart for vector functions.

The Chain Rule

$$\frac{d}{dt}\mathbf{r}(a(t)) = \mathbf{r}'(a(t))a'(t).$$

Let us sketch a proof of this rule. If $a(t)$ is a real-valued function and

$$\mathbf{r}(t) = x(t)\,\mathbf{i} + y(t)\,\mathbf{j} + z(t)\,\mathbf{k}$$

is a vector function, then

$$\mathbf{r}(a(t)) = x(a(t))\,\mathbf{i} + y(a(t))\,\mathbf{j} + z(a(t))\,\mathbf{k}$$

is a vector function whose derivative is found by using the chain rule on each coordinate. The result is

$$\begin{aligned}\frac{d}{dt}\mathbf{r}(a(t)) &= x'(a(t))a'(t)\,\mathbf{i} + y'(a(t))a'(t)\,\mathbf{j} + z'(a(t))a'(t)\,\mathbf{k}\\ &= \mathbf{r}'(a(t))a'(t).\end{aligned}$$

We illustrate some of these rules in the following example.

EXAMPLE 2 Let $\mathbf{r}(t) = t\,\mathbf{i} + (t^3 - 4t)\,\mathbf{j} + (\ln t)\,\mathbf{k}$, $\mathbf{s}(t) = 3\,\mathbf{i} - t^2\,\mathbf{k}$ and $a(t) = 7t^2$. Compute the derivative of (a) $a(t)\mathbf{r}(t)$, (b) $\mathbf{r}(t)\cdot\mathbf{s}(t)$, (c) $\mathbf{r}(2t^4 - t)$.

Solution: We compute each derivative in two ways just to illustrate how the general case might be treated. The first method uses one of the derivative formulas given above whereas the second method expands the expression and then computes the derivative of each coordinate.

a. For the derivative of the product $a(t)\mathbf{r}(t)$ we have

$$\begin{aligned}&\frac{d}{dt}\big(a(t)\mathbf{r}(t)\big)\\ &\quad= a'(t)\mathbf{r}(t) + a(t)\mathbf{r}'(t)\\ &\quad= 14t\big(t\,\mathbf{i} + (t^3 - 4t)\,\mathbf{j} + (\ln t)\,\mathbf{k}\big) + 7t^2\big(\mathbf{i} + (3t^2 - 4)\,\mathbf{j} + (1/t)\,\mathbf{k}\big)\\ &\quad= 21t^2\,\mathbf{i} + (35t^4 - 84t^2)\,\mathbf{j} + (14t\ln t + 7t)\,\mathbf{k}.\end{aligned}$$

For the second method of evaluating this derivative, we perform the multiplication before doing the differentiation to get

$$\begin{aligned}\frac{d}{dt}\left(a(t)\mathbf{r}(t)\right) &= \frac{d}{dt}\left(7t^3\,\mathbf{i} + (7t^5 - 28t^3)\,\mathbf{j} + 7t^2\ln t\,\mathbf{k}\right)\\ &= 21t^2\,\mathbf{i} + (35t^4 - 84t^2)\,\mathbf{j} + (14t\ln t + 7t)\,\mathbf{k}.\end{aligned}$$

The two computations give the same result.

b. Now for the derivative of the dot product of $\mathbf{r}$ and $\mathbf{s}$, we have

$$\begin{aligned}\frac{d}{dt}\left(\mathbf{r}(t)\cdot\mathbf{s}(t)\right) &= \mathbf{r}'(t)\cdot\mathbf{s}(t) + \mathbf{r}(t)\cdot\mathbf{s}'(t)\\ &= (\mathbf{i} + (3t^2 - 4)\,\mathbf{j} + (1/t)\,\mathbf{k})\cdot(3\,\mathbf{i} - t^2\,\mathbf{k})\\ &\quad + \left(t\,\mathbf{i} + (t^3 - 4t)\,\mathbf{j} + (\ln t)\,\mathbf{k}\right)\cdot(0\,\mathbf{i} - 2t\,\mathbf{k})\\ &= 3 - t - 2t\ln t.\end{aligned}$$

In the second method, we compute the dot product before doing the differentiation and get

$$\begin{aligned}\frac{d}{dt}\left(\mathbf{r}(t)\cdot\mathbf{s}(t)\right) &= \frac{d}{dt}\left(\left(t\,\mathbf{i} + (t^3 - 4t)\,\mathbf{j} + (\ln t)\,\mathbf{k}\right)\cdot\left(3\,\mathbf{i} - t^2\,\mathbf{k}\right)\right)\\ &= \frac{d}{dt}(3t + 0 - t^2\ln t)\\ &= 3 - (t + 2t\ln t).\end{aligned}$$

Once again the two methods produce the same result.

c. This is a chain rule problem. Let $a(t) = 2t^4 - t$ so we are supposed to find the derivative with respect to t of $\mathbf{r}(a(t))$. The chain rule gives the derivative as

$$\begin{aligned}\frac{d}{dt}\mathbf{r}(a(t)) &= \mathbf{r}'(a(t))a'(t)\\ &= \left(\mathbf{i} + (3a(t)^2 - 4)\,\mathbf{j} + (1/a(t))\,\mathbf{k}\right)a'(t)\\ &= \left(\mathbf{i} + (3(2t^4 - t)^2 - 4)\,\mathbf{j} + (1/(2t^4 - t))\,\mathbf{k}\right)(8t^3 - 1).\end{aligned}$$

The derivative could be found without the chain rule by expressing the coordinates of $\mathbf{r}(2t^4 - t)$ as

$$\mathbf{r}(2t^4 - t) = (2t^4 - t)\,\mathbf{i} + \left((2t^4 - t)^3 - 4(2t^4 - t)\right)\mathbf{j} + \ln(2t^4 - t)\,\mathbf{k}$$

and performing the differentiation directly. The reader can check that the same result is obtained by this method. ■

Derivative of the Magnitude

If $\mathbf{r}(t)$ is a differentiable vector function of t then its magnitude $|\mathbf{r}(t)|$ is also a differentiable function at the values of t where the magnitude is nonzero as we show just below. To simplify the notation we write $r(t) = |\mathbf{r}(t)|$. Figure 13.2 shows the curve $\mathbf{r}(t) = 3\cos 3t\,\mathbf{i} + (2 + 2\sin 4t)\,\mathbf{j}$ defined for $0 \le t \le 1.65$. Its magnitude $|\mathbf{r}(t)|$ is plotted as a function of t in figure 13.3.

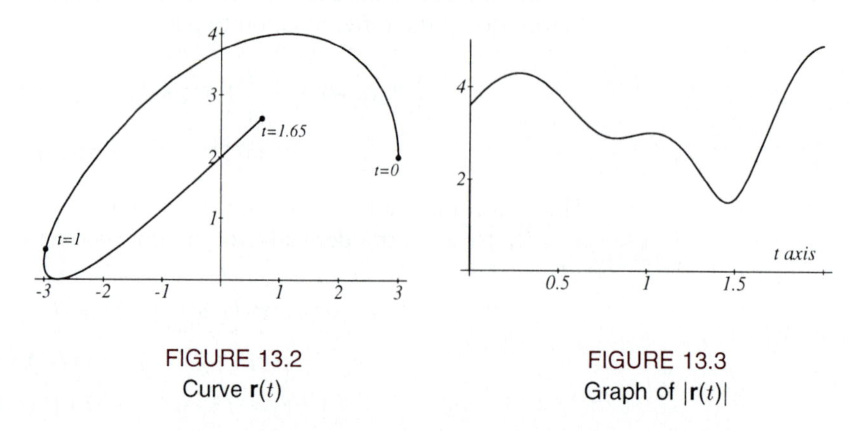

FIGURE 13.2
Curve $\mathbf{r}(t)$

FIGURE 13.3
Graph of $|\mathbf{r}(t)|$

The derivative of $|\mathbf{r}(t)|$ has a very compact expression that can be derived using the rule for the derivative of the dot product. The formula we obtain is

$$r(t)r'(t) = \mathbf{r}(t) \cdot \mathbf{r}'(t). \tag{13.2}$$

To verify this formula we begin with the expression for the square of the magnitude as

$$r(t)^2 = |\mathbf{r}(t)|^2 = \mathbf{r}(t) \cdot \mathbf{r}(t).$$

Take the derivative of the two extreme sides to get

$$2r(t)r'(t) = \mathbf{r}'(t) \cdot \mathbf{r}(t) + \mathbf{r}(t) \cdot \mathbf{r}'(t) = 2\mathbf{r}'(t) \cdot \mathbf{r}(t).$$

After the factor 2 is cancelled, we obtain formula (13.2). If $r(t) \neq 0$, we can solve for $r'(t)$ thus verifying that r is differentiable at t:

$$\frac{d}{dt}|\mathbf{r}(t)| = \frac{\mathbf{r}'(t) \cdot \mathbf{r}(t)}{|\mathbf{r}(t)|} \qquad \text{if} \quad |\mathbf{r}(t)| \neq 0.$$

EXAMPLE 3 Let $\mathbf{r}(t) = 2t\,\mathbf{i} + (1-t)^2\,\mathbf{j}$. Find the derivative of $r(t) = |\mathbf{r}(t)|$.

***Solution*:** From the formula just derived we find

$$\begin{aligned} r(t)r'(t) = \mathbf{r}(t) \cdot \mathbf{r}'(t) &= \left(2t\,\mathbf{i} + (1-t)^2\mathbf{j}\right) \cdot \left(2\,\mathbf{i} - 2(1-t)\,\mathbf{j}\right) \\ &= 4t - 2(1-t)^3. \end{aligned}$$

From the definition of $r(t)$ we have $r(t) = \sqrt{4t^2 + (1-t)^4}$; this function is never 0, so it follows that

$$r'(t) = \frac{\mathbf{r}(t) \cdot \mathbf{r}'(t)}{r(t)} = \frac{4t - 2(1-t)^3}{\sqrt{4t^2 + (1-t)^4}}.$$

Of course one could also compute $r'(t)$ directly differentiating the formula $r(t) = \sqrt{4t^2 + (1-t)^4}$. The same result would be obtained. ■

EXAMPLE 4 If $\mathbf{r}(t) = \cos t\,\mathbf{i} + \sin t\,\mathbf{j}$, show that $\mathbf{r} \cdot \mathbf{r}' = 0$ for all t.

***Solution*:** We have $\mathbf{r} \cdot \mathbf{r} = 1$ for all t so the derivative of the dot product equals 0. The derivative of the dot product also equals $2\mathbf{r}' \cdot \mathbf{r}$ so $\mathbf{r}' \cdot \mathbf{r} = 0$. This shows that the derivative $\mathbf{r}'(t)$ is orthogonal to $\mathbf{r}(t)$ for every t. In this example it would be just as easy to solve the problem directly; we have $\mathbf{r}' = -\sin t\,\mathbf{i} + \cos t\,\mathbf{j}$. One checks that $\mathbf{r} \cdot \mathbf{r}' = 0$. ■

Exercises 13.1

Find $\mathbf{r}'(t)$ and $\mathbf{r}''(t)$ for each of the vector functions.

1. $\mathbf{r}(t) = 3t^4\,\mathbf{i} - (t+2)^2\,\mathbf{j}$

2. $\mathbf{r}(t) = \cos 2t\,\mathbf{i} + \sin 2t\,\mathbf{j} + t\,\mathbf{k}$

3. $\mathbf{r}(t) = (t + \ln t)\,\mathbf{i} - \left(1 - \frac{1}{t}\right)\mathbf{j} + \frac{t}{t+1}\mathbf{k}$

4. $\mathbf{r}(t) = \sin 3t\,\mathbf{i} + \cos 4t\,\mathbf{j} + 2t\,\mathbf{k}$

Find the derivative of $\mathbf{r}(t)$ assuming that $\mathbf{u}$, $\mathbf{v}$, $\mathbf{w}$ and $\mathbf{s}$ are constant vectors.

5. $\mathbf{r}(t) = (\mathbf{w} - t\mathbf{u}) \cdot (\mathbf{s} + 3t\mathbf{v})$

6. $\mathbf{r}(t) = (3t\mathbf{u} + t^2\mathbf{v}) \cdot (t\mathbf{u} - t^3\mathbf{v})$

7. $\mathbf{r}(t) = (\mathbf{u} + t\mathbf{v}) \times (\mathbf{s} + t^2\mathbf{w})$

8. $\mathbf{r}(t) = \Big((t-1)\mathbf{u} - (t+1)\mathbf{v}\Big) \times \Big(t\mathbf{u} + (t-4)\mathbf{v}\Big)$

Find $d\mathbf{r}/dt$ in each of the following.

9. $\mathbf{r}(s) = s^2\,\mathbf{i} - \ln|s+1|\,\mathbf{k}, \quad \text{where } s = t^2 - 1$

10. $\mathbf{r}(p) = 4(p^2 - 1)\,\mathbf{i} + (4 - 3p)\,\mathbf{j}, \quad \text{where } p = (t - 1/t)^3$

11. $\mathbf{r}(s) = s\,\mathbf{i} + \tan s\,\mathbf{j} + \tan 2s\,\mathbf{k}, \quad \text{where } s = \arctan 2t$

12. $\mathbf{r}(p) = p^2[p^3\,\mathbf{i} - p\,\mathbf{j} + (2p+3)^2\,\mathbf{k}]$
    where $p = (t^2 - 3t + 4)$

Sketch the curve $\mathbf{r}(t)$ and the graph of the magnitude $|\mathbf{r}(t)|$.

13. $\mathbf{r}(t) = 4t^2\,\mathbf{i} + t^4\,\mathbf{j}, \ -2 \le t \le 2$

14. $\mathbf{r}(t) = (t+1)^2\,\mathbf{i} + t\,\mathbf{j}, \ -2 \le t \le 2$

15. $\mathbf{r}(t) = \cos 2t\,\mathbf{i} + 4\sin 2t\,\mathbf{j}, \ 0 \le t \le 2\pi$

16. $\mathbf{r}(t) = \Big(2t\,\mathbf{i} + (1 - t^2)\mathbf{j}\Big)/(t^2 + 1), \ -4 \le t \le 4$

17. Let $\mathbf{u}(t) = (t+1)\,\mathbf{i} + 2t\,\mathbf{j}$. Find a positive number δ such that $|\mathbf{u}(t) - \mathbf{i}| < 0.1$ whenever $|t| < \delta$.

18. Let $\mathbf{u}(t) = t^2\,\mathbf{i} - 4t\,\mathbf{j}$ and $\mathbf{v} = \mathbf{i} - 4\,\mathbf{j}$. Find a number $\delta > 0$ such that $|\mathbf{u}(t) - \mathbf{v}| < 0.5$ whenever $|t - 1| < \delta$.

19. Let $\mathbf{u}(t) = (t+1)\,\mathbf{i} + 2t\,\mathbf{j}$. Let a be any fixed number and ϵ any positive number. Find a positive number δ such that $|\mathbf{u}(t) - \mathbf{u}(a)| < \epsilon$ whenever $|t - a| < \delta$.

20. Let $\mathbf{u}(t) = t^2\,\mathbf{i} - 4t\,\mathbf{j}$ and let $\mathbf{v} = \mathbf{i} - 4\,\mathbf{j}$. Let ϵ be any positive number. Find a positive number δ described in terms of ϵ such that $|\mathbf{u}(t) - \mathbf{v}| < \epsilon$ whenever $|t - 1| < \delta$.

21. Let $\mathbf{r}(t) = x(t)\,\mathbf{i} + y(t)\,\mathbf{j} + z(t)\,\mathbf{k}$. Use the definition of the limit of a vector function and the properties of limits of (scalar) functions to prove the assertion made in the text that

$$\lim_{t\to a} \mathbf{r}(t) = \Big(\lim_{t\to a} x(t)\Big)\,\mathbf{i} + \Big(\lim_{t\to a} y(t)\Big)\,\mathbf{j} + \Big(\lim_{t\to a} z(t)\Big)\,\mathbf{k}.$$

22. Let $\mathbf{r}(t)$ be a differentiable vector function such that $|\mathbf{r}(t)| = c$ for some constant c and all t. Show that $\mathbf{r}(t)$ is orthogonal to $\mathbf{r}'(t)$ for all t. Show conversely if $\mathbf{r}(t)$ is orthogonal to $\mathbf{r}'(t)$ for all t, then $|\mathbf{r}(t)|$ is constant.

23. Let $\mathbf{u}(t) = x(t)\,\mathbf{i} + y(t)\,\mathbf{j}$ and $\mathbf{v}(t) = a(t)\,\mathbf{i} + b(t)\,\mathbf{j}$. Prove the derivative formula $(\mathbf{u} \cdot \mathbf{v})' = \mathbf{u}' \cdot \mathbf{v} + \mathbf{u} \cdot \mathbf{v}'$ by expressing $\mathbf{u}$ in terms of x and y, $\mathbf{v}$ in terms of a and b, computing each side of the equation and comparing the results.

24. With $\mathbf{u}$ and $\mathbf{v}$ as in exercise 23, follow the suggestion given there and verify the formula

$$(\mathbf{u} \times \mathbf{v})' = \mathbf{u}' \times \mathbf{v} + \mathbf{u} \times \mathbf{v}'.$$

25. Using Equation (13.2) for the formula of the derivative of $|\mathbf{r}(t)|$ prove the formula

$$\frac{d}{dt}\frac{\mathbf{r}}{|\mathbf{r}|} = \frac{|\mathbf{r}|^2\mathbf{r}' - (\mathbf{r} \cdot \mathbf{r}')\mathbf{r}}{|\mathbf{r}|^3}.$$

26. A proof of a derivative formula is called *coordinate-free* if the formula is derived using the definition $\lim_{h\to 0}[\mathbf{r}(t+h) - \mathbf{r}(t)]/h$ directly without resorting to the use of coordinate functions. Give a coordinate-free proof of each of the following formulas:
    a. $(a\mathbf{r})' = a'\mathbf{r} + a\mathbf{r}'$,
    b. $(\mathbf{r} \cdot \mathbf{s})' = \mathbf{r} \cdot \mathbf{s}' + \mathbf{r}' \cdot \mathbf{s}$.

13.2 Integral of a vector function

If $f(t)$ is a continuous function, then the indefinite integral $\int f(t)\,dt$ is a function whose derivative equals $f(t)$. By analogy, we define the indefinite integral of a vector function to be a vector function whose derivative is the original function; if $\mathbf{r}(t)$ is a vector function of t, then $\int \mathbf{r}(t)\,dt$ is a vector function whose derivative is $\mathbf{r}(t)$. When $\mathbf{r}(t)$ is given in terms of coordinate functions

$$\mathbf{r}(t) = x(t)\,\mathbf{i} + y(t)\,\mathbf{j} + z(t)\,\mathbf{k}, \tag{13.3}$$

then the indefinite integral of $\mathbf{r}$ is found by integrating the coordinate functions:

$$\begin{aligned}\int \mathbf{r}(t)\,dt &= \int \big(x(t)\,\mathbf{i} + y(t)\,\mathbf{j} + z(t)\,\mathbf{k}\big)\,dt \\ &= \Big(\int x(t)\,dt\Big)\,\mathbf{i} + \Big(\int y(t)\,dt\Big)\,\mathbf{j} + \Big(\int z(t)\,dt\Big)\,\mathbf{k}.\end{aligned}$$

We can see the correctness of this method of integration by computing the derivative of the right-hand side and finding that it equals $\mathbf{r}(t)$.

Many of the familiar properties of the indefinite integral of scalar functions carry over to the case of vector functions. We list three of these properties:

$$\int \mathbf{u}(t) + \mathbf{v}(t)\,dt = \int \mathbf{u}(t)\,dt + \int \mathbf{v}(t)\,dt,$$

$$\int a\mathbf{u}(t)\,dt = a\int \mathbf{u}(t)\,dt, \qquad a = \text{constant},$$

$$\int \mathbf{u}'(t)\,dt = \mathbf{u}(t) + \mathbf{c}, \qquad \mathbf{c} = \text{a constant vector}.$$

The Definite Integral of a Vector Function

The Fundamental Theorem of Calculus gives a method of computing the definite integral of a continuous scalar function; if $F'(x) = f(x)$, then $\int_a^b f(x)\,dx = F(b) - F(a)$. This equality is not the definition of the definite integral of $f(x)$ but rather a method by which the definite integral can be computed.

The definite integral of a vector function is defined by using the Fundamental Theorem of Calculus for scalar functions. When $\mathbf{r}(t)$ is given in terms of coordinate functions as in Equation (13.3) we have

$$\begin{aligned}\int_a^b \mathbf{r}(t)\,dt &= \int_a^b x(t)\,\mathbf{i} + y(t)\,\mathbf{j} + z(t)\,\mathbf{k}\,dt \\ &= \mathbf{i}\int_a^b x(t)\,dt + \mathbf{j}\int_a^b y(t)\,dt + \mathbf{k}\int_a^b z(t)\,dt.\end{aligned}$$

The following formulation can be considered as the Fundamental Theorem of Calculus for vector functions.

THEOREM 13.1

Let $\mathbf{r}(t)$ be a continuous vector function for $a \le t \le b$. Then there is a function $\mathbf{u}(t)$ such that $\mathbf{u}'(t) = \mathbf{r}(t)$ and

$$\int_a^b \mathbf{r}(t)\,dt = \mathbf{u}(t)\Big|_a^b = \mathbf{u}(b) - \mathbf{u}(a).$$

The proof is carried out by expressing $\mathbf{r}$ in terms of its coordinate functions and then applying the Fundamental Theorem for scalar functions to each coordinate function. We leave this as an exercise for the reader.

EXAMPLE 1 If $\mathbf{r}(t) = 2(t + 1)\mathbf{i} + \sin t\,\mathbf{j}$, evaluate the integrals

$$\int \mathbf{r}(t)\,dt \quad \text{and} \quad \int_0^2 \mathbf{r}(t)\,dt.$$

Solution: The indefinite integral is found by integrating the coordinate functions:

$$\int \mathbf{r}(t)\,dt = \int (t+1)\mathbf{i} + \sin t\,\mathbf{j}\,dt = (\frac{1}{2}(t+1)^2 + c_1)\mathbf{i} + (-\cos t + c_2)\mathbf{j}$$
$$= \frac{1}{2}(t+1)^2\mathbf{i} - \cos t\,\mathbf{j} + \mathbf{c}$$

where $\mathbf{c} = c_1\mathbf{i} + c_2\mathbf{j}$ is a constant vector.

The definite integral is computed by evaluating the definite integral of the coordinate functions:

$$\int_0^2 (t+1)\mathbf{i} + \sin t\,\mathbf{j}\,dt = \frac{1}{2}(t+1)^2\Big|_0^2\mathbf{i} - \cos t\Big|_0^2\mathbf{j} = 4\mathbf{i} - (\cos 2 - 1)\mathbf{j}.$$ ■

In preparation for use in the next theorem, we give a property relating the dot product and the integral of a vector function. Observe that the integral $\int_a^b \mathbf{r}(t)\,dt$ is a vector and so the dot product $\mathbf{u} \cdot \int_a^b \mathbf{r}(t)\,dt$ is meaningful for any vector $\mathbf{u}$. Also $\mathbf{u} \cdot \mathbf{r}(t)$ is a scalar function so $\int_a^b \mathbf{u} \cdot \mathbf{r}(t)\,dt$ is the integral of a continuous scalar function. The next theorem relates these two integral expressions.

THEOREM 13.2

Let $\mathbf{u}$ be a constant vector and $\mathbf{r}(t)$ a continuous vector function for $a \le t \le b$. Then $\mathbf{u} \cdot \int_a^b \mathbf{r}(t)\,dt = \int_a^b \mathbf{u} \cdot \mathbf{r}(t)\,dt$.

Proof

We give the proof in two dimensions; the proof in three dimensions is similar. Let $\mathbf{r}(t) = x(t)\mathbf{i} + y(t)\mathbf{j}$ and $\mathbf{u} = p\mathbf{i} + q\mathbf{j}$. Then

$$\mathbf{u} \cdot \int_a^b \mathbf{r}(t)\,dt = (p\mathbf{i} + q\mathbf{j}) \cdot \left(\int_a^b x(t)\mathbf{i} + y(t)\mathbf{j}\,dt\right)$$
$$= (p\mathbf{i} + q\mathbf{j}) \cdot \left(\left(\int_a^b x(t)\,dt\right)\mathbf{i} + \left(\int_a^b y(t)\,dt\right)\mathbf{j}\right)$$
$$= p\left(\int_a^b x(t)\,dt\right) + q\left(\int_a^b y(t)\,dt\right)$$
$$= \int_a^b px(t) + qy(t)\,dt = \int_a^b \mathbf{u} \cdot \mathbf{r}(t)\,dt,$$

which proves the theorem. □

Next we use this to prove an important inequality satisfied by the integral of a vector function.

THEOREM 13.3

Let $\mathbf{r}(t)$ be a continuous vector function for $a \leq t \leq b$. Then

$$\left|\int_a^b \mathbf{r}(t)\,dt\right| \leq \int_a^b |\mathbf{r}(t)|\,dt. \tag{13.4}$$

Proof

We begin by reminding the reader of the basic inequality relating the dot product and the length of vectors. If **u** and **v** are vectors then, using the fact that $|\cos\theta| \leq 1$, we have

$$|\mathbf{u}\cdot\mathbf{v}| = |\mathbf{u}|\,|\mathbf{v}|\,|\cos\theta| \leq |\mathbf{u}|\,|\mathbf{v}|,$$

where θ is the angle between **u** and **v**.

Now let

$$\mathbf{u} = \int_a^b \mathbf{r}(t)\,dt$$

so that **u** is a constant vector. In the next step we use the inequality $|\int_a^b f(x)\,dx| \leq \int_a^b |f(x)|\,dx$ that holds for any scalar function f (because $f(x) \leq |f(x)|$). Because **u** is a constant vector, we have

$$|\mathbf{u}|^2 = |\mathbf{u}\cdot\mathbf{u}| = \left|\mathbf{u}\cdot\int_a^b \mathbf{r}(t)\,dt\right| = \left|\int_a^b \mathbf{u}\cdot\mathbf{r}(t)\,dt\right|$$

$$\leq \int_a^b |\mathbf{u}\cdot\mathbf{r}(t)|\,dt \leq \int_a^b |\mathbf{u}|\,|\mathbf{r}(t)|\,dt = |\mathbf{u}|\int_a^b |\mathbf{r}(t)|\,dt.$$

If $|\mathbf{u}| \neq 0$ then we can cancel $|\mathbf{u}|$ and obtain the inequality we want. If $|\mathbf{u}| = 0$ then inequality (13.4) is true because the left side equals 0 while the right side is nonnegative as it is the integral of a nonnegative function. □

Riemann Sums

The expression of the definite integral of a scalar function as a Riemann sum was given in chapter 5 as

$$\int_a^b f(x)\,dx = \lim_{n\to\infty}\sum_{i=1}^{n} f(w_{i,n})(x_{i,n} - x_{i-1,n}), \tag{13.5}$$

where $P_n : a = x_{0,n} < x_{1,n} < \cdots < x_{n,n} = b$ is a partition of $[a,b]$, the numbers $w_{i,n}$ lie on $[x_{i-1,n}, x_{i,n}]$ and the sequence of norms $\|P_n\|$ has limit 0.

There is a corresponding expression for the definite integral of a vector function as a limit of Riemann sums. Let P_n be a partition of the interval $[a,b]$ over which the vector function $\mathbf{r}(t)$ is continuous. Suppose we have

$$P_n : a = t_{0,n} < t_{1,n} < \cdots < t_{n,n} = b.$$

Assume that $\lim_{n\to\infty}\|P_n\| = 0$ and that some number $w_{i,n}$ is selected on each interval $[t_{i-1,n}, t_{i,n}]$. Then we have

Definite Integral of a Vector Function

$$\int_a^b \mathbf{r}(t)\,dt = \lim_{n\to\infty} \sum_{i=1}^{n} \mathbf{r}(w_{i,n})(t_{i,n} - t_{i-1,n}).$$

To verify that the vector form follows from the scalar form of the theorem, we write $\mathbf{r}(t) = x(t)\,\mathbf{i} + y(t)\,\mathbf{j}$ (in the two dimensional case). We then have

$$\begin{aligned}
\int_a^b \mathbf{r}(t)\,dt &= \mathbf{i}\int_a^b x(t)\,dt + \mathbf{j}\int_a^b y(t)\,dt \\
&= \mathbf{i}\left(\lim_{n\to\infty}\sum_{i=1}^{n} x(w_{i,n})(t_{i,n} - t_{i-1,n})\right) + \mathbf{j}\left(\lim_{n\to\infty}\sum_{i=1}^{n} y(w_{i,n})(t_{i,n} - t_{i-1,n})\right) \\
&= \lim_{n\to\infty}\sum_{i=1}^{n}\big(x(w_{i,n})\,\mathbf{i} + y(w_{i,n})\,\mathbf{j}\big)(t_{i,n} - t_{i-1,n}) \\
&= \lim_{n\to\infty}\sum_{i=1}^{n} \mathbf{r}(w_{i,n})(t_{i,n} - t_{i-1,n}).
\end{aligned}$$

The proof of the formula in the three-dimensional case requires only minor modification. This formula is used later in this chapter to give an integral formula for the length of a curve in two or three dimensions.

Exercises 13.2

Evaluate the integrals.

1. $\int 3t^4\,\mathbf{i} - (t+2)^2\,\mathbf{j}\,dt$

2. $\int \sin 4t\,\mathbf{i} - 2\cos 2t\,\mathbf{j} + t^2\,\mathbf{k}\,dt$

3. $\int_0^1 (1-t)\,\mathbf{i} + (1-t^2)\,\mathbf{j}\,dt$

4. $\int_\pi^{2\pi} \sin t\,\mathbf{i} + \sin 2t\,\mathbf{j} + \sin 4t\,\mathbf{k}\,dt$

5. $\int_0^{\pi/2} (\sin t\,\mathbf{i} - \cos t\,\mathbf{j})\cdot(\mathbf{i} + \mathbf{j})\,dt$

6. $\int_0^1 |(1-t^2)\,\mathbf{i} + 2t\,\mathbf{j}|\,dt$

For each value of $\mathbf{c}$ and function $\mathbf{r}(t)$, compute the two expressions

$$\int_0^1 \mathbf{c}\cdot\mathbf{r}(t)\,dt \quad \text{and} \quad \mathbf{c}\cdot\int_0^1 \mathbf{r}(t)\,dt.$$

7. $\mathbf{c} = 2\,\mathbf{i} + 3\,\mathbf{j}$, $\mathbf{r}(t) = t^2\,\mathbf{i} + (1-t)\,\mathbf{j}$

8. $\mathbf{c} = 4\,\mathbf{i} - 3\,\mathbf{j}$, $\mathbf{r}(t) = (3-t)^2\,\mathbf{i} + (4-t)^2\,\mathbf{j}$

9. $\mathbf{c} = \sin p\,\mathbf{i} + \cos p\,\mathbf{j}$, $\mathbf{r}(t) = \sin t\,\mathbf{i} + \cos t\,\mathbf{j}$

10. $\mathbf{c} = e^{-1}\,\mathbf{i} + e\,\mathbf{j}$, $\mathbf{r}(t) = e^t\,\mathbf{i} + e^{-t}\,\mathbf{j}$

Find the function $\mathbf{r}(t)$ from the given data.

11. $\mathbf{r}'(t) = t\,\mathbf{i} - t^2\,\mathbf{j} + 4\,\mathbf{k}$, $\mathbf{r}(0) = \mathbf{i}$

12. $\mathbf{r}'(t) = \cos t\,\mathbf{i} - 4\sin t\,\mathbf{j}$, $\mathbf{r}(0) = \mathbf{i} + \mathbf{j}$

13. $\mathbf{r}'(t) = 2e^{-t}\,\mathbf{i} - e^{2t}\,\mathbf{k}$, $\mathbf{r}(0) = \mathbf{i}$

14. $\mathbf{r}'(t) = (e^t - e^{-t})\,\mathbf{i} - t^2\,\mathbf{j} + (e^t + e^{-t})\,\mathbf{k}$, $\mathbf{r}(0) = \mathbf{i}$

15. Let $\mathbf{r}(t)$ be a continuous vector function for t on the interval $I = [a, b]$ and for every t on I let

$$\mathbf{u}(t) = \int_a^t \mathbf{r}(s)\,ds.$$

Prove $\mathbf{u}'(t) = \mathbf{r}(t)$. Conclude that every continuous vector function has an antiderivative.

16. Let $\mathbf{r}(t)$ be a continuous vector function that satisfies $|\mathbf{r}(t)| = 2$ for all t on $[0, 4]$. Prove $|\int_0^4 \mathbf{r}(t)\,dt| \leq 8$.

17. Let g be a continuous scalar function on $I = [a, b]$. Prove

$$\left|\int_a^b \big(\cos g(t)\,\mathbf{i} + \sin g(t)\,\mathbf{j}\big)\,dt\right| \leq b - a.$$

18. Express the integral $\int_0^2 t\,\mathbf{i} + (4t^2 + 1)\,\mathbf{j}\,dt$ as a limit of Riemann sums.

19. Express the integral $\int_1^3 4t^2\,\mathbf{i} + (2t - 1)\,\mathbf{j} + t^3\,\mathbf{k}\,dt$ as a limit of Riemann sums in two different ways by making different choices of points on the subintervals for the evaluation of the function.

20. Let $\mathbf{r}(t) = t\,\mathbf{i} + t^2\,\mathbf{j}$. Evaluate the limit

$$\lim_{n\to\infty} \sum_{i=1}^{n} \mathbf{r}\left(\frac{i}{n}\right)\left(\frac{1}{n}\right)$$

by interpreting it as a definite integral.

21. Let $\mathbf{r}(t) = (1-t)\,\mathbf{i} + 2t\,\mathbf{j} + (t^2+4)\,\mathbf{k}$. Evaluate the limit

$$\lim_{n\to\infty} \sum_{j=1}^{n} \mathbf{r}\left(\frac{n+2j}{n}\right)\left(\frac{2}{n}\right)$$

by interpreting it as a definite integral.

13.3 Curves in parametric and vector form

We consider two methods for describing curves in two and three dimensions. We have used functions $y = f(x)$ to describe curves in two dimensions but the graph of a function is a somewhat special curve in that each vertical line meets the graph of a function in at most one point. A curve as familiar as a circle is not the graph of a function. We now introduce two methods, closely related, to describe very general curves: position vectors and parametric equations can be used to represent curves more general than the graph of a function. Moreover the same notions carry over to three dimensions with almost no changes.

Suppose a point moves in three-dimensional space along a curve so that its position at time t is the point $P = (x(t), y(t), z(t))$ where $x(t)$, $y(t)$ and $z(t)$ are real-valued functions of t. The vector from the origin to the point P is a vector function of t defined by

$$\mathbf{r}(t) = x(t)\,\mathbf{i} + y(t)\,\mathbf{j} + z(t)\,\mathbf{k}$$

called the *position vector* of the moving particle. The path of the particle is a curve traced out by the terminal point of $\mathbf{r}(t)$ as t varies. We call $\mathbf{r}(t)$ a *vector representation* of the curve. If the path of the particle lies in the xy plane and the position at time t is the point $(x(t), y(t))$ then the position vector is $\mathbf{r}(t) = x(t)\,\mathbf{i} + y(t)\,\mathbf{j}$. In either case we refer to the vector function $\mathbf{r}(t)$ as a representation of the curve. We illustrate these ideas with some two-dimensional examples.

EXAMPLE 1 Describe the path of the moving particle whose position at time t is $(\cos t, \sin t)$.

Solution: The position vector $\mathbf{r}(t)$ has terminal point $(\cos t, \sin t)$ and initial point at the origin for any t, so the position vector is

$$\mathbf{r}(t) = \cos t\,\mathbf{i} + \sin t\,\mathbf{j}.$$

The terminal point of $\mathbf{r}(t)$ lies on the circle with equation $x^2 + y^2 = 1$. For $t = 0$, the particle is at $(1, 0)$ and, as t increases, the point at $(\cos t, \sin t)$ moves counterclockwise around the circle returning to $(1, 0)$ when $t = 2\pi$. Thus the path is the unit circle traversed in the counterclockwise direction as t increases. If t ranges over all real numbers, the circle is traced infinitely many times. When t is restricted to an interval of length 2π, such as $[0, 2\pi]$ or $[-\pi, \pi]$, the circle is traced out just once. One point is covered twice as t ranges over the intervals because $\mathbf{r}(0) = \mathbf{r}(2\pi)$ and $\mathbf{r}(-\pi) = \mathbf{r}(\pi)$. If we do not want even that much overlap, then the interval should be restricted to $[0, 2\pi)$ or $(-\pi, \pi]$ for example.

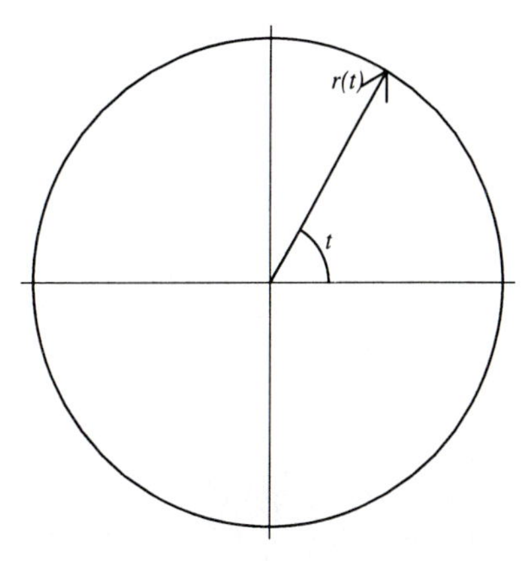

FIGURE 13.4
Position vector $\mathbf{r}(t) = \cos t\,\mathbf{i} + \sin t\,\mathbf{j}$

■

EXAMPLE 2 A particle moves along the parabola $y = 4x^2$ so that its velocity in the x direction at time t is $6t$. If it starts at $(0, 0)$ at time $t = 0$, find the position vector of the moving particle at time t and find the velocity in the y direction.

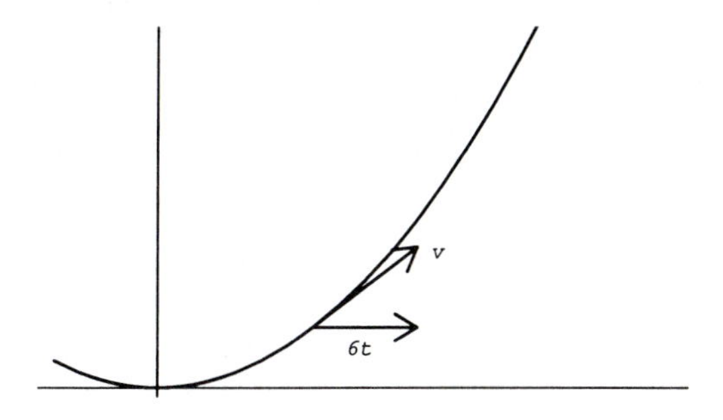

FIGURE 13.5
Velocity tangent to $y = 4x^2$

Solution: The position at time t given by the position vector $\mathbf{r}(t) = x(t)\,\mathbf{i}+y(t)\,\mathbf{j}$ and the particle is at the point $(x(t), y(t))$ at time t. We are told that the particle is on the parabola so $y(t) = 4x(t)^2$ for all t. In addition the velocity in the x direction is $dx/dt = 6t$. To find the function $x(t)$ we integrate once and obtain $x(t) = 3t^2 + c$ for some constant c. At time $t = 0$, we have $x(0) = 0$ because the particle is at the origin at time $t = 0$. Thus $0 = x(0) = 3(0)^2 + c$ and so $c = 0$. Thus $x(t) = 3t^2$ and $y(t) = 4x(t)^2 = 4 \cdot 9t^4$. It follows that the position vector is $\mathbf{r}(t) = 3t^2\,\mathbf{i} + 36t^4\,\mathbf{j}$. The velocity in the y direction is the rate of change of the y coordinate, namely the derivative of $36t^4$. Thus the velocity in the y direction at time t is $4 \cdot 36t^3 = 138t^3$. ■

Parametric Equations

If a two dimensional position vector $\mathbf{r}(t) = x(t)\,\mathbf{i} + y(t)\,\mathbf{j}$ traces out a curve, the equations

$$\begin{cases} x = x(t), \\ y = y(t), \end{cases}$$

are called parametric equations of the curve. A three-dimensional position vector $\mathbf{r}(t) = x(t)\,\mathbf{i} + y(t)\,\mathbf{j} + z(t)\,\mathbf{k}$ traces out a curve consisting of all points (x, y, z) with coordinates given by the parametric equations

$$\begin{cases} x = x(t), \\ y = y(t), \\ z = z(t). \end{cases}$$

The circular path traced out in the first example above has parametric equations

$$\begin{cases} x = \cos t, \\ y = \sin t. \end{cases}$$

In a two-dimensional case such as this one, the rectangular form of the equation of the curve is found by eliminating t from the two equations and expressing the two equations as a single equation relating x and y. Care must be taken to consider the range of t and the values assumed by the two functions $x(t)$ and $y(t)$. When t is eliminated from the pair of equations, there might be additional points (x, y) satisfying the single equation that were not on the original curve. Here is an example to illustrate this point.

EXAMPLE 3 Describe the path traced by the position vector $\mathbf{r}(t) = 3\cos t\,\mathbf{i} - 2\sin t\,\mathbf{j}$ as t ranges over the interval $[0, \pi]$.

Solution: The parametric equations of the path are

$$\begin{cases} x = 3\cos t, \\ y = -2\sin t. \end{cases}$$

We can eliminate the parameter t by using the equation

$$1 = \cos^2 t + \sin^2 t = \left(\frac{x}{3}\right)^2 + \left(\frac{y}{-2}\right)^2.$$

Thus every point of the path lies on the ellipse with equation

$$1 = \left(\frac{x}{3}\right)^2 + \left(\frac{y}{2}\right)^2.$$

The entire ellipse is not traced out however because of the restricted range on t. At $t = 0$ we see the point on the curve is

$$(x, y) = (3\cos 0, -2\sin 0) = (3, 0).$$

As t increases toward π, $x = 3\cos t$ decreases from 3 to -3 and $y = -2\sin t$ varies from 0 to -2 at $t = \pi/2$ and then back to 0 at $t = \pi$. Thus only the lower half of the ellipse is traced out by the position vector with t on the interval $[0, \pi]$. ■

Parametric Equations of an Ellipse

An ellipse or part of an ellipse is traced out by the position vector

$$\mathbf{r}(t) = a\cos t\,\mathbf{i} + b\sin t\,\mathbf{j} \tag{13.6}$$

for nonzero constants a and b. The parametric equations

$$\begin{cases} x = a\cos t, \\ y = b\sin t, \end{cases}$$

determine an ellipse whose equation is found by eliminating the parameter t (as in example 3, just above) to get

$$1 = \left(\frac{x}{a}\right)^2 + \left(\frac{y}{b}\right)^2.$$

The entire ellipse is traced out if t ranges over an interval of length 2π such as $[0, 2\pi]$ or $[-\pi, \pi]$. If t ranges over a shorter interval, only part of the ellipse is covered by the moving point.

We incorporate the parametric equations of a two-dimensional ellipse into the next example of a curve in three dimensions.

EXAMPLE 4 Describe the path having position vector $\mathbf{r}(t) = 4\cos t\,\mathbf{i} + 6\sin t\,\mathbf{j} + t\,\mathbf{k}$ as t ranges over the interval $[0, 2\pi]$.

Solution: The path has parametric equations

$$\begin{cases} x = 4\cos t, \\ y = 6\sin t, \\ z = t. \end{cases}$$

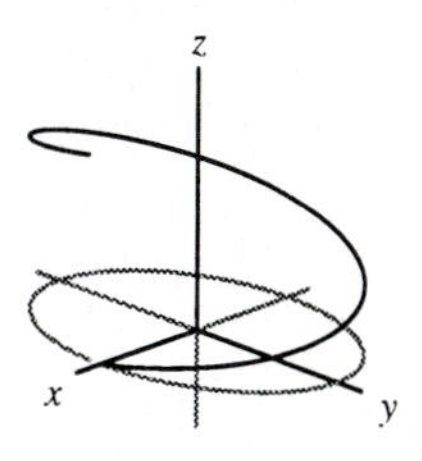

FIGURE 13.6
Spiral lying over an ellipse

To describe the path, we begin by examining its projection into the xy plane. The projection of a point on the curve is found by setting the z coordinate equal to 0; a point (x, y, z) projects to the point $(x, y, 0)$. Thus the projection has parametric equations

$$\begin{cases} x = 4\cos t, \\ y = 6\sin t, \\ z = 0, \end{cases}$$

which is a curve in the xy plane lying on the ellipse with rectangular equation $(x/4)^2 + (y/6)^2 = 1$. Because the range of t is $[0, 2\pi]$, $4\cos t$ varies from -4 to $+4$ and $6\sin t$ varies from -6 to $+6$ and the entire ellipse is traversed by the moving point. Moreover at $t = 0$, the position vector is $\mathbf{r}(0) = 4\,\mathbf{i}$; as t increases, the corresponding point on the projected curve moves in the counterclockwise direction once around the ellipse and returns to the starting point when $t = 2\pi$.

A point on the original three-dimensional curve lies directly above its projection on the ellipse. The height above the xy plane is the z coordinate, namely $z = t$. As t increases, the point on the curve spirals upward around the z-axis always staying directly above the ellipse in the xy plane. The final position $\mathbf{r}(2\pi)$ is 2π units above the initial position $\mathbf{r}(0)$. ■

Here is another example.

EXAMPLE 5 Describe the curve having parametric equations

$$\begin{cases} x = \dfrac{3(t^2-1)}{5(t^2+1)}, \\ y = \dfrac{4(t^2-1)}{5(t^2+1)}, \\ z = \dfrac{2t}{(t^2+1)}. \end{cases} \tag{13.7}$$

Solution: The curve appears difficult to describe because of the somewhat complicated rational functions that define the coordinates. Let us begin with some observations to indicate that we can say something about this curve. Each coordinate function is bounded; that is for any value of t we see that $|x| \leq 1$, $|y| \leq 1$, $|z| \leq 1$. Thus for any value of t, the corresponding point on the curve stays within the box with corners at $(\pm 1, \pm 1, \pm 1)$. Having seen this, we might be inclined to consider a more specific relations between the coordinates. With a bit of algebraic manipulation, we find

$$x^2 + y^2 + z^2 = \frac{9(t^2-1)^2}{25(t^2+1)^2} + \frac{16(t^2-1)^2}{25(t^2+1)^2} + \frac{4t^2}{(t^2+1)^2} = 1.$$

Thus every point on the curve is also on the sphere of radius 1 having center at the origin. One further observation can be made. The projection of the points on the curve to the xy plane trace out a curve given parametrically by

$$\begin{cases} x = \dfrac{3(t^2-1)}{5(t^2+1)}, \\ y = \dfrac{4(t^2-1)}{5(t^2+1)}, \end{cases}$$

which is a part of the line $y = (4/3)x$ in the xy plane passing through the origin. Every point on the curve lies directly above or below the line $y = (4/3)x$. The plane determined by the z-axis and the line $y = (4/3)x$ intersects the unit sphere in a circle that contains the path. The path does not cover the entire circle. The line $y = (4/3)x$ meets the unit sphere at the two points $(3/5, 4/5, 0)$ and $(-3/5, -4/5, 0)$ but the first of these points is not on the curve. This can be seen by showing that the equations

$$x = \frac{3}{5} = \frac{3(t^2-1)}{5(t^2+1)},$$
$$y = \frac{4}{5} = \frac{4(t^2-1)}{5(t^2+1)},$$

cannot be solved for t. We leave it as a nontrivial exercise for the reader to show that all other points on this circle are on the path (13.7). (This can be done by checking the intervals in the x, y and z directions covered by the curve.) ∎

EXAMPLE 6 Find a vector function $\mathbf{r}(t)$ that represents the entire ellipse $4x^2 + y^2 = 16$.

Solution: There are many solutions to this problem; we give only two. By taking $x = 2\cos t$ and $y = 4\sin t$ we see that

$$4x^2 + y^2 = 4(2\cos t)^2 + (4\sin t)^2 = 16(\cos^2 t + \sin^2 t) = 16.$$

Thus $\mathbf{r}(t) = 2\cos t\,\mathbf{i} + 4\sin t\,\mathbf{j}$ is one vector function tracing out the ellipse. Note that this choice traces the ellipse in the counterclockwise direction. Other choices can be found by replacing t with ωt for any nonzero constant ω. A more significant alternative comes by taking $x = 2\sin t$ and $y = 4\cos t$ and the vector function $\mathbf{r}(t) = 2\sin t\,\mathbf{i} + 4\cos t\,\mathbf{j}$. Notice that this vector function traces the ellipse in the clockwise direction. ■

The Cycloid

Next we study a two-dimensional curve that appears in a number of applications. The *cycloid* is the path traced by a fixed point on the circumference of a circle as the circle rolls without slipping along a straight line. As an illustration, imagine a reflector attached to the outer edge of a moving bicycle wheel as the bike and rider pass through the beam of your headlights at night. If the reflector is on the rim of the wheel, the path traced by the reflector is a cycloid. Let us find the parametric equations for this curve.

EXAMPLE 7 Suppose the wheel of a bicycle has radius R and a point on the outside edge of the wheel is marked with a reflector that is initially on the ground at time $t = 0$. If the cycle moves with a constant velocity v find the position of the reflector at time t and thereby determine the parametric equations of the cycloid curve.

Solution: The motion of the reflector can be described in two steps: At time t the center of the wheel has moved parallel to the ground a distance vt and the wheel has rotated through a certain angle θ. The position at time t is the same as if the wheel were first rotated with its center fixed and then, with no further rotation, the center of the wheel is moved to its position at time t.

We first describe the rotational motion. The position vector of a moving particle on a circle of radius R having its center at the origin is given in terms of the trigonometric functions. If the marked point is initially at $R(\cos\alpha\,\mathbf{i} + \sin\alpha\,\mathbf{j})$ and then the circle is rotated through an angle θ, the position is

$$\mathbf{r}_1 = R\cos(\alpha + \theta)\,\mathbf{i} + R\sin(\alpha + \theta)\,\mathbf{j}.$$

If the center is initially at $R\mathbf{j}$ then the position of the marked point after rotation through θ is

$$\mathbf{r}_2 = R\cos(\alpha + \theta)\,\mathbf{i} + R\sin(\alpha + \theta)\,\mathbf{j} + R\,\mathbf{j}.$$

Next we describe the horizontal motion of the circle. The velocity is a constant v so after time t the position of the center is $\mathbf{r}_3 = vt\,\mathbf{i} + R\,\mathbf{j}$. Addition of $vt\,\mathbf{i}$ to any point of the circle gives the position after the horizontal motion. Hence the position of the reflector is

$$\mathbf{r} = \mathbf{r}_2 + vt\,\mathbf{i} = \big(R\cos(\alpha + \theta) + vt\big)\,\mathbf{i} + \big(R\sin(\alpha + \theta) + R\big)\,\mathbf{j}.$$

This must be simplified because θ and t are related. A positive angle of rotation θ produces motion in the negative $\mathbf{i}$ direction. The horizontal distance vt traveled by the center equals the circumference of the part of the circle that travels along the ground; it has length $R\theta$ because θ is the central angle swept out. So we have the relation $R\theta = -vt$. Using the parameter θ rather than t makes the expression simpler. The result is

$$\mathbf{r} = \mathbf{r}(\theta) = R(-\theta + \cos(\alpha + \theta))\,\mathbf{i} + R(1 + \sin(\alpha + \theta))\,\mathbf{j}.$$

In this specific example, the reflector started out on the ground while the center was $(0, R)$. Thus the initial angle α at time $t = 0$ was $\alpha = -\pi/2$. Using the identities

$$\cos(\theta - \frac{\pi}{2}) = \sin\theta, \qquad \sin(\theta - \frac{\pi}{2}) = -\cos\theta,$$

we obtain

$$\mathbf{r} = R(-\theta + \sin\theta)\,\mathbf{i} + R(1 - \cos\theta)\,\mathbf{j}.$$

The parametric equations of this cycloid are

$$\begin{cases} x = R(-\theta + \sin\theta), \\ y = R(1 - \cos\theta). \end{cases}$$

If the position of the reflector is wanted as a function of time t, one substitutes $\theta = vt/R$ into these equations. ■

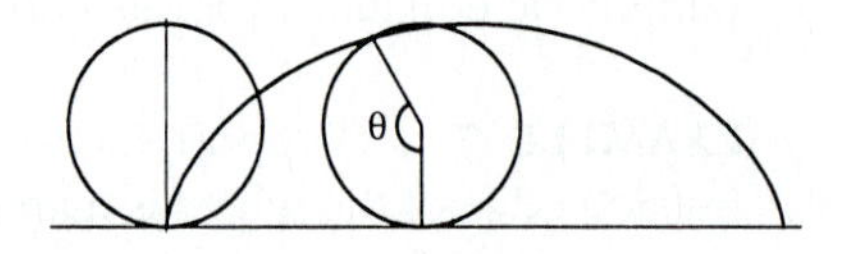

FIGURE 13.7
Cycloid with parameter θ

In the example, the equations obtained for the cycloid have x decreasing with increasing θ and thus the moving wheel travels to the left with increasing θ. By replacing θ with $-\theta$ we obtain the equations

$$\begin{cases} x = R(\theta - \sin\theta), \\ y = R(1 - \cos\theta), \end{cases}$$

of the cycloid in which x increases with increasing θ.

Different Parametric Representations

A curve given parametrically always has many different parametric representations. For example a curve given by

$$\begin{cases} x = x(t) \\ y = y(t) \end{cases} \qquad 0 \le t \le 4$$

can also be represented by the equations

$$\begin{cases} x = x(2s + 2) \\ y = y(2s + 2) \end{cases} \qquad -1 \le s \le 1$$

by making the change of variable $t = 2s + 2$.

A much more general change of variable takes the form $t = g(s)$ for some function g; the curve has parametric equations

$$\begin{cases} x = x(g(s)), \\ y = y(g(s)). \end{cases}$$

Some restrictions must be placed on g. For example if t ranges over the interval $[a, b]$ then the domain of g must be restricted to an interval $[c, d]$ such that for each number t with $a \leq t \leq b$, there is one and only one number s such that $c \leq s \leq d$ and $g(s) = t$. This condition is satisfied if we assume g is continuous, increasing and satisfies $g(c) = a$, $g(d) = b$. In this case we say the change of variable $t = g(s)$ preserves the direction of the curve. If g is continuous, decreasing and satisfies $g(c) = b$, $g(d) = a$ then we say the change of variable $t = g(s)$ reverses the direction of the curve.

Here is a change of variable that reverses the direction of the curve. If t ranges over the interval $[a, b]$ for any $a < b$, then we let $t = a + b - s$. As t ranges from a to b, s ranges from b to a. In other words, s covers the same interval as t but in the reverse direction. Thus the curve

$$\begin{cases} x = x(t) \\ y = y(t) \end{cases} \qquad a \leq t \leq b$$

is the same curve as the one given by

$$\begin{cases} x = x(a + b - s) \\ y = y(a + b - s) \end{cases} \qquad a \leq s \leq b$$

except that in the second case the curve is traversed in the direction opposite to the direction traversed in the first representation.

Here is an example of a change of parameter that is more complicated and slightly more difficult to discover.

EXAMPLE 8 Show that the ellipse with parametric equations $x(\theta) = 3\cos\theta$ and $y(\theta) = 4\sin\theta$ for $0 \leq \theta < 2\pi$ can be parameterized with rational functions as

$$X(t) = \frac{3(1 - t^2)}{1 + t^2}, \qquad Y(t) = \frac{8t}{1 + t^2} \qquad -\infty < t < \infty.$$

Determine if the two curves are identical.

***Solution*:** The rectangular form of the ellipse with parametric equations $x(\theta) = 3\cos\theta$ and $y(\theta) = 4\sin\theta$ is

$$\left(\frac{x}{3}\right)^2 + \left(\frac{y}{4}\right)^2 = 1.$$

The substitution of $x = X(t)$ and $y = Y(t)$ shows that this equation is satisfied so that the curve described as a rational function of t is part of the ellipse given in terms of θ. Now we determine if every point on the ellipse can be expressed as $(X(t), Y(t))$. If we try to solve the equations

$$X(t) = \frac{3(1 - t^2)}{1 + t^2} = 3\cos\theta,$$
$$Y(t) = \frac{8t}{1 + t^2} = 4\sin\theta,$$

we find

$$t^2(1 + \cos\theta) = 1 - \cos\theta,$$
$$t^2\sin\theta - 2t + \sin\theta = 0,$$

The first of these equations cannot be solved if $\cos\theta = -1$ so that value must be omitted. Assume that $\cos\theta \neq -1$. Then from the first equation we obtain

$$t^2 = \frac{1 - \cos\theta}{1 + \cos\theta}$$

and substituting for t^2 in the second equation leads to

$$t = \frac{\sin\theta}{1 + \cos\theta} = \tan\left(\frac{\theta}{2}\right).$$

Thus every point on the ellipse except $(-3, 0)$, the point where $\cos\theta = -1$, is on the curve given parametrically by $(X(t), Y(t))$. The point $(-3, 0)$ is obtained by taking the limit as $t \to \infty$. ■

Exercises 13.3

Express the curves given as vector equations or parametric equations in rectangular coordinates by equations involving only x and y.

1. $x = 2t + 1,\ y = t - 2$

2. $x = 11 - 3t,\ y = 5t - 2$

3. $\mathbf{r}(t) = (2t+1)\,\mathbf{i}+(t-2)\,\mathbf{j}$

4. $\mathbf{r}(t) = (1-3t)\,\mathbf{i}+(5t+1)\,\mathbf{j}$

5. $x = t + 1,\ y = t^2 + 1$

6. $\mathbf{r} = \sin 4t\,\mathbf{i} + 3\cos 4t\,\mathbf{j}$

7. $x = 2\cos t,\ y = 3\sin t$

8. $\mathbf{r} = -3\cos 2t\,\mathbf{i}+6\sin 2t\,\mathbf{j}$

Sketch the curves given by the vector function or by the parametric equations.

9. $\mathbf{r}(t) = \mathbf{i} + t(2\mathbf{i} - \mathbf{j})$

10. $\mathbf{r}(t) = \mathbf{i} + \mathbf{j} - 3t\,\mathbf{i} + 6t\,\mathbf{j}$

11. $\mathbf{r}(t) = \mathbf{i} + t^2(2\mathbf{i} - \mathbf{j})$

12. $x(t) = 1+t^2,\ y(t) = 3-t$

13. $x(t) = 2 + 4t,\ y(t) = -1 + t$

14. $x(t) = -2 + 5t,\ y(t) = 6 + t$

15. $\mathbf{r}(t) = t(\mathbf{i} + \mathbf{j}) + t^2(\mathbf{i} - \mathbf{j})$

16. $\mathbf{r}(t) = \mathbf{i} + (t^2 + 3)(2\mathbf{i} - \mathbf{j})$

Decide if the two curves are the same, if one is part of the other or if they are unrelated. Select a suitable domain for the functions if none is given.

17. $\mathbf{r}_1(t) = \cos 2t\,\mathbf{i} + \sin 2t\,\mathbf{j},\ 0 \le t \le \pi,$
$\mathbf{r}_2(p) = \sin p\,\mathbf{i} + \cos p\,\mathbf{j},\ 0 \le p \le 2\pi$

18. $C_1 : x(t) = \sin t,\ \ y(t) = 2 + 4\sin t,$
$C_2 : x(p) = 2 + 2p,\ y(p) = 10 + 8p$

19. $C_1 : x(t) = \tan t,\ \ y(t) = 2\sec t,$
$C_2 : x(p) = \sqrt{p-1},\ y(p) = 2\sqrt{p}$

20. $\mathbf{r}_1(t) = 2\sin t(\mathbf{i} + \mathbf{j}) + 2\cos t(\mathbf{i} - \mathbf{j}),$
$\mathbf{r}_2(p) = 2\sqrt{2}(\sin p\,\mathbf{i} + \cos p\,\mathbf{j})$

21. $C_1 : x = x(t),\ y = y(t),\ 0 \le t,$
$C_2 : x = x(4p^2),\ y = y(4p^2),\ -\infty < p < +\infty$

22. $C_1 : x = x(2t - 1),\ y = y(2t - 1),\ 0 \le t,$
$C_2 : x = x(1 - p),\ y = y(1 - p),\ p < 2$

Find a parameterization of the form

$$\begin{cases} x = x(t) \\ y = y(t) \end{cases} \qquad 0 \le t \le 1$$

for each curve C. Note the restriction on t.

23. C is the straight line from $(1, 1)$ to $(3, 4)$.

24. C is the line segment from $(0, a)$ to $(b, 0)$.

25. C is the upper half of the circle of radius 4 with center at the origin.

26. C is the part of the parabola $y = x^2$ from $(0, 0)$ to $(4, 16)$.

27. C is part of the parabola $y = x^2 - 4x$ from $(0, 0)$ to $(4, 0)$.

28. Show that the position vector

$$\mathbf{r}(t) = (e^t + e^{-t})\,\mathbf{i} + (e^t - e^{-t})\,\mathbf{j}$$

traces out one branch of the hyperbola $x^2 - y^2 = 4$. Explain how to modify this parametrization to obtain a vector representation of the other branch of the hyperbola.

29. Find a vector representation, using exponential functions, of each branch of the hyperbola

$$\frac{x^2}{a^2} - \frac{y^2}{b^2} = 1.$$

[See exercise 28 for the case $a = b = 2$.]

30. Let $\mathbf{u} = (\mathbf{i} + \mathbf{j})/\sqrt{2}$ and $\mathbf{v} = (\mathbf{i} - \mathbf{j})/\sqrt{2}$. The curve

$$\mathbf{r}(t) = 3\cos t\,\mathbf{u} + 2\sin t\,\mathbf{v}$$

is an ellipse but not in standard position. Sketch the curve and also find the parametric equations of the curve and then eliminate the parameters t and obtain the rectangular equations of the curve.

31. Let $\mathbf{r}(t) = \left(t(2\mathbf{i} + \mathbf{j}) + t^2(\mathbf{i} - 2\mathbf{j})\right)/\sqrt{5}$. Then $\mathbf{r}(t)$ is the vector equation of a rotated parabola. Sketch the curve and find both the parametric and rectangular equations of the curve.

Find a vector function representing the curve given in rectangular coordinates.

32. $x^2 + 16y^2 = 1$

33. $4x^2 + 12y^2 = 9$

34. $y = 4x^2$

35. $y^2 = x$

36. $x^2 + (y + 1)^2 = 4$

37. $y = f(x)$

38. A reflector is attached to the spoke of a bicycle wheel of radius R at a distance a from the center, $a \le R$. If the bicycle travels along the x-axis at a constant velocity v, find the position of the reflector at time t. Assume that at $t = 0$ the reflector is at (a, R) and the center of the wheel is at $(0, R)$.

39. The wheel of a railroad car has radius R and has a cover of radius $R + h$ with $h > 0$ so that the outer edge of the cover rides below the point where the wheel contacts the rail. A spot is marked on the outside edge of the cover at a distance $R + h$ from the center. Find the parametric equations of the spot as the rail car moves. Assume the rail is the x-axis and the spot starts at time 0 at the point $(0, -h)$, its lowest point. Use the parametric equations to determine if there is any time when the spot on the rim is moving in the opposite direction of the car.

40. A bicycle wheel of radius R with center at the origin turns with constant angular velocity ω. An insect walks along a spoke of the wheel with constant linear velocity v from the center of the wheel to the rim. Find the position, with respect to the center of the wheel, of the insect at time t.

41. Repeat exercise 40 except that now assume the bicycle is moving along the x-axis and it is required to find the position of the insect with respect to some fixed point, say the initial position of the insect.

13.4 Tangents and normals to curves

Suppose $\mathbf{r}(t) = x(t)\mathbf{i} + y(t)\mathbf{j} + z(t)\mathbf{k}$ is the position vector representing a curve and that the coordinate functions are differentiable on the interval over which t ranges. How do we find the tangent line to the curve at some particular point $\mathbf{r}(p)$? We proceed in the same manner that was used to find the slope of the tangent line to the graph of a function; we consider a line through the terminal point P of $\mathbf{r}(p)$ and a nearby point Q on the curve. For small values of h, $Q = \mathbf{r}(p + h)$ is a nearby point.

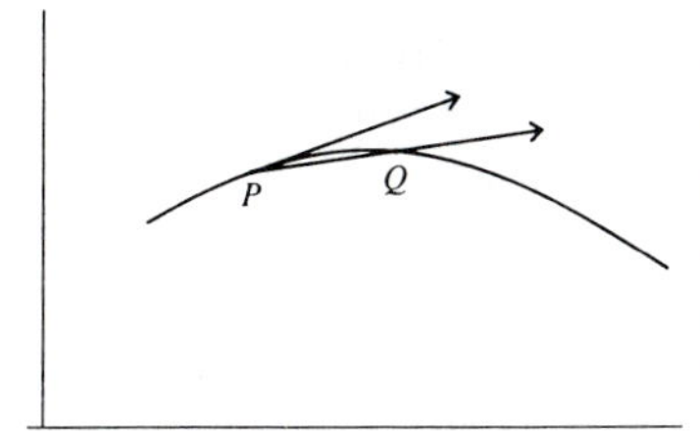

FIGURE 13.8

The vector

$$\overrightarrow{PQ} = \mathbf{r}(p + h) - \mathbf{r}(p)$$

from P to Q is "close" to the tangent line for small values of h and any scalar multiple of $\mathbf{r}(p + h) - \mathbf{r}(p)$ is a parallel vector. In particular

$$\frac{\mathbf{r}(p + h) - \mathbf{r}(p)}{h} \qquad \textbf{(13.8)}$$

is parallel to $\overrightarrow{PQ}$ and is close to the tangent line at the point where $t = p$. As h decreases toward 0, vector (13.8) approaches a vector tangent to the path at $t = p$. There is an exception that must be made here. It could happen that $\mathbf{r}'(p) = 0$ in which case no conclusion can be drawn about the direction of $\mathbf{r}'(p)$ inasmuch as the zero vector does not have a specific direction.

We summarize these facts in the following.

Tangent Vector

Suppose $\mathbf{r}(t)$ is a differentiable function at $t = p$ and $\mathbf{r}'(p) \neq 0$. Then the vector $\mathbf{r}'(p)$ is parallel to the tangent line to the curve $\mathbf{r}(t)$ at the point with position vector $\mathbf{r}(p)$.

EXAMPLE 1 Find the tangent vector to the cycloid $\mathbf{r}(t) = (t - \sin t)\mathbf{i} + (1 - \cos t)\mathbf{j}$. (This cycloid corresponds to the path of a point on the rim of a circle of radius 1 moving to the right on the x-axis with velocity 1, as described earlier.)

Solution: The tangent vector is the derivative of the position function at points where the derivative is nonzero. In this case we have

$$\mathbf{r}'(t) = (1 - \cos t)\,\mathbf{i} + \sin t\,\mathbf{j}.$$

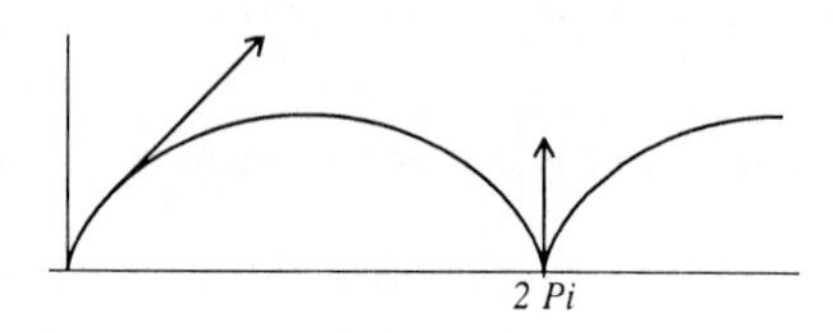

FIGURE 13.9
Two tangents on a cycloid

The derivative is zero when $\sin t = 0$ and $\cos t = 1$. This happens when $t = 0$ or more generally at $t = 2\pi n$ for any integer n. The tangent line to the curve is parallel to $\mathbf{r}'(t)$ except possibly at the points where $t = 2\pi n$. An examination of the graph of the cycloid suggests that there is a vertical tangent line at these points. The proof that this is so is left to the exercises following this section. ■

Unit Tangent

Suppose a curve C is represented by the vector function

$$\mathbf{r}(t) = x(t)\,\mathbf{i} + y(t)\,\mathbf{j} + z(t)\,\mathbf{k}$$

with t ranging over some interval I (possibly infinite). Suppose further that the function $\mathbf{r}(t)$ is twice differentiable and that $\mathbf{r}'(t)$ is nonzero for all values of t on I. Then $\mathbf{r}'(t)$ is a nonzero tangent vector to the curve at the point $(x(t), y(t), z(t))$. If we divide the tangent vector by its length we obtain a unit tangent vector

$$\mathbf{T}(t) = \frac{\mathbf{r}'(t)}{|\mathbf{r}'(t)|}. \tag{13.9}$$

We give some properties in the following theorem.

THEOREM 13.4

Suppose that $\mathbf{r}(t)$ is twice differentiable and $\mathbf{r}'(t)$ is nonzero for all t on an interval I. Then the unit tangent vector $\mathbf{T}(t)$ given by Equation (13.9) is differentiable at each t on I and its derivative $\mathbf{T}'(t)$ is orthogonal to $\mathbf{T}(t)$ for each t in I.

Proof

The denominator $|\mathbf{r}'(t)|$ is nonzero so the derivative of $\mathbf{T}$ can be computed by the quotient rule. We now show that the derivative of $\mathbf{T}$ is orthogonal to $\mathbf{T}$. From the fact that $\mathbf{T}(t)$ is a unit vector for each t we have $\mathbf{T}(t)\cdot\mathbf{T}(t) = 1$. Take derivatives in this equation to get

$$\mathbf{T}(t)\cdot\mathbf{T}'(t) + \mathbf{T}'(t)\cdot\mathbf{T}(t) = 0.$$

It follows that $2\mathbf{T}'(t) \cdot \mathbf{T}(t) = 0$ and finally $\mathbf{T}'(t) \cdot \mathbf{T}(t) = 0$. This proves that $\mathbf{T}'(t)$ is orthogonal to $\mathbf{T}(t)$. □

Principal Normal

For those values of t for which $\mathbf{T}'(t) \neq 0$, we can form the unit vector

$$\mathbf{N}(t) = \frac{\mathbf{T}'(t)}{|\mathbf{T}'(t)|}.$$

The vector $\mathbf{N}(t)$ is called the *principal normal vector.* It is orthogonal to $\mathbf{T}(t)$ and so $\mathbf{N}(t)$ is normal to the curve. The line through $(x(t), y(t), z(t))$ and parallel to $\mathbf{N}(t)$ is called the *principal normal line* to the curve at the point. In the case of a two-dimensional curve, there is at most one normal line, the line perpendicular to the tangent line but in three dimensions there are infinitely many normal lines to a curve at a point. They all lie in the plane orthogonal to the tangent line.

It might happen that $\mathbf{T}'(t)$ is zero for some t; when $\mathbf{T}'(t) = 0$ then $\mathbf{N}(t)$ is not defined by the above formula. There may or may not be a uniquely defined normal line in this case. When we refer to $\mathbf{N}(t)$ in the following discussion, we always understand that we mean t is restricted to values where $\mathbf{N}(t)$ is defined.

EXAMPLE 2 Let the curve C have representation $\mathbf{r}(t) = 2t\,\mathbf{i} - 3t^2\,\mathbf{j} + 3t^3\,\mathbf{k}$ for all numbers t. Find the unit tangent $\mathbf{T}$ and the principal normal $\mathbf{N}$ for general t and then find the unit tangent and principal normal at the point where $t = 1$.

Solution: We first differentiate the position vector to get

$$\mathbf{r}'(t) = 2\,\mathbf{i} - 6t\,\mathbf{j} + 9t^2\,\mathbf{k}.$$

Then

$$|\mathbf{r}'(t)| = \sqrt{4 + (-6t)^2 + (9t^2)^2} = \sqrt{(2 + 9t^2)^2} = 2 + 9t^2.$$

For the unit tangent vector we obtain

$$\mathbf{T}(t) = \frac{\mathbf{r}'(t)}{|\mathbf{r}'(t)|} = \frac{2\,\mathbf{i} - 6t\,\mathbf{j} + 9t^2\,\mathbf{k}}{2 + 9t^2}.$$

The principal normal is found by dividing the derivative of $\mathbf{T}$ by its length. First compute the derivative $\mathbf{T}'(t)$. By applying the quotient rule we obtain

$$\mathbf{T}'(t) = \frac{-36t\,\mathbf{i} - (12 - 54t^2)\,\mathbf{j} + 36t\,\mathbf{k}}{(2 + 9t^2)^2}.$$

The length of $\mathbf{T}'$ is

$$\frac{\sqrt{(-36t)^2 + (12 - 54t^2)^2 + (36t)^2}}{(2 + 9t^2)^2} = \frac{6}{2 + 9t^2}.$$

Finally we have

$$\mathbf{N}(t) = \frac{(2 + 9t^2)\mathbf{T}'(t)}{6} = \frac{-6t\,\mathbf{i} - (2 - 9t^2)\,\mathbf{j} + 6t\,\mathbf{k}}{2 + 9t^2}.$$

Thus we have $\mathbf{T}(t)$ and $\mathbf{N}(t)$ for general t. The point on the curve corresponding to $t = 1$ has position vector $\mathbf{r}(1) = 2\,\mathbf{i} - 3\,\mathbf{j} + 3\,\mathbf{k}$. The unit tangent and principal normal at this point are

$$\mathbf{T}(1) = \frac{2\,\mathbf{i} - 6\,\mathbf{j} + 9\,\mathbf{k}}{11} \quad \text{and} \quad \mathbf{N}(1) = \frac{-6\,\mathbf{i} + 7\,\mathbf{j} + 6\,\mathbf{k}}{11}.$$

A check shows $\mathbf{T}(1) \cdot \mathbf{N}(1) = 0$ so they are indeed orthogonal. As a check on the calculations we verify that $\mathbf{T}(t)$ and $\mathbf{N}(t)$ are orthogonal for every t by forming the dot product:

$$\begin{aligned}\mathbf{T}(t)\cdot\mathbf{N}(t) &= \frac{2\mathbf{i}-6t\,\mathbf{j}+9t^2\,\mathbf{k}}{2+9t^2}\cdot\frac{-6t\,\mathbf{i}-(2-9t^2)\mathbf{j}+6t\,\mathbf{k}}{(2+9t^2)}\\ &= \frac{-12t-6t(-2+9t^2)+9t^2(6t)}{(2+9t^2)^2}=0\end{aligned}$$

verifying what we knew ought to be true in general. ■

Direction of the Tangent and Normal

If P is a point on the differentiable curve C, there are two unit tangent vectors to C at P. If $\mathbf{T}$ is one choice then $-\mathbf{T}$ is the other. Similarly, if $\mathbf{N}$ is a unit normal to the curve at P then $-\mathbf{N}$ is another unit normal. If the curve has a vector representation given by $\mathbf{r}(t)$ with $a \le t \le b$ then the unit tangent vector

$$\mathbf{T}(t) = \frac{\mathbf{r}'(t)}{|\mathbf{r}'(t)|}$$

is only one of the two possible unit tangent vectors at $\mathbf{r}(t)$. It could happen that the same curve is represented by another vector function $\mathbf{F}(u)$ for $c \le u \le d$. The point P on the curve has position vector $\mathbf{r}(t_0) = \mathbf{F}(u_0)$ for some t_0 and some u_0. The unit tangent vector using $\mathbf{F}$ is defined by

$$\mathbf{T}_1(u) = \frac{\mathbf{F}'(u)}{|\mathbf{F}'(u)|}$$

and gives a unit tangent $\mathbf{T}_1(u_0)$ at P. This tangent might be equal to $\mathbf{T}(t_0)$ or it might equal the negative of $\mathbf{T}(t_0)$. Let us illustrate the possibilities with an example.

EXAMPLE 3 The arc of the parabola $y = x^2$ for $-1 \le x \le 1$ has the vector representations

$$\begin{aligned}\mathbf{r}(t) &= t\,\mathbf{i} + t^2\,\mathbf{j} &\quad -1 \le t \le 1,\\ \mathbf{F}(u) &= -u\,\mathbf{i} + u^2\,\mathbf{j} &\quad -1 \le u \le 1.\end{aligned}$$

Compute the unit tangent vectors at every point of the curve using the two vector representations.

Solution: The derivative of each position vector divided by the length gives the two unit tangent functions

$$\mathbf{T}(t) = \frac{\mathbf{i}+2t\,\mathbf{j}}{\sqrt{1+4t^2}} \qquad \mathbf{T}_1(u) = \frac{-\mathbf{i}+2u\,\mathbf{j}}{\sqrt{1+4u^2}}.$$

The point $P = (a, a^2)$ has position vector given as $\mathbf{r}(a)$ and $\mathbf{F}(-a)$, so the respective unit tangent vectors are

$$\mathbf{T}(a) = \frac{\mathbf{i}+2a\,\mathbf{j}}{\sqrt{1+4a^2}} \qquad \mathbf{T}_1(-a) = \frac{-\mathbf{i}-2a\,\mathbf{j}}{\sqrt{1+4a^2}}.$$

Thus the direction of the unit tangent vector depends on the particular parameterization of the curve and is not determined by any property of the curve itself. Note that the path is traced from left to right by $\mathbf{r}(t)$ as t ranges from -1 to $+1$, whereas the path is traced from right to left by $\mathbf{F}(u)$ as u ranges from -1 to $+1$. Thus the unit tangent points roughly in the direction of the motion. ■

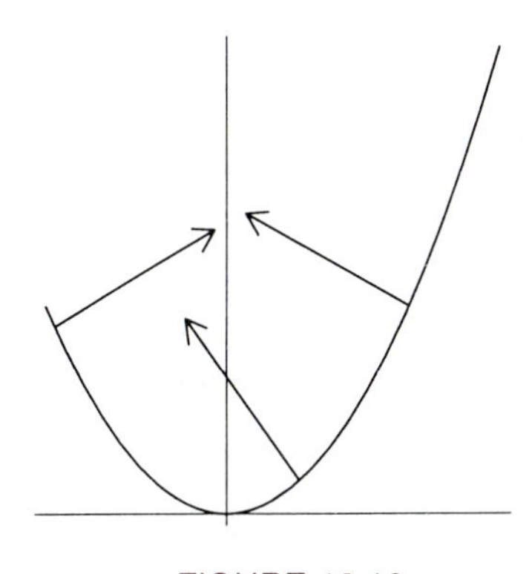

FIGURE 13.10
Unit normals to $y = x^2$

The normal vector depends less on the choice of parameterization. The vector **N** is parallel to the derivative of **T** and points in the direction in which **T** is changing. For a small value of $|h|$, **N** points nearly in the direction of $\mathbf{T}(t+h) - \mathbf{T}(t)$, which is the vector from the terminal point of $\mathbf{T}(t)$ to the terminal point of $\mathbf{T}(t+h)$. If $\mathbf{r}(t)$ is your position at time t as you walk along the curve and if the curve causes you to turn to the right at $\mathbf{r}(t)$, then the unit tangent is turning to the right and the principal normal points to the right. Thus the principal normal points in the direction of the bend in the curve.

Exercises 13.4

For each curve find the unit tangent and the principal normal at the indicated point and verify that the two vectors are orthogonal.

1. $\mathbf{r}(t) = 3t\,\mathbf{i} + 2t^2\,\mathbf{j},\ t = 1$

2. $\mathbf{r}(t) = (t^2 - 4)\,\mathbf{i} + (3t + 2)\,\mathbf{j},\ t = 0$

3. $\mathbf{r}(t) = e^t\,\mathbf{i} + e^{-t}\,\mathbf{j},\ t = 0$

4. $\mathbf{r}(t) = (t + \sin t)\,\mathbf{i} + \cos t\,\mathbf{j},\ t = \pi$

5. $\mathbf{r}(t) = \left(2t\,\mathbf{i} + (1 - t^2)\,\mathbf{j}\right)/(1 + t^2),\ t = 1$

6. $\mathbf{r}(t) = t\,\mathbf{i} + t^2\,\mathbf{j} + t^3\,\mathbf{k},\ t = 0$

A rocket travels along a curved path while its engine is running but travels in a straight line with the engine off. The straight line path is tangent to the curved path if the engine is shut down in midflight. Answer the following two questions based on this premise.

7. The position of the rocket under power is

$$\mathbf{r}(t) = (50 + 100/t^2)\,\mathbf{i} + (160t - 40/t)\,\mathbf{j} + 4t\,\mathbf{k}.$$

If the engine is shut down at $t = 10$, what is the position of the rocket for all t with $t \geq 10$?

8. The position of the rocket at time t is

$$\mathbf{r}(t) = (2t + 1)\,\mathbf{i} + (2 + 4/t)\,\mathbf{j} + (10 + \ln t)\,\mathbf{k}.$$

At what time should the engine be shut off to have the rocket "coast" into the space station located at $33\,\mathbf{i} + (13 + 2\ln 2)\,\mathbf{k}$?

9. Find the points of the curve $\mathbf{r}(t) = t\,\mathbf{i} + (4 + t^2)\,\mathbf{j}$ where $\mathbf{r}(t)$ and $\mathbf{r}'(t)$ are perpendicular. Find the points where $\mathbf{r}(t)$ and $\mathbf{r}'(t)$ are parallel.

10. Let $a(t)$ be any differentiable scalar function and **u** and **v** any constant vectors. Prove that the curve $\mathbf{r}(t) = a(t)\mathbf{u} + \mathbf{v}$ is part of or all of a straight line. Compute the unit tangent $\mathbf{T}(t)$ to this curve.

11. Sketch the curve with equation $\mathbf{r}(t) = t^2\,\mathbf{i} + t^3\,\mathbf{j}$ for $|t| \leq 1$. Note that Theorem 13.4 does not apply at $t = 0$. Does this curve have a tangent at $\mathbf{r}(0)$?

12. The two circles $\mathbf{r}_1(t) = \cos t\,\mathbf{i} + \sin t\,\mathbf{j}$ and $\mathbf{r}_2(u) = \cos u\,\mathbf{j} + \sin u\,\mathbf{k}$ have points in common. Find the tangents to each curve at the common points.

13. Let $\mathbf{r}(t) = x(t)\,\mathbf{i} + y(t)\,\mathbf{j}$ be the vector form of a curve C in the xy plane and suppose that the graph of the function $f(x)$ coincides with C. If the functions $x(t)$ and $y(t)$ are differentiable with respect to t and $f(x)$ is differentiable with respect to x prove the formula

$$\frac{df(x)}{dx} = \frac{dy/dt}{dx/dt} \quad \text{when} \quad dx/dt \neq 0.$$

14. The cycloid with parametrization given by $\mathbf{r}(t) = (t + \sin t)\,\mathbf{i} + (1 - \cos t)\,\mathbf{j}$ has a vanishing derivative at $t = 0$. Argue that the curve has a vertical tangent at $\mathbf{r}(0)$. [Hint: If $\mathbf{r}' = u\,\mathbf{i} + v\,\mathbf{j}$, then the slope of the tangent line at $t = 0$ is the common value of the one sided limits $\lim_{t\to 0^+} v/u$ and $\lim_{t\to 0^-} v/u$.]

15. Suppose the curve $\mathbf{r}(t)$ has the property that the unit tangent vector $\mathbf{T}(t)$ is constant, $\mathbf{T}(t) = \mathbf{u}$ for all t. Prove that the curve $\mathbf{r}(t)$ is a straight line or part of a straight line.

16. The cubic curve $y = x^3$ has the vector representation $\mathbf{r}(t) = t\,\mathbf{i} + t^3\,\mathbf{j}$. Use this to calculate the unit tangent vector $\mathbf{T}(t)$ and the principal normal vector $\mathbf{N}(t)$. Discuss the direction of $\mathbf{N}(t)$ on the intervals $-1 < t < 0$ and $0 < t < 1$. What happens to **N** at $t = 0$?

17. Prove that $d\mathbf{T}/dt = 0$ (and hence the normal $\mathbf{N}(t)$ to the curve $\mathbf{r}(t)$ is undefined) at any point where $\mathbf{r}''(t) = 0$. [Hint: Compute the derivative of **T** in terms of **r**, its derivatives and absolute values of these quantities.]

13.5 Arc length in two and three dimensions

In section 6.3 we defined the length of the graph of a function $y = f(x)$, $a \leq x \leq b$ as the least upper bound of the chord approximations and saw how to compute this length as the integral of $\sqrt{1 + f'(x)^2}$ over the interval $[a, b]$. We use the same ideas now to compute the length of a curve given parametrically in either two or three dimensions. Even in the two-dimensional case we gain some new information inasmuch as we obtain a formula for the length of a curve that might not the graph of a function.

We begin by stating the arc-length formula in several forms and then we give an indication of the proof. Recall that a scalar function $f(t)$ is called smooth if the derivative $f'(t)$ is continuous. For a vector function the definition of smooth is similar but slightly more restrictive.

DEFINITION 13.2

Smooth Vector Function

A vector function $\mathbf{r}(t)$ is called *smooth* over an interval I if its derivative $\mathbf{r}'(t)$ is continuous and *nonzero* at every point t in I.

The condition that the derivative is nonzero ensures that the curve traced out by $\mathbf{r}(t)$ has a tangent line at every point and thus eliminates some pathology. Let $\mathbf{r}(t)$ be a smooth vector function for t on the interval $[a, b]$. Then the length L of the curve traced out by $\mathbf{r}(t)$ as t varies over the interval $[a, b]$ is

Arc-Length Formula

$$L = \int_a^b |\mathbf{r}'(t)|\, dt.$$

If $\mathbf{r}(t) = x(t)\mathbf{i} + y(t)\mathbf{j}$ is a smooth curve in the plane, the length L of the curve traced out as t varies over $[a, b]$ is

Arc Length in Two Dimensions

$$L = \int_a^b \sqrt{x'(t)^2 + y'(t)^2}\, dt.$$

If $\mathbf{r}(t) = x(t)\mathbf{i} + y(t)\mathbf{j} + z(t)\mathbf{k}$ is a smooth curve in the three-dimensional space, the length L of the curve traced out as t varies over $[a, b]$ is

Arc Length in Three Dimensions

$$L = \int_a^b \sqrt{x'(t)^2 + y'(t)^2 + z'(t)^2}\, dt.$$

We prove these formulas are correct. We first recall some theoretical points. The length of the curve traced out by $\mathbf{r}(t)$ for t on $[a, b]$ is the least upper bound of the chord approximations. The meaning of chord approximation is the same as in section 6.3. Start with a partition $P : a = t_0 < t_1 < \cdots < t_n = b$ and consider the chords connecting the adjacent points $\mathbf{r}(t_0), \ldots, \mathbf{r}(t_n)$ on the curve. The sum of the lengths of these chords is the chord approximation corresponding to the partition P. The vector from the terminal point of $\mathbf{r}(t_{i-1})$ to the terminal point of $\mathbf{r}(t_i)$ is the vector $\mathbf{r}(t_i) - \mathbf{r}(t_{i-1})$ and so the chord approximation corresponding to P is

$$L_P = \sum_{i=1}^{n} |\mathbf{r}(t_i) - \mathbf{r}(t_{i-1})|. \tag{13.10}$$

Let $s(t)$ be the length of the arc from $\mathbf{r}(a)$ to $\mathbf{r}(t)$ for any t on $[a, b]$. Our goal is to show

$$s(t) = \int_a^t |\mathbf{r}'(u)|\, du. \tag{13.11}$$

We accomplish this by proving $s'(t) = |\mathbf{r}'(t)|$ for all t on $[a, b]$. The proof is not just one step and requires some preparation.

As the first step we prove

$$s(t) \le \int_a^t |\mathbf{r}'(u)|\, du. \tag{13.12}$$

Let P be the partition of $[a, b]$ as above and L_P the chord approximation in Equation (13.10). We have

$$\mathbf{r}(t_i) - \mathbf{r}(t_{i-1}) = \int_{t_{i-1}}^{t_i} \mathbf{r}'(u)\, du$$

and so

$$|\mathbf{r}(t_i) - \mathbf{r}(t_{i-1})| = \left| \int_{t_{i-1}}^{t_i} \mathbf{r}'(u)\, du \right| \le \int_{t_{i-1}}^{t_i} |\mathbf{r}'(u)|\, du.$$

It follows

$$L_P = \sum_{i=1}^{n} |\mathbf{r}(t_i) - \mathbf{r}(t_{i-1})| \le \sum_{i=1}^{n} \int_{t_{i-1}}^{t_i} |\mathbf{r}'(u)|\, du = \int_a^b |\mathbf{r}'(u)|\, du.$$

Thus the integral on the right is an upper bound for all the chord approximations. But $s(b)$ is the least upper bound for all the chord approximations so we have

$$s(b) \le \int_a^b |\mathbf{r}'(u)|\, du.$$

If we replace b with any point t on the interval and repeat this argument with the interval $[a, t]$ in place of $[a, b]$, then we obtain inequality (13.12).

If u and v satisfy $a \le u \le v \le b$ then the length of the arc from $\mathbf{r}(a)$ to $\mathbf{r}(v)$ is the sum of the lengths of the arc from $\mathbf{r}(a)$ to $\mathbf{r}(u)$ and from $\mathbf{r}(u)$ to $\mathbf{r}(v)$. This additivity property of arc length requires some proof. Because the property is quite intuitive, we leave it to the exercises following this section.

Now take t on $[a, b]$ and a positive h such that $t + h \le b$. Then we have

$$|\mathbf{r}(t + h) - \mathbf{r}(t)| \le s(t + h) - s(t) \le \int_t^{t+h} |\mathbf{r}'(u)|\, du.$$

The first inequality holds because the straight line distance from $\mathbf{r}(t)$ to $\mathbf{r}(t + h)$ is less than or equal to the length of the path between these two points. The second inequality follows from inequality (13.12) applied to the interval $[t, t + h]$.

Now divide each term by the positive number h to get the inequalities

$$\left|\frac{\mathbf{r}(t+h)-\mathbf{r}(t)}{h}\right| \leq \frac{s(t+h)-s(t)}{h} \leq \frac{1}{h}\int_t^{t+h} |\mathbf{r}'(u)|\,du.$$

Take the limit as h approaches 0. The first two terms have limits $|\mathbf{r}'(t)| \leq s'(t)$. The limit of the term with the integral is the derivative of the function $F(t) = \int_a^t |\mathbf{r}'(u)|\,du$ and equals $|\mathbf{r}'(t)|$ by the Fundamental Theorem of Calculus. It follows that $|\mathbf{r}'(t)| \leq s'(t) \leq |\mathbf{r}'(t)|$ and so $|\mathbf{r}'(t)| = s'(t)$ as we wished to prove.

Since $s(a) = 0$, we have

$$s(t) = \int_a^t s'(t)\,dt = \int_a^t |\mathbf{r}'(t)|\,dt,$$

which completes the proof of arc-length formula.

The Arc-Length Function

We summarize what has just been proved. Let $\mathbf{r}(t)$ be a smooth function representing a curve in either two or three dimensions for $a \leq t \leq b$. Let $s(t)$ be defined by

$$s(t) = \int_a^t |\mathbf{r}'(u)|\,du.$$

Then $s(t)$ is a function whose value at t is the length of the path from the terminal point of $\mathbf{r}(a)$ to the terminal point of $\mathbf{r}(t)$.

The derivative of s is $s'(t) = |\mathbf{r}'(t)|$; when $\mathbf{r}$ is expressed in terms of its coordinate functions, the derivative of s is

$$\frac{ds}{dt} = \sqrt{\left(\frac{dx}{dt}\right)^2 + \left(\frac{dy}{dt}\right)^2 + \left(\frac{dz}{dt}\right)^2}.$$

In the two-dimensional case the dz/dt term is omitted. This shows the validity of the integral formulas given at the beginning of the section.

Here are some computations using these integral formulas.

EXAMPLE 1 Find the length of the straight line segment $\mathbf{r}(t) = \mathbf{u} + t\mathbf{v}$ for $a \leq t \leq b$ with $\mathbf{u}$ and $\mathbf{v}$ fixed vectors.

Solution: Using the vector form of the arc-length formula, we obtain

$$L = \int_a^b |\mathbf{r}'(t)|\,dt = \int_a^b |\mathbf{v}|\,dt = |\mathbf{v}|(b-a).$$

Of course this is not a surprising answer, but it does illustrate the formula. ■

EXAMPLE 2 Find the length of one arch of the cycloid having position vector $\mathbf{r}(t) = (t - \sin t)\,\mathbf{i} + (1 - \cos t)\,\mathbf{j}$. The length is the distance traveled by a point on the rim of a wheel having radius 1 as it makes one revolution moving along the ground.

Solution: The derivative of the position function is

$$\mathbf{r}'(t) = (1 - \cos t)\,\mathbf{i} + \sin t\,\mathbf{j},$$

and its length is

$$|\mathbf{r}'(t)| = \sqrt{(1-\cos t)^2 + \sin^2 t} = \sqrt{2 - 2\cos t}.$$

The length of one arch is

$$L = \sqrt{2}\int_0^{2\pi} \sqrt{1 - \cos t}\, dt.$$

This integral can be evaluated exactly with the use of a trigonometric identity:

$$2\sin^2\theta = 1 - \cos 2\theta \quad \text{or} \quad \sqrt{2}|\sin\theta| = \sqrt{1 - \cos 2\theta}.$$

We take $2\theta = t$ and make the substitution. Care must be taken because of the absolute value sign produced by the square root. One arch of the cycloid is traced out as t varies from 0 to 2π. On this interval, $\sin(t/2)$ is nonnegative, so the absolute value signs may be removed and we have

$$L = \sqrt{2}\int_0^{2\pi} \sqrt{2}\sin(t/2)\, dt = 8.$$

■

Here is an example of a curve in three dimensions whose length is surprisingly easy to compute.

EXAMPLE 3 Find the length of the spiral having parametric equations

$$\begin{cases} x = 4\cos t, \\ y = 4\sin t, \\ z = at, \end{cases}$$

for any positive constant a as t ranges over the interval $[0, b]$.

Solution: The application of the integral formula gives the length as

$$L = \int_0^b \sqrt{x'(t)^2 + y'(t)^2 + z'(t)^2}\, dt = \int_0^b \sqrt{4^2\sin^2 t + 4^2\cos^2 t + a^2}\, dt$$

$$= \int_0^b \sqrt{4^2 + a^2}\, dt = b\sqrt{4^2 + a^2}.$$

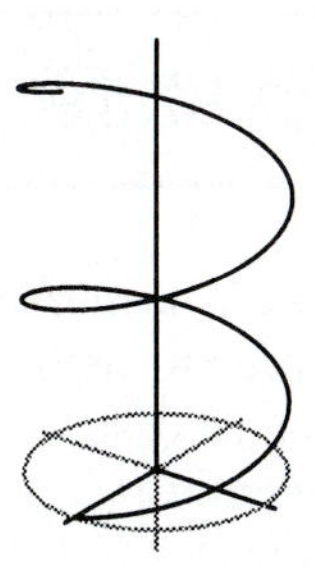

FIGURE 13.11
Spiral over a circle in the xy plane

The integral would not be so easy to evaluate if the projection of the curve into the xy plane were an ellipse in place of a circle. ■

Exercises 13.5

Find the length of each curve.

1. $\mathbf{r}(t) = t(2\mathbf{i} + 3\mathbf{j} - \mathbf{k}),\ 0 \le t \le 4$
2. $\mathbf{r}(t) = t^2(\mathbf{i} - 2\mathbf{j} - 2\mathbf{k}) + (3\mathbf{i} - 4\mathbf{k}),\ 0 \le t \le 4$
3. $\mathbf{r}(t) = 2t^2\,\mathbf{i} + t^3\,\mathbf{j},\ 0 \le t \le 1$
4. $\mathbf{r}(t) = e^t \cos t\,\mathbf{i} + e^t \sin t\,\mathbf{j},\ 0 \le t \le 2\pi$
5. $\mathbf{r}(t) = 2\sin t\,\mathbf{i} - 2\cos t\,\mathbf{j} + 3t\,\mathbf{k},\ 0 \le t \le \pi$
6. $\mathbf{r}(t) = 3t^2\,\mathbf{i} + t^3\,\mathbf{j} + 6t\,\mathbf{k},\ 0 \le t \le 1$
7. $\mathbf{r}(t) = \frac{1}{2}t^2\,\mathbf{i} + \ln(2t)\,\mathbf{j} + \sqrt{2}t\,\mathbf{k},\ \ 1 \le t \le 2$
8. $\mathbf{r}(t) = t\,\mathbf{i} + (2\sqrt{2}/3)t^{3/2}\,\mathbf{j} + (t^2/2)\,\mathbf{k},\ \ 0 \le t \le 1$

Find the length of the curve $\mathbf{r}(t)$ in three dimensions and compare it with the length of the projection of the curve into the xy plane.

9. $\mathbf{r}(t) = 3\cos t\,\mathbf{i} + 3\sin t\,\mathbf{j} + t\,\mathbf{k},\ \ 0 \le t \le 2\pi$
10. $\mathbf{r}(t) = a\cos t\,\mathbf{i} + a\sin t\,\mathbf{j} + t\,\mathbf{k},\ \ 0 \le t \le 2\pi,\ a = \text{constant}$
11. For two positive constants a and b find the ratio of the lengths of the two curves

$$\mathbf{r}_1(t) = 4\cos 2t\,\mathbf{i} + 4\sin 2t\,\mathbf{j} + at\,\mathbf{k}, 0 \le t \le \pi,$$
$$\mathbf{r}_2(t) = 4\cos 2t\,\mathbf{i} + 4\sin 2t\,\mathbf{j} + bt\,\mathbf{k}, 0 \le t \le \pi.$$

12. The curve $\mathbf{r}(t) = R(\cos 2\pi mt\,\mathbf{i} + \sin 2\pi mt\,\mathbf{j}) + t\,\mathbf{k}$ is a model for a coiled spring of radius R that has m complete turns for each unit of height. Which spring requires more material: a spring of radius 6 with 4 turns per unit height or a spring of radius 8 with 3 turns per unit height?
13. Let $\mathbf{r}(t)$ be a smooth curve for t on $[a, b]$ and let $g(s)$ be a differentiable function having a positive derivative for s on $[c, d]$ and satisfying $g(c) = a$, $g(d) = b$. Then the curve $\mathbf{u}(s) = \mathbf{r}(g(s))$ for s on $[c, d]$ is the same as the curve $\mathbf{r}(t)$. Show that the formula for the length of the curve gives the same result when applied to either representation; that is, $\int_a^b |\mathbf{r}'(t)|\,dt = \int_c^d |\mathbf{u}'(s)|\,ds$.
14. Show how to derive the formula $\int_a^b \sqrt{1 + f'(x)^2}\,dx$ for the length of the graph of a function $f(x)$ from the formula for the length of a parametric curve in two dimensions.
15. If a curve C has a representation $\mathbf{r}(t) = x(t)\,\mathbf{i} + y(t)\,\mathbf{j} + z(t)\,\mathbf{k}$ and if C' is the curve obtained by projecting C into one of the coordinate planes, then intuitively we expect that the length of C is greater than or equal to the length of C'. Prove this is true at least for smooth curves.
16. Prove the additivity property of the definition of arc length. If $\mathbf{r}(t)$ is a smooth curve for t on $[a, b]$ and if $a \le u \le b$ then the length of the arc from $\mathbf{r}(a)$ to $\mathbf{r}(b)$ equals the sum of the lengths of the arcs from $\mathbf{r}(a)$ to $\mathbf{r}(u)$ and from $\mathbf{r}(u)$ to $\mathbf{r}(b)$. [Hint: The arc length is the least upper bound of the set of chord approximations. For any chord approximation L_1 of the arc from $\mathbf{r}(a)$ to $\mathbf{r}(b)$ corresponding to the partition $a = t_0 < t_1 < \cdots < t_n = b$ there is a chord approximation L_2 obtained by adding the point u to this partition and then L_2 is the sum of two chord approximations over the intervals $[a, u]$ and $[u, b]$ and $L_1 \le L_2$.]

13.6 Curvature

Curvature is a measure of the bend in a curve at a point. It takes some care to give a precise definition of the notion of curvature. Let us first consider a two-dimensional case $\mathbf{r}(t) = x(t)\,\mathbf{i} + y(t)\,\mathbf{j}$. We see later that some pathology is avoided if we assume $\mathbf{r}(t)$ is twice differentiable and the first derivative is nonzero. The tangent vector $\mathbf{r}'(t)$ to the curve changes in length and direction as t varies. The change in direction of the tangent gives some measure of the bend of the curve; the change of length of the tangent is related to the speed with which the moving point moves along the path and has no particular relation to the bend of the curve. Thus instead of $\mathbf{r}'(t)$, we consider the unit tangent $\mathbf{T}(t) = \mathbf{r}'(t)/|\mathbf{r}'(t)|$. As t increases, $\mathbf{T}(t)$ changes direction but does not change length. The change in $\mathbf{T}$ is a measure of the bend of the curve. It is tempting to guess that the curvature should be computable from the rate of change of $\mathbf{T}$; but we must ask for the rate of change of **T** with respect to what? If we were to use the rate of change of **T** with respect to t to compute the curvature, then this would not be a property of only the curve but rather a property depending on the particular parameterization of the curve. For example suppose we parameterize a curve as $\mathbf{r}(t)$ for $0 \leq t \leq 1$ and also as $\mathbf{u}(w) = \mathbf{r}(w^2)$ for $0 \leq w \leq 1$ where $w^2 = t$. The unit tangent vector **T** can be expressed as a function of the parameter t or of the parameter w and there is no reason why the derivative of **T** with respect to t should equal the derivative with respect to w even when we evaluate at a point where $w^2 = t$. (See the exercises following this section for a specific example.) The definition of curvature is given by using one special parameterization of the curve. You may ask how one is to select a special parameterization from the infinitely many different parameterizations of a given curve? The answer is that we use the geometry of the curve (its length in particular) to select a parameterization, called the *arc-length parameterization,* that can be described for very general curves. Then we take the length of the derivative of **T** with respect to that parameter as the of curvature.

Arc-Length Parameterization

Let C be a curve in two or three dimensions and let $\mathbf{r}(t)$ be a vector representation of C. Select some initial point $\mathbf{r}(t_0)$ on the curve and for any $t \geq t_0$, let $s(t)$ be the length of the curve from $\mathbf{r}(t_0)$ to $\mathbf{r}(t)$. Thus

$$s(t) = \int_{t_0}^{t} |\mathbf{r}'(u)|\, du. \tag{13.13}$$

For $t \leq t_0$ we define $s(t)$ as the negative of the length of the curve from t to t_0. In this case the value of $s(t)$ is also given by the integral formula (13.13). Notice that $s(t)$ is an increasing function for all t because it is expressed as an integral of a positive function. Thus for any s_0, the equation $s(t) = s_0$ has at most one solution t. For any number s_0, there is at most one point on the curve at a distance s_0 from the starting point. We also regard s as a parameter with which we associate a unique point on the curve; let $\mathbf{g}(s)$ be the unique point of the curve at a distance s from $\mathbf{r}(t_0)$. Thus for some coordinate functions we express the curve as

$$\mathbf{g}(s) = x(s)\,\mathbf{i} + y(s)\,\mathbf{j} + z(s)\,\mathbf{k} \quad \text{and} \quad \mathbf{r}(t) = \mathbf{g}\big(s(t)\big)$$

We emphasize that $\mathbf{g}(s)$ with parameter s represents the same curve as that represented by $\mathbf{r}(t)$ with parameter t. The main difference is that the parameter s has a geometric meaning directly related to the curve whereas the parameter t is a dummy variable that has no particular connection to the curve (as in the case where t represents time). The parameter s depends on the choice of initial point $\mathbf{r}(t_0)$; a different choice of initial point gives a different parameter but one that differs from s only by a constant. (See the exercises following this section.) Now we give an example where the parameter s is easy to determine.

EXAMPLE 1 Let $\mathbf{r}(t) = \cos 2\pi t\,\mathbf{i} + \sin 2\pi t\,\mathbf{j}$ for $0 \leq t < 1$. Thus $\mathbf{r}(t)$ represents the circle of radius 1 with center at the origin. Find the parametric representation with arc length as the parameter.

***Solution*:** We first select an initial point on the curve from which to compute distances. Let $t_0 = 0$ so that $\mathbf{r}(0) = \mathbf{i}$. Compute $s(t)$ as

$$s(t) = \int_{t_0}^{t} |\mathbf{r}'(u)|\, du = \int_0^t |-2\pi \sin 2\pi u\,\mathbf{i} + 2\pi \cos 2\pi u\,\mathbf{j}|\, du = 2\pi \int_0^t du = 2\pi t.$$

As t ranges over $[0, 1)$, s ranges over $[0, 2\pi)$. Thus the arc-length parameterization of the circle is found by substituting s for $2\pi t$:

$$\mathbf{g}(s) = \cos s\,\mathbf{i} + \sin s\,\mathbf{j}, \quad 0 \le s < 2\pi.$$

In many examples, finding the arc-length parameterization is more complicated. ■

Now we turn to the definition of curvature again. The definition of curvature takes into account the rate of change of the unit tangent vector with respect to distance measured along the curve. Because s measures the distance along the curve, $d\mathbf{T}/ds$ is a vector that measures the instantaneous rate of change of the unit tangent vector with respect to distance along the curve. The curvature is defined to be the magnitude of this vector.

DEFINITION 13.3

Curvature

Let $\mathbf{r}(s)$ for s on some interval I be a parametric representation of a curve C in two or three dimensions. Assume that the second derivative $\mathbf{r}''(s)$ exists and the first derivative $\mathbf{r}'(s)$ is nonzero for all s on I. We assume that s is the arc-length parameter for C. Then the *curvature* at the point $\mathbf{r}(s)$ is given by the function

$$\kappa = \kappa(s) = \left|\frac{d\mathbf{T}}{ds}\right|$$

where $\mathbf{T}(s)$ is the unit tangent vector to C at $\mathbf{r}(s)$.

The computation of the curvature is usually done by applying the chain rule rather than actually finding the parameterization of the curve in terms of the arc-length parameter. If a curve is given as $\mathbf{r}(t)$ for some arbitrary parameter t and if $\mathbf{T}(t) = \mathbf{T}$ is the unit tangent vector, then the curvature is computed by the chain rule as

$$\kappa = \left|\frac{d\mathbf{T}}{ds}\right| = \left|\frac{d\mathbf{T}}{dt}\right|\left|\frac{ds}{dt}\right|^{-1} = \left|\frac{d\mathbf{T}}{dt}\right|\left|\mathbf{r}'(t)\right|^{-1}.$$

We illustrate with a circle, a case for which the computations work out nicely.

Curvature of a Circle

EXAMPLE 2 Find the curvature function for the circle $\mathbf{r}(t) = R\cos\omega t\,\mathbf{i} + R\sin\omega t\,\mathbf{j}$ where $\omega \neq 0$ is a constant.

Solution: The unit tangent is the derivative of $\mathbf{r}(t)$ divided by its length; in this case we have

$$\mathbf{T}(t) = \frac{\mathbf{r}'(t)}{|\mathbf{r}'(t)|} = -\sin\omega t\,\mathbf{i} + \cos\omega t\,\mathbf{j}.$$

We find $d\mathbf{T}/ds$ by the chain rule

$$\frac{d\mathbf{T}}{dt} = \frac{d\mathbf{T}}{ds}\frac{ds}{dt}.$$

We find easily that

$$\frac{d\mathbf{T}}{dt} = -\omega \cos \omega t\, \mathbf{i} - \omega \sin \omega t\, \mathbf{j}.$$

The term ds/dt is computed from the general formula (13.13) with the result

$$\frac{ds}{dt} = |\mathbf{r}'(t)| = |-R\omega \sin \omega t\, \mathbf{i} + R\omega \cos \omega t\, \mathbf{j}| = R|\omega|.$$

We combine the equations obtained so far to get

$$\frac{d\mathbf{T}}{ds} = \left(\frac{d\mathbf{T}}{dt}\right)\left(\frac{ds}{dt}\right)^{-1} = \frac{-\omega \cos \omega t\, \mathbf{i} - \omega \sin \omega t\, \mathbf{j}}{R|\omega|}.$$

The curvature is the magnitude of this vector so we finally have

$$\kappa = \frac{1}{R}.$$

The curvature of a circle of radius R is the constant $1/R$; the greater the radius, the smaller the curvature. This agrees with our intuitive idea that the curvature should measure the bend in the curve. A circle with a very large radius has much less bend per unit length than a circle with a very small radius. That the curvature is the same at every point of a circle should be no surprise because the "bend" in the circle is the same at every point by the symmetry of the circle. ■

The general formula for the curvature of a curve given in vector form can be computed using the same procedure as in example 2, but the details might be more complicated, of course. Suppose $\mathbf{r}(t)$ is a twice differentiable vector function representing a curve. The curvature function for the curve is

$$\begin{aligned} \kappa &= \left|\frac{d\mathbf{T}}{ds}\right| = \left(\frac{ds}{dt}\right)^{-1}\left|\frac{d\mathbf{T}}{dt}\right| \\ &= |\mathbf{r}'(t)|^{-1}\left|\frac{d}{dt}\frac{\mathbf{r}'(t)}{|\mathbf{r}'(t)|}\right| = \left|\frac{|\mathbf{r}'(t)|\mathbf{r}''(t) - \mathbf{r}'(t)|\mathbf{r}'(t)|'}{|\mathbf{r}'(t)|^3}\right|. \end{aligned} \tag{13.14}$$

This equation can be put into a slightly simpler form. The derivative of $|\mathbf{r}'(t)|$ is computed by differentiating the equation $\mathbf{r}'(t) \cdot \mathbf{r}'(t) = |\mathbf{r}'(t)|^2$ to get

$$|\mathbf{r}'(t)|' = |\mathbf{r}'(t)|^{-1}\mathbf{r}'(t) \cdot \mathbf{r}''(t).$$

Substituting this into Equation (13.14) we obtain

$$\kappa = \frac{\left||\mathbf{r}'|^2\mathbf{r}'' - \mathbf{r}'(\mathbf{r}' \cdot \mathbf{r}'')\right|}{|\mathbf{r}'|^4}. \tag{13.15}$$

If $\mathbf{r}(t) = x(t)\,\mathbf{i} + y(t)\,\mathbf{j}$, then the curvature can be expressed in terms of x and y and their derivatives with respect to t:

$$\kappa = \frac{|x'y'' - x''y'|}{\left[(x')^2 + (y')^2\right]^{3/2}}. \tag{13.16}$$

If a curve is given in rectangular form $y = f(x)$, we can use the parameter $x = t$ and $y = f(t)$ to express the curve in vector form as

$$\mathbf{r}(x) = x\,\mathbf{i} + f(x)\,\mathbf{j}.$$

Then the derivative with respect to t is the derivative with respect to x; the rectangular form for the curvature function of the graph of $y = f(x)$ is

$$\kappa = \frac{|y''|}{\left[1 + (y')^2\right]^{3/2}}. \qquad \textbf{(13.17)}$$

EXAMPLE 3 Compute the curvature at each point of the parabola $y = x^2$ and show that the curvature is greatest at the vertex.

Solution: Using formula (13.17) for the curvature, we find

$$\kappa = \frac{2}{\left[1 + (2x)^2\right]^{3/2}}.$$

The curvature is greatest when the denominator is smallest, namely when $x = 0$. ■

Another Formula for Curvature

Here is another expression for the curvature of a curve C that lies in the xy plane. Let C be given by the two-dimensional vector equation $\mathbf{r}(s)$ with s the arc-length parameter. Let $\mathbf{T}(s)$ be the unit tangent vector and let ϕ be the angle from the positive x-axis to $\mathbf{T}$. Thus $\mathbf{T} \cdot \mathbf{i} = \cos\phi$. We assume the vector function $\mathbf{r}$ is sufficiently well-behaved so that ϕ is a continuous and differentiable function of s. This means that the change in the direction of $\mathbf{T}$ is a differentiable function of the arc length. It should be intuitively clear that the rate of change of the direction of the tangent is closely related to the curvature. We make this connection precise.

THEOREM 13.5

Curvature in Terms of ϕ

With the notation as just described, the curvature function of the curve $\mathbf{r}(s)$ is given by

$$\kappa(s) = \left|\frac{d\phi}{ds}\right|.$$

Proof

The definition of ϕ as the angle from the x-axis to $\mathbf{T}$ and the fact that $\mathbf{T}$ is a unit vector imply that we can write

$$\mathbf{T}(s) = \cos\phi\,\mathbf{i} + \sin\phi\,\mathbf{j}.$$

Then we differentiate using the chain rule to get

$$\frac{d\mathbf{T}}{ds} = (-\sin\phi\,\mathbf{i} + \cos\phi\,\mathbf{j})\frac{d\phi}{ds}.$$

Then we get the curvature as the absolute value of this expression

$$\kappa = \left|\frac{d\mathbf{T}}{ds}\right| = |-\sin\phi\,\mathbf{i} + \cos\phi\,\mathbf{j}|\left|\frac{d\phi}{ds}\right| = \left|\frac{d\phi}{ds}\right|. \qquad \square$$

We can give an application of this formula to prove a fact that is geometrically very intuitive. We show that a curve of constant (nonzero) curvature must be a part of a circle.

THEOREM 13.6

Plane Curves of Constant Curvature

Let $\mathbf{r}(s)$ be a differentiable vector function describing a curve in the xy plane with s the arc-length parameter varying over an interval I. Assume further that the unit tangent $\mathbf{T}(s)$ is differentiable for s on I. If the curvature function $\kappa(s)$ is a constant $\kappa \neq 0$ for all s on I, then the curve lies on a circle having radius $1/\kappa$.

Proof

The description of κ in terms of the angle ϕ from the x-axis to $\mathbf{T}$ gives $|d\phi/ds| = \kappa$, which is constant and positive. Because $\mathbf{T}$ and ϕ are continuous and differentiable functions of s we must have either $d\phi/ds = \kappa$ for all s on I or $d\phi/ds = -\kappa$ for all s on I. (The continuity is used to ensure that $d\phi/ds$ does not have one sign over part of the interval and the other sign on another part of the interval.) We assume it is the first case; the second case is treated in the same way. The equation $d\phi/ds = \kappa$ implies

$$\phi = \kappa s + c, \qquad c = \text{constant.}$$

The tangent vector $\mathbf{T}$ now has the form

$$\mathbf{T} = \cos\phi\,\mathbf{i} + \sin\phi\,\mathbf{j} = \cos(\kappa s + c)\,\mathbf{i} + \sin(\kappa s + c)\,\mathbf{j}.$$

From the derivative $d\mathbf{r}/ds = \mathbf{T}$ we obtain, for a constant a on I,

$$\begin{aligned}\mathbf{r}(s) - \mathbf{r}(a) &= \int_a^s \mathbf{T}(u)\,du \\ &= \int_a^s \cos(\kappa u + c)\,\mathbf{i} + \sin(\kappa u + c)\,\mathbf{j}\,du \\ &= \left(\frac{1}{\kappa}\right)\left(-\sin(\kappa s + c)\,\mathbf{i} + \cos(\kappa s + c)\,\mathbf{j}\right) + \mathbf{v},\end{aligned}$$

for a constant vector $\mathbf{v}$. If we write $R = 1/\kappa$ and combine the constant vectors $\mathbf{r}(a)$ and $\mathbf{v}$ into one constant $\mathbf{u}$ then

$$\mathbf{r}(s) = \mathbf{u} + R\big(-\sin(\kappa s + c)\,\mathbf{i} + \cos(\kappa s + c)\,\mathbf{j}\big),$$

which shows that $\mathbf{r}(s)$ lies on the circle of radius R and center at $\mathbf{u}$. □

Exercises 13.6

Find the arc-length parameterization of the following curves using $\mathbf{r}(0)$ as initial point.

1. $\mathbf{r}(t) = 2t\,\mathbf{i} + (t + 1)\,\mathbf{j}$
2. $\mathbf{r}(t) = -3t\,\mathbf{i} + (3t - 4)\,\mathbf{j}$
3. $\mathbf{r}(t) = t^2\,\mathbf{i} + 3t^2\,\mathbf{j} - 4t^2\,\mathbf{k}$
4. $\mathbf{r}(t) = (2 - 3t^3)\,\mathbf{i} + t^3\,\mathbf{j} + (1 - 4t^3)\,\mathbf{k}$
5. $\mathbf{r}(t) = R\cos at\,\mathbf{i} + R\sin at\,\mathbf{j}$, $0 \le at < 2\pi$, $aR \neq 0$
6. $\mathbf{r}(t) = \cos(\pi t + 1)\,\mathbf{i} + \sin(\pi t + 1)\,\mathbf{j}$, $0 \le \pi t + 1 < 2\pi$
7. $\mathbf{r}(t) = \sin 2t\,\mathbf{i} + \cos 2t\,\mathbf{j} + 3t\,\mathbf{k}$, $0 \le t \le \pi$
8. $\mathbf{r}(t) = t\,\mathbf{i} - \sin t\,\mathbf{j} + \cos t\,\mathbf{k}$, $0 \le t \le 8\pi$

Find the curvature function for each curve.

9. $\mathbf{r}(t) = 2t\,\mathbf{i} + (t + 1)\,\mathbf{j}$
10. $\mathbf{r}(t) = 3t\,\mathbf{i} + (3t^2)\,\mathbf{j}$
11. $\mathbf{r}(t) = t\,\mathbf{i} + e^{-t}\,\mathbf{j}$
12. $\mathbf{r}(t) = R(\sin t\,\mathbf{i} - \cos t\,\mathbf{j})$
13. $\mathbf{r}(t) = \ln t\,\mathbf{i} - t\,\mathbf{j}$
14. $\mathbf{r}(t) = t\,\mathbf{i} + (t - t^2)\,\mathbf{j}$
15. The graph of the equation $x^{2/3} + y^{2/3} = 1$ is called the *hypocycloid* and has a parametrization $\mathbf{r}(t) = \cos^3 t\,\mathbf{i} + \sin^3 t\,\mathbf{j}$. Find the curvature function and determine the points where the curvature has an extreme value.
16. Show that the curvature of an ellipse ($\mathbf{r}(t) = a\cos t\,\mathbf{i} +$

$b \sin t\,\mathbf{j})$ has its extreme values on the x- or y-axis. If $a > b$ where is the maximum curvature?

17. Let $\kappa(t)$ be a continuous and nonnegative function on the interval $[a, b]$. Show that there is a curve $\mathbf{r}(t) = x(t)\,\mathbf{i} + y(t)\,\mathbf{j}$, $a \le t \le b$, whose curvature function is κ. [Hint: Define $g(t) = \int_a^t \kappa(t)\,dt$ and set

$$x(t) = \int_a^t \cos g(u)\,du \quad y(t) = \int_a^t \sin g(u)\,du.]$$

The *circle of best fit* (also called the *circle of curvature* or the *osculating circle*) to a curve $\mathbf{r}(t)$ at a point $P = \mathbf{r}(c)$ on the curve is a circle that is tangent to the curve at P, has the same curvature as the curve at P and lies on the same side of the tangent as the curve. Find the circle of best fit for each curve at the given point. Make a sketch showing the curve and the circle. Note that the center of the circle of best fit lies on the normal to the curve at the given point and has radius $1/\kappa$.

18. $\mathbf{r}(t) = t\,\mathbf{i} + t^2\,\mathbf{j}$, $P = \mathbf{r}(0)$

19. $\mathbf{r}(t) = t\,\mathbf{i} + t^2\,\mathbf{j}$, $P = \mathbf{r}(1)$

20. $y = e^x$, $P = (0, 1)$

21. $y = 1/x$, $P = (1, 1)$

22. $y = e^x + e^{-x}$, $P = (0, 2)$

23. $y = e^{x/a} + e^{-x/a}$, $P = (0, 2)$, $a > 0$

24. If $\mathbf{r}(t)$ is a smooth curve and $\mathbf{T}(t)$, $\mathbf{N}(t)$ and $\kappa(t)$ are the unit tangent, principal normal and curvature functions show that the circle of best fit at $\mathbf{r}(a)$ has the vector equation

$$\mathbf{u}(\theta) = \mathbf{r}(a) + \mathbf{N}(a)/\kappa(a) + (\cos\theta\,\mathbf{i} + \sin\theta\,\mathbf{j})/\kappa(a).$$

[This is useful for plotting the graph of a curve and a circle of best fit.]

25. The Folium of Descartes is the graph of $x^3 + y^3 = 3xy$. Find the curvature at the point $(3/2, 3/2)$. Inspect an accurate graph and offer a guess as to whether or not the maximum curvature occurs at this point.

26. For the graph of $x^{2/n} + y^{2/n} = 2$, with $n > 2$, find the curvature at $(1, 1)$.

27. Let $s = s(t)$ be the arc-length parameter for a curve $\mathbf{r}(t)$ using initial point $\mathbf{r}(t_0)$ and let $S = S(t)$ be the arc-length parameter for the same curve using initial point $\mathbf{r}(t_1)$. Show that s and S are related by $S(t) = s(t) + c$ for some constant c. Express c as an integral of $|\mathbf{r}'(t)|$.

28. If a curve has constant curvature equal to 0, show the curve lies on a straight line.

29. Let $a(t)$, $b(t)$, $c(t)$ and $d(t)$ be polynomials of degree at most one and let C be the curve with parametric equations

$$x(t) = \frac{a(t)}{b(t)}, \quad y(t) = \frac{c(t)}{d(t)}.$$

Show that t cannot be the arc-length parameter for C unless C is a straight line. [Hint: If t is the arc-length parameter, then for some constant p we have

$$t = \int_p^t \sqrt{x'(u)^2 + y'(u)^2}\,du.$$

Differentiate the integral to get $1 = \sqrt{x'(t)^2 + y'(t)^2}$ for all t. Use the fact that a polynomial of degree one has the form $mt + n$ for some real numbers m and n and prove the latter equation can only be solved when x' and y' are constant functions. From this conclude C is a straight line.] [Comment: This problem suggests that a curve expressed using the arc-length parameter must have fairly complicated functions in its description. It can be proved (with considerable effort) that a curve cannot be expressed in terms of the arc-length parameter using rational functions for the coordinates unless the curve is a straight line. The reader is invited to try this exercise for rational functions that equal the quotient of polynomials of small degree but be warned that the case for general rational functions is very difficult and beyond the material presented in this text.]

13.7 Curves in three dimensions

A curve in two dimensions can only bend toward or away from its tangent vector at a point. A curve in three dimensions has the freedom of bending toward or away from the tangent vector and also above or below it. To be more precise, there is bending in the plane of the unit tangent and the principal normal and also bending above or below this plane. We describe the bending by using the derivatives of the unit tangent $\mathbf{T}$, the principal normal $\mathbf{N}$ and the derivative of the unit vector $\mathbf{B} = \mathbf{T} \times \mathbf{N}$, which is orthogonal to the plane of $\mathbf{T}$ and $\mathbf{N}$. The vector $\mathbf{B}$ is called the *binormal* to the curve.

The curvature κ of a curve $\mathbf{r}(t)$ in three dimensions is defined exactly as in the two dimensional case by Equation (13.15), namely

$$\kappa = \kappa(s) = \left|\frac{d\mathbf{T}}{ds}\right|.$$

If we express the functions in terms of t and let *primes* denote the derivative with respect to t, the formula for curvature becomes

$$\kappa = \frac{\left||\mathbf{r}'|^2\mathbf{r}'' - \mathbf{r}'(\mathbf{r}' \cdot \mathbf{r}'')\right|}{|\mathbf{r}'|^4}. \tag{13.18}$$

The *principal normal* is the unit vector

$$\mathbf{N} = \frac{d\mathbf{T}}{dt}\left|\frac{d\mathbf{T}}{dt}\right|^{-1}.$$

One easily verifies, just as in the two dimensional case, that **T** and **N** are orthogonal. We define a third vector **B** called the *binormal*, which is normal to the plane of **T** and **N**, by

$$\mathbf{B} = \mathbf{T} \times \mathbf{N}.$$

Then **B** is also a unit vector and the triple **T**, **N**, **B** form a right-handed system that moves along the curve in such a way that **T** is tangent and **N** is normal to the curve and **B** is orthogonal to the plane of **T** and **N**. The rate of change of each vector with respect to length of the arc can be expressed in a combination of formulas called the *Frenet formulas.*

THEOREM 13.7

The Frenet Formulas

Let $\mathbf{r} = \mathbf{r}(t)$ be a twice differentiable smooth curve and let s denote the arc-length parameter of the curve. The the unit tangent, principal normal and binormal satisfy the equations

$$\frac{d\mathbf{T}}{ds} = \kappa\mathbf{N}, \quad \frac{d\mathbf{B}}{ds} = \tau\mathbf{N}, \quad \frac{d\mathbf{N}}{ds} = -\kappa\mathbf{T} - \tau\mathbf{B},$$

where κ is the curvature and τ is a scalar function.

The scalar function $\tau = \tau(t)$ whose existence is proved in this theorem is called the *torsion* of the curve.

Proof

For the first formula we apply the definitions and the chain rule. We know that $s = s(t)$ is the length of the arc from some given point and so it is an increasing function. Thus ds/dt is a nonnegative function. Thus

$$\begin{aligned}\frac{d\mathbf{T}}{ds} &= \left(\frac{ds}{dt}\right)^{-1}\frac{d\mathbf{T}}{dt} = \left(\frac{ds}{dt}\right)^{-1}\left|\frac{d\mathbf{T}}{dt}\right|\mathbf{N}\\ &= \left|\left(\frac{ds}{dt}\right)^{-1}\frac{d\mathbf{T}}{dt}\right|\mathbf{N} = \left|\frac{d\mathbf{T}}{ds}\right|\mathbf{N}\\ &= \kappa\mathbf{N}.\end{aligned}$$

The proof of the other two formulas uses three ideas: The first is that **T**, **N**, **B** form a right-handed system of mutually orthogonal unit vectors and so

$$\mathbf{T} \times \mathbf{N} = \mathbf{B}, \qquad \mathbf{N} \times \mathbf{B} = \mathbf{T}, \qquad \mathbf{B} \times \mathbf{T} = \mathbf{N}.$$

The second idea is that the cross product of two vectors is orthogonal to each of the two vectors. The third idea is that a vector function having constant length is orthogonal to its derivative, as we have seen several times now.

Differentiate both sides of the equation $\mathbf{N} \times \mathbf{B} = \mathbf{T}$ with respect to s to get

$$\frac{d\mathbf{N}}{ds} \times \mathbf{B} + \mathbf{N} \times \frac{d\mathbf{B}}{ds} = \frac{d\mathbf{T}}{ds} = \kappa \mathbf{N}. \tag{13.19}$$

Now we examine each term. The vector $d\mathbf{N}/ds$ is orthogonal to $\mathbf{N}$ so it lies in the plane determined by $\mathbf{T}$ and $\mathbf{B}$. The cross product of $d\mathbf{N}/ds$ and $\mathbf{B}$ is orthogonal to the plane containing these two vectors; thus the cross product is parallel to $\mathbf{N}$ and we put $(d\mathbf{N}/ds) \times \mathbf{B} = a\mathbf{N}$. From Equation (13.19) we obtain

$$\mathbf{N} \times \frac{d\mathbf{B}}{ds} = \kappa \mathbf{N} - a\mathbf{N} = (\kappa - a)\mathbf{N}.$$

This equation implies that $\mathbf{N} \times (d\mathbf{B}/ds)$ is parallel to $\mathbf{N}$. By properties of the cross product, it must also be orthogonal to $\mathbf{N}$ and so

$$\mathbf{N} \times \frac{d\mathbf{B}}{ds} = \mathbf{0}.$$

(Here we reason that only the zero vector can be simultaneously parallel to and orthogonal to a given vector.) The cross product of two vectors is zero only when they are parallel so $\mathbf{N}$ and $d\mathbf{B}/ds$ are parallel. There is a scalar function τ satisfying

$$\frac{d\mathbf{B}}{ds} = \tau \mathbf{N}.$$

This proves the second equation stated in the theorem. To prove the third, we differentiate the equation $\mathbf{N} = \mathbf{B} \times \mathbf{T}$ to get

$$\begin{aligned}\frac{d\mathbf{N}}{ds} &= \frac{d\mathbf{B}}{ds} \times \mathbf{T} + \mathbf{B} \times \frac{d\mathbf{T}}{ds} = \tau \mathbf{N} \times \mathbf{T} + \mathbf{B} \times \kappa \mathbf{N} \\ &= -\tau \mathbf{B} - \kappa \mathbf{T}.\end{aligned}$$

The proof of the Frenet formulas is complete. □

The computation of **T**, **N**, **B** can often be rather complicated but for some special curves the computation is manageable. The circular helix lying over a circle in the xy plane is one case where the computations can be made with little trouble.

EXAMPLE 1 Find the vectors **T**, **N**, **B** for the curve given by

$$\mathbf{r}(t) = R\cos t\,\mathbf{i} + R\sin t\,\mathbf{j} + at\,\mathbf{k}$$

for some positive constants R and a and then use these computations to determine the curvature and the torsion functions of the curve.

Solution: We first compute ds/dt because that term occurs in the remaining formulas. We have

$$\frac{ds}{dt} = |\mathbf{r}'(t)| = |-R\sin t\,\mathbf{i} + R\cos t\,\mathbf{j} + a\,\mathbf{k}| = \sqrt{R^2 + a^2} = \lambda,$$

showing that ds/dt is constant. We use the Greek letter lambda (λ) to denote this constant. Notice that in this example we have

$$\frac{d\mathbf{u}}{ds} = \left(\frac{ds}{dt}\right)^{-1}\frac{d\mathbf{u}}{dt} = \frac{1}{\lambda}\frac{d\mathbf{u}}{dt}$$

for any differentiable function $\mathbf{u}$.

Next we find $\mathbf{T}$ as

$$\mathbf{T} = \frac{\mathbf{r}'(t)}{|\mathbf{r}'(t)|} = \frac{-R\sin t\,\mathbf{i} + R\cos t\,\mathbf{j} + a\,\mathbf{k}}{\lambda}.$$

We apply the chain rule to compute $\kappa\mathbf{N}$ and $\mathbf{N}$ by using the first Frenet formula $d\mathbf{T}/ds = \kappa\mathbf{N}$ with the result

$$\kappa\mathbf{N} = \frac{d\mathbf{T}}{ds} = \left(\frac{ds}{dt}\right)^{-1}\frac{d\mathbf{T}}{dt} = \frac{1}{\lambda}\frac{d\mathbf{T}}{dt} = \frac{1}{\lambda}\frac{-R\cos t\,\mathbf{i} - R\sin t\,\mathbf{j}}{\lambda}.$$

We can determine the curvature function and $\mathbf{N}$ from this equation because

$$\kappa = |\kappa\mathbf{N}| = \frac{R}{\lambda^2} = \frac{R}{R^2 + a^2}$$

and it follows that

$$\mathbf{N} = (1/\kappa)(\kappa\mathbf{N}) = -\cos t\,\mathbf{i} - \sin t\,\mathbf{j}.$$

Next we obtain $\mathbf{B}$ from the definition $\mathbf{B} = \mathbf{T} \times \mathbf{N}$:

$$\mathbf{B} = \frac{-R\sin t\,\mathbf{i} + R\cos t\,\mathbf{j} + a\,\mathbf{k}}{\lambda} \times (-\cos t\,\mathbf{i} - \sin t\,\mathbf{j})$$
$$= \frac{a\sin t\,\mathbf{i} - a\cos t\,\mathbf{j} + R\,\mathbf{k}}{\lambda}.$$

To find the torsion function we use the second Frenet formula and evaluate

$$\tau\mathbf{N} = \frac{d\mathbf{B}}{ds} = \frac{1}{\lambda}\frac{d\mathbf{B}}{dt}$$
$$= \frac{1}{\lambda}\frac{d}{dt}\frac{(a\sin t\,\mathbf{i} - a\cos t\,\mathbf{j} + R\,\mathbf{k})}{\lambda}$$
$$= \frac{1}{\lambda}\frac{1}{\lambda}(a\cos t\,\mathbf{i} + a\sin t\,\mathbf{j}).$$

By comparing this with the value of $\mathbf{N}$ computed just above, we find

$$\tau = \frac{-a}{\lambda^2} = \frac{-a}{R^2 + a^2}.$$

Thus the torsion is also constant. ■

REMARK: The three vectors $\mathbf{T}$, $\mathbf{N}$ and $\mathbf{B}$ form a moving coordinate frame on a given curve and so the three vectors and their derivatives tell us something about how the curve changes. If the curve lies in a plane, say the xy plane, then $\mathbf{T}$ and $\mathbf{N}$ also lie in the xy plane and $\mathbf{B}$ is a constant unit vector orthogonal to that plane. In particular the curve lies in the plane of $\mathbf{T}$ and $\mathbf{N}$. For a three-dimensional curve, the plane containing $\mathbf{T}(t)$ and $\mathbf{N}(t)$ is tangent to the curve and one may think of "a very small part" of the curve near $\mathbf{r}(t)$ as as lying in that plane (in reality, there might be only one point of the curve in the plane of $\mathbf{T}$ and $\mathbf{N}$). Because $\mathbf{B}$ is the unit normal to that plane, the change of $\mathbf{B}$ measures the change of orientation of the plane; the change of $\mathbf{B}$ with respect to distance s along the curve is given by the third Frenet formula as $\tau\mathbf{N}$. The absolute value of this vector is $|\tau|$ and so the absolute value of the torsion gives a measure of the twisting of the curve above and below the moving tangent plane. This is analogous to the curvature, which by definition is the absolute value of the derivative of $\mathbf{T}$ with respect to distance along the curve. The curvature measures the bending of the curve in the plane of $\mathbf{T}$ and $\mathbf{N}$ and the torsion measure the twisting of the curve above and below the plane of $\mathbf{T}$ and $\mathbf{N}$.

Curves with Common Normals

Consider the following geometric problem. Suppose C_1 is a curve with a principal normal at every point. Can there be another curve C_2 that has a normal at every point lying in the same line as a normal to C_1? First consider a rather obvious example. If C_1 is a circle then any circle C_2 having the same center as C_1 has the property described in the problem. The normal to C_1 lies along the radius of the circle and hence is parallel to the normal to C_2 at the corresponding point.

We give a solution to this problem in two dimensions that makes use of many of the ideas developed in this chapter.

In the case of a curve C_1 lying in the xy plane, there are many curves C_2 having normals in common with C_1. We describe all of them.

THEOREM 13.8

Plane Curves with Common Normals

Let C_1 be a curve in the xy plane with twice differentiable vector representation $\mathbf{r}(s)$, with s the arc-length parameter. Let $\mathbf{N}(s)$ be the principal normal. Suppose C_2 is a differentiable curve in the xy plane with the property that the normal line to C_1 at any point of C_1 meets C_2 and is normal to C_2. Similarly assume that every normal to C_2 meets C_1 along a normal to C_1. Then C_2 has the vector representation $\mathbf{r}_2(s) = \mathbf{r}(s) + a\,\mathbf{N}(s)$ for some constant a. Conversely every normal to a curve given by $\mathbf{r}_2(s) = \mathbf{r}(s) + a\,\mathbf{N}(s)$ is parallel to a normal to C_1.

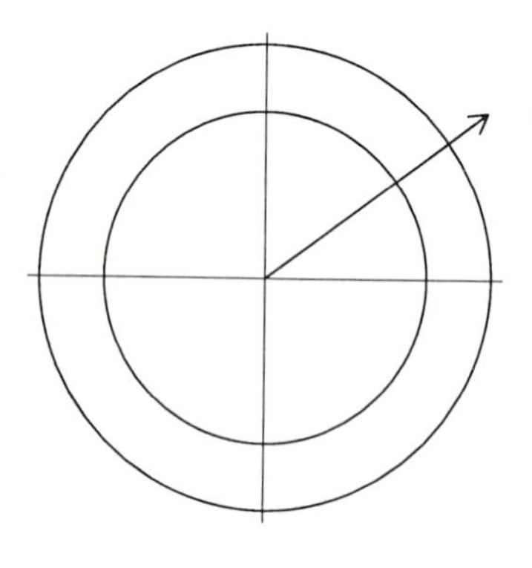

FIGURE 13.12
Concentric circles share normals

Proof

Let C_2 be a curve such that every normal line to C_1 is a normal to C_2 and conversely. To reach a point on C_2, we start from the origin and move to a point on C_1, that is to $\mathbf{r}(s)$, and then move along the normal $\mathbf{N}(s)$ some distance, say $a(s)$, to reach a point on C_2. We can assume the distance function $a(s)$ is continuous. The assumptions about normals ensure that every point of C_2 is obtained this way so C_2 has the vector representation

$$\mathbf{r}_2(s) = \mathbf{r}(s) + a(s)\mathbf{N}(s). \tag{13.20}$$

Note that s need not be the arc-length parameter for the second curve. We begin the analysis of this situation by showing that $a(s)$ is a constant. Because $\mathbf{N} \cdot \mathbf{N} = 1$, we have

$$a(s)^2 = a(s)\mathbf{N} \cdot a(s)\mathbf{N} = \big(\mathbf{r}_2(s) - \mathbf{r}(s)\big) \cdot \big(\mathbf{r}_2(s) - \mathbf{r}(s)\big).$$

Differentiate this to get

$$\begin{aligned} \frac{da(s)^2}{ds} &= 2(\mathbf{r}_2(s) - \mathbf{r}(s)) \cdot (\mathbf{r}_2'(s) - \mathbf{r}'(s)) \\ &= 2a(s)\mathbf{N}(s) \cdot (\mathbf{r}_2'(s) - \mathbf{r}'(s)). \end{aligned}$$

The derivative $\mathbf{r}'(s)$ is tangent to C_1 and hence is perpendicular to $\mathbf{N}(s)$

because $\mathbf{N}(s)$ is normal to C_1. Thus $\mathbf{N}\cdot\mathbf{r}'(s) = 0$. Similarly $\mathbf{r}_2'(s)$ is tangent to C_2 and hence is perpendicular to $\mathbf{N}(s)$ because $\mathbf{N}(s)$ is also normal to C_2 by assumption. Thus $\mathbf{N}\cdot\mathbf{r}'(s) = 0$ and it follows that the derivative of $a(s)^2$ is 0. Thus $a(s)^2$ is constant and $a(s)$ is constant.

Hence C_2 has the vector representation stated in the theorem. Now we show that the curve described by $\mathbf{r}_2$ in Equation (13.20) with $a(s) = a$ constant has its normals in common with C_1.

The tangent to C_2 at $\mathbf{r}_2(s)$ is given by the derivative

$$\mathbf{r}_2'(s) = \frac{d}{ds}(\mathbf{r}(s) + a\,\mathbf{N}(s)) = \mathbf{T}(s) - a\kappa\mathbf{T}(s) = (1 - a\kappa)\mathbf{T}.$$

We have used the Frenet formulas to get this equation. This shows that the tangent to C_2 is a scalar multiple of the unit tangent $\mathbf{T}$ to C_1. Thus the two curves have parallel tangents at corresponding points and so their normals are also parallel at corresponding points. □

We do not claim that the curves given by $\mathbf{r}_2(s) = \mathbf{r}(s) + a\,\mathbf{N}(s)$ have nonzero tangents. It is possible that at some points $\mathbf{r}_2$ has a 0 tangent vector and hence no principal normal exists at that point.

The curves $\mathbf{r} + a\mathbf{N}$ obtained from $\mathbf{r}$ by adding a constant multiple of the normal appear to be parallel to the first curve in an intuitive sense. However our intuition can be somewhat misleading. We consider the example of a parabola and some "parallel" curves. The results seem surprising.

EXAMPLE 2 Let $\mathbf{r}(t) = 4t\,\mathbf{i} + t^2\,\mathbf{j}$ and let $\mathbf{N}$ be the principal normal. Sketch the three curves $\mathbf{r}(t)$, $\mathbf{r}(t) + 4\mathbf{N}(t)$ and $\mathbf{r}(t) + 12\mathbf{N}(t)$ and comment on the difference of the shapes.

Solution: Figure 13.13 shows the curve $\mathbf{r}(t)$ with a few normal lines of length 4. The curve $\mathbf{r}(t) + 4\mathbf{N}(t)$ passes through the endpoints of these normal lines and the curve has roughly the same shape as the parabola.

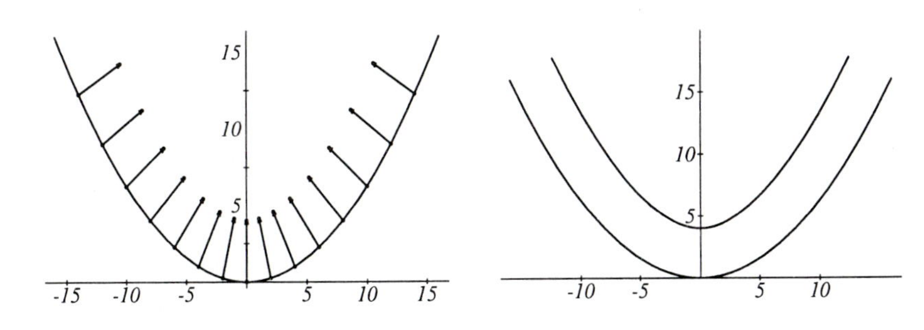

FIGURE 13.13
Curve with normals

FIGURE 13.14
Curve and parallel curve

Figure 13.15 shows $\mathbf{r}(t)$ with a few normal lines of length 12. Some of these normal lines intersect each other. The curve $\mathbf{r}(t) + 12\mathbf{N}(t)$ passes through the endpoints of the normals; the x coordinate of the points on the curve is not an increasing function of t; x changes direction at two points as can be seen in figure 13.16 showing the curves.

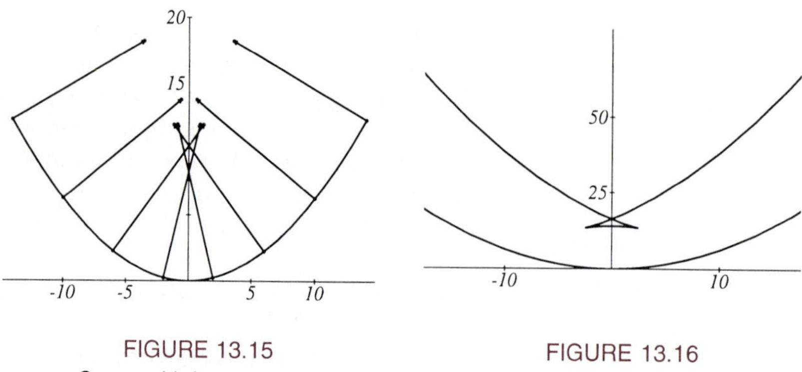

FIGURE 13.15
Curve with long normals

FIGURE 13.16
Long normals produce cusps

The curve $\mathbf{r} + 12\mathbf{N}$ has two cusps (or "corners") that can be determined as the points where $\mathbf{r} + 12\mathbf{N}$ has a zero derivative and so the tangent not exist. See the exercises following this section. ■

Parallel Curves in Three Dimensions

Suppose we consider the same problem in three dimensions. Unlike the case in two dimensions, for a given curve C_1 there may not be any other curve C_2 for which the two share common principal normals at every point. We do not give an analysis of the situation but simply give an example.

EXAMPLE 3 Show that the circular helix $\mathbf{r}(t) = R\cos t\,\mathbf{i} + R\sin t\,\mathbf{j} + at\,\mathbf{k}$, with R and a constants, shares a common normal at every point with the circular helix $\mathbf{r}_2(t) = R_2\cos t\,\mathbf{i} + R_2\sin t\,\mathbf{j} + bt\,\mathbf{k}$.

Solution: The unit tangent and principal normal to $\mathbf{r}$ are computed to be

$$\mathbf{T} = \frac{R}{\sqrt{a^2 + R^2}}\left(-\sin t\,\mathbf{i} + \cos t\,\mathbf{j} + \frac{a}{R}\mathbf{k}\right)$$
$$\mathbf{N} = -\cos t\,\mathbf{i} - \sin t\,\mathbf{j}.$$

Notice that the principal normal does not involve either of the two constants a or R. Thus when the normal $\mathbf{N}_2$ for $\mathbf{r}_2$ is computed, we find $\mathbf{N} = \mathbf{N}_2$ so the two curves have the same principal normal at $\mathbf{r}(t)$ and $\mathbf{r}_2(t)$ for each t. ■

In this example, the curve $\mathbf{r}$ has constant curvature and constant torsion. For any curve C_1 having constant (nonzero) curvature but nonconstant (and continuous) torsion function, there is no curve C_2 that shares common normals with C_1 at every point. This is quite difficult to prove and we will not attempt to do so. In case you wonder if a curve with constant curvature and nonconstant torsion really does exist, we mention that not only do such curves exist but even much more is true. Given any positive continuous function $\kappa(t)$ and any continuous function $\tau(t)$, there exists a curve $\mathbf{r}$ having $\kappa(t)$ as its curvature function and $\tau(t)$ as its torsion function. We give no proof because one would require a study of solutions of systems of differential equations and would take us beyond the material of this text.

Exercises 13.7

Compute the vector functions **T**, **N** and **B** for each given **r**(t).

1. $\mathbf{r}(t) = 4\cos t\,\mathbf{i} - 8t\,\mathbf{j} - 4\sin t\,\mathbf{k}$

2. $\mathbf{r}(t) = at\,\mathbf{i} - b\cos t\,\mathbf{j} - b\sin t\,\mathbf{k}$

3. $\mathbf{r}(t) = \dfrac{1-t^2}{1+t^2}\,\mathbf{i} + \dfrac{2t}{1+t^2}\,\mathbf{j}$

4. If $\mathbf{r}(t)$ is a curve for which the torsion function τ is the zero function, argue that the curve lies in a plane orthogonal to **B**.

5. If $\mathbf{r}(t)$ is a twice differentiable curve for which the binormal $\mathbf{B}(t)$ is constant, then the plane determined by $\mathbf{T}(t)$ and $\mathbf{N}(t)$ is the same for all values of t. If **u** is a given unit vector, find an example of a curve having constant binormal $\mathbf{B}(t) = \mathbf{u}$.

6. Let $\mathbf{r}(t) = x(t)\,\mathbf{i} + y(t)\,\mathbf{j} + z(t)\,\mathbf{k}$ be a differentiable curve that lies in the sphere consisting of all points (x, y, z) satisfying $x^2 + y^2 + z^2 = 1$. Show that $\mathbf{r}'(t)$ is orthogonal to $\mathbf{r}(t)$ for every t.

7. Suppose $\mathbf{r}(t)$ is a three times differentiable curve. Verify the following formulas for the curvature and torsion:
$$\kappa = \frac{|\mathbf{r}' \times \mathbf{r}''|}{|\mathbf{r}'|^3}, \qquad \tau = \frac{\mathbf{r}' \times \mathbf{r}'' \cdot \mathbf{r}'''}{|\mathbf{r}' \times \mathbf{r}''|^2}.$$

8. Let $\mathbf{r}(t) = (\cos t + t\sin t)\,\mathbf{i} + (\sin t - t\cos t)\,\mathbf{j}$, the vector equation of the involute of a circle. Compute the curvature function κ at every point and determine where the curvature is a maximum.

9. Let $\mathbf{r}(t) = 10\cos t\,\mathbf{i} + 5\sin t\,\mathbf{j}$ and let $\mathbf{N}(t)$ be the principal normal. Then the curve **r** is an ellipse and **N** is inward pointing. Consider the related curve given by $\mathbf{u}(t) = \mathbf{r}(t) - \mathbf{N}(t)$. Show that **u** is *not* an ellipse. [Hint: If it were, then argue by symmetry that it would have a vector form $\mathbf{r}_1(w) = a\cos w\,\mathbf{i} + b\sin w\,\mathbf{j}$ for some a and b. Show this is impossible.]

10. Let $\mathbf{r}(t)$ be a three times differentiable curve and $\mathbf{N}(t)$ the principal normal. Assume $\mathbf{r}'(t) \neq 0$ so that the curve has a unit tangent $\mathbf{T}(t)$ at every point. Assume also that $\mathbf{N}(t)$ is defined for every t. Consider a new curve $\mathbf{u}(t) = \mathbf{r}(t) + a\,\mathbf{N}(t)$ for a positive constant a. Prove
$$\mathbf{u}'(t) = \Big(1 - a\kappa(t)\Big)\,s'(t)\mathbf{T}(t)$$
where $s(t)$ is the arc-length function. Thus argue that $\mathbf{u}'(t) = 0$ for those t where the circle of best fit to $\mathbf{r}(t)$ has radius equal to a. [Hint: Use the Frenet formulas.]

11. Let $\mathbf{r}(t)$ and $\mathbf{u}(t)$ be as in exercise 10. Let $s(t)$ be the length of the curve **r** from $\mathbf{r}(0)$ to $\mathbf{r}(t)$ and let $S(t)$ be the length of the curve **u** from $\mathbf{u}(0)$ to $\mathbf{u}(t)$. Show that $s(t) - S(t)$ is a nonnegative function that can be expressed in terms of a, s' and the curvature of **r**.

12. Let $\mathbf{r}(t) = 4t\,\mathbf{i} + t^2\,\mathbf{j}$ and let $\mathbf{N}(t)$ be the principal normal to the parabola at $\mathbf{r}(t)$. For a constant a let $\mathbf{u}(t) = \mathbf{r}(t) + a\,\mathbf{N}(t)$. Find the points where the curve $\mathbf{u}(t)$ has cusps. These are detected as the points where $\mathbf{u}'(t) = 0$. [Hint: Use exercise 11.]

13.8 Motion along a curve

If a particle moves along a curve in either two or three dimensions, its location with respect to the origin is described by the position vector
$$\mathbf{r}(t) = x(t)\,\mathbf{i} + y(t)\,\mathbf{j} \quad \text{or} \quad \mathbf{r}(t) = x(t)\,\mathbf{i} + y(t)\,\mathbf{j} + z(t)\,\mathbf{k}.$$
The velocity $\mathbf{v}(t)$ of the particle at time t is the rate of change of position with respect to time. Thus
$$\mathbf{v}(t) = \mathbf{r}'(t) = \frac{d\mathbf{r}}{dt}.$$
The acceleration $\mathbf{a}(t)$ of the particle is the rate of change of velocity with respect to time. Thus
$$\mathbf{a}(t) = \mathbf{v}'(t) = \mathbf{r}''(t).$$
The velocity and acceleration are vectors with a direction and magnitude.

We have already seen that $\mathbf{r}'(t)$ is the tangent vector to the curve so we conclude that the velocity vector is tangent to the path of the motion of the particle.

The acceleration vector can be interpreted using Newton's law which states that the force $\mathbf{F}$ acting on the particle equals the mass times the acceleration vector (provided suitable units are used). Thus the force vector is a positive scalar multiple of the acceleration vector. If the force acting on a particle is zero, then the acceleration is zero and the velocity is constant. When the path of the particle is curved (that is, not a straight line) the velocity is not constant because the velocity is parallel to the tangent vector. If the velocity is not constant, there is a nonzero acceleration and hence a nonzero force. Thus a particle moving in a curved path must be subject to some nonzero force. Similarly, a particle moving in a straight line but with nonconstant velocity (the velocity vector changes length but not direction) must be subject to a nonzero force.

Speed

The magnitude of the velocity is called the *speed*, denoted by $v(t)$; thus

$$v(t) = |\mathbf{v}(t)| = |\mathbf{r}'(t)|.$$

This is called the speed because it equals the rate of change of distance along the curve with respect to time. We see this as follows: The distance traveled by the particle from time $t = a$ to time t is

$$s(t) = \int_a^t |\mathbf{r}'(u)|\, du = \int_a^t |\mathbf{v}(u)|\, du.$$

The rate of change of distance with respect to time is

$$v(t) = s'(t) = |\mathbf{r}'(t)| = |\mathbf{v}(t)|$$

by the Fundamental Theorem of Calculus.

EXAMPLE 1 A ball is thrown into the air and falls to the ground. After the ball is released, the only force acting on it is the constant force of gravity, which we denote by $\mathbf{F} = -mg\,\mathbf{k}$ where g is the gravitational constant and m is the mass of the ball. Find the position vector $\mathbf{r}(t)$ that gives the position of the ball at time t. Assume that the position at time $t = 0$ is given as $\mathbf{r}(0) = \mathbf{r}_0$ and the velocity at time $t = 0$ is given as $\mathbf{v}(0) = \mathbf{v}_0$.

Solution: By Newton's law we have $\mathbf{F} = m\mathbf{a} = m\mathbf{r}''$ and so $-g\,\mathbf{k} = \mathbf{r}''$. Because $\mathbf{r}' = \mathbf{v}$ we have $\mathbf{r}'' = \mathbf{v}'$ and

$$\mathbf{v}' = -g\,\mathbf{k};$$

thus $\mathbf{v}(t) = -gt\,\mathbf{k} + \mathbf{u}$ for some constant vector $\mathbf{u}$. The condition at time $t = 0$ gives

$$\mathbf{v}_0 = \mathbf{v}(0) = -g(0)\,\mathbf{k} + \mathbf{u} = \mathbf{u}.$$

Thus $\mathbf{u} = \mathbf{v}_0$ and $\mathbf{v}(t) = -gt\,\mathbf{k} + \mathbf{v}_0$. Next we use this equation to find $\mathbf{r}$; we have

$$\mathbf{r}'(t) = \mathbf{v}(t) = -gt\,\mathbf{k} + \mathbf{v}_0.$$

It follows that $\mathbf{r}(t) = -(1/2)gt^2\,\mathbf{k} + \mathbf{v}_0 t + \mathbf{u}_0$ for some constant $\mathbf{u}_0$. Evaluate at time $t = 0$ to get

$$\mathbf{r}_0 = \mathbf{r}(0) = -(1/2)g \cdot 0\,\mathbf{k} + 0 \cdot \mathbf{v}_0 + \mathbf{u}_0 = \mathbf{u}_0.$$

Thus we finally obtain

$$\mathbf{r}(t) = -\frac{gt^2}{2}\,\mathbf{k} + t\mathbf{v}_0 + \mathbf{r}_0$$

where $\mathbf{v}_0$ is the velocity vector at time $t = 0$ and $\mathbf{r}_0$ is the position vector at time $t = 0$. ■

Let us take a special instance of example 1.

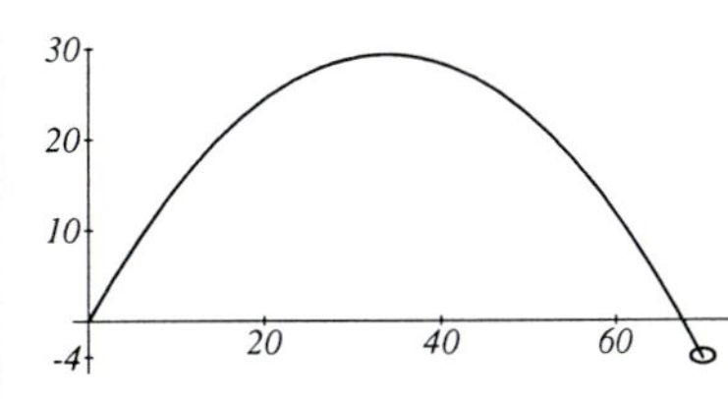

FIGURE 13.17
Path of the ball

EXAMPLE 2 A football is kicked with an initial velocity of 50 ft/sec at an angle of $\pi/3$ (60 degrees) above the horizontal. The ball impacts with the kicker's foot at a point 2 ft above the ground. Using $g = 32\,\text{ft/sec}^2$, find the position of the ball at time t for t ranging from the time of impact until the ball returns to the ground.

Solution: We select the coordinate system so that the origin is at the point of impact, the z-axis is perpendicular to the ground and the initial direction of the ball is in the xz plane. Then the initial velocity is

$$\mathbf{v}_0 = 50\big(\cos(\pi/3)\,\mathbf{i} + \sin(\pi/3)\,\mathbf{k}\big) = 25\,\mathbf{i} + 25\sqrt{3}\,\mathbf{k}.$$

The velocity vector at time t is $\mathbf{v}(t) = -32t\,\mathbf{k} + \mathbf{v}_0$. The position vector at time t is

$$\mathbf{r}(t) = -16t^2\,\mathbf{k} + t\mathbf{v}_0 = 25t\,\mathbf{i} + (25\sqrt{3}t - 16t^2)\,\mathbf{k}.$$

The function $\mathbf{r}(t)$ gives the position of the ball during the time that the only force acting on it is the force of gravity, that is from the time the ball leaves the kicker's foot until the time the ball strikes the ground. We assume the ground is horizontal so that ground level is $z = -2$ because we took the origin at the point where impact occurred. The ball hits the ground when the coefficient of $\mathbf{k}$ is -2; namely when $-2 = 25\sqrt{3}t - 16t^2$. This quadratic equation has only one positive solution, $t_0 \approx 2.75175$. At this time the position is $\mathbf{r}(t_0) \approx 68.7939\,\mathbf{i} - 2\,\mathbf{k}$ so the ball hits the ground about 23 yards from the point of the kick. ■

Uniform Circular Motion

If a particle moves around the circle of radius R with constant angular velocity ω then its position vector can be written as

$$\mathbf{r}(t) = R\cos\omega t\,\mathbf{i} + R\sin\omega t\,\mathbf{j}$$

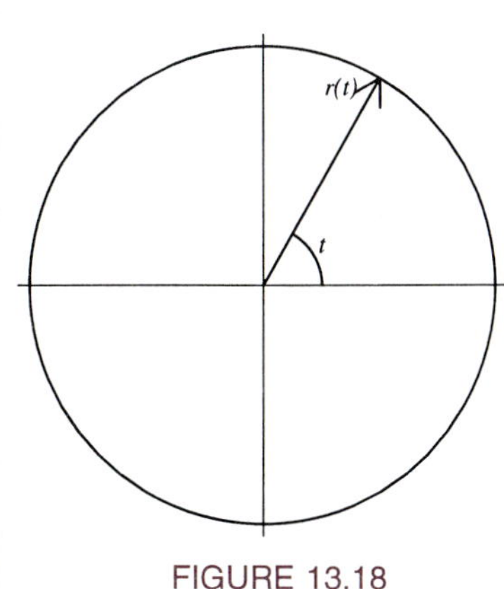

FIGURE 13.18

assuming that the particle is at $(R, 0)$ at time $t = 0$.

For $\omega > 0$, the motion is counterclockwise and one complete revolution is made in $2\pi/\omega$ time units. The velocity and acceleration vectors are found by differentiating the position vector to get

$$\mathbf{v}(t) = \mathbf{r}'(t) = \omega R(-\sin\omega t\,\mathbf{i} + \cos\omega t\,\mathbf{j}),$$
$$\mathbf{a}(t) = \mathbf{r}''(t) = -\omega^2 R(\cos\omega t\,\mathbf{i} + \sin\omega t\,\mathbf{j}).$$

The speed is $v(t) = |\mathbf{v}(t)| = |\omega|R$, a constant. The velocity, of course, is not constant because it is tangent to the path. The velocity vector has constant length but nonconstant direction. Note that the acceleration is a negative constant multiple of the position vector. Thus the force exerted on the particle to keep it in uniform circular motion must be directed along the position vector but opposite in direction. The magnitude of the acceleration is proportional to the square of the angular velocity. In this example the force is orthogonal to the velocity at all time because $\mathbf{v}(t) \cdot \mathbf{a}(t) = 0$.

Decomposition of Acceleration

If $\mathbf{r}(t)$ is the position of a moving particle at time t, and if the path is not a straight line, then some force must be acting on the particle to cause it to move along the path it follows. According to Newton's law the force $\mathbf{F} = \mathbf{F}(t)$ satisfies

$$\mathbf{F} = m\mathbf{r}'' = m\mathbf{a},$$

where $\mathbf{a}$ is the acceleration and m is the mass of the particle. The force is a constant multiple of the acceleration so a description of the force is obtained from the acceleration or the acceleration is obtained from the force, depending on what is given. We show how to express the acceleration in terms of $\mathbf{T}$, $\mathbf{N}$ and $\mathbf{B}$ making use of the curvature and torsion functions. We use the Frenet formulas and the equation $ds/dt = |\mathbf{r}'| = s'$ to express the derivative $\mathbf{r}''$ in terms of derivatives with respect to s. Start with the equation $\mathbf{r}' = |\mathbf{r}'|\mathbf{T} = s'\mathbf{T}$ (that comes from the definition of $\mathbf{T}$) and differentiate with respect to t to get

$$\begin{aligned}\mathbf{r}'' &= \frac{d}{dt}(s'\mathbf{T}) = s''\mathbf{T} + s'\frac{d\mathbf{T}}{dt}\\ &= s''\mathbf{T} + s'\left(\frac{ds}{dt}\right)\frac{d\mathbf{T}}{ds} = s''\mathbf{T} + (s')^2\kappa\,\mathbf{N}.\end{aligned} \tag{13.21}$$

Thus the acceleration is a linear combination of $\mathbf{T}$ and $\mathbf{N}$ and hence $\mathbf{r}''$ lies in the plane of $\mathbf{T}$ and $\mathbf{N}$. The coefficients of $\mathbf{T}$ and $\mathbf{N}$ in Equation (13.21) are important in describing the acceleration. We study them now in more detail.

Components of Acceleration

For any vector $\mathbf{u} = x\,\mathbf{i} + y\,\mathbf{j} + z\,\mathbf{k}$, the coordinates can be computed by

$$x = \mathbf{u}\cdot\mathbf{i} \qquad y = \mathbf{u}\cdot\mathbf{j} \qquad z = \mathbf{u}\cdot\mathbf{k}$$

as is verified by carrying out the dot products. The property depends on the fact that the three vectors $\mathbf{i}$, $\mathbf{j}$ and $\mathbf{k}$ are mutually orthogonal unit vectors. Thus the vector can be expressed as

$$\mathbf{u} = (\mathbf{u}\cdot\mathbf{i})\,\mathbf{i} + (\mathbf{u}\cdot\mathbf{j})\,\mathbf{j} + (\mathbf{u}\cdot\mathbf{k})\,\mathbf{k}.$$

The vectors $\mathbf{T}$, $\mathbf{N}$ and $\mathbf{B}$ are also mutually orthogonal unit vectors so any vector $\mathbf{u}$ can be expressed in the form $\mathbf{u} = a\mathbf{T} + b\,\mathbf{N} + c\,\mathbf{B}$; the constants a, b and c can be computed using the orthogonality of $\mathbf{T}$, $\mathbf{N}$ and $\mathbf{B}$ to obtain

$$a = \mathbf{u}\cdot\mathbf{T}, \qquad b = \mathbf{u}\cdot\mathbf{N}, \qquad c = \mathbf{u}\cdot\mathbf{B};$$

we obtain the analogous formula for the expression of $\mathbf{u}$ in terms of these three vectors, namely,

$$\mathbf{u} = (\mathbf{u}\cdot\mathbf{T})\mathbf{T} + (\mathbf{u}\cdot\mathbf{N})\,\mathbf{N} + (\mathbf{u}\cdot\mathbf{B})\,\mathbf{B}$$

for any vector $\mathbf{u}$.

If $\mathbf{r}$ is the position vector of a moving point and $\mathbf{a}$ is its acceleration vector, then $\mathbf{a}$ can be expressed as

$$\mathbf{a} = (\mathbf{a}\cdot\mathbf{T})\mathbf{T} + (\mathbf{a}\cdot\mathbf{N})\,\mathbf{N} + (\mathbf{a}\cdot\mathbf{B})\,\mathbf{B}.$$

For brevity, the coefficients are denoted by $a_T = \mathbf{a}\cdot\mathbf{T}$, $a_N = \mathbf{a}\cdot\mathbf{N}$, $a_B = \mathbf{a}\cdot\mathbf{B}$ and are called the *components of acceleration.* The scalars are defined by the equation

$$\mathbf{a} = a_T\mathbf{T} + a_N\,\mathbf{N} + a_B\,\mathbf{B}.$$

The computations in Equation (13.21) show that

$$a_T = s'', \qquad a_N = \kappa(s')^2, \qquad a_B = 0.$$

The tangential component of acceleration is the time derivative of $s' = |\mathbf{v}|$ and so a_T depends on the rate of change of the speed. The normal component of acceleration is the curvature times the square of the speed.

The components of acceleration can be felt by the passengers in a train or in an automobile as the vehicle travels around a curve. We take a sample problem in the next example.

EXAMPLE 3 A car maintains a constant speed of 30 miles/hr (= 44 ft/sec) as it travels along the curve in a road. Describe the force on the car and how it is felt by the passengers.

Solution: The position of the car at time t is given by the vector $\mathbf{r}(t)$ and the force on the car is a positive constant times the acceleration $\mathbf{r}''(t)$. The general formulas are simplified in this example because the speed is constant; that is, $|\mathbf{r}'| = ds/dt = 44$ (using feet and seconds as the units). It follows that the velocity vector is $\mathbf{r}' = 44\mathbf{T}$. The acceleration is

$$\begin{aligned}\mathbf{r}'' &= \frac{d}{dt}(44\mathbf{T}) = \left(\frac{ds}{dt}\right)\frac{d}{ds}(44\mathbf{T}) \\ &= 44(44\kappa\,\mathbf{N}) = 44^2\kappa\,\mathbf{N}.\end{aligned}$$

Thus the force is a constant times the curvature in the direction of the principal normal. We account for this force by considering the friction between the tires of the car and the road. The passengers inside the car would travel in a straight line if no force were applied. A passenger traveling along the identical path as the car, feels the force between herself and the seats of the car. If the car seats are very slippery (so there is little friction) there is not enough force to keep the passenger moving along the same path as the car. The deviation from the car's path usually results in the passenger feeling as if she is thrown toward the car door (i.e., opposite to the direction of the normal). The magnitude of this force depends on the *square* of the speed. Thus a small increase in speed can produce a large increase in the force in the normal direction. The magnitude of the force is also proportional to the curvature; a very sharp curve has a large curvature and results in a large component of force in the normal direction. ■

The computation of a_T and a_N can be simplified sometimes by using the following ideas. We have already seen that $a_T = ds/dt = d|\mathbf{v}|/dt$ so this can be determined without finding $\mathbf{T}$ first. Starting from the equation $\mathbf{a} = a_T\mathbf{T} + a_N\mathbf{N}$ we obtain

$$|\mathbf{a}|^2 = \mathbf{a}\cdot\mathbf{a} = a_T^2 + a_N^2$$

and so

$$a_N = \sqrt{|\mathbf{a}|^2 - a_T^2}.$$

Note that the curvature κ is never negative (by definition) and the term $(s')^2$ is never negative so $a_N = \kappa(s')^2$ is the nonnegative square root of $|\mathbf{a}|^2 - a_T^2$. This method of computation eliminates the need for determining the curvature. In fact it can be used in some cases to determine the curvature more simply than by using the definition.

EXAMPLE 4 The *involute of a circle* is the path traced by the endpoint of a string that is unwinding from a circular spool. Find the position vector and then compute the components of acceleration for this moving point.

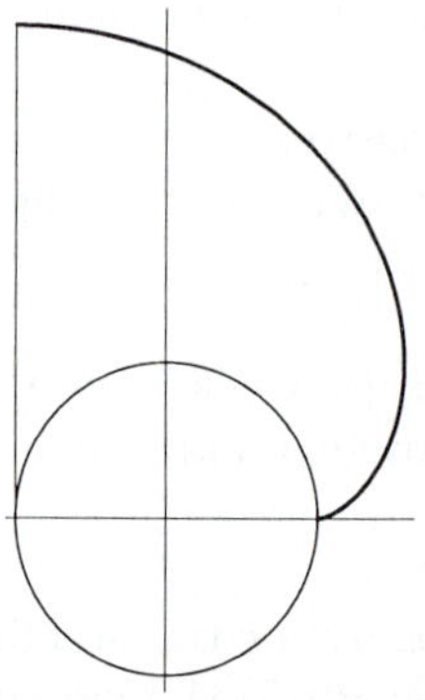

Solution: We assume the spool is a circle of radius 1 and the string is unwound with unit angular velocity. Thus at time t the point of contact of the string and the circle is given by the vector $\mathbf{u} = \cos t\,\mathbf{i} + \sin t\,\mathbf{j}$. The string is tangent to the circle and so is parallel to the tangent vector $\mathbf{n} = \sin t\,\mathbf{i} - \cos t\,\mathbf{j}$. The distance from the moving end to the point of contact with the spool equals the length of the arc from which the string was unwound, namely t. Thus the position vector is

$$\mathbf{r}(t) = \mathbf{u} + t\mathbf{n} = (\cos t + t\sin t)\mathbf{i} + (\sin t - t\cos t)\mathbf{j}.$$

FIGURE 13.19
Involute of a circle

By taking derivatives we find

$$\mathbf{r}' = \mathbf{v} = t\cos t\,\mathbf{i} + t\sin t\,\mathbf{j}$$
$$\mathbf{r}'' = \mathbf{a} = (\cos t - t\sin t)\mathbf{i} + (\sin t + t\cos t)\mathbf{j}.$$

Now we find a_T by taking the derivative of $|v| = t$ to get $a_T = 1$. To find a_N, we first compute

$$|\mathbf{a}|^2 = (\cos t - t\sin t)^2 + (\sin t + t\cos t)^2 = t^2 + 1.$$

Then

$$a_N = \sqrt{|\mathbf{a}|^2 - a_T^2} = \sqrt{t^2 + 1 - 1} = t.$$

We assume here that $t \geq 0$. From these computations we obtain

$$\mathbf{a} = a_T\mathbf{T} + a_N\mathbf{N} = \mathbf{T} + t\mathbf{N}.$$

One could determine the curvature from these computations. That problem is left to the exercises. ■

Exercises 13.8

If $\mathbf{r} = \mathbf{r}(t)$ is the position of a moving particle, compute the speed and the tangential and normal components of acceleration.

1. $\mathbf{r} = 5\sin \pi t\,\mathbf{i} + 5\cos \pi t\,\mathbf{j}$ **2.** $\mathbf{r} = 3t\,\mathbf{i} + (2t^2 - 1)\mathbf{j}$

3. $\mathbf{r} = 2\cos at\,\mathbf{i} + 2\sin at\,\mathbf{j} + 3t\,\mathbf{k}$

4. $\mathbf{r} = a\cos 2t\,\mathbf{i} + a\sin 2t\,\mathbf{j} + bt\,\mathbf{k}$

5. $\mathbf{r} = t\,\mathbf{i} + (t^2/2)\mathbf{j} + (t^3/3)\mathbf{k}$

6. $\mathbf{r} = e^t\sin t\,\mathbf{i} + e^t\cos t\,\mathbf{j} + e^t\,\mathbf{k}$

Find the position vector of a moving particle having the given acceleration and initial data.

7. $\mathbf{a}(t) = 4t\,\mathbf{i}$, $\mathbf{v}(0) = 3\,\mathbf{j}$, $\mathbf{r}(0) = -\mathbf{i} + \mathbf{j}$

8. $\mathbf{a}(t) = (t^2 + 1)\mathbf{i} + 3t^3\,\mathbf{j}$, $\mathbf{v}(0) = \mathbf{i} + \mathbf{j}$, $\mathbf{r}(0) = 2\mathbf{j}$

9. $\mathbf{a}(t) = 4\cos t\,\mathbf{i} - 4\sin t\,\mathbf{j}$, $\mathbf{v}(0) = 4\,\mathbf{j}$, $\mathbf{r}(0) = -4\,\mathbf{i}$

10. $\mathbf{a}(t) = \cos 2t\,\mathbf{i} + 3\sin 2t\,\mathbf{j}$, $\mathbf{v}(0) = \mathbf{i}$, $\mathbf{r}(0) = -4\,\mathbf{i}$

11. Describe in words the motion of the point having position vector $\mathbf{r}(t) = 10\sin 5t\,\mathbf{i} - 10\cos 5t\,\mathbf{j}$. Include the starting point, direction of rotation and the number of revolutions in unit time.

12. As a car moves with constant speed around a curve having constant curvature, the driver feels a force seemingly pushing him to the outside of the path. If the speed is initially 55 mph, how much should the speed be reduced to cut in half the force felt by the driver?

13. Prove the statement: If a particle moves with constant speed along a path, the magnitude of the force on the particle is some constant times the curvature of the path. [Hint: Constant speed means that $ds/dt = |\mathbf{r}'(t)|$ is constant; the force is a constant times $\mathbf{r}''(t)$. Use the Frenet formulas to express the curvature.]

14. If the force acting on a moving particle is perpendicular to the direction of motion (at every point) prove that the particle moves with constant speed.

15. If $\mathbf{r}(t)$ is the position vector of a moving point, $\mathbf{v}(t)$, $\mathbf{a}(t)$ the velocity and acceleration of the point, show that the curvature of the path can be expressed as

$$\kappa = |\mathbf{v} \times \mathbf{a}|\left|\frac{ds}{dt}\right|^{-3}.$$

[Hint: Express $\mathbf{v}$ and $\mathbf{a}$ in terms of $\mathbf{T}$ and $\mathbf{N}$.]

16. Prove that there is a function $f(x) = ax^4 + bx^3 + cx^2$, with a, b, c constants, whose graph passes through $(0,0)$ and $(1,1)$ and having curvature function $\kappa(x)$ that satisfies $\kappa(0) = 1/4$ and $\kappa(1) = 1/8$. Note that the actual computation of the constants can be more difficult than simply proving that such constants exist. [Remark: A curve satisfying $\kappa(0) = \alpha$ and $\kappa(1) = \beta$ for arbitrary positive α and β is used as a "transition curve" over $[0, 1]$ to connect a curve to the left of the origin with a curve to the right of $(1, 1)$ in such a way that the curvature of the resulting curve continuous.]

13.9 Planetary motion

The position and motion of the planets was a matter of great interest and mystery in the late sixteenth century. Tycho Brahe, a Danish astronomer, began making records of the positions of the planets and continued for twenty years. Johannes Kepler made detailed analysis of Brahe's data and formulated three laws of planetary motion, one of which states that the path of a planet is an ellipse with the sun at one focus. His laws were empirical and were not justified by any general principals of physics. One of the first triumphs of the invention of calculus was the theoretical explanation of Kepler's laws of planetary motion by Isaac Newton approximately 40 years after Kepler had stated his principles. Newton based the derivation on his Universal Law of Gravitation that can be stated as follows:

Newton's Law of Gravitation

The force exerted by one body on another acts along the line between the two bodies and is directly proportional to the product of the masses of the two bodies and inversely proportional to the square of the distance between them.

We express this law in vector notation as follows. If M is the mass of the sun and m is the mass of the planet and if $\mathbf{r}$ is the vector from the sun to the planet then the force exerted by the sun on the planet is

$$\mathbf{F} = -\frac{GMm}{|\mathbf{r}|^2}\frac{\mathbf{r}}{|\mathbf{r}|} \qquad \textbf{(13.22)}$$

where the constant G is the same for all planets. Note that $\mathbf{r}/|\mathbf{r}|$ is a unit vector in the direction of $\mathbf{r}$ and $|\mathbf{r}|^2$ is the square of the distance between the two bodies.

A force of the form $\mathbf{F} = A\mathbf{r}$ with A a scalar function is called a *central force.* In the case that the magnitude of $\mathbf{F}$ is inversely proportional to the square of the magnitude of the position, $|\mathbf{F}| = \text{constant}/|\mathbf{r}|^2$ is called an *inverse square central force.* Thus Newton's law of gravitation states that the force exerted on a planet by the sun is an inverse square central force.

We show how to use this law to derive the equation for the path followed by a planet around the sun. We make some assumptions. For example we assume that the sun remains fixed. In principle, the planet exerts a force on the sun but the effect is is negligible. The error is about the same as the error in assuming that the earth remains still when a tennis ball is thrown into the air and then falls back to the ground. In principle the earth attracts the tennis ball and the tennis ball attracts the earth but when we calculate the motion of the ball, we can safely assume the earth does not move. A second assumption is that we ignore the effect of other planets, moons, asteroids, etc. The sun so dominates the solar system, that the effect on one planet of other bodies is space is extremely small and is ignored.

We take our coordinates so that the center of the sun is the origin. The direction of the coordinate axes will be selected later to simplify some of the computations. Let $\mathbf{r} = \mathbf{r}(t)$ be the position of the planet at time t. By Newton's relation between acceleration and force we have $\mathbf{F} = m\mathbf{r}''$. When this is combined with Equation (13.22) and the mass m is cancelled we have

$$\mathbf{r}'' = -\frac{GM}{|\mathbf{r}|^3}\mathbf{r}. \tag{13.23}$$

The solution of this differential equation is the heart of the problem. It has been the object of study since Newton's time (about 1760). We employ several unmotivated steps to find a solution. The solution probably seems rather mysterious to the reader seeing this for the first time but it helps to understand that there are many ways to solve this equation. We profit from the efforts of many mathematicians and physicists to distill a method that is particularly elegant.

Consider the vector function $\mathbf{w} = \mathbf{r} \times \mathbf{r}'$ and evaluate its derivative:

$$\begin{aligned}\frac{d}{dt}\mathbf{w} &= \frac{d}{dt}\mathbf{r} \times \mathbf{r}' = \mathbf{r}' \times \mathbf{r}' + \mathbf{r} \times \mathbf{r}'' \\ &= 0 + \mathbf{r} \times \left(-\frac{GM}{|\mathbf{r}|^3}\mathbf{r}\right) = \left(-\frac{GM}{|\mathbf{r}|^3}\right)(\mathbf{r} \times \mathbf{r}) \\ &= 0.\end{aligned}$$

Here we have used the facts $(\mathbf{r} \times \mathbf{r}) = 0 = (\mathbf{r}' \times \mathbf{r}')$. Thus the derivative of the function $\mathbf{w}$ is 0 and it follows that $\mathbf{w}$ is a constant vector.

Because $\mathbf{w}$ is constant and orthogonal to the plane containing $\mathbf{r}$ and $\mathbf{r}'$, it follows that $\mathbf{r}$ is always in the same plane. In other words, the path of the planet lies in the unique plane that passes through the center of the sun and is orthogonal to the constant vector $\mathbf{w}$. If we select the z-axis to lie along $\mathbf{w}$ then the path of the planet is in the xy plane.

And now for the next step, we compute the derivative of the unit vector pointing along $\mathbf{r}$. We have

$$\frac{d}{dt}\frac{\mathbf{r}}{|\mathbf{r}|} = \frac{(\mathbf{r} \cdot \mathbf{r})\mathbf{r}' - (\mathbf{r} \cdot \mathbf{r}')\mathbf{r}}{|\mathbf{r}|^3}. \tag{13.24}$$

To get this form we apply the quotient rule and then the rule for the derivative of $|\mathbf{r}|$ that is found by differentiating the equation $|\mathbf{r}|^2 = \mathbf{r} \cdot \mathbf{r}$. We have the following formula for the cross product of three vectors: $\mathbf{u} \times (\mathbf{v} \times \mathbf{w}) = (\mathbf{u} \cdot \mathbf{w})\mathbf{v} - (\mathbf{u} \cdot \mathbf{v})\mathbf{w}$ so the numerator of Equation (13.24) can be written as a triple product. The result is

$$\frac{d}{dt}\frac{\mathbf{r}}{|\mathbf{r}|} = \frac{\mathbf{r} \times (\mathbf{r}' \times \mathbf{r})}{|\mathbf{r}|^3}.$$

We compute the derivative of $\mathbf{r}' \times \mathbf{w}$ making use of the fact that $\mathbf{w} = \mathbf{r} \times \mathbf{r}'$ is a constant vector:

$$\begin{aligned}\frac{d}{dt}(\mathbf{r}' \times \mathbf{w}) &= \mathbf{r}'' \times \mathbf{w} = \mathbf{r}'' \times (\mathbf{r} \times \mathbf{r}')\\ &= -\frac{GM}{|\mathbf{r}|^3}\mathbf{r} \times (\mathbf{r} \times \mathbf{r}') = \frac{GM}{|\mathbf{r}|^3}\mathbf{r} \times (\mathbf{r}' \times \mathbf{r})\\ &= GM\frac{\mathbf{r} \times (\mathbf{r}' \times \mathbf{r})}{|\mathbf{r}|^3}\\ &= GM\frac{d}{dt}\frac{\mathbf{r}}{|\mathbf{r}|}.\end{aligned}$$

From this equation we conclude

$$\frac{d}{dt}\left(\mathbf{r}' \times \mathbf{w} - GM\frac{\mathbf{r}}{|\mathbf{r}|}\right) = 0.$$

Only a constant vector function has a 0 derivative so there is a constant vector $\mathbf{c}$ such that

$$\mathbf{r}' \times \mathbf{w} - GM\frac{\mathbf{r}}{|\mathbf{r}|} = \mathbf{c}. \tag{13.25}$$

We get further simplification if we dot this last equation with $\mathbf{r}$, but before doing so we recall the scalar triple product identity that implies

$$\mathbf{r} \cdot \mathbf{r}' \times \mathbf{w} = \mathbf{r} \times \mathbf{r}' \cdot \mathbf{w} = \mathbf{w} \cdot \mathbf{w} = |\mathbf{w}|^2.$$

Now we dot both sides of Equation (13.25) with $\mathbf{r}$ to obtain

$$\begin{aligned}\mathbf{r} \cdot \mathbf{c} &= \mathbf{r} \cdot \mathbf{r}' \times \mathbf{w} - GM\frac{\mathbf{r} \cdot \mathbf{r}}{|\mathbf{r}|}\\ &= |\mathbf{w}|^2 - GM|\mathbf{r}|.\end{aligned}$$

Using the expansion of the dot product and rearranging the terms, this equation becomes

$$|\mathbf{r}|(GM + |\mathbf{c}|\cos\theta) = |\mathbf{w}|^2,$$

where θ is the angle between $\mathbf{r}$ and $\mathbf{c}$. This equation is actually a solution of the problem because we can now describe the vector $\mathbf{r}$ in terms of the constant vector $\mathbf{c}$ and the angle θ. We now select our coordinate system so that the positive x-axis points in the direction of $\mathbf{c}$. A unit vector that makes an angle θ with $\mathbf{c}$ is $\mathbf{u} = \cos\theta\,\mathbf{i} + \sin\theta\,\mathbf{j}$. Then $\mathbf{r}$ is the vector

$$\mathbf{r} = |\mathbf{r}|\mathbf{u} = |\mathbf{r}|(\cos\theta\,\mathbf{i} + \sin\theta\,\mathbf{j}) = |\mathbf{w}|^2\frac{\cos\theta\,\mathbf{i} + \sin\theta\,\mathbf{j}}{GM + |\mathbf{c}|\cos\theta}.$$

This vector equation has the form

$$\mathbf{r} = \frac{a\cos\theta\,\mathbf{i} + a\sin\theta\,\mathbf{j}}{1 + e\cos\theta}, \tag{13.26}$$

where $a = |\mathbf{w}|^2/GM$ and $e = |\mathbf{c}|/GM$ are nonnegative constants and θ is a parameter that measures the angle from the positive x-axis to the position vector $\mathbf{r}$. The curve described by this equation is one of the conic curves as we see in a moment. The particular conic represented by Equation (13.26) depends on the choice of constants. Let us first consider some trivial cases.

If $a = 0$ then $\mathbf{r} = 0$ for all t and this is the case in which the planet and the sun coincide.

If $e = 0$ then $\mathbf{r} = a\cos\theta\,\mathbf{i} + a\sin\theta\,\mathbf{j}$ and the path is a circle of radius a.

In all remaining cases we assume that a and e are positive constants. The path turns out to be an ellipse, parabola or hyperbola depending on $0 < e < 1$, $e = 1$ or $e > 1$; only the case $0 < e < 1$ corresponds to the motion of a planet in our solar system. When $e \geq 1$, there are values of θ that make the denominator equal to 0 and so the planet moves "infinitely" far away from the sun as θ approaches such a value.

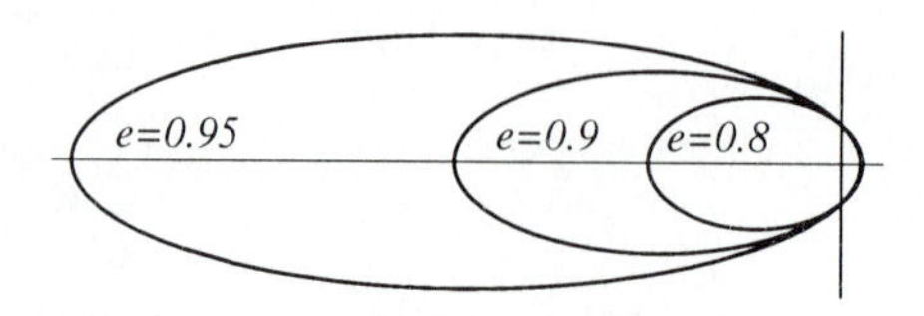

FIGURE 13.20
Orbits with different e

THEOREM 13.9

Suppose a planet of mass m moves under the effect of an inverse square central force $\mathbf{F} = -GMm\,\mathbf{r}/|\mathbf{r}|^3$. If the position vector $\mathbf{r}$ is given by Equation (13.26) for positive constants a and e then the path P is all or part of the curve given in rectangular coordinates as follows:

i. If $0 < e < 1$, P is an ellipse with equation

$$\left(\frac{x+h}{p}\right)^2 + \left(\frac{y}{q}\right)^2 = 1, \qquad h = \frac{ae}{1-e^2},\ p = \frac{a}{1-e^2},\ q = \frac{a}{\sqrt{1-e^2}};$$

ii. If $e = 1$, P is a parabola with equation $y^2 = a^2 - 2ax$;

iii. If $e > 1$, P is one branch of the hyperbola with equation

$$\left(\frac{x+h}{p}\right)^2 - \left(\frac{y}{q}\right)^2 = 1, \qquad h = \frac{ae}{1-e^2},\ p = \frac{a}{e^2-1},\ q = \frac{a}{\sqrt{e^2-1}}.$$

Proof

The conversion to rectangular coordinates is based on the equation $\mathbf{r} = x\,\mathbf{i} + y\,\mathbf{j}$ and the relations

$$|\mathbf{r}| = \sqrt{x^2+y^2} \quad \text{and} \quad x = |\mathbf{r}|\cos\theta.$$

Take absolute values in Equation (13.26) to get $|\mathbf{r}||1 + e\cos\theta| = a$. For $0 < e \leq 1$ we have $|1 + e\cos\theta| = 1 + e\cos\theta$ so

$$a = |\mathbf{r}|(1 + e\cos\theta) = |\mathbf{r}| + e|\mathbf{r}|\cos\theta = \sqrt{x^2+y^2} + ex. \qquad \textbf{(13.27)}$$

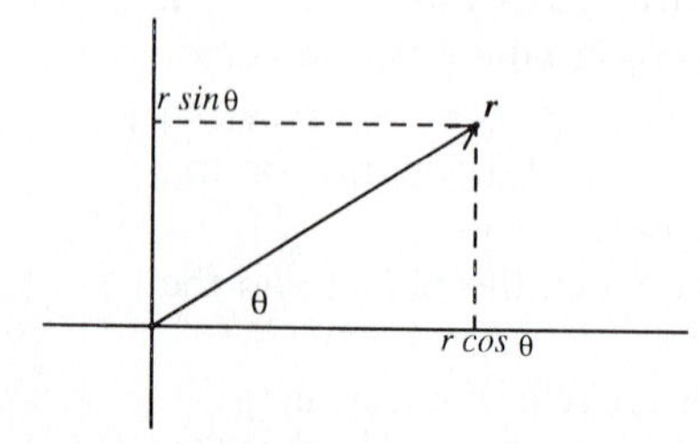

FIGURE 13.21

By rearranging terms and squaring one produces the equation

$$(1-e^2)x^2 + 2aex + y^2 = a^2. \qquad \textbf{(13.28)}$$

In case $e = 1$, we immediately get the equation of statement ii. For $0 < e < 1$, the term $1 - e^2$ is positive and so $\sqrt{1-e^2}$ is a real number. By completing the square in Equation (13.28) the form of the equation given in statement i can be obtained.

In case $e > 1$ we restrict θ to lie on an interval so that $1 + e\cos\theta > 0$. Then Equation (13.27) is still valid but now $1 - e^2$ is a negative number so $\sqrt{e^2 - 1}$ is a real number. Completing the square leads to the equation given in part iii. □

Kepler's first law states that the orbit of a planet is an ellipse. One deduces from Theorem 13.9 just proved that the orbit is one of the conic sections. In case the conic is a parabola or a hyperbola, the planet makes only one pass around the sun and then moves away forever. None of the planets exhibit that kind of behavior. Other celestial objects behave in this way however.

Kepler's Second and Third Laws

We do not give a complete verification of the other two Kepler laws but we state them and briefly indicate how they are derived. The first goal is to prove that $|\mathbf{r}|^2\theta'$ is constant.

Becausee the motion of the planet is in a plane, we use a rotating frame of unit vectors to describe the motion. In the description of the position vector $\mathbf{r}$ given by Equation (13.8), the angle θ is the angle between the position vector and some fixed vector that we selected to lie along the positive x-axis. We use unit vectors $\mathbf{u}$ in the direction of $\mathbf{r}$ and a perpendicular unit vector $\mathbf{v}$ defined as follows:

$$\mathbf{u} = \cos\theta\,\mathbf{i} + \sin\theta\,\mathbf{j}, \qquad \mathbf{v} = -\sin\theta\,\mathbf{i} + \cos\theta\,\mathbf{j}.$$

Note that $\mathbf{v} = d\mathbf{u}/d\theta$ and $-\mathbf{u} = d\mathbf{v}/d\theta$. So we have the following formulas for the time derivatives:

$$\frac{d\mathbf{u}}{dt} = \frac{d\theta}{dt}\frac{d\mathbf{u}}{d\theta} = \theta'\mathbf{v},$$
$$\frac{d\mathbf{v}}{dt} = \frac{d\theta}{dt}\frac{d\mathbf{v}}{d\theta} = -\theta'\mathbf{u}.$$

We now express $\mathbf{r}$ and its derivatives in terms of $\mathbf{u}$ and $\mathbf{v}$. Starting from $\mathbf{r} = r\mathbf{u}$ (recall that we write $|\mathbf{r}| = r$ for short) and differentiating with respect to t we obtain

$$\begin{aligned}\mathbf{r}' &= r'\mathbf{u} + r\theta'\mathbf{v}\\ \mathbf{r}'' &= (r''\mathbf{u} + r'\theta'\mathbf{v}) + (r'\theta'\mathbf{v} + r\theta''\mathbf{v} - r(\theta')^2\mathbf{u}\\ &= (r'' - r(\theta')^2)\mathbf{u} + (r\theta'' + 2r'\theta')\mathbf{v}.\end{aligned}$$

The assumption that the force is central implies that $\mathbf{r}'' = (\text{constant})\mathbf{u}$ or that $\mathbf{r}'' \cdot \mathbf{v} = 0$. From the last equation just above we see that $\mathbf{r}'' \cdot \mathbf{v} = (r\theta'' + 2r'\theta')$ and so $r\theta'' + 2r'\theta' = 0$. Note that

$$\frac{d}{dt}(r^2\theta') = r^2\theta'' + 2rr'\theta' = r(r\theta'' + 2r'\theta') = 0$$

and so $r^2\theta' = H$ is constant as time varies. This fact is the basis for proving the second and third Kepler laws. The second law states:

Kepler's Second Law

The position vector from the sun to a moving planet sweeps out an area that is the same per unit time at any part of the orbit.

One can relate the expression $r^2\theta'$ to the area swept out by the position vector; the fact that this expression is constant implies that the area swept out in unit time is the same for every part of the orbit.

Kepler's third law relates the length of the major axis of the orbit to the period. The period is the length of time required to complete one orbit. Let T be the time required for one orbit. Thus for the planet earth, $T = 365.25$ days (approximately). We bring T into the previous equations by making a change of variable in the simple integral

$$TH = \int_0^T H\,dt.$$

We use the equation $H = r^2 d\theta/dt$ or $H\,dt = r^2\,d\theta$. As t varies from 0 to T, θ varies from 0 to 2π (provided we begin with $\theta = 0$ at time $t = 0$, which we may do). Making use of Equation (13.8) to evaluate r^2, we obtain

$$TH = \int_0^T H\,dt = \int_0^{2\pi} r^2\,d\theta = \int_0^{2\pi} \frac{a^2\,d\theta}{(1 + e\cos\theta)^2}.$$

The last integral is a rational function of $\cos\theta$ and can be evaluated by the methods indicated in section 8.6. The result is

$$T = \frac{2\pi a^2}{H(1 - e^2)^{3/2}}.$$

Next we need the length of the major axis of the path. The length of the position vector $\mathbf{r}$ is the the distance from the focus of the ellipse to the planet. The major axis is the line through the two foci and the length of the major axis is the sum of the minimum value of r and the maximum value of r. From the equation $r = a/(1 + e\cos\theta)$ we seen the minimum value of r is $a/(1 + e)$ and the maximum value is $a/(1 - e)$. The length of the major axis is

$$A = \frac{a}{1 + e} + \frac{a}{1 - e} = \frac{2a}{1 - e^2}.$$

Substituting this into the equation for T, we obtain

$$T^2 = 4\pi^2 \frac{a}{H^2} \frac{a^3}{(1 - e^2)^3} = 4\pi^2 \frac{a}{H^2} A^3.$$

We now show that the constant a/H^2 does not depend on the planet but only on G and M.

We express $\mathbf{r}$ and $\mathbf{r}'$ in terms of $\mathbf{u}$ and $\mathbf{v}$ and then compute $\mathbf{w}$ as

$$\mathbf{w} = \mathbf{r} \times \mathbf{r}' = r\mathbf{u} \times (r'\mathbf{u} + r\theta'\mathbf{v}) = r^2\theta'(\mathbf{u} \times \mathbf{v}) = H(\mathbf{u} \times \mathbf{v}).$$

Because $\mathbf{u}$ and $\mathbf{v}$ are orthogonal vectors of length 1, their cross product also has length 1. Thus $H = |\mathbf{w}|$. The constant a was defined above as $a = |\mathbf{w}|^2/GM$, and so we have

$$\frac{a}{H^2} = \frac{1}{GM}.$$

Thus

$$T^2 = \frac{\pi^2}{2GM} A^3$$

where A is the length of the major axis of the orbit, G is the gravitational constant and M is the mass of the sun. Kepler had stated this third law as

Kepler's Third Law

The square of the period of a planet's orbit is proportional to the cube of the length of the major axis and the constant of proportionality is the same for all planets.

Although we have presented the theoretical verification of Kepler's laws for planets orbiting the sun, the computations are valid for the motion of any body under the influence of an inverse square force field that moves in an elliptical orbit. For example the laws apply to the motion of a man-made satellite that orbits the earth or the moon orbiting the earth.

EXAMPLE 1 A satellite is placed into orbit around the earth so that its orbit is (nearly) circular. If the time for one orbit is 90 minutes, how high above the center of the earth is the satellite? It is necessary to know the value of the gravitational constant, $G = 6.67 \times 10^{-11}$ in a system of units using kilograms, meters and seconds. The mass of the earth is $m = 5.98 \times 10^{24}$ kg.

Solution: We neglect the effect of all particles in the universe except that of the earth on the satellite. If D is the distance of the satellite above the center of the earth, the the major axis of the satellite is $A = 2D$ and the relation $T^2 = \frac{\pi^2}{2Gm}A^3$ can be solved for D to give

$$D = \left(\frac{T^2GM}{4\pi^2}\right)^{1/3}.$$

Substituting the numerical values of the constants gives $D \approx 6650$ km. We can compare this to the radius of the earth at the equator, which is approximately 6380 km, to conclude that the satellite is roughly 270 km or 168 miles above the surface of the earth. ■

Exercises 13.9

Give an accurate plot of the following vector curves.

1. $\mathbf{r}(\theta) = \dfrac{4\cos\theta\,\mathbf{i} + 4\sin\theta\,\mathbf{j}}{2 + \cos\theta}$

2. $\mathbf{r}(\theta) = \dfrac{5\sin\theta\,\mathbf{i} - 5\cos\theta\,\mathbf{j}}{4 - 2\cos\theta}$

Convert the equations from polar form to rectangular form and identify the curve.

3. $r = 4/(5 + \cos\theta)$ **4.** $r = 1/(3 + 2\cos\theta)$

5. $r = 6/(8 - 3\sin\theta)$ **6.** $r = 12/(4 - 4\cos\theta)$

7. Let $\mathbf{r}$ be the position of a particle acted on by any force. Show that the angle α between the velocity vector $\mathbf{r}'$ and $\mathbf{r}$ satisfies $\cos\alpha = |\mathbf{r}|'/|\mathbf{r}'|$.

8. Let $\mathbf{r}$ be the position of a particle acted on by an inverse square central force. Show that the velocity vector is perpendicular to the position vector only at the times when $|\mathbf{r}|$ is a (local) maximum or a minimum. [Hint: Use exercise 7.]

9. A comet is observed to make its nearest approach to the sun at a distance D and at this time it has speed V. Find the condition in terms of D and V that the path of the comet is an ellipse. Given that this is the case, find the farthest distance between the comet and the sun. Assume that the effects of other planets on the comet is negligible compared to the effect of the sun.

14 Partial Derivatives

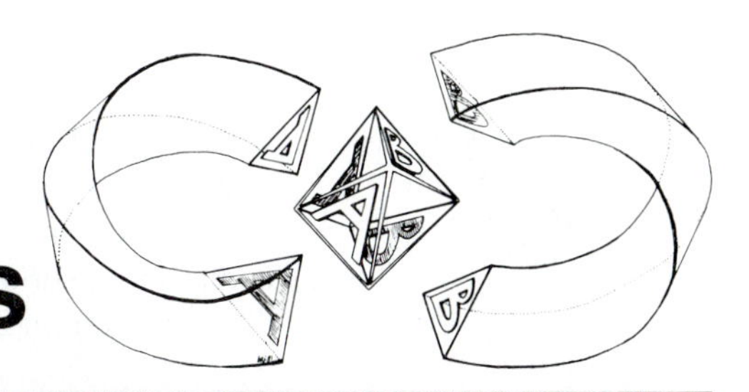

In this chapter we extend many of the ideas developed for functions of one variable to functions of several variables. In place of $y = f(x)$ we now study functions $z = f(x, y)$ or $w = g(x, y, z)$ of two or three variables. Curves are replaced by surfaces, derivatives are replaced by partial derivatives and new concepts are introduced as an aid to understanding how functions of several variables behave and how they are used. We consider the problem of finding the maximum and minimum values of a function; they correspond to the location of extreme points of the surface $z = f(x, y)$. Just as the derivative of a function of one variable is used in finding the extreme values of $h(x)$, the partial derivatives of $f(x, y)$ or $g(x, y, z)$ are used to locate the extreme values of f or g. The gradient of a function of several variables is introduced and is shown to be useful in a number of types of problems such as finding the direction from a point in which a function increases most rapidly. The Lagrange multiplier method for finding extrema of functions subject to side conditions is given a thorough treatment with motivation from geometric considerations. Concepts are illustrated with an extended example that considers the problem of selecting the path to ski down a hill to reach the bottom in the fastest possible time.

14.1 Surfaces in three dimensions

The study of a function $y = f(x)$ is facilitated by the study of its graph, a curve in the two-dimensional plane. The study of functions of two variables $z = f(x, y)$ is enhanced by the study of surfaces. In this section we give examples of some surfaces in xyz space.

Cylinders

Begin with a curve in the yz plane, for example the parabola $z = y^2$. Through every point, $(0, a, a^2)$ of this parabola consider the line parallel to the x-axis passing through the point. This line can be described as the set of all (x, a, a^2) with x any real number. The collection of all lines, one for each point of the

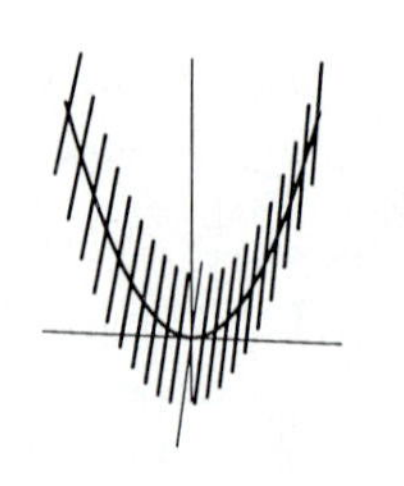

FIGURE 14.1
Cylinder generated by the curve $z = y^2$

parabola, is a surface. A point (x, y, z) is on this particular surface if and only if $z = y^2$. There is no restriction on x. This surface is called a *cylinder* generated by the curve $z = y^2$. To be more specific we should also indicate that the lines through the curve are taken parallel to the x-axis. The x- axis is called *directing line* of the surface. We say that the surface is the graph of the equation $z = y^2$ since a point (x, y, z) is on the surface if and only if the coordinates satisfy the equation.

More generally if L is a line and C is a curve lying in a plane that does not contain L, a surface can be obtained by taking the lines parallel to L and passing through the points of C. This is the surface *generated by* C with *directing line* L.

EXAMPLE 1 Discuss the graph of the equation $x^2 + y^2 = 4$ representing a surface in three dimensions.

Solution: A point (x, y, z) lies on the graph of the equation if its coordinates satisfy the given equation. The z coordinate does not appear in the equation so z is arbitrary. The point $(x, y, 0)$ is on the graph if the point (x, y) satisfies $x^2 + y^2 = 4$, that is if the point is on the circle in the xy plane of radius 2 with center at the origin. The graph of the equation is a right circular cylinder of radius 2. It is the surface generated by the circle of radius 2 with center at the origin in the xy plane and having the z-axis as its directing line. ■

Vector Representation of a Cylinder

The cylinder $z = y^2$ can be represented in vector form. We think of moving from the origin to reach a point on the cylinder in two steps. We first write a vector from the origin to some point on the curve $y\,\mathbf{j} + z\,\mathbf{k} = y\,\mathbf{j} + y^2\,\mathbf{k}$ and then move along a vector $x\,\mathbf{i}$ parallel to the x-axis to reach the point $x\,\mathbf{i} + y\,\mathbf{j} + y^2\,\mathbf{k}$. The vector function given by

$$\mathbf{r}(x, y) = x\,\mathbf{i} + y\,\mathbf{j} + y^2\,\mathbf{k}$$

is a vector representation of the cylinder $z = y^2$.

The cylinder with equation $x^2 + y^2 = 4$ can be represented by the vector function

$$\mathbf{r}(\theta, z) = 2\cos\theta\,\mathbf{i} + 2\sin\theta\,\mathbf{j} + z\,\mathbf{k}$$

because every vector in the xy plane from the origin to a point on the circle $x^2 + y^2 = 4$ has the form $2\cos\theta\,\mathbf{i} + 2\sin\theta\,\mathbf{j}$ for some number θ.

More generally suppose that a curve is given lying in a plane and L is a line not in the plane of the curve. How do we represent the cylinder generated by the curve having L as directing line? One way is to first find a vector representation of the curve, say $\mathbf{r}(t)$, and then take a vector $\mathbf{u}$ parallel to L. To reach a point in the required surface we move from the origin to $\mathbf{r}(t)$ and then in the direction of $\mathbf{u}$ some arbitrary distance. The vector function

$$\mathbf{U}(s, t) = \mathbf{r}(t) + s\mathbf{u}$$

represents the surface as t ranges over domain of $\mathbf{r}$ and s ranges over all real numbers.

Here is a slightly more complicated example in which the curve is not in one of the coordinate planes.

EXAMPLE 2 Describe a parabola lying in the plane containing the origin and the two points $(1, 1, 0)$ and $(1, -1, 1)$ and then find a vector representation of the cylinder generated by this parabola with directing line parallel to the vector $5\mathbf{i} + 7\mathbf{k}$.

Solution: Let $\mathbf{u} = (\mathbf{i} + \mathbf{j})/\sqrt{2}$ and $\mathbf{v} = (\mathbf{i} - \mathbf{j} + \mathbf{k})/\sqrt{3}$. Then $\mathbf{u}$ and $\mathbf{v}$ are unit vectors, orthogonal to each other and the plane $s\mathbf{u} + t\mathbf{v}$ contains the points $(1, 1, 0)$ and $(1, -1, 1)$ as the terminal points of $\sqrt{2}\mathbf{u}$ and $\sqrt{3}\mathbf{v}$ respectively. For an example of a parabola lying in this plane we take the curve $\mathbf{r}(t) = t\mathbf{u} + t^2\mathbf{v}$. (Many other choices are possible.) One vector form of the required cylinder is then

$$\begin{aligned}\mathbf{U}(s, t) &= \mathbf{r}(t) + s(5\mathbf{i} + 7\mathbf{k}) = t\mathbf{u} + t^2\mathbf{v} + s(5\mathbf{i} + 7\mathbf{k}) \\ &= \left(\frac{t}{\sqrt{2}} + \frac{t^2}{\sqrt{3}} + 5s\right)\mathbf{i} + \left(\frac{t}{\sqrt{2}} - \frac{t^2}{\sqrt{3}}\right)\mathbf{j} + \left(\frac{t^2}{\sqrt{3}} + 7s\right)\mathbf{k}.\end{aligned}$$

Many other correct solutions are possible. ■

Surfaces of Revolution

A *surface of revolution* is the surface obtained by revolving a curve around a fixed line. For example if $f(x)$ is a positive function for $a \leq x \leq b$ then a surface is obtained by revolving a curve $y = f(x)$ around the x-axis.

Starting with the straight line $y = x/2$ for $x \geq 0$ and revolving the curve around the x-axis, we obtain a cone.

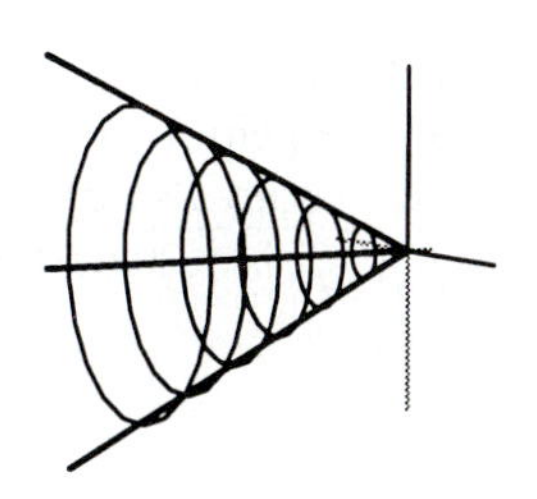

FIGURE 14.2
Cone with generator $2y = x$

We can find the equation satisfied by the coordinates of a point on a surface of revolution. Suppose the curve $y = f(x)$ is revolved around the x-axis to produce a surface S. Consider the plane parallel to the yz plane that crosses the x-axis at some particular x_1. Then the intersection of the surface S and this plane is a circle of radius $f(x_1)$. Another point on this intersection has coordinates (x_1, y, z) where the distance from (x_1, y, z) to $(x_1, 0, 0)$ is the same as the distance from $(x_1, f(x_1), 0)$ to $(x_1, 0, 0)$. In other words the coordinates of (x_1, y, z) satisfy the equation

$$y^2 + z^2 = f(x)^2. \tag{14.1}$$

Conversely if the coordinates of (x, y, z) satisfy Equation (14.1), then the distance from (x, y, z) to $(x, 0, 0)$ equals the distance from $(x, f(x), 0)$ to $(x, 0, 0)$ and so the point (x, y, z) is on the surface of revolution obtained by revolving the graph of $y = f(x)$ around the x-axis.

A similar formula can be derived for the surface obtained by revolving a curve in one of the other coordinate planes about an axis. For example the equation of the surface obtained by revolving a curve $z = g(y)$ around the y-axis is

$$z^2 + x^2 = g(y)^2.$$

EXAMPLE 3 Write the equation of the solid of revolution obtained by revolving the upper half ellipse with equation $y = \sqrt{9 - 4x^2}$ around the x-axis.

Solution: Using $f(x) = \sqrt{9 - 4x^2}$ the formula for the surface is obtained from Equation (14.1) and reads

$$y^2 + z^2 = 9 - 4x^2.$$

This surface is an example of an *ellipsoid*. ■

EXAMPLE 4 Describe the surface having equation $4x^2 + 4y^2 - 64z^2 = 400$.

Solution: If the equation is written as $x^2 + y^2 = 16z^2 + 100$ then it has the same form as Equation (14.1) except that y, z and x in Equation (14.1) are replaced by x, y and z, respectively. The equation may be expressed in the form $x^2 + y^2 = f(z)^2$ where $f(z)^2 = 16z^2 + 100$. There are two obvious choices for the function $f(z)$, namely

$$f_1(z) = \sqrt{16z^2 + 100} \quad \text{and} \quad f_2(z) = -\sqrt{16z^2 + 100}.$$

If we use the first choice and revolve the curve $x = f_1(z)$ around the z-axis we obtain a surface having equation

$$x^2 + y^2 = f_1(z)^2 = 16z^2 + 100,$$

which is the equation we are studying. If we use the second choice f_2, exactly the same equation and the same surface is obtained.

We could also approach this by considering the curve in the yz plane given by $y = \sqrt{16z^2 + 100}$ and revolving it around the z-axis.

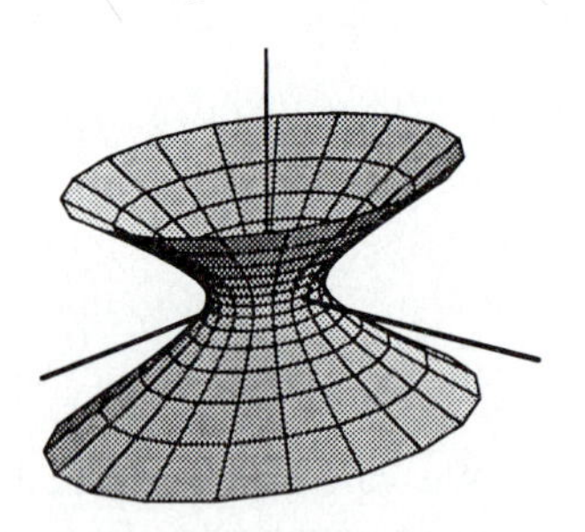

FIGURE 14.3
Revolved hyperbola

■

Vector Form of a Surface of Revolution

The vector form for a surface obtained by revolving the graph of $y = f(x)$ around the x-axis is found as follows. The circle of radius 1 in the yz plane has vector form $\cos\theta\,\mathbf{j} + \sin\theta\,\mathbf{k}$ for $0 \leq \theta \leq 2\pi$ and the circle of radius $f(x)$ has the form

$$f(x)(\cos\theta\,\mathbf{j} + \sin\theta\,\mathbf{k}).$$

The intersection of the surface of revolution with the plane parallel to the yz plane passing through $(x, 0, 0)$ is a circle of radius $f(x)$. Starting from the origin move along the x-axis to reach $x\,\mathbf{i}$ and then move to any point on that circle to reach an arbitrary point of the surface of revolution; the vector equation is

$$\mathbf{U}(x, \theta) = x\,\mathbf{i} + f(x)(\cos\theta\,\mathbf{j} + \sin\theta\,\mathbf{k}).$$

EXAMPLE 5 Find the vector form of the surface of revolution obtained by rotating the curve $y = z^2$ in the yz plane about the y-axis.

Solution: Because we revolve around the y-axis, we express z in terms of y as $z = \sqrt{y}$. The intersection of the the surface of revolution and plane parallel to the xz plane crossing the y-axis at $(0, y, 0)$ is the circle $\sqrt{y}(\cos\theta\,\mathbf{i} + \sin\theta\,\mathbf{k})$ and the equation of the surface is

$$\mathbf{U}(y, \theta) = y\,\mathbf{j} + \sqrt{y}(\cos\theta\,\mathbf{i} + \sin\theta\,\mathbf{k}).$$

We can use from this form to see that no new points are obtained by considering the second curve $z = -\sqrt{y}$. If we set $\phi = \theta + \pi$ then the function

$$\mathbf{V}(y, \phi) = y\,\mathbf{j} + \sqrt{y}(\cos\phi\,\mathbf{i} + \sin\phi\,\mathbf{k})$$

represents the same surface as $\mathbf{U}$. But now $\cos\phi = \cos(\theta + \pi) = -\cos\theta$ and $\sin\phi = -\sin\theta$. Then

$$\mathbf{V}(y, \phi) = y\,\mathbf{j} - \sqrt{y}(\cos\theta\,\mathbf{i} + \sin\theta\,\mathbf{k}),$$

which is the vector form of the surface obtained by revolving $z = -\sqrt{y}$ around the y-axis. ■

Surfaces in Parametric Form

When a surface is given in vector form it can be regarded as a *parametric form* just as we have seen earlier for curves given either in vector or parametric form. In the two examples given in this section, cylinders and surfaces of revolution, the vector equation of the surface has the form

$$\mathbf{U}(s, t) = X(s, t)\,\mathbf{i} + Y(s, t)\,\mathbf{j} + Z(s, t)\,\mathbf{k}$$

for some functions X, Y and Z, each depending on two parameters s and t. Thus a point (x, y, z) is on the surface if there are values of s and t such that

$$\begin{cases} x = X(s, t), \\ y = Y(s, t), \\ z = Z(s, t). \end{cases}$$

These equations are called *parametric equations for the surface.* Just as the case for curves, the given surface might have many different sets of parametric equations that describe it.

For the example of a surface of revolution generated by the graph of $y = f(x)$ the parametric equations of the surface are

$$\begin{cases} x = s, \\ y = f(s)\cos t, \\ z = f(s)\sin t. \end{cases}$$

The cylinder generated by the graph of $y = f(x)$ in the xy plane and having directing line parallel to the z-axis has parametric equations

$$\begin{cases} x = s, \\ y = f(s), \\ z = t. \end{cases}$$

Exercises 14.1

Describe in words the graphs in xyz space of the equations, give a rough sketch of the surface. Then find a vector representation for each surface.

1. $x^2 + z^2 = 4$

2. $3y^2 + 12z^2 = 27$

3. $x = 4z^2$

4. $2x^2 = 3z - 2y^2$

5. $x = 8y^2 + 4$

6. $y^2 + z^2 = -3x$

7. $x^2 + y^2 = z/2$

8. $x^2 + z^2 = y^2$

9. $x^2 + y^2 - z^2 = 1$

10. $y^2(x^2 + z^2) = 1$

11. Find the equation of the surface obtained by revolving the ellipse in the yz plane having equation $(ay)^2 + (bz)^2 = (ab)^2$ around the y-axis.

12. Find the equation of the paraboloid of revolution obtained by revolving the parabola $y = x^2$ around the y-axis.

Identify the surfaces having the following vector representations as cylinders or surfaces of revolution and give a rough sketch. Indicate a directing line or an axis of revolution as is appropriate.

13. $\mathbf{U}(s, t) = 4t\,\mathbf{i} + t^2\,\mathbf{j} + s\,\mathbf{k}$

14. $\mathbf{W}(s, t) = s\,\mathbf{i} + (2t + 1)\,\mathbf{j} + t^2\,\mathbf{k}$

15. $\mathbf{V}(s, t) = s\,\mathbf{i} + s^2(\cos t\,\mathbf{j} + \sin t\,\mathbf{k})$

16. $\mathbf{U}(s, t) = 3s\cos t\,\mathbf{i} + 3s\sin t\,\mathbf{k} + s\,\mathbf{j}$

17. $\mathbf{V}(s, t) = s\,\mathbf{i} + t\,\mathbf{j} + 3\,\mathbf{k}$

18. $\mathbf{W}(s, t) = s\,\mathbf{i} - 6\,\mathbf{j} + t\,\mathbf{k}$

Describe in words and then give a rough sketch of the surfaces having the following parametric equations.

19. $x = t, \quad y = s, \quad z = s^2 + t^2$ for $0 \le s, t \le 1$

20. $x = 2t, \quad y = 4t, \quad z = s + 1$ for $t \ge 0$ and $s \ge 0$

14.2 Quadric surfaces

A quadric surface is one whose coordinates satisfy a polynomial equation of the form

$$ax^2 + by^2 + cz^2 + dxy + exz + fyz + gx + hy + kz + m = 0$$

for some constants $a, b, c, d, e, f, g, h, k$ and m. In the simpler case of curves in two dimensions, a quadric has the form

$$ax^2 + by^2 + cxy + dx + ey + f = 0.$$

By completing the square and translating the origin we know that the curve can be put in one of the standard forms for a conic curve, and the graph is a parabola, an ellipse or a hyperbola. The case of quadric surfaces in xyz space presents more options and some slightly more complicated standard forms. After completing the square and translating the origin, the equation can be simplified considerably. We consider some of the possibilities. The case of a *cone* is left to the exercises following this section.

Traces

To get some idea of the shape of the surface we consider the intersection of the surface with planes, usually planes parallel to one of the coordinate planes. The intersection of a surface and a plane is usually a curve. Of course the intersection might be empty or consist of a single point or even several points. The curve obtained by intersecting a surface and a plane is called the *trace* of the surface in the plane. If the plane is parallel to one of the coordinate planes, we simply call the curve a *trace of the surface* without specific reference to the particular plane.

The Ellipsoid

$$\frac{x^2}{a^2} + \frac{y^2}{b^2} + \frac{z^2}{c^2} = 1.$$

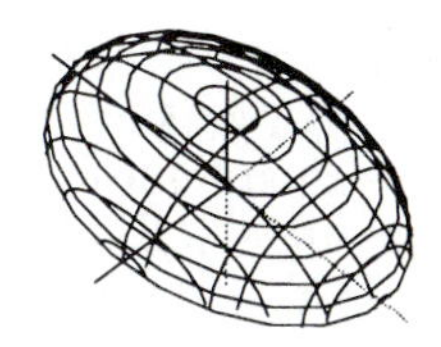

FIGURE 14.4
Traces of an ellipsoid

The graph of the equation of the ellipsoid is confined to the region $|x| \leq a$, $|y| \leq b$, $|z| \leq c$ assuming that a, b and c are positive constants. The traces parallel to any coordinate plane are ellipses. If we cut the surface with the plane $z = p$ for $|p| < c$ we get a curve whose coordinates are (x, y, p) with x and y satisfying the equation

$$\frac{x^2}{a^2} + \frac{y^2}{b^2} = 1 - \frac{p^2}{c^2},$$

which is the equation of an ellipse. Similarly the intersection of the ellipsoid with a plane $x = q$ or $y = r$ is an ellipse.

If any two of the three constants a, b and c are equal, the ellipsoid is a surface of revolution. If all three constants are equal the ellipsoid is a sphere.

Hyperboloid of One Sheet

$$\frac{x^2}{a^2} + \frac{y^2}{b^2} - \frac{z^2}{c^2} = 1.$$

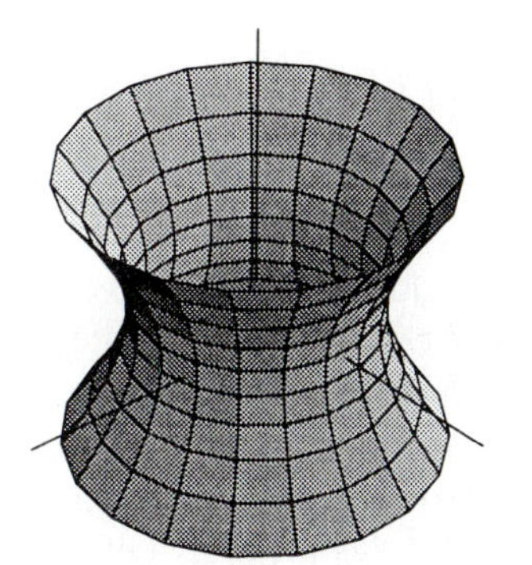

FIGURE 14.5
Hyperboloid of one sheet

For any value $z = p$ there is a point (x, y, p) on the graph and the intersection of the surface with the plane $z = p$ is an ellipse with equation

$$\frac{x^2}{a^2} + \frac{y^2}{b^2} = \frac{p^2}{c^2} + 1.$$

Note that as p increases the ellipse gets larger. For its traces parallel to the xy plane, the hyperboloid of one sheet has ellipses that increase in size as the intersecting plane moves up the positive z-axis or moves down the negative z-axis.

Intersection of the surface with a plane of the form $x = q$ produces a hyperbola with equation

$$\frac{y^2}{b^2} - \frac{z^2}{c^2} = 1 - \frac{p^2}{a^2}.$$

This is a hyperbola for which y is unrestricted but z must satisfy $|z| \geq 1 - (p/a)^2$.

The name of the surface is meant to suggest that the surface has a single "sheet" so that one can trace a path from one point of the surface to any other point without ever leaving the surface. The next example illustrates a different phenomenon.

Hyperboloid of Two Sheets

$$\frac{z^2}{c^2} - \frac{y^2}{b^2} - \frac{x^2}{a^2} = 1.$$

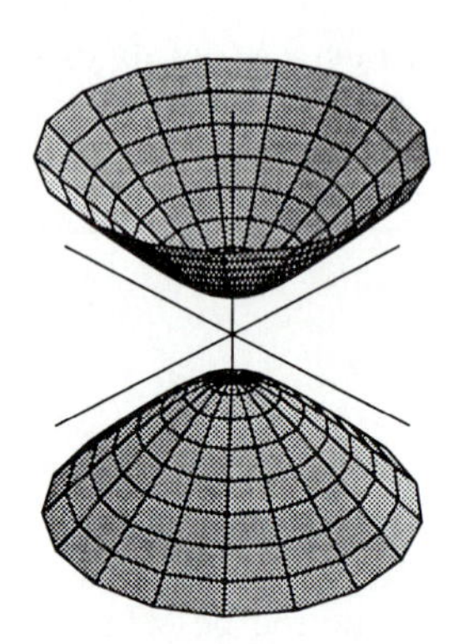

FIGURE 14.6
Hyperboloid of two sheets

If we select some plane $z = q$ and determine its intersection with the surface we get a curve with equation that can be written as

$$\frac{y^2}{b^2} + \frac{x^2}{a^2} = \frac{q^2}{c^2} - 1.$$

The left-hand side of the equation is nonnegative so if there is a point (x, y, q) in the surface, we must have $(q/c)^2 - 1 \geq 0$ or equivalently, $|q| \geq c$. Thus the z coordinate of points on the surface either less than or equal to $-c$ or greater than or equal to c. This surface has two sheets unlike the previous example. It is not possible to connect every pair of points in the surface by a path that lies entirely in the surface. For example there is no path in the surface connecting $(0, 0, c)$ and $(0, 0, -c)$.

The trace of the surface at $z = q$ is an ellipse that increases in size as q increases along the positive z-axis or decreases along the negative z-axis. The trace of the surface in the plane $x = p$ is a hyperbola with equation

$$\frac{z^2}{c^2} - \frac{y^2}{b^2} = \frac{p^2}{c^2} + 1.$$

The traces in the planes $y = r$ are also hyperbolas.

The Elliptic Paraboloid

$$\frac{x^2}{a^2} + \frac{y^2}{b^2} = cz.$$

FIGURE 14.7
Elliptic paraboloid

We assume the constant c is positive so the values of z must be nonnegative. (For negative c, z is also negative.) For any $p > 0$, the intersection of the surface and the plane $z = p$ is an ellipse and the size of the ellipse increases as p increases. The trace of the surface in the plane $x = q$ is a parabola opening upward. The same holds for the traces in the planes $y = r$.

Figure 14.7 shows the graph of the equation $x^2 + y^2 = z$ for $|x| \leq 2$ and $|y| \leq 3$.

The Hyperbolic Paraboloid

$$\frac{x^2}{a^2} - \frac{y^2}{b^2} = cz.$$

This is the most interesting of the quadric surfaces because of the variety of the different traces. Assume c is a positive constant. We first consider the trace of the surface in the plane $z = p$. For a positive value p, the trace is a hyperbola

$$\frac{x^2}{a^2} - \frac{y^2}{b^2} = cp \qquad \textbf{(14.2)}$$

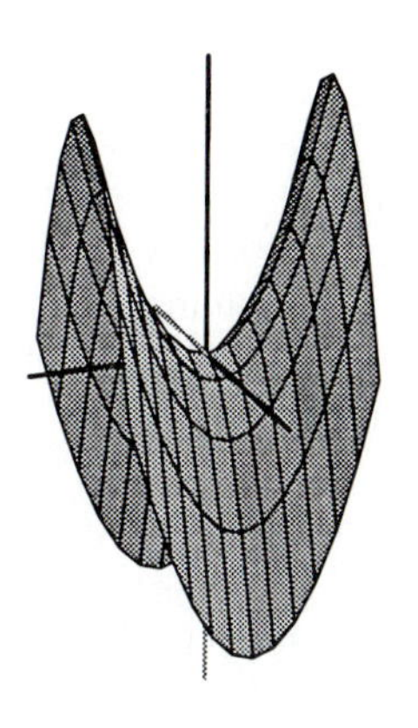

FIGURE 14.8
$z = x^2 - y^2$
Hyperbolic paraboloid

that opens along the x-axis. When p is negative, (14.2) is the equation of a hyperbola that opens along the y-axis. When $p = 0$, (14.2) is the equation of two straight lines $(x/a) = \pm(y/b)$. Thus we have a surface that contains a straight lines.

The trace in the plane $x = q$ is a parabola that opens downward, whereas the trace in a plane $y = r$ is a parabola that opens upward.

Vector Representation of Quadric Surfaces

There are many different vector representations of all or part of the quadric surfaces mentioned above. We do not attempt a thorough discussion but will be content to mention some vector representations that employ trigonometric functions. The methods in section 8.6 may be used to discover rational parametrizations of some of these surfaces. Parametrizations of surfaces are often useful when using a computer to display surfaces. The surfaces might be easier to describe parametrically than as graphs of functions. For example consider a hyperboloid of two sheets that opens upward along the positive z-axis and symmetrically opens downward along the negative z-axis. The rectangular equation has the form $z^2 = 1 + x^2/a^2 + y^2/b^2$ for some constants a and b. For a fixed z, the trace of the surface is an ellipse; we have seen how to parametrize an ellipse using trigonometric functions in the form $p\cos t\,\mathbf{i} + q\sin t\,\mathbf{j}$ where the constants determine the shape of the ellipse. If we introduce an additional parameter to allow the ellipse to grow as the parameter changes we could write $x = ar\cos t$ and $y = br\sin t$. Then the equation of the hyperboloid has the form

$$\begin{aligned} z^2 &= 1 + \frac{x^2}{a^2} + \frac{y^2}{b^2} \\ &= 1 + \frac{r^2a^2\cos^2 t}{a^2} + \frac{r^2b^2\sin^2 t}{b^2} \\ &= 1 + r^2. \end{aligned}$$

Thus we obtain the vector representation

$$H(r,t) = ar\cos t\,\mathbf{i} + br\sin t\,\mathbf{j} \pm \sqrt{r^2+1}\,\mathbf{k}$$

for a hyperboloid of two sheets. The double sign is necessary to account for both positive and negative square roots of $1 + r^2$. This is only one example of many parametrizations that are possible.

Each of the following vector functions of two parameters s and t corresponds to a quadric surface (a, b and c are constants):

A. $\mathbf{E}(s,t) = a\cos t\cos s\,\mathbf{i} + b\sin t\cos s\,\mathbf{j} + c\sin s\,\mathbf{k}$,

B. $\mathbf{H_1}(s,t) = a\cos t\sec s\,\mathbf{i} + b\sin t\sec s\,\mathbf{j} + c\tan s\,\mathbf{k}$,

C. $\mathbf{H_2}(s,t) = a\sin s\tan t\,\mathbf{i} + b\cos s\tan t\,\mathbf{j} + c\sec t\,\mathbf{k}$,

D. $\mathbf{EP}(s,t) = as\cos t\,\mathbf{i} + bs\sin t\,\mathbf{j} + s^2\,\mathbf{k}$,

E. $\mathbf{HP}(s,t) = as\sec t\,\mathbf{i} + bs\tan t\,\mathbf{j} \pm s^2\,\mathbf{k}$.

We leave it for the reader to verify that $\mathbf{E}(s,t)$ represents an ellipsoid, $\mathbf{H_1}(s,t)$ represents a hyperboloid of one sheet, $\mathbf{H_2}(s,t)$ represents a hyperboloid of two sheets, $\mathbf{EP}(s,t)$ represents an elliptic paraboloid and $\mathbf{HP}(s,t)$ represents a hyperbolic paraboloid. In the last case there are really two equations corresponding to the two choices of sign; this double sign arrangement is convenient because z can take on both positive and negative values.

There are many alternate parametrizations of quadric surfaces. The restricted range of some functions used has the implication that only part of the surface is represented and some modification is needed to obtain the remaining part of the surface. This is seen in the double sign in E for example. The hyperbolic functions are often useful for alternate parametrizations. The functions

$$\sinh t = \frac{e^t - e^{-t}}{2} \quad \text{and} \quad \cosh t = \frac{e^t + e^{-t}}{2}$$

satisfy the equation $\cosh^2 t - \sinh^2 t = 1$. Thus the substitutions $x = a\cosh t$ and $y = b\sinh t$ satisfy

$$\frac{x^2}{a^2} - \frac{y^2}{b^2} = 1.$$

The following vector functions are alternate parametrizations of B, C and E above:

B$'$. $\mathbf{H_1}(s,t) = a\cos t\cosh s\,\mathbf{i} + b\sin t\cosh s\,\mathbf{j} + c\sinh s\,\mathbf{k}$,

C$'$. $\mathbf{H_2}(s,t) = a\sin s\sinh t\,\mathbf{i} + b\cos s\sinh t\,\mathbf{j} \pm c\cosh t\,\mathbf{k}$,

E$'$. $\mathbf{HP}(s,t) = as\cosh t\,\mathbf{i} + bs\sinh t\,\mathbf{j} \pm s^2\,\mathbf{k}$.

Exercises 14.2

Identify the surfaces represented by the following equations. Give a rough sketch that shows the general orientation and enough detail to identify it as one of the five quadric surfaces whose figures are sketched in this section.

1. $x^2 + y^2 + z^2 = 4$

2. $x^2 + 4y^2 + 16z^2 = 64$

3. $x^2 + y^2 - z^2 = 4$

4. $x^2 + 9y^2 - 81z^2 = 81$

5. $x^2 - 4y^2 - 16z^2 = 64$

6. $x^2 - y^2 + z^2 = 4$

7. $x^2 - 9y^2 + 81z^2 = 81$

8. $x^2 + y^2 - z^2 = 4$

9. $-x^2 + 4y^2 - 16z^2 = 64$

10. $z = x^2 - y^2$

11. $2z = x^2 - y^2/8$

12. $x = z^2 - y^2$

13. $8y = 2x^2 - z^2$

14. $z + 5 = x^2 - y^2$

15. $z = (x-3)^2 - (y-2)^2$

16. $\left(\frac{x-2}{2}\right)^2 + \left(\frac{y+1}{4}\right)^2 + z^2 = 1$

17. A surface given by an equation $z^2 = (x/a)^2 + (y/b)^2$ is called a *cone*. Verify that the cone has elliptical traces when intersected with certain planes parallel to the coordinate planes but the traces parallel to the other coordinate planes are hyperbolas or straight lines.

18. Generate a surface in the following way. Let C be the circle in the plane $z = 4$ consisting of all points $(x, y, 4)$ for which $x^2 + y^2 = 1$. Take all half-lines that begin at the origin and pass through a point of C. This collection of half-lines is a surface. Find the equation of this surface.

19. Find a vector representation of the cone $z^2 = (x/a)^2 + (y/b)^2$ in the general form $\mathbf{U}(s,t) = sm(t)\,\mathbf{i} + sn(t)\,\mathbf{j} + s^2\,\mathbf{k}$ for some functions $m(t)$ and $n(t)$

Give a rough sketch of the cones having the following equations. Show a typical trace and the orientation of the cone.

20. $z^2 = 2x^2 + y^2$

21. $x^2/4 = y^2/9 + z^2/9$

22. $2y^2 = 4x^2 + z^2$

23. $(x-2)^2 = (y-3)^2 + z^2$

24. Verify that each of the vector functions described in A–E, B$'$, C$'$ and E$'$ above does represent one of the quadric surfaces as indicated.

25. Determine if the entire quadric surface having rectangular coordinates is represented by the corresponding vector equation in A–E$'$ above.

26. A surface obtained by revolving a curve $y = f(x)$ around the x-axis has equation $y^2 + z^2 = f(x)^2$. Similar equations are obtained for surfaces that are obtained by revolving curves around the y-axis or the z-axis. Determine which of the surfaces in Exercises 1-16 are surfaces of revolution and give the curve that is revolved when appropriate.

27. Determine the rectangular equation of the surface having vector representation

$$W(r,t) = 2r\sin t\,\mathbf{i} + 5r\cos t\,\mathbf{j} + r^2\,\mathbf{k}.$$

28. Determine the rectangular equation of the surface having vector representation

$$V(r,t) = 2r\cos t\,\mathbf{i} - 3r\sin t\,\mathbf{j} + 5r\,\mathbf{k}.$$

29. Determine the rectangular equation of the surface having vector representation

$$W(r,t) = (r^2+4)\,\mathbf{i} + r\sin t\,\mathbf{j} - 3r\cos t\,\mathbf{k}.$$

14.3 Functions of several variables

At the beginning of the text we considered functions of one variable, $f(x)$ where x lies in some domain and $f(x)$ is a real number. Later we considered vector functions $\mathbf{r}(t)$ where t was a real number but the value of the function $\mathbf{r}(t)$ was a vector. We also studied vector functions of two variables in the form $\mathbf{r}(s,t) = s\mathbf{u} + t\mathbf{v}$ to describe a plane containing the vectors $\mathbf{u}$ and $\mathbf{v}$ and we saw some vector valued functions $\mathbf{U}(s,t)$ of two variables in the previous sections of this chapter. Now we consider functions $f(x,y)$ and $f(x,y,z)$ of two or three independent variables. The values of f are real numbers.

Many of the statements we make apply equally well to real-valued functions of any number of variables $f(x_1, x_2, \ldots, x_n)$.

Just for the record, let us give the definition and a few examples of functions of several variables.

DEFINITION 14.1

Function

Let D be a set of points in the xy plane or in xyz space. A function f with domain D is a rule that assigns to each point $\mathbf{p}$ of D a unique real number denoted by $f(\mathbf{p})$.

In the definition, if $\mathbf{p} = (x,y)$ then we write $f(\mathbf{p}) = f(x,y)$, whereas if $\mathbf{p} = (x,y,z)$ then we write $f(\mathbf{p}) = f(x,y,z)$.

The reader can easily write down many examples of functions of several variables using familiar functions of one variable; for example:

$$f(x,y) = e^x(x^2+5y^3), \qquad g(x,y) = \sin(xy) + x\cos(y^2+x),$$
$$h(x,y,z) = xy + xz + yz, \qquad F(x,y,z) = \ln(x^2+y^2+z^2).$$

The largest domain on which these functions can be defined is all of the xy plane for f and g and all of xyz space for h. The domain of F is all xyz space with the origin omitted because the logarithm is not defined at 0.

EXAMPLE 1 The hyperboloid of two sheets with equation

$$\frac{z^2}{5^2} - \frac{x^2}{4^2} - \frac{y^2}{2^2} = 1$$

has one sheet above and one sheet below the xy plane. We can see because $z^2 \geq 25$ for every (x,y,z) on the surface and so either $z \geq 5$ or $z \leq -5$. For any point $(x,y,0)$ in the xy plane, find the height to the point directly above it in the surface.

***Solution*:** The point directly above $(x,y,0)$ in the surface is (x,y,z) where x, y and z satisfy the equation of the surface. Solve the equation for z and use the fact that $z \geq 0$ to get

$$H(x,y) = 5\sqrt{1 + \frac{x^2}{4^2} + \frac{y^2}{2^2}},$$

which is the distance from $(x, y, 0)$ up to the point directly above it in the surface. ■

EXAMPLE 2 A pile of sand on the ground covers a circular area of radius 10 units and is cone shaped with the highest point of the cone 5 units above the center of the base. For each point $p = (x, y)$ of the base find the height H of the sand directly above p.

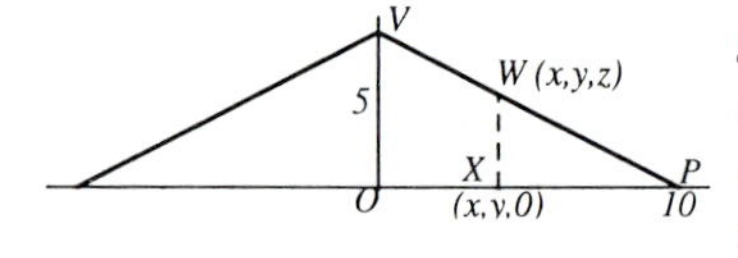

FIGURE 14.9

Solution: Draw the line from the origin through (x, y) and out to the edge of the pile. If O is the origin, X is (x, y), P the point where OX meets the edge of the pile, W is the point directly above X on the surface of the pile, and V is the highest point of the pile, the triangle OPV is similar to XPW and so

$$|XW| = \frac{|XP| \cdot |OV|}{|OP|} = \frac{(10 - \sqrt{x^2 + y^2}) \cdot 5}{10}.$$

Thus the height function is the function of two variables given by $H(x,y) = (10 - \sqrt{x^2 + y^2})/2$. ■

Graphs of Functions

A function of two variables determines a surface in the following way. If a function $f(x, y)$ is defined for all (x, y) in some domain D in the xy plane then the surface determined by f is the set

$$S = \{(x, y, z) : (x, y) \in D, \quad \text{and} \quad z = f(x, y)\}.$$

The surface S is the *graph of* $f(x, y)$. We use several variants of terminology such as f determines the surface $z = f(x, y)$ or S is the graph of $z = f(x, y)$.

EXAMPLE 3 Describe the graph of the function $f(x, y) = 2x - 3y + 12$.

Solution: The function is defined for all (x, y) and the equation $z = 2x - 3y + 12$ is the equation of a plane. The plane crosses the coordinate axes at the points $(-6, 0, 0)$, $(0, 4, 0)$ and $(0, 0, 12)$. ■

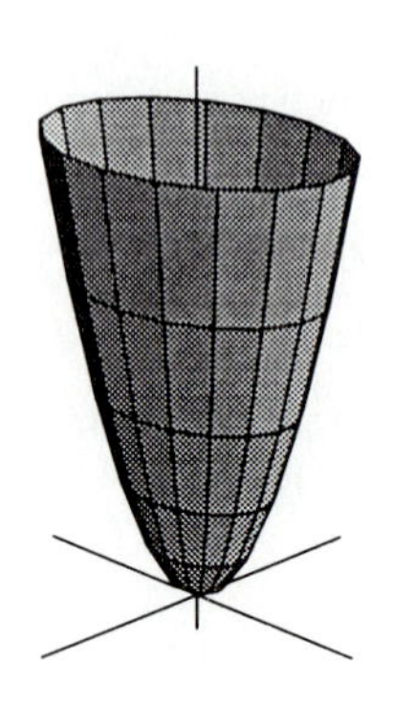

FIGURE 14.10

EXAMPLE 4 Describe the surface defined by the function $z = f(x, y) = 2x^2 + y^2$.

Solution: The equation $z = f(x, y) = 2x^2 + y^2$ describes an elliptic paraboloid passing through the origin and opening upward with the positive z-axis as its central axis although it is not a surface of revolution. The solutions of the equations $f(x, y) = c$ for a positive constant c correspond to the intersection of the surface with the plane $z = c$. In this case that intersection is an ellipse. ■

A surface S is determined by a function if and only if for each point (x, y, z) in S there is no other point having the same first two coordinates; that is if (x, y, z_1) and (x, y, z_2) are points in S then $z_1 = z_2$. When S has this property then we can define a function by the rule $f(x, y) = z$ whenever (x, y, z) is a point of S. Conversely if $f(x, y)$ is a function defined for all (x, y) in some set D then the surface of f consists of the points $(x, y, f(x, y))$ for all (x, y) in D. Clearly there can be no two points with the same xy coordinates because the definition of function is such that for each (x, y) in the domain of f there is only one number assigned by f to that point.

Level Curves and Contours

Let S be the graph of $f(x, y)$. The intersection of S with a plane $z = p$ is called a *contour curve* of f and its projection to the xy-plane is called a *level curve* of height p of f. The level curve consists of the points $(x, y, 0)$ for which $f(x, y) = p$.

EXAMPLE 5 Plot the level curves of height 1, 2, 3 and 4 for the function $f(x, y) = 2 - x + y^2$.

Solution: For a general height p the level cure of height p consists of the point in the xy plane whose coordinates satisfy $2 - x + y^2 = p$ or $x = 2 - p + y^2$. We may now plot these for the indicated values of p.

The graph of the function f has the shape indicated in figure 14.12.

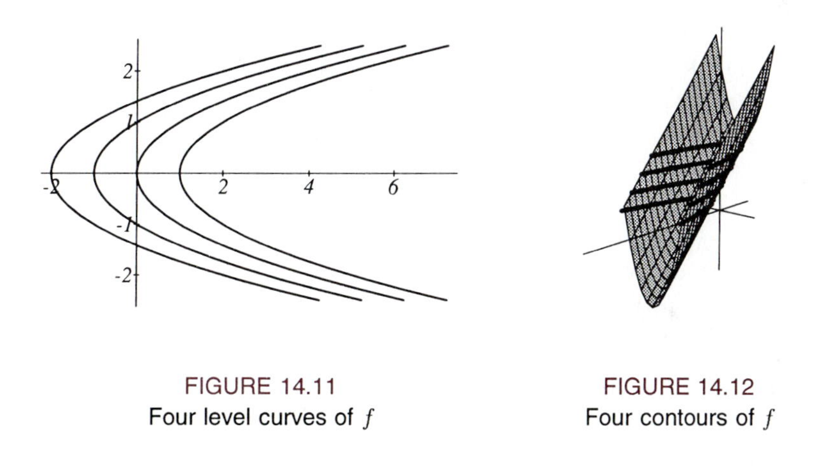

FIGURE 14.11
Four level curves of f

FIGURE 14.12
Four contours of f

■

The level curves of f give a way of representing in two dimensions the surface $z = f(x, y)$ in three dimensions.

Level Surfaces

It is not possible to represent the graph of a function $f(x, y, z)$ of three variables in the usual way because the graph would require consideration of the points $(x, y, z, f(x, y, z))$ in a four-dimensional space. By analogy with level curves for a function of two variables we can consider the *level surfaces* of f. A level surface for f of height c is the surface in xyz space defined by equation

$$f(x, y, z) = c.$$

Level surfaces arise in practical applications as illustrated in the following.

EXAMPLE 6 Suppose a solid ball of radius R is centered at the origin and represents a laboratory experiment in which the ball is subjected to a heat source so that the temperature at a point (x, y, z) inside the ball is given by a function $f(x, y, z)$. Then the level surfaces of f are the regions inside the ball having constant temperature. For example, if

$$f(x, y, z) = 212 - 15\sqrt{x^2 + y^2 + z^2},$$

then the region within the ball where the temperature is 182 is the level surface defined by

$$182 = f(x, y, z) = 212 - 15\sqrt{x^2 + y^2 + z^2}.$$

This is equivalent to the equation $4 = x^2 + y^2 + z^2$, which is the equation of the sphere of radius 2. All the surfaces of constant temperature are spheres in this example. ■

REMARK: The value of the numerical exercises in the following set will be greatly enhanced if the student has available a computer with a three-dimensional graphics package and uses it to picture and study the graphs.

Exercises 14.3

Sketch the surface of each function. Label the axes and relevant points like the points where the surface crosses an axis and highlight the curves where the surface meets one of the coordinate planes.

1. $g(x, y) = 4 - x^2 - y^2$

2. $h(x, y) = 4x - 3y$

3. $f(x, y) = 1 + xy$

4. $F(x, y) = x + y^2$

5. Examine the list of quadric surfaces and decide which can be the graph of a function. Find the function when possible. [A complete answer to this problem requires many cases if different orientations of each possible surface are taken into account. Just consider the different equations as given and also with different permutations of the variables; that is, both $x^2 - y^2 = z$ and $z^2 - x^2 = y$ are equations of quadric surfaces with different permutations of the variables.]

Sketch the level curves of the function f at the given heights p and then sketch the graph of the function over the region containing the corresponding contours.

6. $f(x, y) = 16 - x^2 - y^2$, $p = 0, 5, 12, 16$

7. $f(x, y) = 10 - 2x^2 - y^2$, $p = 0, 1, 6, 10$

8. $f(x, y) = x^2 - y^2$, $p = -4, -1, 0, 1, 4$

9. $f(x, y) = x - y^2$, $p = -4, -1, 0, 1, 4$

Find examples $z = f(x, y)$ that have the following level curves.

10. Circles with center at $(1, 0)$

11. Ellipses with major axis twice the minor axis

12. Hyperbolas centered at $(0, 0)$ and crossing the x-axis

Identify the level curves and sketch a representative sample of them.

13. $f = \ln(x^2 + y^2)$

14. $f = \dfrac{x^2}{x^2 + y^2}$

15. $f = (x + y)^2$

16. $f = e^{-xy}$

Describe the level surfaces for the following functions.

17. $f = x - y + z$

18. $f = z^2 - 4x^2 + y^2$

19. $f = (2x)^2 + (3y)^2 + z^2$

20. $f = (x - 1)^2 + y^2 + (z + 2)^2$

21. For a given function f, give a short explanation of why two different level curves (or two different level surfaces) have no points in common.

14.4 Partial derivatives

The derivative of a function of one variable is one of the two fundamental concepts in calculus (the integral is the other). Now that we have introduced functions of several variables, it is natural to look for the analogue of the derivative. A function $f(x, y)$ does not have a derivative in the same sense as does a function of one variable. If we think of the derivative as a measure of the change in the function, then the change in $f(x, y)$ can occur by allowing x to vary, by allowing y to vary or by allowing both x and y to vary. Rather than attempting to define a derivative that measures all these possible changes, we take a simpler course. When the variable y is held fixed, then $f(x, y)$ is a function of x only and its derivative is defined in the usual way for a function of one variable. We call this a *partial derivative of f with respect to x* and denote it $f_x(x, y)$. In the same way if we hold x fixed, then $f(x, y)$ is a function of y only so the derivative with respect to y is called the *partial derivative of f with respect to y* and denote it by $f_y(x, y)$. In a similar way a function of three (or more) variables has a partial derivative with respect to any one of its variables defined by holding all the remaining variables fixed.

Let $f(x, y)$ be a given function of two variables. The partial derivative of f with respect to x at the point (x, y) is the limit

$$f_x(x, y) = \lim_{h \to 0} \frac{f(x + h, y) - f(x, y)}{h}$$

provided the limit exists. Similarly the partial derivative of f with respect to y at the point (x, y) is the limit

$$f_y(x, y) = \lim_{h \to 0} \frac{f(x, y + h) - f(x, y)}{h}$$

provided this limit exists.

Just as we used several notations for derivatives of functions of one variable, there are several notations for partial derivatives. Let $z = f(x, y)$. Then each of the symbols represents the partial derivative of f with respect to x:

$$f_x, \qquad \frac{\partial f}{\partial x}, \qquad \frac{\partial z}{\partial x}, \qquad \frac{\partial f(x, y)}{\partial x}, \qquad \frac{\partial}{\partial x} f(x, y).$$

Similarly the derivative of f with respect to y can be written as any of

$$f_y, \qquad \frac{\partial f}{\partial y}, \qquad \frac{\partial z}{\partial y}, \qquad \frac{\partial f(x, y)}{\partial y}, \qquad \frac{\partial}{\partial y} f(x, y).$$

If we deal with a function of three variables, $g(x, y, z)$ then the partial derivatives with respect to x, y or z are defined just as in the previous case. We only write down the partial derivative of g with respect to z:

$$\frac{\partial g}{\partial z} = \lim_{h \to 0} \frac{g(x, y, z + h) - g(x, y, z)}{h}$$

provided the limit exists.

Let us compute some derivatives.

EXAMPLE 1 Compute the partial derivatives of $f(x, y) = x^3 - 5y^2$ and then find $f_x(2, 3)$ and $f_y(2, 3)$.

Solution: When computing f_x, treat y as a constant so that

$$f_x(x, y) = \frac{\partial}{\partial x}(x^3 - 5y^2) = \frac{\partial}{\partial x}(x^3) - \frac{\partial}{\partial x}(5y^2) = 3x^2 + 0 = 3x^2.$$

Now we compute f_y treating x as a constant:

$$f_y(x, y) = \frac{\partial}{\partial y}(x^3 - 5y^2) = \frac{\partial}{\partial y}(x^3) - \frac{\partial}{\partial y}(5y^2) = 0 - 10y = -10y.$$

Evaluating at a specific point $(x, y) = (2, 3)$, we have

$$f_x(2, 3) = 3 \cdot 2^2 = 12 \qquad \text{and} \qquad f_y(2, 3) = -10 \cdot 3 = -30. \qquad \blacksquare$$

EXAMPLE 2 Compute the partial derivatives $f_x(x, y)$ and $f_y(x, y)$ for the function $f(x, y) = (2x + 5y^2)e^{3x}$.

Solution: To compute f_x we treat y as a constant and differentiate with respect to x. The product rule is required so we write out the steps:

$$\begin{aligned} f_x(x, y) &= \frac{\partial}{\partial x}\left[(2x + 5y^2)e^{3x}\right] \\ &= \left[\frac{\partial}{\partial x}(2x + 5y^2)\right] e^{3x} + [2x + 5y^2]\frac{\partial}{\partial x}e^{3x} \\ &= [2]e^{3x} + [2x + 5y^2]3e^{3x} \\ &= (2 + 6x + 15y^2)e^{3x}. \end{aligned}$$

Now to compute the derivative with respect to y we treat x as constant and obtain

$$\begin{aligned} f_y(x, y) &= \frac{\partial}{\partial y}\left[(2x + 5y^2)e^{3x}\right] \\ &= \left[\frac{\partial}{\partial y}(2x + 5y^2)\right] e^{3x} + [2x + 5y^2]\frac{\partial}{\partial y}e^{3x} \\ &= [10y]e^{3x} + [2x + 5y^2] \cdot 0 \\ &= 10ye^{3x}. \end{aligned}$$

$\blacksquare$

In the single variable case, $f'(a)$ is interpreted ad the rate of change of f at $x = a$. Similarly $f_x(a, b)$ is the rate of change of f in the x direction at (a, b) and $f_y(a, b)$ is the rate of change of f in the y direction at (a, b).

Higher Derivatives

On the assumption that all relevant partial derivatives exist, the partial derivative with respect to x of a function has partial derivatives with respect both x and y. Similarly the partial derivative with respect to y has derivatives with respect to x and y. Thus a function of two variables has four possible second derivatives. They are

$$f_{xx}(x, y) = \frac{\partial}{\partial x}f_x(x, y) = \frac{\partial^2 f}{\partial x^2}, \qquad f_{yy}(x, y) = \frac{\partial}{\partial y}f_y(x, y) = \frac{\partial^2 f}{\partial y^2},$$

$$f_{xy}(x, y) = \frac{\partial}{\partial y}f_x(x, y) = \frac{\partial^2 f}{\partial y \partial x}, \quad f_{yx}(x, y) = \frac{\partial}{\partial x}f_y(x, y) = \frac{\partial^2 f}{\partial x \partial y}.$$

Take note of the order of the differentiation: f_{xy} means first differentiate f with respect to x and then with respect to y. In practice the order is usually not important because, as we prove later, $f_{xy} = f_{yx}$ whenever the second partial derivatives of f are continuous.

Partial derivatives of order higher than the second are illustrated with the following notation. Suppose $w = f(x, y, z)$. Then

$$\frac{\partial^3 w}{\partial x\,\partial y\,\partial y} = \frac{\partial^3}{\partial x\,\partial y\,\partial y} f(x, y, z) = f_{yyx}(x, y, z),$$

$$\frac{\partial^4 w}{\partial x\,\partial z\,\partial y\,\partial x} = \frac{\partial^4}{\partial x\,\partial z\,\partial y\,\partial x} f(x, y, z) = f_{xyzx}(x, y, z).$$

EXAMPLE 3 Let $f(x, y) = \ln(x^2 + y^2)$. Compute the second partial derivatives of f and then observe that

$$\frac{\partial^2 f}{\partial x^2} + \frac{\partial^2 f}{\partial y^2} = 0 \quad \text{and} \quad f_{xy}(x, y) = f_{yx}(x, y).$$

Solution: We compute the first derivatives:

$$f_x(x, y) = \frac{2x}{x^2 + y^2}, \qquad f_y(x, y) = \frac{2y}{x^2 + y^2}.$$

Next we compute the partial derivative with respect to x of each of these functions:

$$f_{xx}(x, y) = \frac{-2x^2 + 2y^2}{(x^2 + y^2)^2}, \qquad f_{yx} = \frac{-4xy}{(x^2 + y^2)^2}.$$

Finally we compute the partial derivatives of f_x and f_y with respect to y:

$$f_{xy} = \frac{-4xy}{(x^2 + y^2)^2}, \qquad f_{yy} = \frac{2x^2 - 2y^2}{(x^2 + y^2)^2}.$$

It is now evident that $f_{xx} + f_{yy} = 0$ and that $f_{xy} = f_{yx}$ at all points where the functions are defined; that is, for all $(x, y) \neq (0, 0)$. ■

Exercises 14.4

Find the first partial derivatives with respect to each variable for the following functions.

1. $f = 2x^3 - 4xy^2 + y^4$

2. $f = 7x^2 - x^3y^2 - 3x$

3. $f = \sin(2x + 3y)$

4. $f = xe^{x+y}$

5. $g = r\cos\theta$

6. $g = r\sin\theta\cos\phi$

7. $f = e^y(\cos x + \sin x)$

8. $f = \dfrac{x^2 - y^2}{x^2 + y^2}$

9. $f = x^3 + y^3 + z^3 + xyz$

10. $f = (x^n + y^n)e^{xyz}$

11. $g = e^{-r}\cos\theta + e^r\sin\theta$

12. $g = r\sin\theta + r^{-1}\cos\phi$

13. $f = \dfrac{x - z}{y - z}$

14. $f = \dfrac{xyz - 1}{x + y + z}$

Compute all the second partial derivatives of the following functions.

15. $f = 3x - 5y + 2x^2 + 11xy - 3y^2$

16. $f = ax + by + cx^2 + hxy + ky^2$

17. $f = \arctan(y/x)$

18. $f = \dfrac{\cos(x + y)}{\cos(x - y)}$

19. $f = ax + by + cz + hx^2 + ky^2 + mz^2 + nxy + pxz + qyz$

20. $f = \dfrac{xyz}{x + y + z}$

21. $f = \ln(x^2 + y^2 + z^2)$

The partial differential equation

$$\frac{\partial^2 u}{\partial x^2} + \frac{\partial^2 u}{\partial y^2} = 0$$

involving an unknown function $u = u(x, y)$ is called *Laplace's equation* in two variables. Determine which of the following functions are solutions of Laplace's equation.

22. $ae^x\sin y + be^x\cos y$

23. $\arctan(x/y)$

24. $3x^2 + 4xy + 3y^2$

25. $4x^3 - 3x^2y + 12xy^2 - y^3$

A string is stretched along the x-axis and held fixed at each end. The string is given a (small) displacement and released so that it vibrates in the xy plane. At time t, the shape of the string is described by a curve $u = u(x, t)$. If suitable conditions are satisfied, the function $u(x, t)$ satisfies the *one-dimensional wave equation*

$$\frac{\partial^2 u}{\partial t^2} = a^2\frac{\partial^2 u}{\partial x^2},$$

where a is a constant that depends on the string and

initial conditions. Show that each of the following functions satisfies the one-dimensional wave equation.

26. $u = 4\cos(x + at)$

27. $u = \dfrac{(x + at)^2 - 1}{(x + at)^2 + 1}$

28. $u = \sin nx \cos ant \quad (n \text{ constant})$

29. $u = g(x + at)$ [$g(x)$ is a twice differentiable function]

30. The area of a parallelogram is $A = A(a, b, \theta) = ab\sin\theta$ if two adjacent sides have lengths a and b and θ is the angle between them. Compute the rate of change of A with respect to a and with respect to θ. Compute the values $A(10, 15, \pi/4)$, $A(10.1, 15, \pi/4)$ and $A(10, 15, 1.01\pi/4)$ and determine which change, the increase of a or θ by 1%, produces the greatest increase in area. Could this have been predicted by computing $A_a(10, 15, \pi/4)$ and $A_\theta(10, 15, \pi/4)$?

14.5 Limits and continuity

The definition of partial derivatives required the use of limits but only the familiar limits of functions of one variable. Now we extend the notion of limit to functions of several variables. The intuitive notion of the limit of $f(x, y)$ as the point (x, y) approaches (a, b), written as

$$\lim_{(x,y)\to(a,b)} f(x, y) = L,$$

is that the values $f(x, y)$ are close to L for all (x, y) sufficiently close to (a, b). As usual, the terms "close" and "sufficiently close" must be made precise as in the following more formal definition.

DEFINITION 14.2

Limit of a Function

The function $f(x, y)$ has a limit L at the point (a, b), written as

$$\lim_{(x,y)\to(a,b)} f(x, y) = L,$$

if for every $\epsilon > 0$ there is some number $\delta > 0$ such that whenever

$$0 < \sqrt{(x - a)^2 + (y - b)^2} < \delta$$

then

$$|f(x, y) - L| < \epsilon.$$

This can be put into less technical words by saying that for every positive number ϵ, there is a circle of some positive radius δ such that the distance between L and the value of $f(x, y)$ at any point on the interior of the circle, other than the center, is at most ϵ. The definition depends on the notion of "closeness" for points (x, y) in the plane. Thus the point (x, y) is "close" to (a, b) if (x, y) lies inside a small circle of positive radius δ with center at (a, b).

The formal properties that were discussed or proved about limits of a function of one variable carry over to the case of a function of two variables. For example the limit of the sum, product or quotient of two functions is exactly what you would expect. We state the facts as a theorem but we do not give a proof.

THEOREM 14.1

Let $f(x, y)$ and $g(x, y)$ be functions such that the limit equations

$$\lim_{(x,y)\to(a,b)} f(x, y) = L, \qquad \lim_{(x,y)\to(a,b)} g(x, y) = M$$

both hold. Then the following limits exist and have the indicated values:

A. $\lim_{(x,y)\to(a,b)} f(x, y) + g(x, y) = L + M,$

B. $\lim_{(x,y)\to(a,b)} f(x, y)g(x, y) = LM,$

C. $\lim_{(x,y)\to(a,b)} f(x, y)/g(x, y) = L/M,$ provided $M \neq 0$.

Continuity

The concept of continuity extends from functions of one-variable to functions of several variables with little change in the wording.

DEFINITION 14.3

Continuity of a Function of Two Variables

The function $f(x, y)$ is continuous at the point (a, b) if

i. $f(a, b)$ is defined;

ii. $\lim_{(x,y)\to(a,b)} f(x, y)$ exists and is equal to $f(a, b)$.

Here is an example where we prove continuity holds at a point.

EXAMPLE 1 Let $f(x, y)$ be the function defined by

$$f(x, y) = \begin{cases} \dfrac{x(x^2 - y^2)}{x^2 + y^2} & \text{for } (x, y) \neq (0, 0) \\ 0 & \text{for } (x, y) = (0, 0). \end{cases}$$

Prove that f is continuous at $(0, 0)$.

Solution: The idea is to show that f is small at all points near the origin. We first observe that $|x^2 - y^2| \leq x^2 + y^2$ as an easy application of the Triangle Inequality. We now have the estimate

$$|f(x, y)| = \left| \frac{x(x^2 - y^2)}{x^2 + y^2} \right| \leq |x| \frac{|x^2 - y^2|}{x^2 + y^2} \leq |x|.$$

If the distance of (x, y) from the origin is less than some positive number δ, then, in particular, $|x| < \delta$ and it follows that $|f(x, y)| \leq |x| < \delta$. To show continuity of f at the origin, it is necessary to find a δ corresponding to any given ϵ such that the conditions in the definition hold. If we are given a positive number ϵ then we select $\epsilon = \delta$. Then we get $|f(x, y) - f(0, 0)| < \epsilon$ whenever $\sqrt{x^2 + y^2} < \delta$. Thus

$$\lim_{(x,y)\to(0,0)} f(x, y) = 0 = f(0, 0),$$

which proves $f(x, y)$ is continuous at the origin. ∎

The continuity of f at (a, b) is often used in the following way. We have some functions $u = u(x, y)$ and $v = v(x, y)$ for which it is known that

$$\lim_{(x,y)\to(p,q)} u(x, y) = a \qquad \text{and} \qquad \lim_{(x,y)\to(p,q)} v(x, y) = b.$$

The continuity of f then permits the evaluation

$$\lim_{(x,y)\to(p,q)} f(u, v) = f\Big(\lim_{(x,y)\to(p,q)} u, \lim_{(x,y)\to(p,q)} v\Big) = f(a, b).$$

The reasoning is essentially the same as in the one-variable case.

Continuity of $f(x, y)$ at (a, b) imposes somewhat more severe restrictions on the behavior of f than in the single-variable case. The restrictions come from the greater flexibility with which a point (x, y) can approach (a, b). For example let us permit (x, y) approach $(0, 0)$ along any of the straight lines $y = mx$ for m fixed. If f is continuous at $(0, 0)$, then for any m, the limits

$$\lim_{(x,y)\to(0,0)} f(x, y) = \lim_{x\to 0} f(x, mx)$$

all have the same value, namely $f(0, 0)$. The approach to $(0, 0)$ along straight lines is only one of many possible paths of approach all of which lead to the same limit for a continuous function. We illustrate with a specific example.

EXAMPLE 2 Show that the function

$$f(x, y) = \frac{xy}{x^2 + y^2} \qquad (x, y) \neq (0, 0)$$

does not have a limit at $(0, 0)$.

Solution: At points where $y = mx$, we have

$$f(x, y) = f(x, mx) = \frac{x \cdot mx}{x^2 + m^2x^2} = \frac{m}{1 + m^2}.$$

This shows that f has a constant value on the straight lines through the origin and that value is not 0 except when $m = 0$. Thus as (x, y) approaches $(0, 0)$ along the line $y = x$, f has a limit $1/2$ $(m = 1)$, whereas if (x, y) approaches $(0, 0)$ along the line $y = -x$, f has a limit $-1/2$ $(m = -1)$. Any circle of positive radius (δ) with center at $(0, 0)$ has points at which f has the value $1/2$ and points at which f has the value $-1/2$. Thus we cannot have all the values of f be within $0.25 = \epsilon$ of any particular L. Hence the limit does not exist. ■

Functions of Three Variables

The notions of limit and continuity for a function of three variables is easily obtained by imitating what has been given for the case of two variables. In the definition of limit, the restriction $0 < \sqrt{(x - a)^2 + (y - b)^2} < \delta$ is replaced by

$$0 < \sqrt{(x - a)^2 + (y - b)^2 + (z - c)^2} < \delta,$$

which is the condition that (x, y, z) lies inside the sphere of radius δ and center (a, b, c) but $(x, y, z) \neq (a, b, c)$.

Continuity of f at (a, b, c) is succinctly described by the condition

$$\lim_{(x,y,z)\to(a,b,c)} f(x, y, z) = f(a, b, c).$$

The reader should have no difficulty in formulating a statement defining continuity of a function of any number of variables based on the statement for functions of two or three variables.

Linear Approximation Theorem

The Mean Value Theorem for $f(x)$ states that when $f'(x)$ exists on (a, b) and f is continuous on $[a, b]$ then for every x in $[a, b]$ we have $f(x) - f(a) = f'(c)(x - a)$ for some c on $[a, b]$. This statement is not in a form that suggests the correct analogue for functions of several variables. We give a restatement of this theorem in a form that can be generalized to functions of several variables. Consider the function

$$g(h) = \begin{cases} \dfrac{f(x+h) - f(x)}{h} - f'(x) & \text{if } h \neq 0 \\ 0 & \text{if } h = 0. \end{cases}$$

Then $g(h)$ is continuous at $h = 0$ (because $f'(x)$ exists) and we have a formula

$$f(x + h) - f(x) = f'(x)h + g(h)h.$$

For $h \neq 0$ the formula is essentially the definition of g whereas for $h = 0$ both sides are 0. One can interpret this equation as saying that not only is the difference $(f(x + h) - f(x)) - f'(x)h = g(h)h$ small (close to 0) but in fact this quantity divided by h is also small ($g(h)$ approaches 0 as h approaches 0). The function of h defined by $f(x) + f'(x)h$, for fixed x is a linear function of h that approximates $f(x+h)$. The statement that the difference between these is $g(h)h$ where $g(h) \to 0$ as $h \to 0$ is called the Linear Approximation Theorem (LAT) for f.

The extension of the LAT to several variables that we are about to give is analogous to this last formulation of the LAT for a single variable function. It plays a crucial role in the development of basic differentiation rules. For example, we will apply the next theorem to determine the proper form of the chain rule for functions of several variables.

THEOREM 14.2

Linear Approximation Theorem in Two Variables

Let $f(x, y)$ be defined in some rectangular region R containing (a, b). If the first partial derivatives f_x and f_y of $f(x, y)$ are continuous at (a, b) then there exist functions $g_1(h, k)$ and $g_2(h, k)$ such that

$$f(a + h, b + k) - f(a, b) = f_x(a, b)h + f_y(a, b)k + g_1(h, k)h + g_2(h, k)k \tag{14.3}$$

for all (h, k) sufficiently near $(0, 0)$. The functions g_1 and g_2 have the additional property

$$\lim_{(h,k)\to(0,0)} g_1(h, k) = \lim_{(h,k)\to(0,0)} g_2(h, k) = 0.$$

Proof

The point (a, b) is fixed and (x, y) is a variable point. We have written conditions using the variables h and k defined by

$$h = x - a \qquad k = y - b$$

to emphasize the view that we are examining the behavior of f near (a, b) and the variables h and k measure how far (left-right or up-down) (x, y) is from (a, b). We rewrite the left side of Equation (14.3) as a sum of two terms, one measuring the change with x varying and the other with y varying:

$$f(a+h,b+k)-f(a,b)=[f(a+h,b+k)-f(a,b+k)]+[f(a,b+k)-f(a,b)].$$

Each term enclosed in square brackets can be regarded as the change in a function as one variable changes. This is a context in which the MVT for functions of one variable can be applied.

If the MVT is applied to $f(x, b+k)$ we obtain a number c_1 (depending on x and k) such that c_1 is between $x = a+h$ and a and

$$\frac{f(a+h,b+k)-f(a,b+k)}{h}=f_x(c_1,b+k).$$

Similarly the second term can be expressed as

$$\frac{f(a,b+k)-f(a,b)}{k}=f_y(a,c_2)$$

for some c_2 between $y = b+k$ and b. Take notice that as $h \to 0$ we have $c_1 \to a$ and that as $k \to 0$ we have $c_2 \to b$. Now substitute the approximations for the terms in square brackets to obtain

$$\begin{aligned} f(a+h,b+k)-f(a,b+k)&=f_x(c_1,b+k)h\\ f(a,b+k)-f(a,b)&=f_y(a,c_2)k. \end{aligned}$$

Now to get the form stated in the theorem, we set

$$\begin{aligned} g_1(h,k)&=f_x(c_1,b+k)-f_x(a,b),\\ g_2(h,k)&=f_y(a,c_2)-f_y(a,b). \end{aligned}$$

By substitution, one shows the equation stated in the theorem is valid. It remains to show that g_1 and g_2 have the limit properties. We have

$$\lim_{(h,k)\to(0,0)}(c_1,b+k)=(a,b)$$

and so, using the continuity of f_x at (a, b), we find

$$\lim_{(h,k)\to(0,0)} g_1(h,k)=\lim_{(h,k)\to(0,0)}(f_x(c_1,b+k)-f_x(a,b))=f_x(a,b)-f_x(a,b)=0.$$

In the same way one obtains $\lim_{(h,k)\to(0,0)} g_2(h,k) = 0$ and the proof is complete. □

EXAMPLE 3 Let $f(x, y) = x^2 - 2xy$. Find functions g_1 and g_2 whose existence is asserted by the LAT for f for an arbitrary point (a, b).

Solution: We have

$$f_x(x,y)=2x-2y \qquad \text{and} \qquad f_y(x,y)=-2x.$$

We follow the outline of the proof of the LAT for f by writing

$$\begin{aligned} f(a+h,&b+k)-f(a,b)\\ &=[f(a+h,b+k)-f(a,b+k)]+[f(a,b+k)-f(a,b)]\\ &=[(a+h)^2-2(a+h)(b+k)-a^2+2a(b+k)]\\ &\qquad+[a^2-2a(b+k)-a^2+2ab]\\ &=[2ah+h^2-2h(b+k)]+[-2ak]\\ &=[(2a-2b)+h-2k]h+[-2a]k\\ &=f_x(a,b)h+f_y(a,b)k+(h-2k)h+0\cdot k. \end{aligned}$$

If we set $g_1(h, k) = (h - 2k)$ and $g_2(h, k) = 0$, then all the requirements stated in the LAT are satisfied.

Note that there are other choices for g_1 and g_2. We could also take $g_1(h,k) = h$ and $g_2(h,k) = -2h$ so that we still have $g_1(h,k)h + g_2(h,k)k = (h-2k)h$ with both g_1 and g_2 approaching 0 as h and k approach 0. ■

Differentiable functions

A function for which the conclusion of the Linear Approximation Theorem holds is called a *differentiable* function. Thus $f(x,y)$ is differentiable at (a,b) if $f_x(a,b)$ and $f_y(a,b)$ exist and

$$f(a+h,b+k) - f(a,b) = f_x(a,b)h + f_y(a,b)k + g_1(h,k)h + g_2(h,k)k$$

for all (h,k) sufficiently near $(0,0)$ and for some functions g_1 and g_2 with the property

$$\lim_{(h,k)\to(0,0)} g_1(h,k) = \lim_{(h,k)\to(0,0)} g_2(h,k) = 0.$$

Thus, by the LAT, a function with continuous first partial derivatives at (a,b) is differentiable at (a,b). Take note that the terminology for functions of two variables differs from the corresponding terminology for functions of one variable where we said $f(x)$ is differentiable if the derivative $f'(x)$ exists. For functions of several variables, the mere existence of the partial derivatives is not sufficient for the conclusion of the LAT to hold.

Another difference between the one-variable and two-variables cases is the relation between the existence of derivatives and the continuity of the function. For a function g of one variable, the existence of g' at a point implies the continuity of g at that point. We will see in the exercises that the partial derivatives f_x and f_y of a function of two variables can exist at a point even though f is not continuous at that point. However the stronger condition of differentiability does imply continuity as we now prove.

THEOREM 14.3

Differentiability Implies Continuity

If $f(x,y)$ is differentiable at (a,b) then f is continuous at (a,b).

Proof

To show the continuity of f at (a,b) it is necessary to show that $\lim_{(x,y)\to(a,b)} f(x,y) = f(a,b)$. Let $h = x - a$ and $k = y - b$. By the differentiability assumption there exist functions g_1 and g_2 such that

$$f(a+h,b+k) = f(a,b) + f_x(a,b)h + f_y(a,b)k + g_1(h,k)h + g_2(h,k)k$$

and $\lim_{(h,k)\to(0,0)} g_1(h,k) = \lim_{(h,k)\to(0,0)} g_2(h,k) = 0$. By taking the limit as $(h,k) \to (0,0)$ in this equation we see

$$\lim_{(x,y)\to(a,b)} f(x,y) = \lim_{(h,k)\to(0,0)} f(a+h,b+k) = f(a,b)$$

because each of the terms $f_x(a,b)h$, $f_y(a,b)k$, $g_1(h,k)h$, $g_2(h,k)k$ has limit 0. This proves the continuity of f at (a,b). □

Mixed Partial Derivatives

We have mentioned earlier that the mixed partial derivatives f_{xy} and f_{yx} are equal at every point where the first partial derivatives and the mixed partial derivatives are continuous. We prove this fact with several applications of the MVT for functions of one variable.

THEOREM 14.4

Equality of Mixed Partial Derivatives

If the first partial derivatives f_x and f_y and the second partial derivatives f_{xy} and f_{yx} of a function f are continuous in a rectangular region R, then $f_{xy}(x, y) = f_{yx}(x, y)$ at all points of R.

Proof

Let (a, b) be a point in R and let h and k be (small) numbers such that the three points $(a + h, b + k)$, $(a + h, b)$ and $(a, b + k)$ are also in R. Then

$$\begin{aligned}
f_{xy}(a, b) &= \lim_{k\to 0} \frac{1}{k}\left(f_x(a, b + k) - f_x(a, b)\right)\\
&= \lim_{k\to 0} \frac{1}{k}\left(\lim_{h\to 0} \frac{f(a + h, b + k) - f(a, b + k)}{h}\right.\\
&\qquad \left. - \lim_{h\to 0} \frac{f(a + h, b) - f(a, b)}{h}\right)\\
&= \lim_{k\to 0}\lim_{h\to 0}\left(\frac{f(a + h, b + k) - f(a, b + k) - f(a + h, b) + f(a, b)}{hk}\right).
\end{aligned}$$

The analogous expression for $f_{yx}(a, b)$ has the same term to the right of the limits but the order of the limits is reversed; that is $\lim_{h\to 0}\lim_{k\to 0}$ replaces $\lim_{k\to 0}\lim_{h\to 0}$. It is not obvious that the same result is obtained by taking the limits in different orders. In fact, there are cases in which the order is important and different results are obtained by reversing the order. Hence we must make use of the continuity to verify the equality in this case.

Let D and F be the functions

$$\begin{aligned}
D(h, k) &= f(a + h, b + k) - f(a, b + k) - f(a + h, b) + f(a, b),\\
F(u) &= f(a + h, u) - f(a, u)
\end{aligned}$$

so that $D(h, k) = F(b + k) - F(b)$. The continuity of f_y implies the continuity of $F'(u)$ so the Linear Approximation Theorem can be applied to F to obtain

$$D(h, k) = F'(u_1)k = (f_y(a + h, u_1) - f_y(a, u_1))k$$

for some u_1 between b and $b + k$. For later use we note that as k approaches 0, u_1 approaches b. Now the continuity of f_{yx} allows the application of the Mean Value Theorem to this last difference to obtain

$$f_y(a + h, u_1) - f_y(a, u_1) = f_{yx}(v_1, u_1)h$$

for some v_1 between a and $a + h$. Note that v_1 approaches a as h approaches 0. Combine the last two equations to get

$$D(h, k) = f_{yx}(v_1, u_1)hk.$$

Now we repeat this line of reasoning by starting with a function that holds the second variable fixed in place of the first variable as done by F. Let

$$G(v) = f(v, b + k) - f(v, b)$$

so that $D(h, k) = G(a + h) - G(a)$. Apply the MVT to G to get

$$D(h, k) = G'(v_2)h = (f_x(v_2, b + k) - f_x(v_2, b))h$$

for some v_2 between a and $a + h$. One final application of the MVT yields

$$f_x(v_2, b + k) - f_x(v_2, b) = f_{xy}(v_2, u_2)k$$

for some u_2 between b and $b + k$. Thus

$$D(h, k) = f_{xy}(v_2, u_2)hk.$$

Equate the two expressions for $D(h, k)$ to conclude

$$f_{yx}(v_1, u_1)hk = f_{xy}(v_2, u_2)hk.$$

Now cancel the hk and take the limit as h and k approach 0. The continuity of the mixed partial derivatives implies that the limit exists and equals each of the two expressions $f_{yx}(a, b) = f_{xy}(a, b)$. This proves the theorem. □

In the exercises following this section we give an example to show that the continuity assumptions in the theorem cannot be disregarded.

For functions satisfying continuity conditions on all relevant partial derivatives, the order of differentiation can be interchanged. For example

$$f_{xyx} = f_{yxx} = f_{xxy}$$

at all points where these three derivatives and the derivatives f_x, f_y, f_{xy}, f_{yx} of f are continuous. A similar property holds for functions of more than two variables. If $w = f(x, y, z)$ is a function with continuous partial derivatives of the first and second orders, then

$$\frac{\partial^2 w}{\partial x \partial y} = \frac{\partial^2 w}{\partial y \partial x}, \qquad \frac{\partial^2 w}{\partial z \partial y} = \frac{\partial^2 w}{\partial y \partial z}, \qquad \frac{\partial^2 w}{\partial x \partial z} = \frac{\partial^2 w}{\partial z \partial x}.$$

Exercises 14.5

1. Let $g(x, y) = (x^2 - y^2)/(x^2 + y^2)$ for $(x, y) \neq (0, 0)$. Show that there is no way to define $g(0, 0)$ so that g is continuous at $(0, 0)$.

2. Let $F(x, y) = x^2y^2/(x^2+y^2)$ for $(x, y) \neq (0, 0)$. Show that there is a constant c such that the definition $F(0, 0) = c$ makes F continuous at $(0, 0)$.

Show that the following limits do not exist.

3. $\displaystyle\lim_{(x,y)\to(0,0)} \frac{xy}{x^2 + y^2}$

4. $\displaystyle\lim_{(x,y)\to(0,0)} \frac{x}{\sqrt{x^2 + y^2}}$

5. Show that the limit

$$\lim_{(x,y)\to(0,0)} \frac{x^n y^m}{(x^2 + y^2)^k}$$

exists if $0 \leq k < n, m$.

Evaluate the partial derivatives and verify the equality as required.

6. For $f(x, y) = e^{-xy}/(x^2 + y^2)$ verify by direct computation that $f_{xy} = f_{yx}$ at every point where f is defined.

7. For $f(x, y) = \ln(x^2 + y^2) - 4x^2y^2$ verify by direct computation that $f_{xy} = f_{yx}$ at every point other than $(0, 0)$.

8. For $g(x, y, z) = xy^2z^2 + x^2y^2z + x^2yz^2$ verify by direct computation that $g_{xyz} = g_{zyx} = g_{xzy}$ at every point (x, y, z).

9. For $g(x, y, z) = x \sin yz + y \sin xz$ verify by direct computation that $g_{xyz} = g_{xzy} = g_{yxz}$ at every point (x, y, z).

10. Let f be the function defined by $f(x, y) = xy/(x^2 + y^2)$ if $(x, y) \neq (0, 0)$ and $f(0, 0) = 0$. We have seen that f is not continuous at $(0, 0)$. Show that the first derivatives f_x and f_y exist at every point including $(0, 0)$. [In contrast to the single variable theorem, the existence of the partial derivatives for a function of two variables does not imply continuity of the function.]

11. Let $f(x, y)$ be defined by

$$f(x,y) = \begin{cases} \dfrac{xy(x^2 - y^2)}{x^2 + y^2} & \text{for } (x,y) \neq (0,0) \\ 0 & \text{for } (x,y) = (0,0). \end{cases}$$

Show that

$$f_x(0,y) = \lim_{h \to 0} \frac{f(h,y) - f(0,y)}{h} = -y$$

and

$$f_y(x,0) = \lim_{k \to 0} \frac{f(x,k) - f(x,0)}{k} = x.$$

Conclude from these computations that $f_{xy}(0,0) = -1$ and $f_{yx}(0,0) = +1$. Notice that $f_{xy}(0,0) \neq f_{yx}(0,0)$. Explain what part of the hypothesis of Theorem 14.4 fails for this function.

For a function $f(x, y)$ that is differentiable at a point (a, b) the function df is defined by $df(h, k) = f_x(a, b)h + f_y(a, b)k$ and is called the *total differential* of f at (a, b). The total differential is a function of the numbers h and k that measures the distance from the point (a, b) where the partial derivatives are evaluated. By the LAT we have

$$f(a+h, b+k) - f(a,b) = df(h,k) + g_1(h,k)h + g_2(h,k)k$$

where g_1 and g_2 have a limit 0 as $(h, k) \to (0, 0)$. For $|h|$ and $|k|$ small, the expressions $g_1(h, k)h$ and $g_2(h, k)$ are usually very small (product of two small numbers) so that $f(a, b) + df(h, k)$ often gives a good approximation of the values of $f(a + h, b + k)$ at points near to (a, b). Use this idea to find approximate values of the functions below at the indicated points. If you have a calculator or computer available, compute the values $f(a + h, b + k)$ on your device and compare them with the approximations given by $f(a, b) + df(h, k)$.

12. For $f(x, y) = x^3y^2$ approximate $f(1.1, 2.05)$ with $f(1, 2) + df(0.1, 0.05)$.

13. For $f(x, y) = \sin x \cos y$ approximate $f(\pi + 0.05, 0.05)$ with $f(\pi, 0) + df(0.05, 0.05)$.

14. For $f(x, y) = e^x \ln(x^2 + y^2)$ approximate $f(0.05, 0.95)$ with $f(0, 1) + df(0.05, -0.05)$.

15. For $f(x, y) = e^{(x^2 - y^2)} \sin(xy\pi/2)$ approximate the value $f(1.05, 0.95)$ with $f(1, 1) + df(0.05, -0.05)$.

The expression $f(a + h, b + k) - f(a, b) \approx df(h, k)$ is not useful in case the differential of f at (a, b), df, is identically 0. Show that this is the case in each example below and then give an estimate by direct computation of the maximum and minimum difference $f(a + h, b + k) - f(a, b)$ for $|h| \leq 0.1$ and $|k| \leq 0.1$.

16. $f(x, y) = 1 - x^2 - y^2, \quad (a, b) = (0, 0)$

17. $f(x, y) = 4 - 2x^2 + 3y^2, \quad (a, b) = (0, 0)$

18. $f(x, y) = 12 - 3x^3 - 4y, \quad (a, b) = (0, 0)$

19. $f(x, y) = 1 - 2x^4 - y^3, \quad (a, b) = (0, 0)$

20. $f(x, y) = x^4(y^2 + 1) + y^4(x^2 + 1) - x^2 - y^2, \quad (a, b) = (0, 0)$

21. If $f(x, y)$ is continuous at (a, b), prove that it is *separately continuous* at (a, b). This means

$$\lim_{x \to a} f(x, b) = f(a, b)$$

and

$$\lim_{y \to b} f(a, y) = f(a, b).$$

22. Show that the function $f(x, y) = xy/(x^2 + y^2)$ is separately continuous at $(0, 0)$ but it is not continuous there. [See exercise 21 and example 2 of this section.]

14.6 The chain rule

The next step in the study of differentiation is the chain rule. In the case of a function of a single variable $f(x)$, the chain rule is used to evaluate the derivative of f with respect to a variable t when $x = x(t)$. It reads

$$\frac{d}{dt} f(x) = f'(x) \frac{dx}{dt}.$$

We rewrite this using the notation of partial derivatives as

$$\frac{d}{dt} f(x) = f_x(x) \frac{\partial x}{\partial t} \qquad \mathbf{(14.4)}$$

to be suggestive of the chain rule we are about to develop for functions of several variables.

Now suppose that f is a function of two or more variables and each variable is a function of t and perhaps other variables as well. To be specific let us consider the case of a function f of two variables x and y each of which is a function of t, $x = x(t)$ and $y = y(t)$. The derivative of f with respect to t must take into account the fact that as t changes, both variables x and y change and thus produce a change in f.

The formula we derive below in this case is

$$\frac{\partial}{\partial t} f(x, y) = f_x(x, y)\frac{\partial x}{\partial t} + f_y(x, y)\frac{\partial y}{\partial t}.$$

Let us apply it to an example before we give the proof.

EXAMPLE 1 Let $f(x, y) = x^3 y$ and let $x = x(t) = \sin t$, $y = y(t) = 2t^3$. Then f is a function of t. Find the derivative of f with respect to t.

Solution: We compute some derivatives first:

$$f_x(x, y) = 3x^2 y, \qquad \frac{dx}{dt} = \cos t,$$

$$f_y(x, y) = x^3, \qquad \frac{dy}{dt} = 6t^2.$$

The formula just above gives

$$\begin{aligned}\frac{\partial}{\partial t} f(x, y) &= 3x^2 y \cos t + x^3(6t^2)\\ &= 6t^3 \sin^2 t \cos t + 6t^2 \sin^3 t.\end{aligned}$$

In this case, we can compute the derivative by first expressing f in terms of t as

$$f(x(t), y(t)) = (\sin^3 t)(2t^3).$$

The derivative with respect to t can be computed directly and it is found to agree with the value obtained by the chain rule. ■

A slightly more general situation is one in which f is given as a function of x and y and each of x and y is a function of several variables, say u and v. Then f is a function of u and v and we ask for the derivatives of f with respect to u and v. Here is the statement of the general rule and its proof.

THEOREM 14.5

The Chain Rule

If $f(x, y)$ is a differentiable function of x and y and if both x and y are differentiable functions of u and v, then the partial derivatives of f with respect to u and v exist and are given by the formulas

$$\frac{\partial f}{\partial u} = \frac{\partial f}{\partial x}\frac{\partial x}{\partial u} + \frac{\partial f}{\partial y}\frac{\partial y}{\partial u},$$

$$\frac{\partial f}{\partial v} = \frac{\partial f}{\partial x}\frac{\partial x}{\partial v} + \frac{\partial f}{\partial y}\frac{\partial y}{\partial v}.$$

Proof

By definition of the partial derivative of f with respect to u we have

$$\frac{\partial}{\partial u} f(x(u,v), y(u,v)) = \lim_{h \to 0} \frac{f(x(u+h,v), y(u+h,v)) - f(x(u,v), y(u,v))}{h}. \tag{14.5}$$

We use the differentiability assumptions and the LAT several times to write the right-hand side in a form where the limit can be evaluated. We hold u and v constant and write $x = x(u,v)$ and $y = y(u,v)$. There exist functions g_1, g_2, g_3, g_4 such that

$$\begin{aligned}
x(u+h,v) &= x(u,v) + x_u(u,v)h + g_1(h)h, \\
y(u+h,v) &= y(u,v) + y_u(u,v)h + g_2(h)h, \\
f(x+H, y+K) &= f(x,y) + f_x(x,y)H + f_y(x,y)K \\
&\quad + g_3(H,K)H + g_4(H,K)K.
\end{aligned}$$

Each of the functions g_i has limit 0 as the variables on which the functions depend approach 0.

When u changes to $u + h$, we denote the change in x by H so that

$$H = x(u+h,v) - x(u,v) = x_u(u,v)h + g_1(h)h;$$

similarly we denote the change in y by K so that

$$K = y(u+h,v) - y(u,v) = y_u(u,v)h + g_2(h)h.$$

Now we substitute into the right side of Equation (14.5) to get

$$\begin{aligned}
\frac{\partial f}{\partial u} &= \lim_{h \to 0} \frac{f(x+H, y+K) - f(x,y)}{h} \\
&= \lim_{h \to 0} \frac{f_x(x,y)H + f_y(x,y)K + g_3(H,K)H + g_4(H,K)K}{h} \\
&= \lim_{h \to 0} \left(f_x(x,y)\frac{H}{h} + f_y(x,y)\frac{K}{h} + g_3(H,K)\frac{H}{h} + g_4(H,K)\frac{K}{h} \right).
\end{aligned} \tag{14.6}$$

The remaining limits involving h can be evaluated. We have

$$\begin{aligned}
\lim_{h \to 0} \frac{H}{h} &= \lim_{h \to 0} \frac{x_u(u,v)h + g_1(h)h}{h} \\
&= \lim_{h \to 0} x_u(u,v) + g_1(h) = x_u(u,v).
\end{aligned}$$

Similarly $\lim_{h \to 0}(K/h) = y_u(u,v)$. Furthermore for $j = 3$ or 4 and using the fact that $H \to 0$ and $K \to 0$ as $h \to 0$ we have

$$\lim_{(h,k) \to (0,0)} g_j(H,K) = \lim_{(H,K) \to (0,0)} g_j(H,K) = 0.$$

Now substitute the value of these limits into expression (14.6) for $\partial f/\partial u$ to get

$$\frac{\partial}{\partial u} f(x,y) = f_x(x,y)x_u(u,v) + f_y(x,y)y_u(u,v).$$

This is the chain rule as stated in the theorem. The formula for $f_v(x,y)$ is derived in the analogous way. □

The chain rule for functions of three variables takes the analogous form. If f is a function of x, y and z and each of these variables is a function of u and perhaps other variables as well, then the derivative of f with respect to u is given by the rule

$$\frac{\partial f}{\partial u} = \frac{\partial f}{\partial x}\frac{\partial x}{\partial u} + \frac{\partial f}{\partial y}\frac{\partial y}{\partial u} + \frac{\partial f}{\partial z}\frac{\partial z}{\partial u}. \tag{14.7}$$

EXAMPLE 2 Let $f(x, y) = x \sin y + y \cos x$ and suppose that $x = \sqrt{u^2 + v^2}$ and $y = 3v + uv$. Compute the partial derivative of f with respect to v at the point where $(u, v) = (0, 1)$.

Solution: We begin by computing the partial derivatives of f with respect to x and y and the partial derivatives of x and y with respect to v:

$$f_x(x, y) = \sin y - y \sin x, \qquad f_y(x, y) = x \cos y + \cos x,$$
$$x_v(u, v) = \frac{v}{\sqrt{u^2 + v^2}}, \qquad y_v(u, v) = 3 + u.$$

The general formula for $\partial f/\partial v$ is

$$\begin{aligned}\frac{\partial}{\partial v} f(x, y) &= f_x(x, y)x_v(u, v) + f_y(x, y)y_v(u, v)\\ &= (\sin y - y \sin x)\left(\frac{v}{\sqrt{u^2 + v^2}}\right) + (x \cos y + \cos x)u.\end{aligned}$$

Now we evaluate everything at $(u, v) = (0, 1)$ to get

$$x(0, 1) = 1, \qquad y(0, 1) = 3,$$
$$x_v(0, 1) = 1, \qquad y_v(0, 1) = 3,$$
$$\begin{aligned}\frac{\partial f}{\partial v}(x(0, 1), y(0, 1)) &= f_x(1, 3) \cdot 1 + f_y(1, 3) \cdot 3\\ &= (\sin 3 - 3 \sin 1) + (\cos 3 + \cos 1) \cdot 3.\end{aligned}$$

■

EXAMPLE 3 Let $f(x, y, z) = x^2 + y^2 + z^2$ and let

$$x = r \sin\theta \cos\phi, \quad y = r \sin\theta \sin\phi, \quad z = r \cos\theta.$$

Compute the three partial derivatives $\partial f/\partial r$, $\partial f/\partial\theta$, $\partial f/\partial\phi$ and then evaluate each at $(r, \theta, \phi) = (2, \pi/2, \pi)$.

Solution: We apply the chain rule for a function of three variables:

$$\begin{aligned}\frac{\partial f}{\partial r} &= \frac{\partial f}{\partial x}\frac{\partial x}{\partial r} + \frac{\partial f}{\partial y}\frac{\partial y}{\partial r} + \frac{\partial f}{\partial z}\frac{\partial z}{\partial r}\\ &= 2x \sin\theta \cos\phi + 2y \sin\theta \sin\phi + 2z \cos\theta,\\ \frac{\partial f}{\partial \theta} &= \frac{\partial f}{\partial x}\frac{\partial x}{\partial \theta} + \frac{\partial f}{\partial y}\frac{\partial y}{\partial \theta} + \frac{\partial f}{\partial z}\frac{\partial z}{\partial \theta}\\ &= 2xr \cos\theta \cos\phi + 2yr \cos\theta \sin\phi - 2zr \sin\theta,\\ \frac{\partial f}{\partial \phi} &= \frac{\partial f}{\partial x}\frac{\partial x}{\partial \phi} + \frac{\partial f}{\partial y}\frac{\partial y}{\partial \phi} + \frac{\partial f}{\partial z}\frac{\partial z}{\partial \phi}\\ &= -2xr \sin\theta \sin\phi + 2y \sin\theta \cos\phi + 2zr \cdot 0.\end{aligned}$$

Now to evaluate these partial derivatives at the indicated values of r, θ and ϕ we evaluate x, y and z first:

$$x = x(2, \pi/2, \pi) = -2, \qquad y = y(2, \pi/2, \pi) = 0, \qquad z = z(2, \pi/2, \pi) = 0.$$

The value of the partial derivatives at this point is

$$\left.\frac{\partial f}{\partial r}\right|_{(r,\theta,\phi)=(2,\pi/2,\pi)} = 2(-2)(1)(-1) + 0 + 0 = 4,$$

$$\left.\frac{\partial f}{\partial \theta}\right|_{(r,\theta,\phi)=(2,\pi/2,\pi)} = 2(-2)(0)(-1) + 0 + 0 = 0,$$

$$\left.\frac{\partial f}{\partial \phi}\right|_{(r,\theta,\phi)=(2,\pi/2,\pi)} = -2(-2)(1)(0) + 0 + 0 = 0.$$

The chain rule can be avoided by expressing f in terms of r, θ and ϕ to get

$$f = (r\sin\theta\cos\phi)^2 + (r\sin\theta\sin\phi)^2 + (r\cos\theta)^2 = r^2.$$

Thus we find directly that $f_r = 2r$ and $f_\theta = f_\phi = 0$. This would have been discovered by expressing the partial derivatives of f_θ and f_ϕ in terms of r, θ and ϕ only. ■

EXAMPLE 4 A parallelepiped (a cube, except that not all edges need have the same length) has sides of length x, y and z and volume $V = xyz$. The face with sides x and y has diagonal of length $u = \sqrt{x^2 + y^2}$; the face with sides y and z has diagonal of length $v = \sqrt{y^2 + z^2}$; the diagonal passing through the interior of the solid has length $w = \sqrt{x^2 + y^2 + z^2}$. One can express x, y and z in terms of u, v and w so that V is a function of u, v and w. Compute the partial derivatives of V with respect to u.

Solution: To compute V_u by the chain rule, we compute $V_x = yz$, $V_y = xz$, $V_z = xy$. The three partial derivatives x_u, y_u and z_u are also needed. To compute these we first express x, y and z in terms of u, v and w. With a bit of manipulation we find:

$$x = \sqrt{w^2 - v^2}, \quad y = \sqrt{u^2 + v^2 - w^2}, \quad z = \sqrt{w^2 - u^2}.$$

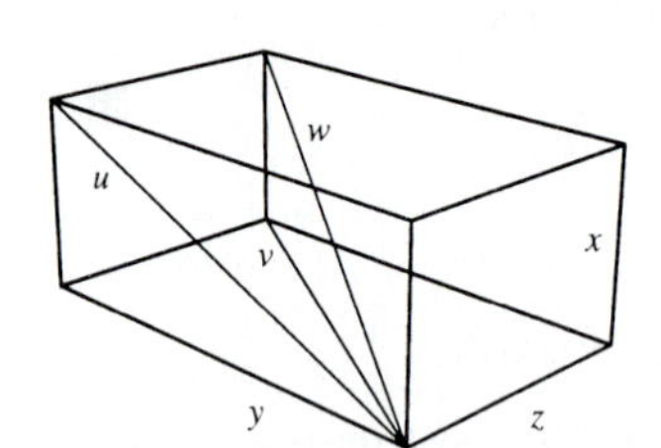

FIGURE 14.13

Now the partial derivatives of V can be written down explicitly:

$$\begin{aligned}\frac{\partial V}{\partial u} &= \frac{\partial V}{\partial x}\frac{\partial x}{\partial u} + \frac{\partial V}{\partial y}\frac{\partial y}{\partial u} + \frac{\partial V}{\partial z}\frac{\partial z}{\partial u}\\ &= yz \cdot 0 + xz\frac{u}{\sqrt{u^2 + v^2 - w^2}} - xy\frac{u}{\sqrt{w^2 - u^2}},\\ \frac{\partial V}{\partial v} &= \frac{\partial V}{\partial x}\frac{\partial x}{\partial v} + \frac{\partial V}{\partial y}\frac{\partial y}{\partial v} + \frac{\partial V}{\partial z}\frac{\partial z}{\partial v}\\ &= -yz\frac{v}{\sqrt{w^2 - v^2}} + xz\frac{v}{\sqrt{u^2 + v^2 - w^2}} + xy \cdot 0,\\ \frac{\partial V}{\partial w} &= \frac{\partial V}{\partial x}\frac{\partial x}{\partial w} + \frac{\partial V}{\partial y}\frac{\partial y}{\partial w} + \frac{\partial V}{\partial z}\frac{\partial z}{\partial w}\\ &= yz\frac{w}{\sqrt{w^2 - v^2}} - xz\frac{w}{\sqrt{u^2 + v^2 - w^2}} + xy\frac{w}{\sqrt{w^2 - u^2}}.\end{aligned}$$

It would be possible to express these partial derivatives entirely in terms of u, v and w if necessary by substituting for x, y and z the expressions given above. ■

Implicit Derivatives

In section 3.6 we discussed how the derivative of an implicit function could be found using the chain rule for a function of one variable. We take up the same problem again but now use the chain rule for functions of several variables to determine the derivative of an implicit function.

Suppose $y = g(x)$ is a differentiable function and it is known that

$$f(x, y) = c$$

for some constant c and all x (on some interval). We differentiate both sides of this equation with respect to x, taking into account the fact that y is a function of x to get

$$f_x(x, y)\frac{\partial x}{\partial x} + f_y(x, y)\frac{\partial y}{\partial x} = \frac{\partial c}{\partial x} = 0.$$

The notation $\partial y/\partial x$ is just another notation for dy/dx; we solve for this, taking into account the fact that $\partial x/\partial x = 1$ and get

$$\frac{dy}{dx} = -\frac{f_x(x, y)}{f_y(x, y)}. \tag{14.8}$$

Let us apply this computation in an example.

EXAMPLE 5 Let $y = g(x)$ be a differentiable function of x such that $x^2 + 2xy - 3y^2 = 4$ for all x. Compute the general formula for dy/dx and then evaluate this derivative at the points where $x = 2$.

Solution: Let $f(x, y) = x^2 + 2xy - 3y^2$ so that

$$f_x(x, y) = 2x + 2y, \qquad f_y(x, y) = 2x - 6y.$$

Formula (14.8) yields

$$\frac{dy}{dx} = -\frac{f_x(x, y)}{f_y(x, y)} = -\frac{2x + 2y}{2x - 6y}.$$

Now to evaluate this at the points where $x = 2$, it is necessary to know the value of $y = g(2)$. Any information about the value of y at $x = 2$ must be deduced from the equation

$$f(2, y) = 2^2 + 2 \cdot 2y - 3y^2 = 4 \qquad \text{or} \qquad y(4 - 3y) = 0.$$

Thus there are two possible values of $y = g(2)$; either $y = 0$ or $y = 4/3$. We record the value of y' in each case:

$$\text{If} \quad y = 0, \quad \text{then} \quad y' = -\frac{2(2) + 2(0)}{2(2) - 6(0)} = -1;$$

$$\text{If} \quad y = \frac{4}{3}, \quad \text{then} \quad y' = -\frac{2(2) + 2(4/3)}{2(2) - 6(4/3)} = \frac{5}{3}.$$

■

The idea we just used to compute the derivative of an implicitly defined function of one variable can be extended to an implicit function of several variables. Suppose $z = g(x, y)$ is a differentiable function that satisfies

$$f(x, y, z) = c \tag{14.9}$$

for some differentiable f and constant c and all x and y in some region. We evaluate the partial derivatives of z with respect to x and y by taking the partial derivative of both sides of Equation (14.9). For example

$$f_x(x, y, z)\frac{\partial x}{\partial x} + f_y(x, y, z)\frac{\partial y}{\partial x} + f_z(x, y, z)\frac{\partial z}{\partial x} = \frac{\partial c}{\partial x} = 0. \tag{14.10}$$

We assume that x and y are independent variables so that $\partial y/\partial x = 0$ (because y

is held constant as x varies) and, of course, $\partial x/\partial x = 1$. Thus we can use Equation (14.10) to solve for $\partial z/\partial x$ and obtain

$$\frac{\partial z}{\partial x} = -\frac{f_x(x, y, z)}{f_z(x, y, z)}.$$

Similarly, we obtain

$$\frac{\partial z}{\partial y} = -\frac{f_y(x, y, z)}{f_z(x, y, z)}.$$

Let us apply these formulas in an example.

EXAMPLE 6 Find the partial derivatives $\partial z/\partial x$ and $\partial z/\partial y$ if $z = g(x, y)$ and satisfies the equation

$$xy^2 + xyz - z^2y - 3zx^2 - 8 = 0.$$

Solution: Let $f(x, y, z) = xy^2 + xyz - z^2y - 3zx^2 - 8$. The formulas for the derivatives of z with respect to x and y give

$$\frac{\partial z}{\partial x} = -\frac{f_x(x, y, z)}{f_z(x, y, z)} = -\frac{y^2 + yz - 6zx}{xy - 2zy - 3x^2},$$

$$\frac{\partial z}{\partial y} = -\frac{f_y(x, y, z)}{f_z(x, y, z)} = -\frac{2xy + xz - z^2 - 3x^2}{xy - 2zy - 3x^2}.$$

■

Some Applications of the Chain Rule

The chain rule can be used to describe the rate of change of a function of several variables. As examples we consider a variation on the classic box problem and a related rate problem that involves more free variables than the one-variable case that was considered in section 4.10.

EXAMPLE 7 A rectangular box has height x, width y, and depth z. The height, width and depth change at a constant rate of a units per second. How fast is the volume of the box changing in general and in the specific instant when $x = 4$, $y = 8$ and $z = 12$?

Solution: The volume is $V = xyz$ and its rate of change with respect to time is found by the chain rule. We require the rate of change with respect to time of the three variables; this is given as $dx/dt = dy/dt = dz/dt = a$. By the chain rule we have

$$\frac{\partial V}{\partial t} = V_x\frac{\partial x}{\partial t} + V_y\frac{\partial y}{\partial t} + V_z\frac{\partial z}{\partial t} = yza + xza + xya.$$

At the given instant, we have

$$\left.\frac{dV}{dt}\right|_{(x,y,z)=(4,8,12)} = (8)(12)a + (4)(12)a + (4)(8)a = 176a.$$

Note that the expression $yz + xz + xy$ is half the surface area of the box so this argument shows that when each of the sides increases at the same rate, the volume increases at a rate proportional to the surface area. ■

EXAMPLE 8 A right circular cone with base of radius r and height h has volume $V = \frac{1}{3}\pi r^2 h$. Suppose the radius and height change with time according to the rules $dr/dt = -1$ and $dh/dt = 4$ for $0 \leq t \leq 10$. Thus the radius is decreasing and the height is increasing as time t increases. Is the volume increasing or decreasing at the time when $r = 50$ and $h = 50$ What relation between r and h indicates that the volume is increasing?

Solution: The volume is increasing at time t if the derivative dV/dt is positive at t and similarly the volume is decreasing at time t if the derivative is negative. We compute using the chain rule,

$$\frac{dV}{dt} = V_r \frac{\partial r}{\partial t} + V_h \frac{\partial h}{\partial t} = \frac{1}{3}(2rh)(-1) + \frac{1}{3}r^2(4).$$

The assertion that the volume is increasing when $r = 50$ and $h = 50$ is equivalent to

$$\frac{1}{3}[(2 \cdot 50 \cdot 50)(-1) + (50)^2(4)] > 0.$$

This inequality is satisfied so the volume is increasing. Thus the volume for general r and h is increasing when $dV/dt > 0$; in view of the equation for the derivative, this is equivalent to

$$4r^2 > 2rh \qquad \text{or} \qquad 2r > h.$$

There are many different functions $r = r(t)$ and $h = h(t)$ that satisfy $dr/dt = -1$ and $dh/dt = 4$. One could write down all the possibilities and then determine in each case when V is increasing or decreasing. The use of the chain rule allows a uniform treatment without the case-by-case analysis. ■

Exercises 14.6

Compute f_u and f_v for each function given.

1. $f(x, y) = xy - 2x^2 - 3y^2$, $x = u + 4v$, $y = 3u - 7v^2$

2. $f(x, y) = x^2 - 3y^2 - 3x^4y^3$, $x = uv$, $y = 2u^2 - 5v^2$

3. $f(x, y) = \sin(x + 2y)$, $x = 2u - 3v$, $y = v - u$

4. $f(x, y) = \ln(x^2 + xy + y^2)$, $x = u/v$, $y = v/u$

5. $f(x, y) = \sqrt{x^2 + y^2}$, $x = u \sin v$, $y = u \cos v$

6. $f(x, y) = e^{-xy}$, $x = \ln(u^2 + v^2)$, $y = \ln(u/v)$

Find $\partial w/\partial u$, $\partial w/\partial v$ and $\partial w/\partial t$ for each case.

7. $w = xy + yz + zx + xyz$, $x = u \cos t$, $y = u \sin t$, $z = vt$

8. $w = x^3 + y^3 + z^3 - 3xyz$, $x = uv \cos t$, $y = uv \sin t$, $z = -uv$

9. $w = xy^2 + yz^2 + zx^2$, $x = u + v + t$, $y = uv + vt + ut$, $z = uvt$

10. $w = xe^{yt} + ye^{xt}$, $x = (\ln u)/t$, $y = (\ln v)/t$

For functions $u = u(x, y)$ and $v = v(x, y)$ the equations

$$u_x = v_y \quad \text{and} \quad u_y = -v_x$$

are the Cauchy-Riemann equations that play a central role in the study of functions of a complex variable. Verify that the following pairs of functions satisfy the Cauchy-Riemann equations.

11. $u = x^2 - y^2$, $v = 2xy$

12. $u = x^3 - 3xy^2$, $v = 3x^2y - y^3$

13. $u = e^x \cos y$, $v = e^x \sin y$

14. $u = (e^y + e^{-y}) \cos x$, $v = (e^{-y} - e^y) \sin x$

Use the chain rule for functions of several variables to determine the derivative dz/dt or $\partial z/\partial t$ when $z = g(t)$ or $z = g(u, t)$ and z is related to u and t by the given equation.

15. $z^3 + t^3 - 3zt = 1$

16. $4zt - z^2 - t^2 = 12$

17. $zt + z^2 + t^2 = \sin zt$

18. $\sin z \cos t = \dfrac{2z - 1}{2t - 1}$

19. $ztu + zt + tu + zu = z^2 + t^2 + u^2$

20. If $z = f(x, y)$, $x = r \cos t$ and $y = r \sin t$ prove that

$$\left(\frac{\partial z}{\partial r}\right)^2 + \frac{1}{r^2}\left(\frac{\partial z}{\partial t}\right)^2 = \left(\frac{\partial z}{\partial x}\right)^2 + \left(\frac{\partial z}{\partial y}\right)^2.$$

21. If p and v are independent variables, $H = 3uv^2 + pv$ and $u = f(p, v)$, find an expression for $\partial H/\partial v$.

22. Let $w = f(x, y, z)$ be a differentiable function and define $g(u) = f(ux, uy, uz)$ for some fixed x, y, z. Find $g'(u)$ in terms of the partial derivatives of f.

23. Let $\mathbf{r}(t) = x(t)\,\mathbf{i} + y(t)\,\mathbf{j} + z(t)\,\mathbf{k}$ be the position vector of a particle moving through space. Let $T(x, y, z)$ be the temperature of the air at the point (x, y, z). What is the rate of change with respect to time t of the air temperature surrounding the particle $\mathbf{r}(t)$ as it moves through space?

24. Let $\mathbf{r}(t) = 3\cos t\,\mathbf{i} + 2\sin t\,\mathbf{j} + 2t\,\mathbf{k}$ be the position vector of a particle moving through space. Let $T(x, y, z) = 212 - 33\sqrt{(x-2)^2 + (y-1)^2 + z^2}$ be the temperature of the air at the point (x, y, z). What is the rate of change with respect to time t of the air temperature surrounding the particle $\mathbf{r}(t)$ as it moves through space?

25. A right circular cylinder has a radius and altitude that vary with time. At a certain instant, the altitude is increasing at 0.5 feet per second and the radius is decreasing at a rate 0.2 feet per second. How fast is the volume changing if at this time the radius is 20 feet and the altitude is 60 feet?

26. How fast is the surface area (sides plus top and bottom) changing of the cylinder in exercise 25?

27. An airplane flying is directly east at a rate of 300 miles per hour and climbing at the rate of 20 miles per hour. How fast is the distance between the airplane and an observer on the ground changing at the instant when the plane is 5 miles east of the observer and the plane is at an altitude of 6 miles.

28. Rework exercise 27 with all the same data except that we now ask for the rate of change of the distance between the observer and the airplane at the instant when the plane is 5 miles north of the observer rather than 5 miles east of the observer. [Now the plane is moving across the field of vision of the observer instead of directly away from the observer.]

14.7 The gradient

We introduce a vector function associated with a function f of two or three variables. The *gradient of* f is denoted by ∇f and is defined by the equation

$$\nabla f(x, y, z) = f_x(x, y, z)\,\mathbf{i} + f_y(x, y, z)\,\mathbf{j} + f_z(x, y, z)\,\mathbf{k}$$

at all points (x, y, z) where the partial derivatives exist. If f is a function of only x and y then

$$\nabla f(x, y) = f_x(x, y)\,\mathbf{i} + f_y(x, y)\,\mathbf{j}.$$

For example the gradient of the function $f(x, y) = x^2 - 3xy + 7y^2$ is $\nabla f(x, y) = (2x - 3y)\,\mathbf{i} + (14y - 3x)\,\mathbf{j}$; the gradient of the function $G(x, y, z) = xyz - 3x - 11y$ is $\nabla G(x, y, z) = (yz - 3)\,\mathbf{i} + (xz - 11)\,\mathbf{j} + xy\,\mathbf{k}$.

The gradient is used in a number of problems such as finding the tangent plane to a surface, the normal to a space curve and the directional derivative of a scalar function. Let us begin with an application to curves in two dimensions. The next theorem shows that the gradient $\nabla f(a, b)$ is normal to the level curve $f(x, y) = 0$ at the point (a, b) provided that the point is on the curve; that is $f(a, b) = 0$.

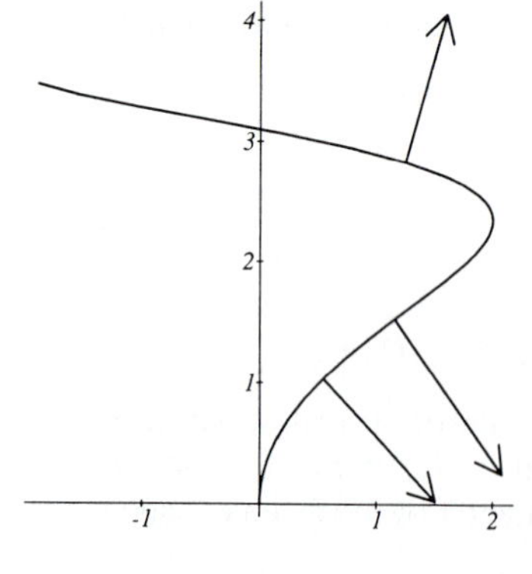

FIGURE 14.14
∇f normal to $f = 0$

THEOREM 14.6

Gradient is Normal to the Curve

Let $f(x, y) = 0$ be the equation of a curve C in the xy plane and let $\mathbf{r}(t)$ be a differentiable position vector that traces out a part of C containing the point (x_0, y_0) corresponding to $\mathbf{r}(t_0) = x_0\,\mathbf{i} + y_0\,\mathbf{j}$. Assume C has a nonzero tangent $\mathbf{r}'(t_0)$ at this point. If $f_x(x, y)$ and $f_y(x, y)$ exist at (x_0, y_0), then the gradient $\nabla f(x_0, y_0)$ is normal to C at (x_0, y_0).

Proof

First we write $\mathbf{r}(t) = X(t)\,\mathbf{i} + Y(t)\,\mathbf{j}$ with differentiable functions $X = X(t)$ and $Y = Y(t)$ and note that

$$f(X(t), Y(t)) = 0 \qquad \textbf{(14.11)}$$

because the position vector lies on the curve C determined by $f(x, y) = 0$. Notice that the vector $\mathbf{r}'(t)$ is tangent to the curve. We now differentiate both sides of Equation (14.11) with respect to t using the chain rule to get

$$f_x(X(t), Y(t))X'(t) + f_y(X(t), Y(t))Y'(t) = 0.$$

If t_0 is a value of t such that

$$\mathbf{r}(t_0) = X(t_0)\,\mathbf{i} + Y(t_0)\,\mathbf{j} = x_0\,\mathbf{i} + y_0\,\mathbf{j}$$

then we have

$$0 = f_x(x_0, y_0)X'(t_0) + f_y(x_0, y_0)Y'(t_0) = \nabla f(x_0, y_0) \cdot \mathbf{r}'(t_0).$$

This shows that the gradient is orthogonal to the tangent vector $\mathbf{r}'(t_0)$ and so the gradient is normal to the curve. □

The key idea in the above argument is that the chain rule can be expressed as a dot product. The particular parameterization $\mathbf{r}(t)$ of part of the curve plays no real role except to serve as a convenient tool so that the chain rule can be brought into the computation.

EXAMPLE 1 Let E be the ellipse with equation $16 = 3x^2 + 4y^2$. The point $(2, 1)$ is on E; find the normal to the curve at this point.

Solution: Let $f(x, y) = 3x^2 + 4y^2 - 16$ so that $f(x, y) = 0$ is the equation of the ellipse and $f(2, 1) = 0$ verifying that $(2, 1)$ is a point on the curve. By the previous theorem, $\nabla f(2, 1)$ is a normal vector to E at $(2, 1)$. We evaluate the gradient of f to get

$$\nabla f(x, y) = f_x(x, y)\,\mathbf{i} + f_y(x, y)\,\mathbf{j} = 6x\,\mathbf{i} + 8y\,\mathbf{j}.$$

Thus $\nabla f(2, 1) = 12\,\mathbf{i} + 8\,\mathbf{j}$ is normal to E at $(2, 1)$. ■

Curves in Explicit Form

When the curve is given in the explicit form $y = g(x)$ we consider the function $f(x, y) = y - g(x)$ so the curve is given by the equation $f(x, y) = 0$. Then the gradient of f is

$$\nabla f(x, y) = -g'(x)\,\mathbf{i} + \mathbf{j}$$

and this is seen to be normal to the curve by the following reasoning. The slope of the tangent to the curve at (x, y) is $g'(x)$ and a vector parallel to the tangent is then $\mathbf{i} + g'(x)\,\mathbf{j}$. Clearly this is orthogonal to $\nabla f(x, y)$.

Three-Dimensional Gradient

There is a three-dimensional version of the previous theorem in which f is replaced by a function of three variables.

THEOREM 14.7

Let S be a surface with equation $F(x, y, z) = 0$ and let $P_0 = (x_0, y_0, z_0)$ be a point of S where F is differentiable. Suppose $\mathbf{r}(t) = X(t)\mathbf{i} + Y(t)\mathbf{j} + Z(t)\mathbf{k}$ is a differentiable curve lying in the surface S and passing through P_0 when $t = t_0$ and having a nonzero tangent at that point. If ∇F exists and is nonzero at P_0, then $\nabla F(x_0, y_0, z_0)$ is normal to the curve $\mathbf{r}$.

Proof

The proof runs along the same lines as the two-variable version. Because the curve lies in the surface, we have $F(X(t), Y(t), Z(t)) = 0$; differentiate with respect to t using the chain rule to get

$$0 = F_x(X, Y, Z)X' + F_y(X, Y, Z)Y' + F_z(X, Y, Z)Z' = \nabla F(X, Y, Z) \cdot \mathbf{r}'.$$

This shows $\nabla F(X, Y, Z)$ is orthogonal to the tangent $\mathbf{r}'$ to the curve and so the gradient is normal to the curve. □

EXAMPLE 2 Let S be the ellipsoid $(x-1)^2 + 2(y+3)^2 + 4(z-2)^2 = 7$. Find a normal vector to any curve lying in the surface at the point $(2, -2, 1)$.

Solution: Let $F(x, y, z) = (x-1)^2 + 2(y+3)^2 + 4(z-2)^2 - 7$ so the equation of the ellipsoid is $F(x, y, z) = 0$. Note that $F(2, -2, 1) = 0$ so that $(2, -2, 1)$ is indeed a point on the ellipsoid. The gradient of F is

$$\nabla F(x, y, z) = 2(x-1)\mathbf{i} + 4(y+3)\mathbf{j} + 8(z-2)\mathbf{k}$$

and this vector is normal to any curve in the surface passing through (x, y, z). In particular

$$\nabla F(2, -2, 1) = 2\mathbf{i} + 4\mathbf{j} + 8\mathbf{k}$$

is such a normal. ■

Notice that in both the examples, no curve $\mathbf{r}$ was mentioned as there was in the theorems. The hypothesis involving the curve $\mathbf{r}$ is a convenient device to detect orthogonality. We know that $\mathbf{r}'$ is a tangent to the curve so a vector orthogonal to $\mathbf{r}'$ is normal to the curve. In the three-dimensional case, we see that $\nabla F(x, y, z)$ is orthogonal to every curve in the surface $F = 0$ through the point (x, y, z). We restate the three-dimensional theorem without the clutter of the curve $\mathbf{r}$, but first we define what we mean by a normal to a surface.

A nonzero vector $\mathbf{u}$ is *normal to a surface* S at a point P in S if $\mathbf{u}$ is orthogonal at P to every curve lying in S and passing through P.

THEOREM 14.8

The Gradient is Normal to the Surface

Let $F(x, y, z)$ be a differentiable function and S the surface with equation $F(x, y, z) = 0$. If $F(x_0, y_0, z_0) = 0$ then the gradient $\nabla F(x_0, y_0, z_0)$ is normal to S at (x_0, y_0, z_0).

This is merely a restatement of Theorem 14.7 along with the definition just given of a normal to a surface.

For future reference we highlight the form of the chain rule that made its appearance in the previous discussion. If f is either a function of two or three variables and $\mathbf{r}(t)$ is a vector function in two or three dimensions, respectively, then we have

$$\frac{d}{dt} f(\mathbf{r}(t)) = \nabla f(\mathbf{r}(t)) \cdot \mathbf{r}'(t). \qquad \textbf{(14.12)}$$

Tangent Planes to a Surface

If a surface is given by an equation $F(x, y, z) = 0$ and $P = (x_0, y_0, z_0)$ is a point on the surface and if $\mathbf{N}$ is a normal to the surface at P, then the *tangent plane* to the surface at P is the plane through P having $\mathbf{N}$ as normal. If the is no normal to the surface at P then there there is no tangent plane to the surface at P. We can find the equation of the plane that is tangent to the surface using the information just obtained about the normal. The vector $\nabla F(x_0, y_0, z_0)$ is normal to the surface and we suppose it is nonzero. Then the tangent to the surface at P has $\nabla F(x_0, y_0, z_0)$ as a normal vector so the vector form of the equation of the tangent plane is

$$((x - x_0)\mathbf{i} + (y - y_0)\mathbf{j} + (z - z_0)\mathbf{k}) \cdot \nabla F(x_0, y_0, z_0) = 0.$$

This can be expressed in rectangular coordinates as

$$a(x - x_0) + b(y - y_0) + c(z - z_0) = 0, \quad \text{where} \quad \nabla F(x_0, y_0, z_0) = a\,\mathbf{i} + b\,\mathbf{j} + c\,\mathbf{k}.$$

EXAMPLE 3 Find the equation of the plane tangent to the ellipsoid

$$x^2 + 5y^2 + 4z^2 = 30$$

at the point $(-3, 1, 2)$.

Solution: The ellipsoid has equation $F(x, y, z) = 0$ with

$$F(x, y, z) = x^2 + 5y^2 + 4z^2 - 30.$$

We have $\nabla F(x, y, z) = 2x\,\mathbf{i} + 10y\,\mathbf{j} + 8z\,\mathbf{k}$.

Thus $\nabla F(-3, 1, 2) = -6\,\mathbf{i} + 10\,\mathbf{j} + 16\mathbf{k}$ is normal to the surface and to the tangent plane at the point $(-3, 1, 2)$. The equation of the plane through $(-3, 1, 2)$ with this normal is

$$-6(x + 3) + 10(y - 1) + 16(z - 2) = 0.$$

■

Tangent Plane to an Explicitly Defined Surface

If a surface is given as $z = f(x, y)$ then we set $F(x, y, z) = z - f(x, y)$ so the surface is also given by the equation $F(x, y, z) = 0$. We have then $F_z(x, y, z) = 1$, $F_x(x, y, z) = -f_x(x, y)$ and $F_y(x, y, z) = -f_y(x, y)$. Thus the plane tangent to the surface $z = f(x, y)$ at the point $(a, b, f(a, b))$ has equation

$$f_x(a, b)(x - a) + f_y(a, b)(y - b) - (z - f(a, b)) = 0.$$

EXAMPLE 4 Find the equation of the plane tangent to the surface $z = 4x^2 - y^2$ at the point $(1, 3, -5)$.

Solution: Evaluate $\partial z/\partial x = 8x$ and $\partial z/\partial y = 2y$ at $(x, y) = (1, 3)$ and substitute into formula (14.12) to get

$$8(x - 1) + 6(y - 3) - (z - (-5)) = 0 \quad \text{or} \quad 8x + 6y - z = 31.$$

■

Here is a slightly more subtle use of the gradient.

EXAMPLE 5 Find the points P in the surface $F(x, y, z) = x^2 + y^2 + 2z^2 - 4 = 0$ where the line from P to $(2, 2, 2)$ is tangent to the surface.

Solution: Let $(u, v, w) = P$ be a point in the surface meeting the conditions. Then the vector from P to $T = (2, 2, 2)$ is in the tangent plane, hence is normal to the gradient $\nabla F(u, v, w)$. This translates to the equation $\overrightarrow{PT} \cdot \nabla F(u, v, w) = 0$ and so

$$\big((u-2)\mathbf{i} + (v-2)\mathbf{j} + (w-2)\mathbf{k}\big) \cdot \big(2u\,\mathbf{i} + 2v\,\mathbf{j} + 4w\,\mathbf{k}\big) = 2u(u-2) + 2v(v-2) + 4w(w-2) = 0.$$

In addition,

$$F(u, v, w) = 0 = u^2 + v^2 + 2w^2 - 4.$$

Subtracting twice this equation from the previous equation gives us the condition

$$4u + 4v + 8w = 8.$$

Thus (u, v, w) is a point on the surface $F = 0$ and on the plane $x + y + 2z = 2$. The intersection of the surface and the plane is a curve that can be explicitly described as follows. We look for a parametric representation of the curve in which z is the parameter. (The choice of z is quite arbitrary.) From the equation of the plane we obtain $x = 2 - 2z - y$. Substitute into the equation of the surface to get

$$0 = (2 - 2z - y)^2 + y^2 + 2z^2 - 4 = 4 - 8z - 4y + 6z^2 + 4yz + 2y^2 - 4.$$

Solve this quadratic for y in terms of z to get

$$y = 1 - z \pm \sqrt{1 + 2z - 2z^2}$$
$$x = 2 - 2z - y = 1 - z \mp \sqrt{1 + 2z - 2z^2}.$$

If you feel more comfortable using another variable as parameter, it would be just fine to write the equations of the curve as

$$\begin{cases} x = x(t) = 1 - t \mp \sqrt{1 + 2t - 2t^2}, \\ y = y(t) = 1 - t \pm \sqrt{1 + 2t - 2t^2}, \\ z = z(t) = t. \end{cases}$$

The ambiguous signs must be selected so that we use either both upper signs or both lower signs. ■

Tangents to Intersections of Surfaces

We give one more application of the gradient. Suppose two surfaces given by $F(x, y, z) = 0$ and $G(x, y, z) = 0$ intersect and the intersection is a differentiable curve. The tangent to this curve is orthogonal to the gradient of F and also orthogonal to the gradient of G at each point of the intersection. The cross product $\nabla F \times \nabla G$ is also orthogonal to each of ∇F and ∇G and so the cross product is parallel to the tangent to the curve of intersection. Thus we can find the direction of the tangent to the curve without actually finding the equation of the curve.

EXAMPLE 6 Find the tangent to the curve of intersection of the two surfaces

$$F(x, y, z) = x^2 - 2y^2 + 3z^2 - 2 = 0,$$
$$G(x, y, z) = 2x^2 + 3y^2 - 2z^2 - 3 = 0$$

at the point $(1, 1, 1)$.

Solution: We first check that $(1, 1, 1)$ is on both surfaces by verifying that $F(1, 1, 1) = G(1, 1, 1) = 0$. Then we compute

$$\nabla F(1, 1, 1) = 2\mathbf{i} - 4\mathbf{j} + 6\mathbf{k},$$
$$\nabla G(1, 1, 1) = 4\mathbf{i} + 6\mathbf{j} - 4\mathbf{k}.$$

It follows that the cross product

$$\nabla F(1, 1, 1) \times \nabla G(1, 1, 1) = 4\mathbf{i} + 32\mathbf{j} + 28\mathbf{k}$$

is tangent to the intersection curve at $(1, 1, 1)$. ■

Exercises 14.7

Compute the gradient of each function and find the points, if any, where the gradient equals 0.

1. $f = x^2 + 2y^3$

2. $f = xe^y$

3. $f = 2x^3 - 4xy^2 + 3y^2$

4. $f = xy - x - y$

5. $f = \cos x \sin y$

6. $f = xe^y - y$

7. $f = \dfrac{x - y}{x + y}$

8. $f = \dfrac{x - y}{x^2 + y^2}$

Use the gradient to find a vector normal to the curve $f = 0$ at the indicated point P and then use this information to find a vector tangent to the curve at P.

9. $f = 2x^2 + 3y^2 - 5, \quad P = (1, -1)$

10. $f = x^2 - 14y + 24, \quad P = (2, 2)$

11. $f = y - x^2 - 3x, \quad P = (0, 0)$

12. $f = 3x + xy - 2y - 2, \quad P = (1, 1)$

Find the equation of the plane tangent to the surface $F(x, y, z) = 0$ or $z = f(x, y)$ at the indicated point P.

13. $z = x^2 - y^2$, $P = (2, -1, 3)$

14. $F(x, y, z) = x^2 + y^2 + z^2 - 12, \quad P = (2, -2, 2)$

15. $z = x^2 + y^2, \quad P = (-2, 1, 5)$

16. $F(x, y, z) = 3x^2 - 4y^2 - z^2 - 7, \quad P = (3, -2, 2)$

17. Show that the plane tangent to the cone $z^2 = x^2/a^2 + y^2/b^2$ at any point other than the origin intersects the cone in a line passing through the origin.

If two curves in the xy plane intersect at a point P, the angle between the curves is defined as the angle between the tangent lines at P. This is equal to the angle between the normal lines to the curves at P. Use this along with the gradient to compute the cosine of the angle between the following pairs of curves at the indicated point.

18. $x^2 + y^2 = 7$, $2x^2 + y^2/4 = 7$, $P = (\sqrt{3}, 2)$

19. $y = x^2$, $y = 8 - x^2$, $P = (2, 4)$

20. $x^2 - y^2 = 8$, $2x + 3y = 9$, $P = (3, 1)$

21. $(x - a)^2 + y^2 = a^2$, $x^2 + y^2 = a^2$, $P = (a/2, a\sqrt{3}/2)$

22. The paraboloid with equation $z = x^2 + 3y^2$ intersects the plane with equation $x + y + z = 6$ in a curve. Find the tangent to that curve at the point $(1, 1, 4)$.

23. The hyperbolic paraboloid $z = x^2 - 4y^2$ and the cone $z^2 = x^2 + 4y^2$ intersect in a curve that passes through the point $P = (1, 0, 1)$. Find the tangent to the curve at P using gradients. Then find a parametric representation of the curve and find a tangent to the curve at P from the parametric representation.

24. The surfaces $z = 4xy$ and $z^2 + 4x^2 - 2y^2 - 2x = 16$ meet in a curve containing the point $(1, 1, 4)$. Find a tangent to the intersection curve at that point.

25. Find all points in the surface $f(x, y, z) = z^2 - xy = 4$ where a tangent to the surface passes through the point $P = (3, 3, 1)$. Describe the set of points as the intersection of the surface and a plane.

26. Show that the Linear Approximation Theorem 14.2 in section 14.5 can be written in the following vector form. If f is a differentiable function, then

$$f(\mathbf{u} + \mathbf{v}) = f(\mathbf{u}) + \nabla f(\mathbf{u}) \cdot \mathbf{v} + G(\mathbf{v})$$

where $\lim_{|\mathbf{v}| \to 0} G(\mathbf{v})/|\mathbf{v}| = 0$.

27. To find the point P on the ellipsoid $f(x, y, z) = x^2 + 4y^2 + 6z^2 - 11 = 0$ that is nearest the point $Q = (a, b, c)$, (where $f(a, b, c) \neq 0$) one could proceed as follows.

The shortest line from Q to the surface is normal to the surface (this will be more believable after the section on Lagrange multipliers) and thus some multiple of the gradient equals $\overrightarrow{PQ}$. Writing $P = (x, y, z)$ and then, for some λ, we have

$$(x\,\mathbf{i} + y\,\mathbf{j} + z\,\mathbf{k}) + \lambda\nabla f(x, y, z) = a\,\mathbf{i} + b\,\mathbf{j} + c\,\mathbf{k}.$$

Use this to express x, y, z in terms of λ and the fact that $f(x, y, z) = 0$ to obtain an equation that determines λ. If convenient, use some calculating device to approximate the solutions of this equation for some specific choices of a, b, c. [The complete solution of this problem, even for some specific values of a, b and c, requires a great deal of computation.]

14.8 Directional derivatives

For a function $f(x, y)$ defined over a region R, the first partial derivative, $f_x(a, b)$ evaluated at a point (a, b), measures the rate of change of the function in the direction of the x-axis. Similarly, $f_y(a, b)$ measures the rate of change in the y direction. It makes good sense to ask for the rate of change of $f(x, y)$ in some other directions. More specifically, given the point (a, b) and a vector $\mathbf{v}$ (that indicates a direction), we ask what is the rate of change of $f(x, y)$ at (a, b) in the direction of $\mathbf{v}$? We refer to this concept as the *directional derivative* of the function; we use the gradient to evaluate the directional derivative. The formulation of the problem can be made in any number of dimensions so we state it in a form that applies equally well to functions of two or three (or any number of) variables.

If f is a function of two variables x and y, then we may write $\mathbf{u} = x\,\mathbf{i} + y\,\mathbf{j}$ and then write $f(\mathbf{u}) = f(x, y)$. Similarly if f is a function of x, y and z, then we may write $\mathbf{u} = x\,\mathbf{i} + y\,\mathbf{j} + z\,\mathbf{k}$ and set $f(\mathbf{u}) = f(x, y, z)$.

DEFINITION 14.4

Directional Derivative

Let $f(\mathbf{u})$ be defined in an open sphere centered at the terminal point of the vector $\mathbf{A}$ and let $\mathbf{v}$ be a vector of unit length. The directional derivative of $f(\mathbf{u})$ at $\mathbf{A}$ in the direction $\mathbf{v}$ is

$$D_{\mathbf{v}} f(\mathbf{A}) = \lim_{t \to 0} \frac{f(\mathbf{A} + t\mathbf{v}) - f(\mathbf{A})}{t},$$

provided this limit exists.

Consider what this limit measures. The numerator $f(\mathbf{A}+t\mathbf{u}) - f(\mathbf{A})$ measures the difference between the values of f at $\mathbf{A}$ and a nearby point $\mathbf{A} + t\mathbf{v}$ that is t units away from $\mathbf{A}$ in the direction of $\mathbf{v}$. This is divided by t to get the average rate of change and then the limit is taken as t approaches 0.

EXAMPLE 1 Let $f(x, y) = x^2 - 3xy$ and let $\mathbf{v} = c\,\mathbf{i} + s\,\mathbf{j}$ be some unit vector.

For any $\mathbf{A} = a\,\mathbf{i} + b\,\mathbf{j}$ we have $\mathbf{A} + t\mathbf{v} = (a + tc)\,\mathbf{i} + (b + ts)\,\mathbf{j}$ and

$$\begin{aligned}\lim_{t\to 0}\frac{f(\mathbf{A}+t\mathbf{v})-f(\mathbf{A})}{t} &= \lim_{t\to 0}\frac{f(a+tc,b+ts)-f(a,b)}{t}\\ &= \lim_{t\to 0}\frac{(a+tc)^2-3(a+tc)(b+ts)-a^2+3ab}{t}\\ &= \lim_{t\to 0}\frac{2atc+t^2c^2-3tcb-3ats+3t^2cs}{t}\\ &= 2ac-3cb-3as.\end{aligned}$$

This direct computation shows that the result depends on (a, b) and on $\mathbf{v} = c\,\mathbf{i}+s\,\mathbf{j}$. Note that by taking $\mathbf{v} = \mathbf{i}$, we have the derivative in the direction of the x-axis evaluated at (a, b), namely $f_x(a, b) = 2a - 3b$. If we take $\mathbf{v} = \mathbf{j}$, then the directional derivative is $f_y(a, b) = -3a$. ■

The observation made at the end of the example is true for any function; that is,

$$D_{\mathbf{i}}f(x,y) = f_x(x,y), \qquad D_{\mathbf{j}}f(x,y) = f_y(x,y).$$

Of course a similar result holds for functions of three variables.

In example 1 we used only the definition of directional derivative to evaluate $D_{\mathbf{v}}f(a, b)$. The next result shows how to use the gradient to evaluate this derivative in the case the function is differentiable.

THEOREM 14.9

Directional Derivative and the Gradient

Let f be a function defined in a region R containing (a, b) and let let $\mathbf{v}$ be a unit vector. If the first partial derivatives $f_x(x, y)$ and $f_y(x, y)$ are continuous in the region R, then

$$D_{\mathbf{v}}f(a,b) = \nabla f(a,b)\cdot\mathbf{v}.$$

Proof

Because the first derivatives are continuous, f is differentiable and we can apply Theorem 14.2 to express the difference of $f(x, y)$ and $f(a, b)$ as

$$f(x,y) - f(a,b) = f_x(a,b)h + f_y(a,b)k + g_1(h,k)h + g_2(h,k)k$$

with $h = x - a$, $k = y - b$ and g_1, g_2 functions having limit 0 as $(h, k) \to (0, 0)$.

Let $\mathbf{v} = c\,\mathbf{i} + s\,\mathbf{j}$. Now use the definition of $D_{\mathbf{v}}f(a, b)$ and this last equation with $h = x - a = tc$ and $k = y - b = ts$ to obtain

$$\begin{aligned}D_{\mathbf{v}}f(a,b) &= \lim_{t\to 0}\frac{f(a+tc,b+ts)-f(a,b)}{t}\\ &= \lim_{t\to 0}\frac{f_x(a,b)tc+f_y(a,b)ts+g_1(tc,ts)tc+g_2(tc,ts)ts}{t}\\ &= \lim_{t\to 0}(f_x(a,b)c+f_y(a,b)s+g_1(tc,ts)c+g_2(tc,ts)s)\\ &= f_x(a,b)c+f_y(a,b)s\\ &= (f_x(a,b)\,\mathbf{i}+f_y(a,b)\,\mathbf{j})\cdot(c\,\mathbf{i}+s\,\mathbf{j})\\ &= \nabla f(a,b)\cdot\mathbf{v}.\end{aligned}$$

Note that the functions g_1 and g_2 have limit 0 because $\lim_{t\to 0}(tc, ts) = (0, 0)$.
□

Theorem 14.9 is valid for functions of three variables with no significant change. We state it without proof.

THEOREM 14.10

Directional Derivative in Three Dimensions

Let f be a function defined in a region R of xyz space containing (a, b, c) and let let $\mathbf{v}$ be a unit vector. If f is differentiable at (a, b), then

$$D_{\mathbf{v}} f(a, b, c) = \nabla f(a, b, c) \cdot \mathbf{v}.$$

Return to example 1 and compute the directional derivative of $f(x, y) = x^2 - 3xy$ in the direction of $\mathbf{v} = c\,\mathbf{i} + s\,\mathbf{j}$ using the gradient as

$$\begin{aligned} D_{\mathbf{v}} f(a, b) &= [f_x(a, b)\,\mathbf{i} + f_y(a, b)\,\mathbf{j}] \cdot \mathbf{v} \\ &= [(2a - 3b)\,\mathbf{i} - 3a\,\mathbf{j}] \cdot [c\,\mathbf{i} + s\,\mathbf{j}] \\ &= 2ac - 3bc - 3bs. \end{aligned}$$

Of course this computation agrees with the result obtained earlier.

EXAMPLE 2 Compute the rate of change of $f(x, y) = x^2 + 3xy + 2y^2$ at $(2, 2)$ in the direction of $\mathbf{v} = \cos\theta\,\mathbf{i} + \sin\theta\,\mathbf{j}$ for any number θ.

Solution: The gradient of f is

$$\nabla f(x, y) = (2x + 3y)\,\mathbf{i} + (3x + 4y)\,\mathbf{j}$$

and $\nabla f(2, 2) = 10\,\mathbf{i} + 14\,\mathbf{j}$. The directional derivative at $(2, 2)$ in the direction of $\mathbf{v}$ is

$$\begin{aligned} D_{\mathbf{v}} f(2, 2) &= \nabla f(2, 2) \cdot \mathbf{v} = (10\,\mathbf{i} + 14\,\mathbf{j}) \cdot (\cos\theta\,\mathbf{i} + \sin\theta\,\mathbf{j}) \\ &= 10\cos\theta + 14\sin\theta. \end{aligned}$$

For some specific computations we record the directional derivative in the directions for which $\theta = 0, \pi/4, \pi/2$. We have

$$\begin{aligned} D_{\mathbf{i}} f(2, 2) &= 10\cos 0 + 14\sin 0 = 10, \\ D_{(\mathbf{i}+\mathbf{j})/\sqrt{2}} f(2, 2) &= 10\cos\pi/4 + 14\sin\pi/4 = 12\sqrt{2}, \\ D_{\mathbf{j}} f(2, 2) &= 10\cos\pi/2 + 14\sin\pi/2 = 14. \end{aligned}$$

■

The Direction of Greatest Change

The examples and discussion above shows that the rate of change of a function at some point depends on the direction from the point. Because the directional derivative of f determines the rate of change of f in a specific direction, one might ask how to find the direction in which f changes most rapidly. For example, the function $f(x, y) = x^2 + 3xy + 2y^2$ was seen to have several different rates of change at the point $(2, 2)$. Of the three directions tested in that example, we saw that $D_{(\mathbf{i}+\mathbf{j})/\sqrt{2}} f(2, 2)$ was greater than either of $D_{\mathbf{i}} f(2, 2)$ or $D_{\mathbf{j}} f(2, 2)$.

We now show that the direction of the greatest rate of change is the direction of the gradient.

THEOREM 14.11

Direction of Greatest Rate of Change

Let f be a differentiable function of two or three variables in a region R containing the point P. The directional derivative of f at P has the largest value in the direction of the gradient $\nabla f(P)$ evaluated at P, provided $\nabla f(P) \neq 0$. The direction in which the directional derivative is least is the negative $-\nabla f(P)$ of the gradient at P.

Proof

Let $\mathbf{v}$ be a unit vector. Then

$$D_{\mathbf{v}} f(P) = \nabla f(P) \cdot \mathbf{v} = |\nabla f(P)|\,|\mathbf{v}| \cos\theta = |\nabla f(P)| \cos\theta,$$

where θ is the angle between $\nabla f(P)$ and $\mathbf{v}$. This last expression is largest when $\cos\theta = 1$. This occurs when the angle between $\nabla f(P)$ and $\mathbf{v}$ is 0; in other words the vectors are parallel and point in the same direction. The directional derivative is smallest when $\cos\theta = -1$. This occurs when the angle between $\nabla f(P)$ and $\mathbf{v}$ is π; that is, the vectors are parallel but point in opposite directions. □

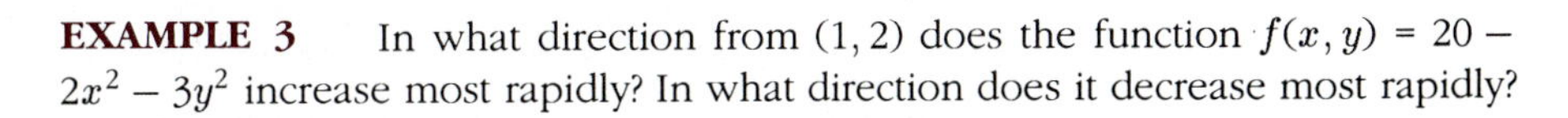

EXAMPLE 3 In what direction from $(1, 2)$ does the function $f(x, y) = 20 - 2x^2 - 3y^2$ increase most rapidly? In what direction does it decrease most rapidly?

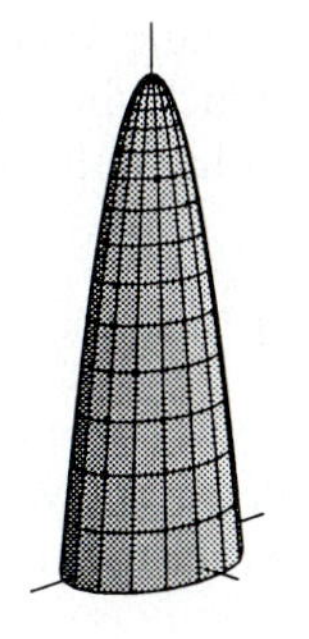

FIGURE 14.15
$z = 20 - 2x^2 - 3y^2$

Solution: The gradient of f at $(1, 2)$ is $\nabla f(1, 2) = -4\mathbf{i} - 12\mathbf{j}$. This is the direction of greatest increase of f when moving from the point $(1, 2)$. (From a rough sketch of the surface we note that the function is generally increasing as we move in the direction toward the origin.) The function is decreasing most rapidly at $(1, 2)$ in the direction of $4\mathbf{i} + 12\mathbf{j}$. ■

Now we consider an extended example that makes significant use of the gradient.

EXAMPLE 4 Which path allows the skier to reach the bottom of the hill in the shortest time?

Solution: To solve this problem, we first need a mathematical model of the hill. We consider only a fairly simple hill that has a lot of symmetry. Imagine that ground level is the xy plane and $f(x, y)$ gives the height of the hill above $(x, y, 0)$. We assume the top of the hill is at $(0, 0, H)$, H is the height of the hill, and the level curves $f(x, y) = c$ are ellipses. One function that meets this criterion is

$$f(x, y) = H - Ax^2 - By^2,$$

where A and B are positive constants. This function might give a reasonable approximation of the height of a hill except for values close to ground level where it would be more accurate to have the hill "flatten out" more gradually. We do not attempt to make this modification and use the function $f(x, y)$ as the description of the hill. The constants A and B can be interpreted as determining the hill along the x or y direction; if $y = 0$ the trace of the hill above the x-axis is the curve $z = H - Ax^2$ so that the horizontal distance from the top of the hill to ground level measured in the x direction is $\sqrt{H/A}$. Similarly the profile of the hill over the y-axis is the curve $z = H - By^2$ and the horizontal distance to ground level in the y direction is $\sqrt{H/B}$.

The skier follows a path so that at time t the position is $(x(t), y(t), z(t))$ that lies entirely in the surface $z = f(x, y)$. It follows that $z(t) = f(x(t), y(t))$ so the z component is determined by the x and y components. It is sufficient to consider the vector $x(t)\,\mathbf{i} + y(t)\,\mathbf{j}$; this is the projection of the skier's path into the xy plane.

From all of the possible paths, how do we select a path that gives the quickest descent? We assume that at each point of the path, the skier is travelling in the direction in which the height of the hill is decreasing most rapidly. This should seem a reasonable assumption because experience would dictate that the faster the decrease in the height of the hill, the greater the velocity of the skier. We do not prove that this assumption produces a path of quickest descent. It is also necessary to assume some starting point so let us assume that (x_0, y_0) is the point in the xy plane above which the skier begins his descent. To keep some equations valid later, we assume that $x_0 \neq 0$ and $y_0 \neq 0$. We know that at any point during the descent, the tangent to the path points in the direction of instantaneous motion and this direction should be the direction in which the height of the hill is decreasing most rapidly.

The tangent to the path has the projection into the xy plane equal to $\mathbf{v}(t) = x'(t)\,\mathbf{i} + y'(t)\,\mathbf{j}$ and the direction from the point (x, y) in which the height of the hill is decreasing most rapidly is $-\nabla f(x, y)$. If two vectors have the same direction, then one is a scalar multiple of the other. Thus

$$x'(t)\,\mathbf{i} + y'(t)\,\mathbf{j} = -M\nabla f(x, y) = 2AMx\,\mathbf{i} + 2BMy\,\mathbf{j}$$

for some M. However the M might not be the same number when we move from one point to another. All we know is that M is positive (because the vectors $\mathbf{v}(t)$ and $-\nabla f(x, y)$ point in the same direction). So we are left with the problem of finding two unknown functions, $x(t)$ and $y(t)$ that satisfy

$$\begin{cases} x'(t) = 2AMx(t), \\ y'(t) = 2BMy(t), \end{cases}$$

for some positive $M = M(x, y)$.

This problem has two equations involving unknown functions and their derivatives. This is a situation frequently encountered in many physical problems. We do not take any time to discuss general problems of this sort because the equations in this problem can be solved without the general theory. The mysterious function M can be eliminated from the equations. Multiply the first equation by Ay and the second by Bx to make the right-hand sides equal. Then we obtain $Byx' = Axy'$ or $Byx' - Ayx' = 0$. This expression might remind you of the numerator of the derivative of a quotient. In the special case $A = B = 1$ we would have $yx' - xy' = 0$ and then $(x/y)' = (yx' - xy')/y^2 = 0$ would imply $x/y = c = \text{constant}$. In this derivation we must assume that $y \neq 0$ because we have divided by y. Notice that the two curves representing the straight lines along one of the axes (either $x = 0$ or $y = 0$) satisfy the differential equation $Byx' - Ayx' = 0$ but we exclude these two as trivial cases and look for others.

To give due consideration to the constants A and B we consider the quotient x^B/y^A and compute the derivative:

$$\begin{aligned} \frac{d}{dt}\frac{x^B}{y^A} &= \frac{y^A Bx^{B-1}x' - x^B Ay^{A-1}y'}{y^{2A}} \\ &= x^{B-1}y^{A-1}\frac{Byx' - Axy'}{y^{2A}} = 0. \end{aligned}$$

It follows that $x^B/y^A = c$, a constant or $x^B = cy^A$. There is a bit of a problem with this solution. We require that x^A and y^B be defined. So to be safe, we should assume $x \geq 0$ and $y \geq 0$ inasmuch as x^A need not be defined for negative x and certain A. We are given that the skier begins at the point (x_0, y_0) so we should be working with the conditions $x_0 > 0$ and $y_0 > 0$. The constant c is determined by the equation $c = {x_0}^B/{y_0}^A$ and we have the nice symmetrical equation of the path

$$\left(\frac{x}{x_0}\right)^B = \left(\frac{y}{y_0}\right)^A.$$

An argument using symmetry serves to find the solutions when (x_0, y_0) is in some quadrant other than the first. The cases in which either $x_0 = 0$ or $y_0 = 0$ are exceptional and do not fall into the general case just covered.

If $A = B$, then the cross sections of the hill are circles; everything is perfectly symmetrical and the solution curve is a straight line $y^A = cx^A$ that can be written as $y = \pm c_1 x$ with $c_1 = |c|^{1/A}$. The fastest path down the hill is any path straight down the hill.

For $A \neq B$, the paths are curves of the form $y = c_1 x^m$ with $m = B/A$ and c_1 a constant. Figures 14.16 and 14.17 show the case $A = 0.5$, $B = 1$ and $c = 1$.

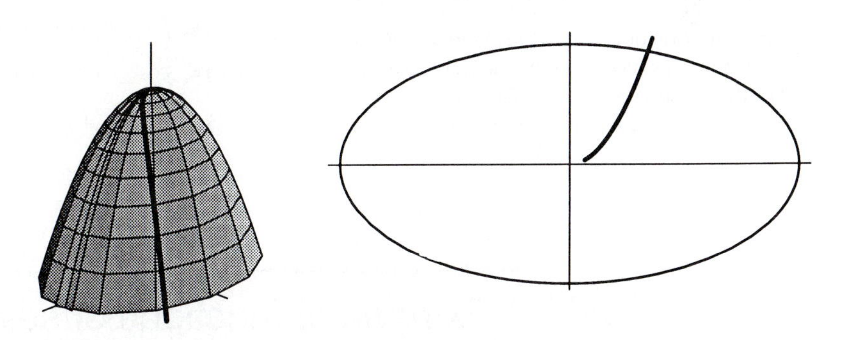

FIGURE 14.16
A fastest path

FIGURE 14.17
Projection of path

Note that the path crosses the level curves at right angles (that is, the curves are orthogonal). To the skier this means that the fastest path is taken by crossing the lines of equal altitude at right angles. ■

Exercises 14.8

Compute the directional derivative $D_{\mathbf{u}}f$ for the given functions, the vectors $\mathbf{u}$ and points P and then also compute $D_{\mathbf{v}}f$ for $\mathbf{v}$ a unit vector in the direction of the gradient of f at P.

1. $f(x, y) = 2x^2 - 5y^2$; $\mathbf{u} = \mathbf{i}$, $(\mathbf{i} + \mathbf{j})/\sqrt{2}$ and $\mathbf{j}$; $P = (3, 1)$

2. $f(x, y) = x^2 + 2xy + 3y^2$; $\mathbf{u} = (3\mathbf{i} + 4\mathbf{j})/5$, $(\mathbf{i} - \mathbf{j})/\sqrt{2}$ and $\mathbf{j}$; $P = (1, 1)$

3. $f(x, y, z) = x^2 + 3y^2 + 4z^2 - 8xyz$; $\mathbf{u} = \mathbf{i}$, $\mathbf{j}$, $\mathbf{k}$ and $(\mathbf{i} + \mathbf{j} + \mathbf{k})/\sqrt{3}$; $P = (1, 1, 1)$

4. $f(x, y, z) = x^k y^m z^n$; $\mathbf{u} = \mathbf{i}$, $\mathbf{j}$, $\mathbf{k}$ and $(\mathbf{i} + \mathbf{j} + \mathbf{k})/\sqrt{3}$; $P = (1, 1, 1)$

5. Find the rate of change of $f(x, y) = 2x^2 - y^2 + 4xy$ at $(1, 1)$ in the direction that $g(x, y) = 4x^3 - 6y^2x$ increases most rapidly.

6. Find the rate of change of $e^{-x}\cos y$ at $(\pi/4, \pi/4)$ in the direction that $e^{-y}\cos x$ decreases most rapidly.

7. A ski hill has the mathematical model given by

$$1000(z - 100)^2 = 400x^2 + 900y^2,$$

where the ground is the xy plane and z is the height of the hill above the point (x, y). If a skier starts at the point on the hill directly above (10, 10) and comes down the hill so that at every point of the path she is moving in the direction of the most rapid decline in the height of the hill, what path does she follow to reach the ground? Express the solution as a curve in the xy plane representing the projection of the skiers path.

8. Show that, for any "hill" $z = f(x, y)$, the path of quickest descent always crosses a level curve orthogonally; that is, the tangent to the path at the point of intersection with a level curve is normal to the tangent to the level curve.

9. The temperature at a point (x, y) in the xy plane is $T(x, y) = a + x^2 - y^2$ for some constant a. If you are standing at the point (1, 1), what path should you take so that the temperature at your feet increases as rapidly as possible?

10. The cross section of a wall is the strip $0 \leq x \leq 1$, $y \geq 0$. The temperature at the point (x, y) inside the wall is given by the function

$$T(x, y) = 100e^{-\pi y} \sin \pi x - 10e^{-2\pi y} \sin 2\pi x.$$

Find the direction from $(1/2, b)$ in which the temperature increases most rapidly and decreases most rapidly. Do the same computation for a point (ϵ, b) where ϵ is a positive, but very small, number.

11. Let $W = W(x, y) = x^2+y^2-xy$ and suppose that each of x and y is a function of t satisfying $x'(0) = a$ and $y'(0) = b$ for constants a and b. Find a condition involving the constants a and b and the values $(x(0), y(0))$ to ensure that W is a decreasing function of t at $t = 0$. Express this condition in terms of the angle between the gradient ∇W and the velocity vector of the position $x(t)\,\mathbf{i} + y(t)\,\mathbf{j}$.

12. The directional derivative $D_{\mathbf{u}}f(x, y)$ is a function of two variables and so one may consider the second directional derivative

$$D^2_{\mathbf{u}}f(x, y) = D_{\mathbf{u}}[D_{\mathbf{u}}f(x, y)].$$

Prove the following formula in which we assume $\mathbf{u} = c\,\mathbf{i} + s\,\mathbf{j}$:

$$D^2_{\mathbf{u}}f(x, y) = f_{xx}(x, y)c^2 + 2f_{xy}(x, y)cs + f_{yy}(x, y)s^2.$$

Using exercise 12, calculate $D^2_{\mathbf{u}}f(x, y)$ at P.

13. $f = x^2 - 7xy + y^2$, $\mathbf{u} = (\mathbf{i} + \mathbf{j})/\sqrt{2}$, $P = (0, 0)$

14. $f = ax^2 + bxy + cy^2$, $\mathbf{u} = c\,\mathbf{i} + s\,\mathbf{j}$, $P = (0, 0)$

15. $f = x^3 + y^3 - 3xy$, $\mathbf{u} = (3\,\mathbf{i} + 4\,\mathbf{j})/5$, $P = (3/2, 3/2)$

16. $f = \ln(x^2 + y^2)$, $\mathbf{u} = (a\,\mathbf{i} + b\,\mathbf{j})/\sqrt{a^2 + b^2}$, $P = (1, 1)$

14.9 Extrema of functions of several variables

Some of the most appealing applications of the theory of the derivative of a function of one variable are to the solution of maximum and minimum value problems. The situation with extrema of a function of several variables is somewhat more complicated but still has many very interesting applications. We begin with the definitions.

DEFINITION 14.5

Local Maximum and Minimum

A function $f(x, y)$ has a local maximum at point (x_0, y_0) if there is a positive number δ such that $f(x, y) \leq f(x_0, y_0)$ for every (x, y) satisfying $(x - x_0)^2 + (y - y_0)^2 < \delta^2$.

A function $f(x, y)$ has a local minimum at point (x_0, y_0) if there is a positive number δ such that $f(x, y) \geq f(x_0, y_0)$ for every (x, y) satisfying $(x - x_0)^2 + (y - y_0)^2 < \delta^2$.

A local maximum or a local minimum is called a *local extremum.* If f has a local maximum at (a, b) that satisfies the stronger condition $f(a, b) \geq f(x, y)$ for all (x, y) in the domain of f then we say f has an *absolute maximum* (a, b). A similar definition applies for an *absolute minimum.*

If the function f is continuous in a rectangular region R defined by a set of inequalities $a \le x \le b$, $c \le y \le d$ then there exists a point (x_m, y_m) in R at which f attains its minimum value. In other words $f(x_m, y_m) \le f(x, y)$ for every (x, y) in R. Similarly there is a point (x_M, y_M) in R at which f attains its maximum value. We give a proof of this statement in the next section. For now we accept these results and take up the problem of finding the points at which f has a minimum or maximum value.

For a function of one variable we saw that a maximum or minimum occurs at a critical point, that is, at a point where either the derivative does not exist or where the derivative is zero. There is the additional possibility that a local extrema could occur at the endpoint of an interval on which the function is defined. We now obtain an analogue for functions of two variables. Suppose that $f(a, b)$ is a local maximum of f. Consider the function of one variable defined by $g(x) = f(x, b)$. The graph of $g(x)$ is the curve obtained by intersecting the surface $z = f(x, y)$ and the plane $y = b$. Because $f(a, b)$ is a local maximum of f, it follows that $g(a)$ is a local maximum of g. Thus a is a critical point for g. This translates to the statement that either f_x does not exist at (a, b) or $f_x(a, b) = 0$. If we repeat this argument with the function $h(y) = f(a, y)$ we find that either f_y does not exist at (a, b) or $f_y(a, b) = 0$. In either case there is the additional possibility that (a, b) is an endpoint of the region on which the function is defined. A completely analogous argument permits us to reach the same conclusion if $f(a, b)$ is a local minimum of f. We summarize these statements as follows.

When we consider a rectangular region R defined by inequalities $p \le x \le q$, $r \le y \le s$, we say that a point (a, b) is an *interior point* of R if $p < a < q$ and $r < b < s$. In other words, an interior point does not lie on the edges of the rectangle.

THEOREM 14.12

Let $f(x, y)$ be defined at all points of a rectangular region R and suppose that (a, b) is an interior point of R. If $f(x, y)$ has either a local maximum or local minimum at (a, b) and if the partial derivatives f_x and f_y exist at (a, b) then $f_x(a, b) = f_y(a, b) = 0$.

Thus the problem of finding the extreme points of f is translated into a two step problem: First find the points where either the first partial derivatives do not exist or where both f_x and f_y vanish. Second, test each of these points to determine if f has a local maximum or local minimum there.

In some cases this second step can be carried out by inspection as we now illustrate.

EXAMPLE 1 Let $f(x, y) = 11 - x^2 + 2x - y^2 - 2y$. Find the local extreme points of f.

Solution: The first partial derivatives are given by

$$f_x(x, y) = -2x + 2 \quad \text{and} \quad f_y(x, y) = -2y - 2.$$

These derivatives exist at every point and are both equal to 0 only at $(x, y) = (1, -1)$. Is this likely to be a local maximum, a local minimum or neither. Let us try a few values. We compute $f(1, -1) = 13$, $f(1, 0) = 12$, $f(0, -1) = 12$ and $f(2, -1) = 12$. This suggests (but does not prove) that $f(1, -1)$ might be a local maximum. So we try to prove that this is the case. We expect to prove that

$$f(1,-1) - f(x,y) \geq 0$$

for all x and y. We have

$$\begin{aligned} f(1,-1) - f(x,y) &= 13 - (11 - x^2 + 2x - y^2 - 2y) \\ &= x^2 - 2x + 1 + y^2 + 2y + 1 \\ &= (x-1)^2 + (y+1)^2 \geq 0. \end{aligned}$$

This gives a proof that f has an absolute maximum at $(1,-1)$ because at all other points the value of f is less than its value at $(1,-1)$. ■

The vanishing of the two partial derivatives is not enough to ensure the existence of a local extreme as the following example shows.

EXAMPLE 2 Discuss the extreme points of the function $f(x,y) = 1 - x^2 + y^2$.

Solution: The partial derivatives $f_x = -2x$ and $f_y = 2y$ vanish only at $(0,0)$. The difference

$$f(0,0) - f(x,y) = x^2 - y^2 = (x+y)(x-y)$$

takes on both positive and negative values in any circle of positive radius having center at the origin. For example, if we move toward the origin along the line $y = 2x$ the above difference is $(3x)(-x) = -3x^2 \leq 0$ whereas if we move to the origin along the line $2y = x$, the above difference is $(3y)(y) = 3y^2 \geq 0$. Thus f has neither a local maximum nor a local minimum at $(0,0)$. The domain of definition of f is the entire xy plane so every point is an interior point and there are no other points where both first partial derivatives vanish. Hence this function has no extreme points. ■

Test for Extrema

We now consider how we can recognize the extreme points of a function. The next theorem may be regarded as the analogue of the second derivative test for functions of one variable. Recall that the second derivative test is the following: Suppose $g(x)$ satisfies $g'(a) = 0$. Then $g''(a) > 0$ implies g has a local minimum at a and $g''(a) < 0$ implies g has a local maximum at a. There is no information if $g''(a) = 0$.

The analogue of the second derivative test for a function $f(x,y)$ is somewhat more complicated partly because there are three second derivatives f_{xx}, f_{xy} and f_{yy} that affect the result. Before stating the second derivative test, we give the plan for attacking this problem. We start with a point (a,b) where $f_x(a,b) = f_y(a,b) = 0$. It is the goal to determine if f has a maximum, a minimum, or neither at (a,b). We subtract the value of f at (a,b) from the values of f at nearby points. If $f(a,b) - f(h,k) \geq 0$ for all (h,k) "near" (a,b) then f has a maximum at (a,b). If the sign is reversed for all (h,k), then f has a minimum. If there are positive and negative differences, then f does not have an extreme value at (a,b).

We parametrize the points near (a,b) by giving the direction θ and distance $t \geq 0$ so the point $(a + t\cos\theta, b + t\sin\theta)$ is "nearby". We try to estimate the difference in the function values at the two points. To simplify the notation, we hold θ fixed and write $c = \cos\theta$ and $s = \sin\theta$. Let

$$F(t) = f(a + tc, b + ts) - f(a,b).$$

Clearly if $F(t) \geq 0$ for all values of t satisfying $0 \leq t < \epsilon$ for some positive ϵ and for all θ then $f(a,b)$ is a relative minimum. Similarly if $F(t) \leq 0$ for all t satisfying $0 \leq t < \epsilon$ and all θ then $f(a,b)$ is a local maximum. The statement of the theorem is simplified slightly if we introduce the *discriminant* function of f. Let

$$D(x,y) = f_{xx}(x,y)f_{yy}(x,y) - f_{xy}^2(x,y).$$

In preparation for the statement and proof of the second derivative test, we require the following fact.

THEOREM 14.13

Let A, B and C be constants, not all zero such that

$$A\cos^2\theta + 2B\cos\theta\sin\theta + C\sin^2\theta > 0 \qquad \textbf{(14.13)}$$

for all θ. Then $AC - B^2 > 0$. If

$$A\cos^2\theta + 2B\cos\theta\sin\theta + C\sin^2\theta < 0$$

for all θ then $AC - B^2 > 0$. If the expression $A\cos^2\theta + 2B\cos\theta\sin\theta + C\sin^2\theta$ assumes both positive and negative values for various choices of θ then $AC - B^2 < 0$.

Proof

For any $(x,y) \neq (0,0)$ we have $x\,\mathbf{i} + y\,\mathbf{j} = \rho(\cos\theta\,\mathbf{i} + \sin\theta\,\mathbf{j})$ for some θ and $\rho = \sqrt{x^2 + y^2} > 0$. If inequality (14.13) holds for all θ then

$$\rho^2(A\cos^2\theta + 2B\cos\theta\sin\theta + C\sin^2\theta) > 0$$
$$Ax^2 + 2Bxy + Cy^2 > 0.$$

By taking $y = 0$ and $x = 1$ we see that $A > 0$. Next complete the squares by first factoring out A to get

$$A\left[x^2 + \frac{2B}{A}xy + \frac{B^2}{A^2}y^2 + \left(\frac{C}{A} - \frac{B^2}{A^2}\right)y^2\right] \geq 0$$
$$A\left[\left(x + \frac{B}{A}y\right)^2 + \left(\frac{AC - B^2}{A^2}\right)y^2\right] \geq 0.$$

Now by taking $x = -By/A$ and $y \neq 0$ we get $AC - B^2 > 0$. This completes the proof in case the first inequality holds for all θ. If the second inequality holds for all θ, we can multiply by -1 to get a form of the first inequality with constants $-A$, $-B$ and $-C$ replacing A, B, C. The conclusion is $(-A)(-C) - (-B)^2 = AC - B^2 > 0$, which is what we wanted to prove in this case also.

Now suppose the function $A\cos^2\theta + 2B\cos\theta\sin\theta + C\sin^2\theta$ assumes both positive and negative values for various choices of θ. If $A \neq 0$ then the completion of squares argument in the preceding case can be used to argue that $AC - B^2 < 0$. Similarly if $C \neq 0$ a completion of squares argument beginning with a removal of a factor C leads to the same conclusion. In the remaining case, we suppose that $A = C = 0$. Because not all three constants are 0, we have $B \neq 0$ and $AC - B^2 = -B^2 < 0$. □

This theorem is used in the proof of the next theorem.

THEOREM 14.14

Second Derivative Test for Extrema

Let $f(x, y)$ have continuous second partial derivatives in a rectangular region R and let (a, b) be an interior point of R such that $f_x(a, b) = f_y(a, b) = 0$. Let $D(a, b) = f_{xx}(a, b)f_{yy}(a, b) - [f_{xy}(a, b)]^2$.

i. If $D(a, b) > 0$ and $f_{xx}(a, b) < 0$ then $f(a, b)$ is a local maximum of f;
ii. If $D(a, b) > 0$ and $f_{xx}(a, b) > 0$ then $f(a, b)$ is a local minimum of f;
iii. If $D(a, b) < 0$ then f does not have an extreme value at (a, b);
iv. If $D(a, b) = 0$ then no conclusion can be drawn.

Proof

We give a proof that assumes f has continuous third partial derivatives. A more complicated argument is needed to give the proof under the slightly weaker hypothesis as stated.

The idea of the proof is to compare the values of f at (a, b) and nearby points. A point near (a, b) can be expressed in terms of vectors as

$$a\,\mathbf{i} + b\,\mathbf{j} + t(\cos\theta\,\mathbf{i} + \sin\theta\,\mathbf{j}).$$

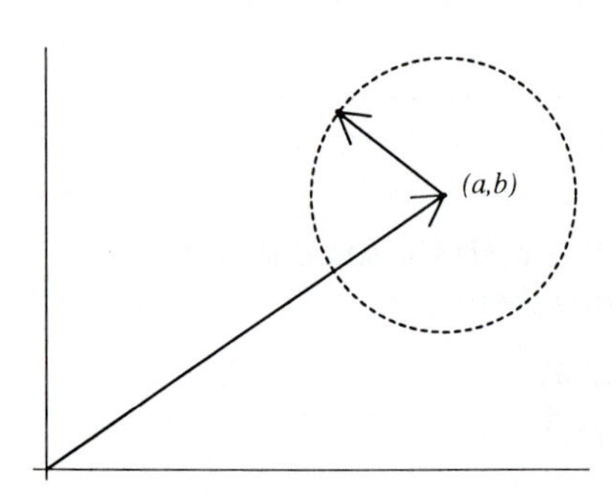

FIGURE 14.18
Vector to points near (a, b)

Let θ be an arbitrary but fixed number and let

$$F(t) = f(a + t\cos\theta, b + t\sin\theta) - f(a, b).$$

If $f(a, b)$ is a local maximum of f, then there is some positive number δ such that $F(t) \leq 0$ for all t satisfying $0 \leq t \leq \delta$ and all θ. Similarly if $f(a, b)$ is a local minimum of f there is a positive δ with $F(t) \geq 0$ for all t with $0 \leq t \leq \delta$ and all θ.

We regard F as a function of a single variable t (because θ is held fixed) and apply Taylor's Theorem to it. We need an explicit expression for first two derivatives of F with respect to t. These are computed by applying the chain rule several times. For simplicity of notation, let $P = (a + t\cos\theta, b + t\sin\theta)$. Then

$$F'(t) = f_x(P)\cos\theta + f_y(P)\sin\theta,$$
$$F''(t) = f_{xx}(P)\cos^2\theta + 2f_{xy}(P)\cos\theta\sin\theta + f_{yy}(P)\sin^2\theta.$$

By Taylor's Theorem (9.5 part B) there is a number z between 0 and t such that

$$F(t) = F(0) + F'(0)t + \frac{1}{2}F''(0)t^2 + \frac{1}{6}F'''(z)t^3.$$

We observe from the definition of F that $F(0) = 0$. For the first derivative we have

$$F'(0) = f_x(a, b)\cos\theta + f_y(a, b)\sin\theta = 0;$$

this equals 0 because we have assumed that the first partial derivatives of f are 0 at (a, b). For brevity we write $A = f_{xx}(a, b)$, $B = f_{xy}(a, b)$ and $C = f_{yy}(a, b)$. Thus we have

$$\begin{aligned} F(t) &= \frac{1}{2}F''(0)t^2 + \frac{1}{6}F'''(z)t^3 \\ &= \frac{1}{2}\left(A\cos^2\theta + 2B\cos\theta\sin\theta + C\sin^2\theta\right)t^2 + \frac{F'''(z)t^3}{6}. \end{aligned} \quad \textbf{(14.14)}$$

We want a condition on t that ensures $F(t)$ has constant sign in a small region near (a, b). The idea is that the sign of a sum $p + q$ is the same as the sign of p if $|p| > |q|$. So we seek a restriction on t to ensure $|F''(0)t^2| > |F'''(z)t^3/3|$.

By assumption on the continuity of the third partial derivatives of f, F''' is continuous and so there is a number M such that $|F'''(t)| < M$ if $0 \leq t \leq 1$ for all θ. (The restriction to $t \leq 1$ is arbitrary; any positive number in place of 1 would do just as well.) Now we want to select a further limitation of t so that

$$0 \leq t \leq k = \frac{|F''(0)|}{M}.$$

Then

$$F(t) = \frac{1}{2}F''(0)t^2 + \frac{1}{6}F'''(z)t^3 \quad \text{and} \quad \frac{1}{2}F''(0)t^2$$

have the same sign. Now suppose that f has a minimum at (a, b). We call this case I. Then there is a positive number δ such that $F(t) \geq 0$ for all t satisfying $0 \leq t \leq \delta$ and for all θ. We can replace δ by any smaller positive number so we may assume that $\delta \leq k$. It follows that $(A\cos^2\theta + 2B\cos\theta\sin\theta + C\sin^2\theta)$ is nonnegative for every θ. From Theorem 14.13, we conclude

$$D(a, b) = f_{xx}(a, b)f_{yy}(a, b) - [f_{xy}(a, b)]^2 = AC - B^2 > 0.$$

Note that in this case the values of $(A\cos^2\theta + 2B\cos\theta\sin\theta + C\sin^2\theta)$ are nonnegative and A is one of these values. Hence $A \geq 0$.

If f has a maximum value at (a, b) we call this case II. Almost the same argument as just given can be applied to show that $D(a, b) > 0$. We need only note that the values of $F(t)$ are nonpositive instead of nonnegative. In this case the values of $(A\cos^2\theta + 2B\cos\theta\sin\theta + C\sin^2\theta)$ are nonpositive so this time $A \leq 0$.

If f has neither a local maximum nor a local minimum at (a, b), we call this case III. Then $F(t)$ must take on both positive and negative values in any small circle centered at (a, b). Thus by Theorem 14.13 we obtain $D(a, b) = AC - B^2 < 0$.

Now we must prove the statements in the conclusion of the theorem. If $D(a, b) > 0$ and $A > 0$, then we cannot have either case II or case III, so f has a local minimum at (a, b).

If $D(a, b) > 0$ and $A < 0$, then we cannot have either case I or case III so f has a local maximum at (a, b).

If $D(a, b) < 0$ then we cannot have case I or case II and so f has neither a relative minimum nor a local maximum at (a, b). □

Classification of Critical Points

A point where both the first partial derivatives of f are zero or one of the first partial derivatives does not exist is called a *critical point.* The extrema of f occur at critical points. The second derivative test provides a means of classifying the critical points as points where f has a local maximum, local minimum or neither of these. A point that is a critical point of f but at which f has neither a local maximum nor a relative minimum is called a *saddle point.* To classify the critical points means to first locate the critical points and then determine if f has a local maximum, a local minimum or a saddle point at each. Let us apply this to an example.

EXAMPLE 3 Locate and classify the critical points of the function

$$f(x, y) = 3y^2 - 3xy^2 - x^3 + 3x^2.$$

***Solution*:** Because f is a polynomial, all its partial derivatives are continuous so the second derivative test can be applied. We first find the critical points. The first partial derivatives set equal to 0 give the following:

$$f_y = 6y - 6xy = 6y(1 - x) = 0, \quad \text{either } y = 0 \text{ or } x = 1$$
$$f_x = -3y^2 - 3x^2 + 6x = 0; \quad \text{when } x = 1,\ y = \pm 1; \quad \text{when } y = 0,\ x = 0 \text{ or } 2.$$

Thus there are four critical points to be tested $(1, \pm 1)$, $(0, 0)$, $(2, 0)$. The second derivatives and discriminant are

$$f_{xx} = -6x + 6, \qquad f_{xy}(x, y) = -6y, \qquad f_{yy}(x, y) = 6 - 6x,$$
$$D(x, y) = (-6x + 6)(6 - 6x) - (-6y)^2 = 36[(1 - x)^2 - y^2].$$

We have $D(1, \pm 1) = -36 < 0$, so the two critical points $(1, \pm 1)$ are saddle points and $f(1, \pm 1) = 2$.

At $(0, 0)$, we have $D(0, 0) = 36 > 0$ and $f_{xx}(0, 0) = 6 > 0$; thus f has a relative minimum at the origin, $f(0, 0) = 0$.

At $(2, 0)$ we have $D(2, 0) = 36 > 0$ and $f_{xx}(2, 0) = -6$; thus f has a relative maximum at $(2, 0)$, $f(2, 0) = 4$.

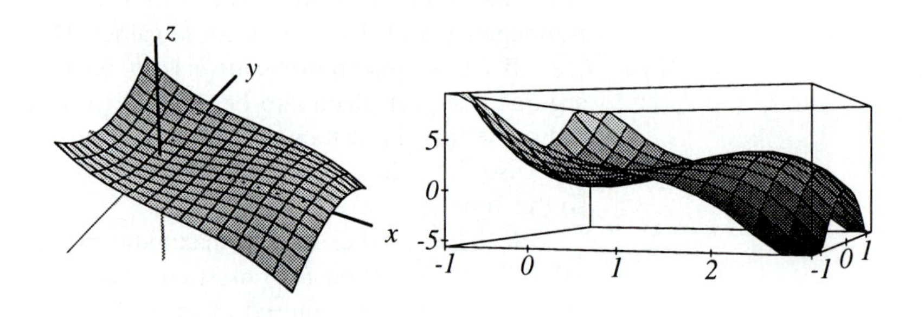

FIGURE 14.19
$z = f(x, y)$

FIGURE 14.20
Another view

The two views of the surface (that resembles a towel in the wind suspended by two corners) might help visualize the two local extreme points. ■

Method of Least Squares

Next we consider again the problem of finding the line of best fit to a given set of points in the plane. We have solved this problem using only the theory of maxima and minima of functions of one variable. We solve it again with the theory of functions of two variables.

EXAMPLE 4 Let (a_1, b_1), (a_2, b_2), ..., (a_n, b_n) be a set of n points in the plane. Let d_i be the vertical distance from (a_i, b_i) to a line $y = mx + b$ with m and b unknown constants. Find m and b that makes the sum of squares $\sum d_i^2$ as small as possible.

Solution: The point on the line having a_i as its x coordinate is $(a_i, ma_i + b)$ so $d_i = |ma_i + b - b_i|$. The sum we are trying to minimize is

$$S(m, b) = \sum_{i=1}^{n} d_i^2 = \sum_{i=1}^{n} (ma_i + b - b_i)^2.$$

The minimum occurs at the points where the first partial derivatives S_m and S_b of S vanish. We compute

$$\begin{aligned} S_m &= \sum_{i=1}^{n} 2a_i(ma_i + b - b_i) \\ &= 2m\sum_{i=1}^{n} a_i^2 + 2b\sum_{i=1}^{n} a_i - \sum_{i=1}^{n} a_i b_i = 0, \end{aligned}$$

$$S_b = \sum_{i=1}^{n} 2(ma_i + b - b_i) = 2m\sum_{i=1}^{n} a_i + 2nb - \sum_{i=1}^{n} b_i = 0.$$

Let $A = \sum a_i$, $B = \sum b_i$, $C = \sum a_i b_i$ and $E = \sum a_i^2$. The equations can be rewritten as

$$Am + nb = B \qquad Em + Ab = C.$$

We can solve this system of linear equations to get

$$\begin{aligned} m &= \frac{nC - AB}{nE - A^2} = \frac{n\sum a_i b_i - (\sum a_i)(\sum b_i)}{n\sum a_i^2 - (\sum a_i)^2} \\ b &= \frac{BE - AC}{nE - A^2} = \frac{(\sum b_i)(\sum a_i^2) - (\sum a_i)(\sum a_i b_i)}{n\sum a_i^2 - (\sum a_i)^2}. \end{aligned}$$

One can argue that we have not divided by zero here so long as the a_i are all different and $n \geq 2$.

We can use the second derivative test to determine if we really have a minimum. The second partial derivatives are

$$S_{mm} = 2E, \qquad S_{bb} = n, \qquad S_{mb} = 2A$$

and the discriminant is a constant $D = 2nE - 4A^2 = 2n\sum a_i^2 - 4(\sum a_i)^2$. With some computation along the line indicated in the exercises following section 4.5, one can show that this constant is nonnegative. ■

Exercises 14.9

Classify the critical points of each function.

1. $f(x, y) = x^2 - 3x - y^2$

2. $f(x, y) = x^2 - 4x + y^2 - 4y - 16$

3. $f(x, y) = x^2 + 2x - y^2 - 4y - 4$

4. $f(x, y) = x^3 + y^2 + 2xy + 4x - 3y - 5$

5. $f(x, y) = \dfrac{x^2y^2 + 4y + 2x}{xy}$

6. $f(x, y) = e^x \cos y$

7. $f(x, y) = y(e^x - 1)$

8. $f(x, y) = x^4 + 2x^2y^2 + y^4$

9. $f(x, y) = x^3 - 3x + 4y^2 - 16y + 1$

10. $f(x, y) = x + \dfrac{1}{x} + y^2$

11. $f(x, y) = (x^2 - 1)^2(y^2 - 1)^2$

12. Find the point on the line $3x + 5y = 8$ nearest the origin.

13. Find the point on the line $3x + 5y = 8$ nearest $(2, -2)$.

14. Find the point in the plane $2x + 4y - 5z = 8$ nearest the origin.

15. Find the point in the plane $2x + 4y - 5z = 8$ nearest the point $(1, 2, 3)$.

16. Show that the box of largest volume that can be inscribed in a sphere of radius R is a cube.

17. Find the shape of a rectangular box of maximum volume with surface area 100 square units. Count all six sides in the surface area.

18. Find the shape of an open-top box of maximum volume with surface area 100 square units. Count only five sides in the surface area.

19. Given a set of n points (a_1, b_1), (a_2, b_2), $\ldots$, (a_n, b_n), find the coordinates of a point (p, q) that minimizes the sum of the squares of the distances to the (a_i, b_i).

20. For a given function $f(x)$, find the constants a and b that minimize the integral

$$F(a, b) = \int_{-1}^{1} [f(x) - a - bx]^2 \, dx.$$

14.10 Extrema with constraints

Problems involving maximum and minimum values of a function of several variables often involve some side conditions that restrict the variables under consideration. Here is an example of a side condition in an extremum problem.

Find the maximum value of $f(x, y)$ when (x, y) is restricted to be on a curve $g(x, y) = 0$. In this problem the variable point (x, y) is restricted to lie on a particular curve (this is the side condition) and we are asked to find the maximum value of a function f subject to this side condition. In some cases such a problem admits a very nice geometric interpretation as we illustrate in the following extended example.

Extreme Values of a Linear Function on a Curve

EXAMPLE 1 Find the extreme values of the function $f(x, y) = 5x + 6y$ at points on the ellipse having vector equation $\mathbf{r}(t) = a \cos t\, \mathbf{i} + b \sin t\, \mathbf{j}$.

Solution: We begin with a geometric interpretation of this problem. A fixed ellipse E is given and we want to find the maximum value of $f(x, y)$ when (x, y) a point on E. We focus attention on the level curves of f. In this particular case f has a simple form; each level curve of f is given by an equation $f(x, y) = c$ (for a constant c) that is the equation of a straight line, written as $f = c$. For any point (x, y) on the line $f = c$, the function f has the constant value c. The family of lines $\{f = c\}$ all have the same slope, $-5/6$ in this case. The maximum value of f at a point of E is the largest number c such that the line $f = c$ has a point in common with the ellipse E. Think of starting with some very large number c so the line $f = c$ is far above or to the right of E and then allow c to decrease. As c decreases, the line $f = c$ moves toward E and for some value of c, the line $f = c$ firsts contact E as a tangent to E. The largest such c is the maximum value of f on E. The smallest c for which $f = c$ touches E is the minimum value of f on E. The key point is that the line $f = c$ is tangent to the ellipse when c is either the maximum or minimum value of f at a point of E.

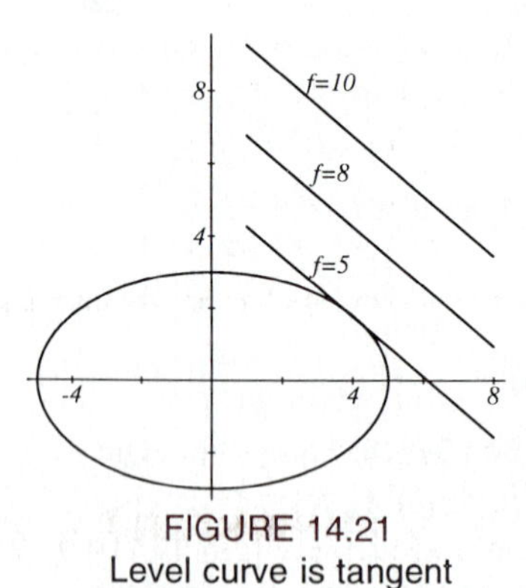

FIGURE 14.21
Level curve is tangent

We have argued that at the point (x_0, y_0) of E where f has a maximum (or a minimum), the line $f(x, y) = c$ with $c = f(x_0, y_0)$ is tangent to the ellipse. Thus one way to determine (x_0, y_0) is to look for the points where the tangent to the ellipse is parallel to the level curves $f(x, y) = c$. In this example all such lines have slope $-5/6$.

The tangent line is parallel to the vector $\mathbf{r}'(t) = -a\sin t\,\mathbf{i} + b\cos t\,\mathbf{j}$ that has slope $(-b\cot t)/a$. The lines $f(x, y) = 5x + 6y = c$ all have slope $-5/6$, so we require

$$\frac{-b\cot t}{a} = -\frac{5}{6} \quad \text{or} \quad \cot t = \frac{5a}{6b}.$$

There are two solutions of this equation with $0 \leq t \leq 2\pi$, one corresponding to the maximum value of f on E and the other corresponding to the minimum value of f on E.

For use in the next idea note that the statement "the line $f(x, y) = c$ is parallel to the tangent vector" can be also stated as "∇f is orthogonal to the tangent vector" in view of the fact that ∇f is orthogonal to the level curve $f(x, y) = c$ and hence to every line L_c.

Now we give a nongeometric solution of this problem using ideas that apply to more complicated functions than the one given in this example.

We write $f(\mathbf{r}(t))$ as shorthand for $f(a\cos t, b\sin t)$. If t_1 is a number such that f has an extreme value at $\mathbf{r}(t_1)$, then the function of one variable $f(\mathbf{r}(t))$ has derivative equal to 0 at $t = t_1$. By the chain rule we have

$$0 = \frac{d}{dt}f(\mathbf{r}(t)) = \nabla f(\mathbf{r}(t)) \cdot \mathbf{r}'(t) = f_x(\mathbf{r}(t))(-a\sin t) + f_y(\mathbf{r}(t))(b\cos t).$$

By evaluating this in our particular case, we have

$$-5a\sin t + 6b\cos t = 0 \quad \text{or} \quad \tan t = \frac{6b}{5a}.$$

There are two values of t on the interval $0 \leq t \leq 2\pi$ that satisfy $\tan t = 6b/5a$, one giving the maximum of f on the ellipse, the other giving the minimum.

We point out that $\nabla f = 5\,\mathbf{i} + 6\,\mathbf{j}$ is a constant vector perpendicular fo the level curves $f(x, y) = c$ and so the points of E at which f takes an extreme value are the points where ∇f is also orthogonal to the tangent vector at E, or more simply where ∇f is orthogonal to E. ■

We now prove the general statement that is suggested by example 1.

THEOREM 14.15

Let $\mathbf{R}(t) = x(t)\,\mathbf{i} + y(t)\,\mathbf{j} = x\,\mathbf{i} + y\,\mathbf{j}$ be the position vector to a curve C such that $\mathbf{R}'(t)$ exists and is nonzero for every t on an interval $a \leq t \leq b$. Let f be a function that is differentiable in a rectangular region containing C. Suppose that when f is evaluated along C, it takes an extreme value at $\mathbf{R}(t_0)$. If $\mathbf{R}(t_0) = x_0\,\mathbf{i} + y_0\,\mathbf{j}$ then $\nabla f(x_0, y_0)$ is orthogonal to the curve C; that is $\nabla f(x_0, y_0)$ is orthogonal to the tangent vector to C, $\mathbf{R}'(t_0)$.

Proof

The function of one variable t defined by $f(\mathbf{R}(t))$ has a maximum or minimum at $t = t_0$. Either the derivative of $f(\mathbf{R}(t))$ does not exist at t_0 or the derivative is 0 at t_0. By assumption the function is differentiable and so its derivative with respect to t is zero at t_0; that is

$$0 = \frac{d}{dt}f(\mathbf{R}(t)) = \nabla f(\mathbf{R}(t)) \cdot \mathbf{R}'(t)$$

at $t = t_0$. This shows that $\nabla f(\mathbf{R}(t_0))$ is perpendicular to $\mathbf{R}'(t_0)$ as required. Using different words, the gradient of f at the point on the curve where f

has an extreme value is orthogonal to the tangent to the curve at the same point. □

EXAMPLE 2 Find the extreme values of $f(x, y) = xy - 7x - 2y$ when (x, y) is a point on the curve $y = x^2$.

Solution: To match the notation of Theorem 14.15, we represent the curve in vector form as $\mathbf{r}(t) = t\,\mathbf{i} + t^2\mathbf{j}$ with $x = t$ and $y = t^2$. The tangent vector is $\mathbf{r}'(t) = \mathbf{i} + 2t\,\mathbf{j}$. The gradient of f is $\nabla f(x, y) = (y - 7)\mathbf{i} + (x - 2)\mathbf{j}$. At an extreme point of f on the curve we have, after expressing ∇f in terms of t,

$$\nabla f(x(t), y(t)) \cdot \mathbf{r}'(t) = 0 = ((t^2 - 7)\mathbf{i} + (t - 2)\mathbf{j}) \cdot (\mathbf{i} + 2t\,\mathbf{j}) = 3t^2 - 4t - 7.$$

The solutions of the quadratic equation are found by the quadratic formula to be $t = -1,\ 7/3$ corresponding to the two points $(-1, 1)$ and $(7/3, 49/9)$. The values of f at the two points are

$$f(-1, 1) = 4 \qquad \text{and} \qquad f(7/3, 49/9) = -98/27.$$

These represent local extreme values of f.

This problem deals with functions that are sufficiently simple that we could reduce it to a single variable problem right from the start. Because we restrict the function to points (x, y) for which $y = x^2$, we can express the function f in terms of x only as

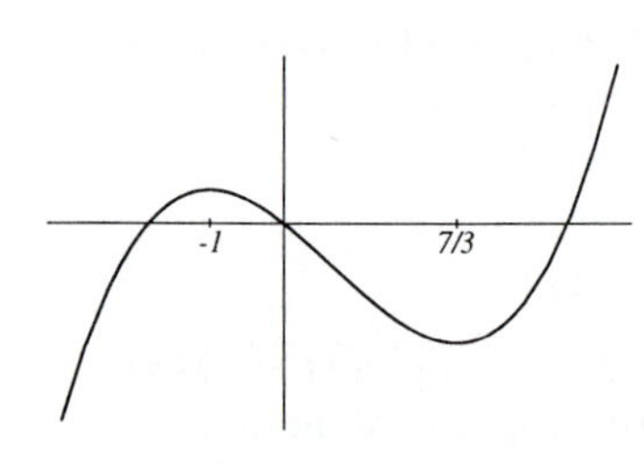

FIGURE 14.22
$y = g(x)$

$$f(x, x^2) = g(x) = xx^2 - 7x - 2x^2 = x^3 - 2x^2 - 7x.$$

We want the local maximum and minimum values of this function. If the derivative is set equal to 0 we get exactly the same quadratic equation as in the preceding method. The graph of $g(x)$ is shown in figure 14.22. ■

EXAMPLE 3 What is the rectangle of maximum area that can be inscribed in the ellipse $b^2x^2 + a^2y^2 = a^2b^2$ with its sides parallel to the coordinate axes?

Solution: Let (u, v) be the coordinates in the first quadrant of one vertex of a rectangle inscribed in the ellipse. The four vertices of the rectangle are $(\pm u, \pm v)$, the area of the rectangle is $A(u, v) = 4uv$ and the variables are constrained so the point lies on the ellipse; thus $b^2u^2 + a^2v^2 = a^2b^2$ for $a > 0$ and $b > 0$. Let us use a trigonometric parametrization of the ellipse so that it is given by

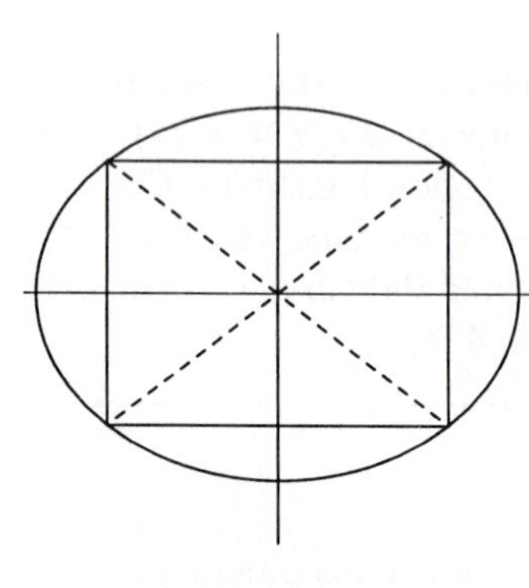
FIGURE 14.23
Maximum rectangle

$$\mathbf{r}(t) = a\cos t\,\mathbf{i} + b\sin t\,\mathbf{j}.$$

The gradient of A is $\nabla A(u, v) = 4v\,\mathbf{i} + 4u\,\mathbf{j}$ and at a point on the curve we have

$$\nabla A(a\cos t, b\sin t) = 4b\sin t\,\mathbf{i} + 4a\cos t\,\mathbf{j}.$$

At an extremum of A on the curve, this gradient is orthogonal to the tangent $\mathbf{r}'(t)$. This gives

$$(-a\sin t\,\mathbf{i} + b\cos t\,\mathbf{j}) \cdot (4b\sin t\,\mathbf{i} + 4a\cos t\,\mathbf{j}) = 4ab(\cos^2 t - \sin^2 t) = 0.$$

Using the identity $\cos^2 t - \sin^2 t = \cos 2t$ we find the solutions of this equation to be

$$t = \frac{\pi}{4}, \frac{3\pi}{4}, \frac{5\pi}{4}, \quad \text{and} \quad \frac{7\pi}{4}$$

for t on the interval $0 \le t \le 2\pi$. These four solutions correspond to the four vertices of the same rectangle. The extremal rectangle has one vertex at $(u, v) = (a/\sqrt{2}, b/\sqrt{2})$ corresponding to $t = \pi/4$. The largest area of a rectangle inscribed in the ellipse is $A_{\text{max}} = 2ab$.

It is interesting to notice that no matter what the shape of the ellipse (that is no matter what the values of a and b), the angle made by the diagonal of the maximum rectangle with respect to the x-axis is $\pi/4$. ■

Lagrange Multipliers

We give a formulation of the previous technique for finding the extrema of a function subject to a side condition that has generalizations to functions of more than two variables and even to more than one side condition. We consider only the simplest situation, however.

THEOREM 14.16

Lagrange Multiplier Method

Suppose the functions $f(x, y)$ and $g(x, y)$ are differentiable and $\nabla g(x, y)$ is continuous in a rectangular region containing (a, b) in its interior. Suppose that (a, b) is a point where f has an extreme value subject to the constraint $g(x, y) = 0$. Then $\nabla f(a, b)$ and $\nabla g(a, b)$ are parallel. In particular, if $\nabla g(a, b) \neq 0$, then there is a scalar λ such that $\nabla f(a, b) = \lambda g(a, b)$.

Proof

If $\nabla g(a, b) = 0$ then $\nabla g(a, b)$ is parallel to every vector so there is nothing more to prove. Assume then that $\nabla g(a, b) \neq 0$ and let $c = g(a, b)$. We have regarded an equation of the form $g(x, y) = c$ as defining a curve (either x is a function of y or y is a function of x). The condition that $\nabla g(a, b) \neq 0$ is enough to ensure that this is correct. More precisely at least one of the partial derivatives g_x or g_y is nonzero at (a, b). Let us assume (just to be specific) that $g_y(a, b) \neq 0$. Then there is a rectangular region R_0 containing (a, b) defined by inequalities $u < x < v$ and $p < y < q$ and a function $y = M(x)$ such that for all x satisfying $u < x < v$, we have $g(x, M(x)) = c$. In other words, g does define y as a function of x at least on some part of the region originally under consideration. Moreover $M(x)$ is differentiable at $x = a$ with derivative given by $M'(a) = -g_x(a, b)/g_y(a, b)$. Thus we have the situation considered in the previous paragraph. Namely we have an extreme value of the function f at a point (a, b) of the curve $y = M(x)$. Thus the gradient of f is orthogonal to the tangent to the curve. The gradient $\nabla g(a, b)$ is also orthogonal to the level curve $g(x, y) = c = g(a, b)$. Thus both $\nabla f(a, b)$ and $\nabla g(a, b)$ are orthogonal to the tangent and so $\nabla f(a, b)$ and $\nabla g(a, b)$ are parallel. Given two parallel vectors, one is a scalar multiple of the other (if they are nonzero). Hence there is a λ such that $\nabla f(a, b) = \lambda g(a, b)$, as we wished to prove. □

One can restate the extremum problem as follows. The points where $f(x, y)$ has an extreme value subject to the side condition $g(x, y) = 0$ are among the points where either $\nabla g(x, y) = 0$ or $\nabla f(x, y) = \lambda \nabla g(x, y)$ for some scalar λ. The scalar λ that appears in this problem is called *a Lagrange multiplier* for the problem.

Lagrange Multipliers with Three Variables

One advantage of stating the theorem about extrema with side conditions in the form of Lagrange multipliers is that the same statement is valid for functions of three variables. Thus suppose we want to find the extreme values of a function $f(x, y, z)$ subject to the side condition that $g(x, y, z) = 0$. If f and g satisfy the appropriate differentiability conditions then the points where f assumes its extreme values are among the points for which either $\nabla g = 0$ or $\nabla f = \lambda \nabla g$ for some scalar λ. We do not give a proof of this extended version but simply illustrate it with an example.

EXAMPLE 4 Find the extreme values of the function $f(x, y, z) = 2x - 3y + z$ subject to the condition $g(x, y, z) = x^2 + y^2 + z^2 - 4 = 0$. Thus we are asking for the extreme values of f taken over all points of the sphere of radius 2 with center at the origin.

***Solution*:** We use the method of Lagrange. We note that

$$\nabla g(x, y, z) = 2x\,\mathbf{i} + 2y\,\mathbf{j} + 2z\,\mathbf{k}$$

and this is never zero on the sphere. According to Lagrange, the extreme values occur at the points on the sphere where $\nabla f = \lambda \nabla g$. This becomes, in our case,

$$\lambda(2x\,\mathbf{i} + 2y\,\mathbf{j} + 2z\,\mathbf{k}) = 2\,\mathbf{i} - 3\,\mathbf{j} + \mathbf{k}.$$

We conclude from this that

$$x\,\mathbf{i} + y\,\mathbf{j} + z\,\mathbf{k} = (2\,\mathbf{i} - 3\,\mathbf{j} + \mathbf{k})/2\lambda,$$

or

$$x = \frac{1}{\lambda}, \quad y = -\frac{3}{2\lambda}, \quad z = \frac{1}{2\lambda}.$$

In addition (x, y, z) must be a point on the sphere (that is $g(x, y, z) = 0$) so that

$$4 = x^2 + y^2 + z^2 = \frac{4 + 9 + 1}{4\lambda^2}.$$

We conclude that $\lambda = \pm\sqrt{7/8}$ and the two extreme values of f occur at the points $\pm(\sqrt{8}/\sqrt{7}, -3\sqrt{2}/\sqrt{7}, \sqrt{2}/\sqrt{7})$ and at these points the value of f is $\pm 14\sqrt{2}/\sqrt{7}$.

This problem can be given a purely geometric solution in the same manner as was done with the lines and ellipses earlier. The equation $g(x, y, z) = 0$ is satisfied by the points on the sphere of radius 2 with center at the origin. The equation $f(x, y, z) = c$ is satisfied by points in a plane. For c sufficiently large, the plane has no points in common with the sphere; as c decreases, the first contact of the plane $f = c$ with the sphere occurs with the plane as a tangent. The vector $2\mathbf{i} - 3\mathbf{j} + \mathbf{k}$ is normal to the plane and this is parallel to the vector from the origin to the point of tangency of the plane and the sphere. The sphere has radius 2 so the vector $2(2\mathbf{i} - 3\mathbf{j} + \mathbf{k})/\sqrt{14}$ of length 2 and its negative are the position vectors to the points of tangency. ■

Exercises 14.10

1. Find the minimum value of $f(x, y) = 4x - 3y - 6$ subject to the condition that $x^2 + y^2 = 1$.

2. Find the minimum value of $f(x, y) = x^2 + y^2$ subject to the condition that $4x - 3y = 6$.

3. Find the extreme values of $f(x, y) = 4x + 7y$ subject to the condition that $3x^2 + 4y^2 = 7$.

4. Find the minimum value of $f(x, y) = 3x^2 + 4y^2$ subject to the condition that $4x + 7y = 11$.

5. Find the minimum value of $f(x, y, z) = 3x^2 + 2y^2 + 4z^2$ subject to the condition that $2x + 4y - 6z = -5$.

6. Find the dimensions of the rectangular parallelepiped of maximum volume that has edges parallel to the coordinate axes and that can be inscribed in the ellipsoid $(x/a)^2 + (y/b)^2 + (z/c)^2 = 1$.

7. Find the minimum distance from $(4, 4, 2)$ to a point on the cone with equation $4x^2 + 4y^2 = z^2$.

8. Find the minimum distance from the origin to a point of the ellipsoid with equation $(x - 1)^2 + (y - 3)^2 + z^2 = 16$.

9. Find the smallest and largest sum $x^4 + y^4$ if $x^2 + y^2 = 1$.

10. Find the smallest and largest product xy if $x^2 + y^2 = 1$.

11. Let $F(x, y, z) = 0$ be the equation of a surface and let $Q = (a, b, c)$ be a point not on the surface. Interpret the method of Lagrange multipliers to argue that the minimum distance from Q to a point in the surface is measured along a normal to the surface.

Let **u** and **v** be orthogonal unit vectors. The graph of the vector equation

$$\mathbf{r}(t) = a \cos t\mathbf{u} + b \sin t\mathbf{v} \qquad \mathbf{(14.15)}$$

is an ellipse with center at the origin and major and minor axes along the directions of **u** and **v**. The rectangular equation of this ellipse can be put in the form $Ax^2 + Bxy + Cy^2 = D$ for some constants A, B, C, D. If we start with this equation in rectangular coordinates, the directions of **u** and **v** can be determined, because the two farthest points on the curve from the origin lie in the direction of either **u** or **v** and the nearest points on the curve to the origin lie along the line through the origin containing the other unit vector. Carry out this idea for each of the following rectangular equations of ellipses and describe the curves in vector form as in Equation (14.15) for suitable **u**, **v** and constants a and b.

12. $5x^2 + 8xy + 4y^2 = 36$

13. $41x^2 - 24xy + 34y^2 = 2$

14.11 Properties of continuous functions

In this rather theoretical section we prove a number of facts about continuous functions: continuity implies uniform continuity on closed rectangles, the implicit and inverse function theorems.

We have seen in chapter 5 that continuity of a function of one variable implies the uniform continuity of the function over a closed interval. We do the two-variable case of this theorem for function continuous on rectangular regions. It follows from this that the function is bounded and attains its minimum and maximum values.

Uniform Continuity

Throughout this section R denotes a rectangular region consisting of all points (x, y) satisfying $a \leq x \leq b$ and $c \leq y \leq d$. For brevity we write

$$R = [a, b; c, d]$$

as a short notation for the rectangle.

Our first goal is to prove that a function that is continuous in R is uniformly continuous in R. The definition of uniform continuity is essentially the same as given in chapter 5 for functions of one variable but we give it again for completeness.

DEFINITION 14.6

Uniform Continuity

The function f is uniformly continuous in R if, for every $\epsilon > 0$, there is a

$\delta > 0$ such that whenever (x, y) and (u, v) are points in R with

$$(x-u)^2 + (y-v)^2 < \delta^2,$$

then

$$|f(x, y) - f(u, v)| < \epsilon.$$

In slightly different words, f is uniformly continuous in R if for every given positive number ϵ, there is a positive number δ with the property that whenever the distance between two points of R is less than δ, then the function values at these two points are within ϵ of each other. This is a somewhat stronger requirement than continuity because continuity is a condition that depends on first selecting a point and then producing the particular δ for the given ϵ. The δ might change as we move from point to point (even with the same ϵ); uniform continuity implies that one δ can be selected that works for all points in R.

The first step in proving that continuity implies uniform continuity is to show a slightly weaker result about the existence of a partition with special properties. Let us set up a notation for partitions of R.

Starting from partitions of the intervals $[a, b]$ and $[c, d]$ such as

$$P_1 : a = x_0 < x_1 < \ldots < x_m = b, \qquad P_2 : a = y_0 < y_1 < \ldots < y_n = b$$

we denote by

$$P : [x_0, x_1, \ldots, x_m; y_0, y_1, \ldots, y_n]$$

the partition of R into subrectangles $R_{ij} = [x_{i-1}, x_i; y_{j-1}, y_j]$. The *norm of the partition* is the largest of the lengths of any side of an R_{ij}; thus the norm of P is the maximum of the lengths

$$x_1 - x_0, x_2 - x_1, \ldots, x_m - x_{m-1}, y_1 - y_0, y_2 - y_1, \ldots, y_n - y_{n-1}.$$

THEOREM 14.17

Let f be continuous in the rectangular region $R = [a, b; c, d]$ and let ϵ be any given positive number. Then there is a partition P of R with the property that whenever (x, y) and (u, v) are points in the same subrectangle,

$$|f(x, y) - f(u, v)| < \epsilon. \qquad \textbf{(14.16)}$$

Proof

We argue by assuming that no such partition exists and then drawing some conclusions from this assumption. Eventually we reach a statement that logically follows from the supposition but is inconsistent with the continuity of f. It follows that our supposition is false and hence the theorem is true.

Suppose that no partition of R exists having the property stated in the theorem for a particular positive number ϵ. Divide R into four subrectangles according to the partition

$$\left[a, \frac{a+b}{2}, b; c, \frac{c+d}{2}, d\right].$$

Now we argue that the theorem cannot be true for each of the four rectangles with the same given ϵ. If it were possible to partition each of the four subrectangles such that inequality (14.16) is true for any two points in same subrectangle of the partition, then we would use those partitions to obtain a partition of R for which the statement of the theorem is true. No such partition of R exists (by assumption) so at least one of the four subrectangles also fails to have a partition with the required property. This is the key step because we now have the theorem failing on a rectangle with sides half the length of the original and we can repeat this to further divide that rectangle into four subrectangles and the theorem must fail on one of them.

We obtain a decreasing sequence of subrectangles $S_k = [a_k, b_k; c_k, d_k]$

$$S_1 \supset S_2 \supset \cdots \supset S_k \supset \cdots$$

each having the property that there is no partition such that inequality (14.16) holds for every pair of points on the same subinterval. In addition have the sides of S_k decreasing by a factor of one-half at each step because

$$b_k - a_k = \frac{b-a}{2^k}, \qquad d_k - c_k = \frac{d-c}{2^k}.$$

Now consider the sequence of left end points

$$a_1 \leq a_2 \leq \cdots \leq a_k \leq \cdots.$$

Every a_k is less than b so $\{a_k\}$ is a bounded nondecreasing sequence. Such a sequence has a least upper bound so there is a smallest number A with $a_k \leq A$ for every k. Because b is an upper bound for the sequence and A is a least upper bound, it follows that $A \leq b$. Similarly there is a smallest number C with $c_k \leq C \leq d$ for all k. The point (A, C) is in R so f is continuous at (A, C). For the ϵ that was given at the start, continuity implies that there is some positive number γ such that

$$|f(A, C) - f(x, y)| < \epsilon/2 \quad \text{provided} \quad (A - x)^2 + (C - y)^2 < \gamma^2.$$

The rectangles $S_k = [a_k, b_k; c_k, d_k]$ have their lower left corners getting close to (A, C) and the side lengths decreasing by a factor of one-half as k increases. For sufficiently large k, the entire rectangle S_k lies inside the circle with center at (A, C) and radius γ. Now we estimate the difference in function values at two points of S_k. Let (x, y) and (u, v) be two points in S_k. Then the two points have distance from (A, C) less than γ so we have

$$\begin{aligned} |f(x, y) - f(u, v)| &= |f(x, y) - f(A, C) + f(A, C) - f(u, v)| \\ &\leq |f(x, y) - f(A, C)| + |f(A, C) - f(u, v)| \\ &< \frac{\epsilon}{2} + \frac{\epsilon}{2} = \epsilon. \end{aligned}$$

Thus we have found a partition of S_k, namely the trivial partition having only one subrectangle, for which the statement of the theorem is true. We had supposed that S_k was selected so that no such partition was possible. Thus our original assumption is incorrect and the theorem holds. □

Now we improve the statement of the previous theorem by removing the reference to a specific partition of R and prove that continuity implies uniform continuity.

THEOREM 14.18

Continuity on *R* Implies Uniform Continuity

If f is a continuous function on the rectangle $R = [a, b; c, d]$ then f is uniformly continuous on R.

Proof

Let ϵ be a positive number. We must prove there is a positive number δ such whenever the distance between two points of R is less than δ then the difference in the values of f at these two points is less than ϵ.

Use Theorem 14.17 with $\epsilon/2$ in place of ϵ to obtain a partition P of R such that inequality (14.16) holds whenever (x, y) and (u, v) are in the same subrectangle of the partition. Now let δ be the norm of P and let (x, y) and (u, v) be two points in R that are at a distance less than δ apart. Then one possibility is that (x, y) and (u, v) lie in a subrectangle of P. In this case the difference in the function values at (x, y) and (u, v) is at most $\epsilon/2$ by the choice of P. In particular inequality (14.16) holds.

If (x, y) and (u, v) are not in the same subrectangle of P, then they must lie in adjacent subrectangles, that is subrectangles having a common point on the boundary. This is so because (x, y) and (u, v) are less than δ apart and all sides of subrectangles of the partition P have length at most δ. Let (p, q) be a point on the straight line connecting (x, y) and (u, v) and also on the boundary of one of the subrectangles of P. Then the pair (x, y) and (p, q) lie in one subrectangle of P and the pair (p, q) and (u, v) lies in a subrectangle. Thus we have

$$\begin{aligned}|f(x, y) - f(u, v)| &= |f(x, y) - f(p, q) + f(p, q) - f(u, v)| \\ &\le |f(x, y) - f(p, q)| + |f(p, q) - f(u, v)| \\ &< \frac{\epsilon}{2} + \frac{\epsilon}{2} = \epsilon.\end{aligned}$$

This proves the theorem. □

Uniform continuity of continuous functions holds for regions more complicated than rectangles. The critical condition that a region should satisfy in order that the analogue of Theorem 14.18 hold is that the region be closed and bounded. "Closed" means roughly that the region includes its boundary and "bounded" means that the region is contained in some sufficiently large circle in the plane. We do not attempt to give more general results but we make use of the results for regions that can be closely approximated by sets of sufficiently small rectangles like the region bounded by an ellipse or other smooth curve.

There are two important properties of uniformly continuous functions that we use often—they are bounded and attain their maximum and minimum values. We prove these only for rectangular regions.

THEOREM 14.19

Uniform Continuity on *R* Implies Bounded

Let f be continuous in a rectangular region R. Then f is bounded on R.

Proof

Because f is continuous on R, it is uniformly continuous. By uniform continuity, there is a number δ such that $|f(x, y) - f(u, v)| < 1$ whenever the distance between (x, y) and (u, v) is less than δ (use $\epsilon = 1$ in the definition of uniform continuity).

Now fix a point (x_0, y_0). We know something about how the values of f change from one point (x_0, y_0) to another point less than δ units away. For any point (p, q) in R, we can estimate how much the function changes from (x_0, y_0) to (p, q) by finding the number of steps of length at most δ are required to walk from (x_0, y_0) to (p, q). Construct the straight line path between (x_0, y_0) and (p, q) that lies entirely in R. Since R is a rectangle, this is possible. The length of the path connecting two points of R has length at most D, the length of the diagonal of R. The straight-line path connecting (x_0, y_0) and (p, q) has length no more than D and is composed of no more than D/δ segments of length at most δ. The most that a value of f changes from one end of a segment to the other is 1 and so the greatest possible change in f from (x_0, y_0) to (p, q) is D/δ. We have then

$$|f(x_0, y_0) - f(p, q)| < \frac{D}{\delta},$$

which implies

$$\frac{D}{\delta} - f(x_0, y_0) < f(p, q) < \frac{D}{\delta} + f(x_0, y_0).$$

Inasmuch as (p, q) was a completely arbitrary point in R and (x_0, y_0) is a fixed point, this proves that f is bounded on R. □

Finally we use this boundedness result to prove that f attains its maximum and minimum values on rectangular regions.

THEOREM 14.20

Let f be continuous on the rectangle R. Then there are points $(x_{\min}, y_{\min})$ and $(x_{\max}, y_{\max})$ in R satisfying

$$f(x_{\min}, y_{\min}) \leq f(x, y) < f(x_{\max}, y_{\max})$$

for every point (x, y) in R. In other words, f takes on its maximum and minimum values.

Proof

Theorem 14.19 implies that f is bounded so, by the Completeness Property of the Real Numbers, there is a smallest number M such that $f(x, y) \leq M$ for all (x, y) in R. We obtain the right half of the inequality in the conclusion of the theorem if we show that M is actually a value of f at some point of R. Let us suppose that this is not the case. In other words assume that $f(x, y) < M$ for all (x, y) in R. Then the function

$$g(x, y) = \frac{1}{M - f(x, y)}$$

is continuous in R. But a continuous function in R is uniformly continuous and hence bounded. Thus there is some number K such that

$$g(x, y) = \frac{1}{M - f(x, y)} \leq K$$

for all (x, y) in R. Using the fact that $M - f(x, y) > 0$ we conclude from this inequality that

$$f(x, y) \leq M - \frac{1}{K}.$$

This statement conflicts with the assumption that M was the least upper bound for the values of f because we have just produced a smaller upper bound, namely $M - 1/K$. The conflict arises because we made the assumption that $M - f(x, y)$ was never equal to 0 and so this assumption cannot stand. It follows that there is some point $(x_{\max}, y_{\max})$ in R satisfying $M = f(x_{\max}, y_{\max})$ and $f(x, y) < f(x_{\max}, y_{\max})$. The proof of the left hand inequality in the conclusion of the theorem is carried out in the analogous way and is left to the reader. □

In the case of functions of one variable, we saw that a function continuous on the closed interval $[a, b]$ must be uniformly continuous but a function that is continuous on a nonclosed interval need not be uniformly continuous. The situation is analogous for functions of two variables.

EXAMPLE 1 Show that the function $f(x, y) = x/y$ is continuous but not uniformly continuous in the region S consisting of all (x, y) satisfying $0 \leq x \leq 2$ and $0 < y \leq 2$.

Solution: The only points where f fails to be continuous are those (x, y) with $y = 0$. Thus f is continuous in S. By using only the definition we show that f is not uniformly continuous in S. Take $\epsilon = 1$. We show there are points in S as close together as you like but where the function values differ by more than 1. Take the point $p_k = (1, 10^{-k})$ for some positive integer k. Then p_k is in S and $f(p_k) = 10^k$. Let δ be any positive number. Select positive integers $k < h$ so that $10^{-k} < \delta$ and $10^{-h} < \delta$. Then the numbers 10^{-k} and 10^{-h} differ by less than δ but the function values $f(p_k) = 10^k$ and $f(p_h) = 10^h$ differ by much more than $\epsilon = 1$. Hence for this ϵ it is not possible to find a δ that meets the conditions required in the definition of uniform continuity. ■

Implicit Function Theorem

Now we take up the the implicit function theorem in the simplest case. It is a case that has been used several times in the text. We refer to this as the simplest case because there are versions for functions of more than two variables but we do not prove these generalizations. In the theorem, we consider a function $g(x, y)$ and the possibility that the equation $g(x, y) = 0$ implicitly defines y as a function of x. We prove that, under suitable restrictions, there is such a function that defines a curve passing through any given point (a, b) for which $g(a, b) = 0$.

To realize the necessity of some of the maneuvers in the following theorem, we consider a simple example. Let $g(x, y) = x^2 + y^2 - 1$ so the graph of $g(x, y) = 0$ is a circle of radius 1 with center at the origin. There is no function $y = M(x)$ whose graph is the entire circle. If we restrict x to the interval $-1 \leq x \leq 1$ and y to the interval $0 \leq y \leq 1$, then there is a function $y = M(x) = \sqrt{1 - x^2}$ such that $g(x, M(x)) = 0$ for all x on $-1 \leq x \leq 1$. The derivative of M exists on the open interval $-1 < x < 1$ and can be computed as $-g_x/g_y$ since $g_y \neq 0$ at the points where $-1 < x < 1$. For a general g, we know much less about the graph and the points where the derivative exists but the assumption of continuity ensures the existence of intervals with similar properties.

THEOREM 14.21

Implicit Function Theorem

Let $g(x, y)$ be a function such that g, g_x and g_y are continuous in some rectangular region R that contains the point (a, b) in its interior. Assume that $g(a, b) = 0$ and $g_y(a, b) \neq 0$. Then

i. there are numbers u, v, p and q such that $u < a < v$ and $p < b < q$ and a function $M(x)$ defined on the interval $u < x < v$ such that $y = M(x)$ is the unique number on the interval $p < y < q$ satisfying $g(x, y) = 0$;

ii. the function M and its derivative dM/dx are continuous on the interval $u < x < v$.

Proof

Because $g_y(a, b) \neq 0$ and g_y is continuous, there must be a square (possibly very small) $\{(x, y) : |x - a| < h,\ |y - b| < h\}$ contained in R on which g_y does not change sign. We assume that g_y is positive on the square. If it were negative, we could replace g with $-g$ and apply the same arguments as those about to be given.

For each x on the interval $a - h < x < a + h$, the function $g(x, y)$ of y has a positive derivative and so is an increasing function of y. Because $g(a, b) = 0$, we must have

$$g(a, b + h) > 0 \qquad \text{and} \qquad g(a, b - h) < 0.$$

Now we use the continuity twice to get a possibly smaller interval around a. There is some positive δ_1 such that $g(x, a + h) > 0$ for $|x - a| < \delta_1$ and similarly a positive δ_2 such that $g(x, a - h) < 0$ for $|x - a| < \delta_2$. If δ is the smaller of δ_1 and δ_2 then we have

$$g(x, b + h) > 0 \quad \text{and} \quad g(x, b - h) < 0, \quad \text{for } x \text{ on } \quad (a - \delta, a + \delta).$$

Now we have the intervals we want. Let $u = a - \delta$, $v = a + \delta$, $p = b - h$ and $q = b + h$. For any x satisfying $a - \delta < x < a + \delta$, the function $g(x, y)$ is an increasing function of y with $g(x, b - h) < 0$ and $g(x, b + h) > 0$. Thus there is a y, depending on x, such that $g(x, y) = 0$ by the Intermediate Value Theorem for functions of one variable. Moreover because $g(x, y)$ is an increasing function of y, there is only one such y for the given x. Let us denote this unique y by $y = M(x)$. Now clearly we have $g(x, M(x)) = 0$. Thus we have the existence of the function M defined for x on some open interval containing a. Notice that $u < M(x) < v$ for every x on I because y was restricted to lie on this interval.

Now we must show that M is continuous and differentiable and that the derivative $M'(x)$ is continuous on the interval $I = (u, v)$. Let x_1 be a point on I. To show that M is continuous at x_1 we must prove that for any positive ϵ there is some positive δ such that $|M(x) - M(x_1)| < \epsilon$ whenever $|x - x_1| < \delta$. We apply the argument made at the beginning of the proof to construct a suitable interval. Replace ϵ with a smaller value if necessary to ensure that $x_1 \pm \epsilon$ are both on I. Let S_ϵ be the square with center $(x_1, M(x_1))$ and sides of length 2ϵ. Then $g(x_1, M(x_1)) = 0$ so by the reasoning applied above, there is a number $\delta > 0$ such that

$$g(x, M(x_1) + \epsilon) > 0 \qquad \text{and} \qquad g(x, M(x_1) - \epsilon) < 0$$

for x on $(x_1 - \delta, x_1 + \delta)$. It follows then for any x satisfying $|x - x_1| < \delta$ the unique y satisfying $g(x, y) = 0$ must lie between $M(x_1) - \epsilon$ and $M(x_1) + \epsilon$. Because this unique y equals $M(x)$, we have $|M(x) - M(x_1)| < \epsilon$ whenever $|x - x_1| < \delta$. Hence M is continuous on I.

To show that M' is continuous we use the Linear Approximation Theorem for g, which asserts that

$$g(x_1 + k_1, y_1 + k_2) - g(x_1, y_1) = g_x(x_1, y_1)k_1 + g_y(x_1, y_1)k_2 + G_1(k_1, k_2)k_1 + G_2(k_1, k_2)k_2$$

where G_1 and G_2 approach 0 as k_1 and k_2 approach 0. Select any number s such that both $x_1 \pm s$ are on I and write $M(x_1 + s) = M(x_1) + \Delta$. For use in just a few lines we note that $\lim_{s \to 0} \Delta = 0$ because M is continuous. We have

$$g(x_1 + s, M(x_1 + s)) = 0 \quad \text{and} \quad g(x_1, M(x_1)) = 0$$

by definition of M. By the LAT, the difference (which equals 0)

$$g(x_1 + s, M(x_1 + s)) - g(x_1, M(x_1)) = g(x_1 + s, M(x_1) + \Delta) - g(x_1, M(x_1))$$

can be expressed as

$$0 = g_x(x_1, M(x_1))s + g_y(x_1, M(x_1))\Delta + G_1(s, \Delta)s + G_2(s, \Delta)\Delta.$$

Save this for a moment and turn to the main issue, that of proving M' exists. The quotient for the derivative of M is

$$\frac{M(x_1 + s) - M(x_1)}{s} = \frac{\Delta}{s} = -\frac{g_x(x_1, M(x_1)) + G_1(s, \Delta)}{g_y(x_1, M(x_1)) + G_2(s, \Delta)}.$$

Now let s approach 0; the terms G_1 and G_2 approach 0, (because $\Delta \to 0$) and the left-hand side approaches $M'(x_1)$, so we have

$$M'(x_1) = -\frac{g_x(x_1, M(x_1))}{g_y(x_1, M(x_1))}.$$

This shows the existence of the derivative $M'(x_1)$ and also that the derivative is continuous because g_x and g_y are continuous by assumption. The proof is complete. □

The hypothesis of the Implicit Function Theorem is not symmetric and neither is the conclusion; there is some favoritism shown for the variable y. We can balance this lack of symmetry by pointing out an alternate conclusion, given an alternate hypothesis. If we keep all assumptions in the theorem except we replace the condition $g_y(a, b) \neq 0$ with the assumption that $g_x(a, b) \neq 0$ then we conclude that there is some function $x = N(y)$ with a continuous derivative such that $g(N(y), y) = 0$ for y on a sufficiently small interval.

To make the hypothesis completely symmetric we could assume $\nabla g(a, b) \neq 0$ and then conclude that either $y = M(x)$ or $x = N(y)$ with the appropriate differentiability conclusions about M or N.

EXAMPLE 2 Show that the relation $g(x, y) = y^3 + x^2 + y - x + 5x^4y = 0$ defines y as a function of x over some open interval with a at its center, for any a. Similarly show that x is defined as a function of y for some small intervals around "most" points.

***Solution*:** Because g is a polynomial, g, g_x and g_y are continuous everywhere. The partial derivative $g_y(x, y) = 3y^2 + 5x^4 + 1$ is positive at every point. Thus for any solution $g(a, b) = 0$, there is an interval $u < a < v$ and a function $M(x)$ such that $y = M(x)$ satisfies $g(x, y) = 0$ for all x with $u < x < v$.

The partial derivative $g_x(x, y) = 2x - 1 + 20x^3y$ takes on both positive and negative values. To invoke the Implicit Function Theorem, we must be near points where g_x is nonzero. The points where $g_x = 0$ form a curve $y = (1 - 2x)/20x^3$. If (a, b) is not on this curve and if $g(a, b) = 0$ then there is an open interval $I = (p, q)$ containing b and a differentiable function N such that $x = N(y)$ and $g(N(y), y) = 0$ for all y on I.

Note that in this example the functions, which are guaranteed to exist by the Implicit Function Theorem, are difficult to write down. There is no easy way to solve the equation

$$y^3 + x^2 + y - x + 5x^4y = 0$$

for x in terms of y or y in terms of x. ■

Inverse Functions

Suppose we consider an equation of the form $y = x^5 - 7x^3 - 1$ or the related function $g(x, y) = y - x^5 - 7x^3 + 1$. Then g and its partial derivatives are continuous and $g_y = 1$ at every point. Thus y can be expressed as a function of x. But we knew that because we started with y expressed as a function of x. That is not very interesting. What about the other variable? We have $g_x(x, y) = 5x^4 - 21x^2$, which equals 0 only at the three points $x = 0$ and $x = \pm\sqrt{21/5}$. Hence we can express x as a function of y, at least on certain intervals that avoid the three zeros of the derivative. That is interesting. Well at least it is something that we did not know immediately at the start. The expression of x as a function of y is given by the inverse function that we discussed in chapter 5. Of course we have to worry a little about the intervals over which the inverse function is defined. Here is the general statement that follows almost immediately from the Implicit Function Theorem.

THEOREM 14.22

Inverse Function Theorem

Let $y = f(x)$ be a continuous function for which the derivative $f'(x)$ is also continuous in an interval (u, v) containing a. Let $f(a) = b$ and suppose that $f'(a) \neq 0$. Then there is an open interval I containing b and an inverse function f^{-1} of f defined on I such that $f(f^{-1}(y)) = y$ for all y on I. Moreover f^{-1} is differentiable on I with derivative given by

$$\frac{d}{dy}f^{-1}(y) = \frac{1}{f'(x)},$$

where $y = f(x)$.

Proof

We apply the Implicit Function Theorem to the function $g(x, y) = y - f(x)$. The hypothesis is satisfied because g, $g_x = -f'(x)$ and $g_y = 1$ are continuous and $g_x(a, b) = -f'(a) \neq 0$. Because $g(a, b) = 0$, there is a differentiable function $x = N(y)$ such that $g(N(y), y) = y - f(N(y)) = 0$ for all y on some open interval containing b. Thus $f(N(y)) = y$. The derivative of N is determined from the Implicit Function Theorem (with the roles of x and y

reversed) as

$$N'(y) = -\frac{g_y(x,y)}{g_x(x,y)} = -\frac{1}{-f'(x)} = \frac{1}{f'(x)}$$

where (x, y) satisfy $g(x, y) = 0$; that is $y = f(x)$. Thus N is the inverse f^{-1} of f on a suitable interval. □

The Inverse Function Theorem as presented here is really not very deep. If f has a continuous derivative on an open interval containing b and $f'(b) \neq 0$, then there is a small interval containing b on which f is differentiable and either increasing or decreasing. It is easy to see that an increasing function (or decreasing function) has an inverse so we have, in either case, that there is an inverse of f that is differentiable on the interval containing b. The version of the Implicit Function Theorem for functions of several variables is not so easy. We give the statement without proof.

THEOREM 14.23

Implicit Functions of Two Variables

Let $G(x, y, z)$ be a continuous and differentiable function on a region R of the form

$$|x - a| < h, \qquad |y - b| < k, \qquad |z - c| < p,$$

for some point (a, b, c). Assume further that the first partial derivatives of G are continuous in R and that $G(a, b, c) = 0$ but $G_z(a, b, c) \neq 0$. Then there are positive numbers H, K, P such that, for any x and y in the region S defined by $|x - a| < H$ and $|y - b| < K$, there is a unique z, denoted by $z = M(x, y)$, satisfying $|z - c| < P$ such that $G(x, y, z) = G(x, y, M(x, y)) = 0$. The function M has continuous first partial derivatives M_x and M_y in the region S.

EXAMPLE 3 Let $G(x, y, z) = x^3 + y^3 + z^3 - 3xyz$ so that $G(0, 1, -1) = 0$ and $G_z(0, 1, -1) = 3 \neq 0$. By the Implicit Function Theorem, there is a rectangular region S around the point $(0, 1)$ and a function $z = M(x, y)$ such that $G(x, y, M(x, y)) = 0$ for all (x, y) in S and the first partial derivatives of M are continuous in S. Compute a general formula for $M_x(x, y)$ and the value $M_x(0, 1)$.

Solution: We set $z = M(x, y)$ and then differentiate both sides of the equation $G(x, y, z) = 0$ with respect to x to get

$$\begin{aligned} 0 &= G_x \frac{\partial x}{\partial x} + G_y \frac{\partial y}{\partial x} + G_z \frac{\partial z}{\partial x} \\ &= (3x^2 - 3yz) \cdot 1 + (3z^2 - 3xy)\frac{\partial z}{\partial x}. \end{aligned}$$

Remember that $z = M(x, y)$, so we solve for M_x to get

$$M_x = \frac{\partial z}{\partial x} = -\frac{(3x^2 - 3yz)}{(3z^2 - 3xy)}.$$

Evaluate at $(0, 1, -1)$ to obtain $M_x(0, 1) = 1$. By a similar computation you can discover that $M_y(0, 1) = -1$. ■

Exercises 14.11

Decide if the functions f are uniformly continuous in the given region S. If the function is uniformly continuous, find its maximum and minimum values and the points of S at which these are attained.

1. $f(x, y) = \frac{1}{x+y}$, $S = \{(x, y) : 0.1 \leq x \leq 1; 0.1 \leq y \leq 1\}$

2. $f(x, y) = \frac{1}{x+y}$, $S = \{(x, y) : 0 \leq x \leq 1; 0 \leq y \leq 1\}$

3. $f(x, y) = \frac{1}{x+y}$, $S = \{(x, y) : 0 < x \leq 1; 0 < y \leq 1\}$

4. $f(x, y) = \frac{1}{x+y}$, $S = \{(x, y) : 0.1 \leq x; 0.1 \leq y\}$

5. If f is uniformly continuous in a region S of the plane, show that f is continuous at every point of S.

6. Suppose S is a *convex* region in the plane; that is if (p, q) and (r, s) are points in S then the straight line segment between these two points also lies in S. Suppose in addition that S is bounded; that is S lies inside a circle of radius r for some sufficiently large r. Prove that a function that is uniformly continuous in S is bounded in S.

In each of the following exercises, use the Implicit Function Theorem to show that some part of the graph of $g(x, y) = 0$ can be given as the graph of $y = M(x)$ for a suitable function M in a region sufficiently near the point (a, b). Compute the derivative $M'(a)$.

7. $g(x, y) = y^3 + y - x^2$, $(a, b) = (0, 0)$

8. $g(x, y) = y + x - x \sin y$, $(a, b) = (0, 0)$

9. $g(x, y) = y + xe^y - 1$, $(a, b) = (1, 0)$

10. $g(x, y) = y^3 + x^3 - 2xy$, $(a, b) = (1, 1)$

11. $g(x, y) = x - \tan y$, $(a, b) = (1, \pi/4)$

12. $g(x, y) = x - \sin y$, $(a, b) = (0, 0)$

In the following exercises, quote the Implicit Function Theorem for two variables to show that the surface defined by the equation $G(x, y, z) = 0$ can be expressed as the graph of a function $z = M(x, y)$ in a region sufficiently near the point (a, b, c). Find the partial derivatives $M_x(a, b)$ and $M_y(a, b)$.

13. $G(x, y, z) = x^3 + y^3 + z^3 - 6xyz + 2$, $(a, b, c) = (1, 2, 1)$

14. $G(x, y, z) = x^4 + y^4 + z^4 - 4x - 4y - 4z + 9$, $(a, b, c) = (1, 1, 1)$

15. $G(x, y, z) = e^{-z} - z + x^2 - 2y^2$, $(a, b, c) = (1, 1, 0)$

16. $G(x, y, z) = x + y + z - 2\cos xyz$, $(a, b, c) = (0, 1, 1)$

15 Integration in Higher Dimensions

In this chapter we take up the integration of functions of several variables. We describe a generalization of the definite integrals of functions of one variable evaluated in chapter 5 to the case of functions of two or three variables. The theory is much richer than the one-variable theory. We not only have to consider a function to be integrated but also we must give careful attention to the region over which the integration is performed. A double integral is used to determine the volume of a solid bounded below by the xy plane and above by a surface given as $z = f(x, y)$. Other applications include the computation of the mass of a plane region having variable density and the computation of moments of plane and solid regions. The concept of a line integral is introduced in this chapter. The line integral has ingredients similar to that of the integral of a function of one variable. Single-variable integration can be regarded as integration along a path that lies on the x-axis, usually an interval, whereas a line integral involves integration along a path that lies on a curve in the plane or even in three-dimensional space. The line integral is used to compute the work done in moving along a path in two or three dimensions in the presence of a variable force acting at each point of the path. Green's Theorem, which relates a double integral over a region to a line integral taken around the boundary of the region, is proved and applied to give a change of variables formula for double integrals.

15.1 Double integrals

Let R be the rectangular region of the plane bounded by the lines $x = a$, $x = b$, $y = c$ and $y = d$. For brevity we write $R = [a, b; c, d]$ to denote this region. Let f be a function that is defined and continuous in the region R. We now define the integral of f over R.

A partition of R is a decomposition of R into subrectangles of the following type. Start with partitions of the intervals $[a, b]$ and $[c, d]$ of the usual type,

$$P_1 : a = x_0 < x_1 < \cdots < x_m = b \qquad P_2 : a = y_0 < y_1 < \cdots < y_n = b,$$

and let

$$P : [x_0, x_1, \ldots, x_m; y_0, y_1, \ldots, y_n]$$

denote the partition of R into subrectangles $R_{ij} = [x_{i-1}, x_i; y_{j-1}, y_j]$. The area of R_{ij} is denoted by ΔR_{ij} and is is given by

$$\Delta R_{ij} = (x_i - x_{i-1})(y_j - y_{j-1}).$$

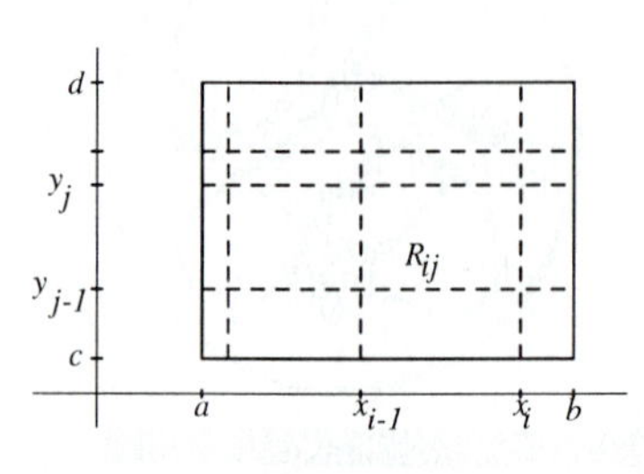

FIGURE 15.1 Subrectangle of a partition

The area A of R is $(b-a)(d-c)$ and clearly the area of R is the sum of the areas of the subrectangles; thus

$$A = \sum_{i=1}^{m} \sum_{j=1}^{n} \Delta R_{ij}.$$

We now form two sums depending on the function f and the partition P of R. First notice that because f is continuous in all of R, it is continuous in R_{ij} and thus has a maximum value M_{ij} and a minimum value m_{ij} on R_{ij}. The lower sum $L(P)$ is defined by taking the sum of the products of the area of R_{ij} times the minimum values of f on R_{ij}; the upper sum $U(P)$ is the sum of the products of the area of R_{ij} and the maximum value of f on R_{ij}. In symbols we have

$$L(P) = \sum_{i=1}^{m} \sum_{j=1}^{n} m_{ij} \Delta R_{ij} \qquad U(P) = \sum_{i=1}^{m} \sum_{j=1}^{n} M_{ij} \Delta R_{ij}.$$

From $m_{ij} \leq M_{ij}$, it follows that $L(P) \leq U(P)$. We prove that there is a unique number that is simultaneously greater than or equal to every lower sum of f over R and less than or equal to every upper sum of f over R. That number is called the *double integral of f over R*.

THEOREM 15.1

Existence of the Double Integral

Let f be a continuous function in the rectangular region R. There is a unique number that is simultaneously an upper bound for all lower sums of f over R and a lower bound for all upper sums of f over R.

Proof

We keep the notation introduced just before the statement of the theorem so that P is a particular partition of R. Because f is continuous on R, it has a minimum value m_0 on R. Thus $m_0 \leq m_{ij}$ for every i and j. It follows that

$$m_0 A = m_0 \sum_{i=1}^{m} \sum_{j=1}^{n} \Delta R_{ij} \leq \sum_{i=1}^{m} \sum_{j=1}^{n} m_{ij} \Delta R_{ij} = L(P).$$

If M_0 is the maximum value of f on R then by similar reasoning we obtain $M_{ij} \leq M_0$ and $U(P) \leq M_0 A$. By combining these inequalities with the fact that $L(P) \leq U(P)$ we find

$$m_0 A \leq L(P) \leq U(P) \leq M_0 A.$$

Notice that the terms $m_0 A$ and $M_0 A$ do not depend on the choice of the partition P so it follows that $M_0 A$ is an upper bound for the set of all lower sums and $m_0 A$ is a lower bound for the set of all upper sums. By the completeness property of the real numbers the set of all lower sums has a least upper bound, which we denote by I_{lower}. Thus $L(P) \leq I_{\text{lower}}$.

Analogously we write I_{upper} for the greatest lower bound of the set of upper sums of f over R. If ϵ is any positive number then $I_{\text{lower}} - \epsilon$ is less than the least upper bound for all lower sums so there is a partition Q'

such that

$$L(Q') > I_{\text{lower}} - \epsilon.$$

Similarly there is a partition Q'' such that

$$U(Q'') < I_{\text{upper}} + \epsilon.$$

By refining the two partitions Q' and Q'' we form a new partition Q that contains all the horizontal and vertical lines determined by either Q' or Q''. By taking a finer partition, the lower sum cannot decrease, so $L(Q') \leq L(Q)$ and the upper sum cannot increase, so $U(Q'') \geq U(Q)$. Thus we have

$$L(Q) \geq L(Q') \geq I_{\text{lower}} - \epsilon, \quad U(Q) \leq U(Q'') \leq I_{\text{upper}} + \epsilon.$$

By combining these appropriately we obtain

$$0 \leq U(Q) - L(Q) \leq I_{\text{upper}} - I_{\text{lower}} + 2\epsilon.$$

This implies that

$$I_{\text{lower}} \leq I_{\text{upper}} + 2\epsilon$$

for every positive number ϵ. It follows that

$$I_{\text{lower}} \leq I_{\text{upper}}.$$

The next goal is to prove these numbers are actually equal. We accomplish this by showing there is a partition P of R for which the difference $U(P) - L(P)$ is smaller than any given positive ϵ.

Let $\epsilon > 0$ be given. Because f is uniformly continuous on R, there is a partition P such that the difference in values of f at points in the same subrectangle is less than ϵ/A. In particular, the difference $M_{ij} - m_{ij}$ is less than ϵ/A. Then

$$\begin{aligned} U(P) - L(P) &= \sum_{i=1}^{m}\sum_{j=1}^{n}(M_{ij} - m_{ij})\Delta R_{ij} \leq \sum_{i=1}^{m}\sum_{j=1}^{n}\left(\frac{\epsilon}{A}\right)\Delta R_{ij} \\ &= \frac{\epsilon}{A}\sum_{i=1}^{m}\sum_{j=1}^{n}\Delta R_{ij} = \frac{\epsilon}{A}A = \epsilon. \end{aligned}$$

Now the inequality

$$L(P) \leq I_{\text{lower}} \leq I_{\text{upper}} \leq U(P)$$

implies

$$0 \leq I_{\text{lower}} - I_{\text{upper}} \leq \epsilon.$$

Because ϵ was an arbitrary positive number, it follows that $I_{\text{lower}} = I_{\text{upper}}$. This proves the theorem. □

Now we formally give a name to the unique number whose existence is assured by this theorem.

DEFINITION 15.1

Double Integral over Rectangular Regions

Let f be continuous in the rectangular region R. The unique number that is both an upper bound for all lower sums of f over R and a lower bound for all upper sums of f over R is called the *double integral* of f over R

and is denoted by the symbol

$$\iint_R f(x, y)\, dR.$$

For the record we repeat one of the inequalities derived above for I_{lower} and I_{upper}: If f is continuous in the rectangular region R of area A and if m_0 and M_0 are the minimum and maximum values of f in R then

$$m_0 A \leq \iint_R f(x, y)\, dR \leq M_0 A. \tag{15.1}$$

We have restricted the region R to be rectangular until now. It is natural to ask for a definition of a double integral of a function over a more general region. The definition can be extended in a rather natural way to regions that are sufficiently nice. What does "nice" mean? We do not give a precise definition. We take up this problem in the next section for regions that we call Type X and Type Y. We proceed to give a definition of a double integral over any bounded region but it is necessary to leave open the question of existence of the integral.

Let S be a bounded region of the xy plane so that S is contained in some rectangle $R = [a, b; c, d]$ bounded by the lines $x = a$, $x = b$, $y = c$ and $y = d$. For any partition $P : [x_0, x_1, \ldots, x_m; y_0, y_1, \ldots, y_n]$ of R, let $P(S)$ denote the collection of subrectangles that contain points of S. We call the set of subrectangles $P(S)$ *a partition of* S. Number the subrectangles that contain points of S in some order as R_1, R_2, ..., R_k. For any rectangle R_i in $P(S)$, let ΔR_i denote its area and let (u_i, v_i) be a point in $S \cap R_i$. The sum

$$\sum_{i=1}^{k} f(u_i, v_i)\Delta R_i$$

is called an *approximating sum* corresponding to the partition $P(S)$. The *norm* of P is the longest side of a subrectangle of P. With all this notation, we now give the definition of the double integral.

If f is a function defined on R, the double integral of f over S is a number I with the property that for any positive number ϵ there is some positive number δ such that

$$\left| I - \sum_{i=1}^{k} f(u_i, v_i)\Delta R_i \right| < \epsilon$$

whenever P is a partition with norm less than δ.

Of course we do not know that the double integral of f over S exists in general. The arguments above showed that the double integral of a continuous function over a rectangular region does exist.

We return to this point in the next section.

Volumes as Double Integrals

Let f be defined and continuous over a region S of the xy plane and suppose that $f(x, y) \geq 0$ at every point (x, y) of S. Then f and S define a region of three-dimensional space

$$V = \{(x, y, z) : (x, y) \in S,\ 0 \leq z \leq f(x, y)\},$$

which we refer to as the *solid defined by* f *over* S. We consider the problem of computing the volume of this solid. First we take the view that V does have a volume and we must find a way to compute it. For example if S is a rectangular region of the xy plane and f is a continuous function on S, then the double integral gives us the volume.

Let us give a plausible argument for this statement.

We use the approximation method that parallels the discussion of the determination of the area bounded by the x-axis and the graph of a positive function $y = g(x)$ in chapter 5. Assume that S is a bounded region. Start with a rectangle S_1 that lies entirely within S. Let m_1 be the minimum value of f on S_1. Then the parallelepiped (= box) with base S_1 and height m_1 lies completely inside the solid so the volume of the parallelepiped is at most equal to the volume of the solid. Let ΔS_1 denote the area of the rectangle S_1 and vol(V) the volume of the solid V. We have

$$m_1 \Delta S_1 \leq \text{vol}(V).$$

Instead of just one rectangle inside S, let us consider a partition $P(S)$ of S and let $S_1, \ldots, S_n$ be the subrectangles of the partition lying within S. For each i, let m_i be the minimum value of f on S_i and ΔS_i the area of S_i. Then the parallelepipeds having base S_i and height m_i lie within the solid V and so

$$I_S(P) = \sum_{S_i \in P(S)} m_i \Delta S_i \leq \text{vol}(V).$$

The amount of the solid left out of the sum $I_S(P)$ can be reduced by taking a partition with smaller norm so we are led to the result that the volume of V is the least upper bound of the sums $I_S(P)$ as P ranges over all partitions of S. Thus

$$\text{vol}(V) = \iint_S f(x, y)\, dS,$$

the volume is the double integral of f over S.

Some computations of volumes are given after we find methods of evaluating double integrals in the next section.

Riemann Sums

It is sometimes convenient to express the double integral of f over S as a limit rather than as a least upper bound of some set of numbers. This was done using Riemann sums in the case of functions of one variable and the analogous result holds in the case of double integrals. Let us discuss it for a rectangular region $R = [a, b; c, d]$. Let $P : [x_0, \ldots, x_m; y_0, \ldots, y_n]$ be a partition of R into subrectangles $R_{ij} = [x_{i-1}, x_i; y_{j-1}, y_j]$. Let (u_{ij}, v_{ij}) be any point in the rectangle R_{ij}. Then the sum

$$RS = \sum_{i=1}^{m} \sum_{j=1}^{n} f(u_{ij}, v_{ij}) \Delta R_{ij}$$

is called a *Riemann sum* of f over R. Different Riemann sums are obtained by making different choices of the points (u_{ij}, v_{ij}) in each R_{ij} and by taking different partitions P. Let m_{ij} and M_{ij} be the minimum and maximum values of f on R_{ij}, so then we have

$$m_{ij} \leq f(u_{ij}, v_{ij}) \leq M_{ij}.$$

It follows that the Riemann sum RS lies between the lower sum and the upper sum corresponding to P:

$$L(P) \leq RS \leq U(P).$$

Suppose we form a sequence of partitions, $P_1, P_2, \ldots$ of R with the property that the norm of P_k approaches 0 as k grows to infinity. Then the difference between the upper and lower sums approaches 0 and, because the Riemann sums for each partition lies between the lower and upper sums, we find that the Riemann sums approach the double integral. We collect these facts in a theorem.

THEOREM 15.2

Riemann Sums

Let $P_1, P_2, \ldots, P_k, \ldots$ be a sequence of partitions of the rectangular region R such that the norm of P_k has limit 0 as k approaches infinity. For any function f that is continuous on R, the double integral of f over R is the limit of the Riemann sums

$$\iint_R f(x, y)\, dR = \lim_{k\to\infty} \sum_{i=1}^{m_k} \sum_{j=1}^{n_k} f(u_{ij}, v_{ij}) \Delta R_{ij}^{(k)},$$

where $R_{ij}^{(k)}$ is a subrectangle in the kth partition P_k and (u_{ij}, v_{ij}) is a point in $R_{ij}^{(k)}$.

We illustrate the use of Riemann sums to evaluate a double integral.

EXAMPLE 1 Let $R = [0, 1; 0, 1]$ and let $f(x, y) = ax + by$ for two constants a and b. Evaluate the double integral of f over R by using Riemann sums.

Solution: We select regular partitions P_k such that all subrectangles are squares with sides of length $1/k$. Thus

$$P_k : \left[0, \frac{1}{k}, \frac{2}{k}, \ldots, \frac{k}{k}; 0, \frac{1}{k}, \frac{2}{k}, \ldots, \frac{k}{k}\right].$$

The subrectangles are described by

$$R_{ij} = \left[\frac{i-1}{k}, \frac{i}{k}; \frac{j-1}{k}, \frac{j}{k}\right]$$

so that $\Delta R_{ij} = 1/k^2$. Thus P_k is a sequence of partitions of R and the norm of P_k is $1/k$, a quantity that approaches 0 as k approaches infinity.

Now we evaluate the Riemann sum that is obtained by selecting the point

$$(u_{ij}, v_{ij}) = \left(\frac{i-1}{k}, \frac{j-1}{k}\right).$$

The term $f(u_{ij}, v_{ij})$ can be evaluated by substitution to get

$$f(u_{ij}, v_{ij}) = a\frac{i-1}{k} + b\frac{j-1}{k} = \frac{1}{k}(ai - a + bj - b).$$

The Riemann sum for the partition P_k and this choice of points is

$$RS_k = \sum_{i=1}^{k} \sum_{j=1}^{k} \frac{1}{k}(ai - a + bj - b)\frac{1}{k^2}.$$

By using the formulas

$$\sum_{j=1}^{k} 1 = k, \qquad \sum_{j=1}^{k} j = \frac{k(k+1)}{2},$$

we simplify the inner sum to get

$$\sum_{j=1}^{k} \frac{1}{k}(ai - a + bj - b)\frac{1}{k^2} = \frac{1}{k^3}\left[(ai - a - b)\sum_{j=1}^{k} 1 + b\sum_{j=1}^{k} j\right]$$
$$= \frac{1}{k^3}\left[(ai - a - b)k + b\frac{k(k+1)}{2}\right].$$

Now substitute this value for the inner sum to compute the remaining sum as

$$RS_k = \sum_{i=1}^{k} \frac{1}{k^3}\left[(ai - a - b)k + b\frac{k(k+1)}{2}\right]$$
$$= \frac{ak}{k^3}\sum_{i=1}^{k} i + \frac{bk(k+1) - 2(a+b)k}{2k^3}\sum_{i=1}^{k} 1$$
$$= \frac{a}{k^2}\frac{k(k+1)}{2} + \frac{b(k+1) - 2a - 2b}{2k^2}k$$
$$= \frac{a}{2} + \frac{b}{2} - \frac{a+b}{2k}.$$

So we have a closed expression for the Riemann sum and we can easily evaluate the limit as k grows to infinity. We have

$$\iint_R (ax + by)\, dR = \lim_{k \to \infty}\left[\frac{a}{2} + \frac{b}{2} - \frac{a+b}{2k}\right] = \frac{a}{2} + \frac{b}{2}.$$

The method of Riemann sums can be used in this example because the function f was particularly simple. This method is usually not practical for complicated functions. ■

We study more efficient ways to evaluate double integrals in the next section.

Exercises 15.1

Use the estimate given in inequality (15.1) to find upper and lower bounds for the value of the following integrals.

1. $\iint_R \sin x \cos y\, dR,\ R = [\pi/4, \pi/2; 0, \pi/4]$

2. $\iint_R e^{-x} \cos(x^2 + y^2)\, dR,\ R = [0, 1; 0, 1]$

3. $\iint_R e^{x^2+y^2} \sqrt{xy}\, dR,\ R = [-2, -1; -4, -3]$

4. $\iint_R \ln(x^2 + 2xy + y^2)\, dR,\ R = [1, a; 1, a],\ a > 1$

Let R be the rectangular region $[0, 1; 0, 1]$ and let P_k be the partition

$$\left[0, \frac{1}{k}, \frac{2}{k}, \ldots, \frac{k}{k}; 0, \frac{1}{k}, \frac{2}{k}, \ldots, \frac{k}{k}\right].$$

5. Let $f(x, y) = 2x + 4y$. Find the lower and upper sums $L(P_2)$ and $U(P_2)$ for f over the region R, and then use the computation of the double integral in the example to compare these sums with the value of the double integral.

6. Let $f(x, y) = x + 3y$. Find the lower and upper sums $L(P_3)$ and $U(P_3)$ for f over the region R, and then use the computation of the double integral in the example to compare these sums with the value of the double integral.

7. Express a Riemann sum corresponding to the partition P_k for the function $f(x, y) = xy$ over R in terms of k and evaluate the limit as $k \to \infty$ to get the exact value of the double integral of f over R.

8. Express a Riemann sum corresponding to the partition P_k for the function $f(x, y) = x^2 - y^2$ over R in terms of k and evaluate the limit as $k \to \infty$ to get the exact value of the double integral of f over R.

9. Using Riemann sums, express as a limit the volume of the solid bounded below by the rectangle $[0, 2; 0, 3]$ and above by the plane $z = 2x + 5y + 2$.

10. Without attempting to evaluate the limit, express as a limit of Riemann sums the volume of the solid bounded by the rectangle $[0, 2; 0, 2]$ in the xy plane and above by the surface $z = 4x^2 + xy$.

11. Find the lower sum and the upper sum for $z = f(x, y) = 4x^2 + xy$ over the square $R = [0, 2; 0, 2]$ corresponding to the partition of R into 16 subsquares with sides of length $1/2$. Take note that $f(x, y)$ is an increasing function of x for any fixed y and an increasing function of y for any fixed x. Thus the maximum and minimum values of f on any square occur at diagonally opposite corners.

15.2 Iterated integrals

In this section we show how to evaluate double integrals over rectangular regions and then consider the problem for more general regions. As a preliminary to this topic, we consider a function defined as an integral with respect to one variable of a function of two variables.

Let f be a continuous function on the rectangle $R = [a, b; c, d]$ and let $F(x)$ be defined as

$$F(x) = \int_c^d f(x, y)\, dy.$$

Let us show that this is a continuous function of x on the interval $[a, b]$. Let ϵ be a given positive number. To prove continuity it is necessary to show that there exists a positive number δ with the property that

$$|F(x + h) - F(x)| < \epsilon \quad \text{whenever} \quad |h| < \delta.$$

Let $\epsilon_1 = \epsilon/(d - c)$. By the uniform continuity of f over R we know there is a positive number δ such that the values of f differ by at most ϵ_1 when f is evaluated at two points having distance apart less than δ. Thus for $|h| < \delta$, we have $|f(x + h, y) - f(x, y)| < \epsilon_1$. Now we have

$$\begin{aligned} |F(x + h) - F(x)| &= \left| \int_c^d f(x + h, y) - f(x, y)\, dy \right| \\ &\leq \int_c^d |f(x + h, y) - f(x, y)|\, dy \\ &\leq \int_c^d \epsilon_1\, dy = \epsilon_1(d - c) = \epsilon. \end{aligned}$$

This proves the continuity of F over $[a, b]$. Thus the integral of $F(x)$ over $[a, b]$ exists and we write it as

$$\int_a^b F(x)\, dx = \int_a^b \left(\int_c^d f(x, y)\, dy \right) dx.$$

The expression on the right is called an iterated integral. There is another iterated integral obtained by starting with the function

$$G(y) = \int_a^b f(x, y)\, dx.$$

Again this function is continuous for y on $[c, d]$ and the integral of G over $[c, d]$ exists. The integral

$$\int_c^d G(y)\, dy = \int_c^d \left(\int_a^b f(x, y)\, dx \right) dy$$

is another iterated integral. Our goal is to show that both the iterated integrals defined here are equal and that they equal the double integral of f over R.

THEOREM 15.3

Let f be continuous on the rectangular region $R = [a, b; c, d]$. Then

$$\iint_R f(x,y)\,dR = \int_a^b \Big(\int_c^d f(x,y)\,dy\Big)\,dx = \int_c^d \Big(\int_a^b f(x,y)\,dx\Big)\,dy.$$

Proof

Let $P : [x_0, \ldots, x_m; y_0, \ldots, y_n]$ be a partition of R and let R_{ij} denote the subrectangle $[x_{i-1}, x_i; y_{j-1}, y_j]$ having area ΔR_{ij}. Let m_{ij} and M_{ij} denote the minimum and maximum values of f on R_{ij}.

For any fixed x on $[x_{i-1}, x_i]$ and every y on $[y_{j-1}, y_j]$, we have

$$m_{ij} \le f(x,y) \le M_{ij}.$$

By integrating with respect to y along the interval $[y_{j-1}, y_j]$ we obtain

$$m_{ij}(y_j - y_{j-1}) \le \int_{y_{j-1}}^{y_j} f(x,y)\,dy \le M_{ij}(y_j - y_{j-1}).$$

We have a system of n inequalities $j = 1, \ldots, n$, holding i fixed, that we intend to sum. First we concentrate on the middle terms. Let $F(x)$ be defined by

$$F(x) = \sum_{j=1}^{n} \int_{y_{j-1}}^{y_j} f(x,y)\,dy = \int_c^d f(x,y)\,dy.$$

For x on $[x_{i-1}, x_i]$ we have

$$\sum_{j=1}^{n} m_{ij}(y_j - y_{j-1}) \le F(x) \le \sum_{j=1}^{n} M_{ij}(y_j - y_{j-1}).$$

Note that the extreme left and right terms in this inequality depend on i explicitly whereas the middle term depends on i because x is restricted to lie on $[x_{i-1}, x_i]$. The inequality gives bounds for $F(x)$ and it follows that the integral of F over $[x_{i-1}, x_i]$ is bounded; namely

$$\Big(\sum_{j=1}^{n} m_{ij}(y_j - y_{j-1})\Big)(x_i - x_{i-1})$$
$$\le \int_{x_{i-1}}^{x_i} F(x)\,dx \le \Big(\sum_{j=1}^{n} M_{ij}(y_j - y_{j-1})\Big)(x_i - x_{i-1}).$$

Now form the sum of the respective terms over $i = 1, \ldots, m$ taking notice of the fact that $\Delta R_{ij} = (x_i - x_{i-1})(y_j - y_{j-1})$ to get

$$\sum_{i=1}^{m}\sum_{j=1}^{n} m_{ij}\Delta R_{ij} \le \int_a^b F(x)\,dx \le \sum_{i=1}^{m}\sum_{j=1}^{n} M_{ij}\Delta R_{ij}.$$

The middle term is an iterated integral, the left term is the lower sum, $L(P)$, and the rightmost term is the upper sum, $U(P)$. We rewrite the previous inequality as

$$L(P) \leq \int_a^b \left(\int_c^d f(x, y)\, dy \right) dx \leq U(P).$$

The iterated integral an upper bound for the set of lower sums and also a lower bound for the set of upper sums. Because the double integral of f over R is the unique number satisfying these two conditions, we have proved

$$\iint_R f(x, y)\, dR = \int_a^b \left(\int_c^d f(x, y)\, dy \right) dx.$$

The statement of the conclusion of the theorem involving the other iterated integral is proved analogously and is left to the reader. □

Let us rework the example from the previous section in which we evaluated a double integral using a limit of Riemann sums.

EXAMPLE 1 Evaluate the double integral

$$\iint_R ax + by\, dR,$$

where R is the rectangle $R = [0, 1; 0, 1]$.

Solution: We express the double integral as an iterated integral:

$$\begin{aligned} \iint_R ax + by\, dR &= \int_0^1 \left(\int_0^1 ax + by\, dy \right) dx \\ &= \int_0^1 \left(axy + \frac{by^2}{2} \right) \Big|_{y=0}^{y=1} dx \\ &= \int_0^1 \left(ax + \frac{b}{2} \right) dx \\ &= \frac{ax^2}{2} + \frac{bx}{2} \Big|_0^1 \\ &= \frac{a}{2} + \frac{b}{2}. \end{aligned}$$

This agrees with the result obtained previously. ■

We now have a technique for evaluating a double integral over a rectangular region; the next step is to consider the problem defining and evaluating the double integral of a continuous function over a more general region. It is not a trivial matter to extend the notion of the double integral to completely general regions so we will be content to discuss nonrectangular regions that are still quite well-behaved, the so-called Type X and Type Y regions that are defined below. Although these conditions are still rather restrictive, many more general regions can be decomposed into subsets of Type X or Type Y and the integration done over each piece by the method we are about to describe.

Some Suitable Regions

We consider a region S that is contained in a rectangular region $R = [a, b; c, d]$ and a function f that is continuous on R. Even though we are interested in defining the integral of f over S, we insist that f be continuous on R rather than just on S. In this way we avoid questions of whether or not f assumes its minimum or maximum values on S or subsets of S and we are confident that f is uniformly continuous. We must make an assumption about S that is satisfied by all the examples of regions to be considered later (and in fact is satisfied by all sets defined by collections of inequalities involving continuous functions). We require that for any subrectangle A in R, the intersection of A and S is a set that has an area. Moreover we suppose that S has an area and the area of S is the sum of the areas of $R_{ij} \cap S$ where the R_{ij} are the subrectangles of a partition of R. These should seem fairly intuitive assumptions and we use these properties of S without further reference.

Now we turn to the definition of upper and lower sums of f over S. Let $P : [x_0, \ldots, x_m; y_0, \ldots, y_n]$ be a partition of R and let $R_{ij} = [x_{i-1}, x_i; y_{j-1}, y_j]$ be a subrectangle of the partition and let

$$S_{ij} = S \cap R_{ij} \qquad \Delta S_{ij} = \text{area of } S_{ij}.$$

Thus $\Delta S_{ij} = 0$ if R_{ij} has no points in S (or even if R_{ij} has only a finite number of points of S). Let m_{ij} and M_{ij} denote the minimum and maximum values of f on the subrectangle R_{ij}. (It would seem more logical to take the minimum and maximum values of f on S_{ij}, but we wish to avoid consideration of the problem of the existence of such extrema. For the regions that we consider below, this distinction is irrelevant.) Now we define the lower sum $L(P(S))$ and upper sum $U(P(S))$ of f over S relative to the given partition P:

$$L(P(S)) = \sum_{i=1}^{m} \sum_{j=1}^{n} m_{ij} \Delta S_{ij},$$

$$U(P(S)) = \sum_{i=1}^{m} \sum_{j=1}^{n} M_{ij} \Delta S_{ij}.$$

Because $m_{ij} \leq M_{ij}$ and $\Delta S_{ij} \geq 0$, it follows that

$$L(P(S)) \leq U(P(S))$$

for every partition P.

Notice that when S is a rectangular region, the lower and upper sums here defined agree with the definition of upper and lower sums given previously for rectangular regions.

The plan is to define the double integral as the least upper bound of all the lower sums and, at the same time, show that this number is the greatest lower bound of the set of upper sums. To do that we need to compare lower sums for different partitions (and upper sums for different partitions). This is accomplished with the use of refinements.

Refinements

Given the two partitions

$$P : [x_0, \ldots, x_m; y_0, \ldots, y_n], \qquad P' : [u_0, \ldots, u_p; v_0, \ldots, v_q]$$

of the same rectangle

$$R = [x_0, x_m; y_0, y_n] = [u_0, u_p; v_0, v_q] = [a, b; c, d],$$

we say that P' is a refinement of P if each x_i is equal to some u_k and each y_j is equal to some v_r. Thus a refinement of P is obtained by adding additional points to the partition of $[a, b]$ or $[c, d]$ determined by P.

We begin the comparison by showing that lower sums do not decrease when we pass from a partition to a refinement of it. We do this by adding one point at a time. Suppose we refine P by the addition of one point x^* between x_{i-1} and x_i for one particular i so we obtain the partition

$$P' : [x_0, \ldots, x_{i-1}, x^*, x_i, \ldots, x_m; y_0, \ldots, y_n].$$

The subrectangle R_{ij} is divided into two subrectangles

$$R'_{ij} = [x_{i-1}, x^*; y_{j-1}, y_j], \qquad R''_{ij} = [x^*, x_i; y_{j-1}, y_j]$$

on which f has minimum values m'_{ij} and m''_{ij} respectively. Because m_{ij} is the minimum of f on all of R_{ij}, it follows that

$$m_{ij} \leq m'_{ij}, \qquad m_{ij} \leq m''_{ij}.$$

The contribution to $L(P(S))$ of the rectangle R_{ij} is $m_{ij}\Delta S_{ij}$; the contribution to $L(P'(S))$ of the two parts of the rectangle R_{ij} is

$$\begin{aligned} m'_{ij}\Delta S'_{ij} + m''_{ij}\Delta S''_{ij} &\geq m_{ij}\Delta S'_{ij} + m_{ij}\Delta S''_{ij} \\ &= m_{ij}(\Delta S'_{ij} + \Delta S''_{ij}) \\ &= m_{ij}\Delta S_{ij}. \end{aligned}$$

By taking sums over all subrectangles of P we obtain $L(P(S)) \leq L(P'(S))$. Of course a similar inequality is obtained if we refine P by the addition of one point to the collection $\{y_0, \ldots, y_n\}$. The inequality extends to more general refinements of P by the addition of one point at a time. If P' is a refinement of P then

$$L(P(S)) \leq L(P'(S)).$$

By an analogous argument one shows that the upper sums do not increase as we pass from a partition to a refinement of the partition. Thus if P' is a refinement of P, then

$$U(P'(S)) \leq U(P(S)).$$

It now follows that every upper sum is an upper bound for every lower sum. To see this let P and Q be any two partitions of R and let P' be a partition that refines both P and Q. Then

$$L(P(S)) \leq L(P'(S)) \leq U(P'(S)) \leq U(Q(S)).$$

Thus $U(Q(S))$ is an upper bound for every lower sum and $L(P(S))$ is a lower bound for every upper sum. It follows that the set of all lower sums of f over S has a least upper bound, denoted as $I_{\text{lower}}(f, S)$ and the set of upper sums has a greatest lower bound, denoted as $I_{\text{upper}}(f, S)$.

Inasmuch as every upper sum is an upper bound for the set of lower sums, it follows that the least upper bound of the lower sums is less than or equal to every upper sum; that is

$$I_{\text{lower}}(f, S) \leq U(P(S))$$

for every partition P. Thus $I_{\text{lower}}(f, S)$ is a lower bound for the set of upper sums and so

$$I_{\text{lower}}(f, S) \leq I_{\text{upper}}(f, S).$$

The last step in the development is to show the equality of I_{lower} and I_{upper}. For this we use the uniform continuity of f in R. Let A denote the area of S and let ϵ be any positive number. The uniform continuity ensures the existence of a number $\delta > 0$ such that

$$|f(x, y) - f(u, v)| < \frac{\epsilon}{A} \quad \text{if} \quad (x - u)^2 + (y - v)^2 < \delta^2.$$

Select a partition P of R such that the diagonal of each subrectangle R_{ij} has length less than δ. Then the difference between the maximum and minimum values of f on a subrectangle is less than ϵ/A. In other words,

$$\begin{aligned} U(P(S)) - L(P(S)) &= \sum\sum(M_{ij} - m_{ij})\Delta S_{ij} \\ &\leq \frac{\epsilon}{A}\sum\sum \Delta S_{ij} = \frac{\epsilon}{A}(\text{area of } S) \\ &= \frac{\epsilon}{A}A = \epsilon. \end{aligned}$$

From this and the inequality

$$L(P(S)) \leq I_{\text{lower}}(f, S) \leq I_{\text{upper}}(f, S) \leq U(P(S))$$

we conclude

$$0 \leq I_{\text{upper}}(f, S) - I_{\text{lower}}(f, S) < \epsilon.$$

Because ϵ was an arbitrary positive number, the difference $I_{\text{upper}} - I_{\text{lower}}$ cannot be positive (if it were a positive number, we could take ϵ equal to half that positive number) so $I_{\text{lower}} = I_{\text{upper}}$.

Double Integrals over Nonrectangular Regions

We now make the expected definition: The double integral of f over S is the unique number that is simultaneously the least upper bound of the set of lower sums of f over S and the greatest lower bound of the set of upper sums of f over S and we denote this number by

$$\iint_S f(x, y)\, dS.$$

The evaluation of double integrals for even mildly complicated sets S might be quite a formidable problem. We show how to evaluate double integrals over certain types of regions by using iterated integrals.

Regions of Type X and Type Y

A region S is said to be of Type X if it consists of all points (x, y) satisfying

$$a \leq x \leq b, \quad \text{and} \quad g_1(x) \leq y \leq g_2(x) \qquad \textbf{(15.2)}$$

where g_1 and g_2 are continuous functions on the interval $[a, b]$. In fact we usually insist that g_1 and g_2 are piecewise differentiable functions without further comment.

Here is the mnemonic device associating the name Type X with the description of the region. Begin by projecting the region to the x-axis. The projection is an interval $[a, b]$ and for every x on this interval, the points (x, y) that lie in the region are described by the condition $g_1(x) \leq y \leq g_2(x)$.

A region is said to be of Type Y if it consists of all points (x, y) satisfying

$$c \leq y \leq d \quad \text{and} \quad h_1(y) \leq x \leq h_2(y). \qquad \textbf{(15.3)}$$

The procedure for describing a region of Type Y is analogous to that just described for regions of Type X. First project the region to the y-axis to obtain the interval $[c, d]$. For each y on the interval, the set of points (x, y) in the region correspond to an interval $[h_1(y), h_2(y)]$ depending on y, over which x ranges.

Of course, some regions cannot be described as either a Type X or a Type Y region. Also, some regions can be described in both ways.

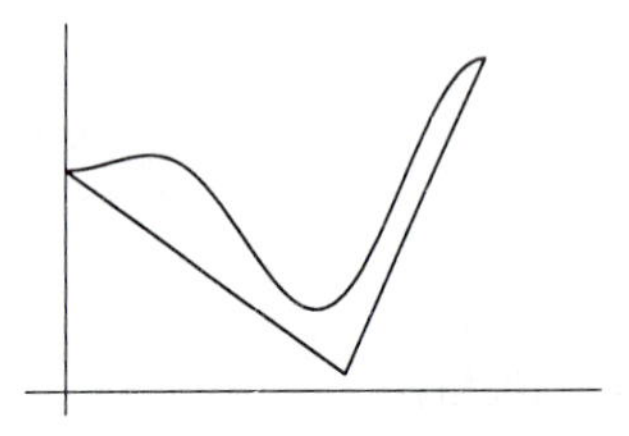

FIGURE 15.2
A Type X region
not Type Y

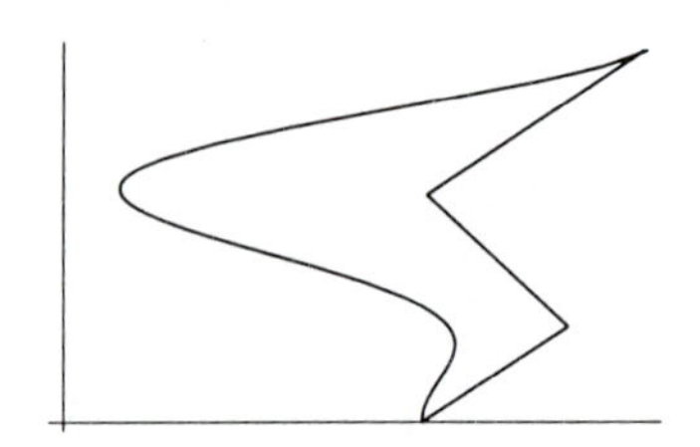

FIGURE 15.3
A Type Y region
not Type X

We show how to evaluate a double integral over a region of Type X or Y as an iterated integral.

Let S be a region of Type X given as in statement (15.2). Assume that R is a rectangle containing S and that f is a continuous function on R. For x on $[a, b]$, let $F(x)$ be defined by

$$F(x) = \int_{g_1(x)}^{g_2(x)} f(x, y)\, dy. \tag{15.4}$$

Then F is continuous on $[a, b]$, as is seen below, and hence the integral $\int_a^b F(x)\, dx$ exists. This is the iterated integral that we intend to prove is equal to the double integral of f over S.

THEOREM 15.4

Double Integrals over Regions of Type X

Let S be a region of Type X as described in statement (15.2) and let f be a function that is continuous in a rectangular region R containing S. Then the double integral of f over S is equal to an iterated integral:

$$\iint_S f(x, y)\, dS = \int_a^b \left(\int_{g_1(x)}^{g_2(x)} f(x, y)\, dy \right) dx.$$

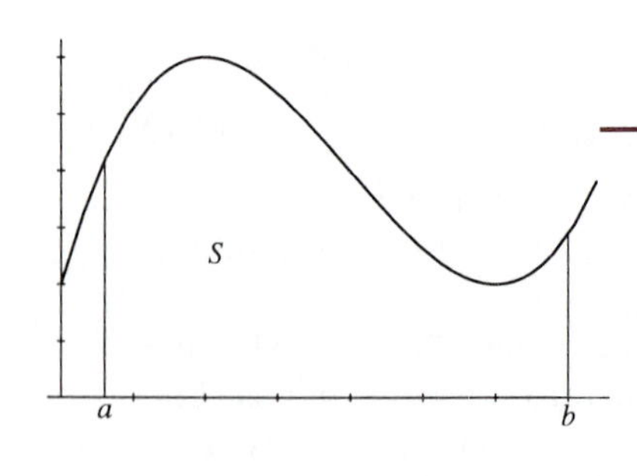

FIGURE 15.4
Simplified region

Proof

We simplify the notation just a bit. If c is a constant and $g_1(x) \le c \le g_2(x)$ for x on some interval, then

$$\int_{g_1(x)}^{g_2(x)} = \int_c^{g_2(x)} - \int_c^{g_1(x)}$$

for all x on the interval. With this idea in mind, we work with a region S given with the following description:

$$S = \{(x, y) : a \le x \le b,\ c \le y \le g(x)\}$$

where $g(x)$ is a differentiable function on $[a, b]$. Once this case is completed, the more general case can be obtained from it.

Let $F(x)$ be defined by Equation (15.4) using $g_1(x) = c$ and $g_2(x) = g(x)$. We first show the continuity of F. Given any positive number ϵ, we must show that some $\delta > 0$ exists so that $|h| < \delta$ implies $|F(x+h)-F(x)| < \epsilon$. We have from the definition of F that

$$\begin{aligned} &F(x+h) - F(x) \\ &= \int_c^{g(x+h)} f(x+h, y)\,dy - \int_c^{g(x)} f(x,y)\,dy \\ &= \int_c^{g(x+h)} f(x+h,y) - f(x,y)\,dy + \int_c^{g(x+h)} f(x,y)\,dy - \int_c^{g(x)} f(x,y)\,dy \\ &= \int_c^{g(x+h)} f(x+h,y) - f(x,y)\,dy + \int_{g(x)}^{g(x+h)} f(x,y)\,dy. \end{aligned} \tag{15.5}$$

To estimate the two integrals, we first apply the MVT to g (which is assumed differentiable on $[a, b]$) to get $g(x+h) = g(x) + g'(z)h$ for some z between x and $x+h$. The second integral approaches 0 as h approaches 0. We see this by applying the Fundamental Theorem of Calculus (writing $u = g(x)$) to get

$$\lim_{h \to 0} \frac{1}{g'(z)h} \int_u^{u+g'(z)h} f(x,y)\,dy = f(x,u);$$

it follows that

$$\lim_{h \to 0} \int_u^{u+g'(z)h} f(x,y)\,dy = 0.$$

Thus for $|h| < \delta_1$ the second integral in Equation (15.5) is less than $\epsilon/2$. To estimate the first integral on the right of Equation (15.5) we use the uniform continuity of f over S to get some δ_2 with the property that the difference in the values of f at two points is less than $\epsilon/2(d-c)$ whenever the distance between the two points is less than δ_2. Then we have

$$\begin{aligned} \left| \int_c^{g(x+h)} f(x+h,y) - f(x,y)\,dy \right| &\le \int_c^{g(x+h)} |f(x+h,y) - f(x,y)|\,dy \\ &\le \int_c^{g(x+h)} \frac{\epsilon}{(d-c)}\,dy \\ &\le \frac{\epsilon}{2(d-c)}(g(x+h) - c) \le \frac{\epsilon}{2}. \end{aligned}$$

Take $0 < \delta \le \delta_1$ and δ_2; for $|h| < \delta$ we have

$$\begin{aligned} &|F(x+h) - F(x)| \\ &\le \left| \int_c^{g(x+h)} f(x+h,y) - f(x,y)\,dy \right| + \left| \int_{g(x)}^{g(x+h)} f(x,y)\,dy \right| \\ &\le \frac{\epsilon}{2} + \frac{\epsilon}{2} = \epsilon. \end{aligned}$$

This proves that F is continuous on $[a, b]$ and is therefore integrable over that interval. The integral of F over $[a, b]$ is the iterated integral that we must show equals the double integral of f over S.

We begin by selecting a partition of R that has some slightly special requirements. Starting from any partition $P : [x_0, \ldots, x_m; y_0, \ldots, y_n]$ of R we add, if necessary, some additional points to the set $\{y_0, \ldots, y_n\}$ as follows. For any i, $1 \le i \le m$, let u_i be the minimum value of $g(x)$ for x on $[x_{i-1}, x_i]$. We want one of the y_k to equal u_i; if this is not the case initially, add the point u_i to obtain a refinement of P. Rather than change notation to indicate a new partition containing all the u_i, we just assume that P already meets this requirement. For each i, let j_i be the index such that $u_i = y_{j_i}$. The minimum and maximum values of f on any subrectangle R_{ij} are m_{ij} and M_{ij}, respectively, so that for (x, y) in R_{ij} we have

$$m_{ij} \leq f(x, y) \leq M_{ij}.$$

Integrate each term with respect to y over the interval $[y_{j-1}, y_j]$ to get

$$m_{ij}(y_j - y_{j-1}) \leq \int_{y_{j-1}}^{y_j} f(x, y)\, dy \leq M_{ij}(y_j - y_{j-1}).$$

Next we sum these inequalities for $1 \leq j \leq j_i$ and note that

$$\sum_{j=1}^{j_i} \int_{y_{j-1}}^{y_j} = \int_c^{y_{j_i}} = \int_c^{g(x)} - \int_{y_{j_i}}^{g(x)}$$

to get

$$\sum_j m_{ij}(y_j - y_{j-1}) \leq \int_c^{g(x)} f\, dy - \int_{y_{j_i}}^{g(x)} f\, dy \leq \sum_j M_{ij}(y_j - y_{j-1}).$$

We can rewrite this as

$$\sum_{j=1}^{j_i} m_{ij}(y_j - y_{j-1}) + \int_{y_{j_i}}^{g(x)} f\, dy \leq F(x) \leq \sum_{j=1}^{j_i} M_{ij}(y_j - y_{j-1}) + \int_{y_{j_i}}^{g(x)} f\, dy.$$

Our next step is to integrate each term over $[x_{i-1}, x_i]$ and then sum over all i. This produces

$$\sum_{i=1}^{m}\sum_{j=1}^{j_i} m_{ij}(y_j - y_{j-1})(x_i - x_{i-1}) + \sum_{i=1}^{m} \int_{x_{i-1}}^{x_i} \left(\int_{y_{j_i}}^{g(x)} f\, dy \right) dx$$

$$\leq \sum_{i=1}^{m} \int_{x_{i-1}}^{x_i} F(x)\, dx = \int_a^b F(x)\, dx$$

$$\leq \sum_{i=1}^{m}\sum_{j=1}^{j_i} M_{ij}(y_{j-1} - y_j)(x_i - x_{i-1}) + \sum_{i=1}^{m} \int_{x_{i-1}}^{x_i} \left(\int_{y_{j_i}}^{g(x)} f\, dy \right) dx.$$

The double sums on the extremes of these inequalities are not quite lower or upper sums because the range of summation does not take into account the region between the lines $y = y_{j_i}$ and $y = g(x)$. However the sum of the integrals in the extreme terms accounts for these missing terms. Let

$$W_i = \int_{x_{i-1}}^{x_i} \left(\int_{y_{j_i}}^{g(x)} f\, dy \right) dx.$$

In the expression for W_i if f is replaced by a constant C then the expression is equal to C times the area of the region bounded by the curve $y = g(x)$ and the the line $y = y_{j_i}$ over the interval $[x_{i-1}, x_i]$. For C take the minimum value of f (and substitute into the left side) and the maximum value of f (and substitute into right side) over the rectangle R_{ij_i+1}. We are then left with

$$\sum_i \sum_j m_{ij} \Delta S_{ij} \leq \int_a^b F(x)\, dx \leq \sum_i \sum_j M_{ij} \Delta S_{ij}.$$

Equivalently we get the iterated integral trapped between a lower sum and an upper sum. By selecting the partition to have upper sum and lower sum less than some given ϵ, we find the difference between the iterated integral and the double integral to be less than the given ϵ. It follows that the iterated integral and the double integral are equal.

To get the result for regions bounded by two curves as defined in inequalities (15.2), we apply the result just proved to separate regions bounded by a straight line and a curve and then combine the calculations. We do not carry out the exercise of describing the details. □

Of course there is a symmetric result for regions of Type Y.

THEOREM 15.5

Double Integrals over Regions of Type Y

Let S be a region of Type Y as described by inequalities (15.3) and let f be a function that is continuous in a rectangular region R containing S. Then the double integral of f over S is equal to an iterated integral:

$$\iint_S f(x, y)\,dS = \int_c^d \left(\int_{h_1(y)}^{h_2(y)} f(x, y)\,dx\right) dy.$$

After all this work, it is time for some examples.

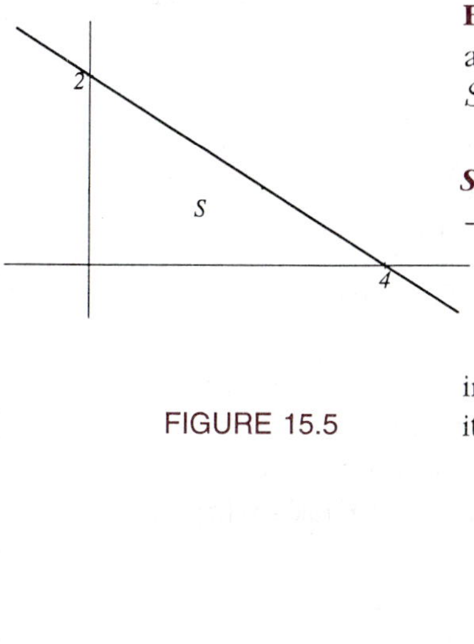

EXAMPLE 2 Let S be the triangular region bounded by the x- and y-axes and the line $x + 2y = 4$. Evaluate the double integral of $f(x, y) = x/(y + 6)^2$ over S.

FIGURE 15.5

Solution: Note first that there is no point in the region having y coordinate -6; thus f is continuous over S.

We can view S as a Type X region with the description

$$S = \{(x, y) : 0 \le x \le 4, \quad 0 \le y \le (4 - x)/2\}$$

in which $g(x) = (4 - x)/2$ is continuous. Then the double integral equals the iterated integral

$$\begin{aligned}
\int_0^4 \int_0^{(4-x)/2} \frac{x}{(y + 6)^2}\,dy\,dx &= \int_0^4 \left(\frac{-x}{y + 6}\right)\Bigg|_{y=0}^{y=(4-x)/2} dx \\
&= \int_0^4 \left(\frac{x}{6} + \frac{2x}{x - 16}\right) dx \\
&= \left(\frac{x^2}{12} + 2x + 32\ln|x - 16|\right)\Bigg|_0^4 \\
&= \frac{28}{3} + 32(\ln 3 - \ln 4).
\end{aligned}$$

The region S can also be viewed as a Type Y region in which we use the description

$$S = \{(x, y) : 0 \le y \le 2, \quad 0 \le x \le 4 - 2y\},$$

where the $h(y)$ in the definition of Type Y region is $h(y) = 4 - 2y$.

If we evaluate the double integral as an iterated integral over a region of

Type Y it becomes

$$\int_0^2 \int_0^{4-2y} \frac{x}{(y+6)^2}\,dx\,dy = \int_0^2 \left(\frac{x^2}{2(y+6)^2}\right)\Bigg|_{x=0}^{x=4-2y} dy$$

$$= \int_0^2 \frac{(4-2y)^2}{2(y+6)^2}\,dy$$

$$= \int_0^2 2 - \frac{32}{y+6} + \frac{128}{(y+6)^2}\,dy$$

$$= \left(2y - 32\ln|y+6| - \frac{128}{y+6}\right)\Bigg|_0^2$$

$$= \frac{28}{3} + 32(\ln 3 - \ln 4).$$

As expected, the two different ways of evaluating the double integral give the same result. ■

The two different ways of viewing the integral in this example are about equally as much work to evaluate. In some cases where a region can be viewed either as Type X or Type Y, one choice might produce substantially easier integration than the other. The next example illustrates this point.

EXAMPLE 3 Evaluate the double integral $\iint_S ye^{x^2}\,dS$ where S is the region bounded by the lines $y = 0$, $x = 1$ and the graph of $y = \sqrt{x}$.

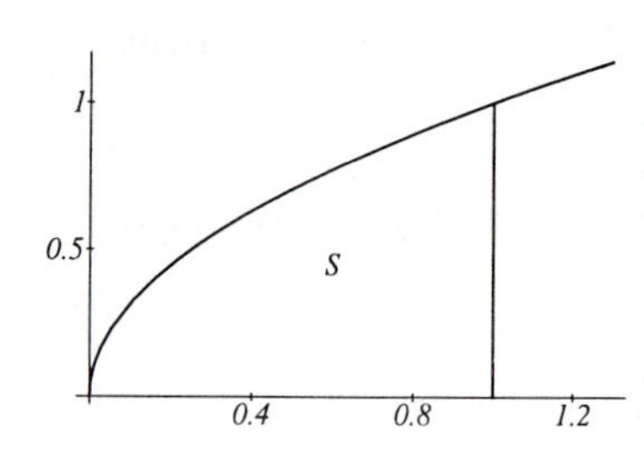

FIGURE 15.6
Type X and Type Y

Solution: The region S can be described as a region of either Type X or Type Y but the integration is not possible in one order.

We try using a Type Y description of S, so that

$$S = \{(x,y) : 0 \le y \le 1, \quad y^2 \le x \le 1\}.$$

Then the iterated integral is

$$\iint_S ye^{x^2}\,dS = \int_0^1 \int_{y^2}^1 ye^{x^2}\,dx\,dy.$$

We cannot get started with the inside integral because there is no elementary function whose derivative is e^{x^2}. Rather than throwing up our hands in frustration, we try describing S as a Type X region.

We have

$$S = \{(x,y) : 0 \le x \le 1, \quad 0 \le y \le \sqrt{x}\}$$

and then the integral is

$$\iint_S ye^{x^2}\,dS = \int_0^1 \int_0^{\sqrt{x}} ye^{x^2}\,dy\,dx$$

$$= \int_0^1 \frac{1}{2}\left(y^2 e^{x^2}\right)\Bigg|_{y=0}^{y=\sqrt{x}} dx$$

$$= \int_0^1 \frac{1}{2}\left(xe^{x^2}\right)dx = \frac{1}{4}e^{x^2}\Bigg|_0^1$$

$$= \frac{e-1}{4}.$$

There is no general way to know that one description will produce easier integration than the other. When faced with difficulties using one description, try the alternate description. ■

In the next example, the integration can be carried out in either order but one choice requires some additional work.

EXAMPLE 4 Express the integral $\iint_S f(x, y)\, dS$ as an iterated integral using descriptions of S as Type X and Type Y when S is the region enclosed by the graphs of $x = y^2$ and $y = 2 - x$.

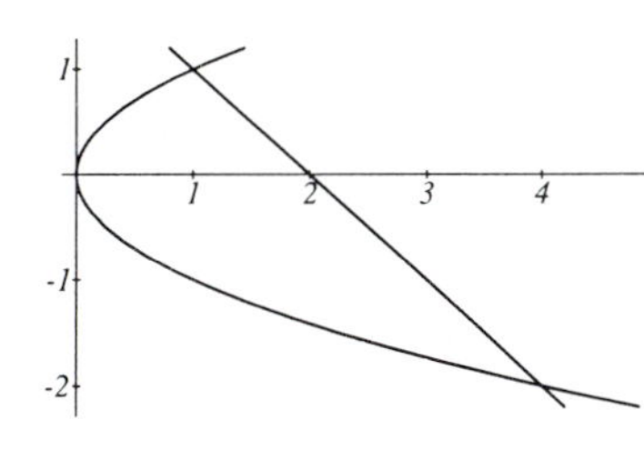

FIGURE 15.7
The region S

Solution: We first sketch the region S (see figure 15.7).
The description of S as a Type Y region is

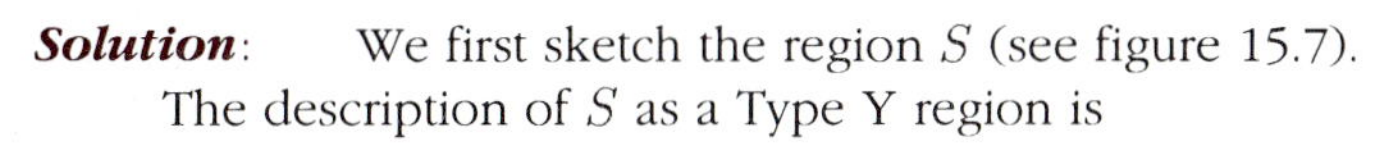

$$S = \{(x, y) : -2 \leq y \leq 1, \quad y^2 \leq x \leq 2 - y\}$$

and the iterated integral is

$$\iint_S f(x, y)\, dS = \int_{-2}^{1} \int_{y^2}^{2-y} f(x, y)\, dx\, dy.$$

The description of S as a Type X region is complicated by the fact that the upper curve requires two formulas for its description as x ranges over $[0, 4]$. Let

$$S_1 = \{(x, y) : 0 \leq x \leq 1, \quad -\sqrt{x} \leq y \leq \sqrt{x}\},$$
$$S_2 = \{(x, y) : 1 \leq x \leq 4, \quad -\sqrt{x} \leq y \leq 2 - x\}.$$

Then S is the union of the two Type X regions S_1 and S_2 so we have

$$\iint_S f(x, y)\, dS = \int_0^1 \int_{-\sqrt{x}}^{\sqrt{x}} f\, dy\, dx + \int_1^4 \int_{-\sqrt{x}}^{2-x} f\, dy\, dx.$$

So we have expressed the integral by iterated integrals using both orders of integration. ■

Exercises 15.2

Evaluate the double integrals over the indicated rectangular regions.

1. $\iint_R x^2 + y^2\, dR,\ R = [0, 2; 0, 3]$

2. $\iint_R 2x - 3y\, dR,\ R = [-2, 5; -1, 3]$

3. $\iint_R x \sin y\, dR,\ R = [0, 2; 0, \pi]$

4. $\iint_R x e^{xy}\, dR,\ R = [1, 3; 1, 2]$

5. $\iint_R y e^{xy}\, dR,\ R = [1, 2; 1, 3]$

6. $\iint_R \dfrac{3x}{1 + y^2}\, dR,\ R = [-1, 1; -2, 2]$

7. $\iint_R e^{(x-y)}\, dR,\ R = [-1, 1; -1, 1]$

8. $\iint_R \sin(x + y)\, dR,\ R = [0, \pi/2; 0, \pi/2]$

Sketch the region S and describe it as a Type X or Type Y region, if possible. Then express the area of the region as an iterated integral of one or both forms if possible.

9. S is the region enclosed by the graph of $y = 3x^2 + 1$ and the straight line $y = 4$

10. S is the region in the first quadrant bounded by the lines $y = x/2$ and $y = 3x - 4$

11. S is the region enclosed by the graphs of $y = 4 - x^2$ and $y = x^2$

12. S is the region common to the two circles with equations $x^2 + y^2 = 4$ and $(x - 2)^2 + y^2 = 4$

13. S is the region that lies inside the circle with equation $x^2 + y^2 = 16$ but outside the circle with equation $x^2 + (y + 5)^2 = 64$

14. S is the region in the half plane $x \geq 0$ bounded by the graph of the hyperbola with equation $x^2 - y^2 = 4$ and the parabola $x = 5 - y^2$

15. S is the region inside the square having three vertices at $(0, -\pi)$, $(0, \pi)$, $(2\pi, \pi)$ and bounded above by the graph of $y = \sin x$ and below by the graph of $y = \cos x$

16. S is the region in the first quadrant bounded by the graphs of $xy = 4$ and $x + y = 5$

Evaluate each double integral as an iterated integral and sketch the region over which the integral is taken.

17. $\iint_S (x^2 - y^2)\,dS$, $S = \{(x, y) : 1 \le x \le 4,\ 0 \le y \le 2x\}$

18. $\iint_S 2x - 3y + 4\,dS$, $S = \{(x, y) : 1 \le y \le 3,\ 1 \le x \le y^2\}$

19. $\iint_S \frac{\sin x}{x}\,dS$, $S = \{(x, y) : 0 \le x \le \pi,\ 0 \le y \le x\}$

20. $\iint_S x \cos y\,dS$,
$S = \{(x, y) : 0 \le x \le \sqrt{\pi/2},\ 0 \le y \le x^2\}$

Each of the following iterated integrals equals a double integral $\iint_S f\,dS$ for some region S. Sketch the regions S and then express the double integral as an iterated integral with the order of integration reversed. In attempting to reverse the order of integration, it might be necessary to subdivide the region and use the sum of two iterated integrals. Finally evaluate the double integral.

21. $\int_0^2 \int_0^{x^2} xy^2\,dy\,dx$

22. $\int_1^3 \int_x^{x^2} \frac{y}{x}\,dy\,dx$

23. $\int_{-1}^3 \int_{x^2}^{2x+3} xy\,dy\,dx$

24. $\int_0^1 \int_y^1 \frac{1}{1 + x^2}\,dx\,dy$

25. $\int_0^R \int_0^{\sqrt{R^2-y^2}} (R^2 - x^2)^{3/2}\,dx\,dy$

26. $\int_0^1 \int_y^1 \sqrt{1 + x^2}\,dx\,dy$

27. Use the expression of a double integral as a limit of Riemann sums to prove the following statement about integrals over a suitably nice region S: If f and g are continuous functions on S and if $f(x, y) \le g(x, y)$ at every point of S then

$$\iint_S f(x, y)\,dS \le \iint_S g(x, y)\,dS.$$

28. Let $S = [a, b; c, d]$ and $f(x, y) = U(x)V(y)$ for functions U and V that are continuous on $[a, b]$ and $[c, d]$ respectively. Express the double integral $\int_S f(x, y)\,dS$ as a product of two integrals of functions of one variable.

29. Let $R = [-a, a; -b, b]$ and let $f(x, y)$ be an odd function of x; that is $f(-x, y) = -f(x, y)$ for (x, y) in R. Prove $\iint_R f(x, y)\,dR = 0$.

30. Let $f(x, y)$ be continuous over the entire xy plane and suppose

$$\iint_R f(x, y)\,dR = 0$$

for every rectangle $R = [a, b; c, d]$. Show that $f(x, y) = 0$ for all (x, y). [Hint: If f is nonzero at some point, then there is some point P where f is positive or negative. By continuity there is some (small but nonzero) rectangle where f is positive throughout the rectangle or f is negative throughout the rectangle. Use this to get a conflict with the assumption about the integrals over rectangles.]

31. If m and M are constants such that $m \le f(x, y) \le M$ for every (x, y) in S and if $A(S)$ is the area of S, show

$$mA(S) \le \iint_S f(x, y)\,dS \le MA(S).$$

32. Prove the Mean Value Theorem for Double Integrals that states: Let S be a region of Type X or Type Y and let f be a function that is continuous on S. Then there is a point (x_0, y_0) in S that satisfies

$$\iint_S f(x, y)\,dS = f(x_0, y_0)A(S)$$

where $A(S)$ is the area of S. [Hint: Let the minimum of f on S be $m = f(x_1, y_1)$ and the maximum of f on S be $M = f(x_2, y_2)$ where the two points (x_i, y_i) are in S. Then there is a continuous path $\mathbf{r}(t) = u(t)\mathbf{i} + v(t)\mathbf{j}$ that lies in S and has $\mathbf{r}(0) = x_1\mathbf{i} + y_1\mathbf{j}$ and $\mathbf{r}(1) = x_2\mathbf{i} + y_2\mathbf{j}$. Thus the function $f(\mathbf{r}(t))A(S)$ is continuous for $0 \le t \le 1$ and takes on a minimum value $mA(S)$ and a maximum value $MA(S)$. Apply the Intermediate Value Theorem for functions of one variable, to conclude that there is some value of t at which the value of the function is equal to the double integral.]

15.3 More volumes using double integrals

We have briefly indicated that the volume of the solid bounded above by the surface $z = f(x, y)$, for a positive function f, and below by a region S of the xy

plane is given by the double integral

$$\iint_S f(x,y)\,dS.$$

If S is either a Type X or a Type Y region, then the volume can be expressed as an iterated integral.

EXAMPLE 1 Determine the volume of tetrahedron bounded by the coordinate planes and the plane with equation

$$\frac{x}{a}+\frac{y}{b}+\frac{z}{c}=1$$

with a, b and c positive constants.

Solution: The solid in question is bounded below by the triangle in the xy plane with sides $x=0$, $y=0$ and $x/a+y/b=1$. We view this as a region of Type X with description

$$S=\{(x,y):0\le x\le a,\quad 0\le y\le b-bx/a\}.$$

The volume is

$$\begin{aligned}V&=\iint_S c\left(1-\frac{x}{a}-\frac{y}{b}\right)dS\\&=\int_0^a\int_0^{b-bx/a}c\left(1-\frac{x}{a}-\frac{y}{b}\right)dy\,dx\\&=c\int_0^a\left(y-\frac{xy}{a}-\frac{y^2}{2b}\right)\Big|_0^{b-bx/a}dx\\&=\frac{bc}{2}\int_0^a 1-\frac{2x}{a}+\frac{x^2}{a^2}\,dx\\&=\frac{abc}{6}.\end{aligned}$$

This formula might be familiar; it is the volume of a right tetrahedron with sides of lengths a, b and c emanating from the right angle. ■

Solids Bounded by Two Surfaces

A solid might be described by some means other than that used in the opening paragraph of this section. For example a solid as familiar as the interior of a sphere requires a description somewhat more subtle than just giving a region of the xy plane and a surface above it. The sphere of radius a with center at the origin can be described by referring to the circle of radius a center at the origin in the xy plane as a base region and then, for each point (x,y) in that circle, restricting the z coordinates to lie between two numbers depending on (x,y). More precisely, the interior of the sphere is

$$B=\{(x,y,z):x^2+y^2\le a^2,-\sqrt{a^2-x^2-y^2}\le z\le\sqrt{a^2-x^2-y^2}.$$

The volume of a solid having a description of this form is easily expressed as a double integral. Let us give the statement in a somewhat more general form.

THEOREM 15.6

Volume Bounded by Graphs of Two Functions

Let S be a region of the xy plane contained in some rectangle and let V be a solid described as

$$V = \{(x, y, z) : (x, y) \in S, \quad b(x, y) \leq z \leq t(x, y)\}$$

for b and t two functions continuous on S (b is the "bottom" function and t is the "top" function). The the volume of V is given by the double integral

$$\iint_S t(x, y) - b(x, y)\, dS.$$

Proof

For any partition P of the rectangle containing S and a subrectangle $S_{ij} = [x_{i-1}, x_i; y_{j-1}, y_j]$, the parallelepiped with base S_{ij} and extending from the bottom, $b(x_i, y_j)$ to the top, $t(x_i, y_j)$ of the solid has volume

$$\Delta S_{ij}\left(t(x_i, y_j) - b(x_i, y_j)\right);$$

by taking the sum over all subrectangles of the partition and then taking the limit over a sequence of partitions whose norm approaches 0, we get on the one hand, the double integral of $t(x, y) - b(x, y)$ over S, and on the other hand, we get the volume of the solid. □

EXAMPLE 2 Find the volume of the region enclosed by the two paraboloids with equations $z = 100 - 4x^2 - y^2$ and $z = 6x^2 + 9y^2$.

Solution: The points lying on both surfaces have coordinates that satisfy both equations. If we eliminate z from the two equations and simplify, then the condition satisfied by (x, y) is $x^2 + y^2 = 10$. Let S be the region of the xy plane bounded by this circle. Then (x, y, z) is a point in the solid bounded by the two surfaces if and only if

$$x^2 + y^2 \leq 10, \quad 6x^2 + 9y^2 \leq z \leq 100 - 4x^2 - y^2.$$

The volume is evaluated as a double integral and as an iterated integral:

$$\iint_S (100 - 4x^2 - y^2) - (6x^2 + 9y^2)\, dS = \int_{-\sqrt{10}}^{\sqrt{10}} \int_{-\sqrt{10-x^2}}^{\sqrt{10-x^2}} 100 - 10x^2 - 10y^2\, dy\, dx.$$

The integration is rather complicated but can be carried out by making suitable substitutions in the single-variable integration. We give the results of a machine-aided computation:

$$\int_{-\sqrt{10-x^2}}^{\sqrt{10-x^2}} 100 - 10x^2 - 10y^2\, dy$$

$$= 200\sqrt{10 - x^2} - 20x^2\sqrt{10 - x^2} - 20\frac{(10 - x^2)^{3/2}}{3}.$$

Then the indefinite integral of this (with respect to x) equals

$$200\left(\frac{x\sqrt{10-x^2}}{2}+5\arcsin\left(\frac{x}{\sqrt{10}}\right)\right)$$
$$-20\left(\frac{x\sqrt{10-x^2}(-10+2x^2)}{8}\right)-250\arcsin\left(\frac{x}{\sqrt{10}}\right)$$
$$-20\left(\frac{x(50-2x^2)\sqrt{10-x^2}}{24}\right)-250\arcsin\left(\frac{x}{\sqrt{10}}\right).$$

The terms can be evaluated at $x=-\sqrt{10}$ and $x=\sqrt{10}$ to yield the final value of 500π for the integral. ■

Volumes by Polar Coordinates

The use of polar coordinates to describe a region in the xy plane can, in some cases, make the integration over the region simpler. The simplest regions of the xy plane, where this applies, are those that are described by a polar rectangle

$$P: \quad a\le r\le b, \qquad \alpha\le\theta\le\beta. \tag{15.6}$$

A more general region that we call a region of Type θr has the description

$$P: \quad \alpha\le\theta\le\beta, \qquad h_1(\theta)\le r\le h_2(\theta) \tag{15.7}$$

for nonnegative functions h_1 and h_2. With a region P having this description, the equations $x=r\cos\theta$ and $y=r\sin\theta$ transform points in the $r\theta$ plane to points in the xy plane. Let R denote the region into which the region P is transformed. In section 15.10 we prove the following change of variable formula that expresses the integral over R (in xy coordinates) as an integral over P (in $r\theta$ coordinates):

$$\iint_R f(x,y)\,dR=\int_\alpha^\beta\int_{h_1(\theta)}^{h_2(\theta)} f(r\cos\theta, r\sin\theta)\,r\,dr\,d\theta.$$

Take notice of the extra factor of r in the integrand. This factor arises as the Jacobian of the transformation, as explained in section 15.10.

Here is a case that illustrates the advantage of polar coordinates.

EXAMPLE 3 A drill of radius 1 is used to bore a hole through a solid ball of radius 4 in such a way that the center of the hole coincides with the center of the sphere. What is the volume of material removed from the ball?

Solution: The equation of the surface of the ball is $x^2+y^2+z^2=16$. Assume the hole is drilled along the z-axis. Then the volume of the solid removed is the part of the ball directly above and below the interior of the circle S with equation $x^2+y^2=1$ in the xy plane. If (x,y) is in S, then (x,y,z) is in the part of the material removed if and only if

$$-\sqrt{16-x^2-y^2}\le z\le\sqrt{16-x^2-y^2}.$$

So the volume of material removed is the double integral expressed as an iterated integral

$$\iint_S \sqrt{16 - x^2 - y^2} - (-\sqrt{16 - x^2 - y^2})\, dS$$

$$= 2\iint_S \sqrt{16 - x^2 - y^2}\, dS$$

$$= 2\int_{-1}^{1}\int_{-\sqrt{1-x^2}}^{\sqrt{1-x^2}} \sqrt{16 - x^2 - y^2}\, dy\, dx$$

We do not attempt to evaluate this integral in rectangular coordinates. The problem is simpler in polar coordinates. The region in the xy plane over which the integration is performed is described by

$$0 \le r \le 1, \quad 0 \le \theta \le 2\pi.$$

The surface is given by the equation $r^2 + z^2 = 4^2$ so that we have

$$b(r) = -\sqrt{16 - r^2} \le z \le \sqrt{16 - r^2} = t(r).$$

The volume of the hole is given by the integral of $t(r) - b(r)$, namely

$$V = \int_0^{2\pi}\int_0^1 2\sqrt{16 - r^2}\, r\, dr\, d\theta$$

$$= \int_0^{2\pi} -\frac{2}{3}(16 - r^2)^{3/2}\Big|_0^1\, d\theta$$

$$= \frac{4\pi}{3}(64 - 15^{3/2}) \approx 24.7358.$$

Let us try to check if this answer is reasonable. The material removed from the sphere fits into a cylinder of radius 1 and height 8 (the diameter of the sphere). Such a cylinder has volume $8\pi \approx 25.1327$. This seems close enough to feel confident that we have the correct number. ■

Exercises 15.3

Find the volume of the solid V described in each problem.

1. $V = \{(x, y, z) : 0 \le x \le 4, \quad 0 \le y \le x, \quad 0 \le z \le 8 - x - y\}$

2. $V = \{(x, y, z) : -2 \le y \le 2, \quad 0 \le x \le \sqrt{4 - y^2}, \quad -2 \le z \le x + y\}$

3. $V = \{(x, y, z) : 1 \le y \le 5, \quad -y \le x \le 5 - y, \quad 0 \le z \le x^2 + y^2\}$

4. $V = \{(x, y, z) : 0 \le x \le 1, \quad 0 \le y \le 2 - x^2, \quad -y \le z \le (x + y)^2\}$

Describe each region as a set of all (x, y, z) satisfying some inequalities in the form illustrated by the first four exercises.

5. V is bounded by the surfaces $z = 0$, $x = 0$, $y^2 = 4 - x$ and $z = y + 2$

6. V is bounded by the cylinders $y = x^2$ and $y = 4 - z^2$

7. V is bounded by the planes $z = 0$, $x + y + z = 1$, $-x - y + z = 1$, $-x + y + z = 1$, and $x - y + z = 1$

8. V is bounded by the surfaces $z = 0$, $x^2 + y^2 = 4$, $z = 2x^3 + 3y^2$

9. V is bounded by the surfaces $z = 0$, $(y - 1)^2 + x^2 = 1$, and $z = 16 - 4y^2 - x^2$

10. V is bounded by the surfaces $z = 0$, $z = x^2 + y^2$ and $x^2 + y^2 = 1$

Change to polar coordinates to evaluate the following integrals.

11. $\iint_R \sqrt{1 + x^2 + y^2}\, dR$, R is the disc $x^2 + y^2 \le a^2$

12. $\iint_R e^{-(x^2+y^2)}\, dR$, R is the disc $x^2 + y^2 \le a^2$

13. $\iint_R x^2\, dR$, R is the half disc $x^2 + y^2 \le 1$, $y \ge 0$

14. $\iint_R \frac{dR}{x^2 + y^2}$, R is the quarter disc $x^2 + y^2 \le a^2$, $x \ge 0$, $y \le 0$

15. $\iint_R (1 - x^2 - y^2)\, dR$, R is the triangle with vertices

(0, 0), (1, 0), (1, 1)

16. Rework example 2 from this section by changing to polar coordinates.

17. A cylindrical hole of radius a is drilled through the center of a sphere of radius R, $R > a$. Express the volume of the solid that remains in terms of the height of the remaining solid. The result is unexpected because it does not involve either the radius of the hole or the radius of the sphere.

18. The indefinite integral $\int e^{-x^2}\,dx$ cannot be expressed as an elementary function so the evaluation of a definite integral $\int_a^b e^{-x^2}\,dx$ must be approximated by numerical methods. However the improper integral $I = \int_0^\infty e^{-x^2}\,dx$ can be evaluated exactly using the following "trick":

$$I^2 = \left(\int_0^\infty e^{-x^2}\,dx\right)\left(\int_0^\infty e^{-y^2}\,dy\right)$$
$$= \iint_R e^{-(x^2+y^2)}\,dR,$$

where R is the first quadrant of the xy plane. Change to polar coordinates, represent R as the region $0 \leq \theta \leq \pi/2$ $0 \leq r < \infty$ and deduce that $I = \sqrt{\pi/8}$.

15.4 Center of mass and moments of inertia

The material in this section is presented as an application of double integrals; the concepts presented here play an important role in the study of mechanics. We do not give the motivation from physics; just a statement of definitions and evaluation of several examples.

We have considered moments and centers of mass of plane lamina in chapter 6. In the examples to be presented in this section, we are able to compute moments and centers of mass for lamina having nonconstant density. We now consider some of these fairly general situations in which the computations can be made with a double integral.

The density function of a solid V can be defined as follows. Let (x, y, z) be a point in a solid V and let $M(\delta)$ be the mass of a cube with side of length δ and center at (x, y, z). The volume of the cube is $v(\delta) = \delta^3$ and the average mass (or mass per unit volume) is the ratio $M(\delta)/v(\delta)$. The limit as $\delta \to 0$ of this ratio is called the density at (x, y, z) and is denoted by $\rho(x, y, z)$. We usually assume that ρ is continuous at every point of V. This might be only an approximation to a real physical situation but it is a good approximation in most cases.

If V is a plane figure covering a region S of the xy plane and having unit thickness, then its volume is numerically the surface area. If the units are such that the thickness is a constant c, then the volume is c times the surface area. We use $c = 1$ in most cases.

Now consider the case in which the density of the figure is given by a function $\rho(x, y)$, which is assumed to be continuous over S. We want to express the total mass as an integral. Partition the region into small subrectangles S_{ij} having area ΔS_{ij} so that the density varies little over S_{ij}. If (x_i, y_j) is a point in S_{ij} then the density is $\rho(x_i, y_j)$ and the mass is $\rho(x_i, y_j)\Delta S_{ij}$. We take the sum of these approximations over all subrectangles of the partition to get an approximation

$$\sum_i \sum_j \rho(x_i, y_j)\Delta S_{ij}$$

of the mass. In the limiting case (as the partitions are taken with norm approaching

0) the mass M is given as the double integral

$$M = \iint_S \rho(x, y)\, dS.$$

Moments

The moment of a point mass about a line is the product of the mass and the distance from the point to the line. The moment of a region S about a line L is first approximated by taking a partition of S into subrectangles S_{ij}, selecting a point (x_i, y_j) in the subrectangle and forming the sum of the products $\rho(x_i, y_j) D_L(x_i, y_j) \Delta S_{ij}$, where $\rho(x_i, y_j)\Delta S_{ij}$ is the approximate mass of the subrectangle and $D_L(x, y)$ is the distance from (x, y) to L. When we pass to the limit of such approximating sums we get the moment M_L of the region S, with density function ρ, about the line L is expressed as the double integral

$$M_L = \iint_S \rho(x, y) D_L(x, y)\, dS.$$

In the special cases where the line L is either the x- or y-axis we get a slightly simplified formula. The moment about the x-axis is

$$M_x = \iint_S y\rho(x, y)\, dS,$$

and the moment about the y-axis is

$$M_y = \iint_S x\rho(x, y)\, dS.$$

Note that the distance from (x, y) to the x-axis is y and the distance from (x, y) to the y-axis is x.

A typical problem dealing with moments is to find the center of mass of a region. Recall that the center of mass is the point where all of the mass of the region could be concentrated and the moment of the point about the x- or y-axis would be the same as for the entire region. If $(\overline{x}, \overline{y})$ is the center of mass for the region S then

$$(\overline{x}, \overline{y}) = \left(\frac{M_y}{M}, \frac{M_x}{M}\right).$$

Thus we compute the center of mass from the integrals giving the mass and the moments about the x- and y-axes:

$$\overline{x} = \frac{\iint_S x\rho(x, y)\, dS}{\iint_S \rho(x, y)\, dS}, \quad \overline{y} = \frac{\iint_S y\rho(x, y)\, dS}{\iint_S \rho(x, y)\, dS}.$$

EXAMPLE 1 Let S be the triangle with vertices $(0, 0)$, $(1, 0)$ and $(0, 2)$ and suppose a plate covers the region S exactly and the density of the plate $\rho(x, y) = k(3 - x)$ for a positive constant k. Find the mass and center of mass of the plate.

Solution: Two sides of the triangle lie along the coordinate axes and the third side is part of the line with equation $2x + y = 2$. The mass is given by

$$\begin{aligned} M &= \iint_S k(3 - x)\, dS = \int_0^1 \int_0^{2-2x} k(3 - x)\, dy\, dx \\ &= k\int_0^1 6 - 8x + 2x^2\, dx = \frac{8k}{3}. \end{aligned}$$

The moment around the x-axis is

$$M_x = \iint_S yk(3-x)\,dS = \int_0^1 \int_0^{2-2x} yk(3-x)\,dy\,dx$$
$$= \int_0^1 \frac{1}{2}(2-2x)^2 k(3-x)\,dy\,dx = \frac{11k}{6}.$$

The moment about the y-axis is

$$M_y = \iint_S x\rho(x,y)\,dS = \int_0^1 \int_0^{2-2x} xk(3-x)\,dy\,dx$$
$$= \int_0^1 (2-2x)xk(3-x)\,dx = \frac{5k}{6}.$$

The center of mass is determined from these computations as

$$(\overline{x}, \overline{y}) = \left(\frac{M_y}{M}, \frac{M_x}{M}\right) = \left(\frac{5/6}{8/3}, \frac{11/6}{8/3}\right) = \left(\frac{5}{16}, \frac{11}{16}\right).$$

If the triangle had uniform density the center of mass would be be $(1/3, 2/3)$ so the center of mass of the nonuniform plate has moved slightly to the left and up from the center of mass of the uniform plate. ■

Moment of Inertia

The moment of inertia of a point mass about a line is the product of the mass and the square of the distance from the line. If S is the region with density function ρ and L is a line, then

$$I_L = \iint_S \rho(x,y) D_L(x,y)^2\,dS$$

is the moment of inertia of the region about the line. Note the line need not be in the plane of the regions S. The concept of moment of inertia plays an important part in the study of rotating objects and in other physical problems.

EXAMPLE 2 Find the moment of inertia about the z-axis of the circular disc bounded by the graph of $x^2 + y^2 = R^2$ if the density function is a constant.

Solution: Let ρ be the constant density. The moment of inertia about the z-axis is

$$I_z = \iint_S (x^2+y^2)\rho\,dS,$$

which can be evaluated by transforming to polar coordinates;

$$I_z = \int_0^{2\pi} \int_0^R r^2 \rho r\,dr\,d\theta = 2\pi\rho\frac{R^4}{4} = \frac{MR^2}{2}$$

where $M = \rho\pi R^2$ is the mass of the disc. ■

The Radius of Gyration

Suppose a plate of mass M has I as its moment of inertia about some line L. If all the mass is concentrated at a point whose distance from L is r then the point mass has Mr^2 as its moment of inertia. The number

$$r = \sqrt{\frac{I}{M}}$$

is called the *radius of gyration* of the plate and is the distance from L at which all the mass could be concentrated to produce the same moment of inertia about L.

Using the example of the uniform circular plate of radius R given just above, the moment of inertia about the z-axis was seen to be $I = MR^2/2$. The radius of gyration is then $r = R/\sqrt{2}$.

Exercises 15.4

Compute the moments around the x-axis and around the y-axis of the following regions R with the given density functions ρ and find the center of mass of R.

1. R is bounded by $y = 11 - x^2$ and $y = 4x - 10$, $\rho = 1$
2. R is bounded by $y = 6x - x^2$ and $y = x$, $\rho = 1$
3. R is bounded by $y = x^3$ and $y = x$ with $x \geq 0$, $\rho = 1$
4. R is bounded by $y = x^2 - 2$ and $y = 0$, $\rho = 1$
5. R is the triangle with verticies $(0, 0)$, $(2, 0)$, $(0, 2)$, $\rho = 2x$
6. R is the triangle with verticies $(0, 0)$, $(2, 0)$, $(0, 2)$, $\rho = x + y$
7. R is the triangle with verticies $(0, 0)$, $(2, 0)$, $(0, 2)$, $\rho = x^2 + y^2$
8. R is the half-disc $x^2 + (y - 1)^2 = 1$, $y \geq 0$, $\rho = xy$
9. Find the center of mass of the triangle with vertices $(0, 0)$, $(a, 0)$ and $(0, b)$ $(a > 0,\ b > 0)$ if the density varies in proportion to the distance from the x-axis. Repeat the problem if the density varies in proportion to the square of the distance from the y-axis.
10. Determine the moment of inertia and the radius of gyration of the rectangular plate with vertices at $(0, 0)$, $(a, 0)$, $(0, b)$ and (a, b) about the x-axis. With no further integration determine the moment of inertia of the plate about the y-axis.
11. Let I_x, I_y and I_z denote the moment of inertia about the x-axis, the y-axis and the z-axis, respectively, of a region in the xy plane. Prove $I_z = I_x + I_y$.
12. A region S is a "washer" in the shape of a circular disc of radius R with a circular hole of radius r removed from its center. Assuming the density in constant, compute the moment of inertia of S about (a) its diameter; (b) a line through the center perpendicular to the plane of the circles.
13. Find the center of mass of the half-ring that lies between the two half-circles, $x^2 + y^2 = a^2$, $x^2 + y^2 = A^2$, $y \geq 0$ and where $a < A$ and $\rho = 1$. How would you interpret the limit as a approaches A of your solution?
14. Let R be a region of the xy plane that completely contains a subregion S. Let R/S denote the region R with S removed (so R now has a hole where S used to be). Show that the moment of R/S about any line is found by subtracting the moment of S from the moment of R (about the same line).
15. Prove the Parallel Axis Theorem that states the following: Let a plate have moment of inertia I about a line L that passes through the center of mass of the plate. If L' is a line parallel to L at a distance d from L then the moment of inertia of the plate about L' is $I + d^2M$, where M is the mass of the plate.
16. If a plate lies in the xy plane and if I_L is the moment of inertia of the plate about a line L that is parallel to the z-axis, show that I_L is minimized when L passes through the center of mass of the plate.

15.5 Surface area

Let a surface S be given by the function $z = f(x, y)$ for all points (x, y) in some region A of the xy plane. We would like to compute the area of the surface S. To do this we follow a familiar pattern of approximating the surface area by first partitioning A into small rectangles and getting an approximation to the "patch" of the surface above each subrectangle of the partition.

Let us describe how we get an approximating patch. Suppose that A is a rectangle and that (u, v) is some point in A. As a first approximation to the surface area, we take a parallelogram in the tangent plane to S at $(u, v, f(u, v))$; the parallelogram in the tangent plane that we select is the one that projects onto A under the projection $(x, y, z) \to (x, y, 0)$. The area of this parallelogram can be determined from some basic data describing A and the tangent plane. Of course this first approximation might not be very close to the area of the surface so we partition A into a large number of small subrectangles and make the same computation of a tangent plane and parallelogram for each subrectangle. Then the sum of the areas of these patches gives an approximation to the surface area. Of course this idea works well only for fairly well-behaved functions f. Recall that the analogous situation for functions of one variable is the computation of the length of the graph of $g(x)$ over an interval. We obtained the formula for the arc length

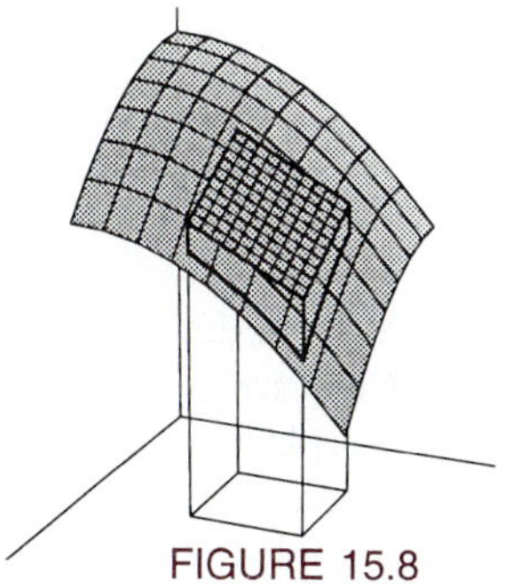

FIGURE 15.8
Patch in tangent plane approximates surface area

$$\int_a^b \sqrt{1 + g'(x)^2}\, dx$$

under the assumption that g' is continuous over the interval. In the case of surface area, we assume that f has continuous first partial derivatives throughout A. Let us set up some appropriate notation. Assume that A is a region contained in a rectangle R lying in the xy plane and suppose that f has continuous first derivatives in the rectangle R. Let

$$P : [x_0, \ldots, x_m; y_0, \ldots, y_n]$$

be a partition of R with subrectangles denoted by $R_{ij} = [x_{i-1}, x_i; y_{j-1}, y_j]$ and let $A_{ij} = A \cap R_{ij}$ be the part of the subrectangle that lies in A. Let ΔA_{ij} denote the area of A_{ij}. (Here we make an assumption about A, namely that it is a sufficiently nice region so that every A_{ij} has an area.) Select a point (u_i, v_j) in A_{ij} and let S_{ij} be a parallelogram in the tangent plane to the surface at $(u_i, v_j, f(u_i, v_j))$ that projects onto R_{ij} and let ΔS_{ij} denote its area. Then the sum

$$\sum_{i=1}^{m} \sum_{j=1}^{n} \Delta S_{ij} \qquad \textbf{(15.8)}$$

is an approximation to the area of the surface S. We take the position that as the partition gets finer, the sum of the areas of the patches get closer and closer to the surface area. We are led to the following natural computation. The area of the surface S is a number L that satisfies the condition: For every positive number ϵ, there is a positive number δ such that

$$\left| \sum_{i=1}^{m} \sum_{j=1}^{n} \Delta S_{ij} - L \right| < \epsilon$$

whenever the norm of the partition P is less than δ.

So we now have a plan for finding the surface area. If the function f is sufficiently well behaved, the surface area can be computed as an integral. We must express the areas ΔS_{ij} in terms of f and hope that an integral turns up as the number approximated by all the sums (15.8).

Area of One Patch

The tangent plane to S at $(u, v, f(u, v))$ has equation

$$f_x(u, v)(x - u) + f_y(u, v)(y - v) - (z - f(u, v)) = 0,$$

and a normal to this plane is the gradient of $f(x, y) - z$

$$\mathbf{N}(x, y) = f_x(x, y)\mathbf{i} + f_y(x, y)\mathbf{j} - \mathbf{k}$$

evaluated at $(x, y) = (u, v)$.

Let $\mathbf{p}$ and $\mathbf{q}$ be two vectors in the tangent plane that form two sides of the parallelogram; then the area of the parallelogram is $|\mathbf{p} \times \mathbf{q}|$. Now we want an expression for the area of the projection of this patch onto the xy plane. To find the projection of $\mathbf{p}$ onto the xy plane, we must subtract from $\mathbf{p}$ its component in the direction of $\mathbf{k}$. That component is $(\mathbf{p} \cdot \mathbf{k})\mathbf{k}$. The two vectors onto which $\mathbf{p}$ and $\mathbf{q}$ project are

$$\mathbf{p}' = \mathbf{p} - (\mathbf{p} \cdot \mathbf{k})\mathbf{k}, \quad \mathbf{q}' = \mathbf{q} - (\mathbf{q} \cdot \mathbf{k})\mathbf{k}. \tag{15.9}$$

The area of the rectangle with sides $\mathbf{p}'$ and $\mathbf{q}'$ is $|\mathbf{p}' \times \mathbf{q}'|$. It is necessary to relate the two areas. The relation between the expressions is clearer if we express them without absolute values.

The vector $\mathbf{p} \times \mathbf{q}$ is the cross product of two vectors in the tangent plane and so the cross product is parallel to the normal $\mathbf{N}$ to the plane. If the area of the patch is $a = |\mathbf{p} \times \mathbf{q}|$, then

$$\mathbf{p} \times \mathbf{q} = \pm a\mathbf{n}, \quad \text{where} \quad \mathbf{n} = \frac{\mathbf{N}}{|\mathbf{N}|}.$$

Similarly, the area $a' = |\mathbf{p}' \times \mathbf{q}'|$ is the length of a vector that is normal to the xy plane; that is

$$\mathbf{p}' \times \mathbf{q}' = \pm a'\mathbf{k}.$$

In particular, we have $a' = \pm\mathbf{p}' \times \mathbf{q}' \cdot \mathbf{k}$. Now we use this and the equations (15.9) to get the relation we need. The following computation uses the equation $\mathbf{r} \times \mathbf{k} \cdot \mathbf{k} = \mathbf{k} \times \mathbf{r} \cdot \mathbf{k} = 0$ for any vector $\mathbf{r}$:

$$\begin{aligned} \pm a' = \mathbf{p}' \times \mathbf{q}' \cdot \mathbf{k} &= \big(\mathbf{p} - (\mathbf{p} \cdot \mathbf{k})\mathbf{k}\big) \times \big(\mathbf{q} - (\mathbf{q} \cdot \mathbf{k})\mathbf{k}\big) \cdot \mathbf{k} \\ &= \mathbf{p} \times \mathbf{q} \cdot \mathbf{k} + 0 \\ &= \pm a\mathbf{n} \cdot \mathbf{k} = \pm a \cos\gamma, \end{aligned}$$

where γ is the angle between $\mathbf{n}$ and $\mathbf{k}$. We have to clear up the ambiguous signs. The vector $\mathbf{n}$ is a unit normal to the surface but could be pointing in either of two directions. Both a and a' are positive because they represent areas thus we should make a choice of unit normal $\mathbf{n}$ so that γ lies between $-\pi/2$ and $\pi/2$ thereby making $\cos\gamma$ nonnegative; that is $\mathbf{n}$ generally points toward the positive z direction (to within $\pi/2$). Such a choice is called the *outer normal* the surface. Thus we have the relation between the approximating patch and its projection written as

$$a = \frac{a'}{\cos\gamma}, \quad \cos\gamma = \mathbf{k} \cdot \mathbf{n}$$

where $\mathbf{n}$ is the unit outer normal. Replacing a' by ΔR_{ij} in the approximating sum (15.8), we find

$$\sum_{i=1}^{m}\sum_{j=1}^{n} \Delta S_{ij} = \sum_{i=1}^{m}\sum_{j=1}^{n} \frac{1}{\cos\gamma}\Delta R_{ij}.$$

(Of course the angle γ depends on i and j and the particular point of the subrectangle at which the normal vector was computed. It would be more precise to write γ_{ij} in place of just γ.) This latter sum is exactly an approximating sum for a double integral over A of the function $1/\cos\gamma$ where γ is determined from the f as explained above. Thus we find the surface area is

$$\iint_A \frac{1}{\cos\gamma}\, dA, \qquad \cos\gamma = \mathbf{n}\cdot\mathbf{k}.$$

Let us interpret these computations and obtain the result more directly expressed in terms of f.

THEOREM 15.7

Surface Area

Let S be the surface given by $z = f(x, y)$ lying over a region A of the xy plane. Assume that f has continuous first partial derivatives. Then the area of S is given by the integral

$$\iint_A \sqrt{1 + f_x(x, y)^2 + f_y(x, y)^2}\, dA.$$

Proof

Let us assume that $f(x, y) > 0$ over S. If this is not the case, we could add a large constant to f to reposition the surface above the xy plane without changing the surface area of the integral in the conclusion. A normal to the surface is found by taking the gradient of the function $F(x, y, z) = z - f(x, y)$; let

$$\mathbf{N}(x, y) = -f_x(x, y)\,\mathbf{i} - f_y(x, y)\,\mathbf{j} + \mathbf{k}.$$

The unit normal pointing in the direction of $\mathbf{N}$ is

$$\mathbf{n} = \frac{\mathbf{N}}{|\mathbf{N}|} = \frac{-f_x\,\mathbf{i} - f_y\,\mathbf{j} + \mathbf{k}}{\sqrt{1 + f_x^2 + f_y^2}},$$

Then

$$\cos\gamma = \mathbf{n}\cdot\mathbf{k} = \frac{1}{\sqrt{1 + f_x^2 + f_y^2}},$$

and so

$$\iint_A \frac{1}{\cos\gamma}\, dA = \iint_A \sqrt{1 + f_x(x, y)^2 + f_y(x, y)^2}\, dA,$$

as we wished to prove. Note that we did make certain that $\cos\gamma$ was nonnegative as the earlier discussion had required. □

The expression $dS = \sqrt{1 + f_x(x, y)^2 + f_y(x, y)^2}\, dA$ is called the *differential of surface area*. The differential of surface area can have different forms if the surface is not given in the rectangular form $z = f(x, y)$.

Let us apply the surface area formula in a few examples.

EXAMPLE 1 The cylinder $x^2 + y^2 = 4$ cuts the plane $ax + by + c = z$ in a closed curve. Find the area it encloses on that plane.

Solution: We are asking for the area of the surface with equation $z = f(x, y) =$

$ax + by + c$ lying over the region A of the xy plane bounded by the circle of radius 2 with center at the origin. The surface area is

$$\iint_A \sqrt{1 + f_x(x, y)^2 + f_y(x, y)^2}\, dA = \iint_A \sqrt{1 + a^2 + b^2}\, dA$$
$$= \sqrt{1 + a^2 + b^2} \iint_A dA = \sqrt{1 + a^2 + b^2}\, 4\pi.$$

We have evaluated the double integral $\iint_A dA = 4\pi$ as the area of the circle A of radius 2. ■

In the next example we exhibit a function with a very unusual property.

EXAMPLE 2 Show that the area of the surface given by the equation

$$f(x, y) = \frac{1}{2}\left[e^{(x+y)/\sqrt{2}} + e^{-(x+y)/\sqrt{2}}\right]$$

over a region S of the xy plane numerically equals the volume of the solid bounded below by S and above by the graph of f.

Solution: It is necessary to compute the function that is integrated to give the surface area but before doing so we make some changes in notation.

The hyperbolic cosine and hyperbolic sine are defined by

$$\cosh u = \frac{e^u + e^{-u}}{2}, \qquad \sinh u = \frac{e^u - e^{-u}}{2}$$

and satisfy the identities

$$\frac{d}{du}\cosh u = \sinh u, \qquad \frac{d}{du}\sinh u = \cosh u, \qquad \cosh^2 u - \sinh^2 u = 1.$$

Notice that $f(x, y) = \cosh[(x + y)/\sqrt{2}]$. Thus

$$1 + f_x^2 + f_y^2 = 1 + \frac{1}{2}\sinh^2\frac{x+y}{\sqrt{2}} + \frac{1}{2}\sinh^2\frac{x+y}{\sqrt{2}}$$
$$= 1 + \sinh^2\frac{x+y}{\sqrt{2}} = \cosh^2\frac{x+y}{\sqrt{2}}.$$

It follows that

$$\sqrt{1 + f_x^2 + f_y^2} = \cosh\frac{x+y}{\sqrt{2}} = f(x, y).$$

We conclude that

$$\int_S \sqrt{1 + f_x^2 + f_y^2}\, dS = \iint_S f(x, y)\, dS,$$

which implies that the surface area of the graph of f over S equals the volume of the solid below the graph of f and over S. ■

Areas of Surfaces in Parametric Form

The formula given in Theorem 15.7 is used to determine the area of a surface that is explicitly given as $z = f(x, y)$. An alternative formula is necessary if the surface is given in parametric form or in vector form. We develop such a formula now.

Suppose a surface S is given parametrically by the equations

$$\begin{cases} x = x(u,v), \\ y = y(u,v), \\ z = z(u,v), \end{cases}$$

where the point (u, v) is restricted to lie in some region R of the uv plane. For simplicity we assume that R is a rectangular region $R = [a, b; c, d]$ although the extension to more general regions of Type U or Type V (the analogue of Type X and Type Y for the xy plane) is equally valid.

Take a partition $P : [u_0, \ldots, u_m; v_0, \ldots, v_n]$ of R and focus on one subrectangle $R_{ij} = [u_{i-1}, u_i; v_{j-1}, v_j]$. Form a small rectangle in the tangent plane to S and use its area as an element in the sum that approximates the surface area.

Let

$$\mathbf{F}(u,v) = x(u,v)\,\mathbf{i} + y(u,v)\,\mathbf{j} + z(u,v)\,\mathbf{k} = x\,\mathbf{i} + y\,\mathbf{j} + z\,\mathbf{k},$$

so that $\mathbf{F}(u, v)$ is the position vector to a point in S. Let $Q = \mathbf{F}(u_{i-1}, v_{j-1})$ be a point in S; the tangent plane to S at Q can be described as follows. The vectors

$$\mathbf{w}_1 = \mathbf{F}_u(u_{i-1}, v_{j-1}) \quad \text{and} \quad \mathbf{w}_2 = \mathbf{F}_v(u_{i-1}, v_{j-1})$$

are tangent to S at Q. We assume they are not parallel vectors so that they can be used to describe the tangent plane by the function

$$\mathbf{v}(u,v) = \mathbf{F}(u_{i-1}, v_{j-1}) + (u - u_{i-1})\mathbf{w}_1 + (v - v_{j-1})\mathbf{w}_2.$$

For every (u, v) in R, $\mathbf{v}(u, v)$ is in the tangent plane to S at Q and $\mathbf{v}(u_{i-1}, v_{j-1}) = Q$. The rectangle $[u_{i-1}, u_i; v_{j-1}, v_j]$ corresponds to a rectangle R'_{ij} in the tangent plane. We take the area of this "patch" as an approximation to the surface area so that

$$\sum_{i=1}^{m}\sum_{j=1}^{n} \Delta R'_{ij}$$

serves as the approximation to the surface area corresponding to the partition P. We must express this sum as the sum of terms multiplied by the area of R_{ij} so that a double integral results as the limit of such sums.

The area of the patch R'_{ij} is the absolute value of the cross product of the two vectors that make two adjacent sides of the parallelogram, namely

$$\begin{aligned} \Delta R'_{ij} &= |(u_i - u_{i-1})\mathbf{w}_1 \times (v_j - v_{j-1})\mathbf{w}_2| \\ &= |\mathbf{F}_u(u_{i-1}, v_{j-1}) \times \mathbf{F}_v(u_{i-1}, v_{j-1})|(u_i - u_{i-1})(v_j - v_{j-1}). \end{aligned}$$

This cross product can be evaluated using the definition of $\mathbf{F}$ but we leave it in this form; notice that $(u_i - u_{i-1})(v_j - v_{j-1}) = \Delta R_{ij}$, the area of R_{ij}. Finally we obtain the approximation to the surface area is

$$\sum_{i=1}^{m}\sum_{j=1}^{n} |\mathbf{F}_u(u_{i-1}, v_{j-1}) \times \mathbf{F}_v(u_{i-1}, v_{j-1})|\Delta R_{ij}.$$

If we take a limit of such sums over a sequence of partitions whose norm approaches 0, the expression should approach what we intuitively accept as the area of the surface. The limit is also a double integral so we have

$$\text{Area of } S = \iint_R |\mathbf{F}_u(u,v) \times \mathbf{F}_v(u,v)|\, dR.$$

Let us compare this formula with that derived previously for the area of a nonparametric surface. Take $u = x$ and $v = y$ and $z = f(u, v) = f(x, y)$. Then

$$\begin{aligned} \mathbf{F}(x, y) &= x\,\mathbf{i} + y\,\mathbf{j} + f(x, y)\,\mathbf{k} \\ \mathbf{F}_x(x, y) &= \mathbf{i} + f_x(x, y)\,\mathbf{k} \\ \mathbf{F}_y(x, y) &= \mathbf{j} + f_y(x, y)\,\mathbf{k}. \end{aligned}$$

The absolute value of the cross product term is

$$\begin{aligned} |\mathbf{F}_x \times \mathbf{F}_y| &= |(\mathbf{i} + f_x\,\mathbf{k}) \times (\mathbf{j} + f_y\,\mathbf{k}) \\ &= |\mathbf{k} - f_x\,\mathbf{i} - f_y\,\mathbf{j}| \\ &= \sqrt{1 + f_x^2 + f_y^2}. \end{aligned}$$

So we recover the previous formula that the area of the surface is given by Theorem 15.7.

Now here is an example of an area computation that would be rather difficult without the use of some parametric representation of the surface.

EXAMPLE 3 Compute the area of the surface of a sphere of radius a.

Solution: The reader may want to try this with the representation $z = \sqrt{a^2 - x^2 - y^2}$ for the upper half sphere. The integration becomes quite cumbersome. Here is a parametric representation of the upper half of the sphere from which the integration is much simpler:

$$\begin{cases} x = a \cos u \cos v, \\ y = a \sin u \cos v, \\ z = a \sin v, \end{cases}$$

where $0 \le u \le 2\pi$ and $0 \le v \le \pi/2$. We restrict v to an interval where $\cos v$ and $\sin v$ are nonnegative; this parametrization sweeps out the upper half of the sphere of radius a. We should take care that we cover no region of the sphere more than once. For fixed v and $0 \le u < 2\pi$, (x, y) traces out a circle of radius $a \cos v$ and then $a \sin v$ is the distance from such a point to the upper half of the sphere. There is some double coverage of points when $u = 0$ and $u = 2\pi$ as well as at $v = 0$. This (as can be shown) does not affect the computation of area because the region of double coverage has area 0. The double coverage of some points could be avoided by dividing the surface into pieces and performing the computation over the separate pieces. The only overlap would be some boundary lines that have area 0.

In keeping with the notation just above, we set

$$\mathbf{F}(u, v) = a \cos u \cos v\,\mathbf{i} + a \sin u \cos v\,\mathbf{j} + a \sin v\,\mathbf{k}$$

and obtain

$$\begin{aligned} \mathbf{F}_u(u, v) &= -a \sin u \cos v\,\mathbf{i} + a \cos u \cos v\,\mathbf{j}, \\ \mathbf{F}_v(u, v) &= -a \cos u \sin v\,\mathbf{i} - a \sin u \sin v\,\mathbf{j} + a \cos v\,\mathbf{k}. \end{aligned}$$

The cross product is

$$\mathbf{F}_u \times \mathbf{F}_v = a^2(\cos u \cos^2 v\,\mathbf{i} + \cos^2 v \sin u\,\mathbf{j} + \cos v \sin v\,\mathbf{k})$$

and the absolute value is

$$\begin{aligned} |\mathbf{F}_u \times \mathbf{F}_v| &= a^2\sqrt{\cos^2 u \cos^4 v + \cos^4 v \sin^2 u + \cos^2 v \sin^2 v} \\ &= a^2|\cos v| = a^2 \cos v. \end{aligned}$$

The absolute value signs can be removed because we have selected an interval for v on which $\cos v$ is nonnegative. Thus the formula for the area of the upper half of the sphere is

$$\iint_R |\mathbf{F}_u \times \mathbf{F}_v|\, dR = a^2 \int_0^{2\pi} \int_0^{\pi/2} \cos v \, dv \, du,$$

which is easily evaluated to give the numerical answer $2\pi a^2$. Because this was the area of only half the sphere, the area of the full sphere of radius a is $4\pi a^2$. ■

The parametric representation of a surface that is initially given in rectangular coordinates can be viewed as a change of variables problem for the double integral that gives the surface area. The change of variables problem for more general double integrals is discussed in section 15.10.

Integrals over a Surface

We now mention the more general notion of an integral of a function over a surface. Let S be a surface given either as $z = f(x, y)$ with (x, y) in a region G of the xy plane or in parametric form. Suppose that F is a function defined at points of a region V in xyz space that contains S. We assume that F is continuous on V. The surface integral of F over S is the limit of approximating sums of the form

$$\sum F(u_i, v_i, w_i)\Delta S_i$$

where (u_i, v_i, w_i) is a point in a "patch" on S corresponding to some partition and ΔS_i is the area of that patch on S. If $dS = \sqrt{1 + f_x^2 + f_y^2}\, dG$ is the element of surface area then the surface integral of F over S is evaluated as

$$\int_S F\, dS = \iint_G F(x, y, z)\sqrt{1 + f_x^2 + f_y^2}\, dG.$$

If the surface is given by a vector function $\mathbf{r}(u, v)$ with (u, v) in a region G of the uv plane then an approximating sum is obtained from a partition of G and the area of a patch is $|\mathbf{r}_u \times \mathbf{r}_v|$. The surface integral of F over S is given by

$$\iint_G F(\mathbf{r}(u, v))|\mathbf{r}_u \times \mathbf{r}_v|\, dG.$$

We give an illustration of this concept in each form.

EXAMPLE 4 A container has the shape of the surface $z = x^2+y^2$ for $x^2+y^2 \leq 4$ but its walls (of negligible thickness) have variable density with density function $\rho(x, y, z) = k|xy|$. Find the mass of the container.

Solution: Let G be the disc in the xy plane bounded by the circle $x^2 + y^2 = 4$ so the surface lies above G. If we partition G into small subrectangles G_{ij} and let (u_i, v_j) be a point in G_{ij} then the small patch of the surface above G_{ij} has mass approximated by $k|u_i v_j|\Delta S_{ij}$ where S_{ij} is the area of the patch in the surface just above G_{ij}. By taking sums over all subrectangles of the partition we get an approximation to the mass that has, in the limiting case, the value given by the surface integral of $k|xy|$ over S. We find it useful to use the symmetry of the problem to eliminate the absolute values. If we partition the region G into the four quadrants, the mass is the same over each quadrant as we see by the symmetry of the container and of the density function. Hence the total mass is four times the mass of the part of the container over the first quadrant G_1. The advantage of doing the calculation over only the first quadrant is that $k|xy| = kxy$

for (x, y) in the first quadrant. We evaluate the surface integral as an iterated integral:

$$\begin{aligned}\int_S k|xy|\,dS &= 4\int_{G_1} kxy\sqrt{1+z_x^2+z_y^2}\,dG \\ &= \int_0^2\int_0^{\sqrt{4-x^2}} kxy\sqrt{1+4x^2+4y^2}\,dy\,dx \\ &= \frac{k}{3}\int_0^2 17^{\frac{3}{2}}x - x\left(1+4x^2\right)^{\frac{3}{2}}\,dx \\ &= \frac{k}{3}\left(\frac{1}{20}+2\cdot 17^{\frac{3}{2}}-\frac{17^{\frac{5}{2}}}{20}\right) \approx 80.65\,k.\end{aligned}$$

■

EXAMPLE 5 The drum of a photocopy machine is right circular cylinder on which an electrostatic charge is placed to attract the toner. In a testing mode, the charge is distributed on the drum at a density inversely proportional to the distance from a point source. If the surface of the drum is represented by the vector function $\mathbf{r}(u, v) = 4\cos u\,\mathbf{i} + 4\sin u\,\mathbf{j} + v\,\mathbf{k}$ with $0 \le u \le 2\pi$, $-7 \le v \le 7$ then the source is located at the point $(8, 0, 0)$. Assume that the only part of the drum that receives any charge is the part that can be "seen" from the source without obstruction from other parts of the drum. Write an integral that represents the total charge on the drum.

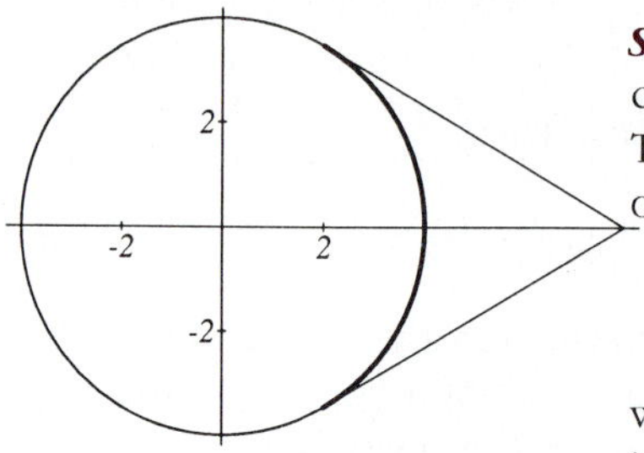

FIGURE 15.9
End-view of cylinder

Solution: The geometry of the configuration implies that the source at $(8, 0, 0)$ can "see" the points $\mathbf{r}(u, v) = 4\cos u\,\mathbf{i} + 4\sin u\,\mathbf{j} + v\,\mathbf{k}$ for which $-(\pi/3) \le u \le \pi/3$. The distance from the point $\mathbf{r}(u, v)$ to the source is $|\mathbf{r}(u, v) - 8\,\mathbf{i}|$ so the amount of charge on a small patch at the point corresponding to (u, v) is

$$k\frac{|\mathbf{r}_u \times \mathbf{r}_v|}{|\mathbf{r}(u, v) - 8\,\mathbf{i}|}\Delta = k\frac{4}{\sqrt{80 - 64\cos u + v^2}}\Delta$$

where k is a constant of proportionality and Δ is the area of the subrectangle in the uv plane. The integral over the surface is then expressed as the iterated integral

$$\int_{-\pi/3}^{\pi/3}\int_{-7}^{7} k\frac{4}{\sqrt{80 - 64\cos u + v^2}}\,dv\,du.$$

The inside integral can be evaluated exactly but the second integral cannot be done exactly. An approximation method gives the numerical result of about $19.017k$. ■

Surface integrals are discussed again when we consider Stokes's Theorem and the Divergence Theorem in chapter 16.

Exercises 15.5

Find an equation in rectangular form for the (parts of) quadric surfaces with the parametric form given in each exercise.

1. $\mathbf{r}(u, v) = pv\sin u\,\mathbf{i} + qv\cos u\,\mathbf{j} + v^2\,\mathbf{k}$

2. $\mathbf{r}(u, v) = a\cos u\sin v\,\mathbf{i} - b\cos u\cos v\,\mathbf{j} + c\sin u\,\mathbf{k}$

3. $\mathbf{r}(u, v) = a\cos u\sin v\,\mathbf{i} + b\cos u\cos v\,\mathbf{j} - c\sin u\,\mathbf{k}$

4. $\mathbf{r}(u, v) = a\sin u\sin v\,\mathbf{i} - b\sin u\cos v\,\mathbf{j} - c\cos u\,\mathbf{k}$

Find the area of the surface S described in each of the following six exercises.

5. S is the surface with equation $z^2 = 2xy$ lying above the rectangle $R = [1, 2; 1, 3]$ in the xy plane

6. S is the surface with equation $z = \dfrac{1}{x+y} - \dfrac{(x+y)^3}{12}$ that lies over the region $1 \leq x \leq 4$, $1 \leq y \leq 3$

7. S is the surface with equation $3z = x^{3/2} + y^{3/2}$ lying over the triangle $0 \leq y \leq 3$, $0 \leq x \leq y$

8. S is the part of the cylinder $x^2 + z^2 = 1$ inside the cylinder $x^2 + y^2 = 1$

9. S is the part of the sphere of radius $2R$ that lies inside a right circular cylinder of radius R situated so that the center of the sphere is on the surface of the cylinder

10. S is the part of the cone $z^2 = x^2 + y^2$ that lies inside the cylinder $(x-1)^2 + y^2 = 1$

11. Let S be the surface of a cylinder given by the equation $z = g(y)$ with g a positive, differentiable function. Show that the area A of S that lies above the rectangle $[a, b; c, d]$ in the xy plane can be computed as follows. Treat $z = g(y)$ as a curve in the yz plane and let L be the length of the arc of the graph of g for $c \leq y \leq d$. Then $A = (b-a)L$.

12. A plane P intersects the xy plane at an angle θ with $0 \leq \theta < \pi/2$. A region A in P projects to a region P' in the xy plane of area 1. What is the area of A?

13. Show that the ellipsoid with rectangular equation

$$\frac{x^2}{p^2} + \frac{y^2}{p^2} + \frac{z^2}{q^2} = 1$$

can be parameterized by the vector equation

$$\mathbf{r}(u, v) = p \cos u \sin v\, \mathbf{i} + p \sin u \sin v\, \mathbf{j} + q \cos v\, \mathbf{k},$$
$$0 \leq u \leq 2\pi, \quad 0 \leq v \leq \pi.$$

Use this to express the surface area of the ellipsoid as a single integral involving p and q. Evaluate the integral in the special case $p = q$.

14. A plane P is given in vector form as $\mathbf{r}(s, t) = s\mathbf{u} + t\mathbf{v}$ for two fixed vectors $\mathbf{u}$ and $\mathbf{v}$. A region R of P is defined by restricting s and t to the intervals $0 \leq s \leq 1$ and $0 \leq t \leq 1$. Suppose the density function for R is defined at the point $\mathbf{r}(s, t)$ as $\rho(s, t) = 2st$. Find the mass of the region by evaluating it as a surface integral.

15. Let $x\,\mathbf{i} + y\,\mathbf{j} = x(t)\,\mathbf{i} + y(t)\,\mathbf{j}$ for $0 \leq t \leq 1$ be a simple closed curve in the xy plane, with $x(t)$ and $y(t)$ differentiable functions of t. For $a > 0$ construct a "cone-like" surface S by drawing a straight line from $(0, 0, 0)$ to $(x(t), y(t), a)$ for $0 \leq t \leq T$. Show that the area of S is given by the single integral

$$\int_0^T \sqrt{a^2[x'(t)^2 + y'(t)^2] + [x'(t)y(t) - x(t)y'(t)]^2}\, dt.$$

Apply this in the case $x = R\cos t$, $y = R\sin t$ to determine the lateral area of a right circular cone of height a having base of radius R.

16. Evaluate the surface integral of the function $x^2 + xy + y^2$ taken over the surface of the cylinder $x^2 + y^2 = 4$ for $0 \leq z \leq 1$.

17. Evaluate the integral of $3x - 5y$ taken over the surface $z = x^2 + 2y^2$ for $0 \leq z \leq 1$.

18. Parametrize the ellipsoid S having equation

$$\frac{x^2}{a^2} + \frac{y^2}{b^2} + \frac{z^2}{c^2} = 1$$

by using parameters u, v, w and relations $x\,\mathbf{i} + y\,\mathbf{j} + z\,\mathbf{k} = au\,\mathbf{i} + bv\,\mathbf{j} + cw\,\mathbf{k}$ express express the surface integral

$$\iint_S f(x, y, z)\, dS$$

as a surface integral over the sphere of radius 1. Check your result by deriving the area of the surface of the ellipsoid given that the area of the sphere of radius 1 is 4π. [Hint: This change of variables does not precisely fit the description in the text because the surface is not given in the form $z = g(x, y)$. Derive a correct formula by computing the area of an approximating "patch" in terms of u and v and then writing the surface integrals as limits of approximations.]

15.6 Triple integrals

Now that the reader has seen the progression of ideas that passed from single integrals to double integrals, the passage to triple integrals should not be too surprising. The context in which a triple integral is defined is this: We have a region V of xyz space and a function $f(x, y, z)$ defined and continuous at every point of V. The triple integral

$$\iiint_V f(x, y, z)\, dV$$

is defined by the following limit. Let V be contained in the parallelogram

$$V_0 = [a, b; c, d; p, q] = \{(x, y, z) : a \leq x \leq b, \quad c \leq y \leq d, \quad p \leq z \leq q\}$$

and let

$$P : [x_0, \ldots, x_m; y_0, \ldots, y_n; z_0, \ldots, z_r]$$

be a partition of V_0 into subparallelepipeds $V_{ijk} = [x_{i-1}, x_i; y_{j-1}, y_j; z_{k-1}, z_k]$ having volume ΔV_{ijk} and let (u_i, v_j, w_k) be a point in $V \cap V_{ijk}$. Then a sum

$$\sum_{i=1}^{m} \sum_{j=1}^{n} \sum_{k=1}^{r} f(u_i, v_j, w_k) \Delta V_{ijk}$$

is called a Riemann sum of f over V corresponding to the partition P. The norm of P is, as usual, the maximum of the sides of the subparallelepipeds V_{ijk}. The triple integral is defined to be the number I, if it exists, such that for every $\epsilon > 0$ there is a $\delta > 0$ such that

$$\left| I - \sum_{i=1}^{m} \sum_{j=1}^{n} \sum_{k=1}^{r} f(u_i, v_j, w_k) \Delta V_{ijk} \right| < \epsilon$$

whenever P is a partition with norm less than δ.

For certain regions V the triple integral exists whenever f is continuous on V. We give the description of regions analogous to Type X and Type Y for double integrals and describe how the triple integral is evaluated as an iterated integral. Because the statements and proofs are analogous to the case of double integrals, no proofs are given.

Regions of Type XY and Type YX

A region in xyz space is called Type XY if its projection to the xy plane is a two dimensional region of Type X. A region of Type XY can be described by first restricting x to lie on an interval, then limiting the choice of y to an interval that depends on the choice of x and finally limiting z to an interval that depends on the choice of x and y. Thus a region of Type XY can be described as

$$V = \{(x, y, z) : a \leq x \leq b, \quad g_1(x) \leq y \leq g_2(x), \quad h_1(x, y) \leq z \leq h_2(x, y)\}.$$

We usually assume that the g_i and h_j are continuous and, at least, piecewise differentiable. When V admits this description, then the triple integral over V is an iterated integral

$$\iiint_V f(x, y, z)\, dV = \int_a^b \int_{g_1(x)}^{g_2(x)} \int_{h_1(x,y)}^{h_2(x,y)} f(x, y, z)\, dz\, dy\, dx.$$

A region of xyz space is of Type YX if its projection to the xy plane is a region of Type Y. Such a region has a description of the form

$$V = \{(x, y, z) : c \leq y \leq d, \quad g_1(y) \leq x \leq g_2(y), \quad h_1(x, y) \leq z \leq h_2(x, y)\}.$$

We give an illustration of regions of these types.

EXAMPLE 1 Evaluate the integral of $2xz$ over the region enclosed by the two paraboloids with equations $z = 12 - 2x^2 - y^2$ and $z = x^2 + 5y^2$.

Solution: The first paraboloid is an inverted "cup" and the second is an upright "cup". To describe the enclosed region, we must find the limitations on

x and y and then for each (x, y) we have the limitation of z by

$$h_1(x, y) = x^2 + 5y^2 \le z \le 12 - 2x^2 - y^2 = h_2(x, y).$$

The surfaces intersect in a curve which, when projected to the xy plane, defines the region over which (x, y) may vary. The intersection consists of the points (x, y, z) that satisfy both equations. Eliminating z from the two equations gives us

$$12 - 2x^2 - y^2 = x^2 + 5y^2$$

or

$$3x^2 + 6y^2 = 12.$$

The region of the xy plane bounded by this ellipse is a region of Type X with the description

$$\left\{(x, y) : -2 \le x \le 2, \quad -\frac{1}{2}\sqrt{4 - x^2} \le y \le \frac{1}{2}\sqrt{4 - x^2}\right\}.$$

When this is combined with the limitations on z, we obtain the description of V as a region of Type XY. Now we evaluate the triple integral as an iterated integral:

$$\begin{aligned}
&\iiint_V 2xz\,dV \\
&= \int_{-2}^{2}\int_{-\sqrt{4-x^2}/2}^{\sqrt{4-x^2}/2}\int_{x^2+5y^2}^{12-2x^2-y^2} 2xz\,dz\,dy\,dx \\
&= \int_{-2}^{2}\int_{-\sqrt{4-x^2}/2}^{\sqrt{4-x^2}/2} xz^2\Big|_{x^2+5y^2}^{12-2x^2-y^2} dy\,dx \\
&= \int_{-2}^{2}\int_{-\sqrt{4-x^2}/2}^{\sqrt{4-x^2}/2} x[(12 - 2x^2 - y^2)^2 - (x^2 + 5y^2)^2]\,dy\,dx \\
&= \int_{-2}^{2}\int_{-\sqrt{4-x^2}/2}^{\sqrt{4-x^2}/2} x[(144 - 48x^2 + 3x^4) - (24 + 6x^2)y^2 - 24y^4]\,dy\,dx.
\end{aligned}$$

The inside integral can be evaluated directly to give a value

$$x\left[(144 - 48x^2 + 3x^4) - (24 + 6x^2)\frac{1}{12}(4 - x^2) - \frac{3}{10}(4 - x^2)^2\right]\sqrt{4 - x^2}.$$

The integration of this function of x can be accomplished with the substitution $u = 4 - x^2$ and $du = -2x\,dx$ to produce the value 0 for the integral. (All that work for 0.) We omit the details of the computation.

The region V could have been also described as the set of all (x, y, z) satisfying

$$-\sqrt{2} \le y \le \sqrt{2}, \quad -\sqrt{4 - 2y^2} \le x \le \sqrt{4 - 2y^2}, \quad h_1(x, y) \le z \le h_2(x, y),$$

which shows it to be a Type YX region. ■

Volume as a Triple Integral

For a region V in xyz space, the triple integral

$$\iiint_V 1\,dV$$

equals the volume of V. If V is described as a Type XY region

$$V = \{(x, y, z) : a \leq x \leq b, \quad g_1(x) \leq y \leq g_2(x), \quad h_1(x, y) \leq z \leq h_2(x, y)\},$$

then the volume of V is computed as the iterated integral

$$\int_a^b \int_{g_1(x)}^{g_2(x)} \int_{h_1(x,y)}^{h_2(x,y)} dz\, dy\, dx.$$

Analogous iterated integrals give the volume if V is described as a Type YX region or even one of the many other variations such as Type XZ or Type ZY, and so on.

EXAMPLE 2 Express the volume as an iterated integral of the region bounded above by the cylinder with equation $z = 8 - x^2$ and below by the plane with equation $y - z = 4$ with y limited by the inequality $y \geq 0$.

Solution: The upper surface is formed by lines parallel to the y-axis meeting the xz plane at points of the plane curve $z = 8 - x^2$. This surface intersects the plane $y - z = 4$ at points (x, y, z) whose coordinates satisfy both equations; eliminating z gives the condition $y = 12 - x^2$. The projection of the region into the xy plane is the set $(x, y, 0)$ satisfying

$$-2\sqrt{3} \leq x \leq 2\sqrt{3}, \quad 0 \leq y \leq 12 - x^2.$$

For any (x, y) in this projection z is restricted by the condition

$$y - 4 \leq z \leq 8 - x^2.$$

Thus the region of xyz space has been described as a Type XY region and the iterated integral giving the volume is

$$\begin{aligned}
&\int_{-2\sqrt{3}}^{2\sqrt{3}} \int_0^{12-x^2} \int_{y-4}^{8-x^2} dz\, dy\, dx \\
&= \int_{-2\sqrt{3}}^{2\sqrt{3}} \int_0^{12-x^2} (8 - x^2) - (y - 4)\, dy\, dx \\
&= \int_{-2\sqrt{3}}^{2\sqrt{3}} (8 - x^2)(12 - x^2) - \frac{1}{2}(12 - x^2)^2 + 4(12 - x^2)\, dx.
\end{aligned}$$

With a little computation, the value of the integral is found to be $768\sqrt{3}/5$. ■

Mass as a Triple Integral

A triple integral can be used to determine the mass of a solid of nonuniform density. The density function of a solid V can be defined as follows. Let (x, y, z) be a point in V and let $M(\delta)$ be the mass of a cube with side of length δ and center at (x, y, z). The volume of the cube is $v(\delta) = \delta^3$ and the average mass (or mass per unit volume) is the ratio $M(\delta)/v(\delta)$. The limit as $\delta \to 0$ of this ratio is called the density and denoted by $\rho(x, y, z)$. We usually assume that ρ is continuous at every point of V.

With $\rho(x, y, z)$ giving the density of a solid at the point (x, y, z) then a very small parallelepiped with center at (x, y, z) and volume ΔV_{ijk} has mass approximately equal to the product $\rho(x, y, z)\Delta V_{ijk}$ and so a sum of such terms gives an approximation to the mass. In the limiting case the mass is given by the triple integral

$$\iiint_V \rho(x, y, z)\, dV.$$

EXAMPLE 3 The density at a point of a certain solid sphere of radius R and center at the origin is proportional to the square of the distance from the center. Write the total mass as an iterated integral.

Solution: For the density function we have $\rho(x, y, z) = c(x^2+y^2+z^2)$ for some constant c. The sphere can be expressed as a region of Type XY as the set of all (x, y, z) satisfying

$$-R \leq x \leq R, \quad -\sqrt{R^2-x^2} \leq y \leq \sqrt{R^2-x^2},$$
$$-\sqrt{R^2-x^2-y^2} \leq z \leq \sqrt{R^2-x^2-y^2},$$

so the total mass is given by

$$\iiint_S \rho\, dS = \int_{-R}^{R} \int_{-\sqrt{R^2-x^2}}^{\sqrt{R^2-x^2}} \int_{-\sqrt{R^2-x^2-y^2}}^{\sqrt{R^2-x^2-y^2}} c(x^2+y^2+z^2)\, dz\, dy\, dx.$$

Because of the symmetry, this integral is best evaluated by expressing everything in polar coordinates, a step that is justified in the section on change of variables. We leave the evaluation by that means until later.

If you have a computer package that does integration, you might find the results in steps as follows:

$$\int_{-\sqrt{R^2-x^2-y^2}}^{\sqrt{R^2-x^2-y^2}} c(x^2+y^2+z^2)\, dz$$
$$= 2c(x^2+y^2)\sqrt{R^2-x^2-y^2} + \frac{2\left(R^2-x^2-y^2\right)^{3/2}}{3}.$$

Denote this expression by $P(x, y)$ and ask the computer to do some more work to get

$$\int_{-\sqrt{R^2-x^2}}^{\sqrt{R^2-x^2}} P(x, y)\, dy = \pi c x^2\left(R^2-x^2\right) + \frac{\pi c\left(R^2-x^2\right)^2}{2}.$$

Finally the last integration yields the value $4c\pi R^5/5$. ■

Moments as Triple Integrals

If V is a solid region in xyz space its moment about a line L can be expressed as a triple integral. Let $\rho(x, y, z)$ be the density function expressing the mass per unit volume at a point and $D_L(x, y, z)$ the distance from (x, y, z) to the nearest point on L. If V is partitioned into "small" parallelepipeds V_{ijk} having volume ΔV_{ijk} and if (x_i, y_j, z_k) is a point in V_{ijk}, then the approximate mass of V_{ijk} is $\rho(x_i, y_j, z_k)\Delta V_{ijk}$ and the moment of that mass about the line L is approximately $D_L(x_i, y_j, z_k)\rho(x_i, y_j, z_k)\Delta V_{ijk}$; if we take the sum over all parallelepipeds of the partition, we obtain an approximation to the moment of V about L. Passing to the limit in the usual way gives the moment of V about L as

$$m_L = \iiint_V D_L(x, y, z)\rho(x, y, z)\, dV. \tag{15.10}$$

The formula is simplified if L is one of the coordinate axes; for example the distance from (x, y, z) to the x-axis is $\sqrt{y^2 + z^2}$ with analogous formulas for distances to the other coordinate axes.

With exactly the same reasoning we can express the moment of V about a plane P; replace the function $D_L(x, y, z)$ in Equation (15.10) by a function $\delta_P(x, y, z)$ that gives the distance from (x, y, z) to the plane P then the moment of V about P is

$$m_P = \iiint_V \delta_P(x, y, z)\rho(x, y, z)\, dV.$$

If P is the yz plane then δ_P is written as δ_{yz} and we find $\delta_{yz}(x, y, z) = x$. The moments about the various coordinate planes are given by the formulas

$$m_{yz} = \iiint_V x\rho(x, y, z)\, dV,$$

$$m_{xz} = \iiint_V y\rho(x, y, z)\, dV,$$

$$m_{xy} = \iiint_V z\rho(x, y, z)\, dV.$$

If $M = M(V)$ denotes the mass of V then the center of mass is the point where all the mass of V can be concentrated so that the point mass has the same moment about the three coordinate planes as the entire solid V. If $(\overline{x}, \overline{y}, \overline{z})$ is the center of mass, then

$$\overline{x}M = m_{yz}, \qquad \overline{y}M = m_{xz}, \qquad \overline{z}M = m_{xy}.$$

We provide a rather straightforward example.

EXAMPLE 4 Let V be the cube $[1, 2; 2, 3; 0, 1]$ having nonconstant density function $\rho(x, y, z) = xyz$. Find the center of mass.

Solution: We compute the mass and the moments as triple integrals as follows:

$$M = \int_1^2 \int_2^3 \int_0^1 xyz\, dz\, dy\, dx = \int_1^2 \int_2^3 \frac{xy}{2}\, dy\, dx = \int_1^2 \frac{5x}{4}\, dx = \frac{15}{8},$$

$$m_{xy} = \int_1^2 \int_2^3 \int_0^1 xyz^2\, dz\, dy\, dx = \int_1^2 \int_2^3 \frac{xy}{3}\, dy\, dx = \int_1^2 \frac{5x}{6}\, dx = \frac{5}{4},$$

$$m_{xz} = \int_1^2 \int_2^3 \int_0^1 xy^2 z\, dz\, dy\, dx = \int_1^2 \int_2^3 \frac{xy^2}{2}\, dy\, dx = \int_1^2 \frac{19x}{6}\, dx = \frac{19}{4},$$

$$m_{yz} = \int_1^2 \int_2^3 \int_0^1 x^2 yz\, dz\, dy\, dx = \int_1^2 \int_2^3 \frac{x^2 y}{2}\, dy\, dx = \int_1^2 \frac{5x^2}{6}\, dx = \frac{35}{12}.$$

The center of mass $(\overline{x}, \overline{y}, \overline{z})$ is determined by these computations to give the values

$$(\overline{x}, \overline{y}, \overline{z}) = \left(\frac{m_{yz}}{M}, \frac{m_{xz}}{M}, \frac{m_{xy}}{M}\right) = \left(\frac{14}{9}, \frac{38}{15}, \frac{2}{3}\right).$$

Just to get some validation of our computations take note that the center of the cube is $(3/2, 5/2, 1/2)$ and this would be the center of mass if the density were constant. The density function is xyz that grows larger as the point moves away from the origin. Thus a small part of the cube far from the origin has greater mass than a part of equal volume nearer the origin. Thus the center of mass is shifted from the center of the cube a small amount in a direction away from the origin. Our computation shows the shift from the center is given by the vector

$$\frac{14}{9}\mathbf{i}+\frac{38}{15}\mathbf{j}+\frac{2}{3}\mathbf{k}-\left(\frac{3}{2}\mathbf{i}+\frac{5}{2}\mathbf{j}+\frac{1}{2}\mathbf{k}\right)=0.0556\,\mathbf{i}+0.0333\,\mathbf{j}+0.1667\,\mathbf{k}.$$

At least we verify that the shift is in the right direction. ■

Exercises 15.6

Evaluate the iterated integrals.

1. $\int_0^1\int_0^2\int_0^3 4\,dx\,dy\,dz$

2. $\int_{-1}^1\int_{-2}^0\int_{-3}^1 x\,dx\,dz\,dy$

3. $\int_0^a\int_0^b\int_0^c (x+y+z)^2\,dy\,dx\,dz$

4. $\int_{-1}^1\int_{-2}^2\int_{-3}^3 xyz^2\,dx\,dy\,dz$

Let V be the Type XY solid defined by (for a constant $p > 1$)

$$V=\{(x,y,z):1\le x\le p,\quad 1\le y\le x,\quad y\le z\le x\}.$$

Evaluate the triple integrals by first expressing each as an iterated integral.

5. $\iiint_V x^2-y^2+z^2\,dV$

6. $\iiint_V \frac{(1+x)(1+y)}{(1+z)^2}\,dV$

7. $\iiint_V xyz\,dV$

8. $\iiint_V \frac{(1+z)^2\,dV}{(1+x)^2(1+y)^2}$

9. Find the volume of the solid consisting of all (x, y, z) with $0\le x\le 1$, $x\le y\le 4x$ and $0\le z\le x^2+y^2$.

10. Find the volume of the solid consisting of the points (x, y, z) satisfying $0\le y\le 2$, $y-1\le x\le y^2+1$ and $0\le z\le x^2+y^2$.

11. Find the volume of the solid bounded by the two paraboloids $z=x^2+y^2$ and $z=2-x^2-y^2$.

12. Find the center of mass of the solid having constant density ρ consisting of the points (x, y, z) satisfying $-1\le x\le 1$, $0\le y\le x$, $0\le z\le 4-x^2+y^2$.

13. Find the center of mass of the tetrahedron of constant density that has vertices $(0,0,0)$, $(a,0,0)$, $(0,b,0)$ and $(0,0,c)$.

14. Find the center of mass of a cube having side of length a and for which the density varies in proportion with the square of the distance from one face.

15. Find the center of mass of a cube having side of length a and for which the density varies in proportion with the square of the distance from one vertex.

Let U be the unit cube $U = [0,1;0,1;0,1]$ having constant density $\rho = 1$.

16. Find the moment of U about the x-axis.

17. Find the moment of U about the xy plane.

18. Find the moment of U about the plane $y = -2$.

19. Find the moment of the sphere of radius a about its diameter and about a plane passing through its center.

20. Let V be a bounded solid region having constant density. Suppose L is a line passing through V with the property that any plane normal to L that intersects V does so in a circular disc with L passing through the center. Prove that the center of mass lies on L.

21. Prove the Mean Value Theorem for triple integrals that states: Let V be a connected solid region in xyz space and f a continuous function defined throughout V. Then there is some point (x_1, y_1, z_1) in V such that

$$\iiint_V f\,dV = f(x_1,y_1,z_1)\,\mathrm{vol}(V).$$

To say V is connected means that between any two points of V there is a continuous curve lying entirely in V.

15.7 Line integrals

We now introduce the notion of a line integral. This concept bears some of the ingredients of the integral of a function of one variable where the integration is done along an interval and also some ingredients of the double integral because the computations are done at points in a plane.

Let C be a curve from an initial point $A = (a, b)$ to a terminal point $B = (c, d)$ and let f be a function defined and continuous in a rectangular region R containing the curve C. The concept of the line integral of f along C is defined by using the now familiar pattern of taking a partition of C, evaluating f at a point of the subarc, multiplying by a length related to the partition, summing these terms and then taking a limit. There is some choice in the selection of the multiplying factor that gives rise to several different forms of the line integral. Let us be more precise.

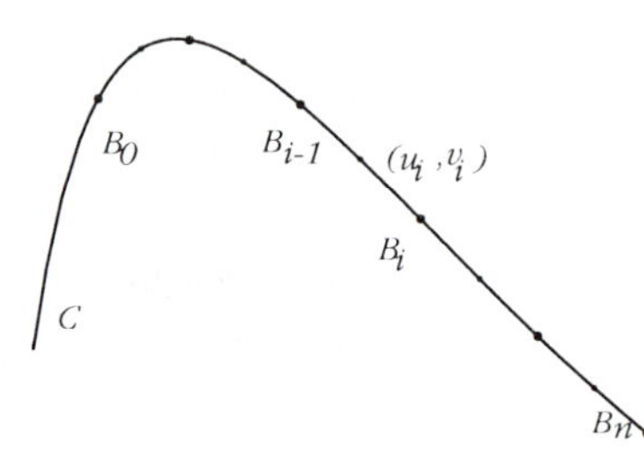

FIGURE 15.10
Partition of a curve

Partition the curve by selecting $n + 1$ points $B_0 = A, B_1, \ldots, B_n = B$ on the curve; let the coordinates of B_i be (x_i, y_i). Let C_i denote the arc of C from B_{i-1} to B_i. On each arc C_i select a point (u_i, v_i) and form the sum of the terms $f(u_i, v_i)(x_i - x_{i-1})$ to get a number depending on the partition and the selection of points on the subarc:

$$L_x = \sum_{i=1}^{n} f(u_i, v_i)(x_i - x_{i-1}). \qquad \textbf{(15.11)}$$

Note that there are other possible choices of the multiplier $(x_i - x_{i-1})$; for example, $(y_i - y_{i-1})$ or $\sqrt{(x_i - x_{i-1})^2 + (y_i - y_{i-1})^2}$ will be selected in a moment to give other line integrals.

Let the *norm of the partition* be the largest distance between two consecutive points B_j and B_{j+1}. We use the phrase *directed curve* to mean a curve for which one endpont has been designated as the initial point and another point designated as the terminal point. The definition of line integral requires the use of directed curves.

DEFINITION 15.2

Line Integral With Respect To x

The line integral of f with respect to x over the directed curve C is a number I with the following property: Keeping the notation introduced just above, for every positive number ϵ, there is a positive number δ such that

$$\left| \sum_{i=1}^{n} f(u_i, v_i)(x_i - x_{i-1}) - I \right| < \epsilon$$

whenever the norm of the partition of C is less than δ. If the number I exists with this property, we denote it by $\int_C f(x, y)\, dx$.

We remark on the form of this definition. The studious reader might have expected that we would have taken m_i and M_i as the minimum and maximum values of f on the arc C_i and then formed lower and upper sums of terms $m_i(x_i - x_{i-1})$ and $M_i(x_i - x_{i-1})$. The natural next step would require that $m_i(x_i - x_{i-1}) \leq M_i(x_i - x_{i-1})$ which, in the past, would follow from $m_i \leq M_i$ and $x_i - x_{i-1} \geq 0$. After this step we would pass to least upper bounds and the like. However this first step does not work because we do not know that $x_i - x_{i-1} \geq 0$. As we move along the curve, some of the x's might turn back. This little complication prompts us to disregard maximum and minimum values of f and use just any value of f on C_i.

We define two other line integrals that differ from the line integral with respect to x in the multiplying factors used in forming the sums in Equation (15.11). We define the line integral with respect to y of f over C as the number that differs from the sum

$$L_y = \sum_{i=1}^{n} f(u_i, v_i)(y_i - y_{i-1}) \qquad \textbf{(15.12)}$$

by no more than ϵ if the norm of the partition is less than some δ depending on ϵ. If such a number exists we denote it by $\int_C f(x, y)\,dy$.

Finally the line integral of f with respect to arc length is denoted by $\int_C f(x, y)\,ds$ and is defined to be the number that differs from all the sums

$$L_s = \sum_{i=1}^{n} f(u_i, v_i)\sqrt{(x_i - x_{i-1})^2 + (y_i - y_{i-1})^2} \qquad \textbf{(15.13)}$$

by no more than any given ϵ if the norm of the partition is less than δ.

Notice what happens in case the curve C is an interval $[a, c]$ on the x-axis. Then the subarcs C_i are just subintervals $[x_{i-1}, x_i]$ and the points $(u_i, v_i) = (u_i, 0)$ are just the coordinates of a partition of $[a, c]$. The sum $\sum f(u_i, 0)(x_i - x_{i-1})$ is just a Riemann sum for the function $f(x, 0)$ over $[a, b]$ and the line integral with respect to x coincides with the usual integral of a function over an interval

$$\int_C f(x, y)\,dx = \int_a^c f(x, 0)\,dx.$$

If C happens to be a vertical line, then all the points (x_i, y_i) have the same first coordinate. Thus the differences $(x_i - x_{i-1})$ are equal to 0 and $L_x = 0$. Thus the line integral with respect to x (of any function) equals 0.

Evaluation of Line Integrals

Now that we have definitions for the three type of line integrals, we need some conditions that ensure their existence and a method of evaluation. The conditions, as expected, depend both on the curve and the function to be integrated.

THEOREM 15.8

Evaluation of Line Integrals

Let C be a curve given in parametric form as

$$C = \{(x, y) : x = x(t), y = y(t) \text{ for } p \le t \le q\}$$

with initial point $A = (x(p), y(p))$ and final point $B = (x(q), y(q))$. Assume that both $x'(t)$ and $y'(t)$ are continuous on the interval $[p, q]$. Let f be a function of (x, y) that is continuous in a rectangular region containing C. Then the line integrals of f over C exist and can be computed as ordinary integrals according to the formulas

$$\int_C f(x, y)\,dx = \int_p^q f(x(t), y(t))x'(t)\,dt,$$

$$\int_C f(x, y)\,dy = \int_p^q f(x(t), y(t))y'(t)\,dt,$$

$$\int_C f(x, y)\,ds = \int_p^q f(x(t), y(t))\sqrt{x'(t)^2 + y'(t)^2}\,dt.$$

Proof

We do not give all the details of the proof but simply give an indication of the argument. A partition of C arises from a partition

$$P : [t_0, t_1, \ldots, t_n]$$

of $[p, q]$ by setting $(x_i, y_i) = (x(t_i), y(t_i))$. The points (u_i, v_i) on the subarc C_i of C correspond to some point t_i^* on the interval $[t_{i-1}, t_i]$ with $(u_i, v_i) = (x(t_i^*), y(t_i^*))$. The sum for the line integral with respect to x has the form

$$\begin{aligned}\sum_{i=1}^{n} f(u_i, v_i)(x_i - x_{i-1}) &= \sum_{i=1}^{n} f(x(t_i^*), y(t_i^*))(x(t_i) - x(t_{i-1})) \\ &= \sum_{i=1}^{n} f(x(t_i^*), y(t_i^*))(x'(t_i^{**})(t_i - t_{i-1})\end{aligned}$$

where t_i^{**} is a point on $[t_{i-1}, t_i]$ that arises by an application of the Mean Value Theorem to the function $x(t)$. The final sum is (almost) a Riemann sum for the integral of $f(x(t), y(t))x'(t)$ over $[p, q]$. There is a subtlety that makes this sum slightly different from a Riemann sum; namely the two points t_i^* and t_i^{**} would be the same in a Riemann sum. The difference here can be made negligible by taking finer and finer partitions but we do not take the time to carry out the ϵ, δ computations to prove that the sums as written approach the integrals as stated in the theorem.

Similar arguments apply to the two remaining type of line integrals. □

Here are some examples of line integral computations.

EXAMPLE 1 Evaluate the line integrals with respect to x, with respect to y and with respect to arc length of the function $f(x, y) = x^2 - xy$ over the straight-line path from $(0, 0)$ to $(1, 2)$.

Solution: We begin with a parametrization of the curve as

$$C = \{(x, y) : x = t, \quad y = 2t, \quad 0 \leq t \leq 1\}.$$

Then at the point $(t, 2t)$ of C, f has the value $f(t, 2t) = t^2 - t(2t) = -t^2$. By applying the formulas in the theorem, we find

$$\begin{aligned}\int_C f(x, y)\, dx &= \int_0^1 (-t^2)(1)\, dt = -\frac{t^3}{3}\Big|_0^1 = -\frac{1}{3}, \\ \int_C f(x, y)\, dy &= \int_0^1 (-t^2)(2)\, dt = -\frac{2t^3}{3}\Big|_0^1 = -\frac{2}{3}, \\ \int_C f(x, y)\, ds &= \int_0^1 (-t^2)\sqrt{5}\, dt = -\sqrt{5}\frac{t^3}{3}\Big|_0^1 = -\frac{\sqrt{5}}{3}.\end{aligned}$$

■

There are instances when we combine the line integrals with respect to x and y; the notation

$$\int_C P(x, y)\, dx + Q(x, y)\, dy$$

stands for the sum of the two line integrals

$$\int_C P(x,y)\,dx \quad \text{and} \quad \int_C Q(x,y)\,dy.$$

EXAMPLE 2 Evaluate the integral $\int_C (x^2+y^2)\,dx + 2xy\,dy$ when C is the arc of the circle of radius 1, center at the origin, from $(1,0)$ to $(1/\sqrt{2}, 1/\sqrt{2})$.

Solution: We use the parametrization $x(t) = \cos t$ and $y(t) = \sin t$ for $0 \le t \le \pi/4$. Then $x'(t) = -\sin t$ and $y'(t) = \cos t$. Now we have

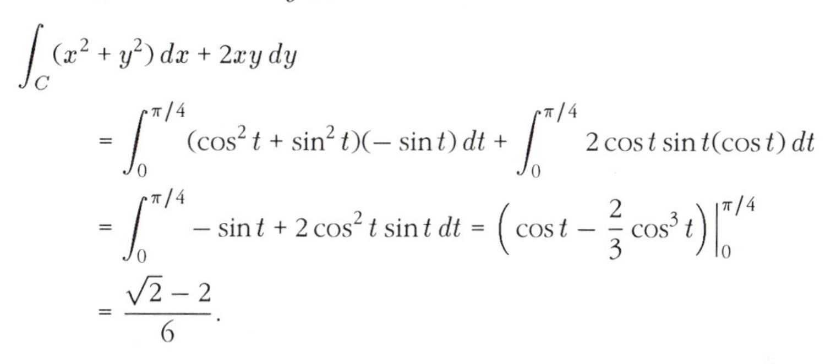

$$\begin{aligned}\int_C (x^2+y^2)\,dx + 2xy\,dy &= \int_0^{\pi/4} (\cos^2 t + \sin^2 t)(-\sin t)\,dt + \int_0^{\pi/4} 2\cos t \sin t(\cos t)\,dt \\ &= \int_0^{\pi/4} -\sin t + 2\cos^2 t \sin t\,dt = \left(\cos t - \frac{2}{3}\cos^3 t\right)\Big|_0^{\pi/4} \\ &= \frac{\sqrt{2}-2}{6}.\end{aligned}$$

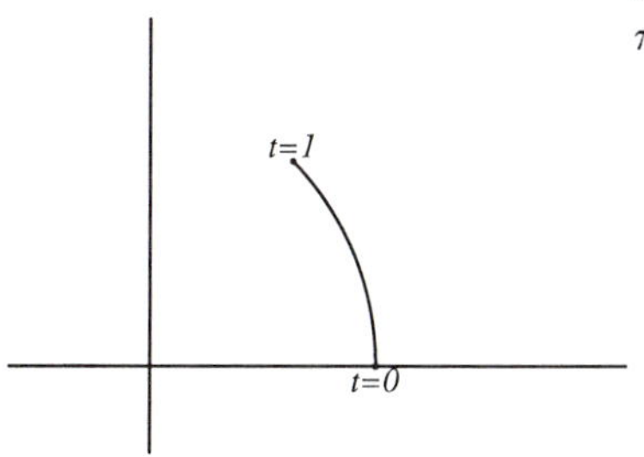

FIGURE 15.11
Path in example 2

■

Work as a Line Integral

The work done by applying a constant force to move an object in a straight line is the product of the force and the distance. If the force is not constant but the motion is still along a straight line, say along the interval from a to b, then the work is the integral

$$\int_a^b F(x)\,dx,$$

where $F(x)$ is the applied force at the point x.

We now consider the computation of the work done in applying a force that varies from point to point while moving an object along a curved path C. Assume that there is a force acting at each point of a region R given by

$$\mathbf{F}(x,y) = u(x,y)\,\mathbf{i} + v(x,y)\,\mathbf{j}.$$

If an object moves along C and if (x,y) is a point on C then the force acting on the object must be tangent to C. Thus the effect of $\mathbf{F}$ must be computed along the tangent to C. Let $\mathbf{T}(x,y)$ be a unit tangent vector to C at (x,y); then the dot product $\mathbf{F}\cdot\mathbf{T}$ represents the component of the force acting on the object. The work done is the line integral

$$\int_C \mathbf{F}\cdot\mathbf{T}\,ds.$$

We see this by getting an approximation to the work; let us take a partition of C by $n+1$ points $P_i = (x_i, y_i)$ very near each other. Then the work done in moving from P_{i-1} to P_i can be approximated by assuming that the force is constant and the path is a straight line. If (u_i, v_i) is a point on C between P_{i-1} and P_i, then the work done along this part of the path is approximately

$$\mathbf{F}(u_i,v_i)\cdot\mathbf{T}(u_i,v_i)\sqrt{(u_i-u_{i-1})^2 + (v_i-v_{i-1})^2},$$

the force at the point taken in the direction of the tangent to the path times the length of the path. If we sum these over a partition of C and then take limits as the norm of the partition approaches 0, the approximation approaches the line integral with respect to arc length of $\mathbf{F}(x, y) \cdot \mathbf{T}(x, y)$.

If the curve is given in vector form, say by the equation $\mathbf{r}(t)$ for $p \le t \le q$ then the unit tangent $\mathbf{T}$ is given by taking the derivative with respect to arc length,

$$\mathbf{T} = \frac{d\mathbf{r}}{ds} = \frac{d\mathbf{r}}{dt}\left(\frac{dt}{ds}\right) = \mathbf{v}\left(\frac{dt}{ds}\right).$$

Here, of course, $\mathbf{v}$ is the velocity when $\mathbf{r}$ is regarded as the position of a moving point. It now follows that $\mathbf{T}\,ds = \mathbf{v}\,dt$ and the work can be expressed in terms of t as

$$\int_C \mathbf{F} \cdot \mathbf{T}\,ds = \int_p^q \mathbf{F} \cdot \mathbf{v}\,dt \tag{15.14}$$

with F expressed in terms of t.

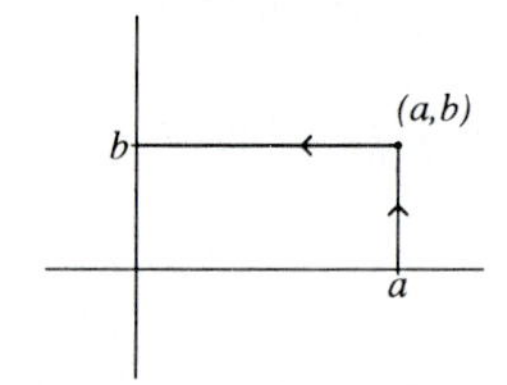

FIGURE 15.12
Path in example 3

EXAMPLE 3 Suppose an object moves under the effect of a variable force

$$\mathbf{F}(x, y) = \frac{x\,\mathbf{i} + y\,\mathbf{j}}{x^2 + y^2}$$

for $(x, y) \neq (0, 0)$ in a region of the first quadrant not containing the origin. Compute the work done in moving the object along the broken-line path from $(a, 0)$ to (a, b) to $(0, b)$ for a and b positive constants.

Solution: The path consists of two segments C_1 and C_2 that we treat separately.

A vector equation for C_1 is $\mathbf{r} = a\,\mathbf{i} + bt\,\mathbf{j}$ for $0 \le t \le 1$. The velocity vector along C_1 is the constant vector $b\,\mathbf{j}$, so the work done moving along C_1 is

$$\begin{aligned}\int_{C_1} \mathbf{F} \cdot \mathbf{T}\,ds &= \int_{C_1} \mathbf{F} \cdot \mathbf{v}\,dt \\ &= \int_0^1 \frac{a\,\mathbf{i} + bt\,\mathbf{j}}{a^2 + b^2t^2} \cdot b\,\mathbf{j}\,dt = \int_0^1 \frac{b^2 t}{a^2 + b^2 t^2}\,dt \\ &= \frac{1}{2}\ln(a^2 + b^2t^2)\Big|_0^1 = \frac{1}{2}[\ln(a^2 + b^2) - \ln a^2].\end{aligned}$$

A vector equation for C_2 is $\mathbf{r} = (a - at)\,\mathbf{i} + b\,\mathbf{j}$ for t ranging from 0 to 1. The velocity vector along C_2 is $-a\,\mathbf{i}$ so the work done along this part of the path is

$$\begin{aligned}\int_{C_2} \mathbf{F} \cdot \mathbf{T}\,ds &= \int_{C_2} \mathbf{F} \cdot \mathbf{v}\,dt \\ &= \int_0^1 \frac{(a - at)\,\mathbf{i} + b\,\mathbf{j}}{a^2t^2 + b^2} \cdot (-a\,\mathbf{i})\,dt = \int_1^0 \frac{a^2(t-1)}{a^2t^2 + b^2}\,dt \\ &= \frac{1}{2}\ln(a^2t^2 + b^2) - \frac{a}{b}\arctan\frac{at}{b}\Big|_0^1 \\ &= \frac{1}{2}[\ln(a^2 + b^2) - \ln b^2] - \frac{a}{b}\arctan\frac{a}{b}.\end{aligned}$$

The total work is the sum of the work on the two legs of the path. ■

Line Integrals in Space

The treatment of line integrals in the plane has a fairly direct extension to line integrals over a path in three-dimensional space. Suppose that B is a curve given in vector form by $\mathbf{r}(t) = x(t)\,\mathbf{i} + y(t)\,\mathbf{j} + z(t)\,\mathbf{k}$ for $a \le t \le b$. A line integral over B is an expression denoted by

$$\int_B P(x,y,z)\,dx + Q(x,y,z)\,dy + R(x,y,z)\,dz,$$

which is defined by taking limits of approximating sums of the form

$$\sum_{i=1}^{m}\sum_{j=1}^{n}\sum_{k=1}^{p} P(x_i - x_{i-1}) + Q(y_j - y_{j-1}) + R(z_k - z_{k-1})$$

where P, Q and R are evaluated at points on the curve between $(x_{i-1}, y_{j-1}, z_{k-1})$ and (x_i, y_j, z_k). The evaluation of the integral expressed in terms of the parameter is carried out exactly as in the two dimensional case to give

$$\int_a^b Px' + Qy' + Rz'\,dt$$

where all functions are expressed in terms of t; that is

$$\begin{aligned} Px' &= P(x(t), y(t), z(t))x'(t), \\ Qy' &= Q(x(t), y(t), z(t))y'(t), \\ Rz' &= R(x(t), y(t), z(t))z'(t). \end{aligned}$$

Exercises 15.7

Evaluate each of the line integrals over the indicated path. In these exercises either x or y can be used as parameter to describe the path.

1. $\int_C (x+2y)^2\,dx$, C is the straight line from $(0,0)$ to $(5,5)$

2. $\int_C xy^2\,dx$, C is the straight line from $(0,0)$ to $(2,5)$

3. $\int_C (x+2y)^2\,dy$, C is the straight line from $(0,0)$ to $(5,5)$

4. $\int_C xy^2\,dy$, C is the straight line from $(0,0)$ to $(2,5)$

5. $\int_C (x+y)\,dx + (x-y)\,dy$, C is the straight line from $(0,0)$ to $(3,6)$

6. $\int_C (x+y)\,dx + (x-y)\,dy$, C is the broken line from $(0,0)$ to $(0,6)$ to $(3,6)$

7. $\int_C (x+y)\,dx + (x-y)\,dy$, C is the broken line from $(0,0)$ to $(3,0)$ to $(3,6)$

8. $\int_C 2xy\,dx + (x^2+y^2)\,dy$, C is the arc of $y = x^2$ from $(0,0)$ to $(2,4)$

9. $\int_C xy^2\,dx + (x^2-y^2)\,dy$, C is the arc of $y^2 = x+1$ from $(0,1)$ to $(3,2)$

10. $\int_C x\,dx + (x+y)\,dy$, C is the arc of $y = x^2$ from $(0,0)$ to $(2,4)$

11. $\int_C x\,dx + y\,dy + z\,dz$, C is the straight line from $(0,0,0)$ to $(1,2,3)$

12. $\int_C y\,dx + z\,dy + x\,dz$, C is the straight line from $(0,0,0)$ to $(1,2,3)$

13. $\int_C (y+z)\,dx + (x+z)\,dy + (x+y)\,dz$, C is the arc of the path $x = 2t$, $y = 1-3t$, $z = t^2+1$ from $(0,1,1)$ to $(2,-2,2)$

14. Find the work done in moving an object along the straight path from $(0,0)$ to $(2,2)$ if the force at (x,y) is $\mathbf{F} = x\,\mathbf{i} + y\,\mathbf{j}$.

15. Find the work done in moving an object along the straight path from $(0,0)$ to $(a,2a)$ if the force at (x,y) is $\mathbf{F} = x\,\mathbf{i} - y\,\mathbf{j}$.

16. Find the work done in moving an object along the straight path from $(0,0,0)$ to $(2,2,2)$ if the force at (x,y,z) is $\mathbf{F} = x\,\mathbf{i} + y\,\mathbf{j} + z\,\mathbf{k}$. Repeat the calculation if the path is the broken line path from $(0,0,0)$ to $(2,0,0)$ to $(2,2,0)$ to $(2,2,2)$.

Evaluate the line integrals. Find a suitable parameter to describe the path if none is already specified.

17. $\int_C \frac{y}{\sqrt{x^2+y^2}}\,dx + \frac{x}{\sqrt{x^2+y^2}}\,dy$, C is the circle parametrized by $x = \cos t$ and $y = \sin t$ for $0 \le t \le 2\pi$

18. $\int_C \frac{-y}{\sqrt{x^2+y^2}}\,dx + \frac{x}{\sqrt{x^2+y^2}}\,dy$, C is the circle parametrized by $x = \cos t$ and $y = \sin t$ for $0 \le t \le 2\pi$

19. $\int_C \frac{x}{\sqrt{x^2+y^2}}\,dx + \frac{y}{\sqrt{x^2+y^2}}\,dy$, C is the circle parametrized by $x = \cos t$ and $y = \sin t$ for $0 \le t \le 2\pi$

20. $\int_C x^2\,dx + xy\,dy$, C is the broken line from $(0,0)$ to

(0, 4) to (4, 4)

21. $\int_C xy^2\,dx + x^2y\,dy$, C the square with vertices $(\pm 1, \pm 1)$ traversed in the counterclockwise direction

22. $\int_C x^2\,dx + xy\,dy + xz\,dz$, C is the broken line from (0, 0, 0) to (0, 4, 0) to (4, 4, 4)

23. $\int_C yz\,dx + xz\,dy + xy\,dz$, C is the broken line from (0, 0, 0) to (1, 2, 3) to (−1, −2, −3) to (0, 0, 0)

15.8 Path-independent line integrals

A line integral of the form $\int_C P\,dx + Q\,dy$ is called *independent of path* if it has the same value for any path C connecting two given points A and B. We always restrict our attention to paths that are differentiable except possibly at a finite number of points. Such a path is called *piecewise differentiable* because it is composed of a finite number of paths each differentiable except possibly at the endpoints. Let us give an example of a path independent integral.

EXAMPLE 1 Show that the value of the integral $\int_C y\,dx + x\,dy$ is the same for all piecewise differentiable paths C from (0, 0) to (1, 2).

Solution: Let C be a path with vector equation $\mathbf{r}(t) = x(t)\,\mathbf{i} + y(t)\,\mathbf{j}$ with x and y differentiable functions of t and $(x(0), y(0)) = (0, 0)$, $(x(1), y(1)) = (1, 2)$. Then

$$\begin{aligned}\int_C P\,dx + Q\,dy &= \int_0^1 \big(y(t)x'(t) + x(t)y'(t)\big)\,dt \\ &= \int_0^1 (x(t)y(t))'\,dt = x(t)y(t)\Big|_0^1 \\ &= x(1)y(1) - x(0)y(0) = 2.\end{aligned}$$

This shows that the line integral has the value 2 no matter what path is used from (0, 0) to (1, 2). The critical point in the computation is that the expression

$$x\,dy + y\,dx = \frac{d(xy)}{dt}dt$$

is precisely the differential of the product xy.

For this particular integral, the argument just given did not depend on the particular initial and terminal points. If we have C given by $\mathbf{r}(t) = x(t)\,\mathbf{i} + y(t)\,\mathbf{j}$ with $\mathbf{r}(0) = a\,\mathbf{i} + b\,\mathbf{j}$ and $\mathbf{r}(1) = c\,\mathbf{i} + d\,\mathbf{j}$ then the same computation as just made shows

$$\int_C P\,dx + Q\,dy = \int_0^1 y(t)x'(t) + x(t)y'(t)\,dt = x(1)y(1) - x(0)y(0) = cd - ab.$$

This is a special instance of a much more general phenomenon that we now investigate. ■

Exact Differentials

Let $f(x, y)$ be a differentiable function over a region S. The expression

$$df = f_x(x, y)\,dx + f_y(x, y)\,dy$$

is called *the differential of* f. The symbols dx and dy are to be regarded as variables independent of x or y; thus df is a function of four independent variables. We have previously discussed differentials in the context of approximations to function values (see the exercises following section 14.5). An expression of the form

$$P(x, y)\,dx + Q(x, y)\,dy$$

is called an *exact differential* if there is some function f such that $df = P(x, y)\,dx + Q(x, y)\,dy$ for all (x, y) in a region.

We make the connection between exact differentials and path-independent line integrals in the next theorem. A line integral $\int_C P(x, y)\,dx + Q(x, y)\,dy$ is independent of path throughout a region R if the value of the integral is the same for any two paths in R from one given point of R to another given point of R.

THEOREM 15.9

Line Integrals of Exact Differentials

Let $P(x, y)$ and $Q(x, y)$ be functions with continuous first partial derivatives throughout a rectangular region R. Suppose that for any two points (a, b) and (c, d) in R, the value of the line integral

$$\int_C P(x, y)\,dx + Q(x, y)\,dy \tag{15.15}$$

does not depend on the choice of differentiable path C from (a, b) to (c, d). Then there is a function f defined and differentiable throughout R such that

$$df = P(x, y)\,dx + Q(x, y)\,dy.$$

Conversely if $P(x, y)\,dx + Q(x, y)\,dy$ is an exact differential defined throughout R, then the line integral (15.15) is independent of path.

Proof

Suppose first that f is defined throughout R and

$$df = P(x, y)\,dx + Q(x, y)\,dy.$$

For any differentiable path C given by $\mathbf{r}(t) = x(t)\,\mathbf{i} + y(t)\,\mathbf{j}$ for $0 \le t \le 1$ lying in R, set $(a, b) = (x(0), y(0))$ and $(c, d) = (x(1), y(1))$. Then

$$\begin{aligned}\frac{d}{dt} f(x(t), y(t)) &= f_x(x, y)\frac{dx}{dt} + f_y(x, y)\frac{dy}{dt}\\ &= P(x(t), y(t))\frac{dx}{dt} + Q(x(t), y(t))\frac{dy}{dt}.\end{aligned}$$

It follows that

$$\begin{aligned}\int_C P(x, y)\,dx + Q(x, y)\,dy &= \int_0^1 \left(f_x(x, y)\frac{dx}{dt} + f_y(x, y)\frac{dy}{dt}\right) dt\\ &= \int_0^1 \frac{df(x(t), y(t))}{dt}\,dt = f(x(t), y(t))\Big|_0^1\\ &= f(x(1), y(1)) - f(x(0), y(0))\\ &= f(c, d) - f(a, b).\end{aligned}$$

This proves one implication of the theorem, namely that the value of a line integral of an exact differential depends only on the endpoints of the path. Now we must show that the independence of path implies the existence of a function f such that $P\,dx + Q\,dy$ is the differential of f. How do we construct f? We use the fact that the line integral between two points of R does not depend on the path but only on the two endpoints and the direction taken from one point to the other. Start with any point (a, b) in R and define f by the rule

$$f(u, v) = \int_C P(x, y)\,dx + Q(x, y)\,dy \quad C \text{ a path from } (a, b) \text{ to } (u, v).$$

The independence of path is important for the definition of f because the value of the integral is the same no matter what path is used to get from (a, b) to (u, v). Only the endpoints are significant, so we write

$$\int_C P(x, y)\,dx + Q(x, y)\,dy = \int_{(a,b)}^{(u,v)} P(x, y)\,dx + Q(x, y)\,dy.$$

Now we must compute the differential of f. We compute the partial derivatives using the definition; temporarily we use (u, v) as the generic point of R so as not to confuse the notation involving the dx and dy behind the integral sign:

$$\begin{aligned} f_u(u, v) &= \lim_{h \to 0} \frac{f(u + h, v) - f(u, v)}{h} \\ &= \lim_{h \to 0} \frac{1}{h}\left[\int_{(a,b)}^{(u+h,v)} P\,dx + Q\,dy - \int_{(a,b)}^{(u,v)} P\,dx + Q\,dy\right] \\ &= \lim_{h \to 0} \frac{1}{h} \int_{(u,v)}^{(u+h,v)} P\,dx + Q\,dy. \end{aligned}$$

We evaluate the integral from (u, v) to $(u + h, v)$ along the horizontal path from the first point to the second. On the horizontal path we have $y = v$, a constant and $dy = 0$. Thus

$$\begin{aligned} f_u(u, v) &= \lim_{h \to 0} \frac{1}{h} \int_{(u,v)}^{(u+h,v)} P\,dx + Q\,dy \\ &= \lim_{h \to 0} \frac{1}{h} \int_{(u,v)}^{(u+h,v)} P(x, v)\,dx \\ &= \lim_{h \to 0} \frac{1}{h} \int_{u}^{u+h} P(x, v)\,dx \\ &= P(u, v) \end{aligned}$$

by the Fundamental Theorem of Calculus (applied to the function $P(x, v)$ of one variable x). Thus

$$f_u(u, v) = P(u, v)$$

at every point (u, v) of R. In the same way we find

$$f_v(u, v) = Q(u, v).$$

Now returning to the (x, y) notation, we have

$$df = f_x(x, y)\,dx + f_y(x, y)\,dy = P(x, y)\,dx + Q(x, y)\,dy$$

as we had hoped to show. By assumption P and Q are continuous in R. Thus f has continuous first derivatives in R so f is continuous and differentiable in R. □

This theorem gives a condition for the integral of $P\,dx + Q\,dy$ to be independent of path (namely the existence of a certain f) but the statement does not provide a test that can be applied directly to P and Q. The next theorem remedies this deficiency when the first derivatives are continuous.

THEOREM 15.10

Test for Exact Differentials

Let $P(x, y)$ and $Q(x, y)$ have continuous first partial derivatives throughout a rectangular region R. Then $P\,dx + Q\,dy$ is an exact differential if and only if $P_y = Q_x$ at every point of R.

Proof

If $P\,dx + Q\,dy$ is an exact differential then there is a function f for which $f_x(x, y) = P(x, y)$ and $f_y(x, y) = Q(x, y)$. By assumption, P and Q have continuous first partial derivatives and so $f_{xy} = f_{yx}$ by the theorem on the equality of mixed partial derivatives. Because $f_{xy} = P_y$ and $f_{yx} = Q_x$ we obtain $P_y = Q_x$ as we wished to prove.

Now suppose $P_y = Q_x$ in R. Once again we must produce a function f having special properties. We select any point (a, b) in R and then, for (x, y) in R, define $f(x, y)$ by the rule

$$f(x, y) = \int_a^x P(x, y)\,dx + \int_b^y Q(a, y)\,dy.$$

Note that both integrals in the definition of f are integrals of functions of one variable along the x-axis or y-axis.

Now we show that f has the properties stated in the conclusion of the theorem. The second integral depends only on y so the derivative of f with respect to x can be computed by the Fundamental Theorem of Calculus:

$$\begin{aligned} f_x(x, y) &= \frac{\partial}{\partial x}\int_a^x P(x, y)\,dx + \frac{\partial}{\partial x}\int_b^y Q(a, y)\,dy \\ &= P(x, y). \end{aligned}$$

The second step is to show that $f_y = Q$. This step requires the equality

$$\frac{\partial}{\partial y}\int_a^x P(x, y)\,dx = \int_a^x P_y(x, y)\,dx, \qquad \textbf{(15.16)}$$

which we is proved below. By using this derivative formula we find

$$\begin{aligned} f_y(x, y) &= \frac{\partial}{\partial y}\int_a^x P(x, y)\,dx + \frac{\partial}{\partial y}\int_b^y Q(a, y)\,dy \\ &= \int_a^x P_y(x, y)\,dx + Q(a, y) = \int_a^x Q_x(x, y)\,dx + Q(a, y) \\ &= Q(x, y)\Big|_{x=a}^{x=x} + Q(a, y) = Q(x, y) - Q(a, y) + Q(a, y) = Q(x, y). \end{aligned}$$

Thus f has the correct properties and the theorem is proved except for the proof of Equation (15.16) to which we now turn. □

THEOREM 15.11

If P has continuous first partial derivatives throughout a rectangular region

R then

$$\frac{\partial}{\partial y}\int_a^b P(x,y)\,dx = \int_a^b P_y(x,y)\,dx,$$

provided (a, y) and (b, y) lie in R.

Proof

By the definition of the partial derivative we find

$$\begin{aligned}\frac{\partial}{\partial y}\int_a^b P(x,y)\,dx &= \lim_{k\to 0}\frac{1}{k}\left[\int_a^b P(x,y+k)\,dx - \int_a^b P(x,y)\,dx\right]\\ &= \lim_{k\to 0}\int_a^b \frac{P(x,y+k)-P(x,y)}{k}\,dx.\end{aligned}$$

We rewrite the term behind the integral sign by applying the MVT to get

$$\frac{P(x,y+k)-P(x,y)}{k} = P_y(x,v) = P_y(x,y) + [P_y(x,v) - P_y(x,y)]$$

for some number v between y and $y + k$. Because P is continuous in R, it is uniformly continuous in R. Given any $\epsilon > 0$ there is a $\delta > 0$ such that

$$|P_y(x,v) - P_y(x,y)| < \epsilon \quad \text{whenever} \quad |v - y| < \delta.$$

Restrict k so that $|k| < \delta$ and then

$$\begin{aligned}\left|\int_a^b [P_y(x,v) - P_y(x,y)]\,dx\right| &\le \int_a^b |P_y(x,v) - P_y(x,y)|\,dx\\ &\le \int_a^b \epsilon\,dx = \epsilon(b-a).\end{aligned}$$

If we take smaller and smaller choices for ϵ, the corresponding δ also gets smaller and force the $|k|$ to get smaller. So we take the limit as $k \to 0$ and find the integral is less than any given positive number; thus

$$\lim_{k\to 0}\int_a^b [P_y(x,v) - P_y(x,y)]\,dx = 0.$$

Now substitute back into the earlier equation to get

$$\begin{aligned}\frac{\partial}{\partial y}\int_a^b P(x,y)\,dx &= \lim_{k\to 0}\int_a^b P_y(x,y) + [P_y(x,v) - P_y(x,y)]\,dx\\ &= \int_a^b P_y(x,y)\,dx + \lim_{k\to 0}\int_a^b [P_y(x,v) - P_y(x,y)]\,dx\\ &= \int_a^b P_y(x,y)\,dx,\end{aligned}$$

as we wished to prove. □

EXAMPLE 2 Show that $(x^2 + y^2)\,dx + 2xy\,dy$ is an exact differential in any rectangular region and find a function f of which it is the differential.

Solution: The functions $P(x,y) = x^2 + y^2$ and $Q(x,y) = 2xy$ have continuous first partial derivatives in the entire xy plane and

$$P_y(x,y) = 2y, \qquad Q_x(x,y) = 2y.$$

Thus $P_y = Q_x$ and the theorem implies that $P\,dx + Q\,dy$ is an exact differential. To find the function f we appeal to the proof of the theorem in which f was defined by the rule (using $a = 0$ and $b = 1$)

$$\begin{aligned} f(x, y) &= \int_0^x P(x, y)\,dx + \int_1^y Q(0, y)\,dy \\ &= \int_0^x x^2 + y^2\,dx + \int_1^y 2(0)y\,dy = \frac{1}{3}x^3 + xy^2\Big|_0^x \\ &= \frac{x^3}{3} + xy^2. \end{aligned}$$

Just for some reassurance let us verify that this computation did produce a correct function:

$$\begin{aligned} f_x(x, y) &= \frac{\partial}{\partial x}\left(\frac{x^3}{3} + xy^2\right) = x^2 + y^2 = P(x, y) \\ f_y(x, y) &= \frac{\partial}{\partial y}\left(\frac{x^3}{3} + xy^2\right) = 2xy = Q(x, y). \end{aligned}$$

Of course there was nothing special about making the choice $a = 0$ and $b = 1$. Any other constants would have produced a correct function (possibly different from this one). ■

We give an alternative method for solving the problem in the previous example; that solution treats the problem from the view of definite integration; now we present a point of view more like indefinite integration.

EXAMPLE 3 Find a function f whose differential is

$$(6x^2y^2 + 2xy + 3)\,dx + (x^2 + 4x^3y + 3y^2)\,dy.$$

Solution: Let $P = 6x^2y^2 + 2xy + 3$ and $Q = x^2 + 4x^3y + 3y^2$. Then

$$P_y = 12x^2y + 2x \quad \text{and} \quad Q_x = 2x + 12x^2y$$

so $P_y = Q_x$ and there is a function f that satisfies $f_x = P$ and $f_y = Q$. Starting with the equation $f_x = P$ we integrate with respect to x, treating y as a constant to get

$$f(x, y) = \int P(x, y)\,dx = \int 6x^2y^2 + 2xy + 3\,dx = 2x^3y^2 + x^2y + 3x + g$$

where g is a constant (with respect to x). The constant of integration must be regarded as a function of y to complete the solution. With $g = g(y)$ and this formula for f we now use the equation $f_y = Q$ to evaluate g:

$$f_y = 4x^3y + x^2 + \frac{\partial g}{\partial y} = Q = x^2 + 4x^3y + 3y^2.$$

It follows that

$$\frac{\partial g}{\partial y} = 3y^2 \quad \text{and} \quad g = g(y) = y^3 + c.$$

Thus f is given by $f(x, y) = 2x^3y^2 + x^2y + 3x + y^3 + c$ with c a constant. ■

Exact Differentials in Three Dimensions

The notion of an exact differential extends in the natural way to functions of three variables. Given $f(x, y, z)$ twice differentiable at all points of some parallelepiped V in xyz space, its differential is

$$df = f_x\,dx + f_y\,dy + f_z\,dz,$$

and we have the equations $f_{xy} = f_{yx}$, $f_{xz} = f_{zx}$, $f_{yz} = f_{zy}$. It follows that an expression

$$P\,dx + Q\,dy + R\,dz$$

can be an exact differential only if the equations $P_y = Q_x$, $P_z = R_x$, $Q_z = R_y$ holds at all points of V. It can be proved that these conditions are also sufficient; this gives the analogue of the two-dimensional test for exactness.

Line Integrals over Closed Paths

There is another way to characterize the line integrals that are independent of path. We need the notion of a *closed path.* A curve given in vector form as $\mathbf{r}(t) = x(t)\,\mathbf{i} + y(t)\,\mathbf{j}$ for $p \le t \le q$ is called a closed path if the initial point and the terminal point coincide; that is $\mathbf{r}(p) = \mathbf{r}(q)$.

THEOREM 15.12

Integrals over a Closed Path

Let $P(x, y)$ and $Q(x, y)$ be continuous in a rectangular region R. The line integrals

$$\int_C P(x, y)\,dx + Q(x, y)\,dy$$

are independent of path in R if and only if the line integral over every closed path in R equals 0;

$$\int_G P(x, y)\,dx + Q(x, y)\,dy = 0 \quad G \text{ a closed path in } R.$$

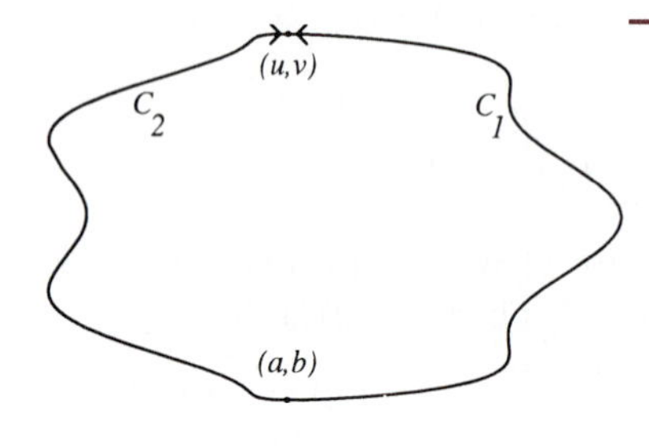

FIGURE 15.13

Proof

The proof uses the simple idea that if C_1 and C_2 are two paths in R each having initial point (a, b) and each having terminal point (u, v) then the path G formed by passing from (a, b) to (u, v) along C_1 and then back to (a, b) by traversing C_2 in the opposite direction is a closed path and

$$\int_G = \int_{C_1} - \int_{C_2}.$$

It follows that the integral along G is 0 if and only if the integral along C_1 equals the integral along C_2. □

By combining this last idea with the previous idea we obtain the following theorem.

THEOREM 15.13

Let P and Q be functions with continuous first partial derivatives throughout the rectangular region R. Then the integral

$$\int_G P\,dx + Q\,dy \tag{15.17}$$

equals 0 for every closed path G in R if and only if $P_y = Q_x$ at every point of R.

Proof

If the integral around every closed path is 0, then the line integral is independent of path and it follows that $P_y = Q_x$ in R. Conversely if $P_y = Q_x$ in R, the integral (15.17) is the integral of an exact differential and is independent of path and so the integral around a closed path is 0. □

In this theorem, the condition that both P and Q have continuous first partial derivatives throughout the rectangular region is important. Consider the following example.

EXAMPLE 4 Let

$$P\,dx + Q\,dy = \frac{-y\,dx + x\,dy}{x^2 + y^2}.$$

Show that the differential of $f(x, y) = \arctan(y/x)$ is $P\,dx + Q\,dy$ but yet

$$\int_G P\,dx + Q\,dy \neq 0$$

when G is the circle of radius 1 with center at the origin.

Solution: It is straightforward to verify that

$$\begin{aligned} f_x &= \frac{1}{1+(y/x)^2}\frac{-y}{x^2} = \frac{-y}{x^2+y^2} = P \\ f_y &= \frac{1}{1+(y/x)^2}\frac{1}{x} = \frac{x}{x^2+y^2} = Q \end{aligned}$$

so that $df = P\,dx + Q\,dy$. It follows that at every point where the partial derivatives of f are continuous we have $f_{xy} = f_{yx}$ or equivalently that $P_y = Q_x$. Now we evaluate the line integral of f around the unit circle. Use the parametric representation $x = \cos t$ and $y = \sin t$ for $0 \leq t \leq 2\pi$; then $P = -\sin t$ and $Q = \cos t$ and we have

$$\int_G P\,dx + Q\,dy = \int_0^{2\pi} [-\sin t(-\sin t) + \cos t\cos t]\,dt = \int_0^{2\pi} dt = 2\pi.$$

The line integral around the closed curve is not zero. This does not conflict with the theorem just proved, however, because the theorem required P and Q to be continuous throughout the region enclosed by the curve. In this case, P and Q are not continuous at the origin. If the line integral were taken around a closed curve that did not enclose the origin, the value would be zero as the reader might wish to verify in some specific cases. ■

Exercises 15.8

Test each differential for exactness and find a function of which it is the differential when such a function exists.

1. $(y-2)\,dx + x\,dy$

2. $3y\,dx + (3x-4y)\,dy$

3. $(8xy^2+2)\,dx + (8x^2y-1)\,dy$

4. $(\sin y + y\sin x)\,dx + (x\cos y - \cos x)\,dy$

5. $(e^{-y} - ye^x)\,dx - (xe^{-y} - e^x)\,dy$

6. $\left(\dfrac{2x^2}{x^2+y^2} + \ln(x^2+y^2)\right) dx + \left(\dfrac{2xy}{x^2+y^2}\right) dy$

Show that each line integral is independent of path and then evaluate the integral.

7. $\int_C (8xy^2+2)\,dx + (8x^2y-1)\,dy$ where C is a path from $(0,0)$ to $(3,4)$

8. $\int_C (3x^2+4xy-2y^2)\,dx + (2x^2-4xy-3y^2)\,dy$, where C is a path from $(0,1)$ to $(3,3)$

9. $\int_C e^{-x}\cos y\,dx + e^{-x}\sin y\,dy$, where C is a path from $(0,\pi/2)$ to $(0,\pi)$

10. $\int_C \left(y\ln(xy+1) + \dfrac{xy^2}{xy+1}\right) dx + \left(x\ln(xy+1) + \dfrac{x^2y}{xy+1}\right) dy$, where C is a path in the first quadrant from $(1,1)$ to $(4,3)$.

11. Let $P(x,y)$ and $Q(x,y)$ be continuous functions in a rectangular region R and let $f(x,y)$ have continuous first partial derivatives in R. Prove that for any simple closed path C lying in R we have

$$\int_C (P+f_x)\,dx + (Q+f_y)\,dy = \int_C P\,dx + Q\,dy.$$

12. Let $P(x,y)$ and $Q(x,y)$ be continuous functions in a rectangular region R and let $f(x,y)$ have continuous first partial derivative in R. Prove that for any continuous path C from (a,b) to (c,d) and lying in R we have

$$\int_C (P+f_x)\,dx + (Q+f_y)\,dy = f(c,d) - f(a,b) + \int_C P\,dx + Q\,dy.$$

13. Let $g(x)$ and $h(y)$ be continuous everywhere and let $P(x,y)$ and $Q(x,y)$ be differentiable functions in a rectangular region R. If C is a simple closed curve lying in R then

$$\int_C (P(x,y)+g(x))\,dx + (Q(x,y)+h(y))\,dy = \int_C P(x,y)\,dx + Q(x,y)\,dy.$$

14. Let $P(x,y)$ and $Q(x,y)$ be continuous functions in a rectangular region R. Show that there is some function $M(x,y)$ such that

$$\int_C P\,dx + Q\,dy = \int_C (P+M)\,dx.$$

[Hint: Consider the previous exercises in this section.]

15. Use Theorem 15.11 and mathematical induction to prove the formula

$$\int_0^x \frac{dt}{(t^2+y)^{n+1}} = \frac{(-1)^n}{n!}\frac{\partial^n}{\partial y^n}\int_0^x \frac{dt}{t^2+y}$$

and then evaluate the right-hand side for $n = 1, 2, 3$. [Comment: This technique is useful when applying the method of partial fractions to evaluate integrals.]

16. Let P and Q have continuous first derivatives throughout a rectangle R in the xy plane. Prove that $P\,dx+Q\,dy$ is an exact differential if and only if $P\mathbf{i}+Q\mathbf{j}$ is the gradient of a function having continuous second partial derivatives throughout R.

17. Suppose we write a position vector as $\mathbf{r} = x\,\mathbf{i} + y\,\mathbf{j}$ and use the notation $d\mathbf{r} = dx\,\mathbf{i} + dy\,\mathbf{j}$. For a differentiable function $f(x,y)$ and a path C given in vector form as $\mathbf{r}(t)$ for $a \le t \le b$ show that the following formula is a correct interpretation of Theorem 15.9:

$$\int_C \nabla f(x,y)\cdot d\mathbf{r} = f(\mathbf{r}(b)) - f(\mathbf{r}(a)).$$

15.9 Green's theorem

The theorems in the previous section imply that a line integral $\int_C P\,dx + Q\,dy$ around every closed path in a region is 0 if the difference $P_y - Q_x$ is 0 throughout the region. In this section we prove a theorem that allows the possibility that

$P_y - Q_x \neq 0$, first proved by the English mathematician and physicist George Green in 1828, which relates the double integral of $P_y - Q_x$ over a region to the line integral of $P\,dx + Q\,dy$ around the closed curve that bounds the region.

Some care is necessary in formulating the statement because we must traverse the closed path in a specific direction. Obviously we cannot expect to get the same result when integrating around a closed curve in two different directions. Also the statement has exceptions when the region has "holes" in it (see the examples given below). We do not attempt to prove, or even state, the theorem in its most general setting. Instead, we restrict our attention to regions of Type X and Type Y and some other regions that can be decomposed into subregions of these types.

We have a notion of the positive direction of traversing a circle. If the circle has its center at the origin and is parametrized by $(\cos t, \sin t)$ then the direction of increasing t is the positive direction. If we "walk around" the circle in the positive direction, the interior of the circle is on our left. We use the same orientation for regions bounded by more complicated curves. For a given region R having a boundary curve that is piecewise differentiable, we select the positive direction of the curve to be that direction that keeps the region (or at least the nearest part of the region) to our left. We also want to consider curves that do not cross themselves except that the initial and terminal points of the curve coincide. We refer to such a curve as *simple* and *closed.* Thus an ellipse is a simple closed curve but a figure eight is not simple. Regions of Type X or Type Y are bounded by simple closed curves. We give the important ideas of the proof but we do only a special case completely.

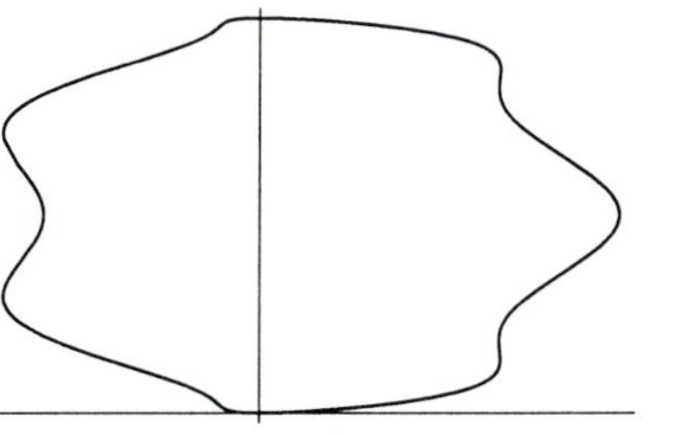

FIGURE 15.14
A simple closed curve

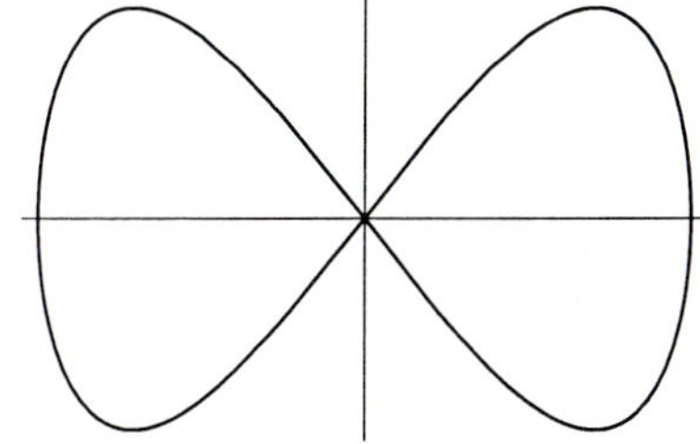

FIGURE 15.15
Closed, but not simple

THEOREM 15.14

Green's Theorem

Let S be a region bounded by a simple closed curve G and let P and Q be functions having continuous first partial derivatives throughout S. Then

$$\iint_S (Q_x - P_y)\,dS = \int_G P\,dx + Q\,dy,$$

where the integration around the curve G is in the positive direction.

Proof

We first assume that the region is both of Type X and of Type Y. In its role as a region of Type X it is described by

$$S = \{(x, y) : a \leq x \leq b, \quad g_1(x) \leq y \leq g_2(x)\}.$$

The boundary of the region is made up of four pieces:

$$\begin{aligned}
G_1 &= \{(x, g_1(x)) : a \leq x \leq b\}, \\
G_2 &= \{(b, y) : g_1(b) \leq y \leq g_2(b)\}, \\
G_3 &= \{(x, g_2(x)) : a \leq x \leq b\}, \\
G_4 &= \{(a, y) : g_1(a) \leq y \leq g_2(a)\}.
\end{aligned}$$

We evaluate the line integral first as a sum of the contribution from P and and then the contribution from Q, and each is evaluated as a sum of line integrals over the four pieces of the path:

$$\int_G P\,dx = \int_{G_1} P\,dx + \int_{G_2} P\,dx + \int_{G_3} P\,dx + \int_{G_4} P\,dx.$$

Over the paths G_2 and G_4, x is constant so $dx = 0$; more precisely in the partition of the path by points (x_i, y_i) we have $x_i - x_{i-1} = 0$ so all approximating sums equal 0. Thus we have only to evaluate the integrals over G_1 and G_3. View these curves as parametrized by the parameter x so we can evaluate the integrals as follows:

$$\begin{aligned}
\int_G P\,dx &= \int_a^b P(x, g_1(x))\,dx + \int_b^a P(x, g_2(x))\,dx \\
&= -\int_a^b [P(x, g_2(x)) - P(x, g_1(x))]\,dx \\
&= -\int_a^b \left(\int_{g_1(x)}^{g_2(x)} P_y(x, y)\,dy\right) dx \\
&= -\iint_S P_y(x, y)\,dS.
\end{aligned} \qquad \textbf{(15.18)}$$

This expresses the first half of the line integral in terms of a double integral of S.

Now use the assumption that the region is also of Type Y with a description

$$S = \{(x, y) : c \leq y \leq d, \quad h_1(y) \leq x \leq h_2(y)\}$$

to evaluate the second part of the line integral. The computations are so similar to those just given that we do not repeat them but simply state the final result in the form

$$\int_G Q\,dy = \iint_S Q_x\,dS. \qquad \textbf{(15.19)}$$

Now we combine the results of equations (15.18) and (15.19) to obtain the statement of Green's Theorem for regions that are both Type X and Type Y.

The extension of Green's theorem to regions that are not both Type X and Type Y is made by the device of subdividing the region or making "cuts". For example the region in the figure is of Type Y but not of Type X. But Green's Theorem is true for this region, as we now indicate.

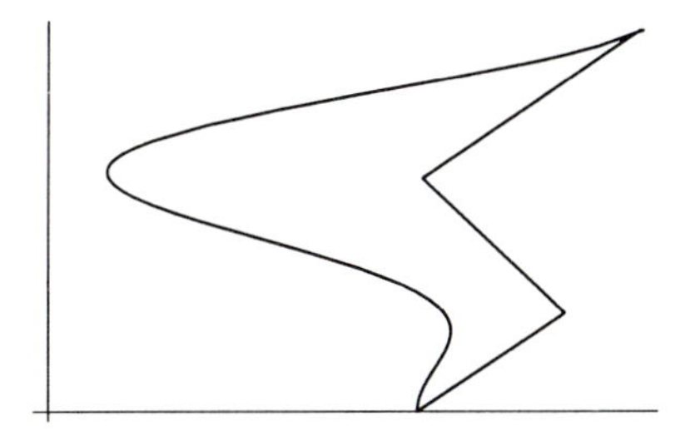

FIGURE 15.16
Type Y but not Type X

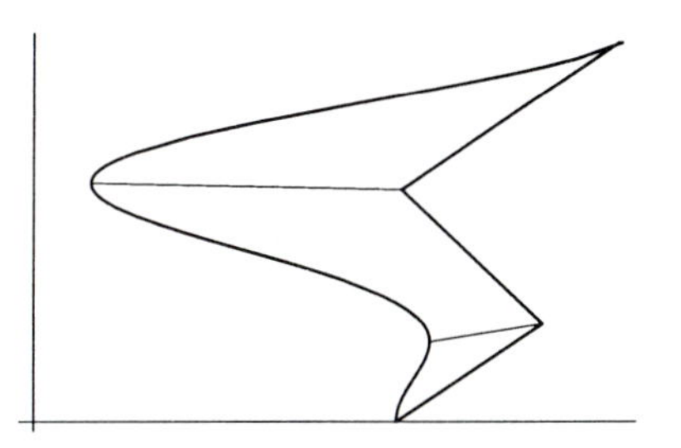

FIGURE 15.17
Subregions are both Type X and Type Y

Make a cut along the lines as shown in figure 15.17; the region is divided into three subregions each of which is of Type X and Type Y. Green's Theorem can be applied to each region separately. The integration is carried out in the positive direction around each boundary. In the sum of the three line integrals, there is an integration around the boundary of the original (uncut) regions plus two integrations along each cut-line, once in the each direction. The two integrations along the cut-lines cancel each other so all that remains is the integral on the boundary of the original region. The double integrals over the subregions add to give the double integral over the entire region.

One can extend Green's Theorem to more general regions that can be decomposed into regions of Type X and Type Y, but we will not attempt a general description of such regions. □

EXAMPLE 1 Use Green's theorem to evaluate the line integral

$$\int_C (5y + (1 - x^2)^{1/3})\,dx + (x - e^{\sin y})\,dy$$

where C is the circle $x^2 + y^2 = 1$ traversed in the positive direction.

Solution: This would be a very complicated line integral to evaluate from first principles but Green's Theorem makes the evaluation relatively easy. Let

$$P(x, y) = 5y + (1 - x^2)^{1/3}, \qquad Q(x, y) = x - e^{\sin y}$$

so that

$$Q_x - P_y = 1 - 5 = -4.$$

The region S bounded by C is the interior of the circle of radius 1 and area π. Thus by Green's Theorem the line integral equals

$$\iint_S Q_x - P_y\,dS = \iint_S -4\,dS = -4\pi.$$

■

EXAMPLE 2 Evaluate the line integral

$$\int_C (x - y)^2\,dx + (x + y)^2\,dy$$

where C is the boundary of the region enclosed by the graphs of $x = y^2$ and $y = x - 2$.

Solution: The region enclosed by S is of Type Y with the description

$$S = \{(x, y) : -1 \leq y \leq 2, \quad y^2 \leq x \leq y + 2\}.$$

We could directly evaluate the line integral by breaking the curve up into two parts and parametrizing them separately and doing the evaluation. Rather than do this, let us apply Green's Theorem. With $P = (x - y)^2$ and $Q = (x + y)^2$ we have

$$Q_x - P_y = 2(x + y) + 2(x - y) = 4x$$

so the line integral equals

$$\iint_S 4x \, dS = \int_{-1}^{2} \int_{y^2}^{y+2} 4x \, dx \, dy = \int_{-1}^{2} 2(y + 2)^2 - 2(y^2)^2 \, dy = \frac{117}{15}.$$ ■

Applications of Green's Theorem

Green's Theorem can be applied to determine the area of a region S bounded by a simple closed curve C. Let P and Q be functions of (x, y) with continuous first partial derivatives throughout the region S and suppose that

$$Q_x(x, y) - P_y(x, y) = 1 \qquad \textbf{(15.20)}$$

at every point of S. Then

$$\int_C P \, dx + Q \, dy = \iint_S Q_x - P_y \, dS = \iint_S dS = \text{ area of } S.$$

Some possible choices of P and Q that satisfy Equation (15.20) are $P = 0$ and $Q = x$ or $Q = 0$ and $P = -y$. Thus we have

$$\int_C x \, dy = -\int_C y \, dx = \text{ area of } S.$$

If a and b are constants such that $a + b = 1$ then the previous equation implies that

$$\int_C ax \, dy - by \, dx = \text{ area of } S. \qquad \textbf{(15.21)}$$

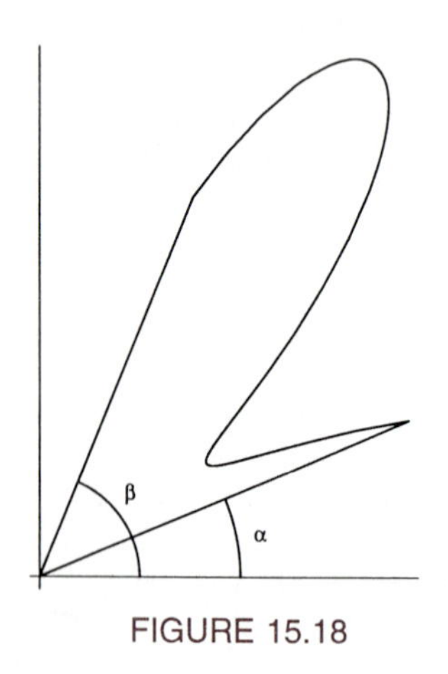

FIGURE 15.18

This last formula with $a = b = 1/2$ can be used to derive a formula for the region bounded by a curve given in *polar coordinates.*

Recall that the polar coordinates (r, θ) of the point (x, y) are determined geometrically as follows: r is the distance from the origin to the point (x, y) so that $r \geq 0$ and θ is the angle from the positive x-axis to the line from the origin to (x, y). The equations relating x, y, r, θ are

$$x = r \cos \theta, \qquad y = r \sin \theta.$$

A curve might be given in *polar form* by an equation $r = f(\theta)$ for $\alpha \leq \theta \leq \beta$ and $f(\theta) \geq 0$ where α and β are fixed numbers.

Suppose a region is given in the form

$$S = \{(r, \theta) : \alpha \leq \theta \leq \beta, \quad 0 \leq r \leq f(\theta)\}.$$

By Equation (15.21) the area of S can be determined as the value of a line integral around the boundary C of S as

$$\frac{1}{2}\int_C x\,dy - y\,dx = \text{area of } S.$$

The boundary C is described in three parts: On the first and third parts, θ is constant and on the second part we have the curve $r = f(\theta)$; more precisely

$$\begin{aligned} C_1 &: \theta = \alpha, \quad 0 \le r \le f(\alpha), \\ C_2 &: \alpha \le \theta \le \beta, \quad r = f(\theta), \\ C_3 &: \theta = \beta, \quad 0 \le r \le f(\beta). \end{aligned}$$

Evaluate the line integral (15.21) with $a = b = 1/2$ on C_1; we first note that on this curve we have

$$\begin{aligned} x &= r\cos\alpha, & dx &= \cos\alpha\,dr, \\ y &= r\sin\alpha, & dy &= \sin\alpha\,dr. \end{aligned}$$

It follows that $x\,dy - y\,dx = r\cos\alpha\sin\alpha\,dr - r\sin\alpha\cos\alpha\,dr = 0$ and

$$\frac{1}{2}\int_{C_1} -y\,dx + x\,dy = 0.$$

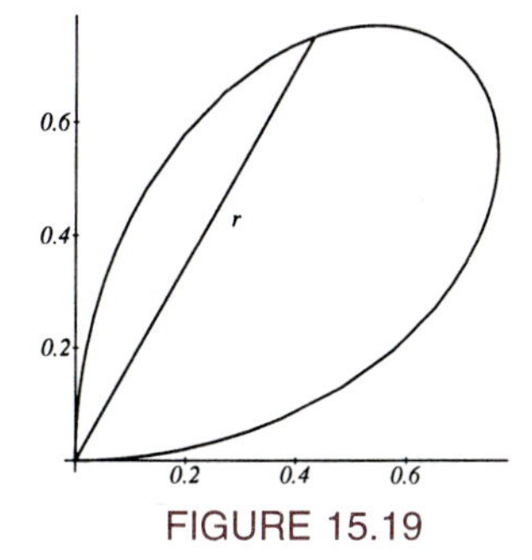

FIGURE 15.19

The analogous computation shows that the integral over C_3 is also 0. The integral over C_2 requires the substitution

$$\begin{aligned} x &= r\cos\theta, & dx &= (-r\sin\theta + r'\cos\theta)\,d\theta, \\ y &= r\sin\theta, & dy &= (r\cos\theta + r'\sin\theta)\,d\theta \end{aligned}$$

from which we get

$$\begin{aligned} -y\,dx + x\,dy &= -r\sin\theta(-r\sin\theta + r'\cos\theta)\,d\theta + r\cos\theta(r\cos\theta + r'\sin\theta)\,d\theta \\ &= (r^2\sin^2\theta + r^2\cos^2\theta)\,d\theta = r^2\,d\theta. \end{aligned}$$

Now we obtain the area of the region S as

$$\begin{aligned} \iint_S dS &= \frac{1}{2}\int_C -y\,dx + x\,dy = \frac{1}{2}\int_{C_2} -y\,dx + x\,dy \\ &= \frac{1}{2}\int_\alpha^\beta r^2\,d\theta. \end{aligned}$$

We have proved the following theorem.

THEOREM 15.15

Areas in Polar Coordinates

Let S be the region bounded by the lines $\theta = \alpha$ and $\theta = \beta$ and the curve $r = f(\theta)$ with f a nonnegative continuous function. Then the area of S is

$$\frac{1}{2}\int_\alpha^\beta r^2\,d\theta.$$

EXAMPLE 3 Find the area of the "loop" in the first quadrant (see figure 15.19) with polar equation $r = \sin 2\theta$.

Solution: A sketch of the region is shown in the figure. The region is described as

$$S = \{(r,\theta) : 0 \le \theta \le \frac{\pi}{2}, \quad 0 \le r \le \sin 2\theta\}.$$

According to the formula for areas in polar coordinates, the area is

$$\frac{1}{2}\int_0^{\pi/2} \sin^2 2\theta \, d\theta = \frac{1}{4}\int_0^{\pi/2} 1 - \cos 4\theta \, d\theta = \frac{\pi}{8}.$$

■

Exercises 15.9

Evaluate the line integrals taken around C in the positive direction using Green's Theorem.

1. $\int_C 2y\,dx - 5x\,dy$, C is the path $x^2 + y^2 = 4$
2. $\int_C 3x\,dx + 4y\,dy$, C is the path $x^2 + y^2 = 9$
3. $\int_C y^2\,dx$, C is the rectangle with vertices $(0,0)$, $(1,0)$, $(1,5)$, $(0,5)$
4. $\int_C 2y\,dx - 5x\,dy$, C is the path $x^2 + y^2 = 4$
5. $\int_C (x^2 + 10xy + 2y^2)\,dx + (5x^2 - xy + y^2)\,dy$, C is the path $(x-1)^2 + y^2 = 1$
6. $\int_C e^x \sin y\,dx + e^x \cos y\,dy$, C is the path $(x-a)^2 + (y-b)^2 = 1$

Express the area enclosed by the curve as a line integral

7. $\mathbf{r}(t) = 4\cos t\,\mathbf{i} + 2\sin t\,\mathbf{j}$, $0 \le t \le 2\pi$
8. $\mathbf{r}(t) = a\cos t\,\mathbf{i} + b\sin t\,\mathbf{j}$, $0 \le t \le 2\pi$

Compute both sides of the equation in Green's Theorem with the given P, Q and region G.

9. $P = xy$, $Q = -2xy$, $G = \{(x,y) : 0 \le x \le 1, 0 \le y \le 2\}$
10. $P = 2xy^3 - 3x^2y$, $Q = 3x^2y^2$, $G = \{(x,y) : 0 \le x \le 1, 0 \le y \le 1 - x\}$

In exercises 11 and 12, G denotes the region inside the circle $x^2 + y^2 = 4$ and outside the circle $x^2 + y^2 = 1$ with the boundary βG oriented in the positive direction after making a "cut" from $(1,0)$ to $(2,0)$ that is traversed twice. Evaluate the integrals around βG using Green's Theorem.

11. $\displaystyle\int_{\beta G} \frac{x\,dx}{x^2+y^2} + \frac{y\,dy}{x^2+y^2}$
12. $\displaystyle\int_{\beta G} \frac{-y^3\,dx}{x^2+y^2} + \frac{xy^2\,dy}{x^2+y^2}$
13. If $(\overline{x}, \overline{y})$ is the center of mass of the region G having uniform density show that

$$2\overline{x}\int_{\beta G} x\,dy = \int_{\beta G} x^2\,dy$$

where βG is the boundary of G. State and prove a similar formula for $\overline{y}$.

15.10 Change of variables

The familiar change of variables formula,

$$\int_a^b f(x)\,dx = \int_c^d f(g(u))g'(u)\,du, \quad g(c) = a, \; g(d) = b,$$

for functions of one variable proved to be very useful for evaluation of certain types of definite integrals. Change of variable formulas for line integrals or double integrals are generally more complicated. One reason for the additional complications is that we must take into account not only the change of the function but also the change of the region of integration. In the one-variable case the change of the function is made by transforming $f(x)$ into $f(g(u))g'(u)$ and the change of the interval $[a,b]$ is taken into account by arranging g so that it transforms the interval $[c,d]$ (or $[d,c]$ if $d < c$) onto $[a,b]$. In the two-variable case we must take into account how a change of variable transforms a region of the uv plane into a region of the xy plane (or the other way around in some cases).

An Extended Example

We begin by looking at a specific transformation to illustrate the type of considerations made for more general transformations. The computations done in this specific case will be repeated in a more general setting below. A careful reading of this section will reveal all the ideas necessary to carry through the proofs later.

Let T be the transformation from the uv plane to the xy plane defined by

$$(x, y) = T(u, v) = (2u + 4v, 4u + 5v). \tag{15.22}$$

Thus T associates with every point in the uv plane a point (x, y) in the xy plane.

In general we say that T is one to one if two different points of the uv plane correspond to two different points in the xy plane. Let us prove that this T is one to one. Suppose (u_1, v_1) and (u_2, v_2) are points in the uv plane such that

$$T(u_1, v_1) = T(u_2, v_2).$$

Then we have

$$\begin{aligned}(2u_1 + 4v_1, 4u_1 + 5v_1) &= (2u_2 + 4v_2, 4u_2 + 5v_2), \quad \text{or} \\ 2u_1 + 4v_1 &= 2u_2 + 4v_2, \\ 4u_1 + 5v_1 &= 4u_2 + 5v_2.\end{aligned}$$

With a bit of manipulation one finds that these equations are satisfied only when $(u_1, v_1) = (u_2, v_2)$. This proves that T is one to one on the entire uv plane, in fact.

Every point in the xy plane equals T applied to some point in the uv plane. To prove this it is only necessary to show that the system of equations

$$\begin{cases} 2u + 4v = x, \\ 4u + 5v = y, \end{cases}$$

has a solution for each choice of x and y. One finds the solution as

$$u = \frac{-5x + 4y}{6}, \qquad v = \frac{2x - y}{3}.$$

Now we examine how T transforms some lines in the uv plane into curves (actually lines) in the xy plane.

Consider the general straight line $v = pu + r$ with p and r constants. A point on this line has the form $(u, v) = (u, pu + r)$ and T applied to that point produces

$$\begin{aligned}(x, y) = T(u, pu + r) &= (2u + 4(pu + r), 4u + 5(pu + r)) \\ &= ((4p + 2)u + 4r, (5p + 4)u + 5r).\end{aligned}$$

This is the parametric representation of a line in the xy plane, namely

$$\begin{cases} x = (4p + 2)u + 4r, \\ y = (5p + 4)u + 5r, \end{cases}$$

where u is regarded as the parameter. For a specific example, the line $v = u + 1$ is transformed by T into the line with parametric representation $x = 6u + 4$ and $y = 9u + 5$. By eliminating the parameter u, we find the rectangular-form equation of this line to be $y - 5 = 3(x - 4)/2$.

Note that the vertical line $u = a$, consisting of points (a, v), is transformed into the line

$$(x, y) = T(a, v) = (2a + 4v, 4a + 5v),$$

which is the parametric representation of the line having rectangular equation $(x - 2a)/4 = (y - 4a)/5$.

A horizontal line $v = c$, consisting of points (u, c), is is transformed into the line

$$(x, y) = T(u, c) = (2u + 4c, 4u + 5c), \qquad \textbf{(15.23)}$$

which is the parametric representation of the line having rectangular equation $(x - 4c)/2 = (y - 5c)/4$.

Now we combine these specific facts about transformations of lines to discuss the transformation of a rectangular region.

Let S be the region of the uv plane bounded by the horizontal and vertical lines $u = 1$, $u = 3$, $v = 1$ and $v = 2$. Then T transforms these into the lines

$$y = 2x - 3, \quad y = 2x - 9, \quad y = (5x + 6)/4, \quad y = (5x + 12)/4.$$

The region and the transformed region are shown in figures 15.20 and 15.21, respectively. It is worthwhile to notice that the areas of the two regions are not the same. The region in the uv plane has area 2 while the transformed region has area 12 (as one can determine with a bit of calculation). The transformed region has area 6 times larger than the original region. The factor 6 will appear in the computations below.

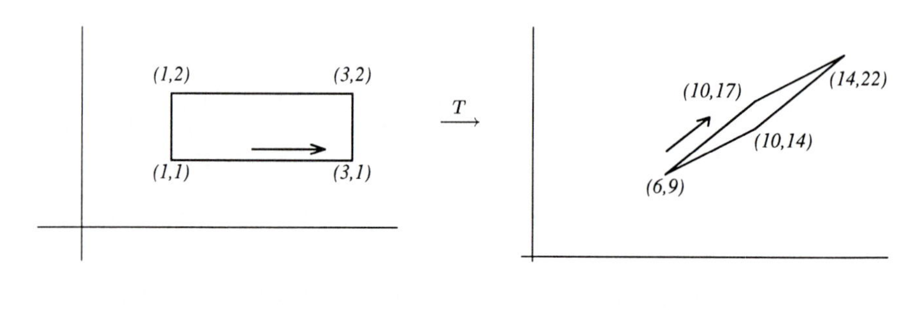

FIGURE 15.20
uv rectangle

FIGURE 15.21
Transformed region

It is also important to pay attention to the orientation. Suppose we traverse the rectangle S in the positive direction; we encounter, in order, the points $(1, 1)$, $(3, 1)$, $(3, 2)$, $(1, 2)$, and back to $(1, 1)$. When T is applied to these points, we obtain $(6, 9)$, $(10, 17)$, $(14, 22)$, $(10, 14)$ and back to $(6, 9)$. This "walk" around the region of the xy plane is in the negative direction. Thus T has reversed the orientation of the boundary.

Now we illustrate the main point. Suppose we want to evaluate a double integral over the region $A = T(S)$ in the xy plane,

$$\int_{T(S)} f(x, y)\, dA,$$

or a line integral

$$\int_C P\, dx + Q\, dy$$

where C is the boundary of $T(S)$. It might be possible to just directly evaluate either integral but we want to illustrate the method of transforming these integrals into a double integral or line integral in the uv plane over the region S or its boundary B.

With this in mind we use a single parameter t to describe the boundary of S. One possibility, for example, would be for the curve $(u(t), v(t))$ to trace out the piece $a \le u \le b$ and $v = c$ for $0 \le t \le 1$. Then for $1 \le t \le 2$, $(u(t), v(t))$

traces out the piece $u = b$ and $c \leq v \leq d$, and so on. So for $0 \leq t \leq 4$, $(u(t), v(t))$ traces out the boundary of S. This parametrization is transferred to the boundary of $T(S)$ by the equations (15.22) so that x and y are functions of t by the rule

$$(x(t), y(t)) = T(u(t), v(t)).$$

When the line integral is evaluated, we express C in terms of t, P and Q as functions of t and dx and dy in terms of t and dt. The familiar result is

$$\int_C P\,dx + Q\,dy = \int_0^4 P(x(t), y(t))x'(t) + Q(x(t), y(t))y'(t)\,dt.$$

To complete the expression, we need an explicit form for $dx = x'(t)dt$ and $dy = y'(t)\,dt$. Actually we do not really want an explicit form but rather a form in terms of u and v. We compute (keeping in mind $x = 2u + 4v$ and $y = 4u + 5v$)

$$\begin{aligned} dx = x'(t)\,dt &= \frac{\partial x(u,v)}{\partial t}\,dt = \left(\frac{\partial x}{\partial u}\frac{\partial u}{\partial t} + \frac{\partial x}{\partial v}\frac{\partial v}{\partial t}\right)dt \\ &= \left(2\frac{du}{dt} + 4\frac{dv}{dt}\right)dt = 2\,du + 4\,dv. \\ dy = y'(t)\,dt &= \frac{\partial y(u,v)}{\partial t}\,dt = \left(\frac{\partial y}{\partial u}\frac{\partial u}{\partial t} + \frac{\partial y}{\partial v}\frac{\partial v}{\partial t}\right)dt \\ &= \left(4\frac{du}{dt} + 5\frac{dv}{dt}\right)dt = 4\,du + 5\,dv. \end{aligned}$$

We can now express the line integral around C in the xy plane in terms of the integral around B in the uv plane:

$$\begin{aligned} \int_C P\,dx + Q\,dy &= \int_B P(T(u,v))(2\,du + 4\,dv) + Q(T(u,v))(4\,du + 5\,dv) \\ &= \int_B \big(2P(T(u,v)) + 4Q(T(u,v))\big)\,du + \big(4P(T(u,v)) + 5Q(T(u,v))\big)\,dv. \end{aligned}$$

We have succeeded in expressing the line integral given in terms of x and y in terms of u and v. The last step is to do the same for a double integral. We use the computations just completed and Green's Theorem to change variables in the double integral.

Suppose we want to evaluate a double integral over $A = T(S)$:

$$\iint_A M(x,y)\,dA.$$

Assume we have some pair of functions $P(x,y)$ and $Q(x,y)$ such that

$$Q_x - P_y = M$$

at every point of A. One possible choice is $P = 0$ and $Q(x,y) = \int M(x,y)\,dx$ (where y is held constant in the integral with respect to x). Then by Green's Theorem, we have

$$\begin{aligned} \iint_A M(x,y)\,dA &= \int_C P(x,y)\,dx + Q(x,y)\,dy \\ &= \int_B \big(2P(T(u,v)) + 4Q(T(u,v))\big)\,du + \big(4P(T(u,v)) + 5Q(T(u,v))\big)\,dv. \end{aligned}$$

The last integral is evaluated by Green's Theorem on S. Let

$$\begin{aligned} p(u,v) &= 2P(T(u,v)) + 4Q(T(u,v)) = 2P(x,y) + 4Q(x,y), \\ q(u,v) &= 4P(T(u,v)) + 5Q(T(u,v)) = 4P(x,y) + 5Q(x,y). \end{aligned}$$

We will apply Green's Theorem in a moment but first, there is one small point that must not be ignored. We assume that the boundary B of S is traversed in the positive direction and we select the parameter so that this occurs. After we apply the transformation T to the curve, we cannot be sure that the resulting path will be traversed in the positive direction. In fact it is not difficult to verify that the parametrization of the boundary of $T(S)$ causes the curve to be traversed in the negative direction. (Just examine the direction of one segment to see for yourself.) Thus an unexpected minus sign creeps into the computation when we apply Green's Theorem, which assumes the boundary is traversed in the positive direction. By Green's Theorem we have

$$-\int_B p(u,v)\,du + q(u,v)\,dv = \iint_S (q_u - p_v)\,dS,$$

so the last step in this long computation is to express $q_u - p_v$ in terms of P, Q, and M. Apply the chain rule to get the expressions

$$\begin{aligned} q_u &= 4\left(P_x\frac{\partial x}{\partial u} + P_y\frac{\partial y}{\partial u}\right) + 5\left(Q_x\frac{\partial x}{\partial u} + Q_y\frac{\partial y}{\partial u}\right) \\ &= 4(P_x(2) + P_y(4)) + 5(Q_x(2) + Q_y(4)), \\ p_v &= 2\left(P_x\frac{\partial x}{\partial v} + P_y\frac{\partial y}{\partial v}\right) + 4\left(Q_x\frac{\partial x}{\partial v} + Q_y\frac{\partial y}{\partial v}\right) \\ &= 2(P_x(4) + P_y(5)) + 4(Q_x(4) + Q_y(5)). \end{aligned}$$

Now subtract and use the definition $Q_x - P_y = M$ to get

$$q_u - p_v = (16 - 10)P_y + (10 - 16)Q_x = -6(Q_x - P_y) = -6M.$$

(It is not just an accident that the Q_y and P_x terms have 0 coefficients in this last computation.) In the last expression of M, we want to evaluate not at (x, y) but at $T(u, v)$, which is the same as (x, y). Combine the results of the last two computations to get

$$\begin{aligned} \iint_{T(S)} M(x,y)\,dA &= \int_C P(x,y)\,dx + Q(x,y)\,dy \\ &= \int_B p(u,v)\,du + q(u,v)\,dv = -\iint_S (q_u - p_v)\,dS \\ &= \iint_S 6M(T(u,v))\,dS \\ &= 6\int_a^b \int_c^d M(2u + 4v, 4u + 5v)\,dv\,du. \end{aligned}$$

The mysterious factor 6 that appears will be explained by placing this in a more general context below. If you follow the computations very closely you will discover that the 6 appears because

$$-6 = \frac{\partial x}{\partial u}\frac{\partial y}{\partial v} - \frac{\partial x}{\partial v}\frac{\partial y}{\partial u},$$

a quantity that is called the *Jacobian of the transformation.* The minus sign was negated by consideration of the orientation of the boundary curve.

Change of Variables in Double Integrals

Now we consider a more general case. We begin with a region A in the xy plane and a double integral

$$\iint_A M(x, y)\, dA$$

that we wish to evaluate by making a transformation of the variables (x, y) to (u, v) where the variables are related by

$$x = f(u, v) \qquad y = g(u, v).$$

The change of variables should be regarded as a geometric transformation T from the uv plane to the xy plane determined by

$$T(u, v) = (x, y) = (f(u, v), g(u, v)).$$

Let S be a region of the uv plane that is transformed by T onto the region A. Let B denote the boundary of S and C the boundary of A. Then T transforms points on B to points on C. We require that T be a one-to-one transformation from S to A; in other words, different points of S are mapped by T to different points of A. In particular, we assume this is true for points on the boundary.

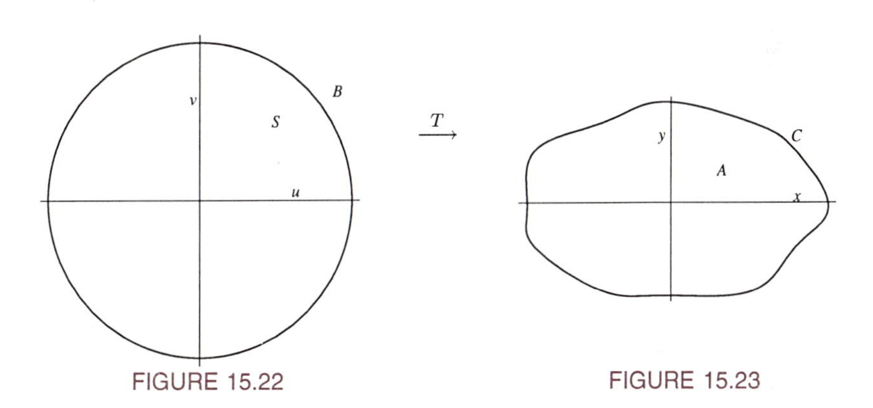

FIGURE 15.22

FIGURE 15.23

We must assume differentiability conditions on f and g so that we may eventually apply Green's Theorem, which requires certain functions to have continuous first partial derivatives. It is sufficient to assume that f and g have continuous first partial derivatives in S.

THEOREM 15.16

Change of Variables

Let $T(u, v) = (f(u, v), g(u, v)) = (x, y)$ be a transformation by which the region S in the uv plane is transformed to the region A in the xy plane in a one to one manner. Assume that f and g have continuous first partial derivatives in S. If M is a continuous function on A then

$$\iint_A M(x, y)\, dA = \pm \iint_S M(T(u, v))J(f, g)\, dS,$$

where

$$J(f, g) = f_u(u, v)g_v(u, v) - f_v(u, v)g_u(u, v).$$

The + sign is used if T preserves the orientation of the boundary of S and the − sign is used if T reverses the orientation of the boundary of S.

Proof

We first outline the proof. Apply Green's Theorem to express the double integral over A as a line integral around the boundary C of A. We then obtain a parametrization of C by applying T to a parametrization of B, the boundary of S, and use it to change variables so the line integral around C is expressed as a line integral around B. Then Green's Theorem is applied to the line integral around B to express it as a double integral over S.

Select P and Q differentiable functions on A that satisfy

$$Q_x - P_y = M.$$

There are usually many choices, for example take $P = 0$ and $Q(x, y) = \int M(x, y)\, dx$ with y held fixed. We have seen earlier that a function Q defined in this way is differentiable with respect to y whenever M is differentiable with respect to y. This represents an additional restriction on M that is not necessary for the validity of the theorem but we will assume it to give the proof we have in mind. By Green's Theorem we express the double integral as a line integral

$$\iint_A M\, dA = \iint_A (Q_x - P_y)\, dA = \int_C P\, dx + Q\, dy. \qquad \textbf{(15.24)}$$

Here C represents the boundary of A traversed in the positive direction. We evaluate the line integral by using a parametric representation obtained from a representation of the boundary of S.

Let $u = u(t)$, $v = v(t)$ be a parametric representation of the boundary B with $c \leq t \leq d$ and assume that as t increases, the point $(u(t), v(t))$ traverses B in the positive direction. Then the curve $T(u(t), v(t))$ is the boundary of A except that we do not know the direction. Let $\epsilon(T) = +1$ if the direction is positive and let $\epsilon(T) = -1$ if it is negative. We use the sign later when we apply Green's Theorem to the line integral around B.

The parametrization of C is given explicitly by

$$x = f(u(t), v(t)), \qquad y = g(u(t), v(t)).$$

We compute $dx = x'(t)\, dt$ and $dy = y'(t)\, dt$ from these equations with the aid of the chain rule:

$$dx = \left(f_u \frac{du}{dt} + f_v \frac{dv}{dt}\right) dt = f_u\, du + f_v\, dv,$$

$$dy = \left(g_u \frac{du}{dt} + g_v \frac{dv}{dt}\right) dt = g_u\, du + g_v\, dv.$$

We now have an expression of the line integral over C in terms of t and also in terms of u and v and a line integral around B:

$$\begin{aligned}
\int_C P\, dx + Q\, dy &= \int_c^d \left(P(T(u, v))\frac{dx}{dt} + Q(T(u, v))\frac{dy}{dt}\right) dt \\
&= \int_B P(T(u, v))(f_u\, du + f_v\, dv) + Q(T(u, v))(g_u\, du + g_v\, dv) \\
&= \int_B (P(T(u, v))f_u + Q(T(u, v))g_u)\, du \\
&\qquad + (P(T(u, v))f_v + Q(T(u, v))g_v)\, dv.
\end{aligned}$$

If we introduce short names for the functions appearing here,

$$p = P(T(u, v))f_u + Q(T(u, v))g_u, \qquad q = P(T(u, v))f_v + Q(T(u, v))g_v$$

then we have

$$\int_C P\,dx + Q\,dy = \int_B p\,du + q\,dv.$$

The last integral is, by Green's Theorem, a double integral over S of the difference $q_u - p_v$. The computation of these derivatives requires the derivatives with respect to u and v of $P(T(u,v)) = P(x,y)$ be determined; that is accomplished by the chain rule. We have

$$\begin{aligned} q_u &= [P_x x_u + P_y y_u] f_v + P f_{uv} + [Q_x x_u + Q_y y_u] g_v + Q g_{vu} \\ &= [P_x f_u + P_y g_u] f_v + P f_{uv} + [Q_x f_u + Q_y g_u] g_v + Q g_{vu}, \\ p_v &= [P_x x_v + P_y y_v] f_u + P f_{uv} + [Q_x x_v + Q_y y_v] g_u + Q g_{vu} \\ &= [P_x f_v + P_y g_v] f_u + P f_{uv} + [Q_x f_v + Q_y g_v] g_u + Q g_{vu}. \end{aligned}$$

Finally the difference simplifies to

$$q_u - p_v = (Q_x - P_y)(f_u g_v - f_v g_u) = M J(f,g).$$

The term $J(f,g) = f_u g_v - f_v g_u$ is called the Jacobian of (f,g) with respect to (u,v). Now we combine all the various computations to get the final string of equalities:

$$\begin{aligned} \iint_A M\,dA &= \epsilon(T) \int_C P\,dx + Q\,dy \\ &= \epsilon(T) \int_B p\,du + q\,dv \\ &= \epsilon(T) \iint_S (q_u - p_v)\,dS \\ &= \epsilon(T) \iint_S M J(f,g)\,dS, \end{aligned}$$

which proves the change of variables formula as stated in the conclusion of the theorem. □

Notation for the Jacobian

The Jacobian $J(f,g)$ can be expressed in the form of a 2×2 determinant: For $f = f(u,v)$ and $g = g(u,v)$, we have

$$J(f,g) = \begin{vmatrix} \dfrac{\partial f}{\partial u} & \dfrac{\partial f}{\partial v} \\ \dfrac{\partial g}{\partial u} & \dfrac{\partial g}{\partial v} \end{vmatrix} = f_u g_v - f_v g_u.$$

As an aid to keep the functions and variables in the correct order, this expression is sometimes written as

$$J(f,g) = \frac{\partial(f,g)}{\partial(u,v)}.$$

Now let us work a few examples to illustrate a change of variables.

EXAMPLE 1 Let A be the diamond-shaped region of the xy plane bounded by the four lines $x + y = \pm 1$ and $x - y = \pm 1$. Evaluate the double integral

$$\iint_A e^{2x+6y}\,dA.$$

Solution: We note that the region is bounded by two pairs of parallel lines so it would be a good idea to try a substitution in which $u = x + y$ and $v = x - y$. This seems like a good idea because the boundary is then described by the four lines $u = \pm 1$ and $v = \pm 1$, which is a good region for integration in the uv plane. Solve the pair of equations for x and y in terms of u and v to get

$$x = \frac{u+v}{2}, \qquad y = \frac{u-v}{2}.$$

The Jacobian of the transformation is

$$J = x_u y_v - x_v y_u = -\frac{1}{2}\frac{1}{2} - \frac{1}{2}\frac{1}{2} = -\frac{1}{2}.$$

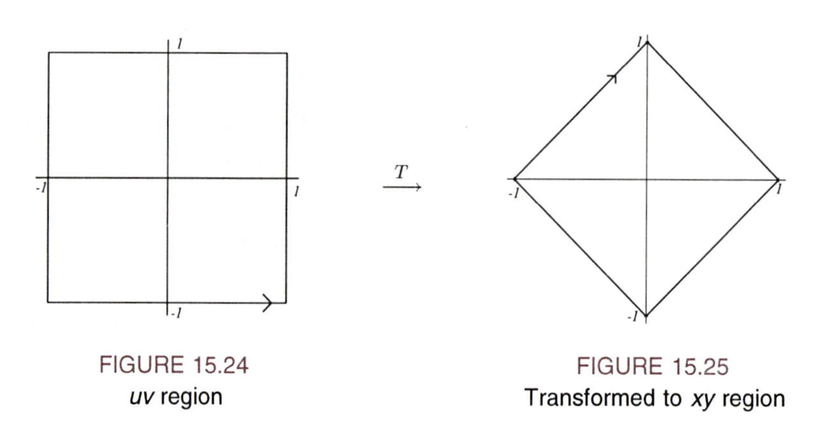

FIGURE 15.24
uv region

FIGURE 15.25
Transformed to *xy* region

Let us check on the orientation of the boundary after transformation by this correspondence. The segment $v = -1$, $-1 \le u \le 1$ is transformed into the segment having parametric representation

$$x = \frac{u-1}{2}, \qquad y = \frac{u+1}{2} \quad \text{for} \quad -1 \le u \le 1.$$

When $u = -1$ we get the point $(x, y) = (-1, 0)$; when $u = 1$ we get the point $(x, y) = (0, 1)$. Thus the curve in the xy plane is traversed in the negative direction when the curve in the uv plane is traversed in the positive direction. Thus a minus sign must be introduced in the final equation. We now have

$$\begin{aligned}\iint_A e^{2x+6y}\, dA &= -\iint_S e^{(u+v)+3(u-v)} J\, dS \\ &= -\int_{-1}^{1}\int_{-1}^{1} e^{4u} e^{-2v} \left(\frac{-1}{2}\right) du\, dv \\ &= \frac{1}{16}(e^4 - e^{-4})(e^2 - e^{-2}).\end{aligned}$$

As a check that we have accounted for all the minus signs correctly we observe that the function e^{2x+6y} is positive everywhere so its integral over a region should be positive, and indeed, the answer above is positive. ∎

EXAMPLE 2 Let A be the region in the xy plane bounded by two hyperbolas with equations $x^2 - y^2 = 1$ and $x^2 - y^2 = 4$ and two lines with equations $x + y = 10$ and $x - y = 10$. Find a change of variable that makes evaluation of an integral

$$\iint_A M(x, y)\, dA$$

somewhat simpler, at least insofar as the region is concerned.

Solution: The region is not of Type X but it is of Type Y but, unfortunately, the boundary is not given by a single formula. We look for a transformation to simplify the region of integration. Because two of the boundary curves have the form $x^2 - y^2$ = constant, we try $u = x^2 - y^2$ so in the uv plane, two boundaries are $u = 1$ and $u = 4$. The choice for v is not so clear, so we try $v = x + y$ because that choice makes a third boundary segment given by $v = 10$. By using the equations

$$u = x^2 - y^2 = (x + y)(x - y) = v(x - y)$$

and $v = x + y$, we can solve for x and y in terms of u and v and get

$$x = \frac{1}{2}\left(v + \frac{u}{v}\right), \qquad y = \frac{1}{2}\left(v - \frac{u}{v}\right).$$

The corresponding region in the uv plane is bounded by the lines $u = 1$, $u = 4$, $v = 10$ and the line

$$10 = x - y = \frac{1}{2}\left(v + \frac{u}{v}\right) - \frac{1}{2}\left(v - \frac{u}{v}\right) = \frac{u}{v},$$

or, more simply, $u = 10v$. So we do get a region bounded by straight lines in the uv plane. The transformation is not continuous where $v = 0$ but the region S does not include any points with $v = 0$ so the transformation functions are differentiable.

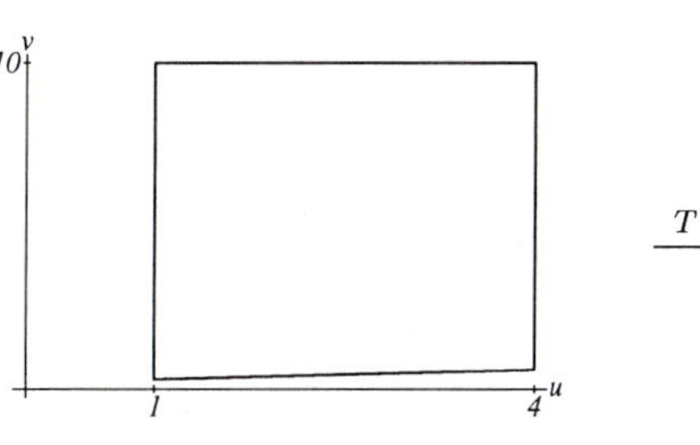

FIGURE 15.26
uv region

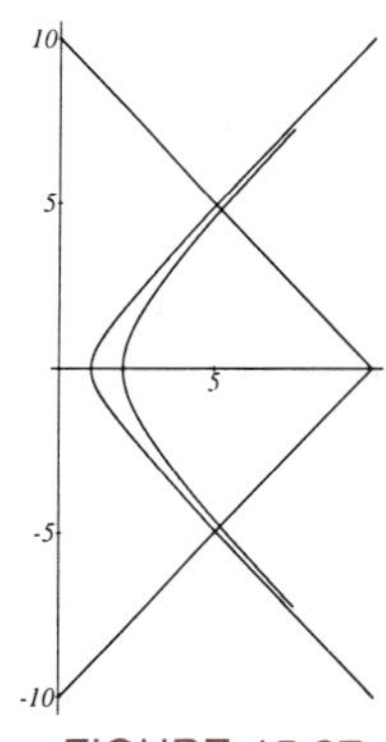

FIGURE 15.27
Transformed to *xy* region

The Jacobian is

$$J = \frac{\partial(x, y)}{\partial(u, v)} = x_u y_v - x_v y_u = \frac{1}{2v},$$

as can easily be computed.

The boundary of S is transformed with the orientation preserved as one checks, for example, by looking at one segment $u = 10v$ where $1/10 \le v \le 4/10$ and increasing v indicates the positive direction. This corresponds to the parametric curve

$$x = \frac{1}{2}\left(v + 10\right), \qquad y = \frac{1}{2}\left(v - 10\right)$$

on which increasing v is also the positive direction. So finally we have

$$\iint_A M(x, y)\, dA = \int_1^4 \int_{u/10}^{10} M\left(\frac{1}{2}\left(v + \frac{u}{v}\right), \frac{1}{2}\left(v - \frac{u}{v}\right)\right) \frac{1}{2v}\, dv\, du.$$

The ease in evaluating of this integral depends on the function M. ■

Polar Coordinates in Double Integrals

The general change of variable formula for double integrals can be applied to the case in which a region of the xy plane is described in polar coordinates. We view the equations

$$x = r\cos\theta, \qquad y = r\sin\theta,$$

as a transformation from the $r\theta$ plane to the xy plane

$$T(r,\theta) = (r\cos\theta, r\sin\theta) = (x,y).$$

If a region R in the xy plane is the transform of a region S in the $r\theta$ plane then a double integral over R can be expressed as a double integral over S. The Jacobian of the transformation is

$$\frac{\partial(x,y)}{\partial(r,\theta)} = x_r y_\theta - x_\theta y_r = r\cos^2\theta + r\sin^2\theta = r.$$

If we suppose that the description of R in terms of polar coordinates is given by

$$\alpha \le \theta \le \beta, \qquad g_1(\theta) \le r \le g_2(\theta),$$

then

$$\iint_R f(x,y)\,dR = \int_\alpha^\beta \int_{g_1(\theta)}^{g_2(\theta)} f(r\cos\theta, r\sin\theta)\,r\,dr\,d\theta.$$

We have already illustrated this change of variable in section 15.3.

Summation of a Special Series

We close this section with an application of the change of variable formula to find the sum of a special infinite series. The series

$$S = \sum_{n=1}^{\infty} \frac{1}{n^2}$$

converges as we can verify by any of several tests; finding the exact value of the sum is rather difficult. We use an indirect method that involves evaluating the integral

$$I = \int_0^1 \int_0^1 \frac{1}{1-xy}\,dx\,dy$$

in two ways. The first way uses the geometric series to obtain

$$\frac{1}{1-xy} = 1 + xy + x^2y^2 + \cdots + x^ny^n + \cdots = \sum_{n=0}^{\infty} x^ny^n.$$

Then the integral I is evaluated as

$$I = \sum_{n=0}^{\infty} \int_0^1 \int_0^1 x^ny^n\,dx\,dy = \sum_0^{\infty} \frac{1}{(n+1)^2} = \sum_1^{\infty} \frac{1}{n^2} = S.$$

The second method for evaluating I is to make a change of variable and then actually obtain the numerical value of I. Use the transformation from the uv

plane to the xy plane given by

$$(x, y) = T(u, v) = \left(\frac{u+v}{\sqrt{2}}, \frac{u-v}{\sqrt{2}}\right).$$

The region that is transformed onto the unit square $[0, 1; 0, 1]$ is a diamond-shaped region that is the union of two triangles

$$R_1 = \{(u, v) : 0 \le u \le \sqrt{2}/2, \quad -u \le v \le u\}$$
$$R_2 = \{(u, v) : \sqrt{2}/2 \le u \le \sqrt{2}, \quad u - \sqrt{2} \le v \le \sqrt{2} - u\}.$$

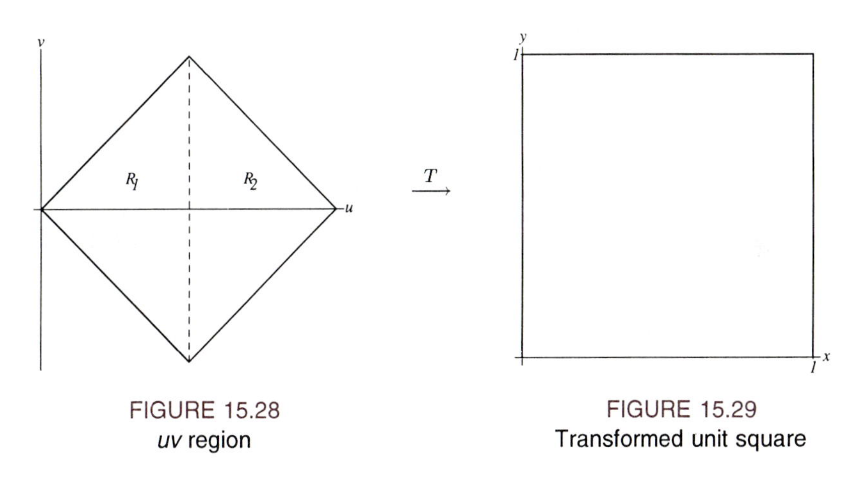

FIGURE 15.28
uv region

FIGURE 15.29
Transformed unit square

The function $1/(1 - xy)$ is given in terms of u and v by

$$a(u, v) = \frac{2}{2 - u^2 + v^2}.$$

The Jacobian of the transformation is found to equal 1 so we have the formula (by the Change of Variables Theorem)

$$I = \iint_{R_1} a(u, v)\, dv\, du + \iint_{R_2} a(u, v)\, dv\, du.$$

So now we separately evaluate the two integrals on the right. For both integrals, the first step is to obtain the antiderivative of $a(u, v)$ when viewed as a function of v only:

$$A(u, v) = \int a(u, v)\, dv = \frac{2}{\sqrt{2 - u^2}} \arctan \frac{v}{\sqrt{2 - u^2}}.$$

Making use of the symmetry of the arctangent function in the form $\arctan(-w) = -\arctan w$, we find

$$\iint_{R_1} a(u, v)\, dv\, du = \int_0^{\sqrt{2}/2} \frac{2}{\sqrt{2 - u^2}} \cdot 2 \cdot \arctan \frac{u}{\sqrt{2 - u^2}}\, du.$$

This last integral looks very complicated but a judicious change of variable makes it easier. [I don't know who first thought of this trick, but it sure is not transparent.] Let

$$\theta = \arctan \frac{u}{\sqrt{2 - u^2}}.$$

For a right triangle with sides u, $\sqrt{2 - u^2}$ and hypotenuse 2, one of the acute angles is θ. We see that

$$2\sin\theta = u, \quad 2\cos\theta\, d\theta = du \quad \text{and} \quad 2\cos\theta = \sqrt{2-u^2}.$$

When $u = 0$, $\theta = 0$; when $u = \sqrt{2}/2$, $\theta = \pi/6$. Now make all the substitutions in the integral to get

$$\begin{aligned}\iint_{R_1} a(u,v)\, dv\, du &= \int_0^{\pi/6} \frac{1}{\cos\theta} \cdot 2\theta \cdot 2\cos\theta\, d\theta \\ &= \int_0^{\pi/6} 4\theta\, d\theta = \frac{\pi^2}{3\cdot 6}.\end{aligned}$$

With the help of some fortunate cancellation, we have the first integral. Now we try the second integral:

$$\iint_{R_2} a(u,v)\, dv\, du = \int_{\sqrt{2}/2}^{\sqrt{2}} \frac{2}{\sqrt{2-u^2}} \cdot 2 \cdot \arctan\frac{\sqrt{2}-u}{\sqrt{2-u^2}}\, du.$$

Note that

$$\frac{\sqrt{2}-u}{\sqrt{2-u^2}} = \frac{\sqrt{2}-u}{\sqrt{(\sqrt{2}-u)(\sqrt{2}+u)}} = \frac{\sqrt{\sqrt{2}-u}}{\sqrt{\sqrt{2}+u}}.$$

Let θ be a variable such that

$$\theta = \arctan\frac{\sqrt{\sqrt{2}-u}}{\sqrt{\sqrt{2}+u}} \quad \text{or} \quad \tan\theta = \frac{\sqrt{\sqrt{2}-u}}{\sqrt{\sqrt{2}+u}}.$$

Then we find (after some manipulation of terms)

$$2\sqrt{2}\cos^2\theta = \sqrt{2}+u, \quad 2\sqrt{2}\sin^2\theta = \sqrt{2}-u, \quad 2\sqrt{2}\sin\theta\cos\theta = \sqrt{2-u^2}.$$

We also get the differential as $du = -4\sqrt{2}\cos\theta\sin\theta\, d\theta$. When $u = \sqrt{2}/2$, $\theta = \pi/6$; when $u = \sqrt{2}$, $\theta = 0$. Putting all these facts together, we find

$$\begin{aligned}\iint_{R_2} a(u,v)\, dv\, du &= \int_{\pi/6}^{0} \frac{4}{2\sqrt{2}\sin\theta\cos\theta} \cdot \theta \cdot (-4\sqrt{2})\cos\theta\sin\theta\, d\theta \\ &= \int_{\pi/6}^{0} -8\theta\, d\theta = \frac{2\pi^2}{3\cdot 6}.\end{aligned}$$

Now combine the two computations to get the remarkable formula

$$\sum_{n=1}^{\infty} \frac{1}{n^2} = \frac{\pi^2}{6}.$$

This series sum has been known for a long time. Leonhard Euler posed the problem (in 1736) of finding the sum of the series of the reciprocals of the cubes of the positive integers. The geometric series argument implies that

$$\sum_{n=1}^{\infty} \frac{1}{n^3} = \int_0^1 \int_0^1 \int_0^1 \frac{1}{1-xyz}\, dx\, dy\, dz. \qquad \textbf{(15.25)}$$

However no one has been able to evaluate the triple integral and so the sum is unknown in the sense that it cannot be expressed in terms of familiar constants. It was proved (by Roger Apéry) in the late-1970s that the sum $\sum 1/n^3$ is not a rational number. [See the article by A. Van der Poorten, Math. Intellegencer, No. 1, 1978-79, pp. 195-203.] Here is your chance for fame (and fortune?). Evaluate the triple integral (15.25).

Exercises 15.10

In each of the next six problems, evaluate the integral $\iint_A xy\,dA$ when A is the parallelogram indicated. Find a change of variables into u, v coordinates that transforms the integral to one taken over a rectangle in the uv plane. Make a sketch and indicate the direction of the boundary and its transform.

1. $0 \le 2x + y \le 4,\ 0 \le -x + 3y \le 1$

2. $-1 \le x + 4y \le 12,\ -2 \le x + y \le 0$

3. $2 \le 2x + 5y \le 10,\ 0 \le 2x + 6y \le 9$

4. $-4 \le -3x + 7y \le 4,\ -1 \le -x + 3y \le 1$

5. $10 \le x + y \le 11,\ 10 \le -x + y \le 11$

6. $-5 \le 2x + 13y \le -4,\ -2 \le x + 6y \le -1$

Evaluate the integral $\iint_S x^2 y\,dS$, where S is the region of the xy plane bounded by the indicated curves. Carry out the evaluation by making a transformation of variables as indicated. Sketch the region S and the region of the uv plane that corresponds to S.

7. S is the region in the first quadrant bounded by the the lines $y\sqrt{3} = x$, $y = x$ and the circles $x^2 + y^2 = 1$, $x^2 + y^2 = 4$ and $T(u, v) = (e^{-v}\cos u, e^{-v}\sin u)$

8. S is bounded by the curves $y = (x/2)^2 - 1$, $y = 16 - (x/8)^2$ and the transformation is $(x, y) = T(u, v) = (2uv, v^2 - u^2)$

9. S is the region in the upper half plane ($y \ge 0$) bounded by the curves $x = 1 - (y/2)^2$, $x = 25 - (y/10)^2$, $x = -4 + (y/4)^2$, $x = -25 + (y/10^2$ and the transformation is $T(u, v) = (u^2 - v^2, 2uv)$

10. S is the roughly triangular shaped region bounded by the curves $x = y$, $x = -2y^2$, $x = -2 + 7y - 4y^2$ (and containing the point $(0.2, 0.3)$) and the transformation is $(x, y) = T(u, v) = (u - 2v^2, -u + v^2)$

Find a suitable transformation $T(u, v) = (x, y)$ that make evaluation of the following integrals easier in the sense that all limits of the u,v integrals are constant and only one double integral is required.

11. $$\int_0^3 \int_{x-3}^{x+4} f(x, y)\,dy\,dx$$

12. $$\int_{-1}^4 \int_{2y+5}^{2y+8} f(x, y)\,dx\,dy$$

13. $$\left(\int_0^1 \int_{-x}^{7x/5} + \int_1^2 \int_{(7x-12)/5}^{-x+12/5}\right) f(x, y)\,dy\,dx$$

14. $$\int_1^2 \int_0^{4x-4} f(x, y)\,dy\,dx + \int_2^3 \int_{4x-8}^{4} f(x, y)\,dy\,dx$$

15. Prove the Chain Rule for Jacobians that states

$$\frac{\partial(x, y)}{\partial(u, v)} \frac{\partial(u, v)}{\partial(s, t)} = \frac{\partial(x, y)}{\partial(s, t)}$$

provided all the functions involved are differentiable.

16. Suppose $T(u, v) = (x, y)$ is a transformation from the uv plane to the xy plane that carries a region U onto a region R. Suppose that $M(s, t) = (u, v)$ is a transformation from the st plane to the uv plane that carries a region Q onto the region U. State a formula that relates an integral $\iint_R f(x, y)\,dR$ to an integral over Q. Assume whatever hypothesis is necessary about the two transformations.

15.11 Triple integrals by spherical coordinates

We have given a the change of variable formula for double integrals. We do not treat the general case for triple integrals but present two special techniques that arise frequently. Polar coordinates for regions of the xy plane have a natural analogue in xyz space where the notion is called cylindrical coordinates. We also describe spherical coordinates, which are convenient for regions having some spherical symmetry.

Cylindrical Coordinates

If the region of integration in three dimensions involves circular symmetry in the

xy plane, cylindrical coordinates might simplify the integration. The change of variable is

$$x = r\cos\theta, \quad y = r\sin\theta, \quad z = z;$$

this is just polar coordinates for the x and y variables and no change for the z variable. The Jacobian for the change from x, y coordinates to r, θ is just r so there is no surprise in the following formula for integration in cylindrical coordinates:

$$\iiint_V f(x, y, z)\, dV = \iiint_V f(r\cos\theta, r\sin\theta, z)\, r\, dz\, dr\, d\theta,$$

where in the integral on the right, V is described in r, θ, z coordinates by inequalities of the form

$$\alpha \leq \theta \leq \beta, \quad g_1(\theta) \leq r \leq g_2(\theta), \quad h_1(r, \theta) \leq z \leq h_2(r, \theta).$$

We give a simple illustration where cylindrical coordinates simplify the integration.

EXAMPLE 1 Find the mass of a cylindrical container of height h and base of radius R if the density at a point (x, y, z) is given by $\rho = 100z^2$.

Solution: The mass is given as the triple integral

$$M = \iiint_V 100z^2\, dV,$$

where V is the region described in rectangular coordinates by

$$V: \; -R \leq x \leq R, \quad -\sqrt{R^2 - x^2} \leq y \leq \sqrt{R^2 - x^2}, \quad 0 \leq z \leq h.$$

The region is somewhat simpler to describe in cylindrical coordinates:

$$V: \; 0 \leq r \leq R, \quad 0 \leq \theta \leq 2\pi, \quad 0 \leq z \leq h.$$

The mass is

$$M = \int_0^{2\pi} \int_0^R \int_0^h 100z^2 r\, dz\, dr\, d\theta = \frac{100h^3R^2\pi}{3}.$$

■

Spherical Coordinates

Now we introduce a system of coordinates that is especially useful in three-dimensional problems having spherical symmetry.

The idea is to locate a point by giving its distance from the origin (ρ) and then giving two angles to uniquely locate the point on the sphere of radius ρ. We saw in chapter 13 that every unit vector in xyz space has the form

$$\mathbf{u} = \mathbf{u}(\theta, \phi) = \sin\phi\cos\theta\,\mathbf{i} + \sin\phi\sin\theta\,\mathbf{j} + \cos\phi\,\mathbf{k}.$$

Any vector $\mathbf{v}$ with initial point at the origin and terminal point at (x, y, z) can be expressed as its length times a unit vector. Thus we have

$$\mathbf{v} = x\,\mathbf{i} + y\,\mathbf{j} + z\,\mathbf{k} = \rho(\sin\phi\cos\theta\,\mathbf{i} + \sin\phi\sin\theta\,\mathbf{j} + \cos\phi\,\mathbf{k})$$

where ρ is the distance from the origin to (x, y, z), ϕ is the angle from the z-axis to $\mathbf{v}$ and θ is the same angle defined in cylindrical coordinates.

We summarize the relationships between the various coordinates and the

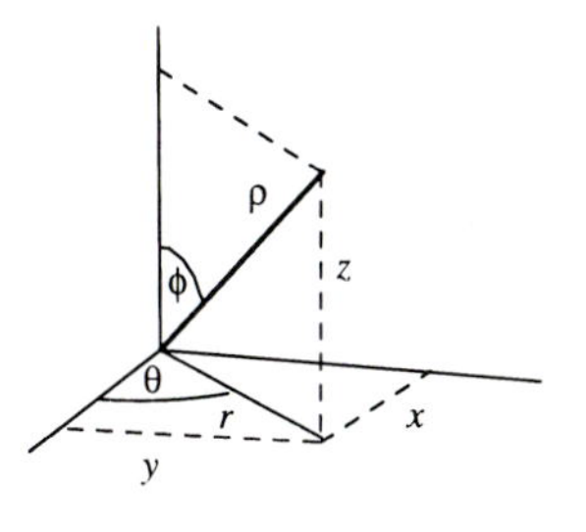

FIGURE 15.30
Spherical coordinates (ρ, θ, ϕ) of (x, y, z)

restrictions on the variables:

$$x = \rho \sin \phi \cos \theta, \quad y = \rho \sin \phi \sin \theta, \quad z = \rho \cos \phi,$$

$$\rho = \sqrt{x^2 + y^2 + z^2}, \quad \theta = \arctan \frac{y}{x}, \quad \phi = \arccos \frac{z}{\rho},$$

$$\rho \geq 0, \qquad 0 \leq \theta < 2\pi, \qquad 0 \leq \phi \leq \pi.$$

The numbers (ρ, θ, ϕ) corresponding to (x, y, z) are the *spherical coordinates* of (x, y, z). Take notice of the order in which these are recorded. These are related to the cylindrical coordinates (r, θ, z) of (x, y, z) by the rules

$$r = \rho \sin \phi, \qquad \theta = \theta, \qquad z = \rho \cos \phi.$$

Surfaces in Spherical Coordinates

The set of all points whose spherical coordinates are $(2, \theta, \phi)$, with no restriction on θ or ϕ, is written as $\rho = 2$. This is a sphere centered at the origin and having radius 2. This example shows the advantage of using spherical coordinates to simplify certain equations of surfaces involving the rectangular expression $x^2 + y^2 + z^2$.

The equation $\theta = \alpha$, a constant, represents a plane containing the z-axis such that the line of intersection with the xy plane makes an angle $\theta = \alpha$ with the positive x-axis.

The equation $\phi = \alpha$ represents a (half) cone with vertex at the origin. For $0 \leq \alpha \leq \pi/2$ the cone opens upward (so it holds water) and for $\pi/2 \leq \alpha \leq \pi$ it opens downward. The cases $\alpha = 0$, $\pi/2$ or π are not usually called cones because in the first and last case, the surface is a half-line and in the remaining case it is the xy plane.

EXAMPLE 2 Express the circular paraboloid $z = 4x^2 + 4y^2$ in spherical coordinates.

Solution: Use the equations for x, y, z in terms of ρ, θ, ϕ and substitute into the equation of the paraboloid to get

$$\begin{aligned} \rho \cos \phi &= 4\rho^2 \sin^2 \phi \cos^2 \theta + 4\rho^2 \sin^2 \phi \sin^2 \theta \\ &= 4\rho^2 \sin^2 \phi(\cos^2 \theta + \sin^2 \theta) \\ &= 4\rho^2 \sin^2 \phi; \end{aligned}$$

more simply,

$$\rho = \frac{\cos \phi}{4 \sin^2 \phi}.$$

■

EXAMPLE 3 Express in rectangular form the surface having spherical equation $\rho = a \cos \phi$, with a constant.

Solution: Multiply the equation by ρ to get $\rho^2 = 2a\rho \cos \phi$. Now substitute from the transformation equations to get

$$x^2 + y^2 + z^2 = 2az.$$

After completing the square, we find this is equivalent to the equation

$$x^2 + y^2 + (z - a)^2 = a^2.$$

Hence the surface is a sphere of radius a with center $(0, 0, a)$. ■

Integration

We turn immediately to the application of spherical coordinates to the evaluation of certain triple integrals. Suppose we wish to evaluate the integral

$$\iiint_V f(x, y, z)\, dV$$

where V is a region of xyz space that can be conveniently described in spherical coordinates. We have used regions of Type XY and Type YX as convenient for rectangular coordinates. By analogy we call a region V Type $\phi\theta$ or Type $\theta\phi$ if V consists of all points with spherical coordinates (ρ, θ, ϕ) satisfying either

$$\text{Type } \theta\phi: \quad a \le \theta \le b, \quad g_1(\theta) \le \phi \le g_2(\theta), \quad h_1(\theta, \phi) \le \rho \le h_2(\theta, \phi), \quad \textbf{(15.26)}$$

or

$$\text{Type } \phi\theta: \quad a \le \phi \le b, \quad g_1(\phi) \le \theta \le g_2(\phi), \quad h_1(\theta, \phi) \le \rho \le h_2(\theta, \phi). \quad \textbf{(15.27)}$$

When V is such a region it would be natural to convert an integral from rectangular form to spherical form but the change of variable must be done with an appropriate scaling factor. In the change from rectangular to cylindrical coordinates, $dx\, dy\, dz$ was replaced by $r\, dr\, d\theta\, dz$. In the conversion to spherical coordinates, the scaling factor replaces $dx\, dy\, dz$ by $\rho^2 \sin\phi\, d\rho\, d\phi d\theta$ or $\rho^2 \sin\phi\, d\rho\, d\theta\, d\phi$, depending on the type description. Here is a statement of the main transformation rule.

THEOREM 15.17

Spherical Coordinates

Let V be a region that is described as Type $\theta\phi$ as in (15.26). Then

$$\iiint_V f(x, y, z)\, dV =$$

$$\int_a^b \int_{g_1(\theta)}^{g_2(\theta)} \int_{h_1(\theta,\phi)}^{h_2(\theta,\phi)} f(\rho \sin\phi \cos\theta, \rho \sin\phi \sin\theta, \rho \cos\phi)\rho^2 \sin\phi\, d\rho\, d\phi\, d\theta.$$

A similar statement holds if V is Type $\phi\theta$.

A proof of this theorem would require the discussion of the Jacobian of a transformation in three dimensions. We considered the Jacobian only in two dimensions (section 15.10) and do not digress now to prove the extension of the theorem. We illustrate it with several examples.

EXAMPLE 4 Compute the volume enclosed by a sphere of radius a.

Solution: The solid bounded by a sphere of radius a is a Type $\theta\phi$ region with description

$$0 \le \theta \le 2\pi, \quad 0 \le \phi \le \pi, \quad 0 \le \rho \le a,$$

so by the theorem, the volume is

$$\iiint_V 1\,dV = \int_0^{2\pi}\int_0^{\pi}\int_0^{a} \rho^2 \sin\phi \, d\rho\, d\phi\, d\theta$$
$$= 2\pi \int_0^a \rho^2(-\cos\phi)\Big|_0^{\pi}$$
$$= \frac{4\pi}{3}\rho^3\Big|_0^a = \frac{4\pi a^3}{3}.$$

The reader should compare this computation with the method by cylindrical coordinates and rectangular coordinates. ■

Here is another example showing the advantage of this change of variable.

EXAMPLE 5 Compute the mass of a hemispherical solid of radius a in which the density is proportional to the kth power ($k \neq -3$) of the distance from the center of the complete sphere.

Solution: The hemisphere V has the description in spherical coordinates

$$V: \quad 0 \le \theta \le 2\pi, \quad 0 \le \phi \le \pi/2 \quad 0 \le \rho \le a.$$

The density at the point (ρ, θ, ϕ) is ρ^k so the total mass is given by the integral

$$M = \int_0^{\pi/2}\int_0^{2\pi}\int_0^{a} \rho^k \rho^2 \sin\phi\, d\rho\, d\theta\, d\phi = \frac{a^{k+3}}{k+3}(2\pi)(-\cos\phi)\Big|_0^{\pi/2} = \frac{2a^5\pi}{5}. \quad ■$$

Exercises 15.11

Transform the integrals in rectangular coordinates to cylindrical or spherical coordinates and evaluate the integrals.

1. $\iiint_V x^2\,dV$, where V is the solid
$-1 \le x \le 1, \quad |y| \le \sqrt{1-x^2}, \quad 0 \le z \le 4$

2. $\iiint_V \sqrt{x^2+y^2+z^2}\,dV$, where V is the solid
$-9 \le x \le 9, \quad |y| \le \sqrt{9-x^2}, \quad 0 \le z \le a$

3. $\iiint_V \frac{dV}{\sqrt{x^2+y^2}}$, where V is the solid
$|x| \le 1,\ 0 \le y \le \sqrt{1-x^2},\ 0 \le z \le \sqrt{1-x^2-y^2}$

4. $\iiint_V \frac{dV}{\sqrt{x^2+y^2}}$, where V is the solid
$0 \le x \le 4,\ 0 \le y \le \sqrt{4-x^2},\ 0 \le z \le \sqrt{16-x^2-y^2}$

5. Find the center of mass of the part of the solid ball of radius a having center at the origin lying in the positive octant (x, y, z all ≥ 0). Assume uniform density.

6. Find the center of mass of the paraboloid given in rectangular coordinates by $z = 3(x^2+y^2)$ and $0 \le z \le 4$. Assume uniform density.

7. Find the volume of the "flat ice-cream cone" bounded below by the cone $\phi = \pi/6$ and above by the sphere $\rho = a$.

8. Compute the moment of inertia about the z-axis of the solid sphere of radius a having center at the origin.

9. Find the volume of the solid bounded above by the surface of the sphere $x^2+y^2+z^2 = 2$ and below by the paraboloid $z = x^2+y^2$.

10. Define a function F by the rule $F(x, y, z)$ is the product of the distance from (x, y, z) to the origin and from (x, y, z) to the z-axis. Evaluate the integral

$$\iiint_V F(x, y, z)\,dV,$$

where V is bounded by the sphere of radius a having its center at the origin.

16 Two Theorems in Vector Calculus

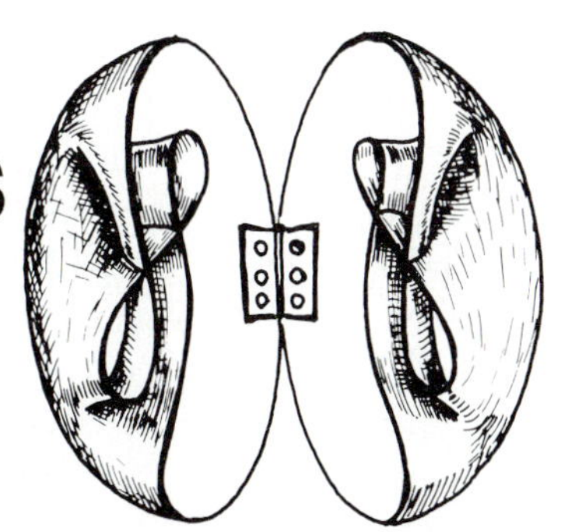

This short chapter is devoted to the proofs of two important theorems that deal with integration and vector functions. Stokes's Theorem relates the line integral of a vector function over a path C to the surface integral of the curl of the vector function taken over a surface having C as its boundary. Stokes's Theorem has many applications to physical problems; for example, it can be used to give an easy proof that the curl of an electrostatic field is zero. The Divergence Theorem relates an integral over a closed surface of the normal component of a vector function to the triple integral of the divergence of the vector function over the region enclosed by the surface. As an application of the Divergence Theorem we show how to compute the divergence of an electrostatic field.

16.1 Oriented surfaces and Stokes's theorem

We have presented Green's Theorem, which related a line integral around a simple closed path in the xy plane to a double integral over the region enclosed by the path. In this section we consider Stokes's Theorem, which has a similar context but is considerably more general. It relates a line integral taken along a closed path in three-dimensional space to a double integral taken over a surface having the path as its boundary. Because many different surfaces have the same boundary, this theorem is considerably more surprising than Green's Theorem. We have discussed integrals of a scalar function taken over a surface in the section on the area of a surface. We extend that discussion to consider now integrals of vector functions over oriented surfaces. In Green's Theorem it was necessary to take the orientation of the simple closed curve into account. The matter is more complicated in three dimensions and we must introduce the notion of an oriented surface along with some additional notions for plane regions.

Stokes's Theorem is of great value in the study of vector calculus. The vector notation provides a compact way of expressing what would otherwise be quite complicated expressions. For this reason we conform to standard practice and formulate Stokes's Theorem in vector notation.

Boundaries and Interiors

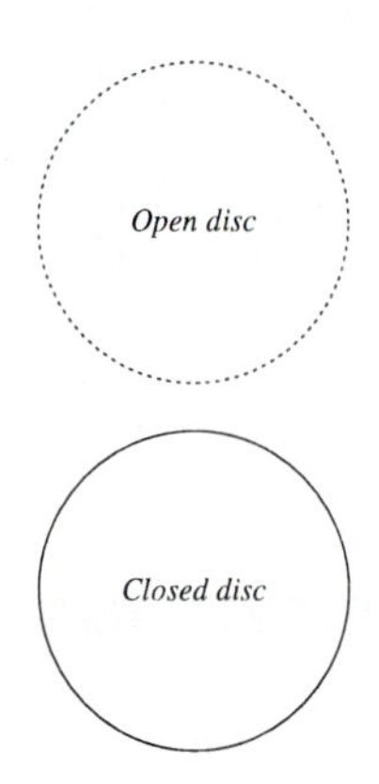

FIGURE 16.1

We have used the notions of open intervals (a, b) and closed intervals $[a, b]$ on the line. We need the analogous notions for two-dimensional regions. An open disc $D(r)$ of radius r and a closed disc $D_1(r)$ are easily defined by inequalities:

$$D(r) = \{(x, y) : x^2 + y^2 < r^2\} \qquad D_1(r) = \{(x, y) : x^2 + y^2 \leq r^2\}.$$

The *boundary* of either disc is the circle B with equation $x^2 + y^2 = r^2$. Note the terminology here. We call B the boundary of $D(r)$ and $D_1(r)$ even though B is part of $D_1(r)$ but not part of $D(r)$. These are discs centered at the origin. An open or closed disc centered at a general point (h, k) is defined analogously. We use the term *interior* to describe the part of the disc that excludes the boundary. Thus $D(r)$ is the interior of $D_1(r)$ and $D(r)$ is also the interior of $D(r)$.

For more general regions in the plane, we use an intuitive notion of boundary and interior. A region G is frequently described by specifying a closed curve as the boundary and including as part of G the region enclosed by the curve. If the curve is a familiar one, such as a circle or ellipse, or one given parametically by "nice" functions, then one has no difficulty deciding if a point is inside the region or outside the region. We restrict our attention to well-behaved regions and be content with a few examples.

A rectangle defined by

$$R[a, b; c, d] = \{(x, y) : a \leq x \leq b; \quad c \leq y \leq d\}$$

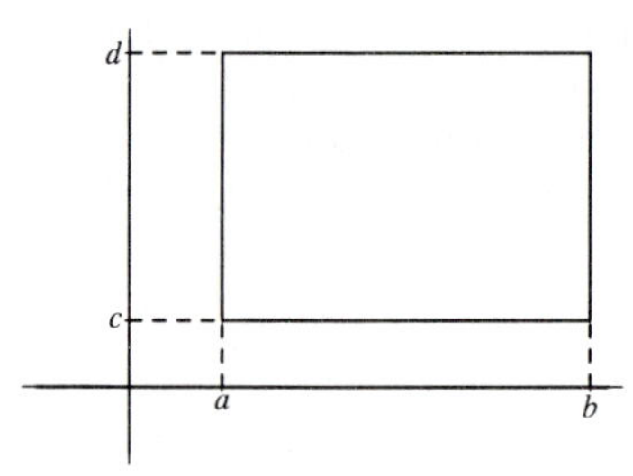

FIGURE 16.2
$R[a, b; c, d]$

is a closed region whose boundary consists of the four line segments and whose interior is the set

$$I[a, b; c, d] = \{(x, y) : a < x < b; \quad c < y < d\}.$$

A more general region can be described by applying a transformation to a rectangular region. Let $f(x, y)$ and $g(x, y)$ be differentiable functions at every point of a rectangle $R[a, b; c, d]$ and let T be the transformation defined by the rule $T(u, v) = (f(u, v), g(u, v))$ for (u, v) in $R = R[a, b; c, d]$. Let G be the set $\{T(u, v) : (u, v) \in R\}$. Then the boundary of G is the set of points $T(u, v)$ where (u, v) is on the boundary of R; the interior of G is the set of all $T(u, v)$ where (u, v) runs over all the interior points of R. These statements are not meant to be obvious but their proof is not be given here.

Oriented Surfaces

The concept of an oriented surface (to be defined) is technically rather subtle but the intuitive purpose is to provide a device to ensure that we can tell one side of the surface from another. There are examples of "one-sided" surfaces that cannot be oriented and to which the following discussion does not apply. The surfaces that we wish to consider are defined in terms of the existence of a parameterization that has some "nice" properties.

DEFINITION 16.1

Regular Surface

Let S be a set in xyz space. S is a regular surface if there is a region G of the uv plane and a continuous vector function

$$\mathbf{r}(u, v) = x(u, v)\,\mathbf{i} + y(u, v)\,\mathbf{j} + z(u, v)\,\mathbf{k}$$

defined and having continuous first partial derivatives throughout some rectangle R containing G such $\mathbf{r}_u \times \mathbf{r}_v$ is nonzero at every point on the interior of G and that $(u, v) \to (x(u, v), y(u, v), z(u, v))$ is a correspondence between the points of G and the points of S that is one to one on the interior of G.

The definition insists that a parametrization is one to one on the interior points but it is allowed to send different boundary points to the same point. As we because see below, it is also useful to consider surfaces that are made up of regular surfaces, pieced together along curves in their boundaries.

DEFINITION 16.2

Orientable Surface

A surface S is orientable if it is a regular surface and there is defined a continuous vector function $\mathbf{n}(x, y, z)$, defined for all (x, y, z) on S, such that $\mathbf{n}(x, y, z)$ is a unit normal to S at (x, y, z).

There are several subtle points in this definition that should be noticed. The set S is required to be a surface with a tangent plane at every point. The vector function $\mathbf{n}$ gives a unit normal to S at each point of S and, most importantly, $\mathbf{n}$ is a continuous function. The fact that S is a regular surface implies there is a parametrization $\mathbf{r}(u, v)$ for (u, v) in G for which $\mathbf{r}_u \times \mathbf{r}_v$ is nonzero at every point of the interior of G. Thus there is a unit normal that varies continuously on the interior of the surface. The condition of orientability requires that the unit normal is defined at every point of the surface, not only at the points corresponding to the interior of G.

We remark that one can compute a unit normal to a surface from a parametric representation if suitable conditions hold. For example suppose that a surface is given by the vector representation

$$\mathbf{r}(u, v) = x(u, v)\,\mathbf{i} + y(u, v)\,\mathbf{j} + z(u, v)\,\mathbf{k},$$

with (u, v) ranging over a region G of the uv plane. Assume that the correspondence is one to one on all of G. If the two derivatives $\mathbf{r}_u$ and $\mathbf{r}_v$ exist and their cross product is nonzero at every point of G, then the function

$$\mathbf{n}(u, v) = \frac{\mathbf{r}_u \times \mathbf{r}_v}{|\mathbf{r}_u \times \mathbf{r}_v|}$$

is a unit normal to the surface. If the derivatives $\mathbf{r}_u$ and $\mathbf{r}_v$ are continuous then $\mathbf{n}$ is continuous and the surface is orientable.

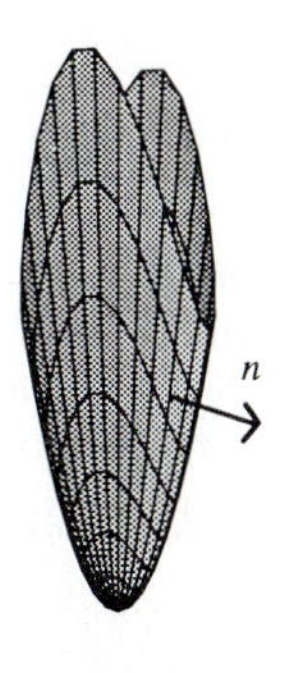

FIGURE 16.3 A normal to the paraboloid

EXAMPLE 1 Show that the paraboloid $z = ax^2 + by^2$, with $0 \le ax^2 + by^2 \le c$, is an orientable surface. Here we assume a, b and c are positive constants.

Solution: Let us first try the most obvious parametrization in which the position vector to the surface is given in rectangular coordinates by

$$\mathbf{r}(x, y) = x\,\mathbf{i} + y\,\mathbf{j} + (ax^2 + by^2)\,\mathbf{k} \qquad 0 \le ax^2 + by^2 \le c.$$

Thus the region G is the elliptic region of the xy plane given by $0 \le ax^2 + by^2 \le c$. This representation is one to one on any region of the xy plane. Then

$$\mathbf{r}_x = \mathbf{i} + 2ax\,\mathbf{k}, \qquad \mathbf{r}_y = \mathbf{j} + 2by\,\mathbf{k}.$$

Clearly these are nonzero vectors. Their cross product is

$$\mathbf{r}_x \times \mathbf{r}_y = (\mathbf{i} + 2ax\,\mathbf{k}) \times (\mathbf{j} + 2by\,\mathbf{k}) = -2ax\,\mathbf{i} - 2by\,\mathbf{j} + \mathbf{k},$$

which is a nonzero vector for every x and y. The unit vector function

$$\mathbf{n} = \frac{-2ax\,\mathbf{i} - 2by\,\mathbf{j} + \mathbf{k}}{\sqrt{1 + 4a^2x^2 + 4b^2y^2}}$$

is continuous so the surface is orientable. The negative of $\mathbf{n}$ is also a continuous unit normal to the surface. For each point (x, y) satisfying $0 \leq ax^2 + by^2 \leq c$ there is one and only one point (x, y, z) in the surface so the surface surface is regular. ■

A Nonorientable Surface

To convince the reader that not every regular surface is orientable, we give an example of a surface, called the Mobius strip, that is not orientable. A model of the Mobius strip can be formed by starting with a rectangular piece of paper, say 1-unit wide and 12-units long, giving one end a half-twist and then joining the two 1-unit edges producing a loop with a twist. The effect of the twist is to make it impossible to distinguish one side from the other (if "one side" makes any sense). To see what we mean by this, take the paper strip and draw a pencil line down the center parallel to the 12-unit sides and continue the line until returning to the starting point. The surprising effect is that there is a line on both "sides" of the paper. If the pencil is held normal to the surface (and if the pencil is one unit in length) then the pencil serves as a unit normal. The normal is moved continuously but on returning for the first time to the starting point (actually on the opposite side of the paper), the normal is pointing in the direction opposite to that it was initially. This is meant to suggest that it is impossible to define a continuous unit normal at every point of the Mobius band. The problem is that at the starting point, it seems that two normals must be defined, but that is not allowed. This surface is the simplest example of a "one-sided" surface.

A Mobius strip can be represented in vector form by the equation

$$\mathbf{r}(u, v) = \left(1 + v\cos\frac{u}{2}\right)\left(\cos u\,\mathbf{i} + \sin u\,\mathbf{j}\right) + v\sin\frac{u}{2}\,\mathbf{k},$$

where $0 \leq u \leq 2\pi$ and $-0.25 \leq v \leq 0.25$. A rendering using this vector representation is shown in figure 16.4.

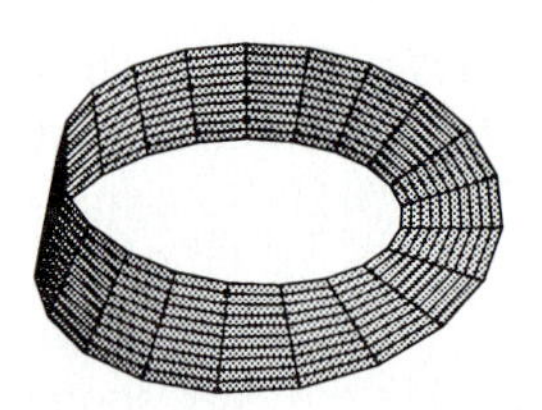

FIGURE 16.4
A Mobius strip

This representation is constructed as follows. Begin with a vector $\mathbf{A} = \cos u\,\mathbf{i} + \sin u\,\mathbf{j}$ from the origin to a point on the unit circle in the xy plane. At the end of that vector we want to insert a line segment that is perpendicular to the xy plane when $u = 0$ but, as u increases, the inserted segment gradually rotates so that when $u = 2\pi$ the segment has rotated through π radians. This is achieved if the angle made by the segment and the xy plane is $u/2$; the segment should project onto the vector $\mathbf{A}$ so it has the shape $p\cos u\,\mathbf{i} + p\sin u\,\mathbf{j} + q\,\mathbf{k}$ for some p and q. If we denote the angle between this vector and the xy plane by ϕ then

$$\cos\phi = \frac{\mathbf{A}\cdot(p\cos u\,\mathbf{i} + p\sin u\,\mathbf{j} + q\,\mathbf{k})}{|p\cos u\,\mathbf{i} + p\sin u\,\mathbf{j} + q\,\mathbf{k}|} = \frac{p}{\sqrt{p^2 + q^2}}.$$

By taking $p = \cos(u/2)$ and $q = \sin(u/2)$ we get $\phi = u/2$ and the vector is of unit length. Multiply that unit vector by v to a obtain vector of length v and then form the sum

$$(\cos u\,\mathbf{i} + \sin u\,\mathbf{j}) + v\left(\cos\frac{u}{2}\cos u\,\mathbf{i} + \cos\frac{u}{2}\sin u\,\mathbf{j} + \sin\frac{u}{2}\,\mathbf{k}\right)$$

to produce the representation of the Mobius strip.

If S is an orientable surface, then at each point of S there are two unit normals, one being the negative of the other. The restriction that the unit normal be continuous means in effect that there are only two choices for the unit normal function on an orientable surface. In some cases there is a (more or less) natural choice of unit normal function. A surface may be *closed* by which we mean that the surface divides three-dimensional space into two disjoint subsets, the points inside and on the surface (a bounded set) and the points outside the surface (an unbounded set); a sphere has this property whereas a plane does not. If S is a closed orientable surface the unit normal that points into the unbounded part of space is called the *exterior* normal. Its negative is called the *interior* normal. For the sphere with equation $x^2 + y^2 + z^2 = R^2$ the unit exterior normal is parallel to the position vector $x\,\mathbf{i} + y\,\mathbf{j} + z\,\mathbf{k}$, which has length R if (x, y, z) is on the sphere. Thus $\mathbf{n} = (x\,\mathbf{i} + y\,\mathbf{j} + z\,\mathbf{k})/R$ is the unit exterior normal and $-\mathbf{n}$ is the interior normal.

Exercises 16.1

1. Show that the graph of $z = x^2 - y^2$ is a regular surface when (x, y) is restricted to lie in any rectangle $[a, b; c, d]$.

2. Show that the graph of $z = 4x^2 - xy + 3y^2$ is a regular surface when (x, y) is restricted to lie in any rectangle $[a, b; c, d]$.

3. Find a continuous vector function $\mathbf{n}$ that gives a unit vector at each point of the surface

$$\{(x, y, z) : -1 \leq x \leq 1,\ 0 \leq y \leq x^2,\ z = x + xy^2 + 4\}.$$

4. Find a continuous vector function $\mathbf{n}$ that gives a unit vector at each point of the ellipsoid with equation $(x - 1)^2 + 4(y - 3)^2 + 9(z - 2)^2 = 25$.

5. Let the surface S be given by the vector equation $\mathbf{r}(u, v) = (u^2 - v^2)\,\mathbf{i} + 2uv\,\mathbf{j} + (u^2 + v^2)\,\mathbf{k}$ with (u, v) restricted to lie in a region G of the uv plane. Show that S is regular provided G does not contain the origin in its interior.

6. Let the surface S be given by the vector equation $\mathbf{r}(u, v) = u\cos v\,\mathbf{i} + u\sin v\,\mathbf{j} + 3u^2\,\mathbf{k}$ with (u, v) restricted to lie in a region G of the uv plane. Show that S is a regular surface so long as G contains no interior points with $u = 0$.

7. Modify the parametric representation of the Mobius strip given in the text to obtain the equation of a strip with two full twists. Similarly obtain a representation of a strip with n twists. Do you think the surface is orientable for any choices of n? [Give a plausible, but not necessarily rigorous, explanation of your answer.]

16.2 The curl and divergence

In this section we introduce an operator ∇, which can be applied to scalar functions to produce a vector function or it can be applied in two different ways to vector functions to produce a scalar function in one case and a vector function in the other. In some sense, all of the definitions of the operation of ∇ are like differentiation operators that can be applied to either scalar or vector functions.

We use the symbolic notation

$$\nabla = \frac{\partial}{\partial x}\mathbf{i} + \frac{\partial}{\partial y}\mathbf{j} + \frac{\partial}{\partial z}\mathbf{k}.$$

This is not a vector but rather a device to aid the memory and facilitate computations. It is called an *operator* because it is applied to functions to produce other functions. We have already used this symbol to define the gradient. If f is a scalar function of (x, y, z) having first partial derivatives then

$$\nabla f = \left(\frac{\partial}{\partial x}\mathbf{i} + \frac{\partial}{\partial y}\mathbf{j} + \frac{\partial}{\partial z}\mathbf{k}\right) f = f_x\,\mathbf{i} + f_y\,\mathbf{j} + f_z\,\mathbf{k}.$$

The properties of the gradient have been studied earlier. The gradient of f may sometimes be written as grad f so that $\nabla f = \operatorname{grad} f$.

Curl of a Vector Function

For a vector function

$$\mathbf{F}(x, y, z) = P(x, y, z)\,\mathbf{i} + Q(x, y, z)\,\mathbf{j} + R(x, y, z)\,\mathbf{k}$$

with P, Q, R differentiable functions, we define the curl of $\mathbf{F}$ to be the vector function

$$\operatorname{curl}\mathbf{F} = (R_y - Q_z)\,\mathbf{i} + (P_z - R_x)\,\mathbf{j} + (Q_x - P_y)\,\mathbf{k}.$$

This definition is rather difficult to remember because of the minus signs so we give a commonly used device that aids the memory.

The curl of a vector function can be expressed using the formalism of the cross product.

If $\mathbf{F} = P\,\mathbf{i} + Q\,\mathbf{j} + R\,\mathbf{k}$ is a vector function of (x, y, z) then, using the determinant notation for the cross product,

$$\nabla \times \mathbf{F} = \begin{vmatrix} \mathbf{i} & \mathbf{j} & \mathbf{k} \\ \dfrac{\partial}{\partial x} & \dfrac{\partial}{\partial y} & \dfrac{\partial}{\partial z} \\ P & Q & R \end{vmatrix}$$

and so

$$\operatorname{curl}\mathbf{F} = \nabla \times \mathbf{F}.$$

Divergence of a Vector Function

One last bit of formalism is used to define the divergence of a vector function. With $\mathbf{F}$ as above, the divergence of $\mathbf{F}$ is

$$\operatorname{div}\mathbf{F} = P_x + Q_y + R_z.$$

Note that this can be expressed as a dot product

$$\operatorname{div}\mathbf{F} = \nabla \cdot \mathbf{F}.$$

EXAMPLE 1 Compute the curl and divergence of the vector function

$$\mathbf{F}(x, y, z) = (x^2 + e^y)\,\mathbf{i} + (\sin z - 3y)\,\mathbf{j} + xyz\,\mathbf{k}.$$

Solution: For the divergence of $\mathbf{F}$ we have

$$\begin{aligned} \nabla \cdot \mathbf{F} &= \left(\frac{\partial}{\partial x}\mathbf{i} + \frac{\partial}{\partial y}\mathbf{j} + \frac{\partial}{\partial z}\mathbf{k}\right) \cdot \left((x^2 + e^y)\,\mathbf{i} + (\sin z - 3y)\,\mathbf{j} + xyz\,\mathbf{k}\right) \\ &= \frac{\partial}{\partial x}(x^2 + e^y) + \frac{\partial}{\partial y}(\sin z - 3y) + \frac{\partial}{\partial z}xyz \\ &= 2x - 3 + xy. \end{aligned}$$

To compute the curl we expand the cross product

$$\nabla\times(x^2+e^y)\mathbf{i}+(\sin z-3y)\mathbf{j}+xyz\,\mathbf{k}=$$

$$\frac{\partial}{\partial x}(\sin z-3y)\mathbf{i}\times\mathbf{j}+\frac{\partial}{\partial x}xyz\,\mathbf{i}\times\mathbf{k}+\frac{\partial}{\partial y}(x^2+e^y)\mathbf{j}\times\mathbf{i}+\frac{\partial}{\partial y}xyz\,\mathbf{j}\times\mathbf{k}$$

$$+\frac{\partial}{\partial z}(x^2+e^y)\mathbf{k}\times\mathbf{i}+\frac{\partial}{\partial z}(\sin z-3y)\mathbf{k}\times\mathbf{j}$$

$$=(xz-\cos z)\mathbf{i}-yz\,\mathbf{j}-e^y\,\mathbf{k}.$$

■

The compact notations ∇f, $\nabla\cdot\mathbf{F}$ and $\nabla\times\mathbf{F}$ suggest certain relations by analogy with the corresponding properties for vectors. We list a few examples.

THEOREM 16.1

Let f be a scalar function with continuous second partial derivatives. Then

$$\nabla\times(\nabla f)=0.$$

Proof

This is suggested by the rule for vectors $\mathbf{u}\times\mathbf{u}=0$ and the proof is carried out by just computing the coefficients of **i**, **j** **k**. For example the coefficient of **i** in $\nabla\times(f_x\,\mathbf{i}+f_y\,\mathbf{j}+f_z\,\mathbf{k})$ is $f_{zy}-f_{yz}$. Because the second partial derivatives of f are continuous, the mixed partial derivatives f_{yz} and f_{zy} are equal at every point. The theorem follows by verifying the similar occurrence for the other components. Reverting to the alternate notation, this theorem states that

$$\text{curl}\,(\text{grad}\,f)=0.$$

□

THEOREM 16.2

Let **F** be a vector function whose second partial derivatives are continuous. Then

$$\nabla\cdot(\nabla\times\mathbf{F})=0.$$

Proof

This rule is suggested by the vector property $\mathbf{u}\cdot(\mathbf{u}\times\mathbf{v})=0$ for any vectors **u** and **v**. The proof is carried out by computing the components and using the equality of the mixed partial derivatives. □

The Curl and Divergence in Applications

The divergence and curl of a vector function arise frequently in the description of physical phenomena. We give just an indication of one such instance.

Suppose fluid is moving through a channel. The channel need not have parallel sides and there might be many sources from which the fluid flows into the channel. Set up a coordinate system to represent the points in the channel and suppose that $\mathbf{v}(x,y,z)$ is the velocity of a particle of fluid located at (x,y,z).

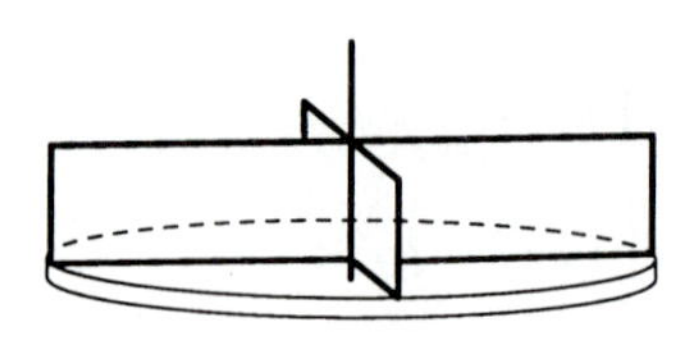

FIGURE 16.5
Inverted paddle wheel

For simplicity we assume that the fluid is in a *steady state*. This means that the velocity vector $\mathbf{v}(x, y, z)$ depends only on the point (x, y, z) and does not change over time. The divergence $\operatorname{div}\mathbf{v}$ has something to do with the tendency for the fluid to accumulate toward or dissipate from the point (x, y, z). The curl of $\mathbf{v}$ has something to do the the rotational tendency produced by the flow. This can be visualized using the image of a *paddle wheel*. We do not mean the type of paddle wheel once used to power a steam boat but rather a very small disc that floats on the surface of the fluid. Vertical "paddles" are slats or boards attached to the disc so that any rotation in the fluid induces a rotation of the disc. We are interested in the behavior of the fluid near a point so the paddle wheel should have a disc of very small radius (in fact we ought to think of a shrinking paddle wheel whose radius tends to zero).

If $\operatorname{curl}\mathbf{v} \neq 0$ the paddle wheel tends to rotate; if $\operatorname{curl}\mathbf{v} = 0$ the paddle wheel will tend not to rotate. Of course this tendency can change from point to point and the situation would be described by evaluating $\operatorname{curl}\mathbf{v}$ at the points in question.

EXAMPLE 2 A thin layer of fluid is flowing across a plane region so that the velocity vector at (x, y) of a particle of the fluid is $\mathbf{v}(x, y) = ax\,\mathbf{i} + ay\,\mathbf{j}$. Compute the divergence and curl of $\mathbf{v}$.

Solution: The divergence is given by

$$\operatorname{div}\mathbf{v} = \nabla \cdot \mathbf{v} = \frac{\partial}{\partial x}(ax) + \frac{\partial}{\partial y}(ay) = 2a.$$

So the divergence is constant. This can be thought of as saying that the area occupied by any portion of the fluid expands at a constant rate if $a > 0$ (and contracts at a constant rate if $a < 0$). The velocity of the fluid at (x, y) is a times the radial vector from the origin to the point. If $a > 0$ then the motion of the fluid is directly away from the origin (and directly toward the origin if $a < 0$).

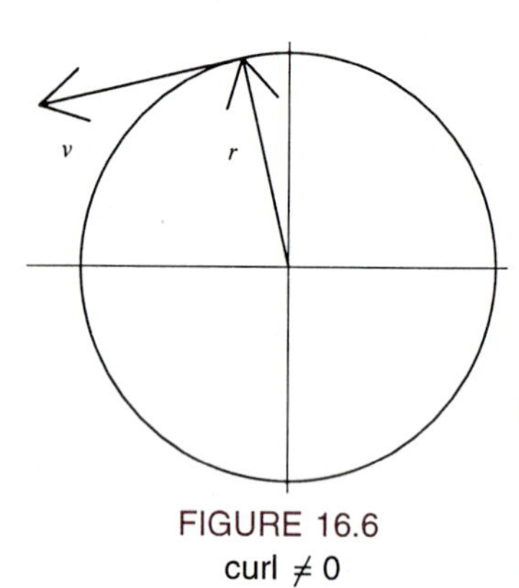

FIGURE 16.6
curl $\neq 0$

The curl of $\mathbf{v}$ is computed using the definition as

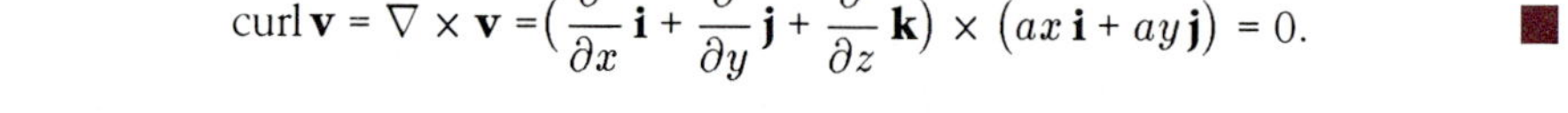

$$\operatorname{curl}\mathbf{v} = \nabla \times \mathbf{v} = \left(\frac{\partial}{\partial x}\mathbf{i} + \frac{\partial}{\partial y}\mathbf{j} + \frac{\partial}{\partial z}\mathbf{k}\right) \times \left(ax\,\mathbf{i} + ay\,\mathbf{j}\right) = 0.$$

■

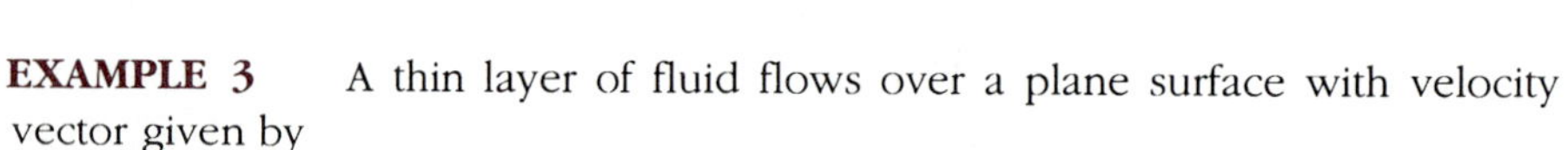

EXAMPLE 3 A thin layer of fluid flows over a plane surface with velocity vector given by

$$\mathbf{v}(x, y) = -y\,\mathbf{i} + x\,\mathbf{j}.$$

Compute the divergence and curl of $\mathbf{v}$.

Solution: The flow is purely rotational since the velocity vector is normal to the position vector. That is if (x, y) lies on the circle of radius R with center at the origin, then the velocity vector $\mathbf{v}(x, y)$ is tangent to the circle pointing in the counterclockwise direction. We find the divergence by

$$\operatorname{div}\mathbf{v} = \frac{\partial(-y)}{\partial x} + \frac{\partial x}{\partial y} = 0.$$

The curl is

$$\operatorname{curl}\mathbf{v} = \begin{vmatrix} \mathbf{i} & \mathbf{j} & \mathbf{k} \\ \frac{\partial}{\partial x} & \frac{\partial}{\partial y} & \frac{\partial}{\partial z} \\ -y & x & 0 \end{vmatrix} = 2\mathbf{k}.$$

The curl is nonzero and is a constant vector pointing in the direction normal to the plane of the rotation. ■

A somewhat more detailed interpretation of the properties of the curl and divergence is given after we prove Stokes's Theorem and the Divergence Theorem.

Stokes's Theorem

To motivate Stokes's Theorem we restate Green's Theorem in vector form. Suppose S is a region of the xy plane and its boundary, denoted by βS, is a simple closed curve. If P and Q are functions defined on S having continuous first partial derivatives then

$$\iint_S (Q_x - P_y)\, dS = \int_{\beta S} P\, dx + Q\, dy,$$

where the integration around the curve βS is in the positive direction.

We rewrite this in vector form by setting $\mathbf{F}(x, y) = P(x, y)\,\mathbf{i} + Q(x, y)\,\mathbf{j}$ so that the line integral can be expressed as

$$\int_{\beta S} P\, dx + Q\, dy = \int_{\beta S} \mathbf{F} \cdot d\mathbf{r}$$

with $\mathbf{r} = x\,\mathbf{i} + y\,\mathbf{j}$ and $d\mathbf{r} = dx\,\mathbf{i} + dy\,\mathbf{j}$. Thus Green's Theorem relates the integral of $\mathbf{F} \cdot d\mathbf{r}$ around the boundary of S to the double integral over S of a function derived from $\mathbf{F}$.

For Stokes's Theorem we consider an orientable surface S and its boundary βS. We relate the line integral of $\mathbf{F} \cdot d\mathbf{r}$ around the boundary to the integral taken over the surface S of a function derived from $\mathbf{F}$. The curl is used in determining the function derived from $\mathbf{F}$.

We give the statement of the theorem in the form suitable for the proof we intend to give. This is not the most general form of Stokes's Theorem but one that gives a good indication of the general case. The principal restriction is in the type of surface over which the integration is performed.

Let S be a regular surface in xyz space given in vector form by a function

$$\mathbf{r}(u, v) = x(u, v)\,\mathbf{i} + y(u, v)\,\mathbf{j} + z(u, v)\,\mathbf{k}$$

where (u, v) runs through all the points of a region G in the uv plane that is bounded by a simple closed curve βG. We assume that $\mathbf{r}$ is a one-to-one correspondence between the points of G and the points of the surface S and that the boundary βG is mapped by $\mathbf{r}$ to the boundary βS of S. We assume that $\mathbf{r}$ has continuous second derivatives throughout G. The boundary βG of G is assumed to have a piecewise parametrization by continuously differentiable functions. Furthermore we assume that the derivatives $\mathbf{r}_u$ and $\mathbf{r}_v$ are continuous and that the cross product $\mathbf{r}_u \times \mathbf{r}_v$ is nonzero so that

$$\mathbf{n} = \mathbf{n}(u, v) = \frac{\mathbf{r}_u \times \mathbf{r}_v}{|\mathbf{r}_u \times \mathbf{r}_v|}$$

is a continuous unit normal to S. Let

$$\mathbf{F} = \mathbf{F}(x, y, z) = P(x, y, z)\,\mathbf{i} + Q(x, y, z)\,\mathbf{j} + R(x, y, z)\,\mathbf{k}$$

be a vector function defined in a parallelepiped containing S and having continuous first derivatives in that region. As usual we write $\mathbf{r} = x\,\mathbf{i} + y\,\mathbf{j} + z\,\mathbf{k}$ and $d\mathbf{r} = dx\,\mathbf{i} + dy\,\mathbf{j} + dz\,\mathbf{k}$.

THEOREM 16.3

Stokes's Theorem

Let the vector function $\mathbf{F}$ have continuous first partial derivatives throughout a solid region that contains a surface S. Assume the surface S is regular with boundary βS and is parametrized as described just above. We have

$$\iint_S (\operatorname{curl} \mathbf{F}) \cdot \mathbf{n}\, dS = \int_{\beta S} \mathbf{F} \cdot d\mathbf{r}$$

where the integration around βS is in the positive direction determined by the parametrization.

Proof

The ideas behind the proof have been used before but the computations are quite lengthy. The theorem relates a line integral of $\mathbf{F} \cdot d\mathbf{r}$ along the boundary βS of S to the integral over the surface S of the product $\operatorname{curl} \mathbf{F} \cdot \mathbf{n}$. The proof is carried out by expressing the line integral in terms of u and v as a line integral around βG, applying Green's Theorem in the uv plane and then interpreting the integral over the region G as an integral over S.

We begin with the line integral. We are assuming all of the notation given just before the statement of the theorem. Let the boundary of G be given parametrically by

$$u = u(t), \qquad v = v(t), \qquad 0 \le t \le 1.$$

Then the boundary βS of S is given parametrically by

$$x = x(u(t), v(t)), \quad y = y(u(t), v(t)), \quad z = z(u(t), v(t)).$$

If t is supressed from the notation, the line integral has the form

$$\begin{aligned} \int_{\beta S} \mathbf{F} \cdot d\mathbf{r} &= \int_{\beta S} P\, dx + Q\, dy + R\, dz \\ &= \int_{\beta G} P(x_u\, du + x_v\, dv) + Q(y_u\, du + y_v\, dv) + R(z_u\, du + z_v\, dv) \\ &= \int_{\beta G} (P x_u + Q y_u + R z_u)\, du + (P x_v + Q y_v + R z_v)\, dv. \end{aligned}$$

To simplify the notation somewhat, let

$$\begin{aligned} p &= P x_u + Q y_u + R z_u, \\ q &= P x_v + Q y_v + R z_v, \end{aligned}$$

so that by application of Green's Theorem we have

$$\int_{\beta S} \mathbf{F} \cdot d\mathbf{r} = \int_{\beta G} p\, du + q\, dv = \iint_G (q_u - p_v)\, dG.$$

Next we rewrite the double integral over G by evaluating the partial derivatives q_u and p_v. We find

$$\begin{aligned} q_u &= \frac{\partial}{\partial u}(Px_v + Qy_v + Rz_v) \\ &= \frac{\partial P}{\partial u}x_v + Px_{vu} + \frac{\partial Q}{\partial u}y_v + Qy_{vu} + \frac{\partial R}{\partial u}z_v + Rz_{vu}, \\ p_v &= \frac{\partial}{\partial u}(Px_u + Qy_u + Rz_u) \\ &= \frac{\partial P}{\partial v}x_u + Px_{uv} + \frac{\partial Q}{\partial v}y_u + Qy_{uv} + \frac{\partial R}{\partial v}z_u + Rz_{vu}. \end{aligned}$$

There are still some terms to which the chain rule must be applied; the derivatives of P, Q, R with respect to u and v can be rewritten as follows:

$$\begin{aligned} P_u &= P_x x_u + P_y y_u + P_z z_u, & P_v &= P_x x_v + P_y y_v + P_z z_v, \\ Q_u &= Q_x x_u + Q_y y_u + Q_z z_u, & Q_v &= Q_x x_v + Q_y y_v + Q_z z_v, \\ R_u &= R_x x_u + R_y y_u + R_z z_u, & R_v &= R_x x_v + R_y y_v + R_z z_v. \end{aligned}$$

Now we substitute these expressions into the description of q_u and p_v and cancel the common terms to obtain

$$\begin{aligned} q_u - p_v = &P_y(y_u x_v - y_v x_u) + P_z(z_u x_v - z_v x_u) \\ &+ Q_x(x_u y_v - x_v y_u) + Q_z(z_u y_v - z_v y_u) \\ &+ R_x(x_u z_v - x_v z_u) + R_y(y_u z_v - y_v z_u). \end{aligned}$$

That is as far as we go with the string of equalities from this end. Next we start with the surface integral in the conclusion of the theorem and show that it is equal to the double integral over G of this last expression.

In the section on integrals over a surface in parametric form we have developed the formula

$$\iint_S f\, dS = \iint_G f\, |\mathbf{r}_u \times \mathbf{r}_v|\, dG,$$

where f is a scalar function and (as in our present case) $\mathbf{r}$ transforms the region G of the uv plane to the surface S in xyz space. In view of the definition of the unit normal $\mathbf{n}$, and using in place of f the scalar function $\operatorname{curl} \mathbf{F} \cdot \mathbf{n}$, we have

$$\iint_S (\operatorname{curl} \mathbf{F} \cdot \mathbf{n})\, dS = \iint_G (\operatorname{curl} \mathbf{F} \cdot \mathbf{r}_u \times \mathbf{r}_v)\, dG.$$

The next, and last step, is to express $\operatorname{curl} \mathbf{F} \cdot \mathbf{r}_u \times \mathbf{r}_v$ in terms of P, Q, R and the derivatives of x, y and z with respect to u and v. For the first term we have

$$\begin{aligned} \mathbf{r}_u \times \mathbf{r}_v &= \begin{vmatrix} \mathbf{i} & \mathbf{j} & \mathbf{k} \\ x_u & y_u & z_u \\ x_v & y_v & z_v \end{vmatrix} \\ &= (y_u z_v - y_v z_u)\mathbf{i} - (x_u z_v - x_v z_u)\mathbf{j} + (x_u y_v - x_v y_u)\mathbf{k} \end{aligned}$$

For the term with the curl we have

$$\operatorname{curl} \mathbf{F} = (R_y - Q_z)\mathbf{i} + (P_z - R_x)\mathbf{j} + (Q_x - P_y)\mathbf{k}.$$

Now by careful inspection of the terms listed, we find

$$\operatorname{curl} \mathbf{F} \cdot \mathbf{r}_u \times \mathbf{r}_v = q_u - p_v.$$

From this we then have

$$\iint_G (\operatorname{curl} \mathbf{F} \cdot \mathbf{r}_u \times \mathbf{r}_v)\, dG = \iint_G (q_u - p_v)\, dG.$$

Replacing each integral by the corresponding expression involving an integral over S or βS from the equalities already derived now gives us the equality stated in the conclusion of the theorem. □

Stokes's Theorem can be applied to surfaces that do not quite satisfy the assumptions made before the proof. If S can be partitioned into a finite number of surfaces each satisfying the assumptions of Stokes's Theorem and such that the pieces overlap only along boundary lines then Stokes's Theorem can be applied over each piece separately. If the subdivision can be done so that each of the overlapping curves is traversed twice, once in each direction, then cancellation occurs and only the integration over the boundary of the entire region remains. We do not attempt to give a careful or complete statement of this process but it is illustrated in the next example and in the exercises.

EXAMPLE 4 Let S be the surface composed of two parts, one is the cylinder of height a given by

$$S_1 = \{(x, y, z) : x^2 + y^2 = 1, \quad 0 \le z \le a\}$$

and the other is the top of the cylinder given by

$$S_2 = \{(x, y, z) : 0 \le x^2 + y^2 \le 1, \quad z = a\}.$$

Let $\mathbf{F}$ be the vector function $\mathbf{F}(x, y, z) = 2y\,\mathbf{i} - 2x\,\mathbf{j} + xyz\,\mathbf{k}$. Verify Stokes's Theorem by computing both integrals directly.

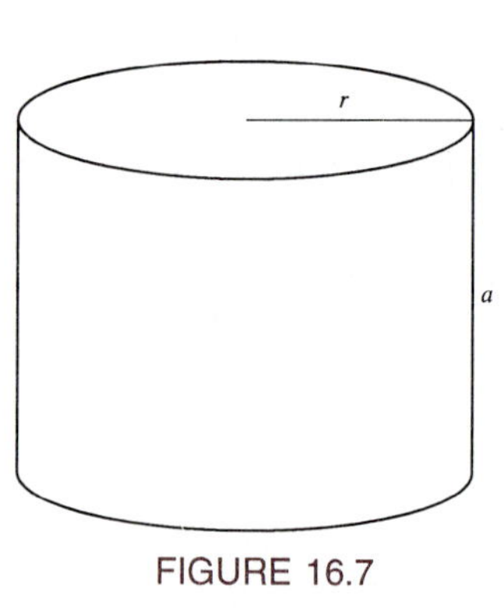

FIGURE 16.7

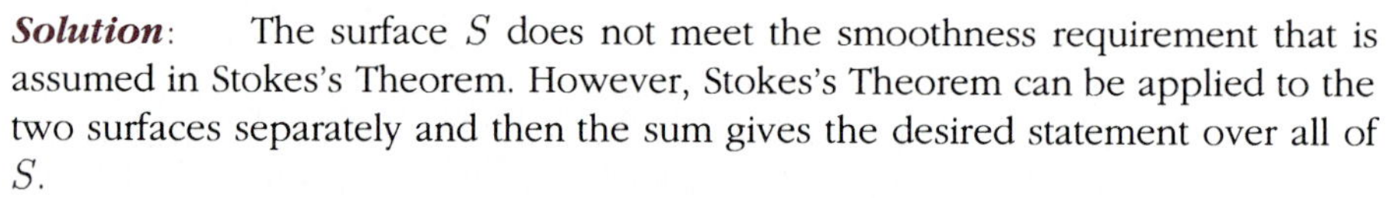

Solution: The surface S does not meet the smoothness requirement that is assumed in Stokes's Theorem. However, Stokes's Theorem can be applied to the two surfaces separately and then the sum gives the desired statement over all of S.

We first write the curl of $\mathbf{F}$ for reference: $\operatorname{curl} \mathbf{F} = xz\,\mathbf{i} - yz\,\mathbf{j} - 4\mathbf{k}$. The unit outer normal to S_1 is a radial vector given by

$$\mathbf{n}_1(x, y, z) = x\,\mathbf{i} + y\,\mathbf{j}, \quad \text{for} \quad x^2 + y^2 = 1.$$

A representation of S_1 for which

$$\mathbf{n}_1 = \frac{\mathbf{r}_u \times \mathbf{r}_v}{|\mathbf{r}_u \times \mathbf{r}_v|}$$

is given by

$$S_1 = \{(x, y, z) : x = \cos u, \quad y = \sin u, \quad z = v, \quad 0 \le u \le 2\pi, \quad 0 \le v \le a\}$$

and $\mathbf{r} = \cos u\,\mathbf{i} + \sin u\,\mathbf{j} + v\,\mathbf{k}$. Then

$$\mathbf{r}_u \times \mathbf{r}_v = (-\sin u\,\mathbf{i} + \cos u\,\mathbf{j}) \times \mathbf{k} = \cos u\,\mathbf{i} + \sin u\,\mathbf{j}.$$

Note that $|\mathbf{r}_u \times \mathbf{r}_v| = 1$. Now we have

$$dS_1 = |\mathbf{r}_u \times \mathbf{r}_v|\, du\, dv = du\, dv.$$

Put the results of these computations together to get

$$\begin{aligned}\iint_{S_1} (\operatorname{curl} \mathbf{F}) \cdot \mathbf{n}_1 \, dS_1 &= \iint_{S_1} (xz\,\mathbf{i} - yz\,\mathbf{j} - 4\,\mathbf{k}) \cdot (x\,\mathbf{i} + y\,\mathbf{j}) \, dS_1 \\ &= \iint_{S_1} z(x^2 - y^2) \, dS_1 \\ &= \int_0^a \int_0^{2\pi} v(\cos^2 u - \sin^2 u) \, du \, dv \\ &= 0.\end{aligned}$$

(Note that by either symmetry or trigonometric identities, one argues that $\int_0^{2\pi} \cos^2 u \, du = \int_0^{2\pi} \sin^2 u \, du$.)

Because S_2 is parallel to the xy plane, a parametrization for which $(\mathbf{r}_u \times \mathbf{r}_v)/|\mathbf{r}_u \times \mathbf{r}_v|$ equals the outer unit normal $\mathbf{k}$ is given by $\mathbf{r}(u, v) = u\,\mathbf{i} + v\,\mathbf{j}$. The differential of surface area is $dS_2 = dv \, du$. The unit outer normal to S_2 is the constant vector $\mathbf{k}$. Then we have

$$\begin{aligned}\iint_{S_2} (\operatorname{curl} \mathbf{F}) \cdot \mathbf{n}_2 \, dS_2 &= \iint_{S_2} (xz\,\mathbf{i} - yz\,\mathbf{j} - 4\,\mathbf{k}) \cdot (\mathbf{k}) \, dS_2 \\ &= \iint_{S_2} -4 \, dS_2 \\ &= -4\pi.\end{aligned}$$

The last integral was evaluated by realizing that it represents (-4) times the area of S_2. Thus we have

$$\iint_S (\operatorname{curl} \mathbf{F}) \cdot \mathbf{n} \, dS = \iint_{S_1} (\operatorname{curl} \mathbf{F}) \cdot \mathbf{n}_1 \, dS_1 + \iint_{S_2} (\operatorname{curl} \mathbf{F}) \cdot \mathbf{n}_2 \, dS_2 = -4\pi.$$

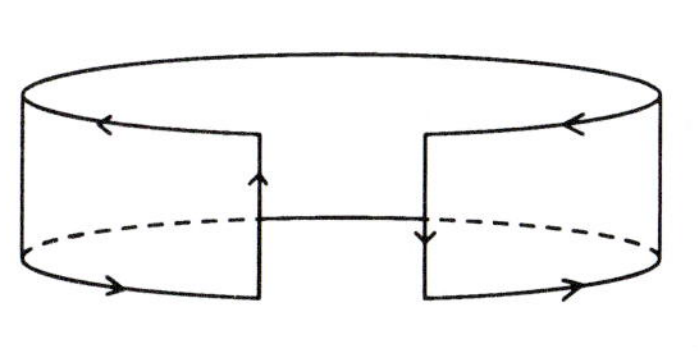

FIGURE 16.8
Side of the cylinder

Now we compute the other side of the equation in Stokes Theorem. It is necessary to discuss the boundary of S_1 first. We can think of cutting the cylinder with a vertical cut parallel to the z-axis, say from $(1, 0, 0)$ to $(1, 0, a)$, and rolling the cylinder flat into a rectangle, in which case the boundary is clear. We traverse it by starting at $(1, 0, 0)$ and walking around the circle at the base in the counterclockwise direction, then walking up the cut and around the circle at the top in the clockwise direction and then down the cut back to the starting point. The integration along this path permits cancellation along the vertical cut because that line is traversed once in each direction.

The integration around the boundary of S_2 is taken around the circle at the top of S_1 but in the clockwise direction; this cancels the part of the integration along the top part of the boundary of S_1. Thus all that remains in the sum of the line integrals over the boundaries of S_1 and S_2 is the line integral over the bottom of S_2, which is the boundary of the original surface S. So we take $\mathbf{r} = \cos t\,\mathbf{i} + \sin t\,\mathbf{j}$ with $0 \le t \le 2\pi$ as the representation of the lower circle bounding S and get

$$\begin{aligned}\int_{\beta S} \mathbf{F} \cdot d\mathbf{r} &= \int_{\beta S} (2y\,\mathbf{i} - 2x\,\mathbf{j} + xyz\,\mathbf{k}) \cdot d\mathbf{r} \\ &= \int_0^{2\pi} (2 \sin t\,\mathbf{i} - 2 \cos t\,\mathbf{j}) \cdot (-\sin t\,\mathbf{i} + \cos t\,\mathbf{j}) \, dt \\ &= \int_0^{2\pi} (-2)(\sin^2 t + \cos^2 t) \, dt \\ &= (-2)(2\pi) = -4\pi.\end{aligned}$$

Thus the conclusion of Stokes's Theorem is verified.

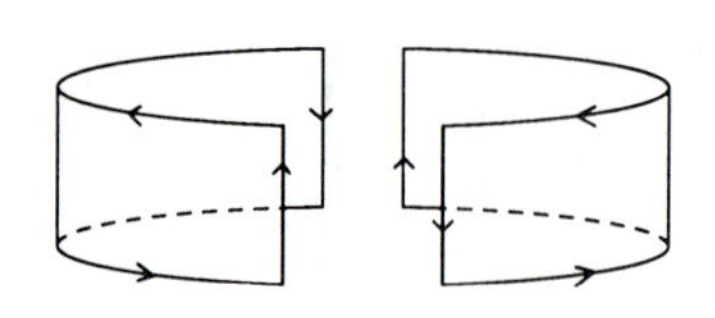

FIGURE 16.9
1:1 parametrization
of cylinder side

Here we have used a slight extension of the version of Stokes's theorem stated above, because our parametrization of S_1 was not one to one on the boundary of the parametrizing rectangle. However it is easy to check that the proof given above works equally well for this case. Alternatively, we could have stayed with the original statement of the theorem and used the device of parametrizing S_1 by using two rectangles rather than one. ■

If we have a smooth surface that is closed, that is the surface encloses a region of space like the surface of a sphere, then there is no boundary of the surface. Thus the integral in the conclusion of Stokes's Theorem dealing with the integral around the boundary must equal 0. This implies the following corollary of Stokes's Theorem.

Corollary 16.1

If S is a smooth closed surface and $\mathbf{F}(x, y, z)$ has continuous first partial derivatives throughout the region enclosed by S, then

$$\iint_S (\text{curl}\, \mathbf{F}) \cdot \mathbf{n}\, dS = 0.$$

Circulation

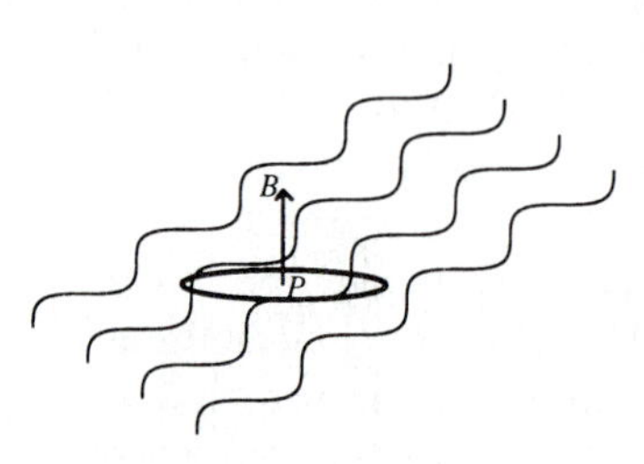

FIGURE 16.10

Suppose we interpret a vector function $\mathbf{F}$ as the velocity of a flowing fluid. We explain how the curl of $\mathbf{F}$ is related to the idea of circulation. Pick a point P in the region through which the fluid is flowing. Fix a plane through P and let S be a circular disc in the plane with P as center and some (small) number δ as radius; let C be the boundary of the disc. Select a unit vector $\mathbf{n}$ that is normal to the disc and such that C is oriented in the positive direction with respect to $\mathbf{n}$. Then Stokes's Theorem implies

$$\iint_S \text{curl}\, \mathbf{F} \cdot \mathbf{n}\, dS = \int_C \mathbf{F} \cdot d\mathbf{r}.$$

The line integral on the right is called the *circulation* of $\mathbf{F}$ at P around C. The integral on the left we interpret as follows. For a continuous function $f(x, y)$ defined over a region S of area $A(S)$, the product

$$\frac{1}{A(S)} \iint_S f(x, y)\, dS$$

is called the average value of f over S. Hence we can say that the circulation of $\mathbf{F}$ around C is the average value of $\text{curl}\, \mathbf{F} \cdot \mathbf{n}$ over the disc. In words, the circulation is the average value of the normal component of the curl of $\mathbf{F}$. By passing to the limit as $\delta \to 0$ we obtain a concept that depends only on the point P and the direction $\mathbf{n}$:

$$\text{curl}\, \mathbf{F} \cdot \mathbf{n} = \lim_{\delta \to 0} \frac{\text{circulation of } \mathbf{F} \text{ around } C}{\pi \delta^2}.$$

In case $\text{curl}\, \mathbf{F} = 0$ then the normal component in any direction is 0 so the circulation is 0 around any disc.

Exercises 16.2

Compute curl **F** and div **F** for each function. [We use $\mathbf{r} = x\,\mathbf{i} + y\,\mathbf{j} + z\,\mathbf{k}$.]

1. $\mathbf{F} = 3x\,\mathbf{i} + 5y\,\mathbf{j}$

2. $\mathbf{F} = 4y\,\mathbf{i} - 7x\,\mathbf{j}$

3. $\mathbf{F} = \dfrac{x\,\mathbf{i} + y\,\mathbf{j}}{x^2 + y^2}$

4. $\mathbf{F} = \dfrac{y\,\mathbf{i} + x\,\mathbf{j}}{x^2 + y^2}$

5. $\mathbf{F} = yz\,\mathbf{i} + xz\,\mathbf{j} + xy\,\mathbf{k}$

6. $\mathbf{F} = xy\,\mathbf{i} + yz\,\mathbf{j} + xz\,\mathbf{k}$

7. $\mathbf{F} = \mathbf{r}/|\mathbf{r}|^2$

8. $\mathbf{F} = \mathbf{r}/|\mathbf{r}|^3$

Use direct calculation and verification by Stokes's Theorem to determine the value of the surface integrals. Here S is the indicated oriented surface with **n** as an upward-pointing normal to S.

9. $\iint_S \operatorname{curl}(y\,\mathbf{i} - x\,\mathbf{j} + z\,\mathbf{k}) \cdot \mathbf{n}\,dS$;

$$S = \{(x, y, z) : -1 \le x \le 1,\ 0 \le y \le \sqrt{1 - x^2},\ z = \sqrt{1 - x^2 - y^2}\}$$

10. $\iint_S \operatorname{curl}(z\,\mathbf{i} + x\,\mathbf{j} + y\,\mathbf{k}) \cdot \mathbf{n}\,dS$; S is the part of the paraboloid $z = 1 - x^2 - y^2$ having $z \ge 0$

11. $\iint_S \operatorname{curl}(xz\,\mathbf{j}) \cdot \mathbf{n}\,dS$; S is the part of the upper half of the sphere $z = \sqrt{9 - x^2 - y^2}$ that lies outside the cylinder $x^2 + y^2 = 4$

12. $\iint_S \operatorname{curl}(-3z\,\mathbf{i} + 2y\,\mathbf{j} + 4x\,\mathbf{k}) \cdot \mathbf{n}\,dS$; S is the plane triangular surface with vertices $(a, 0, 0)$, $(0, b, 0)$ and $(0, 0, c)$ for positive constants a, b and c

Calculate the circulation of **F** around the indicated disc.

13. $\mathbf{F} = -y\,\mathbf{i} + x\,\mathbf{j}$, $C: x^2 + y^2 = 1$

14. $\mathbf{F} = x\,\mathbf{i} + y\,\mathbf{j}$, $C: x^2 + y^2 = 1$

15. $\mathbf{F} = (x - y)\,\mathbf{i} + (x + y)\,\mathbf{j}$, $C: x^2 + y^2 = 1$

16. $\mathbf{F} = (3x - 2y)\,\mathbf{i} + (3x + 2y)\,\mathbf{j}$, $C: x^2 + y^2 = 1$

17. $\mathbf{F} = -y\,\mathbf{i} + z\,\mathbf{j} + x\,\mathbf{k}$; C is the circle of radius 1, center at the origin in the plane normal to $\mathbf{i} - \mathbf{j} - \mathbf{k}$

18. $\mathbf{F} = (z - y)\,\mathbf{i} + x\,\mathbf{j} + (y + z)\,\mathbf{k}$; C is the circle of radius δ, center at the origin in the plane normal to $\mathbf{i} + \mathbf{j} + \mathbf{k}$. What is the circulation at the point $(0, 0, 0)$?

16.3 The divergence theorem

The Divergence Theorem, like Stokes's Theorem and Green's Theorem, equates two integrals of functions related to a vector function; one integral is a triple integral over a solid region of xyz space and the other is a surface integral over the surface of the solid. The solid, its surface and the vector function must be fairly well-behaved. We give the statement of the theorem and then describe some additional assumptions that are made so that a proof can be given without too much pathology creeping in to make the situation complicated.

THEOREM 16.4

The Divergence Theorem

Let V be a solid region in xyz space contained in a parallelepiped H. Assume the boundary βV of V is pieced together from regular surfaces overlapping along boundary curves. Let **F** be a vector function that has continuous first derivatives throughout H. Then

$$\iiint_V \operatorname{div}\mathbf{F}\,dV = \iint_{\beta V} \mathbf{F} \cdot \mathbf{n}\,dS,$$

where **n** is the exterior normal to βV and dS is the differential of the area of the surface βV.

Proof

As with other theorems of this general type, we make some simplifying assumptions about the region V and its boundary βV. The proof is given only in the case that the boundary is a smooth orientable surface. The case for piecewise smooth surfaces requires some modifications and precision that are omitted. Assume further that V can be "cut in half" by a plane $z = c$ into two regions, an upper half and a lower half each of which is a region of Type XY. Similarly we suppose V can be cut by a plane $y = c_2$ into two regions each of Type XZ and that V can be cut by a plane $x = c_3$ into two regions each of Type YZ. For example the interior of a sphere or ellipsoid meets these conditions as do many other solids. We give a proof in this special case only. A proof for somewhat more general regions can be obtained from this case if the region can be decomposed in a suitable way into a finite number of subsolids each of which satisfies the assumptions made here.

Let $\mathbf{F} = P\mathbf{i} + Q\mathbf{j} + R\mathbf{k}$, where P, Q, R are functions with continuous first partial derivatives. If we let $\mathbf{F}_1 = P\mathbf{i}$, $\mathbf{F}_2 = Q\mathbf{j}$ and $\mathbf{F}_3 = R\mathbf{j}$ then

$$\begin{aligned}\mathbf{F} &= \mathbf{F}_1 + \mathbf{F}_2 + \mathbf{F}_3,\\ \operatorname{div}\mathbf{F} &= \operatorname{div}\mathbf{F}_1 + \operatorname{div}\mathbf{F}_2 + \operatorname{div}\mathbf{F}_3,\\ \mathbf{F}\cdot\mathbf{n} &= \mathbf{F}_1\cdot\mathbf{n} + \mathbf{F}_2\cdot\mathbf{n} + \mathbf{F}_3\cdot\mathbf{n}.\end{aligned}$$

If we can prove

$$\iiint_V \operatorname{div}\mathbf{F}_i\, dV = \iint_{\beta V} \mathbf{F}_i\cdot\mathbf{n}\, dS$$

for $i = 1, 2, 3$, then the theorem follows by taking the sum.

Let us work on the evaluation of the integrals for $\mathbf{F}_3$. By assumption V can be described as the union of two solids:

$$\begin{aligned}V_1 &= \{(x,y,z) : a \le x \le b, \quad g_1(x) \le y \le g_2(x), \quad c \le z \le T(x,y)\},\\ V_2 &= \{(x,y,z) : a \le x \le b, \quad g_1(x) \le y \le g_2(x), \quad B(x,y) \le z \le c\},\end{aligned}$$

where $T(x,y)$ and $B(x,y)$ have continuous first partial derivatives and g_1 and g_2 have continuous derivatives. Let U be the region of the xy plane over which V_1 and V_2 are defined; that is,

$$U = \{(x,y) : a \le x \le b, \quad g_1(x) \le y \le g_2(x)\}.$$

Then we compute the integral over V as a sum of the integrals over V_1 and V_2, so the two integrals over V_1 and V_2 should be computed first. We have

$$\begin{aligned}\iiint_{V_1} \operatorname{div}\mathbf{F}_3\, dV &= \iint_U \int_c^{T(x,y)} R_z(x,y,z)\, dz\, dU\\ &= \iint_U [R(x,y,T(x,y)) - R(x,y,c)]\, dU.\end{aligned}$$

For the integral over V_2 we have

$$\begin{aligned}\iiint_{V_2} \operatorname{div}\mathbf{F}_3\, dV &= \iint_U \int_{B(x,y)}^{c} R_z(x,y,z)\, dz\, dU\\ &= \iint_U [R(x,y,c) - R(x,y,B(x,y))]\, dU.\end{aligned}$$

Take the sum of these integrals to get

$$\iiint_V \operatorname{div} \mathbf{F}_3 \, dV = \iint_U [R(x, y, T(x, y)) - R(x, y, B(x, y))] \, dU. \qquad \textbf{(16.1)}$$

It is certainly convenient that the terms involving the constant c have cancelled.

Next we take up the right-hand side of the equation. We see that the upper half of the boundary of V is the surface given by the graph of the function $z = T(x, y)$ for (x, y) in U. The normal to this surface is the gradient of the function $z - T(x, y)$ and so a unit normal is given by

$$\mathbf{n} = \frac{-T_x \mathbf{i} - T_y \mathbf{j} + \mathbf{k}}{\sqrt{T_x^2 + T_y^2 + 1}}.$$

You might question whether this really is the outer normal or whether it should be $-\mathbf{n}$. Inasmuch as this computation is made over the top part of the solid, the outer normal should have a positive component in the direction of $\mathbf{k}$ (think of the outer normal to the top half of a sphere). Thus $\mathbf{n}$ is the correct choice.

Then we have

$$\mathbf{F}_3 \cdot \mathbf{n} = R\mathbf{k} \cdot \frac{-T_x \mathbf{i} - T_y \mathbf{j} + \mathbf{k}}{\sqrt{T_x^2 + T_y^2 + 1}} = \frac{R(x, y, z)}{\sqrt{T_x^2 + T_y^2 + 1}}.$$

We now integrate over the surface βV_1 of V_1, and here we mean the part of the surface of V_1 that is also part of the surface of V. In other words, we do not integrate over the "cut" that we made with the plane $z = c$.

The differential of the surface area is $dS = \sqrt{T_x^2 + T_y^2 + 1} \, dU$ and the value of z at a point (x, y, z) in the surface is $T(x, y)$; thus

$$\iint_{\beta V_1} \mathbf{F}_3 \cdot \mathbf{n} \, dS = \iint_U \frac{R(x, y, T(x, y))}{\sqrt{T_x^2 + T_y^2 + 1}} \sqrt{T_x^2 + T_y^2 + 1} \, dU$$

$$= \iint_U R(x, y, T(x, y)) \, dU.$$

For the analogous computation over the lower half V_2, the surface is given by $z = B(x, y)$ and the outer normal is the gradient of $B(x, y) - z$. The minus sign appears here because this is the lower half of the solid and the outer normal should have a negative z component. Computations like those just given produce the equation

$$\iint_{\beta V_2} \mathbf{F}_3 \cdot \mathbf{n} \, dS = -\iint_U R(x, y, B(x, y)) \, dU.$$

Adding the last two integrals gives the integral over the entire surface of V with the result

$$\iint_{\beta V} \mathbf{F}_3 \cdot \mathbf{n} \, dS = \iint_{\beta V_1} \mathbf{F}_3 \cdot \mathbf{n} \, dS + \iint_{\beta V_2} \mathbf{F}_3 \cdot \mathbf{n} \, dS$$

$$= \iint_U R(x, y, T(x, y)) - R(x, y, B(x, y)) \, dU$$

$$= \iiint_V \operatorname{div} \mathbf{F}_3 \, dV.$$

We have used Equation (16.1) for the last step. This proves the divergence theorem for $\mathbf{F}_3$. Analogous computations give us the result for $\mathbf{F}_1$ and $\mathbf{F}_2$

and then by taking the sum we get the proof for the general **F** as stated in the theorem. □

Interpretation as Outward Flux

The integral $\iint_{\beta V} \mathbf{F}\cdot\mathbf{n}\,dS$ is called the *flux* of **F** out of V or across the surface βV. If we interpret **F** as the velocity vector of a flowing fluid then $\mathbf{F}\cdot\mathbf{n}$ is the component of the velocity pointing outward from the surface βV and the flux is a measure of the amount of fluid flowing out of V through its surface. The Divergence Theorem permits an interpretation of the divergence of **F**. Start with a point P and a solid ball B of radius δ and center at P. The surface of B is a sphere of radius δ. The Divergence Theorem implies that

$$\iiint_B \operatorname{div}\mathbf{F}\,dB = \text{flux of } \mathbf{F} \text{ out of } B.$$

The quotient of the left-hand side by the volume of B is the average value of $\operatorname{div}\mathbf{F}$ over the sphere and so we have

$$\text{average value of div}\,\mathbf{F} \text{ over } B = \frac{\text{flux of } \mathbf{F} \text{ out of } B}{4\pi\delta^3/3}.$$

If we pass to the limit as $\delta \to 0$ we conclude that $\operatorname{div}\mathbf{F}$ is the outward flux per unit volume. (This language is used even when the vector function **F** is not a velocity vector.)

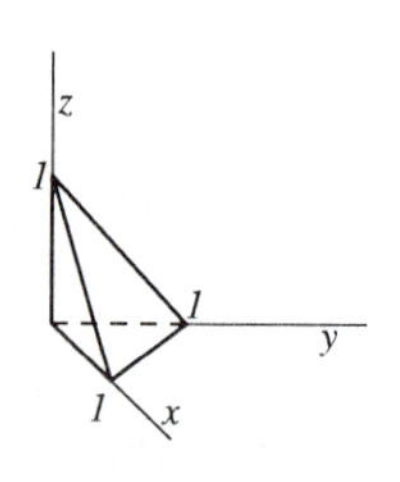

FIGURE 16.11
Tetrahedron

EXAMPLE 1 Compute both integrals in the conclusion of the Divergence Theorem using $\mathbf{F} = x\,\mathbf{i} + 3y\,\mathbf{j} - 2z\,\mathbf{k}$ and V as the tetrahedron with vertices $(0,0,0)$, $(1,0,0)$, $(0,1,0)$ and $(0,0,1)$. In other words, compute the flux of **F** out of V.

Solution: The surface of the tetrahedron is composed of four plane faces: S_1 is in the xz plane, S_2 is in the yz plane, S_3 is in the xy plane and S_4 is in the plane with equation $x + y + z = 1$. We compute the surface integral first; we must integrate $\mathbf{F}\cdot\mathbf{n}$ over each of the four surfaces S_i where **n** is the outward normal to the appropriate S_i. The outward pointing normal to S_1 is $\mathbf{n} = -\mathbf{j}$ and

$$\mathbf{F}\cdot\mathbf{n} = (x\,\mathbf{i} + 3y\,\mathbf{j} - 2z\,\mathbf{k})\cdot(-\mathbf{j}) = -3y.$$

On the surface S_1, we have $y = 0$ so

$$\iint_{S_1} \mathbf{F}\cdot\mathbf{n}\,dS = \iint_{S_1} -3y\,dS = 0.$$

In the same way we find

$$\iint_{S_2} \mathbf{F}\cdot\mathbf{n}\,dS = \iint_{S_3} \mathbf{F}\cdot\mathbf{n}\,dS = 0.$$

The unit normal to S_4 is $(\mathbf{i} + \mathbf{j} + \mathbf{k})/\sqrt{3}$. On S_4 we have $z = 1 - x - y$ so that the differential of surface area is

$$dS = \sqrt{1 + x_z^2 + z_y^2}\,dy\,dx = \sqrt{1 + (-1)^2 + (-1)^2}\,dy\,dx = \sqrt{3}\,dy\,dx.$$

So finally we have

$$\begin{aligned}\iint_{S_4} \mathbf{F}\cdot\mathbf{n}\,dS &= \iint_{S_4} (x\,\mathbf{i} + 3y\,\mathbf{j} - 2z\,\mathbf{k})\cdot\frac{\mathbf{i}+\mathbf{j}+\mathbf{k}}{\sqrt{3}}\,dS \\ &= \iint_{S_4} \frac{x+3y-2z}{\sqrt{3}}\,dS = \iint_{S_4} \frac{x+3y-2z}{\sqrt{3}}\sqrt{3}\,dy\,dx \\ &= \int_0^1\int_0^{1-x} x + 3y - 2(1-x-y)\,dy\,dx = \frac{1}{3}.\end{aligned}$$

Thus the surface integral in the conclusion of the Divergence Theorem has the value $1/3$.

To compute the triple integral we first record the divergence of $\mathbf{F}$:

$$\operatorname{div}\mathbf{F} = \left(\frac{\partial}{\partial x}\mathbf{i} + \frac{\partial}{\partial y}\mathbf{j} + \frac{\partial}{\partial z}\mathbf{k}\right)\cdot(x\,\mathbf{i} + 3y\,\mathbf{j} - 2z\,\mathbf{k}) = 1 + 3 - 2 = 2.$$

The solid region over which the triple integral is evaluated is a Type XY region with description

$$\{(x,y,z) : 0 \le x \le 1,\quad 0 \le y \le 1-x,\quad 0 \le z \le 1-x-y\}.$$

The triple integral is now easily expressed as an iterated integral

$$\iiint_V \operatorname{div}\mathbf{F}\,dV = \int_0^1\int_0^{1-x}\int_0^{1-x-y} 2\,dz\,dy\,dx$$

and the value is computed to be $1/3$ as expected. ■

Electrostatics and the Divergence Theorem

FIGURE 16.12

We give an extended example that illustrates one of the ways the divergence theorem is used in the study of electrostatics, the study of how charged particles interact with each other. We will not explain the basics of electrostatics; it is sufficient to have in mind the idea that particles can carry an electric charge that could be positive or negative. Two charged particles exert a force on each other. Two charges with the same sign (both positive or both negative) repel each other; two charges with opposite signs attract. In either case, each exerts a force on the other that is proportional to the product of their respective charges and inversely proportional to the square of the distance between them and the force vector is parallel to the line between the two particles. It has been found useful to consider the notion of an *electric field* created by a single point charge in order to study the effect on other charges.

If point charge q is placed in space it creates an electric field that affects every other point charge in space. The field denoted by $\mathbf{E}$ is a vector function of (x, y, z). The value $\mathbf{E}(x, y, z)$ is used as follows: If a point charge q' is placed at (x, y, z), then the force exerted on it by q is $q'\mathbf{E}(x, y, z)$. By experiment, $\mathbf{E}$ is determined to be

$$\mathbf{E}(x,y,z) = q\frac{\mathbf{r}}{r^3},$$

where $\mathbf{r}$ is the vector from the position of q to (x, y, z) and $r = |\mathbf{r}|$.

The notion of *flux* is used in the study of electrostatics and in many other different applications. If $\mathbf{F}$ is any (differentiable) vector function, like the electric field $\mathbf{E}$ for example, the flux of $\mathbf{F}$ across a smooth surface S is the surface integral

$$\iint_S \mathbf{F}\cdot\mathbf{n}\,dS,$$

where $\mathbf{n}$ is a unit normal to S. More precisely this surface integral should be called the *flux of* $\mathbf{F}$ *across* S *in the direction of* $\mathbf{n}$.

The Divergence Theorem permits us to use an alternative expression to compute the flux across a smooth surface S that encloses a solid region of xyz space. In the case in which $\mathbf{F}$ denotes the velocity vector of a fluid flowing in a steady state, the flux across the surface is related to the rate at which the fluid is leaving or entering the enclosed region. For electrostatics, the flux has a somewhat more abstract meaning having a more subtle physical interpretation, but is still a very useful concept. In view of the Divergence Theorem, the flux of $\mathbf{F}$ across a closed smooth surface S bounding a solid region G is

$$\iiint_G (\operatorname{div}\mathbf{F})\, dG.$$

We first compute the divergence of the electric field $\mathbf{E}$ and then apply the divergence theorem to draw some conclusions. To simplify the notation we assume for now that the point charge is at the origin so that the vector $\mathbf{r}$ from the charge to the point (x, y, z) is simply $\mathbf{r} = x\,\mathbf{i} + y\,\mathbf{j} + z\,\mathbf{k}$. Because $\mathbf{E} = q\mathbf{r}/r^3$, we can express $\mathbf{E}$ in terms of x, y, z as

$$\mathbf{E} = \frac{q}{r^3}(x\,\mathbf{i} + y\,\mathbf{j} + z\,\mathbf{k}), \quad r = (x^2 + y^2 + z^2)^{1/2}.$$

For the divergence we first observe that $r_x = x/r$, $r_y = y/r$, $r_z = z/r$ and we have

$$\begin{aligned}
\nabla \cdot \mathbf{E} &= q\left(\frac{\partial}{\partial x} x r^{-3} + \frac{\partial}{\partial y} y r^{-3} + \frac{\partial}{\partial z} z r^{-3}\right) \\
&= (r^{-3} - 3x^2 r^{-5}) + (r^{-3} - 3y^2 r^{-5}) + (r^{-3} - 3z^2 r^{-5}) \\
&= 3r^{-3} - 3(x^2 + y^2 + z^2) r^{-5} = 3r^{-3} - 3(r^2) r^{-5} \\
&= 0.
\end{aligned}$$

Now suppose that $\mathbf{E}$ is the electric field produced by a point charge that is not inside the region G bounded by the smooth close surface S. Then $\mathbf{E}$ is continuous and differentiable at every point of G and the Divergence Theorem gives the flux across the surface S as the integral of the divergence, which is 0; hence the flux across S is 0.

Now suppose the point charge is inside the region G enclosed by S. The previous computation does not apply because $\mathbf{E}$ is not continuous at every point inside G. Remember that $\mathbf{r}$ is the vector from the charge to the point (x, y, z) and we divide by the length of this vector. Thus $\mathbf{E}$ is undefined at the origin (assuming still that the charge is at the origin). We overcome this by a useful device. Surround the point charge by a sphere R with center at the charge and with radius a that is small enough so that the entire sphere is inside G. Let G_0 be the region consisting of all points inside G but outside of R. Then $\mathbf{E}$ is continuous and differentiable at every point of G_0. If $\mathbf{n}$ is the outer normal to G and $\mathbf{n}'$ is the outer normal to the sphere R (meaning that $\mathbf{n}'$ points away from the origin) then $-\mathbf{n}'$ is an outer normal to part of the boundary of G_0. The Divergence Theorem can be applied to the region G_0 to obtain that the flux out of G_0 is 0 because $\nabla \cdot \mathbf{E} = 0$ on G_0. Then (and here is the key step)

$$0 = \iint_S \mathbf{E} \cdot \mathbf{n}\, dS + \iint_R \mathbf{E} \cdot (-\mathbf{n}')\, dR$$

or

$$\iint_S \mathbf{E} \cdot \mathbf{n}\, dS = \iint_R \mathbf{E} \cdot \mathbf{n}'\, dR.$$

So the flux out of G is the same as the flux out of a small sphere of radius a with center at the location of the charge. The flux out of the sphere can be computed from the definitions as follows: The unit normal to the sphere is a unit vector parallel to $\mathbf{r}$ so $\mathbf{n}' = \mathbf{r}/r$. Then

$$\mathbf{E}\cdot\mathbf{n}' = q\frac{\mathbf{r}}{r^3}\cdot\frac{\mathbf{r}}{r} = \frac{qr^2}{r^4} = \frac{q}{r^2}.$$

Because $\mathbf{r}$ is a position vector to a sphere of radius a, we have $|\mathbf{r}| = r = a$. Substitute into the integral that defines the flux to get

$$\text{flux out of } R = \iint_R \mathbf{E}\cdot\mathbf{n}'\,dR = \frac{q}{a^2}\iint_R dR = \frac{q}{a^2}4\pi a^2 = 4\pi q.$$

So we see that a point charge q produces a flux $4\pi q$ out of any surface enclosing the charge. [Of course this does not apply if the charge is actually on the surface.]

The equation

$$\iint_S \mathbf{E}\cdot\mathbf{n}\,dS = 4\pi q$$

is called *Gauss's law* and is of fundamental importance in electrostatics. We have derived it for the field corresponding to a point charge q but by additivity it applies to any smooth closed surface S where q is the sum of all the point charges enclosed by S that determine the field $\mathbf{E}$. That is if $q_1, \cdots, q_n$ are point charges enclosed by a surface S and if $\mathbf{E}$ is the field produced by these charges then

$$\iint_S \mathbf{E}\cdot\mathbf{n}\,dS = 4\pi(q_1 + \cdots + q_n) = 4\pi q \qquad \mathbf{(16.2)}$$

where q is the total charge enclosed by S. If a continuous distribution of charge is enclosed by S then a charge density function (charge per unit volume) would be used.

Stokes's Theorem and Gauss's Law

We illustrate how a form of Gauss's law can be derived from Stokes's Theorem. This will show how arguments are often made using the surface and volume integrals. Let V be a solid region bounded by a smooth closed surface S and let $\rho(x, y, z)$ be the charge density at (x, y, z). [To define ρ, take the solid ball B_ϵ of radius ϵ with center at (x, y, z), divide the total charge within B_ϵ by the volume of B_ϵ; the limit as $\epsilon \to 0$ is the value $\rho(x, y, z)$.] When a finite number of point charges lie inside the surface, Gauss's law is obtained by adding the charges. If the charge is distributed continuously (as opposed to discretely) then the continuous version of Equation (16.2) reads

$$\iint_S \mathbf{E}\cdot\mathbf{n}\,dS = 4\pi\iiint_V \rho\,dV.$$

The integral of ρ over the solid region replaces the sum of the point charges.

The Divergence Theorem implies

$$\iint_S \mathbf{E}\cdot\mathbf{n}\,dS = \iiint_V \operatorname{div}\mathbf{E}\,dV$$

and so we conclude

$$\iiint_V \operatorname{div}\mathbf{E}\,dV = 4\pi\iiint_V \rho\,dV.$$

Just because the triple integrals of two functions over a region are the same, we cannot conclude that the functions are equal. However, we have more than the equality for just one region. This equation holds for all V (bounded by smooth surfaces). This additional information is enough to conclude that the functions are equal (see exercises after this section) and we conclude

$$\operatorname{div} \mathbf{E} = 4\pi\rho.$$

Exercises 16.3

Compute each side of the equation in the conclusion of the Divergence Theorem for the vector functions $\mathbf{F}$ and the solid regions V indicated. In cases where the region has a piecewise smooth boundary, it might be necessary to evaluate separate surface integrals over each smooth piece.

1. $\mathbf{F} = 5x\,\mathbf{i} - 3y\,\mathbf{j} + 2z\,\mathbf{k}$, V is the solid tetrahedron with vertices $(0,0,0)$, $(3,0,0)$, $(0,1,0)$ and $(0,0,2)$

2. $\mathbf{F} = xy\,\mathbf{i} + yz\,\mathbf{j} + 2xz\,\mathbf{k}$, V is the solid tetrahedron with vertices $(0,0,0)$, $(1,0,0)$, $(0,4,0)$ and $(0,0,2)$

3. $\mathbf{F} = x\,\mathbf{i} - 2y\,\mathbf{j} + 3z\,\mathbf{k}$, V is the solid ball with center at $(0,0,0)$ and radius 3

4. $\mathbf{F} = x^2\,\mathbf{i} + y^2\,\mathbf{j} + z^2\mathbf{k}$, V is the solid ball with center at $(0,0,0)$ and radius 1

5. $\mathbf{F} = y\,\mathbf{i} - 2z\,\mathbf{j} + 3x\,\mathbf{k}$, V is the solid ball with center at $(0,0,0)$ and radius 3

6. $\mathbf{F} = y\,\mathbf{i} - 2z\,\mathbf{j} + 3x\,\mathbf{k}$, V is the upper half of the solid ball with center at $(0,0,0)$ and radius 3 so its boundary consists of the upper half sphere $x^2+y^2+z^2=9$, $z \geq 0$ and the disc in the xy plane consisting of all $(x,y,0)$ with $x^2+y^2 \leq 9$

7. Compare the flux of $\mathbf{F} = 4xy\,\mathbf{i} + y\,\mathbf{j} + 5xz\,\mathbf{k}$ out of the sphere $x^2+y^2+z^2 = R^2$ to the flux of $\mathbf{F}$ out of the cube $\{(x,y,z) : |x| \leq R,\ |y| \leq R,\ |z| \leq R\}$.

Compute the electrostatic field $\mathbf{E}$ produced by the given configuration of point charges and then determine the force acting on a charge of 1 unit placed at $(0,0,2)$.

8. A positive charge of q units placed at $(-1,0,0)$ and a positive charge of q units placed at $(1,0,0)$

9. A charge of $-q$ units placed at $(-1,0,0)$ and a charge of q units placed at $(1,0,0)$

10. Positive charges of magnitude q placed at $(2,0,0)$, $(0,2,0)$ and $(-\sqrt{2},-\sqrt{2},0)$

11. Negative charges of equal magnitude q placed at the vertices of the square in the xy plane having sides of length 1 and center at the origin.

12. A certain distribution of charges produces an electrostatic field $\mathbf{E} = ayz\,\mathbf{i}+axz\,\mathbf{j}+axy\,\mathbf{k}$, where a is a nonzero constant. Find the total charge enclosed by the upper half sphere $x^2+y^2+z^2=b^2$ ($z \geq 0$) and the disc $x^2+y^2 \leq b^2$ in the xy plane. [Hint: Use Gauss's law and the Divergence Theorem.]

13. A distribution of charges produces an electrostatic field $\mathbf{E} = ax\,\mathbf{i} + ay\,\mathbf{j}$. Find the total charge enclosed by the half sphere described in exercise 12.

14. Let V be a solid with a smooth boundary S. Use the Divergence Theorem to express the volume of V as an integral over the surface S.

15. Suppose f and g are continuous functions in some parallelepiped P and $\iiint_V f\,dV = \iiint_V g\,dV$ for every solid region V bounded by a closed smooth surface contained in P. Prove $f(x,y,z) = g(x,y,z)$ at all points interior to P. [Hint: Let $V(\epsilon)$ be the solid ball with center at (x_1,y_1,z_1) and radius ϵ. Use the Mean Value Theorem for Triple Integrals, which implies that

$$\iiint_{V(\epsilon)} f\,dV = f(x_0,y_0,z_0)\operatorname{vol} V(\epsilon)$$

for a point $(x_0,y_0,z_0) \in V(\epsilon)$, to show

$$f(x_1,y_1,z_1) = \lim_{\epsilon\to 0}\frac{1}{\operatorname{vol} V(\epsilon)}\iiint_{V(\epsilon)} f\,dV.$$

Conclude $f(x_1,y_1,z_1) = g(x_1,y_1,z_1)$ for every point (x_1,y_1,z_1) in the interior of P.]

Appendix A

Answers to selected exercises

Chapter 1

Section 1.1
page 4

1. Any number greater than or equal zero and less than five. **3.** Any number less than five or greater than twenty-five. **5.** The empty set; no numbers satisfy both inequalities. **7.** Any number greater than negative five and less than or equal to negative three, and any number greater than or equal to three and less than five or ant number greater than fifty. **9.** $\{x : -10 < x < 10\}$ **11.** $\{x : x = 1, 2, 3, \ldots, 100\}$ **13.** $\{2n : n$ is an integer $>$ two hundred fifty$\}$

Section 1.2
page 7

1. $[-5, 3]$ **3.** $[-8, -2] \cup [6, 12]$ **5.** $x < -3/2$ or $x > 9/2$ **7.** $-5 \leq x \leq 11$ **9.** $x \geq 3$ or $x \leq 0$ **11.** $0 \leq x \leq 4$

Section 1.3
page 11

1. The rectangular region with corners at $(1, -1)$, $(1, 1)$, $(7, 1)$, $(7, -1)$, including all boundary lines **3.** The semi-infinite strip bounded on the left and right by the lines $x = -1$ and $x = 1$ and bounded at the bottom by the x-axis; the left and right boundary lines are in the region, the bottom line is not **5.** The rectangular region with corners at $(-1, -2)$, $(-1, 2)$, $(1, 2)$, $(1, -2)$; the top and bottom boundary lines are included, the left and right boundary lines are not included **7.** The semi-infinite strip bounded below by the line $y = -2$ and on the left and right by the lines $x = -2$ and $x = 4$; the bottom line is in the regions, the left and right boundary lines are not **9.** The parallelogram with corners at $(-1, -2)$, $(-1, 0)$, $(1, 2)$, $(1, 0)$ including all boundary lines **11.** The disk bounded by the circle of radius 1 with center at $(1, 0)$ **13.** The region in the first quadrant above the graph of $xy = 1$, all of the second quadrant and the region of the third quadrant above the graph of $xy = 1$ **15.** All points in the infinite strip bounded on the left by the line $x = -2\sqrt{2}$ and on the right by the line $x = 2\sqrt{2}$, including both lines **17.** the vertical line $x = 1$ **19.** 6 **21.** Arbitrarily large; take $u = 10^{-k}$ and $v = 10^{-k-1}$ large k **23.** Think of four points on the line and distances between them

Section 1.4
page 14

1. $3y+2x = 8$ **3.** $y = 2x+2$ **5.** $\big((3+\sqrt{39})/10, (-1+3\sqrt{39})/10\big)$, $\big((3-\sqrt{39})/10, (-1-3\sqrt{39})/10\big)$ **7.** No such points **9.** $(1/2, 2)$ **11.** $(127/129, -116/129)$

Section 1.5
page 19

1. Circle with center at $(0, 2)$ passing through $(0, 0)$ **3.** Circle with center at $(-1/3, 1/4)$ and radius $5/12$ **5.** on, on, on, outside **7.** vertex $= (0, 0)$, focus $= (0, 1)$ **9.** $V = (1/8, -1/8)$, $F = (1/8, -3/32)$ **11.** $V = (1/4, 7/8)$, $F = (1/4, 1)$

13. $V = (-4, -2)$, $F = (-15/4, -2)$ **15.** Upper half of the circle of radius 4 and center at $(0, 0)$ **17.** Parabola with vertex at $(3, 10)$, opening down, crossing the x-axis at $3 \pm \sqrt{10}$ **21.** The sequence of parabolas approach the straight line $y = 2x - 1$ as the limiting position

Section 1.6
page 26

1. An ellipse through $(\pm 2, 0)$ and $(0, \pm 1)$ **3.** An ellipse through $(\pm 5, 0)$ and $(0, \pm 5/2)$ **5.** A hyperbola crossing the x-axis at $(\pm 5, 0)$ with asymptotes $x = \pm 2y$ **7.** A hyperbola crossing the x-axis at $(\pm 1, 0)$ with asymptotes $x = \pm 2y$ **9.** $(x-1)^2 + 3(y-1)^2 = 36$, ellipse with center $(1, 1)$ **11.** $5x^2 - 4(y+1)^2 = 28$, hyperbola with center $(0, -1)$ opening left and right **13.** $(x-3)^2 + (y - 7/2)^2 = 193/16$, circle with center at $(3, 7/2)$ and radius $\sqrt{193}/4$ **15.** *Inside* for 1,3,4 and *on* the curve for 2 **17.** Points in common if and only if $m^2 < b^2/a^2$

Section 1.7
page 28

1. $x^2 + y^2 = 2x$; circle with center at $(1, 0)$ and radius 1 **3.** $(x^2 + y^2)^{3/2} = 2axy$ **5.** The positive x-axis **7.** Circle of radius b with center at the origin **9.** $(3x-1)^2 + 12y^2 = 4$, ellipse with center $(1/3, 0)$ **11.** $4x^2 + (3y+1)^2 = 4$, ellipse with center $(0, -1/3)$

Section 1.8
page 31

1. There is at least one car on I 94 having either no driver or having more than one driver **3.** There is at least one real number whose square is less than or equal to zero **5.** At least one car purchased in Wisconsin has no sales tax certificate. **7.** False; try $x = 1$ **9.** True; use $x = -4$

Chapter 2

Section 2.1
page 37

1. 0, −8, −16 **3.** $\sqrt{st}$, $|w|$, 2 **5.** $-12x - 8$ **7.** $0, -320, -4^5 \cdot 46$ **9.** $x^2 + 10$, OCAP* **11.** $x^2(x^2+9)$ **13.** $(x^2 - 5x - 2)/(x-2)$, OCAP **15.** domain = all real numbers; range = all numbers ≥ 8 **17.** domain = all $x \neq 0$; range = all positive numbers **19.** domain = all numbers except $x = 1$;range = all numbers except $y = 1$ **21.** domain = all numbers except $x = 1/3$; range = all numbers except $y = 1$ **23.** $1/(x-1)^2(x-2)^2(x-3)^2$, OCAP

Section 2.2
page 40

1. Line through the origin lying in the second and fourth quadrants, OCAP **3.** Half-line starting at $(0, 5)$ and passing through $(5, 0)$, OCAP **5.** Straight line through $(2/3, 0)$ and $(0, -2)$ **7.** Parabola opening down, vertex at $(7/3, 25/3)$ and crossing the x-axis at $2/3$ and 4 **9.** A curve crossing the x-axis at 0, 1, 2 and increasing without bound for $x > 2$ **11.** A parabola and a line crossing at $((1-\sqrt{5})/2, (3-\sqrt{5})/2)$ and $((1+\sqrt{5})/2, (3+\sqrt{5})/2)$ **13.** Two adjacent parabolas, having vertices at $(0, 0)$ and $(2, 0)$, and crossing at $(1, 1)$ **15.** Graph is composed of three half-lines, the line from $(-3, 0)$ through $(0, 3)$; the line from $(-3, 0)$ through $(-4, 1)$, and the line from $(0, 3)$ through $(-1, 4)$ **17.** The graph rises as it is traced from left to right

Section 2.3
page 42

1. $1, -7/25, (3-2t)/t^2$ **3.** $(1-2x)/(x+1)^2, (-2t-5)/(t+4)^2, (-6w-1)/(3w+2)^2$ **5.** All x except $x = -1$ **7.** $f(g(x)) = 3(3x-7)^2$, $g(f(x)) = 3(3x^2) - 7$

*OCAP = other correct answers possible

9. $f(g(x)) = (x-1)/(2x-1)$, $g(f(x)) = -1/x$ **11.** $f(g(x)) = |x|$, $g(f(x)) = x$, if $x \le 1$ and $= 2-x$, if $x \ge 1$

Section 2.4
page 46

1. $|f(x)| \le 47$, $-33 \le f(x) \le 19$ **3.** $|f(x)| \le 90$, $-90 \le f(x) \le 0$ **5.** $|f(x)| \le 10$, $15/4 \le f(x) \le 10$ **7.** $|f(x)| \le 1$, $1/37 \le f(x) \le 1$ **9.** $|f(x)| \le 3/2$, $1/3 \le f(x) \le 3/2$,(with advanced methods you can get $1/3 \le f(x) \le 1.113$) **11.** $|f(x)| \le 47/5$, $-47/5 \le f(x) \le -47/257$

Section 2.5
page 52

7. $L = 3$ **9.** limit $= 0$ **11.** limit $= 1/2$ **13.** $|f(x) - f(1)| \le 3$, OCAP **15.** $|f(x) - f(1)| \le 1/4$, OCAP **17.** $|f(x) - f(-3)| \le 28/15$, OCAP **21.** A square with side $P/4$ has maximum area

Section 2.6
page 62

1. Limit exists and $= 3$ **3.** Limit exists and $= 4/3$ **5.** Limit exists and $= 0$ **7.** Limit exists and $= -2$ **9.** Limit exists and $= 0$ **15.** limit $= 0$ if $a \ge 0$ and $= a$ if $a \le 0$ **17.** Limit exists and $= 0$ **19.** Limit exists and $= -1$ **21.** $f(x) = 5$ for $x < 3$, $f(x) = 4x - 17$ for $3 \le x \le 5$, $f(x) = -3$ for $x > 5$, OCAP

Section 2.7
page 68

1. 0 **3.** 0 **5.** $+\infty$ **7.** $-\infty$ **9.** $+\infty$ **11.** $1/2$ **12.** 1 **18.** $y = 5x$ **19.** $y = Ax + B - AD$

Section 2.8
page 72

1. $4\sqrt{2} \approx 5.65685$ **3.** $6R$ **5.** $4\sqrt{3}R$

Section 2.9
page 77

1. 10 **13.** 7 **15.** 0 **17.** 2 **19.** 0 **21.** 0 **23.** $2Rn\sin(\pi/n)$

Chapter 3

Section 3.1
page 83

1. All $x \ne 3$ **3.** All $x \ne \pm 2$ **5.** $x \ge -\sqrt[3]{5}$ **7.** All $x \ne 1$ **9.** Not continuous at $x = 1$ **11.** Continuous everywhere **13.** Solutions on $[0,1]$ and $[1,2]$ **15.** Solution on $[-1,0]$ and on $[0,1]$ **17.** $[-1,0]$, $[0,1]$, $[5,6]$ **19.** $[-1,0]$, $[1,2]$, $[4,5]$ **21.** Infinite discontinuity at $x = 1$ **23.** Infinite discontinuity at $x = 0$ **25.** Removable discontinuity at $x = 0$ **27.** Infinite discontinuity at $x = 0$ **31.** Let N be the largest integer not exceeding $a/2\pi$. The number of solutions of $\sin x = x/a$ with $x \ge 0$ is $2N + E$ where E is 0, 1 or 2 (depending on a). For a given a, subtle analysis is needed to determine E.

Section 3.2
page 91

1. $15x^2 - 7$ **3.** $32t - 48$ **5.** $5w^4 - 6w^2 + 5$ **7.** $4x^3 - 6x$ **9.** $y = 0$, $y = 4x - 4$, $y = -4x - 4$ **11.** $y = 4$, $y = -4x + 6$, $y = -4\sqrt{2}x + 8$ **13.** $y = -2x + 3$, $y = 2x + 3$ **15.** $x = \pm 1/\sqrt{3}$ **17.** $x = \pm\sqrt{3}$ **19.** $120x^4$ **21.** $240x$ **23.** $g^{(8)}(x) = -8\cdot 7\cdot 6\cdot 5\cdot 4\cdot 3\cdot 2$, $g^{(9)} = 0$ **25.** $(1,1)$ **27.** Equal slopes at (a, a^3) and $(-a, -a^3)$ for every a **29.** $(m,n) = (4,1)$, $(2,2)$ or $(1,4)$ **31.** $q(x,h) = 12x^2 - 12x + 12xh + h^2 - 6h$ **33.** $q(x,h) = 5x^4 + 10x^3h + 10x^2h^2 + 5xh^3 + h^4$

Section 3.3
page 95

1. $(2x-4)(x+3) + (x^2 - 4x)$ **3.** $1 - 1/x^2$ **5.** $(2x+1)(2x^3 + 7x^2 - 1) + (x^2 + x - 1)(6x^2 + 14x)$ **7.** $4u/(1+u^2)^2$ **9.** $(1 - 2u + 6u^2 + 4u^3 - 8u^5 + u^6)/(1 + u - 4u^2 + u^3)^2$ **11.** $(12 - 2x + x^2)/x^4$ **13.** 6 **15.** 7 **17.** $-2(1+x)^{-3}$ **19.** $-24x^{-4}$ **21.** $24(x+1)^{-5}$ **23.** $1\cdot 2\cdot 3\cdots(n-1)\cdot n(1-x)^{n+1}$ **25.** $(-1)^n 1\cdot 2\cdot 3\cdots(n-1)\cdot n[(x-1)^{n+1} + (x+1)^{n+1}]$

27. $f' = 20x^4 + 2 > 0$ for all x **29.** $f' \geq 0$ if $|x| \geq \sqrt{3}$ **35.** $g(x) = 4x$ for $x \leq 1$ and $g(x) = (x+1)^2$ for $x > 1$ (many other correct solutions possible)

Section 3.4 page 98

1. $4\cos x + 3\sin x$ **3.** $3\cos x + 5\sec^2 x$ **5.** 0 **7.** $-4\csc x\cot x - 3\sec x\tan x$ **9.** $1/(\cos x - 1)$ **11.** $x^2(x\cos x - \sin x)^{-2}$ **13.** $-x^2\cot x\csc x + 2x\csc x + x^{-1}(\csc^2 x - \sec^2 x) + x^{-2}\csc x\sec x$ **15.** $-\sin x$ **17.** $2(\cos^2 x - \sin^2 x)$ **19.** $-2(\cos^2 x - \sin^2 x)$ **21.** $\cot^2 x\csc x + \csc^3 x$ **23.** $2\cot x\csc^2 x$ **25.** $y^{(4)} = y$

Section 3.5 page 103

1. $24(3x-7)^{-9} + 24(3x-7)^7$ **3.** $(6x+4)/(1+x)^3 + 1/2\sqrt{x}(x+1)^{3/2}$ **5.** $\cos(4x+1)\sin(4x+1)[24\sin(4x+1) - 32]$ **7.** $8(2x-5)^3$ **9.** $-4(4-x^2)x$ **11.** 0 **13.** $16(x-5)(x+3)^{-3}$ **15.** $2x\cos(x^2+\pi)$ **17.** $-5\sin 5t$ **19.** $2(x+1)\cos 2x + \sin 2x$ **21.** $-12\sec^2(1-3x)\tan^3(1-3x)$ **23.** $3(x+\sin x)^2(1+x+\cos x+x^2\sin x - 2\sin x\cos x)(1+x\cos x)^{-2}$ **25.** $-B, 0$ **27.** 0 **29.** $a^4\sin ax$ **31.** $-9(4x^2+x)^4(8x+1)^2\sin(4x^2+x)^3 + 6(4x^2+x)(70x^2+20x+1)\cos(4x^2+x)^3$ **33.** $y = a\sin x + b\cos x$, a, b constants **35.** $y = a\sin 3x + b\cos 3x$, a, b constants

Section 3.6 page 106

1. $y' = -4/3y^2$ **3.** $y' = (\sqrt{y^2 - x^2 + 1} + x)/y$ **5.** $y' = \left(2/\sqrt{2(x+y)} - 1/\sqrt{x}\right)\left(1/\sqrt{y} - 2/\sqrt{2(x+y)}\right)^{-1}$ **7.** $y' = 1/5x^{2/3}(1+x^{1/3})^{2/5}$ **9.** $y' = x/6(x^2+4)^{2/3}$ **11.** $4/5$ **13.** $7/10$ **15.** $y'' = -(3y^2+x^2)/9y^3$ **17.** $y'' = 6y/(2-3y^2)^3$ **19.** $y'' = -3/4$ at $(1,1)$ **21.** $y'' = 0$ at $(\pi/4, \pi/4)$ **23.** $y' = -y/x$ and $y' = x/y$ **25.** $y' = 2xy/(x^2-y^2)$ and $y' = (y^2-x^2)/2xy$ **27.** $q(x) = +\sqrt{1-x^2}$ for x on $[-1, a_1]$, $[a_2, a_3]$, $\cdots$ and $q(x) = -\sqrt{1-x^2}$ for x on (a_1, a_2), (a_3, a_4), $\cdots$ **29.** $f(0) = 2$, $f'(0) = 0$, $f''(0) = 2$, $f'''(0) = 0$, $f^{(4)}(0) = 6$, $f^{(5)}(0) = 0$, notice that $y^{(5)} = (x^5 + 10x^3 + 15x)y$

Chapter 4

Section 4.1 page 114

1. critical points: 0 (max), $4/3$ (min) **3.** critical points: $-1-\sqrt{5}$ (min), $-1+\sqrt{5}$ (max) **5.** No critical points on $(1, 10)$ **7.** critical points: $-\pi$ (max), 0 (min), π (max) **9.** critical points: $-5\pi/4$, $-\pi/4$, $3\pi/4$, $7\pi/4$, (neither max nor min) **11.** $x = 1/2$ **13.** $x = -2+\sqrt{3}$ **15.** $x = 1/2$ **17.** $x = -2+\sqrt{3}$ **19.** If $u < v < w$ are solutions of $p(x) = ax^2 + bx + c = 0$ then (by Rolle's theorem) there are numbers s and t ($u < s < v < t < w$) such that $p'(s) = p'(t) = 0$. but $p'(x) = 2ax + b = 0$ has at most one solution.

Section 4.2 page 117

1. $f(1) - f(-1) = f'(c)(1-(-1)) < 2$ and $f(-1) = 3$ so $f(1) < 5$ **3.** $f(2) - f(-2) = f'(c)\cdot 4 \geq 24$ and $f(2) = 199$ so $f(-2) \leq 175$ **5.** For $u < v$, $f(v) - f(u) = f'(c)(v-u)$ and so $|f(v) - f(u)| \leq |v - u|$ **7.** See #5 **9.** $c = \pm 2/\sqrt{3}$ **11.** $c = 0$ or $\pm\sqrt{5/2}$ **13.** $c = 0$ **15.** Suppose $f(a) = f(b) = 0$. Then $0 - 0 = f(b) - f(a) = f'(c)(b-a)$. Since $f'(c) > 0$, it follows $a = b$. **17.** $f(x) - f(1) = f(x) = \frac{1}{c}(x-1)$ for some c, $1 < c < x$. Consider the extreme values of c.

Section 4.3 page 121

1. Decreasing on $[-\sqrt{2}, \sqrt{2}]$, increasing on $(-\infty, -\sqrt{2}]$ and $[\sqrt{2}, \infty)$ **3.** increasing on $(-\infty, \infty)$ **5.** Decreasing on $(-1, 0)$ and $(0, 1)$, increasing on $(-\infty, -1]$ and $[1, \infty)$ **7.** Decreasing on $(-\infty, 0]$, increasing on $[0, \infty)$ **9.** Increasing on $\left[(-3-\sqrt{5})/2, (-3+\sqrt{5})/2\right]$, decreasing elsewhere **11.** Decreasing on $\left[(15-\sqrt{201})/120, (15+\sqrt{201})/120\right]$, increasing elsewhere **13.** $f(x) = x^2 + c$, c constant **15.** $f(x) = x^3/3 + x^2/2 + c$, c constant **17.** The correct conclusion is that $\sin^2 x$ and $-\cos^2 x$ differ by a constant $[\sin^2 x - (-\cos^2 x) = 1]$ **19.** Let $f(x) = \sin x - x\cos x$;

note $f(0) = 0$; show $f'(x) \geq 0$ for x on $[0, \pi]$ **21.** By MVT, $f(x) = f(x) - f(0) = x/(1 + c^2)$ with $0 < c < x$; consider the extreme values of $x/(1 + c^2)$.

Section 4.4
page 124

1. 4 **3.** 3 **5.** $\lim_{x \to 0^+} = \infty$, $\lim_{x \to 0^-} = -\infty$ **7.** 5/3 **9.** $f'(2)$ **11.** $1/3 \cdot 2^{-2/3}$ **13.** 1 **15.** $k = 2$, limit $= -4/3$ **17.** $k = 1$, limit $= 1/3$ **21.** $\lim_{x \to 0} \frac{x - \sin x}{x \sin x} = 0$ **23.** $\lim_{t \to 0^+} \frac{\sin t^2}{t} = 0$ **25.** $\lim_{t \to 0+} \frac{\sqrt{1+t}-1}{t} = \frac{1}{2}$ **27.** $\lim_{t \to 0+}[(1 + t)^{1/3} - 1]/t^{1/3} = 0$

Section 4.5
page 131

1. No local or abs max; abs min is $f(0) = 4$ **3.** abs min at $\pm 1/\sqrt{6}$, local max at $x = 0$, no abs max **5.** Local max at $(3 - \sqrt{3})/3$, local min at $(3 + \sqrt{3})/3$, no abs min or max **7.** local max at 0, abs max at $(15 + \sqrt{33})/12$, local min at $(15 - \sqrt{33})/12$, no abs min **9.** No abs extrema (odd degree), local max at 0, a, local min at $5a/9$ **11.** Domain is $[-2, 2]$, abs min at endpoints, abs max at $x = 0$ **13.** Domain is $[-2, 2]$, abs min at endpoints and$x = 0$, abs max at $\pm\sqrt{8/3}$ **15.** If m is an integer $m > 1$, there is one critical point, $x = b$; if m is odd, $f(b)$ is neither max nor min; if m is even $f(b)$ is an abs min. If $m = 1$, there is no critical point. **17.** $f(x) = x^2$ **19.** $f(x) = -x^3 + 9x^2 - 24x$ $[f'(x) = -3(x - 2)(x - 4)]$ **21.** $N = (b/a)^{1/(n-m)}$ **23.** If $f(c)$ is an absolute minimum then $0 \leq f(c + h) - f(c) = f'(u)h$ for some u between c and $c + h$ and for any number h. Take $h \geq 0$ to conclude $f'(u) \geq 0$ and $\lim_{h \to 0^+} f'(u) = f'(c) \geq 0$. Similarly with $h \leq 0$ get $f'(c) \leq 0$ and so $f'(c) = 0$.

Section 4.6
page 137

1. Normal line: $6y = -x + 6$ **3.** Normal line at (a, a^2): $2ay = -x + a + 2a^3$ [See solution to Ex. 7] **5.** Equilateral triangle with side $R\sqrt{3}$ has max area **7.** The minimum distance from (a, b) to $(x, f(x))$ occurs at a number c satisfying $(c - a) + (f(c) - b)f'(c) = 0$ (assuming no endpoints interfere). The slope of the line from $(c, f(c))$ to (a, b) is $(f(c) - b)/(c - a) = -1/f'(c)$ proving the line is normal to the curve. **9.** 8 **11.** min distance $= p - a$ **13.** $V = 4\pi R^3/3\sqrt{3}$ **15.** When $a \geq \$47.89$ it is least expensive to construct the entire pipeline underwater. **17.** max sum at $(-1/\sqrt{2}, -1/\sqrt{2})$, min sum at $(1/\sqrt{2}, 1/\sqrt{2})$ **19.** (a) Use all the wire to make a circle; (b) Use all the wire to make a square. **21.** $\left(a^{2/3} + b^{2/3}\right)^{3/2}$ **23.** $13y = 9x + 55$ **25.** $7y = x + 28$ **27.** View S_n as a function of x_n only with all other variables held fixed. The function is a quadratic and has a unique absolute minimum. Find the minimum to get $S_n \geq S_{n-1}$.

Section 4.7
page 145

1. $16\pi\sqrt{3}$ **3.** $\sqrt{74.8} \times 93.5/\sqrt{74.8} \approx 8.65 \times 10.81$ **5.** 1.13389 miles west of the intersection **7.** ratio ≈ 0.9515, 20.18° **9.** $v_1/v_2 = 1 = \sin\alpha/\sin\beta$ so $\alpha = \beta$

Section 4.8
page 148

1. Concave on $(-\infty, 0]$, convex on $[0, \infty)$, $(0, 1)$ is an inflection point **3.** Convex everywhere **5.** Convex on $[-2\sqrt{3}/3, 2\sqrt{3}/3]$, inflection points at $x = \pm 2\sqrt{3}/3$, concave elsewhere **7.** Inflection points at $x = n\pi$, n any integer; concave on $[n\pi, (n + 1)\pi]$ if n is an even integer, convex otherwise **9.** Inflection points at $x = n\pi/2$, n any integer; concave on $[n\pi, (2n + 1)\pi/2]$, convex on $[(2n + 1)\pi/2, (n + 1)\pi]$ for any integer n **13.** Apply the analytic definition (4.5) to $f + g$

Section 4.9
page 151

1. $v = 0$ at $t = 0$ and $s(0) = 64$ **3.** $v = 0$ at $t = 1$ and $s(1) = 92$, $s(-1 \pm 3\sqrt{2}/4) = 0$ and $v = \pm 48\sqrt{2}$ **5.** $s'(t) = 0$ at $t = 5$ and $s(5) = 400$ **7.** $s'(t) = 0$ at $t = 1/64$ and $s(1/64) = 500 + 1/256$ **9.** 2.5 sec **11.** Falling time is $G(v_0) = (\sqrt{v_0^2 + 6400} - v_0)/32$; statement is false because $G(2v_0) \neq G(v_0)/2$ **13.** $s(t) = -16(t - w_0)^2 + c$ with $w_0 =$

$v_0/32$ and $c = s_0 - w_0^2$; $s(t_1) = s(t_2)$ implies either $t_1 = t_2$ or $t_1 - w_0 = -(t_2 - w_0)$; it follows that $v(t_1) = \pm v(t_2)$ **15.** acceleration = 1/768 miles/sec^2; 19.2 sec

Section 4.10
page 153

1. (a) $t = 1/4$, (b) $\cos\alpha = 4/5$ **3.** $v_0 \geq \sqrt{g}90$ ft/sec **5.** (70mi/hr = 102.666 ft/sec) max distance if $\alpha = \pi/4$ (45°); max time if $\alpha = \pi/2$ (straight up) **7.** Impact on the hill at $t = 2v_0\cos\alpha[\tan\alpha - \tan\beta]/g$, the x-coordinate at this time is a function of α that has its maximum value when the line making angle 2α with the positive x-axis is perpendicular to the line making angle β.

Section 4.11
page 154

1. $dL/dt = hv/(H-h)$ **3.** When area is A, sides change at the rate of $(2/\sqrt{3}A)^{1/2}$ **5.** 96 × 5280 in^2/year **7.** $d(\text{surface area})/dt = 6r$ **9.** $dx/dt = (1-y)^{-1}dy/dt$, $v_{\max} = -2/9$, $v_{\min} = -2$ **11.** $x(\theta) = \tan\theta$; 8π, 16π **13.** $d\theta/dt = \frac{\cos^2\theta}{1000}\frac{dh}{dt} = \frac{\cos^2\theta}{500}$

Chapter 5

Section 5.1
page 162

1. $\sum_{i=1}^{10}(-1)^{i+1}i$ **3.** $\sum_{k=1}^{9}(-1)^{i+1}\frac{k}{k+2}$ **5.** 112 **7.** 49/20 **9.** 3.9949871... **11.** 1.6349839... **15.** $\frac{1}{2}\sum_{k=1}^{100}\frac{1}{k} - \frac{1}{k+2} = 7625/101 \cdot 102 \approx 0.740148\ldots$ **17.** $n(3n^2+3n-1)(2n+1)(n+1)/30$ **19.** 35,490 **21.** 600,796,660 **23.** $2 - \frac{1}{2^{20}}$ **25.** $\frac{3}{5}\left[\left(\frac{2}{3}\right)^{51} + \left(\frac{2}{3}\right)^{10}\right]$ **27.** $5005 - \left(\frac{1}{2}\right)^9 + \left(\frac{1}{2}\right)^{100}$

Section 5.2
page 167

1. $L = 8$, $U = 24$ **3.** $L = 0.25$, $U = 1.25$ **5.** $L = 0.95$, $U = 1.28333$ **7.** $L = 26581/78440 \approx 0.415244$, $U = 26581/44100 \approx 0.602744$ **9.** $L = 0$, $U = 1/2$ **11.** $L_4 = 0.21875$, $U_4 = 0.46875$; $L_8 = 0.27343$, $U_8 = 0.39843$; $L_{16} = 0.30273$, $U_{16} = 0.36523$ **13.** $L_4 = 0.14062$, $U_4 = 0.39062$; $L_8 = 0.19140$, $U_8 = 0.31640$; $L_{16} = 0.21972$, $U_{16} = 0.28222$ **19.** The diagonal divides a rectangle into two triangles of equal area. On each subinterval, the upper sum overestimates the area by the exact amount that the lower sum under estimates the area. The average equals the area. **21.** If the maximum and minimum values of a function on an interval are equal, the function is constant. The value of the continuous function on two adjacent subintervals must be the same else there would be a jump in the graph.

Section 5.3
page 173

1. $x^4 + x^2 + c$ **3.** $-\cos x + \frac{1}{2}\sin 2x + c$ **5.** $\frac{2}{5}t^5 - \sin t + c$ **7.** $\frac{1}{5}(z-4)^5 + c$ **9.** $\frac{4}{3}z^{3/2} + c$ **11.** $3x^{5/3} - \frac{6}{5}x^{5/2} + c$ **13.** 18 **15.** 2 **17.** $-1 + \pi^5/160$ **19.** 1/5 **21.** $4\sqrt{3} - 4/3$ **23.** $3 \cdot 2^{20/3} + \frac{3}{5}2^{17/2} - \frac{6624}{5}$ **25.** $n(n+2)/2$ **27.** $U_{32} \approx 0.701021$, $L_32 \approx 0.685396$; $U_{1000} - L_{1000} \approx 0.0005$ **29.** [1. 16, 3. 2/3, 7. 1/2, 9. $1/\pi$, 11.1/3, 13. 1/4] **31.** 9

Section 5.4
page 179

1. The largest δ is the positive root of $3h^2 + 12h - 0.2$ which is ≈ 0.01659 **5.** Follow the hint to prove this important theorem

Section 5.5
page 183

1. midpoint approximation = 2 **3.** $\lim_{n\to\infty}\sum_{i=1}^{n}\left(\frac{3i}{n}+2\right)\frac{1}{n} = \frac{7}{2}$ **5.** $\lim_{n\to\infty}\sum_{i=1}^{n}2\left(\frac{2i}{n}\right)^2\frac{2}{n} =$

$\frac{16}{3}$ **7.** $\lim_{n\to\infty}\sum_{i=1}^{n} 2\left(\left(\frac{2i}{n}\right)^2 + \left(\frac{2i}{n}\right) + 2\right)\frac{2}{n} = \frac{26}{3}$ **9.** $\int_0^1 (2+x)^3\,dx$ **11.** $\int_0^1 x^3\,dx$ **13.** $\int_0^1 x^a\,dx$

Section 5.6 page 186

1. 44 **3.** $10a - 4b$ **5.** 20 **7.** 41/6 **9.** 1/6 **11.** 1/20 **13.** $2x - 3$ **15.** $-(x-1)/(x+1)$ **17.** $3x^5/(x^3+2)$ **19.** $(2x+1)\sin(x^2+x)$ **21.** $(x+1)(x+2)$ **23.** $4x$ **25.** $g(b) - g(a) = (b-a)f(c)$ (for some c) since $g'(x) = f(x)$ **27.** $2/\pi$, $2/\pi$

Section 5.7 page 192

1. 16 **3.** 72 **5.** $-2+\pi^2/4$ **7.** $\sqrt{2}$ **9.** $(m^2+12)^{3/2}/6$ **11.** 1/2 **13.** 1/2 **15.** 125/6 **17.** 32/3 **19.** Use $x = au$, $dx = a\,du$ **21.** $8\sqrt{2}$

Section 5.8 page 196

1. $x^3/3 - x^{-1} + c$ **3.** $2x^{1/2}+c$ **5.** $\frac{2}{7}x^{7/2}+\frac{6}{5}x^{5/2}-\frac{2}{3}x^{3/2}+c$ **7.** $-2/w - 3w + w^3/3 + c$ **9.** $x^2/2 - 4x + c$ **11.** $-x^3/3 + 4x + c$ **13.** $x^3/6 + x^2/2 + c_1x + c_2$ **15.** $-\cos 2t + c_1t + c_2$ **17.** $-x^4/4 + x - 3/4$ **19.** $x^2 - 2\sin 2x + 4$ **21.** $x^3/6 + x^2/2 + 4$ **23.** $f''(x) = -c_1\cos x - c_2 \sin x = -f(x)$

Section 5.9 page 200

1. $u = 4x+5$, $(4x+5)^4/16 + c$ **3.** $u = 20 - 3x$, $-\frac{2}{9}(20-3x)^{3/2} + c$ **5.** $u = x^3 - 3$, $-(x^3-3)^{-6}/18+c$ **7.** $u = 1+x^2$, $(1+x^2)^{n+1}/(2n+2)+c$ **9.** $u = z^4-3$, $-(z^4-3)^{-5}/20+c$ **11.** $u = a + x^{m+1}$, $2(a+x^{m+1})^{3/2}/(3m+3)+c$ **13.** $u = x^2+2x$, $\frac{1}{2}\sin(x^2+2x)+c$ **15.** $u = 4x$, $-\frac{1}{4}\cos 4x + c$ **17.** $u = 6 - \pi x$, $-\frac{1}{\pi}\sin(6-\pi x)$ **19.** 31/3 **21.** 1/4 **23.** $191{,}087{,}351/(2^{14}\cdot 3^8 \cdot 5^6) \approx 3.55676 \times 10^{-6}$ **25.** $(5^{n+1}-1)/(2n+2)$ **27.** even **29.** neither **31.** even **33.** $g(x) = \big(g(x)+g(-x)\big)/2 + \big(g(x)-g(-x)\big)/2$ **35.** $\int_{-a}^{0} f(x)/, dx = -\int_a^0 f(-u)\,du = \int_0^a f(u)\,du$ (making the substitution $x = -u$, $dx = -du$ for the first equality)

Section 5.10 page 203

1. $9\alpha/2$ **3.** $\pi^2/4$ **5.** π **7.** $27\pi/2$ **9.** $\pi/8$ **11.** The area is an integral of $g(\cos\theta)/2$ over an interval of length 2π and $m^2 \le g(\cos\theta)^2 \le M^2$.

Chapter 6

Section 6.1 page 209

1. 57π **3.** $32\pi/5$ **5.** $184\pi/15$ **7.** $4\pi/3$ **9.** 18π **11.** $512\pi/15$ **13.** 1500 ft^3 **15.** $\pi/2$ **17.** $\pi - \pi/a$; π **19.** $1408\pi/3$ in^3 **21.** $A^2H/3$ **23.** $ab(h_1+h_2+h_3+h_4)/4$ **25.** $h + R = 5/2$ and h is a solution of $4h^3 - 20h^2 + 25h - 7.03125 = 0$; $h \approx 1.36534$ and box height ≈ 27.2318 in

Section 6.2 page 212

1. 5/6 **3.** 16/3 **5.** 5625 **7.** $\pi\rho R^2h^2/2$ **9.** $25600\rho\pi$ **11.** $GMm(a^{-1} - b^{-1})$, GMm/a

Section 6.3 page 216

1. $3\sqrt{10}$ **3.** 32/3 **5.** 123/32 **7.** 915/16 **9.** $6a$ **11.** Show $|f'(x)| \le 1$ and so $\sqrt{1+f'(x)^2} \le \sqrt{2}$ and the arc length is at most $\sqrt{2}$ which is the length of the straight line from $(0,0)$ to $(1,1)$.

Section 6.4 page 221

1. $3\pi\sqrt{10}h^2$ **3.** $\pi(10\sqrt{10}-1)/27$ **5.** $\left((1+9(1+a)^4)^{3/2} - 40\sqrt{10}\right)\pi/27$ **7.** $(13\sqrt{13}-27)\pi/2$ **9.** 3π **11.** Use the surface area integral for $f(x) = r$

(constant) and $0 \le x \le h$ **13.** $2\pi(2^{7/2} - 4)/3$ **15.** $6\pi a^2/5$

Section 6.5
page 229

1. 1 **3.** $-1/7$ **5.** $1/27$ **7.** mass 10 at $(3, -1)$ **9.** $x^* = 0$, $y^* = 4R/3\pi$ **11.** Center of mass on the vertical axis of symmetry $(6\pi + 28)/(3\pi + 24)$ units above the base **13.** $x^* = -2/15$, $y^* = 0$ **15.** Write the integrals representing the area, volume and y-coordinate of the center of mass and compute the expressions **17.** Express all quantities in terms of integrals and compare **19.** $(0, 4R/3\pi)$ **21.** Surface of sphere is $4\pi R^2$, length $= \pi R$, center of mass is $(0, 2R/\pi)$

Chapter 7

Section 7.1
page 236

1. $2x - 3 = 2u - 3 \Rightarrow x = u$, $f^{-1}(x) = (x+3)/2$ **3.** $x^3 + 5 = u^3 + 5 \Rightarrow x^3 = u^3 \Rightarrow x = u$, $f^{-1}(x) = (x - 5)^{1/3}$ **5.** $f(x) = f(u)$; if $x < 0$ then $f(x) = x < 0$ so $f(u) < 0$ and $f(u) = u$; if $x > 0$ then $u > 0$ and $f(x) = 3x = f(u) = 3u \Rightarrow x = u$, $f^{-1}(x) = x$ for $x < 0$ and $f^{-1}(x) = x/3$ for $x \ge 0$ **7.** $1/x$ is one-to-one on $(0, \infty)$ and equals its own inverse. **9.** One-to-one on the set of all $x \neq a$ with inverse $a(x + 1)/(x - 1)$ **11.** $f(g(a)) = f(g(b)) \Rightarrow g(a) = g(b)$ (because f is one-to-one) $\Rightarrow a = b$ (because g is one-to-one) **13.** If $f = f^{-1}$, the graph of f is symmetric about the line $y = x$; $f(x) = 1/x$ is a nontrivial example **15.** Further hint: If $0 \le u < v < w \le 1$ and $f(u) \le f(w) < f(v)$, then apply the Intermediate Value Theorem to conclude there is a c, $u \le c \le v$ with $f(c) = f(w)$ **17.** By Ex. 15, either f is increasing and $f' > 0$ or f is decreasing and $f' < 0$

Section 7.2
page 242

1. $4\ln 2 + 2\ln 3 = 4.970$ **3.** $(\ln 3)/2 = 0.5495$ **5.** $\ln(2/3) = \ln 2 - \ln 3 = -0.406$ **7.** Domain = all real numbers; range = all $y \ge 0$ **9.** Domain = all x with $|x| > 1$; range = all $y \neq 1$ **11.** $\ln 5 \approx 2\ln 2 + \sum_{i=1}^{4} 1/(16 + i) \approx 1.603$ **13.** For a single interval $[a, b]$, $U - L$ is the area of a rectangle with diagonal = line from $(a, 1/a)$ to $(b, 1/b)$. This line is above the graph so the true area is smaller than the average of L and U

Section 7.3
page 247

1. $2x/(x^2 - 4)$ **3.** $\ln(x - 1) + 1$ **5.** $2/(x^2 - 1)$ **7.** $4x/(x^2 + 4) - 2x/(x^2 + 9) + (4 - 2x)/(x^2 - 1)$ **9.** $3\ln(x^2 - 1) - 2\ln(x^2 + 4) + (x - 1)\big(6x/(x^2 - 1) + 4x/(x^2 + 4)\big)$ **11.** $f(x)\left[\frac{3x^2+1}{x^3+x+1} - \frac{2x-4}{x^2-4x+12} + 2x\cot(x^2-1) + \tan x\right]$ **13.** $\ln|x+4| + c$ **15.** $\ln|x^4+1| + c$ **17.** $-\ln|3 + \cos x| + c$ **19.** $1 + \ln 2$ **21.** $1/2$ **23.** $\ln|\sin x| + c$ **25.** $\lim_{x\to 0^+} \ln x/(1/x) = 0$ **27.** $\lim_{x\to\infty} \ln\big(1/(1 + 1/x)\big) = 0$ **29.** $-1/2$ **31.** $\lim_{x\to\infty} p'(x)/p(x) = 0$ because $\deg p' < \deg p$ **33.** 0

Section 7.4
page 250

1. 3 **3.** $2/3$ **5.** $1/(3x + 1)$ **7.** $\sqrt{x}$ **9.** 0 **11.** (a) $\ln 1^x = x\ln 1 = 0 = \ln 1$ so $1^x = 1$; (b) $\ln a^0 = 0\ln a = 0 = \ln 1$ so $a^0 = 1$ **13.** Suppose $a^u < a^v$. Since $\ln x$ is an increasing function, $u\ln a < v\ln a$. Consider $0 < a < 1$ and $1 < a$ separately. **19.** $x(2\ln 2 + \ln 3) = 0$, $x = 0$ only solution **21.** $(e^x - 1)^2 = 0$, $x = 0$ only solution **23.** $x = 0$ or $x = \ln 2/\ln 3$ **25.** $f(x) = e^{-x} - x$ has $f(0) > 0$ and $f(1) < 0$ so there is a solution to $f(x) = 0$ with $0 < x < 1$

Section 7.5
page 254

1. $4e^{4x}$ **3.** $e^{2x}(1 + 2x)$ **5.** $e^{\sqrt{x}}/2\sqrt{x} - e^{x/2}/2$ **7.** $-(4e^{-3x} - x^3)^{-2/3}(4e^{-3x} + x^2)$ **9.** $-e^x \sin e^x$ **11.** 1 **13.** $\ln a^x = x\ln a$ and also $\ln e^{x\ln a} = (x\ln a)\ln e = x\ln a$; thus $a^x = e^{x\ln a}$ **15.** $(\ln 10)10^x - (\ln 10)10^{-x}$

17. $3^{2x+3}(3+2(\ln 3)(3x-1))$ **19.** $\frac{1}{2}e^{x^2}+c$ **21.** $\ln|1+e^x|+c$ **23.** $\ln|1+e^x|+c$ **25.** $-\cos e^x+c$ **27.** $(u=1+e^x)$, $2\sqrt{1+e^x}+c$ **29.** $f'(x)=f(x)\left[\frac{2}{x^2+1}-\frac{1}{x^2}\ln(x^2+1)\right]$ **31.** $f'(x)=f(x)\left[\frac{1}{x}\ln(e^x+1)+e^x\ln x/(e^x+1)\right]$ **33.** $\frac{\pi}{2}(1-\frac{1}{a^2})$

Section 7.6
page 259

1. $T(t)=10e^{-3t}$ **3.** $T(t)=2+4e^{2t}$ **5.** $G(t)=10e^{-t/10}$ **7.** $G(t)=-400+400e^{-t/10}$ **9.** $G(t)=400-400e^{t/10}$ **11.** 4500 **13.** 246.2 years **15.** 3 days **17.** 2.4 hours **19.** If $k<0$ then $\lim_{t\to\infty}T(t)=T_0/k$; $t>(\ln 101T_0-\ln(100(k^2T(0)-kT_0))$

Section 7.7
page 264

1. e **3.** 1 **5.** 1 **7.** $e^{3/4}$ **9.** See Figure 7.18 for half the inequality; use an upper sum for the other half **13.** The 6.75% account pays \$69.29; the 7% account pays \$71.23 **15.** Apply L'Hôpital's rule m times to get ∞ as the limit

Section 7.8
page 271

1. $\pi/3$ **3.** 0 **5.** $-\pi/4$ **7.** $5\pi/6$ **9.** 1.4706 **11.** 1.5608 **13.** $\sqrt{1-x^2}$ **15.** $1/x$ **17.** $1/\sqrt{1+x^2}$

Section 7.9
page 274

1. $2/\sqrt{1-(1+2x)^2}$ **3.** $-1/|x-1|\sqrt{(x-1)^2-1}$ **5.** $-a/(x^2+a^2)$ **7.** $1/\sqrt{a^2-x^2}$ **9.** $e^{\arcsin x}/\sqrt{1-x^2}$ **11.** $-(5-2x)/|5x-x^2|\sqrt{(5x-2x^2)^2-1}$ **13.** $\pi/4$ **15.** $\arcsin(x/5)+c$ **17.** $\frac{1}{2}\arcsin(x/2)+c$ **19.** $\arctan(t-3)+c$ **21.** $\arcsin\left(\frac{x+3}{\sqrt{19}}\right)+c$ **23.** $x-\arctan x+c$ **25.** $\pi/8\sqrt{6}$ **27.** $\pi/2$

Section 7.10
page 280

1. 1/2 **3.** 1 **5.** $\ln 2$ **7.** ∞ **9.** $-\infty$ **11.** 2 **13.** $-\infty$ **15.** $\pi/2$ **17.** For $x\geq 2$, $e^{-x}/(x^2-1)<e^{-x}$; converges **19.** For $x\geq 2$, $(x^3-x^2+x+1)^{-1}<2/x^3$; converges **21.** Compare with $x^{-1/2}$; diverges **23.** Compare with e^{-x}; converges **25.** Compare with $1/x^2$; converges **27.** π

Chapter 8

Section 8.1
page 284

1. $\frac{1}{8}(\ln|x+4|-\ln|x-4|)+c$ **3.** $-\frac{1}{8}(\ln|x+7|-\ln|x-1|)+c$ **5.** $\frac{1}{2}(x+1)\sqrt{x(x+2)}-\frac{1}{2}\ln|x+1+\sqrt{x(x+2)}|+c$ **7.** $\frac{1}{6}(\ln|x|-\ln|x+6|)+c$ **9.** $\frac{1}{2}\left(x\sqrt{9-x^2}+9\arcsin(x/3)\right)+c$ **11.** $\frac{1}{2}\left(x\sqrt{9+x^2}+9\ln|x+\sqrt{x^2+9}|\right)+c$ **13.** $x\sqrt{3-x^2}+3\arcsin(x/\sqrt{3})+c$ **15.** $16\sqrt{2}+16\ln(1+\sqrt{2})\approx 36.72939$ **17.** $\sqrt{2x-x^2}+\ln|x-1|-\ln|1+\sqrt{2x-x^2}|+c$ **19.** $\frac{1}{2}\sqrt{x^2-8x}-2\,\mathrm{arcsec}\left(\frac{x-4}{4}\right)+c$ **21.** $[u=1/x]$; $-\ln|u+\sqrt{u^2-1}|+c$ **23.** $\pi/6-\sqrt{3}/4\approx 0.09059$ **25.** $\sqrt{2}a=\sin\left(\frac{\pi}{6}-a\sqrt{2-4a^2}\right)$

Section 8.2
page 288

1. $(\sin 2x)/4-(x\cos 2x)/2+c$ **3.** $\frac{1}{2}x^2\ln x-\frac{1}{4}x^2+c$ **5.** $x(\ln x)^2-2x\ln x+2x+c$ **7.** $\frac{1}{15}(6x^2-16x+64)\sqrt{x+2}+c$ **9.** $-\frac{1}{w}(1+\ln w)+c$ **11.** $\frac{1}{2}(x\sin(\ln x)-x\cos(\ln x)+c$ **13.** $(1+x)^{100}\left(101x^2+100x-1\right)/(101\cdot 102)+c$ **15.** $(3t-4)^{12}\left[\frac{t^3}{15}-\frac{16t^2}{15\cdot 14}+\frac{128t}{3\cdot 13\cdot 14\cdot 15}+\frac{128}{27\cdot 13\cdot 14\cdot 15}\right]+c$ **17.** $\frac{1}{9}(x+1)^3(-1+3\ln|x+1|)+c$ **19.** $\frac{1}{2}(x+\sin x\cos x)+c$ **21.** $(1+x^4)^{-3/2}/6-(1+x^4)^{-1/2}/2+c$ **23.** $(1+e^{\pi})/2$ **29.** $e^{ax}(a\sin bx-b\cos bx)/(a^2+b^2)+c$ **31.** $(m\cos mx\cos nx+n\sin mx\sin nx)/(n^2-m^2)$

Section 8.3
page 293

1. $-\frac{1}{8}(3+6x+6x^2+4x^3)e^{-2x}+c$ **3.** $(x^4-4x^3+10x^2-20x+21)e^x+c$ **5.** $\frac{1}{32}(8x^2+12x+5)e^{4x}+c$ **7.** $\frac{1+x}{8(5+2x+x^2)}-\frac{1}{16}\arctan(2/(x+1))+c$ **9.** $-\cos x+\frac{2}{3}\cos^3 x-\frac{1}{5}\cos^5 x+c$

11. $-\frac{1}{6}\cos^6 x+\frac{1}{8}\cos^8 x+c$ **13.** $-\frac{1}{9}\sin^3 3x+\frac{1}{15}\sin^5 3x+c$ **15.** $\ln|\cos x|+\frac{1}{2}\sec^2 x+c$ **17.** $-\frac{1}{2}x^{m-1}e^{-x^2}+\frac{1}{2}(m-1)\int x^{m-2}e^{-x^2}\,dx$, $-(1+x^2+x^4/2)e^{-x^2}+c$ **19.** Let $I_{2n+1}=\int_0^{\pi/2}\sin^{2n+1}x\,dx$; then $I_{2n+1}=\frac{2n}{2n+1}I_{2n-1}$ and $I_{2n+1}=\frac{2\cdot 4\cdots(2n-2)(2n)}{1\cdot 3\cdots(2n-1)(2n+1)}$

Section 8.4 page 305

1. $1/4(x-4)-1/4x$ **3.** $1/(x+1)-1/x+1/(x-1)$ **5.** $\frac{15}{13(x-3)}+\frac{20-2x}{13(x^2+4)}$ **7.** $\frac{7}{25(x-2)}-\frac{2}{5(x-2)^2}+\frac{7x-1}{25(x^2+1)}$ **9.** $\frac{11+4x}{17(x^2-2x+5)}-\frac{2(1+2x)}{17(x^2+4)}$ **13.** $\frac{1}{(x-2)^2}-\frac{2}{x-2}+\frac{1}{(x-1)^2}+\frac{2}{(x-1)}$ **15.** $\frac{1}{10(x-1)^2}-\frac{19}{200(x-1)}+\frac{1}{20(x+1)^2}+\frac{17}{200(x+1)}+\frac{x+3}{100(x^2+9)}$ **17.** $p(x)=p(1)+p'(1)(x-1)+p''(1)(x-1)^2/2+p'''(1)(x-1)^3/6$ so $p(x)=2x^3-6x^2+4x+1$ **19.** $p(x)=\sum_{k=0}^{n}\frac{s_k}{k!}(x-a)^k$

Section 8.5 page 309

1. $\frac{1}{4}(\ln|x-4|-\ln|x|)+c$ **3.** $\ln|x^2-1|-\ln|x|+c$ **5.** $\frac{1}{13}\left(10\arctan(x/2)+15\ln|x-3|-\ln|x^2+4|\right)+c$ **7.** $\frac{1}{50}\left[\frac{20}{x-2}-2\arctan x-14\ln|x-2|+7\ln(x^2+1)\right]+c$ **9.** $\frac{1}{34}\left[15\arctan(\frac{x-1}{2})-2\arctan(x/2)-4\ln(x^2+4)+2\ln|x^2-2x+5|\right]+c$ **11.** $\frac{1}{2}\ln\left(\frac{x-1}{x+1}\right)+c$ **13.** $-\ln|x-1|+3\ln|x|-2\ln|x+1|+c$ **15.** $-\frac{1}{4}\left[-2\arctan x+2\ln|x-1|-\ln(x^4+1)\right]+c$ **17.** $\frac{1}{4}\left[\arctan x-2\ln|x-1|+\ln(x^2+1)-\frac{1}{x-1}-\frac{1}{x^2+1}\right]+c$ **19.** $\frac{x-1}{3(x^2+x+1)}-\arctan x+\frac{8}{3\sqrt{3}}\arctan\left(\frac{2x+1}{\sqrt{3}}\right)+c$ **21.** $\left[\frac{-1}{2(x+1)}-\frac{1}{4}\ln|x+1|+\frac{1}{4}\ln|x+3|\right|_0^1\approx 0.1486$ **23.** $\pi/2\sqrt{3}+(\ln 3)/6$ **25.** After long division, the rational function $r(x)=p(x)/(x-a)^m$ is the sum of a polynomial and a rational function in which the degree of the numerator is at most $m-1$. Assume that the degree of $p(x)$ is at most $m-1$ and that $p(x)$ does not have $(x-a)$ as a factor. Then the integral of $r(x)$ is a rational function if and only if $m\geq 2$ and the degree of $p(x)$ is at most $m-2$

27.

n	4	8	12	16
$\int$	−0.25352	−0.26217	−0.2650	−0.26641

29.

n	4	8	12	16
$\int$	0.01944	0.01769	0.0171	0.01681

Section 8.6 page 317

1. $\frac{1}{2}\left[x^2-x\sqrt{x^1-1}+\ln|x+\sqrt{x^2-1}|+c\right.$ **3.** $\frac{x}{2}\sqrt{x^2+4}-2\ln|x+\sqrt{x^2+4}|+c$ **5.** $\int\frac{-4(1-t^2)^n t\,dt}{(1+t^2)^{n+1}(1+2t-t^2)}$ **7.** $-\frac{1}{2}\ln|3-2\sin\theta|+c$ **9.** $\frac{\theta}{5}-\frac{2}{5}\ln|\cos\theta-2\sin\theta|+c$ **11.** $-\frac{\theta}{2}+\frac{2}{\sqrt{3}}\arctan\left(\sqrt{3}\tan(\theta/2)\right)+c$ **13.** For $m=2k$, k an integer, $(\sqrt{p(x)})^m=p(x)^k$; for $m=2k+1$, $(\sqrt{p(x)})^m=p(x)^k\sqrt{p(x)}$ **15.** $\frac{2t}{t^2-1}+\ln|(t-1)/(t+1)|+c$, $t=\left(\frac{x-1}{x+1}\right)^{1/2}$ **17.**

n	4	8	16	32	64	128
$\int$	2.14185	2.64161	2.8916	3.0166	3.0791	3.11034

Section 8.7 page 321

1. $\frac{x}{2}\sqrt{4-x^2}+2\arcsin(x/2)+c$ **3.** $-\frac{x}{2}\sqrt{25-x^2}+\frac{25}{2}\arcsin(x/2)+c$ **5.** $x/\sqrt{1-9x^2}+c$ **7.** $\frac{x}{2}\sqrt{a^2-x^2}+\frac{a^2}{2}\arcsin(x/a)+c$ **9.** $\arcsin(x/3)+c$ **11.** $\frac{1}{2}\left[\ln|\sec\theta+\tan\theta|+\sec\theta\tan\theta\right]+c$ **13.** $-\frac{1}{3}\sec^3\theta+\frac{1}{5}\sec^5\theta+c$ **15.** $\frac{1}{8}\left[\ln|\sec\theta-\tan\theta|-\sec\theta\tan\theta+\sec^3\theta\tan\theta\right]+c$

Chapter 9

Section 9.1 page 335

1. $1/2-x/4+x^2/8-x^3/16+x^4/32-x^5/64$ **3.** $1-2x^2+2x^4/3-4x^6/45$ **5.** Let $u=x-\pi/3$; $\sqrt{3}/2+u/2-\sqrt{3}u^2/4-u^3/12+\sqrt{3}u^4/48+u^5/240-\sqrt{3}u^6/1440+u^7/10080$ **7.** Let $u=x-1$; $16u^3+32u^4+24u^5+8u^6+u^7$ **9.** $2+x/4-x^2/64+x^3/512-5x^4/16384$ **11.** $x-x^2/2+x^3/6-x^4/12$ **13.** $x+x^3/6$ **15.** $-x-x^3-x^5-x^7-x^9$ **17.** Use $e^2\approx 9$ to bound the nth derivative at 2; $|e^2-P_n(2)|<9\cdot 2^{n+1}/(n+1)!$; this is $<10^{-2}$ with

$n = 10$ **19.** $|\sin(22/7) - P_n(22/7)| \le (22/7)^{n+1}/(n+1)! = E(n)$; $E(12) \approx 0.00047$, $P_{12}(22/7) = P_{13}(22/7) = -0.00124$ **21.** $\ln(0.95) = \ln(1 - 0.05)$; $|\ln(1 - 0.05) - P_2(0.05)| < (0.95)^{-3}(0.05)^3/3! \approx 0.000024$ **23.** $P_2(x) = 2 + x/32 - 3x^2/4096$; $|17^{1/4} - P_2(1)| < 2.56 \times 10^{-5}$ **25.** $x^4/384 < e^{-x} - (1 - x + x^2/2 - x^3/6) < x^4/24$ on $[0, 4]$ **27.** $0 \le \cos 2x - (1 - 4x^3/3)| \le x^5/5!$ on $[0, \pi/4]$ **29.** The coefficient of x^k in the Taylor polynomial for $g(x) = f(-x)$ is $g^{(k)}(0)/k! = (-1)^k f^{(k)}(0)/k!$ which is also the coefficient of x^k in $P_n(-x)$ **31.** See remark in answer to Ex. 29

Section 9.2
page 339

1. monotone increasing, unbounded **3.** not monotone, $|a_n| \le 1$ **5.** monotone increasing, $|b_k| < 1$ **7.** monotone increasing, unbounded **9.** not monotone, bounded; $c_{j+1} < c_j$ if $j > 10^3$ **11.** monotone decreasing and $0 < d_i \le d_1$ **13.** monotone increasing, $0 \le f_n < 1$ **15.** monotone increasing, $0 \le a_n \le 1$ **17.** $|c_n| < B \Rightarrow |a_n| = |(-1)^n c_n| < B$ **19.** $\lim_{x \to \infty} x^{1/x} = 1 \Rightarrow a_n = n^{1/n}$ is bounded; $a_n^n = n$ is not bounded **21.** $g_{n+1}/g_n < 1$ because $a_1 > \cdots > a_n \Rightarrow 1/(a_1 \cdots a_n) < (1/a_{n+1})^n$

Section 9.3
page 344

1. $-2/3$ **3.** 1 **5.** ∞ **7.** 0 **9.** 1 **11.** 0 **13.** e **15.** 0 **17.** ∞ **19.** Imitate the proof in example 4; first show $a_1 < a_2$ and then $a_k < u \Rightarrow a_{k+1}^2 = 1 + a_k < 1 + u = u^2$, etc. **21.** Let u be the positive solution of $u^2 = 3 + u$; compute$a_1 = 2$, $a_2 = \sqrt{5} \approx 2.23606$, $a_3 \approx \sqrt{5.223606} \approx 2.2882$, $a_4 \approx \sqrt{5.2882} \approx 2.29961$ so $a_1 < a_2 < a_3 < a_4 < u$; imitate the method of Example 4. **23.** If $\lim a_n = L$ then, for any positive ϵ there is an N such that $|a_k - L| < \epsilon$ for all $k > N$. Use $a_n \ge f(x) \ge a_{n+1}$ for x on $[n, n+1]$ to conclude $|f(x) - L| < \epsilon$ for $x > N$ **23.** Let N_n be the number of sides of T_n; $N_1 = 3$ and at every step, one side gets replaced by 4 sides; $N_n = 3 \cdot 4^{n-1}$. Each side of T_n has length $1/3^{n-1}$ so the perimeter of T_n is $3(4/3)^{n-1}$ which approaches ∞ as n increases. The area of T_{n+1} is the area of T_n plus N_n times the area of an equilateral triangle with side $1/3^{n-1}$; the area approaches a finite limit $53\sqrt{3}/20$

Chapter 10

Section 10.1
page 354

1. $f^{(n)} = (-1)^n n!/(1 + x)^{n+1}$; converges on $(-1/2, 1/2)$ but even on $(-1, 1)$ **3.** $f^{(2k)} = (-1)^k 2^{2k} \sin 2x$, $f^{(2k+1)} = (-1)^k 2^{2k+1} \cos 2x$; converges for every x **5.** $f^{(n)} = e^{1+x}$; converges for every x **7.** $f^{(n)} = 4 \cdot 5^n e^{5x-3}$; converges for every x **9.** $f^{(n)} = (-1)^n/(2 + x)^{n+1}$; $\sum_{k=0}^{\infty}(-1)^k x^k/2^{k+1} k!$ **11.** $f^{(n)} = 3^n n!/(1 - 3x)^{n+1}$; $\sum_{k=0}^{\infty}(3x)^k$ **13.** $f = 1/(x + 1) - 1/(x + 2)$; $\sum_{k=0}^{\infty}(-1)^k(1 - 2^{-k-1})x^k$ **15.** $f = -\frac{3}{2}\frac{1}{x+1} + \frac{5}{2}\frac{1}{x+3}$; $\sum a_k x^k$ with $a_k = (-1)^k(-3/2 + 5/6 \cdot 3^k)$ **17.** For $0 < c < 1$, $\ln c < 0$ and $\lim_{n \to \infty} \ln c^n = \lim_{n \to \infty} n \ln c = -\infty$; follows $\lim c^n = 0$ **21.** $|\ln 2 - P_n(1)| < 1/(n + 1)$; take $n + 1 > 10{,}000$ **23.** 7/9 **25.** 1 **27.** 176/225 **29.** 10471/3333 **31.** 0.16666... **33.** 2.571428571428... **35.** 30 ft; $h + 2h\rho + 2h\rho^2 + \cdots = h(1 + \rho)/(1 - \rho)$

Section 10.2
page 361

1. $r = 1$; $|r^k/k| < 1$; converges absolutely on $(-1, 1)$ **3.** $r = 1$; $|r^{2k}/(4 + k)| < 1$; converges absolutely on $(-1, 1)$ **5.** For a given x, take $r > x$, $N > 3r$ and $B = max\{(3r)^k/k! : 0 \le k \le N\}$; then $|(3r)^k/k!| \le B$; converges absolutely for every x **7.** $r = \sqrt{5}$; $|r^{2k}/5^k(k^2 + 1)| < 1$; converges absolutely on $(-\sqrt{5}, \sqrt{5})$ **9.** $r = 2$; $|r^k/(2^k + k)| < 1$; converges absolutely on $(-2, 2)$ **11.** $r = 200$; $|r^k/200^k k(k + 1)(k + 2)| < 1$; converges absolutely on $(-200, 200)$ **13.** By nth term test, $\lim a_n = 0$ so the sequence a_n is bounded; with $r = 1$ we have

$|a_n| = |a_n r^n| < B$ so series converges absolutely on $(-1,1)$ **15.** sum $= 2$; $a_n = s_n - s_{n-1} = 1/n(n+1)$ **17.** Convergence $\Rightarrow \lim a_k = 0$ so $|a_n| < 1$ for all except possibly a finite number of terms which we may ignore. Then $0 \le a_n^2 \le a_n$ and $\sum a_n^2 \le \sum a_n$, both converge

Section 10.3
page 369

"rc" = "radius of convergence" **1.** Ratio text implies rc = 1 **3.** Ratio test implies rc = 1 **5.** Ratio test implies rc $= \infty$ **7.** Ratio test implies rc $= 1/4$ **9.** $(-2,2)$; converges if $-1 < cx < 1$ or x on $(-1/c, 1/c)$ **11.** rc = 1 by the ratio test since $\lim_{x\to\infty} f(x+1)/f(x) = 1$ for any polynomial $f(x)$ **13.** $\lim_{n\to\infty} a_n = A$ implies a_n is bounded (all but a finite number of the a_n are within 1 of A by definition of limit) so the Bounded Terms Test (with $r = 1$) implies $\sum a_n x^n$ converges on $(-1,1)$ **15.** Let $p(t) = At^m + a_1 t^{m-1} + \cdots + a_m$ with $A \neq 0$ and $q(t) = t^m + b_1 t^{m-1} + \cdots + b_m$; then $\sum p(k)x^k/q(k)$ converges absolutely on $(-1,1)$ but does not converge at ± 1

Section 10.4
page 377

1. $f = \sum_0^\infty (-1)^k x^k/k!$, $f' = \sum_0^\infty (-1)^k x^{k-1}/(k-1)! (= -f)$; $\int_0^x f = \sum_1^\infty (-1)^k x^{k+1}/(k+1)!$ **3.** $f = \sum_1^\infty (-1)^{k+1} x^k/k5^k$, $f' = \sum_1^\infty (-1)^{k+1} x^{k-1}/5^k$; $\int_0^x f = \sum_1^\infty (-1)^{k+1} x^{k+1}/k(k+1)5^k$

Let $G(x) = \sum_{i=0}^\infty x^i = 1/(1-x)$ and then $G'(x) = \sum_1^\infty kx^{k-1} = -1/(1-x)^2$.

5. $x(G'-1) = x^2(2-x)/(1-x)^2$ **7.** $x^2G' = x^2/(1-x)^2$ **9.** $\sum_1^\infty k(x^k - x^{k-1}) = (x-1)\sum_1^\infty kx^{k-1} = (x-1)(G'-1) = x(x-2)/(1-x)$ **11.** $F = \int_0^x \sum_1^\infty (-1)^{k+1} t^{2k-2}/(2k)!\, dt = \sum_1^\infty (-1)^{k+1} x^{2k-1}/(2k)!(2k-1)$; converges everywhere **13.** $F = \int_0^x \sum_1^\infty (-1)^{k+1} t^{k-1}/k\, dt = \sum_1^\infty (-1)^{k+1} t^k/k^2$; rc = 1 **15.** $P_2(x) = 1 - \frac{x^2}{4}$; $P_4(x) = 1 - \frac{x^2}{4} + \frac{x^4}{16}$; $P_6(x) = 1 - \frac{x^2}{4} + \frac{x^4}{16} - \frac{x^6}{36\cdot 64}$ **17.** If $f(x) = 0$ for all x on $(-r,r)$, then $f^{(n)}(x) = 0$ for every positive integer n and all x on the interval. From the power series, we see that $f^{(n)}(0) = n!a_n$ and so $a_n = 0$ for all n. **19.** If $f(x) = \sum_{k=0}^\infty a_k x^k$ then $f(x) - f(-x) = \sum_{k=0}^\infty [a_k x^k - a_k(-x)^k] = \sum_{m=1}^\infty 2a_{2m-1}x^{2m-1}$. For an even function, this difference is 0 and every $a_{2m-1} = 0$. **21.** $(\cos x - 1)/x^2 = -\frac{1}{2!} + \frac{x^4}{4!} - \cdots$; $c = -1/2$ **23.** $c = 1$

Section 10.5
page 386

1. $-(1 + x^2 + x^4 + \cdots + x^{2n} + \cdots$, for $|x| < 1$ **3.** $\sum_{k=0}^\infty a_k x^k$, $a_k = \frac{2}{3} - (-1)^k \frac{1}{6\cdot 2^k}$ **5.** $\sum_{k=0}^\infty a_k (x-3)^k$, $a_k = -\frac{1}{2}(-1)^k[2^{-k} - 2\cdot 3^{-k} + 4^{-k}]$ **7.** $\frac{1}{16}\sum_{k=0}^\infty \left(2k^2 + 8k + 7 + (-1)^k\right)x^{2k}$ **9.** $\sum_{k=0}^\infty (-1)^k (t-\pi)^{2k+1}/(2k+1)!$ **11.** $\ln(x+4) = \ln 5 - \sum_{k=1}^\infty (-1)^k x^k/k5^k$ **13.** $e^x \cos x = 1 + x - x^3/3 - x^4/6 - x^5/30 + \cdots$; $e^x \cos x = 1 - x + x^3/3 - x^4/6 + x^5/30 + \cdots$ **15.** The A row is for the n terms of the $\ln(1+x)$ series at $x = -0.7$; the B row is for n terms of the $\ln[(1+x)/(1-x)]$ series at $x = -7/13$:

n	4	6	8	10	12
A	−1.11936	−1.17258	−1.19155	−1.19886	−1.20181
B	−1.20286	−1.20391	−1.20397	−1.20397	−1.20397

17. $(\cos x)^p = 1 - px^2/2 + (3p^2 - 2p)x^4/24 - (15p^3 - 30p^2 + 16p)x^6/720 + \cdots$ **19.** The series will converge for $|x| < 1$ if $1 \le R$ and for $|x| < R$ when $0 < R < 1$ **21.** $x + x^2 + 3 + \frac{5}{6}(x^3 + x^4) + \frac{101}{120}(x^5 + x^6) + \cdots$ **23.** $1 - x + \frac{1}{2}(x^2 - x^3) + \frac{13}{24}(x^4 - x^5) + \frac{389}{720}(x^6 - x^7) + \cdots$ **25.** $[a_0 + (a_0 + a_2)x^2 + (a_0 + a_2 + a_4)x^4 + \cdots] + [a_1 x + (a_1 + a_3)x^3 + (a_1 + a_3 + a_5)x^5 + \cdots]$ **27.** $(e^x + e^{-x})/2$ **29.** $(e^x + e^{-x})/2x$ **31.** $1 - x/2 + x^2/12 - x^4/720 + x^6/30240 - x^8/1209600 + \cdots$

33. $(1-x)f(x) = a_0 + \sum_{k=0}^{\infty}(a_{k+1} - a_k)x^k$; at $x = 1$ the nth partial sum is $s_n = a_{n+1}$ which has limit A

Section 10.6
page 393

1. $(e^x + e^{-x} - 2\cos x)/x^2 = 2+$ terms with higher powers of x; lim $= 2$ **3.** $\left[(1 - t + t^2/2 - \cdots) - (1 - t/2 + t^2/4 + \cdots)\right]/(t - t^2/2 + \cdots) \to -1/2$ as $t \to 0$ **5.** 2 **7.** $\frac{1}{2}f''(0)$ **9.** $L_1(x) = a_0(1-x)$; $L_2(x) = a_0(1 - 2x + x^2/2)$; $L_3(x) = a_0(1 - 3x + 3x^2/2 - x^3/6)$; $L_4(x) = a_0(1 - 4x + 3x^2 - 2x^3/3 + x^4/24)$ **11.** $y = a_0(1 + x + x^2 + x^3 + \cdots) = a_0/(1-x)$ **13.** Substitution of the series into the differential equation produces $a_1 = 0$ and $a_{k+1} = -a_{k-1}/(k+1)(k+2)$ which leads to $a_{2k} = (-1)^k a_0/(2k+1)!$ and $a_{2k+1} = 0$

Section 10.7
page 396

1. Diverges by the integral test **3.** Converges by the ratio test **5.** Diverges by the integral test **7.** Diverges by the integral test **9.** $a_{k+1}/a_k = c^{2k+1}$ which has limit 0, if $|c| < 1$, 1 if $c = 1$,$\pm\infty$ if $|c| > 1$ **11.** $\int_n^{\infty} x^{-4}\,dx = 1/3n^3 < 9 \times 10^{-6}$ implies $n \geq 34$.

Section 10.8
page 402

1. Conditionally conv.; sum of the first 10 terms **3.** Divergent by nth term test **5.** Convergent by the alternating series test **7.** $|x| < 5$, divergent at $x = \pm 5$ by nth term test **9.** $|x| < 5^{1/3}$, divergent at both endpoints by nth term test **11.** Convergent if $|x| < 1$, divergent at $x = \pm 1$ by nth therm test **13.** Convergent if $0 < x \leq 2$; divergent at $x = 0$ by the integral test; convergent at $x = 2$ by the alternating series test **15.** convergent for all x by the ratio test

Section 10.9
page 407

1. $P_3 = x + x^3/6$; $P_5 = x + x^3/6 + x^5/5!$; $P_7 = x + x^3/6 + x^5/5! + x^7/7!$ **3.** $|\sinh x - P_3(x)| < 3 \times 10^{-4}$ if $|x| < 0.3$ **5.** The derivation by power series depends on the identity $\sum_{m=0}^{h} \frac{1}{(2m)!}\frac{1}{(2h+1-2m)!} = \frac{2^{2h}}{(2h+1)!}$ **9.** $x = a\sinh t$, $dx = a\cosh t$, $\sqrt{a^2 + x^2} = a\cosh t$ and the integral equals $x\sqrt{a^2+x^2} + \frac{a^2}{2}\ln(x + \sqrt{a^2 + x^2}) + c$ **11.** $(0.9)^{1/6} = (1 - 0.1)^{1/6} = 1 - \frac{1}{6}(0.1) + \frac{1}{2}\frac{1}{6}\frac{-5}{6}(0.1)^2 - \frac{1}{3!}\frac{1}{6}\frac{-5}{6}\frac{-11}{6}(0.1)^3 + \cdots \approx 0.982635\ldots$ **13.** $(17)^{1/4} = (16+1)^{1/4} = 2\left(1 + \frac{1}{16}\right)^{1/4} \approx 2\left[1 + \frac{1}{4}\frac{1}{16} + \frac{1}{2!}\frac{1}{4}\frac{-3}{4}\frac{1}{16^2} + \frac{1}{3!}\frac{1}{4}\frac{-3}{4}\frac{-7}{4}\frac{1}{16^3} + \frac{1}{4!}\frac{1}{4}\frac{-3}{4}\frac{-7}{4}\frac{-11}{4}\frac{1}{16^4}\right] \approx 2.0305431$

Chapter 11

Section 11.1
page 420

1. $[45/32, 46/32]$; $f(91/64) = 0.021728$ **3.** $[41/32, 42/32]$; $f(83/64) = -0.02124$ **5.** $[40/32.41/32]$; $f(83/32) = 0.02728$ **7.** $[43/32, 44/32]$; $f(87/64) = 0.02346$ **9.** $x_0 = 2$, $x_1 = 1.5$, $x_2 = 1.4167$, $x_3 = 1.41422$, $x_4 = 1.41421$ **11.** $x_0 = 1$, $x_1 = 1.3333$, $x_2 = 1.30303$, $x_3 = 1.30278$, $x_4 = 1.30278$ **13.** $x_0 = 1$, $x_1 = 1.3333$, $x_2 = 1.26389$, $x_3 = 1.25993$, $x_4 = 1.25992$ **15.** $x_0 = 1$, $x_1 = 1.5$, $x_2 = 1.36842$, $x_3 = 1.35339$, $x_4 = 1.35321$ **17.** $x_0 = 5$, $x_1 = -1$, $x_2 = 5$, $x_3 = -1$; $x_0 = 4$, $x_1 = 1.51515$, $x_2 = 2.00515$, $x_3 = 2.0000$

21.

	n	1	2	3	4	5	6	7	8
$c = 1.9$	c_n	1.71	1.21	0.25	−0.19	0.22	−0.18	0.21	−0.16
$c = 1.99$	c_n	1.97	1.91	1.74	1.29	0.38	−0.23	0.28	−0.21
$c = 2$	c_n	2	2	2	2	2	2	2	2
$c = 2.01$	c_n	2.03	2.09	2.28	2.93	5.63	26	654	427,720
$c = 2.1$	c_n	2.31	3.026	6.131	31.46	958.3	917,350	8×10^{11}	7×10^{23}

23.

	n	1	2	3	4	5	6	7	8
$c = 1$	c_n	0	0	0	0	0	0	0	0
$c = 0$	c_n	0	0	0	0	0	0	0	0
$c = -1$	c_n	-2	-12	-1872	-6×10^9	-10^{29}	-10^{88}	-10^{265}	-10^{795}

25. $\lim_{n\to\infty} c_n = 0$ because $c_{n+1} = c_n/2$

Section 11.2
page 432

1.

n	2	4	6	8
MP_n	0.7906	0.7867	0.78598	0.78572
T_n	0.775	0.78279	0.78424	078474

3.

n	2	4	6	8
MP_n	0.87788	0.86961	0.86814	0.86763
T_n	0.84559	0.86173	0.86465	0.86567

5. Use $P_4(x) = 1 - x^2/2 + x^4/24$ to get an approximation to the integral = 0.47942 with an error $\leq 10^{-4}$ **7.** $\int_0^x \operatorname{arctan} t/t\, dt = x - x^3/9 + x^5/25 - \cdots$ is an alternating series for which a partial sum serves to approximate with an error less than the first ommitted term; use $x - x^3/9\Big|_0^{0.5} \approx 0.4872$ with an error $\leq 10^{-4}$

Section 11.3
page 438

1.

n	2	4	6	8	10
S_n	6.41667	6.40104	6.40021	6.40007	6.40003

3.

n	2	4	6	8	10
S_n	0.16797	0.16675	0.16668	0.16667	0.16667

5.

n	2	4	6	8
S_n	0.785392	0.785398	0.785398	0.785398

7.

n	2	4	6	8
S_n	0.5493626	0.5493100	0.5493069	0.5493064

9. The square root function is evaluated $2n + 1$ times and is multiplied by 4, n times; by 2, $n - 1$ times; by 1, 2 times. If E is the maximum error in each evaluation, the accumulated error in forming the sum of the terms is (at most) $(4n + 2(n - 1) + 2)E = 6nE$. This sum is multiplied by $(b - a)/6n$ producing an error of at most $(b - a)E$ (which does not get smaller with increasing n). The Simpson's rule error decreases with increasing n.

Chapter 12

Section 12.1
page 444

1. $\sqrt{2}$ **3.** $\sqrt{14}$ **5.** $z = 3$ **7.** $y = -2$ **9.** inside **11.** inside **13.** $(x - 1)^2 + (y - 2)^2 + (z - 3)^2 = 9$ **15.** $(x - a)^2 + (y - 4)^2 + (z - b)^2 = a^2 + b^2 + 16$ **17.** $(x - 2)^2 + (y - b)^2 + (z - r)^2 = r^2$, b, r constant. **19.** Yes (distance between centers is less than the sum of the two radii) **21.** Yes **23.** Yes, if $|c| \leq 1/\sqrt{2}$

Section 12.2 page 440

1. $[2, 1]$, $\sqrt{5}$ **3.** $[3, -1]$, $\sqrt{10}$ **9.** $[-3, 1]$, $\sqrt{10}$ **11.** $[13, 0]$ 13 **13.** $2\sqrt{2}(\cos(\pi/4)\mathbf{i} + \sin(\pi/4)\mathbf{j})$ **15.** $2(\cos(5\pi/6)\mathbf{i} + \sin(5\pi/6)\mathbf{j})$ **17.** $5(\cos 0.6435\,\mathbf{i} + \sin 0.6435\,\mathbf{j})$ **19.** $10(\cos 0.6435\,\mathbf{i}+\sin 0.6435\,\mathbf{j})$ **21.** $(r, \theta) = (2, \pi/4)$, $(2.\pi/3)$, $(5, 0.6435)$, $(5, -0.6435)$ **23.** All numbers from 0 to 2 (inc.)

Section 12.3 page 452

1. $[2, 1, -2]$, 3 **3.** $[3, -1, 0]$, $\sqrt{10}$ **5.** $-5\mathbf{i} + 2\mathbf{j} - \mathbf{k}$, $\sqrt{30}$ **7.** $-6\mathbf{i} + 21\mathbf{j} + 9\mathbf{k}$, $3\sqrt{62}$ **9.** $a = -4$ **11.** $[-2, 0, -1]$, $[4, 2, -5]$ **13.** $t = \pm 1/\sqrt{14}$ **15.** $t = \pm\sqrt{3/13}$ **17.** $(\rho, \theta, \phi) = (3\sqrt{2}, 0, \pi/4)$ **19.** $(\rho, \theta, \phi) = (\sqrt{2}, \pi/4, 0)$ **21.** $(\rho, \theta, \phi) = (2\sqrt{3}, \pi/4, \pi/4)$ **23.** (a) single point on the z-axis, 2 units above the xy-plane; (b) sphere of radius 1; (c) the entire xy-plane

Section 12.4 page 458

1. 2 **3.** 0 **5.** -2 **7.** $c = 0$ **9.** $-1/3$ **11.** $\mathbf{u} = \mathbf{i}$, $\mathbf{v} = \mathbf{j}$, $\mathbf{w} = \mathbf{k}$ **13.** $(\mathbf{i} + \mathbf{j})/\sqrt{2}$ **15.** $-(\mathbf{i} + \mathbf{j} + \mathbf{k})/3$, $-(\mathbf{i} - \mathbf{j} - \mathbf{k})/3$ **17.** $(\mathbf{i} + \mathbf{j})/2 + (-\mathbf{i} + \mathbf{j})/2$ **19.** $-4(\mathbf{i}+3\mathbf{j})/5+3(3\mathbf{i}-\mathbf{j})/5$ **21.** Either $\mathbf{u}$ is orthogonal to $\mathbf{v}$ or $\mathbf{u} = \mathbf{v}$. **25.** $\pm 2\pi/3$ **27.** $\mathbf{u} = a\mathbf{i}+b\mathbf{j}+c\mathbf{k} \Rightarrow \mathbf{u}\cdot\mathbf{i} = a$, $\mathbf{u}\cdot\mathbf{j} = b$, $\mathbf{u}\cdot\mathbf{k} = c$ and $\mathbf{u} = (\mathbf{i}\cdot\mathbf{i})\mathbf{i}+(\mathbf{i}\cdot\mathbf{j})\mathbf{j}+(\mathbf{i}\cdot\mathbf{k})\mathbf{k}$

Section 12.5 page 467

1. $\mathbf{j} + \mathbf{k}$ **3.** $-2\mathbf{i} + \mathbf{j} + \mathbf{k}$ **5.** 0 **7.** 2 **9.** 2 **11.** 0 **13.** $3\mathbf{i} + 6\mathbf{k}$ **15.** 7 **17.** 7/2 **19.** 2 **21.** 2 **23.** 2 **25.** $(1, 2, 3)$, $(2, 3, 8)$, $(3, 5, 5)$, $(5, 6, 7)$, $(4, 6, 10)$, $(6, 7, 12)$, $(7, 9, 9)$, $(8, 10, 14)$; vol = 16 **27.** Maximum volume = $2\sqrt{19}$ **29.** Either $\mathbf{v} = 0$ or $\mathbf{v}$ is parallel to $\mathbf{u}$. **31.** $\pm|\mathbf{v}|^2$ **33.** $|(a\mathbf{i} + b\mathbf{j}) \times (c\mathbf{i} + d\mathbf{j})| = |ad - bc|$ = the absolute value of the determinant

Section 12.6 page 473

1. $t(\mathbf{i} + 2\mathbf{j} + \mathbf{k})$, $\{x = t,\ y = 2t,\ z = t\}$, $2x = y = 2z$ **3.** $(-\mathbf{i} - \mathbf{j}) + t(\mathbf{i} + 2\mathbf{j} - \mathbf{k})$; $\{x = -1+t, y = -1+2t, z = -t\}$; $x+1 = z = -(y+1)/2$. **5.** $2\mathbf{i}-4\mathbf{j}-\mathbf{k})+t(2\mathbf{i}+4\mathbf{j}+6\mathbf{k})$; $\{x = 2 + 2t,\ y = -4 + 4t,\ z = -1 + 6t\}$; $(x - 2)/2 = (y + 4)/4 = (z + 1)/6$. **7.** $(0, -5/2, 17/6)$, $(5, 0, 11/3)$,$(-17, -11, 0)$ **9.** $(2, 0, 17/5)$,$(2, 17, 0)$, no point on the yz-plane and the line **11.** $(1, -6, 9)$ **13.** parallel **15.** $3\sqrt{61}/\sqrt{46}$ **17.** $7/\sqrt{2}$ **19.** $|\mathbf{u} - \mathbf{v} - \frac{(\mathbf{u}-\mathbf{v})\cdot\mathbf{w}}{\mathbf{w}\cdot\mathbf{w}}\mathbf{w}|$ **21.** Paths cross at $(-1, -3)$ but they reach that point at different times **23.** $\mathbf{r}(t) = t(-3\mathbf{i} + 2\mathbf{j} - \mathbf{k})$ **25.** Assume $ab \neq 0$; the line has vector form $\mathbf{r}(t) = [0, -c/b] + t[1/a, -1/b]$; apply Eq. (12.15)

Section 12.7 page 483

1. $2x + 2y - 3z = 0$ **3.** $2x - y - 3z = -1$ **5.** $x + y + z = 0$ **7.** $x + y + z = 1$ **9.** $2x - y + z = 3$ **11.** $s[1, 1, 0] + t[0, 1, 2]$,(many other correct solutions are possible **13.** $[3, 1, 0] + s[2, 4, 0] + t[0, 3, 2]$ **15.** $3/\sqrt{6}$ **17.** $1/\sqrt{29}$ **19.** $\mathbf{r}(t) = t[1, 7, 5]$ **21.** $\mathbf{r}(x) = [3/8, 7/4, -1/8] + x[1, 0, 2]$ **23.** $-\sqrt{3}/2\sqrt{7}$ **25.** $c = 1$; no **27.** $5/\sqrt{3}$ **29.** $(0, 0, m/c)$ is in the plane; $\mathbf{u} = [x_0, y_0, x_0 - m/c]$, $\mathbf{w} = [a, b, c]$ and $D = |(\mathbf{u}\cdot\mathbf{w}/\mathbf{w}\cdot\mathbf{w})\mathbf{w}|$

Chapter 13

Section 13.1 page 491

1. $12t^3\mathbf{i} - 2(t + 2)\mathbf{j}$; $36t^2\mathbf{i} - 2\mathbf{j}$ **3.** $(1 + \frac{1}{t})\mathbf{i} - \frac{1}{t}\mathbf{j} + \frac{1}{(t+1)^2}\mathbf{k}$; $-\frac{1}{t^2}\mathbf{i} + \frac{2}{t^3}\mathbf{j} - \frac{2}{(t+1)^3}\mathbf{k}$ **5.** $-\mathbf{u}\cdot\mathbf{s} + 3\mathbf{w}\cdot\mathbf{v} - 6t\mathbf{u}\cdot\mathbf{v}$ **7.** $\mathbf{v}\times\mathbf{s} + 3t^2\mathbf{v}\times\mathbf{w} + 2t\mathbf{u}\times\mathbf{w}$ **9.** $4t(t^2 - 1)\mathbf{i} - (2/t)\mathbf{k}$ **11.** $\frac{2}{4t+1}\mathbf{i} + 2\mathbf{j} + \frac{4+16t^2}{1-4t^2}\mathbf{k}$ **13.** The graph of $\mathbf{r}(t)$ is the parabola $16y = x^2$; $|r(t)| = t^2\sqrt{16 + t^4}$ **15.** $\mathbf{r}(t)$ covers the ellipse $x^2 + (y/4)^2 = 1$ two times **17.** Any positive $\delta \leq 1/10\sqrt{5}$ will do. **19.** Any positive $\delta \leq \epsilon/\sqrt{5}$ will do. **23.** $(\mathbf{u}\cdot\mathbf{v})' = (ax + by)' = a'x + ax' + b; y + by' = \mathbf{u}'\cdot\mathbf{v} + \mathbf{u}\cdot\mathbf{v}'$ **25.** Differentiate using the quotient rule and the formula $|\mathbf{r}|' = \mathbf{r}\cdot\mathbf{r}'/|\mathbf{r}|$

Section 13.2
page 495

1. $\frac{3}{5}t^5\,\mathbf{i} - \frac{1}{3}(t+2)^3\,\mathbf{j}$ **3.** $\frac{1}{2}\mathbf{i} + \frac{2}{3}\mathbf{j}$ **5.** 0 **7.** $3t - \frac{3}{2}t^2 + \frac{2}{3}t^3$ **7.** $\frac{13}{6}$ **9.** $\sin p + \sin(1-p)$ **11.** $\frac{1}{2}(2+t^2)\,\mathbf{i} - \frac{1}{3}t^3\,\mathbf{j} + 4t\,\mathbf{k}$ **13.** $(3 - 2e^{-t})\,\mathbf{i} + \frac{1}{2}(1 - e^{2t})\,\mathbf{k}$ **15.** Apply the Fundamental Theorem of Calculus to each coordinate. **17.** The integral is $\le \int_a^b |\cos g(t)\,\mathbf{i} + \sin g(t)|\,dt = \int_a^b 1\,dt = b - a$ **19.** $\lim_{n\to\infty}\sum_{i=1}^n \mathbf{r}(1 + \frac{2i}{n})\frac{2}{n}$; $\lim_{n\to\infty}\sum_{i=1}^n \mathbf{r}(1 + \frac{2(i-1)}{n})\frac{2}{n}$ **21.** $\int_1^3 \mathbf{r}(t)\,dt = -2\,\mathbf{i} + 8\,\mathbf{j} + (50/3)\,\mathbf{k}$

Section 13.3
page 504

1. $x - 2y = 5$ **3.** $x = 2y + 5$ **5.** $y = x^2 - 2x + 2$ **7.** $(x/2)^2 + (y/3)^2 = 1$ **9.** Straight line through $(1,0)$ and $(3,-1)$. **11.** Half-line from $(1,0)$ through $(3,-1)$. **13.** $x = 4y + 6$ **15.** A parabola "rotated" through $\pi/4$ (rectangular equation $x^2 + y^2 - 2x + 2y + 2xy = 0$ **17.** $C_1 = C_2$ **19.** C_2 has $x \ge 0$ so C_2 is part of C_1 **21.** $C_1 = C_2$ but every point of C_2 is covered twice in the given parametric form **23.** $x = 1 + 2t,\ y = 1 + 3t$ for $0 \le t \le 1$ **25.** $x = 4\cos\pi t,\ y = 4\sin\pi t$ for $0 \le t \le 1$ **27.** $x = 4t,\ y = 16t^2 - 8t$ **29.** $x = a(e^t + e^{-t})/2,\ y = b(e^t - e^{-t})/2$ **31.** $x = (2t + t^2)/\sqrt{5},\ y = (t - 2t^2)/\sqrt{5}$; $4x^2 - x\sqrt{5} + 4xy + 2\sqrt{5}y + y^2 = 0$ **33.** $\frac{3}{2}\cos t\,\mathbf{i} + \frac{\sqrt{3}}{2}\sin t\,\mathbf{j}$ **35.** $t^2\,\mathbf{i} + t\,\mathbf{j}$ **37.** $t\,\mathbf{i} + f(t)\,\mathbf{j}$ **39.** Assume the car moves to the right with constant velocity v so the center of the wheel is (vt, R) at time t and the wheel has turned through an angle $\theta = vt/R$; if ϕ is the measure of the angle in the conventional way (from the positive x-axis counter clockwise) then $\phi - \theta = -\pi/2$; $\mathbf{r}(t) = ((R+h)\sin\theta + Rvt)\,\mathbf{i} + (R - (R+h)\cos\theta)\,\mathbf{j}$; $\mathbf{v}(t) = v[(1 + \frac{R+h}{R}\cos\theta)\,\mathbf{i} + \frac{R+h}{R}\sin\theta\,\mathbf{j}]$; there are values of θ that make the x-coordinate of the velocity negative and so the spot does move opposite to the direction of the car at some times. **41.** $\mathbf{r}(t) = (R\omega t + vt\cos\omega t)\,\mathbf{i} - vt\sin\omega t\,\mathbf{j}$

Section 13.4
page 509

1. $\mathbf{T}(t) = \frac{3}{\sqrt{9+16t^2}}\,\mathbf{i} + \frac{4t}{\sqrt{9+16t^2}}\,\mathbf{j}$, $\mathbf{T}(1) = (3/5)\,\mathbf{i} + (4/5)\,\mathbf{j}$; $\mathbf{N}(t) = \{-4t\sqrt{(9+16t^2)^{-1}}\,\mathbf{i} + 3\sqrt{(9+16t^2)^{-1}}\,\mathbf{j}\}$, $\mathbf{N}(1) = [-4/5, 3/5]$ **3.** $\mathbf{T}(t) = (e^t\,\mathbf{i} - e^{-t}\,\mathbf{j})/\sqrt{e^{2t} + e^{-2t}}$; $\mathbf{T}(1) = (e\,\mathbf{i} - e^{-1}\mathbf{j})/\sqrt{e^2 + e^{-2}}$; $\mathbf{N}(t) = (e^t\,\mathbf{i} + e^{-t}\,\mathbf{j})/\sqrt{e^{2t} + e^{-2t}}$; $\mathbf{N}(1) = (e\,\mathbf{i} + e^{-1}\mathbf{j})/\sqrt{e^2 + e^{-2}}$. **5.** $\mathbf{T}(1) = -\mathbf{j}$, $\mathbf{N}(1) = -\mathbf{i}$ **7.** $\mathbf{r}(10) + (t - 10)\mathbf{v}(10) = ((265 - t)\,\mathbf{i} + (802t - 40)\,\mathbf{j} + 20t\,\mathbf{k})/5$ **9.** $\mathbf{r}\cdot\mathbf{r}' = 0$ at $t = 0$; $\mathbf{r}$ and $\mathbf{r}'$ are parallel at $t = \pm 2$ **11.** $y = x^{3/2}$ which has a derivative at $x = 0$ **13.** $(x(t), y(t)) = (x, f(x))$ implies $x = x(t)$ and $y(t) = f(x(t))$; $y' = f'(x)x'$ by the chain rule **15.** $\mathbf{r}' = |\mathbf{r}'|\mathbf{T} = |\mathbf{r}'|\mathbf{u}$, $\mathbf{u}$ and $\mathbf{v}$ constant; integrate with respect to t and find the curve is part of the line $\mathbf{w}(s) = s\mathbf{u} + \mathbf{v}$ **17.** $d\mathbf{T}/dt = (|\mathbf{r}'|^2\mathbf{r}'' - \mathbf{r}'(\mathbf{r}''\cdot\mathbf{r}'))/|\mathbf{r}'|^3$ which equals 0 wwhen $\mathbf{r}'' = 0$

Section 13.5
page 514

1. $4\sqrt{14}$ **3.** $61/27$ **5.** $\pi\sqrt{13}$ **7.** $\frac{3}{2} + \ln 2$ **9.** $2\pi\sqrt{10},\ 6\pi$ **11.** $\ell_1/\ell_2 = \sqrt{64 + a^2}/\sqrt{64 + b^2}$ **13.** The derivative of $\mathbf{r}(g(s))$ is $\mathbf{r}'(g(s))g'(s)$; substitute $t = g(s)$ in the integral **15.** The arc lengths satisfy $\int_a^b \sqrt{(x')^2 + (y')^2}\,dt \le \int_a^b \sqrt{(x')^2 + (y')^2 + (z')^2}\,dt$

Section 13.6
page 519

1. $t = s/\sqrt{5}$ **3.** $t^2 = s/\sqrt{26}$ **5.** $t = s/|aR|$ **7.** $t = s/\sqrt{13}$ **9.** $\kappa = 0$ **11.** $\kappa(t) = e^{2t}/(1 + e^{2t})^{3/2}$ **13.** $\kappa(t) = |t|/(1 + t^2)^{3/2}$ **15.** $\kappa(t) = 2/3|\sin 2t|$, κ is a min at $t = \pi/4 + n\pi$, κ has no maximum **19.** $\kappa(1) = 2/5^{3/2}$, center $(-4, 7/2)$, radius $1/\kappa(1)$ **21.** $\kappa(1) = 1/\sqrt{2}$, center $(2,2)$, radius $1/\kappa(1)$ **23.** $\kappa(0) = 1/\sqrt{2}$, center $(2,2)$, radius $1/\kappa(0)$ **25.** Use implicit differentiation to get $\kappa = 16/3\sqrt{2}$ at $(3/2, 3/2)$ **27.** $S(t) - s(t) = \int_{t_1}^{t_0} |\mathbf{r}'(t)|\,dt$

Section 13.7
page 527

1. $\mathbf{T} = (-\sin t\,\mathbf{i} - 2\mathbf{j} - \cos t\,\mathbf{k})/\sqrt{5}$, $\mathbf{N} = -\cos t\,\mathbf{i} + \sin t\,\mathbf{k}$, $\mathbf{B} = (-2\sin t\,\mathbf{i} + \mathbf{j} - 2\cos t\,\mathbf{k})/\sqrt{5}$ **3.** $\mathbf{T} = (-2t\,\mathbf{i} + (1-t^2)\mathbf{j})/(1+t^2)$, $\mathbf{N} = ((t^2-1)\mathbf{i} - 2t\,\mathbf{j})/(1+t^2)$, $\mathbf{B} = \mathbf{k}$ **5.** Let $\mathcal{P}$ be a plane through the origin and normal to $\mathbf{u}$; take a circle in $\mathcal{P}$ with center at the origin **7.** These are two identities that follow from applications of the Frenet formulas; the second is quite complicated and subtle **9.** Let $\mathbf{r}(t) - \mathbf{N}(t) = x(t)\,\mathbf{i} + y(t)\,\mathbf{j}$; if this curve is an ellipse, then the symmetry of the construction indicates it must coincide with an ellipse $a\cos u\,\mathbf{i} + b\sin u\,\mathbf{j}$ for some constants a and b; but then $g(t) = (x(t)/a)^2 + (y(t)/b)^2 = 1$. Evaluate $g(0)$, $g(\pi/4)$, $g(\pi/2)$ to find this is impossible **11.** $\int_0^t |s'|\,|1 - a\kappa(t)|\,dt$

Section 13.8
page 532

1. $a_T = 0$, $a_N = 5\pi^2$, speed $= 5\pi$ **3.** $a_T = 0$, $a_N = 2a^2$, speed $= \sqrt{9+4a^2}$ **5.** $a_T = (t+2t^3)/\sqrt{1+t^2+t^4}$, $a_N = \sqrt{1+4t^2+t^4}/\sqrt{1+t^2+t^4}$, speed $= \sqrt{1+t^2+t^4}$ **7.** $\mathbf{v}(t) = 2t^2\,\mathbf{i} + 3\,\mathbf{j}$; $\mathbf{r}(t) = \frac{2}{3}t^3\,\mathbf{i} + 3t\,\mathbf{j} - \mathbf{i} + \mathbf{j}$ **9.** $\mathbf{v}(t) = 4(\sin t\,\mathbf{i} + \cos t\,\mathbf{j})$; $\mathbf{r}(t) = -4(\cos t\,\mathbf{i} - \sin t\,\mathbf{j})$ **11.** The path is a circle of radius 10, center at the origin, $(-10, 0)$ is the starting point, the motion is clockwise and $5/2\pi$ revolutions are made in unit time **13.** $\mathbf{F} = m\mathbf{r}'' = m(s''\mathbf{T} + (s')^2\kappa\mathbf{N}) = m(s')^2\kappa\mathbf{N}$ since $s' = 0$ **15.** $|\mathbf{v}\times\mathbf{a}| = |\mathbf{r}'\times\mathbf{r}''| = |s'\mathbf{T}\times(s''\mathbf{T} + (s')^2\kappa\mathbf{N})| = |s'|^3\kappa|\mathbf{T}\times\mathbf{N}| = |s'|^3$

Section 13.9
page 539

1. Rectangular equation is $9(x+4/3)^2/64 + 3y^2/16 = 1$ **3.** Ellipse with center at $(-1/6, 0)$, extreme points at $(-1/6, \pm 2/\sqrt{6})$ and $(-1/6 \pm 5/6, 0)$ **5.** Differentiate $|\mathbf{r}|^2 = \mathbf{r}\cdot\mathbf{r}$ **9.** $\mathbf{r}$ given by (13.25); find $D = a/(1+e)$; $V = D|\theta'|$; use the equation from the derivation of Kepler's third law that says $r^2\theta' = H = \sqrt{aGM}$; obtain $0 < e = -1 + DV^2/GM < 1$ if the path is an ellipse

Chapter 14

Section 14.1
page 546

(Note that many different vector representations of a surface may be correct.) **1.** Right circular cylinder, radius 2, y-axis is axis of symmetry, $\mathbf{r}(s,t) = 2\cos t\,\mathbf{i} + s\,\mathbf{j} + 2\sin t\,\mathbf{k}$ **3.** Parabolic cylinder formed with a parabola in the xz-plane and lines parallel to the y-axis, $\mathbf{r}(s,t) = 4s^2\,\mathbf{i} + t\,\mathbf{j} + s\,\mathbf{k}$ **5.** A cylinder with directing line parallel to the z-axis generated by the parabola; $\mathbf{r}(s,t) = (8t^2+4)\,\mathbf{i} + t\,\mathbf{j} + s\,\mathbf{k}$ **7.** A surface of revolution; revolve the parabola $y^2 = z/2$ around the z-axis, $\mathbf{r}(s,t) = s\cos t\,\mathbf{i} + s\sin t\,\mathbf{j} + 2s^2\,\mathbf{k}$. **9.** A surface of revolution; revolve the hyperbola $y^2 - z^2 = 1$ around the z-axis, $\mathbf{r}(s,t) = (\sqrt{1+s^2}(\cos t\,\mathbf{i} + \sin t\,\mathbf{j}) + s\,\mathbf{k}$ **11.** $a^2x^2 + a^2y^2 + b^2z^2 = a^2b^2$ **13.** A cylinder generated by the parabola $y = (x/4)^2$ with directing line parallel to the z-axis **15.** The parabola $y = x^2$ is revolved around the x-axis **17.** The plane with equation $z = 3$ **19.** Part of the surface of revolution: $z^2 = y$ is revolved around the z-axis (restricted by $0 \le x, y \le 1$

Section 14.2
page 550

1. Sphere of radius 2, center at $(0,0,0)$ **3.** Hyperboloid of one sheet not intersecting the z-axis **5.** Hyperboloid of two sheets not intersecting the yz-plane **7.** Hyperboloid of one sheet not intersecting the y-axis **9.** Hyperboloid of two sheets not intersecting the xz-plane **11.** Hyperbolic paraboloid sitting astride the x-axis **13.** hyperbolic paraboloid (side view) **15.** hyperbolic paraboloid with center $(3,2,0)$ **17.** For $x = c \neq 0$ the trace is a hyperbola; for $x = 0$, the trace consists of two straight lines **19.** $U(s,t) = sa\cos t\,\mathbf{i} + sb\sin t\,\mathbf{j} + s^2\,\mathbf{k}$ **21.** Surface of revolution: revolve the line $3x = 2y$ around the x-axis **23.** Cone with vertex at $(2,3,0)$ obtained by revolving the line $x - 2 = y - 3$ around the x-axis

25. Yes for all except E, C and E' (equations with the double sign) where two equations must be used **27.** $(x/2)^2 + (y/5)^2 = z^2$ **29.** $y^2 + (z/3)^2 = x - 4$

Section 14.3
page 554

1. Meets xy-plane in the hyperbola $xy = -1$, meets the xz-plane in the line $t\mathbf{i} + \mathbf{k}$, meets the yz-plane in the line $t\mathbf{j} + \mathbf{k}$ **3.** meets xy-plane in the circle $x^2 + y^2 = 4$, meets the xz-plane in the parabola $z = 4 - x^2$, meets the yz-plane in the parabola $z = 4 - y^2$ **5.** The surfaces whose equations involve z but not z^2 are graphs of functions $z = f(x, y)$ **7.** Concentric ellipses with center at $(0, 0)$; the last is just a point **9.** Parabolas obtained by shifting $x = y^2$ to the right and left **11.** $f(x, y) = x^2 + (y/2)^2$ or $(x/2)^2 + y^2$ **13.** Circles with center $(0, 0)$ **15.** Straight lines with slope -1 **17.** Planes normal to $\mathbf{i} - \mathbf{j} + \mathbf{k}$ **19.** Ellipsoids, center at $(0, 0, 0)$ with ratio x-intercept:y- intercept:z-intercept = 3:2:6 **21.** If (a, b) is on the level curve $f(x, y) = p$ and on the level curve $f(x, y) = q$ then $p = f(a, b) = q$ since (a, b) satisfies both equations

Section 14.4
page 557

1. $f_x = 6x^2 - 4y^2$, $f_y = -8xy + 4y^3$ **3.** $f_x = 2\cos(2x + 3y)$, $f_y = 3\cos(2x + 3y)$ **5.** $g_r = \cos\theta$, $g_\theta = -r\sin\theta$ **7.** $f_x = e^y(-\sin x + \cos x)$, $f_y = f$ **9.** $f_x = 3x^2 + yz$, $f_y = 3y^2 + xz$, $f_z = 3z^2 + xy$ **11.** $g_r = -e^{-r}\cos\theta + e^r\sin\theta$, $g_\theta = -e^{-r}\sin\theta + e^r\cos\theta$ **13.** $f_x = 1/(y - z)$, $f_y = (x - z)/(y - z)^2$, $f_z = (y - x)/(x - z)^2$ **15.** $f_{xx} = 4$, $f_{xy} = f_{yx} = 11$, $f_{yy} = -6$ **17.** $f_{xx} = 2xy/(x^2 + y^2)^2$, $f_{xy} = (y^2 - x^2)/(x^2 + y^2)^2$, $f_{yy} = -f_{xx}$ **19.** $f_{xx} = 2h$, $f_{xy} = f_{yx} = n$, $f_{yy} = 2k$, $f_{xz} = p$, $f_{zy} = f_{yz} = q$, $f_{zz} = 2m$ **21.** $f_{xx} = 2(y^2 + z^2 - x^2)/(x^2 + y^2 + z^2)^2$, $f_{xy} = f_{yx} = 4xy/(x^2 + y^2 + z^2)^2$, other 2nd derivatives obtained by symmetry of the variables **23.** yes **25.** no

Section 14.5
page 565

1. $\lim g(x, mx) = (1 - m^2)/(1 + m^2)$, no single limit at $(0, 0)$ **3.** approach $(0, 0)$ along line $y = mx$ to get different limits **5.** Let $(x, y) = (r\cos t, r\sin t)$ to see $\lim = 0$ **7.** $f_{xy} = -16xy - 4xy/(x^2 + y^2)^2 = f_{yx}$ **9.** $g_{xyz} = \cos yz - yz\sin yz + \cos xz - xz\cos xz$ **11.** f_{xz} is not continuous at $(0, 0)$ **13.** -0.05 **15.** 1.2 **17.** max = 0.03, min = -0.02 **19.** max = $f(0, -0.1) = 0.001$; min = $f(0.1, 0.1) = -0.0012$

Section 14.6
page 573

1. $f_u = (y - 4x)1 + (x - 6y)3$, $f_v = (y - 4x)4 + (x - 6y)(-14v)$ **3.** $f_u = 7\cos(7u + 4v - 14v^2)$; $f_v = (4 - 28v)\cos(7u + 4v - 14v^2)$ **5.** $f_u = |u|/u$; $f_v = 0$ **7.** $w_u = (y + z + yz)\cos t + (x + z + xz)\sin t + (y + x + xy)(0)$, $w_v = (y + x + xy)(t)$, $w_t = -(y + z + yz)u\sin t + (x + z + xz)u\cos t + (y + x + xy)(v)$ **9.** $w_u = (y^2 + 2xz)1 + (2xy + z^2)(v + t) + (2zy + x^2)vt$, $w_v = (y^2 + 2xz)1 + (2xy + z^2)(v + t) + (2zy + x^2)ut$, $w_t = (y^2 + 2xz)1 + (2xy + z^2)(u + v) + (2zy + x^2)uv$ **11.** $u_x = v_y = 2x$, $u_y = -v_x = -2y$ **13.** $u_x = v_y = e^x\cos y$, $u_y = -v_x = -e^x\sin y$ a15$z_t = (z - t^2)/(z^2 - t)$ **17.** $-(z + 2t - z\cos zt)/(t + 2z - t\cos zt)$ **19.** $z_t = -(zu + z + u - 2t)/(tu + t + u - 2z)$ **21.** $3v^2 f_v(p, v) + 6uv + p$ **23.** $dT/dt = \nabla T(x, y, z)\cdot\mathbf{r}'(t)$ **19.** -280π **21.** $1620/\sqrt{61}$ mph

Section 14.7
page 579

1. $\nabla f = 2x\,\mathbf{i} + 6y^2\,\mathbf{j}$; $(0, 0)$ **3.** $\nabla f = (6x^2 - 4y^2)\mathbf{i} + (6y - 8xy)\mathbf{j}$; $(0, 0)$, $(3/4, \pm 3\sqrt{6}/8)$ **5.** $\nabla f = -\sin x\sin y\,\mathbf{i} + \cos x\cos y\,\mathbf{j}$, $\nabla f = 0$ at points $(n\pi, (2m + 1)\pi/2)$ and $((2m + 1)\pi/2, n\pi)$ for any integers n and m **7.** $\nabla f = 2(y\,\mathbf{i} - x\,\mathbf{j})/(x + y)^2$; ∇f never 0 (but undefined at $(0, 0)$ **9.** $4\mathbf{i} - 6\mathbf{j}$, tangent vector = $6\mathbf{i} + 4\mathbf{j}$ **11.** $-3\mathbf{i} + \mathbf{j}$, tangent vector = $\mathbf{i} + 3\mathbf{j}$ **13.** $4(x - 2) + 2(y + 1) - (z - 3) = 0$ **15.** $-4(x + 2) + 2(y - 1) - (z - 5) = 0$ **17.** Equation of the tangent plane to the cone at (p, q, r) is $p(x - a)/a^2 + q(y - q)/b^2 - r(z - r) = 0$ and this passes through the origin **19.** $-15/17$ **21.** $1/2$ **23.** Tangent line is parallel to $(2\mathbf{i} - \mathbf{k}) \times (2\mathbf{i} + 2\mathbf{k}) = -6\mathbf{j}$;

one parametrization of the intersection is $\mathbf{r}(u) = \frac{\sqrt{1+u^2}}{1-u^2}\,\mathbf{i} + \frac{u}{2}\frac{\sqrt{1+u^2}}{1-u^2}\,\mathbf{j} + \frac{1+u^2}{1-u^2}\,\mathbf{k}$ and $\mathbf{r}(0) = (1, 0, 1)$ and $\mathbf{r}'(0) = (1/2)\mathbf{j}$ (which agrees) **25.** The intersection of the surface with the plane $3x + 3y - 2z + 8 = 0$ **27.** The equation for λ is $a^2(1 + 2\lambda)^{-2} + 4b^2(1 + 8\lambda)^{-2} + 6c^2(1 + 12\lambda)^{-2} = 11$

Section 14.8
page 585

1. 18, $4\sqrt{2}$, -10, $6\sqrt{14}$ **3.** -6, -2, 0, $2\sqrt{10}$ **5.** $4/\sqrt{5}$ **7.** $(x/10)^9 = (y/10)^4$ **9.** direction $\mathbf{i} - \mathbf{j}$ **11.** Need $\nabla W \cdot (x'\,\mathbf{i} + y'\,\mathbf{j}) < 0$ so angle between $\pi/2$ and π **13.** -5 **15.** $117/25$

Section 14.9
page 593

1. Saddle point $= (3/2, 0)$ **3.** Saddle point $= (-1, -2)$ **5.** Local min at $(2, 1)$ **7.** Saddle point at $(0, 0)$ (14.14 does not apply; inspect the sign of the function in each quadrant **9.** Saddle points at $\pm(-1, \sqrt{2})$; local max at $(-1, -\sqrt{2})$; local min at $(1, \sqrt{2})$ **11.** Local max at $(0, 0)$; global min at every $(\pm 1, y)$ and $(x, \pm 1)$ **13.** $(52/17, -4/17)$ **15.** $(71/45, 142/45, 70/45)$ **17.** Cube with side $\sqrt{50/3}$ **19.** $p = (\sum a_i)/n$, $q = (\sum b_i)/n$

Section 14.10
page 598

1. $f_{min} = -11$ at $(-4/5, 3/5)$ **3.** Extreme values $\pm\sqrt{211}\sqrt{21}$ **5.** $165/121$ at $(-2/11, -6/11, 9/122)$ **7.** Minimum distance $= (10\sqrt{2} - 11)/2\sqrt{10}$ **9.** Smallest sum $= 1/2$; largest sum $= 1$ **11.** The minimum occurs at a point (x, y, z) in the surface where $2(x-a)\mathbf{i} + 2(y-b)\mathbf{j} + 2(z-c)\mathbf{k}$ is parallel to the normal $\nabla F(x, y, z)$ **13.** The curve is given parametrically as $a\cos\theta\mathbf{u} + b\sin\theta\mathbf{v}$ with $a = 1/5$, $b = \sqrt{2}/5$, $\mathbf{u} = (4\mathbf{i} - 3\mathbf{j})/5$, $\mathbf{v} = (3\mathbf{i} + 4\mathbf{j})/5$

Section 14.11
page 609

1. max $= 5$ at $(0.1, 0.1)$; min $= 0.5$ at $(1, 1)$ **3.** Pick any positive δ and take any k with $10^{-k} < \delta$. Take $(x_1, y_1) = (10^{-(k+1)}, 10^{-(k+1)})$ and $(x_2, y_2) = (10^{-2(k+1)}, 10^{-2(k+1)})$; then (x_1, y_1) is within δ of (x_2, y_2) but $|f(x_1, y_1) - f(x_2, y_2)| > 10^k$ which can be larger than any give positive number **5.** Take (a, b) in S and $\epsilon > 0$. There is a $\delta > 0$ satisfying the definition (14.6) of uniform continuity. Thus $|f(x, y) - f(a, b)| < \epsilon$ when $(x - a)^2 + (y - b)^2 < \delta^2$ and thus the definition of continuity at (a, b) is satisfied **7.** $M'(0) = 0$ **9.** $M'(0) = 1$ **11.** $M'(1) = -1/2$ **13.** $M_x(1, 2) = 1$; $M_y(1, 2) = 2/3$ **15.** $M_x(1, 1) = 1$; $M_y(1, 1) = -2$

Chapter 15

Section 15.1
page 617

1. $\pi^2/8 \leq$ integral $\leq \pi^2/4$ **3.** $\sqrt{3}e^{10} \leq$ integral $\leq 2\sqrt{2}e^{18}$ **5.** $L(P_2) = 3/2$, $U(P_2) = 9/2$, exact value $= 3$ **7.** $L(P_k) = \left(\frac{1}{2} - \frac{1}{2k}\right)^2$; lim $= 1/4$ **9.** Vol $= \lim_{k\to\infty} \sum_{i=0}^{k-1} \sum_{j=0}^{k-1} \left[\frac{2i}{k} + \frac{15j}{k} + 2\right] \frac{6}{k^2} = 69$ **11.** $L(P) = 65/4$; $U(P) = 129/4$

Section 15.2
page 629

1. 26 **3.** 4 **5.** $\frac{1}{2}(e^6 - e^2) - e^3 + e$ **7.** $(e - e^{-1})^2$ **9.** Type X and Type Y; 6 **11.** Best done as Type X; $16\sqrt{2}/3$ **13.** Type X; area $= \frac{3\sqrt{119}}{2} - 64 \arcsin \frac{3\sqrt{119}}{80} + 16 \arcsin \frac{3\sqrt{119}}{40} \approx 4.72185$ **15.** $\int_{\pi/4}^{3\pi/4} \int_{\cos x}^{\sin x} dy\, dx = \sqrt{2}$ **17.** $(1 - 16^2)/6$ **19.** 2 **21.** $\int_0^4 \int_{\sqrt{y}}^2 xy^2\, dx\, dy = 32/3$ **23.** $\left(\int_0^1 \int_{-\sqrt{y}}^{\sqrt{y}} + \int_1^9 \int_{(y-3)/2}^{\sqrt{y}}\right) xy\, dx\, dy = 160/3$ **25.** $\int_0^R \int_0^{\sqrt{R^2-x^2}} (R^2 - x^2)^{3/2}\, dy\, dx = 8R^5/15$ **27.** For a partition into subrectangles S_{ij} and (u_i, v_j) in S_{ij} we have $f(u_i, v_j) \leq g(u_i, v_j)$ and $f(u_i, v_j)\Delta S_{ij} \leq g(u_i, v_j)\Delta S_{ij}$. Take sums and pass to the limit **29.** For a regular partition of

$[-a, a]$, points may be selected in the subrectangles so that whenever (u_i, v_j) is one selected point, so is $(-u_i, v_j)$. Then every Riemann sum is 0.

Section 15.3
page 634

1. 32 **3.** 280 **5.** $-2 \leq y \leq 2$, $0 \leq x \leq 4 - y^2$, $0 \leq z \leq y + 2$, vol $= 16/3$ **7.** $0 \leq |x| \leq 1, 0 \leq |y| \leq 1-|x|, 0 \leq z \leq 1-|x|-|y|$; or write this without absolute values as a union of four solids corresponding to the four quadrants in which (x, y) may lie **9.** $-1 \leq x \leq 1$, $1 - \sqrt{1 - x^2} \leq y \leq 1 + \sqrt{1 - x^2}$, $0 \leq z \leq 16 - x^2 - 4y^2$ **11.** $\int_0^{2\pi} \int_0^a r\sqrt{1 + r^2}\, dr\, d\theta = \frac{2\pi}{3}[(1 + a^2)^{3/2} - 1]$ **13.** $\int_0^{\pi} \int_0^1 r^2 \cos^2 \theta\, r\, dr\, d\theta = \pi/8$ **15.** $\int_0^{\pi/4} \int_0^{\sec\theta} (1 - r^2) r\, dr\, d\theta = 1/6$ **17.** height $= 2h$, volume remaining $= 4\pi h^3/3$

Section 15.4
page 638

1. $M = 500/3$, $M_x = -4000/3$, $M_y = -1000/3$; $(-2, -8)$ **3.** $M = 1/4$, $M_x = 2/21$, $M_y = 2/15$; $(8/15, 8/21)$ **5.** $M = 8/3$, $M_x = 4/3$, $M_y = 8/3$; $(1, 1/2)$ **7.** $M = 8/3$, $M_x = 32/15 = M_y$; $(4/5, 4/5)$ **9.** $M = \rho ab^2/6$, $M_x = \rho ab^3/12$, $M_y = \rho a^2b^2/24$, $(a/4, b/2)$; $(a/2, b/4)$ **13.** $(0, 4(A^2 + aA + a^2)/(3\pi(A + a))$; $\lim_{a\to A} = (0, 2A/\pi)$ is the center of mass of the circumference of the half-circle of radius A

Section 15.5
page 646

1. $(x/p)^2 + (y/q)^2 = z$ **3.** $(x/a)^2+(y/b)^2+(z/c)^2 = 1$ **5.** $2\sqrt{2}(2-3\sqrt{2}-4\sqrt{3}+5\sqrt{6})/3$ **7.** $(32 + 100\sqrt{10} - 2 \cdot 7^{5/2})/15 \approx 5.92961$ **9.** $8R^2(\pi - 2)$ (top and bottom included) **11.** Area $S = \int_a^b \left(\int_c^d \sqrt{1 + g'(y)^2}\, dy\right) dx = \int_a^b L\, dx = L(b - a)$ **13.** $\int_0^{\pi} \int_0^{2\pi} p|\sin v|\sqrt{p^2 \cos^2 v + q^2 \sin^2 v}\, du\, dv$; for $p = q$ values is $4\pi p^2$ **15.** slant height of cone is $L = \sqrt{R^2 + a^2}$ and lateral area is πRL **17.** 0

Section 15.6
page 653

1. 24 **3.** $abc[2(a^2+b^2+c^2)+3(ab+ac+bc)]$ **5.** $2p^5/15-p^4/3+p^3/6+p^2/6-p/6+1/30$ **7.** $(p^6 - 1)/48 - (p^4 - p^2)/16$ **9.** 6 **11.** $2\pi/3$ **13.** $(a/4, b/4, c/4)$ **15.** If the square is given by $0 \leq x, y, z \leq a$ and the mass is proportional to the square of the distance from the origin, then the center of mass is $(7a/12, 7a/12, 7a/12)$ **17.** 1/2 **19.** 0; 0 (argue by symmetry--no integration needed)

Section 15.7
page 659

1. 375 **3.** 375 **5.** 9/2 **7.** 9/2 **9.** 557/30 **11.** 7 **13.** −5 **15.** $-3a^2/2$ **17.** 0 **19.** 0 **21.** 0 **23.** 0

Section 15.8
page 668

1. Differential of $xy - 2x$ **3.** Differential of $4x^2y^2 + 2x - y$ **5.** Not exact **7.** 566 **9.** 1 **11.** $\int_C f_x\, dx + f_y\, dy = 0$ because $f_x\, dx + f_y\, dy$ is exact **13.** $g(x)\, dx + h(y)\, dy$ is an exact differential, so apply Exercise 11 **15.** Assuming $y > 0$: $n = 1$, $-y^{-1/2} \arctan(xy^{-1/2})$; $n = 2$, $-\frac{1}{4}y^{-3/2} \arctan(xy^{-1/2}) - \frac{x}{4y(x^2+y)}$; $n = 3$, $\frac{1}{8}y^{-5/2} \arctan(xy^{-1/2}) + \frac{1}{12}\frac{x}{y}\frac{1}{x^2+y} - \frac{1}{6xy^2} + \frac{1}{(x^2+y)^2}$. For $y < 0$ analogous formulas are obtainted involving $\sqrt{-y}$ and $\ln|x \pm \sqrt{-y}|$

Section 15.9
page 674

1. -28π **3.** −25 **5.** 0 **7.** If $a + b = 1$, area $= \int_C -by\, dx + ax\, dy = 16\pi$ **9.** −5 **11.** 0 **13.** $y^* = \frac{1}{2}\int_{\beta G} y^2\, dx / \int_{\beta G} y\, dx$

Section 15.10
page 687

In Exercises 1 to 6, more than one correct answer is possible.
1. $T(u, v) = \big((3u - v)/7, (u + 2v)/7\big)$, preserves orientation, $R = [0, 4; 0, 1]$ **3.** $T(u, v) = \big((5v - 4u)/2, u - v\big)$, reverses orientation, $R = [2, 10; 0, 9]$ **5.** $T(u, v) = \big((u - v)/2, (u + v)/2\big)$, preserves orientation, $R = [10, 11; 10, 11]$ **7.** T preserves

orientation, $R = [2, 3; -\ln 2, 0]$; $\left(\frac{3\sqrt{3}}{24} - \frac{2\sqrt{2}}{24}\right)$ $\frac{31}{3}$ **9.** T preserves the orientation, $R = [1, 5; 2, 5]$; -60480 **11.** $T(u, v) = (u, u+v)$, $\int_0^3 \int_{x-3}^{x+4} f\, dy\, dx = \int_0^3 \int_{-3}^{4} f(u, u+v)\, dv\, du$ **13.** $T(u, v) = (5(u+v)/12, (7v - 5u)/12)$, preserves orientation; transformation converts the integral to one taken over $R = [0, 12/5; 0, 12/5]$

Section 15.11
page 691

1. $\int_0^4 \int_0^{2\pi} \int_0^1 r^3 \cos^2\theta\, dr\, d\theta\, dz = \pi$ **3.** $\int_0^\pi \int_0^1 \int_0^{\sqrt{1-r^2}} dz\, dr\, d\theta = \pi^2/4$ **5.** $(3a/8, 3a/8, 3a/8)$ **7.** $\pi a^3(2 - \sqrt{3})/3$ **9.** $\pi(8\sqrt{2} - 7)/6$

Chapter 16

Section 16.1
page 697

1. $\mathbf{r}(x, y) = x\,\mathbf{i} + y\,\mathbf{j} + (x^2 - y^2)\mathbf{k}$; $\mathbf{r}_x \times \mathbf{r}_y = 2x\,\mathbf{i} + 2y\,\mathbf{j} + \mathbf{k} \neq 0$ **3.** $\mathbf{A}(x, y) = (2x^2 + y^2)\,\mathbf{i} + 2xy\,\mathbf{j} - \mathbf{k}$ is normal to the surface and never 0; $\mathbf{n} = \mathbf{A}/|\mathbf{A}|$ **5.** $\mathbf{r}_u \times \mathbf{r}_v = 4(u^2 + v^2)\,\mathbf{i} - 8uv\,\mathbf{j} + 4(u^2 + v^2)\,\mathbf{k} \neq 0$ if $(u, v) \neq (0, 0)$ **7.** $\mathbf{r}(u, v) = (1 + v\cos(nu/2))(\cos u\,\mathbf{i} + \sin u\,\mathbf{j}) + v\sin(nu/2)\mathbf{k}$ ($|v| \leq 1$, $0 \leq u \leq 2\pi$) is the vector representation of a "Mobius strip" with n twists. The representation is one-to-one on the interior of the uv-rectangle but not on the boundary since $\mathbf{r}(0, v) = \mathbf{r}(2\pi, (-1)^n v)$. One can show $\mathbf{r}_u \times \mathbf{r}_v \neq 0$. Then $\mathbf{n} = \mathbf{r}_u \times \mathbf{r}_v / |\mathbf{r}_u \times \mathbf{r}_v|$ is a unit normal and to be continuous, it is necessary that $\mathbf{n}(0, v) = \mathbf{n}(2\pi, (-1)^n v)$; this happens only for n even

Section 16.2
page 707

1. 0; $3\,\mathbf{i} + 5\,\mathbf{j}$ **3.** $\frac{-4xy}{(x^2+y^2)^2}\,\mathbf{k}$ **5.** 0;0 **7.** 0; $1/|\mathbf{r}|^2$ **9.** $-\pi$ **11.** $-4\pi\sqrt{5}$ **13.** 2π **15.** 2π **17.** 3π

Section 16.3
page 714

1. 4 **3.** $8\pi/3$ **5.** 0 **7.** Out of the sphere: $4\pi R^3/3$; out of the cube: $8R^3$ **9.** $-2q\,\mathbf{i}/5\sqrt{5}$ **11.** $-8q\,\mathbf{k}/5\sqrt{5}$ **13.** $ab^3/3$

Index

PARTIAL FRACTION DECOMPOSITION

52. $\dfrac{1}{(x-a)(x-b)} = \dfrac{1}{a-b}\left[\dfrac{1}{x-a} - \dfrac{1}{x-b}\right]$

53. $\dfrac{x}{(x-a)(x-b)} = \dfrac{1}{a-b}\left[\dfrac{a}{x-a} - \dfrac{b}{x-b}\right]$

54. $\dfrac{1}{(x-a)(x-b)(x-c)} = \dfrac{1}{(b-a)(c-a)(x-a)} + \dfrac{1}{(a-b)(c-b)(x-b)} + \dfrac{1}{(a-c)(b-c)(x-c)}$

55. $\dfrac{1}{(x-a)^2(x-b)} = \dfrac{1}{(a-b)^2}\left[\dfrac{a-b}{(x-a)^2} - \dfrac{1}{x-a} + \dfrac{1}{x-b}\right]$

56. $\dfrac{1}{(x-a)^2(x-b)^2} = \dfrac{1}{(a-b)^3}\left[\dfrac{a-b}{(x-a)^2} - \dfrac{2}{x-a} + \dfrac{a-b}{(x-b)^2} - \dfrac{2}{x-b}\right]$

57. $\dfrac{1}{(x-a)(x^2+bx+c)} = \dfrac{1}{a^2+ab+c}\left[\dfrac{1}{(x-a)} - \dfrac{x+a+b}{x^2+bx+c}\right]$

58. $\dfrac{1}{(x-a)p(x)} = \dfrac{1}{p(a)}\dfrac{1}{x-a} + \dfrac{h(x)}{p(x)}, \quad h(x) = -\dfrac{1}{p(a)}\dfrac{p(x)-p(a)}{x-a}, \quad p(a) \neq 0$

INFINITE SERIES

59. $\dfrac{1}{1-x} = 1 + x + x^2 + \cdots + x^n + \cdots \qquad |x| < 1$

60. $e^x = 1 + x + \dfrac{x^2}{2!} + \cdots + \dfrac{x^n}{n!} + \cdots$

61. $\sin x = x - \dfrac{x^3}{3!} + \cdots + (-1)^n \dfrac{x^{2n+1}}{(2n+1)!} + \cdots$

62. $\cos x = 1 - \dfrac{x^2}{2!} + \cdots + (-1)^n \dfrac{x^{2n}}{(2n)!} + \cdots$

63. $\ln(1+x) = x - \dfrac{x^2}{2} + \dfrac{x^3}{3} - \cdots + (-1)^{n+1}\dfrac{x^n}{n} + \cdots \qquad |x| < 1$

64. $\ln\left(\dfrac{1+x}{1-x}\right) = 2\left(x + \dfrac{x^3}{3} + \dfrac{x^5}{5} + \cdots + \dfrac{x^{2n+1}}{2n+1} + \cdots\right) \qquad |x| < 1$

65. $(1+x)^a = 1 + ax + \dfrac{a(a-1)x^2}{2!} + \cdots + \dfrac{a(a-1)\cdots(a-n+1)x^n}{n!} + \cdots \qquad |x| < 1$

66. $\arctan x = x - \dfrac{x^3}{3} + \dfrac{x^5}{5} - \cdots + (-1)^n \dfrac{x^{2n+1}}{2n+1} + \cdots \qquad -1 < x \leq 1$

FINITE SUMS

67. $\displaystyle\sum_{k=1}^{n} k = \frac{n(n+1)}{2}$

68. $\displaystyle\sum_{k=1}^{n} k^2 = \frac{n(2n+1)(n+1)}{6}$

69. $\displaystyle\sum_{k=1}^{n} k^3 = \frac{n^2(n+1)^2}{4}$